III-Nitride, SiC and Diamond Materials for Electronic Devices

MATERIALS RESEARCH SOCIETY
SYMPOSIUM PROCEEDINGS VOLUME 423

III-Nitride, SiC and Diamond Materials for Electronic Devices

Symposium held April 8-12, 1996, San Francisco, California, U.S.A.

EDITORS:

D. Kurt Gaskill
Naval Research Laboratory
Washington, D.C., U.S.A.

Charles D. Brandt
Northrop Grumman Science and Technology Center
Pittsburgh, Pennsylvania, U.S.A.

Robert J. Nemanich
North Carolina State University
Raleigh, North Carolina, U.S.A.

MRS
MATERIALS RESEARCH SOCIETY
PITTSBURGH, PENNSYLVANIA

This work was supported in part by the Army Research Office under Grant Number ARO: DAAH04-96-1-0084. The views, opinions, and/or findings contained in this report are those of the author(s) and should not be construed as an official Department of the Army position, policy, or decision, unless so designated by other documentation.

This work was supported in part by the Office of Naval Research under Grant Number ONR: N00014-96-1-0249. The United States Government has a royalty-free license throughout the world in all copyrightable material contained herein.

Single article reprints from this publication are available through University Microfilms Inc., 300 North Zeeb Road, Ann Arbor, Michigan 48106

CODEN: MRSPDH

Published by:

Materials Research Society
9800 McKnight Road
Pittsburgh, Pennsylvania 15237
Telephone (412) 367-3003
Fax (412) 367-4373
Website: http://www.mrs.org/

Library of Congress Cataloging in Publication Data

III-Nitride, SiC, and diamond materials for electronic devices : symposium held April 8–12, 1996, San Francisco, California, U.S.A. / editors, D. Kurt Gaskill, Charles D. Brandt, Robert J. Nemanich
p. cm—(Materials Research Society symposium proceedings ; v. 423)
Includes bibliographical references and index.
ISBN 1-55899-326-6
1. Electronics—Materials—Congresses. I. Gaskill, D. Kurt II. Brandt, Charles D. III. Nemanich, Robert J. IV. Series: Materials Research Society symposium proceedings ; v. 423.
TK7871.I53 1996
621.3815'2—dc20 96-38385
CIP

Manufactured in the United States of America

CONTENTS

Preface xvii

Materials Research Society Symposium Proceedings xviii

PART I: DEVICE TECHNOLOGIES

Devices

*SiC Semiconductor Applications - An Air Force Perspective 3
Laura S. Rea

*SiC Power Devices 9
T.P. Chow and M. Ghezzo

*Growth of High Quality (In,Ga,Al)N/GaN Heterostructure Materials and Devices by Atmospheric Pressure MOCVD 23
S.P. DenBaars, S. Keller, B.P. Keller, Y.F. Wu, D. Kapolnek, and U.K. Mishra

Investigation of an NEA Diamond Vacuum Microtriode Array 33
C.W. Hatfield, G.L. Bilbro, A.S. Morris, P.K. Baumann, B.L. Ward, and R.J. Nemanich

Tetrahedral Amorphous Carbon Thin Film Transistors 39
F.J. Clough, B. Kleinsorge, W.I. Milne, and J. Robertson

Theoretical Prediction of Zinc Blende Phase GaN Avalanche Photodiode Performance Based on Numerically Calculated Electron and Hole Impact Ionization Rate Ratio 45
J. Kolnik, I.H. Oguzman, K.F. Brennan, R. Wang, and P.P. Ruden

OBIC Measurements on 6H-SiC Schottky Diodes 51
K. Rottner, A. Schöner, M. Frischholz, and R. Helbig

Defect Dominant Junction Characteristics of 4H-SiC p^+/n Diodes 57
J. Scofield, M. Dunn, K. Reinhardt, Y.K. Yeo, and R. Hengehold

Fabrication of Highly Oriented, Smooth Diamond Films on Silicon for Electronic Devices 63
S.D. Wolter, T.H. Borst, P. Gluche, W. Ebert, A. Vescan, and E. Kohn

*Invited Paper

Radiative Recombination Rates in GaN, InN, AlN and Their Solid Solutions 69
A.V. Dmitriev and A.L. Oruzheinikov

Pyroelectric Effect in Wurtzite Gallium Nitride 75
A.D. Bykhovski, V.V. Kaminski, M.S. Shur, Q.C. Chen, and M.A. Khan

CVD Diamond Wires as X-ray Detectors 81
C. Manfredotti, F. Fizzotti, M. Galetto, A. Lo Giudice, D. Margherita, C. Ongaro, P. Polesello, and E. Vittone

Critical Materials, Device Design, Performance and Reliability Issues in 4H-SiC Power UMOSFET Structures 87
A.K. Agarwal, R.R. Siergiej, S. Seshadri, M.H. White, P.G. McMullin, A.A. Burk, L.B. Rowland, C.D. Brandt, and R.H. Hopkins

High-Temperature Switching Characteristics of 6H-SiC Thyristor 93
K. Xie, J.R. Flemish, T. Burke, W.R. Buchwald, and J.H. Zhao

Characterization of 4H-SiC MOS Capacitors by a Fast-Ramp Response Technique 99
T.S. Sudarshan, V.P. Madangarli, G. Gradinaru, C.C. Tin, R. Hu, and T. Isaacs-Smith

Characterization of 4H-SiC JFETs for Use in Analog Amplifiers Capable of 723K Operation 105
J.B. Casady, D.C. Sheridan, W.C. Dillard, and R.W. Johnson

High Electric Field Breakdown of 4H-SiC p-n Junction Diodes 111
C.C. Tin, V. Madangarli, R. Hu, E. Luckowski, J. Casady, T. Isaacs-Smith, G. Gradinaru, T.S. Sudarshan, and R.W. Johnson

Metallizations

Improved Nickel Silicide Ohmic Contacts to n-Type 4H and 6H-SiC Using Nichrome 119
E.D. Luckowski, J.R. Williams, M.J. Bozack, T. Isaacs-Smith, and J. Crofton

Chemical and Structural Characterization of the Ni-Ti Alloy/ 6H-SiC Contacts 125
M. Levit, I. Grimberg, B.-Z. Weiss, and M. Eizenberg

Low Resistance Ohmic Contact on *p*-type GaN Grown by Plasma-Assisted Molecular Beam Epitaxy 131
Myung C. Yoo, J.W. Lee, J.M. Myoung, K.H. Shim, and K. Kim

Approaches to High Temperature Contacts to Silicon Carbide 137
J.M. Delucca and S.E. Mohney

Characterization of Zirconium - Diamond Interfaces 143
P.K. Baumann, S.P. Bozeman, B.L. Ward, and R.J. Nemanich

Ohmic Contacts to *n*-Type 6H-SiC Without Post-Annealing 149
Tokuyuki Teraji, Shiro Hara, Hideyo Okushi, and Koji Kajimura

Etching

Photothermally Assisted Dry Etching of GaN 157
R.T. Leonard and S.M. Bedair

Dry Etch Damage in InN, InGaN and InAlN 163
S.J. Pearton, J.W. Lee, J.D. MacKenzie, C.B. Vartuli, S.M. Donovan, C.R. Abernathy, R.J. Shul, F. Ren, and J.R. Lothian

Optical Spectroscopy of Dry-Etched GaN-Nanostructures 169
H. Zull, J. Müller, J. Koeth, F. Kieseling, and A. Forchel

Photocurrent Sensitivities, Surface Color, and Auger Spectroscopy of Silicon Carbide (SiC) by Photoelectro-Chemical (PEC) Etching 175
D.M. Collins, G.L. Harris, and Naiqun Chen

Implantation

Si Implantation and Annealing of GaN for n-Type Layer Formation 183
B. Molnar, A.E. Wickenden, and M.V. Rao

p- and n-Type Implantation Doping of GaN with Ca and O 189
J.C. Zolper, R.G. Wilson, S.J. Pearton, and R.A. Stall

Ion Beam Synthesis of Tungsten Implantation into 6H-Silicon Carbide at Elevated Temperatures 195
Hannes Weishart, W. Matz, and W. Skorupa

TEM Structure Characterization of Ti/Al and Ti/Al/Ni/Au Ohmic Contacts for n-GaN 201
S. Ruvimov, Z. Liliental-Weber, J. Washburn, K.J. Duxstad, E.E. Haller, Z.-F. Fan, S.N. Mohammad, W. Kim, A.E. Botchkarev, and H. Morkoc

BETA Silicon Carbide pn Junction Diodes 207
J.C. Coleman, G.L. Harris, and D.B. Poker

PART II: CRYSTAL AND FILM GROWTH

Bulk and Bulk-Like Crystal Growth

Analysis of the Sublimation Growth Process of Silicon Carbide Bulk Crystals 215
R. Eckstein, D. Hofmann, Y. Makarov, St. G. Müller, G. Pensl, E. Schmitt, and A. Winnacker

Gallium Nitride Thick Films Grown by Hydride Vapor Phase Epitaxy 221
R.J. Molnar, P. Maki, R. Aggarwal, Z.L. Liau, E.R. Brown, I. Melngailis, W. Götz, L.T. Romano, and N.M. Johnson

Optimization of Reactor Geometry and Growth Conditions for GaN Halide Vapor Phase Epitaxy 227
S.A. Safvi, N.R. Perkins, M.N. Horton, A. Thon, D. Zhi, and T.F. Kuech

Thermodynamic Analysis and Growth Characterization of Thick GaN Films Grown by Chloride VPE Using $GaCl_3/N_2$ and HN_3/N_2 233
Heon Lee, Masaaki Yuri, Tetsuzo Ueda, and James S. Harris, Jr.

High Resolution X-ray Diffraction Analysis of Gallium Nitride Grown on Sapphire by Halide Vapor Phase Epitaxy 239
R.J. Matyi, D. Zhi, N.R. Perkins, M.N. Horton, and T.F. Kuech

Structural Characterization of Thick GaN Films Grown by Hydride Vapor Phase Epitaxy 245
L.T. Romano, R.J. Molnar, B.S. Krusor, G.A. Anderson, D.P. Bour, and P. Maki

Film Growth

***Silicon Carbide CVD Approaches Industrial Needs** 253
R. Rupp, P. Lanig, J. Völkl and D. Stephani

***Kinetics Approach to the Growth of Cubic Boron Nitride** 265
C.A. Taylor, II and Roy Clarke

The Impact of Pregrowth Conditions and Substrate Polytype on SiC Epitaxial Layer Morphology 275
A.A. Burk, Jr., L.B. Rowland, G. Augustine, H.M. Hobgood, and R.H. Hopkins

*Invited Paper

Nucleation and Growth of Oriented Diamond Films on Nickel Substrates ... 281
P.C. Yang, W. Liu, D.A. Tucker, C.A. Wolden, R.F. Davis, J.T. Glass, J.T. Prater, and Z. Sitar

Nucleation and Growth of Gallium Nitride Films on Si and Sapphire Substrates Using Buffer Layers ... 287
N.R. Perkins, M.N. Horton, D. Zhi, R.J. Matyi, Z.Z. Bandic, T.C. McGill, and T.F. Kuech

Seeding with a Diamond Suspension for Growth of Smooth Polycrystalline Diamond Surfaces ... 293
I. St. Omer, T. Stacy, E.M. Charlson, and E.J. Charlson

Properties of Tetrahedral Amorphous Carbon Deposited by a Filtered Cathodic Vacuum Arc ... 299
M. Chhowalla, C.W. Chen, B. Kleinsorge, J. Robertson, G.A.J. Amaratunga, and W.I. Milne

X-ray Diffraction Analysis of Strain and Mosaic Structure in (001) Oriented Homoepitaxial Diamond Films ... 305
W. Brock Alexander, Pehr E. Pehrsson, David Black, and James E. Butler

TEM and PL Characterisation of MBE-Grown Epitaxial GaN/GaAs ... 311
Yan Xin, P.D. Brown, C.B. Boothroyd, A.R. Preston, C.J. Humphreys, T.S. Cheng, C.T. Foxon, A.V. Andrianov, and J.W. Orton

Chemical Beam Epitaxy of $GaN_xP1\text{-}_x$ Using a N Radical Beam Source ... 317
N.Y. Li, D.H. Tomich, W.S. Wong, J.S. Solomon, and C.W. Tu

Improved Aluminum Nitride Thin Films Grown by MOCVD From Tritertiarybutylaluminum and Ammonia ... 323
T. Metzger, E. Born, R. Stömmer, W. Rieger, R. Dimitrov, D. Lentz, H. Angerer, O. Ambacher, and M. Stutzmann

Properties of Homoepitaxially MBE-Grown GaN ... 329
T. Suski, J. Krueger, C. Kisielowski, P. Phatak, M.S.H. Leung, Z. Liliental-Weber, A. Gassmann, N. Newman, M.D. Rubin, E.R. Weber, I. Grzegory, J. Jun, M. Bockowski, S. Porowski, and H.I. Helava

Lattice-Matched InAsN(X=0.38) on GaAs Grown by Molecular Beam Epitaxy ... 335
Y.C. Kao, T.P.E. Broekaert, H.Y. Liu, S. Tang, I.H. Ho, and G.B. Stringfellow

Growth and Characterization of AlInGaN/InGaN Heterostructures 341
J.C. Roberts, F.G. McIntosh, M. Aumer, V. Joshkin, K.S. Boutros, E.L. Piner, Y.W. He, N.A. El-Masry, and S.M. Bedair

Growth and Characterization of AlGaN/GaN Heterostructures with Multiple Quantum Wells by PAMBE 347
K.H. Shim, J.M. Myoung, O.V. Gluschenkov, C. Kim, K. Kim, M.C. Yoo, S. Kim, D.A. Turnbull, and S.G. Bishop

Origin of High-Conductivity Layer Near the Surface in As-Grown Diamond Films 353
Sadanori Yamanaka, Kazushi Hayashi, Hideyo Okushi, and Koji Kajimura

GaN Film Growth by a Supersonic Arcjet Plasma 359
M.A. Cappelli, A.E. Kull, K. Schwendner, H. Lee, S.J. Harris, Jr., and J. Mroczkowski

Particle-Assisted Oriented Deposition of Diamond Thin Films 365
Dong-Gu Lee and Rajiv K. Singh

Deposition of Adherent Diamond Films on Sapphire Substrates 371
Donald R. Gilbert and Rajiv K. Singh

GaN Growth by Nitrogen ECR-CVD Method 377
K.H. Chen, C.H. Chao, T.J. Chuang, Y.J. Yang, L.C. Chen, C.K. Chen, Y.F. Huang, C.H. Yang, H.Y. Lin, I.M. Chang, and Y.F. Chen

High Quality p-Type GaN Films Grown by Plasma-Assisted Molecular Beam Epitaxy 385
J.M. Myoung, C. Kim, K.H. Shim, O. Gluschenkov, K. Kim, and M.C. Yoo

Influence of the Growth Atmosphere on the Properties of AlN Grown by Plasma-Assisted Pulsed Laser Deposition 391
T. Ogawa, M. Okamoto, Y. Mori, and T. Sasaki

Laser Wavelength Dependent Properties of BN Thin Films Deposited by Laser Ablation 397
Y. Suda, T. Nakazono, K. Ebihara, and K. Baba

Preparation and Characterization of Wurtzitic GaN Single Crystals in Nano and Micro Scale 403
San Yu, Hongdong Li, Haibin Yang, Dongmei Li, Haiping Sun, and Guangtian Zou

Growth of Crystalline Quality SiC on Thin and Thick Silicon-on-Insulator Structures 409
F. Namavar, P. Colter, A. Cremins-Costa, C-H. Wu, E. Gagnon, D. Perry, and P. Pirouz

Growth and Characterization of Polycrystalline Diamond Thin Films on Porous Silicon by Hot Filament Chemical Vapor Deposition 415
S. Mirzakuchaki, E.J. Charlson, E.M. Charlson, T. Stacy, F. Shahedipour, and H.W. White

Reconstruction of Nonpolar GaN Surfaces 421
J. Elsner, M. Haugk, and Th. Frauenheim

Stability and Reconstruction of β-SiC (100) Surfaces 427
R. Gutierrez and Th. Frauenheim

A New, Nearly Single-Domain Surface Structure of Homoepitaxial Diamond (001) Films 433
Yalei Kuang, Naesung Lee, Andrzej Badzian, Teresa Badzian, and Tien T. Tsong

Computer Simulation of Surface Diffusion of Silicon and Carbon Adatoms on SiC(001) 439
Q.A. Bhatti, G.J. Moran, and C.C. Matthai

High Temperature Gas Phases Reactions of Trimethylgallium with Ammonia and Trimethylamine 445
A. Thon, S.A. Safvi, and T.F. Kuech

Atomic Force Microscopy Study of GaN-Buffer Layers on SiC(0001) by MOCVD 451
Dongsup Lim, Dongjin Byun, Gyeungho Kim, Ok-Hyun Nam, In-Hoon Choi, Dalkeun Park, and Dong-Wha Kum

TEM Analysis of the Observed Phases During the Growth of Oriented Diamond on Nickel Substrates 457
W. Liu, P.C. Yang, D.A. Tucker, C.A. Wolden, R.F. Davis, J.T. Glass, J.T. Prater, and Z. Sitar

Part III: DEFECTS, DOPANTS AND CHARACTERIZATION

Defects and Structural Properties

***Theory of Defects, Doping, Surfaces and Interfaces in Wide Gap Nitrides** 465
J. Bernholc, P. Boguslawski, E.L. Briggs, M. Buongiorno Nardelli, B. Chen, K. Rapcewicz, and Z. Zhang

***Structural Defects and Their Relationship to Nucleation of GaN Thin Films** 475
Weida Qian, Marek Skowronski, and Greg S. Rohrer

*Invited Paper

Effect of Si Doping on the Structure of GaN 487
Zuzanna Liliental-Weber, S. Ruvimov, T. Suski, J.W. Ager, III, W. Swider, Y. Chen, Ch. Kisielowski, J. Washburn, I. Akasaki, H. Amano, C. Kuo, and W. Imler

ESR and LESR Studies in CVD Diamond 495
C.F.O. Graeff, E. Rohrer, C.E. Nebel, M. Stutzmann, H. Güttler, and R. Zachai

Electron Paramagnetic Resonance Studies of HF-CVD Diamond Films 501
B. Ramakrishnan, D.J. Keeble, H. Rodrigo, and A. Kulkarni

Electrical and Optical Investigation of the Position of Vanadium Related Defects in the 4H and 6H SiC Bandgaps 507
J.R. Jenny, M. Skowronski, W.C. Mitchel, S.R. Smith, A.O. Evwaraye, H.M. Hobgood, G. Augustine, and R.H. Hopkins

Electron Scattering by Native Defects in III-V Nitrides and Their Alloys 513
L. Hsu and W. Walukiewicz

Characterization of Deep Level Defects in 4H and 6H SiC Via DLTS, SIMS and MeV E-Beam Irradiation 519
J.P. Doyle, M.O. Aboelfotoh, M.K. Linnarsson, B.G. Svensson, A. Schöner, N. Nordell, C. Harris, J.L. Lindström, E. Janzén, and C. Hemmingsson

Deep Level Defects in Mg-Doped GaN 525
Gyu-Chul Yi and Bruce W. Wessels

DLTS and CV Analysis of Doped and N-Implanted GaN 531
D. Haase, M. Schmid, A. Dörnen, V. Härle, H. Bolay, F. Scholz, M. Burkard, and H. Schweizer

The Relationship Between Micropipes and Screw Dislocations in PVT Grown 6H-SiC 539
Jennifer Giocondi, Gregory S. Rohrer, Marek Skowronski, V. Balakrishna, G. Augustine, H.M. Hobgood, and R.H. Hopkins

Characterization of Defect Structures in 3C-SiC Single Crystals Using Synchrotron White Beam X-ray Topography 545
W. Huang, M. Dudley, and C. Fazi

Structure of the 2H-AlN/6H-SiC Interface 551
P. Vermaut, P. Ruterana, G. Nouet, and H. Morkoc

Extrapolation of Critical Thickness of GaN Thin Films from Lattice Constant Data Using Synchrotron X-ray 557
Chinkyo Kim, I.K. Robinson, Jaemin Myoung, Kyuhwan Shim, Kyekyoon Kim, and Myung-Cheol Yoo

***Ex Situ* and *In Situ* Methods for Complete Oxygen and Non-Carbidic Carbon Removal from $(0001)_{Si}$ 6H-SiC Surfaces** 563
Sean W. King, Mark C. Benjamin, Richard S. Kern, Robert J. Nemanich, and Robert F. Davis

High Temperature Surface Degradation of III-V Nitrides 569
C.B. Vartuli, S.J. Pearton, C.R. Abernathy, J.D. MacKenzie, J.C. Zolper, and E.S. Lambers

Morphology of Optically Transparent Cubic Silicon Carbide Prepared by Chemical Vapor Deposition 575
Michael W. Russell, Jaime A. Freitas, Jr., and James E. Butler

Doping and Electrical Properties

Dopant Incorporation Efficiencies of SiC Crystals Grown on {1100}-Face 583
Naohiro Sugiyama, Atsuto Okamoto, Toshihiko Tani, and Nobuo Kamiya

Low Volume Resistivity Chemical Vapor Deposited Boron Doped Polycrystalline Thin Diamond Film Growth on Sapphire 589
Hassan Golestanian, S. Mirzakuchaki, E.J. Charlson, T. Stacy, and E.M. Charlson

Activation of Acceptors in Mg-Doped, p-Type GaN 595
W. Götz, N.M. Johnson, J. Walker, and D.P. Bour

Electrical Characterization of Magnesium-Doped Gallium Nitride Grown by Metalorganic Vapor Phase Epitaxy 601
J.W. Huang, H. Lu, J.G. Cederberg, I. Bhat, and T.F. Kuech

The Influence of Hydrogen Plasma Passivation on Electrical and Optical Properties of AlGaN Samples Grown on Sapphire 607
A.Y. Polyakov, M. Shin, S.J. Pearton, M. Skowronski, D.W. Greve, and J.A. Freitas

Theoretical Study of Hydrogen in Cubic GaN 613
Stefan K. Estreicher and Djordje M. Maric

Role of Hydrogen and Hydrogen Complexes in Doping of GaN 619
Jörg Neugebauer and Chris G. Van de Walle

Diffusion of Hydrogen in 6H Silicon Carbide 625
M.K. Linnarsson, J.P. Doyle, and B.G. Svensson

Electronic Structure of Beryllium, Magnesium and Silicon Impurity in Cubic Gallium Nitride 631
V.A. Gubanov, E. Pentaleri, C.Y. Fong, and B.M. Klein

Impurity Conduction in n-Type 4H-SiC 637
A.O. Evwaraye, S.R. Smith, W.C. Mitchel, and M.D. Roth

Factors Influencing the Electrical and Optical Properties of AlGaN Layers on Sapphire 643
M. Shin, A.Y. Polyakov, M. Skowronski, D.W. Greve, R.G. Wilson, and J.A. Freitas

Influence of Forced Diffusion of Boron on Electrical Conductivity of Diamond Films 649
T. Sung, G. Popovici, M.A. Prelas, R.G. Wilson, Kim Bigelow, J. Chacon, and S.K. Loyalka

Measurement of the Activation Energy in Phosphorous Doped Polycrystalline Diamond Thin Films Grown on Silicon Substrates by Hot Filament Chemical Vapor Deposition 655
S. Mirzakuchaki, H. Golestanian, E.J. Charlson, and and T. Stacy

Theory and Realization of Two-Layer Hall Effect Measurement Concept for Characterization of Epitaxial and Implanted Layers of SiC 661
Adolf Schöner, Kurt Rottner, and Nils Nordell

Formation and High Frequency CV-Measurements of Aluminum/Aluminum Nitride/6H Silicon Carbide Structures 667
C.-M. Zetterling, K. Wongchotigul, M.G. Spencer, C.I. Harris, S.S. Wong, and M. Östling

Optical and Field Emission Properties

Electronic Structure and Temperature Dependence of Excitons in GaN 675
I.A. Buyanova, J.P. Bergman, W. Li, B. Monemar, H. Amano, and I. Akasaki

Optical Properties and Morphology of GaN Grown by MBE on Sapphire Substrates 681
E. Tuncel, D.B. Oberman, H. Lee, T. Ueda, and J.S. Harris, Jr.

A Study of Electroluminescent Emission from CVD Diamond 687
C. Manfredotti, F. Fizzotti, E. Vittone, P. Polesello, F. Wang, R. Schwarz, and S. Grebner

Cathodoluminescence Studies of Bound Excitons and Near Band Gap Emission Lines in Boron-and Phosphorus-Doped CVD-Diamonds 693
H. Sternschulte, T. Albrecht, K. Thonke, R. Sauer, M. Grießer, and M. Grasserbauer

Characterisation of Diamond-Like Carbon by Raman Spectroscopy and Optical Constants 699
C. Moßner, P. Grant, H. Tran, G. Clarke, D.J. Lockwood, H.J. Labbé, B. Mason, and R. Berriche

Thermal and Optical Admittance Spectroscopy Studies of Defects in 15R-SiC 705
S.R. Smith, A.O. Evwaraye, and W.C. Mitchel

Local-Field and Exchange-Correlation Effects in Optical Spectra of Wide-Band-Gap Semiconductors 711
V.I. Gavrilenko and F. Bechstedt

Photoluminescence Properties of CVD Diamond Excited by Ultraviolet Synchrotron Radiation 717
Jaihyung Won, Akimitsu Hatta, Toshimichi Ito, Takatomo Sasaki, and Akio Hiraki

Surface Chemical Effects on the Optical Properties of Thin Nanocrystalline Diamond Films 723
A.V. Khomich, V.I. Polyakov, P.I. Perov, V.P. Varnin, I.G. Teremetskaya, V.G. Balakirev, and E.D. Obraztsova

Raman Spectroscopy Investigation of $(SiC)_{1-x}(AlN)_x$ Layers Formed by Ion Implantation in 6H-SiC 729
D.R.T. Zahn, T. Werninghaus, M. Thümer, J. Pezoldt, and V. Heera

Photoreflectance Study of GaN Film Grown by Metalorganic Chemical Vapor Deposition 735
K. Yang, R. Zhang, Y.D. Zheng, L.H. Qin, B. Shen, H.T. Shi, Z.C. Huang, and J.C. Chen

Thermomodulated Reflectivity Spectra of GaN/Sapphire Epilayer 741
Y. Li, Y. Lu, H. Shen, M. Wraback, M. Schurman, L. Koszi, and R.A. Stall

Optical Characterization of GaN Films Grown on (0001) Sapphire Substrate 747
K. Yang, R. Zhang, Y.D. Zheng, L.H. Qin, B. Shen, H.T. Shi, Z.C. Huang, and J.C. Chen

Optical and Structural Properties of α-$Si_{1-x}C_x$ Films 753
Zhizhong Chen, Kai Yang, Rong Zhang, Hongtao Shi, and Youdou Zheng

Surface Photovoltage Effects in Photoemission from Diamond Surfaces 759
C. Bandis and B.B. Pate

Investigation of the Field Emission Current from Polycrystalline Diamond Films 765
J.W. Glesener and A.A. Morrish

Field Emission and Band Bending Considerations from High-Quality NEA Diamond 771
C. Bandis, B.B. Pate, W. Phillips, M.A. Plano, M.D. Moyer, and M.A. Moreno

Electron Field Emission from Diamond-Like Carbon 777
J. Robertson, S.R.P. Silva, G.A.J. Amaratunga, and W.I. Milne

Author Index 783

Subject Index 789

PREFACE

This volume contains papers presented at the 1996 MRS Spring Meeting during the symposium on "III-Nitride, SiC and Diamond Materials For Electronic Devices." The meeting was held in San Francisco, California, from April 8-12. The symposium involved over 140 papers that included invited presentations and contributed oral, poster and late news posters, as well as a panel discussion held mid-week. This symposium differed from several previous MRS symposia on wide bandgap semiconductors in that the emphasis was specifically on materials aspects related to electronic properties and devices. The proceedings volume is organized much as the meeting, but with poster and oral presentations mixed according to the session topics. Solid advances were reported in the growth techniques of all three materials groups. Contributions demonstrated the critical importance of surfaces, interfaces, doping, defects, and impurities. Reports showed potential device applications ranging from new high-frequency, high-power all solid-state devices to unique cold cathode electronic devices. While the results presented here demonstrate real promise for a wide range of new solid-state devices that are not feasible with current production materials, substantial progress in materials research is necessary to fulfill the real potential of these applications. A lively panel discussion was held in the middle of the conference which focused on several critical issues related to the electronic potential of the three materials.

We gratefully acknowledge Rockwell International, Kobe Steel, USA and Northrop Grumman for financial support of the panel discussion. We wish to thank all of the contributors and participants who made the symposium so successful. We appreciate the efforts of the invited speakers for setting the stage for the program. We also thank the sessions chairs for coordinating and leading the discussion, and we express our gratitude to the panel moderator and members for stimulating such a lively session. We appreciate the efforts of the meeting chairs, Thomas F. Kuech, Clifford L. Renschler, and Chuang Chuang Tsai, and the MRS staff for their efforts in making the whole meeting a success. We gratefully acknowledge the financial support provided by the following organizations: Defense Advanced Research Projects Agency, Air Force Wright Laboratory, Army Research Office, and Office of Naval Research.

D. Kurt Gaskill
Charles D. Brandt
Robert J. Nemanich

June, 1996

Materials Research Society Symposium Proceedings

Volume 395— Gallium Nitride and Related Materials, F.A. Ponce, R.D. Dupuis, S.J. Nakamura, J.A. Edmond, 1996, ISBN: 1-55899-298-7

Volume 396— Ion-Solid Interactions for Materials Modification and Processing, D.B. Poker, D. Ila, Y-T. Cheng, L.R. Harriott, T.W. Sigmon, 1996, ISBN: 1-55899-299-5

Volume 397— Advanced Laser Processing of Materials—Fundamentals and Applications, R. Singh, D. Norton, L. Laude, J. Narayan, J. Cheung, 1996, ISBN: 1-55899-300-2

Volume 398— Thermodynamics and Kinetics of Phase Transformations, J.S. Im, B. Park, A.L. Greer, G.B. Stephenson, 1996, ISBN: 1-55899-301-0

Volume 399— Evolution of Epitaxial Structure and Morphology, A. Zangwill, D. Jesson, D. Chambliss, R. Clarke, 1996, ISBN: 1-55899-302-9

Volume 400— Metastable Phases and Microstructures, R. Bormann, G. Mazzone, R.D. Shull, R.S. Averback, R.F. Ziolo, 1996 ISBN: 1-55899-303-7

Volume 401— Epitaxial Oxide Thin Films II, J.S. Speck, D.K. Fork, R.M. Wolf, T. Shiosaki, 1996, ISBN: 1-55899-304-5

Volume 402— Silicide Thin Films—Fabrication, Properties, and Applications, R. Tung, K. Maex, P.W. Pellegrini, L.H. Allen, 1996, ISBN: 1-55899-305-3

Volume 403— Polycrystalline Thin Films: Structure, Texture, Properties, and Applications II, H.J. Frost, M.A. Parker, C.A. Ross, E.A. Holm, 1996, ISBN: 1-55899-306-1

Volume 404— *In Situ* Electron and Tunneling Microscopy of Dynamic Processes, R. Sharma, P.L. Gai, M. Gajdardziska-Josifovska, R. Sinclair, L.J. Whitman, 1996, ISBN: 1-55899-307-X

Volume 405— Surface/Interface and Stress Effects in Electronic Materials Nanostructures, S.M. Prokes, R.C. Cammarata, K.L. Wang, A. Christou, 1996, ISBN: 1-55899-308-8

Volume 406— Diagnostic Techniques for Semiconductor Materials Processing II, S.W. Pang, O.J. Glembocki, F.H. Pollack, F.G. Celii, C.M. Sotomayor Torres, 1996, ISBN 1-55899-309-6

Volume 407— Disordered Materials and Interfaces, H.Z. Cummins, D.J. Durian, D.L. Johnson, H.E. Stanley, 1996, ISBN: 1-55899-310-X

Volume 408— Materials Theory, Simulations, and Parallel Algorithms, E. Kaxiras, J. Joannopoulos, P. Vashishta, R.K. Kalia, 1996, ISBN: 1-55899-311-8

Volume 409— Fracture—Instability Dynamics, Scaling, and Ductile/Brittle Behavior, R.L. Blumberg Selinger, J.J. Mecholsky, A.E. Carlsson, E.R. Fuller, Jr., 1996, ISBN: 1-55899-312-6

Volume 410— Covalent Ceramics III—Science and Technology of Non-Oxides, A.F. Hepp, P.N. Kumta, J.J. Sullivan, G.S. Fischman, A.E. Kaloyeros, 1996, ISBN: 1-55899-313-4

Volume 411— Electrically Based Microstructural Characterization, R.A. Gerhardt, S.R. Taylor, E.J. Garboczi, 1996, ISBN: 155899-314-2

Volume 412— Scientific Basis for Nuclear Waste Management XIX, W.M. Murphy, D.A. Knecht, 1996, ISBN: 1-55899-315-0

Volume 413— Electrical, Optical, and Magnetic Properties of Organic Solid State Materials III, A.K-Y. Jen, C.Y-C. Lee, L.R. Dalton, M.F. Rubner, G.E. Wnek, L.Y. Chiang, 1996, ISBN: 1-55899-316-9

Volume 414— Thin Films and Surfaces for Bioactivity and Biomedical Applications, C.M. Cotell, A.E. Meyer, S.M. Gorbatkin, G.L. Grobe, III, 1996, ISBN: 1-55899-317-7

Volume 415— Metal-Organic Chemical Vapor Deposition of Electronic Ceramics II, S.B. Desu, D.B. Beach, P.C. Van Buskirk, 1996, ISBN: 1-55899-318-5

MATERIALS RESEARCH SOCIETY SYMPOSIUM PROCEEDINGS

Volume 416— Diamond for Electronic Applications, D. Dreifus, A. Collins, T. Humphreys, K. Das, P. Pehrsson, 1996, ISBN: 1-55899-319-3

Volume 417— Optoelectronic Materials: Ordering, Composition Modulation, and Self-Assembled Structures, E.D. Jones, A. Mascarenhas, P. Petroff, R. Bhat, 1996, ISBN: 1-55899-320-7

Volume 418— Decomposition, Combustion, and Detonation Chemistry of Energetic Materials, T.B. Brill, T.P. Russell, W.C. Tao, R.B. Wardle, 1996 ISBN: 1-55899-321-5

Volume 420— Amorphous Silicon Technology—1996, M. Hack, E.A. Schiff, S. Wagner, R. Schropp, M. Matsuda, 1996, ISBN: 1-55899-323-1

Volume 421— Compound Semiconductor Electronics and Photonics, R.J. Shul, S.J. Pearton, F. Ren, C.-S. Wu, 1996, ISBN: 1-55899-324-X

Volume 422— Rare Earth Doped Semiconductors II, S. Coffa, A. Polman, R.N. Schwartz, 1996, ISBN: 1-55899-325-8

Volume 423— III-Nitride, SiC, and Diamond Materials for Electronic Devices, D.K. Gaskill, C. Brandt, R.J. Nemanich, 1996, ISBN: 1-55899-326-6

Volume 424— Flat Panel Display Materials II, M. Hatalis, J. Kanicki, C.J. Summers, F. Funada, 1996, ISBN: 1-55899-327-4

Volume 425— Liquid Crystals for Advanced Technologies, T.J. Bunning, S.H. Chen, W. Hawthorne, N. Koide, T. Kajiyama, 1996, ISBN: 1-55899-328-2

Volume 426— Thin Films for Photovoltaic and Related Device Applications, D. Ginley, A. Catalano, H.W. Schock, C. Eberspacher, T.M. Peterson, T. Wada, 1996, ISBN: 1-55899-329-0

Volume 427— Advanced Metallization for Future ULSI, K.N. Tu, J.W. Mayer, J.M. Poate, L.J. Chen, 1996, ISBN: 1-5899-330-4

Volume 428— Materials Reliability in Microelectronics VI, W.F. Filter, J.J. Clement, A.S. Oates, R. Rosenberg, P.M. Lenahan, 1996, ISBN: 1-55899-331-2

Volume 429— Rapid Thermal and Integrated Processing V, J.C. Gelpey, M. Öztürk, R.P.S. Thakur, A.T. Fiory, F. Roozeboom, 1996, ISBN: 1-55899-332-0

Volume 430— Microwave Processing of Materials V, M.F. Iskander, J.O. Kiggans, E.R. Peterson, J.Ch. Bolomey, 1996, ISBN: 1-55899-333-9

Volume 431— Microporous and Macroporous Materials, R.F. Lobo, J.S. Beck, S. Suib, D.R. Corbin, M.E. Davis, L.E. Iton, S.I. Zones, 1996, ISBN: 1-55899-334-7

Volume 432— Aqueous Chemistry and Geochemistry of Oxides, Oxyhydroxides, and Related Materials J.A. Voight, B.C. Bunker, W.H. Casey, T.E. Wood, L.J. Crossey, 1996, ISBN: 1-55899-335-5

Volume 433— Ferroelectric Thin Films V, S.B. Desu, R. Ramesh, B.A. Tuttle, R.E. Jones, I.K. Yoo, 1996, ISBN: 1-55899-336-3

Volume 434— Layered Materials for Structural Applications, J.J. Lewandowski, C.H. Ward, W.H. Hunt, Jr., M.R. Jackson, 1996, ISBN: 1-55899-337-1

Volume 435— Better Ceramics Through Chemistry VII—Organic/Inorganic Hybrid Materials, B. Coltrain, C. Sanchez, D.W. Schaefer, G.L. Wilkes, 1996, ISBN: 1-55899-338-X

Volume 436— Thin Films: Stresses and Mechanical Properties VI, W.W. Gerberich, H. Gao, J-E. Sundgren, S.P. Baker 1996, ISBN: 1-55899-339-8

Volume 437— Applications of Synchrotron Radiation to Materials Science III, L. Terminello, S. Mini, D.L. Perry, H. Ade, 1996, ISBN: 1-55899-340-1

Prior Materials Research Society Symposium Proceedings available by contacting Materials Research Society

Part I

Device Technologies

Devices

SiC Semiconductor Applications - An Air Force Perspective

L.S. Rea
U.S. Air Force Wright Laboratory Materials Directorate
3005 P St. Ste. 6, WPAFB, Ohio 45433-7707, reals@ml.wpafb.af.mil

ABSTRACT

The Department of Defense (DoD) is investing in the development of Silicon Carbide (SiC) for a wide range of applications. Over the past year, SiC technology has demonstrated excellent device performance results for power devices, high temperature electronic devices and microwave devices. The materials growth and processing technology for SiC is now at a level of sufficient maturity to support substantial device development efforts. While there is still considerable materials and device research required for SiC to achieve it's full potential, the fundamental technology has been proven for several critical applications. A perspective on some Air Force device performance requirements will be presented. The status of SiC materials development, material limits to advances in device performance and issues relating to supporting technology will also be discussed.

INTRODUCTION

The U.S. Air Force is developing advanced semiconductor materials to meet requirements for next generation systems. "Conventional" semiconductors, such as silicon and gallium arsenide, won't be able to provide performance for many high temperature, high power applications. SiC-based electronic devices are being developed to address some of these military applications. Much work is being performed and sponsored by other U.S. government agencies, including DARPA, the Navy, the Army and NASA. Many of the applications for SiC discussed here are quite similar to applications being pursued by these other agencies and services. This paper will focus, however, on Air Force applications.

Three primary system application areas will be presented: Power Components, High Temperature Sensors and High Frequency Applications. There are other technical areas where SiC (and other wide bandgap semiconductors) offer advantages, such as for UV sensors and optoelectronics, but the majority of the development efforts within the Air Force are presently focused on these three primary applications. Each will be discussed in turn.

Other wide bandgap semiconductors are beginning to show great promise for dramatic improvements in device performance. There has been a tremendous increase in the research efforts on gallium nitride-based devices. These materials have already demonstrated superior

Mat. Res. Soc. Symp. Proc. Vol. 423

optical performance over SiC , and will possibly push the performance envelope of RF devices beyond that obtainable with SiC. These materials are, however, much less mature than SiC. For some relatively near-term applications needed by the Air Force, SiC technology is becoming mature enough to be considered as a candidate for these requirements. For non-optoelectronic applications, the electronic and thermal properties of SiC are sufficient to meet many device performance requirements. The mobility and breakdown field strengths are high enough, and most importantly, SiC substrates are available. This will allow SiC devices (on SiC substrates) to perform closer to theoretical limits than devices built by heteroepitaxy. While there are still further improvements required in the materials technology, the level of technical risk has decreased to where it is reasonable to pursue some specific system insertions.

POWER COMPONENTS

Very high power, high temperature devices will enable development of the "More Electric Aircraft". This is part of a tri-service, multi-agency "More Electric Initiative" [1]. The goals of this initiative are to reduce the total system weight, while increasing reliability and performance. Reduction (or elimination) of active cooling and hydraulic actuation, as well as a reduction in part counts are planned. To accomplish these goals, very high power, high temperature electronics will be required to support the power management and distribution system.

One great advantage of SiC for power components is in efficiency. SiC device efficiency has been modeled at over 99% [2]. The best performance for a silicon device will be at 95% or less. This increase in efficiency will result in a five fold reduction in electronics heat load. This will enable the reduction or complete elimination of the cooling systems required for the Power Management and Distribution (PMAD) system, flight actuator, stores system and the engine control electronics [1].

To accomplish this level of performance, devices must possess the following characteristics: high voltage (600-1000 V) with high conductance, low switching loss and high current density (1000 A/cm^2). These devices must operate up to 350 °C and must be highly reliable. For electric motor and generator control applications, motor controller devices operational to 350°C are required. These include highly efficient inverter/converter circuits (95-98% efficient) and 50-200 hp electric drives. To handle the power, solid state circuit breakers (greater than 1000 V) must also be developed.

The most critical materials limitation for power device development is the requirement for large area (greater than 1 cm^2), low (zero) defect density bulk and epitaxial materials. Since current density scales with device area, the only practical means to achieve the high current densities required is to produce larger devices. This doesn't mean that large diameter, defect-free wafers are required, only that wafers exhibiting large, defect-free regions must exist. Large, defect-free wafers will be necessary to improve device yields as the device production technology matures. Dopant uniformity is also an issue. Another area requiring improvement relates to the oxide/dielectric layers. These layers must exhibit both high temperature reliability (at 350°C) and high breakdown field strength (greater than 10 MV/cm). Low resistance contacts, especially p-type, with high temperature stability are also required.

HIGH TEMPERATURE SENSORS

Distributed electronics, which can be considered a "local engine network" [3] are being developed to replace hydraulic actuation. Again, this will lead to reduced system weight, complexity and maintenance. Very high temperature sensors, capable of performing some signal conversion functions, are required to accomplish this goal. In particular, pressure, temperature and position sensors capable of performing A/D, D/A and low signal amplification at the sensor must be produced and integrated into the aircraft control system. These components must function reliably at temperatures above 350°C. Some specific near-engine sensor requirements call for operation at temperatures near 650°C. High temperature packaging and stable metallizations for these temperature ranges will also be necessary.

The limitations being encountered in the development of these devices, namely undesirable threshold voltage variations and inadequate mobilities, relate primarily to epitaxial materials growth. As reported by Krishnamurthy et al [4], undesired aluminum (Al) contamination from p-type substrates (or epitaxial layers) into an n-type SiC layer can result in threshold voltage variations. In addition, the electron mobility of the n-type layers are also reduced by the presence of Al [4]. Stable p-type dopants, possibly alternatives to Al, with large area uniformity will be necessary. Also, dopant predictability and reproducibility must be achieved before devices can be produced in quantity. Other materials limitations include the reliability of the oxide layers required for many of the device designs currently under development.

HIGH FREQUENCY, HIGH POWER DENSITY RF

Efforts are underway to replace vacuum tubes with solid state components. In general, moving to solid state components (based on GaAs and InP) will improve system safety and maintainability. SiC-based solid state devices offer the possibility of further size and weight reductions due to higher power densities, reduced cooling requirements and higher temperature and voltage operation. Specific RF and microwave system requirements potentially benefited by SiC device development involve frequencies between 200 MHz and 12 GHz (UHF through X-Band). Potential applications are in the areas of communications, radar (airborne and ground-based) and electronic warfare [5]. Power requirements for these devices are for 50-1000 W (pulsed) and at least 200 W (continuous wave, CW). High efficiency (power added efficiency, PAE, greater than 40%) and high linearity are also essential.

Materials limitations which are slowing device development include small substrate diameters, high defect levels and unstable electronic properties. Cost effective, production-worthy devices will require substrates at least three inches in diameter, with fewer than 10 micropipes/cm^2, and less than 1000 dislocations/cm^2. Uniform, stable, semi-insulating material, with resistivities greater than 1E6 ohm-cm must also be available, in particular, to support the development of high power MESFET structures. Other requirements for successful SiC RF products include uniform, large area epitaxial growth, ion implantation to enable planar circuit development and stable p-type contacts.

SUMMARY

While examples of SiC power devices, high temperature sensors and RF-devices have been successfully demonstrated, there are still fundamental materials problems which must be resolved for this technology to reach its full potential. The Air Force will continue to emphasize uniform epitaxial growth, stable p-type dopants and process control. Bulk substrate materials are becoming consistently better, but larger diameters are necessary to assure technological acceptance by the device community. Other materials requirements include research on oxides/dielectrics to produce low interface state densities (SiC/dielectric interface) and high temperature reliability, and research on metal contacts which remain stable at high temperatures and current densities.

ACKNOWLEDGMENTS

The author gratefully acknowledges the advice, guidance and technical contributions of John King, Bill Mitchel, Steve Przybylko and Kitt Reinhardt from the Air Force Wright Laboratory supporting the preparation of this manuscript, and the patience and guidance of Kurt Gaskill of the Naval Research Laboratory in the final preparation of this paper.

REFERENCES

1. K. C. Reinhardt, J. D. Scofield and W.C. Mitchel, in Proceedings from the Workshop on High Temperature Electronics for Vehicles, edited by G. Khalil, H. Singh and T. Podlesak (Army Research Laboratory Technical Report, April, 1995) pp. 73-79.

2. K. C. Reinhardt, private communication.

3. S. J. Przybylko in the Proceedings from the 29th Joint Propulsion Conference and Exhibit (AIAA/SAE/ASME/ASEE, June, 1993), pp. 1-9.

4. V. Krishnamurthy, D.M. Brown, E. Downey, J. Kretchmer and D. Larkin, "Characterization of 6H SiC Epitaxial Layers Grown Using Site-Competition Epitaxy", to be published in Transactions, 3rd High Temperature Electronics Conference, (June, 1996) Albuquerque, NM.

5. J. King, private communication.

SiC POWER DEVICES

T.P. Chow* and M. Ghezzo**

*Rensselaer Polytechnic Institute, Troy, NY 12180-3590, chow@unix.cie.rpi.edu
**General Electric Corporate Research and Development, Schenectady, NY 12301, ghezzo@crdgw2.crd.ge.com

ABSTRACT

The present status of SiC high-voltage power switching devices is reviewed. The figures of merits that have been used for unipolar and bipolar devices to quantify the intrinsic performance improvement over silicon are presented. Analytical and numerical modeling and simulations to estimate the BV and device choice are described. The active area and termination design of trenched-gate MOS power transistors, together with an integrated process for their fabrication, is presented. The progress in high-voltage power device experimental demonstration is described. The material and process technology issues that need to be addressed for SiC device commercialization are discussed. Finally, the impact of SiC power devices on motor drive systems is estimated.

INTRODUCTION

While silicon is the dominant semiconductor of choice for high-voltage power device applications, wide bandgap semiconductors, particularly SiC, have attracted much attention recently because they are projected to have much better performance than silicon [1-5]. Compared to silicon, these wide bandgap semiconductors, SiC and GaInN can be categorized into one group while diamond and AlN into another because the former has bandgaps of 2-3.5 eV and the latter 5.5-6.5 eV. On the other hand, the Group IV or IV-IV semiconductors have indirect bandgaps whereas the Group III-nitrides are direct. In particular, various polytypes of SiC has superior physical properties which offer a wider energy bandgap (2 to 3 times), a higher electric breakdown field (8-10 times), a higher thermal conductivity (3 times), a larger saturated electron drift velocity (2-2.5 times), when compared to silicon (See Table I). Hence, SiC power devices can operate at a higher temperature, a higher frequency, and a lower power loss, together with a less stringent heat sinking requirement. In this paper, the unipolar and bipolar figures of merit for power devices will be described. The analytical and numerical modeling analyses used for SiC are presented. The design of MOS SiC device structures, including termination, and processes will then be discussed. Also, the outstanding material and processing issues that need to be overcome to commercial SiC devices will be pointed out. Finally, the impact of SiC technology into motor drive systems on efficiency will be estimated.

FIGURES OF MERIT

To quantify the intrinsic improvement of replacing silicon with wide bandgap semiconductors, several different figures of merit (FM), which are proposed specifically for unipolar and bipolar high-voltage power devices [1,2,4,5], have been used. It should be noted that the avalanche electric field is the most important physical material parameter in all of these figures of merit, followed by thermal conductivity and carrier mobility. The unipolar figures of merit for the major wide bandgap semiconductors is shown in Table II. A low frequency FM (like BM [1]) shows a

Mat. Res. Soc. Symp. Proc. Vol. 423 © 1996 Materials Research Society

Table I: Physical properties of important semiconductors for high-voltage power devices.

Material	E_g eV	ε_r	μ_n cm^2/V.s	μ_p cm^2/V.s	E_c 10^6 V/cm	v_{sat} 10^7 V/cm	λ W/cm	n_i cm^{-3}
Si	1.1	11.8	1350	450	0.3	1.0	1.5	1.5×10^{10}
Ge	0.66	16.0	3900	1900	0.1	0.5	0.6	2.4×10^{13}
GaAs	1.4	12.8	8500	400	0.4	2.0	0.5	1.8×10^{6}
GaN	3.39	9.0	900	50	3.3	2.5	1.3	1.9×10^{-10}
3C-SiC	2.2	9.6	100	50	1.2	2.0	4.5	6.9
4H-SiC	3.26	10	720^a 650^c	50	2.0	2.0	4.5	8.2×10^{-9}
6H-SiC	2.86	9.7	370^a 50^c	50	2.4	2.0	4.5	2.4×10^{-5}
diamond	5.45	5.5	1900	1500	5.6	2.7	20	2.4×10^{-27}

Note: *a* — mobility along a-axis, *c* — mobility along c axis

Table II: Normalized unipolar figures of merit of important semiconductors for high-voltage power devices[4].

Material	JM	KM	Q_{F1}	Q_{F2}	BM (Q_{F3})	BHFM
Si	1	1	1	1	1	1
Ge	0.03	0.2	0.06	0.02	0.2	0.3
GaAs	7.1	0.5	5.2	7	16	11
GaN	756	1.6	563	6220	653	78
3C-SiC	64	1.6	112	446	37	11
4H-SiC	178	4.6	387^a 350^c	2580^a 2330^c	129^a 117^c	23^a 21^c
6H-SiC	256	4.7	334^a 45^c	2670^a 360^c	111^a 15^c	17^a 2^c
diamond	2540	32	54860	1024000	4110	470

Note: *a* — values calculated using mobility along a-axis, *c* — values calculated using mobility along c axis

10-100 times improvement over silicon due to the fact that a reduced drift length needed to support the required voltage rating results in a lower ON-resistance. By contrast, a better high-frequency FM (like BHFM) is due to a reduced capacitance from decreased active area [2]. Comparison of wide bandgap power devices with silicon counterparts is complicated by the fact that these devices can be grouped into two distinct groups - those having an odd number of junctions and those having an even number. Junction rectifiers and thyristors belong to the former group while the bipolar junction transistor belongs to the latter. Consequently, in the ON state, junction rectifiers and thyristors have a turn-on ("knee") voltage which is about 70% of the semiconductor bandgap but the BJT does not have such a turn-on knee due to junction drop cancellation between the emitter-base and base-collector junctions in saturation. While the second group of devices can use the unipolar figures of merit, for the first group, the high turn-on voltage (1.5 to 2.8V) for the wide bandgap semiconductors results in a higher forward drop, when compared to silicon (with a 0.7V turn-on voltage), for low-frequency applications where conduction loss dominates. However, with increasing switching frequency, the reduced lifetime in the drift layer to improve the switching power loss causes a rapid increasing forward drop.

Hence, a bipolar figure of merit using the frequency (f_{min}) at which the total power loss silicon device becomes equal to that of the wide bandgap semiconductors has been defined [5]. Another point that needs to be pointed out is that only indirect semiconductors are considered because only their minority carrier lifetimes are sufficiently long (>1μs) to enable conductivity modulation of the drift layer. Tables III and IV shows the various power loss components and this FM for a 1000V pin junction rectifier and IGBT (a dominant MOS-controlled bipolar power transistor), respectively, and we can see that SiC polytypes and diamond have a lower total power loss when the switching frequencies exceed 30 kHz [5]. Similar improvement has been estimated for thyristor and GTO [5].

Table III: A bipolar figure of merit applied to PiN rectifier (Calculated at $J_F = 100\,A/\text{cm}^2$, $V_{BR} = 1000\text{V}$, $\tau_{n0} = 1.0\mu s$, $\tau_{p0} = 0.1\mu s$) [5].

Name	N_d (cm^{-3})	W_{N^-} (μm)	V_F (V)	J_{end} (A/cm^2)	J_{off} (A/cm^2)	t_s (ms)	t_{f1} (s)	t_{f2} (ms)	t_{tot} (s)	E_{off} (mJ)	f_{min} (kHz)
Si	1.3×10^{14}	100	0.88	9.7	2×10^{-5}	7	0.81	0	0.82	54	-
Ge	4.4×10^{13}	200	0.47	8.3	8×10^{-2}	7.6	1.8	0	1.8	62	0
3C-SiC	3.8×10^{15}	16.7	1.98	42.8	2×10^{-15}	4.4	0.2	2.2	0.2	20	1.62
6H-SiC	1.6×10^{16}	8.33	2.65	46.2	3×10^{-21}	10.0	0.14	4.6	0.14	5.5	1.82
Diam	1.2×10^{17}	2.3	5.24	93.4	5×10^{-44}	2.6	0.02	10	0.029	1.1	4.12

Table IV: A bipolar figure of merit applied to IGBT (Calculated at $J_F = 100\,A/\text{cm}^2$, $V_{BR} = 1000\text{V}$) [5].

Name	N_d (cm^{-3})	W_{N^-} (μm)	τ_{n0} (μs)	V_F	α_{PNP}	W_R (μm)	J_{off} (A/cm^2)	τ_B (ms)	E_{off} (mJ)	f_{min} (kHz)
Si	1.3×10^{14}	100	1.0	1.2	0.14	6.0	2.0×10^{-5}	0.285	2.55	-
Ge	4.4×10^{13}	200	0.95	0.63	0.14	12.0	8.5×10^{-2}	0.302	2.70	<190
3C-SiC	3.8×10^{15}	16.7	0.37	2.74	0.14	1.0	5.4×10^{-15}	0.058	0.55	38.4
6H-SiC	1.6×10^{16}	8.33	0.15	2.97	0.14	0.5	1.8×10^{-20}	0.016	0.15	36.9
Diam	1.2×10^{17}	2.3	0.0011	5.04	0.14	0.14	5.0×10^{-44}	6.0×10^{-4}	5.0×10^{-4}	75.3

Modeling and Simulations

Analytical modeling and numerical simulations have been used for SiC device performance projection. The material parameters and their variation with temperature used in these simulations have been carefully assessed and extracted from available experimental data. These include impact ionization coefficients, bandgap energy, carrier mobility and dopant energy levels. In particular, the bandgap variation with temperature of 6H-SiC is obtained by a two-parameter fit to experimental data. Impact ionization coefficients of electrons and holes (α_n and α_p respectively) as a function of electric field were also estimated by using the conventional two-parameter fit of the best available experimental data [6]. To simplify our analysis and obtain an approximate design curve, we have approximated [7] the ionization coefficients by a single ionization coefficient using a power law (following Fulop's approach used for Si [8]) and obtain the ideal breakdown voltage (BV_{pp}) and depletion layer width at breakdown (W_{pp}) as a function of background doping as shown in Fig. 1. For example, at a doping concentration of 10^{16} cm^{-3}, a BV_{pp} of 2810 V and a W_{pp} of 17.4 μm were obtained, compared to a BV_{pp} of 2700V obtained using the exact α's. Furthermore, the effective avalanche field estimated for doping concentration of 10^{15} to 10^{17} cm^{-3} is the range 2.5 x 10^6 V/cm, close to the experimental value of 3 x 10^6 V/cm. Also, in contrast to silicon, the breakdown voltage of 6H-SiC is expected to decrease when temperature increases [9]. While this analytical curve is very useful in estimating the drift length required, more accurate two- or three- dimensional numerical simulations solving the continuity equations are used to assess the device performance. However, even these simulations are only approximate due to a lack of basic material parameters (like Auger coefficients) and their orientation and polytype dependence for SiC.

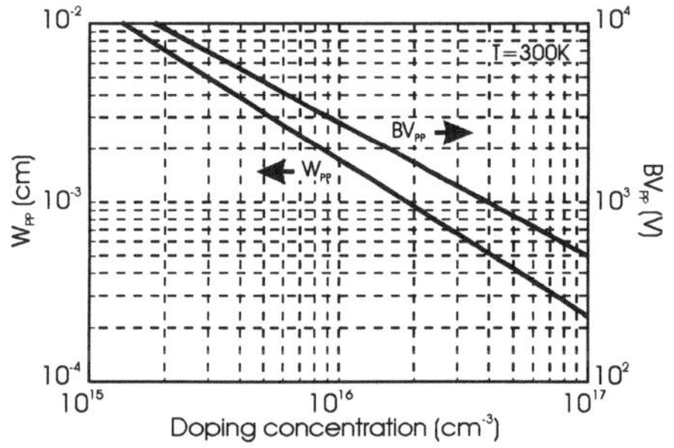

Figure 1. Breakdown voltage of parallel-plane, one-sided abrupt junction (BVpp) and its depletion layer width at breakdown (Wpp) for 6H-SiC at 300K [7].

The choice of whether a unipolar device (like a MOSFET) or a bipolar device (like an IGBT) should be used for a particular application depends on the blocking voltage and operating temperature. For silicon, for blocking voltage above 300V, the IGBT is preferred over the power MOSFET due to a lower drift layer resistance resulting from conductivity modulation. For SiC, the reduction in ON-resistance makes IGBT uncompetitive below 1000V. However, at higher blocking voltages or operating temperatures or in applications where a high overcurrent (sometimes as high as 10 times the ON-state current) condition may exist, the IGBT can perform better than the power MOSFET. More on this will be discussed in the next section.

DEVICE STRUCTURES AND PROCESSES

Active Area Design

Silicon power MOS devices almost exclusively use a Double-diffused MOS (DMOS) structure in which the channel is formed between the difference in junction depths of the p-body and n+ source along the MOS surface. DMOS in SiC is difficult due to slow diffusion of most n- and p-type dopants at or below 1200°C. By contrast, the UMOS in which the channel p body layer can be grown epitaxially after the growth of the n drift layer can be formed by subsequent n+ source implantation, vertical trench etching and gate oxidation. Since the SiC UMOS structure can be realized with present process technology, its use in high-voltage power MOSFET and

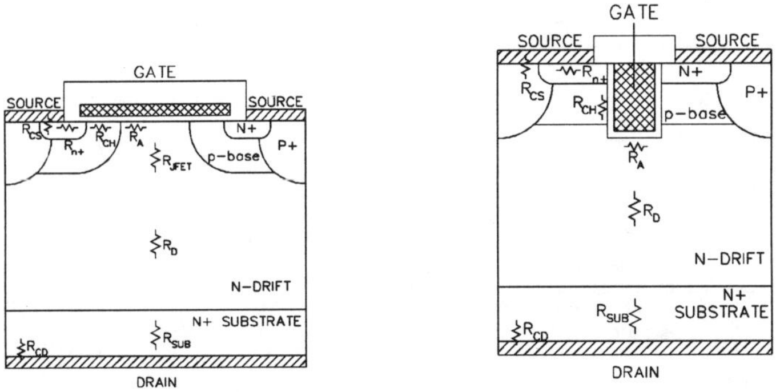

Figure 2. Schematic Cross-section of high-voltage DMOS and UMOS FET and IGBT.

IGBT has been experimentally demonstrated [10,11]. However, care must be taken in the device and process design so as to avoid high electric field at the trench corners. To illustrate this point, the breakdown voltage of a UMOS FET as a function of mesa width for various trench depths is shown in Fig. 3. For deeper trenches, a very narrow mesa width (<3μm) is needed to provide charge shielding and hence higher BV.

In the on-state, it has been found that a 1000V SiC power UMOSFET (with a fairly aggressive pitch spacing of 15 μm) is expected to have an ON-resistance about 100 times better than Si at room temperature as well as at 400°C. In contrast, despite its conductivity modulation, SiC IGBT's are not always better than the SiC MOSFET's due to a higher turn-on voltage [3]. However, at higher current densities (>500A/cm^2) and higher operating temperatures (>200°C), SiC IGBT's can have a lower forward drop than SiC power MOSFET (see Figs. 4 and 5). Furthermore, the minority carrier lifetime in the drift region needs to be sufficiently high (>0.3 μs) so that the drift region drop is minimized.

Termination

Besides the active area design, the termination design is critical to realize the optimal blocking voltage of a power switch. For a DMOS FET or IGBT, a lightly doped, implanted RESURF region can be used to reduce the surface electric field so as to obtain blocking voltages close to an one-dimensional, ideal junction breakdown voltage. (See [12] for a design example of this.) While the placement (by implantation, for example) of the proper space charge would yield the optimal BV, one usually do not know *a priori* the actual amount of active charge (vs. the amount implanted). Hence, the exact dosage needs to be determined empirically. For UMOS devices, a three-step termination design, shown in Fig. 6, which removes part of the p-body epi layer by vertical plasma etching, are explored and they are compatible with the UMOS process, which is used to fabricate the vertical MOS channels. This termination technique closely resembles that of angle beveling in that a certain amount of dopant space charge is removed from the more heavily

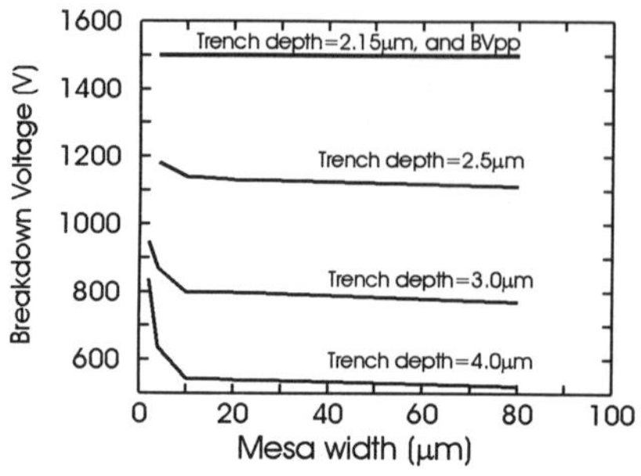

Figure 3. Avalanche breakdown voltage of UMOS structure as a function of mesa width and trench depth [11].

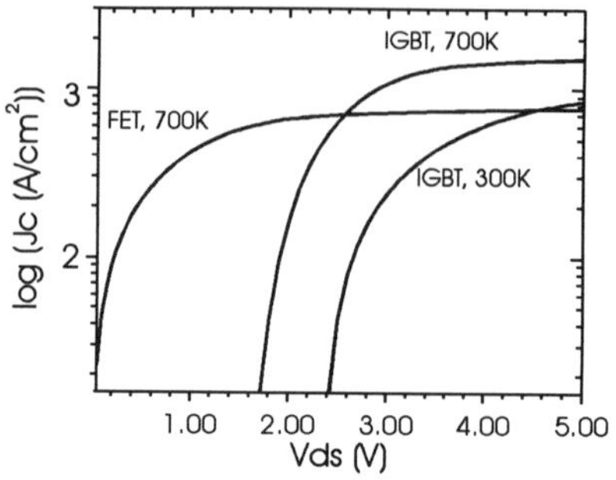

Figure 4. Calculated I-V characteristics of 1000V UMOS FET and IGBT at room temperature and 700K [7].

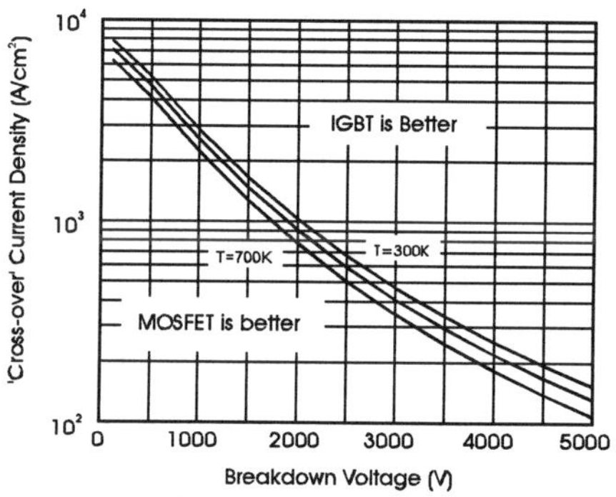

Figure 5. Calculated 'cross-over' current densities of 6H-SiC power UMOS FETs and IGBTs as a function of forward blocking voltage and temperature (τ_{n0}=0.5μs, τ_{p0}=50ns).

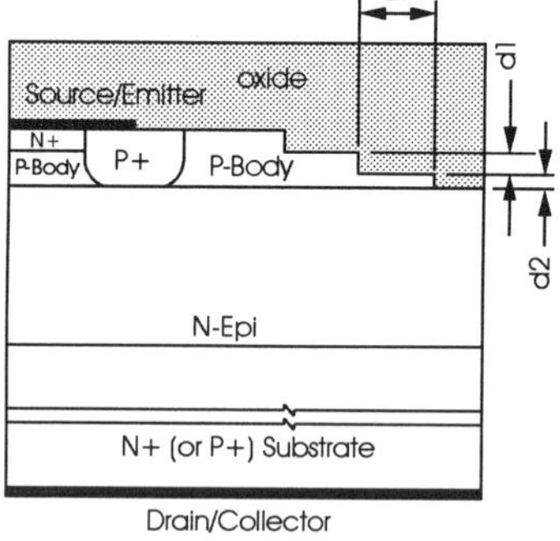

Figure 6. A termination structure for UMOS FET and IGBT [7].

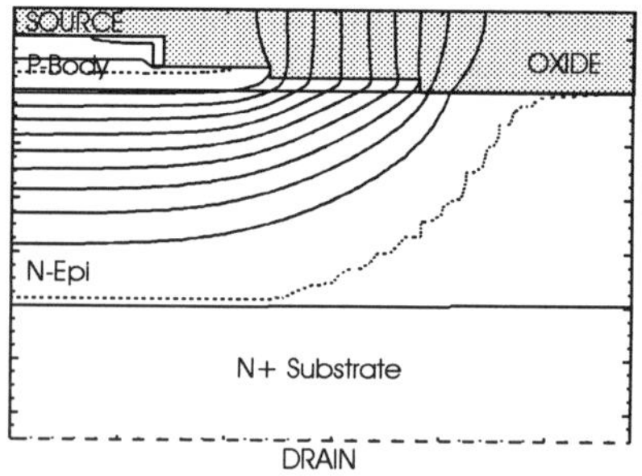

Figure 7. Equipotential contour of SiC termination design at breakdown [7].

doped side so as to spread out the total electric field across the junction. In our three-step termination design, the step region nearest to the pn junction should be totally depleted when BV is reached. (Such a case is depicted in Fig. 7.) The key design parameters are the length and height of each step. By varying the step heights, near-optimal blocking voltages have been obtained with reasonable process latitudes, as shown in Figs. 8 and 9.

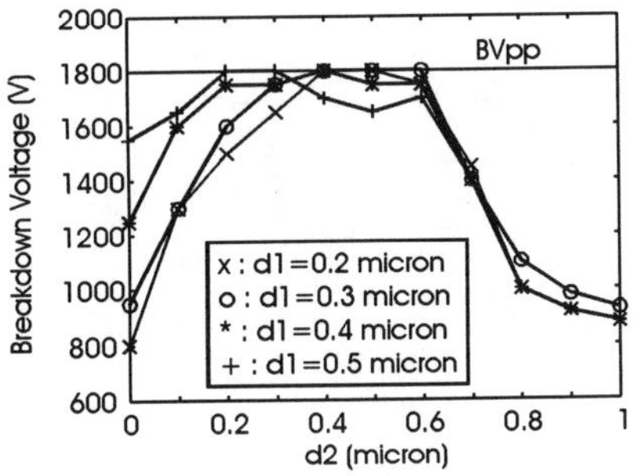

Figure 8. Breakdown voltage of the termination structure as a function of d1 and d2 [7].

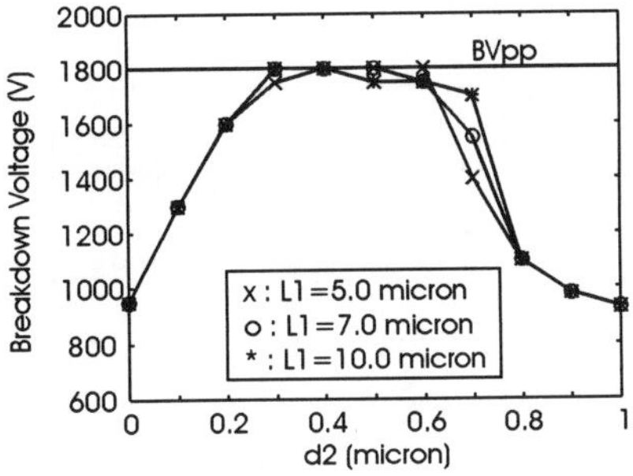

Figure 9. Breakdown voltage of the termination structure as a function of d2 and L1 [7].

Switching Characteristics

Due to the absence of minority carriers, the ultimate switching speed of the power MOSFET is the carrier relaxation time in semiconductor, which is about 1ps. Practical silicon power MOSFET switching times are usually limited by gate resistance and capacitance and package inductance and typically range from 1-10ns. By contrast, minority carrier storage and its removal slows than the turn-off time of the Si IGBT, which is larger than 100ns. SiC IGBT's can be faster than the silicon counterparts due to a thinner drift layer and less minority carrier storage.

Integrated UMOS Process

In silicon UMOS processes [13,14], the most important processing steps are the vertical trench etching and gate poly-Si planarization. The trench etching conditions determine the exact trench shape and carrier mobility along the trench sidewalls and the gate poly-Si planarization can greatly affect gate-to-source shorts and hence device yield. Most of the SiC UMOS devices reported [10] are non-planar because the gate electrode is not self-aligned and planarized. Recently, we have developed a self-aligned, poly-Si gate UMOS process and implemented it on 6H-SiC to fabricate UMOS FET and IGBT with the three-step termination described earlier [11]. Highlights of this process flow are depicted in Fig. 10. This fully planarized process resembles that of a state-of-art silicon UMOS process in using anisotropic etching to define vertical trenches, which are subsequently oxidized and filled with a conformal layer of CVD poly-Si, and successive oxidation and etch-back to planarize the poly-Si. On the other hand, unlike any Si UMOS process, the gate oxidation is performed after the N+ source and P+ short implantation due to the high (>1200°C) dopant activation temperature. Furthermore, different barrier metals need to be used for Ohmic contacts to the N+ and P+ implanted regions. Other process enhancements, such as a thicker oxide on the trench bottoms, can be further incorporated into this basic process.

SiC UMOS FET/IGBT Process Summary

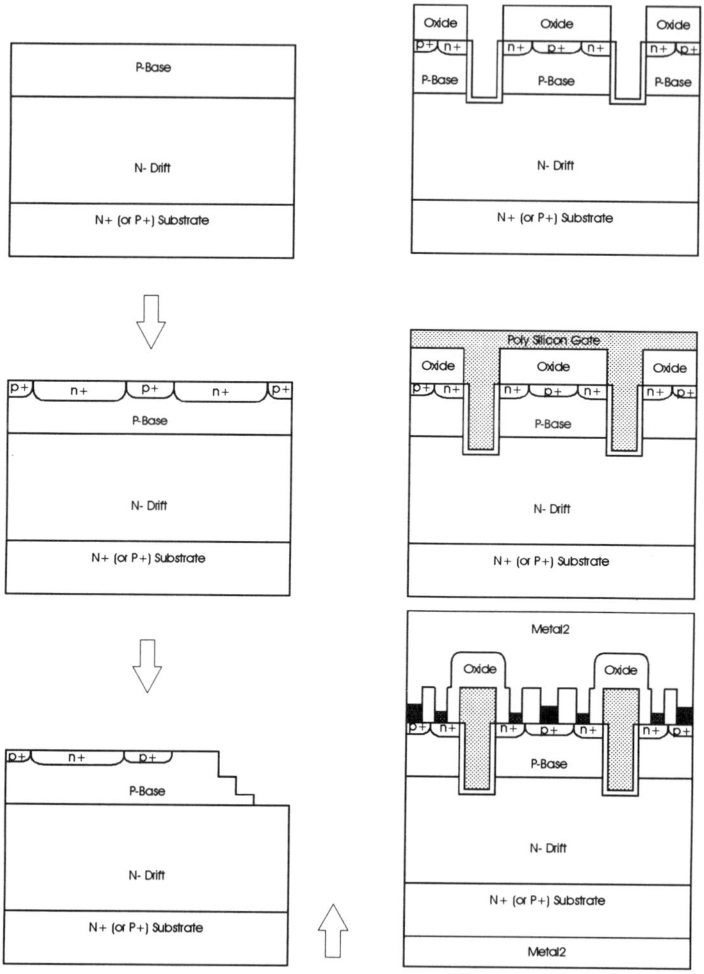

Figure 10. SiC UMOS FET/IGBT process flow.

EXPERIMENTAL DEVICES

Many 6H- and 4H-SiC test devices of BV higher than 100V have already been demonstrated. These are summarized in Table V. Junction rectifiers as high as 4500V has been reported for 6H-SiC [15]. High-quality 3C-SiC junction rectifiers with a BV of 350V have also been grown on 6H-SiC substrates [16]. Schottky rectifiers of BV as high as 1100V for 6H-SiC [17] and 4H-SiC [18,19] . A npn BJT with an epitaxially grown base has been demonstrated in 6H-SiC, having a $\beta \sim 10$ and a blocking voltage of 200V [20]. 6H- and 4H-SiC vertical power UMOSFET's of up to 300V and 1A has been fabricated and characterized but their ON-resistance is higher than optimal [10,20]. The highest BV (450V) for a SiC FET was reported for a substrate-gate, lateral JFET on 6H-SiC [21]. 6H- and 4H-SiC gate-triggered thyristors reported have BV ranging from 100 to 375V and current levels as high as 1.8A [10,20,22]. Recently, we have also fabricated UMOSFET and IGBT [11] with BV up to 200V and current up to 1mA using the fully planarized UMOS process described above. The experimental and simulated I-V characteristics of one of

Table V: A list of SiC power devices that have been experimentally demonstrated.

Device Type	Polytype	Power Ratings	Features	Developer
Rectifier				
Schottky	6H-SiC	550V, Al/Ti	Guard-Ring Termination	Fuji Electric, 1995
	6H-SiC	1100V, Ni		U. of Cincinnati, 1995
	4H-SiC	1100V, Ti	C-Face	Kyoto Univ. 1995
	4H-SiC	1000V, Ti		N. Carolina State U., 1995
Junction	6H-SiC	4500V, 20mA	$V_F \sim 6V$	Linköping Univ. 1995
	6H-SiC	2000V, 1mA		NASA, 1993
	3C-SiC	350V	On 6H-SiC substrate	NASA, 1994
Transistors:				
MOSFET	6H-SiC	60V, 125mA	UMOS, $38m\Omega - cm^2$	Cree, 1993
	4H-SiC	150V, 150mA	UMOS, $33m\Omega - cm^2$	Cree, 1995
JFET	6H-SiC	450V, 12mA	Lateral, Substrate Gate	N. Carolina State U., 1995
	4H-SiC	170V, 50mA	Lateral, Buried Gate	Ioffe Institute, 1995
BJT	6H-SiC	200V, 20mA	$\beta \sim 10$, $126m\Omega - cm^2$	Cree, 1993
IGBT	6H-SiC	200V, 1mA	Self-Aligned UMOS	RPI/GE, 1996
Thyristors:				
Thyristor	6H-SiC	100V, 20mA	Gate Triggered	Cree, 1993
	4H-SiC	375V, 1A	Gate Triggered	Cree, 1995
	6H-SiC	100V, 1.8A	Gate Turn-Off	ARL, 1995

our UMOS IGBT at 250°C are shown in Fig. 11. A very low channel mobility (<5% of the bulk value) needs to be used for data fitting.

MATERIAL AND PROCESSING CHALLENGES

At present, commercial SiC substrates of 6H and 4H polytypes up to 2 inches in diameter are available and custom n- and p-type epitaxial layers up to 10 μm can be ordered. For high voltage devices, total epitaixal layer thickness of at least up to 30μm with acceptable surface flatness and doping uniformity and minimum compensation is needed. To minimize parasitic substrate resistance and maximum carrier concentration, a doping of 10^{19} cm^{-3} would be desired. The

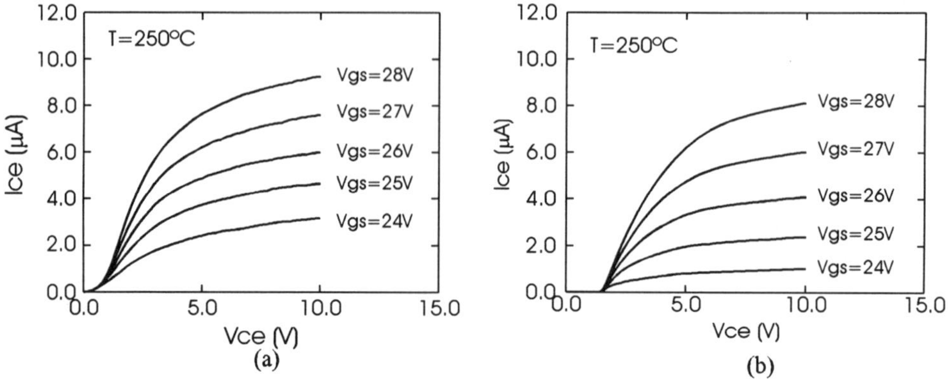

Figure 11. Experimental (a) and simulated (b) current characteristics of 6H-SiC UMOS IGBT at 250°C [11].

most severe structural defect is the micropipes but its density has been substantially improved (to a recently reported value of less than 10 micropipes/cm^2 [23]). A micropipe density of less than 1/cm^2 is needed to realize devices of current ratings larger than 100A with reasonable yield. 3C-SiC can be heteroepitaxially grown on silicon substrates but due to a large (18%) lattice mismatch, the interfacial region is full of defects and poor film qualities often results. High-quality 3C-SiC film growth on 6H-SiC substrates has been demonstrated [16] but careful control on surface and growth conditions need to be performed to prevent nucleation of other polytypes. AlN substrates are not commercially available but heteroepitaxial hexagonal AlN films can be grown on sapphire substrates. SiC is the only compound semiconductor on which a high-quality SiO_2 film can be grown, much like silicon. While MOS interfacial properties of thermally grown oxide on lightly doped n-type 6H-SiC can be controlled to the extent that a fixed oxide charge (Q_f) less than 10^{11}/cm^2 and interface state density less than 10^{11}/cm^2-eV can be obtained, those on p-type appears to be much worse [24,25] and its cause has been attributed to either aluminum [24] or carbon [26] at the SiO_2/SiC interface. High surface state densities also degrade termination design by enhancing surface electric field crowding. Also, the effect of RIE on the MOS parameters has not been investigated. Furthermore, high-temperature reliability of gate oxide is also of major concern [27]. While dopant diffusion in SiC is very slow when compared to silicon, planar, ion-implanted n+/p and p+/n junctions, using nitrogen and aluminum or boron respectively, have been demonstrated, despite a relatively high implantation/activation temperature (>1200°C) and small percentage (<20%) of dopant activation [28-32]. However, its forward and reverse I-V characteristics is inferior to either silicon implanted junctions or epitaxially grown SiC junctions, probably due to defects introduced during implantation. Ohmic contacts of reasonable contact resistivities ($<10^{-4}$ Ω-cm^2) have been made to SiC. Usually, Ni is used on n-type contacts and Al is used for p-type contacts. A relatively high (around 1000°C) sintering step is performed for contact formation. Formation of p-type contacts is more difficult because a high surface hole concentration is often hard to achieve due to the deepness of the acceptor levels (>180meV).

SYSTEM EFFICIENCY

The performance enhancement projected for SiC MOSFET will be further demonstrated in using a 15A motor drive system as a test case. Efficiency due to device power dissipation has been shown [33] to be drastically reduce when SiC-based devices replace conventional silicon devices. In Fig. 12, the system efficiency (due to device power loss) calculated analytically using a self-consistent model [34]) is shown as a functional of frequency at room temperature for Si and SiC devices respectively. It can be seen that the SiC devices dissipate much less power and operate at a higher (>10 times) switching than the Si counterparts, despite a much larger voltage rating. These efficiency figures demonstrate that SiC devices can be used to reduce energy consumption and relax the heat sinking requirements of the package.

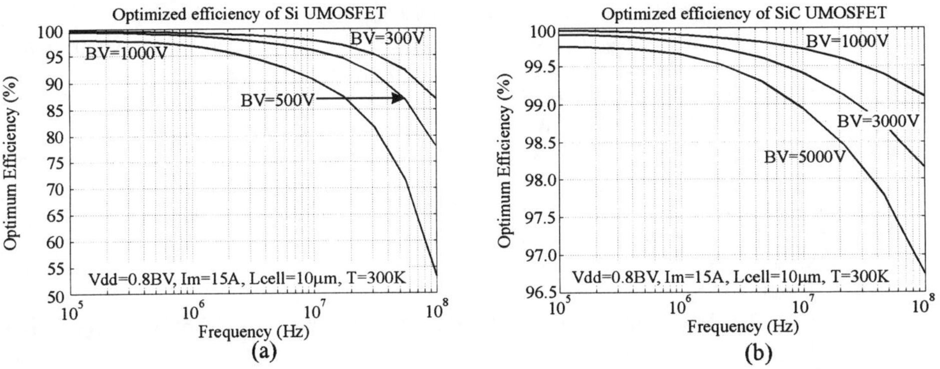

Figure 12. The optimum system efficiency of Si and SiC devices as a function of switching frequency [34].

SUMMARY

We have reviewed the present status of SiC devices for high-voltage power electronics applications. In particular, we have presented the figures of merit, modeling and simulation approaches, the design of device structures and processes, major device achievements, material and processing challenges and system efficiency impact of SiC power devices. With the advance of material, process and device technology, we expect to see SiC power devices commercially by the year 2000.

ACKNOWLEDGMENTS

We would like to thank N. Ramungul, R. Tyagi-Sodhi, J. Kretchmer, and W. Hennessy for technical assistance and acknowledge financial support from Army TACOM under contract #DAAE07-94-C-R066 and the Office of Naval Research under a MURI contract.

REFERENCES

1. K. Shenai, R.S. Scott, and B.J. Baliga., IEEE Trans. Electron Devices **36**, 1811 (1989).
2. B.J. Baliga, IEEE Electron Device Lett. **10**, 455 (1989).
3. M. Bhatnagar and B.J. Baliga, IEEE Trans. Electron Devices,

4. T.P. Chow and R. Tyagi, IEEE Trans. Electron Devices **41**, 1481 (1994).
5. A. Bhalla and T.P. Chow, Paper 6.13, Proc. 6th International Symp. Power Semiconductor Devices and ICs, pp.287-292, May 31-Jun 2 (1994).
6. A.S. Kyuregyan and S.N. Yurkov, Sov. Phys. Semicond., **23**, 1126 (1989).
7. N. Ramungul, R. Tyagi, A. Bhalla, T.P. Chow, M. Ghezzo, J. Kretchmer, and W. Hennessy, Int'l Conf. on SiC and Related Materials - ICSRCM-95, Tech. Dig., 322, 1995 (unpublished).
8. W. Fulop, Solid-State Electron. **10**, 39 (1967).
9. A.O. Konstantinov, in Amorphous and Crystalline Silicon Carbide III, edited by G.L. Harris, M.G. Spencer, and C.Y.-W. Yang (Srpinger-Verlag, Berlin Heidelberg, 1992), p.213.
10. J.W. Palmour, S.T. Allen, and D.G. Waltz, Int'l Conf. on SiC and Related Materials - ICSRCM-95, Tech. Dig., 319, 1995 (unpublished).
11. N. Ramungul, T.P. Chow, M. Ghezzo, J. Kretchmer, and W. Hennessy, to be presented at the 54th Device Research Conference, 1996 (unpublished).
12. T.K. Wang, T.P. Chow, D.M. Brown, and M. Ghezzo, Proc. 4th International Symposium on Power Semiconductor Devices and ICs, pp.303-308, May 19-21, 1992.
13. H.-R. Chang, R.D. Black, V.A.K. Temple, W. Tantraporn, and B.J. Baliga, IEEE Trans. Electron Devices **ED-34**, 2329 (1987).
14. S. Matsumoto, T. Ohno, H. Ishii, and H. Yoshino, IEEE Trans. Electron Devices, **41**, 814 (1994).
15. O. Kordinar, J.P. Bergman, A. Henry, E. Janzen, S. Savage, J. Andre, L.P. Ramberg, U. Lindefelt, W. Hermansson, and K. Bergman, Appl. Phys. Lett., **11**, 1561 (1995)
16. P.G. Neudeck, D.J. Larkin, J.E. Starr, J.A. Powell, C.S. Salupo, and L.G. Matus, IEEE Trans. Electron Devices **41**, 826 (1994).
17. J.N. Su and A.J. Steckl, Int'l Conf. on SiC and Related Materials - ICSRCM-95, Tech. Dig., 597, 1995 (unpublished).
18. A. Itoh, T. Kimoto, H. Matsunami, Int'l Conf. on SiC and Related Materials - ICSRCM-95, Tech. Dig., 559, 1995 (unpublished).
19. R. Raghunathan, D. Alok, and B.J. Baliga, IEEE Electron Device Letters, **16**, 226 (1995).
20. J.W. Palmour, J.A. Edmond, and C.H. Carter, presented at the 51st Device Research Conference, 1993 (unpublished).
21. D. Alok and B.J. Baliga, Int'l Conf. on SiC and Related Materials - ICSRCM-95, Tech. Dig., 307, 1995 (unpublished).
22. K. Xie, J.H. Zhao, J.R. Flemish, T. Burke, W.R. Buchwald, G. Lorenzo, and H. Singh, IEEE Electron Device Letters, **17**, 142 (1996).
23. J.W. Palmour, presented at the Workshop on High-Temperature Electronics, Fort Mommouth, NJ, 1995 (unpublished).
24. D.M. Brown, M. Ghezzo, J. Kretchmer, E. Downey, J. Pimbley, and J. Palmour, IEEE Trans. Electron Devices, **41**, 618 (1994).
25. J.N. Shenoy, G.L. Chindalore, M.R. Melloch, J.A. Cooper, J.W. Palmour, and K.G. Irvine, J. Electron. Mater. **24**, 303 (1995).
26. G. Pensl, Int'l Conf. on SiC and Related Materials - ICSRCM-95, Tech. Dig., 585, 1995 (unpublished).
27. V. Krishnamurthy, D.M. Brown, M. Ghezzo, J. Kretchmer, W. Hennessy, E. Downey, and G. Michon, in Silicon Carbide and Related Compounds, edited by M.G. Spencer, R.P. Devaty, J.A. Edmond, M. Asif Khan, R. Kaplan, and M. Rahman (Institute of Physics Publishing, Bristol and Philadelphia, 1994), pp.483-486.

28. J.A. Edmond, K. Das, and R.F. Davis, J. Appl. Phys., **63**, 922 (1988).
29. M. Ghezzo, D.M. Brown, E. Downey, J. Kretchmer, W. Hennessy, D.L. Polla, and H. Bakhru, IEEE Electron Device Lett., **13**, 639 (1992).
30. M. Ghezzo, D.M. Brown, E. Downey, and J. Kretchmer, Appl. Phys. Lett., **63**, 1206 (1993).
31. N. Ramungul, V. Khemka, R. Tyagi, T.P. Chow, M. Ghezzo, P.G. Neudeck, J. Kretchmer, W. Hennessy, and D.M. Brown, Int'l Conf. on SiC and Related Materials - ICSRCM-95, Tech. Dig., 561, 1995 (unpublished).
32. P.M. Shenoy and B.J. Baliga, IEEE Electron Device Lett., **16**, 454 (1995).
33. B.J. Baliga, Proc. IEEE, **82**, 1112 (1994).
34. N. Ramungul, T.P. Chow, D.A. Torrey, M. Ghezzo, R.D. King, J. Kretchmer, and W. Hennessy, First Int'l Conf. on All Electric Combat Vehicle (AECV), Conf. Proc., 211, 1995 (unpublished).

GROWTH OF HIGH QUALITY (In,Ga,Al)N/GaN HETEROSTRUCTURE MATERIALS AND DEVICES BY ATMOSPHERIC PRESSURE MOCVD

S. P. DENBAARS*, S. KELLER, B. P. KELLER, Y.F. WU, D. KAPOLNEK*, and U. K. MISHRA
Electrical & Computer Engineering and *Materials Departments, University of California, Santa Barbara, CA 93106, U.S.A.

ABSTRACT

Using atmospheric pressure MOCVD we have obtained high quality InGaN/GaN and AlGaN/GaN heterostructure materials and devices. For nominally undoped 4 µm thick GaN films, we obtained 300 K mobilities of 780 cm^2/Vs and an unintentional background impurity level of $n_{300K} = 6*10^{16}\ cm^{-3}$. For InGaN/GaN heterostructures we have obtained direct band-edge transitions with FWHM as narrow as 7.9 nm (59 meV) for 50Å thick $In_{0.16}Ga_{0.84}N$ quantum wells at 300K, which is the among the best reported values. The quantum wells display energy shifts towards shorter wavelength with decreasing well thickness, and the shift agrees with predicted quantum effects. These materials have been incorporated into InGaN single quantum well LEDs that emit at 450 nm. In addition AlGaN/GaN heterostructure materials have been incorporated into HFETs and MODFETs. Gate-drain breakdown voltage well exceeding 100 V, and extrinsic transconductance g_m of up to 140 mS/mm were realized in the MODFET.

INTRODUCTION

Recently, the Group-III Nitride based semiconductors have emerged as the leading material for short wavelength optoelectronic devices operating from the UV to blue wavelengths. Furthermore, the group III-nitrides are distinguished by their high thermal conductivity and physical hardness, making them suitable for high temperature, high power electronic devices applications. The majority of the InGaN based p-n junction light emitting diodes (LEDs) are employing impurity related transition for blue and green emission.[1,2,3] Recently, direct bandgap emission in the blue-green spectral region has been obtained using high In content in single quantum well (SQW) LEDs and lasers[4]. However, there are few reports on the optimum growth condition for InGaN which has proven to be difficult to grow due to the high volatility of indium at normal nitride growth temperatures. Understanding the growth of ultra-thin InGaN/GaN quantum wells by metal organic chemical vapor deposition (MOCVD) is of extreme importance in improving the properties of these devices. In this paper we will discuss aspects of the MOCVD growth of high optical quality InGaN quantum wells and quantum well LEDs. In addition, we present preliminary results on GaN/AlGaN HFETs.

Mat. Res. Soc. Symp. Proc. Vol. 423

EXPERIMENTAL

Epitaxial films were grown on c-plane sapphire in a two-flow atmospheric pressure MOCVD reactor (Thomas Swan, Ltd.). Trimethylgallium (TMGa), trimethylindium (TMIn) and ammonia were used as precursors. After annealing the substrates at 1050 °C, a 190 Å thick GaN nucleation layer was deposited at 600 °C. The temperature was then raised to 1080 °C to grow GaN of 1.4 to 2.4 μm thickness. The growth of InGaN was investigated in the temperature range between 650 and 850 °C. The 0.1 to 0.2 μm thick InGaN layers were grown on top of the thick GaN layers. The input flows of TMGa and TMIn were varied from 1 to 13 μmol/min and 1.9 to 24 μmol/min, respectively. The ammonia flow was kept constant at 0.12 mol/min.

RESULTS

For nominally undoped 4 μm thick GaN films, we obtained 300 K mobilities of 780 cm^2/Vs ($n_{300K} = 6*10^{16}$ cm^{-3}), which is among the highest values reported for this carrier concentration range. Surface nitridation of the sapphire substrate was found to highly influence the GaN film mobility and double crystal x-ray diffraction linewidth. Double crystal x-ray rocking curves display a minimum FWHM of 40 arcsec or 300 arcsec for 2 micron thick GaN films depending on the nitridation procedure. Surprisingly the best mobilities are routinely obtained from the samples with broader DCXRD linewidths. On these samples dislocation densities as low as $4*10^8$ cm^{-2} for GaN on Sapphire substrates were observed in cross-sectional TEM measurements. The films are characterized by a band edge to deep level luminescence intensity ratio of higher than 1300 for photoluminescence measurements (220 mW/cm^2) performed at 22 K. Even at excitation levels as low as 2.2 mW/cm^2, the 300 K PL is dominated by the near band edge emission . The FWHM of the (002) x-ray reflection was 180 arcsec. A relatively low dislocation density was obtained from cross-sectional transmission electron microscopy analyses and amounted to $4x10^8$ cm^{-2}.[5]

TABLE I. FWHM of symmetric and asymmetric XRD GaN reflections and dislocation densities for 1.2 μm thick films grown on sapphire with different NH_3 pretreatments

NH_3 Preflow	Symmetric (002) (FWHM/ arcsec)	Asymm(102) (FWHM/ arcsec)	Dislocations (cm-2)
60 s	269"	413"	$4*10^8$
400 s	40"	740"	$2*10^{10}$

The growth of InGaN has to be performed at temperatures below 850 °C because of the high volatility of indium at common GaN growth temperatures of above 1000 °C. But even on InGaN layers grown at temperatures below 800 °C, In droplet formation was observed.[6] Fig. 1a shows the 300 K PL spectra of three InGaN films of varying composition. Sample (a) was deposited at 800 °C with a TMGa and TMIn flow of 3.8 and 10 μmol/min, respectively. Sample (b) was grown at 760 °C with a TMGa and TMIn flow of 3.8 and 7.2 μmol/min, respectively, and sample (c) at 700 °C with a TMGa flow of 1.9 μmol/min and TMIn flow of 4.8 μmol/min. The InGaN films were Si doped with a Si_2H_6 flow of 0.7 nmol/min. The spectra are dominated by the recombination of the donor bound excitons. No luminescence related to deep levels was apparent, up to a wavelength of 800 nm. The luminescence intensity decreases slightly with decreasing growth temperature and increasing indium content of the layers. The emission wavelength increased from 375 nm (a) to 389 and 419 nm for samples (b) and (c), respectively, corresponding to indium compositions of x^s_{In} = 0.04,

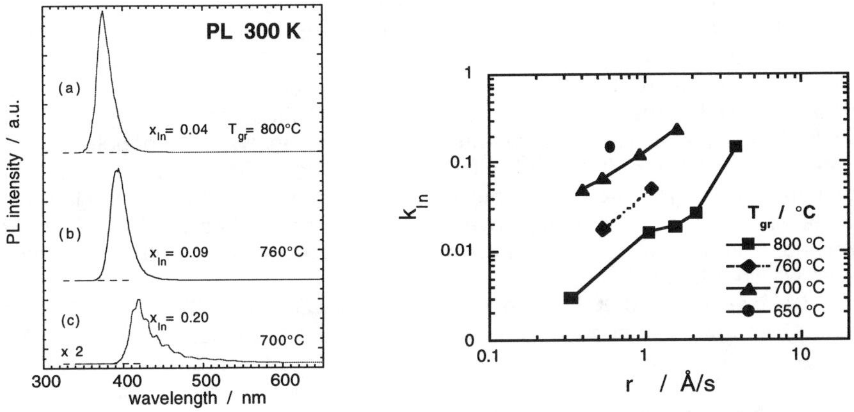

(a) (b)

Fig. 1 . (a) 300 K PL spectra of InGaN films grown at (a) 800° C (b) 760°C and (c) 700°C (excitation density 220 mW/cm²). (b) Dependence of the relative indium segregation coefficient, *kIn*, on the growth rate determined at different growth temperatures.

(a), x^s_{In} = 0.09, (b), and x^s_{In} = 0.2, (c)[7] Fig. 1b shows the dependence of the relative indium incorporation coefficient,

$$k_{In} = x^S_{In} * f_{TMGa} / (1-x^S_{In}) * f_{TMIn} ,$$

on the growth rate of InGaN for samples grown at different growth temperatures. Besides the expected increase of k_{In} with decreasing growth temperature[8], we observe an increase of k_{In} with increasing growth rate at a given temperature. This indicates that the incorporation of indium is limited by the evaporation of indium species from the surface. The tendency for evaporation decreases at lower temperatures and/or increasing growth rate, when the indium species become trapped by the growing layer. $In_{0.16}Ga_{0.84}N$ single quantum wells were embedded in InGaN of graded composition. First a 250 Å thick InGaN:Si layer with an In composition increasing from 4.5 to 6.5 % In was grown on a 2.4 µm thick GaN film by slowly lowering the growth temperature from 760 to 750 °C. The temperature was then ramped to 700 °C for the growth of the undoped $In_{0.16}Ga_{0.84}N$ quantum well. After accomplishing the quantum well growth, the temperature was raised again to 760 °C and a 10 Å thick $In_{0.05}Ga_{0.95}N$ layer was grown. TMGa and TMIn were injected into the reactor at a flow rate of 1.9 and 3.7 µmol/min, respectively, throughout the whole InGaN layer growth. The layer structure was accomplished with the deposition of a 0.1 µm thick GaN cap at 1050 °C. The bulk layers were analyzed by double crystal x-ray diffraction using the (002) reflection of GaN and InGaN. Room temperature photoluminescence (PL) measurements were performed using the 325 nm line of an He-Cd laser operating at a pump level of 1 mW. The thickness of the QWs was determined from transmission electron microscopy calibration of 100 Å thick InGaN quantum wells embedded in GaN.

Fig. 2 shows the 300 K PL spectra of four $In_{0.16}Ga_{0.84}N$ SQWs of a thickness between 17 and 50 Å embedded in the graded $In_xGa_{1-x}N$ barriers. The graded barrier was chosen since SQWs embedded directly in GaN showed only weak luminescence under the excitation conditions applied in this study. The reason for this behavior is presently under investigation. The emission wavelength of 409 nm measured for the 50 and 34 Å thick wells is the same as for InGaN bulk layers grown under these conditions. The peak wavelength shifted towards lower values at a thickness below 30 Å. With decreasing thickness, the intensity of the QW luminescence decreased by a factor of about 4. The loss of intensity with thinner well width results from the increased penetration of the electron wave function into the barriers, causing a decreased probability of carrier recombination in the well. Additionally, the influence of nonradiative carrier recombination due to interface related defects increases. The FWHM of the QW luminescence increased from 7.9 nm for the 50 Å thick well to about

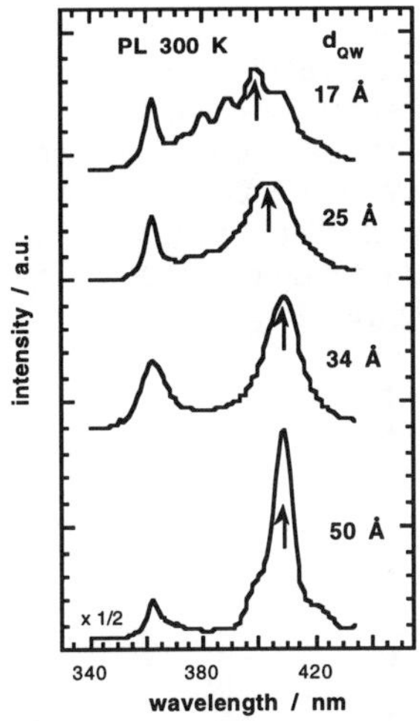

Fig. 2. 300 K PL spectra of $In_{0.16}Ga_{0.84}N$ SQWs of different thickness.

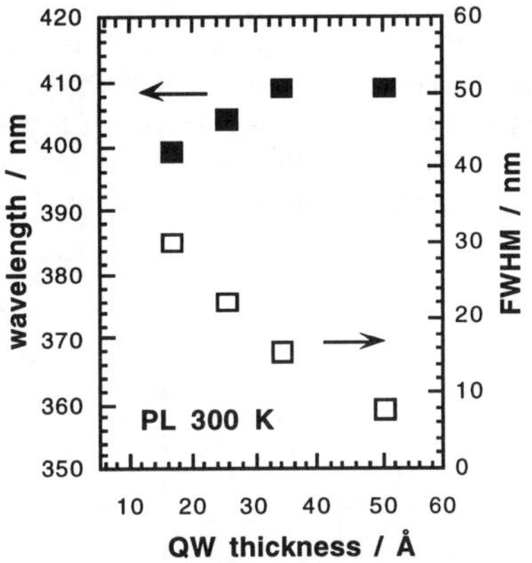

Fig. 3. 300 K PL emission wavelength and FWHM of the SQWs.

30 nm for the 17 Å thick well (Fig. 3). The increase the FWHM of the QW luminescence with decreasing thickness is thought to originate from interface fluctuations, together with band filling effects caused by the silicon doping of the barrier material.[9] The large effective electron mass in the nitrides together with the graded barriers result in quantum effects only being observed for a thickness less than 30 Å for this QW composition. In contrast to the results obtained for GaN QWs embedded in AlGaN,[10] we did not observe any remarkable strain induced shifts of the emission wavelength of the quantum wells. Transmission electron microscope investigations indicated a relatively high stacking fault density in the InGaN layer, which may be responsible for the high degree of relaxation of the quantum wells.[11]

OPTICAL DEVICES (InGaN /GaN DH LED)

For the LED structure indium contents exceeding 25% were explored in order to shift the wavelength into the range of greater eye sensitivity. In the LED structure the first layer consisted of 1 μm undoped GaN deposited by applying a TMGa flow rate of 51 μmol/min followed by 3 μm silicon doped GaN, grown with a TMGa flow of 13 μmol/min and a Si_2H_6 flow of 1.2 nmol/min. The growth of the 20 nm thick $In_{0.27}G_{0.73}N$ active layer was performed at 830 °C. A 20 nm $Al_{0.15}Ga_{0.85}N$ cap was grown immediately after the quantum well. The contact layer consisted of a 0.5 μm thick Mg doped p-GaN cap layer deposited at 1080 °C with TMGa and Cp_2Mg flows of 13 and 1.3 μmol/min, respectively, resulting in a magnesium concentration in the solid of $1*10^{20}$ cm^{-3}, measured by secondary ion mass spectroscopy (SIMS). After the growth , the samples were annealed in nitrogen at 730 °C for 15 min. The LED process was accomplished by evaporating a 10 nm Ni / 10 nm Au current spreading layer, and a p-type contact consisting of 20 nm Ni / 200 nm Au. After chlorine reactive ion etching (RIE) of the mesa, 200 nm Al was evaporated as n-contact metal. Figure 4 shows the electroluminescence spectra of the LED at forward currents of 10, 20 and 100 mA. The peak electroluminescence (EL) wavelength shifts from green to blue at increasing forward currents. We believe the low current spectra is caused by deep levels in our high indium containing ternary alloys and can be further improved by optimizing

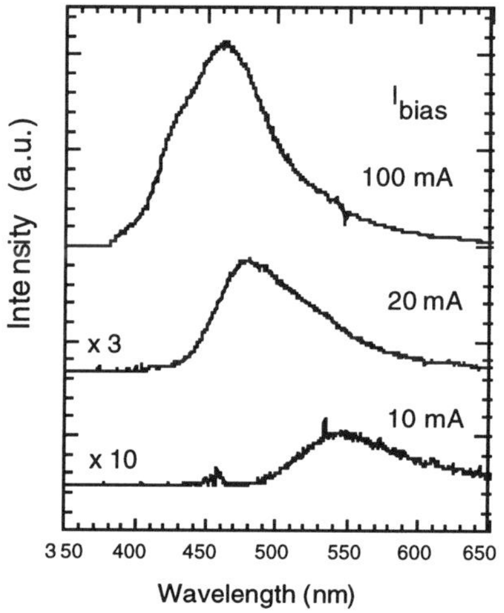

Fig. 4 Electroluminescence of GaN/InGaN double heterostructure LED for different forward currents.

growth conditions. At higher forward currents, the spectrum is dominated by band-edge related luminescence around 450 nm. The current - voltage (I - V) characteristics of the DH LED exhibits a forward voltage of 3.6V at 20 mA. A series resistance of 55 Ω was measured for these diodes.

ELECTRONIC DEVICES (HFETS AND MODFETS)

Owing to its large bandgap, high electron saturation velocity and large conduction band offset, the $GaN/Al_xGa_{1-x}N$ system is particularly attractive for high power HEMT devices. Early 0.25 μm gate-length HFETs[12] showed promise (f_t ~ 20 GHz, f_{max} ~ 77 GHz). Performance was limited by the low transconductance (g_m ~ 28 mS/mm) resulted from low sheet charge density and poor ohmic contact resistance. Aktas et al.[13] demonstrated MODFETs with an improved g_m of 120 mS/mm. We have further improved on these earlier studies and achieved GaN HFETs and MODFETs with a gate to drain diode breakdown voltage well in excess of 100 V, transconductances as high as 140 mS/mm, and a full channel current > 300 mA/mm. The transistor channels were grown on the c-plane sapphire substrate with Channel thickness of 0.4 μm for the HFETs and 0.3 μm for the MODFETs. They are both unintentionally doped with a background electron concentration around $4x10^{16}$ cm^{-3}. Gate barrier of the HFETs was an undoped 300 Å $Al_{0.15}Ga_{0.85}N$ layer. It was modulation-doped for the MODFETs with a 30 Å spacer, a 150 Å doped layer (N_d ~ $3X10^{18}$ cm^{-3}) and a 120 Å undoped cap. Mesa isolation was done with Cl_2 RIE. Source and drain ohmic scheme was Ti/Al annealed at 650 C° , while gate metalisation was E-beam evaporated Au. Gate lengths were 1.5 mm and 1 mm for the HFETs and the MODFETs respectively. On wafer TLM pattern measurement yielded a contact resistance of 2.5 W-mm for the HFETs and 3.0 W-mm for the MODFETs.

The GaN/AlGaN MODFET exhibited the highest external transconductances of approximately 140 mS/mm as shown in figure 5. Gate to drain diode I-V characteristics of a representative MODFET with gate-drain separation of 2 mm showed a turn-on at 1.6 V while the reverse breakdown is beyond the limit of the HP 4145B Semiconductor Parameter Analyzer, with a leakage current density of 0.6 pA/mm^2 at -100 V. Large voltage I-V characteristics of the same MODFET was recorded up to V_{ds} = 60 V. With its full channel current of > 300 mA/mm, a RMS class

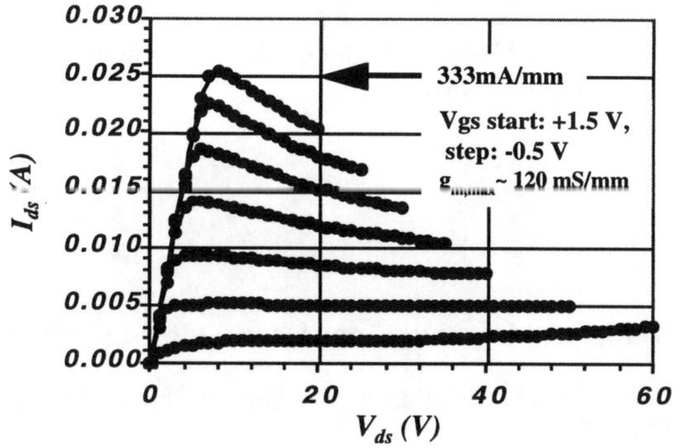

Figure 5. GaN/$Al_{0.15}Ga_{0.85}$N MODFET exhibiting high extrinsic transconductance and a high >300mA/mm channel conductance.

A amplifier power output over 2 W/mm is possible. I-V characteristics of another MODFET with 150 mA/mm channel current was obtained with channel voltage up to 100 V without inducing damage. The maximum on-stage voltage is thermally limited. The HFETs demonstrated similar high voltage capability and channel current density as the MODFETs with a lower knee voltage as the result of the lower ohmic contact resistance. Initial microwave measurement indicated $f_t \sim 5$ GHz and $f_{max} \sim 11$ GHz at $V_{ds} = 20$ V for both the 1.5 μm gate-length HFETs and the 1 μm gate-length MODFETs, lower than expected. The reason is currently under investigation.

CONCLUSIONS

In conclusion, narrow and bright band edge related luminescence was observed for InGaN SQWs grown at 700 °C. The quantum wells display energy shifts towards shorter wavelength with decreasing well thickness. The increase in the emission energy was accompanied by an increase in the full width at half maximum and a decrease in the intensity of the QW luminescence. The FWHM of the QW luminescence of the 50 Å thick well was as narrow as 7.9 nm (59 meV). InGaN/GaN double heterostructure LEDs showed blue luminescence visible in bright room light. In addition, we have successfully fabricated GaN HFETs and MODFETs with very high breakdown voltage and large transconductance. In particular, a gate-drain breakdown voltage well exceeding 100 V, g_m up to 120 mS/mm, channel current density > 300 mA/mm were realized in the same MODFET.

ACKNOWLEDGMENTS

The authors would like to thank H. Masui for the support in growth and A. C. Abare for modeling of quantum wells. The authors are grateful for the support of the National Science Foundation, the Army Research Office, Hughes Research Laboratories, and a UC Micro Grant from Rockwell International.

REFERENCES

1. N. Nakamura, T. Mukai, M. Senoh; Appl. Phys. Lett. **64** (1994) 1687

2. I. Akasaki, H. Amano, H. Murakami, M. Sassa, H. Kato, K. Manabe; J. Cryst. Growth **128** (1993) 379

3. M.A. Kahn, Q. Chen, R.A. Skogman, J.N. Kuznia; Appl. Phys. Lett. **66** (1995) 2046

[4] S. Nakamura, M. Senoh, S. Nagahama, N. Iwasa, T. Yamada, T. Matshushita, H. Kiyoku, and Y. Sugimoto, Jpn. J. Appl. Phys. **35** (1996) L74-L76.

[5] D. Kapolnek, X.H. Wu, B. Heying, S. Keller, B.P. Keller, U.K. Mishra, S.P. DenBaars, and J.S. SPeck, Appl. Physics. Lett. **67** (1995) 1541.

[6] M. Shimizu, K. Hiramatsu and N. Sawaki, J. Cryst. Growth **145** (1994) 209

[7] K. Osamura, S. Naka, Y. Murakami; J. Appl. Phys. **46** (1975) 3432

[8] S. Nakamura, J. Cryst. Growth **145** (1994) 911

[9] D.F. Welch, G.W. Wicks, and L.F. Eastman, Appl. Phys. Lett. **46** (1985) 991

[10] S. Krishnankutty, R.M. Kolbas, M.A. Kahn, J.N. Kuznia, J.M. van Hove and D.T. Olson, J. Electron. Mat. **21** (1992) 609

[11] X.H. Wu, S. Keller, B.P. Keller, U.K. Mishra, S.P. DenBaars, and J.S. Speck, to be published in Appl Physics. Lett.

[12] M. Asif Khan, Michael S. Shur, John N. Kuznia, Q. Chen, Jin Burm and William Schaff, Appl. Phys. Lett. **66** (1995) 27

[13] O. Aktas, Wkim, Z. Fan, F. Stengel, A. Botchkarev, A. Salvador, B. Sverdlov, S.N. Mohammad and Morkoc, IEDM, Tech. Digest, 205, Dec.10-13,1995

INVESTIGATION OF AN NEA DIAMOND VACUUM MICROTRIODE ARRAY

C.W. HATFIELD*, G.L. BILBRO*, A.S. MORRIS*, P.K. BAUMANN**, B.L. WARD**, and R.J. NEMANICH**
*ECE Department, North Carolina State University, Raleigh, NC 27695
**Dept. of Physics, North Carolina State University, Raleigh, NC, 27695

ABSTRACT

The properties and characteristics of vacuum microtriodes based on NEA diamond surfaces were modelled. Specifically, an NEA diamond vacuum microtriode array was investigated using electrical measurements, electron optics software, and microwave circuit simulation. Data for emission current versus applied voltage for various anode-to-cathode distances for diamond NEA surfaces was analyzed and various parameters were extracted. Electron optics software was used to determine Fowler-Nordheim and space-charge-limited DC I-V characteristics for each microtriode. Microwave circuit simulation was done to determine the behavior of arrays of these vacuum microtriodes in an RF amplifier circuit.

INTRODUCTION

In this work, a set of simulations are presented for diamond NEA emission vacuum microelectronic devices, in which electrons are emitted from a diamond cathode into vacuum and are collected at an anode spaced a few micrometers away. Unlike traditional thermionic vacuum tubes, which require a heated cathode in order to obtain significant current densities, these cathodes would be fabricated by growing a thin film of polycrystalline diamond on silicon and then using appropriate surface treatments to obtain a cold cathode with a negative electron affinity (NEA) surface. Much work has been previously reported concerning current emission from diamond surfaces [1, 2, 3, 4, 5].

Vacuum microtriode structures have been investigated utilizing a molybdenum or silicon cathode shaped in the form of a conic tip. Measurements have been reported for arrays of these types of structures [6, 7, 8]. In the simulations done in this work, the difficulties involved in fabricating these tips is avoided by assuming that the emitting cathode surfaces are flat. The usefulness of these types of flat-emitter vacuum microtriodes in high-frequency RF circuits was recognized by Eastman [9].

We simulated the DC I-V characteristics of the vacuum microtriodes using EGN2, a commercially available electron optics simulation program [10]. First, a set of simulations were done using parameters extracted from recent diamond NEA emission data. It is hoped that, eventually, for NEA diamond surface emission, space-charge-limited current (SCLC) will be obtained (i.e., the electric field at the cathode surface will drop from a large value to approximately zero). In order for the microtriodes examined in this work to operate in the SCLC regime, emitted current denisities must exceed 100 A/cm^2. To investigate the usefulness of these microtriodes for microwave circuits, simulations for these structures assuming SCLC were performed.

From the SCLC results, a large-signal circuit model for a 1 $\times$ 1 mm^2 array of each microtriode was developed. This model was implemented in Microwave Harmonica, a commercially available microwave circuit simulation program [11]. The circuit simulator was used to predict the performance for each microtriode array in a microwave amplifier circuit. The results of these simulations indicate that diamond NEA vacuum microtriodes are promising devices for high-power, high-frequency microwave circuits.

Mat. Res. Soc. Symp. Proc. Vol. 423 © 1996 Materials Research Society

METHODS

Diamond NEA Emission Data

Typical diamond NEA emission data was taken from our research group and used as a basis for Fowler-Nordheim (F-N) emission simulations. This data indicates that diamond surfaces for which the presence of NEA has been confirmed by photoemission measurements show a F-N dependence of current with anode voltage and cathode-to-anode spacing. That is, the emitted current has the form

$$I = k\left(\frac{\beta V_a}{d}\right)^2 \exp\left(\frac{-6530d\phi^{3/2}}{\beta V_a}\right) \quad (1)$$

where V_a is the anode voltage (V), d is the anode-to-cathode spacing (μm), ϕ is the barrier height of the NEA surface (eV), β is the field-enhancement factor, and k is a data-fitting parameter (with units of $\mu A \cdot \mu m^2/V^2$).

Specifically, the data used in these simulations was originally reported by Bozeman, et al [12]. This data was obtained from a 3.2-μm-thick chemical vapor deposited (CVD) diamond film grown on silicon. This film had a 1.1×10^{18} cm^{-3} boron concentration (confirmed by SIMS measurements), a (110) texture surface morphology, and a Raman FWHM of 9.4 cm^{-1}. The current was measured by placing a 2-mm-diameter cylindrical platinum anode with a flat end near the diamond film surface (2.6 to 19.8 μm).

Analysis of the data indicated that this film exhibited a threshold electric field of 28±2 V/μm, i.e., 0.1 μA of current were obtained when 28±2 V/μm appeared at the sample surface. With the assumption that $\beta = 1$ (the surface is perfectly flat, so no field enhancement has taken place), it was found that a reasonably good fit to the data could be obtained using $\phi = 0.16 \pm 0.04$ eV and k = 387.1 $\mu A \cdot \mu m^2/V^2$. These values of ϕ, β, and k were used in the subsequent F-N DC I-V simulations.

Device Geometries

A vacuum microtriode structure originally proposed by Eastman [9] is shown in Fig. 1. Electrons are emitted into vacuum by a cathode, pass by 0.3×0.3 μm^2 metal grid electrodes

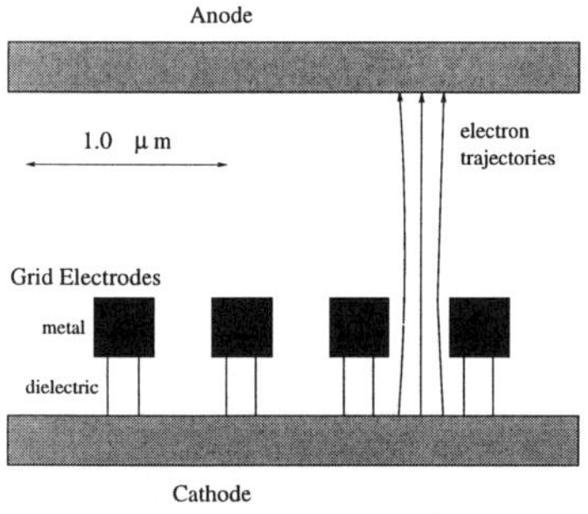

Figure 1: Vacuum microtriode structure proposed by Eastman [9].

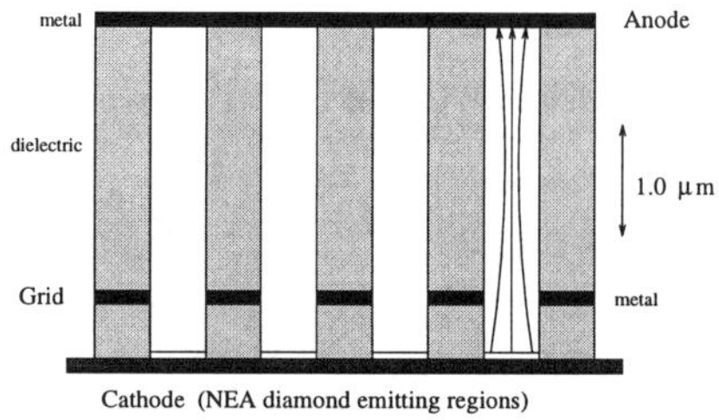

Figure 2: Vacuum microtriode structure proposed by our group.

(which are spaced 0.3 μm apart and sit on 0.15×0.3 μm^2 SiO_2 supports), and proceed on an additional 1.0 μm to an anode.

The electrode geometry of a vacuum microtriode of our own design is shown in Fig. 2. In this structure, electrons are emitted from 0.5-μm-wide flat NEA diamond surfaces, pass by 0.5×0.1 μm grid electrodes (which are spaced 0.5 μm apart and sit on 0.5×0.45 μm^2 SiO_2 supports), and proceed an additional 2.5 μm to the anode (supported by 0.5×2.45 μm^2 SiO_2 supports).

F-N-Type Emission DC I-V Simulations

DC I-V simulations were done for the two microtriode geometries using the EGN2 electron optics simulator [10]. The NEA emission data described above was implemented using the derived parameters β,ϕ, and k to calculate F-N current emission. To maximize the F-N current flowing from cathode to anode, the voltage on the anode should be as large as possible. However, there is a limit to the electric field that can exist in the device; dielectric breakdown and other detrimental effects can occur. For this reason, the maximum electric field allowed in the F-N current simulations was 100 V/μm. This means that no more than $\pm$30 V can be applied to the grid electrodes in the Eastman device, and no more than $\pm$45 V can be applied to the grid electrodes in our proposed structure, when the cathode is grounded.

An example of a F-N emission simulation for the Eastman device is shown in Fig. 3. Note that these device structures can be analyzed in half-space due to their symmetry, and this improves computational efficiency. The Eastman device was biased with 30 V on the control grid and 130 V on the anode. For a 1 $\times$ 1 mm^2 array of triodes, this resulted in

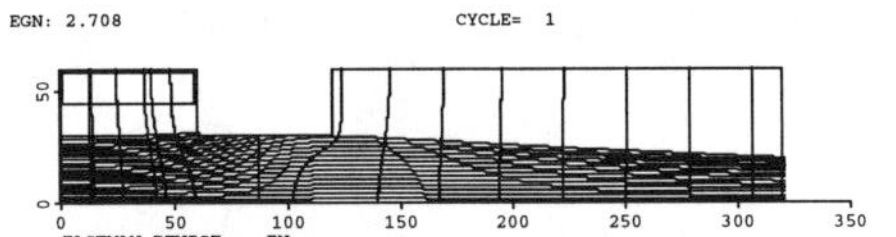

Figure 3: Fowler-Nordheim current emission simulation for Eastman structure.

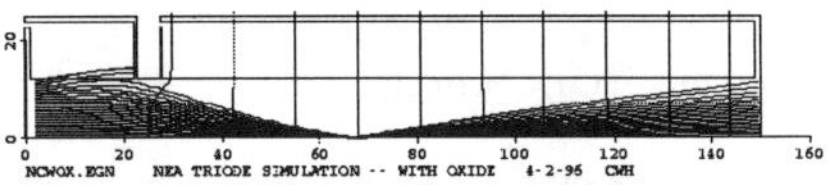

Figure 4: Space-charge-limited current simulation for our proposed structure.

a DC anode current of 1.34 mA and a DC grid current of 0.676 mA. Thus, the DC power dissipation at the anode is 0.174 W and the dissipation at the grid is 20.3 mW. At this bias point, simulation of anode current modulation by grid voltage indicated a transconductance (g_m) of 0.242 mS and an anode resistance (r_a) of 137 kΩ.

For our proposed structure, the microtriode was biased with 45 V on the control grid and 290 V on the anode. For a 1 $\times$ 1 mm^2 array, the DC anode current was 5.48 mA and the DC grid current was 0.514 mA. Thus, the DC power dissipation at the anode is 1.59 W and the disspation at the grid is 23.1 mW. At this bias point, simulation of current modulation indicated a g_m of 0.658 mS and an r_a of 78.1 kΩ.

SCLC DC I-V Simulations

In the SCLC regime, the vacuum microtriode anode current is given by [13]

$$I_A = K(V_A + \mu V_G)^{3/2} \qquad (2)$$

where K is the perveance (A/V$^{3/2}$) and μ ($= r_a \times g_m$) is the amplification factor. We approximated the grid current by way of the expression

$$I_G = FK\mu^{3/2}(V_A + \mu V_G)^{3/2} \qquad (3)$$

where F is determined by device simulations.

For a 1 $\times$ 1 mm^2 array of the Eastman device, simulations yielded an amplification factor (μ) of 171, a perveance (K) of 1.75 mA/V$^{3/2}$, and a grid current constant (F) of 3.09 $\times$ 10^{-4}. An example of a SCLC simulation for our proposed structure is shown in Fig. 4. For a 1 $\times$ 1 mm^2 array, simulations yielded a μ of 56.4, a K of 9.90 mA/V$^{3/2}$, and an F of 2.75 $\times$ 10^{-4}. These simulations lead to an anode current vs. anode voltage and grid voltage characteristics for arrays of the Eastman device and our proposed structure as shown in Figures 5 and 6, respectively.

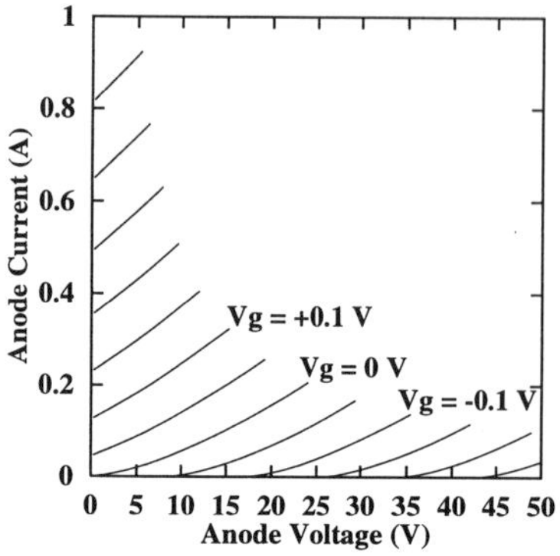

Figure 5: SCLC DC I-V characteristics for Eastman structure array.

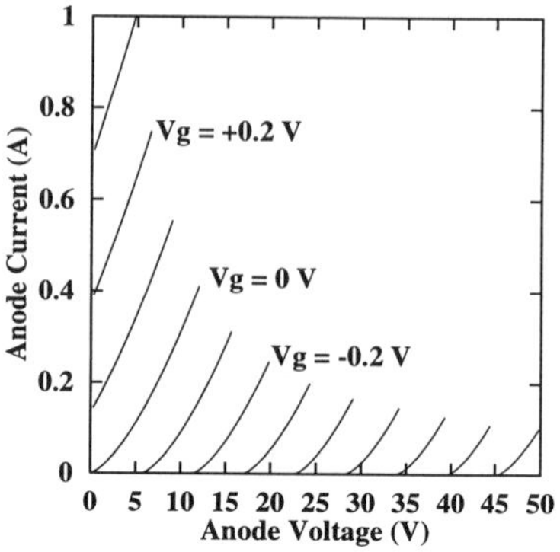

Figure 6: SCLC DC I-V characteristics for an array of our proposed structure.

Microwave Circuit Simulations

Microwave circuit simulations were performed using Microwave Harmonica, a commercially available software package. This program is capable of both linear and nonlinear circuit analysis by way of harmonic balance calculations.

To perform an accurate nonlinear simulation of each microtriode array, a large-signal model was developed, as shown in Fig. 7. This large-signal model contains two current

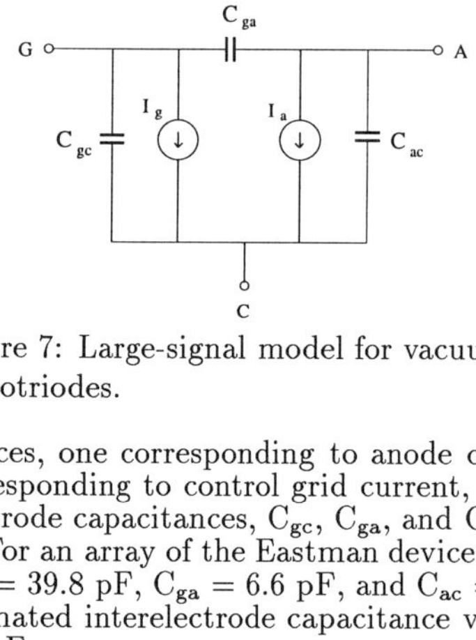

Figure 7: Large-signal model for vacuum microtriodes.

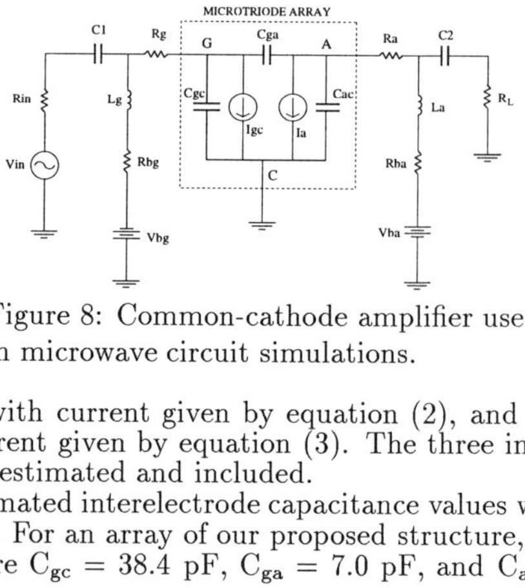

Figure 8: Common-cathode amplifier used in microwave circuit simulations.

sources, one corresponding to anode current, with current given by equation (2), and one corresponding to control grid current, with current given by equation (3). The three interelectrode capacitances, C_{gc}, C_{ga}, and C_{ac}, were estimated and included.

For an array of the Eastman device, the estimated interelectrode capacitance values were C_{gc} = 39.8 pF, C_{ga} = 6.6 pF, and C_{ac} = 2.8 pF. For an array of our proposed structure, the estimated interelectrode capacitance values were C_{gc} = 38.4 pF, C_{ga} = 7.0 pF, and C_{ac} = 1.5 pF.

An array of each type of microtriode was simulated in a common-cathode amplifier circuit, as shown in Fig. 8, in order to determine each array's usefulness as an amplifier component for frequencies rnging from 10 MHz up to 10 GHz. The simulation assumed low-resistance DC and RF voltage sources and low-resistance array package leads. The load resistance was set equal to the anode resistance of the array at the quiescent point [14].

For the Eastman array circuit simulation, a load of 70 Ω was used. The array was biased with 19.2 V on the anode and 0.0547 V on the grid. A 0.15-V magnitude RF input signal was used. As shown in Fig. 9, this array delivered about 1.1 W of power to the load at low

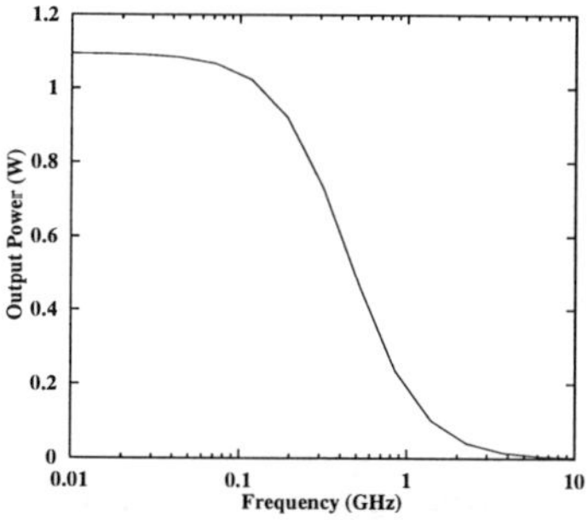

Figure 9: Eastman array output power vs. frequency with a 70-Ω load.

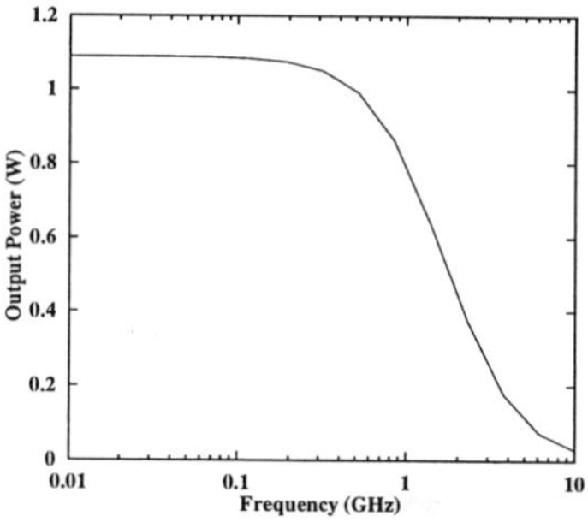

Figure 10: Output power for our proposed array vs. frequency with a 24-Ω load.

frequency and about 0.2 W of power at 1 GHz.

For our proposed device array circuit simulation, a load of 24 Ω was used. The array was biased with 11.0 V on the anode and 0.0345 V on the grid. A 0.23-V magnitude RF input signal was used. As shown in Fig. 10, the array delivered about 1.1 W of power to the load at low frequency and about 0.8 W of power at 1 GHz.

RESULTS AND DISCUSSION

The results obtained for the F-N emission simulations in this work show what type of vacuum microtriode device performance could be expected using typical NEA diamond cold cathodes as they exist today. Obviously, these devices require large anode voltages (130 to 290 V) and suffer from small anode currents (1 to 6 mA), small transconductances (0.2 to 0.7 mS) and large grid leakage currents (0.5 to 0.7 mA). Obviously, these are not practical devices for high-power, high-frequency microwave applications.

When the SCLC DC I-V characteristics for the Eastman device and our proposed structure are compared, it is seen that although Eastman's structure has an amplification factor three times as high (i.e., the grid voltage exerts more control over the grid current), the perveance for our structure is more than five times higher. It should be noted that some of the electron trajectories in the simulations of our proposed device entered and passed through the dielectric layers, indicating that dielectric charging of the SiO_2 surface may need to be taken into account in future simulations.

The microwave circuit simulations reveal that both microtriode geometries suffer from the Miller effect. That is, the capacitance between the grid and the anode is a feedback capacitor, and circuit analysis for these triodes shows that the total capacitance which appears at the grid input is the sum of the grid-to-cathode capacitance plus the grid-to-anode capacitance times the voltage gain of the array [15]. This phenomenon severely reduces gain at higher frequencies. To overcome this problem, a screen grid electrode could be included in the device design, inserted between the control grid and the anode. This would essentially eliminate the control-grid-to-anode capacitance and increase the output power at higher frequencies.

This work has shown the feasibility of using NEA diamond vacuum microtriodes for microwave circuits, but the device geometry and circuit design presented are not optimized for any particular application. Additional work needs to be done to determine optimal designs for these vacuum microtriodes and circuits containing them for particular microwave circuit applications, such as cellular phone or radar subsystems.

CONCLUSIONS

In this work, typical diamond NEA emission data was used to predict the performance of two different vacuum microtriode arrays. It was concluded that with currently attainable current emission densities, useful devices for high-power and high-frequency RF circuits are not attainable. Simulations were then done for the vacuum microtriode structures assuming that emission from the NEA diamond surface would allow the devices to operate in the space-charge-limited regime. These device and circuit simulations indicate that these vacuum microtriode structures operating in the SCLC regime have potential for use in high-power and high-frequency microwave circuits.

ACKNOWLEDGMENTS

This work was supported by the North Carolina State University and the Office of Naval Research under contract no. N00014-92-J-1477. The authors also appreciate the software support provided by Electron Optics Simulations (Los Altos, CA) and Compact Software (Paterson, NJ).

REFERENCES

[1] C. Wang, A. Garcia, D.C. Ingram, M. Lake, and M.E. Kordesch, Electron. Lett. **27**, 1459 (1991).

[2] N.S. Xu, R.V. Latham, and Y. Tzeng, Electron. Lett. **29**, 1596 (1993).

[3] Z. Feng, I.G. Brown, and J.W. Ager, J. Mater. Res. **10**, 1585 (1995).

[4] W. Zhu, G.P. Kochanski, S. Jin, and L. Seibles, J. Appl. Phys. **78**, 2707 (1995).

[5] M.W. Geis, J.C. Twichell, J. Macaulay, and K. Okano, Appl. Phys. Lett. **67**, 1328 (1995).

[6] C.A. Spindt, C.E. Holland, A. Rosengreen, and I. Brodie, IEDM 93, 749 (1993).

[7] G. Gammie, R. Kozlowski, R. Mallavarpu, and A.Palevsky, IEDM 93, 753 (1993).

[8] D. Palmer, H.F. Gray, J. Mancusi, D. Temple, C. Ball, J.L. Shaw, and G.E. McGuire, J. Vac. Sci. Technol. B **13**, 150 (1995).

[9] L.F. Eastman in *Proceedings of the Second International Conference on Vacuum Microelectronics*, edited by R.E. Turner, Bath, U.K., July 24-26, 1989, 189.

[10] *EGN2*, Electron Optics Simulations, 451 Patrick Way, Los Altos, CA 94022.

[11] *Microwave Harmonica*, Compact Software, 483 McLean Blvd. & Corner of 18th Avenue, Paterson, NJ 07504.

[12] S.P. Bozeman, P.K. Baumann, B.L. Ward, M.J. Powers, J.J. Cuomo, R.J. Nemanich, and D.L. Dreifus, "Electron Emission Measurements from CVD Diamond Surfaces." To be published in Diamond and Related Materials, 1996.

[13] H.J. Reich, *Theory and Applications of Electron Tubes* (McGraw-Hill Book Company, New York, 1944).

[14] W.B. Nottingham, Proc. I.R.E. **29**, 620 (1941).

[15] A.S. Sedra and K.C. Smith, *Microelectronic Circuits* (Saunders College Publishing, Philadelphia, 1991).

TETRAHEDRAL AMORPHOUS CARBON THIN FILM TRANSISTORS

F.J. CLOUGH *, B. KLEINSORGE, W.I. MILNE and J. ROBERTSON
Cambridge University Engineering Department, Trumpington Street, Cambridge, CB2 1PZ, U.K.
* Emerging Technologies Research Centre, DeMontfort University, The Gateway, Leicester, LE1 9BH, U.K.

ABSTRACT

This paper describes the design and fabrication of a carbon based thin film transistor (TFT). The active layer is formed from a novel form of amorphous carbon (a-C) known as tetrahedrally bonded amorphous carbon (ta-C) which can be deposited at room temperature using a filtered cathodic vacuum arc (FCVA) technique. In its 'as grown' condition, ta-C is p-type and the devices described here, produced using undoped material, exhibit p-channel operation.

INTRODUCTION

The flat panel display market is currently dominated by active matrix addressed liquid crystal displays (AMLCDs) for which the preferred TFT switching elements are manufactured from either hydrogenated amorphous silicon (a-Si:H) or polycrystalline silicon (poly-Si). However, the temperatures required to produce these materials make them incompatible with the desire to use low cost plastic substrates. Further, a-Si:H devices continue to suffer from gate bias induced instabilities thought to be associated with hydrogen motion [1] and the production of poly-Si films, uniformly over large areas, remains problematical.

Our solution is to investigate the potential of diamond-like carbon (DLC) as an alternative material for the active layer in the LCD pixel switches. Although a number of carbon-based TFTs have previously been demonstrated they have either been produced from high temperature (> 800°C) CVD diamond [2] or have used more diverse forms of carbon, such as polymeric p-phenelyne vinylene (PPV) [3] or carbon C_{60} [4], for which encapsulation is required to ensure satisfactory operation.

DLC is an amorphous form of carbon which contains a large fraction of sp^3 bonding. There are two basic types of DLC, hydrogenated and unhydrogenated. Hydrogenated amorphous carbon (a-C:H) is usually prepared by the plasma-enhanced CVD (PECVD) technique and its properties can be varied by altering the dc 'self bias' on the electrode supporting the substrate. A more novel type of unhydrogenated a-C can be produced using a FCVA technique. This material can be deposited at room temperature and, when deposited at an ion energy of 100 eV, has a bandgap of approximately 2.5 eV, contains negligible hydrogen (< 0.5 at %) and is up to 85% sp^3 bonded [5]. As grown the material is p-type with a conductivity activation energy of approximately 0.35 eV. N-type doping can be achieved by the addition of nitrogen to the reaction vessel during film deposition [6].

This paper will describe the design, manufacture and performance of TFTs made using undoped ta-C as the active layer. Some investigation of the mechanisms responsible for the observed FET operation is described.

EXPERIMENT

Device Design

It has been shown by Davis et al [7] that, although bulk ta-C films have a high sp^3 content, there is always a surface layer, rich in sp^2 bonding, present. This is illustrated in Figure 1 which shows a bright field TEM image of a ta-C film. The film was grown in two stages between which it was exposed to the atmosphere for a period of 10 minutes. The silicon substrate is shown on

Mat. Res. Soc. Symp. Proc. Vol. 423 © 1996 Materials Research Society

the extreme left of the picture and five distinct layers can be identified by Fresnel contrast. The lighter regions marked A, C and E correspond to regions of lower density which, using nanometre spatial resolution techniques, have been shown to be sp^2 rich [7]. The graphitic layer at the top surface is of the order of 1 - 2 nm and 3 - 5 nm at the bottom interface with the substrate. Such layers will lead to the pinning of the Fermi Level and prevent transistor action. The device structure and the fabrication process must therefore allow this layer to be removed.

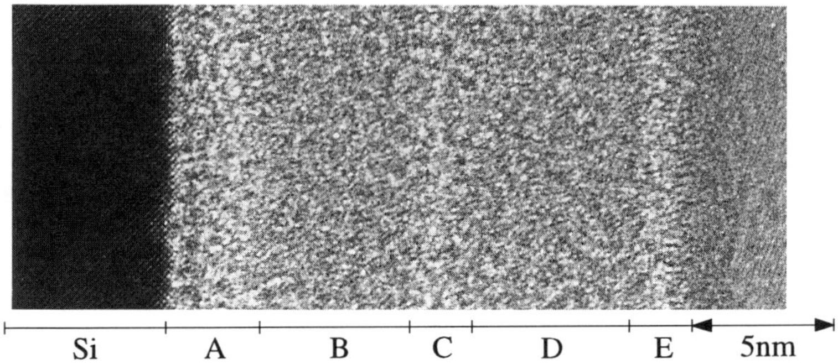

Figure 1. Cross-sectional bright field TEM of a ta-C double layer on silicon [7].

From the several TFT structures available a coplanar type was chosen for the final device. This configuration has several advantages: the graphitic layer on the top surface is thinner and the least contaminated; the top layer can be easily accessed to allow etching off of the sp^2 layer and; the subsequent deposition of the gate insulator layer can be effected in the same pump down sequence thereby minimising the chance of the sp^2 layer reforming and reducing potential contamination problems at this critical interface.

Device Fabrication

A lift -off process was used to pattern a 50 nm thick layer of undoped ta-C into 1.2 x 0.6 mm islands on a Corning 7059 glass substrate. Aluminium was then deposited by thermal evaporation and patterned to define the source and drain contacts. The graphitic layer on the top surface of the ta-C islands was then etched away in an N_2O plasma and a 400 nm layer of PECVD silicon nitride was then immediately deposited to form the TFT gate insulator. An aluminium top gate contact completed the process. A detailed description of the device fabrication is given in [8] and a layout of the completed planar structure is shown in Figure 2.

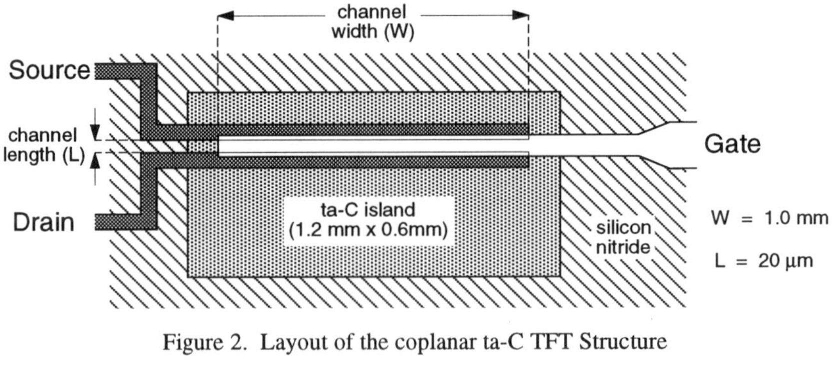

Figure 2. Layout of the coplanar ta-C TFT Structure

Results

A typical transfer characteristic for a device produced using undoped ta-C as the active layer is shown in Figure 3. This device operates in the p-channel mode. As the bias on the gate is ramped from + 100 V to - 100 V the drain current increases by 1 to 2 orders. For all devices tested the gate leakage is less than a few picoamps.

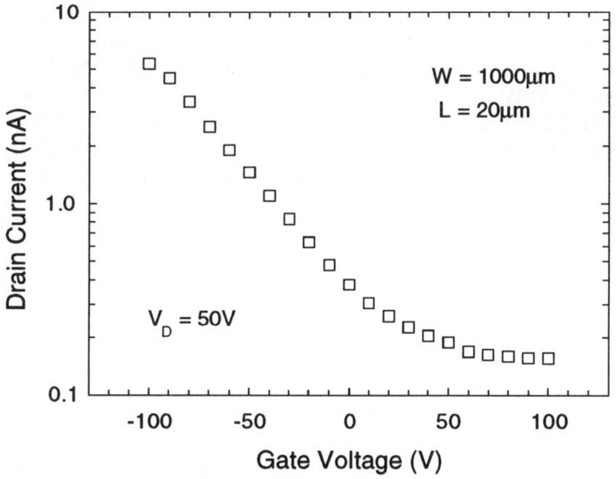

Figure 3. Transfer Characteristic for p-channel ta-C TFT

Figure 4 shows a series of drain characteristics for the same device. The drain current is clearly modulated by the applied gate voltage. However, no saturation in the drain characteristics has been observed, even for drain voltages as high as 100 V. The low drain current at low drain bias may be attributed to the poor ohmicity of the aluminium source and drain contacts to ta-C layer. We are currently investigating other contact materials in order to eliminate this effect.

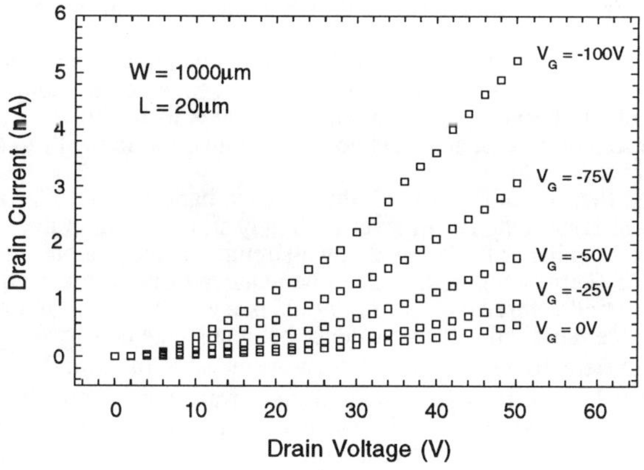

Figure 4. Drain Characteristics for a typical p-channel ta-C TFT

DISCUSSION

The TFTs operate by accumulation in the p-channel enhancement mode. This is to be expected as the undoped, 'as grown' ta-C is strongly p-type. The increase in the drain current as the gate voltage in made increasingly more negative is associated with the field induced movement of the Fermi level closer to the valence band. Similarly, the reduction in drain current as the gate bias is ramped positively is consistent with the movement of the Fermi level towards midgap.

This is confirmed by the data presented in Figure 5 which shows the conductivity activation energy (E_A) as a function of applied gate bias. When there is no gate bias applied the measured activation energy of 0.35 eV is consistent with the p-type nature of undoped ta-C. As the gate voltage is increased negatively the activation energy decreases. This is in agreement with the observed rise in the TFT drain current and corresponds to the field induced shifting of the Fermi level closer to the valence band in the ta-C layer at the ta-C/Silicon nitride interface. As the gate bias is increased more negatively E_A eventually saturates at around 0.26 eV.

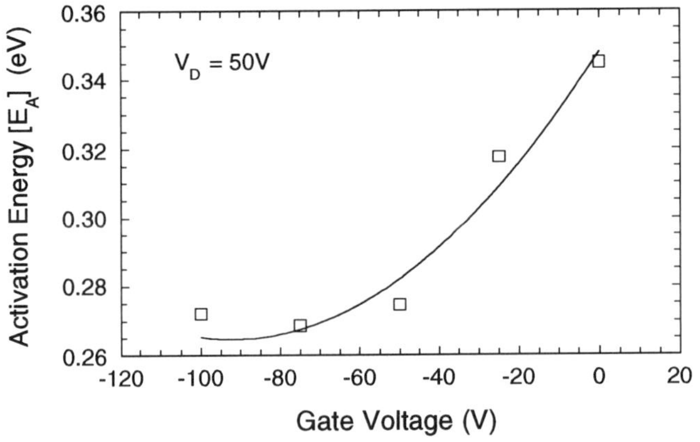

Figure 5. Conductivity Activation Energy versus Gate Voltage.

The slow switching speed and high 'Off' current (when compared to a-Si:H TFTs) can be attributed to the large density of deep or localised states which are known to exist in a-C films [9]. Movement of the Fermi level requires these states to be filled or emptied and large gate fields are therefore required. In addition, hopping conduction in the deep states is responsible for the high 'Off' current.

It is suggested that, in undoped ta-C, the Valence band tail is wide (again in comparison to device quality a-Si:H) and the Fermi level is already situated within the Valence band tail states. Thus, at fields compatible with the breakdown strength of the silicon nitride gate insulator, only small shifts in the Fermi level position can be achieved due to the high tail state density. Our observation of a small saturated shift of 0.09 eV is therefore due to the movement of the Fermi level deeper into the rapidly increasing valence band tail state density.

The very small field effect mobilities which we measure in these devices, of the order 10^{-5} to 10^{-6} $cm^2V^{-1}s^{-1}$ [8], indicate that extended state transport is not the dominant conduction mechanism. Hopping conduction in the valence band tail states is the most likely mechanism. The gate field dependence of the hopping current can be attributed to the movement of the Fermi level closer to the valence band where the defect density is higher.

Further comparison with the a-Si:H system suggests that improvements in TFT 'On' state performance may be obtained by operating the device in n-channel mode. The field effect mobility of electrons in a-Si:H is higher and the conduction band tail is sharper both of which are responsible for a the superior performance of n-channel a-Si:H TFTs. A significant reduction in the deep state density of the ta-C films will be required to improve the 'Off' state performance of the ta-C devices.

CONCLUSIONS

Low temperature, glass compatible, amorphous carbon based TFTs, operating in the p-channel enhancement mode, have been produced. Gate control of the drain current has been observed. The measured hole mobility is too low for practical applications. Future work will concentrate on optimisation of the material in order to steepen the valence band tail and reduce the deep state density. Efforts are also continuing to produce n-channel devices from nitrogen doped ta-C layers. It is hoped that the electron mobility and the shape of the conduction band tail will be more conducive to high performance ta-C TFT operation.

ACKNOWLEDGEMENTS

The authors are grateful to K.M. Knowles of the Department of Materials Science and Metallurgy, Cambridge University for the TEM picture and the information relating to it.

REFERENCES

1. M.J. Powell, C. Van Berkel, A.R. Franklin, S.C. Deane and W.I. Milne, Phys Rev B **45**, 4160 (1992).
2. G.S. Gildenblatt, S.A. Grot, C.W. Hatfield and A.R. Badzian, IEEE Electron Device Letts. **12**, 37 (1991).
3. K. Pichler, C.P. Jarrett, R.H. Friend, B. Ratier and A.Moliton, J. Appl. Phys. **77**, 3523 (1995).
4. R.C. Haddon, A.S. Perel, R.C. Morris, T.T.M. Palstra, A.F. Hebard and R.M. Fleming, Appl. Phys. Letts. **67**, 121 (1995).
5 V.Veerasamy, G.A.J.Amaratunga, W.I.Milne and P.J.Fallon, J. Non. Crystalline Solids 164-**166**, 1111, 1993.
6. V.S. Veerasamy, J. Yuan, G.A.J. Amaratunga, W.I. Milne, K.W. Gilkes, M.Weiler and L.M. Brown, Phys Rev B, **48**, 17954 (1993).
7. C.A. Davis, K.M. Knowles and G.A.J. Amaratunga, Surface and Coatings Technology, **76 - 77**, 316 (1995).
8. F.J. Clough, W.I. Milne, B. Kleinsorge, J. Robertson, G.A.J. Amaratunga and B.N. Roy, Electronics Letters **32**, 498 (1996).
9. V.S. Veerasamy, G.A.J. Amaratunga, C.A. Davis, W.I. Milne, P.Hewitt and M. Weiler Solid State Electronics **37**, 319 1994.

Theoretical Prediction of Zinc Blende Phase GaN Avalanche Photodiode Performance Based on Numerically Calculated Electron and Hole Impact Ionization Rate Ratio

J. Kolnik, I. H. Oguzman, K. F. Brennan, R. Wang* and P. P. Ruden*
School of ECE, Georgia Tech, Atlanta, GA 30332, kolnik@celdec1.mirc.gatech.edu
*Dept. of Electrical Engineering, University of Minnesota, Minneapolis, MN 55455

ABSTRACT

In this paper, we present the first calculations of the electron and hole initiated interband impact ionization rate in zinc blende phase GaN as a function of the applied electric field strength. The calculations are performed using an ensemble Monte Carlo simulator including the full details of the conduction and valence bands along with a numerically determined, wave-vector dependent interband ionization transition rate determined from an empirical pseudopotential calculation. The first four conduction bands and first three valence bands, which fully comprise the energy range of interest for device simulation, are included in the analysis. It is found that the electron and hole ionization rates are comparable over the full range of applied electric field strengths examined. Based on these calculations an avalanche photodiode, APD, made from bulk zinc blende GaN then would exhibit poor noise and bandwidth performance. It should be noted however, that the accuracy of the band structure employed and the scattering rates is presently unknown since little experimental information is available for comparison. Therefore, due to these uncertainties, it is difficult to unequivocally conclude that the ionization rates are comparable.

INTRODUCTION

The III-Nitride materials are presently attracting great interest for numerous device applications owing to their relatively large band gap energies [1]. The energy gap of these materials can be spectrally matched for blue and ultraviolet emission and detection [2]. Additionally, the large band gap of the III-Nitrides, particularly GaN, offers the potential for ultra-high temperature and radiation hard electronics. In many of these device applications, particularly short channel field effect devices and avalanche photodiode, APD, detectors, high electric fields exist which can lead to carrier multiplication through impact ionization. In short channel devices it is often important to avoid carrier multiplication while in APD devices multiplication provides gain which can be exploited for low light level detection. In any case, it is important to know at what electric field strengths impact ionization will occur and to know more about the nature of the ionization breakdown, i.e., whether the carrier ionization rates are comparable, etc. Knowledge of the ionization rates ratio enables the determination of the noise and bandwidth properties of an APD, which in turn dictate its usefulness in most detector applications [3,4].

Spectral matching to blue and ultraviolet wavelengths coupled with the high quantum efficiency of most of the III-Nitride materials makes them attractive candidates for new, solid state detectors. To date, GaN based detectors have been made [5-7]. Photoconductive UV sensors made on insulating GaN [5,6] as well as p-n junction detectors have been demonstrated [7]. However, there is no information presently available about GaN based detectors utilizing impact ionization such as APDs. As is well known, optimal performance of an APD, as

Mat. Res. Soc. Symp. Proc. Vol. 423

measured in terms of low noise and high bandwidth operation, occurs under single carrier ionization conditions [4]. Though few materials are known to exhibit single carrier-type ionization, satisfactory noise performance can still be attained provided the ionization rate of the secondary carrier species is very much less, by more than an order of magnitude, than the primary carrier species ionization rate.

It is the purpose of this paper to present the first determination of the electron and hole ionization rates in bulk zinc blende phase GaN. The rates are determined theoretically using an ensemble Monte Carlo calculation. The Monte Carlo simulators for the electrons and holes contain the full details of the conduction and valence bands covering the full energy range of interest. Results are presented for the electron and hole ionization rates, average carrier energy and originating band as a function of applied electric field strength.

MODEL DESCRIPTION

The calculations are performed using ensemble Monte Carlo simulators for electron and hole transport which include the full details of the first four conduction and first three valence bands respectively. These bands cover the full energy range of interest. The band structure is obtained from an empirical pseudopotential calculation [8] using an expansion set of 113 reciprocal lattice vectors. All of the relevant phonon scattering mechanisms, as well as ionized impurity scattering are included into the simulators. The parameters used to determine the electron scattering rates, i.e., phonon energies, dielectric constants, deformation potential constants, etc. have been reported in reference [8]. In the hole simulator, owing to the very anisotropic nature of the valence bands, the phonon scattering rates are all calculated numerically by integrating over the actual pseudopotential band structure and incorporated into the simulations following the approach of Hinckley and Singh [9]. In both the electron and hole simulators, the high energy phonon scattering rates are assumed to be dominated by deformation potential scattering. In this regime, the phonon scattering rate is calculated through use of a time dependent perturbation theory expansion by integrating the transition rate over the final, numerically determined, density of states including collision broadening effects [10].

The wavevector dependent interband impact ionization transition rate is calculated numerically using the empirical pseudopotential band structure. The transition rate is determined by integrating Fermi's golden rule for a two-body, screened Coulomb interaction over the possible final states using a numerically generated wavevector dependent dielectric function [11] and pseudowavefunctions. The usual first order umklapp processes as well as those processes arising from the pseudowavefunction expansion as described by Sano and Yoshii [12] are included in the evaluation of the direct and exchange terms which comprise the full matrix element for the transition. The wavefunctions are determined throughout the entire Brillouin zone using the transformation properties of the irreducible wedge for the zinc blende symmetry [12]. Free carrier screening is also added, and for simplicity, is treated in the Debye approximation with a carrier concentration of 10^{17} cm^{-3} at 300 K. The integration is performed using the Monte Carlo method by sampling over several million final states. The wavevector dependent transition rate for both the electrons and holes is calculated for 1419 initial states which is then incorporated into the Monte Carlo simulators.

For simplicity, the bands are labeled in the present simulators using a basic ordering scheme. The lowest energy points are assigned to the first band, either conduction or valence bands, which are labeled #1. The next lowest energy points are assigned to band #2, etc. Though band crossing effects can sometimes be important, for the purposes of identifying the bands, they are neglected. In the case of zinc blende phase GaN, the number of band crossings is relatively small anyway. Nevertheless, we investigated the effect of including band crossings in the calculations. It was found that the calculated results for the ionization rates do not change significantly if the band crossing effects are accounted for or not during the course of the simulation.

RESULTS

As mentioned above, presently there is no available information, either experimental or theoretical, about the interband impact ionization coefficients in bulk GaN. The calculations presented here are the first reported estimates of these rates. The calculated electron and hole impact ionization rates in bulk zinc blende phase GaN as a function of inverse applied electric field are shown in Figure 1. As can be seen from Figure 1, the rates are nearly the same at the highest electric field strength, 3.0 MV/cm, and below. Only at the lowest field strength, 1.0 MV/cm, does the hole ionization rate appear to be significantly higher than the electron rate. Some caution should be exercised however in interpreting the lower field results. At 1.0 MV/cm the ionization counts are too few to provide good statistical estimates. Therefore, there is a high degree of uncertainty about the level to which the ionization rates differ in this range.

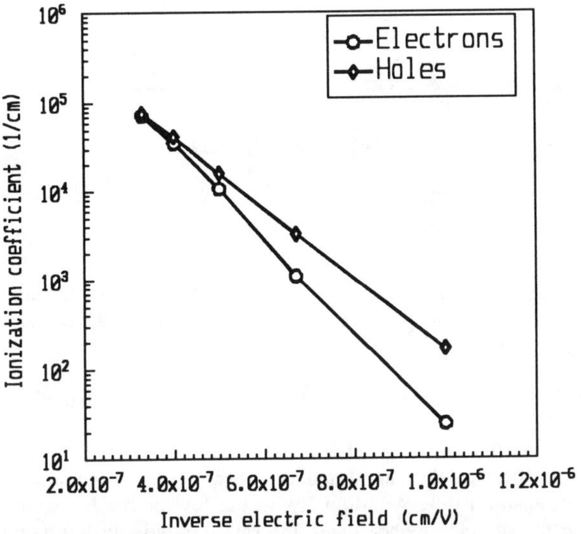

Figure 1: Calculated electron (squares) and hole (diamonds) impact ionization coefficients as a function of inverse applied electric field in bulk zinc blende phase GaN.

The percentage of ionization events originating out of each band for electron initiated events is plotted in Figure 2 as a function of electric field strength. In zinc blende phase GaN, the first four conduction bands dominate the carrier transport dynamics. As can be seen from Figure 2, most of the electron initiated ionization events originate out of the second conduction band. At higher electric fields the third conduction band begins to contribute significantly to the ionization rate. Neither the first nor the fourth conduction bands contribute much to the ionization rate at the field strengths examined here.

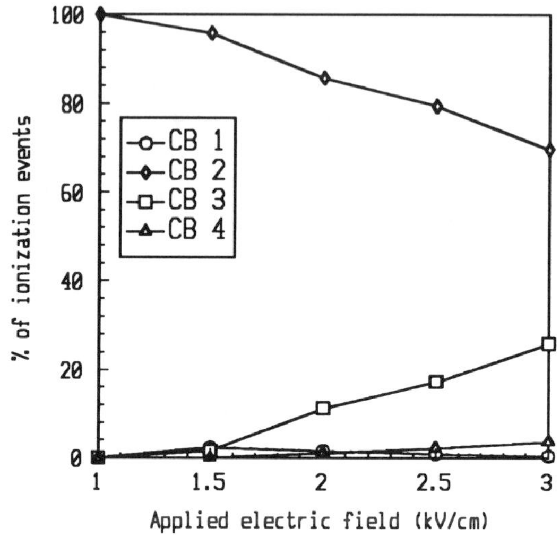

Figure 2: Percentage of electron initiated impact ionization events arising from each of the first four conduction bands as a function of applied electric field.

The case for hole initiated ionization is very different. In this case all of the ionization events originate out of the third valence band. The third valence band contributes nearly 100% of the ionization events over the full range of applied electric fields. This is due to the fact that there are no points in the first valence band and very few points in the second valence band which have a nonzero transition rate. Additionally, at the few points for which the transition rate does not vanish in the second valence band, the rate is relatively low compared with the rate from the third valence band. Subsequently, the third valence band dominates the hole initiated ionization process.

Finally, the quantum yield as a function of applied electric field strength for both electron and hole initiated impact ionization is plotted in Figure 3. The quantum yield is defined as the average number of impact ionization events caused by a high energy injected carrier until its kinetic energy relaxes below the ionization threshold energy through scattering and/or ionization events. The quantum yield provides an excellent means of determining the relative "hardness" of the threshold, i.e., the degree to which a carrier will survive to energies much higher than the ionization threshold energy. As can be seen from Figure 3, the quantum yield increases at a substantially greater rate for hole initiated ionization than for electron initiated ionization. Notice, that the quantum yield reaches 1 for hole initiated ionization at a much lower energy than for electron initiated ionization indicating a much harder threshold energy in the valence band. This implies that the electrons survive to much higher energies on average compared with the holes prior to suffering an impact ionization event. Consequently, the threshold energy for hole initiated ionization is significantly "harder" than that for electron initiated ionization in bulk zinc blende phase GaN.

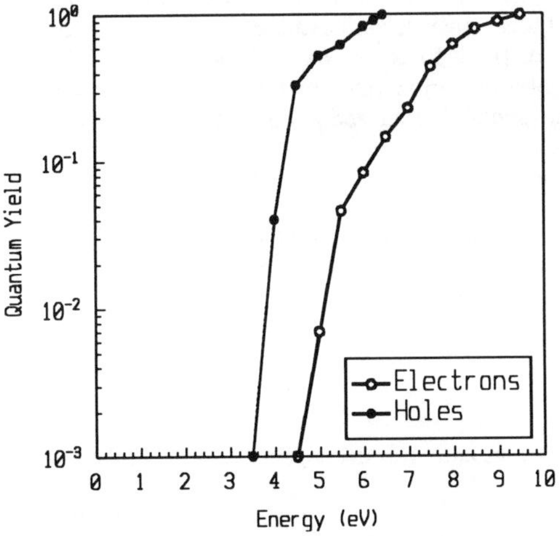

Fig. 3: Calculated quantum yield for both electrons and holes as a function of energy of the carriers.

CONCLUSIONS

In this paper, we present the first determination of the electron and hole impact ionization rates in bulk zinc blende phase GaN. The rates are determined theoretically from ensemble Monte Carlo simulators. The Monte Carlo calculations include the full details of the first four conduction and first three valence bands derived from an empirical pseudopotential calculation. Additionally, the wavevector dependent impact ionization transition rate is determined using a numerically generated dielectric function [11] and pseudowavefunctions. It is found that the carrier ionization rates are nearly identical at high electric field strength, ~3 MV/cm, becoming somewhat different at lower field strengths. Though there appears to be a difference in the electron and hole ionization rates at ~1.0 MV/cm, due to poor statistics, this difference cannot as yet be considered reliable.

It is further found that the threshold for hole initiated ionization in bulk zinc blende phase GaN is significantly "harder" than the corresponding threshold for electron initiated ionization. As a result, the electrons will survive to higher energies, on average, before suffering an impact ionization event than the holes.

It should be further noted that the calculations rely to some extent on the knowledge of several parameters, such as phonon energies, deformation potentials, etc, which are presently not well known. In addition, we have found that the ionization transition rate is highly sensitive to the calculated band structure and can vary significantly depending upon the details of the particular bands used. Presently, the accuracy of our band structure at high energies is unknown since little experimental data at these energies are available to provide a comparison. Therefore, due to these uncertainties, it is difficult to unequivocally conclude, based on these calculations alone, that the electron and hole ionization rates in bulk zinc blende GaN are nearly equal. Further work, principally experimental, is required to confirm these predictions.

ACKNOWLEDGEMENTS

This work was sponsored in part by the National Science Foundation through a collaborative grant made to Georgia Tech (ECS-9313635) and to the University of Minnesota (ECS-9408479).

REFERENCES

1. S. N. Mohammad, A. A. Salvador, and H. Morkoc, Proceedings of the IEEE, **83**, 1306 (1995).
2. S. Nakamura, T. Mukai, and M. Senoh, Appl. Phys. Lett, **64**, 1687 (1994).
3. R. J. McIntyre, IEEE Trans. Electron Dev., **ED-13**, 164 (1966).
4. M. C. Teich, K. Matsuo, and B. E. A. Saleh, **QE-22**, 1184 (1986).
5. M. A. Khan, J. N. Kuznia, D. T. Olson, J. M. Van Hove, M. Blasingame, L. F. Leitz, Appl. Phys. Lett., **60**, 2917 (1992).
6. M. A. Khan, Microwave Journal, **36**, 69 (1993).
7. Q. Chen, M. A. Khan, C. J. Sun, J. W. Yang, Electronics Lett., **31**, 1781 (1995).
8. J. Kolnik, I. H. Oguzman, K. F. Brennan, R. Wang, P. P. Ruden and Y. Wang, J. Appl. Phys., **78**, 1033 (1995).
9. J. M. Hinckley and J. Singh, Phys. Rev. B, **41**, 2912 (1990).
10. Y. C. Chang, D. Z.-Y. Ting, J. Y. Tang and K. Hess, Appl. Phys. Lett., **42**, 76 (1983).
11. R. Wang, P. P. Ruden, J. Kolnik, I. Oguzman, and K. F. Brennan, Proceedings of the MRS, Fall, 1995.
12. N. Sano and A. Yoshii, Phys. Rev. B, **45**, 4171 (1992).

OBIC MEASUREMENTS ON 6H-SiC SCHOTTKY DIODES

K. ROTTNER *, A. SCHÖNER *, M. FRISCHHOLZ **, R. HELBIG **
* IMC, P.O. Box 1084, S-16421 Stockholm-Kista, Sweden
** Institut für Angewandte Physik, Universität Erlangen, Staudtstr. 7, D-91058 Erlangen, Germany

ABSTRACT

The OBIC (Optical Beam Induced Current) technique is a powerful method to investigate the electric field distribution of p-n junctions in SiC. In a previous work we found strong indications for the presence of a high density of negative surface charge in n-type SiC. In order to study samples of both conductivity types under similar conditions we prepared Schottky contacts on n-type and p-type 6H-SiC CVD epitaxial layers.

OBIC measurements show an extension of the depletion region of several hundreds of microns from the edge of the contact on n- and p-type samples, thus interconnecting diodes on an area up to several mm^2. Our results imply that there is no fixed surface charge but a high density of both acceptor- and donorlike surface states leading to a dependence of the net surface charge on the Fermi energy, in which case the sign of the surface charge reverses from negative on n-type material to positive on p-type 6H-SiC.

INTRODUCTION

Recent progress in the growth of single crystal SiC wafers as well as CVD layers has made it possible to produce devices with a blocking capability of up to 4.5 kV [1]. The ability of supporting high electric fields is limited by the onset of avalanche breakdown, which can occur both within the interior regions of the device and at the edges. The resulting high electric surface fields in SiC power devices are one of the major problems to be solved to further improve device performance. A careful design of the edge termination of power devices is necessary to raise the breakdown voltage to the upper bound provided by the analysis of the semi-infinite junction.

Another factor that comes into play is the passivation of the surface of discrete power devices which is in fact closely related to the problem of edge termination. The passivation induced surface charges strongly influence the electric field in the device regions lying close to the surface and a proper choice of the passivant can lower the surface electric field considerably. The higher surface fields of SiC power devices make surface passivation even more critical in comparison to Si power devices.

A previous study [2] on SiC p^+-n mesa diodes showed that the experimental breakdown voltage exceeded the value expected from numerical studies regardless of the specific edge termination design (mesa diodes with and without multiple floating field rings). OBIC measurements revealed a substantial expansion of the depletion layer to a value ten times as large as that provided by the numerical analysis, thus decreasing the surface electric field and in consequence increasing the maximum achievable breakdown voltage. These results implied the presence of a negative surface charge with a concentration of $\geq 3\times10^{12}$ cm^{-2}. Surface charge variations will result in irregularly shaped depletion layers which itself will cause premature breakdown. Locally reduced surface charge concentrations may result in increased electric surface fields in this spot. On the opposite high surface charge concentrations may lead to the incorporation of remote defects like pinholes in the active device area due to an increased depletion width. The origin of the surface charge is still unknown and it is not yet possible to control its precise concentration.

Mat. Res. Soc. Symp. Proc. Vol. 423

Therefore an optimized junction termination design including the effect of this charge is not possible to date and the efforts should be directed towards removing the charge.

In this study we further investigated the questions that originated from those findings about origin and character of the surface charge by studying Schottky diodes on n- and p-type 6H-SiC with respect to conductivity type, doping concentration and surface processing. The use of Schottky contacts instead of p-n mesa diodes simplified the processing since no RIE process and no oxide formation were involved. Furthermore it was then possible to study the as-grown SiC surface.

EXPERIMENT

The samples used in this work were CVD grown p- and n-type epilayers with thickness ranging from 10 μm to 30 μm grown on the Si-face of 3.5° off-oriented 6H-SiC substrates from Cree. The doping levels of the epilayers were in the range of $2x10^{14}$ cm^{-3} to $1x10^{16}$ cm^{-3}. On top of the epilayers Schottky barrier contacts were formed by electron beam evaporation of Al on n-type SiC and Ti on p-type SiC, respectively. The contacts had a diameter of 200 μm and were arranged in an array with a spacing of 300 μm. For the large area backside ohmic contact we used Al. In order to study the influence of surface processing on the depletion layer width several samples were etched using a standard RIE process ($CF_4/H_2/Ar$ plasma) to remove approximately 1 μm of the CVD layer prior to the formation of Schottky contacts.

A mapping of the depletion layer was accomplished by measuring the spatial variation of the photocurrent in the device. The principle configuration for this technique is illustrated in Fig. 1. In order to generate electron-hole pairs we used monochromatic light from a 100 W mercury arc lamp in combination with interference filters (bandwidth 5 nm) for the UV mercury lines. An aperture was homogeneously illuminated and reimaged onto the sample surface using a microscope with a long working distance objective. For 3.4 eV photons the penetration depth ($1/\alpha$, with the absorption coefficient α) in 6H-SiC is about 14 μm [3]. The spotsize was approximately 15 μm. In order to separate the photocurrent from the dark current and for increased sensitivity the light beam was mechanically chopped (f = 70 Hz) and a lock-in technique was used. A more detailed description of the set-up can be found in [4].

Within the space charge region the charge carriers are separated by the electric field before they can recombine. At low injection levels and for doping levels much larger than the intrinsic carrier concentration the photocurrent outside the space charge region will decrease roughly

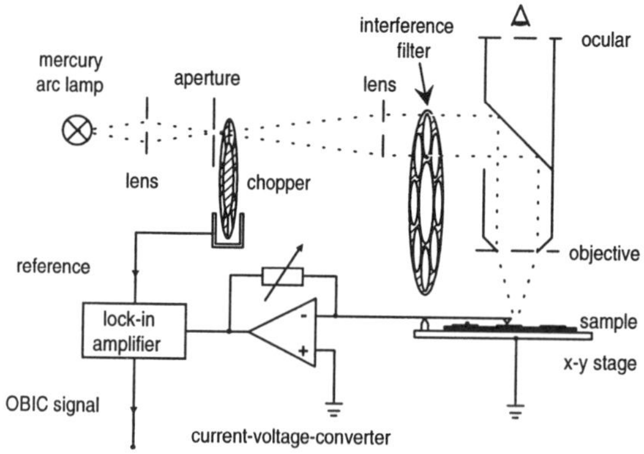

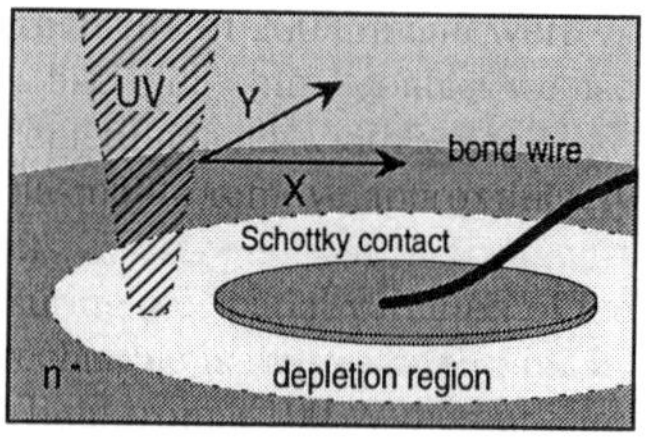

Fig. 1 Schematic view of the OBIC set-up (left) and principle of scanning the depletion region of a Schottky contact (magnified section of the sample surface, right).

exponentially according to the minority carrier diffusion length. Thus the OBIC images can be directly related to the expansion of the depletion layer at the surface of the device.

RESULTS

The OBIC mapping for a low N-doped n-type sample (doping concentration $2x10^{14}$ cm^{-3}) is shown in Fig. 2 in grayscale encoding, where white color represents the maximum photocurrent and black the minimum photocurrent, respectively. Superimposed is a drawing of the contact array. The depletion layer of the device under test extends over the whole area that was investigated without a significant decrease in the signal except for the area under the opaque metal contacts. This has a large impact on the result of any electrical measurements since the actual device area is much larger than the area covered by the metallisation. Furthermore the OBIC image indicates that all diodes are interconnected and it is not possible to single out a specific diode for electrical characterization. This result implies that the presence of a negative surface charge is not distinctive for mesa diodes where process steps as RIE and oxidation are involved, but is also present at the as-grown surface.

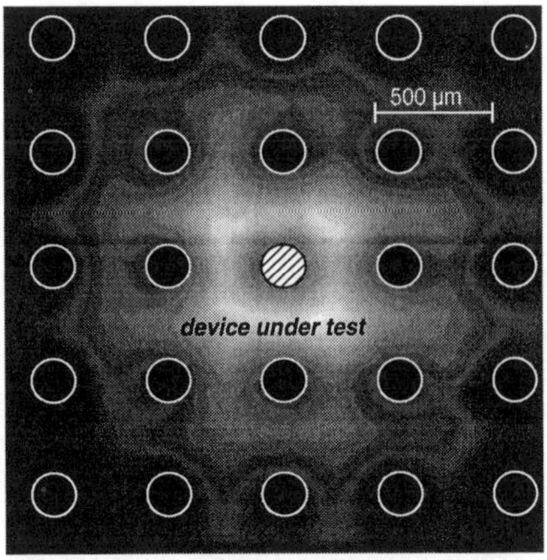

Fig. 2 OBIC image of the depletion region of a reverse biased Schottky contact on n-type 6H-SiC ($N_D = 2x10^{14}$ cm^{-3}, irradiation wavelength 365 nm, reverse bias 3 V)

Fig. 3 illustrates the influence of doping concentration on the extension of the space charge region. A magnified section of the OBIC image of Fig. 2 is compared to an image of a higher doped sample (doping concentration $1x10^{16}$ cm^{-3}). In the latter case the space charge region extends only 220 µm from the contact edge with a well defined perimeter. Nevertheless this

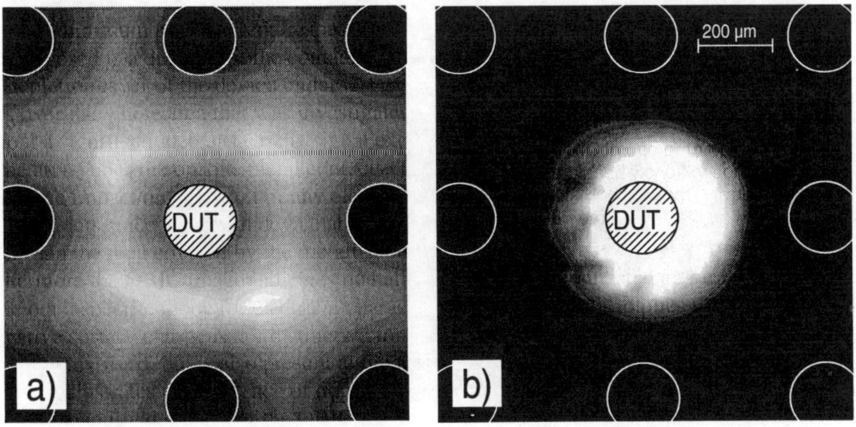

Fig. 3 Dependence of the depletion layer width on the doping concentration (irradiation wavelength 365 nm, reverse bias 3 V).
a) doping concentration $2x10^{14}$ cm^{-3}
b) doping concentration $1x10^{16}$ cm^{-3}.

value is still more than ten times larger than expected from theory. A brief estimate of the diffusion length of 6H-SiC from mobility and lifetime data reported in literature [1,5,6] for the particular doping levels yields a difference in diffusion lengths by a factor of 5. From the experimentally observed decay of the photocurrent with distance from the contact edge we found the diffusion length for the low doped sample to be 50 times larger than for the high doped sample (Fig. 4). This result does not agree with the sole assumption of carrier transport by lateral diffusion. A more practical assumption to explain above results is the formation of an inversion channel.

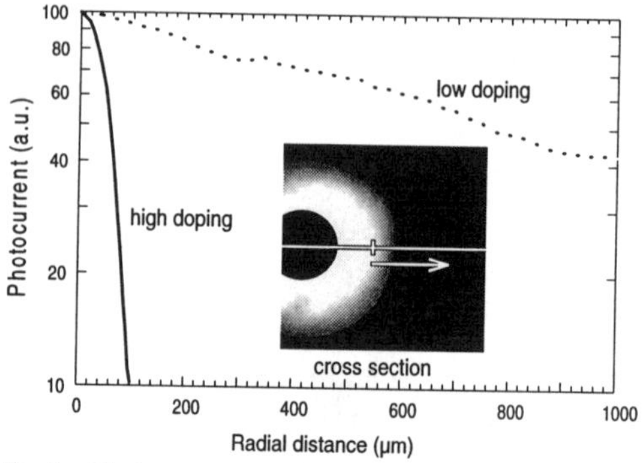

Fig. 4 Photocurrent as function of distance from the depletion layer boundary

In order to study the process dependence of the negative surface charge we selected the RIE process, which is a characteristic process in the fabrication of SiC device structures and has been one of the process steps used for p-n mesa diodes investigated previously. The process dependence of the expansion of the depletion layer width is shown in Fig. 5. The RIE etched sample has a markedly smaller extent of the electric field compared to the as-grown sample. But in spite of the removal of 1 µm thick portion of the CVD layer the outspread of the space charge region is still noticeable which means that the surface charge was retained and could not be entirely removed.

As additional confirmation of the assumption of a negative surface charge we prepared Schottky contacts on p-type CVD epilayers. In this case we expected to be able to demonstrate a decrease in depletion width. Yet in contrast to our expectation, for p-type samples a similar behavior as for n-type samples was observed. This is demonstrated in Fig. 6a where three diodes

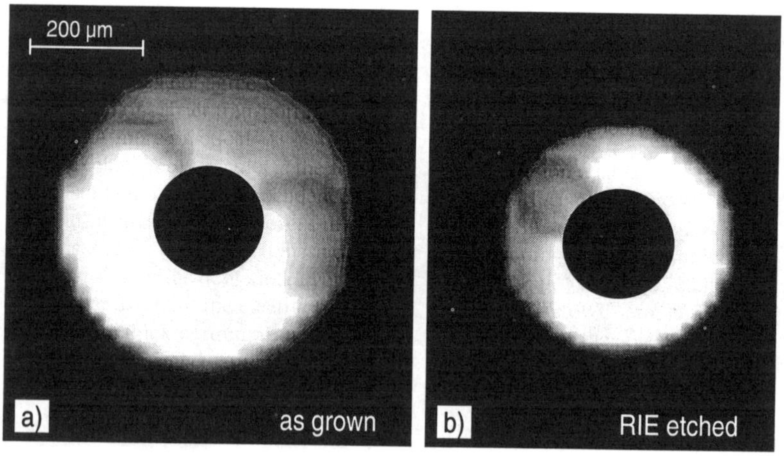

Fig. 5 Process dependence of depletion layer width (irradiation wavelength 365 nm, reverse bias 3 V).
a) n-type ($N_D = 5x10^{15}$ cm^{-3}) 6H-SiC, as-grown surface
b) after removal of 1 µm using a RIE ($CH_4/H_2/Ar$) process.

adjacent to the diode under test exhibit an increase in signal in their proximity. The OBIC signal is likewise enhanced in-between the diodes but to a lesser extent. Outside the region encircled by this four diodes the photocurrent is reduced rapidly with distance. An enlarged depletion region width for p-type 6H-SiC however can only be explained if one assumes a positive surface charge.

In order to establish the fact that the photocurrent observed in the OBIC measurement originates indeed from charge carrier separation due to the presence of an electric field with a drift component directed to the device under test we performed three-terminal measurements on the contact configuration as seen in Fig. 6. For that purpose we applied a high reverse bias voltage of 100 V to one of the adjoining Schottky contacts to extract all charge carriers generated in its vicinity. As a consequence the photocurrent as measured in the device under test should be significantly reduced in the proximity of the biased contact. The experimental result as presented in Fig. 6b apparently shows the expected behavior for the biased and the most distant contact with regard to the diode under test while the depletion layer for the remaining two contacts is barely affected by the applied bias voltage. This result can readily be interpreted in terms of the electric field sensitive character of the OBIC method.

CONCLUSIONS

The markedly increased depletion layer width at the surface of SiC Schottky contacts can be explained by the presence of surface charge. The fact that this surface charge is effective for n- as well as p-type material indicates that there exist acceptor-like surface states along with donor-like surface states. Assuming an arbitrary distribution of these surface states within the bandgap the Fermi energy determines the net charge of the surface charge. This simple model renders a

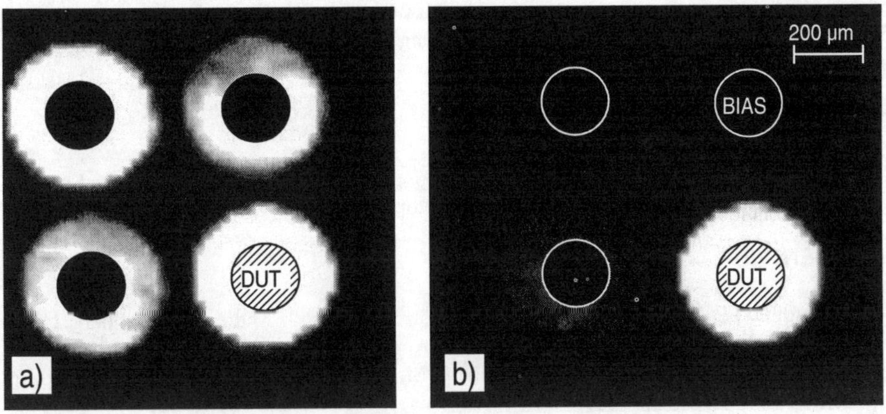

Fig. 6 OBIC image of the depletion region of a Schottky contact on p-type 6H-SiC (irradiation wavelength 365 nm, reverse bias 3 V).
The experimental configuration for three terminal measurements is shown to the right.

a) voltage was applied to diode "DUT"
b) an additional reverse bias voltage was applied to diode "BIAS"

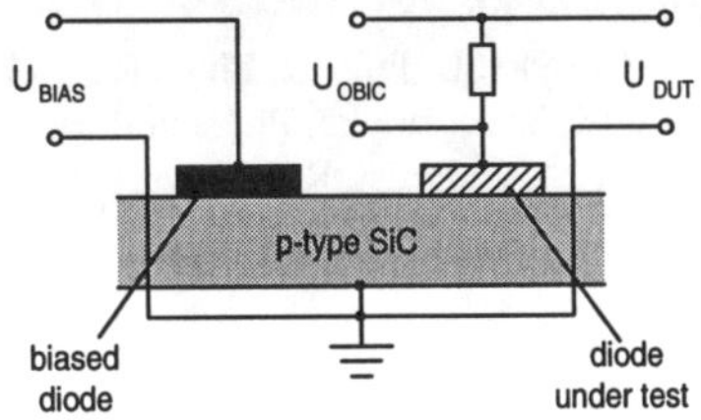

negative net charge for n-type SiC and a positive net charge for p-type SiC, respectively. Space charge neutrality results in a band bending at the semiconductor surface (Fig. 7). This is supported by the trend observed for variation of the doping level.

The formation of a space charge region at the surface due to surface charge increases the effective device area and correspondingly accounts for the experimentally observed leakiness of Schottky barrier rectifiers. On extremely low doped material the extension of the depletion layer over several millimeters without a significant decrease in photocurrent is indicative of an inversion channel since the photocurrent dependency on the distance cannot be understood in terms of carrier transport by diffusion alone.

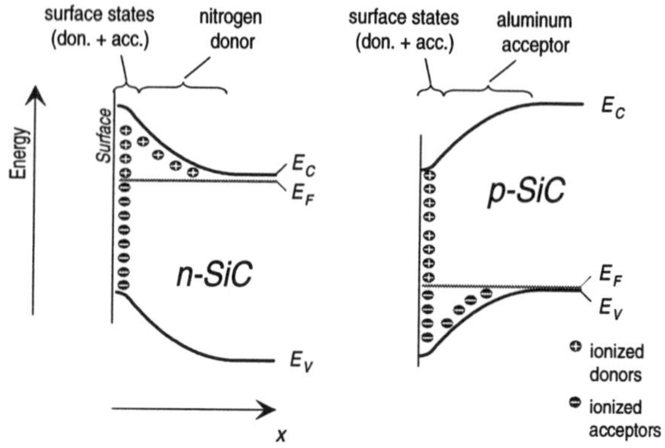

Fig. 7 Resulting band diagram for the 6H-SiC surface including the effect of acceptor- and donorlike surface states on n- and p-type SiC.

While the specific origin of the surface charge is not yet clear, it is suggested by the fact that the surface charge has been retained even after the removal of several μm of the CVD layer the presence of a surface charge is not just a contamination effect but of more fundamental nature. Further work must be directed to identify the origin of the surface charge in order to be able to control it. The seemingly ubiquitous presence of a surface charge has a high impact on the actual design of junction termination and an understanding of its physics will be important in addressing other passivation related problems.

ACKNOWLEDGMENTS

The authors gratefully acknowledge financial support by Asea Brown Boveri and the Swedish National Board for Industrial and Technical Development (NUTEK).

REFERENCES

1. O. Kordina, J.P. Bergman, A. Henry, E. Janzén, S. Savage, J. André, L.P. Ramberg, U. Lindefelt, W. Hermansson, and K. Bergman, Appl. Phys. Lett. **67** (11), 1561 (1995).
2. K. Rottner, S. Savage, J. André, L. Ramberg, M. Frischholz, and R. Helbig in Proc. 6th Int. Conf. on Silicon Carbide and Related Materials - 1995 , Sept. 18-21, 1995, Kyoto, Japan.
3. M. Frischholz, Th. Mandel, R. Helbig, A. Hammerschmidt, and G. Schmidt, Appl. Surface Science **65/66**, 784 (1993).
4. W.J. Choyke, L. Patrick, Phys. Rev. **172**, 769 (1968).
5. G. Pensl, W.J.Choyke, Physica B **185**, 264 (1993).
6. M. Schadt, G. Pensl, R.P. Devaty, W.J. Choyke, R. Stein, and D. Stephani, Appl. Phys. Lett. **65** (24), 3120 (1994).

DEFECT DOMINANT JUNCTION CHARACTERISTICS OF 4H-SiC p^+/n DIODES

J. Scofield,* M. Dunn,* K. Reinhardt,** Y.K. Yeo,* and R. Hengehold*
*Air Force Institute of Technology, Wright Patterson AFB OH 45433
**Wright Laboratory, Wright-Patterson AFB OH 45433

ABSTRACT

Forward and reverse current-voltage (I-V-T) measurements of MOCVD grown 4H-SiC p^+/n diodes are compared to classical recombination-generation theory over the temperature range of 100 to 750 K. The forward bias data indicate that the I-V characteristics of the well-behaved devices follow a classical recombination dominant transport mechanism. Ideality factors were determined to be in the range of 1.85 to 2.09, and the forward activation energy found to be E_A=1.56 eV compared to a nearly ideal value of 1.6 eV. A majority of the devices tested under forward bias conditions were, however, found to exhibit significant leakage current components due to tunneling at forward biases of up to 2.2 V for turn-on voltages in the 2.5 to 3.0 range. Deep level transient spectroscopy (DLTS) was also performed on the diode structures over the same wide temperature range, and the results were correlated to those obtained from reverse I-V-T and C-V-T characterization. Deep level defects at energies between 200 and 856 meV were identified from the DLTS data, and these levels are believed to be responsible for the tunneling dominant current conduction. Intrinsic deep levels, common to all devices tested, are emphasized and suggested as possible reverse bias tunneling paths for breakdown to explain the lack of an avalanche mechanism in all of the 4H-SiC diodes tested.

INTRODUCTION

As device performance requirements increase and geometry innovations continue, existing Si technologies are becoming increasingly limited by intrinsic material properties. With the theoretical limits of performance being closely approached by many present day devices, especially for high power and high temperature applications, the need for viable alternatives is becoming imperative. Wide-bandgap semiconductors which show negligible intrinsic generation and leakage current, large saturated carrier velocities, an increase in electric field breakdown, and in many instances excellent thermal conductivity may satisfy the demand for greater performance for high temperature and power applications.

Due to the recent availability of large area substrate material, α-SiC has attracted renewed attention as one of the most promising materials for high-power, high-temperature, and high-frequency devices. Ultraviolet photodiodes,[1] junction devices,[2] numerous field effect structures,[3] and most notably blue and violet light emitting diodes[4] have all been demonstrated to various levels of efficiency and performance. Concurrently, continued improvements to substrate quality has significantly reduced the macroscopic structural defect density such as micropipes and inclusions toward reasonable levels. However, difficulties in processing and the need for better quality base material continue to be the primary impediments to the realization of the full performance potential of SiC electronic devices. For example, poor oxide-semiconductor interfacial quality results in degraded MOSFET transconductance and high-voltage performance, and minority carrier lifetime limited bipolar device characteristics.

In this paper, we report results on the electrical characterization of the junction characteristics and current conduction mechanisms of 4H-SiC p^+/n diode structures. Forward

Mat. Res. Soc. Symp. Proc. Vol. 423 © 1996 Materials Research Society

and reverse current-voltage-temperature (I-V-T), capacitance-voltage-temperature (C-V-T), deep level transient spectroscopy (DLTS), electroluminescence (EL), and reverse breakdown measurements were performed to correlate junction characteristics with defects in the material.

EXPERIMENTAL

The SiC diodes used in this study were mesa isolated structures, off-axis MOCVD grown on the (0001) surface of n-conductivity, 4H polytype wafers. Asymmetric abrupt p^+/n junctions were obtained by doping Al and N in p- and n-type regions to $\sim 1x10^{19}$ and $5x10^{15}$ cm^{-3}, respectively. Al and Ni standard ohmic contact metalliztions were utilized on the p-and n-type surfaces, respectively.

Temperature dependent I-V measurements were performed in a closed cycle He refrigerated cryostat, which includes a heated stage enabling measurements to temperatures in excess of 750 K during a single scan. Fully automated computer controlled instrumentation and data acquisition capabilities allow the performance of a wide range of detailed electrical characterizations including C-V and several variations of DLTS.

RESULTS AND DISCUSSION

Forward biased current conduction characteristics of the 4H-SiC diodes were first compared to the classical Sah-Noyce-Schockley[5] (SNS) recombination-generation theory. Typical forward biased I-V data taken on the 4H-SiC diodes are illustrated in Fig. 1 for both a well-behaved and a leaky diode with a significant shunting characteristic. Solid lines in the figure are numerical fits to the classical SNS equation,[6]

$$I = I_{diff} \exp(qV/A_1kT) + I_{rec} \exp(qV/A_2kT) + I_0 \exp(BV), \qquad (1)$$

where I is the total forward bias dark current and V is the applied junction voltage. The first and second terms account for diffusion and recombination mechanisms, and the third term is included to account for current conduction due to tunneling[6], with B an empirical fitting parameter. As can be seen from the figure, the 4H-SiC diodes exhibit classical forward bias characteristics, including the dominance of recombination over ideal diffusion conduction in the well-behaved devices as expected for large bandgap (E_G) materials. An average value for I_{rec}/I_{diff} obtained from several ideal diodes was in the range of 10^{25}. Typical values of the diode ideality factors are illustrated next to their respective conduction regions in the figure. The A_2 factor corresponding to recombination in the space charge region (SCR) is indicative of recombination via traps near the center of the bandgap.

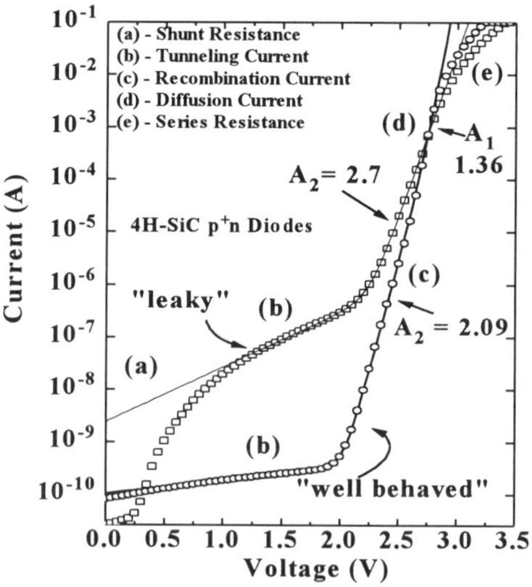

Figure 1. Forward I-V characteristics for near ideal (o) and strongly shunted ( ) diodes.

Although it is difficult to obtain the diffusion ideality factor, A_1, due to the dominance of recombination currents at lower voltages and due to series resistance effects at larger forward biases, it was observed to be consistently greater than the ideal Schockley value of one. This dominant recombination processes[7] along with the large number of diodes exhibiting shunting behavior are the first indications as to the presence of significant concentrations of deep traps and/or recombination centers in these devices. In fact, the majority of devices characterized during the present study revealed that the shunted type leaky diodes were found to be over *70%* of the randomly selected devices from the 1.125" wafer.

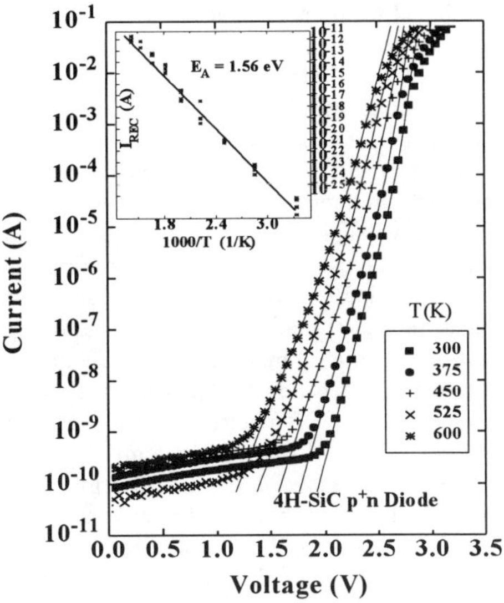

Figure 2. Forward I-V-T series for good device with recombination-activation Arrhenius plot inset.

Figure 2 illustrates a typical temperature dependent series of I-V data curves for a well-behaving diode along with the solid line fits to the diffusion and recombination components of conduction. Diodes without significant leakage or shunt characteristics were found to have recombination to diffusion ratios which agreed closely with the predicted temperature dependence, $I_{rec}/I_{diff} \propto T^{-1/2}\exp(E_G/2kT)$, over the range of 300 to 725 K. The activation energy for the $A_2 \approx 2$ current component can be obtained from an Arrhenius plot of $\ln(I_{rec})$ vs 1000/T using the best linear fit to the expression $I_{rec} \propto T^{5/2}\exp(-E_A/kT)$. The results of these analyses for several diodes are shown in the inset of Fig. 2. The obtained value of 1.56 eV is in agreement with statistical theory[5] which predicts $E_A=E_{G0}/2$, where $E_{G0}=E_G(T=0)$.

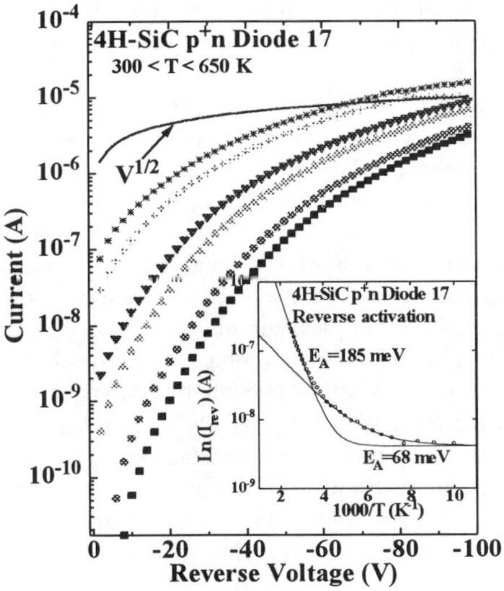

Figure 3. Temperature dependent series of reverse I-V measurement data with activation plot inset.

Figure 3 displays a set of typical temperature dependent reverse bias I-V curves for diodes exhibiting a significant tunneling component in the forward I-V characteristic. Included in the figure is the theoretical thermal generation current, $I_{rev} \propto V^{1/2}$, curve expected for abrupt doping profiles.[8] In wide bandgap materials, it is predicted that the thermally generated SCR component (I_{rev}) will be preeminent over bulk-diffusion and surface-perimeter leakage currents,[8] which are neglected in the analysis. As seen from Fig. 3, the trend is exponential for the mid and larger reverse bias voltages rather than the ideal square root dependence, which can be attributed to the neglected perimeter and shunt reverse leakage components for the larger and smaller bias ranges, respectively. Similar behaviors have been previously reported for $GaInP_2$[9] and InGaAs[9] diodes. The inset of Fig. 3, showing the Arrhenius plot of the reverse activation energy, illustrates the typical exponential dependence upon temperature of the reverse leakage current [$I_{rev} = C_1\exp(-E_{A1}/kT) + C_2\exp(-E_{A2}/kT)$]. The activation energies listed on the plot are characteristic of the dominant generation centers controlling reverse current at the selected bias and measured temperature. The energies obtained in this manner were relatively insensitive to the bias voltage chosen as long as the shunting regions of small bias were avoided. The 185 meV generation center was the most consistently observed in all the devices tested, and was also detected using DLTS methods. The shallower 68 meV level is attributed to the nitrogen donor levels in the 4H material, since calculations of the free carrier concentration at equilibrium over this temperature range support a Fermi level effect associated with these shallow levels. Temperature dependent capacitance measurements corroborate this effect.

A range of reverse activation energy levels were obtained from the large number of devices tested. The results predominantly indicated the presence of shallower generation centers in the 4H material, while a few energy levels were in the 700 meV range. This observation is most likely due to three conditions in the material. First, for the reverse I-V-T method, which at best will resolve only two or three levels of reasonable energetic separation, the presence of numerous levels in the forbidden gap results in decreasing accuracy and doubtful validity. Secondly, the temperature range of the measurements (300 - 600 K) will not result in significant thermal emission from deep centers greater than ~1.1 eV as the Fermi level is pushed toward its intrinsic position. Thus, generation centers located more deeply in the gap will not contribute to the thermally stimulated reverse leakage currents. Finally, if there are numerous levels present in the material, with energy states spanning the energy gap, there may exist a significant probability for reverse bias tunneling through a series of trap levels. This may occur until a sufficiently shallow level is subsequently occupied, which has a greater thermal emission probability than tunneling to the resonant band or to another intermediate level. The DLTS and reverse breakdown data presented in the following paragraphs support the role of all these defect related mechanisms in the 4H-SiC diode characteristics.

DLTS characterization of the diodes subjected to I-V-T analysis was typically performed using classical constant-voltage (CVDLTS) methods. In some instances in which the trap concentration exceeded $0.1N_d$, the constant-capacitance (CCDLTS) mode of operation was utilized. It should be noted that although thermal scans are accomplished to over 750 K, the emission rate [$e_n = N_C \langle v \rangle \sigma \exp(-E_t/kT)$] limitation will restrict observation of deep levels to within only ~$E_G/3$ of either bandedge. Thus a large portion of the bandgap will remain uncharacterized using thermal generative techniques such as DLTS. Nevertheless, several deep-level defects were consistently recorded across the matrix of samples, and the typical DLTS rate window plots of eight deep level majority-electron and minority-hole trapping centers are shown in Fig. 4. The concentrations of defects labeled A-H in Fig. 4 are large enough to affect device junction characteristics. They range from $0.01N_d$ for level A to $0.3N_d$ for level F, where the shallow donor doping level N_d is $5x10^{15}$ cm^{-3}. These deep traps with ionization energies over the range of 200 to 856 meV are implicated in the shunting and leakage currents seen in the forward I-V characteristics of a majority of devices. The majority carrier electron traps shown in Fig. 4 have been observed in all diodes exhibiting significant tunneling leakage currents. Furthermore, these defects have characteristically large capture cross sections ranging from $5x10^{-17}$ to $6x10^{-13}$ cm^2. When these defects occur in large concentrations, the probability of transitions through these centers will significantly increase, and may dominate junction characteristics. Levels A, B, and C located at energies of E_C-192, 286, and 340 meV, respectively, have all been implicated as dominant reverse bias generation centers from I-V-T analysis. When considering the large number of traps observed, although only considering roughly one third of the 4H-SiC bandgap, the possibility of tunneling through several undetected intermediate levels prior to thermal emission from the A, B, or C levels becomes more probable. It should also be noted that several additional deep level electron and hole trapping centers were detected, but they were consistently found to exist in much lower concentrations. Table I enumerates the energy levels, ΔE, capture cross sections, σ, and trap concentrations, N_t, obtained primarily from DLTS measurements with superscripts indicating corroborating I-V-T data. C-V-T measurements were taken on several devices at 1 MHz and 1 kHz source frequencies. In addition to the detected levels illustrated in Fig. 4, Table I lists 4 additional deep level centers which were all detected in the n-type diode base of a number of other samples using DLTS methods. The centers labeled e1, e2, and e3 are electron traps, and h1 is a hole trapping center. These defects were found to be present in much lower concentrations than those of Fig. 4, as seen in Table I, and in comparison will have negligible effects on device I-V behavior.

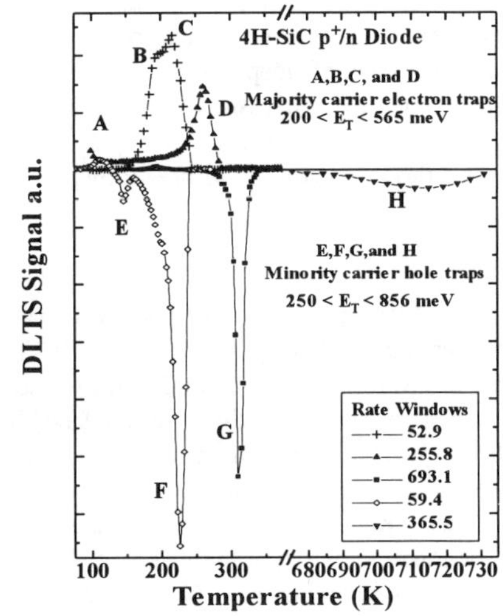

Figure 4. DLTS rate window spectra for several large concentration traps implicated in the shunt leakage currents of the p+/n 4H-SiC diodes.

Additionally, minority carrier hole traps labeled F and G in Fig. 4 have also been observed in diodes with shunting forward I-V characteristics and in diodes exhibiting significant leakage and tunneling currents. However, these are present only in diodes which exhibited a bright green (~2.4 eV) defect electroluminescence. Field dependent DLTS emission rate measurement results and the extremely large capture cross section identify an acceptor nature of the F-defect, which make it a potential dominant recombination center in the n-type material. This fact together with the presence of this center in all diodes exhibiting a bright green luminescence lead us to conclude that the 620 meV center is the activator responsible for the bright green electroluminescence. The deep energy level, large cross section, and high concentration of this hole trap results in a large recombination probability and short radiative lifetime associated with this center, which may explain the lack of a violet N-Al donor-acceptor-pair radiative recombination in these diodes.

Characterization of the reverse breakdown mechanism in these devices was marginally successful due to the destructive failure of the diodes when biased into the breakdown region.

TABLE I. Deep level trap summary of DLTS detected defects in the 4H-SiC p^+/n diodes

Trap	A†	B†	C†	D	E*	F*	G†*	H*	e1	e2	e3	h1*
ΔE	200	285	340	564	246	620	730	856	585	611	536	350
σ	2E-17	4E-17	1E-16	6E-13	1E-18	2E-12	4E-12	1E-18	2E-14	5E-20	3E-21	1E-19
N_t	0.01	0.06	0.09	0.07	0.03	0.3	0.25	0.03	0.002	0.002	0.003	0.002

*-Minority carrier hole traps. †-Observed also using I-V-T technique. Energies in meV, cross sections in cm^2, and trap concentrations normalized to the shallow donor $5x10^{15}$ cm^{-3} value.

Diodes with well-behaved I-V characteristics were consistently able to sustain an average 475 V reverse bias before catastrophic breakdown. Leaky devices, characterized by high concentrations of deep defect centers, were seen to breakdown at a much lower 95 V average. An avalanche mechanism was not observed in any of the devices tested. Temperature dependent trends were characterized by a negative coefficient for reverse breakdown up to 423 °C, although once a diode was biased to breakdown, it was no longer useful for further testing.

SUMMARY

Recently grown p^+/n 4H-SiC rectifiers have been studied to determine current conduction mechanisms and deep levels using I-V-T and DLTS. Roughly 30% of the measured diodes showed well-behaved forward and reverse I-V characteristics. However, the majority of diodes tested were found to exhibit shunted I-V behavior which obeyed a tunneling conduction model. DLTS measurements of the devices revealed the presence of numerous deep trapping centers, several of which are implicated in the large reverse saturation currents observed in many devices. Reverse breakdown measurements were characterized by the lack of an avalanche mechanism, a negative temperature coefficient, and the catastrophic failure of each device after biasing to breakdown. The negative temperature coefficient is indicative of reverse bias tunneling through allowed states in the bandgap[7]. The detection of several deep level defects with significant concentrations support this supposition. Furthermore, the large number of DLTS detected centers observed within roughly $E_G/3$ of the bandedges suggests the possibility of additional undetected deeper levels in the material. The sources of the defects observed in these devices are unknown. However, the presence of structural flaws such as micropipes and polytype inclusions in hexagonal SiC are well-known and thus, dislocations, vacancies, grain boundaries, etc. may be closely associated with the defects detected in this study. Additionally, support for this link to crystalline flaws may be related to the existence of extremely high concentrations for several defects. That is, the unintentional introduction of atomic impurities, at the concentrations observed for several of the defects, is unlikely in high quality epitaxial CVD reactors. Thus, if lattice imperfections are in fact dominating junction characteristics, scaling the device areas to dimensions appropriate for applications such as high power will only exacerbate the problems. The results here indicate that while the improvements to material technology have been significant with steady progression, there continue to exist significant concentrations of structural and impurity related defects. This highlights the need for additional study of the defects in these materials, especially their sources, and continued attention and improvements to material quality.

REFERENCES

1. D. Brown, E. Downey, M. Ghezzo, J. Kretchmer, R. Saia, Y. Liu, J. Edmond, G. Gati, J. Pimbley, and W. Schneider, IEEE Trans. Elec. Dev. **40**, 325 (1993).
2. G. Kelner, S. Binari, M. Shur, K. Sleger, J. Palmour, and H. Kong, Mat. Sci. Egr. **B11**, 121 (1992).
3. J. Palmour, H. Kong, and C. Carter, in Trans. 1st Int. High Temperature Electronics Conf, Albuquerque NM, p. 491 (1991).
4. B. Vishnevskaya, V. Dmitriev, I. Kovalenko, L. Kogan, Y. Morozenko, V. Rodkin, A. Syrkin, B. Tsarenkov, and V. Chelnokov, Fiz. Tekh. Poluprovodn. **22**, 664, (1988) [Sov. Phys. Semicond. **22** (4), 414 (1988)].
5. C.T. Sah, R.N. Noyce, and W. Schockley, Proc. IRE **45**, 1228 (1957).
6. H.J. Hovel, in Semiconductors and Semimetals, edited by R.K. Willardson and A.C. Beer, (Academic, New York, 1975), Vol 11 p. 246.
7. S.M. Sze, in Physics of Semiconductor Devices, 2nd ed. (John Wiley and Sons, 1981) p. 92.
8. K. Reinhardt, Y. Yeo, and R. Hengehold, J. Appl. Phys. **77**, 5763 (1995).
9. G. Shaw, S. Messenger, R. Walters, and G. Summers, J. Appl. Phys. **73**, 7244 (1993).

FABRICATION OF HIGHLY ORIENTED, SMOOTH DIAMOND FILMS ON SILICON FOR ELECTRONIC DEVICES

S.D. WOLTER*, T.H. BORST, P. GLUCHE, W. EBERT, A. VESCAN, AND E. KOHN
Department of Electron Devices and Circuits, University of Ulm, Ulm/Donau, Germany, D-89069

*Current address: The Pennsylvania State University, 207 Steidle Bldg., University Park, PA, USA 16802

ABSTRACT

A new bias-enhanced nucleation method based on an AC-bias step to form highly oriented diamond (HOD) nuclei on silicon substrates is presented. The uniformity of the nucleated film and the bias time strongly depended on the substrate temperature and the substrate holder. In our case the shortest bias time and highest nucleation densities were achieved at ~ 850°C while using a graphite susceptor. Following this nucleation enhancement step the diamond films were grown out using conditions employing an α-parameter slightly greater than 2. This ultimately leads to extremely smooth and well-faceted (100) textured HOD films which could be used as substrates for the fabrication of electronic devices.

Schottky diodes with high rectification ratios and high breakdown voltages have been fabricated for the first time via selective growth of the active boron doped diamond layers on these HOD films. Results of the growth procedure and diode performance will be given.

INTRODUCTION

Synthetic diamond is readily used for tribology purposes; however, this unique material also holds promise for commercial application in the microelectronics arena. Although significant progress has been made in the fabrication of diamond over the past decades, the development of large area single crystals has not yet been achieved. Further improvements in diamond synthesis are essential in order for diamond to realize its full potential.

The recent development of a nucleation enhancement process, termed bias-enhanced nucleation (BEN), which is performed *in situ* during microwave plasma chemical vapor deposition led to the discovery of obtaining broad area highly oriented diamond (HOD) on β-SiC[1] and Si(100)[2,3]. The microelectronic character of these films was first investigated by Stoner et al.[4]. Later work by Fox et al.[5] revealed carrier mobilities approaching that of devices fabricated via homoepitaxial diamond growth.

Mat. Res. Soc. Symp. Proc. Vol. 423

The purpose of this paper is to discuss the application of BEN utilizing an AC bias rather than the conventional DC bias source used in previous works[1-5]. It has been observed that a carburization treatment prior to BEN is essential for growing oriented diamond on silicon[3]. Alternating current bias-enhanced nucleation (AC BEN) was observed to form HOD without this additional carburization pretreatment. Subsequent to AC BEN, diamond growth was carried out to form a (100) textured HOD film. Au-Schottky diodes were then fabricated on these surfaces. The fabrication and rectifying character of these devices will be discussed.

EXPERIMENTAL

Diamond deposition was undertaken in an ASTeX™ microwave plasma chemical vapor deposition (MP CVD) chamber using methane and hydrogen as source gases. This growth system has been modified so that a bias potential may be applied to the substrate. One pole of a 50Hz voltage source is connected to the metal casing of the thermocouple which is situated in the graphite susceptor containing the substrate and the other pole is connected to ground. The substrate biasing was only performed as a nucleation enhancement step; no substrate biasing was used during diamond deposition.

Single-sided highly polished silicon(100) wafers up to 3 inches in diameter were entered into the growth chamber as-received. The growth chamber was immediately evacuated by way of a turbomolecular pump to 1×10^{-6} torr. Substrate biasing was performed immediately following a 10 minute hydrogen plasma cleaning step. An AC bias of $150V_{RMS}$ was used which produced a current of 10 to 40mA. This procedure was performed at a substrate temperature ~850°C as measured by a calibrated optical pyrometer. Following the formation of a diamond film determined by use of laser reflection interferometry[6], the bias voltage was terminated and diamond deposition was continued using a two-step process. First, the oriented particles were grown to form diamond octahedra under conditions employing an α-parameter of 3 as shown in Figure 1a. These process conditions were initially used yielding the fastest growth direction normal to the substrate. Once the diamond octahedra had overgrown the randomly oriented diamond (Figure 1b), the substrate temperature was increased to form diamond cubo-octahedra. During this procedure an α-parameter of ~2 was utilized. Continuation of this deposition procedure resulted in a (100) textured surface. The process conditions used in this study are summarized in Table I. A schematic representation of this growth system and further details of the procedure used to obtain these HOD films have been presented previously[7].

Au-Schottky devices were then fabricated onto the intrinsic HOD substrates. A boron doped p^+-ohmic contact layer was deposited on these substrates to ~400nm in thickness. A boron-coated molybdenum rod of ~1mm in diameter was injected into the plasma via an externally controlled manipulator arm and served as the dopant source. Next, a silicon dioxide layer was deposited and circular regions of various sizes ranging from ~50µm to greater than 100µm in diameter were wet etched revealing the underlying p^+- layer. A pulse-doped p- layer was then selectively grown to approximately 400nm in thickness. The pulse doping was undertaken by injecting the boron-coated rod into the plasma for 50 seconds at the beginning of a 25 minute growth duration and quickly withdrawing it away from the plasma. The silicon dioxide was removed and gold contacts were deposited onto the p- layer. An in-depth discussion of this experimental procedure has been presented previously[8].

Figure 1. SEM micrographs of highly oriented diamond formed by AC BEN followed by (a) a short and (b) an extended diamond deposition period.

(a)

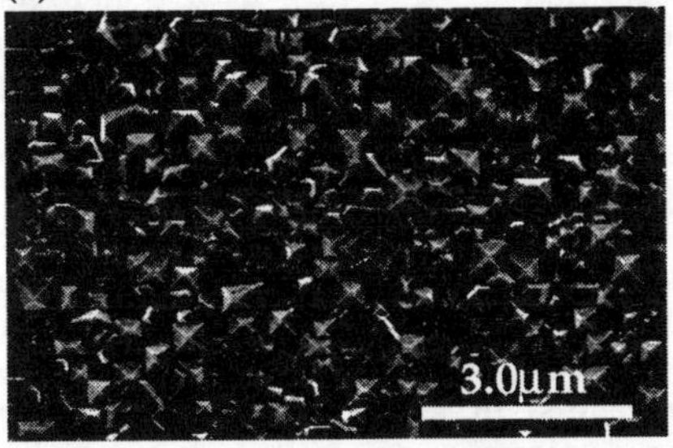

(b)

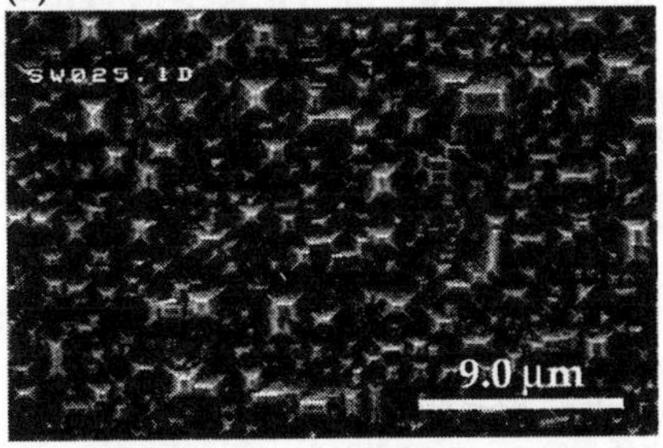

Table I. The processing conditions for alternating current bias-enhanced nucleation (AC BEN) and growth of (100) textured diamond.

Parameters	AC BEN	Diamond deposition
Bias voltage	150VRMS	0
Bias current	10 - 40 mA	0
CH_4/H_2	2%	1.5%
Pressure	~17 Torr	30 Torr
Microwave power	700 Watts	700 Watts
Substrate temperature	approx. 850°C	780 - 850°C

RESULTS AND DISCUSSION

The biasing pretreatment was determined to be optimum at a substrate temperature of 850°C per the experimental conditions used in this study. Any increase or decrease in the substrate temperature over the temperature range of

650°C to 850°C resulted in a longer bias duration to obtain a diamond film. A high percentage of the diamond formed when using AC substrate biasing were also observed to be oriented relative to the Si(100). The diamond(100) is parallel to the silicon(100) and a diamond<110>//silicon<110> relationship exists, which has been observed in previous studies[1-3]. However, there is evidense of misorientation and tilting of the diamond particles relative to the substrate which has been attributed to the large degree of mismatch between the Si/SiC and diamond lattices[9].

The AC bias is acting to enhance diamond nucleation as well as providing a surface conducive to oriented diamond nucleation. It is proposed that the extreme negative voltages of the AC waveform are acting to promote diamond nucleation. Based on past work in this area[10-19], only voltages of approximately >-100volts were found to promote diamond nucleation. Also, the positive and moderate negative voltage portion of the AC waveform is speculated to be acting as a concurrent carburization procedure. This process may be responsible for forming an epitaxial carbide conversion layer onto which the oriented diamond is nucleating as proposed in past work[3].

The diamond films shown in Figures 2a and b were grown out using the two-step process previously discussed. An I-V plot of a 50μm diameter diode is exhibited in Figure 3a. Reverse breakdown voltages in the range of 40 to 50 volts were typically measured for this particular diode size as is evident in this graphic. Point contact reverse breakdown voltages of ~100volts were also measured. A correlation between the size of the diode devices and the reverse breakdown voltages were not obvious which may be due to non-uniform boron doping across the substrate. The breakdown field strength was calculated to be ~$2x10^6$V/cm based on the reverse breakdown voltages and the thickness of the p- layer.

Figure 2. SEM micrographs of (a) a plan view and (b) an inclined view of a highly oriented diamond surface. The p^+- layer, p- layer, and gold contacts are shown on these substrates.

(a) (b)

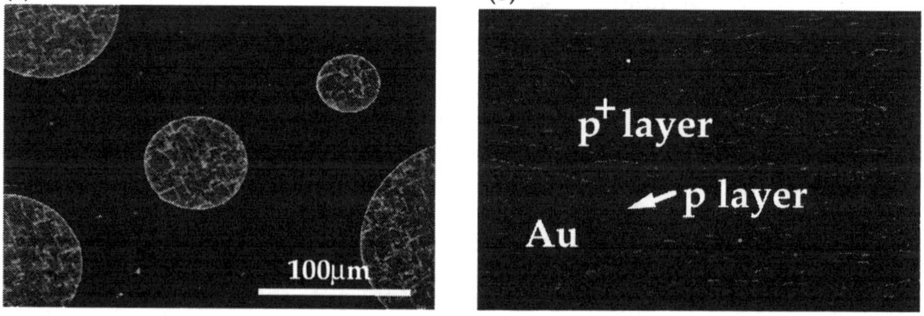

Figure 3. (a) An I-V plot of a 50μm diameter diode fabricated on the intrinsic HOD surface. (b) The temperature dependent I-V character of these diodes.

(a)

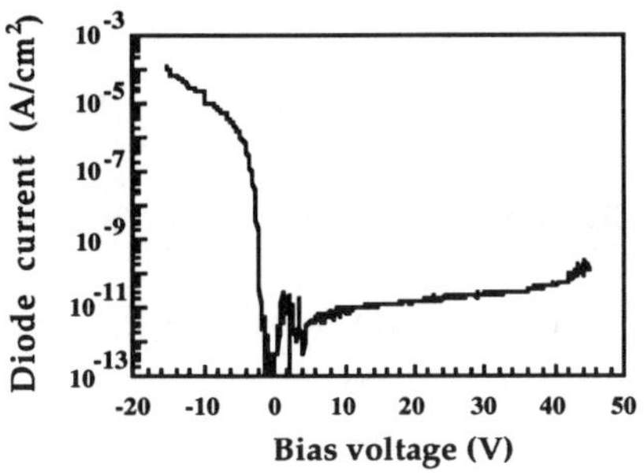

(b)

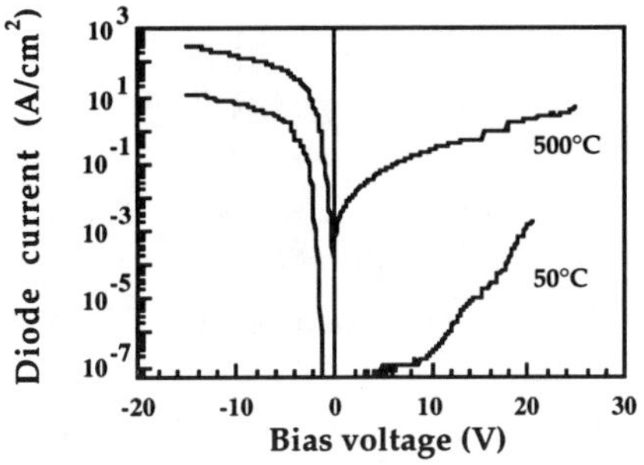

The low forward bias current at -15volts in Figure 3a (i.e. ~10^{-4}A/cm^2) may be attributed to the thin p$^+$- layer. It is speculated that increasing the thickness of this layer and/or increasing the dopant concentration within this layer may lead to an increase in the forward current density. The I-V character of a 50μm diameter diode measured at 50°C and 500°C may be seen as well in Figure 3b. These devices exhibited rectification ratios $I_f/I_r > 10^6$ at 50°C and $>10^2$ at 500°C (measured at +/- 15 volts). A Richardson plot using the data extracted from several temperature dependent I-V plots yielded a linear slope which is indicative of Schottky behavior. The potential barrier measured via the slope of this plot was determined to be ~1.6eV, which is comparable to diodes fabricated by homoepitaxial means.

SUMMARY

Alternating current bias-enhanced nucleation was responsible for forming highly oriented diamond on silicon(100). The AC bias is acting to promote diamond nucleation while concurrently forming a surface conducive to oriented diamond nucleation. Growth of these films were continued employing a two-step process to form a (100) textured surface. Au-Schottky diodes were fabricated using solid boron as the dopant source. Reverse breakdown voltages in the range of 40 to 50volts were measured for 50μm diameter diode structures. When measuring the I-V character of these diodes at 50°C and 500°C, rectification ratios, I_f/I_r, of $>10^6$ and $>10^2$ were observed, respectively. A Richardson plot was constructed using data from the temperature dependent I-V plots and indicated a Schottky barrier height of ~1.6eV.

ACKNOWLEDGMENTS

We gratefully acknowledge the funding which was provided for by the German Ministry of Research and Education (BMDF).

REFERENCES

1. B. R. Stoner and J. T. Glass, Appl. Phys. Lett. **60**, 698 (1992).

2. X. Jiang, C.-P. Klages, R. Zachai, M. Hartweg, and H. -J. Füsser, Appl. Phys. Lett. **62**, 3438 (1993).

3. S. D. Wolter, B. R. Stoner, J. T. Glass, P. J. Ellis, D. S. Buhaenko, C. E. Jenkins, and P. Southworth, Appl. Phys. Lett. **62**, 1215 (1993).

4. B.R. Stoner, Chien-teh Kao, D.M. Malta, R.C. Glass, Appl. Phys. Lett. **62**, 2347 (1993).

5. B.A. Fox, B.R. Stoner, D.M. Malta, P.J. Ellis, R.C. Glass, and F.R. Sivazlian, Diamond and Related Materials **3**, 382 (1994).

6. B.R. Stoner, B.E. Williams, S.D. Wolter, K. Nishimura, and J.T. Glass, J. Mater. Res., 7 (1992) 1.

7. S.D. Wolter, T.H. Borst, A. Vescan, and E. Kohn, submitted for publication in Appl. Phys. Lett., April, 1996.

8. P. Gluche, S.D. Wolter, T.H. Borst, W. Ebert, A. Vescan, and E. Kohn, accepted for publication in Electronic Device Letters, June, 1996.

9. B.R. Stoner, G.H. Ma, S.D. Wolter, W. Zhu, Y.-C. Yang, R.F. Davis, and J.T. Glass, Dia. Rel. Mat. **2**, 142 (1993).

10. S. Yugo, T. Kanai, T. Kimura, and T. Muto, Appl. Phys. Lett. **58**, 1036 (1991).

11. S. Yugo, T. Kanai, and T. Kimura, Dia. Rel. Mat. **1**, 388 (1992).

12. S. Yugo, T. Kimura, and T. Kanai, Dia. Rel. Mat. **2**, 328 (1992).

13. X. Jiang, R. Six, C.-P. Klages, R. Zachai, M. Hartweg, and H.-J. Füßer, Dia. Rel. Mat. **2**, 407 (1992).

14. S. D. Wolter, J. T. Glass, and B. R. Stoner, Thin Solid Films **261**, 4 (1995).

15. B. R. Stoner, G. H-. M. Ma, S. D. Wolter, and J. T. Glass, Phys. Rev. B **45**, 11067 (1991).

16. J. Gerber, S. Sattel, K. Jung, H. Erhardt, and J. Robertson, Dia. Rel. Mat. **4**, 559 (1995).

17. J. Robertson, Dia. Rel. Mat. **4**, 549(1995).

18. R. Beckmann, B. Sobisch, W. Kulisch, and C. Rau, Dia. Rel. Mat. **3**, 555(1994).

19. R. Spitzl, V. Raiko, R. Heiderhoff, H. Gnaser, and J. Engemann, Dia. Rel. Mat. **4**, 563(1995).

RADIATIVE RECOMBINATION RATES IN GaN, InN, AlN AND THEIR SOLID SOLUTIONS

A.V.DMITRIEV, A.L.ORUZHEINIKOV

Department of Low Temperature Physics, Faculty of Physics,
M.V.Lomonosov Moscow State University, Moscow, 119899, Russia.
Fax: (7–095) 932 88 76, E-mail: Dmitriev@lt.phys.msu.su

ABSTRACT

The radiative recombination rates have been calculated for the first time in the wide band gap wurtzite semiconductors GaN, InN and AlN and their solid solutions $Ga_xAl_{1-x}N$ and $In_xAl_{1-x}N$ on the base of existing data on the energy band structure and optical absorption in these materials. We calculated the interband matrix elements for the direct optical transitions between the conductivity band and the valence one using the experimental photon energy dependence of the absorption coefficient near the band edge. In our calculations we assumed that the material parameters of the solid solutions (the interband matrix element, carrier effective masses and so on) could be obtained by a linear interpolation between their values in the alloy components. The temperature dependence of the energy gap was taken in the form proposed by Varshni. The calculations of the radiative recombination rates were performed in the wide range of temperature and alloy compositions.

INTRODUCTION

Nowadays, the nitride semiconductors such as GaN, AlN and InN attract a considerable attention due to their outstanding physical, chemical and mechanical properties and also because of the recent progress in the technology that allowed to produce high quality nitride films with help of MOVPE and MBE (for a recent review, see Ref. [1]). The attractive properties of the nitrides include high heat conductivity, hardness, chemical stability and high luminescence intensity. These wide gap semiconductors are very promising materials for LEDs and semiconductor lasers in wide spectral interval from ultra-violet to green and even orange [2] because their solid solutions may have the energy gap varying from 2 eV in InN to 6.2 eV in AlN.

Important characteristics of materials used in luminescence devices are the rates of different electron–hole recombination processes. However, there is no information at the moment about the intensities of these processes in the nitrides. In this work, we concentrate on the calculation of the radiative recombination rate in GaN, InN, AlN and their binary alloys $Ga_xAl_{1-x}N$ and $In_xAl_{1-x}N$ on the base of the experimental absorption data that is present in the literature [3-7], the calculated electron energy band dispersion laws [8-11] and the temperature dependence of the band gaps in the pure materials [12,13].

ENERGY SPECTRA OF THE NITRIDE SEMICONDUCTORS

We consider more common and popular hexagonal phase of the nitrides. All of them belong to the crystal class C_{6v}. Their conductivity bands are non-degenerate, and their

Mat. Res. Soc. Symp. Proc. Vol. 423

electron states originate from atomic s-functions. In the Γ point of Brilluine zone they transform according to Γ_1, the unite representation of C_{6v}. The valence band is complicated and consists of two branches. One of them transforms according to Γ_1 whereas the other is degenerate and form the two-dimensional representation Γ_6. If spin–orbit interaction is taken into account, Γ_6 farther splits into two bands, Γ_7 and Γ_9 [14]. However, the latter splitting manifest itself only along k_x and k_y, but it equals zero in the very Γ point and along k_z, z being the direction of the hexagonal axis c which usually coincides with the normal to the film. We will neglect this splitting and consider Γ_6 as a degenerate band.

According to the results of Ref. [6,8-10,15,16], the order of levels in the valence band of the nitrides is different: in GaN and InN, Γ_6 branch lies above Γ_1, whereas in AlN it lies below Γ_1.

The symmetry of the electron wave functions in different bands and band branches leads to the following selection rules for the radiative transitions:

- For the transition from Γ_1^c to Γ_1^v, only z- component of the transition matrix element differs from zero, and correspondingly, the emitted photon is polarized along z axis.
- On the contrary, for the transition from Γ_1^c to Γ_6^v, z-component of the transition matrix element equals zero, and the emitted photon polarization is perpendicular to z axis.

RADIATIVE RECOMBINATION RATE: THE METHOD OF CALCULATION

Usually, Shockley–van Roosbroeck formula is used for the calculation of the radiative recombination intensity [18], which allows one to calculate the transition rate if the spectral dependence of the absorption coefficient is known. However, this is not the case in the nitrides where the optical absorption has been measured only in a small vicinity of the band edge, and another method is needed that would allow to express the recombination rate through the material parameters.

A straightforward quantum-mechanical calculation similar to given in Ref. [19] leads to the following expression for the spontaneous radiative recombination rate:

$$R = \frac{ne^2}{m_0^2 c^3 \hbar^2}\, \mathbf{M}^2 \left[\frac{2k_BT}{\pi\hbar^2}\right]^{3/2} \sqrt{\mu_x\mu_y\mu_z}\; E_g \left[1+\frac{3k_BT}{2E_g}\right] e^{-E_g/k_BT}$$

where n is the refractive index, m_0 is free electron mass, $\mu_x = (m_{e,x}^{-1}+m_{h,x}^{-1})^{-1}$ is the reduced carrier mass in x direction, and similar for y and z, and

$$\mathbf{M}^2 = \frac{1}{4\pi}\sum_{\lambda=1}^{2}\int d\Omega_{\mathbf{k}}\, |\mathbf{e}_{\mathbf{k}\lambda}\,\mathbf{P}_{cv}|^2$$

where $\mathbf{e}_{\mathbf{k}\lambda}$ is the polarization direction of the photon with the momentum $\mathbf{k}$, the sum is over two polarizations, the averaging is over all photon momentum directions, and $\mathbf{P}_{cv}$ is the interband transition matrix element at the Γ-point of the Brilluine zone.

Defining the radiative recombination coefficient according to the equality

$$R = B\, np$$

where n and p are carrier concentrations, one comes to the following expression:

$$B = \frac{ne^2}{m_0^2 c^3 \hbar^2}\, \mathbf{M}^2 \left[\frac{2\pi\hbar^2}{k_BT}\right]^{3/2} \frac{1}{(\bar{m}_x\bar{m}_y\bar{m}_z)^{1/2}}\; E_g(T)\left[1+\frac{3k_BT}{2E_g(T)}\right]$$

where $\bar{m}_x = m_{e,x} + m_{h,x}$ and so on.

TEMPERATURE AND ALLOY COMPOSITION DEPENDENCE OF THE ENERGY GAP

To derive the accurate temperature dependence of the radiative recombination rate, one needs to know how the gap varies with the temperature in the materials under consideration. In many semiconductors, including the nitrides, the empiric Varshni formula [17] approximates well the observed temperature dependence of the gap:

$$E_g(T) = E_g(0) - \frac{\gamma T^2}{T + \beta}$$

where γ and β are parameters. Their values for nitride thin films were found out in Ref. [12,13]:

AlN: $E_g(300\mathrm{K}) = 6.026$ eV, $\gamma = 1.799 \times 10^{-3}$ eV/K, $\beta = 1462$ K,
GaN: $E_g(0) = 3.427$ eV, $\gamma = 0.939 \times 10^{-3}$ eV/K, $\beta = 772$ K,
InN: $E_g(300\mathrm{K}) = 1.970$ eV, $\gamma = 0.245 \times 10^{-3}$ eV/K, $\beta = 624$ K.

For the gap value in the binary alloys with the components A and B, we used the expression

$$E_g^{(AB)}(x,T) = \left[x\, E_g^{(A)}(T) + (1-x)\, E_g^{(B)}(T) - d\, x(1-x)\right]$$

with $d = 1.0$ for $Ga_xAl_{1-x}N$ and $d = 2.6$ for $In_xAl_{1-x}N$ [3,21,22].

CALCULATION OF THE INTERBAND MATRIX ELEMENT

To calculate the interband transition matrix element, we made use of the formula for the absorption coefficient that also contains $\mathbf{P}_{cv}$ (see, for example, Ref. [19]):

$$\alpha(\hbar\omega) = b\sqrt{\hbar\omega - E_g}, \qquad b = \frac{2e^2\sqrt{8\mu_x\mu_y\mu_z}}{\hbar^2 m_0^2 ncE_g}\,\frac{1}{2\pi}\sum_{\lambda=1}^{2}\int d\phi\, |\mathbf{e}_{\mathbf{k}\lambda}\, \mathbf{P}_{cv}|^2$$

where the integration is over all polarization directions in the plane perpendicular to the incident photon momentum.

Extracting the coefficient b from the measured frequency dependence of α, one can find $\mathbf{P}_{cv}$ components in the xy plane assuming that the light beam was perpendicular to the film surface. However, this method does not allow us to find $P_{cv,z}$. This is not important for the transitions from $\Gamma_{1,c}$ to $\Gamma_{6,v}$ as in InN and GaN (see selection rules in Sec. 2 above), but in AlN where $\Gamma_{1,v}$ band lies higher than $\Gamma_{6,v}$, this may cause problems. We still have calculated B for AlN, too, assuming that $\Gamma_{1,v}$–$\Gamma_{6,v}$ splitting in AlN is rather small [20], and the hole population of $\Gamma_{6,v}$, and hence the transition probability to this band, may be higher than that in $\Gamma_{1,v}$ due to much higher density of states in $\Gamma_{6,v}$.

The values of $P_{cv,x}=P_{cv,y}$ found this way together with other material parameters used in calculations are listed below:

AlN: $P_{cv,x}=13.4 \times 10^{-20}$ g cm/s, $n=2.15$, $m_e=0.45$, $m_{h,z}=3.68$, $m_{h,xy}=3.78$;
GaN: $P_{cv,x}=6.3 \times 10^{-20}$ g cm/s, $n=2.67$, $m_e=0.22$, $m_{h,z}=2.49$, $m_{h,xy}=2.02$;
InN: $P_{cv,x}=11.5 \times 10^{-20}$ g cm/s, $n=2.1$, $m_e=0.12$, $m_{h,xyz}=0.5$.

The material parameters for the alloys were found using a simple linear interpolation between the values for the alloy components.

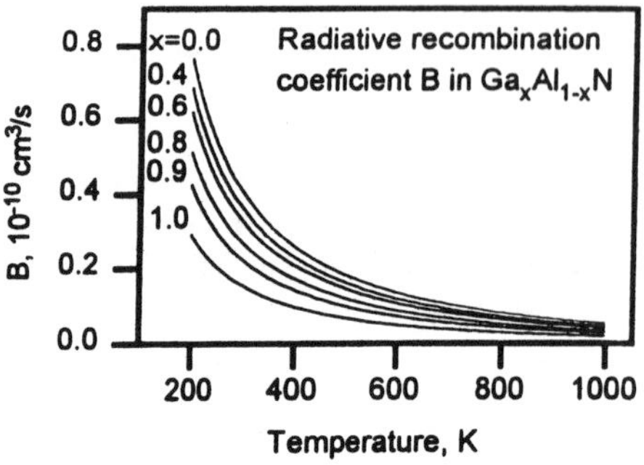

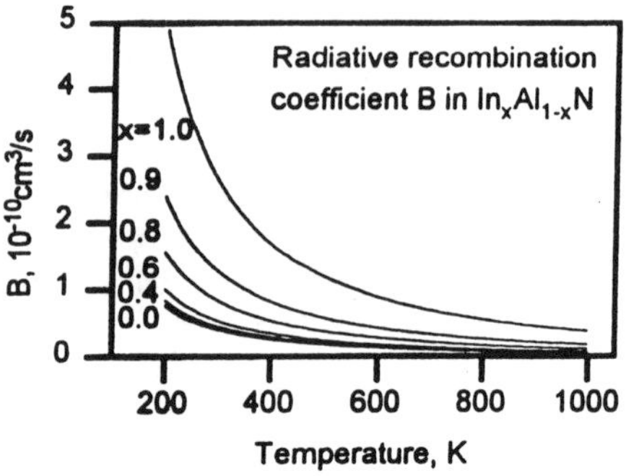

The calculations of the radiative recombination rates were performed in the full range of alloy compositions and in the temperature interval from 200 to 1000K.

The calculated temperature dependence of the radiative recombination coefficient, B, is shown in the figures. B is higher in InN and lower in GaN, taking intermediate values in AlN. For example, at 300K B = (2.7, 0.4 and 0.15) $\times$ 10^{-10}cm^3/s for InN, AlN and GaN, respectively. The corresponding radiative lifetime of non-equilibrium carriers in GaN at T = 300K and $n = 1 \times 10^{18} cm^{-3}$ equals 60 ns.

CONCLUSION

We calculated radiative recombination coefficients for three nitride semiconductors and their binary alloys. To do it, we extracted values of the interband matrix elements from the absorption data. The matrix elements do not differ considerably in these semiconductors, and the difference between the recombination coefficients is connected also with the difference in the band gap values and carrier masses.

ACKNOWLEDGEMENTS

The work was supported in part by Russian Foundation for Fundamental Research and by the International Science Foundation.

REFERENCES

[1] S.Strite, M.E.Lin and H.Morkoç: Thin Solid Films **231** (1993) 197.
[2] S.Nakamura et al.: Jpn. J. Appl. Phys. **34** (1995) 1832; Appl. Phys. Lett. **67** (1995) 1866.
[3] S.Yoshida, S.Misawa and S.Gonda: J. Appl. Phys. **53** (1982) 6844.
[4] R.Dingle and D.D.Sell: Phys. Rev. **B4** (1971) 1211.
[5] P.B.Perry and R.F.Rutz: Appl. Phys. Lett. **33** (1978) 319.
[6] S.Bloom and G.Hakbeke: Physica status solidi (b) **66** (1974) 161.
[7] C.P.Foly and T.M.Tasley: J. Appl. Phys. **59** (1986) 3241.
[8] D.Jones and A.H.Lettington: Solid State Commun. **11** (1972) 701.
[9] A.Rubio, J.L.Corkill et al.: Phys. Rev. **48** (1993) 11810.
[10] C.P.Foly and T.M.Tasley: Phys. Rev. **33** (1986) 1430.

[11] M.Z.Huang and W.Y.Ching: J. Phys. Chem. Solids **46** (1985) 977.
[12] Qixin Guo and A.Yoshida: Jpn. J. Appl. Phys. **33** (1994) 2453.
[13] H.Tessiere, P.Perlin et. al.: J. Appl. Phys. **76** (1994) 2429.
[14] J.L.Birman: Phys. Rev. **114** (1959) 1490.
[15] J.Pastrnak and L.Roskova: Physica status solidi **26** (1968) 591.
[16] S.Bloom: J. Phys. Chem. Solids **32** (1971) 2027.
[17] Y.P.Varshni: Physica **34** (1967) 149.
[18] W.van Roosbroeck and W.Shockley: Phys. Rev. **94** (1954) 1558.
[19] P.T.Landsberg: Recombination in Semiconductors, Cambridge University Press, 1991.
[20] H.Yamashita, K.Fukui, S.Misawa and S.Yoshida: J. Appl. Phys. **50** (1979) 896.
[21] Y.Koide, H.Itoh et al.: J. Appl. Phys. **21** (1987) 4540.
[22] J.C.Phillips: Bonds and Bands in Semiconductors, Academic, New York, 1973.

PYROELECTRIC EFFECT IN WURTZITE GALLIUM NITRIDE

A. D. BYKHOVSKI [(a),(b)], V. V. KAMINSKI [(b)], M. S. SHUR [(a)], Q. C. CHEN [(c)], and M. A. KHAN [(c)]

[(a)] *Department of Electrical Engineering, University of Virginia, Charlottesville, VA 22903-2442, USA, ab4k@virginia.edu*
[(b)] *A. F. Ioffe Physico-Technical Institute, 194021 St. Petersburg, Russia*
(c) *APA Optics, Inc., 2950 N. E. 84th Lane, Blaine, MN 55449, USA*

ABSTRACT

We report on the measurements of the pyroeffect in wurtzite n-type GaN films deposited over basal plane sapphire substrates. The voltage drop between the contacts was measured while the sample was subjected to uniform heating. Our results show that the pyroelectric effect in GaN can be partially attributed to the secondary pyroelectricity, caused by the development of strain in the material due to thermal expansion.

INTRODUCTION

Large piezoelectric constants of GaN and AlN [1] point out possible applications of GaN - based materials in piezoelectric sensors. Due to a wide band gap, these sensors are expected to operate in a broad temperature range and/or in a harsh environment. Recently, we reported [2] on the measurements of the piezoresistive effect in n-type wurtzite GaN films, which confirm these expectations. Our study showed that these samples had a strong dynamic response to an applied force. The gauge factor (GF) of these structures was measured at room temperature for both longitudinal and transverse configurations. The dynamic effect was related to a strong piezoeffect in GaN. The maximum dynamic GF observed was ~ 130 (approximately four times larger than for SiC [3]).

In this paper, we present the results of our experimental and theoretical studies of the pyroelectric effects in wurtzite GaN films in a temperature range from 20 °C to 80 °C. We also propose simple equation describing the pyroeffect under realistic condition of the pyroelectric charge relaxation. This equation relates the pyroelectric coefficients to piezoelectric properties of GaN. Our theory is in good agreement with our experimental data and can be applied to analyze the pyroelectric effect in many different types of materials.

I. EXPERIMENT

GaN layers (3 - 5 μm thick) were deposited over basal plane sapphire substrates, slightly off-axis. The sample geometry is shown in Fig. 1. Except for the center stripe, the GaN film has been completely etched down to the sapphire substrate (see Fig. 1). The standard Hall measurements showed that the electron concentration was close to 5×10^{16} cm^{-3} and the Hall mobility was on the order of 350 cm^2/V-s. A typical sample resistance was close to 2 kΩ.

Mat. Res. Soc. Symp. Proc. Vol. 423 © 1996 Materials Research Society

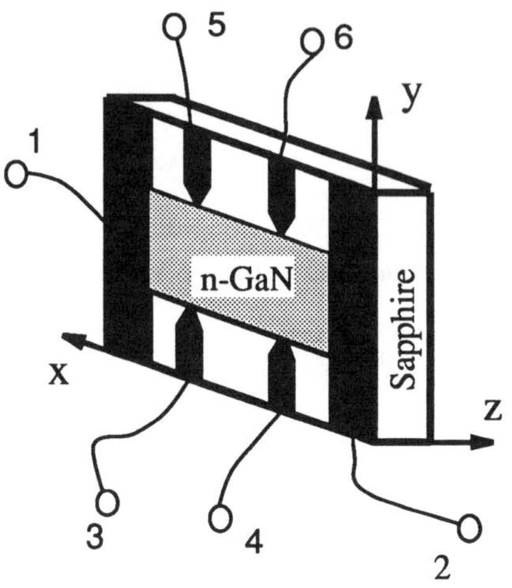

The voltage drop between the contacts 1 and 2 was measured while the sample was subjected to uniform heating. In these experiments, we abruptly changed the ambient temperature and monitored the sample temperature (using a SmS based thermoresistor) and voltage responses as functions of time. The experimental results will be described in Section III, where they will be interpreted using our theory. We conducted hot thermal shock experiments in the range +20 °C - +80 °C using oil as a thermal bath.

Fig. 1. Schematic of GaN sample. 1 - 6 are the contacts to GaN layer.

II. THEORY

The wurtzite GaN is a natural pyroelectric with the pyroelectric polarization in the direction of c-axis. It generates an electric charge in response to heat flow. Under the steady state conditions, the pyroelectric polarization can not be detected because of the compensating charges from the sample and from the ambiance [4]. This is why we study the change in the pyroelectric charge as a function of temperature variation. Both the sample temperature change T(t) - T(0) (which is proportional to the thermal energy stored) and the thermal flux affect the current value of the polarization [5].

In our experiments, the pyroelectric effect can be described by the system of the heat transfer equation, the equation of motion of a viscous fluid, and the equation for pyroelectric polarization (voltage) generation.

Heat Transfer

In our study, the sample was subjected to a uniform heating. The heat transfer was relatively low (due to the high viscosity of oil), and, therefore, it was caused both by free convection and thermal conduction. Since the measurement temperatures were relatively low, we may neglect the radiative transfer. Due to the high thermal conductivity of sapphire substrate, we can neglect the temperature gradients in the sample, and apply the lumped system analysis [6].

Pyroelectric Effect

Our samples had a strong electric response to a heat flow into the sample. In the thermodynamic equilibrium, the effect vanished. Taking into account the response to the heat accumulation and to a secondary pyroelectricity (piezoeffect) [5], the effect can be described by the simple equation which fits well our experimental data:

$$\dot{V} = \alpha_1 \dot{W} + \dot{V}_{piezo} - \frac{V}{\tau_s} \tag{1}$$

In Eq. (1), W is heat accumulated (dissipated) by the sample, and V_{piezo} describes the piezoelectric effect due the non-uniform thermal expansion of the sample [5], τ_s is the relaxation time, α_1 is a constant, and the dots denote the time derivatives. The heat accumulated by the sample is given by

$$W = c_p \rho (T - T_o), \tag{2}$$

where ρ is the sample density, c_p is the sample specific heat, T_0 is an initial sample temperature (in our experiments, it is a room temperature), and the weak dependence $c_p \rho$ on temperature is neglected. The piezoelectric term caused by non-uniform deformation is equal to:

$$V_{piezo} = \alpha_{piezo}(T - T_o) \tag{3}$$

where α_{piezo} is proportional to the effective piezoelectric constant of GaN and to the difference between GaN and sapphire thermal expansion coefficients. Using Eqs. (2), (3), we can write Eq. (1) in the simple form:

$$\dot{V} = \alpha \dot{T} - \frac{V}{\tau_s} \tag{4}$$

where

$$\alpha = \alpha_1 c_p \rho + \alpha_{piezo} \tag{5}$$

The solution of the Eq. (4) with the initial condition

$$V(t_o) = 0 \tag{6}$$

is

$$V = \alpha e^{-\Delta t / \tau_s} \int_0^{t - t_o} \dot{T}(t') e^{-\Delta t' / \tau_s} dt' \quad , t \geq t_o \tag{7}$$

According to Eq. (7), in order to find the voltage generated by the pyroeffect, we need to know the time derivative of sample temperature, $\dot{T}$. We determined $\dot{T}$ from the experimentally measured dependence $T(t)$. We calculate α_{piezo} from Eq. (5) using the theory of the piezoeffect in III-Nitrides [7] and the piezoelectric constants extracted from the data on GaN mobility [2]. In our experiments, α_{piezo} is the main contribution to the pyroeffect, and $\alpha \approx \alpha_{piezo}$. To compare the pyroelectric properties of GaN and another pyroelectric materials, we estimated the pyroelectric voltage coefficient [5] of GaN:

$$P_v = \frac{dF}{dT} \tag{8}$$

where F is the electric field in the sample caused by the piezoelectricity. In order to find the electric field, we solve Poisson's equation with the appropriate boundary conditions taking into account the piezoelectric polarization along the contacts [7]. It yields:

$$P_v = \alpha_{piezo} / 2 r_d \tag{9}$$

where r_d is the Debye length.

III. RESULTS

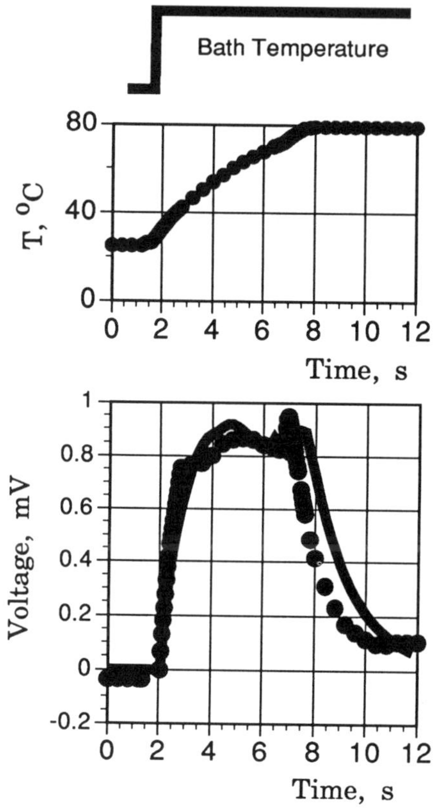

Fig. 2. Upper curve represents the sample temperature, and the lower curve represents the pyroelectric voltage. Circles are the data, solid line is the calculation.
The temperature of the environment (bath) increases from room temperature to 80 ^{o}C.

In Fig. 2, we plotted the results of our hot thermal shock experiments. Notice that in oil bath the heat transfer was relatively small due to the high oil viscosity. In this case, the heat conduction (thermal diffusion) mechanism is important. Indeed, the analysis of the monitored sample temperature increase shows that the heat transfer is due to a mixed diffusion-convection mechanism. As mentioned above, we used the experimental dependence T(t) in order to calculate the generated voltage. Also, our calculation shows that the main contribution to this effect comes from the secondary piezoelectricity. In Fig. 2, we plot the dependence V(t) calculated according to Eq. (7) with the only contribution left due to the piezoeffect. The theory reproduces well the experimental data. Notice that the only fitting parameter in this calculation is the relaxation time (we use τ_s = 1.6 s), and its value does not affect the forward front of the voltage pulse. The value of the pyroelectric voltage coefficient estimated according to Eq. (9) is ~ $3x10^3$ V/mK.

IV. CONCLUSIONS

We performed the experimental and the theoretical studies of the pyroelectric effect in wurtzite n-type GaN grown on basal plane sapphire slightly off (0001) axis. The voltage drop between the contacts was measured while the sample was subjected to uniform heating. Our results show that the pyroelectric effect in GaN can be attributed to the secondary pyroelectricity (piezoeffect), caused by the development of strain in the material due to non-uniform thermal expansion.

ACKNOWLEDGMENTS

This work was supported by the Office of Naval Research STTR contract (project monitor Max Yoder).

REFERENCES

1. J. G. Gualieri, J. A. Kosinski, and A. Ballato, IEEE Trans. Ultrason., Ferroel., and Freq. Control, 41, 53 (1994).

2. A. D. Bykhovski, V. V. Kaminski, M. S. Shur, Q. C. Chen, and M. A. Khan, Appl.. Phys. Lett., 68(6), 818 (1996).

3. J. S. Shor, L. Bemis, and A. D. Kurtz, IEEE Trans. Electron Dev. , 41(5), 661 (1994).

4. D. F. Nelson, Electric, Optic, and Acoustic Interactions in Dielectrics, (John Wiley & Sons, New York, 1979), pp. 244-246.

5. J. Fraden, AIP Handbook of Modern Sensors (AIP, New York, 1993), pp. 90-97.

6. Y. Bayazitolu and M. N. Ozisik, Elements of Heat Transfer (McGraw-Hill Book Company, New York, 1988), pp. 18-22.

7. A. Bykhovski, B. Gelmont, and M. Shur, Appl. Phys. Letters, 63, No 16, 2243 (1993); J. Appl. Phys., 74(11), 6734 (1993).

CVD DIAMOND WIRES AS X-RAY DETECTORS

C.Manfredotti,F.Fizzotti,M.Galetto,A.Lo Giudice,D.Margherita,C.Ongaro,P.Polesello,E.Vittone
Experimental Physics Department, University of Torino and INFN, sezione di Torino, via Giuria 1, 10125 Torino, Italy, manfredotti@to.infn.it

ABSTRACT

Miniature metal-diamond wire detectors have been fabricated by Hot Filament Chemical Vapor Deposition (HFCVD) method and characterized as x-ray dosimeters in the energy range 50 keV-250 keV. Linearity of the response and the sensitivity as a function of energy have been accurately determined in a range routinely used for x-ray radiotherapy. With a Mo wire substrate, a 10 μm thick CVD diamond film can reach sensitivities of the order of 4×10^{-8} A/Gy/min, more than one order of magnitude larger than for a 6 cm^3 standard ionization chamber.

INTRODUCTION

For x-ray detectors, particularly at low energies, diamond has many advantages over other semiconductor materials, as silicon and germanium. Diamond is faster due to larger saturated carrier drift velocity, a factor of 2 greater than silicon, less noisy due to the lower dielectric constant and more resistant to radiation damage (the displacement threshold is approximately 43 eV compared with approximately 20 eV for silicon). Due to its large band gap, a junction is not necessary to reduce the leakage current for proper operation of the detector at room temperature. Therefore, a detector can be made into a simple metal-semiconductor-metal (MSM) device with a mode of operation being that of a photoconductor, both pulsed and DC.

Few papers, however, have been published until now about the use of Chemical Vapor Deposited (CVD) diamond as x-ray detector, both in the field of hard x-rays [1, 2, 3] and for soft x-rays applications [4].

As a matter of fact, diamond is approximately a tissue-equivalent detector (Z=6 against Z=6.7 for the biological tissue) and this advantage can be probably used for the realization of solid state tissue-equivalent gamma cameras, particularly at low x-ray energies (below 30 keV, for instance). Moreover, diamond is also tissue-compatible and environmentally hard, being therefore exceptionally suitable for in-vivo probes for absorbed dose monitoring. For these applications, needle-like shapes are evidently to be preferred. Now, among the various CVD techniques, Hot Filament Chemical Vapour Deposition (HFCVD) turns out to be particularly suitable and simple for obtaining diamond wires and tubes [5, 6]. In this work, we present the results which have been obtained in deposition of CVD diamond onto some kind of metal wires (W, Mo, Ta) in order to prepare x-ray detectors for energies in the interval 50÷250 keV of maximum energy. The substrate metal wire is used in order to include the CVD diamond response in this energy region. For higher x-ray energies, larger Z-value substrates may be needed (Au, Pt, Pb)

EXPERIMENTAL

Our hot filament chemical vapor deposition apparatus, consists of a cylindrical stainless steel chamber (250 mm diameter, 300 mm height) with a viewport. The gas flow is controlled by two mass flow controllers and the pressure is measured by a capacitance manometer. The deposition parameters are listed in table I.

Before deposition, a base pressure of 10^{-6} mbar was maintained for 24 hours.

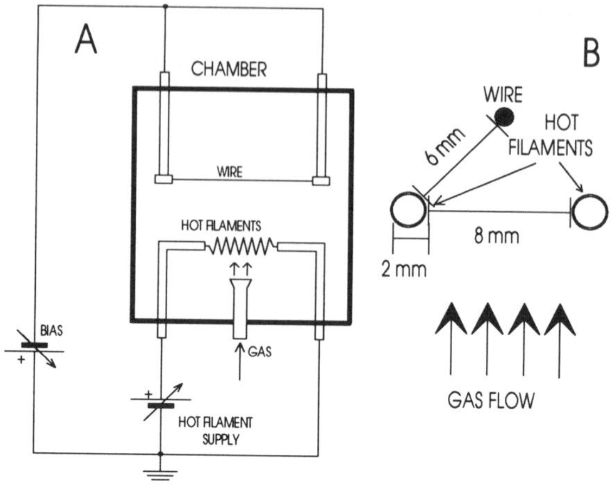

Fig. 1 - HFCVD apparatus used for diamond wire fabrication (A) and the particular arrangement of the two hot filaments in order to obtain a uniform deposition (B)

Fig. 2 - SEM micrograph of a cross section of a diamond deposition (20 μm) on a W wire with a 100 μm diameter at the origin

Two tantalum wires (0.25 mm diameter) are wound into helical shapes with about 2 mm turn diameter. The filaments were heated up to 2300°C as measured by an optical pyrometer (Minolta-Land Cyclops 152). The filaments were located 8 mm apart and 6 mm distant from the substrate which consists of a molybdenum wire (50 mm long and 0.3 mm diameter). In order to increase initial diamond nucleation rates, they were manually abraded with 0.25 μm diamond paste and then washed in acetone.

The temperature of the substrate wire, heated by radiation and conduction from the tantalum filaments, was not measured due to the difficulty in placing a thermocouple through the filament and was estimated to be about 950°C. The tantalum filaments dissociated the gases allowing carbon to deposit on to the surface of the wires in the form of a polycrystalline diamond film whose thickness, as evaluated by scanning electron microscopy, was about 10 μm after 8 hours deposition.

After deposition, and before contacts were applied, the diamond coated wire was annealed at 600°C in an argon environment and rinsed in acetone.

Parameter	
CH_4 flow rate	2.5 sccm
H_2 flow rate	500 sccm
Pressure	15 mbar
Filament temperature	2300°C
Substrate temperature (estimated)	950°C
Deposition Time	8 hours

Table I Growth parameter used for the deposition of diamond film

A metallic thermal evaporation technique was used to coat 4 mm of the diamond surface with successive metals: Ti (40 nm) and Au (100 nm). The thickness of the metallic layers was measured during evaporation with an in-situ thickness monitor. Titanium was used since it easily forms a carbide structure; gold was used to prevent oxidation of the Ti layer. The sample was then rinsed in acetone and annealed at 600°C in a N_2 environment to allow the Ti to form a carbide with diamond.

The electronic properties of the Au/Ti-diamond-Mo structure were investigated by measuring the current-voltage characteristic by means of a mod.617 Keithley electrometer.

RESULTS AND DISCUSSION

The CVD diamond films onto metal wires grow at a reasonable deposition rate (1-2 μm/h) and, with the particular hot filaments arrangement adopted, quite uniformly. The best films, from the point of view of adhesion and thickness uniformity, are those deposited onto tungsten wires: an example is shown in Fig. 2.

Tungsten, however, is relatively fragile, due to the strong carburization; and, moreover, the cleaning procedure before deposition is difficult to apply. This is to some extent disappointing, since Raman spectra indicate the best diamond quality for W deposition. Tantalum giving

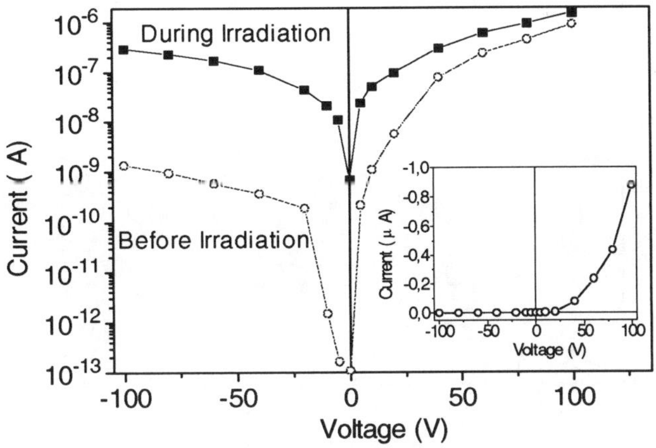

Fig. 3 - I-V characteristics of a Mo-diamond wire (300 μm diameter plus a 10 μm of deposited diamond) before and during an x-ray irradiation of about 150 keV maximum energy. In the insert, the rectifying properties of the diamond wire are displayed on a linear scale

adhesion problems, the choice fell onto Mo, which gave very good electrical characteristics, displaying a strong rectification behavior (see Fig. 3, insert) which allows current intensities lower or of the order of 1 nA below 100 V applied bias. This strong rectification is not well understood at the moment, but it is certainly related to the different kinds of electrodes (Mo, or better, MoC inside, Ti/Au outside).

The current increase corresponding to a dose rate of 10.8 Gy/min is quite noticeable, particularly at bias voltages around or below 20 V (see Fig. 3): however, since in the investigated dose rate interval the S/N ratio was very good, all the measurements reported here will be referred to 100 V bias. After an irradiation corresponding to a dose of 20 Gy, the diamond wire reaches a reproducible starting condition, which can be used for any kind of dose measurement.

Both direct and reverse I-V characteristics are compatible with a Schottky or Poole-Frenkel behavior: but it seems more likely that direct I-V characteristics, at least at larger bias voltages, is due to a Space Charge Limited Current mechanism.

Mo-diamond wires have been subjected to a careful investigation about their response to x-rays in the energy interval from 50 keV up to 250 keV and for x-ray tube current from 6 to 14 mA. The responses in terms of current as a function of the dose rate are always linear (see Fig. 4, which reports the results obtained at a maximum energy of 150 keV, for an applied bias of 100 V). The x-ray tube was filtered by a Be window: a second filtration was set up by placing the wire detector in a box of aluminum 2 mm thick, in order to cut off energies below 20 keV. In the same box, we placed a 6 cm^2 ionization chamber (mod. 1015× Dosimeter Corporation) with a time equivalent response within 10 % from 20 keV to 660 keV, in order to measure directly the dose rate. A further measurement by using much smaller TLD-100 thermoluminescent dosimeters (about 3×3×1 mm^3) gave approximately the same results. Therefore, one can conclude that, at least in a dose rate interval typical for x-ray radiotherapy, and for maximum energies from 50 keV to 250 keV, Mo-diamond wire response is linear. In terms of current, Mo-wire displays a

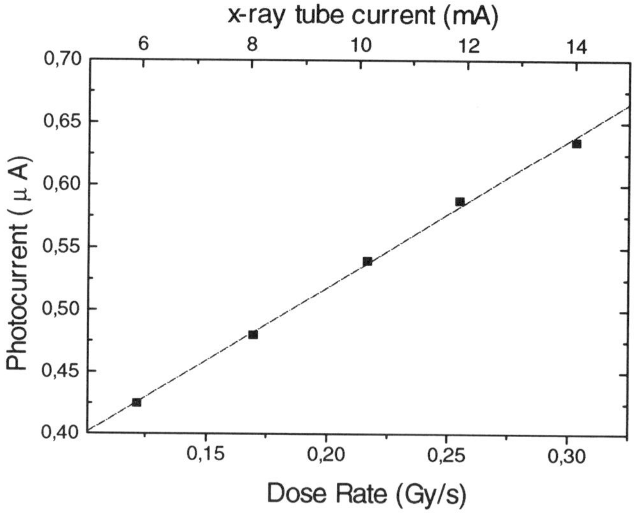

Fig. 4 - Behaviour of the steady-state photocurrent in a Mo-diamond wire, biased at 100 V, as a function of dose rate (Gy/s). X-ray tube voltage is 150 kV, while the current varies from 6 to 14 mA.

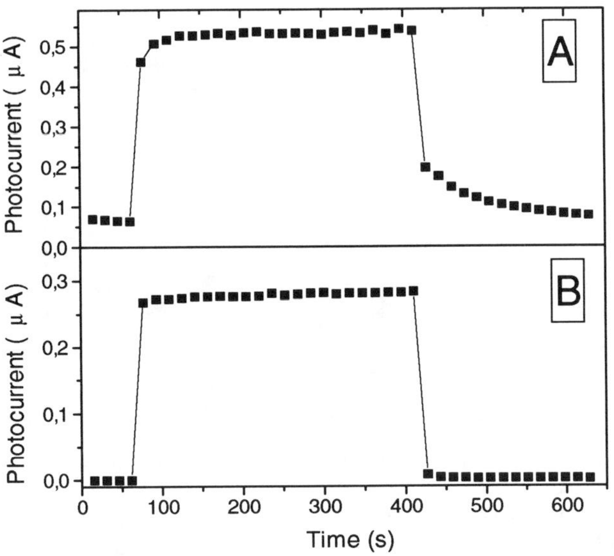

Fig. 5 - Time response of a Mo-diamond wire (A) and of a diamond planar detector (B) as exposed to x-rays (150 kV, 12 mA) for a time of 6 minutes. The diamond planar detector has dimensions of 1 $cm^2 \times 400$ μm. In both cases, bias voltage is 100 V

sensitivity about 14 times larger than the 6 cm^3 ionization chamber placed in the same conditions (7×10^{-12} Ah/R against 5×10^{-13} Ah/R, where Roentgen, R, refers to a ionization density of 2×10^9 electron-ion pair per cm^3 of air).

Time response of Mo-diamond wire is also relatively fast, at least if compared with a very fast (about 1 ns) planar CVD diamond detector in approximately the same conditions (Fig. 5). The long tail of the x-ray beam shutoff is not present in planar detector, where the substrate and a 50 μm thick layer next to the substrate have been cut off, which contained mainly small dimension grains. Deep trapping in this region, which is present in wire detector, and, likely, the presence of large capacitances due to technologically imperfect contact, could be responsible for the observed effect, which seems to be limited to 2 or 3 minutes and which may be likely eliminated by a more careful contact preparation.

Finally, Mo-diamond wire sensitivity (in terms of As/Gy) is reported as a function of x-ray maximum energy in Fig. 6. The sensitivity is decreased as a function of maximum energy mainly because low energy electrons produced by low energy x-rays close to the Mo surface are more effective, because of the short range and of the small thickness of diamond. The results of theoretical calculations, based on Monte Carlo method, are also plotted, and it turns out to be in reasonable agreement with experimental data. The calculations simulate the whole x-ray spectrum impinging on the wire detector, calculate the interaction point, follow the emitted photoelectron according to the particular angular distribution and, finally, evaluate the charge (or the current) created in the diamond film, according to the particular irradiation geometry, which is normal to the wire axis. By comparing with the actual detector response one can obtain the charge collection efficiency of the CVD diamond films, which is of the order of few percent. The

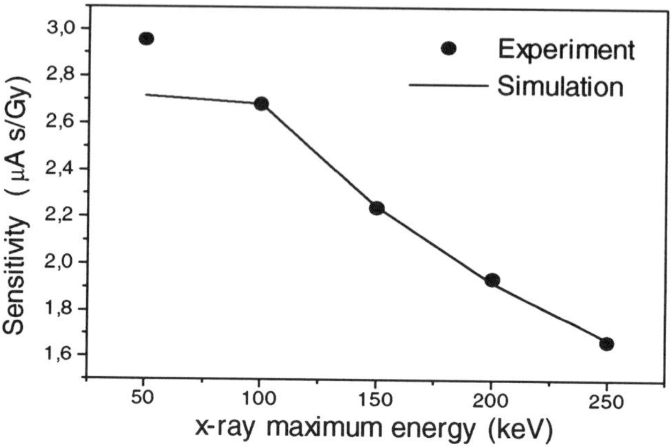

Fig. 6 - Mo-diamond wire detector sensitivity (As/Gy) as a function of x-rays maximum energy, for an x-ray tube current of 10 mA. Bias voltage is 100 V. The results obtained by a simple Monte Carlo calculation are also shown for comparison. A Gray (Gy) corresponds to a dose of 1 J deposited energy in 1 kg of detector material (diamond in this case)

program code will be very useful in order to optimize the wire detector from the point of view both of the geometry and of the kind of substrate material.

CONCLUSION

Miniature diamond wire detector have been fabricated and characterized with respect to medium energy x-ray. The linearity of response and the sensitivity vs. x-ray maximum energy have been measured between 50 keV and 250 keV. In spite of the very small dimensions, the sensitivity is more than one order of magnitude larger than for a 6 cm^2 standard ionization chamber. By varying the wire substrate one can try to increase the sensitivity for higher energies or, also, to make the detector tissue-equivalent for low energies applications. A wire mesh detector may be also fabricated for dose imaging purposes.

REFERENCES

[1] V.F. Dvoryankin, A.A Kudryashov, Yu. Sh. Temirov, A.F. Alexenko, L.L. Bouilov, A.A. Botev and G.A. Sokolina, in Proceedings of "*Diamond Films '95*" 6th European Conference on Diamond and Related Materials, Barcelona, Spain, September1995 in press

[2] F. Foulon, T. Pochet, E. Gheeraert and A. Deneuville, IEEE Trans. Nucl. Sci. **41**, 927 (1994)

[3] V.F. Dvoryankin,A.A Kudryashov,Yu.Sh.Temirov,L.L.Bouilov,G.A.Sokolina,A.F. Alexenko, in Proc. of "*Appl. Diamond Conference 1995*" , Gaithersburg (MA), August 1995, in press.

[4] S. Han private communication

[5] A.A. Morrish, J.W. Glesener, P.E. Pehrson, B. Maruyama nad P.M. Natishan, J. Mat. Sci. **30**, 834 (1995)

[6] P.G. Partridge, M.N.R. Ashfold, P.W. May and E.D. Nicholson, J. Mat. Sci. **30**, 3973 (1995)

CRITICAL MATERIALS, DEVICE DESIGN, PERFORMANCE AND RELIABILITY ISSUES IN 4H-SIC POWER UMOSFET STRUCTURES

A. K. AGARWAL, R. R. SIERGIEJ, S. SESHADRI, M. H. WHITE*, P. G. MCMULLIN, A. A. BURK, L. B. ROWLAND, C. D. BRANDT AND R. H. HOPKINS
Northrop Grumman Science & Technology Center, 1310 Beulah Road, Pittsburgh, PA 15235, Tel. (412) 256-2037, FAX (412) 256-1877, agarwal@mmwcad.pgh.wec.com,
*Lehigh University, Sherman Fairchild Center, Bethlehem, PA 18015.

ABSTRACT

The long-term reliability of gate insulator under high field stress of either polarity presents a constraint on the highest electric field that can be tolerated in a 4H-SiC UMOSFET under on or off condition. A realistic performance projection of 4H-SiC UMOSFET structures based on electric field in the gate insulator (1.5 MV/cm under on-condition and 3 MV/cm under off-condition) consistent with long-term reliability of insulator is provided for the breakdown voltage in the range of 600 to 1500 V. The use of P^+ polysilicon gate allows us to use a higher field of 3 MV/cm in the insulator under off-condition and leads to a higher breakdown voltage as the Fowler Nordheim (FN) injection from the gate electrode is reduced. FN injection data is presented for p type 4H-SiC MOS capacitor under inversion at room temperature and at 325°C. It is concluded that the insulator reliability, and not the SiC, is the limiting factor and therefore the high temperature operation of these devices may not be practical.

INTRODUCTION

Extremely attractive performance projections have been made for 4H-SiC UMOSFET structure for power electronic applications claiming 100x better performance than silicon power devices [1]. These projections are made on the basis of 10x higher breakdown field in SiC compared to silicon. However, a closer look reveals that the limitations of the gate insulator such as SiO_2 will prevent us from exploiting the full potential of these devices. Significant progress has been reported towards fabricating these devices [2,3]. The results from CREE Research are particularly encouraging although the specific on-resistance of 17.5 $m\Omega\text{-}cm^2$ has been obtained with positive gate bias far exceeding the practical limit of 1.5 MV/cm electric field in the insulator.

INSULATOR ISSUES

Electric Field in the Gate Insulator Under On-Condition

The electric field in the insulator overlapping the N^+ region under on-condition (Fig. 1), E_{i+}, should not exceed 1.5 MV/cm at room temperature and should be much less at elevated temperature. The energy band diagram under these conditions is shown in Fig. 2. Since the barrier for electrons in the conduction band of N^+ 4H-SiC is only 2.7 eV [4] compared to 3.15 for Si, FN tunneling of electrons will dictate the maximum field consistent with the long-term time dependent dielectric breakdown (TDDB) of the insulator at a given temperature. In silicon MOS devices, this field is kept at about 2 MV/cm. Due to the lower barrier in SiC, 1.5 MV/cm at room temperature is more appropriate. This then places a limit on maximum positive gate bias that can be applied under on-condition and on the doping of the p-region due to the need for minimizing the threshold voltage.

Mat. Res. Soc. Symp. Proc. Vol. 423 © 1996 Materials Research Society

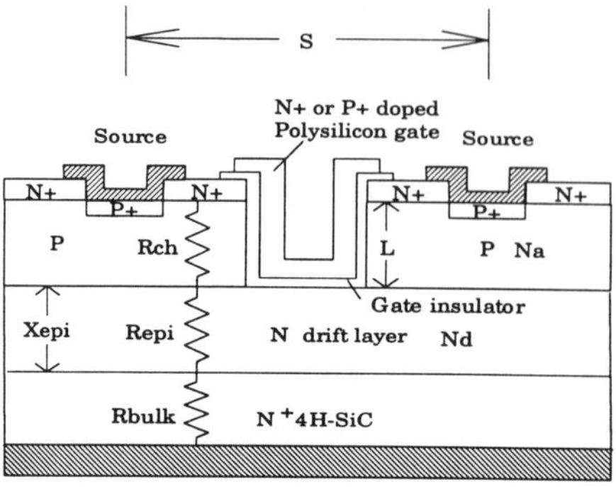

Fig. 1 A schematic cross-section of a unit cell of the 4H-SiC UMOS showing all the active layers. In a device, there are N cells each of width W and pitch S.

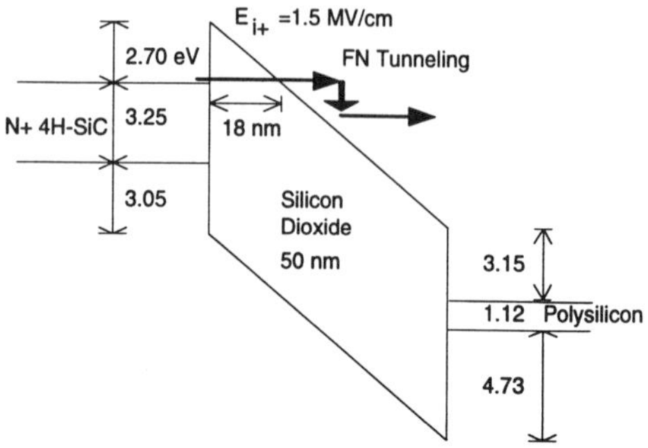

Fig. 2 Energy band diagram of N^+ 4H-SiC/SiO_2/Polysilicon system under on-condition

In order to study the FN tunneling, metal-oxide-semiconductor (MOS) capacitors were made on the silicon face of the p-type 4H-SiC by pyrogenic oxidation at 1050°C for six hours followed by 1 hr. anneal in Argon and 1.5 hr. anneal in oxygen at 950°C. Approximately 460 Å of silicon nitride was deposited on top of the thermal oxide and gold dots were evaporated. The backside of the wafer was cleaned by reactive ion etching the oxide and the nitride and 5000 Å of Al layer was thermally evaporated. Finally, the wafer was annealed in forming gas at 450°C for 1 Hr. FN tunneling measurements were made by stepping a positive bias on the top electrode, creating an inversion layer at each bias step by illuminating with a UV source for 1 minute, and then measuring the current in the dark. The results are shown in Fig. 3. The composite insulator breaks down at about 11-12 MV/cm electric field at room temperature and about 5 MV/cm at 325°C. The FN plots indicate much higher current density at a given field in SiC as compared to silicon. The current density in SiC increases much more rapidly with temperature as compared to silicon indicating the presence of enhanced thermionic field emission due to lower barrier in SiC/SiO_2 system.

Electric Field in the Gate Insulator Under Off-Condition

The electric field in the insulator over the bottom corner of the trench, E_{i-}, under off condition with maximum voltage on the drain should not exceed about 2 MV/cm to avoid FN tunneling of electrons from the conduction band of N^+ ploysilicon gate into the insulator at room temperature. This limit may be raised to about 3 MV/cm if P^+ doped polysilicon gate is employed due to about 1 V increase in the barrier (Fig. 4) [5]. A 2-D simulation using MEDICI software from TMA indicates the field crowding in the insulator near the bottom corner. Taking into account the ratio of dielectric constants of SiO_2 and SiC (3.9/10), the maximum field in SiC at the SiC/SiO_2 interface in the bottom flat region of the trench should not exceed about 0.75 MV/cm. Assuming some charge sharing with the adjacent channel [6], this limit may be raised to about 1 MV/cm at room temperature. Again, this limit should be lowered at elevated temperatures. As the breakdown field of 4H-SiC is about 3 MV/cm, it is clear that in devices such as UMOS or IGBT where gate insulator is subject to the high field stress under off-condition, the design and performance is limited by the insulator reliability issues rather than intrinsic SiC properties. These reliability issues become more severe at elevated temperatures and thus it is difficult to take advantage of low leakage currents in SiC at high operating temperatures. ***Thus the expected reduction in cooling requirements due to higher junction temperatures will not be available in such devices.***

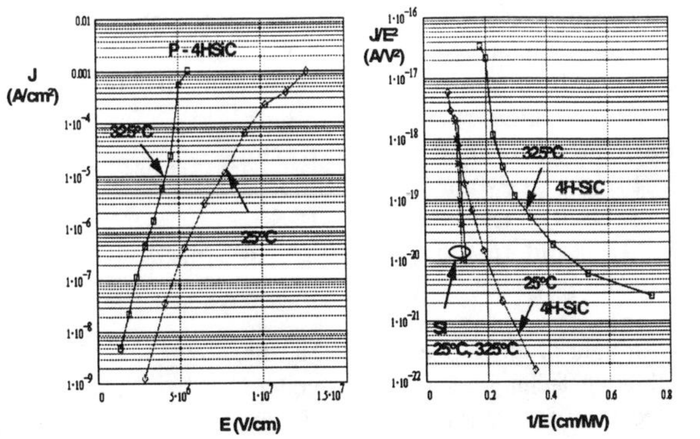

Fig. 3 FN tunneling data on p-4HSiC MOS capacitors under positive bias indicates that much higher injection current densities at 325°C are obtained in 4H-SiC as compared to silicon due to the lower barrier in the SiC/SiO_2 system. The electric field stress in SiC UMOS structures therefore needs to be reduced below that for silicon for long-term reliability.

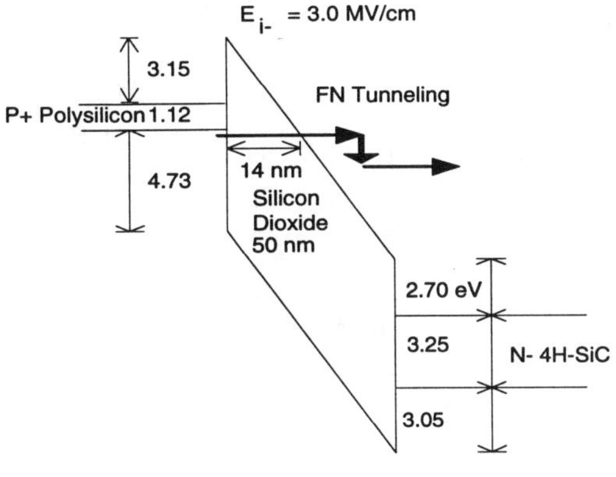

Fig. 4 Energy band diagram of N 4H-SiC/SiO_2/P+ Polysilicon system under off-condition

<u>High Interface State Density and Fixed Charge</u>

High interface state density at the p-SiC/SiO_2 interface and high fixed charge in the gate insulator have been the cause for a very poor effective inversion layer electron mobility of about 10 cm^2/Vs at room temperature [2]. Several good ideas have come forward such as the use of ozone cleaning to remove graphitic carbon from the surface [7] and the growth of thermal oxide at the lowest practical temperature (1050°C) followed by an anneal at an even lower temperature (900°C) [8]. Even with these techniques, the interface state density remains high ($\sim 1\times10^{11}$ $cm^{-2}.eV^{-1}$) in moderately Al-doped layers. Fig. 5 compares high frequency CV curves for MOS samples fabricated as described earlier with or without ozone cleaning prior to thermal oxidation. Ozone cleaning clearly leads to low fixed charge and reduced interface trap density. With further improvements in processing, an inversion layer electron mobility of about 50 cm^2/Vs is not at all unreasonable given the bulk electron mobility of about 600 cm^2/Vs parallel to the c-axis.

<u>Performance Projection:</u>

Considering these factors, and assuming that about 80% of the breakdown voltage is available, the specific on-resistance at room temperature for UMOSFET's has been plotted in Fig. 6 along with the inversion layer resistance and the drift layer resistance for a range of breakdown voltages. **It should be emphasized that even with all the limitations of the gate insulator, the SiC UMOSFET is a very practical and viable choice in this voltage range due to its ease of control and higher switching frequency than regenerative devices such as GTO's.** It is interesting to note that whereas the silicon devices are completely dominated by the drift layer resistance, the 4H-SiC based devices have a potential of being dominated by the channel resistance if surface electron mobility in the inversion layer is much less than 50 cm^2/Vs. **Also, whereas silicon IGBTs gain enormously by cutting down the drift layer resistance by means of conductivity modulation, 4H-SiC IGBTs stand to gain very little in this voltage range.**

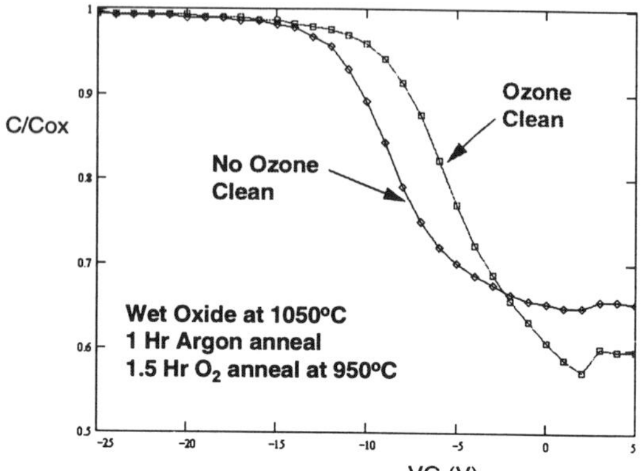

Fig. 5 Ozone cleaning reduces fixed charge and interface trap density in a MOS capacitor formed by thermal oxidation of p-type 4H-SiC.

In light of the above discussion, the following technological goals need to be reached for achieving a specific on-resistance in the range of 30-60 mΩ-cm^2 for a 500-2000 V 4H-SiC UMOSFET:

1. High quality gate dielectric that can withstand a stress of at least 1.5 MV/cm under positive gate bias and at least 3 MV/cm under negative gate bias on a long term basis.

2. High quality gate dielectric/p-SiC interface along the sidewalls leading to an inversion layer mobility of about 50 cm^2/Vs. This implies low density of fast and slow interface traps, $D_{it} < 5x10^{10}$ $cm^{-2}.eV^{-1}$ and low fixed charge in the gate insulator, $Q_i < 1x10^{11}$ cm^{-2}.

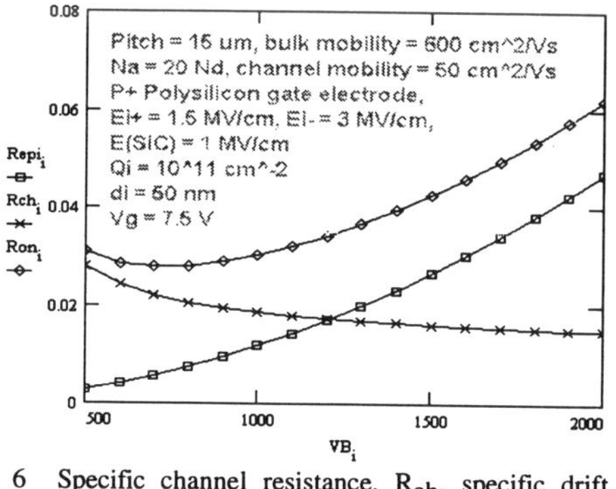

Fig. 6 Specific channel resistance, R_{ch}, specific drift layer resistance, R_{epi}, and specific on-resistance in Ohm.cm^2 for SiC 4H-UMOSFETs at room temperature as a function of breakdown voltage. The fact that R_{ch} dominates means that there is no advantage of using SiC IGBT in this voltage range.

EXPERIMENTAL RESULTS

UMOSFETs were made on the Si-face of 4H-SiC substrates. The device had a periphery of 0.3 cm, side-wall thermal gate oxide thickness of about 100 nm, cell pitch of 29 µm, drift layer of 5 µm doped at about $1x10^{16}$ cm^{-3}, p-type channel layer doped at $1x10^{17}$ cm^{-3} and a channel length of 2 µm. The minimum width of the trench was 3 µm. The gate oxide was grown in three different ways:

Thermal Oxide

Thermal oxide grown at 1150°C wet for 4 hrs followed by 30 min. Ar anneal. Phosphorous doped N^+ polysilicon was used as the gate electrode. The I-V characteristics are shown in Fig. 7. The channel current showed activation with temperature reaching a maximum at 250°C thus indicating large number of interface traps at the thermal oxide/p-SiC sidewall interface. The current density was 18.5 A/cm^2 at 3 V forward voltage corresponding to a specific on resistance of about 162 $m\Omega$-cm^2 at 250°C. The maximum positive gate bias was 18 V to keep the positive field in the side-wall gate oxide below 1.5 MV/cm. It is clear that the on-resistance is dominated by the inversion layer resistance due to the poor quality of the gate oxide/p-SiC interface along the vertical sidewall (inversion layer electron effective mobility, $\mu_{n,ch}$~4 cm^2/Vs). The breakdown voltage was about 100 V for this device. ***The main problem with this approach is that the thermal oxide grown on the silicon face on the bottom of the trench is too thin (~300 Å) whereas it is too thick (~1500 Å) on the side-walls.*** The thin oxide on the trench-floor and the field-crowding at the bottom corners of the trench limit the breakdown voltage to about 100 V.

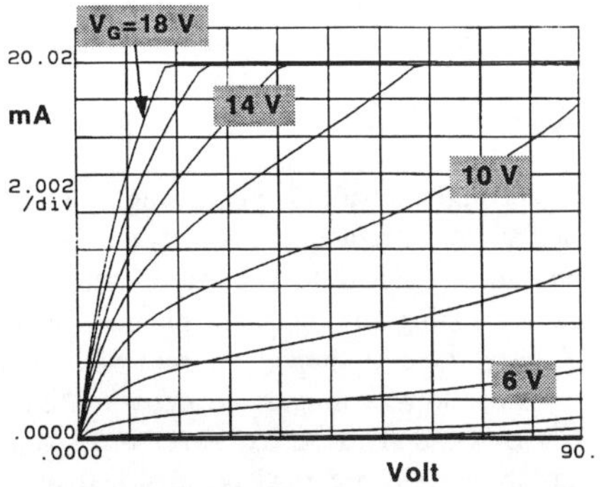

Fig. 7 Experimental I-V characteristics at 250°C for the thermal oxide case.

Deposited and oxidized amorphous Si gate oxide

In order to avoid the anisotropy in gate oxide thickness, the gate oxide was produced by depositing 620Å of amorphous silicon and oxidizing it at 1150°C, wet for 12 min. such that some thermal oxide was grown on the SiC surface after complete oxidation of the deposited amorphous silicon layer. This was followed by 30 min. Ar anneal at 1150°C. A 4000 Å thick layer of phosphorous doped N^+ polysilicon layer was deposited for gate electrode. The I-V characteristics at 300°C are shown in Fig. 8. The current density is too low and shows activation up to 400°C indicating a higher interface state density compared to the pure thermal oxide case. The breakdown voltage is increased to about 150 V due to somewhat thicker oxide on the trench floor as compared to the pure thermal oxide case.

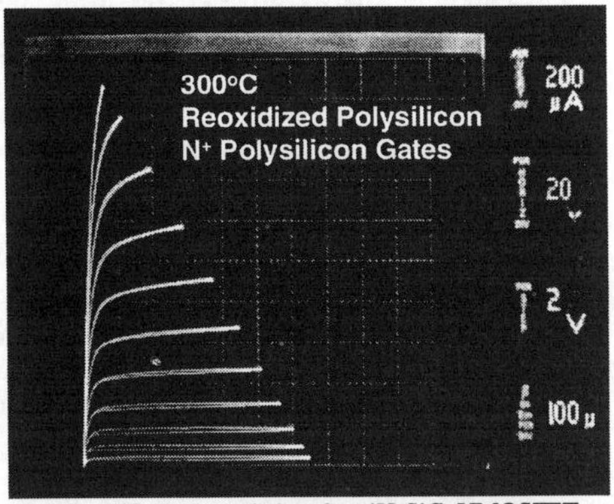

Fig. 8 I-V characteristics for 4H-SiC UMOSFET at 300°C for oxidized amorphous silicon gate insulator.

LPCVD oxide and P^+ doped polysilicon gate

A 700 Å of LPCVD oxide was deposited followed by an anneal in dry O_2 at 1000°C and 30 min. Ar anneal. Boron doped polysilicon gate was used to improve the breakdown voltage up to about 250 V at 300°C as shown in Fig. 9. The

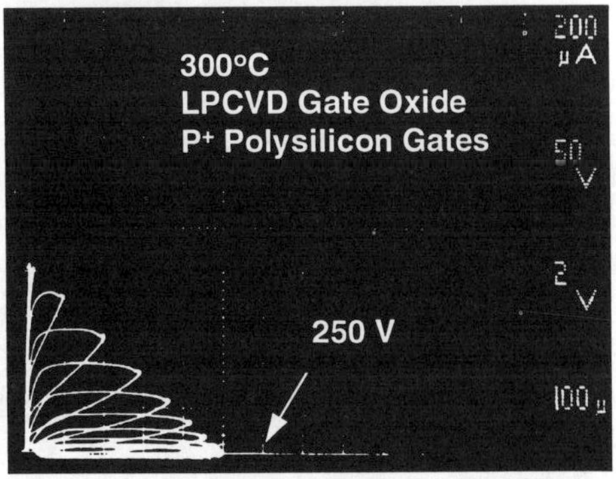

Fig. 9 I-V characteristics for 4H-SiC UMOSFET at 300°C for deposited LPCVD oxide based gate insulator. Use of P^+ polysilicon gate improves the breakdown voltage.

current was in between cases 1 and 2 indicating that interface state density is higher than the thermal oxide but lower than the oxidized amorphous silicon case.

CONCLUSIONS

The above results are encouraging and point towards several problems that need to be solved such as low interface trap density and fixed insulator charge at the gate insulator/4H-SiC side-wall interface, appropriate edge-termination, smaller cell-pitch down to 6-10 μm (to increase the charge sharing among the neighboring cells), and ways to round the sharp corners at the bottom of the vertical U-shaped grooves which tend to concentrate the electric field with the resulting low breakdown voltages. Moreover, the reliability of gate oxides or alternative insulators such as AlN or Oxide/Nitride/Oxide (ONO) stacks needs to be established at high temperature under appropriate electric field polarity and magnitude. Although, thick oxide can be grown on the carbon-face, the interface state density at the SiC/SiO_2 tends to be about one order of magnitude higher than that on the silicon-face. This by itself is not a problem as no current flows through this region but net positive charge in the interface traps under off-condition (n-SiC drift region depleted under the trench) can seriously degrade the breakdown voltage and is a serious long-term reliability issue. It may be better to grow a very thin high quality thermal oxide layer on the silicon-face and on the side-walls and follow it with a deposited and annealed LPCVD oxide to achieve almost a uniform layer of oxide on the sides and bottom of the U-grooves.

ACKNOWLEDGMENT

This work was supported in part by US Air Force (WPAFB, Dayton Ohio) and Office of Naval Research, Arlington, Virginia. Authors wish to thank Mr. Clarence Severt of Wright Laboratory and Dr. Alvin M. Goodman of ONR for helpful technical discussions. The entire AESL staff at Northrop Grumman STC is acknowledged for their contributions to this work.

REFERENCES

1. M. Bhatnagar and B. J. Baliga, "Comparison of 6H-SiC, 3C-SiC and Si power devices," IEEE Trans. on Electron Devices, 40, pp. 645-655, 1993.
2. J. W. Palmour and L. A. Lipkin, "High temperature power devices in silicon carbide," Trans. Second International High Temperature Electronics Conference, p. XI-3, June 1994.
3. N. Tokura, K. Hara, T. Miyajima, H. Fuma and K. Hara, "Current-Voltage and capacitance-voltage characteristics of metal/oxide/6H-silicon carbide structure," Jpn. J. Appl. Phys., 34, pp. 5567-5573, 1995.
4. V. V. Afanas ev, M. Bassler, G. Pensl, E. Stein von Kamienski and M. J. Schulz, "Band offsets and electronic structure of SiC/SiO_2 interfaces," unpublished.
5. Dr. Alvin M. Goodman, private communication.
6. Prof. T. Paul Chow, private communication.
7. Afanas ev, M. Bassler, G. Pensl and M. J. Schulz, "Energy distribution of electron states at SiC/SiO_2 interfaces," IEEE Semiconductor Interface Specialist Conference, Dec. 1995.
8. L. A. Lipkin and J. W. Palmour, "Improved oxidation procedures for reduced SiO_2/SiC defects," IEEE Semiconductor Interface Specialist Conference, Dec. 1995.

High-Temperature Switching Characteristics of 6H-SiC Thyristor

K. Xie, J.R. Flemish, T. Burke, W.R. Buchwald and J.H. Zhao*
US Army Research Laboratory, Physical Sciences Directorate, AMSRL-PS-DB, Fort Monmouth, NJ 07703-5000
* ECE Department, Rutgers University, Piscataway, NJ 08855

ABSTRACT

A 280 V 6H-SiC thyristor has been fabricated and characterized. The switching characteristics of the SiC thyristor were investigated over a temperature range from 23 °C to 400 °C, with a switched current density of 4900 A/cm^2 being observed under pulse bias condition. The thyristor has shown a dV/dt of 400 V/ms. Both the turn-on time and turn-off time increase significantly at 400 °C. The thyristor forward breakover voltage decreases by only 5% when the operating temperature is increased from 23 °C to 400 °C.

INTRODUCTION

SiC is a highly promising semiconductor for high-temperature and high-power electronics applications due to its wide bandgap, high breakdown field and high thermal conductivity [1,2]. These superior properties would result in a high power density (100 times higher than Si counterpart) and low thermal budget for SiC based power systems. It is suggested, under the ideal conditions, that the SiC based MOSFETs are the choice of future power devices because of its many ideal properties, such as low forward drop, high blocking voltage and high frequency operation [3]. However, in reality, many problems are hindering the development of SiC based MOSFETs. The quality of thermal oxide on p-type SiC is still far from acceptable for n-channel MOSFETs applications. The channel mobility of SiC UMOSFETs is around 20 cm^2/V.s, orders magnitude less than that of the Si counterpart [4]. The reliability of SiO_2 at high temperature under high field has proven to be a serious problem even in Si technology [5]. The maximum oxide field is limited to less than 2MV/cm when operating temperature is increased to 350 °C. However, in order to utilized the high breakdown field of SiC, the electric field on the oxide film on SiC will be around 5 MV/cm due to the low dielectric constant of the oxide. It is obvious that such high fields will pose serious problem if the SiC based MOSFETs are to be operated at high temperature. In Si power technology, thyristor structures have been used for high-voltage and high-current applications due to their low forward drop and high current handling capability [6]. Advanced MOS-controlled Si thyristors (MCT) offers additional advantage of voltage gate control. It is expected that SiC thyristors will operate reliably under high-voltage, high-current and high temperature conditions by avoiding the problems associated with oxide based devices. Furthermore, because of the lack of high fields on the oxide, a SiC based thyristor with MOS controlled turn-off could be more easily developed. Thus, of the various SiC power devices demonstrated, the SiC based thyristor is the most promising power device for ultra high-voltage and high-current applications, including high power motor drives for electric vehicle, aircraft engine controls and utilities. In our earlier work, we have reported a high-current and high-temperature SiC thyristor with a blocking voltage of 100 V [7]. In this work we report the switching characteristics of a 280 V 6H-SiC thyristor operated at temperature up to 400 °C.

EXPERIMENT

The 6H-SiC pnpn thyristor structure used in this work is illustrated in the Figure 1. The structure consists of a 0.8 μm p-type base layer grown on an n$^+$ 6H-SiC substrate, a 7.5 μm undoped n$^-$type blocking layer and a 0.5 μm Al doped p$^+$ layer. The doping level in the n$^-$ type blocking layer was designed to be 8x10^{15}/cm^3 for a blocking voltage of 400 V. The gate contact was placed on the n-type base layer. The gate recess and device isolation etching were accomplished by

Mat. Res. Soc. Symp. Proc. Vol. 423 © 1996 Materials Research Society

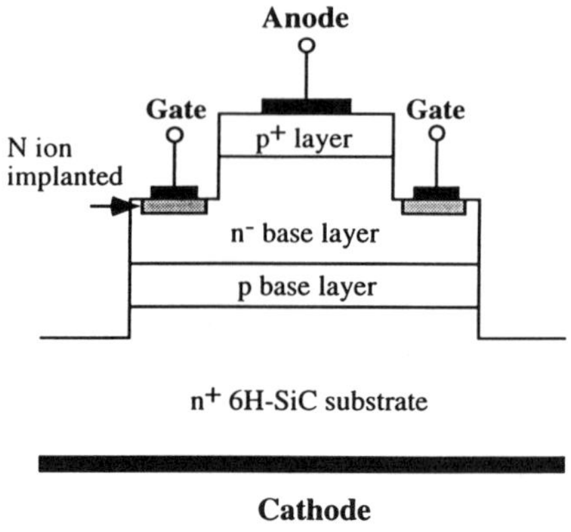

Figure 1. Cross-section of the 6H-SiC thyristor.

employing a newly developed electron cyclotron resonant plasma etch process for SiC [8,9]. The device has an anode diameter of 210 μm and a mesa diameter of 320 μm. The anode area of $3.46\text{x}10^{-4}$ cm^2 was used to calculate the device current density. The high defect density in SiC (100's micropipes/cm^2) is a limiting factor in scaling up device dimensions and hence in increasing maximum device current [10]. Nitrogen was implanted at room temperature in the gate area for the formation of the gate ohmic contact. The implantation depth was 0.25 μm deep with an ion density of $1\text{x}10^{20}/cm^3$. The nitrogen ions were activated at 1350 °C in vacuum for 30 minutes. Al was used as the p-type ohmic contact and Ni was used as the n-type ohmic contact. Both Al and Ni contacts were alloyed at 950 °C in argon for 5 minutes. The specific contact resistances, obtained from transmission line model (TLM) measurements, were found to be 2- $4\text{x}10^{-4}$ ohm-cm^2 for the Ni contacts and 1-$2\text{x}10^{-3}$ ohm-cm^2 for the Al contacts. The Ni and Al contacts were coated with Ti/Au after annealing. The current-voltage and capacitance-voltage measurements are performed to check anode emitter junction and doping level of n^- base layer. The capacitance-voltage measurements indicates that doping level in the n^- base region of SiC thyristor is in the range of 5.5 $\text{x}10^{15}/cm^3$, lower than the design value of $8\text{x}10^{15}/cm^3$. The n^- base layer thickness of 7.5 μm is confirmed by SIMS measurements. The switching performance of the SiC thyristor was characterized at temperatures ranging from 23 °C to 400 °C.

RESULTS AND DISCUSSIONS

The forward blocking voltage of the thyristor was found to be 280 V. This value is consistent with the punchthrough voltage determined from the n^- base layer doping level and thickness, indicating the device was operated in a punchthrough mode. The device leakage current is less than $3\text{x}10^{-5}$ A/cm^2 at 250 V at room temperature. The gate current needed to turn on the thyristor is about 6 times that of the 100 V SiC thyristor at the bias voltages corresponding to the same undepleted n^- base layer width in both thyristors. The higher gate current required in the 280 V thyristor is largely due to the increase in doping level in the n^- base region. It is known that the decrease in base resistance leads to an increase in gate current [5]. It is also expected that a lower base resistance improves the thyristor's dV/dt capability. This is consistent with the observation that the 280 V thyristor has shown a dV/dt of 400 V/μs, a significant improvement over our earlier result of 15 V/μs for the 100 V SiC thyristor.

The switching characteristics of the SiC thyristor were investigated over temperatures from 23 °C to 400 °C. Figure 2 shows a switching waveform at 23 °C with a device current of 1.05 A (equivalent to a current density of 3000 A/cm^2) under gate pulse triggering. The rise time increases from 30 ns at room temperature to 105 ns at 400 °C as shown in Fig. 3. The delay time and rise time are plotted in Fig. 4 as a function of temperature. It is seen that both delay time and rise time increased as the operation temperature is increased. The delay time is related to the carrier transit process which ends with the collapse of space charge region in the thyristor after applying the gate current and is greater than the sum of the carrier base transit times. The rise time has been related to the excess charge build-up process in the base regions after the sum of current gain reaches unity and can be approximated by the geometric mean of the carrier base transit times based on the charge control model [11]. Since the carrier transit time is determined by the base layer thickness and mobility, fast rise times at room temperature could be attributed to the narrow base width in the SiC thyristor while reduction in mobility at high temperature is largely responsible for the observed increase in delay time and rise time.

The turn-off performance of the thyristor was investigated using a double-pulse circuit. The turn-off time t_q is defined as the minimum time required before a second voltage can be applied to the device without switching the device back to the on-state. It is the time required for the excess carriers stored in the base regions to be reduced to below a certain level through recombination processes. It is found that the turn-off time, t_q, is less than 250 ns at room temperature and increases to more than 2 ms at 400 °C as shown in Fig.5. The significant increase in turn-off time by a factor of 8 is most likely due to the increase in carrier lifetimes at high temperature. The effect of temperature on carrier lifetime has been studied by several researchers [12,13] and has been reported that the carrier lifetime increases with increasing temperature. This observation is significant considering the fact that one of the major applications of SiC thyristors is at high temperature. The increase in carrier lifetime at high temperature would lead to the reduction in the on-state resistance and offset the effect of reduction in carrier mobility. However, the carrier lifetime increase as a function of temperature would result in decrease in the thyristor's switching speed and increase in turn-off power loss.

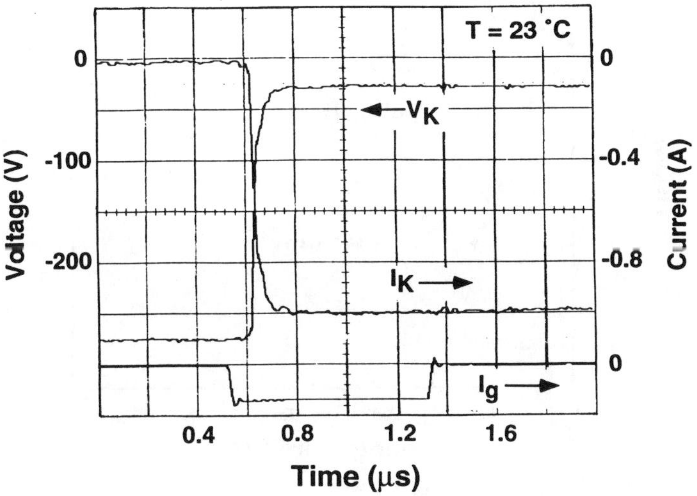

Figure 2. Turn-on switching waveform of the 6H-SiC thyristor with a gate pulse triggering at room temperature. I_g = 5.8 mA/div.

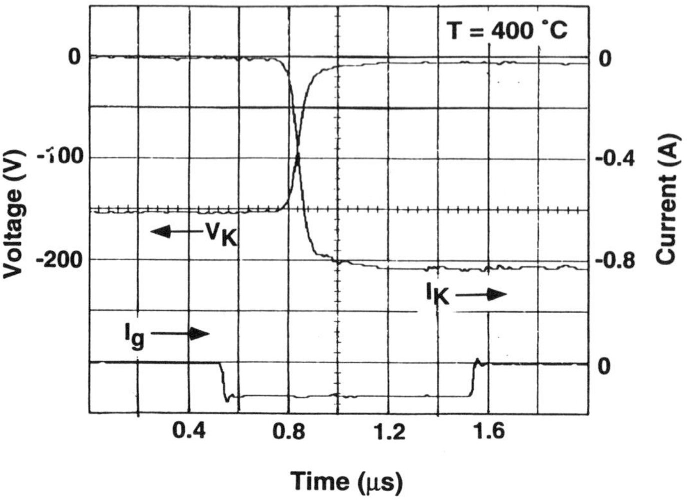

Figure 3. Turn-on switching waveform of the 6H-SiC thyristor with a gate pulse triggering at 400°C. I_g = 5.8 mA/div

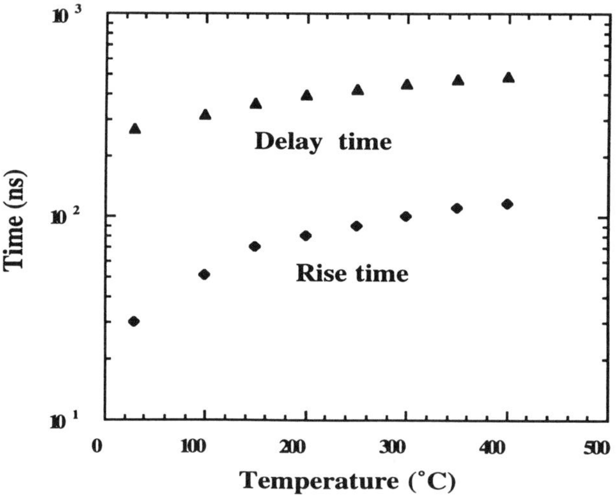

Figure 4. Temperature dependence of the delay time and rise time of the 6H-SiC thyristor during turn-on process.

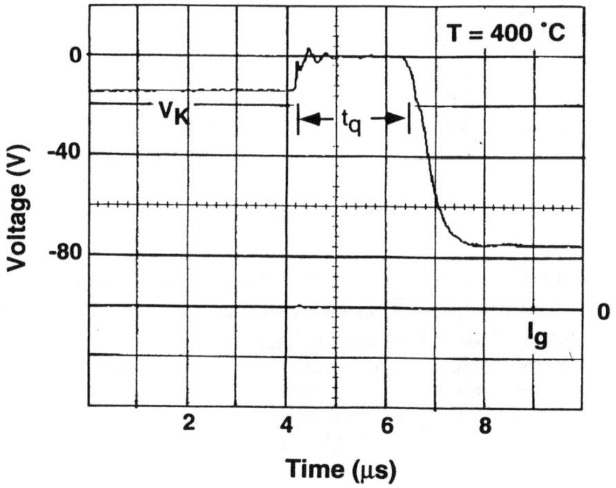

Figure 5. Turn-off time measurement of the 6H-SiC thyristor at 400 °C by a double-pulse circuit. The t_q increases from less than 250 ns at room temperature to over 2 μs at 400 °C.

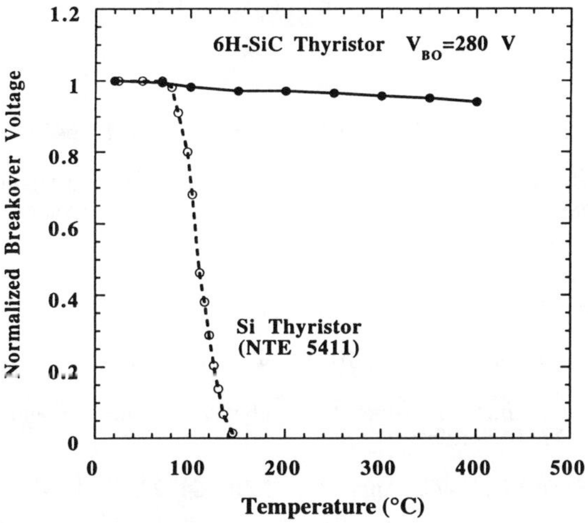

Figure 6. Forward breakover voltage vs temperature for the 6H-SiC thyristor. The data of a conventional Si thyristor (NTE 5411) is included as a comparison.

The thyristor forward breakover voltage decrease by only 5 % and reverse blocking voltage decrease by 8% when the operating temperature is increased from 23 °C to 400 °C, as shown in Fig.6. This performance illustrates the advantage of SiC over Si for high-temperature operation, as the typical maximum rated temperature for Si thyristors are limited to approximately 120 °C. The on-state resistance of this thyristor is dominated by the p-type contact resistance as in the case of our 100 V thyristor [7]. The high p-type contact resistance, which is partly due to the relatively low Al doping level (low $10^{18}/cm^3$) in the p^+ emitter layer, prevents the meaningful study of the temperature dependence of on-state resistance. The maximum current density of 4900 A/cm^2 tested in this thyristor was limited by the high voltage drop in the p-type contact which eventually cause the destruction of the ohmic contacts. Even under this condition, a power density of 1.2 MW/cm^2 is achieved in this device.

CONCLUSIONS

In conclusion, we have demonstrated a SiC thyristor which blocks 280 V and conducts up to 4900 A/cm^2 switched current with fast turn-on and turn-off characteristics. The high temperature performance was investigated over temperature range of 23 °C to 400 °C. While excellent thermal stability was observed in the devices voltage blocking capability, turn-on time and turn-off time were found to increase at high temperature. A low contact resistance of p-type ohmic contact is critical to the SiC thyristor operation. This work demonstrates the potential of SiC thyristors for high-temperature electronics and power control applications.

ACKNOWLEDGMENTS

The authors wish to express their appreciation to Ms. M. Caleca and Prof. H. Du of Stevens Institute of Technology for the high temperature annealing of N implantation.

REFERENCES

1. P.A. Ivanov and V.E. Chelnokov, Semiconductors, **29**, 1003 (1996).

2. H . Morkoc, S. Strite, G.B. Gao, M.E. Lin, B. Sverdlov, and M. Burns, J. Appl. Phys., **76**, 1363 (1994).

3. M. Bhatnagar and B.J. Baliga, IEEE Trans. Electron Devices, **40**, 645 (1993).

4. J.W. Palmour, C.H. Carter,Jr, C.E. Weitzel, and K.J. Nordquist, in Diamond, SiC and Nitride Wide Bandgap Semiconductors, edited by C.H. Carter, Jr et al. (Mater. Res. Soc. Proc. 339, Pittsburgh, PA, 1994), pp. 138-144.

5. J.S. Suehle, P. Chaparala, C. Messick, W.M. Miller, and K.C. Boyko, IEEE Inter. Reliability Physics Proc., (1994), pp. 120-125.

6. B.J. Baliga, Modern Power Devices, (Wiley, New York, 1987).

7. K. Xie, J.H. Zhao, J. Flemish, T. Burke, W. Buchwald, G. Lorenzo, and H. Singh, IEEE Electron Device Lett., **17**, 142 (1996).

8. J. Flemish, K. Xie and J.H. Zhao, Appl. Phys. Lett., **64**, 2315 (1994).

9. K.Xie, J.R. Flemish, J.H. Zhao, W. Buchwald and L. Casas, Appl. Phys. Lett., **67**, 368 (1995).

10. P.G. Neudeck and J.A. Powell, IEEE Electron Device Letters, **15**, 63 (1994).

11. S.K. Ghandhi, "Semiconductor Power devices", (John Wiley & Sons, 1978).

12. R. Bassett, Solid State Electron., **12**, 385 (1969).

13. P. Wilson, Solid State Electron., **10**, 145 (1967).

CHARACTERIZATION OF 4H-SiC MOS CAPACITORS BY A FAST-RAMP RESPONSE TECHNIQUE

T.S. Sudarshan, V. P. Madangarli, and G. Gradinaru*,
C. C. Tin, R. Hu, and T.Isaacs-Smith**
* ECE Department, University of South Carolina, Columbia, SC 29208
** Dept. of Physics, Auburn University, Alabama, AL 36849
sudarsha@charlie.ece.sc.edu; cctin@physics.auburn.edu

ABSTRACT

The current vs voltage characteristics of 4H-SiC MOS capacitors under deep depletion are obtained by a fast ramp response technique so as to obtain the maximum field that can be applied to the MOS structure without the failure of either the semiconductor or the oxide layer. The experiments on n-type 4H SiC wafers having a 5 µm thick epilayer of 10^{15} - 10^{16} cm^{-3} doping concentration and an oxide layer 1200Å - 1500Å thick, indicate the significant influence of the oxide quality and defects in the semiconductor on the nature of the current response during accumulation and deep depletion measurements. The effect of the conductivity of the oxide layer is reflected clearly in the current response, even though classical C-V measurements do not indicate any abnormality. Apart from obtaining the maximum breakdown fields of the semiconductor and the oxide, the fast-ramp response technique provides useful information about the generation processes associated with defects in the MOS structure.

INTRODUCTION

The past few years have seen tremendous improvements made in the growth and processing techniques related to SiC, making high power electronic devices using this wide bandgap semiconductor a reality. Various groups have reported on the fabrication of high voltage pn junction diodes and MOSFETs [1 - 4] using either 6H or 4H SiC. The breakdown voltage of SiC MOS devices is a very important parameter which determines the maximum limits to which the device can be stressed without failure. Considering that one of the major advantages of a wide bandgap semiconductor such as SiC is its suitability for high power devices operating at high voltages, it is essential to characterize the high field properties of the semiconductor *vis-a-vis* what is theoretically expected. Also, the breakdown voltage and leakage current of the oxide layer needs to be characterized in order to make reliable MOS devices. In this context the characterization of a simple MOS capacitor structure by the fast-ramp response technique provides an elegant and inexpensive method to qualify the device worthiness of a MOS structure.

EXPERIMENT

The fast-ramp response technique presented in this paper is capable of providing information regarding the quality of the semiconductor and the oxide layer. The technique involves the measurement of the current response in the accumulation and deep depletion regimes of a MOS capacitor structure while applying a rapid voltage ramp pulse [5]. The measurements yield information about the conductivity and breakdown strength of the oxide layer, the maximum depth of the depletion layer before avalanche breakdown, the maximum field that can be applied across the semiconductor before avalanche generation is observed, etc. The experimental set-up used for the ramp response measurements is shown below.

Mat. Res. Soc. Symp. Proc. Vol. 423

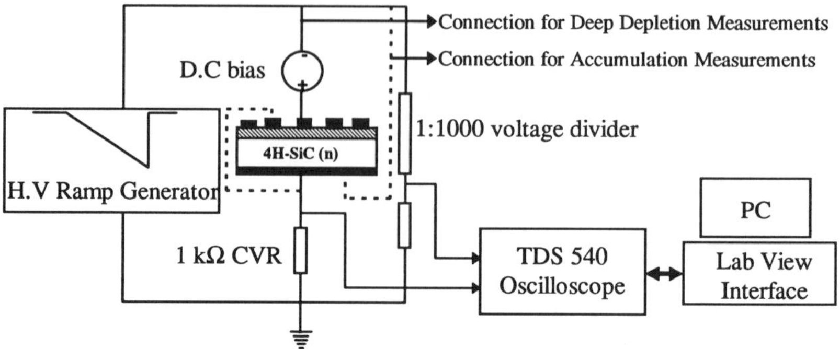

Fig.1: Experimental set-up for fast-ramp response measurements on MOS capacitors.

The experimental system consists of a high voltage ramp generator which is capable of providing a negative voltage pulse ranging from 10 V to 10 kV, at variable slopes from 10 V/ms to 100 V/ms and pulse duration in the range of 0.1ms to 100 ms. Since the ramp generator is designed with independent control over the slope as well as the pulse duration, it is possible to study the variation in the MOS electrical characteristics as a function of either of the above parameters. The experimental sequence for the characterization of MOS structures performed in the present case involves the application of a fixed slope pulse with varying pulse duration so that the peak applied voltage increases with increasing pulse duration. The slope of the ramp voltage is chosen such that it is rapid enough to avoid any thermal generation in the semiconductor.

The MOS capacitor structures used in the present experiments were fabricated on n-type research grade 4H - SiC substrates with ~ 10^{18} cm^{-3} doping concentration and a 5µm epilayer of ~ 10^{15} to 10^{16} cm^{-3} concentration. A thermal oxide of 1200Å - 1800Å thickness was grown on top of the epilayer by wet oxidation at 1150°C in a quartz furnace. Al gate contacts of different diameters (0.1mm to 1.0 mm) were defined on the oxide layer by photolithography. All the MOS structures were at first characterized using room-temperature high frequency C-V measurements at 100 kHz under perfect dark condition. The MOS structures were then tested using the ramp response technique, first in the accumulation mode and then in the deep depletion mode.

As indicated in Fig.1, for the accumulation measurements the negative high voltage pulse is applied to the bottom ohmic contact of the capacitor (for the n-type semiconductor), while the Al gate is connected to the ground via a 1kΩ current viewing resistor (CVR). In this circuit configuration the applied voltage appears entirely across the oxide layer, and hence the characteristics of the oxide layer can be determined. Since the applied ramp voltage has a constant slope (dV/dt), for a fixed oxide capacitance (C_{ox}), the measured current should be ideally a constant value corresponding to the displacement current ($C_{ox}dV/dt$). In case of a poor quality oxide, with significant leakage current, an additional conduction current proportional to the oxide conductivity will be superimposed on the measured displacement current. Hence by comparing the measured total current with the theoretical displacement current that is expected (since C_{ox} is known from the C-V measurements), the oxide conductivity for poor quality oxides can be determined.

The MOS structures with no leakage through the oxide were further tested in the deep depletion mode. In this configuration, as shown in Fig.1, the negative ramp voltage is applied to the gate of the MOS structure in series with a fixed positive D.C. bias while the ohmic contact is connected to the ground through the 1kΩ CVR. The fixed D.C. bias ensures that the MOS structure is at first in accumulation until the ramp voltage magnitude exceeds the D.C. voltage.

Thus in this mode of measurement the MOS structure is swept from accumulation to deep depletion as the ramp voltage increases. Hence the measured current decays from a peak amplitude corresponding to $C_{ox}dV/dt$, as the total displacement current reduces with decrease in depletion capacitance. In the case of ideal MOS structures with good quality semiconductor and oxide layer, the measured current continually decreases with the increase in the ramp voltage, until the field in the depletion region exceeds the breakdown strength of the semiconductor resulting in the breakdown of the semiconductor, which will be reflected as a sudden spike in the measured current. In such a situation any further increase in the ramp voltage will not result in any more growth in the depletion layer, and in fact the measured current will then settle back to the displacement current magnitude corresponding to the oxide capacitance since the semiconductor depletion region will be short-circuited by the breakdown process. On the other hand, in practical MOS structures, the presence of defects in the semiconductor (such as micro-pipes or other kinds of defects common in SiC) will result in an increase in the measured current much before the intrinsic breakdown limit of the semiconductor is reached. Since thermal generation is avoided, we believe that the increase in current is due to local avalanche processes associated with defects in the semiconductor. By measuring the gate voltage at which the current increase is observed (termed as the avalanche voltage, V_{av}), and knowing the epilayer doping concentration and oxide thickness, the depletion width corresponding to 'V_{av}' as well as the field in the oxide and semiconductor can be calculated. Since the measurements are performed in a pulsed regime, even after breakdown, the MOS structure is not permanently damaged and the pre-breakdown current - voltage characteristics can be obtained again at lower voltages. Thus the ramp response technique provides a simple effective method to determine the maximum operating voltage limit of a MOS structure before going through the elaborate steps of fabricating a complete MOS device.

RESULTS

The high frequency C-V characteristics of two different MOS capacitor structures and their corresponding current (I) vs voltage (V) characteristics in accumulation mode (using the ramp response technique) are presented in Fig.2.

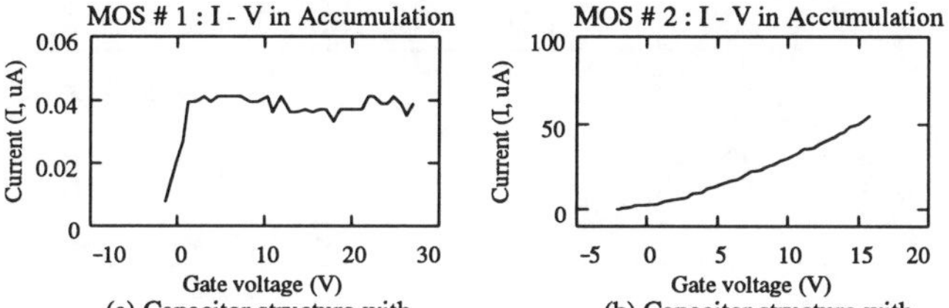

(a) Capacitor structure with "good" quality oxide

(b) Capacitor structure with "bad" quality oxide

Fig.2: High frequency (100kHz) C-V characteristics and I - V characteristics in accumulation using ramp response technique for two different SiC MOS structures.

As observed in Fig.2 while both the MOS structures indicated very good C-V characteristics, the I - V characteristics obtained using the ramp response technique clearly demonstrate that the capacitor structure MOS#2 had significant leakage through the oxide when tested in the accumulation mode while MOS#1 with the good quality oxide did not indicate any measurable leakage. In fact, from the I - V characteristics shown in Fig.2(b), the oxide conductance is calculated to be ~ 3 μS for the "bad" quality oxide.

Typical current and voltage waveforms for a "good" quality MOS structure under deep depletion mode is presented in Fig.3. Note that the current waveform indicates the decay in the displacement current with the increase in the ramp voltage as the depletion capacitance is reduced. In Fig.3 the current and voltage waveforms are shown as a function of time as well as current vs voltage characteristics. While Fig.3(a) shows the situation with no breakdown (peak gate voltage ~ 90 V), Fig.3(b) shows the situation when the same test dot suffered breakdown at V_{av} ~ 113 V.

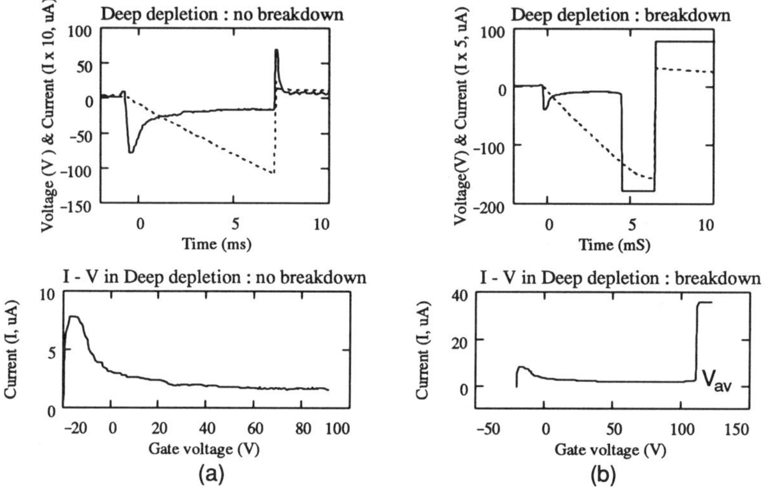

Fig.3: Typical current and voltage waveforms using ramp response technique in deep depletion for a "good" quality MOS structure.

In certain situations it is possible to observe the beginning of the avalanche processes leading to the increase in the current, even before the complete breakdown occurs. A typical I-V characteristics showing the onset of avalanche initiation is depicted in Fig.4.

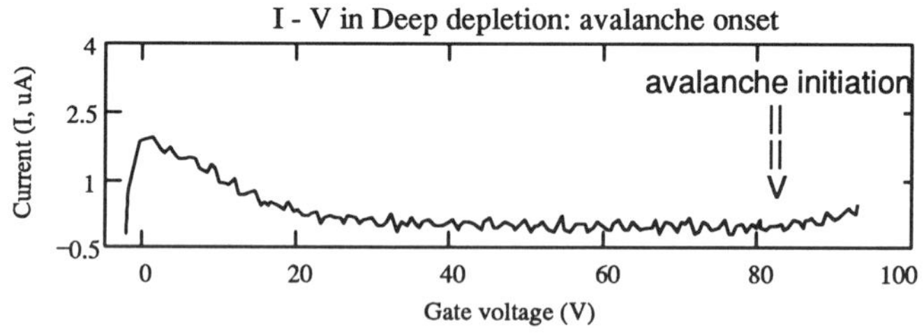

Fig.4: I-V characteristics showing the onset of avalanche in a 4H-SiCMOS capacitor.

Table I summarizes the breakdown data obtained from the deep depletion measurements performed on many (> 50) different test dots of various diameters.

Table I. Avalanche data of 4H SiC MOS capacitors

Diameter d (mm)	Nepi ($10^{16}cm^{-3}$)	t_{ox} (Å)	V_{av} (V)	W_d (μm)	E_{ox} (MV/cm)	E_s (kV/cm)
0.1	0.991	1285	182	4.19	1.96	749.53
0.2	0.886	1279	174	4.34	1.81	694.89
0.4	1.11	1329	137	3.37	1.76	675.01
0.5	4.47	1303	113	1.37	2.88	1104.0
1.0	1.67	1285	60	1.69	1.33	508.99

Nepi : Epilayer doping concentration (cm^{-3}); t_{ox}: oxide thickness (Å); V_{av} :gate voltage at which breakdown or current increase is observed; W_d :depletion width corresponding to V_{av} ; E_{ox} : field in oxide at V_{av} ;E_s : maximum field in semiconductor at V_{av} .

One of the main observations from the breakdown data in Table I is that the maximum field in the semiconductor at which avalanche generation associated with defects is observed lies in a wide range between 500 - 1100 kV/cm for the present MOS structures. This value is below the expected intrinsic breakdown strength of SiC which is in the range of 2 - 3 MV/cm . Since the depletion layer widths calculated in Table I is less than the epilayer thickness of 5 μm, and the field across the oxide at V_{av} is much less than the breakdown strength of the oxide (ideally ~ 10 MV/cm) the breakdown is clearly within the SiC epilayer. Also, a clear dependence of the gate diameter on the avalanche initiation voltage (V_{av}) is observed, with the highest voltage obtained for the smallest gate diameter and the lowest voltage for the largest gate diameter. The dependence of V_{av} on gate diameter is understandable since the probability of including defects in the semiconductor increases with increase in diameter of the gate. But, it has to be noted that the maximum field in the semiconductor at V_{av} do not show any clear correlation with test dot diameter, since the variation in the doping concentration significantly affects the magnitude of the field through the variations in the depletion width.

CONCLUSIONS

The effect of increased conductivity of the oxide layer in some test dots on a given wafer is clearly observed in the current variation under fast-ramp response measurements on the 4H SiC MOS capacitors, even though classical C V measurements do not indicate any abnormality. The maximum field in the semiconductor at which local avalanches associated with defects in the semiconductor are observed lies in the range of 500 - 1100 kV/cm, which is lower than the expected theoretical breakdown limits for SiC. Normal pre-breakdown current and voltage characteristics can be obtained at lower voltages even after breakdown, indicating that the pulse mode measurements in the ramp response technique do not result in a permanent damage of the MOS structure. The avalanche initiation voltage (V_{av}) shows clear dependence on the MOS capacitor dimension, with the highest voltage obtained for the smallest diameter of the gate.

ACKNOWLEDGEMENTS

This work was supported by BMDO/IST and managed by ONR. We are grateful to Dr. D. Duston for his continued interest and support of our work.

REFERENCES

1. P.G. Neudeck, D.J. Larkin, C.S. Salupo, J.A. Powell, and L. G. Matus in Silicon Carbide and Related Materials, edited by M. G. Spencer, R. P. Devaty, J. A. Edmond, M. Asif Khan, R. Kaplan, and M. Rahman, (Proc. of the Fifth Conference on Silicon Carbide and Related Materials, Washington, DC, November, 1993) pp. 475 - 478.
2. V. Krishnamurthy, D. M. Brown, M. Ghezzo, J. Kretchmer, W. Hennessy, E. Downey, and G. Michon, ibid., pp. 483 - 486.
3. J. W. Palmour, J. A. Edmond, H.S. Kong, and C. H. Carter, Jr., ibid., pp. 499 - 502.
4. O. Kordina, J. P. Bergman, A. Henry, E. Janzen, S. Savage, J. Andre, L. P. Ramberg, U. Lindefelt, W. Hermansson, and K. Bergman, Appl. Phys. Lett. **67** (11), 1561 - 1563, (1995).
5. G. Gradinaru, Ph.D Thesis, Polytechnic Institute of Bucharest, Romania, 1989.

CHARACTERIZATION OF 4H-SiC JFETS FOR USE IN ANALOG AMPLIFIERS CAPABLE OF 723K OPERATION

J.B. CASADY *, D.C. SHERIDAN, W.C. DILLARD, and R.W. JOHNSON
NASA Center for Commercial Development of Space, Electrical Engineering Department, 200 Broun Hall, Auburn University, AL 36849-5201, * jbcasady@eng.auburn.edu

ABSTRACT

Discrete, buried-gate 4H-SiC JFETs (W/L = 1 mm/5 µm) were packaged and characterized at temperatures ranging from 290 K to 773 K for use in a hybrid, 4H-SiC analog amplifier. A contaminated passivation oxide was found to limit high-temperature operation initially, but upon removal of the passivation layer the devices demonstrated stable operation to 773 K with adequate amplification (A_V greater than 200 V/V) up to 673 K. From the 16 devices tested, a peak extrinsic saturated transconductance (g_{msat}) of 27.1 mS/mm was measured at 308 K, corresponding to a channel mobility of 400 cm^2/V·s, excluding significant series resistance effects.

INTRODUCTION

Silicon Carbide (SiC) is a wide-bandgap semiconductor well-known for its attractive electrical properties of wide-bandgap, high saturated electron velocity, high thermal conductivity, and high breakdown strength [1-3]. These four material properties give SiC an advantage over other semiconductors such as Si and GaAs in high-temperature, high-power, high-frequency, and high-radiation applications. Of the almost 200 known polytypes of SiC, currently only two are commercially available in bulk wafer form, 4H-SiC and 6H-SiC. While trade-offs exist between the two polytypes, 4H-SiC is generally preferred for analog applications because of its higher and more isotropic electron mobility. Some key electronic properties (bandgap (E_g), thermal conductivity (Θ_K), breakdown strength (E_{crit}), saturated velocity (v_{sat}), electron mobility (μ_n) and intrinsic carrier concentration (n_i)) of 4H-SiC, Si and GaAs are listed in Table 1. Although traditional silicon-based electronics utilize Complementary Metal Oxide Semiconductor (CMOS) logic, Junction Field Effect Transistors (JFETs) have advantages over MOS devices in high-temperature, high-radiation, and even high-power applications in that they do not depend on a gate dielectric such as SiO_2 which often degrades in the aforementioned environments. Here, device performance of 4H-SiC JFETs from 294 K to 723 K and limited packaging issues are examined.

TABLE I: Key Electronic Properties of Si, 4H-SiC, and GaAs

Property	Silicon [3]	4H-SiC [3]	GaAs [4]	Comments
E_g (eV)	1.16	3.26	1.5	T < 5 K
Θ_K (W/cm•K)	1.5	3.7	0.5	300 K
E_{crit} (MV/cm)	0.25	2.2	0.4	$10^{17}<N<10^{18} cm^{-3} at\ T=300\,K$
$v_{sat(e^-)}$ ($\times10^6$ cm/s)*	10	20	8	(overshoot in GaAs)
μ_n (cm^2/V•s)	1400	1000	9200	$T = 300\,K,\ 10^{14} < N < 10^{15}\,cm^{-3}$
n_i (cm^{-3}) at 300 K	$1x10^{+10}$	$5x10^{-9}$	$2.1x10^{+6}$	

*at an electric field of 200 kV/cm

Mat. Res. Soc. Symp. Proc. Vol. 423

EXPERIMENT

Sixteen 4H-SiC buried-gate JFETs, purchased from Cree Research, Inc., were characterized in terms of standard device parameters, such as turn-off voltage (V_T), output resistance (r_{DS}), gate-to-source and gate-to drain capacitances (C_{GS} and C_{GD}, respectively). The devices had a channel width (W) of 1 mm, channel length (L) of 5 μm, gate area of 1.05 x 10^{-3} cm^{-2}, and a channel thickness (a) (set by reactive ion etching) of 0.20 to 0.22 μm. Other parameters provided by the manufacturer include a pn junction built-in potential (V_{bi}) of 2.85 V at 300 K and a nitrogen-doped (n-type) epitaxially-grown channel layer with a doping level of approximately 1.0 x $10^{17}cm^{-3}$. A cross-section of the device is shown in Figure 1. The p-type epitaxial layer is provided to effect charge depletion of the channel. A heavily doped n^+ epitaxial layer is provided to reduce contact resistance to the nickel based ohmic contacts with gold overlayer. Nickel-based ohmic contacts to SiC have been reported elsewhere [5]. The surface was passivated with SiO_2, although this layer was removed prior to testing as detailed below.

The devices were packaged on alumina ceramic chip carriers which had gold thick-film conductor traces and pads screen printed on them to facilitate gold-gold wirebonding and brazing. The bare die were attached by brazing the die using either Au-In (82%/18%) or Au-Ge (88%/12%) preforms in a forming gas (95%H_2/5%N_2) ambient for two minutes. Use of the Au-In braze allowed for higher-temperature operation (up to 773 K), but the high brazing temperature (923 K) degraded the metal contacts in many of the JFETs and prevented operation or wirebonding in these cases. To increase the successfully packaged die yield, Au-Ge brazing (at 683 K) was used for most devices, although this limited the temperature of operation to below the Au-Ge eutectic temperature of 636 K [6].

RESULTS

DC Characterization of 4H-SiC JFETs

Device characteristics of the JFET are described by the equations below taken from Sze [7] which were derived under the assumptions of the gradual-channel approximation, a long channel device, abrupt depletion layer, and constant mobility. Equation (1) describes the built-in potential

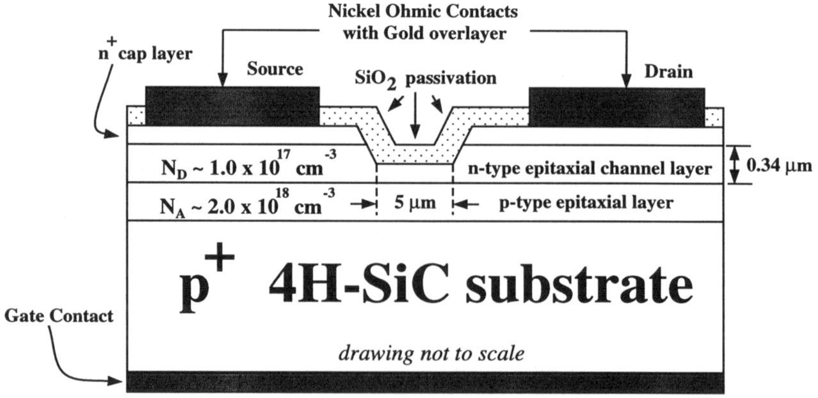

Figure 1: Device cross-section of 4H-SiC buried-gate JFET purchased from Cree Research, Inc.

of an abrupt pn junction in terms of electron charge (q), Boltzmann's constant (k), intrinsic carrier concentration (n_i), temperature (T), and the doping levels of the p and n regions (N_A and N_D, respectively.)

$$V_{bi} = \frac{kT}{q}\ln\left(\frac{N_A N_D}{n_i^2}\right) \approx \frac{kT}{q}\ln\left(\frac{N_D}{n_i}\right) \text{ for a } p^+n \text{ jxn.} \quad (1)$$

Equation (2) gives the saturated drain-to-source current (I_{Dsat}) as a function of relevant device parameters, where V_P is the pinchoff voltage defined by Equation (3). The term ε_S is the semiconductor permittivity (10.0 x (8.854×10^{-14}F/cm^2) for 6H-SiC).

$$I_{Dsat} = \left[\left(\frac{W}{L}\right)\frac{\mu(qn)^2 a^3}{6\varepsilon_S}\right]\left[1 - 3\left(\frac{V_G + V_{bi}}{V_P}\right) + 2\left(\frac{V_G + V_{bi}}{V_P}\right)^{\frac{3}{2}}\right] \quad (2)$$

$$V_P = \left(\frac{qna^2}{2\varepsilon_S}\right) \equiv \text{voltage at which channel is pinched off.} \quad (3)$$

The turn-off voltage (V_T) where the device is turned off is defined by Equation (4) in terms of V_{bi} and V_P. Ideally V_T is controlled primarily through N_D and a, making uniform epilayer doping and thickness critical. Using the manufacturer's supplied values for a, N_D, and V_{bi}, the calculated V_T should range between -0.4 V and -1.5 V at 300 K. Finally, the extrinsic saturated transconductance (g_{msat}) is given in Equation (5). The channel mobility can be extracted from Equation (5) using the values for channel doping and thickness given above.

$$V_T = V_{bi} - V_P \equiv \text{turn} - \text{off voltage} \quad (4)$$

$$g_{msat} = g_{max}\left(1 - \sqrt{\frac{V_G + V_{bi}}{V_P}}\right) \text{ where } g_{max} \equiv \left(\frac{qN_D a\mu W}{L}\right). \quad (5)$$

Effect of Passivation Layerv Removal

In a typical as-received device, the extrinsic, saturated transconductance (g_{msat}) at 0 V gate bias (V_{GS} = 0) was measured to be 6.2 mS/mm at 290 K, falling to 2.27 mS/mm at 673 K. Small-signal voltage gain (A_V) ranged from 232 to 278 V/V up to 473 K for a 10 V drain-to-source bias and 0 V gate-to-source bias. Parasitic drain-to-source leakage current was found to be responsible for degrading A_V. I_{DSS} (I_{DSsat} at V_{GS} = 0 V) fell from ~8.4 mA at 294 K to ~ 5 mA at 723K. The source of the parasitic leakage was identified primarily as leakage through the surface passivation layer. This was confirmed by etching off the surface oxide with buffered oxide etch (BOE) solution of 6:1 H_2O:HF for 15-20 s, and re-testing several devices. Figure 2 illustrates device output characteristics of a typical device at 673 K before and after the removal of the surface passivation. The gate voltage was stepped from -1.5 V to + 1.5 V in 0.5 V increments. Reasons for the dramatically altered DC characteristics of the devices after removal of the passivation oxide could include large amounts of mobile and/or fixed charge within the passivation oxide as a result of Na^+, K^+ or heavy metal contamination. The presence of Na in similar, virgin SiC has been shown elsewhere [8].

Figure 2b illustrates the typical output characteristics of an unpassivated 4H-SiC JFET at 307

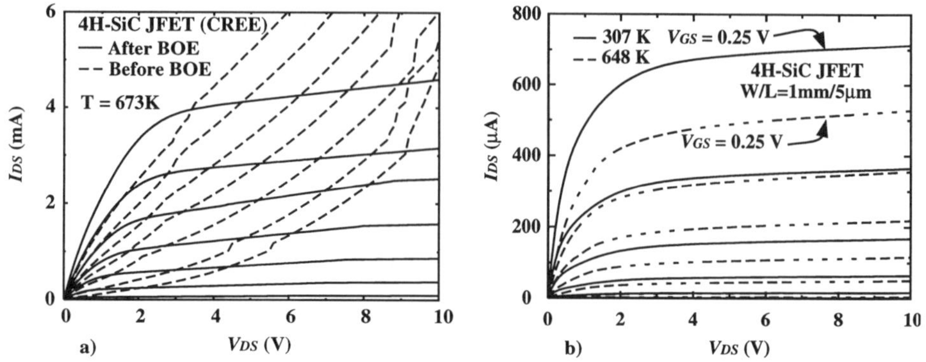

Figure 2a: Output characteristics of a 4H-SiC JFET at 673 K with passivation oxide (before BOE) and without passivation oxide (after BOE). V_{GS} stepped from -1.5 V to 1.5 V in 0.5 V increments.

Figure 2b. Output characteristics of packaged, unpassivated 4H-SiC JFET (J3) at 307 K and 648 K in a dark, shielded, nitrogen ambient. V_{GS} stepped from -1.5 V to +0.25 V in 0.25 V increments.

K and 648 K. These results exhibit the expected decrease in I_{DS} at the higher temperatures from decreased mobility as a result of acoustic phonon scattering. The output resistance (r_{DS}) was fairly constant over the temperature range, in contrast to the passivated devices' output as illustrated in Figure 2a. Typical average values for r_{DS} were 200 kΩ at 300 K and 300 kΩ at 623 K at a DC bias of V_{GS} = 0 V and V_{DS} = 10 V. V_T, g_m, and other DC parameters also were fairly stable with temperature after removing the passivation layer on the devices. V_T at 300 K was recorded before and after etching the surface oxide in four devices and was found to shift by an average of +0.8 V to +1.0 V. V_T ranged from -1.3 V to -2.6 V prior to BOE treatment, and ranged from -0.3 V to -1.7 V afterwards at 300 K in four random devices. V_T for a typical, randomly selected device before and after removal of the surface passivation layer from 300 K to 773 K is shown in Figure 3a. I_{DSS} also decreased by a substantial amount, from an average of 7 mA at 300 K before etching off the surface passivation to 0.8 mA afterwards. This decrease was not surprising since the shift in V_T allowed less conduction at V_{GS} = 0 V for the unpassivated devices. V_T was fairly constant up to 623 K where increased pn junction leakage currents began to contribute. The increased pn junction leakage currents contribute to the increased sub-V_T leakage from drain-to-source as shown by Figure 3b. Increased pn junction leakage is expected since pn junction saturation current is directly proportional to n_i^2, which is strongly temperature-dependent.

Temperature effects on other device parameters

I_{DSsat} and g_{msat} both decreased with increasing temperature as expected from decreasing electron mobility. Aside from μ and the mildly temperature-dependent n, V_{bi} is the remaining first-order temperature-dependent factor in the expression for g_{msat} shown in Equation (5). I_{DSsat} also decreases with increasing temperature, much for the same reasons as discussed for g_{msat}. Figure 4 illustrates I_{DS} and g_m as a function of V_{GS} and temperature. From 300 K to 623 K, a 33% to 50% reduction in g_{msat} was typically observed. A peak g_{msat} of 27.1 mS/mm was measured for one device at 308 K with V_{DS} of 10 V, corresponding to a μ of 400 cm^2/V·s as calculated from Equation (5) while ignoring substantial series resistance effects (approximately 20 kΩ), and using

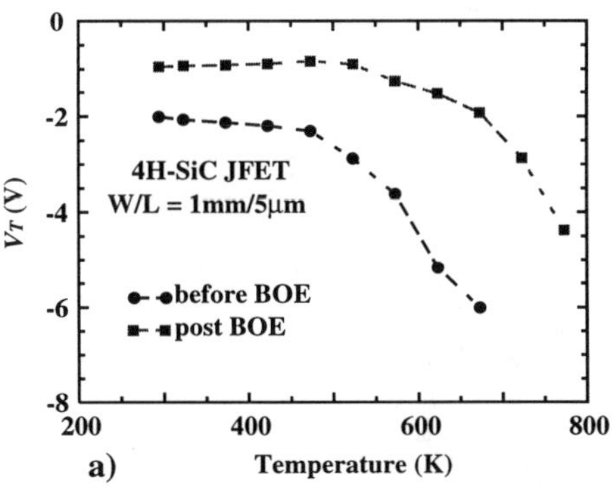

Figure 3a: Turn-off voltage of a 4H-SiC JFET versus temperature before and after removal of contaminated passivation oxide.

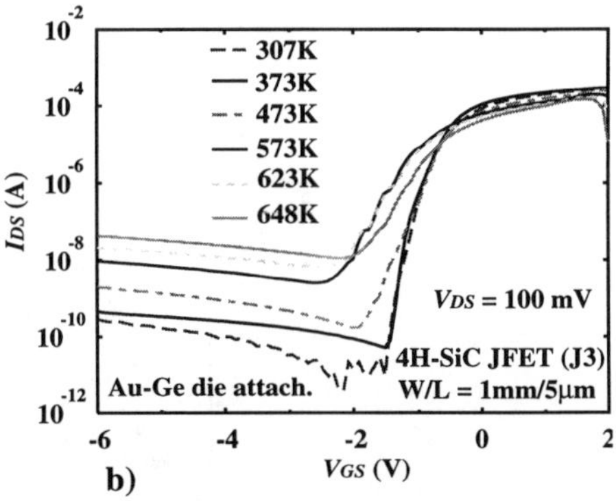

Figure 3b. Linear transfer characteristics illustrating sub-V_T drain-to-source current (I_{DS}) at different temperatures.

the values given for a, N_D, and V_{bi} by the manufacturer. Peak g_{msat} was measured at a V_{DS} of 10 V and a V_{GS} of 2.6 V. The average μ of the 16 devices at 308 K was 250 cm^2/V·s, with a range from 200 to 400 cm^2/V·s observed. Channel μ's were approximately 50% to 80% of expected μ, but are reasonable given the large series resistance of the device and significant defect density in the SiC. Minimizing series resistance and material defect density should increase device performance.

The unbiased C_{GS} and C_{DS} were measured to be in the range of 86 to 55 pF (73 pF average) for all 16 devices at 300 K (f = 100 kHz). Although C_{GS} and C_{DS} increased moderately to an average value of 100 pF at 623 K, the increase may be due to increased capacitance associated with the electrical connections inside the oven.

CONCLUSIONS

Device characteristics such as μ, V_T, leakage currents from drain-to-source and other DC parameters were analyzed quantitatively to 773 K. Reliability studies, improving the quality of the passivation layer and die attach remain the only technological barriers to successful implementation of 4H-SiC JFET's in analog amplifiers capable of 773 K operation. The use of Si_3N_4, oxynitrides and other dielectrics on SiC are currently being evaluated as potential solutions for the passivation. Development of an improved Au-In brazing process (possibly in a N_2 ambient) which does not degrade the ohmic contact metallization should allow sustained operation of packaged devices at 773 K. The actual circuit design, fabrication, and performance of a hybrid amplifier utilizing these components will be presented at a later date [9]. Results obtained with these 4H-SiC JFETs, aside from the contaminated oxide passivation, are consistent with other SiC JFETs characterized elsewhere [10-15], which have demonstrated 1000 hour reliability at 573 K [15].

ACKNOWLEDGMENTS

The authors would like to acknowledge financial support of the NASA Center for Commercial Development at Auburn University under grant NAGW-1192, and the Alabama Microelectronics Science & Tech. Ctr. at Auburn University. The authors would also like to thank Dr. J.W. Palmour at Cree Research, Inc. for valuable input.

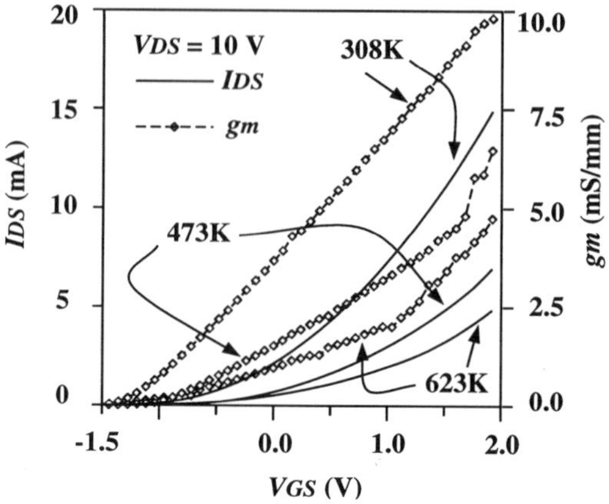

Figure 4: Small-signal, extrinsic saturated transconductance (g_{msat}) and I_{DS} of a typical unpassivated 4H-SiC JFET (L = 5 μm) versus temperature.

REFERENCES

1. H. Morkoc, S. Strite, G.B. Gao, M.E. Lin, B. Sverdlov, and M. Burns, J. Appl. Phys., **76**, 1363 (1994).
2. P.G. Neudeck, J. Elect. Mat., **24**, 283 (1995).
3. J.B. Casady and R.W. Johnson, Solid-St. Elect., in press (1996).
4. O. Madelung, in *Semiconductors: Group IV Elements and III-V Compounds*, P. Poerschke, Chief Ed., (Springer-Verlag, Berlin, 1991), p. 101-113.
5. J. Crofton, P.G. McMullin, J.R. Williams, and M.J. Bozack, J. Appl. Phys., **77**, 1317 (1995).
6. *Constitution of Binary Alloys*, M. Hansen and K. Anderko, Eds., (McGraw-Hill, New York, 1958), p. 206.
7. S.M. Sze, *Physics of Semiconductor Devices*, 2nd ed. (Wiley Press, New York, 1981), p. 318-319.
8. S. Sridevan, P.K. McLarty and B.J. Baliga, Elect. Dev. Lett., **17**, 136 (1996).
9. W.C. Dillard, J.B. Casady, R.C. Jaeger, D.C. Sheridan and R.W. Johnson, to be presented at 3rd Int'l. High Temp. Elect. Conf. (HiTEC), Albuquerque, NM, June 9-14, (1996).
10. K. Dohnke, R. Rupp, D. Peters, J. Völkland, and D. Stephani, in *Silicon Carbide and Rel. Mat., Proc. of 5th Int'l. Conf.*, M.G. Spencer, R.P. Devaty, J.A. Edmond, M.A. Khan, R. Kaplar, and M. Rahman, Eds., (Inst. Phys. Conf. Ser. No. 137, IOP Publishing, Ltd., 1994), p. 625.
11. M.M. Anikin, A.A. Lebedev, M.G. Rastegaeva, N.S. Savkina, A.M. Strel'chuk, A.L. Syrkin and V.E. Chelnokov, in *Silicon Carbide and Rel. Mat., Proc. of 5th Int'l. Conf.*, M.G. Spencer, R.P. Devaty, J.A. Edmond, M.A. Khan, R. Kaplar, and M. Rahman, Eds., (Inst. Phys. Conf. Ser. No. 137, IOP Publishing, Ltd., 1994), p. 597.
12. J.W. Palmour, H.-S. Kong and C.H. Carter, Jr., Int'l. Sem. Device Research Symp., (1991), p. 491.
13. P.G. Neudeck, J.B. Petit and C.S. Salupo, in Trans. 2nd Int'l. High Temp. Elect. Conf. (HiTEC), Session X, (1994), p. 23.
14. F.B. McLean, J.M. McGarrity, C.J. Scozzie, C.W. Tipton, and W.M. Delancey, IEEE Trans. Nuc. Sci., **41**, 1884, (1994).
15. C.J. Scozzie, C.W. Tipton, W.M. DeLancey, J.M. McGarrity, and F.B. McLean, Proc. of 32nd Int'l. Reliability Physics Conf., (IEEE Press, 1994), pp. 351-358.

HIGH ELECTRIC FIELD BREAKDOWN OF 4H-SiC PN JUNCTION DIODES

C.C. TIN*, V. MADANGARLI†, R. HU*, E. LUCKOWSKI*, J. CASADY‡, T. ISAACS-SMITH*, G. GRADINARU†, T.S. SUDARSHAN†, AND R.W. JOHNSON‡
*Department of Physics, Auburn University, AL 36849, cctin@physics.auburn.edu
† Department of Electrical and Computer Engineering, University of South Carolina, Columbia, SC 29208
‡ Department of Electrical Engineering, Auburn University, AL 36849

ABSTRACT

We have characterized the high electric field breakdown process of several epitaxial 4H-SiC p-n structures with oxide passivation. The breakdown voltage was found to be dependent on the size of the diode structures as well as their proximity to any structural defects. The time dependence of the breakdown process was also measured to determine the characteristics of the breakdown mechanism. This time dependence measurement provides an indication of the quality of the diode structures. Both soft and abrupt breakdown mechanisms were observed showing the influence of defects on the high field behavior of the diode structures. Measurements done with and without the use of Fluorinert fluid did not show any difference in the breakdown voltage indicating that surface flashover breakdown mechanism did not play a major role in the avalanche breakdown process.

INTRODUCTION

Silicon carbide (SiC), with its wide bandgap and high breakdown electric field, is an excellent material for the fabrication of high-voltage and high-power devices such as diodes, transistors and thyristors. The high thermal conductivity and wide bandgap of SiC also enable SiC-based devices to operate at extremely high temperatures. Spurred by the early commercial availability of 6H-SiC, many studies [1-5] have been made on the fabrication of power devices on this polytype. However, the wider bandgap and smaller degree of mobility anisotropy of 4H-SiC make it a superior material than 6H-SiC for high power and high temperature applications.

The potential significance of 4H-SiC as the material of choice for the fabrication of devices designed to operate under high voltage and high temperature conditions warrant the study of the high electric field behavior of power devices based on this material. High electric field characterization is necessary in defining the limiting factors in the high field and high power applications of 4H-SiC power devices. The relatively high density of defects in 4H-SiC substrates and epilayers can drastically affect device performance causing premature breakdown of high power 4H-SiC devices.

In this paper, we discuss the high field characterization of 4H-SiC p-n junction diodes in the reverse bias regime using the fast voltage ramp technique in order to determine the maximum reverse breakdown strength of the junction as well as the nature of the breakdown. By monitoring the time evolution of the breakdown mechanism we were able to characterize the behavior of the devices in the various phases up to the limit of avalanche breakdown.

EXPERIMENT

The 4H-SiC epilayers were grown by chemical vapor deposition on the silicon face of 3-4° off-axis n^+-doped 4H-SiC substrates purchased from Cree Research, Inc. The gases used were about 135 ppm of propane and about 90-270 ppm of silane in about 2.2 slm of hydrogen gas. Nitrogen from 2% nitrogen in hydrogen gas mixture and aluminum from trimethylaluminum were used as the n-

Mat. Res. Soc. Symp. Proc. Vol. 423 © 1996 Materials Research Society

and p-type dopants respectively. The 5 mm x 5 mm substrates were degreased by immersing in trichloroethylene, acetone, methanol and deionized water. The samples were then treated with buffered-oxide-etch solution to remove any surface oxide layer before being loaded into the growth chamber. Epitaxial n- and p-type epilayers were grown at atmospheric pressure at about 1450-1500°C. Circular device structures with diameter ranging from 50-300 μm were then patterned on the samples by a two-mask photolithographic process. The first mask enabled the formation of mesa structures by reactive-ion-etching using pure NF_3 gas. The p-n junctions were then passivated by an oxide layer of about 600 Å thick which was grown by thermal oxidation at about 1100°C. This oxide layer was then patterned using the second mask followed by treatment with buffered-oxide-etch (BOE) solution to expose the top surface of the mesa structures for subsequent metallization. The metallization step consisted of sputtering ~ 2000 Å Ni on the back followed by annealing at 950°C for 5 minutes to form the back ohmic contact. The front contact was made by sputtering ~ 2000 Å AlTi and annealed at 1150°C for 10 s. A schematic drawing of the diode is shown in figure 1.

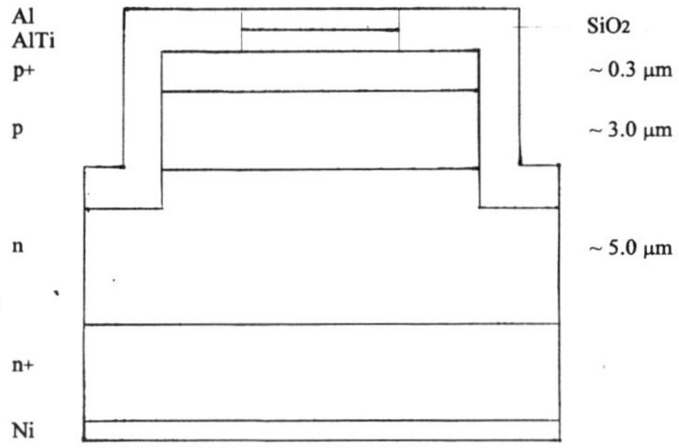

Figure 1. 4H-SiC p-n diode mesa structure with oxide passivation. Doping concentration (cm^{-3}): n = ~$8x10^{15}$; p = ~$5x10^{17}$; p^+ = ~$5x10^{19}$; n^+ = ~$5x10^{18}$.

For high electric field measurements, the p-n junction diode was connected in a reverse bias mode to a specially constructed high voltage ramp generator, which is capable of supplying a negative going ramp voltage pulse of variable slope as well as variable pulse duration. The experiment involved applying a single ramp voltage pulse which gradually increased the bias across the device while the current through the device was being monitored until breakdown current appeared. The advantage of this technique is that breakdown mechanisms due to field-induced thermal effects can be excluded from the study.

RESULTS

The time resolved voltage and current characteristics corresponding to some typical diode structures are shown in figure 2. The broken line in each graph is the ramp voltage pulse. The actual slope and pulse width of the ramp voltage can be deduced from the figure. Since the applied ramp voltage has a constant slope (dV/dt), the measured current has a displacement current component proportional to the capacitance of the depletion region at the reverse biased junction, along with a leakage current component due to the generation processes in the device. The combination of these two components produces the characteristic current profile as shown in figure 2. The characteristic current profile can be explained as follows. At low voltages, the current is observed to decay from an initial peak value (labeled A in figure 2a) in a somewhat exponential manner as expected from the reduction in the displacement current when the depletion capacitance is reduced. When the voltage is increased, the current begins to increase near the end of the ramp voltage (labeled B in figure 2a) due to the generation current dominating over the decaying displacement current. At even higher voltages, the device suffers avalanche breakdown with a large increase in the current.

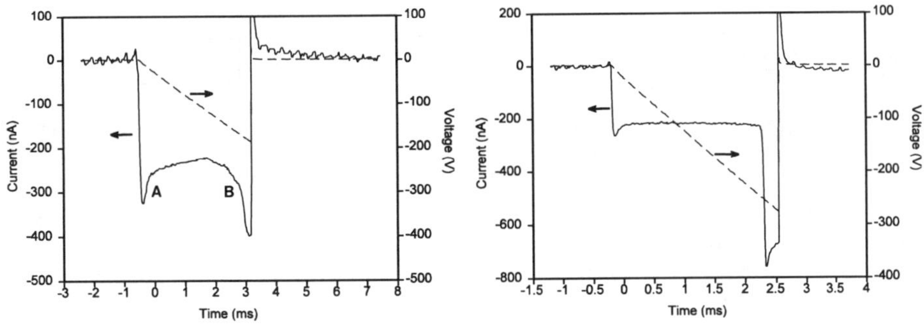

(a) (b)

Figure 2. Current response of diodes with (a) soft and (b) abrupt avalanche breakdown.

The breakdown can be characterized as either a *soft* breakdown with a gradual increase of the current or an *abrupt* breakdown with a rapid increase of the current. The observed avalanche breakdown is not catastrophic, and the diode structure indicates normal behavior when the applied voltage is reduced, which is the advantage of this technique. However, in our study, the measurements were made both with and without catastrophic breakdowns in order to observe the catastrophic breakdown effect on the device structures. Figure 3 shows the CCD camera images of a typical diode before and after catastrophic breakdown. Contact damage is evident in the diode structure after catastrophic breakdown.

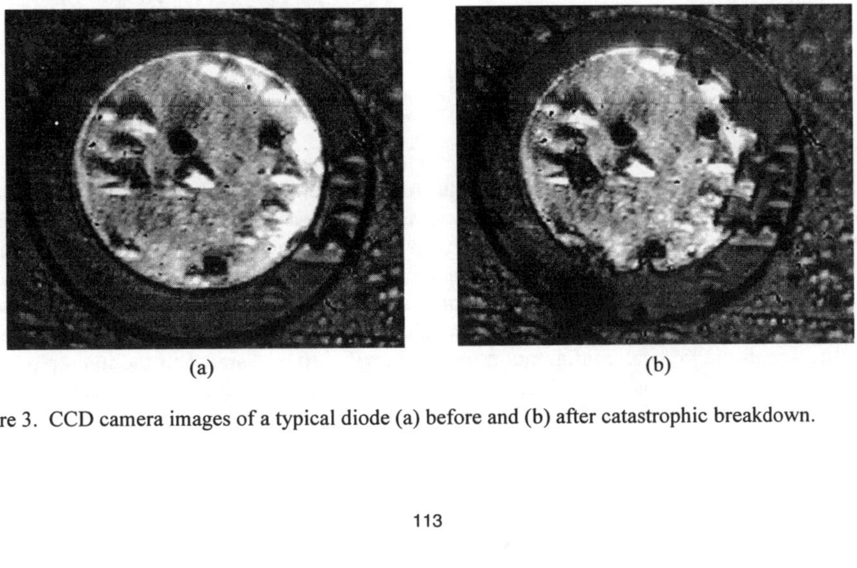

(a) (b)

Figure 3. CCD camera images of a typical diode (a) before and (b) after catastrophic breakdown.

The average breakdown voltages corresponding to different diameters of the diode structures are shown in table I. The average value was obtained from five different diodes of same diameter. It can be seen that there is a direct correlation between the breakdown voltages and the diode size. This could be attributed to the number of defects enclosed within the diode structure, with more defects in a larger size diode than in a smaller size diode.

Table I. Average breakdown voltage (V_B) as function of diode diameter.

Diode Diameter (μm)	V_B (Volts)	Diode Diameter (μm)	V_B (Volts)
300	50	100	100
200	75	75	150
150	95	50	250

A separate study using identically sized diodes on the same substrate showed that the breakdown voltage of a diode was related to its proximity to any visible crystal defects. The breakdown voltages for the diodes shown in figure 4 are tabulated in table II. It can be seen that the best diodes in this case (2 and 8) with the highest breakdown voltages are those which are well separated from any visible defects.

Table II. Breakdown voltage, V_B, in volts as function of temperature (within parenthesis).

Diode	V_B
1	105 (300 K); 70 (600 K)
2	>250 (300 K); >250 (600 K); 150 (900 K)
3	120 (300 K)
4	235 (300 K)
5	150 (300 K)
6	23 (300 K)
7	20 (600 K)
8	>220 (600 K)
9	40 (600 K)
10	35 (900 K)
11	130 (900 K)
12	7 (900 K)

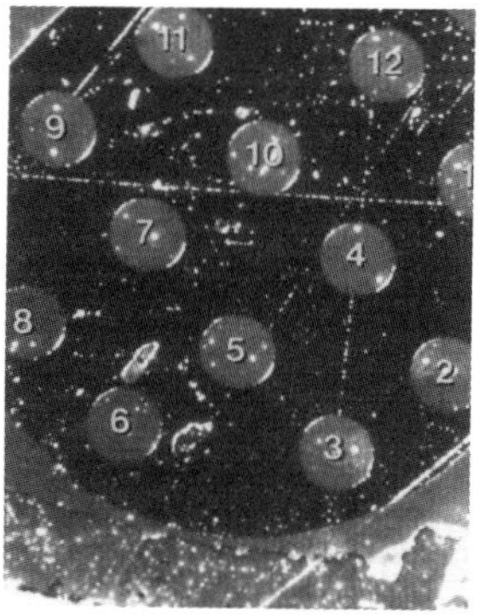

Figure 4. Position of diodes of table II.

The breakdown voltages obtained by DC measurements were typically lower than those obtained by this ramp voltage pulse technique possibly due to thermal effects. The devices also deteriorated more rapidly with increasing leakage current after repeated measurements using the DC technique whereas the characteristics obtained using the ramp voltage pulse method were more reproducible.

The breakdown voltage values are much smaller than the theoretically expected values. Besides the possibility of crystal defects, such as impurities, stacking faults, and dislocations, some other explanations contributing to the low breakdown voltages should also be considered: (i) since the test structues are in the vertical configuration with the applied field parallel to the micropipes, the presence of micropipes could be a detrimental factor, and (ii) surface breakdown due to imperfect edge terminations and surface defects. The latter is usually manifested in the form of *soft* avalanche breakdown. It has been shown by several researchers that edge termination using a high resistive layer is effective in increasing the breakdown voltage of SiC p-n junction [4]. The high resistive layer is produced by the formation of defects induced by ion-implantation. Another factor causing a premature breakdown could be the phenomenon of surface flashover but this is unlikely because measurements done with and without Fluorinert fluid, as shown in figure 5, showed no substantial difference in the breakdown voltage.

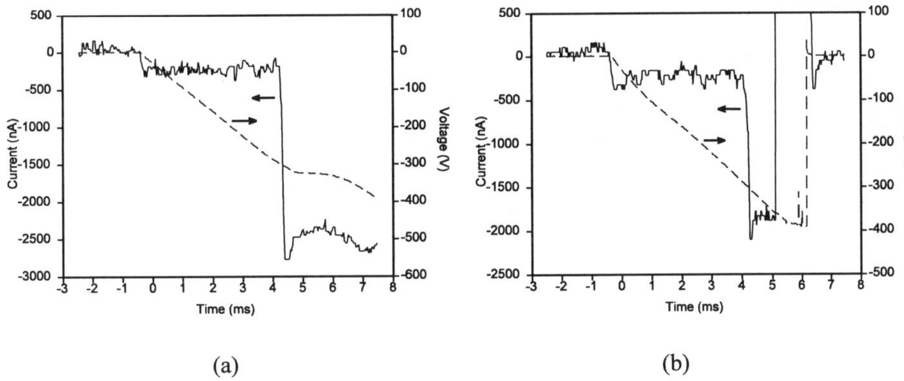

Figure 5. Current responses of a diode measured with (a) and without (b) Fluorinert fluid.

CONCLUSIONS

The effects of high electric field on 4H-SiC p-n diodes have been characterized using the fast voltage ramp technique. The breakdown voltage was found to be directly correlated with the diameter of the diode and its proximity to any crystalline defect. The breakdown voltages were found to be smaller than the theoretical value which could be due to presence of micropipes, imperfect edge terminations and surface defects.

ACKNOWLEDGMENT

This research was supported in part by the Center for Commercial Development of Space Power and Advanced Electronics, located at Auburn University, with funds from NASA Grant NAGW-1192-CCDS-AD, Auburn University and the Center's industrial partners, and by Auburn University's Research Start-up Funds.

REFERENCES

1. L.G. Matus, J.A. Powell, and C.S. Salupo, Appl. Phys. Lett. **59**, 1770 (1991).
2. M. Bhatnagar, P.K. McLarty, and B.J. Baliga, IEEE Electron Device Lett. **13**, 501 (1992).
3. T. Kimoto, T. Urushidani, S. Kobayashi, and H. Matsunami, IEEE Electron Device Lett. **14**, 549 (1993).
4. D. Alok, B.J. Baliga, and P. McLarty, IEEE Electron Device Lett. **15**, 394 (1994).
5. J.W. Palmour, J.A. Edmond, H.S. Kong, and C.H. Carter, Jr., Physica **B 185**, 461 (1993).

Part I

Device Technologies

Metallizations

IMPROVED NICKEL SILICIDE OHMIC CONTACTS TO N-TYPE 4H AND 6H-SIC USING NICHROME

E.D. LUCKOWSKI*, J.R. WILLIAMS*, M.J. BOZACK*, T. ISAACS-SMITH*, J. CROFTON**
*Physics Department, Auburn University, AL 36849, williams@physics.auburn.edu
**Department of Physics, Murray State University, Box 9, Murray, KY 42071

ABSTRACT

Results are reported for ohmic contacts formed on n-type 4H and 6H-SiC using nichrome (80/20 weight percent Ni/Cr). In comparison to contacts formed on 6H-SiC using pure Ni, the electrical characteristics of these NiCr contacts are similar (~ 1E-5 Ω-cm^2 for moderately doped material), and composite Au/NiCr contacts exhibit good stability during long-term anneals (~ 2500 hr) at 300 C without the requirement of a diffusion barrier layer between the ohmic contact layer and the Au cap layer. The use of NiCr also results in success rates near 100% for direct wire bonding to the Au cap layers.

INTRODUCTION

Pure Ni is widely used for the fabrication of ohmic contacts to SiC.[1-6] The nickel silicide layers (e.g., Ni_2Si) which are formed following short, high temperature anneals (~ 1000 C/1 min) exhibit low specific contact resistances and excellent physical stability for long-term anneals at temperatures up to 500 C. However, for wire bonding applications with Au cap layers, pure nickel silicide contacts exhibit two problems which must be overcome. Gold cap layers often do not adhere well to the silicide contact layers, and, for high temperature applications, an intermediate diffusion barrier layer is required to prevent interdiffusion between the nickel silicide contact layer and the Au cap layer. These problems are eliminated when the ohmic contacts are formed using NiCr rather than pure Ni. Results are reported herein for the electrical and physical characterization of ohmic contacts to 4H and 6H-SiC formed using an alloy of composition 80/20 weight percent Ni/Cr (~ 3.5/1 atomic percent).

EXPERIMENTAL PROCEDURE AND RESULTS

Nichrome ohmic contacts were fabricated on 0.5 μm n-type epitaxial layers supplied by Cree Research, Inc. on Si-faced, off-axis, research grade 4H and 6H-SiC p-type substrates. Details of the fabrication process[5,6] and electrical characterization procedures using the linear transmission line model (TLM)[7] have been presented previously. Both thin (~ 50 nm) and thick (~ 150-200 nm) metal layers were sputter deposited onto clean epitaxial layers, and following deposition, all samples were annealed for 3 min at 1100 C in a vacuum ambient of 10^{-6} torr. As can be seen in Figure 1, the total resistance (including the specific contact resistance) between two ohmic contacts decreases with increasing annealing temperature. An optimized anneal cycle reacts the NiCr with the SiC and produces a low specific contact resistance, while simultaneously ensuring that carbon does not accumulate on the surface of the contact layer in concentrations large enough to prevent good adhesion between the contact layer and an Au cap layer. Additionally, the anneal cycle increases the atomic concentration of Cr relative to Ni at the surface of the contact layer[1] which further promotes adhesion of the Au cap layer and also limits interdiffusion between the contact layer and the cap layer.

Mat. Res. Soc. Symp. Proc. Vol. 423 © 1996 Materials Research Society

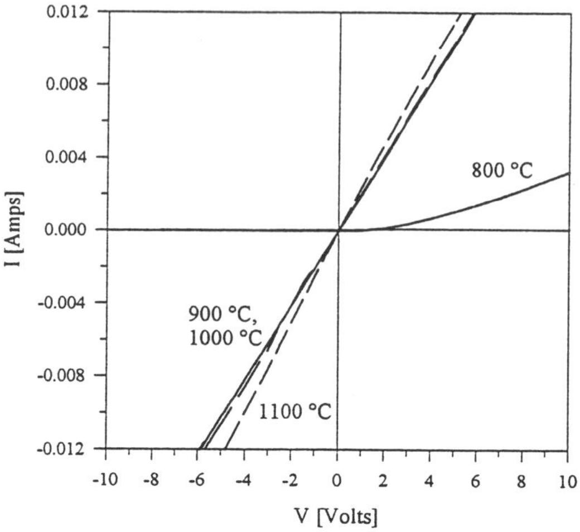

FIG. 1. Current as a function of voltage between two NiCr ohmic contacts on n-type 4H-SiC (4.8E17 cm^{-3}) following three minute anneals at temperatures between 800 and 1100 C.

Anneals were performed using a graphite strip heater, and the anneal temperature was measured using an optical pyrometer positioned to view a 5 mm x 5 mm area of the strip heater immedialely adjacent to the 5 mm x 5 mm SiC samples. Each sample was fabricated with two TLM patterns, each pattern having ten 200 x 200 μm^2 ohmic contacts separated by gaps of constant width and variable lengths which increase from 4 μm to 20 μm (see Figure 2).

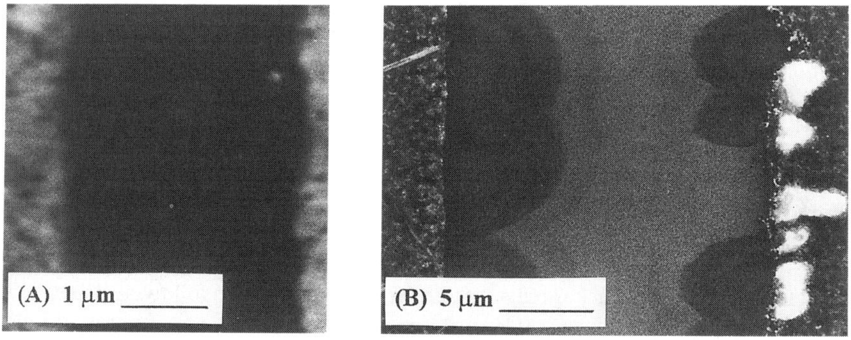

FIG. 2. SEM micrographs of the SiC surface between adjacent ohmic contacts in TLM patterns for (A) sample 187 and (B) sample 155 following 3 min anneals at 1100C.

Four contacts in each TLM pattern on sample 155 had the structure shown in Figure 2. None of the other NiCr samples had these features; however, we have observed similar structure in TLM patterns with ohmic contacts formed by high temperature annealing following sputter deposition of Ni_2Si and AlTi on n- and p-type SiC, respectively. We suspect that the white areas in the ohmic contact near the edge of the gap are associated with carbon segregation during the anneal cycle; however, our attempts to confirm this supposition using Auger spectroscopy have not been successful. We have not been able to isolate the white areas from the rest of the contact with our AES electron beam.

The results of specific contact resistance measurements for several 4H and 6H-SiC samples are shown in Table I. Different doping concertrations refer to different wafers, and repeated doping concentrations indicate different 5 mm x 5 mm samples from a given wafer. The sheet resistance values listed in Table I were measured using Van der Pauw patterns fabricated on the samples with the TLM patterns. In all cases, sheet resistances determined using the Van der Pauw patterns agreed within ±10% with those determined from the TLM calculations.

Table I. Summary of Specfic Contact Resistance Meaurements

Sample No.	N_D (cm^{-3})	R_{SHEET} (Ω/sq)	ρ_C (Ω-cm^2)
155 (6H)	3.2E17	3708	1.2E-6 (4 gaps) 8.9E-6 (8 gaps)
			1.3E-5 (4 gaps) 9.5E-6 (8 gaps)
209 (6H)	3.2E17	3478	2.5E-6
			4.2E-6
211 (6H)	1.4E18	1975	9.1E-5
			6.5E-5
187 (4H)	4.8E17	1923	1.6E-5
			1.4E-5
188 (4H)	4.8E17	1870	1.0E-4
			1.5E-4
185 (4H)	1.3E19	97	1.2E-5
			1.6E-5

Variations in the measured contact resistances for sample 155 result from the number of pairs of contacts used to determine ρ_C. For the remaining samples, the sheet resistances are generally in good agreement and decrease with increasing doping concentration as expected. However, the specific contact resistances do not depend strongly on the doping concentrations. For research grade epi layers using Hg probe profiling, we have measured doping concentrations which vary typically by factors of 2-3 over the surface of 3.3 cm diameter wafers. Variations in doping concentration coupled with variations in epi layer thickness may account for the results presented in Table I. Investigations of the dependence of ρ_C on N_D are ongoing and will take the form of improved statistics generated by characterizing a number of additional samples which are currently being fabricated.

Physical analyses using RBS and AES techniques have been performed for both thin and thick NiCr layers (~ 50 nm and 200 nm, as deposited) following 3 min anneals at 1100 C. Results for a thin sample are shown in Figure 3. Both the AES and RBS spectra show carbon, Cr, Ni and Si at the front surface of the contact layer, with increasing relative concentrations of Ni and Si as one moves away from the surface towards the contact layer-SiC interface. The RBS technique is relatively less sensitive to the prescnce of carbon, and in the RBS simulation, carbon is assumed to be uniformly distributed throughout the contact layer. However, front surface carbon can be observed as a small peak in the spectrum just to the right of the solid line simulation near channel number 120. For the channel number ranges 250-270 and 330-350, deviations between the RBS data and the simulation are due to the surface roughness of the ohmic contact layer.

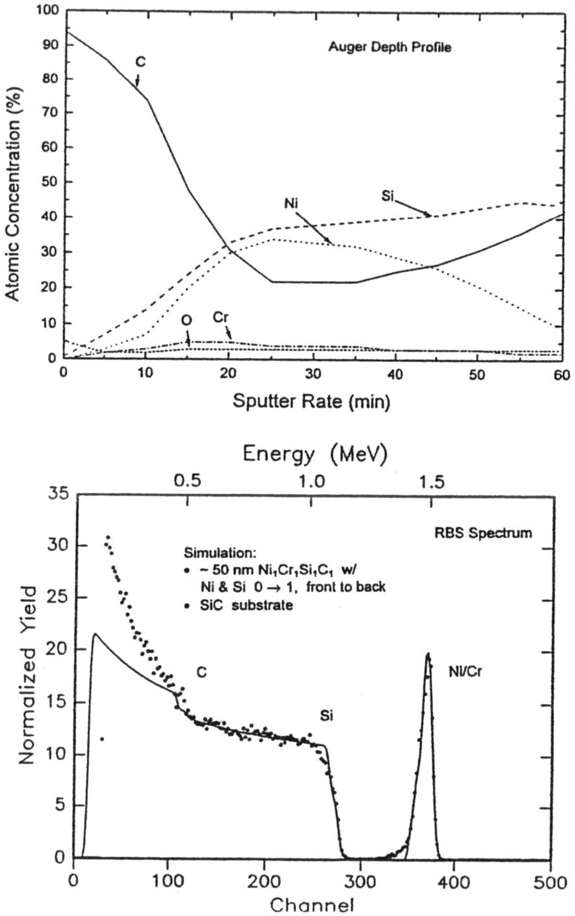

FIG. 3. Auger and Rutherford backscattering spectra for a 50 nm as deposited NiCr layer following a 3 min anneal at 1100 C. (————) RUMP simulation[8].

An Auger depth profile for an annealed, thick NiCr layer (~ 200 nm, as deposited) is shown in Figure 4. In comparison to the depth profile of Figure 3, notice that Cr, Ni and Si are not present at the surface of the contact layer. Instead, the surface is covered with a relatively thick layer of pure carbon. These results are confirmed in RBS spectra which are not shown for this sample. Possible sources of the carbon in the contact layers are incorporation during sputter deposition of the NiCr films, incorporation from the carbon strip heater during annealing, and accumulation during annealing as the NiCr reacts with the SiC epi layer. Rutherford backscattering and AES

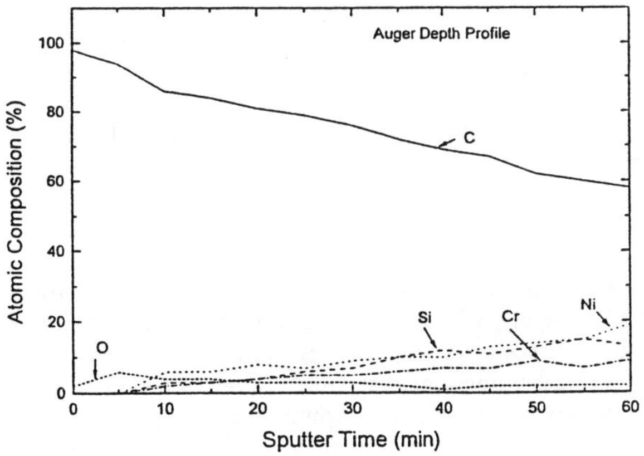

FIG.4. Auger spectra for a 200 nm as deposited NiCr film following a 3 min, 1100 C anneal.

analyses for samples annealed for times between 1 and 6 minutes have eliminated the strip heater as a source, and AES depth profiles for unannealed NiCr layers show that very little carbon is incorporated in the metal film during sputter deposition. Therefore, our supposition is that residual carbon from the NiCr-SiC interaction diffuses rapidly through the interaction region and accumulates at surface of the contact layer. Results for different NiCr layer thicknesses indicate that the accumulation process is source limited rather than diffusion rate limited. Success rates for direct wire bonding to Au cap layers (~ 200 nm) sputter deposited over the contact layers are very nearly 100% for thin NiCr layers compared to perhaps 20% for thick layers. We attribute these poor results for the thick layers to the excessive carbon buildup on the contact layer surfaces.

Results for long-term anneals of Au/NiCr and Au/Ni samples are shown in Figure 5.

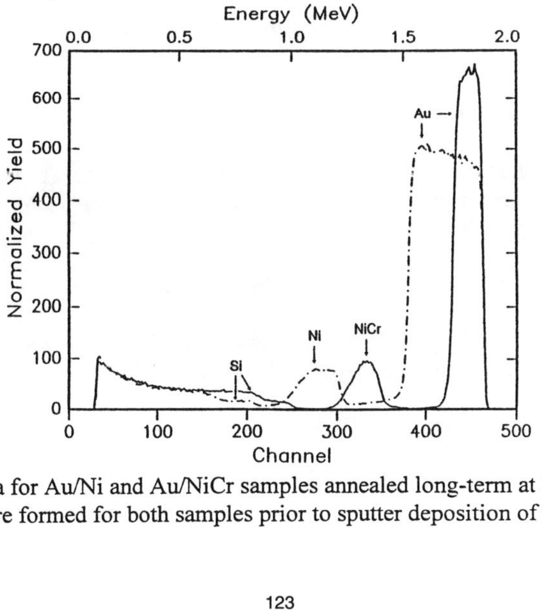

FIG. 5. RBS spectra for Au/Ni and Au/NiCr samples annealed long-term at 300 C and 10^{-6} torr. Ohmic contacts were formed for both samples prior to sputter deposition of the Au cap layers.

The gold cap layers were sputter deposited following short, high temperature anneals used to produced ohmic contacts for both samples. For the Ni sample, interdiffusion between the Au and the nickel silicide contact layer is apparent after 990 hrs (channels 310-360), while the Au and NiCr contact layers show essentially no mixing after 2500 hrs. As discussed previously, the relatively larger Cr concentration at the surface of the NiCr contact layer acts as a barrier to suppress interdiffusion.

CONCLUSION

Nichrome (80/20 weight percent Ni/Cr) can be used to fabricate low resistance ohmic contacts to moderately doped 4H and 6H-SiC. Perhaps as a result of the high annealing temperature (1100 C), no strong dependence on doping was observed for the specific contact resistance. This result is currently under further investigation. Gold cap layers adhere well to thin NiCr contact layers which have relatively larger surface concentrations of Cr compared to Ni and which do not accumulate excessive carbon at the layer surfaces during the anneal cycle which forms the ohmic contacts. As a result, success rates for direct wire bonding to the Au cap layers are very much improved. The higher concentration of Cr near the surface of the contact layer also suppresses interdiffusion between the Au cap layer and the contact layer. The composite Au/NiCr contact is stable at elevated temperatures without the requirement of an intermediate diffusion barrier layer between the Au and NiCr layers.

This work was supported by the Center for Commercial Development of Space and Advanced Electronics, located at Auburn University, under NASA Grant NAGW-1192-CCDS-AD.

REFERENCES

1. J. Crofton, J.M. Ferrero, P.A. Barnes, J.R. Williams, M.J. Bozack, C.C. Tin, C.D. Ellis, J.A. Spitznagel and P.G. McMullin, in Amorphous and Crystalline Silicon Carbide IV, edited by C.Y. Yang, M.M. Rahman and G.L. Harris (Springer, Berlin, 1992) p. 176.

2. C.T. Gardner, J.A. Cooper, M.R. Melloch, J.W. Palmour and C.H. Carter, ibid., p. 338.

3. S. Adams, C. Severt, J. Lenord, S. Liu and S.R. Smith, Trans. Second Int'l. High temperature Electronics Conference, edited by D.B. King and F.V. Thome (1994) p. XIII-9.

4. A.J. Stekl, J.N. Su, P.H. Yih, C. Yuan and J.P. Li, Amorphous and Crystalline Silicon Carbide V, edited by M.G. Spencer, R.P. Devaty, J.A. Edmond, A. Kahn, and M.M. Rahman (IOP Conf. Series No. 137, 1994) p. 653.

5. J. Crofton, P.G. McMullin, J.R. Williams and M.J. Bozack, J. Appl. Phys. **77(3)**, 1317 (1995).

6. J. Crofton, E.D. Luckowski, J.R. Williams, T. Isaacs-Smith, M.J. Bozack and R Siergiej, Proc. Sixth Int'l. Conf. on SiC and Related Materials, in press.

7. H.H. Berger, Solid State Electronics **15**, 145 (1972).

8. L.R. Doolittle, Nucl. Instrs. Meths. in Phys. Res. **B9**, 344 (1985).

CHEMICAL AND STRUCTURAL CHARACTERIZATION OF THE Ni-Ti ALLOY/ 6H-SiC CONTACTS

M. LEVIT, I. GRIMBERG, B. -Z. WEISS, AND M. EIZENBERG

Department of Materials Engineering, Technion - Israel Institute of Technology, Haifa 32000, Israel

ABSTRACT

Ni-Ti alloy is a promising candidate for high-temperature contact metallization for SiC electronic devices. In the present study Ni-Ti alloy thin films (100 nm) of two different compositions ($Ni_{90}Ti_{10}$ and $Ni_{50}Ti_{50}$) were coevaporated on 6H-SiC substrate. Interfacial reactions, microstructure, compositional changes, and phase formation were investigated as functions of heat-treatments in the range of 400-800 °C. The study was carried out using Auger electron spectroscopy, x-ray diffraction, and analytical transmission electron microscopy. In the case of the $Ni_{90}Ti_{10}$ alloy the interaction was found to begin at 450 °C. Ni and C are the dominant diffusing species. The reaction zone is divided into three layers. In the first layer, adjacent to the SiC substrate, the presence of a Ni-rich silicide, Ni_2Si, and C precipitates, was observed. The second layer is composed mainly of TiC, while the third - of Ni_2Si. In the case of the $Ni_{50}Ti_{50}$ alloy the interaction began at 800 °C. Carbon is the dominant diffusant. The reaction zone is divided into two layers. The first, next to the substrate layer is composed of epitaxially grown TiC and the second - of Ni_3Ti_2Si compound. A thin (~5 nm) amorphous discontinuous layer was found at the TiC/SiC interface. Factors controlling phase formation in the Ni-Ti/SiC system are discussed.

INTRODUCTION

Silicon carbide has received recently increasing attention as a wide-bandgap semiconductor material, due to significant advances in its growth and processing technology. SiC is mechanically and chemically stable, and is characterized by high-breakdown electric field and high electron saturation drift velocity [1-8]. Additional attractive features of SiC are its capabilities to be doped both p- and n-type, and to be oxidized, producing electrically insulating SiO_2 on its surface [5]. All these properties make SiC a suitable material for high-temperature, optoelectronic, high-power, and high-frequency applications [1,2].

In the present work the interaction of $Ni_{90}Ti_{10}$ and $Ni_{50}Ti_{50}$ alloy thin films with n-type 6H-SiC single crystal was investigated. Both Ni and Ti films form ohmic contacts on n-type SiC [3,6]. Ni reacts with SiC and forms Ni-silicide (Ni_2Si) and graphite, the latter either uniformly distributed inside the silicide or as a surface layer [4,5]. The reaction between Ti and SiC results in the formation of a two-layer structure consisting either of TiC and the Ti-rich silicide, Ti_5Si_3, or of TiC with free Si as a top layer [7,8]. Thus, bearing in mind the high affinities of Ni to Si and of Ti to C, the formation of a thermally stable contact consisting of Ni-silicide and Ti-carbide should be expected in the interaction between Ni-Ti alloy thin film and SiC.

EXPERIMENT

$Ni_{90}Ti_{10}$ and $Ni_{50}Ti_{50}$ alloy thin films of a thickness of 100 nm were e-gun coevaporated on n-type 6H-SiC (0001) substrates. In order to prevent possible subsequent oxidation and/or contamination of the alloy film, a protective layer of W (~10 nm) was evaporated on top of each sample. During the deposition the pressure was kept at ~10^{-7} torr and the substrate temperature at -100 °C. The composition of the as-deposited samples was verified by Rutherford backscattering spectrometry.

The samples were annealed in a vacuum furnace (~10^{-7} torr) at temperatures 400-800 °C for different durations. Interfacial reactions, phase formation, microstructure, and composition were studied by Auger electron spectroscopy (AES), x-ray diffraction (XRD), and transmission

Mat. Res. Soc. Symp. Proc. Vol. 423 © 1996 Materials Research Society

electron microscopy (TEM) combined with energy dispersive x-ray spectroscopy (EDS).

RESULTS AND DISCUSSION

$Ni_{90}Ti_{10}$ alloy

An XRD spectrum taken from the as-deposited $Ni_{90}Ti_{10}$ alloy thin film shows only Ni and W peaks [Fig. 1(a)]. The peaks are broad, indicating small grain size and/or the presence of microstresses in the as-deposited film. The Ni unit cell parameter calculated on the basis of XRD data is smaller than reported for pure Ni, which suggests the formation of a solid solution and/or the presence of tensile strain. The AES depth profile shows a homogeneous distribution of Ni and Ti [Fig. 1(b)]. The concentration of impurities (C and O) is negligible. TEM studies display fine grains of ~10 nm in the Ni-Ti layer. Some of the grains exhibit a columnar morphology, with the columns oriented perpendicular to the substrate. The selected area electron diffraction (SAED) patterns are composed of spot rings corresponding to randomly oriented nanocrystalline Ni grains.

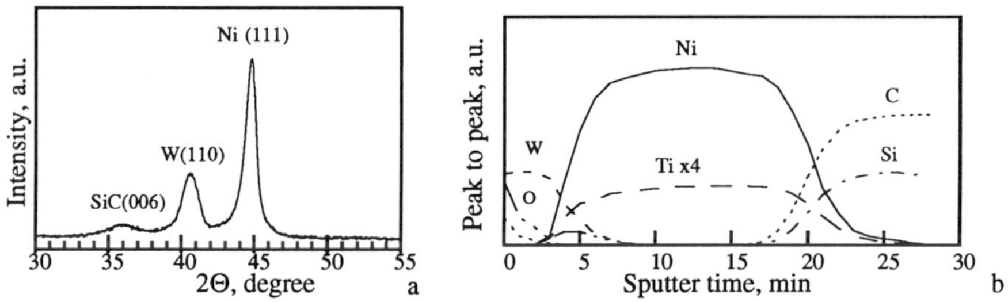

FIG. 1. (a) XRD pattern taken from the as-deposited $Ni_{90}Ti_{10}$ sample. (b) AES depth profile of this sample.

The $Ni_{90}Ti_{10}$ alloy begins an interaction with 6H-SiC at 450 °C. After annealing for 1/2 h, peaks of a Ni-rich silicide, $Ni_{31}Si_{12}$, (also designated as Ni_5Si_2 [9]) and, possibly, traces of TiC are seen in the XRD pattern in addition to peaks of Ni and W. The Auger depth profile of this sample confirms the start of the interfacial reactions. The mobility of Ni and C is evident, but no phase separation could be detected. After annealing at 600 °C for 1/2 h the peaks of Ni disappear from the XRD spectrum [Fig. 2(a)], which shows strong peaks of $Ni_{31}Si_{12}$, as the major phase, as well as peaks of Ni_2Si, of W, and traces of TiC. The AES depth profile of this sample [Fig. 2(b)] indicates that Ni, Si and C, are the mobile species, but C seems to diffuse faster than Ni and Si. Ti and W (top layer) do not diffuse. The reaction zone can be roughly divided into two layers (disregarding the W top layer). The first layer, adjacent to the SiC substrate, is characterized by the presence of Ni, Si, and C. The Ni and Si concentration profiles suggest that Ni-rich silicide have been formed. Since Ni does not form stable carbides, the precipitation of C in this layer should be assumed. It was confirmed by the analysis of the Auger line shape of C (KLL) in this region, which showed features characteristic of elementary C. Because of the possible amorphization of graphite during the sputtering, it is impossible in Auger experiments to distinguish between the originally amorphous C and graphite. An absence of graphite peaks in the XRD spectra suggests that C precipitates in the first layer are either amorphous or dispersed (nano sized) graphite particles.

The second layer, near the W top layer, is approximately three times thinner than the first one and consists of Ti, C, and Ni. The C and Ti lines follow each other, indicating the formation of TiC in this layer. The shape analysis of the C (KLL) Auger line in this region also implies that C is present in the carbide form. By following the Ni line in the AES depth profile [Fig. 2(b)] the formation of a Ni shoulder is observed beyond the TiC region in the second layer of the reaction zone. It is the presence of TiC that drastically slows down Ni diffusion.

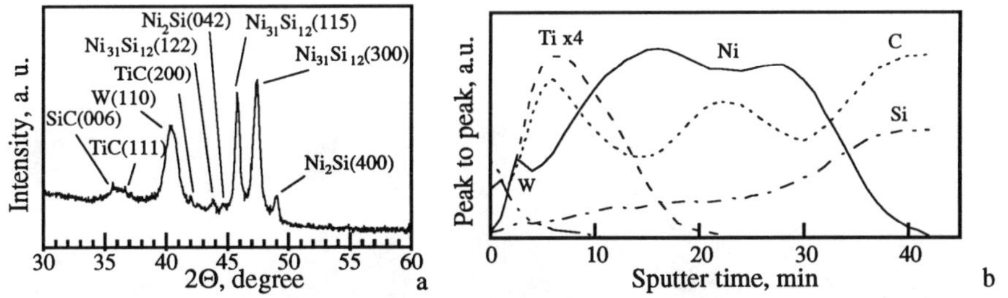

FIG. 2. (a) XRD pattern taken from the $Ni_{90}Ti_{10}$ sample annealed at 600°C for 1/2 h. (b) AES depth profile of this sample.

An XRD pattern taken after annealing at 800 °C for 1/2 h shows that $Ni_{31}Si_{12}$ transforms to Ni_2Si. No silicides richer in Si (namely, NiSi and $NiSi_2$), as revealed on a Si substrate [10], were observed. The peaks of TiC in the XRD spectrum became more pronounced. The spectrum also shows peaks of W, indicating that the top W layer is still present. The AES depth profile from this sample is similar to that taken after annealing at 600 °C, however, the presence of the Ni shoulder behind the TiC layer becomes more evident. Due to their higher mobility at 800 °C, Si atoms reach Ni behind TiC, and a Ni-rich silicide seems to form in this region. Ti and W (the top layer) are still immobile. The AES depth profile suggests the formation of a three-layer structure in the reaction zone. The first layer, adjacent to the SiC substrate, consists of a Ni-rich silicide, Ni_2Si, (XRD data), and C precipitates; the second layer is mainly TiC; and the third - a thin layer of Ni-rich silicide, which judging from the Ni/Si ratio (peak-to-peak height), is probably Ni_2Si.

The results of the TEM studies are consistent with those of AES and XRD. Structures of the reaction zone observed after annealing at 600 °C and at 800 °C are similar and consist, as previously stated, of three layers. The phase composition of each layer was determined by means of SAED patterns and quantitative EDS analysis. The only difference was that while after annealing at 600°C the first layer is mostly composed of $Ni_{31}Si_{12}$, after 800 °C it is Ni_2Si. Fig. 3(a) shows the $Ni_{90}Ti_{10}$ sample annealed at 800°C for 1/2 h. The first layer, adjacent to the SiC substrate, ~120 nm thick, is basically composed of columnar as well as equiaxed grains of Ni_2Si. The small regions (10-20 nm) with bright contrast, seen in this layer, are attributed to the C precipitates. It is its nano size, low atomic scattering factor of C, and small amount, which probably makes graphite undetectable by XRD technique [Fig. 2(a)]. The second layer, ~60 nm thick, is mainly composed of nano TiC grains, while the third (~20 nm thick) contains Ni_2Si grains according to the EDS point analysis. The SAED pattern taken from a grain in the first (Ni_2Si + C) layer [Fig. 3(b)] unambiguously points to the Ni_2Si structure. Since the selected area aperture in this case covered part of the substrate and part of the second layer, dot rings belonging to TiC and several dots of the SiC substrate could be seen.

The interaction between $Ni_{90}Ti_{10}$ alloy film and 6H-SiC was found to begin at 450 °C, when only Ni atoms are mobile. At the first stage of the interaction Ni diffuses toward SiC and may slightly penetrate into SiC through interstitial sites and lattice imperfections, "softening" the SiC lattice [4]. In the second stage Ni begins to react with SiC, and Ni-rich silicide is formed at the interface. Ni-silicide growth is provided by the diffusion of Ni through the silicide. The C atoms, released during silicide formation, diffuse through the Ni-silicide toward the Ti, partly precipitating in the silicide layer. When C atoms reach the Ti, TiC begins to form, adjacent to the Ni-silicide region. The growing TiC layer acts as a diffusion barrier to Ni, and part of the Ni atoms are therefore confined to the outer region of TiC [Fig. 2(b)]. At temperatures higher than 600°C, when the mobility of Si atoms increases, they attain this region and a thin layer of Ni-rich silicide is formed, leading to the three-layer structure in the reaction zone [Fig. 3(a)].

Thermodynamic analysis of the $Ni_{90}Ti_{10}$/SiC interaction [11] shows that the reaction characterized by the largest negative enthalpy change is: $4.5Ni + 0.5Ti + 1.8SiC \rightarrow 0.9Ni_5Si_2 + 0.5TiC + 1.3C$. This is the reason why the Ni_5Si_2 ($Ni_{31}Si_{12}$) silicide appears first. After Ni_5Si_2

is formed, the formation of Ni_2Si becomes thermodynamically favorable, while the formation of silicides richer in Si, namely Ni_3Si_2, NiSi, and $NiSi_2$, is unfavorable [11]. These results are in good agreement with reported Ni-Si-C ternary phase diagram [12].

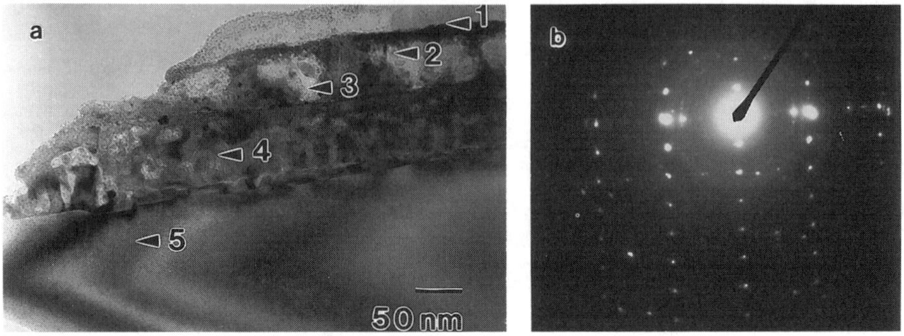

FIG. 3. (a) TEM micrograph taken from the $Ni_{90}Ti_{10}$ sample annealed at 800 °C for 1/2 h. (1) W top layer, (2) Ni_2Si layer, (3) TiC layer, (4) Ni_2Si + C layer, and (5) SiC substrate. (b) SAED pattern taken from a Ni_2Si grain in the Ni_2Si + C layer. Zone axis: [031]. The dot rings belong to TiC.

The observed phase separation in the reaction zone into a layer of Ni_2Si and a layer of TiC could be beneficial in high-temperature SiC metallization. Ni_2Si has low contact resistance and thermal stability when serving as a contact with a n-type SiC substrate [3]. On the other hand, TiC is known to be a diffusion barrier in VLSI technology. The formation of such a structure, consisting of an ohmic Ni-silicide contact followed by a layer of a TiC diffusion barrier, promises high-temperature stability, important for SiC electronic devices.

$Ni_{50}Ti_{50}$ alloy

An XRD spectrum taken from the as-deposited $Ni_{50}Ti_{50}$ sample presents only broad peaks of W [Fig. 4(a)], indicating an amorphous state of the as-deposited alloy film. The AES depth profile of this sample shows a homogeneous distribution of Ni and Ti, and negligible concentration of impurities [Fig. 4(b)]. In order to achieve a crystalline chemically-ordered NiTi structure, long-range diffusion and high atomic mobility are required, which cannot be obtained at -100 °C. On the other hand, in the case of the $Ni_{90}Ti_{10}$ composition, the formation of a crystalline phase requires only short-range atomic movements in order to establish close-packed (Ni-fcc) structure.

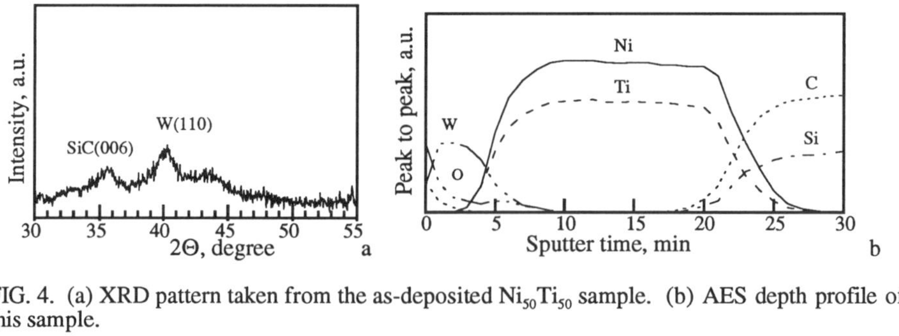

FIG. 4. (a) XRD pattern taken from the as-deposited $Ni_{50}Ti_{50}$ sample. (b) AES depth profile of this sample.

After annealing at 600 °C for 1/2 h no changes could be detected in the AES depth profile compare to the as-deposited state. However, XRD analysis indicates the formation of the NiTi intermetallic, which probably retards reactions of Ni and Ti with SiC. After annealing at 800 °C for 1/2 h [Fig. 5(a)] the peaks of NiTi disappeared, and those of TiC and of the Ni_3Ti_2Si ternary compound are clearly seen (in addition to peaks of W). The AES depth profile of this sample [Fig. 5(b)] indicates that all components (except of W) are mobile, but C seems to be the dominant diffusant. A well defined two-layer structure (disregarding the W top layer) is observed in the reaction zone. The first layer, adjacent to SiC, contains C and Ti, whose lines follow each other, indicating the formation of TiC. In the second layer Ni, Ti, and Si are present. According to the concentration profiles, Ni-Ti-Si ternary phase with composition close to Ni_3Ti_2Si may be situated here.

TEM analysis completely agrees with AES and XRD observations. Fig. 6(a) shows the $Ni_{50}Ti_{50}$ sample annealed at 800 °C for 1/2 h. The reaction zone is well separated into two layers (without the W top layer). The first layer, adjacent to the substrate is ~50 nm thick, and consists of columnar TiC grains, according to SAED pattern analysis. Electron diffraction patterns also indicate that TiC grows epitaxially on the SiC substrate ($(111)_{TiC} \parallel (0001)_{SiC}$, $\langle 1\bar{1}0\rangle_{TiC} \parallel \langle 11\bar{2}0\rangle_{SiC}$). The lattice mismatch between $(111)_{TiC}$ and $(0001)_{SiC}$ is ~1% [9]. The second layer of the reaction zone is ~80 nm thick and is composed of equiaxed grains. As follows from SAED and EDS point analysis these grains belong to the Ni_3Ti_2Si ternary compound.

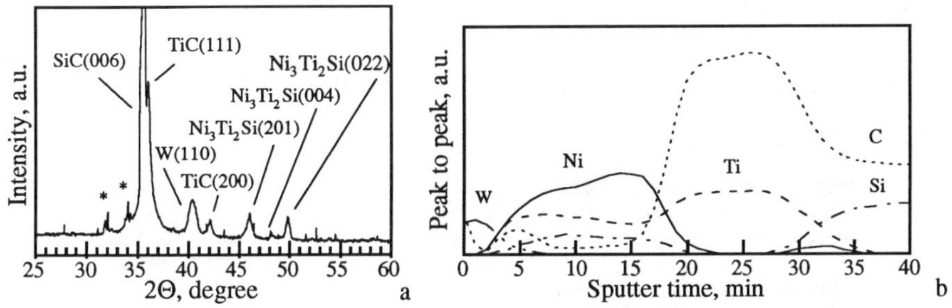

FIG. 5. (a) XRD pattern taken from the $Ni_{50}Ti_{50}$ sample annealed at 800 °C for 1/2 h. Peaks marked by asterisks produced by $W_{L\alpha}$ radiation. (b) AES depth profile of this sample.

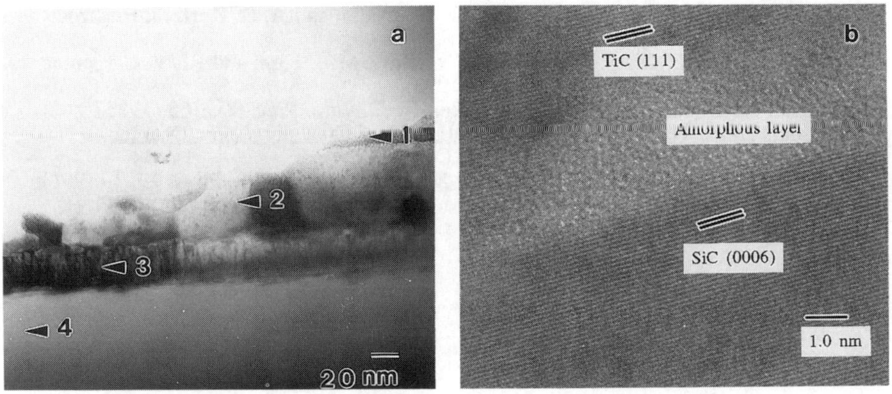

FIG. 6. (a) TEM micrograph taken from the $Ni_{50}Ti_{50}$ sample annealed at 800 °C for 1/2 h. (1) W top layer, (2) Ni_3Ti_2Si layer, (3) TiC layer, and (4) SiC substrate. (b) TiC/SiC interface at higher magnification. (1) TiC layer, (2) amorphous layer, and (3) SiC substrate.

TEM micrographs taken at higher magnifications [Fig. 6(b)] clearly show the epitaxial relationship between TiC and 6H-SiC. In addition, discontinuous amorphous layer up to ~5 nm thick is seen at the TiC/SiC interface. This layer most probably consists of a mixture of C and Si (AES and EDS data), however its precise composition could not be determined.

CONCLUSIONS

The interaction of $Ni_{90}Ti_{10}$ and $Ni_{50}Ti_{50}$ alloy thin films with 6H-SiC single crystal was studied at temperatures between 400 °C and 800 °C. The $Ni_{90}Ti_{10}$/SiC interaction began at 450 °C with the formation of a Ni-silicide, $Ni_{31}Si_{12}$, followed by Ni_2Si as the temperature was raised above 600 °C. Ni and C were found to be the dominant diffusing species. After annealing at 800 °C for 1/2 h a well defined three-layer phase separation was observed. In the first layer, adjacent to the SiC substrate, C precipitates, and a Ni-rich silicide, Ni_2Si, is formed. The latter is also found in the thin outermost layer, while the intermediate layer consists mainly of TiC. Thermodynamic analysis indicates that the structure formed is stable in contact with SiC.

In the case of $Ni_{50}Ti_{50}$ alloy the interaction with 6H-SiC began at 800 °C. C was found to be the dominant diffusant. The reaction zone is divided into two layers. The first, next to the substrate layer, is composed of epitaxially grown TiC and the second - of Ni_3Ti_2Si ternary compound. A thin amorphous discontinuous layer was observed at the TiC/SiC interface.

ACKNOWLEDGMENTS

The authors gratefully acknowledge the support of the Israel Academy of Sciences and Humanities, and Technion V. P. R. Fund - S. and N. Grand Research Fund.

REFERENCES

1. J. H. Edgar, J. Mater. Res. **7**, 235 (1992).
2. D. Alok, P. K. McLarty, and B. J. Baliga, Appl. Phys. Lett. **64**, 2845 (1994).
3. J. Crofton, J. M. Ferrero, P. A. Barnes, J. R. Williams, M.J . Bozack, C. C.Tin, C. D. Ellis, J. A. Spitznagel, and P. G. McMullin, in 4th. Int. Conf. Amorphous and Crystalline SiC, Santa Clara, CA, Oct. 10-11, 1991.
4. M. V. Zeller, J. J .Bellina, N. Saha, J. Filar, J. Hargraeves, and H. Will, Mat. Res. Soc. Symp. Proc. **97**, 283 (1987).
5. I. Ohdomari, S. Sha, H. Aochi, T. Chikyow, and S. Suzuki, J. Appl. Phys. **62**, 3747 (1987).
6. J. S. Shor, R. A. Weber, L. G. Provost, D. Goldstein, and A. D. Kurtz, J. Electrochem. Soc. **141**, 579 (1994).
7. L. M. Porter, R. C. Glass, R. F. Davis, J. S .Bow, M. J. Kim, and R. W. Carpenter, Mat. Res. Soc. Symp. Proc. **282**, 471 (1993).
8. J. J .Bellina, Jr. and M. V. Zeller, Mat. Res. Soc. Symp. Proc. **97**, 265 (1987).
9. P. Villars and L. D. Calvert, Pearson's Handbook of Crystallographic Data for Intermetallic Phases, (ASM, Metals Park, OH, 1985).
10. M. Levit, I. Grimberg, B. -Z. Weiss, and M. Eizenberg, J. Appl. Phys. **79**, 1 (1996).
11. M. Levit, I. Grimberg, and B. -Z. Weiss, and M. Eizenberg, accept. J. Appl. Phys.
12. Y. M. Basin, V. M. Kuznetsov, V. T. Markov, and L. S. Guzei, Russ. Metall. **4**, 197 (1988).

Low Resistance Ohmic Contact on *p*-type GaN Grown by Plasma-Assisted Molecular Beam Epitaxy

Myung C. Yoo*, J.W. Lee*, J.M. Myoung**, K.H. Shim**, and K. Kim**
* Photonics Semiconductor Laboratory, Samsung Advanced Institute of Technology, P.O. Box 111, Suwon 440-600, Korea, mcyoo@saitgw.sait.samsung.co.kr
** Department of Electrical and Computer Engineering, University of Illinois at Urbana-Champaign, Urbana, IL, USA

ABSTRACT

Ohmic contacts on p-type GaN have been investigated. High quality GaN epilayers on c-plane sapphire were prepared using plasma-assisted molecular beam epitaxy that utilized an inductively coupled rf nitrogen plasma source and solid source beams. The resulting film thickness and the doping concentration of the grown samples were in the range of 0.7-1.35 μm and 10^{18} -10^{20}/cm^3, respectively. The metallization consisted of high work function metal bi-layers which included a combinations of 25 nm-thick Ni, Ti, Pt and/or Cr and 200 nm-thick Au on the highly p-doped GaN in a transmission line model pattern. Ohmic contacts were formed by alloying the bi-layers using rapid thermal annealing (RTA) at temperatures in the range of 300-700 °C for 1 min under nitrogen ambient. Current-voltage measurements showed that the specific contact resistance was as low as 1.2×10^{-4} Ω–cm^2 for the sample having 1.4×10^{20}/cm^3 *p*-type doping concentration with a Cr/Au contact annealed at 500 °C for 1 min by RTA. Judging from the scanning Auger microscopy results and the glancing angle x-ray diffraction analysis, this resistance is attributed to Cr diffusion into the GaN layer.

INTRODUCTION

Rapid progress in the GaN-based optoelectronic research has resulted in commercialization of blue/green light emitting diodes (LEDs) [1-2]. More recently, Nakamura et al. of Nichia Chemical have achieved InGaN multi-quantum well (MQW) structure blue/violet lasing under pulsed current injection [3]. Although continuous wave (cw) lasing operation has not yet been demonstrated, this pioneering work is certainly a great step toward the development of III-V nitride semiconductor lasers. To improve the device performance and to realize cw lasing, there are, however, a number of issues that have to be addressed. Among these, p-type ohmic contact is considered one of the most important. According to the I-V characteristics of the Nichia laser, the turn-on voltage of a Ni/Au metal contact was 34 V which could very well be due to the poor p-type contact properties. To ensure reliable laser operation it is, therefore, very important that a low resistive p-type ohmic contacts be developed.

GaN epilayers grown with metalorganic chemical vapor deposition (MOCVD), in many cases, have shown better crystal quality compared with those grown with molecular beam epitaxy (MBE). However, p-type doping concentrations of the MOCVD-grown GaN are known to be lower than those of MBE because of the difficulty in removing the dangling hydrogen bond on the GaN surface [4]. Therefore, post-growth annealing process is essential in MOCVD

Mat. Res. Soc. Symp. Proc. Vol. 423 © 1996 Materials Research Society

to activate the p-type dopants in the GaN crystal [5], while no such procedure is necessary in MBE [6].

It is well known that the as-grown GaN films exhibit n-type nature. This is attributed to the vacancy formation due to improper material incorporation between Ga and N atoms. As a result, forming a metal ohmic contact on the n-type GaN is known to be relatively easier than that on the p-type. Consequently, most work on metal contact on GaN has concentrated on n-type ohmic contact [7,8]. In addition, due to the high work function of GaN (7.5 eV), it is very difficult to find appropriate metals to form an ohmic contact on the p-type GaN. In this paper, we present the results of ohmic contacts fabricated on p-type GaN epilayers grown by plasma-assisted molecular beam epitaxy (PAMBE) using Cr/Au metal contacts.

EXPERIMENT

The PAMBE system used for the growth of GaN epilayers consisted of an inductively coupled rf nitrogen plasma source developed at the University of Illinois, group III solid sources, and Mg and Si dopant sources. After thermal desorption, sapphire substrates were exposed to nitrogen plasma for 20 min. Prior to GaN growth, a 20nm-thick AlN buffer layer was grown on sapphire (0001) at 550 °C. After that, the GaN epilayers were grown at 700 °C. The p-doping was effected using high-purity Mg (6N) as the dopant. A detailed description of the growth system and the film growth procedures can be found elsewhere [9-11].

The transmission line model (TLM) patterns were fabricated on the grown films by the conventional lift-off process. The size of metal pad was 100 x 200 μm and the distances between each metal pad were varied from 5 to 30 μm. After the photolithographic process, ohmic contacts were fabricated by alloying three different combinations of metal layers, such as Cr/Au, Pt/Au, and Ti/Pt/Au, by electron beam evaporation. Deposition conditions for each of the metal layers are listed in Table I.

Table I. Deposition conditions of metal layers for TLM pattern

Metal layers	Layer thicknesses (nm)	Deposition rates (nm/s)
Cr/Au	20/ 300	3/5
Pt/Au	20/ 300	1/5
Ti/Pt/Au	20/ 80/ 300	2.5/1/5

To investigate the effect of annealing on the ohmic characteristics of the metal contacts, RTA was performed under a nitrogen ambient for 1 min. The I-V characteristic was measured by increasing the annealing temperature from 300 ^{0}C to 700 ^{0}C in a TLM pattern.

The contact resistances of all the samples were measured by using the I-V curves in the range of - 10V and + 10V. For heavily p-doped samples, the resistances were measured by the linear fitting method in the +2V and -2V range. Finally, scanning Auger microscopy (SAM) and glancing angle x-ray diffraction (XRD) were used to study the properties of the interface between the metal contact and the p-GaN epilayer. For the SAM analysis, a peak-to-peak signal for each element was detected using 3.5-kV, 500-mA sputtering power at every 30 sec. In the XRD analysis, a $\theta - 2\theta$ scan was performed between 10 and 90 degrees to identify the metal-GaN compound. Intensities and 2θ positions of each peak were indexed using ASTM data.

RESULTS AND DISCUSSION

The p-type GaN samples used for this work were grown by Myoung et al. at the University of Illinois using a PAMBE system. A detailed analysis of the structural, electrical, and optical properties of these samples is provided elsewhere [11]. Two p-type GaN samples, with hole concentrations of 1.4 x 10^{20} cm-^{3} (p^{+} GaN) and 2.7 x 10^{18} cm-^{3} (p^{-} GaN), were used for the present work. The film thicknesses of the samples were 1. 5 μm and 0.6 μm, respectively.

Figure 1 shows an I-V characteristic of as-fabricated Cr/Au metal contact on p^{+}-GaN with 5-μm spacing. As can be seen in the figure p^{+}-GaN sample exhibits ohmic characteristic even without post fabrication annealing treatment. It is also noted that the lowest specific contact resistance of 1.2 x 10 $^{-4}$ ohm-cm^{2} is obtained after annealing at 500 °C for 1 min, which we believe, is the lowest specific contact resistance ever reported for a Cr/Au ohmic contact on p-type GaN.

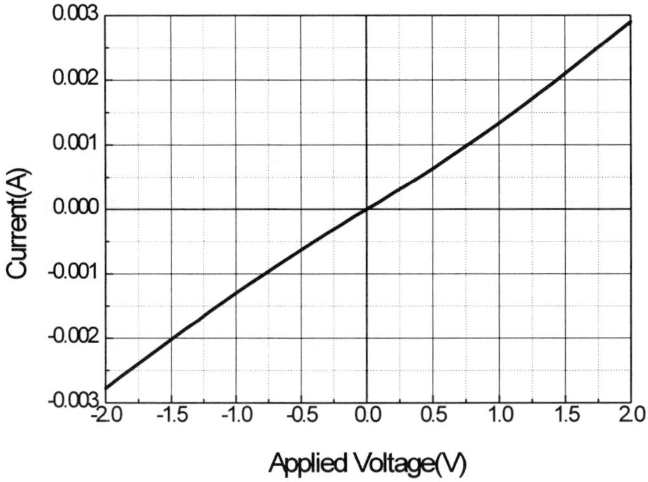

Fig. 1. I-V characteristic of as-fabricated Cr/Au metal contact on p^{+}-GaN with 5-μm spacing.

The effect of p-type doping concentration and contact material on the specific contact resistance is shown in Fig. 2. In the case of p^{+}-GaN, the measured sample shows good ohmic characteristic in the entire annealing temperature range, while the p^{-}-GaN shows ohmic behavior only after annealing at 500 °C or 700 °C . This figure indicates that, regardless of the contact metal, the specific contact resistance of the p^{+}-GaN is always lower than that of the p^{-}-GaN. It is believed that this is due to easy tunneling of electrons in the highly p-doped samples facilitated by the lowering of the contact barrier height.

The effect of annealing temperature on the specific contact resistance of the Cr/Au contact on p^{+}-GaN sample is shown in Fig. 3. Keep in mind that the Cr/Au metal contact exhibits ohmic characteristics without annealing. Fig. 3 shows that as the annealing temperature increases, the specific contact resistance gradually decreases. This decrease could be attributed to the enhanced diffusion of Cr atoms from the metal bilayer to the GaN epilayer at higher temperatures. One may explain this phenomenon by using the same analogy as in the case of a Ti/Al metal contact to n-type GaN, in which low ohmic resistance of a Ti/Al contact results from the reaction between Ti and N to form TiN. This causes surface depletion of N in the heavily doped GaN layer at the interface, hence enabling easy electron tunneling [8].

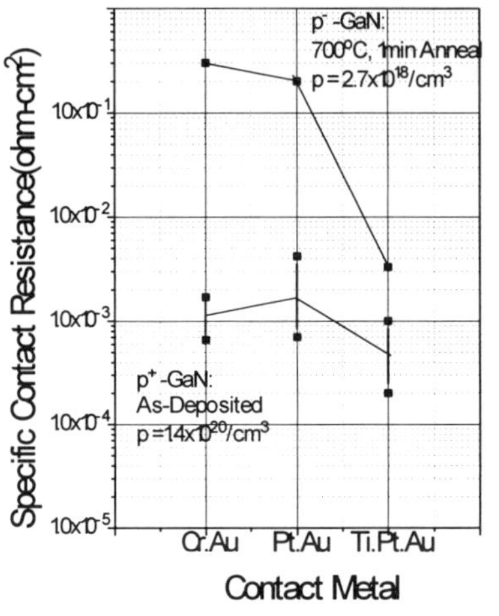

Fig. 2. Effects of doping concentration and metal contact on the specific contact resistance.

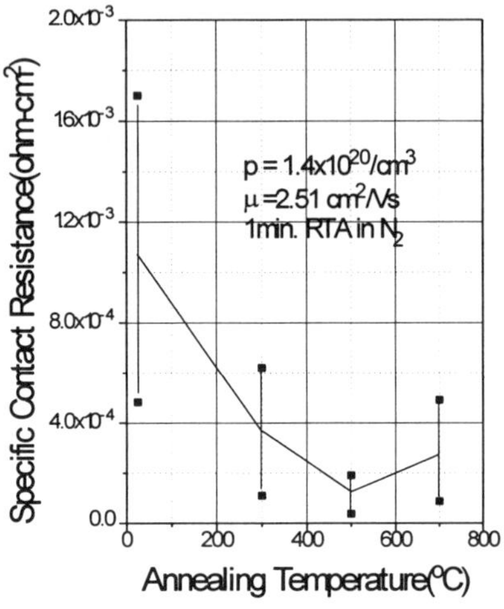

Fig. 3. Change of specific contact resistance as a function of annealing temperature in a Cr/Au metal contact on p^+-GaN.

To verify Cr-diffusion into the GaN epilayer, SAM analysis was performed. In Fig. 4, the results of the SAM analysis of a Cr/Au metal contact after RTA at 700 °C for 1 min are shown. These data clearly indicate that the Cr atoms were diffused into the GaN epilayer. Therefore, it is reasonable to conclude that the diffused Cr atoms formed a Cr-GaN compound at the metal/GaN interface. This result is further confirmed by the glancing angle x-ray diffraction equipped with the thin-film attachment option. The x-ray diffraction of $\theta - 2\theta$ scan showed strong peak intensities at 41.85, 64.68, and 77.55 degrees, which were indexed as a Cr_2GaN. Judging from the SAM and x-ray diffraction analyses, the Cr-GaN compound formed by the diffusion of Cr into the GaN layer may be the main reason why the Cr/Au metal contact shows lower specific contact resistance compared with other metal contacts. However, further study employing analytical transmission electron microscopy (TEM) or scanning transmission electron microscopy (STEM) will be necessary to fully identify the phase of the Cr-GaN compound formed at the interface.

An additional point to be noted from Fig. 3 is that as the annealing temperature is increased from 500 °C to 700 °C, the specific contact resistance increases twofold in magnitude. This means that only up to a certain temperature range, the doping concentration increases at the metal/semiconductor interface with increasing annealing temperature. At higher annealing temperatures, however, it is possible that the doping concentration decreases at the interface due to higher diffusion rate. As a result, dopant atoms are depleted at the interface, thus increasing the number of metal ions that cannot act as active dopants, and causing stress buildup in the crystal lattice acting as a scattering center for the carrier movement and resulting in higher contact resistance [12].

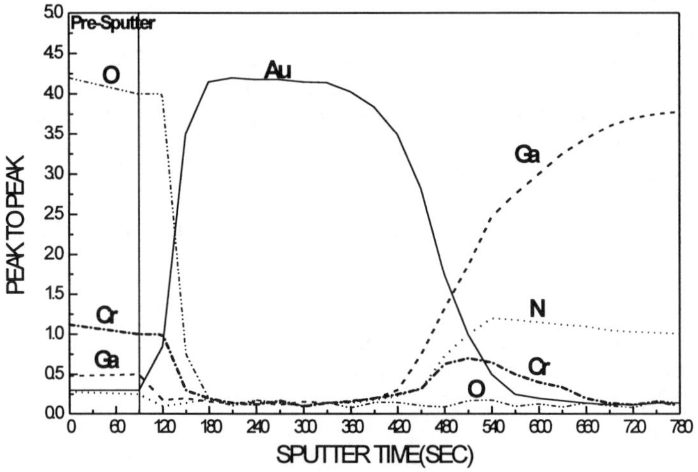

Fig. 4 Scanning Auger data of Cr/Au contact after 700 °C annealing for 1 min in N_2 ambient.

CONCLUSIONS

Ohmic contacts on p-type GaN have been investigated using Cr/Au metal contacts. A low-resistivity Cr/Au ohmic contact of 1.2×10^{-4} Ω–cm^2 was achieved on a p-type GaN after annealing at 500 °C for 1 min using RTA. To our knowledge, this is the lowest-resistivity ohmic contact achieved on a p-type GaN epilayer ever reported. The effect of doping concentration showed that the heavily doped p-type GaN exhibited lower specific contact resistance compared to a sample with lower p-type doping. As the annealing temperature increased, the specific contact resistance decreased. The reason for this lower specific contact resistance of the Cr/Au contact compared with other metal contacts may be due to Cr diffusion into the GaN layer resulting in formation of a Cr_2GaN compound at the interface. However, further study utilizing analytical characterization tools such as analytical transmission electron microscopy (TEM) or scanning transmission electron microscopy (STEM), is needed to identify a very thin layer of Cr-GaN phase formed at the metal/GaN interface.

REFERENCES

1. S. Nakamura, M. Senoh, N. Iwasa, and S. Nagahama, Jpn. J. Appl. Phys. **34**, L797 (1995).
2. S. Nakamura, M. Senoh, N. Iwasa, S. Nagahama, T. yamada, and T. Mukai, Jpn. J. Appl. Phys. **34**, L 1332 (1995).
3. S. Nakamura, M. Senoh, S. Nagahama, N. Iwasa, T. yamada, T. Matsushita, H. Kiyoku and Y. Sugimoto, Jpn. J. Appl. Phys. **35**, L 74 (1996).
4. B. Goldenberg, J.D. Zook, and R.J. Ulmer, Appl. Phys. Lett. **62**, 381 (1993).
5. S. Nakamura, T. Mukai, and M. Senoh, Jpn. J. Appl. Phys. **30**, L1998 (1991).
6. R.J. Molnar, R. Singh, and T.D. Moustakas, Appl. Phys. Lett. **66**, 268 (1995).
7. J.S. Foresi, and T.D. Moustakas, Appl. Phys. Lett. **62**, 2859 (1993).
8. M.E. Lin, Z. Ma, F.Y. Huang, Z.F. Fan, L.H. Allen, and H. Morkoc, Appl. Phys. Lett. **64**, 1003 (1994).
9. M.C. Yoo, T.I. Kim, K. Kim, K.H. Shim, and J.T. Verdeyen, Optical and Quantum Electronics **27**, 427 (1995).
10. K. Kim, M.C. Yoo, K.H. Shim, and J.T. Verdeyen, J. Vac. Sci. Technol. **B11**, 1942 (1993).
11. Z. Yang, L.K. Li, and W.I. Wang, Appl. Phys. Lett. **67**, 1686 (1995).
12. J.M. Myoung, C.K. Kim, K.H. Shim, O.V. Gluschenkov, K. Kim, and M.C. Yoo, Meter. Res. Soc. Symp. Proc. April 1996 (this volume).
13. R.E. Williams, *Gallium Arsenide Processing Techniques*, 2nd ed. (Artech House, Inc., Dedham, MA, 1985), p 236-238.

APPROACHES TO HIGH TEMPERATURE CONTACTS TO SILICON CARBIDE

J. M. DELUCCA and S. E. MOHNEY
Department of Materials Science and Engineering, The Pennsylvania State University, University Park, PA 16802, jmd14@psu.edu, mohney@psu.edu

ABSTRACT

Metallurgical reactions between contacts and SiC can alter the electrical characteristics of the contacts, either beneficially or detrimentally. Simultaneously, consumption of the underlying SiC epilayer takes place. During prolonged operation at elevated temperature, contacts that are not in thermodynamic equilibrium with SiC may continue to react with it. For this reason, interest in thermally stable carbide and silicide contacts to SiC has been growing. To select appropriate carbides or silicides for further study, however, knowledge of the transition metal-silicon-carbon (TM-Si-C) phase equilibria is required. A significant body of literature on the TM-Si-C systems exists and should therefore be examined in the context of electronic applications. In this paper, phase equilibria for representative TM-Si-C systems are presented, trends in these systems with respect to temperature and position of the metal in the periodic table are discussed, and attractive carbide and silicide contacts and processing schemes for thermally stable contacts are highlighted.

INTRODUCTION

Silicon carbide, with its wide band gap and excellent thermal stability, has been targeted for use in high temperature electronics. Many researchers have examined contacts to SiC [1]; however, fewer publications describe the effect of aging at elevated temperatures. During prolonged operation at elevated temperature, contacts that are not in thermodynamic equilibrium with SiC may continue to react with it. Consequently, a contact that initially displays acceptable electrical behavior may eventually change due to continued interfacial reaction. Of course, the rate at which the reaction occurs is governed by the system kinetics and is therefore strongly dependent upon temperature.

Silicide and carbide contacts have attracted attention for use in thermally stable contacts because the driving force for reaction may be reduced or eliminated [2–6]. However, good thermal stability does not ensure optimal electrical characteristics. In many cases, high doping levels and/or solid state reactions appear to be necessary for the formation of ohmic contacts with low resistivities, and it has been observed that ohmic contact formation often requires anneals near 1000°C [1]. Thus, a contact that undergoes interfacial reaction with SiC during annealing, but reacts completely so that the driving force for interfacial reaction is exhausted during the anneal, could also be attractive from the point of view of long-term thermal stability.

In either of the scenarios described above, knowledge of the relevant multi-component phase equilibria for the contact system of interest could prove useful. Therefore, isothermal phase diagrams for the transition metal (TM)-Si-C systems have been compiled from the literature. Particularly useful sources of information include references [7–11]. Since phase diagrams have been reported for most of the TM-Si-C systems only at temperatures well in excess of 1000°C, the relevance of the high temperature diagrams for lower temperature processing or operation is also considered.

DISCUSSION

The TM-Si-C systems show strong trends with respect to position in the period table. In fact, the transition metals can be grouped into those that form binary and/or ternary carbides and those that do not form any carbides. As shown in Fig. 1, the binary carbides form in the early transition metal systems. Ternary phases are reported for the TM-Si-C systems outlined in Fig. 1. Trends in the phases in thermodynamic equilibrium with SiC are also evident based on the position of the metal in the periodic table. For the purposes of discussion, the Ta-Si-C system is

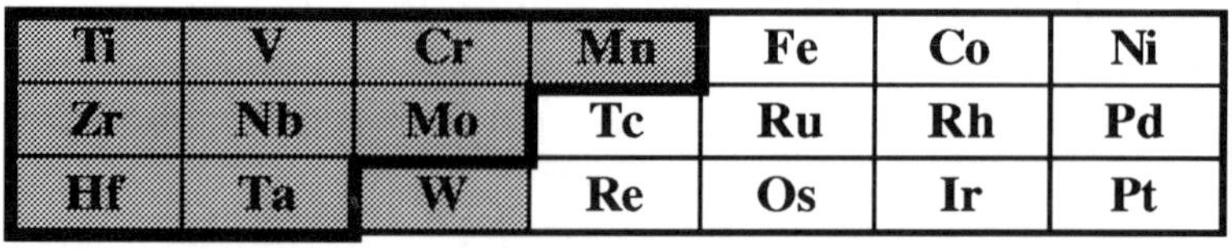

FIGURE 1. Trends in the TM-Si-C systems. The shaded area indicates systems with binary carbides. The bold outline encloses systems with one or more ternary phases. Silicides exist in all of these systems.

chosen as a representative of the early transition metals. The Ni-Si-C and Co-Si-C systems are chosen to represent the late transition metals.

The effect of temperature on the phase equilibria was also examined. Two approaches were taken to determine the applicability of the high temperature isotherms (typically reported between 1000° and 1727°C) for temperatures of interest either for operation or processing of high temperature devices (300–1100 °C). The first approach involved examination of the effect of temperature on the phase diagrams when isotherms for more than one temperature were available in the literature. The second approach involved the use of thermochemical data to estimate the lower temperature phase equilibria. A combination of both approaches was preferred but not always possible due to lack of diagrams or data. The results suggest that phases in thermodynamic equilibrium with SiC will often remain unchanged over a wide temperature range. This observation is not very surprising since the entropies of formation for solid phases from solid elements are generally quite small; so the Gibbs energies of formation, which govern the phase equilibria, are often not a strong function of temperature.

The early transition metal systems are characterized by the presence of silicides, carbides and ternary phases. The Ta-Si-C ternary diagram has been investigated experimentally at 1000 °C [12] and 1727 °C [9]. The diagram for 1000 °C is shown in Fig. 2. Despite the large difference in temperature, the phases in thermodynamic equilibrium with SiC are the same at both 1000 °C and 1727 °C: $TaSi_2$, TaC (at the C-rich end of its range of homogeneity), Si, and C. There is a slight discrepancy between the two diagrams in the Ta-rich portion; however, this portion of the diagram from the study at 1727 °C must be questioned since it does not agree with the most recent Si-Ta binary phase diagram [13]. An estimate of the phase equilibria at 25 °C using thermodynamic data [12–17] once again predicts the same tie-lines to SiC, despite the fact that the ternary phase $Ta_5Si_3C_{1-x}$ was omitted from the calculation due to a lack of thermodynamic data. Consistent with the Ta-Si-C phase diagrams and thermodynamic calculations, Edmond *et al.* [2] found that $TaSi_2$ contacts exhibited good thermal stability as ohmic contacts to n-SiC with electrical characteristics that remained constant for at least 8 hours at 400 °C.

The tie-lines between both C-rich TaC and $TaSi_2$ are characteristic of the early transition metal systems. In fact, for the group IVB and VB metals, all of the metal disilicides and monocarbides are in thermodynamic equilibrium with SiC, at least over a portion of their ranges of homogeneity. In some systems, other silicides and/or ternary phases also exhibit tie-lines to SiC. The metal carbides are particularly interesting since some of these phases are metallic with low bulk resistivities [18] and have the potential for epitaxial growth on SiC (which has been realized for TiC [4,5]).

In moving across the periodic table toward the late transition metals, the absolute value of the free energies of silicide formation pass through a minimum near group VIIB and increase again for the late transition metals [10]. On the other hand, the carbides become increasingly less stable. Thus, in the center of the table, fewer intermetallic phases exist than for the early transition metals, and those that do are less stable. Nevertheless, with the exception of Re at high temperature [10,19], none of the transition metals are found to be in thermodynamic equilibrium with SiC.

Moving again to the right in the periodic table, the stability of the silicides increases for the late transition metals. There are no equilibrium carbide phases under the conditions of interest. In

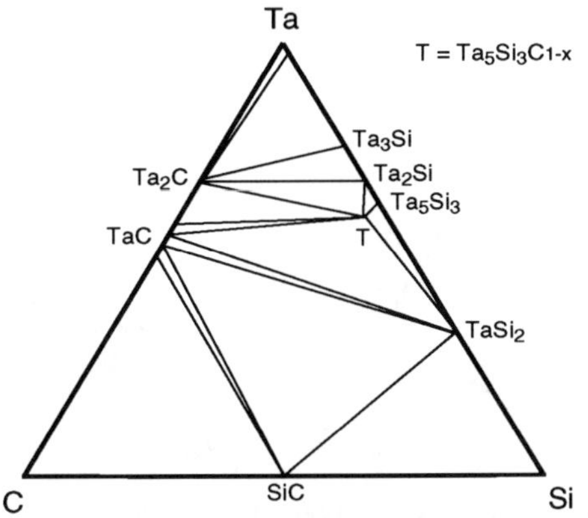

FIGURE 2. Experimental Ta-Si-C ternary isotherm at 1000°C [12].

this study, the Co-Si-C and Ni-Si-C systems were chosen to represent the late transition metals.

As shown in the experimentally determined phase diagrams reproduced in Fig. 3 [11, 20], three cobalt silicides occur at 1000°C and six nickel silicides occur at 900°C. Four of the nickel silicides (Ni_2Si, Ni_3Si_2, NiSi, and $NiSi_2$) and two of the cobalt silicides (CoSi and $CoSi_2$) are in thermodynamic equilibrium with SiC. It should be noted that in contrast to silicide contacts to Si, in which only the most Si-rich silicide is in thermodynamic equilibrium with the semiconductor, more than one silicide can be in thermodynamic equilibrium with SiC. These isotherms from the literature match those calculated by the authors for 25 °C using available thermodynamic data [14–16,21,22].The work of Crofton et al. [23] is also consistent with the Ni-Si-C phase diagrams. They found that a Ni film deposited on n-type SiC and annealed at 950°C formed Ni_2Si along with C. Thus, this contact was annealed until thermodynamic equilibrium had been achieved. The resulting ohmic contact was reported to be chemically and electrically stable at 650°C for many hundreds of hours.

The ternary diagrams of Fig. 3 are typical of those of the late transition metals, as determined at much higher temperatures, primarily by Searcy and Finnie [10]. Interestingly, Searcy also discusses [24] how many of the transition metal disilicides, such as $MoSi_2$, should prove particularly oxidation resistant, which might be useful for high temperature processing and service.

SUMMARY

Table I lists intermetallic phases known to be in thermodynamic equilibrium with SiC, as compiled from isothermal phase diagrams reported in the literature. Some properties relevant to contact formation are also included. Work functions are listed since it has been reported that when metals are deposited on clean SiC surfaces, there is a partial correlation between the work function and the barrier height [1]. Melting points and resistivities are also given.

Based on thermodynamic calculations and the data in Table I, it seems reasonable to conclude that the tie-lines to SiC in these systems will not change dramatically with temperature, though some exceptions can be found in Table I. Of course, changes in the binary phase diagrams or in the existence of ternary phases as the temperature is changed will also necessarily be reflected in the ternary diagrams. However, the abundant condensed phase equilibria data obtained at very high temperatures should at least be a very good starting point in estimating the phase diagrams at typical processing and operation temperatures for SiC devices, even though these temperatures are lower. Besides the direct deposition of phases in thermodynamic equilibrium with SiC, it should also be possible to form thermally stable contacts by reaction at very high temperatures where the kinetics permit complete and timely reaction. Then, at the somewhat less elevated temperatures

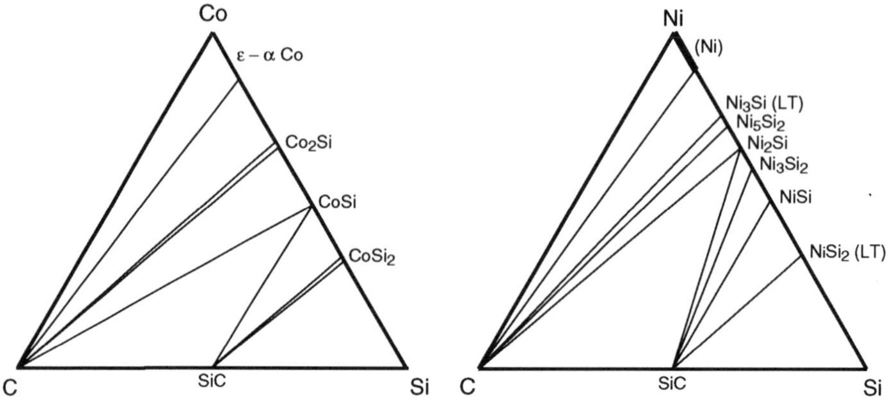

FIGURE 3. Experimental Co-Si-C (1000°C) and Ni-Si-C (900°C) ternary isotherms [12].

required for device operation, the same equilibrium will still hold for many contact systems. By selecting appropriate compositions and thicknesses of initially deposited contact layers, it should also be possible to limit consumption of the underlying SiC epilayer by choosing reactive contacts that will reach thermodynamic equilibrium before too much SiC is consumed. The achievement and confirmation of this equilibria, however, requires adequate annealing time and knowledge of the phases in thermodynamic equilibrium with SiC.

Table I. Intermetallic phases reported to be in equilibrium with SiC. The reported equilibrium temperatures and properties of the intermetallic are provided.

Phase	**Temperatures of Isotherms (°C)**	**Melting Point (°C)**	**Work Function (eV) ***	**Bulk Resistivity (μΩ-cm) ***
TiC	1100[27],1200[28] 1227[29], 1727[9]	3067[17]	3.53[25]	62[18]
$TiSi_2$	1100[27], 1200[28], 1227[29], 1727[9]	1500[15]	3.95–4.18[26]	13–123[18]
Ti_3SiC_2	1100[27], 1200[28] 1227 [29]	-	-	-
ZrC	1300[28,30], 1727[9]	3420[17]	3.20[25]	50[18]
ZrSi	1300[28,30]	2095[15]	-	49[26]
$ZrSi_2$	1300[28,30], 1727[9]	1620[15]	3.8[25], 3.95–4.57[26]	35–161[18]
HfC	1300[28]	3950[17]	3.47[25]	39[18]
$HfSi_2$	1300[28]	2142[15]	4.51[26]	45–70[18]
VC	1000[12]	2700[17]	3.85[25]	78[18]
VSi_2	1000[12]	1680[15]	3.2[26]	9.5–66.5[18]
V_5Si_3C	1000[12]	-	-	-
NbC	1300[28], 1727[9]	3600[17]	4.02[25]	43[18]
$NbSi_2$	1300[28], 1727[9]	1940[15]	4.19[25], 4.18[26]	6.3–100[26]
TaC	1000[12], 1727[9]	4000[17]	3.93[25]	22[18]
$TaSi_2$	1000[12], 1727[9]	2040[15]	4.31[25], 4.71[26]	8.5–80[26]
Ta_5Si_3C	1000[12]	-	-	-

Cr_3C_2	1400[31], 1600[32]	1830[17]	-	-
CrSi	1400[31], 1600[32]	1475[15]	3.47[25]	92–143[26]
$CrSi_2$	1400[31], 1600[32]	1490[15]	3.55–3.74[26]	7–6670[18]
Cr_5Si_3C	1400[31], 1600[32]	-	-	-
Mo_5Si_3	1200[33]	2180[15]	-	45.9[26]
$MoSi_2$	1200[33], 1400[34], 1600[35], 1727[9]	2020[15]	3.87[25], 4.73–6.0[26]	21.5–100[18]
$Mo_{24}Si_{15}C_3$ †	1200[33],1400[34], 1600[35],1727[9]	-	-	-
WC	1727[9]	2747[17]	3.60[25]	20[18]
W_5Si_3	1727[9]	2320[15]	-	-
WSi_2	1727[9]	2160[15]	3.9[25],4.62–6[26]	12.5–70[18]
MnSi	1000[11]	1275[15]	-	200[26]
$MnSi_{1.75}$	1000[11]	1152[15]	-	1073–4550[26]
Mn_5Si_3C	1000[11]	-	-	-
Mn_8Si_2C	1000[11]	-	-	-
Re	1600[10]	3180[18]	4.74–5.1[18]	19[18]
Re_2Si	1600[10]	1810[15]	-	-
ReSi	1600[10]	1880[15]	-	736[26]
$ReSi_2$	1000[11], 1600[10]	1980[15]	3.94[26]	7000[26]
Fe_5Si_3	970[11]	1060[8]	-	170–199[26]
FeSi	970[11]	1410[15]	-	240–287[26]
$FeSi_2$	970[11]	1220[15]	-	$350–10^6$[18]
RuSi	1340[10]	1730[15]	-	-
Ru_2Si_3	1340[10]	1710[15]	-	-
$OsSi_2$	1350[10]	1640[15]	-	-
CoSi	1000[11]	1460[15]	-	30[18],86–168[26]
$CoSi_2$	1000[11]	1326[15]	-	18–68[18]
RhSi	1170[10]	1452[15]	-	-
IrSi	1340[10]	1408[37]	-	-
Ir_2Si_3	1340[10]	-	-	-
$IrSi_2$	1340[10]	-	-	-
$IrSi_3$	1340[10]	1260[37]	-	460–580[26]
Ni_2Si	850[36], 900[20]	1318[15]	-	20–24[26]
Ni_3Si_2	850[36], 900[20]	845[15]	-	79[26]
NiSi	850[36], 900[20]	992[15]	-	29[18], 14–50[26]
$NiSi_2$	850[36], 900[20]	993[15]	-	34–125[26]

* Bulk resistivities and work functions are those reported near room temperature.

† Crystallographic data reported only for $Mo_{24}Si_{15}C_3$ ternary phase [7]. It is assumed that reports of this approximate composition in other studies are actually reports of the same ternary phase.

ACKNOWLEDGMENTS

This work was supported by the Center for Commercial Development of Space Power and Advanced Electronics, located at Auburn University, with funds from NASA Grant NAGW-1192-CCDS-AD and the center's industrial partners. The authors also wish to thank Xin Lin for assistance in preparing the table and figures.

REFERENCES

1. L.M. Porter and R.F. Davis, *Mater. Sci. Eng.* **B34**, 83 (1995) and references therein.
2. J.A. Edmond, J. Ryu, J.T. Glass, and R.F. Davis, *J. Electrochem. Soc.* **135**, 359 (1988).
3. L.G. Matus, J.A. Powell, and J.B. Petit, *NASA Tech. Mem.* 104398, June 1991.
4. J.D. Parsons, G.B. Kruaval, and A.K. Chaddha, *Appl. Phys. Lett.* **65**, 2075 (1994).
5. A.K. Chaddha, J.D. Parsons, and G.B. Kruaval, *Appl. Phys. Lett.* **66**, 760 (1995).
6. M.M. Anikin, M.G. Rastegaeva, A.L Syrkin, and I.V. Chuiko, in *Amorphous and Crystalline Silicon Carbide III*, ed. G.L. Harris, M.G. Spencer, and C.Y. Yang, Springer, Berlin, 1992, p. 183.
7. *Handbook of Ternary Alloy Phase Diagrams*, P. Villars, A Prince, and H. Okamoto, ASM International, Materials Park, OH, 1995.
8. *Journal of Phase Equilibria,* ASM International, Materials Park, OH (refer to end-of-year indices).
9. L. Brewer and O. Krikorian, *J. Electrochem. Soc.*, New York **103**, 38 (1956).
10. A.W. Searcy and L.N. Finnie, *J. Am. Ceram. Soc.* **45**, 268 (1962).
11. J.C. Schuster, *Int. J. Refract. Met. Hard Mater.* **12**, 173 (1993-1994).
12. J.C. Schuster, *J. Chim. Phys.* **90**, 373 (1993).
13. M.E. Schlesinger, *J. Phase Equilibria* **15**, 90 (1994).
14. *Materials Thermochemistry*, O. Kubaschewski, C.B. Alcock, P.J. Spencer, Pergamon Press, New York, 1993, pp. 258-323.
15. *Properties of Metal Silicides*, ed. by K. Maex and M. van Rossum, Inspec, London, 1995, pp. 31-57.
16. *Cohesion in Metals: Transition Metal Alloys*, F.R de Boer, R. Boom, W.C.M. Mattens, A.R. Miedema, and A.K. Niessen, North-Holland Physics Publishing, The Netherlands, 1989.
17. *Refractory Materials* (Vol. 2), E.K. Storms, Academic Press, New York, 1967.
18. *Silicides for VLSI Applications*, S.P. Murarka, Academic Press, Orlando, 1983, pp. 29-78.
19. J.S. Chen, E. Kolawa, M. A. Nicolet, L. Baud, C. Jaussaud, R. Madar, and C. Bernard, *J. Appl. Phys.* **75**, 897 (1994).
20. Y.M. Basin, V.N. Kuznetsov, V.T. Markov, and L.S. Guzei, *Russ. Met.* **4**, 197 (1988).
21. K. Ishida, T. Nishizawa, and M.E. Schlesinger, *J. Phase Equilibria* **12**, 578 (1991).
22. P. Nash and A. Nash, *Bull. Alloy Phase Diagrams* **8**, 6 (1987).
23. J. Crofton, P.G. McMullin, J.R. Williams, and M.J. Bozack, *J. Appl. Phys.* **77**, 1317 (1995).
24. A.W. Searcy, *J. Am. Ceram. Soc.* **40**, 431 (1957).
25. *Handbook of Thermionic Properties,* V.S. Fomenko, ed. G.V. Samsonov., Plenum Press Data Division, New York, 1966, pp. 93-105.
26. *VLSI Electronics Microstructure Science* (Vol. 6), ed. N. G. Einspruch, Academic Press, New York, 1983, pp. 415-23.
27. W.J.J. Wakelkamp, F. J.J. van Loo, and R. Metselaar, *J. Europ. Ceram. Soc.* **8**, 135 (1991).
28. E. Rudy, *Compendium of Phase Diagram Data*, Ternary Phase Equilibria in Transition Metal-Boron-Carbon-Silicon Systems, Part V, AFML-TR-65-2, AD 689843, (1969).
29. M. Touanen, F. Teyssandier, and M. Ducarroir, *J. Mater. Sci. Let.* **8**, 98 (1989).
30. C.E. Brukl, Technical Report, AD 489752, *US Clearinghouse Fed. Sci. Tech. Inform.*, AD, 1966, pp. 37-8.
31. P.W. Pellegrini, B.C. Giessen, and J.M. Feldman, *J. Electrochem. Soc.* **119**, 535 (1972).
32. M.I. Gasik and P.A. Em, Izvestiya Vysshikh Uchebnykh Zavedenii, *Chernaya Metallurgiya* **19**, 93 (1976).
33. F. J.J. van Loo, F.M. Smet, G.D. Rieck, G. Verspui, *High Temp. High Pres.* **14**, 25 (1982).
34. J.M. Guiot, *Silicates Industriels* **31**, 363 (1966).
35. H. Nowotny, E. Parthé, R. Kieffer, and F. Benesovsky, Monatsh. Chem. **85**, 255 (1954).
36. R.C.J. Schiepers, F.J.J. van Loo, and G. de With, *J. Am. Ceram. Soc.* **71**, C284 (1988).
37. H. Okamoto, *J. Phase Equilibria* **16**, 473 (1995).

CHARACTERIZATION OF ZIRCONIUM - DIAMOND INTERFACES

P.K. BAUMANN, S.P. BOZEMAN, B.L. WARD, R.J. NEMANICH
Department of Physics, North Carolina State University, Raleigh, NC 27695-8202

ABSTRACT

Thin Zr films were deposited on natural single crystal diamond (100) substrates by e-beam evaporation in ultra-high vacuum (UHV). Before metal deposition the surfaces were cleaned by UHV anneals at either 500°C or 1150°C. Following either one of these treatments a positive electron affinity was determined by means of UV photoemission spectroscopy (UPS). Depositing 2Å of Zr induced a NEA on both surfaces. Field emission current - voltage measurements resulted in a threshold field (for a current of 0.1 μA) of 79 V/μm for positive electron affinity diamond surfaces and values as low as 20 V/μm for Zr on diamond.

INTRODUCTION

Negative electron affinity (NEA) surfaces based on wide bandgap semiconductors like diamond could enable the development of cold cathodes. So far n-type doping of diamond remains a major challenge. The electron affinity of a semiconductor corresponds to the energy difference between the vacuum level and the conduction band minimum. Usually the conduction band minimum lies below the vacuum level resulting in a positive electron affinity surface. For wide bandgap semiconductors like diamond the conduction band minimum is likely to be close to the vacuum level. For a NEA surface the electrons from the conduction band minimum have sufficient energy to overcome the surface potential and can be emitted into vacuum. Various surface treatments can induce or inhibit a NEA on diamond surfaces [1-6]. Such treatments include plasma cleaning as well as annealing in ultra high vacuum (UHV). Subsequent to a wet chemical etch the diamond surfaces are terminated by oxygen. This chemisorbed oxygen layer forms a surface dipole that results in an increase in the surface workfunction. It is found that such a surface exhibits a positive electron affinity. For the diamond (100) surface an anneal to 900°C - 1050°C or a H-plasma clean results in a NEA and a 2x1 reconstructed, oxygen free surface [3, 5, 6]. The different threshold temperatures are related to different wet chemical pre-treatments [3]. However for all these treatments a positive electron affinity and a 2x1 reconstruction are observed following a 1150°C anneal [6]. The diamond (100) surface has been proposed to be terminated by a monohydride subsequent to a 900°C - 1050°C anneal or a H-plasma exposure. Ab initio calculations for the 2x1 reconstructed surface predict a NEA for a monohydride terminated surface and a positive electron affinity for an adsorbate free surface [5].

Depositing a few Å of a metal like Ti, Ni, Co or Cu on diamond can induce a NEA [7, 8, 9, 10, 3]. The presence of a NEA or positive electron affinity has been correlated with different structures of the metal - diamond interface. Metal - diamond structures with a NEA have been found to exhibit lower Schottky barrier heights than those with a positive electron affinity. Schottky barrier height measurements have been reported for metals deposited on (100), (111) and polycrystalline diamond surfaces [3, 7-20]. No significant dependence of the Schottky barrier height on the metal work function has been found. Lower Schottky barrier heights have been reported for metal films deposited on adsorbate free surfaces than for surfaces terminated by species such as hydrogen or oxygen.

Photoemission spectroscopy is found to be a very sensitive method to distinguish between a NEA or positive electron affinity. Electrons get photoexcited from the valence band into states in the conduction band and can quasi thermalize to the conduction band minimum. Secondary electrons from the conduction band minimum appear as a sharp peak at the low kinetic energy end of photoemission spectra for NEA surfaces [14, 21]. While photoemission spectroscopy determines the emission properties of the surface itself, field emission data reflect a combination of carrier injection, transport and emission processes. In this study diamond (100) surfaces have been cleaned by anneals to 500°C or 1050°C. Thin Zr films were deposited on these diamond substrates. The surface properties were analyzed before and after Zr deposition.

Mat. Res. Soc. Symp. Proc. Vol. 423 © 1996 Materials Research Society

EXPERIMENTAL DETAILS

Several natural p-type single crystal semiconducting diamond (100) substrates were used in this study. To remove nondiamond carbon and metal contaminants an electrochemical etch has been employed. Details of this technique have been described earlier [22, 3]. The samples were then loaded into a UHV system consisting of several interconnected chambers featuring capabilities for annealing, metal deposition, ARUPS and AES. Two different in vacuo cleaning processes were used to study the effect of surface treatment on the characteristics of the zirconium - diamond interface. One procedure involved annealing the wafers to 500°C for 10 minutes. And the other involved a 1150°C anneal for 10 minutes. The base pressure in the annealing chamber was ~ 1×10^{-10} Torr and rose to ~ 8×10^{-10} Torr and ~7×10^{-9} Torr during the anneals, respectively. Subsequent to the anneal a Zr e-beam evaporator was employed to deposit 2 Å thick films. A quartz crystal monitor was used to measure the thickness. During deposition the pressure was ~ 2×10^{-9} Torr. Following each annealing - and deposition step, UPS and AES were used to characterize the samples.

The presence of a Zr film was confirmed by means of AES. AFM images of the diamond wafers clearly display linear groves with a depth of ~ 20 Å. This surface structure is due to polishing the samples with diamond grit. No island structures were observed in AFM measurements after 2 Å of deposition, indicating a uniform 2D layer.

A discharge lamp was employed to excite HeI (21.21 eV) radiation to facilitate the photoemission and a 50 mm hemispherical analyzer with an energy resolution of 0.15 eV was used to detect the emitted electrons. To overcome the workfunction of the analyzer a bias of 2 V was applied to the sample. It was therefore possible to detect the low energy electrons emitted from the NEA surface as a sharp peak at the low energy end of UPS spectra. The position of this feature corresponds to the energy position of the conduction band minimum, E_C. Electrons emitted from E_C appear at $E_V + E_G$ in the spectra, where E_V is the energy of the valence band maximum and E_G the bandgap energy. Furthermore, electrons from E_V get photoexcited to an energy level at $E_V + h\nu$ in the conduction band and are obviously detected at $E_V + h\nu$ in UPS spectra. This corresponds to the high kinetic energy end of the spectra. Therefore the spectral width for a NEA surface is $h\nu - E_G$. Using the value of $h\nu = 21.21$ eV for HeI radiation and $E_G = 5.47$ eV for the bandgap of diamond, a spectral width of ~ 15.7 eV is obtained. For a surface with a positive electron affinity the low energy cutoff is determined by the vacuum level. This results in a smaller value for the spectral width as compared to the case of a NEA.

Photoemission spectra that exhibit features from both the semiconductor and the metal can be used to determine the Schottky barrier height Φ_B (Fig. 1).

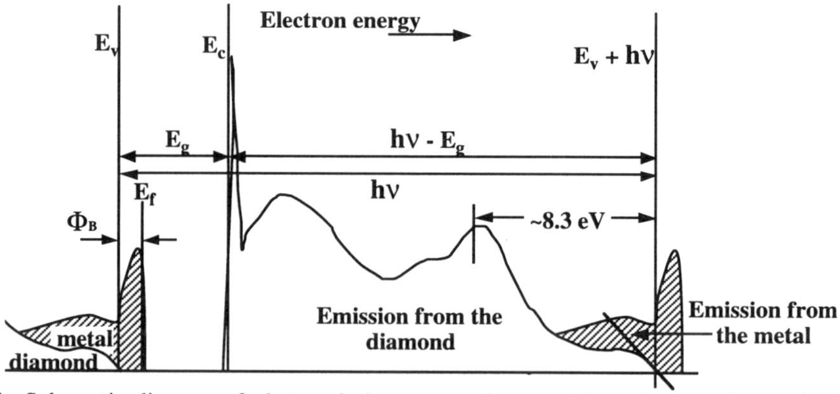

Fig. 1 Schematic diagram of photoemission spectra for metal deposited on diamond. The Schottky barrier height Φ_B is determined from the difference between the position of the valence band edge of diamond E_V and the metal Fermi level E_F.

This method is only suitable for metal films with thicknesses equal to or less than the electron mean free path (≤ 5 Å). For p-type semiconductors like diamond, Φ_B corresponds to the difference between the position of the valence band edge, E_V, of the semiconductor and the Fermi level of the metal, E_F. But the relatively weak onset of emission at E_V may, however, be obscured by the metal Fermi level even for metal layers thinner than the mean free path. As an independent method E_V can be referenced to some strong features in the diamond spectrum before metal deposition. In our case a feature positioned at 8.3 eV below E_V has been chosen. For a NEA the position of the low energy turnon (which corresponds to E_C) can also be used as a reference to find E_V (which is the high energy turnon of the spectrum). The difference between E_C and E_V has to be $h\nu - E_G$.

The base pressure in the field emission chamber was ~ 2 x 10^{-7} Torr. To determine the current - voltage characteristics a bias of 0 to 1100V was applied between the sample and a 2 mm diameter platinum anode with a rounded tip. The distances were varied between 2 and 20 μm.

RESULTS AND DISCUSSION

AES spectra of the as-loaded samples clearly exhibited oxygen peaks. Following a 500°C anneal the oxygen feature was only slightly reduced. Upon heating the samples to 1150°C oxygen could no longer be detected by means of AES. By employing UPS a positive electron affinity of $\chi \cong 1.4$ eV and of $\chi \cong 0.7$ eV were observed for the crystals annealed to 500°C and 1150°C, respectively. These values correspond to previous results [5, 9] Oxygen chemisorbed to diamond is expected to induce a stronger surface dipole and therefore cause an increase in the workfunction compared to a clean surface. Our results are consistent with this. Depositing 2Å of Zr onto oxygen terminated diamond (100) samples resulted in a NEA (Fig. 2a). A Schottky barrier height of $\Phi_B \cong 0.9$ eV was determined by UPS. Subsequent to deposition of 2Å of Zr on clean diamond (100) surfaces a NEA was observed and the spectrum shifted ~ 0.3 eV toward lower energies (Fig. 2b). A lower Schottky barrier height of $\Phi_B \cong 0.7$ eV was measured. This structure exhibited emission even below the conduction band minimum E_C. A summary of these results is shown in Table 1. The electron affinity of a p-type semiconductor following metal deposition is given by equation (1) [21].

$$\chi = (\Phi_M + \Phi_B) - E_G \tag{1}$$

Using the bandgap of diamond $E_G = 5.47$ eV, the workfunction of Zr $\Phi_M = 4.05$ eV and the measured Schottky barrier heights one can calculate electron affinities of $\chi \cong -0.52$ eV for the oxygen terminated surface and $\chi \cong -0.72$ eV for the clean surface. These results are in agreement with the detection of metal induced NEA's by means of UPS. The emission detected below E_C for the clean surface is consistent with the calculated value of $\chi \cong -0.72$ eV. The fact that no emission below E_C was observed for Zr on oxygen terminated diamond may be due to a different interface structure.

This simple workfunction model has been used successfully to explain NEA or positive electron affinity effects for systems like Ti or Ni layers on diamond (111) surfaces [7, 8] and Co or Cu films on diamond (100) surfaces [9, 10]. It has been found that Ni deposited on Ar plasma cleaned diamond (111) substrates induced a NEA. An Ar plasma or a 950°C anneal result in a (111) surface free of adsorbates [4]. In comparison, a positive electron affinity and a larger Schottky barrier were observed for thin Ni films on (111) surfaces terminated by hydrogen. Theoretical results of the Ni - diamond interface have been reported by Erwin and Pickett [24-27] and Pickett, Pederson and Erwin [28]. The most stable configuration for clean (111) and (100) surfaces resulted in a Schottky barrier height of less than 0.1 eV. Lambrecht calculated the Schottky barrier height for copper on diamond (111) surfaces. A value of less than 0.1 eV for a clean surface and greater than 1.0 eV for a surface terminated with hydrogen were found [29]. These results indicate that the Schottky barrier height depends on the interface termination. Lower values for the Schottky barrier height and a greater likelihood of inducing a NEA are expected for metals deposited on clean surfaces than on non adsorbate free surfaces. The Schottky barrier heights reported in this study for Zr on diamond are in agreement with this. It is significant that a metal induced NEA was observed for deposition of Zr on both clean and oxygen terminated surfaces. Previously a positive electron affinity was found for Ti on a oxygen terminated diamond (100) surface [7]. Ti has a workfunction only 0.3 eV higher than that of Zr. So far only metal diamond interfaces with low workfunction metals like Cs have been reported to exhibit a NEA for non adsorbate fee diamond surfaces [30].

Field emission measurements were performed on diamond samples and on Zr films deposited on oxygen terminated and clean diamond surfaces. The emission threshold field was defined for a current of 0.1 μA. Thresholds between 20 and 79 V/μm were measured. For a summary see Table 1. Values of the same magnitude have been reported previously for diamond samples [31, 32]. As an example I-V curves for the diamond surface are shown in Fig. 3. The results from the field emission measurements have been compared to the Fowler - Nordheim equation [33]:

$$I = k\left(\frac{V}{d}\right)^2 \exp\left(\frac{-6530d\varphi^{3/2}}{V}\right) \tag{2}$$

I is the current in amps, V is the bias in volts, d is the distance between the sample and the anode in microns, k is a constant and φ is the effective barrier height in eV. The field enhancement factor has been neglected since the surfaces have been found to be essentially flat by means of AFM. By fitting the field emission data to equation (2) the effective barrier heights φ were obtained as shown in Table 1. Depositing Zr onto both oxygen terminated and clean diamond (100) surfaces improves the emission properties. The best results were obtained for Zr on the clean surface. Both UPS and field emission measurements show these trends consistently.

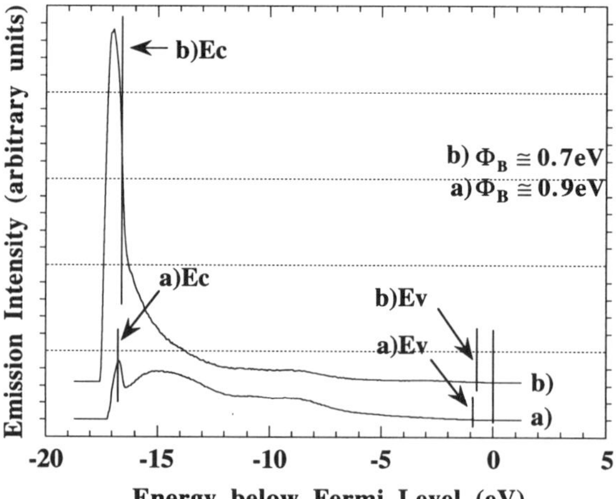

Fig. 2 UV photoemission spectra of 2Å of Zr deposited on a diamond (100) surface annealed to a) 500°C b) 1150°C. Metal induced NEA's are observed upon deposition of Zr for both a) and b). For b) emission below E_C is detected.

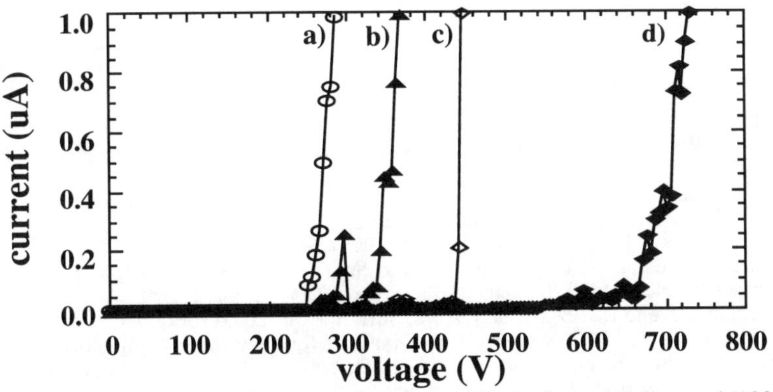

Fig. 3 Field emission current - voltage curves for a type IIb single crystal diamond (100) sample. Distances between the sample and the anode: a) 3.3 μm, b) 4.3 μm, c) 5.6 μm, d) 8.5 μm.

Sample	UPS		Field Emission Threshold (V/μm)	Barrier Height (eV)
C(100)	after 500°C anneal PEA, $\chi \cong 1.4$ eV	after 1150°C anneal PEA, $\chi \cong 0.7$ eV	79 ± 7	0.23 ± 0.01
Zr/C(100) oxygen	NEA, $\chi < 0$, $\Phi_B \cong 0.9$ eV		49 ± 4	0.20 ± 0.01
Zr/C(100) clean	NEA, $\chi < 0$, $\Phi_B \cong 0.7$ eV		20 ± 3	0.09 ± 0.01

Table 1 Results of electron emission measurements. PEA: positive electron affinity, NEA: negative electron affinity. The averages and standard deviations of the field emission measurements at different distances are shown as the field emission threshold and the barrier height. The threshold current is 0.1 μA.

CONCLUSIONS

Thin films of Zr were grown on diamond (100) surfaces annealed to 500°C or 1150°C. Using UPS the diamond samples were found to exhibit a positive electron affinity after either one of the heat treatments. Metal induced NEA's were observed for Zr deposition on either one of these surfaces. For Zr on clean surfaces emission even below E_C was detected. Depositing Zr on diamond reduced both the field emission threshold and the effective barrier height in a manner consistent with the UPS results.

ACKNOWLEDGMENTS

This work was supported by the Office of Naval Research (Contract No. N00014-92-J-1477).

REFERENCES

1. F.J. Himpsel, D.E. Eastman, P. Heimann and J.F. van der Veen, Phys Rev. B **24**, 7270 (1981).

2. B.B. Pate, M.H. Hecht, C. Binns, I. Lindau and W.E. Spicer, J. Vac. Sci. Technol. **21**, 364 (1982).
3. P.K. Baumann, T.P. Humphreys and R.J. Nemanich, in Diamond, SiC and Nitride Wide Bandgap Semiconductors, edited by C.H. Carter, G. Gildenblat, S. Nakamura, R.J. Nemanich, (Mater. Res. Soc. Proc. **339**, Pittsburgh, PA, 1994) 69-74.
4. J. van der Weide and R.J. Nemanich, Appl. Phys. Lett. **62** (1993) 1878.
5. J. van der Weide, Z. Zhang, P.K. Baumann, M.G. Wensell, J. Bernholc and R.J. Nemanich, Phys. Rev. B **50** (1994) 5803.
6. P.K. Baumann and R.J. Nemanich, Proc. of the 5th European Conference on Diamond, Diamond-like and Related Materials, edited by P.K. Bachmann, I.M. Buckley-Golder, J.T. Glass, M. Kamo: J. Diamond Rel. Mat., **4** (1995) 802.
7. J. van der Weide and R.J. Nemanich, J. Vac. Sci. Technol. B **10** (1992) 1940.
8. J. van der Weide and R.J. Nemanich, Phys. Rev. B, **49** (1994) 13629.
9. P.K. Baumann and R.J. Nemanich, Appl. Surf. Sci., accepted for publication.
10. P.K. Baumann and R.J. Nemanich, in Diamond for Electronic Applications, edited by C. Beetz, A. Collins, K. Das, D. Dreifus, T. Humphreys, P. Pehrsson, (Mater. Res. Symp. Soc. Proc. MRS, Pittsburgh, PA 1996), accepted for publication.
11. P.K. Baumann, T.P. Humphreys, R.J. Nemanich, K. Ishibashi, N.R. Parikh, L.M. Porter and R.F. Davis, Proc. of the 4th European Conference on Diamond, Diamond-like and Related Materials, edited by P.K. Bachmann, I.M. Buckley-Golder, J.T. Glass, M. Kamo: J. Diamond Rel. Mat. **3** (1994) 883.
12. C.A. Mead and T.C. McGill, Phys. Lett. **58A** (1976) 149.
13. F.J. Himpsel, D.E. Eastman and J.F. van der Veen, J. Vac. Sci. Technol. **17** (1980) 1085.
14. F.J. Himpsel, P. Heimann and D.E. Eastman, Sol. State Commun. **36,** 631 (1980).
15. J.W. Glesener, A.A. Morrish and K.A. Snail, J. Appl. Phys. **70** (1991) 5144.
16. M.W. Geis, D.D. Rathman, D.J. Ehrlich, R.A. Murphy and W.T. Lindley, IEEE Electron Device Lett. **8** (1987) 341.
17. H. Shiomi, H. Nakahata, T. Imai, Y. Nishibayashi and N. Fujimori, Jpn. J. Appl. Phys. **28** (1989) 758.
18. T. Tachibachi, B.E. Williams and J.T. Glass, Phys. Rev. B **45** (1992) 11975.
19. M.C. Hicks, C.R. Wronski, S.A. Grot, G.S. Gildenblat, A.R. Badzian, T. Badzian and R. Messier, J. Appl. Phys. **65** (1989) 2139.
20. S.A. Grot, S. Lee, G.S. Gildenblat, C.W. Hatfield, C.R. Wronski, A.R. Badzian, T. Badzian and R. Messier, J. Mater. Res. **5** (1990) 2497.
21. B.B. Pate, W.E. Spicer, T. Ohta and I. Lindau, J. Vac. Sci. Technol. **17** (1980) 1087.
22. M. Marchywka, P.E. Pehrsson, S.C. Binari and D. Moses, J. Electrochem. Soc., **140**, No. 2 (1993) L19.
23. E.H. Rhoderick and R.H. Williams, Metal-Semiconductor Contacts Clarendon, Oxford, (1988).
24. S.C. Erwin and W.E. Pickett, Surf. Coat. Technol. **47** (1991) 487.
25. S.C. Erwin and W.E. Pickett, Solid State Commun. **81** (1992) 891.
26. W.E. Pickett and S.C. Erwin, Phys. Rev. B **41** (1990) 9756.
27. W.E. Pickett and S.C. Erwin, Superlatt. Microsruct. **7** (1990) 335.
28. W.E. Pickett, M.R. Pederson and S.C. Erwin, Mater. Sci. Eng. B **14** (1992) 87.
29. W.R.L. Lambrecht, Physica B **185** (1993) 512.
30. M.W. Geis, J.C. Twichell, J. Macaulay, K. Okano, Appl. Phys. Lett. 67 (1995) 1.
31. W. Zhu, G.P Kockanski, S. Jin and L. Siebels, J. of Appl. Phys., in press
32. S.P. Bozeman, P.K. Baumann, B.L. Ward, M.J. Powers, J.J Cuomo, R.J. Nemanich and D.L Dreifus, Proc. of the 6th European Conference on Diamond, Diamond-like and Related Materials, edited by P.K. Bachmann, I.M. Buckley-Golder, J.T. Glass, M. Kamo: J. Diamond Rel. Mat. (1996), accepted for publication.
33. R. Gomer, Field Emission and Field Ionization, Cambridge, MA, (1961).

OHMIC CONTACTS TO *n*-TYPE 6H-SiC WITHOUT POST-ANNEALING

TOKUYUKI TERAJI*, SHIRO HARA, HIDEYO OKUSHI, and KOJI KAJIMURA*
Electrotechnical Laboratory, 1-1-4 Umezono, Tsukuba, Ibaraki 305, Japan

ABSTRACT

We formed titanium Ohmic contacts to *n*-type 6H-SiC epitaxial layer by reducing the Schottky barrier heights. The barrier heights were reduced enough to form the Ohmic contacts by releasing the Fermi level from pinning through making atomically-flat surfaces. The current transport by thermionic emission was dominant at the Ti/SiC interface. Since the Ti contacts were formed without post-annealing, surfaces of the Ti electrodes were flat and homogeneous maintaining as-deposited structures. Contact resistivity was $(6\pm1)\times10^{-3}$ Ω-cm^2, which is comparable to that of the annealed Ni contact formed on the SiC epitaxial layer with the same carrier concentration.

INTRODUCTION

An Ohmic contact is usually formed by heavy impurity doping to a semiconductor by which a tunneling current is increased due to a thinning of a depletion layer. However, it is difficult to apply this conventional method to SiC because high impurity doping to SiC is difficult. Edmond *et al.*[1] reported that high temperature over 1800 °C is required for making of a heavily doped layer with carrier concentration of over 10^{18} cm^{-3} by thermal diffusion. Another impurity doping method by ion implantation also has a problem of a low electrical dopant activity of about 5 % as estimated by Ruff *et al.*[2]

Ni electrodes annealed at high temperatures are widely used as Ohmic electrodes to *n*-type 6H-SiC crystal instead of the heavily doped electrodes. Even in the Ni electrodes, there are some significant problems. The high-temperature annealing typically at around 1000 °C restricts device fabrication processes and makes morphology of electrodes rougher. Therefore, fabrication of Ohmic contacts without heavy impurity doping and post-annealing at high temperatures is one of the key techniques to develop electronic devices using SiC.

In this paper, we propose a new technique for Ohmic contact formation without heavy impurity doping and post-annealing. We also demonstrate an Ohmic contact formation to *n*-type 6H-SiC epitaxial layer using this new technique.

Mat. Res. Soc. Symp. Proc. Vol. 423

NEW TECHNIQUE FOR OHMIC FORMATION

According to the Schottky model[3], an Ohmic contact to an n-type semiconductor is in principle formed when the work function of a metal ϕ_m is smaller than the electron affinity of the semiconductor χ_s, where the Schottky barrier height ϕ_{bn} is negative. However, this method is of hardly any use[4] because high ϕ_{bn} is present at a practical interface even in the case of a low ϕ_m. It is known to be Fermi level pinning that the dependence of ϕ_{bn} on ϕ_m is weak. Our strategy for Ohmic formation is (1) to release the Fermi level from pinning and (2) to use a metal with a low (high) work function to n-type (p-type) semiconductors. The Ohmic contacts are formed without heavy impurity doping and post-annealing.

In the metal/semiconductor system with the Fermi level pinning, ϕ_{bn} is expressed as,

$$\phi_{bn} = S^{\circ}\phi_m + C \qquad (1)$$

where C is a constant. The Fermi level is perfectly released from pinning when S° is 1, whereas it is perfectly pinned when S° is 0. S° of practical interfaces has values within these two limits. Kurtin *et al.*[5] found that semiconductors have their own value. For example, S° of both Si and GaAs is around 0.1 and that of SiC is 0.25.

In 1947, Bardeen[6] proposed that the Fermi level is pinned due to surface states. Owing to the uncertain relationship between the surface states and interface states which pin the Fermi level, only a few experimental chemical trends to suggest the validity of his concept have been observed. Aoki *et al.*[7] reported that the hydrogen termination of dangling bonds on diamond surfaces using hydrogen plasma is effective to improve the Fermi level pinning, where S° improved from around zero to around 0.5. Fan *et al.*[8] found that S° of GaAs increases from 0.14 to 0.53 by sulfur termination. These results indicate that the termination of dangling bonds is effective to reduce interface states.

Recently, many studies concerning hydrogen termination on Si surfaces have been reported. Immersion of Si (111) crystal in a pH-modified buffered HF solution[9] and in boiling water[10] is reported to produce nearly ideal surfaces terminated with monohydrides.

It is known that the two chemical treatments make the surface with atomical flatness as well as mono-hydrogenation. The atomically-flat surface is effective to reduce the density of surface states because such a surface is formed by reducing the density of steps. Advanced oxidation technique was reported to make an atomically-flat Si/SiO_2 interface.[11] In our present study, oxidation followed by HF etching treatment and boiling water treatment were used to make surface with low state density enough for flat band.

The advantages of the present method are as follows. Ohmic contacts are formed without post-annealing and heavy impurity doping. The concept of the method is applicable to other semiconductor materials. By combining this method with the heavy doping method, contact resistivities are more improved.

EXPERIMENT

Three n-type 6H-SiC (0001) epitaxial samples were used to form Ti electrodes. Carrier concentration N_D estimated from capacitance-voltage (C-V) measurements was around 5×10^{17} cm^{-3}. The samples were degreased followed by dipping into 5% HF solution to remove a thin native oxide layer on these surfaces. Two samples were oxidized in a quartz tube furnace. The oxidized layer with a thickness of around 10 nm was etched by dipping in 5% HF solution to remove a surface layer with low crystallinity. Then, one of the two samples was immersed in boiling water for 10 min and rinsed in deionized water. Titanium was deposited onto the three samples at the same time using an E-gun system with a base pressure of around 3×10^{-8} Torr. Electrodes with diameter of 300 µm and thickness of 500 nm were formed by evaporation on the samples through a metal mask. Each sample has over 20 electrodes. During the deposition, the temperature of the samples was maintained below 100 °C. In order to measure the Schottky barrier heights of the Ti contacts, magnesium was thermally evaporated as back contacts. Contact resistivity ρ_c is estimated by measuring the resistance between two electrodes as changing distance of the Ti electrodes. This method is based on a concept of the transmission line model[12] to measure an exact contact resistance. Other metals with different work functions such as Mo and Ni were also evaporated on the SiC surfaces with the same procedures as those of Ti electrodes. Conventional annealed Ni electrodes were fabricated by annealing at 1000 °C for 60 min in pure Ar.

RESULTS

Typical current-voltage (I-V) characteristics of Ti electrodes on the sample with the 5% HF treatment and with the oxidation/etching treatment are shown in Fig. 1. Ti electrodes of 5% HF treatment show rectification properties with ϕ_B of 0.68 eV. After the oxidation/etching treatment, ϕ_B decreases to 0.50 eV. The electrodes with immersing in boiling water show Ohmic property with excellent linearity, where I-V characteristics were measured between two Ti electrodes. Average contact resistivity ρ_c of the Ti electrodes is $(6\pm1)\times10^{-3}$ Ω-cm^2, which is comparable to that of the conventional annealed Ni Ohmic electrodes.

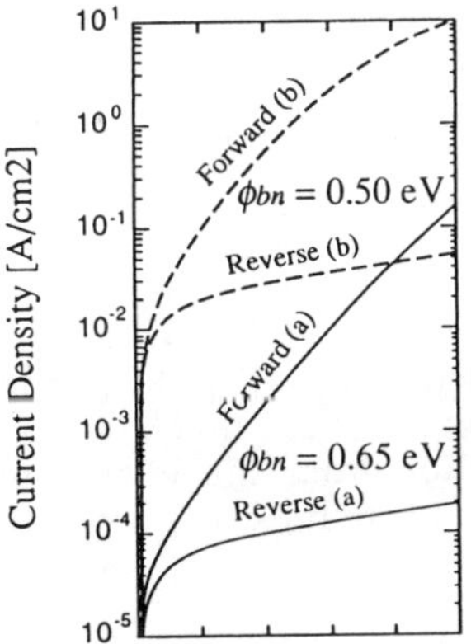

Figure 1. *I-V* characteristics of the Ti electrodes.
(a) 5% HF treatment
(b) oxidation/etching treatment

Figure 2 shows typical optical microscope images of the present Ti electrode and the conventional annealed-Ni electrode. The surface of the Ti electrode is flat, whereas that of the Ni electrode is inhomogeneous because of the high temperature post-annealing at 1000 °C.

Figure 3 shows the Schottky barrier height as a function of the metal work function for different surface treatments. $S°$ values for the three surface treatments are 0.2 for the 5% HF treatment, 0.8 for the oxidation/etching treatment and 1.0 for the boiling water treatment. In other words, the release of Fermi level from pinning increases with proceeding the surface treatment.

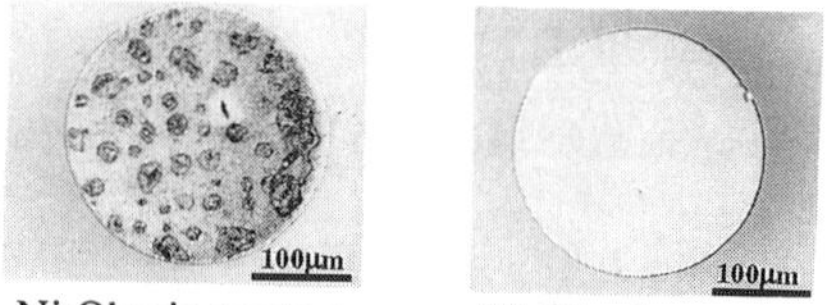

Ni Ohmic contact annealed — Ti Ohmic contact As-deposited

Figure 2. Optical microscope images of the surface of the conventional Ni electrode and the present Ti electrode.

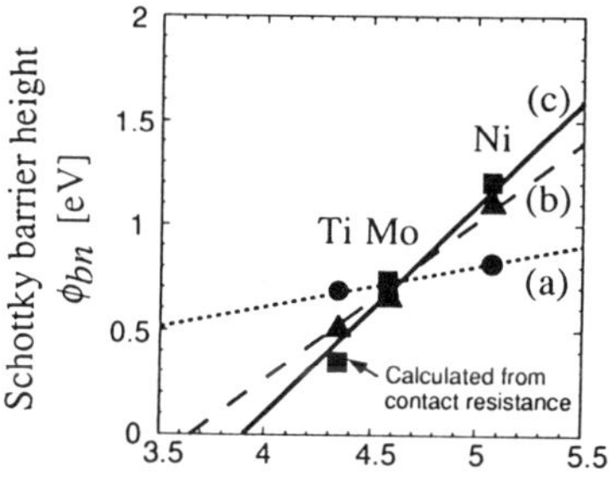

Figure 3. Schottky barrier height as a function of metal work function.
(a) 5% HF treatment
(b) oxidation/etching treatment
(c) boiling water treatment

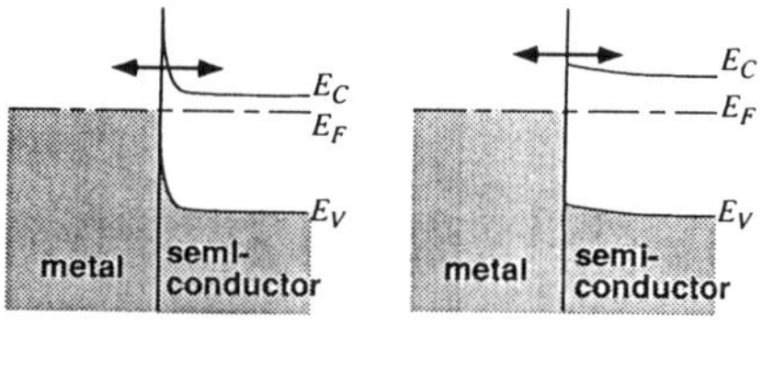

Figure 4. Current transport mechanisms of Ohmic contacts.
(a) heavily doped contact
(b) low barrier contact

DISCUSSION

We discuss the current transport mechanism of the Ti Ohmic contacts. Conventional Ohmic contacts are formed by heavy doping, where the tunneling current is increased due to the decrease in depletion layer thickness as shown in Fig. 4 (a). At this interface, field emission (FE) transport is dominant as explained by Yu[13] and ρ_c is decreased with increasing carrier concentration. Figure 4 (b) shows a band diagram of the lightly doped Ohmic contacts in pinning-free condition. The current transport of the contacts is dominated by the thermionic emission (TE). ρ_c is reduced by decreasing the Schottky barrier height and is essentially independent of N_D. Rhoderick *et al.*[14] explained that there is a gradual transition from the FE to the TE. A dominant transport is determined by a parameter n.

$$n = \left(\frac{qE_{00}}{kT}\right)\coth\left(\frac{qE_{00}}{kT}\right) \quad (2)$$

$$E_{00} = \frac{h}{4\pi}\left(\frac{N_D}{m^* \varepsilon_s}\right) \quad (3)$$

where ε_s is permittivity of the semiconductor, m^* is effective mass, and h is Planck constant. The current transport is the TE when $n = 1$, whereas the current transport is a combination of the TE and the FE when $n > 1$. Figure 5 shows calculated n values for n-type 6H-SiC crystal as a function of the carrier concentrations. Since the carrier concentration we used is 5×10^{17} cm^{-3}, n is estimated to be unity. Therefore, we concluded that the tunneling current of the Ti Ohmic contacts is negligible. ρ_c in TE range is expressed by Yu[13] as

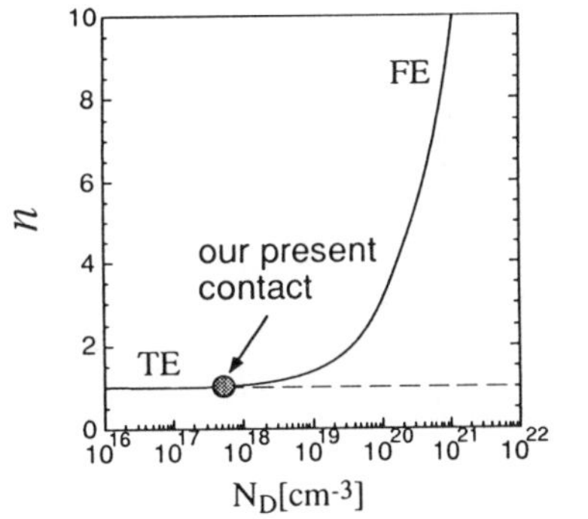

Figure 5. n value as a function of carrier concentration.

$$\rho_C = \frac{k}{qA^*T}\exp\left(\frac{q\phi_B}{kT}\right) \quad (4)$$

where k is the Boltzmann constant and A^* is the Richardson constant. From Eq. (4), we can estimate ϕ_B from ρ_c obtained experimentally when TE is dominant. Estimated ϕ_B of the Ti Ohmic contacts is 0.35 eV.

We discuss next the formation mechanism of the pinning released surface. The surface structure of SiC(0001) is similar to that of Si(111) surface in terms of Si terminated (111) surfaces. Dihydrides and trihydrides on the Si(111) surface may induce interface states because these atoms are unstable. Several types of dislocations, steps and kinks also induce the surface states. Therefore, the removal of these defects is important to make a flat band. In order to reduce the density of dihydrides and trihydrides on the SiC surface, we applied the advanced termination techniques established on the Si(111) surface to the SiC(0001) Si-face. Watanabe *et al.*[15] reported that a Si(111) surface is etched anisotropically in boiling water, leading to the formation of a surface terminated by monohydrides. The monohydride-terminated surfaces are stable compared with surfaces terminated by dihydride or tryhydride because each top Si atom has three covalent bonds with underlying Si atoms.

The density of steps and kinks is reduced by making an atomically-flat surface. Ohishi *et al.*[11] found that oxidation of Si(111) surface at a low pressure of oxygen makes atomically-flat Si/SiO_2 interface. The atomically-flat interface is appeared to be formed by suppressing an oxidation rate to the vertical direction due to lowering a partial pressure of the oxygen. Since the oxidation rate of Si-face of SiC is about 100 times smaller than that of Si, the SiC/SiO_2 interface will also become atomically-flat. In fact, the significant change in the degree of the Fermi level pinning by the oxidation/etching treatment indicates that an atomically-flat surface forms, leading to reducing the density of the surface states and the resultant interface states.

CONCLUSIONS

We formed Ti Ohmic contacts without high impurity doping and post-annealing to *n*-type 6H-SiC epitaxial layer by decreasing the Schottky barrier height. The barrier height is reduced enough to form Ohmic contacts by releasing the Fermi level at the interface from pinning using the oxidation/etching as well as the monohydride termination technique established on Si (111) surfaces. The current transport by the thermionic emission was dominant and the contact resistivity was $(6\pm1)\times10^{-3}$ Ω-cm^2.

REFERENCES

* Also with Faculty of Materials Science, The University of Tsukuba, 1-1-1 Tennoudai, Tsukuba, Ibaraki 305, Japan

1. J. A. Edmond, H. J. Kim, and R. F. Davis in *Rapid Thermal Processing*, edited by T. O. Sedgwick, T. E. Seidel, and B. Y. Tsaur (Mater. Res. Soc. Proc. **52**, Pittsburgh, PA, 1986) pp. 157-164.
2. M. Ruff, H. Mitlehner and R. Helbig, IEEE Trans. Electron. Devices. **41**, 1040 (1994).
3. W. Schottky, Naturwissenschaften, **26**, 843 (1938).
4. M. S. Tyagi, *Metal-Semiconductor Schottky Barrier Junctions and Their Applications*, edited by B. L. Sharma (Plenum Press, New York, 1984), p. 55.
5. S. Kurtin, T. C. McGill, and C. A. Mead, Phys. Rev. Lett. **22**, 1433 (1969).
6. J. Bardeen, Phy. Rev. **71**, 717 (1947).
7. M. Aoki and H. Kawarada, Jpn. J. Appl. Phys. **33** L708 (1994).
8. J. F. Fan, H. Oigawa and Y. Nannichi, Jap. J. Appl. Phy. **27**, L2125 (1988).
9. G. S. Higashi, Y. J. Chabal, G. W. Trucks, and Krishnan Raghavachari, Appl. Phys. Lett. **56**, 656 (1990)
10. S. Watanabe, M. Shigeno, N. Nakayama, and T. Ito, Jpn. J. Appl. Phys. **30**, 3575 (1991).
11. K. Ohishi and T. Hattori, Jpn. J. Appl. Phys. **33**, L675 (1994).
12. H. H. Berger, Solid State. Electron. **15**, 145 (1972).
13. A. Y. C. Yu, Solid State. Electron. **13**, 239 (1970).
14. E. H. Rhoderick and R. H. Williams, *Metal-Semiconductor Contacts, 2nd ed.* (Clarendon Press, Oxford, 1988) p. 113.
15. S. Watanabe, Y. Sugita, Surf. Sci. **327**, 1 (1995).

Part I

Device Technologies

Etching

PHOTOTHERMALLY ASSISTED DRY ETCHING OF GaN

R. T. Leonard* and S. M. Bedair**, North Carolina State University
*Department of Materials Science and Engineering
**Department of Electrical and Computer Engineering

ABSTRACT

Photoassisted dry etching of GaN in HCl by 193 nm ArF excimer laser is developed as a potential alternative process to eliminate the ion damage and surface roughness which occur in etching techniques that involve an energetic ion beam impinging the surface. A directed stream of HCl etchant with background pressure of ~ 5 x 10^{-4} Torr, sample surface temperature between 200 to 400 °C, and laser fluence of 10 to 20 mJ/ pulse combine to produce etching. The photoassisted etching reaction under these process conditions is thermal in nature, with activation energy near 1.2 kcal/ mol. Increases in laser fluence results in increase of etch rate, but the surface also becomes rougher. Distinct etch features can be produced with smooth surfaces at expense of etch rate.

INTRODUCTION

The ability to process devices from III-Nitride materials, especially GaN, depends heavily on the development of controlled etching and masking techniques for these materials. Removing material requires some type of etchant species (reactive or unreactive, as in Ar ion milling) and a means to provide kinetic energy for desorption of etch products and/ or to enhance the etching reaction. Means to provide the necessary energy can be in the form of heat, biased ion beam, kinetic energy transfer, or photons. The masking selects where the material is removed, but the masking material must be compatible with the etching process. Since the ultimate goal is to uniformly etch only beyond the boundaries of the mask, the etching processes must also have a non-crystallographic anisotropic character to transfer the masked pattern.

Both dry and wet techniques can be used for device processing. Dry, plasma etching techniques in the form of reactive ion etching (RIE) [e.g. 1, 2], magnetron electron cyclotron resonance (ECR) etching [3, 4], and chemically assisted ion beam etching (CAIBE) [5, 6] have been very popular for etching the III-N materials. Although these techniques have high etch rates ranging between 200 to 3500 Å/ min, they rely on bombardment of an electrically biased surface with ions that can produce damage, degrading the optical and electrical properties of devices, unless used at low ion energies [7, 8]. Wet etches are ineffective due to the III-N materials resistance to chemical attack, with low etch rates around 20 Å/ min [9] for GaN. Conventional wet etches also have the drawback of isotropic etching, leading to undercutting of the mask; however, these etches can be effective for surface cleaning [10].

Photoassisted etching techniques using both wet and dry etchant species are being developed as alternatives to plasma etching III-N materials to potentially reduce the damage associated with ion beams. Photoassisted wet etching of undoped n-type GaN has also been shown effective in producing etch rates as high as 4000 Å/ min using (1:3) 45% KOH and 400Å/ min using (1:10) HCl; H_2O coupled with a He-Cd (325 nm) laser [11]. The etch features were also reported to be smooth and distinct, with the etch mechanism suggested as photoelectrochemical in nature. Recently we have reported the photoassisted dry etching of GaN [12] using an ArF excimer laser (193 nm) and 100% HCl etchant gas at ~ 0.5 mTorr, producing distinct sidewalls and smooth etched features with an etch rate of ~ 80 Å/ min. Results from so called dark etching, without laser interaction, have shown that no appreciable etching occurs. Also, no appreciable etching is observed with raising the sample temperature to ~ 400 °C under the same HCl conditions, again, without laser radiation. Both HCl and UV radiation were necessary to produce etching of GaN.

Mat. Res. Soc. Symp. Proc. Vol. 423 © 1996 Materials Research Society

In this paper we will present the effect of surface temperature, HCl pressure, and laser fluence on both etch rate and etched surface morphology.

EXPERIMENT

The system used in this study has been previously described [12], and consists of a load locked ultra high vacuum chamber (base pressure $3x10^{-9}$ Torr) equipped with an internal shower head doser for delivering HCl, with the gas flux coincident with laser radiation and the GaN material. GaN samples grown in-house [13] epitaxially on (0001) sapphire and purchased from Cree Research, Inc. (undoped n-type carrier concentrations < $1x\ 10^{17}$ /cm^3) were used in this study. The experiments were performed with no thermal bias to the substrate at ambient temperatures near 23 °C, surface heating due to laser between ~ 200-300 °C, and HCl (99.999% purity) pressures ~ 0.5 to 1 mTorr. An ArF (193 nm) excimer laser was used at 30 Hz and 26 kV to produce laser fluences which were attenuated using non-coated optic flats and all energy losses were measured by a pyroelectric Joule meter. The laser power was measured at the entrance to the chamber, taking into account the loss due to the laser port at the system. The laser beam was focused from the original 1 x 2 cm size at the laser exit to approximately 3 x 6 mm at the sample.

Samples were masked using a contact mask made of tungsten wire physically placed over the surface. After etching, the samples were characterized by optical microscopy with Nomarski differential interference contrast, stylus profilometry, and scanning electron microscopy (SEM) to determine the etch rate and the quality of surface/ sidewall morphology. The etch rate was determined by dividing the etch depth determined by profilometry by the time to pulse the laser (dependent on the repetition rate). Typical number of pulses/ sample was ~ 25,000 taking approximately 14 minutes.

RESULTS AND DISCUSSION

A typical etch profile is shown in Fig. 1. Etching conditions for this sample were ~ 0.5 mTorr HCl with laser fluence of ~ 19 mJ/ pulse. The sidewalls are very distinct and the etch depth for this sample is between ~ 0.6 to 1 μm deep. Non-uniformity in the etch depths on both sides of the mask is due to the non-uniformity of the laser beam itself: the intensity profile of the beam is not a perfect "top hat" shape, but is more rounded in dimensions; thus, the intensity of the laser is not the same for all regions on the sample.

Typical surface morphology of etched regions produced by increasing laser fluences are seen in Fig. 2. Figure 2(a) shows a very slightly textured etched surface compared to the masked unetched surface under low laser fluences of ~ 10 mJ/pulse. Figure 2(b) reveals that at higher

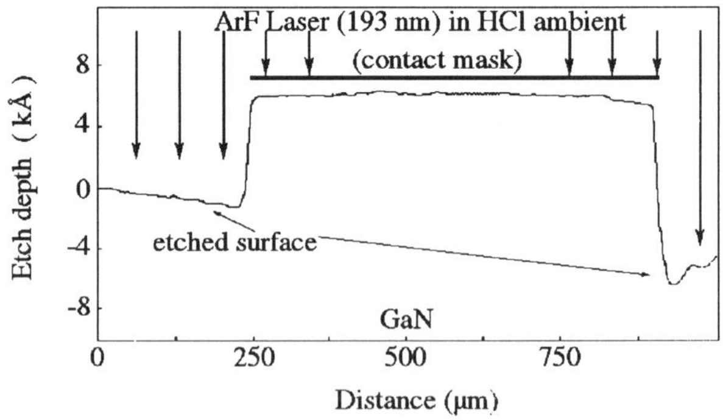

Fig. 1. Typical etch profile measured by profilometry.

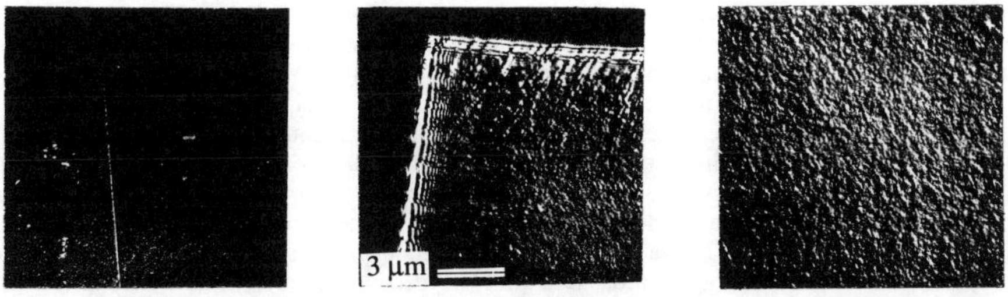

(a) (b) (c)

Fig. 2. Nomarski optical microscopy, 400x observation of etched surface morphology with increasing laser fluence : (a) 10 mJ/ pulse; (b) 12 mJ/ pulse; (c) 19 mJ/ pulse.

laser fluence, 12 mJ/ pulse, the surface becomes more heavily textured. Finally in Fig. 2(c), the surface is roughly textured corresponding to the highest laser fluence used in the study, 19 mJ/ pulse, and is the same sample as in Fig. 1. The edges between the etched and unetched regions of Fig. 2(b) show some diffraction effects due to the incomplete contact masking of the samples. Preliminary work using photolithographed Ni masks on these GaN samples indicate that with a well adhered mask, these specific diffraction effects can be reduced. SEM results have revealed cracks in the Ni mask possibly due to the differences in thermal expansion between GaN and Ni. Because of this, the sidewall morphology cannot be determined due to poor mask adhesion. A more suitable mask is being considered to reduce these effects.

The effect of laser fluence on etch rate is shown in figure 3(a). The temperature rise associated with laser—material interaction has been determined by many authors [e.g. 14] by solution of the heat equation and assumption of a beam shape. Using known thermal constants for

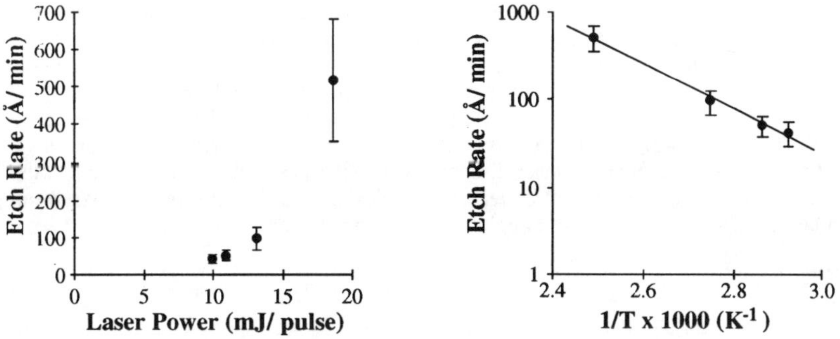

(a) (b)

Fig. 3. (a) Etch rate vs. laser fluence for photoassisted etching of GaN; (b) Arrhenius plot of the etch rate vs. the calculated surface temperature rise with increasing laser fluences.

GaN, a temperature rise between 300 to 400 °C can be associated with the laser power density hitting the sample, assuming all of the laser fluence is transformed to heat. Fig. 3(b) relates the laser fluence to surface temperature, resulting in an Arrhenius plot of etch rate vs. the inverse temperature related to the intensity of the laser radiation. Note that the temperature rise is limited to the surface region only, with the heat dissipating between the 30 Hz, 18 nsec pulses. The error bars included in the graph reveal the effect of a non-uniform radiation source.

The results in Fig. 3(b) suggest that the photoassisted etching of GaN is thermally enhanced by the laser—material interaction in this HCl pressure range. Line fitting techniques through these points indicate a residuals squared value of ~ 0.99, confirming the Arrhenius form of the etch rate vs. 1/T. The activation energy calculated from these results is ~ 1.2 kcal/ mol, higher than the activation energy of 0.65 kcal/ mol for reactive fast atom beam etching (r-FABE) determined by Tanaka [15]. Possible differences in these two activation energies could be due to the different methods of removing the etch products, and considering the effect of the energetic beam in both cases. For the reactive FABE, no reason is given for the temperature dependence of the etch rate or mechanism. The energetic fast atom beam could be adding the necessary energy to further desorb the etch products, thus lowering the activation energy. In photothermally assisted etching, the laser energy is doing all of the work: both the surface is heated and the etch products desorb due to the laser irradiation.

The observations for photoassisted etching of GaN are similar to those for GaAs. Brewer [16] and Tejedor [17] both reported on the photoassisted etching of GaAs with CF_3Br/ 193 nm and Cl_2/ 193 nm, respectively, and showed that with increasing the laser fluence, the etch rate did increase, and the surface morphology roughened with laser fluences above 30 - 35 mJ/ cm^2. They suggested the laser—material interaction produced physical ablation or at least laser enhanced desorption of the GaAs. By comparing the results of Figs. 2 and 3 to that of the GaAs results, similar etching behavior (and possibly mechanism) is occurring.

Compared with the wet photoelectrochemical etching reaction described by Minsky [11], at this time there is no evidence supporting photoelectrochemical etching with the dry etching process conditions used here. This does not mean, however, that they do not exist, since the laser energy (6.4 eV >> E_g^{GaN} ~ 3.4 eV) is more than sufficient for the production of the electron—hole pairs involved in the photoelectrochemical reaction.

The mechanism for photoassisted dry etching is not well understood. The surface roughness resulting with increasing laser fluence suggests that there is some non-uniform interaction of the material with the etching energy of the laser and reactivity of the process gas. Ablation has been suggested to explain the roughness in the photoassisted etching of GaAs, but for GaN, whether or not true ablation occurs, it is obvious that some reactive decomposition of the surface is occurring. The cause of the surface roughness could result from such etching decomposition if the etch products were agglomerating from incomplete desorption, creating local roughness that a uniform etching process could not preferentially remove. For this gas—material interaction, it is expected that the laser dissociates the HCl gas molecules into the etching species. In this case, increasing the amount of reactant species, and/ or increasing the surface temperature would be necessary to remove these agglomerated etch products.

CONCLUSION

We have shown that for the process conditions, the photoassisted etching reaction is thermal in nature, with the activation energy ~ 1.2 kcal/ mol. Increases in the laser fluence results in the increase of etch rate, however, the surface also becomes increasingly rougher. Distinct etch features can be produced with this technique with smooth surfaces at the expense of etch rate. With changes in the process conditions, we hope to improve the surface morphology. Extension of this research to different III-N materials is already underway to determine the effect of photoassisted etch damage and will be reported on in the future.

ACKNOWLEDGMENTS

The authors would like to acknowledge the help of F. G. McIntosh, J. C. Roberts, and the support of the Office of Naval Research University Research Initiative project.

REFERENCES

1. S. J. Pearton, F. Ren, T. R. Fullowan, A. Katz, W. S. Hobson, U. K. Chakrabarti, and C. R. Abernathy, Mater. Chem. Phys. **32**, 215 (1992).

2. M. E. Lin, Z. F. Fan, Z. Ma, L. H. Allen, and H. Morkoç, Appl. Phys. Lett. **64**, 887 (1994)

3. S. J. Pearton, Mater. Sci. Eng. **B27**, 61 (1994).

4. S. J. Pearton, C. R. Abernathy, and F. Ren, Appl. Phys. Lett. **64**, 2294 (1994).

5. A. T. Ping, C. Youtsey, and I. Adesida, J. Electron. Mater. **24**, 229 (1995).

6. A. T. Ping, I. Adesida, and M. Asif Khan, Appl. Phys. Lett. **67**, 1250 (1995).

7. M. Heinbach, J. Kaindl, and G. Franz, Appl. Phys. Lett. **67**, 2034 (1995).

8. S. J. Pearton, J. W. Lee, J. D. MacKenzie, and C. R. Abernathy, Appl. Phys. Lett. **67**, 2329 (1995).

9. S. J. Pearton, C. B. Vartuli, R. J. Shul, and J. C. Zolper, Mater. Sci. Eng. **B31**, 309 (1995).

10. S. W. King, L. L. Smith, J. P. Barnack, J. H. Ku, J. A. Christman, M. C. Benjamin, M. D. Bremser, R. J. Nemanich, and R. F. Davis, to be published in Fall 1995 MRS Proceedings.

11. M. S. Minsky, M. White, and E. L. Hu, Appl. Phys. Lett. **68**, 1531 (1996).

12. R. T. Leonard and S. M. Bedair, Appl. Phys. Lett. **68**, 794 (1996).

13. K. S. Boutros, F. G. McIntosh, J. C. Roberts, S. M. Bedair, E. L. Piner, and N. A. El-Masry, Appl. Phys. Lett. **68**, 1856 (1996).

14. S. M. Bedair and H. P. Smith, Jr., J. Appl. Phys. **40**, 4776 (1969).

15. H. Tanaka, F. Shimokawa, T. Sasaki, and T. Matsuoka, OPTOELECTRONICS—Devices and Technologies **6**, 150 (1991).

16. P. Brewer, S. Halle, and R. M. Osgood, Jr., Appl. Phys. Lett. **45**, 475 (1984).

17. P. Tejedor and F. Briones, Mat. Res. Soc. Symp. Proc. **201**, (1991).

DRY ETCH DAMAGE IN InN, InGaN AND InAlN

S. J. Pearton[1], J. W. Lee[1], J. D. MacKenzie[1], C. B. Vartuli[1], S. M. Donovan[1] and C. R. Abernathy[1], R. J. Shul[2], F. Ren[3] and J. R. Lothian[3]

[1]University of Florida, Gainesville, FL 32611
[2]Sandia National Laboratories, Albuquerque NM 87185
[3]Lucent Technologies, Bell Laboratories, Murray Hill NJ 07074

ABSTRACT

Damage introduction in III-V nitrides during dry etching can be simulated by exposing the samples to pure Ar plasmas for study of the physical (ion-bombardment) effects. Changes in conductivity of InN, $In_{0.5}Ga_{0.5}N$ and $In_{0.5}Al_{0.5}N$ layers exposed to Ar plasmas under both Electron Cyclotron Resonance and reactive ion etching conditions have been measured as a function of rf power, pressure and exposure time. The combination of high microwave and high rf powers produces large increases (10–10^4 times) in sheet resistance of the nitrides, but conditions more typical of real etching processes (rf power < 150W) do not change the bulk electrical properties. The nitrides are more resistant to damage introduction than other III-V semiconductors. The removal of damage-related traps occurs with an activation energy of ~2.7eV. High ion currents during ECR etching can produce substantial conductivity changes, whereas the lower currents under RIE conditions do not affect the nitrides. It is difficult to avoid preferential loss of N in the near-surface of these materials, which leads to leakage currents in rectifying metal contacts deposited on these surfaces.

INTRODUCTION

The wide bandgap InGaAlN alloys are attracting interest for use in blue/UV light emitters and detectors, and for field effect transistors capable of high temperature operation.[1–3] The chemical stability of all of the binary, ternary and quaternary components of this materials system has demanded that dry etching be employed for patterning of device structures. A variety of different plasma chemistries have proven successful, inducing CH_4/H_2, BCl_3/Ar, Cl_2/H_2, $Cl_2/CH_4/H_2/Ar$ and $SiCl_4$ [4-7]. A consistent feature with all of these mixtures is that the III-V nitride etch rates are considerably slower than for more conventional compound semiconductors such as GaAs and InP under the same conditions.[5,7,8] In the reactive ion etching (RIE) mode etch rates for GaN up to ~1100Å/min have been obtained,[6] but higher ion density techniques like magnetron or Electron Cyclotron Resonance (ECR) produce rates up to ~4000Å/min at lower pressures and dc self biases[(7,8)]. It is also clear that the etch rates are invariably limited by sputter removal of the etch products[(4)], and therefore the study of ion-induced damage into the nitrides is a necessary step in optimizing device fabrication processes.

We have examined ion-induced changes in conductivity of InN, InGaN and InAlN under both ECR and RIE etching conditions. In both cases we employed pure Ar discharges in order to simulate the ion bombardment received by the nitrides during plasma etching.

Mat. Res. Soc. Symp. Proc. Vol. 423 © 1996 Materials Research Society

EXPERIMENTAL

The InN, $In_{0.5}Ga_{0.5}N$ and $In_{0.5}Al_{0.5}N$ were grown on semi-insulating GaAs substrates by Metal Organic Molecular Beam Epitaxy at 500°C in an Intevac Gas Source Gen II system.[9] Triethylgallium, trimethylindium and dimethyl ethylamine alane were transported by the carrier gas, while the nitrogen flux was provided by an ECR source (Wavemat MPDR 610) operating at 2.45 GHz and 200W forward power. The films were auto-doped n-type to levels of $\sim 1.5 \times 10^{20} cm^{-3}$ for InN and $In_{0.5}Ga_{0.5}N$ and $\sim 5 \times 10^{19} cm^{-3}$ for $In_{0.5}Ga_{0.5}N$. The shallow donor responsible for this autodoping in InN and related alloys has not been definitively identified at this point, but is thought to be associated with native defects such as nitrogen vacancies.[10] The sample thickness were typically 0.3-0.4μm. Details of the nitride properties have been given previously.[11]

Ohmic contacts to the four corners of 5x5 mm^2 samples were produced by alloying HgIn eutectic at 400°C for 3min. The sheet resistance of the layers was measured before and after exposure to Ar discharges using the Van der Pauw method. The plasma exposures were performed in a Plasma-Therm SLR 770 system employing an ECR source (Astex 4400) and separate rf biasing (13. 56 MHz) of the sample position.

RESULTS AND DISCUSSION

Figure 1(a) shows the ratio of the post-Ar plasma exposed sheet resistances to that of the untreated control samples for the three nitrides investigated. In this case the ECR source was utilized to provide a very high ion density incident on the samples. There is littler change in resistance of the InGaN with increasing rf power until the highest value (450W). The InN shows a similar trend but is more susceptible to damage introduction than the InGaN. This is consistent with the greater bond strength of GaN relative to InN, but this trend does not hold for the third material InAlN. In this case there is a significant change in resistance even at 150W rf. The doping in the InAlN is somewhat lower than that in the other two nitrides which probably accounts for much of the higher sensitivity to plasma-induced changes. The mechanism for the resistance increases is the introduction of deep acceptor states that trap the conduction electrons, and are not thermally ionized at room temperature. This mechanism has been well-established from dry etch damage studies on other III-V semiconductors [13,14] and also from implant isolation work.[14]

In comparison experiments on heavily doped InGaP we found that all of the conditions represented in figure 1(a) produced very large ($>10^4$) resistance changes, even at low rf power (150W). The nitrides are considerably more resistant to damage introduction than more conventional III-V materials, as expected from their mechanical and chemical stability.

Turning to Figure 1(b), it is seen that RIE conditions create significantly smaller changes in resistance in all three materials at the same rf power levels as those used in Figure 1(a). Even very high rf powers produce essentially no change in resistance of InN and InGaN, while InAlN shows severe changes above 250W rf. One would expect to see the onset of resistance increases earlier in this lower-doped material, since if essentially the same number of deep traps are introduced in different nitrides the effect will become evident first in the lowest doped sample.

The time dependence of resistance changes under ECR conditions (1000W, plus 250W of rf power) is shown in Figure 2. These would be typical of a situation in which one selectively etched through one type of nitride layer and stopped at an underlying layer. All three materials show a rapid increase in resistance under these conditions, suggesting that over-etch times should be kept to a minimum in device fabrication processes. We also investigated the pressure dependence of resistance changes under ECR conditions. Figure 3 shows that higher pressure (at least up to 10 mTorr) produce more degradation – this tracks the ion current in our system, although at even higher pressures the resistance changes become less significant due to a induced ionization efficiency.

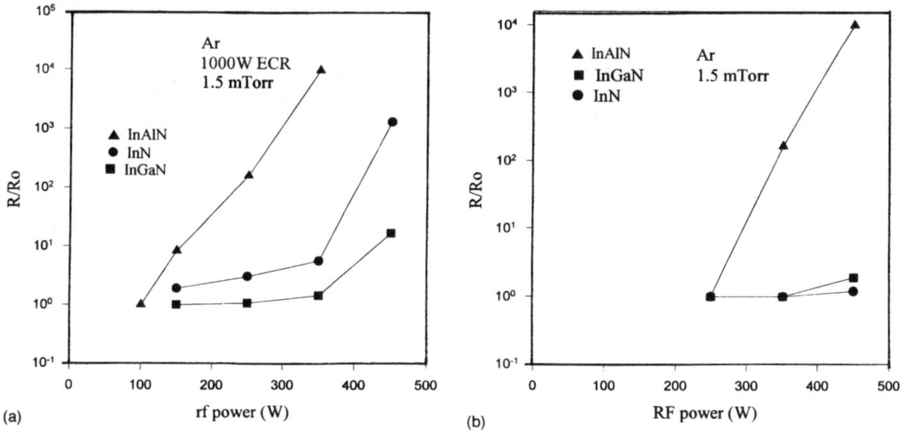

FIG. 1. (a) Increase in sheet resistance relative to the initial value (R_0) of InN, InGaN, and InAlN layers exposed to 1000 W ECR Ar discharges for 1 min, as a function of additional rf power on the electrode. (b) Increase in sheet resistance of nitrides exposed to Ar discharges for 1 min under RIE conditions, as a function of rf power on the electrode.

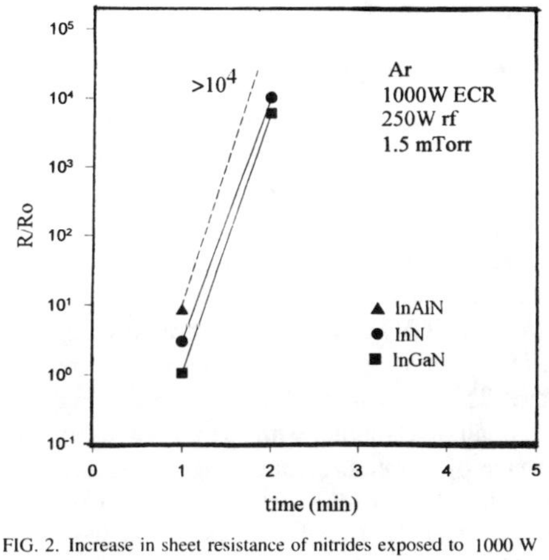

FIG. 2. Increase in sheet resistance of nitrides exposed to 1000 W ECR Ar discharges with 250 W of rf power, as a function of the exposure time.

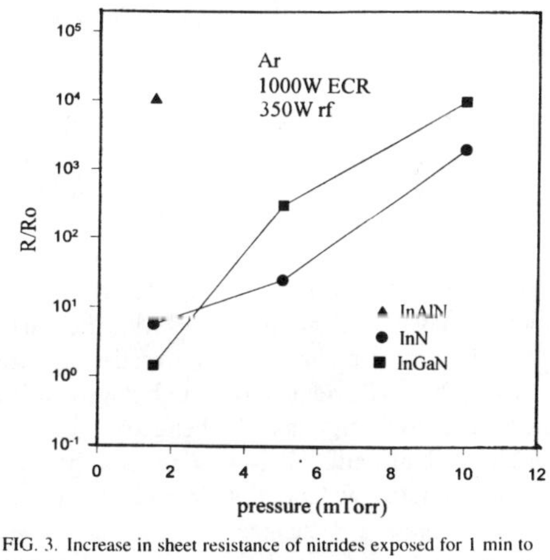

FIG. 3. Increase in sheet resistance of nitrides exposed for 1 min to 1000 W ECR Ar discharges with 350 W of rf power, as a function of the process pressure.

A key question is of course the thermal stability of the plasma-induced damage. Figure 4 shows the results of isochronal anneals (1min) on the resistance of either InN exposed to a 1000W (ECR), 30W rf discharge, or InGaN exposed to a 450W rf plasma. In both cases recovery of the initial conductivity starts above 500°C, and is essentially complete at

650°C. Assuming the recovery mechanism is annealing of the deep acceptor states by diffusion, we can estimate an activation energy of around 2.74eV from the relation

$$R/Ro = 1-\exp[-t\nu\exp(-E_d/kT)] \qquad (1)$$

where t=60 sec, ν the attempt frequency was assumed to be ~$10^{14}S^{-1}$ and T is the absolute anneal temperature. The recovery was apparent over a broader temperature range than observed for single-step processes and indicates that several deep level states are involved with slightly different activation energies. The value estimated above therefore represents an average of these individual energies.

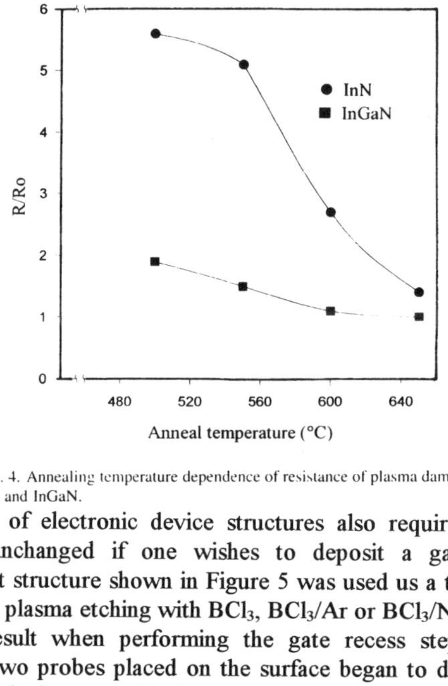

FIG. 4. Annealing temperature dependence of resistance of plasma damaged InN and InGaN.

Dry etching of electronic device structures also requires that the near-surface stoichiometry be unchanged if one wishes to deposit a gate contact for channel modulation. The test structure shown in Figure 5 was used us a test vehicle. A gate mesa was formed by ECR plasma etching with BCl_3, BCl_3/Ar or BCl_3/N_2.

A typical result when performing the gate recess step was that the current measured between two probes placed on the surface began to decrease as the $In_XAl_{1-X}N$ contact layer was etched away, with the current voltage (I-V) characteristic transforming from pure ohmic character to a rectifying nature. However we noticed that the I-V curve always showed evidence of a high degree of leakage. The I-V curves were more rectifying with the BCl_3/N_2 plasma chemistry, but remained almost ohmic with either BCl_3 or BCl_3/Ar. Normally this is an indication of the presence of dry etch-induced changes to the semiconductor surface. For example it is extremely difficult to form a rectifying contact on a dry etched InP surface because the typical Cl_2-or CH_4/H_2-based plasma chemistries employed tend to lead to preferential loss of P, leaving a thin In-rich layer which is strongly n-type and which produces an ohmic I-V characteristic for any subsequently deposited metal.

The etched surface is clearly more In-rich, as shown in Table 1 where we have listed the ratio of raw counts for In/N. While BCl_3/N_2 produces less enrichment than pure BCl_3, there is clearly preferential loss of nitrogen during the dry etch step, and this leads to the presence of a defective layer that prevents achievement of acceptable rectifying contacts.

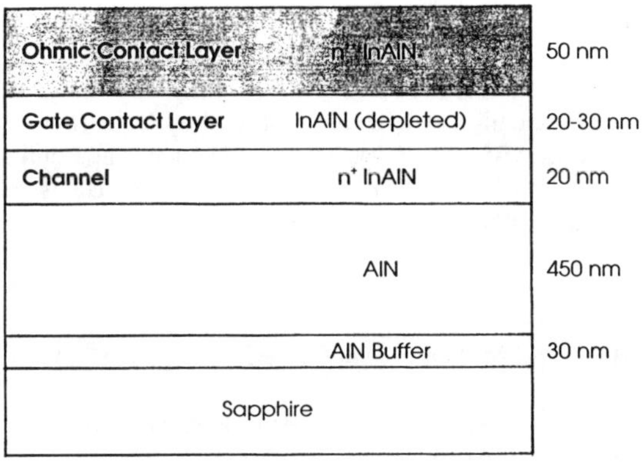

Figure 5 Epitaxial layer structure of InAlN.

Table 1. In/N ratio measured in raw counts from AES analysis of dry etched InAlN samples. No correction for relative sensitivity factor was applied.

Sample	In/N
as-grown	2.86
BCl_3 etch	3.64
BCl_3/N_2 etch	3.49
BCl_3/N_2 + wet etch	3.28

From the I-V measurements we believe this defective layer is probably strong n-type, in analogy with the situation with InP described earlier. At this stage there is no available wet etch solution for InAlN that could be employed to remove the non-stoichiometric layer in the type of clean-up step commonly used in other III-V materials. Other possible solutions to this problem include use of a higher Al concentration in the stop layer, which should be more resistant to nitrogen loss, or employment of a layer structure that avoids the need for gate recess. Addition of steps to enhance group III removal during dry etching, such as addition of CH_4 or heating of the sample, are currently under investigation.

In conclusion, the III-V nitrides InN, InGaN and InAlN can sustain major changes in conductivity as a result of exposure to very high Ar^+ ion fluxes under high power ECR and rf conditions. For plasma conditions more typical of real etching processes (i.e. rf power ≤ 150W) the changes are much less significant and in fact the nitrides are more resistant to damage than conventional III-V semiconductors. The plasma-induced damage is annealed out in the range 550-600°C. InAlN is found to be susceptible to preferential loss of nitrogen during even very low damage ECR etching processes, leading to degraded properties for subsequently deposited rectifying contacts. This behavior is similar to that of InP (and InGaAs).

ACKNOWLEDGMENTS

The work at UF is partially funded by grants from the Division of Materials Research of the National Science Foundation, a URI administrated by ONR (#N 00014-92-J-1895) and an AASERT grant from ARO (Dr. J. M. Zavada). The work at Sandia is supported by DOE under contract no. DE-AC04-94AL85000. The work at UF is performed in the Microfabritech facility whose staff is gratefully acknowledged.

REFERENCES

1. S. Nakamura, T. Mukai and M. Senoh, Appl. Phys. Lett. **64** 1687 (1994)

2. S. C. Binari, L. B. Rowland, W. Kruppa, G. Kelner, K. Doverspike and D. K. Gaskill, Electron. Lett. **30** 1248 (1994)

3. M. A. Khan, A. Bhattarai, J. N. Kuznia and D. T. Olsen, Appl. Phys. Lett. **63** 1214 (1993)

4. I. Adesida, A. Mahajan, E. Andideh, M. A. Khan, D. T. Olsen and J. N. Kuznia, Appl. Phys. Lett. **63** 2777 (1993)

5. S. J. Pearton, C. R. Abernathy and F. Ren, Appl. Phys. Lett. **64** 2294 (1994)

6. M. E. Lin, Z. Fan, Z. Ma, L. H. Allen and H. Morkoe, Appl. Phys. Lett. **64** 887 (1994)

7. R. J. Shul, S. D. Kilcoyne, M. Hagerott-Crawford, J. E. Parmeter, C. B. Vartuli, C. R. Abernathy and S. J. Pearton, Appl. Phys. Lett. **66** 1761 (1995)

8. G. F. McLane, L. Casas, S. J. Pearton and C. R. Abernathy, Appl. Phys. Lett. **66** 3328 (1995)

9 C. R. Abernathy, J. Vac. Sci. Technol. A **11** 889 (1993)

10. T. L. Tansley and R. J. Egan, Phys. Rev. B **45** 10942 (1993)

11. C. R. Abernathy, J. D. Mackenzie, S. R. Bharatan, K. S. Jones and S. J. Pearton, Appl. Phys. Lett. **66** 1632 (1995)

12. S. J. Pearton, F. Ren, C. R. Abernathy, W. S. Hobson, T. R. Fullowan, R. Esaqui and J. R. Lothian, Appl. Phys. Lett. **61** 586 (1992)

13. K. L. Seaward and N. J. Moll, J. Vac. Sci. Technol. B **10** 46 (1992)

14. S. J. Pearton, Mat. Sci. Eng. **4** 313 (1990).

OPTICAL SPECTROSCOPY OF DRY-ETCHED GaN-NANOSTRUCTURES

H. ZULL, J. MÜLLER, J. KOETH, F. KIESELING, A. FORCHEL
Technische Physik, University of Würzburg, Am Hubland, D-97074 Würzburg, Germany
e-mail: zull@wpfx32.physik.uni-wuerzburg.de

ABSTRACT

We have realized GaN wire gratings with periods down to 80 nm and with wire widths down to 26 nm. GaN layers of good structural and optical quality with thicknesses down to 100 nm were grown by electron cyclotron resonance assisted MBE (ECR-MBE). The grating structures were fabricated by high resolution electron beam lithography and electron cyclotron enhanced reactive ion etching (ECR-RIE) using Cl_2/Ar as etching gas. The optical properties of the GaN structures were investigated by photoluminescence (PL) spectroscopy in the temperature range between 40 K and 110 K. The wire patterns show intense excitonic photoluminescence and only a small dry etched induced degradation of the quantum efficiency even for the narrowest wires.

INTRODUCTION

III-V nitride materials are attracting a great deal of attention because they permit to realize visible and near UV light emitting diodes (LED's) and semiconductor lasers due to their large direct band gap energies [1]. Their chemical stability, high thermal conductivity and high melting temperature also make them suitable for high temperature electronic and photonic device applications. Efficient light emitting diodes and room temperature laser diodes have been realized recently [2-4]. For further development of more sophisticated optoelectronic devices, patterning techniques for micro- and nanostructures must be developed. Wet chemical etching techniques have been found to be impractical due to low etch rates [5]. Dry etching techniques are more suitable, because etch rates of some 100 nm/min can be achieved. Recent studies have shown the influence of the etching parameters on the etch rate, the chemical composition and morphology of the etched surfaces [6-9]. To our knowledge no studies on the fabrication and optical characterization of GaN-nanostructures have been published up to now.

EXPERIMENT

For the purpose of the lateral patterning, GaN layers were grown using ECR plasma assisted molecular beam epitaxy (Eiko-100-MBE, Astex CECR) on (0001) sapphire substrates. Using a modified ECR-aperture with a reduced aperture area (6 holes with diameter of 1 mm) we have obtained very flat surfaces suitable for high resolution patterning. During the growth the ECR source was operated at a microwave power of 65 W. The growth started with the deposition of an AlN buffer layer of 30 nm at a substrate temperature of 550°C. Then the GaN epilayer with thicknesses between 100 nm and 600 nm were grown at substrate temperatures between 640°C - 900°C. The layers were characterized by high resolution x-ray diffractometry. They were all single crystalline and the rocking curves showed values of $\Delta\theta$ in the range of 9-10 arcmin which are very good values for such thin layers.

Mat. Res. Soc. Symp. Proc. Vol. 423

For the fabrication of sub 100 nm GaN structures the samples were patterned with 30 nm thick Cr masks. The masks were defined by high resolution electron beam lithography in a 100 nm thick layer of spin coated polymethylmethacrylate (PMMA 950K) and a lift off process. The mask patterns were transfered into the GaN by electron cyclotron resonance (ECR) enhanced reactive ion etching (RIE) using Cl_2/Ar (1:9) as etching gas. The etch conditions including plasma self-bias voltage, chamber pressure, microwave input power, and gas mixture were optimized regarding the etch rates and the surface morphology.

Figure 1 shows the dependence of the etch rate on the Cl_2-content (left) and the self-bias voltage (right). As the Cl_2-content is increased, the etch rate increases rapidly up to a Cl_2-content of 10% and thereafter the etch rate saturates. From this it can be inferred that that surface of the Gan layer is essentially covered by Cl_2 atoms at a Cl_2-content of 10%. During the etching process the Ar^+ ions cause two effects. They remove Ga and N from the sample surface due to physical sputtering and provide energy which promotes reactions between Ga and Cl_2 to form $GaCl_x$ products, which are volatile.

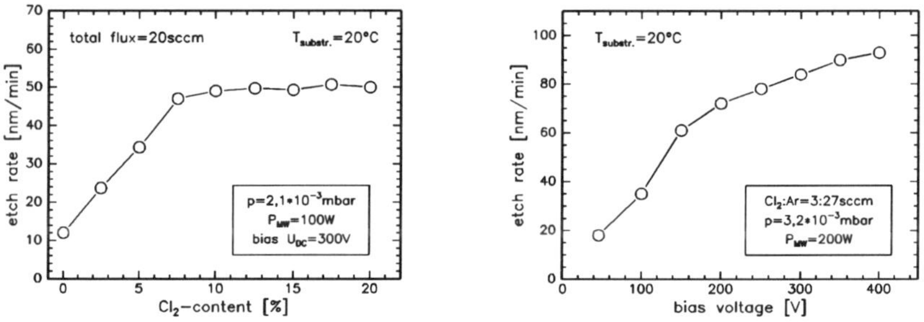

Fig. 1: Etch rate of GaN dependent on the Cl-content (left) and self-bias voltage (right).

By increasing the bias voltage from 40 V to 400 V the etch rate can be varied continuously from ~20 nm/min to ~95 nm/min, while the good surface morphology is preserved. The near-linear increase with increasing bias voltage indicates that sputter-enhanced removal of the etch products is still a limiting factor under these conditions. For the etching of the nanostructures we used a total flow of about 30 sccm (Cl_2:Ar=3:27 sccm) resulting in a chamber pressure of about $3{,}2 * 10^{-3}$ mbar. For low ion damage and reasonable etch rates the self-bias voltage was chosen 200 V. The patterns were etched completely through the GaN and AlN layers down to the sapphire substrates. Finally, the Cr masks were removed in a solution of HCl.

Figure 2 shows an SEM micrograph of small wires with width of 40 nm. The period is approximately 80 nm. This period corresponds to that of a first order distributed feedback (DFB) grating for GaN room temperature emission. The rounded top of the wires indicates that the Cr masks were eroded at the edges during the etching process most likely due to physical sputtering by Ar^+ ions. The structures show good sidewall morphology and reasonable homogeneity over the structured region.

For the optical characterization the samples were mounted in a cryostat which allowed a spatially resolved investigation of the emission at different temperatures. An Ar^+ ion laser was used as excitation source operating at a wavelength of about 335 nm. The luminescence was spectrally resolved by a high resolution spectrometer and detected by a liquid nitrogen cooled CCD camera.

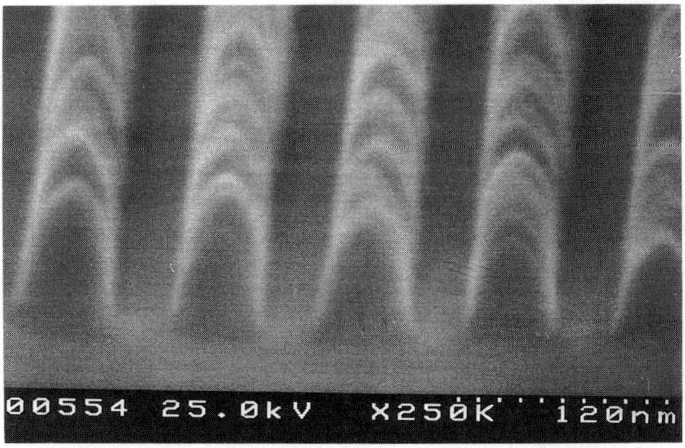

Fig. 2: SEM micrograph of optically active GaN wires obtained by Cl_2/Ar ECR-RIE.

RESULTS

The properties of the GaN layers were investigated by photoluminescence (PL) spectroscopy. Figure 3.a) displays luminescence spectra of two GaN samples at T=10 K, which were grown at different substrate temperatures. The emission line (BX) at E~3.473 eV of the GaN layer grown at T_g=640°C is assigned to the recombination of an exciton bound to a neutral donor (I_2-line). In comparison the high temperature sample (T_g=810°C) emitts free exciton (FX) luminescence with a linewidth of about 8.5 meV at E~3.489 eV. In contrast to the low temperature sample no yellow luminescence is observable for the GaN layer grown at T_g=810°C. In Figure 3.b) the temperature dependence of the peak position is plotted. The solid

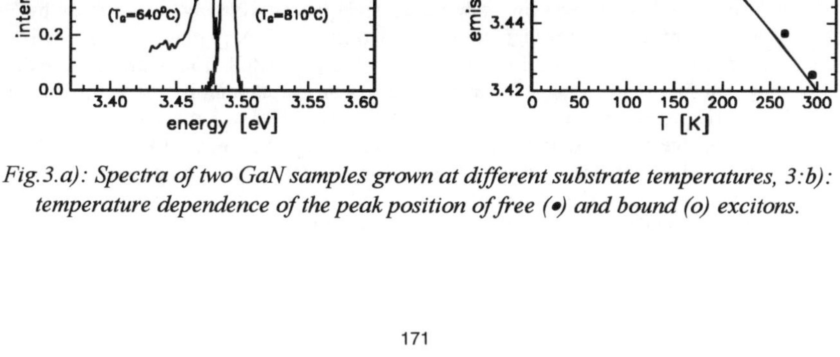

Fig.3.a): Spectra of two GaN samples grown at different substrate temperatures, 3:b): temperature dependence of the peak position of free (•) and bound (o) excitons.

line corresponds to the temperature dependence of free excitons in a very thick GaN layer [10]. The shift of the emission lines with increasing temperature clearly shows the recombination of free (•) and bound (o) excitons. Due to strain effects the free exciton line in our structures (600 nm thick GaN) is shifted to higher energy compared to the emission measured on a thick sample. Thin GaN layers (≤ 1 μm) grown on sapphire substrates experience a compressive strain in plane of the layer, shifting up the free exciton energy by typically about 10 meV [11-13]. Therefore the observed (FX) emsission energy of E~3.489 eV compare favorably with earlier data on free excitons in unstrained thick GaN samples [14,15].

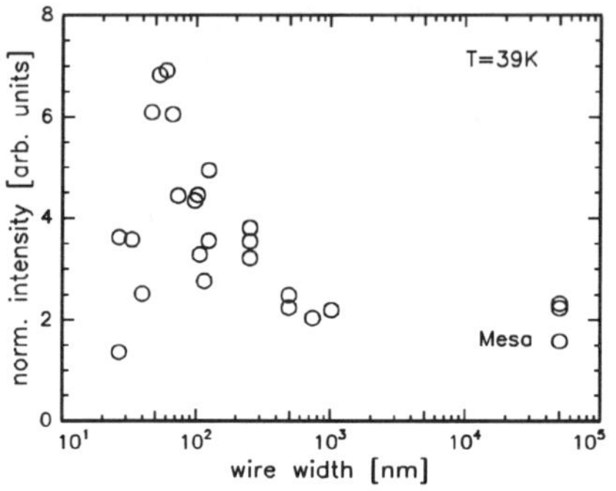

Fig. 4: Wire width dependence of the normalized photoluminescence intensity.

Wires with widths down to 26 nm show intense excitonic photoluminescence up to a temperature of 110 K. This indicates that the ion induced damage of the employed etching process is small.

Figure 4 shows the dependence of the normalized intensity versus wire width for a temperature of 39 K. In going from 1 μm wide wires to wires with widths of about 100 nm the normalized intensity increases by a factor of 5. This may be due to an increased contribution of the edges of the structures to the PL-emission. We observe a maximum of the luminescence intensity for wire widths of about 100 nm, which exceeds the normalized emission intensity of large two dimensional structures. For further decreased wire widths the emission intensity is seen to decrease. We assign the maximum of the PL-intensity at a wire width of about 100 nm to the combined influence of the optical excitation and nonradiative recombination via the etched sidewalls.

CONCLUSIONS

In conclusion, we have grown optically active GaN layers with smooth surfaces by ECR-MBE, which were appropriate for high resolution patterning. GaN grating structures with wire widths down to 26 nm and periods down to 80 nm were fabricated. The patterns were defined by electron beam lithography and a lift off process. For the etching of the nanostructures a suitable ECR-RIE process based on Cl_2/Ar was developed. The structures were characterized by photoluminescence spectroscopy in the temperature range between 40 K and 110 K. Even the smallest wires show intense excitonic luminescence signals in the whole temperature range. The intensity dependence on the wire width is assigned to the combined influence of the optical excitation and nonradiative recombination at the etched sidewalls.

ACKNOWLEDGEMENTS

The financial support of this work by the Bavaria Science Foundation is gratefully acknowledged. We thank S. Kuhn for expert technological assistance.

REFERENCES

1. S. Strite, and H. Morkoc, J. Vac. Sci. Technol. B **10**, 1237 (1992) and references therein.
2. S. Nakamura, T. Mukai, and S. Senoh, Appl. Phys. Lett. **64**, 1687 (1994).
3. R. P. Vaudo, I.D. Goepfert, T.D. Moutsakas, D.M. Beyea, T.J. Frey, and K. Meehan, J. Appl. Phys. **79**, 2779 (1996).
4. R.L. Aggarwal, P.A. Maki, R.J. Molnar, Z.-L. Liau, and I. Melngailis, J. Appl. Phys. **79**, 2148 (1996).
5. T.L. Chu, J. Electrochem. Soc. **118**, 1200 (1971).
6. S.J. Pearton, C.R. Abernathy, and F. Ren, Appl. Phys. Lett. **64**, 2294 (1994).
7 G.F. McLane, L. Casas, S.J. Pearton, and C.R. Abernathy, Appl. Phys. Lett **66**, 3328 (1995).
8. H. Lee, D.B. Obermann, and J.S. Harris Jr., Appl. Phys. Lett. **67**, 2329 (1995).
9. L. Zhang, J. Ramer, J. Brown, K. Zheng, L.F. Lester, and S.D. Hersee, Appl. Phys. Lett. **68**, 367 (1996).
10. W. Shan, T.J. Schmidt, X.H. Yang, S.J. Hwang, and J.J. Song, Appl. Phys. Lett. **66**, 985 (1995).
11. B. Monemar, J.P. Bergmann, H. Amano, I. Akasaki, T. Detchprohm, K. Hiramatsu, and N. Sawaki, International Symposium on Blue Laser and Light Emitting Diodes, Chiba, Japan (1995)
12. C.I. Harris, B. Monemar, H. Amano, and I. Akasaki, Appl. Phys. Lett. **67**, 840 (1995)
13. W. Shan, T.J.Schmidt, X.H. Yang, S.J. Hwang, and J.J. Song, Appl. Phys. Lett. **66**, 985 (1995)
14. R. Dingle, D.D. Sell, S.E. Stokowski, and M. Ilegems, Phys. Rev. B **4**, 1211, (1971)
15. B. Monemar, Phys. Rev B **10**, 676 (1974)

PHOTOCURRENT SENSITIVITIES, SURFACE COLOR, AND AUGER SPECTROSCOPY OF SILICON CARBIDE (SiC) BY PHOTOELECTRO-CHEMICAL (PEC) ETCHING

D. M. COLLINS*, G. L. HARRIS*, NAIQUN CHEN*
*Materials Science Research Center of Excellence, School of Engineering,Howard University, Washington, DC 20059, collins@negril.msrce.howard.edu, gharris@msrce.howard.edu

ABSTRACT

This novel technique involves the ultraviolet (UV) lamp-assisted photoelectrochemical etching of n-type 3C- and 6H-SiC for device processing. In order for this method to be effective, the UV light must be able to enhance the production of holes in the SiC during the etching process thus providing larger currents with light from the photocurrents generated than those currents with no light. Otherwise dark methods would be used as in the case of p-type 3C-SiC.

Experiments have shown that the I/V characteristics of the SiC-electrolyte interface reveal a minimum etch voltage of 3 V and 4 V for n- and p-type 3C-SiC, respectively. Hence, it is possible for etch-stops to occur. From Auger spectroscopy, an oxide formation is present on n-type 3C-SiC where after etching a yellowish layer corresponds to a low silicon to carbon (Si/C) ratio with large photocurrents and a white layer corresponds to a high Si/C ratio with small photocurrents. P-type 3C-SiC shows a grayish or silver layer with a high Si/C ratio and a green layer with a low Si/C ratio. Additionally, n-type 6H-SiC shows a brown or blue layer with a minimum etch voltage of 3 V.

The colors of the etched regions of SiC represent layers that have some degree of porousity formed by the electrochemical process. As a result, the photo-excitation allows control of the porousity and changes the electrical properties of SiC. The nature of these porous layers with its increased resistivity can lead to the formation of devices.

INTRODUCTION

Silicon Carbide (SiC) is an attractive semiconductor material for high speed, high power density, and high temperature device applications. An instrumental process in the fabrication of semiconductor devices is the ability to etch in a highly controlled and selective manner for direct patterning techniques. In this paper, we will report on photoelectrochemical etching of SiC including n-type 3C and 6H, p-type 3C and pn junctions. The etch rate varies from 0.67 to 1.5 μm/min at voltages from 3-4 volts.

The most significant advantage of electrochemical etching is the selective etch stop for different dopant types. Harris and Shor have shown the selectivity of etch stops for pn junctions.[1,4]

EXPERIMENT

The UV lamp setup of the Photo-electrochemical (PEC) system utilized a lamp housing made of aluminum to contain a 1000 Watt mercury (Hg) short arc bulb. The light source had the visible part of the spectrum filtered out to emit UV light.[5]

The electrolytic cell was manufactured out of Teflon with a hole area cut through the center. The purpose of this hole was to expose the sample to the light and to the electrolyte. A spring loaded screw kept the sample in place and also ensured that the container was leak tight. The anode makes contact with the surface of the sample. The cathode, made out of platinum, was dipped into the solution. An digital electrometer measured current flow in the circuit.

The data was collected and measured by a series of on/off light measurements. The non-illumination of UV light on a sample to be etched was performed at incremental voltages to measure the current and likewise for the illumination of UV light. A digital electrometer is hooked to the ground of the power supply to introduce an ammeter into circuit to measure the

Mat. Res. Soc. Symp. Proc. Vol. 423 © 1996 Materials Research Society

current through the wafer and solution. The positive terminal of the power supply is connected to SiC wafer by its ohmic contact (anode) in the Teflon electrochemical cell. The negative terminal of the power supply is connected to a platinum needle (cathode) immersed in the electrolyte.

RESULTS

The PEC process was used to etch n-type cubic and 6H SiC, p-type cubic and cubic pn junctions. Experiments have shown that the I/V characteristics of the SiC-electrolyte interface reveal a minimum etch voltage of 3 V and 4 V for n- and p-type 3C-SiC, respectively. Additionally, n-type 6H-SiC shows a minimum etch voltage of 3 V.

For current-voltage measurements, dark and light currents were measured. The photocurrent equation is the current with UV light minus the current without UV light.. Auger spectroscopy determines the chemical composition of elements in sample and the resulting Si/C ratio chosen by measuring the appropriate peak to peak scaling values of Si and C respectively with the ratio of 3 being a nominal value for SiC.

From Auger spectroscopy, an oxide formation is present on n-type 3C-SiC where after etching a yellowish layer corresponds to a low silicon to carbon (Si/C) ratio with large photocurrents and a white layer corresponds to a high Si/C ratio with small photocurrents. P-type 3C-SiC shows a grayish or silver layer with a low Si/C ratio and a green layer with a high Si/C ratio. Additionally, n-type 6H-SiC shows a brown or blue layer with a minimum etch voltage of 3 V. The blue layer suggests a low Si/C ratio where more Si removal means a carbon rich sample and the brown layer suggests a high Si/C ratio which means more C removal.

The colors of the etched regions of SiC represent layers that have some degree of porousity formed by the electrochemical process. As a result, the photo-excitation allows control of the porousity and changes the electrical properties of SiC. The nature of these porous layers with its increased resisitivity can lead to the formation of devices.

We have summarized the results in the following table:

MATERIAL	AUGER RATIO Si/C	MINIMUM volts required	COMMENTS
p-type 3C	8/0.25, Hi---Pure Si	4	grey/green porous layers
n-type 3C	4/3	3	white/yellow porous layers
n-type 6H	6/2	3	brown/blue porous layers
3C pn	1/3---Lo, 9/3---Hi	3	white/yellow porous; stop etching

The PEC etching of p-type 3C-SiC was the simplest to do among the applications. The sample studied for p-type 3C is labeled as a15. P-type layers of 3 μm or greater were used. Some results in etching p-type 3C-SiC included a minimum etch voltage of 4 volts and etch rates as high as 0.67 μm/min. An important observation was that the light and dark currents were similar. This means that no Ultraviolet (UV) light is needed to etch p-type and the photocurrents were near zero. Dark methods of etching p-type would be appropriate because hole formation during the PEC process is generated easily in p-type material.

Other interesting findings were the change in colors of the SiC in the etched regions. For instance, when the color of the SiC layer was green, there was minimal etching and there was a low Si/C ratio. However, when the color of the SiC layer appeared gray or silver, there was a high Si/C ratio and the etched area that remained after the experiment was pure Si. That is, etching of p-type produced depths toward the Si substrate. Additionally, it is very easy to etch p-type 3C-SiC.

For sample a15, figure 1 shows the photocurrent graph with a minimum etch voltage of 4 Volts. Figure 2 shows a high Si/C ratio of pure Si allowing that minimal C present does not contribute to a ratio calculation.

The most important experimental category of PEC etching is applying this process to n-type 3C-SiC for hole production which is the most basic objective of this project. The challenge of etching n-type 3C proved to be more difficult than p-type and the UV light was critical in this application. The sample studied for n-type 3C is labeled as a2. N-type layers of 5 μm or greater were used.

Some results in etching n-type 3C-SiC included minimum etch voltages of 2 to 3 volts and etch rates as high as 1.4 μm/min. Other interesting findings were the change in colors of the SiC in the etched regions. For instance, when the color of the SiC layer was white, the photocurrents were smaller and there was a high Si/C ratio. However, when the color of the SiC layer was yellow, the photocurrents were larger and there was a low Si/C ratio. For sample a2, figure 3 shows the photocurrent graph with a minimum etch voltage of 3 Volts. Figure 4 shows a low Si/C ratio with an oxide component formed and the color of the layer is yellow.

Before PEC etching began, the 6H sample labeled as 6H3 was etched by KOH for 4 hours to the determine the C face which is smoother than the Si face with some defects or etch pits. Likewise, the photocurrent graph figure 5 shows cut on voltage of 3 Volts. Figure 6 shows a high Si/C ratio and the color of the layer is brown.

The samples studied for pn junctions in 3C-SiC are labeled as a12, and a1 where the n-layer was on the top of the p-layer. However, PEC can be applied to layer in reverse order as well. P-type layers of 3 μm and n-type layers of 5 μm were typically used. For sample a12, figure 7 shows the photocurrent graph with a minimum etch voltage of 3 Volts. Figure 8 shows a low Si/C ratio and the color of the layer is yellow. For sample a1, figure 9 shows the photocurrent graph with a minimum etch voltage of 3 Volts also. Figure 10 shows a high Si/C ratio and the color of the layer is white. The Auger plots show that pn junction exhibit similar characteristics to n-type layers where the white layer corresponds to a high Si/C ratio and the yellow layer to a low Si/C ratio. It also exhibits similar characteristics to p-type layers whereas the white layer has small photocurrents at low voltages and the yellow layer has small photocurrents at high voltages.

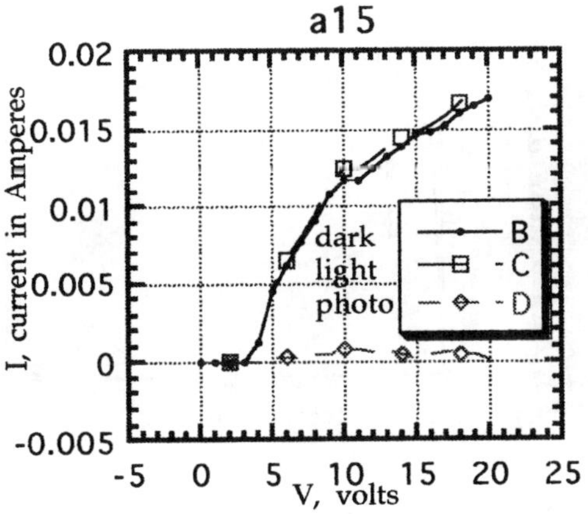

Figure 1: Photocurrent graph for a15 p-type SiC

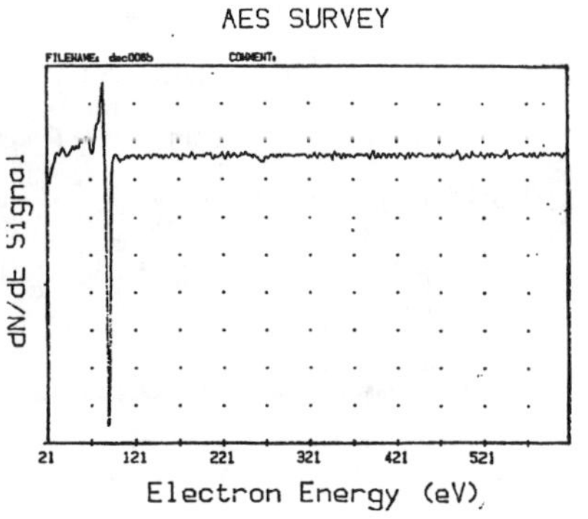

Figure 2: Auger plot for a15 p-type SiC

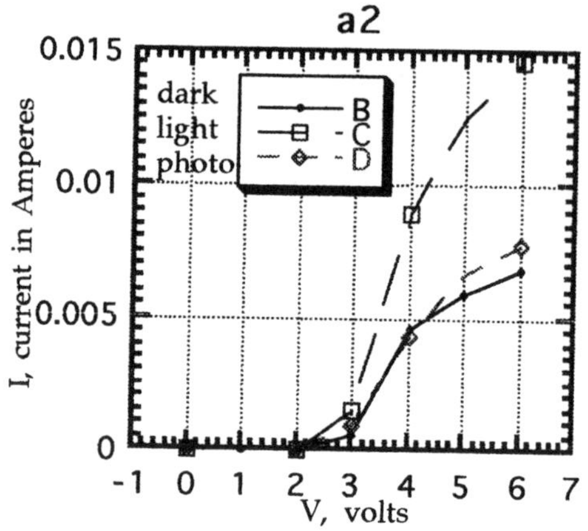

Figure 3: Photocurrent graph for a2 n-type SiC

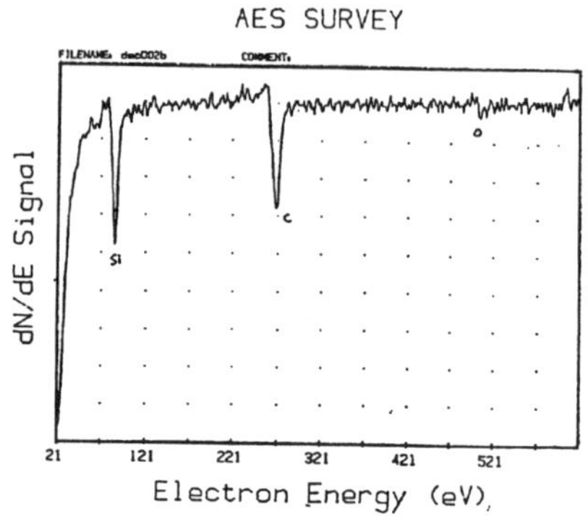

Figure 4: Auger plot for a2 n-type SiC

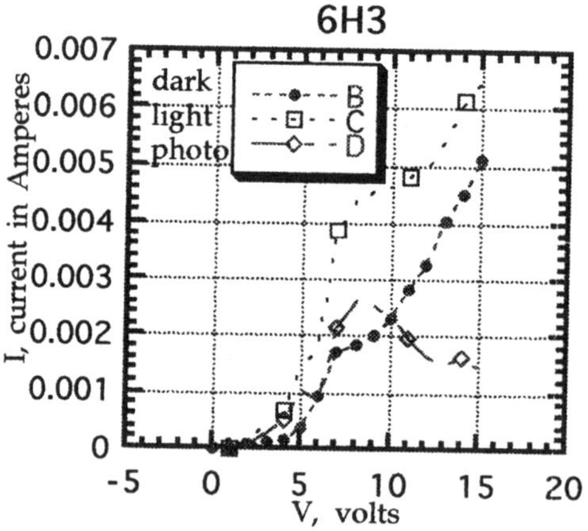

Figure 5: Photocurrent graph for 6H3 6H-SiC

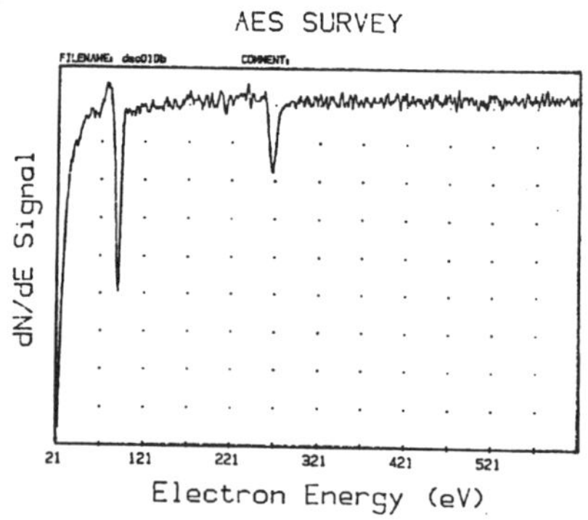

Figure 6: Auger plot for 6H3 6H-SiC

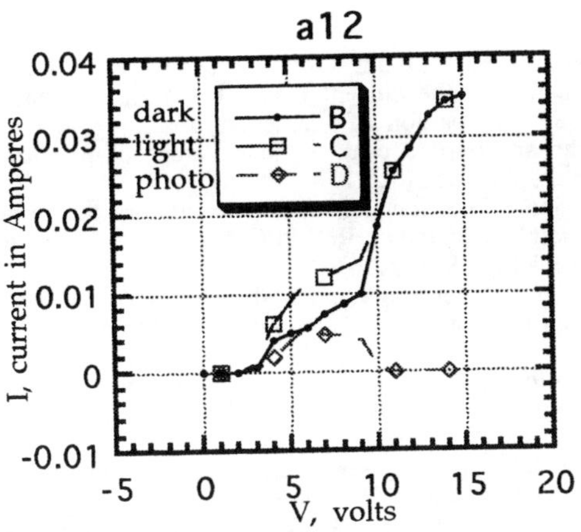

Figure 7: Photocurrent graph for a12 pn layer SiC

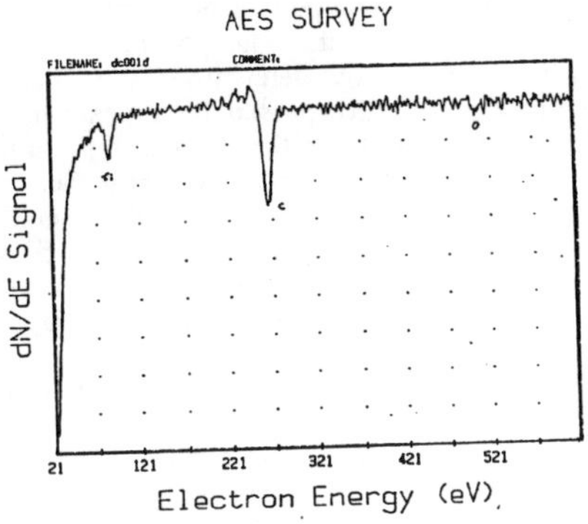

Figure 8: Auger plot for a12 pn layer SiC

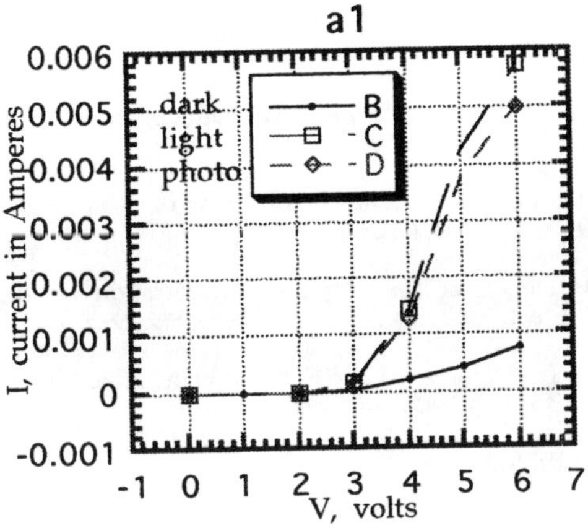

Figure 9: Photocurrent graph for a1 pn layer SiC

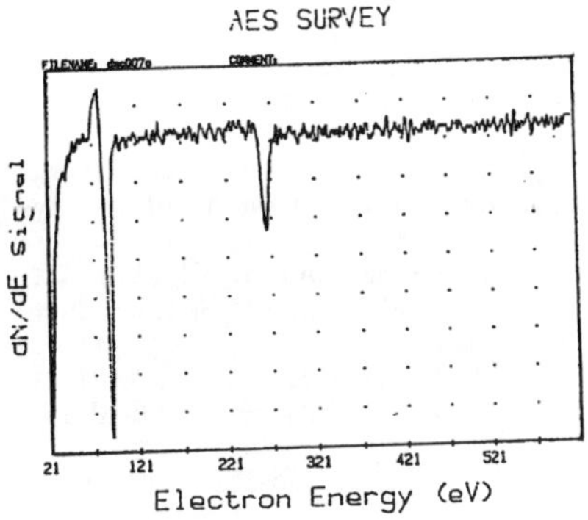

Figure 10: Auger plot for a1 pn layer SiC

It is possible to realize dopant etch-stops of pn junctions with PEC etching of SiC where the n-type semiconductor can be etched. Because etching occurs at higher potentials in p-type semiconductors, etching of n-type should stop at the p-type due to smaller voltages necessary for cut on. Thicker samples and consistent etch rates using the appropriate photocurrent should be able to make consistent etch-stop mechanisms in pn junctions.

CONCLUSIONS

For current-voltage measurements, dark current (no ultraviolet light) and currents with light were measured. The photocurrent equation is the current I(with UV light) - I(dark). Auger spectroscopy determines the chemical composition of elements in sample and the resulting Si/C ratio chosen by measuring the appropriate peak to peak values of Si and C respectively with the ratio of 3 being a nominal value for SiC.

In p-type 3C, no UV light was needed. The light and dark currents are the same. Green colored layers reveal minimum etch depths and thus a low Si/C ratio. Gray or silver color layers remaining after etch is pure Si with a high Si/C ratio.

In n-type 3C, white layers have smaller photocurrents and high Si/C ratios, closer to SiC. The yellow layers have larger photocurrents and low Si/C ratios, preferential to Si removal.

In n-type 6H, the blue layer suggests a low Si/C ratio where more Si removal means a carbon rich sample and the brown layer suggests a high Si/C ratio which means more C removal which is similar to n-type 3C.

PN junctions (n- on top of p-layer) in 3C exhibit similar characteristics to the n-type layers where the white layer corresponds to a high Si/C ratio and the yellow layer corresponds to a low Si/C ratio. It also exhibits similar characteristics to p-type layers where the white layer corresponds to the small photocurrents at low voltages and the yellow layer corresponds to the small photocurrents at high voltages.

ACKNOWLEDGEMENTS

The authors wish to acknowledge NASA headquarters and their support of CSTEA, the NASA Lewis Research Center (NAG 3-431), National Science Foundation (R11-8714676) and NSF-GEE.

REFERENCES

1. J.S. Shor,, R. M. Osgood, and A. D. Kurtz. “Photoelectrochemical conductivity selective etch stops for SiC.” Applied Physics Letters, 60 (Feb 1992), 1001-1003.

2. J.S. Shor,, and R. M. Osgood. “Broad-Area Photoelectrochemical Etching of n-type Beta-SiC.” Journal of Electrochemical Society, 140 (Aug 1993), L123-L125.

3. P. Allongue, and S. Blonkowski. "Influence of the Doping Concentration on the Electrochemical Etching of Semiconductors." Electrochimica Acta, 38 (1993), 889-895.

4. G. L. Harris,, and K. Fekade. "A Technique for Delineation of PN Junction in Cubic Silicon Carbide" Journal of Electrochemical Society, 135 (Feb 1988), 405-407.

5. D. M. Collins, Thesis, Howard University (1995).

Part I

Device Technologies

Implantation

Si IMPLANTATION AND ANNEALING OF GaN FOR n-TYPE LAYER FORMATION

B. MOLNAR *, A. E. WICKENDEN *, M. V. RAO **
*Naval Research Laboratory, Washington, DC, 20375-5347, molnar@estd.nrl.navy.mil
** ECE Department, George Mason University, Fairfax, VA 22030

ABSTRACT

High dose Si has been implanted into MOCVD grown high resistivity and n-type GaN in the 26-500°C temperature range. The implant activation varies widely (30 ->100%) depending on, what energy level is assigned to the Si, the implantation and annealing temperatures, and the quality of the substrate. The usable maximum temperature for activation is limited by the severe decomposition of the GaN. After 1050°C 15s RTA Ga liquid droplet formation has been observed by SEM. This decomposition changes the surface morphology but did not introduce measurable change in the electrical properties up to 1150°C /120s RTA.

INTRODUCTION

Ion implantation is an attractive technique for selective area doping and isolation. Recently high resistive regions have been created in n-or p-type GaN by N implantation and a subsequent annealing [1,2]. Successful p- and n-type doping of GaN by implantation has been also reported [2,3]. In the reported study room temperature Si implantation has been used for n-type doping of GaN with 93% activation of the Si after 1100°C/10s RTA. However, the derived high activation of the Si is based on using 62 meV Si ionization energy. Recently Hall effect measurements on Si doped n-type GaN indicated that the Si ionization energy is between 12 and 17 meV [4].

In the present study, high resistance (>$5x10^6$ ohm/sq), or n-type GaN substrates have been used for Si implantation doping. The Si implant activation results have been evaluated using different high resistance[HR], or n-type GaN layers. In order to gain insight into how the implant temperature controls the electrical activation, the implantations were performed in the temperature range of RT to 500°C. The influence of the annealing (temperature and time) on the GaN decomposition and the effect of decomposition on the GaN electrical properties have been investigated.

EXPERIMENT

The high resistance and n-type GaN layers used in this study were grown at 1040°C on (0001) sapphire substrates by organometallic vapor phase epitaxial (OMVPE) technique at two different laboratories. Van der Pauw type Hall measurements have been used for electrical evaluation of the n-type layers.

Ion implantation was performed in the RT-500°C temperature range. Multiple energy ion implantation has been used in order to create 500 -700 nm thick, uniformly doped layers. The post implantation anneals were performed using either rapid thermal anneal (RTA), or

Mat. Res. Soc. Symp. Proc. Vol. 423 © 1996 Materials Research Society

conventional furnace annealing (FA) in N_2 or forming gas ambient. During anneal the GaN was placed either on a Si_3N_4 surface or on an another GaN wafer in close contact proximity [5].

RESULTS

Since the implantation doping requires high temperature annealing, we have investigated the influence of different annealing conditions in the temperature range 700-1150°C on the surface morphology and electrical properties of the as-grown GaN. The GaN thermal decomposition has been attributed to Ga liquid droplet formation [6,7]. We have correlated the effect of the Ga droplet formation on the electrical properties of the GaN.

The surface of as-grown GaN used in this study is smooth. The morphology of the unprotected GaN surface did not change up to an RTA temperature of 1000°C. After 1050°C/15s anneal, few Ga liquid droplets were observed on the GaN surface as a consequence of N evaporation . As the time or the temperature of the anneals increased, the number of droplets also increased. The Ga liquid droplets formed after 1150°C/15s RTA are shown in Fig. 1(a). The Ga droplets could be removed from the GaN surface by soaking the wafer in HCl. As shown in Fig. 1(b) after HCl soaking the SEM picture revealed holes in the locations where Ga droplets were previously present. The presence of the Ga droplets, however, did not change the integrated electrical properties as determined by the Hall measurements.

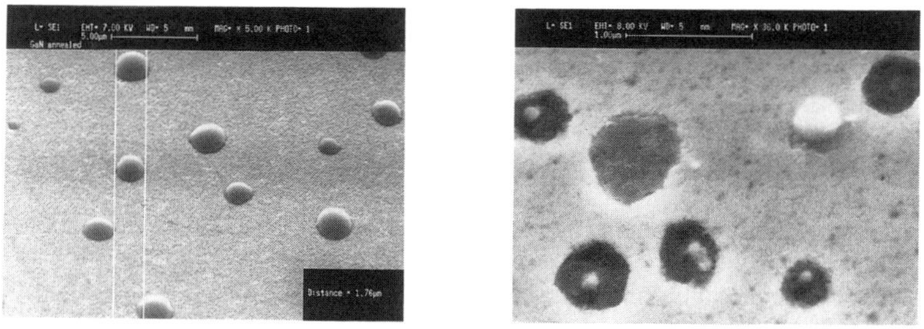

(a) (b)

Fig. 1. SEM micrographs of GaN surface. The samples were annealed at 1150°C/15s. (a) annealed surface, (b) HCl soaked after anneal.

A further increase in the temperature or the time of the anneal, however, leads to coagulation of the Ga-rich liquid droplets. The appearance of Ga coagulation marks the end of the useful range of annealing. Under the coagulated areas the dissolution of the underlying GaN layer has been observed.

The Ga droplet formation could be decreased only by suppressing the nitrogen evaporation. The close contact covering of the GaN surface with another GaN piece or with Si_3N_4 surface during the anneal, reduced the N evaporation. The surface morphologies of GaN heat-treated at 1000°C for 30 min. with and without GaN proximity capping are shown in Fig. 2. As shown in Fig. 2(b) severe deterioration of the surface was observed in the area not covered by the GaN proximity cap. The surface morphology of the areas covered with the GaN proximity cap is smooth, with only few Ga droplets. For the close contact covered areas the sheet resistivity and the mobility values are the same as those in the as-grown material.

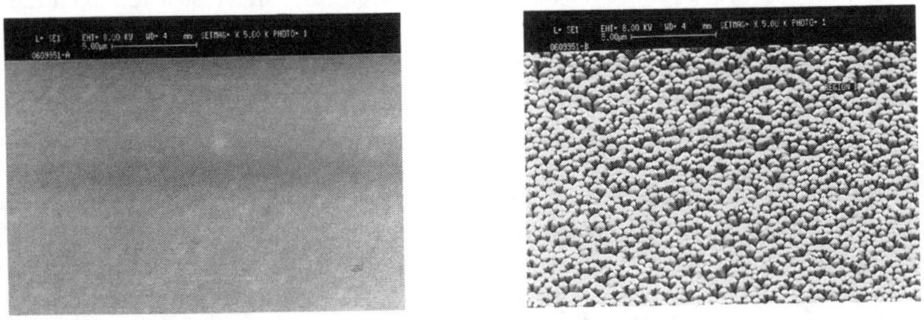

(a) (b)

Fig. 2. SEM micrographs of GaN surface annealed at 1000°C for 30 min. (a) Covered with GaN, (b) without GaN proximity capping.

Our morphological study indicates that the GaN could be heat treated without Ga coagulation for an annealing temperature as high as 1150°C using proximity capping if the anneal duration does not exceed 120s. Therefore, the Si activation RTA has been performed only up to this limit.

First we present our implantation results on high resistance GaN layers. The implantation temperature was 300°C. The Si implant schedule used was 50 keV/4x10^{13} cm^{-2}, 120 keV/1x10^{14} cm^{-2} and 280 keV/3x10^{14} cm^{-2}. These implants created a 600 nm thick surface layer with 10^{19}/cm^{-3} Si concentration. Neither the implanted nor the un-implanted part of the substrate has shown conduction after an RTA at 1000°C for 120s. The Si implanted part begins to show conduction after 1100°C RTA. After 1150°C/120s anneal Hall measurement was possible on both implanted and un-implanted areas.

The measured sheet resistance and mobility values from three different Si implanted samples cut from the same wafer and their average values (<>) are given in the last four rows of Table I. To find the Si implant activation the average sheet carrier concentration $<n_s>$

Table I: Measured and calculated electrical parameters of Si -implanted GaN after 1150°C/120s RTA. For Si activation calculation different E_d's have been used.

E_d (meV)	Θ	(ρ_s)	(cm^2/Vs)	Si activation	n_s, (cm^{-2})
15	0.9	650 (cal)	38 (cal)	48%	2.5x10^{14}
26	0.9	704 (cal)	55 (cal)	75%	1.6x10^{14}
62	0.9	1830 (cal)	85 (cal)	> 100%	4x10^{13}
-	-	627 (meas)	47 (meas)		2x10^{14}
-	-	793 (meas)	57 (meas)		1.4x10^{14}
		1330 (meas)	60 (meas)		8x10^{13}
		<916>	<55>		<1.2x10^{14}>

was compared with the sheet carrier concentrations, derived from the SIMS Si distribution. The ionization energy (E_d) of the Si donor is required for converting the SIMS Si distribution into the electron distribution. For Si dopant the E_d values are reported between 15-62 meV [2,4]. Using 15, 26 and 62 meV values for E_d we have derived three RT carrier profiles from the SIMS Si profile. For calculations of the ρ_s and μ_s values the electron Hall mobility curves of Chin et al. [8] have been used. The measured mobilities for the implanted films are lower than those of epitaxially doped films with the same Si concentration. The calculated mobilities were in agreement with the measured mobilities using the 0.9 compensation ratio. In Table I the calculated n_s and μ_s values are given for different ionization energies and a compensation ratio of 0.9. The percentage activation of the Si varies from 48 to >100%, depending on E_d values used.

We have also observed conduction on the un-implanted part of the HR GaN after 1150°C/120s anneal. We have measured sheet resistance in the range of 0.2-2x10^5 ohms/sq and mobility in the range of 5-20 cm^2/Vs in this material. The error caused by not considering the anneal induced conduction in the above calculation of the Si implant activation is only a few percentage points.

Using HR GaN grown by another laboratory the influence of implantation temperature on electrical properties of the material was investigated. The Si was implanted either at RT or at 500°C . After the 1150°C/120s anneal, using an E_d = 26 meV for Si, the calculated Si activation was 35% for the 500°C implant and only 1% for the RT implant. The difference in the Si activation for different high resistance GaN materials indicate the influence of the background impurities. The higher activation for the 500°C implant compared to the RT implant is due to less lattice damage of the former.

We have also examined the effect of Si implantation on unintentionally doped, n-type GaN. The net carrier concentration of this 1 µm thick GaN was 4x$10^{17}/cm^3$ and the Hall mobility was 130 cm^2/Vs. The same multiple energy implant schedule given before was used. The RT implantation damage removed carriers in the surface layer and hence the layer became highly resistive. After 1000°C/15s anneal the carrier concentration at the surface was still too low for reliable Hall measurements. Only after annealing at 1100°C for 15s or at 900°C for 1 hr we were able to make Hall measurements.

The Si implantation into n-type layer, however, creates two differently conducting parallel layers, the implanted layer and the un-implanted layer. Therefore, a simple Hall measurement gives information for the combined layers. The results on the combined layers are summarized in Table II.

Table II. Hall measurement results for the combined two layers after annealing RT Si implanted n-type GaN with n = 4x10^{17} cm^{-3} background doping.

T{C}	t {s}	ρ_T{ohm/sq}	μ_T {cm^2/Vs}
900	3600	2.8x10^3	40
1100	15	3x10^3	40
1150	15	1x10^3	60
1150	120	950	110

In the two layer structures the total sheet resistance, ρ_T, will be dominated by the sheet resistance of the more conducting layer. Apparently, even after annealing at 1100°C for 15s or at 900°C for 1 hr ρ_T, is still dominated by the un-implanted layer conduction. However, after

1150°C/120s annealing, the implanted layer's conduction begins to show its influence. This indicates an activation of the RT implanted Si, but no activation could be reliably derived.

CONCLUSIONS

In this paper the results of Si implantation doping of MOCVD grown GaN has been presented. Because of the practical need of highly doped regions for contacts, we have only investigated the high dose Si implantation. Such an implantation introduces substantial damage. The ion damage removes carriers from the n-type GaN. By annealing at temperatures > 1000°C the damage could be decreased and the Si activation initiated. The maximum temperature of the anneal, however, is limited by the decomposition of the GaN. The decomposition related Ga liquid formation could be reduced using proximity anneal. We could maximize the Si activation with an 1150°C/120s RTA and still limit the Ga liquid formation. We could determine the activations using high resistance GaN layers in which the parallel conduction of the substrate does not exist. The activation varies widely depending on what energy level is assigned to the Si. We have seen improvement in the activation by raising the temperature of the implantation. The 1150°C/120s anneal alone, has decreased the sheet resistance to $2x10^5$ ohm/sq in HR GaN. The origin of this conduction is presently under investigation.

ACKNOWLEDGMENT

The work at George Mason University is supported by the Office of Naval Research (Dr. Park) under Grant # N00014-95-1-1004.

REFERENCES

1. S.C. Binari, H.B. Dietrich, G. Kelner, L.B. Rowland, K. Doverspike and D.K. Wickenden, J. Appl. Phys. 78, p. 3008 (1995).

2. S.J. Pearton, C.B. Vartuli, J.C. Zolper, C. Yuan and R.A. Stall, Appl. Phys. Lett. 67, p. 1435 (1995).

3. M. Rubin, N. Newman, J.S. Chan, T.C. Fu and J.T. Ross, Appl. Phys. Lett. 64, p. 64 (1994).

4. W. Gotz, N.M. Johnson, D.P. Bour, C.Chen, H. Liu, C. Kuo, and W. Imler, Mat. Res. Soc. Symp. Proc. Vol. 395, to be published.

5. B. Molnar, Appl. Phys. Lett. 36, p. 927 (1980).

6. R. Groh, G. Gerey, L. Bartha and J.I. Pankove, Phys. Stat. Sol. (a) 26, p. 353 (1974).

7. Y. Moromito, J. Electrochem. Soc. 121, p. 13839 (1974).

8. V.W.L. Chin, T.L. Tansley and T. Osotchan, J. Appl. Phys. 75, p. 7386 (1994).

P- AND N-TYPE IMPLANTATION DOPING OF GaN WITH Ca AND O

J. C. ZOLPER,[a] R. G. WILSON,[b] S. J. PEARTON,[c] and R. A. STALL[d]

[a]Sandia National Laboratories, Albuquerque, NM 87185-0603

[b]Hughes Research Laboratory, Malibu CA 90265

[c]University of Florida, Depart. of Materials Science and Engineering, Gainesville, FL 32611

[d]Emcore Corp., Somerset, NJ 0887

ABSTRACT

III-N photonic devices have made great advances in recent years following the demonstration of doping of GaN p-type with Mg and n-type with Si. However, the deep ionization energy level of Mg in GaN (~160 meV) limits the ionized of acceptors at room temperature to less than 1.0% of the substitutional Mg. With this in mind, we used ion implantation to characterize the ionization level of Ca in GaN since Ca had been suggested by Strite [1] to be a shallow acceptor in GaN. Ca-implanted GaN converted from n-to-p type after a 1100 °C activation anneal. Variable temperature Hall measurements give an ionization level at 169 meV. Although this level is equivalent to that of Mg, Ca-implantation may have advantages (shallower projected range and less straggle for a given energy) than Mg for electronic devices. In particular, we report the first GaN device using ion implantation doping. This is a GaN junction field effect transistor (JFET) which employed Ca-implantation. A 1.7 μm JFET had a transconductance of 7 mS/mm, a saturation current at 0 V gate bias of 33 mA/mm, a f_t of 2.7 GHz, and a f_{max} of 9.4 GHz. O-implantation was also studied and shown to create a shallow donor level (~25 meV) that is similar to Si. SIMS profiles of as-implanted and annealed samples showed no measurable redistribution of either Ca or O in GaN at 1125 °C.

INTRODUCTION

The III-V nitride-containing semiconductors InN, GaN, and AlN and their ternary alloys are attracting renewed interest for application to visible light emitters [2,3] and as the basis for high-power or high-temperature electronics [4,5]. Their attractive material properties include bandgaps ranging from 1.9 eV (InN) to 6.2 eV (AlN), an energy gap (E_g(GaN) = 3.39 eV) close to the short wavelength region of the visible spectrum, high breakdown fields, high saturation drift velocities and relatively high carrier mobilities [6,7].

A primary reason for the recent advances in III-N based photonic devices was the demonstration of p-type doping of GaN during MOCVD growth followed by a dehydrogenation anneal to activate the Mg acceptors [8,9]. Unfortunately, the ionization level of the Mg acceptor in GaN is approximately 150 to 165 meV [9] thus limiting the electrically active acceptors at room temperature to ~ 0.3% of the substitutional Mg. The other common column II acceptors used in III-V semiconductors, namely Be, Zn, and Cd, are reported to have still larger ionization energies than Mg, therefore limiting their effectiveness as p-type dopant species [6]. The demonstration of an acceptor in GaN with a smaller ionization level than Mg would contribute to enhancing the p-type conductivity and reducing resistive losses in p/n junction photonic devices. Along these lines, Strite [1] has presented a theoretical argument for Ca to be a shallower acceptor in GaN than Mg based on d-state electron relaxation effects in GaN and the lack of d-state electrons in Ca. One goal of this paper is to test this theoretical hypothesis experimentally by implanting Ca in GaN and measuring the electrical ionization level. The redistribution of Ca in GaN after high temperature annealing is also studied.

Mat. Res. Soc. Symp. Proc. Vol. 423

A second goal of this work is to examine alternative n-type dopants in GaN. N-type doping of GaN during growth [10,11] or by ion implantation [12] has primarily been done using Si which is reported to have a donor ionization level between ~25 and ~60 meV [11,13]. O is of particular interest as a possible alternative n-type dopant due to its position next to N in the periodic table and its suspected role as a background impurity in as-grown GaN [14]. In fact, studies of O introduction in GaN during growth have shown O to act as a donor [14,15]; however, we are not aware of any reports prior to this work on the electrical characterization of O-implanted GaN or on the redistribution properties of O at high temperatures.

We report here on the ion implantation and electrical activation of Ca as an acceptor and O as a donor in GaN. Variable temperature Hall measurements were used to estimate the ionization levels for both dopants. The thermal stability of both species is assessed using secondary ion mass spectrometry (SIMS) profiles of as-implanted and annealed samples. We also reported the first III-N device fabricated with ion implantation doping - a GaN junction field effect transistor.

EXPERIMENTAL

The GaN layers used in the implant doping experiments were 1.5 to 2.0 µm thick grown on c-plane sapphire substrates by metalorganic chemical vapor deposition (MOCVD) in a multiwafer rotating disk reactor at 1040 °C with a ~20 nm GaN buffer layer grown at 530 °C [16]. The GaN layers were unintentionally doped, with background n-type carrier concentrations $\leq 5x10^{16}$ cm^{-3}. ^{40}Ca or ^{16}O ions were implanted at a dose of $5x10^{14}$ cm^{-2} at energies of 180 or 70 keV, respectively, to place the ion peak roughly 100 nm from the surface. One Ca-sample was also implanted with ^{31}P (130 keV, $5x10^{14}$ cm^{-2}) to study the effect of co-implantation, which has been shown to be required to achieve p-type conduction for Mg-implantation in GaN [12]. SIMS samples were prepared with the same implant conditions except ^{18}O isotopes were used to reduced interference problems with background ^{16}O and improve the sensitivity of the SIMS analysis for O. Samples were annealed for 10 to 15 s in flowing N_2 in a SiC coated graphite susceptor between 900 and 1150 °C to study the electrical activation and redistribution of the dopant species.

RESULTS AND DISCUSSION

Figure 1 is the electrical activation data for Ca-implanted GaN with and without the P co-implantation [17]. An unimplanted and annealed sample is included for comparison. Both the Ca-only and the Ca+P samples convert from n-type to p-type after a 1100 °C anneal. This is slightly higher than the temperature required to achieve p-type conduction in Mg+P implanted GaN [12] and may be the result of more implantation induced damage associated with the heavier Ca-ion. The acceptor activity is seen to continue to increase after annealing at 1150 °C with the P co-implanted sample having a 43% lower sheet resistance and a correspondingly higher sheet hole concentration (1.57 versus $1.14x10^{12}$ cm^{-2}). Under the same annealing conditions, the unimplanted sample remains n-type with a slight decrease in sheet resistance that may result from the creation of additional N-vacancies or the depassivation of other n-type impurities. The Hall hole mobility of the Ca-only and Ca+P samples was ~7 cm^2/Vs after a 1150 °C anneal which is in the range reported for epitaxial Mg-doped GaN [8,9]. Figure 2 is an Arrhenius plot of the sheet hole concentration of the 1150 °C annealed Ca-implanted sample along with similar data for a Mg+P implanted sample with the same dose annealed at 1100 °C [13]. A least squares fit to the data gives an ionization level for Ca of 169 meV and 171 meV for Mg+P in GaN. The higher hole concentration in the Ca-sample may be the result of a more optimum annealing temperature for this sample or a difference in the background, compensating, impurity concentrations.

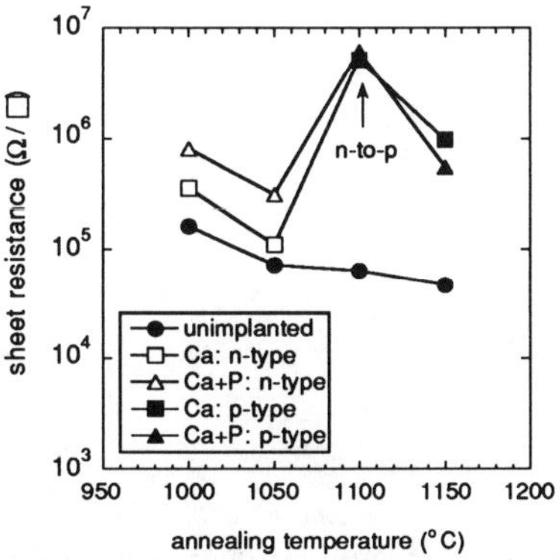

Fig 1. Sheet resistance versus annealing temperature for Ca-implanted (180 keV, 5e14 cm^{-2}) GaN either with or without a P co-implant. Both samples converted from n-to-p type after a 1100 °C anneal.

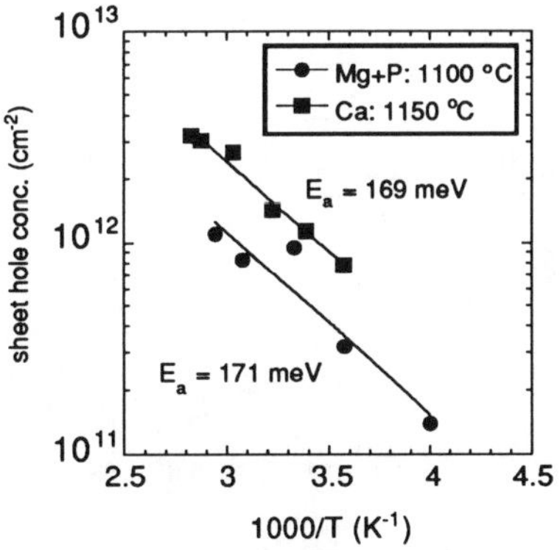

FIG 2. Arrhenius plot of the sheet hole concentration for Mg(+P) and Ca implanted GaN. Both samples have an acceptor ionization energy of ~ 170 meV.

Figure 3 is an Arrhenius plot of the resistance/temperature product of O-implanted GaN annealed at 1050 °C along with data for an unimplanted and annealed (1100 °C) GaN sample. For n-type conduction, an Arrhenius plot of the resistance/temperature product is thought to be more appropriate to account for the potential presence of two band conduction in GaN [18]. O is seen to have an ionization level of 28.7 meV. A similar analysis of Si-implanted GaN yielded an ionization energy of 29 meV [13]. For this ionization energy, 33% of the active donors will be ionized meaning only 3.6% of the implanted O-ions (n_S=5.9x10^{12} cm^{-2}) are activated. The low activation of O may be the result of the lighter O-ion not creating sufficient lattice damage, and therefore N-vacancies, for the O to occupy a substitutional N-site. This situation may be improved in the future by using a co-implantation scheme. The low apparent O activation may also be explained by the existence of a second deep level for O in GaN that is associated with an oxygen complex. If this were the case, the electrons in the deep level would remain unionized at room temperature and not contribute to the measured electron density. Note that the unimplanted and annealed material has an activation energy for conduction of 335 eV.

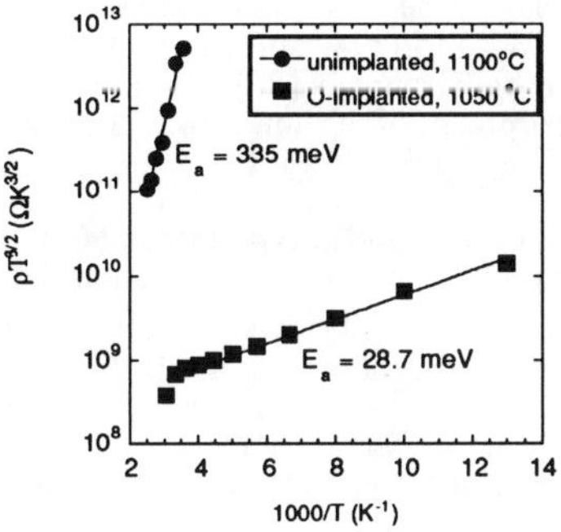

FIG 3. Arrhenius plot of the resistance/temperature product for unimplanted and O-implanted GaN.

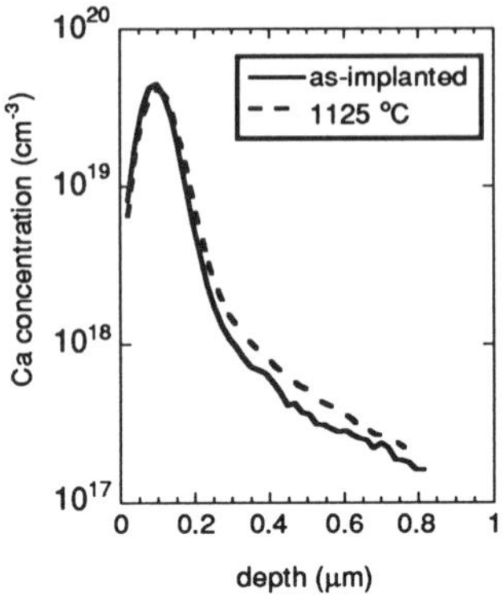

FIG 4. SIMS profiles for as-implanted and annealed (1125 °C, 15 s) Ca (180 keV, $5x10^{14}$ cm^{-2}) in GaN.

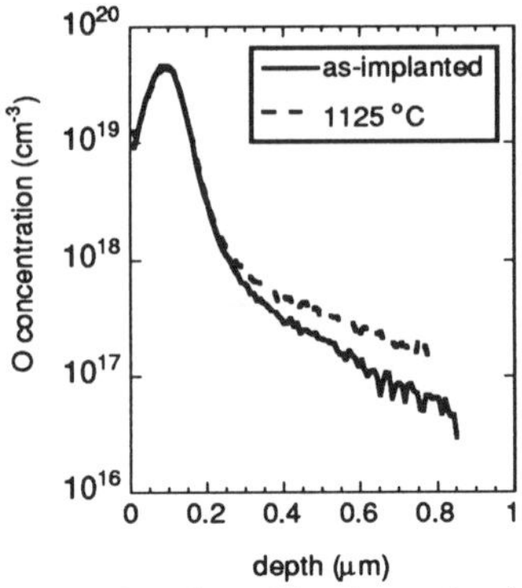

FIG 5. SIMS profiles for as-implanted and annealed (1125 °C, 15 s) O (70 keV, $5x10^{14}$ cm^{-2}) in GaN.

Turning to the thermal stability of Ca and O in GaN, Fig. 4 contains the SIMS profiles for as-implanted and annealed (15 s, 1125 °C) Ca in GaN. Within the resolution of the SIMS measurement (~20 nm), no measurable redistribution has occurred even for this high temperature anneal. This is similar to the redistribution reported for implanted Mg in GaN [19]. By accounting for the exponential dependence for hole ionization [$p \propto p_o exp(E_a/kT)$ where E_a = 169 meV], we estimated that only 0.14 % of the Ca-ions will be ionized at room temperature. Therefore, the peak ionized hole concentration is estimated to be $6.2x10^{16}$ cm^{-3} based on the peak Ca concentration of $4.44x10^{19}$ cm^{-3} in Fig. 4. Although this hole concentration is of the same order as the initial background free electron concentration, the actual acceptor concentration is equal to the Ca-concentration in Fig. 4, since we estimate 100% activation of the implanted Ca ions as acceptors based on a 169 meV ionization energy [17]. Therefore, the unionized Ca-acceptors will compensate the initial background donors and thus allow p-type conduction to be achieved. Figure 5 shows the SIMS profiles for implanted ^{18}O in GaN either as-implanted or annealed at 1125 °C. Here again no measurable redistribution is observed near the peak of the distribution. This is in contrast to dramatic redistribution reported for S, another column VI species, that displayed significant redistribution even after a 600 °C anneal [20]. Based on a conservative estimate of the resolution of the SIMS measurement of 20 nm, an upper limit of $2.7x10^{-13}$ cm^2/s is estimated for the diffusivity of Ca and O in GaN at 1125 °C.

With the demonstration of n- and p-type implantation doping, we have now applied this technology to the fabrication of the first GaN junction field effect transistor (JFET) [21]. Figure 6 shows I_{DS}/V_{DS} data for a ~1.7x50 μm^2 GaN JFET made with a Ca p-type implant and Si n-type implants as represented in the schematic in Fig 7. This device had a maximum transconductance of 7.5 mS/mm at V_{gs} = -2 V and a I_{DS} of 33 mA/mm at 0 V gate bias. The frequency performance of this device was very respectable with a f_t = 2.7 GHz and f_{max} = 9.4 GHz. These metrics are comparable to reported values for epitaxial GaN MESFETs [22].

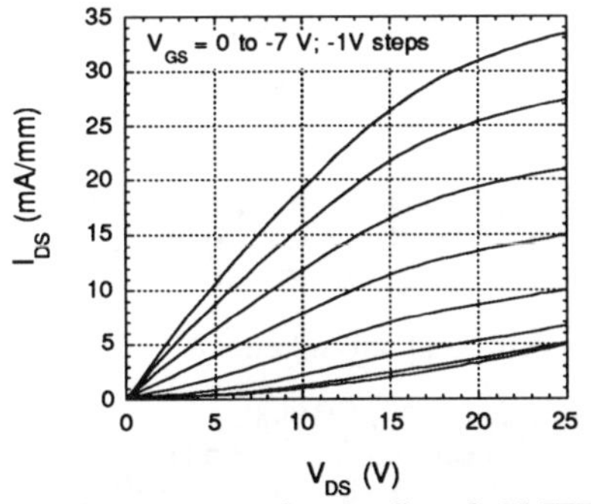

FIG 6: I_{DS} vs V_{DS} for the first GaN JFET. The gate length is ~1.7 μm.

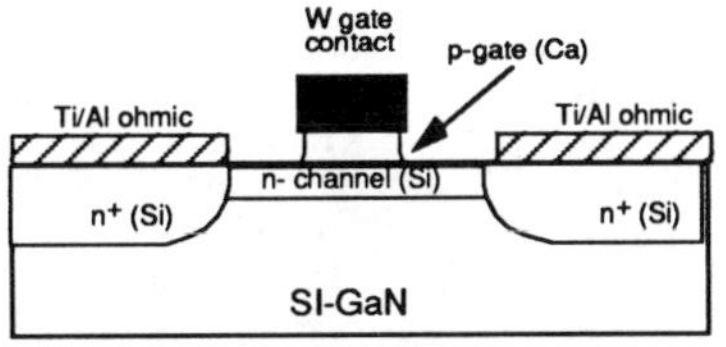

FIG 7. Schematic of all ion implanted GaN JFET using Ca (p-type) and Si (N-type) ion implantation.

CONCLUSION

Ion implantation can be expected to play an enabling role in advanced devices based on III-Nitride materials. In this paper, we have reported results for implantation doping of GaN with Ca for p-type and O for n-type conduction. Ca is seen to have an estimated ionization energy of 170 meV which is similar to that of Mg, the only other acceptor with $E_a < 200$ meV. Both Ca and O where shown to display no measurable redistribute for an RTA at 1125 °C. The first GaN JFET was also report which also represents the first III-Nitride based device fabricated with ion implantation doping.

Acknowledgments: The portion of this work performed at Sandia National Laboratories was supported by the U.S. Department of Energy under contract # DE-AC04-94AL85000. The work at Hughes was supported by ARO (Dr. J. M. Zavada). The work at the University of Florida is partially supported by a National Science Foundation grant (DMR-9421109) and a University Research Initiative grant from ONR (N00014-92-5-1895). The work at EMCORE was supported by BMDO-IST managed by M. Yoder at ONR. The technical support of J. Escobedo at Sandia with implantation and annealing and the support of the MicroFabritech facility at UF is greatly appreciated.

References

[1] S. Strite, Jpn. J. Appl. Phys. **33**, L699 (1994).
[2] S. Nakamura, T. Mukai, and M. Senoh, Appl. Phys. Lett., **64**, 1687 (1994).
[3] I. Akasaki, H. Amano, M. Kito, and K. Hiramatsu, J. Lumin. **48/49**, 666 (1991).
[4] S. C. Binari, L. B. Rowland, W. Kruppa, G. Kelner, K. Doverspike, and D. K. Gaskill, Electronics Lett., **30**, 1248 (1994).
[5] M. A. Khan, J. N. Kuznia, D. T. Olsen, W. J. Schaff, J. W. Burm, and M. S. Shur, Appl. Phys. Lett., **65**, 1121 (1994).
[6] S. Strite and H. Morkoç, J. Vac. Sci. Technol. B **10**, 1237 (1992).
[7] T. P. Chow and R. Tyagi, IEEE Trans. Electron. Dev., **41**, 1481 (1994).
[8] H. Amano, M. Kito, K. Hiramatsu, and I. Akasaki, Jap. J. Appl. Phys. **28** L2112 (1989).
[9] S. Nakamura, T. Mukai, M. Senoh, and N. Iwasa, Jap. J. Appl. Phys. **31** L139 (1992).
[10] S. Nakamura, T. Mukai, and M. Senoh, Jpn. J. Appl. Phys. **31**, 2883 (1992).
[11] J. G. Kim, A. C. Frenkel, H. Liu, and R. M. Park, Appl. Phys. Lett. **65** 91 (1994).

[12] S. J. Pearton, C. B. Vartuli, J. C. Zolper, C. Yuan, R. A. Stall, Appl. Phys. Lett. **67**, 1435 (1995).
[13] J. C. Zolper, M. Hagerott Crawford, S. J. Pearton, C. R. Abernathy, C. B. Vartuli, C. Yuan, and R. A. Stall, J. Elec. Mat. **25**, 839 (1996).
[14] B-C. Chung and M. Gershenzon, J. Appl. Phys. **72**, 651 (1992).
[15] H. Sato, T. Minami, E. Yamada, M. Ishii, and S. Takata, J. Appl. Phys. **75,** 1405 (1994).
[16] C. Yuan, T. Salagaj, A. Gurary, P. Zawadzki, C. S. Chern, W. Kroll, R. A. Stall, Y. Li, M. Schurman, C.-Y. Hwang, W. E. Mayo, Y. Lu, S. J. Pearton, S. Krishnankutty, and R. M. Kolbas, J. Electrochem. Soc. **142**, L163 (1995).
[17] J. C. Zolper, R. G. Wilson, S., J. Pearton, and R. A. Stall, Appl. Phys. Lett. **68**, 1945 (1996).
[18] R. J. Molnar, T. Lei, and T. D. Moustakas, Appl. Phys. Lett. **62**, 72 (1993).
[19] J. C. Zolper, M. Hagerott Crawford, A. J. Howard, S. J. Pearton, C. R. Abernathy, C. B. Vartuli, C. Yuan, R. A. Stall, J. Ramer, S. D. Hersee, R. G. Wilson, Conf Proc. MRS, Fall 1995, symposium AAA (Materials Research Society, Pittsburgh, PA, in press).
[20] R. G. Wilson, S. J. Pearton, C. R. Abernathy, and J. M. Zavada, Appl. Phys. Lett. **66**, 2238 (1995).
[21] J. C. Zolper, R. J. Shul. A. G. Baca, R. G. Wilson, S., J. Pearton, and R. A. Stall, Appl. Phys. Lett. **68**, 2273 (1996).
[22] S. C. Binari, L. B. Rowland, G. Kelner, W. Kruppa, H. B. Dietrick, K. Doverspike, and D. K. Gaskill, in *Proceedings of 1994 International Symposium on Comp. Semiconductors,* San Diego, CA Sept. 1994 (Institute of Physics, Bristol, UK, 1995) pp. 492-496.

ION BEAM SYNTHESIS BY TUNGSTEN IMPLANTATION INTO 6H-SILICON CARBIDE AT ELEVATED TEMPERATURES

HANNES WEISHART, W. MATZ AND W. SKORUPA
Forschungszentrum Rossendorf e.V., Institut für Ionenstrahlphysik und Materialforschung, Postfach 510119, D-01314 Dresden, Germany

ABSTRACT

We studied high dose implantation of tungsten into 6H-silicon carbide in order to synthesize an electrically conductive layer. Implantation was performed at 200 keV with a dose of 1×10^{17} W^+cm^{-2} at temperatures of 90°C and 500°C. The samples were subsequently annealed either at 950°C or 1100°C. The influence of implantation and annealing temperatures on the reaction of W with SiC was investigated. Rutherford backscattering spectrometry (RBS), x-ray diffraction (XRD) and Auger electron spectroscopy (AES) contributed to study the structure and composition of the implanted layer as well as the chemical state of the elements. The implantation temperature influences the depth distribution of C, Si and W as well as the damage production in SiC. The W depth profile exhibits a bimodal distribution for high temperature implantation and a customary gaussian distribution for room temperature implantation. Formation of tungsten carbide and silicide was observed in each sample already in the as-implanted state. Implantation at 90°C and annealing at 950°C lead to crystallization of W_2C; tungsten silicide, however, remains amorphous. After implantation at 500°C and subsequent annealing at 1100°C crystalline W_5Si_3 forms, while tungsten carbide is amorphous.

INTRODUCTION

Silicon carbide is an auspicious material for high-temperature semiconductor devices because of its superior properties such as high thermal conductivity, high electron velocity and wide bandgap [1 , 2]. High-temperature devices, however, are in need of reliable metallization since their working temperature can be as high as 700°C. A proper ohmic contact should have a low resistivity, good adhesion to the underlying SiC, as well as high chemical stability at elevated temperatures for more than 1000 hours. Several attempts have been made to develop such metallization on 6H-SiC [3 - 10]. Contacts that could meet all of the above-mentioned requirements are either layered structures consisting of a reactive metal to ensure adhesion, a diffusion barrier and a low resistivity metal, or highly temperature stable low resistivity materials. Promising materials of the latter class are tungsten silicide and tungsten carbide.

Fabrication of W-based contact layers was achieved up to now by (i) deposition of pure tungsten on SiC and subsequent annealing [5 , 11] and (ii) sputter deposition, subsequent ion beam mixing and final annealing of W/Si multilayers [12]. However, the first technique requires annealing temperatures in excess of 1000°C and, moreover, produces an inhomogenous mixture of polycrystalline tungsten silicide, tungsten carbide and tungsten. The latter needs too much expenditure to form a homogeneous layer of tungsten silicide. The reaction of a deposited W layer with crystalline SiC was investigated earlier [13 , 14] using Auger electron spectroscopy. Already after deposition, little amounts of tungsten carbide and silicide were found, which formed at defects within the W/SiC-interfacial region. Baud et al. showed that annealing at 950°C is necessary to maintain further the reaction between SiC and W [1 1]. We have shown earlier that synthesis of tungsten carbide and silicide in SiC is possible at low temperatures by using ion

Mat. Res. Soc. Symp. Proc. Vol. 423

implantation of W into SiC [15]. This technique introduces at the same time defects and tungsten into SiC and thus facilitates the formation of tungsten compounds in the implanted region. However, annealing temperatures of 1500°C would be necessary to recrystallize the amorphized SiC after room-temperature (RT) implantation [16 , 17]. High temperature implantation may reduce the damage introduced into SiC.

Ion beam synthesis of Ti compounds in SiC has been studied by Vardiman [18]. An analysis of the phases in the implanted and annealed SiC revealed that (i) crystallization is temperature- and composition-dependent and (ii) formation would not occur under equilibrium conditions. He has employed thermodynamic and kinetic factors for explanation of these observations.

The present work reports results on high dose ion implantation of W^+ at an elevated temperature. These results are compared to those obtained earlier by RT implantation [15]. The objective of the investigation is to gain an insight into the influence of both implantation and annealing temperature on the depth distribution of elements, the phase formation as well as the properties of W-implanted 6H-SiC.

EXPERIMENT

Ion implantation of W^+ was performed into wafers of single crystal n-type 6H-SiC from Cree Research, Inc. at a dose of $1\times10^{17}cm^{-2}$ using an energy of 200 keV. The mean projected range for this energy, according to TRIM91, is 51 nm. The substrates were implanted at 500°C and subsequently underwent an anneal for 30 min in Ar either at 950°C or at 1100°C. Analysis of the W profile and stoichiometry of the synthesized layers was made by Rutherford backscattering spectrometry (RBS) of 1.2 MeV $^4He^+$. Auger electron spectroscopy (AES) in combination with sputter depth profiling during sample rotation was performed using a 3 keV Ar^+ beam at an incidence angle of 54° to the surface normal. The AES spectra of crystalline Si, 6H-SiC, WC and WSi_2 were taken as references to characterize the chemical state of the elements. X-ray diffraction (XRD) with gracing incidence was employed to determine the crystalline phases.

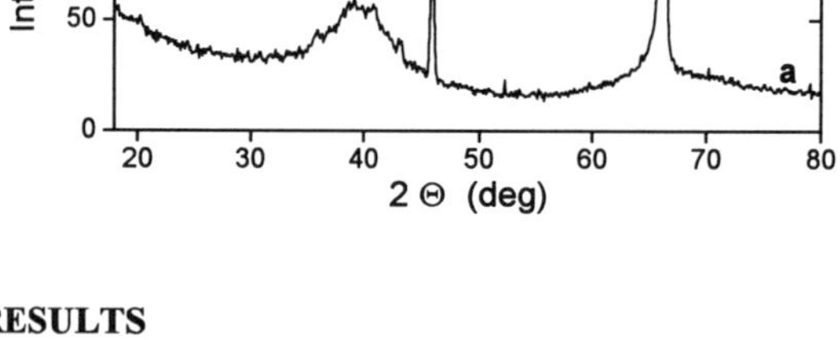

Fig. 1 X ray diffraction pattern of 6H-SiC implanted with W^+ at 500°C. The as implanted sample shows a broad peak around 40° indicating a beginning phase formation (a). The sample annealed at 1100°C shows distinct peaks of crystalline W_5Si_3 (b). The observed SiC peaks are not due to direct Bragg reflection but due to Umweganregung. The W_5Si_3 peaks are marked with arrows.

RESULTS

X-ray diffraction performed on the sample with the lower implantation temperature indicates that in the as implanted state no crystalline phase exists in the near-surface region. Annealing at 950°C causes crystallization of W_2C. No evidence of crystalline WC, WSi_2 or W_5Si_3 is found. Figure 1a shows an x-ray diffraction spectrum of the sample implanted at 500°C in the as

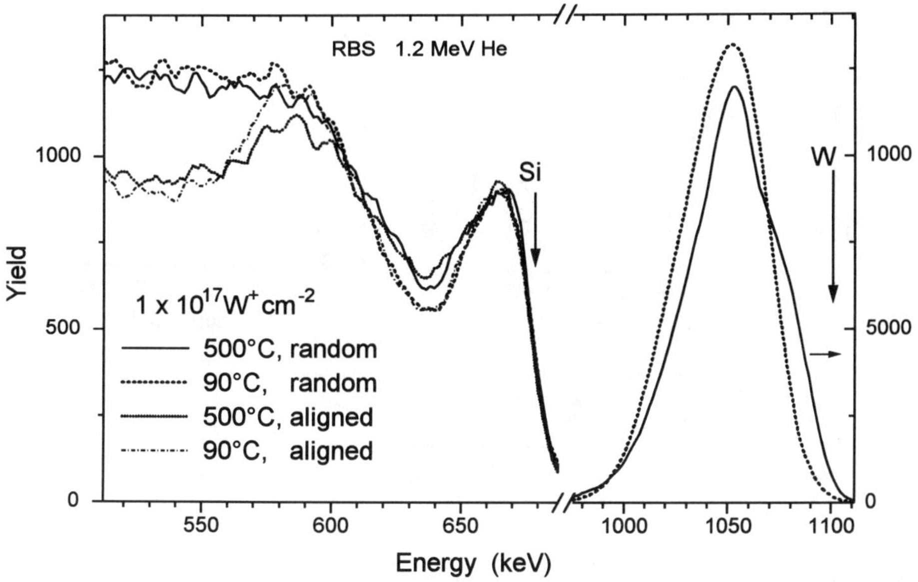

Fig. 2 Channeled and random RBS spectra of SiC samples implanted with 1×10^{17} W^+ cm^{-2} at 200keV.
The implantation temperatures are 90°C and 500°C. The surface edges of Si and W are marked with arrows.
The aligned spectra indicate that implantation at 500°C produces less damage in the SiC.

implanted state. Two distinct SiC peaks, originating from Umweganregung, as well as a broad peak around 40° are observable. Annealing at 1100°C results in crystallization of W_5Si_3 as can be seen from the peaks between 27° and 46° in the spectrum of Fig. 1b.

The RBS channeled and random spectra of the as implanted samples for the implantation temperatures of 90°C and 500°C are depicted in Fig. 2. The W peak for the lower implantation temperature is higher and more narrow. The tungsten peak for the implantation temperature of 500°C shows a small shoulder on its right edge, i.e. in direction to the surface. A comparison of the aligned spectra at an energy of 570 keV reveals that the sample implanted at 90°C has a sharper crystalline/amorphous interface in the SiC. The difference between aligned and random spectra indicate a less pronounced damage for the sample implanted at the higher temperature.

Auger depth profiling of the samples implanted at 500°C reveals a bimodal tungsten distribution. This double peak is clearly visible in Fig. 3a for the unannealed sample. Both the carbon and the silicon concentration show a deep minimum at the second (the deeper) tungsten peak, which is positioned slightly above the depth of the mean projected range. The first tungsten peak is reflected by a minimum only in the carbon concentration. After annealing, the W peaks have decreased in height, while the amount of tungsten has increased beyond the deeper peak. A comparison of the C and the Si line shapes with reference spectra of 6H-SiC, WSi_2 and WC demonstrates clearly the depth dependent chemical states of Si, W and C. Formation of tungsten carbide and tungsten silicide already occurs in the as implanted sample within the implantation zone. Both the C and the Si line shapes indicate a preferential bonding of W with C and Si, respectively, at the expense of SiC within the two tungsten rich layers. Annealing reduces the amount of tungsten silicide and tungsten carbide in the first W peak.

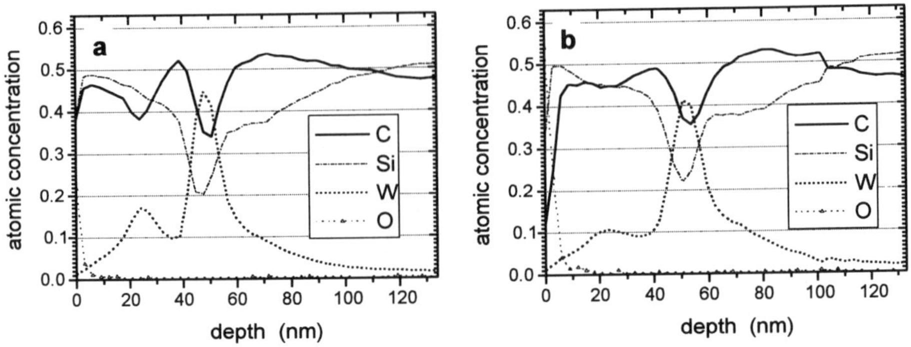

Fig. 3 Auger sputter depth profiles of SiC samples implanted with $1\times10^{17}cm^{-2}$ W^{+} at 500°C. Concentration profiles of carbon, silicon, tungsten and oxygen for the as implanted sample (a) and the sample annealed at 1100°C for 30min (b).
The tungsten profile shows two peaks, while carbon and silicon profiles show minima at that depths. Annealing leads to smoothing of the concentration profiles. The heights of the tungsten peaks are reduced, while the tungsten concentration beyond the second peak increases.

DISCUSSION

Implanting at higher temperatures reduces the damage introduced into SiC. This result can clearly be seen from a comparison of the RBS aligned spectra around 580 keV in Fig. 2. The W double peak found by AES in the samples implanted at 500°C cannot be resolved by RBS. Only a shoulder appears on the right side of the tungsten signal depicted in Fig. 2. The insufficient resolution of RBS is also the reason for the lower height of the tungsten profile compared to that of the sample implanted at 90°C. This phenomenon of an apparently lower W signal is caused by a tungsten rich layer of a thickness beyond the resolution limit of the detector.

The W profile of the 90°C implant displays a broad single peak [15]. The W profile of the 500°C implantation has two peaks, with the second peek being much more pronounced. The diffusion of tungsten is enhanced by the higher implantation temperatures. This fact is obvious in Fig. 2 from the increased tungsten concentration at energies around 980 keV after 500°C implantation. The enhanced mobility of atoms at elevated implantation temperatures may be the reason for the development of two W peaks instead of a single gaussian peak; however, as yet the basic driving force cannot be reasonably accounted for. An explanation can be given in accordance with the results of Vardiman [18]. He found an initially gaussian like Ti peak developing into a double peak structure after annealing at 850°C. Vardiman suggested a preferential nuclei formation of a ternary phase at the location of the steepest initial Ti concentration gradient. Assuming an enhanced erosion of the surface at elevated implantation temperatures, one can expect that the tungsten profile will move towards the surface. If one shifts the profile depicted in Fig. 3 for 15 nm to the right then the minimum between the two peaks will coincide with the mean projected range of 51 nm. In this case the W profile could indeed originate from an initially gaussian distribution, which subsequently transformed to a double peak structure during implantation. First nuclei of tungsten carbide and tungsten silicide might have formed at the location of the steepest initial W concentration gradient. The enhanced atom mobility at elevated implantation temperatures may lead to an Ostwald ripening process, so that the nuclei which form later in the region of the initial W maximum concentration shrink and disappear. Evidence for this nucleation process is the narrow structure of the second W peak as well as the XRD spectrum of the as implanted sample. The indistinct peak centered around 40° is more pronounced in the case of 500°C implantation than in the case of 90°C implantation. This fact indicates the presence of a larger number of microcrystallites or larger sized crystallites in the sample.

The concentration profiles of samples annealed at 950°C remain unchanged, as reported earlier [15]. However, Fig. 3b demonstrates that the elements diffuse during annealing at 1100°C. Consequently, the undulating concentration profiles present after implantation are smoothed. Tungsten seems to diffuse into the bulk. Conversely, carbon seems to move towards the surface. Auger spectra show that the loss of W is accompanied by a loss of tungsten carbide and silicide.

Annealing the samples above 950°C leads to crystallization of new phases. The XRD results indicate crystallization of W_2C for the 90°C implant and of W_5Si_3 for the 500°C implant. The absence of a distinct plateau in the W peak of the RBS spectra, however, indicates the presence of a mixture of several tungsten containing phases. This assumption is verified by the AES depth profiles, which prove the existence of tungsten silicide as well as tungsten carbide in all samples.

According to Vardiman who reported that each phase has its own, composition dependent crystallization temperature, we believe W_2C to be the energetically more favored phase to crystallize during annealing at 950°C under sufficiently high carbon supersaturation in the matrix. The carbon concentration of the sample implanted at 90°C never drops below a value of 40 at% [15]. Some amount of free carbon may therefore exist in the implanted region. The critical radius for W_2C nuclei in such a matrix is much larger than for nuclei which form in a matrix of less excess carbon. The sample implanted at 500°C exhibits a loss of carbon in the tungsten rich regions at 20 nm and 50 nm depth, as is clearly visible in Fig. 3. Thus, formation of W_5Si_3 during annealing appears to be favoured.

Resistance measurements using the four point probe technique on the sample implanted at 500°C and annealed at 1100°C were performed after removal of the oxidized SiC overlayer. A mean value for the sheet resistance of $R_s = 125\ \Omega/\square$ was found. From RBS the thickness of the W containing layer was estimated to be 54 nm. We then obtain a resistivity of $\rho = 675\ \mu\Omega$cm, which is only by 19% higher than the resistivity of the sample implanted at 90°C [15].

CONCLUSIONS

Implantation of W^+ in 6H-SiC leads to the formation of amorphous tungsten carbide and tungsten silicide. Subsequent annealing results in crystallization of one of these phases. Analysis by XRD and AES show that crystalline W_2C mixed with amorphous W_xSi_y and SiC is formed after implantation at 90°C and annealing at 950°C. Crystalline W_5Si_3 mixed with amorphous W_xC and SiC is formed after implantation at 500°C and annealing at 1100°C. The reason for dominant crystallization of tungsten silicide in such a sample may be due to a lack of free carbon in the tungsten rich regions. The resistivity of this layer is $\rho = 675\ \mu\Omega$cm which is only by one order of magnitude higher than that for single crystalline tungsten silicide [19].

The diffusivity of W is increased by elevated implantation temperatures. Such temperatures produce a bimodal W distribution, which cannot reasonably be explained, yet. A conceivable reason may be Ostwald ripening of nuclei which form preferentially at the steepest concentration gradient of an initial gaussian distribution.

ACKNOWLEDGMENTS

The authors want to thank Dr. J. Schöneich for the implantation, Mrs. E. Quaritsch and Dr. H. Reuther for the AES and Drs. M. Voelskow and M. Mäder for the RBS measurements. We are especially indebted to Dr. R. Yankov for fruitful discussions.

REFERENCES

1. W.E. Nelson, F.A. Halden and A. Rosengreen, J. Appl. Phys. **37,** 333 (1966).
2. G. Pensl and R. Helbig, Festkörperprobleme/Advances in Solid State Physics **30**, 133 (1990).
3. J.B. Petit and M.V. Zeller in Wide Band Gap Semiconductors, edited by T.D. Moustaka, J.I. Pankove, and Y. Hamakawa (Mater. Res. Soc. Proc. **242,** Pittsburgh, PA, 1992) pp. 567-572.
4. R.C. Glass, L.M. Spellman, and R.F. Davis, Appl. Phys. Lett. **59,** 2868 (1991).
5. M.M. Anikin, M.R. Rastegaeva, A.L. Syrkin, and I.V. Chuiko, Springer Proceedings in Physics **56,** 183 (1992).
6. J. Crofton, J.M. Ferrero, P.A. Barnes, J.R. Williams, M.J. Bozack, C.C. Tin, C.D. Ellis, J.A. Spitznagel, and P.G. McMullin, Springer Proceedings in Physics **71**, 176 (1992).
7. V.A. Dmitriev, K. Irvine, M. Spencer, and G. Kelner, Appl. Phys. Lett. **64,** 318 (1994).
8. A.K. Chaddha, J.D. Parsons, and G.B. Kruaval, Appl. Phys. Lett. **66**, 760 (1995).
9. T. Uemoto, Jpn. J. Appl. Phys. **34,** L7 (1995).
10. J. Crofton, P.G. McMullin, J.R. Williams, and M.J. Bozack, J. Appl. Phys. **77**, 1317 (1995).
11. L. Baud, C. Jaussaud, R. Madar, C. Bernard, J.S. Chen, and M.A. Nicolet, Mat. Sci. Eng. **B29,** 126 (1995).
12. H. Zhang, PhD thesis, Friedrich Alexander Universität Erlangen, 1990.
13. K.M.Geib, C. Wilson, R.G. Long, and C.W. Wilmsen, J. Appl. Phys. **68,** 2796 (1990).
14. C. Jacob, S. Nishino, M. Mehregany, and P. Pirouz, in Silicon Carbide and Related Materials (Institute of Physics Publishing, Bristol and Philadelphia, 1994) ch.3 p.247.
15. H. Weishart, J. Schöneich, H.-J. Steffen, W. Matz, and W. Skorupa in Beam-Solid Interactions for Materials Synthesis and Characterization, edited by D.E. Luzzi, T.F. Heinz, M. Iwaki, and D.C. Jacobson (Mater. Res. Soc. Proc. **354**, Pittsburgh, PA, 1995) pp. 177-182; Nucl. Instr. Meth. **B10532**, xxx (1996).
16. C.J. McHargue and J.M. Williams, Nuc. Instr. Meth. **B80/81,** 889-94 (1993).
17. V. Heera, R. Kögler, W.Skorupa, and J. Stoemenos, Appl. Phys. Lett. **67**, 1999 (1995).
18. R.G. Vardiman, Materials Science and Engineering **A177,** 209 (1994).
19. S.P. Murarka, in Silicides for VLSI Applications (Academic Press, Orlando, 1983) p. 30.

TEM STRUCTURE CHARACTERIZATION OF Ti/Al and Ti/Al/Ni/Au OHMIC CONTACTS FOR n-GaN

S. Ruvimov, Z. Liliental-Weber and J. Washburn
Materials Sciences Division, Lawrence Berkeley National Laboratory, University of California, MS 62-203, Berkeley, CA 94720

K.J. Duxstad and E.E. Haller
Department of Materials Science and Mineral Engineering, University of California, and Lawrence Berkeley National Laboratory, Berkeley, CA 94720

Z.-F. Fan, S.N. Mohammad, W. Kim, A.E. Botchkarev, and H. Morkoç
Materials Research Laboratory and Coordinated Science Laboratory, University of Illinois at Urbana-Champaign, IL 61810

ABSTRACT

Transmission electron microscopy has been applied to characterize the structure of Ti/Al and Ti/Al/Ni/Au ohmic contacts on n-type GaN (~10^{17} cm^{-3}) epitaxial layers. A thin polycrystalline cubic TiN layer epitaxially matched to the (0001) GaN surface was detected at the interface with the GaN substrate. This layer was studied in detail by electron diffraction and high resolution electron microscopy. The orientation relationship between the cubic TiN and the GaN was found to be: $\{111\}_{TiN}//\{00.1\}_{GaN}$, $[110]_{TiN}//[11.0]_{GaN}$, $[112]_{TiN}//[10.0]_{GaN}$. The formation of this cubic TiN layer results in an excess of N vacancies in the GaN close to the interface which is considered to be the reason for the low resistance of the contact.

INTRODUCTION

Nitride-based electronics and optoelectronics requires high reliability of metal contacts for n- and p-GaN because the performance of devices strongly depends on contact resistance. Recent success in fabrication of such devices as visible light-emitting diodes (LEDs)[1-3] and metal-semiconductor field-effect transistors (MESFETs)[4] evidences that GaN and related compounds are very promising for optoelactronics applications and, in particular, for high temperature applications. Nitrides are known to have an excellent thermal conductivity, high saturation velocity and large breakdown field.[5] Thus, a development of a low resistance ohmic contact for GaN is of great practical importance. A few attempts to achieve good ohmic contacts on GaN epilayers have been reported recently.[5-8] Most of these metallization schemes include Ti as a first metal layer and reactive ion etching (RIE) of the GaN surface prior to the metal deposition. Low resistance contacts to GaN were formed as a result of these processes, even without annealing (e.g. 6.5×10^{-5} Ω cm^2 in the case of a Ti/Ag contact[6]). Similar results were reported for W and WSi_x contacts.[8] The reason for the low resistivity of these contacts is still under discussion. It is resonable to assume that the low resistivity might be related to the interaction of Ti or W with the GaN. As a result, the formation of TiN or WN layers on the GaN surface and, associated with them, an excess of N vacancies near the interface is a possible cause for the low contact resistivity. However, until the present study no direct evidence of the formation of such a layer or its structure has yet been reported.

The structure and morphology of the Ti/Al and Ti/Al/Ni/Au thin composite layers which provide ohmic contacts on n-type GaN (~10^{17} cm^{-3}) epitaxial layers are the focus for the present study. Here, we report results on the metallurgy of GaN/Ti/Al and GaN/Ti/Al/Ni/Au contacts which form a thin cubic TiN layer on the GaN surface.

Mat. Res. Soc. Symp. Proc. Vol. 423 © 1996 Materials Research Society

EXPERIMENTAL

3 μm-thick n-GaN layers were grown by molecular beam epitaxy (MBE) on c-plane Al_2O_3.[7] The GaN layers contained two sublayers, a 2 μm thick Si-doped layer with a carrier concentration of about $5x10^{18}$ cm^{-3} and a 1μm thick undoped n-GaN layer on top with a carrier concentration of about 10^{17} cm^{-3}. Before the metal deposition, the GaN surface was exposed to reactive ion etching (RIE). The composite metal layers were either Ti/Al (20 nm/100nm) or Ti/Al/Ni/Au (15nm/220nm/40nm/50nm). Details of metallization schemes are reported elsewhere.[5,7] A rapid thermal anneal (RTA) at 900 °C for 30 s in an N_2 environment provides the lowest resistance contact of about $8x10^{-6}$ Ω cm^2 and $8.9x10^{-8}$ Ω cm^2 for the Ti/Al and Ti/Al/Ni/Au contacts, respectively.

Transmission electron microscopy (TEM), high resolution electron microscopy (HREM) and electron diffraction have been applied to study the structure of the contacts. Cross-sections were prepared for the TEM study by dimpling followed by ion milling. TEM studies were carried out on Topcon 002B and JEOL 200CX microscopes operated at 200 kV. Energy dispersive X-ray (EDX) spectra were taken from different metal sublayers to study the metal interdiffusion after thermal annealing.

RESULTS AND DISCUSSION

<u>Ti/Al contact</u>

Typical cross-sectional TEM images of a Ti/Al contact are shown in Figs. 1 before (a) and after (b) thermal annealing at 900 °C for 30 s.

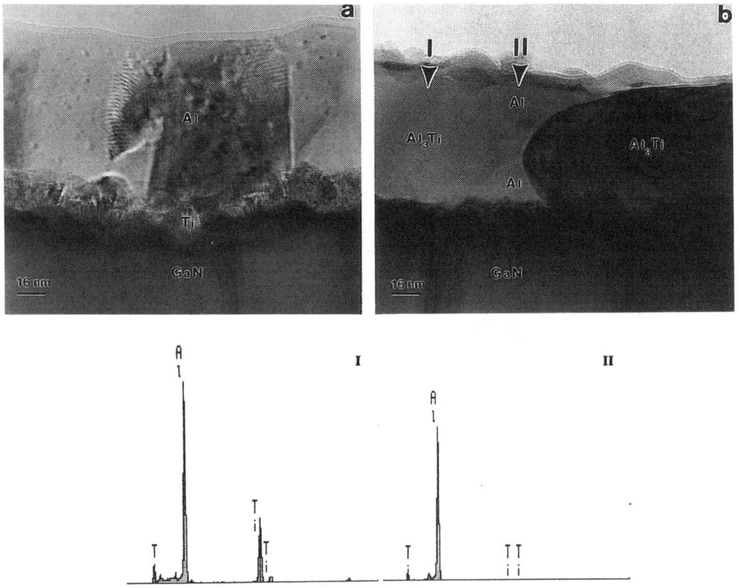

Fig. 1, a-b. Bright field TEM micrographs of Ti/Al contact in cross-section before (a) and after (b) annealing. EDX spectra from regions I and II of micrograph (b) are shown on the bottom.

Both as-deposited layers are textured polycrystalline material with c-planes of the metal grains mostly parallel to the c-plane of GaN.The Al-Ti and Ti-GaN interfaces are very rough as a result of a rough surface morphology of the GaN due to the reactive ion etching prior to metal deposition. Annealing does not change the surface morphology of the GaN significantly, but leads to metal interdiffusion and alloying. The thickness of the metal composite layer decreases after annealing due to the interdiffusion and formation of alloys with higher densities than that of pure Al. In the case of the Ti/Al composite layer, RTA results in the formation of one polycrystalline layer which consists of large $Al_{3+x}Ti$ (x<1) and Al grains with the c-plane almost parallel to that of GaN.

Ti/Al/Ni/Au contact

Similar results were obtained for the Ti/Al/Ni/Au contact (see Fig.2). TEM bright-field micrographs (a) and (b) of Fig.2 represent structure of the metal composite before and after thermal annealing. All of the as-deposited layers are also textured polycrystalline with rough interfaces. In contrast with the Ti/Al contact where no significant interdiffusion was detected by EDX between the Ti and Al in the as-deposited sample, the formation of a Al-Ni alloy was observed in the Ti/Al/Ni/Au metal composite just after Au deposition.

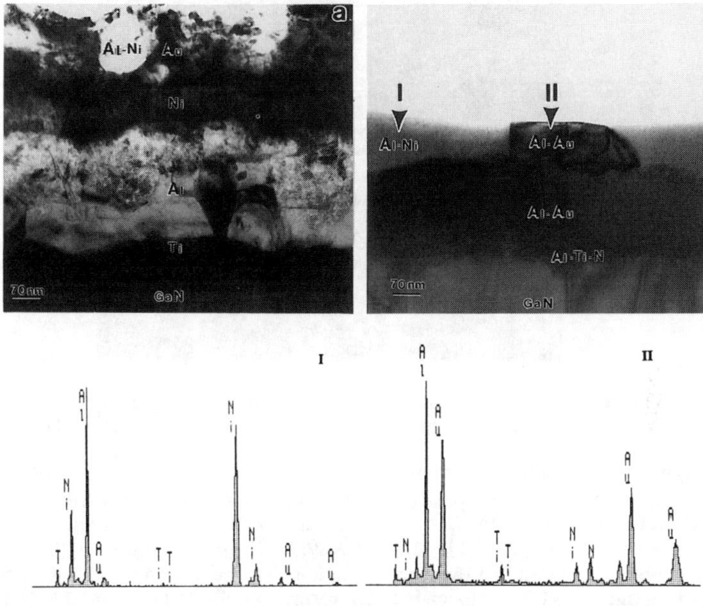

Fig. 2, a-b. Bright field TEM micrographs of Ti/Al/Ni/Au contact in cross-section before (a) and after (b) annealing. EDX spectra from grains I and II of Fig. (b) are shown on the bottom.

Annealing leads to further metal interdiffusion and alloying. The thickness of the metal composite layer decreases after annealing due to the interdiffusion and formation of alloys with higher densities than that of pure Al. For the Ti/Al/Ni/Au composite, annealing leads to the formation of at least three polycrystalline layers. The top layer consists mostly of an Al-Ni alloy with some grains of an Al-Au alloy. The middle layer is formed mostly by an Al-Au alloy with some Al-Ni grains. This is consistent with the high diffusivity of Au and shows a strong interaction between Al and the other species (Au, Ni, Ti). The bottom layer contains mostly Al-Ti and Ti-N alloys, but other species are also present in the layer. An EDX of the GaN interface indicates the presence of 61.5 at.% Ti, 26.2 at.% Al, 6.8 at.% Ga, 3.8 at.% Au, and 1.6 at.% Ni. In fact all of these species are present in various proportions over the metal composite after annealing.

Structure of Ti-GaN interfacial region

The interfacial region between the metal composite layers and the GaN is very complex and also polycrystalline. Both HREM and electron diffraction establish the presence of a textured polycrystalline layer of cubic TiN on the GaN surface. A thin TiN layer was detected by electron diffraction at the Ti-GaN interface even for as-deposited samples. The thickness of this TiN layer increases to 5 nm (Fig.3a) after annealing of the Ti/Al contact and to 10-15 nm (Fig.3b) for the Ti/Al/Ni/Au contact.

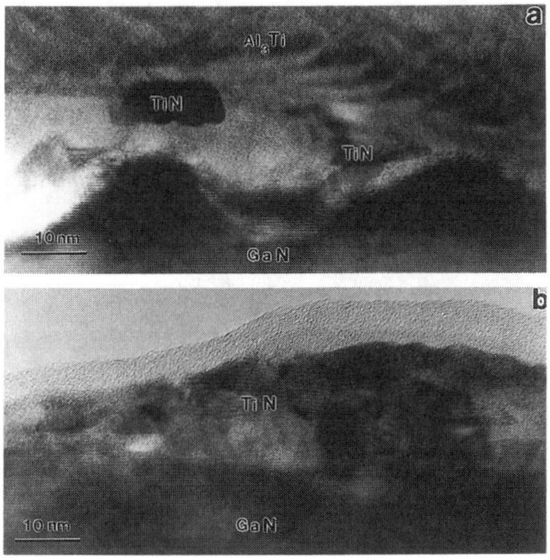

Fig.3, a-b. Low magnification HREM images of the interfacial area near the GaN surface of Ti/Al (a) and Ti/Al/Ni/Au (b) metal composites in cross-section after RTA.

The high resolution electron micrographs in Figs.3b and 4 show the interfacial region between the metal Ti/Al/Ni/Au composite and the GaN after annealing at 900 °C for 30 s. The orientation relationship between the TiN layer and the GaN was determined from the electron diffraction pattern as: $\{111\}_{TiN}//\{00.1\}_{GaN}$, $[110]_{TiN}//[11.0]_{GaN}$, $[112]_{TiN}//[10.0]_{GaN}$.

The TiN grains are often in a twinned orientation. The mismatch in the lattice parameters of TiN and GaN is about 5.6-6.0%. This leads to the formation of relatively small grains of about 5-20 nm in lateral size which are slightly misoriented to each other and with respect to the GaN. The interface between the TiN grains and the GaN in Fig.4 is quite abrupt, but the GaN region near the interface contains many structural defects which are possibly N vacancy complexes. The accumulation of N vacancies may lead to the formation of voids or protrusions at the TiN-GaN interface (see Fig.4).

The resistivity of both the Ti/Al and Ti/Al/Ni/Au contacts decreases drastically after annealing, but their lowest values differ by two orders of magnitude. A thin TiN layer was found at the interface between the metal composite and the GaN in both cases but the TiN is twice as thick in the case of the four metal contact. This may be the result of the specific structure of the composite layer formed after annealing. In the Ti/Al/Ni/Au contact, Au diffuses through the Ni layer and forms an Au-Al alloy while an Al-Ni alloy layer resides at the surface. The Au-Al layer underneath may prevent further diffusion of Ni toward the GaN surface. On the other hand, formation of the Au-Al alloy might prevent outdiffusion of Ti into the Al layer to form an Al-Ti alloy.[7]

Fig.4. HREM micrograph of the interfacial area near the GaN surface of Ti/Al/Ni/Au contacts in cross-section after RTA.

Role of RIE

There are several possible explanations for the reduction of contact resistivity by RIE. First, it provides a method to clean the surface and remove any oxide layer.[7] Second, it increases the GaN surface roughness due to the preferential etching of dislocated material and, hence, increases the contact area. Third, RIE causes radiation damage near the GaN surface and, hence, increases the point defect density in a thin GaN region near the surface. As a result, it makes diffusion of Ti into the GaN more rapid, even during the metal deposition. Thus, the formation of a good contact is expected even before annealing. Indeed, a low resistance contact to GaN of $6.5 \times 10^{-5}\ \Omega\ cm^2$ was reported for the Ti/Ag metallization[6] without annealing. The role of dislocations in GaN in the formation of a good contact is beyond the scope of this paper, but this role appears to be considerable. Indeed, diffusion is enhanced along dislocation cores so that dislocations might be considered as pipes for metal atom migration.

The dislocation density at the top of the GaN layer is as high as 5×10^9. This high dislocation density is associated with the formation of small angle grain boundaries. Preferential etching of GaN at end points of dislocations leads to the formation of grooves on the GaN surface during RIE. The defect distribution over the thickness of the GaN layer is the following: rather high at the interface of the GaN with the AlN buffer, but sharply decreasing over a distance of 0.2 μm toward the layer surface.

Low resistivity metal contacts for n-GaN

Both RIE-induced damage of the GaN and formation of the TiN phase can lead to a supersaturation of N vacancies within the near interface area.[5] For example, formation of a donor-rich surface layer has been detected by Fonash et al.[9] after ion irradiation of the surface. Jenkins & Dow[10] have shown that N vacancies in GaN act as donors. This led Z. Fan et al.[7] to suggest that a heavily doped TiN-GaN interfacial area causes bending of the GaN conduction band sufficient for tunneling. Another type of low resistance ohmic contact is the low barrier Schottky contact[11,12] which might be associated with intermediate or graded band-gap interface material. Because the band gap of AlN is larger than that of GaN, ternary Al-Ti-N compounds or their mixture might have a band gap which is close to that of GaN and, hence, provide such a low-barrier Schottky contact.[5] EDX shows the presence of a certain amount of Al in the interfacial area near the GaN surface. However, the formation of ternary or quaternary Ga-Ti-Al-N compounds has not been detected by TEM in this study. This may be because of the small amount of such material. On the other hand, success in the development of a low resistance contact using the Ti/Ag[6] or W[8] metallization schemes suggests that TiN and WN formation and, associated with it, an excess of N vacancies in the GaN under the contact is the most important factor for achieving low contact resistivity.

CONCLUSIONS

In conclusion, TEM shows the formation of a thin polycrystalline cubic TiN layer at the metal composite-GaN interface. The orientation relationship between the cubic TiN layer and the GaN was found to be: $\{111\}_{TiN}//\{00.1\}_{GaN}$, $[110]_{TiN}//[11.0]_{GaN}$, $[112]_{TiN}//[10.0]_{GaN}$. The formation of this TiN layer and, associated with it, an excess of N vacancies in the GaN under the contact is the most likely explanation for the low resistance of the Ti/Al and Ti/Al/Ni/Au contacts.

ACKNOWLEDGEMENT

The TEM portion of this work was supported by BMDO/IST and administrated by ONR contract number N00014-95-1-1265. K.J.D and E.E.H acknowledge the support of the Director, Office of Energy Research, Office of Basic Energy Sciences, Materials Science Division of the U.S. Department of Energy under Contract No. DE-AC03-76SF00098. The research at the University of Illinois at Urbana-Champaign was supported by grants from ONR, AFOSR and BMDO. The use of the facilities of the National Center of Electron Microscopy is gratefully acknowledged. The authors wish to thank Wendy Swider for TEM sample preparation and C.J. Echer for his assistance with EDX measurements.

REFERENCES

1. S.N. Mohammad, A. Salvador, and H. Morkoç, Proc. IEEE 83, 1306 (1995).
2. H. Amano, M. Kito, X. Hiramatsu, and I. Akasaki, Jpn. J. Appl. Phys. 28, L2112 (1989).
3. S. Nakamura, T. Mukai, and M. Senoh, Jpn. J. Appl. Phys. 30, L1998 (1991).
4. M.A. Khan, A.R. Bhattarai, J.N. Kuznia, and D.T. Olson, Appl. Phys. Lett. 63, 1214 (1993).
5. M.E. Lin, Z. Ma, F.Y. Huang, Z.F. Fan, L.H. Allen, and H. Morkoç, Appl. Phys. Lett. 64, 1003 (1994).
6. J.D. Guo, C.I. Lin, M.S. Feng, F.M. Pan, G.C. Chi, C.T. Lee, Appl. Phys. Lett. 68, 235 (1996).
7. Z. Fan, S.N. Mohammad, W. Kim, O. Aktas, A.E. Botchkarev, and H. Morkoç, Appl. Phys. Lett. 68, (1996) in press.
8. C.B. Vartuli, S.J. Pearton, C.R. Abernathy, J.D. MacKenzie, J.R. Shul, M.I. Lovejoy, J.C. Zopler, A.G. Baca, M.H. Crawford and S.P. Kilcoyne, MRS Proc., v. 421 (1996) in press
9. S. Fonash, S.Ashok, and R. Singh, Appl. Phys. Lett. 39, 423 (1981).
10. D.W Jenkins & J.D. Dow, Phys. Rev. B 39, 3317 (1989).
11. L.H. Allen, L.S. Hung, K.L. Kadavanach, J.R. Phllips, and J.W. Mayer, Appl. Phys. Lett. 51, 326 (1987).
12. Z. Ma, L.H. Allen, B. Blanpain, Q.Z. Hong, J.W. Mayer, and C.J. Palmstrom, Mat. Res. Soc. Symp. Proc. 230, 131 (1991)

BETA SILICON CARBIDE PN JUNCTION DIODES

J.C. COLEMAN *, G.L. HARRIS *, D.B. POKER **+
*Materials Science Research Center of Excellence, School of Engineering, Howard University Washington, DC 20059, count@negril.msrce.howard.edu, gharris@msrce.howard.edu
**Surface Modification and Characterization Research Facility++, Oak Ridge National Laboratory, Oak Ridge, TN 37381, dbp@ornl.gov.

ABSTRACT

Beta silicon carbide (3C-SiC) diodes have been fabricated using **ion implantation** as the selective doping technique. Previous work on 3C-SiC diodes have exhibited properties such as low reverse breakdown voltages and high ideality factors. Also, 6H and 4H SiC diodes have been reported. This paper studies a different procedure to produce better 3C-SiC diodes for use in the electronics industry. Current versus voltage, capacitance versus voltage and **temperature versus voltage** tests were conducted on the devices.

Isolation between devices is a prominent concern when building integrated circuits. Proton bombardment is the preferred planar process for forming isolation regions in gallium arsenide (GaAs) due to the lack of a stable native oxide. Hydrogen and boron in GaAs have exhibited good electrical isolation between devices. This paper investigates using **proton bombardment** to form isolation regions in 3C-SiC. Cubic SiC samples are implanted with a variety of implant doses, ranging from 1×10^{14} to 1×10^{15} ions / cm^2, and implant energies ranging from 150 to 300 keV. **Hall measurement** tests were performed to study the characteristics of the implanted material.

INTRODUCTION

SiC is a wide bandgap semiconductor that has found a variety of applications in the electronics industry. The bandgap of SiC is larger than both Si and GaAs, thus devices made from this material will require a higher temperature to reach the intrinsic region of the material. As a result, SiC is attractive for high temperature electronics. SiC has a higher thermal conductivity than both Si and GaAs, which allows for good heat dissipation and higher packing densities. SiC has a higher saturation electron drift velocity than Si and GaAs. Furthermore, due to SiC's high breakdown electric field high power switching devices and high power MOSFET's have been fabricated.[1,2] From a optical standpoint, SiC has been used in making blue LED's and other photonic devices. Several polytypes exist for this material such as 2H, 4H, 6H, and 3C. These polytypes represent the growth structure of the SiC atoms. 3C-SiC has the highest electron mobility of the SiC polytypes.

Proton bombardment is performed using ion implantation to create damage layers inside a particular substrate. Inside the damage layers are traps in the conduction band which capture electrons, hence decreasing the conductivity of the material. Isolation between diodes is essential for more complex integrated circuits. Proton isolation has been used in GaAs circuits.[3]

EXPERIMENT

The 3C-SiC was grown at the Materials Science Research Center of Excellence (MSRCE) at Howard University using a horizontal chemical vapor deposition (CVD) reactor. The sample was 12 microns (μm) thick. A growth rate of 2.7 microns per hour was obtained at

---

+ The submitted manuscript has been authored by a contractor of the U.S. Government under contract No. DE-AC05-96OR22464. Accordingly, the U.S. Government retains a nonexclusive, royalty-free license to publish or reproduce the published form of this contribution, or allow others to do so, for U.S. Government purposes.

++ Managed by Lockheed Martin Energy Research Corporation for the Division of Materials Science, U.S. Department of Energy under contract DE-AC05-96OR22464.

a pressure of 200 Torr, while the growth temperature was 1385 °C.[4] A Hall measurement was performed on a piece of the material to find the donor concentration (N_d) and Hall mobility. N_d was 2.43 x 10^{16} cm^{-3} and the Hall mobility was 388.9 cm^2 / V-sec . The 3C-SiC was unintentionally doped n-type.

A 300 keV ion implantor was chosen as the selective method of doping the SiC bulk material. At an implant energy of 170 keV, the dose was 1 x 10^{14} ions / cm^2 and the projected range was .4092 microns. Boron triflouride (BF_3) was used to dope part of the SiC sample p-type. For proton isolation, a extrion ion accelerator was used to perform the H^+ and Ne implants. The implant doses for both proton sources into the 3C-SiC samples were 1 x 10^{14}, 4 x 10^{14}, 5 x 10^{14}, and 1 x 10^{15} ions / cm^2. The implant energies were 150, 160, and 170 keV for the hydrogen, while the nitrogen was 150, 200, and 300 keV. Moreover, the isolation dose for the devices were 2.41 x 10^{15} ions / cm^2 and a implant energy of 120 keV. All of the proton isolation work was done at the Oak Ridge National Laboratory.

Proton isolation of the 3C-SiC pn junctions requires a proton mask to protect the patterns on the p-layer from the hydrogen implants. The mask consists of the following layers: [1] 800 Å of Cr, [2] 1000 Å of Ni, [3] 1 micron of Au. The projected range of the hydrogen into the SiC is 9677 Å. A Lectroless Prep and Lectroless Au 2000 solution was used to deposit the 1 micron of Au, while the Cr and Ni metals were deposited using an electron-beam evaporator. This particular mask was chosen because Au is a good trap of H^+ ions.

The 3C-SiC diodes were fabricated using a bi-layer resist process consisting of poly-methyl methacrylate (PMMA) and co-polymer (PMMA/MAA). Two quartz masks sets were used for the p and n layer ohmic contact patterns for the diodes. The ohmic contacts for the p layer was 2500 Å of Al, while the n layer was 800 Å of Mo on top of 700 Å of Ni. The contacts were alloyed in a A&G Associates 410 Heatpulse system for 30 seconds at 950 °C. The gas ambient was argon. Both contacts were deposited using a e-beam evaporator. Standard fabrication techniques were used to make these devices.[5] The mesa areas on these samples range from 6.25 x 10^{-6} cm^2 to 5.62 x10^{-5} cm^2.

From PN junction theory, several parameters can be deduced such as the built-in potential (V_{bi}) and depletion region width. All of the theoretical calculations are for an abrupt diode. For this research, the theoretical value of V_{bi} was 2 Volts. The value for the depletion region width was .9388 μm. The ideal diode equation is used to give a representation of the current versus voltage in the device. However, several conditions of applied voltage and temperature exist where the ideal diode equation fails to adequately represent physical devices. One case is the reverse voltage breakdown (V_{br}). V_{br} is a general formula taken from Sze.[6] The theoretical V_{br} was 87 volts for the diodes. Moreover, there are forward-bias deviations such as the ideality factor (n). This number indicates whether diffusion or generation-recombination is dominant within the depletion region of the device. The value for previous 3C-SiC diodes has been 3.3.[7,8,9] However, the ideality factor should be between 1 and 2. Other important values for the diode, like the abrupt nature of device and N_d, are found by looking at capacitance versus voltage data.

RESULTS

The experimental results from the 3C-SiC diodes fabricated have surpassed previous work done in the area of 3C-SiC diodes. Several diodes were characterized from two different chips. Current versus voltage tests have exhibited breakdown voltages in excess of 20 volts for these devices. The maximum V_{br} obtained for the diodes was 60 Volts. This voltage is the highest seen in a 3C-SiC diode on a Si substrate to this author's knowledge. The average breakdown voltage was 30 volts for the diodes measured. The ideality factor was 1.05. This number is one of the lowest reported in 3C-SiC diodes. The built-in potential was found to be 1.4 volts. This value was obtained by focusing on the diode in the forward-bias region, where the current starts to rise. The I_o was also calculated to be 6.017 x 10^{-14} A. Figure 1 shows the DC I-V characteristic of one of the diodes. All current versus voltage measurements on the devices were done at room temperature.

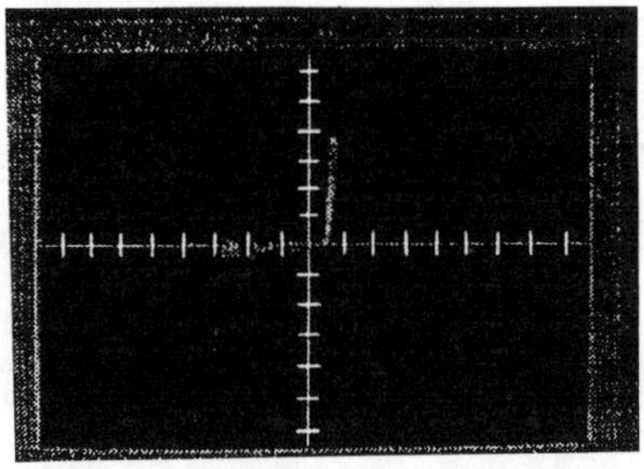

Figure 1 - 3C-SiC Diode With Voltage (20V/div) Horizontally, Current (.5mA/div) Vertically

Temperature versus voltage tests suggest that V_{br} increases as the temperature increases. This positive temperature coefficient demonstrates that avalanche multiplication of carriers was responsible for the junction breakdown voltage. To this author's knowledge, this is the first time that avalanche multiplication has been seen in 3C-SiC. Three different diodes were tested. No microplasmas have been observed on these devices, thus proving that the voltage breakdown is in the bulk region of the device. Another experiment performed was to examine the series resistance versus temperature. The series resistance decreased as the temperature increased. As a result, it does not contribute to the positive temperature coefficient of the devices. Futhermore, a theoretical paper on avalanche stability in wide bandgap semiconductors has predicted that a positive temperature coefficient should be obtained in 3C-SiC.[10] From the above paper, donor carrier concentration is an important parameter in determining a positive or negative temperature coefficient. Figure 2 represents the temperature versus voltage results for one device.

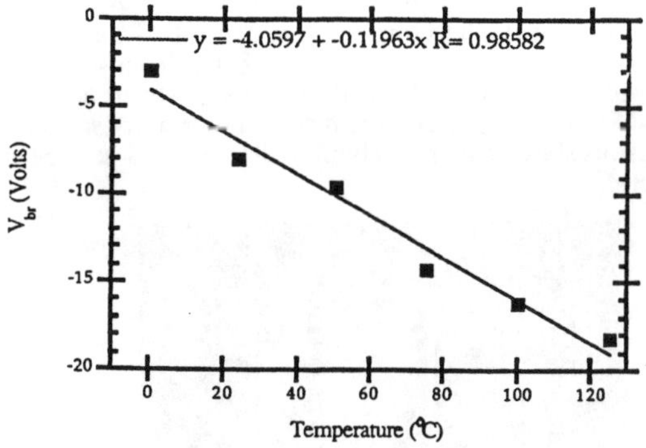

Figure 2 - Temperature Versus Voltage Data For 3C-SiC Diode

Capacitance tests were performed on the diodes using a Hewlett-Packard 4061A Semiconductor / Component Test system. The frequency was set at 12.4 MHz with a oscillation voltage of 1.1 volts. The capacitance reading on the sample was -93.04 degrees, indicating that the readings were accurate. Voltages were changed from -2.75 volts to 0 volts in increments of -.25 volts. The area of the diode measured was 1.546 x 10^{-4} cm^2. The $1/C^2$ versus V data revealed the abrupt nature of the diode. From the same data, the actual built-in potential was 1.4 volts. This value agreed with the information from the current versus voltage tests. Also, the N_d was found to be 1 x 10^{16} cm^{-3}. This result is very close to the donor concentration from the Hall measurement performed on the SiC material. Figure 3 shows the $1/C^2$ versus V data for a diode with an area of 3.125 x 10^{-5} cm^2.

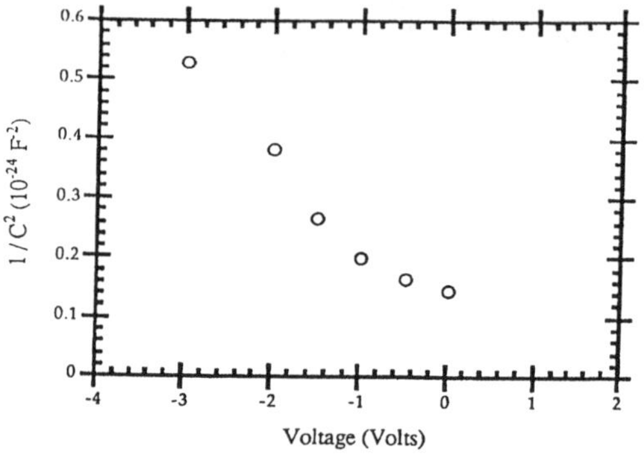

Figure 3 - 1 / C^2 Versus V For 3C-SiC Diode

Both mesa to mesa and gate to mesa isolation were achieved in the samples using protons. The DC current versus voltage results before the proton bombardments on the mesa to mesa isolation sample was 15 and -15 volts. The 3C-SiC samples implanted with hydrogen and neon were characterized by performing Hall measurements. The SiC was removed from the Si substrate before taking the measurements. The resistivity dropped for the implant doses of 1 x 10^{14}, 4 x 10^{14}, and 5 x 10^{14} ions / cm^2 for both the H^+ and Ne 3C-SiC samples. The resistivity slightly increases at the highest implant dose of 1 x 10^{15} ions / cm^2. Figure 5 is the DC I-V characteristics of the device before the proton bombardments, while Figure 4 shows the same device after the implants. Figure 6 shows the resistivity versus dose measurements for hydrogen in 3C-SiC and Figure 7 represents the resistivity versus dose measurements for the Ne in 3C-SiC. These results show that proton bombardment could be used as a viable isolation technique for SiC devices.

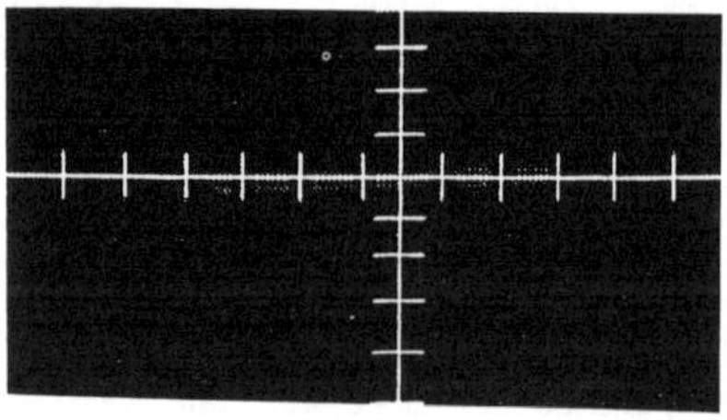

Figure 4 - DC I-V Characteristics After Proton Implants
Current (.5mA/div), Voltage (10V/div)

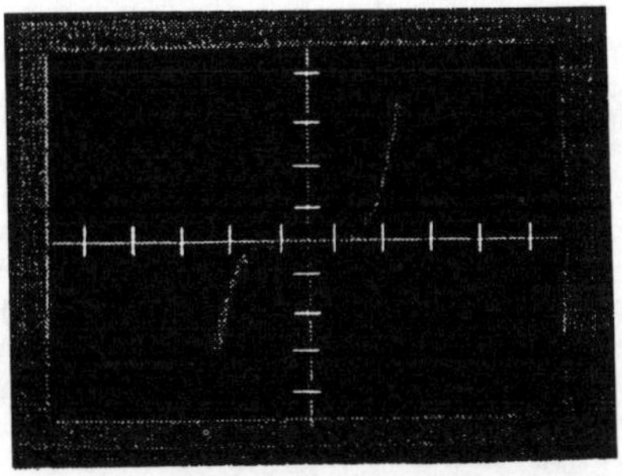

Figure 5 - DC I-V Characteristics Before Proton Implants
Current (.5mA/div), Voltage (10V/div)

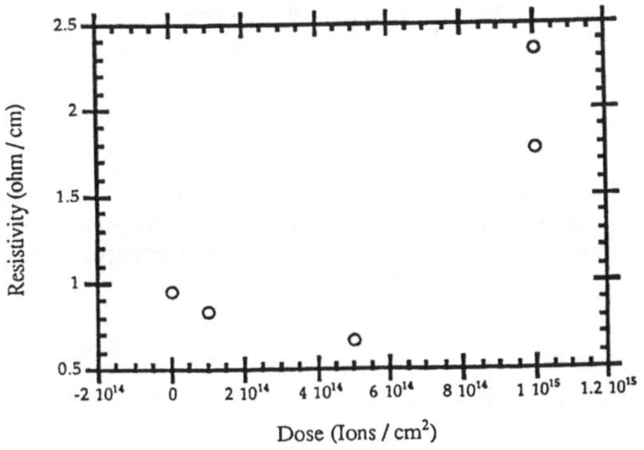

Figure 6 - 3C-SiC Sample Proton Bombarded With Hydrogen

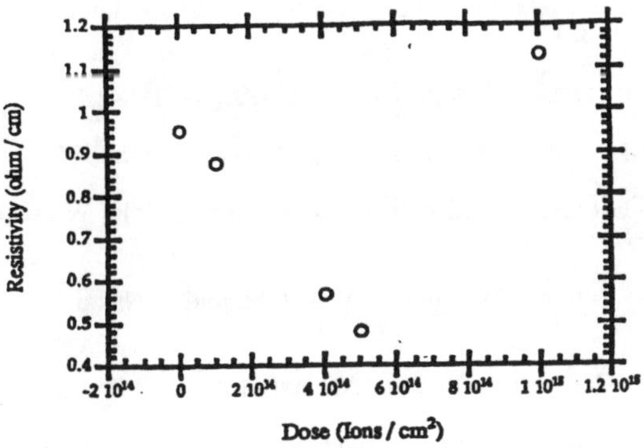

Figure 7 - 3C-SiC Sample Proton Bombarded With Neon

CONCLUSIONS

A unique technique for fabrication of 3C-SiC diodes was explored. The current versus voltages results show improvements from previous work in the areas of ideality factor, turn-on voltage, and reverse breakdown voltages. The ideality factor for the diodes were as low as 1.1. 1.4 volts was the turn-on voltage for the devices, while the maximum breakdown voltage was 60 volts at room temperature. Temperature versus voltage experiments have demonstrated avalanche multiplication in 3C-SiC. No microplasmas are evident on the surface of the device and a positive temperature coefficient was obtained for different devices. The series resistance decreased as the temperature increases. Capacitance versus voltage data had revealed the abrupt nature of the devices, along with expressing values for N_d and V_{bi}. The value of N_d was in th 10_{16} range. This result was similar to the measurement from the Hall measurement, thus proving the accuracy of the findings. Moreover, the Vbi from C-V data was 1.4 volts. This value is the same as that found from the I-V measurements. In the area of proton isolation, 3C-SiC samples and devices have been isolated using hydrogen. Resistivity versus dose measurements have been done for protons implanted in 3C-SiC. The initial results showed the protons acting as acceptor traps. The resistivity started to increase at the highest dose of 1 x 10^{15} ions / cm^2.

Future work includes characterization of more 3C-SiC chips to gather more data on the positive temperature coefficient and to set new reverse breakdown voltage levels. For proton bombardment, several experiments need to be performed such as high dose proton implants,. and temperature effects in the proton damaged 3C-SiC samples.

ACKNOWLEDGMENTS

My sincere thanks to Crawford E. Taylor at M.S.R.C.E. for the 3C-SiC material and Dr. Galina Kelner at the Naval Research Laboratory for the plasma etching. This research was supported by NASA-Lewis (Grant # NAG3-1406), Naval Research Laboratory (Grant # EEC-9116823), and NSF (Grant # N00014-94-C-2018).

REFERENCES

1. J.W. Palmour, L.A. Lipkin, 2nd International High Temperature Electronics Conference, Volume 1, XI-3 - XI-8, (1994).

2. S. Sheng, X. Tang, M.G. Spencer, P.Zhou, and K. Wongchotical , 3rd International High Temperature Electronics Conference, (1996).

3. D.C. D'Avanzo, IEEE Transaction on Electron Devices, Volume ED-29, No. 7 pp.1051-1058, (1982).

4. C.E. Taylor, M.E. Thesis, Howard University (August 1994).

5. J.C. Coleman, M.E. Thesis, Howard University (December 1995).

6. S.M. Sze, Physics of Semiconductors (Second Edition), John Wiley & Sons, Inc., New York, 1981, p. 104.

7. K. Furukawa, A. Uemoto, M. Shigeta, A. Suzuki, and S. Nakajima, Applied Physics Letters 48 (22), (1986).

8. J.A. Edmond, K. Das, and R.F. Davis, Journal of Applied Physics 63 (3), (1988).

9. R.E. Avilia, J.J. Kopanski, and C.D. Fung, Journal of Applied Physics 62 (8), (1987).

10. C.A. Fazi, private communication.

Part II

Crystal and Film Growth

Bulk and Bulk-Like Crystal Growth

ANALYSIS OF THE SUBLIMATION GROWTH PROCESS OF SILICON CARBIDE BULK CRYSTALS

R. ECKSTEIN *, D. HOFMANN *, Y. MAKAROV **, St.G. MÜLLER *, G. PENSL ***, E. SCHMITT * and A. WINNACKER *
* Materials Science Department 6, University of Erlangen-Nürnberg, Martensstr. 7,
** Fluid Mechanics Institute, University of Erlangen-Nürnberg, Cauerstr. 4,
*** Applied Physics Institute, University Erlangen-Nürnberg, Staudtstr. 7,
D-91058 Erlangen, Germany

ABSTRACT

Experimental and numerical analysis have been performed on the sublimation growth process of SiC bulk crystals. Crystallographic, electrical and optical properties of the grown silicon carbide (SiC) crystals have been evaluated by various characterization techniques. Numerical models for the global simulation of SiC bulk growth including heat and mass transfer and chemical processes are applied and experimentally verified.

INTRODUCTION

The present quality of SiC bulk crystals is far from satisfying all the needs of future SiC device applications in high power and high temperature electronics and blue/uv emitting nitride-based III-V optoelectronis. On the one hand SiC substrates are still very defective. Although remarkable progress has been made recently in defect reduction [1], SiC wafers contain still a high number of defects like micropipes, precipitates, dislocations and inhomogeneities (stress and dopant distribution) [2-4]. On the other hand important aspects with regard to economic manufacturing of SiC substrates by the sublimation process has to be improved considerably to achieve higher growth rates and longer crystals with an increase of wafer diameter from 1.25 inch to 4 inch. Research and development activies in these directions require basic understanding of the SiC bulk growth process and defect generation mechanisms.

Despite the technological and commercial relevance of SiC there exist a surprisingly small number of experimental studies which are devoted to the quantitative evaluation of the relation of boundary conditions of SiC bulk growth, e.g. temperature field, system pressure etc., and crystal quality in growth geometries relevant for the preparation of SiC substrates with diameter $d \geq 1$ inch and length $l \geq 10$ mm [5,6]. For considerable smaller crystal geometries or epitxial growth several authors have presented basic experimental studies [7-10]. Numerical modelling which has already shown to be an effective tool for the anaysis of the growth of semiconductors from the melt and its optimization [11] is only at the beginning and marginaly used [1], partly because of the abscence of relevant physical models.

In this paper first results on experimental and numerical analysis of the SiC bulk growth process are introduced. The status of SiC growth activities at the Materials Science Department of the University of Erlangen-Nürnberg is reviewed.

EXPERIMENT

SiC crystals habe been prepared in both vertical (seed located at the bottom of the graphite container) and inverted (seed at the top) configurations by physical vapor transport according to the modified Lely technique (for details see e.g. Ziegler et al. [12] and Tairov [13]). SiC growth runs has been perfomed both in resistively and inductively heated furnace facilities. In table I typical growth conditions are summarized. For seeding Acheson and Lely crystals have been used. SiC powder which is commercially available (Norton) or synthesized in our laboratory was employed as source material. Nominally undoped and n-type nitrogen doped SiC-boules have been grown.

A special crystal/vapor (s/v)-interface demarcation technique has been developed. for the quantitative determination of the dependance of the growth rate and s/v-interface shape from the growth parameters temperature distribution, pressure and source-seed distance. This method is

Mat. Res. Soc. Symp. Proc. Vol. 423

based on the artificial generation of striations in the growing crystal by intermittently introducing nitrogen gas into the reactor. Temperatures have been monitored by 4 pyrometers at the pedestal of the seed and axially along the periphery of the graphite growth container. Prior to temperature measurements the pyrometers have been calibrated at temperatures $T \leq 1800$ °C with thermocouples. At higher temperatures $T \leq 2500$ °C the melting point of different metals, e.g. Pt, Rh, Nb, has been taken as reference. After growth the boules have been cut both perpendicular and along the (0001)/(000$\underline{1}$)-growth direction and the wafers have been polished on both sides for removal of surface damage. For heat transfer analysis the power consumption of the resitively heated growth facility was measured in dependence of the temperature of the growth container.

System pressure p	10 mbar $\leq p \leq$ 50 mbar
Growth temperature T	2150 °C $\leq T \leq$ 2300 °C
Distance seed-source Δz	3 cm $\leq \Delta z \leq$ 5 cm
Axial temperature gradient dT/dz	15 K/cm $\leq dT/dz \leq$ 35 K/cm
Seed orientation	(0001) or (000$\underline{1}$)

Table I: Typical process conditions during SiC sublimation growth

The SiC crystal quality has been evaluated by different characterization techniques: Hall effect, defect etching in molten KOH, stress birefringence microscopy, glow discharge mass spectroscopy (GDMS) and photoluminescence (PL) mapping. For the quantitative determination of the growth rate the crystal length d was measured as distance between artificial introduced striations and the top of the seed and correlated to the elapsed process time t. The growth rates R were calculated as slopes $\partial d/\partial t$ from d(t)-diagrams.

RESULTS

Figure 1 shows a typical as grown 4H-SiC single crystal and a 6H- and a 4H-SiC wafer with a diameter of 25..30 mm. Typical lengths of the crystals are between 10 and 20 mm.

Fig. 1: 1" 4H-SiC bulk single crystal and 6H- and 4H-SiC substrates [orientation (000$\underline{1}$), doping level $1 \cdot 10^{18}$ cm^{-3} $\leq N_D - N_A \leq 5 \cdot 10^{18}$ cm^{-3}] prepared at the Materials Science Dept., Univ. Erlangen-Nürnberg by the modified Lely technique.

From interface demarcation experiments it was possible to determine the crystallization rate R of the crystals as 0.5 mm/h $\leq$ R $\leq$ 2 mm/h. Electronic properties of the doped crystals have been evaluated by Hall effect to n = 1-5$\cdot 10^{18}$ cm^{-3} and ρ = 0.01-0.04 Ωcm. The residual impurity concentrations of relevant elements detected by GDMS are 0.1 ppm$_w$ (B), 0.002 ppm$_w$ (V), $\leq$ 0.01 ppm$_w$ (Al) and 0.1 ppm$_w$ (Ti). The dislocation densities vary between 1500 cm^{-2} $\leq$ EPD $\leq$ 3000 cm^{-2}. Micropipe density was determined by optical microscopy in stress birefringence mode. Fig. 2 shows typical stress patterns of micropipes having a core diameter of 0.5 µm $\leq$ x $\leq$. 20 µm. The micropipe density (MPD) was found to be inhomogeneously distributed along the wafer in a range of 30 cm^{-2} $\leq$ MPD$\leq$ 300 cm^{-2}.

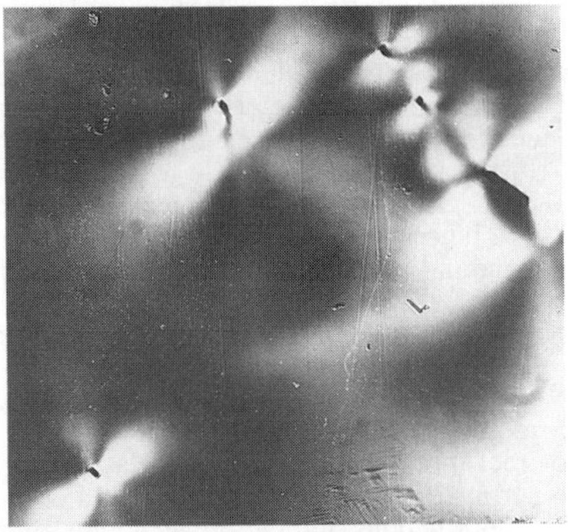

Fig. 2: Micropipes in a (000$\underline{1}$)-4H-SiC wafer visualized microscopically in stress birefringence mode by the surrounding strain field patterns.

PL mapping has been conducted to reveal the uniformity of the crystallographic modification and dopant incorporation. Figure 3a shows a segment of integral PL variations due to 15R polytype formation during growth. An uniformity of the crystallographic modification can be

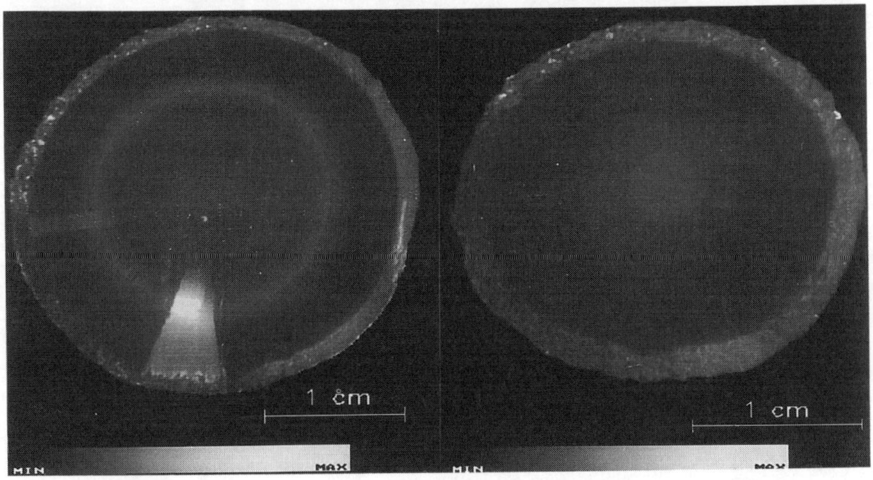

(a) (b)

Fig. 3: Mapping of integral photoluminescence intensity across 1" (000$\underline{1}$) 6H-SiC wafers: (a) Sample with a segment of PL variation due to 15R polytype formation; (b) Sample with uniform crystallographic modification. Circular PL-variations in both wafers originate from increased nitrogen incorporation due to interface demarcation.

observed in the SiC wafer shown in figure 3b. Additional PL-variations which are present in the center (fig. 3b) and circulary (fig. 3a) originate from increased dopant incorporation due to the application of the interface demarcation technique.

MODELLING AND ITS EXPERIMENTAL VERIFICATION

Numerical models are presently elaborated using codes developed at the Fluid Mechanics Institute of the University Erlangen-Nürnberg (e.g. FASTEST) and commercial codes (e.g. FIDAP), respectively for the global simulation of the SiC bulk growth process by the sublimation technique [6]. The axisymmetric models take principally into account all occuring heat generation and transfer mechanisms (radiation, conduction and convection) both for resistively and inductively heated growth reactors. Global simulation is performed by prescribing power distribution at the resistance furnace having two heating segments or power and frequency at the primary side of the medium frequency (MF) motor generator. The radiative heat transfer calculation takes into account the view factor evaluation inside the growth reactor. Mass transport by convection and diffusion in the inert gas (He or Ar) of the following species of SiC sublimation is calculated: Si, Si_2C, SiC_2 and SiC. Chemical data are mainly taken from the publication of Drowart et al. [14]. The respective concentration fields in the growth cell result from consistent sets of equations taking into account the proceeding chemical processes, evaporation from the source, their chemical interaction with the graphite crucible and the deposition at the SiC seed. The underlying considerations and the material properties essential for a realistic simulation of the SiC bulk growth process will be subject of a forthcoming paper (e.g. thermal conductivity of single crystalline SiC and effective conductance of SiC powder have been determined experimentally at growth relevant temperatures).

Figure 4 shows a comparison of measured and globally simulated heating power versus process temperature monitored at the periphery of the graphite crucible. For the establishment of typical growth conditions $T \approx 2500K - 2600K$ considerable power inputs of 25 kW-28 kW are necessary. There is a very good correspondence of experimental results and model predictions.

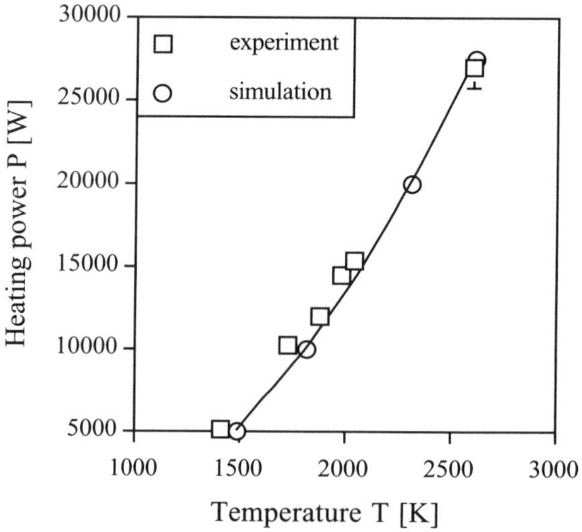

Fig. 4: Comparison of measured and calculated power versus process temperature in the resistively heated SiC growth apparatus.

A functional relationship of the power P = const. $T^3\Delta T$ (ΔT is the temperature difference between crucible and the water cooled reactor wall) is found which indicates radiation dominated heat losses in the temperature range investigated [15].

Figure 5 shows a comparison of the variation of measured and simulated crystal length with process time for a SiC growth run with specific geometrical and thermal boundary conditions. As input parameters temperature distributions have been measured on the periphery of the graphite crucible. From fig. 5 it is obvious that the observed dependence of crystal length versus time does not follow a straight line, i.e. during this process conditions the growth rate was varying considerably. The numerical simulation reflects quite well the experimentally observed growth behavior. Further analysis show that the decrease of the growth velocity with process time is caused by heat transfer limitations under the established process conditions.

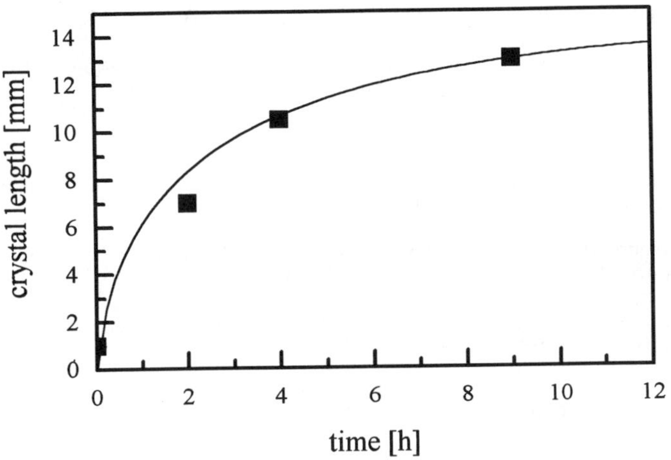

Fig. 5: Comparison of measured (symbols) and calculated (line) SiC crystal length versus process time.

CONCLUSIONS

First experimental and numerical analysis on the SiC bulk growth process have been performed. It is shown that growth velocity can alter considerably during a process run. One reason for this behavior is due to heat transfer limitations when process conditions for high rate SiC growth are established. Global modeling of heat transfer shows a good correspondence of measured and simulated power consumption in our resistively heated SiC growth reactor. Further investigations to verify experimentally our models for SiC bulk sublimation growth are necessary, especially under growth regimes when mass transfer and/or kinetic factors limit crystallization from the vapor.

The SiC wafer quality which has been attained recently is promising suggesting that in the near future further material improvement and increase of crystal diameter and length can be achieved. Numerical modeling and experimental studies on basic growth behavior in the SiC growth system will actively support this goal. A model approach which is under development for the prediction of the generation of parasitic phases (graphitization, Si-droplets) [16] is expected to reveal additional strategies for the optimization of the SiC bulk growth process.

ACKNOWLEDGMENTS

This work is supported by the German Ministery of Education and Research BMBF and Siemens AG, Erlangen under contract nr. 03M2746B0. The contribution of cand.-Ing. M. Kölbl and cand.-Ing. W. Schoirer within their diploma thesis to SiC growth and characterization are gratefully acknowledged. Numerical studies on heat transfer have been performed partially by Dr. T. Jung Inc., Schnaittach, Germany.

REFERENCES

[1] V.F. Tsvetkov, S.T. Allen, H.S. Kong, C.H. Carter, Inst. Phys. Conf. Ser., IOP Publishing, Bristol (U.K.) **142,** 17 (1996)

[2] M. Tuominen, R. Yakimova, R.C. Glass, T. Tuomi, E. Janzen, J. Crystal Growth **144**, 267 (1994)

[3] R.F. Davis, Inst. Phys. Conf. Ser., IOP Publishing Ltd., Bristol (U.K.) **137**, 1 (1994)

[4] S. Wang, M. Dudley, C. Carter,Jr., D. Asbury, C. Fazi in Applications of Synchroton Radiation Techniques to Materials Science edited by D.L. Perry, N.D. Shinn, R.L. Stockbauer, K.L.DAmico, L.J.Terminello (Mat. Res. Soc. Symp. Proc. **307**, Pittsburgh, PA, 1993) 249

[5] D.L. Barrett, J.P. McHugh, H.M. Hobgood, R.H. Hopkins, P.G. McMullin, R.C. Clarke, W.J. Choyke, J. Crystal Growth **128**, 358 (1993)

[6] D. Hofmann, M. Heinze, A. Winnacker, F. Durst, L. Kadinski, P. Kaufmann, Y. Makarov, M. Schäfer, J. Crystal Growth **146**, 214 (1995)

[7] L. Garcon, A. Rouault, M. Anikin, C. Jaussaud, R. Madar, Materials Science and Engineerimg B **29**, 90 (1995)

[8] Y.A. Vodakov, E.N. Mokhov, M.G. Ramm, A.D. Roenkov, Crystal Research and Technology **14**, 729 (1979)

[9] Y.M. Tairov, V.F. Tsvetkov, S.K. Lilov, G.K. Safaraliev, J. Crystal Growth **36**, 147 (1976)

[10] T. Kaneko, J. Crystal Growth **128**, 354 (1993)

[11] G. Müller, A. Ostrogorsky, Convection in Melt Growth, Handbook of Crystal Growth **2**, 709 edited by D.T.J. Hurle (Elsevier Science Publ., Amsterdam 1994)

[12] G. Ziegler, P. Lanig, D. Theis, C. Weyerich, IEEE Transactions on Electron. Devices **30**/4, 277 (1983)

[13] Y.M. Tairov, Materials Science and Engineering B **29**, 83 (1995)

[14] J. Drowart, G. DeMaria, M.G. Inghram, J. Chem. Phys. **41**, 1015 (1958)

[15] J.P. Holman, Heat Transfer, (McGraw-Hill, New York, 1986)

[16] S.Y. Karpov, Y.N. Makarov, M.S. Ramm, Inst. Phys. Conf. Ser., IOP Publishing, Bristol (U.K.) **142,** 177 (1996)

GALLIUM NITRIDE THICK FILMS GROWN BY HYDRIDE VAPOR PHASE EPITAXY

R.J. Molnar, P. Maki, R. Aggarwal, Z.L. Liau, E.R. Brown and I. Melngailis
Massachusetts Institute of Technology, Lincoln Laboratory, Lexington, MA 02173-9108

W. Götz, L.T. Romano and N.M. Johnson
Xerox Palo Alto Research Center, Palo Alto, CA 94304

ABSTRACT

Gallium nitride (GaN) thick films (to 150 μm) have been deposited by hydride vapor phase epitaxy (HVPE). These films are unintentionally doped n-type ($n = 1\text{-}2 \times 10^{17}\ cm^{-3}$ at 300 K) and exhibit structural and electronic properties which are comparable with the best reported for GaN films grown by organometallic vapor phase epitaxy. Additionally, these properties are found to be uniform over 2-in diameter films grown on sapphire substrates. The use of either a GaCl or ZnO surface pretreatment has been found to substantially enhance the nucleation density, resulting in improved surface morphology and film properties, even though it appears that the ZnO film is thermochemically desorbed early on in the growth. Dislocation densities as low as $\sim 5 \times 10^7\ cm^{-2}$ have been attained for films 40 μm thick. Homoepitaxial overgrowths both by electron-cyclotron-resonance plasma enhanced molecular beam epitaxy and OMVPE proceed in a straightforward manner, essentially replicating the defect structure of the HVPE GaN film.

INTRODUCTION

The development of low-defect, controllably doped III-V nitride heterostructures has led to the realization of high power light emitting diodes (LED's) in the yellow to ultraviolet[1-3] and, more recently, injection laser diodes.[4] However, a number of important issues preclude the CW operation of injection lasers and severely inhibit the economical commercialization of these devices. Specifically, epitaxial growth irreproducibility and film inhomogeneity inflict significant yield problems in LED production. In addition, current crowding effects are expected to be predominant in sapphire-substrate based injection lasers with underlying conductive layers <10 μm thick. Vertical-conduction, SiC-substrate based devices currently suffer from cracking problems related to tensile thermal stress in the nitride epilayer, as well as interfacial conduction barriers which increase the generation of heat. Instability and inhomogeneity in the growth appear to be exacerbated by minor process perturbations during the heteronucleation of the nitride film. On the other hand, several groups, including ours, have found that the homoepitaxial growth on preexisting nitride epilayers proceeds in a far more controllable manner, with the nitride overlayer essentially replicating the defect structure of the starting layer.[5] Thus HVPE appears to be a cost-effective way of generating low defect, high uniformity thick films for use as substrates for nitride-based device overgrowths. These films were characterized by reflection high energy electron diffraction (RHEED), x-ray diffraction (XRD), atomic force microscopy (AFM), temperature-dependent Hall effect measurements, deep level transient spectroscopy (DLTS) and 2 K photoluminescence (PL).

EXPERIMENTAL

The films in this study were grown in a vertical HVPE reactor described previously.[6,7] In order to ensure the uniform heteronucleation of the GaN layer, two different procedures have been employed: a GaCl pretreatment[7,8] or an RF-sputtered ZnO buffer-layer technique.[6,9] These techniques are briefly described below.

GaCl Pretreatment

For this case, we followed a procedure modified from that reported by Naniwae et al.[8] A bare sapphire wafer is solvent degreased followed by etching in a 3:1 sulphuric:phosphoric acid solution and deionized water rinse. The wafer is then loaded into the reactor and heated to the growth temperature in a hydrogen stream for 20 minutes. This is followed by a surface nitridation for 20 minutes in a 20% NH_3 stream. Afterwards, the NH_3 is purged from the reactor

Mat. Res. Soc. Symp. Proc. Vol. 423 © 1996 Materials Research Society

and the substrate is exposed to a 0.5-0.75% GaCl stream for 10 minutes. After this step, the flows are adjusted for film growth (with the wafer isolated from the growth stream at an isothermal point in the backflow tube) and then the GaN growth is initiated by reintroducing the wafer to the growth position.

ZnO Pretreatment

Here the sapphire wafers are cleaned as before and loaded into an RF diode sputtering system (Perkin-Elmer 2400), pumped to a base pressure $<10^{-7}$ T and then coated with a 0.1-1.0 μm thick ZnO film deposited under the following conditions (parenthetical values represent empirical optimizations):

$T_{substrate}$	:	Ambient
Target Diameter	:	5-in.
RF Power	:	125-200 W (200 W)
Pressure	:	5-20 mT (20 mT)
% O_2	:	0-100 % (12.5 %) (Balance is Ar).

These polycrystalline ZnO films exhibit a high degree of structural orientation as indicated by the RHEED patterns shown in Figure 1. XRD studies on these films yield a FWHM for the θ-rocking curve of ~55 arcmin and for the θ/2θ scan a FWHM of ~20 arcmin. Prior to loading in the HVPE reactor, the ZnO coated wafer is ultrasonically cleaned in methanol to remove any surface contaminants as well as particulates from the sputtering process. It is then introduced to the reactor, preheated to the growth temperature in the backflow tube and then introduced to the growth stream to initiate growth.

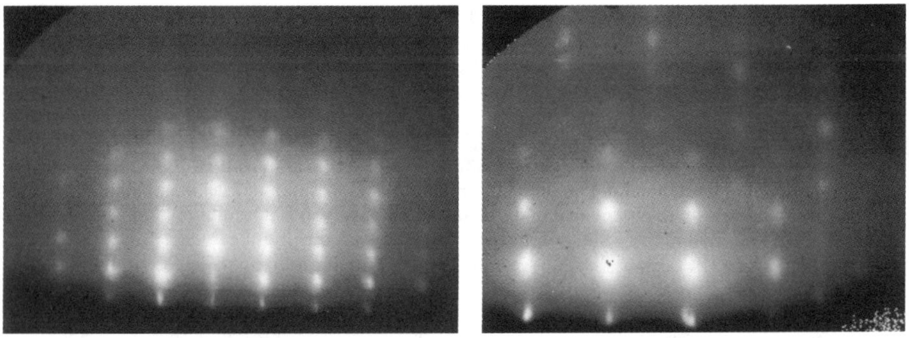

<11$\bar{2}$0> **<1$\bar{1}$00>**

Figure 1 - RHEED patterns along (a) <1$\bar{1}$00> and (b) <11$\bar{2}$0> azimuths for sputter-deposited ZnO.

For all the films in this study, the gas flow parameters during growth were as follows:

HCl over Ga	:	5 sccm
GaCl Diluent	:	100 sccm (H_2)
	:	300 sccm (N_2)
NH_3	:	770 sccm
NH_3 Diluent	:	100 sccm (N_2)
Main Carrier	:	3000 sccm (He or N_2)
HCl in MainCarrier	:	30 sccm.

Under these conditions, we observe from cross-sectional TEM that the ZnO layer is absent from the substrate/film interface and is therefore believe to be thermochemically desorbed in the early stages of growth.[10] Subsequently, we do not observe a strong dependence of the ZnO layer thickness on the GaN film quality, as observed previously.[9]

Film Characterization

Hall measurements were conducted in the temperature range from 80 to 500 K. The magnetic field was 17.4 kG. Samples of 5 x 5 mm size were cut from the wafers and metal dots were vacuum evaporated in the four corners to obtain electrical contacts in the Van der Pauw geometry. The contacts exhibited ohmic current-voltage characteristics over the entire temperature range of the measurement. The DLTS measurement system used in this study was described previously.[11] The measurement was conducted in the temperature range between 80 and 470 K with 1 K increments. For the PL measurements, the samples were mounted in a cryostat and immersed in pumped, liquid He to achieve a sample temperature of ~2 K. The PL spectra were excited with the 325 nm line of a 30 mW HeCd laser with an incident power density of ~1 W / cm^2. The resolution of the spectra was 0.05 nm.

RESULTS and DISCUSSION

The GaN films are generally highly transparent and show little morphological features, with the exception of a narrow (1 mm wide) raised bead around the periphery and randomly distributed hexagonal growth hillocks (~1-5 % surface coverage). These hillocks appear to be related to substrate preparation and have been substantially reduced over previous reports.[7] AFM images, one of which is shown in Figure 2 for a 40 μm thick film, reveals the presence of ~5 Å terraces on the growth surface which are ~75 nm wide. This is indicative of a step-flow mode of growth and the terrace height corresponds to what one would expect for monoatomic steps (c=5.19 Å). Growth rates are typically ~12-15 μm/h and thicknesses up to 40 μm have been grown without thermally induced cracking observable by optical microscopy. Above this value or for cases where surface preparation is nonoptimal, growth striae and/or thermal cracking (usually originating in the sapphire) can be observed. XRD studies yield a θ-rocking curve FWHM of 4.8-7.2 arcmin for the samples measured, which is comparable to reports for high-quality OMVPE material.

Electrical transport measurements as a function of temperature taken in optimized films grown by both pretreatment methods and with varying film thicknesses are shown in Figure 3.

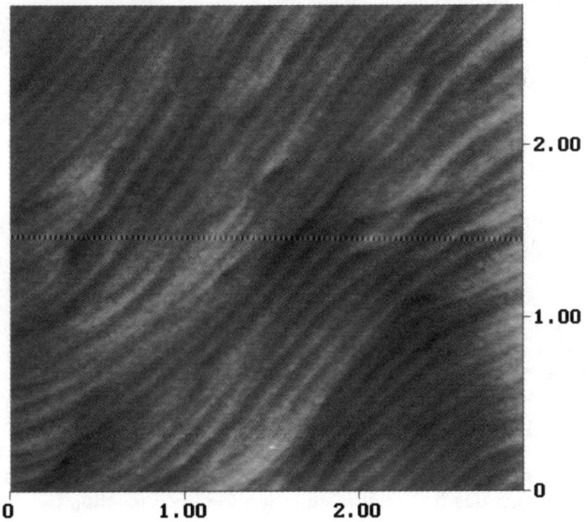

Figure 2 - AFM micrograph of GaN growth surface for 40 μm thick GaN film grown with ZnO pretreatment.

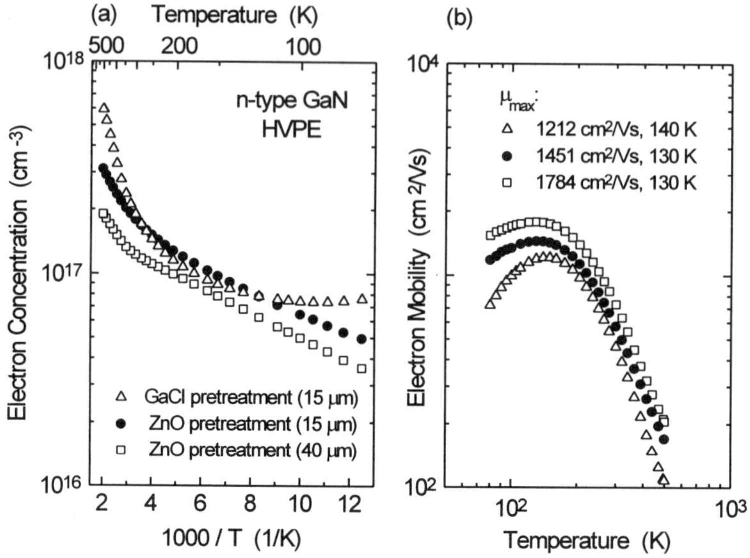

Figure 3 - (a) Electron concentration vs reciprocal temperature and (b) electron mobility vs temperature for unintentionally doped, n-type. Data are shown for three films; one was grown with the GaCl pretreatment (15 μm thick) and the other two were grown with the ZnO pretreatment (15 and 40 μm thick).

The temperature dependence of the electron concentrations is well explained by the presence of Si donors and an additional deeper donor, the nature of which is unclear at the moment. For the 40 μm thick sample, the carrier concentration does not deviate from this Arrhenius behavior at low temperatures, as is typically observed for GaN films doped at this level.[12] This may be the result of low concentrations of compensating defects in these films. The deeper level is found to vary in a range of 100-200 meV from sample to sample. As can be seen these samples exhibit high electron mobility, particularly in consideration of the concentration of residual donors (μ=768 $cm^2/V\cdot s$, n=1.4 x 10^{17} cm^{-3} at 293 K and μ=1784 $cm^2/V\cdot s$, n=6.8 x 10^{16} cm^{-3} at 130 K, for a 40 μm thick ZnO pretreated sample), which is another indication of minimal compensation in these films. The data of Figure 3 also indicate that the average transport properties of these films improve with film thickness. This is consistent with defect annihilation observed by TEM studies,[10] which found dislocation densities of 5 x 10^7 cm^{-2} after 40 μm of growth. Additionally, the homogeneity of these films is excellent as evidenced by the low variation in mobility and carrier concentration over a 2-in. substrate.[7]

Deep level defects were investigated for the sample in Figure 3 with the GaCl pretreatment by DLTS measurements, and the results are shown in Figure 4. Schottky barriers were formed on the GaN surface by evaporating Au contacts. The spectrum exhibits the signature of three deep levels which are labeled DLN_1, DLN_2 and DLN_3. DLN_2 appears as a shoulder of the DLN_3 peak and, therefore, no analysis was attempted. The deep levels DLN_1 and DLN_3 are characterized by activation energies for electron emission to the conduction band of (0.21±0.02) eV and (0.65±0.03) eV, respectively (inset). The concentration of these levels is $(1.2\pm0.5)\times10^{14}$ cm^{-3} and $(4.5\pm0.5)\times10^{15}$ cm^{-3}. Due to the surface sensitive nature of DLTS measurements, one can conclude that the density of deep level defects is comparatively low after ~15 μm of growth.

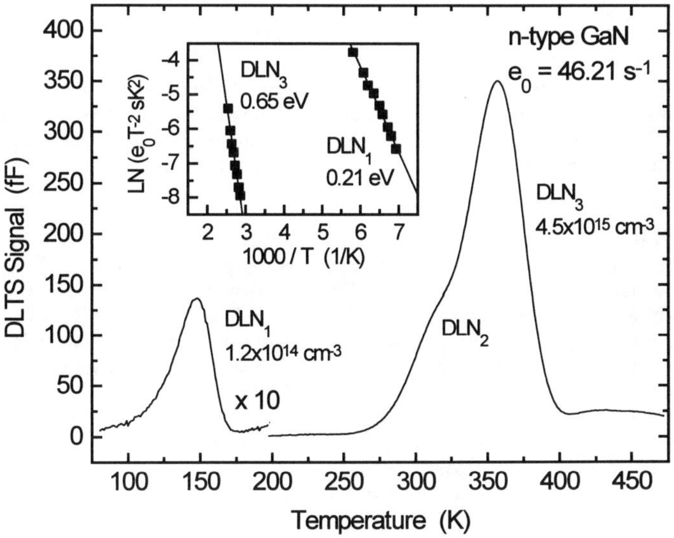

Figure 4 - DLTS spectrum for GaN grown by HVPE with GaCl pretreatment.

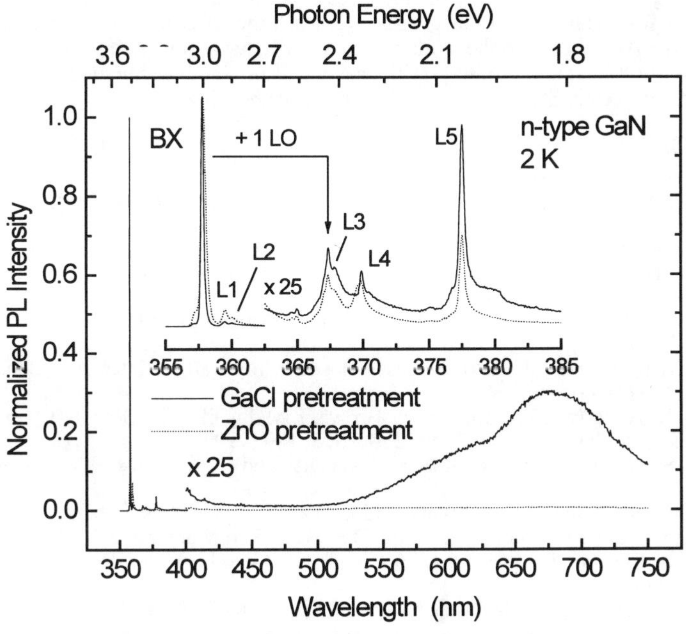

Figure 5 - Normalized PL spectra of GaN films with different substrate pretreatments. The low energy portion of the spectrum is magnified for both spectra by a factor of 25. The inset shows the near-bandedge region of the spectra with a portion of the spectra magnified. PL lines are labeled BX and L1 to L5. The first phonon replica of the BX line is indicated by an arrow.

The 2 K PL spectra of these films, shown in Figure 5, exhibit intense near-bandedge emission associated with donor-bound excitons at 3.468 eV and FWHM = 2.42-5 meV. A high-energy shoulder is believed to be due to the free exciton. The FWHM of the donor-bound excitonic peak does not appear to be correlated with the films' electrical and structural quality, as the highest mobility, lowest-defect-density sample also showed the broadest exciton peak (~5 meV). We are not able to detect any yellow emission centered at 2.2 eV, although weak red emissions centered at 2.0 and 1.8 eV are detected in some of the samples grown using the GaCl pretreatment.

CONCLUSIONS

GaN films have been grown by HVPE uniformly over large (2-in. diameter) areas. The film nucleation on the sapphire substrates has been enhanced by utilizing either a GaCl or ZnO pretreatment. While we believe that the ZnO is thermochemically desorbed early in the growth, this ZnO layer does drastically modify the sapphire surface properties and enhance nucleation, resulting in superior film morphology and electrical properties. The structural, optoelectronic, and electrical properties compare with state-of-the-art OMVPE-grown films. Additionally, these films have been grown to thicknesses of 40 μm without optically visible thermally induced cracking. Consistent with the structural characterization of these films, we find that the film quality continues to improve with thickness, even for thicknesses > 15μm. The high growth rate, comparatively large thickness and low defect density demonstrates that HVPE is viable for the growth of low-defect GaN thick films for use as a substrate for device overgrowths.

ACKNOWLEDGMENTS

The authors are grateful to D. Hovey for fine glasswork and useful discussions and to J. Daneu and B.S. Krusor for technical assistance. The work at Lincoln Laboratory is supported by the Department of the Air Force, and the work at Xerox PARC was supported by ARPA (agreement # MDA972-95-3-008). Opinions, interpretations, conclusions, and recommendations are those of the authors and not necessarily endorsed by the United States Air Force.

REFERENCES

[1] I. Akasaki, H. Amano, M. Kito and K. Hiramatsu, J. Lumin. **48&49**, 666 (1991).
[2] S. Nakamura, T. Mukai and M. Senoh, Appl. Phys. Lett. **64**, 1687 (1994).
[3] S. Nakamura, M. Senoh, N. Iwasa and S. Nagahama, Jpn. J. Appl. Phys. **34**, L797 (1995).
[4] S. Nakamura, M. Senoh, S. Nagahama, N. Iwasa, T. Yamada, T. Matsushita, H. Kiyoku and Y. Sugimoto, Jpn. J. Appl. Phys. **35**, L74 (1996).
[5] P. Maki , R.J. Molnar, R.L. Aggarwal, Z.L. Liau and I. Melngailis, Mater. Res. Soc. Symp. Proc. **395**, 919 (1996).
[6] R.J. Molnar, K.B. Nichols, P. Maki, E.R. Brown and I. Melngailis, Mater. Res. Soc. Symp. Proc. **378**, 479 (1995).
[7] R.J. Molnar, R. Aggarwal, Z.L. Liau, E. R. Brown, I. Melngailis, W. Götz, L.T. Romano and N.M. Johnson, Mater. Res. Soc. Symp. Proc. **395**, 189 (1996).
[8] K. Naniwae, S. Itoh, H. Amano, K. Itoh, K. Hiramatsu and I. Akasaki, J. Cryst. Growth **99**, 381 (1990).
[9] T. Detchprohm, K. Hiramatsu, H. Amano and I. Akasaki, Appl. Phys. Lett. **61**, 2688 (1992).
[10] L.T. Romano, B.S. Krusor, G.A. Anderson, D.P. Bour, R.J. Molnar and P. Maki, *this volume.*
[11] W. Götz, N.M. Johnson, R.A. Street, H. Amano and I. Akasaki, Mater. Res. Soc. Symp. Proc. **378**, 491 (1995)
[12] R.J. Molnar, T. Lei and T.D. Moustakas, Appl. Phys. Lett. **62**, 72 (1993).

OPTIMIZATION OF REACTOR GEOMETRY AND GROWTH CONDITIONS FOR GaN HALIDE VAPOR PHASE EPITAXY

S.A. Safvi*, N.R. Perkins**, M. N. Horton** , A. Thon*, D. Zhi**, T.F. Kuech*'**
*Department of Chemical Engineering, University of Wisconsin, Madison, WI 53706.
**Materials Science Program, University of Wisconsin, Madison, WI 53706.

ABSTRACT

A numerical model of an experimental gallium nitride horizontal vapor phase epitaxy reactor is presented. The model predicts the flow, concentration profiles, and growth rates. The effects of flowrate variation and geometry on the growth rate, growth uniformity and crystal quality were investigated. Numerical model predictions are compared to experimentally observed values. Parasitic gas phase reactions between group III and group V sources and deposition of material on the wall are shown to lead to reduced overall growth rates and inferior crystal quality. A low ammonia concentration is correlated to deposition of polycrystalline films. An optimum HVPE growth process requires selection of reactor geometry and operating conditions to minimize parasitic reactions and wall deposition while providing a uniform reactant distribution across the substrate.

INTRODUCTION

The nitrides of gallium, aluminum, and indium have great potential for applications in electronics and optoelectronic devices due to their wide bandgap range and stability at high temperatures. These nitrides have a direct bandgap ranging from 1.9 eV for InN to 6.3 eV for AlN. GaN with its bandgap of 3.4 eV is particularly suitable for making devices operating in blue to ultraviolet range [1]. The bandgap can be engineered by growing ternary alloys of these nitrides. Zhang et al. [1] measured the bandgap of $Al_xGa_{1-x}N$ using room temperature optical transmission and absorption spectroscopy. By changing the solid composition, a tunable direct bandgap ranging from that of pure GaN to that of pure AlN was obtained.

The heteroepitaxial growth of thin GaN films on sapphire leads to defects arising from lattice mismatch and difference in thermal expansion coefficient. The development of GaN substrates is likely to be a key advance in nitride epitaxial technology, making it feasible to grow homoepitaxial thin GaN films [2]. A promising route for the development of GaN substrates is the heteroepitaxial growth of GaN films by rapid growth techniques, such as halide vapor phase epitaxy (HVPE), followed by the in situ etch removal of the initial substrate to leave a free-standing GaN film. The HVPE technique has been used previously to grow thick layers of GaAs [3], GaN [2, 4, 5], and InP [6]. Mochizuki et. al. [7] have studied the direct reaction between AsH_3 and surface adsorbed GaCl in order to understand the growth chemistry involved in HVPE of GaAs. However, no comparable GaN based studies have been reported.

The development of predictive models of the HVPE process can substantially reduce the time and cost associated with reactor optimization and scale-up by minimizing the required experimental trial and error. It would also aid development of an improved understanding of the HVPE process.

In this study we will describe a two-zone hot wall reactor used to grow thick HVPE GaN films. The emphasis will be on studying the effect of process and geometric parameter variation on film thickness, uniformity and material properties. Experimental results will be compared to

Mat. Res. Soc. Symp. Proc. Vol. 423 © 1996 Materials Research Society

computational predictions. The effect of local gas phase concentrations on film properties will be discussed.

REACTOR MODEL & GROWTH STUDIES

The experiments were carried out in an atmospheric pressure quartz reactor which has been presented earlier[8]. The reactor has three separate concentric inlets for the reaction gases. A mixture of N_2 and HCl is introduced through the central tube, N_2 through the middle annular region, and NH_3 through the outer annular region. The reactor is divided into two separate temperature zones of 850 and 1050 °C. In the first reaction zone, operated at 850 °C, Ga metal is reacted with HCl gas (typical flow ~ 30 sccm) to yield GaCl and H_2 reaction products. The extent of reaction, obtained from decrease in the mass of Ga metal, was in the range of 50-70%. These reaction products are transported to the second zone through the central tube. In the second zone, typically maintained at 1050 °C, a high flowrate N_2 buffer and NH_3 were introduced from the middle and outer annular regions, respectively. The NH_3/HCl ratio was typically 30:1. Two inch diameter (0001) sapphire substrates prepared with a standard solvent degrease are used for the HVPE deposition. For the base case, a N_2 buffer flowrate of 4.5 slm, a NH_3 flowrate of 882 sccm, and, in the central tube, a mixture of 30 sccm HCl and 150 sccm N_2 was employed.

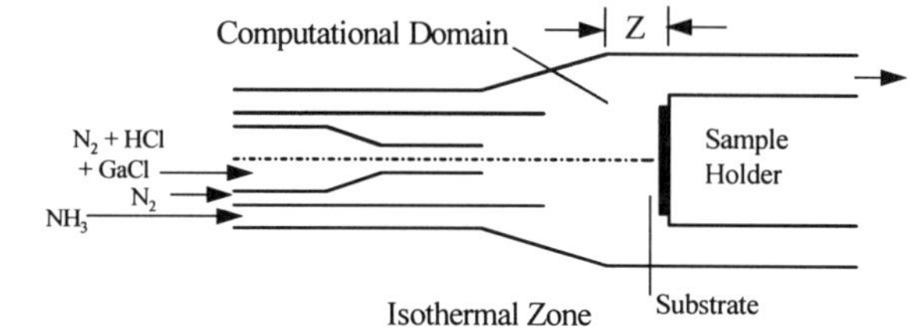

Figure 1: Schematic of the horizontal HVPE reactor. The shaded region defines the computational domain.

A schematic of the computational domain is shown in Figure 1. The variable Z shown in Figure 1 represents the relative axial distance of the sample holder from the inlet. Only the second temperature zone of the reactor, which is of interest for process optimization, was modeled. Cylindrical coordinates have been used. The fundamental equations of continuity, momentum, and species conservation are used to describe the system [9]. With the assumption of no variation in circumferential direction, the flow and concentrations are obtained in two dimensions. The properties of the gas mixture are determined at any point using the concentrated species and applying ideal mixing rules.

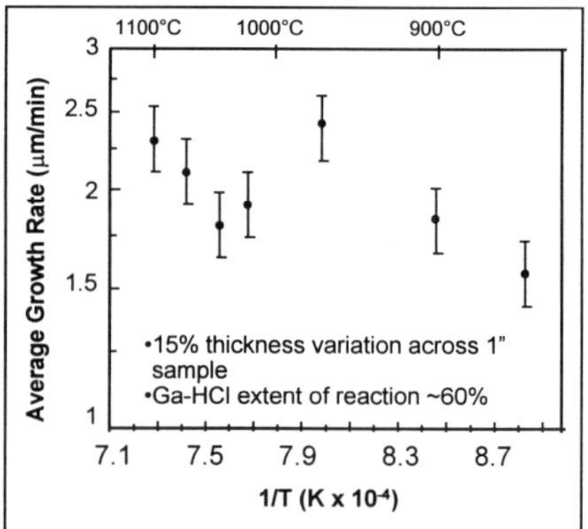

Figure 2: Plot of average growth versus reciprocal of reactor temperature.

Physical and Transport Properties of Gaseous Species: Experimental values of the viscosity of nitrogen [10] and ammonia [11] were fitted to equations as a function of temperature and are listed elsewhere [12]. Binary diffusion coefficients of gas phase species were either obtained from literature or estimated from their Lennard-Jones parameters [13, 14]. The values of the binary diffusion coefficient and Lennard-Jones parameters for all gaseous species used in this study are listed elsewhere [12]. Linear mixing rules were applied to determine local gas phase properties.

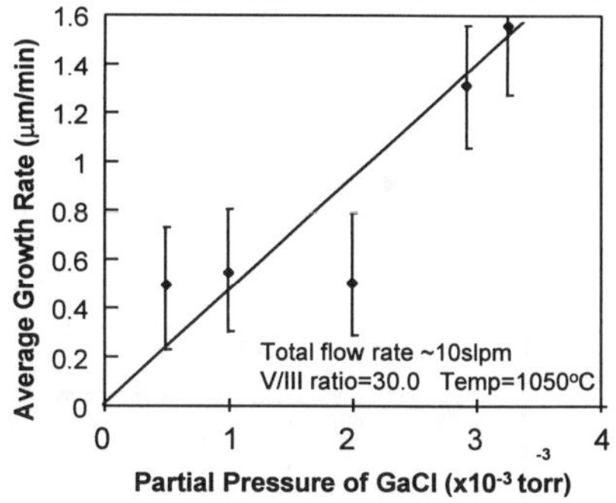

Figure 3: Plot of average growth rate versus GaCl partial pressure at 1050 °C.

Numerical Solution: The system of partial differential equations describing flow and mass transfer was solved using Galerkin's finite-element method [15]. A typical mesh consisted of 1400 trapezoidal elements with the mesh being denser in those parts of the domain where steeper gradients existed. The system of nonlinear algebraic equations obtained after the application of Galerkin's technique was solved by using Newton's method. The system of equations and numerical methodology is presented in detail elsewhere [11]. The computations were performed on a Cray C90 supercomputer.

RESULTS AND DISCUSSION

The first series of experiments were performed in the same reactor with a tilted substrate holder, not shown. The purpose of these experiments was to determine the effect of temperature and GaCl concentration on growth rate. Figure 2 shows a graph of growth rate over a range of temperature. The data indicates that the growth rate at a temperature of 1050 °C is mass transport limited. Subsequent experiments were performed at this temperature because the crystalline quality of the films grown at the higher temperatures in our reactor was better as gauged by x-ray analysis. Figure 3 shows the effect of variation of GaCl partial pressure on the growth rate. The rate of increase in growth rate is the same as the rate of increase of GaCl partial pressure indicating that the growth is limited by transport of the gallium containing species to the growth front. Thus a convection-diffusion model of the reacting species would be a good approximation of the actual process.

In the following set of experiments, a different sample holder that has the growth face perpendicular to the central axis of the reactor was used (Figure 1). This geometric arrangement made the growth system axisymmetric and hence easy to model and compare with experiments.

In experiments where the sample holder was kept close to the inlet, the resulting samples have a dark polycrystalline patch in the center of the wafer, with a clear single crystalline film at the edges. This polycrystalline patch decreased in size and eventually disappeared as the substrate was moved further away from the inlet (Z increases in Figure 1). The patch could be caused due to a deficiency of NH_3 (and hence low effective V/III ratio) near the center of the wafer as NH_3 is introduced from the outermost annular inlet to reduce prereaction. When the substrate is far from the inlet, the ammonia stream is well mixed with

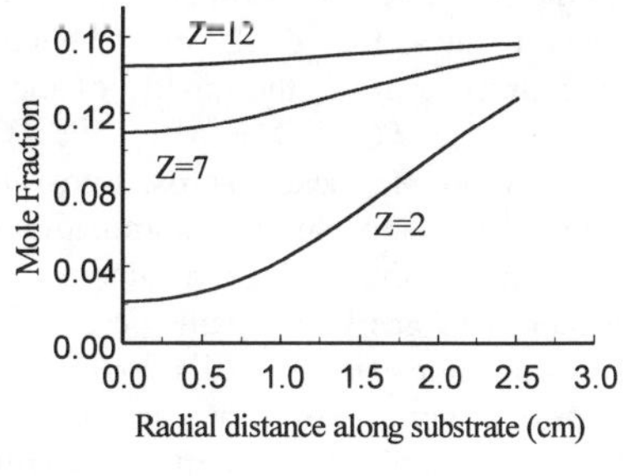

Figure 4: Ammonia concentration across the substrate for different substrate positions.

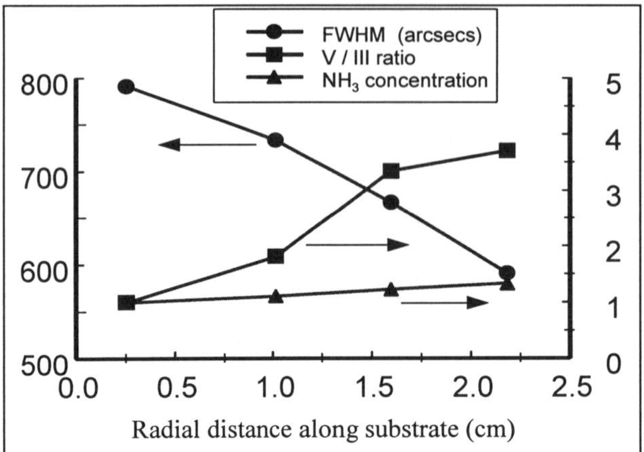

Figure 5: Plot of dimensionless NH_3 concentration, dimensionless V/III ratio, and FWHM from TCXRD (arcsecs) at various points across the substrate.

the other two inlet streams and there is a uniform concentration of ammonia over the substrate at the growth front. Figure 4 shows variation of NH_3 concentration over the substrate for different substrate positions. At a distance of Z=7 cm, the film obtained was single crystalline across the wafer. Figure 5 shows a plot of NH_3 concentration and V/III ratio, made dimensionless with respect to their values at the center of the substrate, across the substrate. The corresponding full width of the triple crystal x-ray diffraction (TCXRD) rocking curve at half maximum ranged from 791 arcsecs at the center to 590 arcsecs at the edge of the wafer. The quality of the film improves with increasing values of ammonia concentration and local V/III ratio. We would expect this trend to reverse after a certain limit as high ammonia concentrations would lead to increased parasitic reactions as discussed later.

The growth uniformity improved with increasing substrate-inlet distance as shown in Figure 6. This is expected as the reactant gases have more time to diffuse and become more uniformly distributed over the substrate. A lowering in overall growth rate with increasing substrate-inlet distance, as shown in Figure 6, could be due to loss of precursors because of surface reaction on the walls (deposition on reactor walls) and gas phase parasitic reactions. The former effect is incorporated in the model but the latter is not, and may explain the predicted growth rates being lower than those experimentally observed. This discrepancy in predicted and observed growth rates increases with increasing substrate-inlet distance. For the process conditions as in the base case, Figure 7 shows a comparison of predicted and experimentally obtained growth rates for Z=12 cm. Though the uniformity is good, the quality of the films, examined by TCXRD, is poor. The TCXRD showed double peaks across most of the sample, probably due to delamination of the film from the substrate upon cooling. Products of gas phase parasitic reactions or from reactions on the walls could adversely interfere with the film growth on the substrate. Additionally, particles coming off the walls may be transported to the film in the absence of thermophoresis effects in the isothermal reactor.

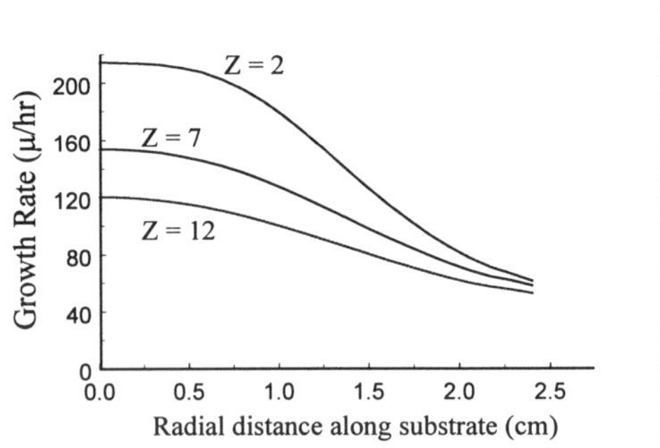

Figure 6: Predicted growth rate variation across the substrate for substrate position of Z =2, 7, and 12 cm.

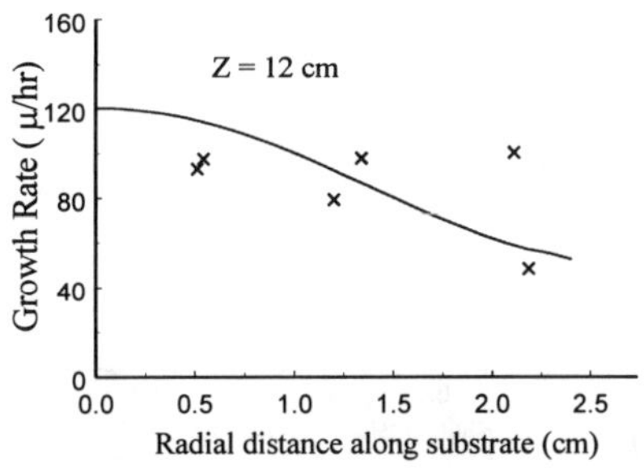

Figure 7: Variation of Growth Rate across the substrate for Z= 12 cm. (x) are the experimental values.

The coupled gas phase and surface reaction between GaCl and NH_3 has been studied by V. Ban using mass spectrometry [16]. The simplest gas phase reaction would lead to the formation of GaN monomer with eventual particulate formation and the resulting loss of reactants at the growth front. In the first zone (at 850 °C) formation of GaCl in the presence of excess HCl is favored. In the second zone (1050 °C), the presence of excess ammonia shifts the equilibrium such that the reduction of GaCl is favored. Both prereaction and wall effects could lead to the same results: lowered growth rates and inferior crystal quality. In order to isolate the effect of gas phase parasitic reactions, we performed some experiments at a sample holder close to the inlet (Z= 4.5 cm) where the effects of the reactions at the wall on the substrate film growth are minimized. All process parameters are kept the same as in the base case, except for the N_2 buffer flow rate through the central annulus. For an N_2 buffer flow of 2.25 and 4.5 slm, computations show that the overall growth rate does not change appreciably, indicating that the walls have little effect on the growth process. Experiments, in contrast, reveal that the growth rates are significantly reduced. Figure 8 and 9 show the predicted and experimentally obtained growth rates for a N_2 buffer flow of 2.25 and 4.5 slm, respectively. Lowering the buffer flowrate increases the residence time in the reactor and reduces the barrier to the mixing of GaCl and NH_3. This gas phase mixing accelerates the pre-reaction between the two precursors. Increasing the residence time increases the extent of this reaction. These results indicate that an optimum buffer flowrate of N_2 should be high enough that precursor prereactions are low, but low enough to allow diffusion of precursors so that they are uniformly distributed across the substrate.

CONCLUSIONS

A numerical model of the HVPE GaN reactor has been presented. Growth rates were predicted and compared to experimentally obtained values for different substrate positions. For substrate positions close to the inlet, non-uniform distribution of precursors across the substrate results in a non-uniform growth rate and film properties. A polycrystalline patch was obtained at the center of the wafer which grew smaller as the substrate was moved further away from the inlet. The crystalline quality of the film was found to improve with increasing local concentration of ammonia and the V/III ratio at the substrate. For substrate positions far from the inlet, films with uniform growth rate were obtained but the crystalline quality was poor. A N_2 buffer between the group III and group V precursor was necessary to limit parasitic

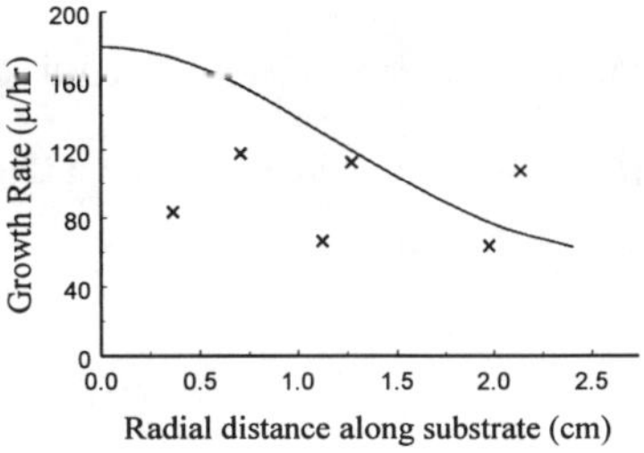

Figure 8: Comparison of computationally and experimentally (x) observed growth rates for a N_2 buffer flow of 2.25 slm, and Z=4.5 cm.

prereactions. The quality of the films and the growth rate was sensitive to the buffer flowrates.

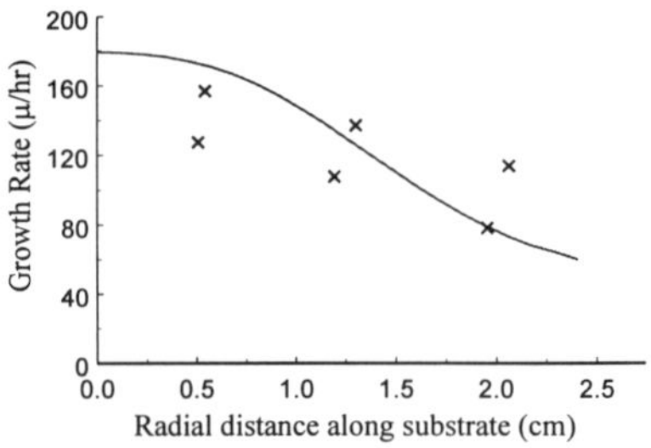

Figure 9: Comparison of computationally and experimentally (x) observed growth rates for a N_2 buffer flow of 4.5 slm, and Z=4.5 cm.

ACKNOWLEDGMENTS

The work was supported by ARPA - URI on Visible Light Emitters. The supercomputing time was provided by DOD.

REFERENCES

1. X. Zhang, P. Kung, A. Saxler, D. Walker, T.C. Wang, and M. Razeghi, Appl. Phys. Lett. **67** (12), 1745 (1995).
2. T. Detchprohm, K. Hiramatsu, N. Sawaki, and I. Akasaki, J. Crystal Growth **137** p. 170-174 (1994).
3. D.W. Shaw, J.Crystal Growth **31** (1975) 130.
4. H. Maruska, and J. Tietjen, Appl. Phys. Lett. **15** p.327 (1969).
5. O. Lagerstedt and B. Monemar, J. Appl. Phys., **45** p. 2266-2272 (1974).
6. R. Beccard, S. Beuven, K. Heime, R. Schmald, H. Jurgensen, P. Harde, and M. Schlak, J. Crystal Growth **121** (1992) 373.
7. Y. Mochizuki, A. Usui, S. Handa and T. Takada, J. Crystal Growth **148** (1995) 96-105.
8. N.R. Perkins, M.N. Horton and T.F. Kuech, MRS Proceedings, 1995.
9. R.B. Bird et al., W.E. Stewart and E.N. Lightfoot, Transport Phenomenon, Wiley, New York (1960)
10. R.H. Perry & C.H. Chilton, Chemical Engineer's Handbook, 5th ed., McGraw-Hill, New York (1977).
11. R.C. Reid, J.M. Prausnitz and T.K. Sherwood, The Properties of Gases and Liquids, McGraw-Hill, New York (1977).
12. S.A. Safvi and T.F. Kuech, Manuscript in preparation.
13. T.J. Mountziaris, S. Kalyanasundram and N.K. Ingle, J. Cryst. Growth, **131**, 283 (1993).
14. J.O. Hirschfelder C.F. Curtiss and R.B. Bird, Molecular Theory of gases and Liquids, Wiley, New York (1967).
15. G. Strang & G. Fix, An Analysis of the Finite Element Method, Prentice Hall, Englewood Cliffs, NJ (1973).
16. V. Ban, J. Electrochem Soc., **119**, 761 (1972).

Thermodynamic Analysis and Growth Characterization of thick GaN films grown by Chloride VPE using $GaCl_3/N_2$ and NH_3/N_2

Heon Lee*, Masaaki Yuri, Tetsuzo Ueda and James S. Harris, Jr.
Solid State Electronics Laboratory, Stanford University, Stanford, CA 94305

ABSTRACT

Thermodynamic calculations were carried out on chloride transport vapor phase epitaxy of GaN using $GaCl_3/N_2$ and NH_3/N_2. At typical growth temperature and gas flow rates, both GaN formation and gas phase etching reactions of GaN are thermodynamically favored. Under thermodynamic equilibrium, most ammonia should decompose to nitrogen and hydrogen gases and gas phase etching of GaN occurs by HCl. From experimental measurements, less than 10% of the incoming ammonia decomposes and under this condition, GaN formation from $GaCl_3/N_2$ and NH_3/N_2 is thermodynamically favored. Higher V/III ratios give a larger driving force for GaN fromation. These calculations match our experimental results. Experimentally, we have optimized the growth conditions of GaN. High crystalline quality thick GaN films (10 ~15μm) were grown on c-Al_2O_3. The GaN films show band edge emission dominated PL at both room temperature and 77 K. Only one set of diffraction peaks from $(10\bar{1}2)$ planes with 60° spacing in the φ-scan of X-ray diffraction are observed. This indicates that the GaN films grown on c-Al_2O_3 are single crystalline. Typical growth rates were about 10 ~15μm/hr and typical Hall mobility values of GaN films were in the range of 3 to 40 $cm^2/Vsec$.

INTRODUCTION

Vapor phase epitaxy (VPE) of GaN has a long history and has been widely used because of its high growth rates (up to a few tens of μm/hr) and lack of carbon incorporation into the film. In conventional chloride transport VPE of GaN, the Ga source is gallium monochloride (GaCl), which is stable only at temperatures above 600°C and is produced by the reaction of liquid gallium with hydrogen chloride (HCl) gas [1~5]. The supply of GaCl is controlled by the Ga temperature and the flow rates of the HCl and H_2 carrier gas. In this work, we used gallium trichloride ($GaCl_3$) as the Ga source. This is the first report of GaN growth using $GaCl_3$ as a precursor. Ammonia (NH_3) and nitrogen (N_2) gases were used as the nitrogen source and carrier gas, respectively. H_2 and HCl gases were not used. Several researchers have reported VPE growth of GaAs using $GaCl_3$ as the initial source, however it was reduced to the monochloride before reaching the reaction zone. Rubinstein and Myers [6] used $GaCl_3$, but it was reduced to a monochloride by reacting with metallic Ga. Hasegawa[7,8] and other researchers also used $GaCl_3$. $GaCl_3$ reacted with the hydrogen to reduce to GaCl before crystal synthesis.

In this paper, we calculate the thermodynamic driving force for all possible reactions inside the reactor and analyze the equilibria between these reactions in order to clarify whether the growth of GaN from $GaCl_3/N_2$ and NH_3/N_2 is predicted from thermodynamic equilibria reactions. In all cases, the standard state is taken as the pure substance at one atmosphere. All thermodynamic data used in this paper is from 'Thermochemical data of pure substance' tabulated by Ihsan Barin[9].

GROWTH SYSTEM DESCRIPTION

An open flow horizontal quartz tube with single zone hot wall furnace was used as the reactor. The crystalline $GaCl_3$ was melted and heated up to 100°C in a stainless steel cell. At this temperature, the vapor pressure of $GaCl_3$ is about 20 mm Hg. The $GaCl_3$ vapor was transported to the reactor by nitrogen carrier gas. Nitrogen gas was chosen in order to dilute the ammonia and to increase the total gas flow rate. The supply of $GaCl_3$ was controlled by both the $GaCl_3$ cell temperature and the flow rate of N_2 carrier gas. The typical N_2 flow rate was 0.1 slm. The actual $GaCl_3$ vapor supply was about 0.003 slm under the above conditions. During the process, 1.5 - 2.0 slm of NH_3 and 1.0 ~ 1.5 slm of N_2 were introduced into the reactor. The resulting V/III ratio

Mat. Res. Soc. Symp. Proc. Vol. 423

was about 500 to 700. The growth temperature was varied over the range of 930 ~ 1050°C. Atmospheric pressure was maintained during the growth. Fig. 1. shows a schematic diagram of the VPE system to grow GaN using $GaCl_3/N_2$ and NH_3/N_3.

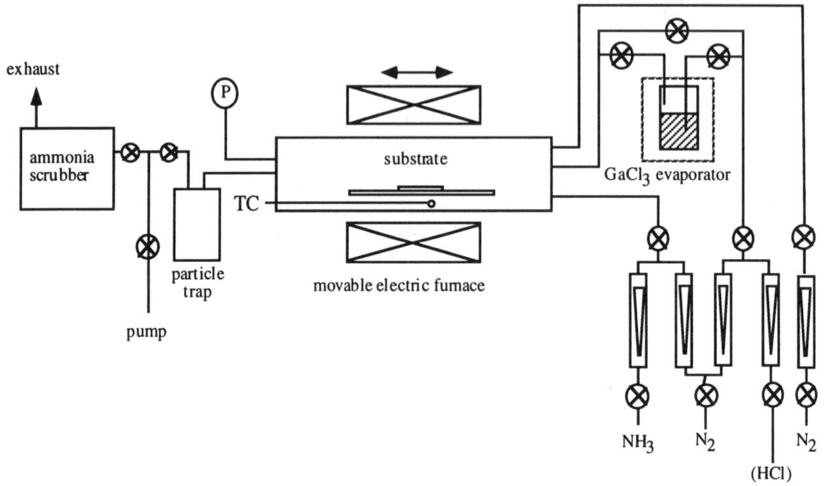

Figure 1. Schematic Apparatus of GaN chloride vapor phase epitaxy system

THERMODYNAMIC ANALYSIS

The nitrogen source, NH_3, is not stable at high temperature, and will thermally decompose to nitrogen and hydrogen gases.

$NH_3(g) = (1\text{-}x)\ NH_3(g) + x/2\ N_2(g) + 3x/2\ H_2(g)$ reaction (1)

where x is the extent of the reaction. The equilibrium value for the extent of the reaction, (x_{eq}), at 1300 K (a typical growth temperature) is about 0.9995. Hence most of the ammonia decomposes to nitrogen and hydrogen at that temperature in thermodynamic equilibrium. However, it is not easy to reach thermodynamic equilibrium in such an open flow system. We can assume that the actual value of x is much lower than its equilibrium value (x_{eq}). The experimentally measured value of x is about 0.1 or less under typical growth parameters[4]. The actual value of x is dependent on NH_3 temperature, NH_3 partial pressure, residence time and surface condition. Both NH_3 partial pressure and residence time can be changed by varying the NH_3 and N_2 flow rates. The H_2 gas, produced from the thermal decomposition of ammonia, will promote the reduction of $GaCl_3$ to GaCl. Even though the direct reduction of $GaCl_3$ to GaCl is not thermodynamically favorable, the hydrogen reduction of $GaCl_3$ to GaCl can occur, spontaneously. Let η be the extent of reaction for hydrogen reduction of $GaCl_3$ to GaCl. The reaction can be written as:

$GaCl_3(g) + H_2(g) = GaCl(g) + 2\ HCl(g)$ reaction (2)

Thus, at a given temperature and extent of the reaction for ammonia decomposition (x), the equilibrium extent of the reaction (η_{eq}) of reaction [2] can be obtained. For typical growth temperatures (1100 K, 1200 K and 1300 K), the relationship between x and η_{eq} is shown in Figure 2. At these temperatures, η_{eq} is almost unity for all values of x larger than 0.01. This implies that most of the $GaCl_3$ will be converted to GaCl at these temperatures under thermodynamic equilibrium. GaCl vapor, produced by reaction (2) reacts with the remaining NH_3 to form GaN.

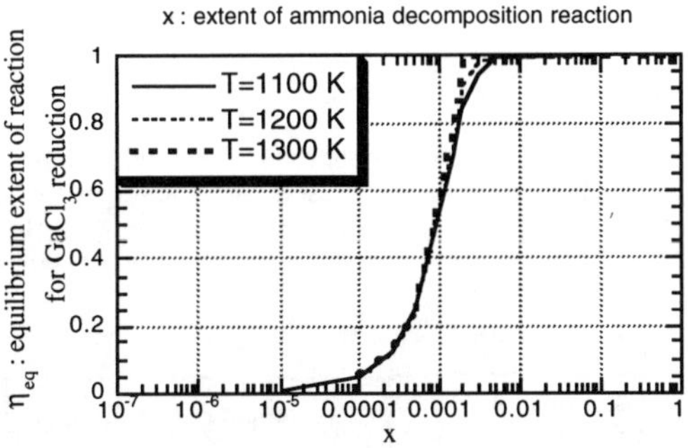

Figure 2. Equilibrium extent of reaction (η_{eq}) for the reduction of $GaCl_3$ with respect to x (extent of reaction) for thermal decomposition of ammonia at various temperatures

Figue 3 shows the equilibrium pressure of any gaseous species inside the reactor. The calculation is based on thermodynamic equilbrium of reactions from (1) to (5) and materials conservation inside the reactor. The activity of GaN was assumed as an unity.

$GaCl + NH_3 = GaN + HCl + H_2$ reaction (3)

$GaCl_3 + NH_3 = GaN + 3\ HCl$ reaction (4)

$GaN + HCl = GaCl + 0.5H_2 + 0.5N_2$ reaction (5)

$P_{total} = 1\ atm = P_{NH3} + P_{N2} + P_{H2} + P_{GaCl3} + P_{GaCl} + P_{HCl}$

$3N^o_{GaCl3} = 3N_{GaCl3} + N_{GaCl} + N_{HCl}$ [Cl]

$3N^o_{NH3} = 3N_{NH3} + 2N_{N2} + N_{HCl}$ [H]

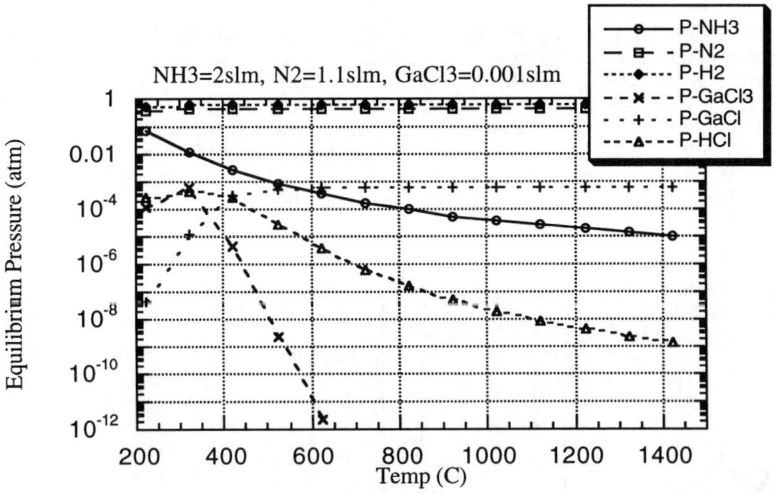

Figure 3. Equilibrium Partial Pressures of gaseous species inside the reactor.

Under thermodynamic equilibrium at typical growth temperature, almost all the ammonia, $GaCl_3$ and HCl gases are consumed by reactions (1), (2) and (5), respectively and GaN formation is not thermodynamically favorable. The driving force for GaN deposition (del Ga) under thermodynamic equilibrium can be defined as: del Ga = $N_{GaCl3} + N_{GaC} - N^{o}_{GaCl3}$, where N stands for the number of molecules per unit itme and volume. If del Ga > 0, gas phase etching of GaN is favored and if del Ga < 0, GaN deposition will occur. The values of del Ga under various x (extent of reaction for ammonia decomposition) over typical growth temperatures are shown in Figure 4. As x increases, GaN formation is depressed and etching of GaN (reaction (5)) is promoted. By adding nitrogen to dilute the ammonia and increase gas flow velocity, thermal decomposition of ammonia (reaction (1)) is reduced, thus the value of x is decreased.

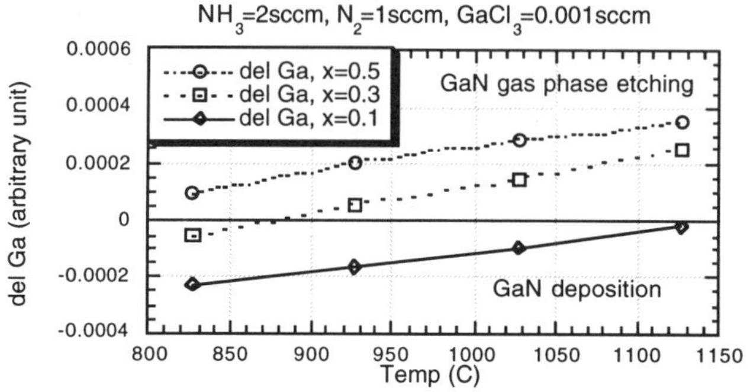

Figure 4. Effects of x (extent of reaction for ammonia decomposition) on the driving force of GaN deposition(del Ga).

EXPERIMENTS AND RESULTS

Growth of GaN using the above $GaCl_3$ precursor and reactions was investigated under a variety of growth conditions. (0001) sapphire substrates were degreased by organic solvents and chemically etched with a hot HCl + H_2PO_4 (3:1) solution in order to remove polishing damage and smooth the surface. After loading the substrates, the reactor was pumped and purged with high purity nitrogen five times. The substrates were heated to the growth temperature under a NH_3 / N_2 ambient. According to XPS analysis, the sapphire surface was nitridized during this heating phase. After reaching the growth temperature, $GaCl_3$ was supplied to the reactor using nitrogen carrier gas. No low temperature buffer layer was used in these experiments.

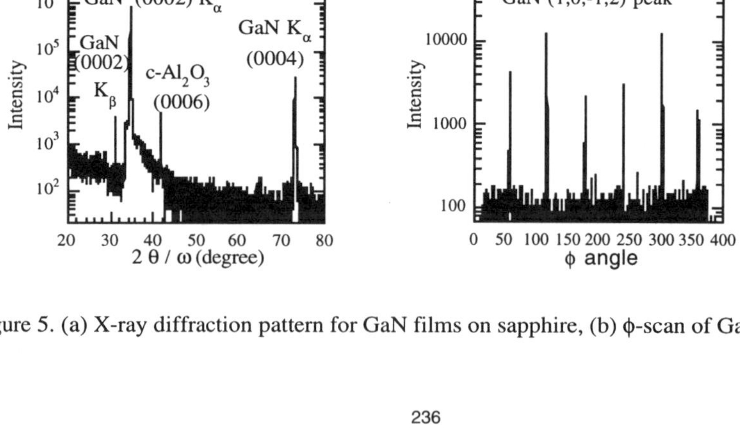

Figure 5. (a) X-ray diffraction pattern for GaN films on sapphire, (b) φ-scan of GaN $(10\bar{1}2)$ plane

X-ray diffraction was used to determine the crystallinity and the orientation of the GaN films. Fig. 5a shows the $2\theta/\omega$ X-ray diffraction pattern for thick GaN films on sapphire. Only (000l) type plane peaks are observed. This implies that the GaN film and sapphire substrate basal planes are parallel to each other, i.e. GaN [0001] // sapphire [0001]. Fig. 5b shows the ϕ-scan of the GaN $(10\bar{1}2)$ plane to determine the in-plane orientation. Only one set of reflection peaks from the $(10\bar{1}2)$ plane is observed with 60° spacing. This indicates that the GaN films grown on sapphire are single crystal. The angular position of the GaN $(10\bar{1}2)$peaks are well matched to the sapphire $(11\bar{2}3)$ peaks. This means the GaN crystal is rotated 90° around the c-axis to minimize the lattice mismatch strain, i.e. GaN $[10\bar{1}0]$ // sapphire $[11\bar{2}0]$.

In order to characterize the optical properties of GaN films, photoluminescence (PL) measurements were done at both room temperature and 77 K. A 10 mW He-Cd laser operating at 3.82 eV was used as the excitation source. At room temperature, only band edge emission near 3.40eV was observed. At 77 K, the spectrum consisted of several peaks, including an exciton bound to a neutral donor (I_2), an exciton bound to a neutral acceptor (I_1) and donor-acceptor pairs. The spectrum is dominated by I_2 emission near 3.47 eV. Fig. 6 shows typical PL spectra at both temperatures. The band edge emission peak had a FWHM of 60 meV at room temperature and the exciton bound to donor emission peak had a FWHM of 25 meV at 77 K. Very weak deep level luminescence near 2.1 eV was also observed at both room temperature and 77 K. According to the PL spectra, the thick GaN films grown by VPE using $GaCl_3/N_2$ and NH_3/N_2 have good crystalline quality.

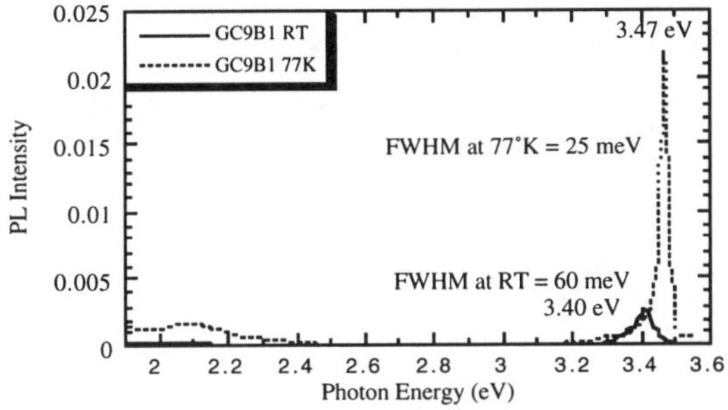

Figure 6. Photoluminescence spectra of GaN film at both room temperature and 77 K

Scanning Electron Microscopy was used to characterize the surface morphologies of GaN films grown on c-sapphire substrates. Commonly, it is very difficult to achieve a very smooth surface over a large area in chloride VPE growth of GaN, due to its high deposition and gas phase etching rates. The localized flux uniformity and growth temperature play key roles in determining the surface morphologies of GaN films. In this experiment, we fixed V/III ratio at approximately 700. The position of the sapphire substrate (1 inch distant from the NH_3/N_2 nozzle) and the configuration of the $GaCl_3/N_2$ and NH_3/N_2 nozzles were also fixed to optimal positions.

The effect of the growth temperature on surface morphology of GaN films was investigated. For GaN films grown below 950°C, shown in Figure 7a, the surface was covered by GaN crystallites (islands) with an approximate size of a few microns. These islands did not coalesce and there are holes all the way to the substrate. The $2\theta/\omega$ scan of X-ray diffraction show only (000l) type plane peaks with broad FWHM. The PL spectra show only band edge emission at room temperature which implies that each GaN crystallite has good crystal quality. According to the X-ray ϕ-scan, several different in-plane orientations were observed. This means each crystallite has a different in-plane orientation.

As the growth temperature is increased to 975°C, (Figure 7b), the crystallites coalesce and the surface became smoother, however, grooves and microholes which have about 50 ~ 60° inclined sidewalls were observed. The films grown in this temperature regime have narrower X-ray diffraction and photoluminescence linewidths. The surface morphologies of films grown at similar temperatures (965°C) with a 3 times longer growth time are shown in Figure 7c. A terrace structure with microholes at the edge of the terraces is observed, although no individual GaN islands are observed. This confirms that the film has grown laterally, rather than three dimensionally. From the ϕ-scan of the GaN $(10\bar{1}2)$ peak, the film is single crystal. Figure 7d shows the surface morphology of a GaN film grown at 995°C, which has the smoothest and flattest surface. At the boundary of two laterally grown islands, thin shallow grooves are observed. Higher growth temperature and longer growth time appear to give better surface morphologies.

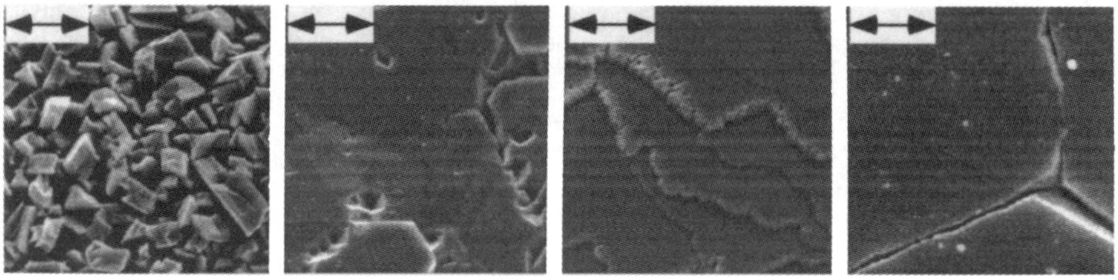

(a) (b) (c) (d)

Figure 7. Surface Morphology of GaN film, (a) grown at 950°C for 1 hour, scale bar = 7.5μm, (b) grown at 975°C for 1 hour, scale bar = 15μm, (c) grown at 965°C for 3 hours, scale bar = 12.5μm, (d) grown at 995°C for 1 hour, scale bar = 3.75μm

According to Hall measurements, GaN films have high n-type background conductivity up to $10^{19}/cm^3$. The electron mobility at room temperature is in the range of 3 to 40 cm^2/Vs.

Conclusion

Thermal decomposition of ammonia and hydrogen reduction of $GaCl_3$ to GaCl were thermodynamically favored under our growth temperature and mass flow conditions. Under thermodynamic equilibrium conditions, gas phase etching of GaN by HCl was preferred over GaN deposition. From experimental measurements, thermal decomposition of ammonia never reached to its equilibrium and only less than 10% of ammonia can be decomposed. The GaN films grown under optimized conditions are single crystalline and show very sharp and strong band edge related emission without any obvious defects related emission in their PL spectra. Higher growth temperature (>975°C) and longer growth time are very important to obtain smoother surface morphologies of GaN films.

REFERENCES

1. J. S. Hwang, A. V. Kuznetsov, S. S. Lee, H. S. Kim, J. G. Choi and P. J. Chong, J. Cryst. Growth 142(1994) 5
2. H. Tsuchiya, T. Okahisa, F. Hasegawa, H. Okumura and S. Yoshida, Jpn. J. Appl. Phys 33(1994) 1747
3. H. P. Maruska, W. C. Rhines and D. A. Stevenson, Mat. Res. Bull. 7(1972) 777
4. M. Illegems, J. Cryst. Growth 13/14(1972) 360
5. M. Illegems and H. C. Montgomery, J. Phys. Chem. Solids, 34(1973) 885
6. M. Rubenstein and E. Myers, J. Electrochem. Soc. 113(1966) 365
7. A. Koukito, F. Hasegawa and H. Seki, Jpn. J. Appl. Phys. 27(1988) L1546
8. F. Hasegawa, H. Yamaguchi and K. Katayama, Jpn. J. Appl. Phys. 27(1988) L1546
9. I. Barin, Thermochemical Data of Pure Substance (VCH, Weinheim, FRG, 1989)

HIGH RESOLUTION X-RAY DIFFRACTION ANALYSIS OF GALLIUM NITRIDE GROWN ON SAPPHIRE BY HALIDE VAPOR PHASE EPITAXY

R.J. MATYI[1], D. ZHI[2], N.R. PERKINS[2], M.N. HORTON[2], and T.F. KUECH[2,3]
[1]Dept. of Materials Science and Engineering, University of Wisconsin, Madison, WI 53706
[2]Materials Science Program, University of Wisconsin, Madison, WI 53706
[3]Dept. of Chemical Engineering, University of Wisconsin, Madison, WI 53706

ABSTRACT

We report a structural analysis of GaN layers with thicknesses ranging from 10 μm to 250 μm which have been grown on sapphire substrates by halide vapor phase epitaxy (HVPE). The effect of growth rate during HVPE growth has also been examined. The growth was performed using GaCl and ammonia as reactants; growth rates in excess of 90 μm/hr have been achieved. The structural characteristics of these layers have been performed with high resolution x-ray diffractometry. Longitudinal scans parallel to the GaN [0002] direction, transverse scans perpendicular to the [0002], and reciprocal space maps of the total diffracted intensity have been obtained from a variety of GaN layers. The transverse scans typically show broad rocking curves with peak breadths of several hundreds of arcseconds. In contrast, the longitudinal scans (or "θ/2θ scans") which are sensitive only to strains in the GaN layers (and not their mosaic distributions) showed peak widths that were at least an order of magnitude smaller and in some cases were as narrow as 16 arcseconds. These results suggest that the defect structure of the GaN layers grown by HVPE is dominated by a dislocation-induced mosaic distribution, with the effects of strain in these materials being negligible in comparison.

INTRODUCTION

Halide vapor phase epitaxy (HVPE) offers the advantage of a high deposition rate during the epitaxial growth of GaN. This high growth rate offers the prospect of a reduction in the dislocation density in the top device region of the grown layer by producing a layer of sufficient thickness (tens to hundreds of microns) so that misfit dislocations and other defects generated during growth will be restricted to the vicinity of the heterointerface. The availability of a GaN layer with a defect-free top layer may also be important to the subsequent growth of either homo-epitaxial GaN or heteroepitaxial structures containing AlN or InN.

The need to reduce or at least control the grown-in defect density is, of course, dictated by the fact that heteroepitaxial GaN grown on common substrates such as SiC or sapphire typically exhibit a high density of structural defects. The relatively large difference in lattice parameter between GaN and these substrate materials typically results in a dislocation density of 10^{10} cm^{-2} or higher [1]. The difference in the coefficient of thermal expansion between GaN and its substrate can lead to the generation of strains and (in the worst case) cracking in the epitaxial layer as it is cooled from the growth temperature. The problems with thermal expansion mismatch are expected to be particularly severe in GaN due to the relatively high temperatures that are typically used in its epitaxial growth. Finally, when growth is performed on sapphire substrates, the difference in crystallographic symmetry between the substrate and the hexagonal GaN wurtzite phase can lead to the generation of antiphase domains.

In the current work we have used a variety of high resolution x-ray diffraction methods to monitor the evolution of the defect structure in GaN grown on sapphire substrates by HVPE. While methods of x-ray diffraction have a long and successful history for the characterization of

Mat. Res. Soc. Symp. Proc. Vol. 423 © 1996 Materials Research Society

defects in epitaxial semiconductor materials, most of these analyses have utilized double crystal rocking curves as their main probe of the structure. Due to the fact that the double crystal method employs a wide-open detector, information is lost due the integration of the diffracted intensity over an angular range corresponding to the acceptance angle of the detector. A more fundamental measurement is performed in a triple crystal diffraction analysis, where an analyzer crystal conditions the beam diffracted by the sample before it strikes the detector. With the use of highly perfect monochromator and analyzer crystals, both the directions and magnitudes of the incident and diffracted wavevectors ($\vec{S}_0/\lambda$ and $\vec{S}/\lambda$, respectively) are well defined. By suitable manipulation of the sample crystal orientation and the diffracted beam direction, it is possible to record the diffracted intensity in any direction in reciprocal space or to map out the intensity distribution around a reciprocal lattice point [2].

Prior x-ray diffraction analyses of GaN grown by MOCVD [3] and MBE [4] have examined the effects of these growth methods on the GaN defect structure. The goal in the current study was to use methods of high resolution x-ray diffraction (particularly triple crystal analyses) to identify the effects growth conditions on the defect structure of HVPE-grown GaN.

EXPERIMENTAL

The HVPE growth was performed using a reactor system that has been described elsewhere [5]. Briefly, HCl is reacted with molten Ga at 850°C to form GaCl; the GaCl is then transported to the substrate (*c*-plane sapphire in this case) where it is reacted with NH_3 at 1030°C to 1050°C to form GaN. The NH_3:HCl ratio is 30:1; the total gas flow rate was typically 10 liter/minute, with an ammonia mole fraction of 0.03 to 0.09. Under these conditions, relatively high growth rates (up to 2 μm/minute) can be achieved. No buffer layers were used; nevertheless, single crystal GaN growth is always obtained.

X-ray diffraction analyses were performed using a Bede 150 double crystal diffractometer that has been modified for high resolution triple crystal diffraction studies [2]. A Rigaku RU200 rotating anode x-ray generator supplied the CuKα x-rays used in this work. Three main types of analyses were performed: (1) maps of the diffuse scattered intensity about the 0002 reciprocal lattice point; (2) transverse scans recorded by rocking the 0002 point with a fixed analyzer crystal; and (3) longitudinal scans which were recorded by moving the analyzer crystal at twice the angular velocity of the sample crystal. In essence, the reciprocal space map shows the details of the intensity distribution around a reciprocal lattice point, while the transverse and longitudinal scans give "cross-sections" of the intensity in directions that are, respectively, perpendicular and parallel to the [0002] direction. Note that the longitudinal scan is analogous to the "θ-2θ" scan that is familiar to workers in powder diffractometry. Further x-ray diffraction analyses include the recording of asymmetric ($10\bar{1}4$) double crystal rocking curves (using a wide open detector), and determining the dependence of the intensity on the scattering angle $q_{10\bar{1}0}$ in a transverse scan.

RESULTS

1. Reciprocal space maps

Figure 1 illustrates a map of the diffracted intensity about the 0002 point for a 125 μm GaN layer grown by HVPE at 125 μm/hour. The Figure shows contours of the logarithm of the intensity, with four contours per decade and the minimum contour level representing $10^{0.25}$ counts/sec. The data were collected by scanning the sample crystal at a fixed analyzer crystal

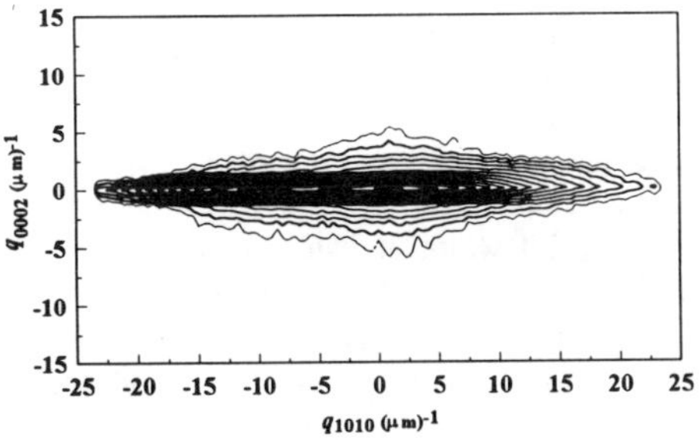

Figure 1. 0002 reciprocal lattice map for HVPE GaN (125 μm thickness) grown at 125 μm/hour.

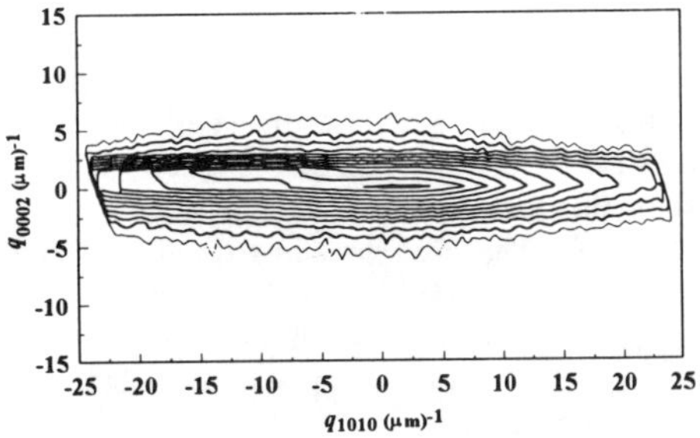

Figure 2. 0002 reciprocal lattice map for HVPE GaN grown at a reduced growth rate of about 30 μm/hour

setting and then incrementing the analyzer crystal between scans; this resulted in the sampling of a trapezoidal region of reciprocal space. The angular deviations of the sample and analyzer crystals from the exact Bragg condition θ_B (denoted α and β, respectively) were converted to orthogonal reciprocal ($q_{0002}, q_{10\bar{1}0}$)space coordinates using the relations

$$q_{0002} = \frac{\beta \cos\theta}{\lambda}; \; q_{10\bar{1}0} = \frac{(2\alpha - \beta)\sin\theta}{\lambda} \tag{1}$$

The Figure shows an isotropic distribution of intensity in reciprocal space, with the intensity being distributed as a "disc" perpendicular to [0002]. Unlike reciprocal space scans from epitaxial materials with higher levels of structural perfection, in this case we do not observe the so-called crystal truncation rod (also known as the "surface streak") which represents the trace of the dynamical reflectivity of the crystal. The absence of this feature suggests that the crystal is diffracting in a purely kinematic fashion due to the presence of structural defects and the absence of long range structural coherence.

Figure 2 shows a reciprocal space map from a GaN sample of similar thickness (about 100 μm) grown by HVPE at a rate of about 30 μm/hour. The intensity is clearly distributed over a larger region of reciprocal space than seen in Figure 1; in fact, it extends beyond the limits of the trapezoidal region of reciprocal space that was sampled in the x-ray data collection. As expected, no evidence for a dynamical surface streak is detected. The Figure suggests that a decrease in growth rate is responsible for an increase in the defect density in the GaN layer.

2. Effect of film thickness:

Table I shows the effect of layer thickness on the observed diffraction behavior from a series

Table I. Effect of HVPE GaN layer thickness on diffraction peak breadths

Thickness (μm)	Growth rate (μm/hour)	(0002) transverse scan	(0002) longitudinal scan	($10\bar{1}4$) rocking curve
250	133	908"	22"	1141"
125	110	635"	16"	640"
22	129	888"	30"	1271"

of GaN samples grown at similar rates (125 to 133 μm/hour). Examination of the reciprocal space scans in Figures 1 and 2 demonstrate that since the intensity distribution is relatively isotropic an acceptable estimate of the extent of the scattering about the 0002 point can be obtained from transverse and longitudinal triple crystal scans. The (0002) transverse scans clearly show that the breadth of the intensity distribution is at a minimum for the layer with a thickness of 125 μm. The peak breadths listed in Table I are the full width at half maximum (FWHM) determined by linear interpolation of the flanks of the reflection profile; although less accurate than values obtained by full profile fitting or from (for instance) an integral breadth, we consider this approach to be acceptable to illustrate the effect of growth parameters on the diffraction characteristics.

We have also recorded conventional double crystal rocking curves using the asymmetric $(10\bar{1}4)$ reflection. It has recently been suggested that the use of the asymmetric reflections should increase the sensitivity of an x-ray analysis to the presence of defects [3]. In our study we see the same trend found from the (0002) transverse peak breadth, although the differences between the samples are somewhat more pronounced.

The longitudinal scans shown in Table I show a much narrower FWHM than found in the corresponding (0002) transverse scans. Remembering that these data are analogous to the θ-2θ scan in a powder diffractometer, the breadths of the longitudinal scans are listed in terms of the angular range through which the sample crystal was rotated during the scan (θ rotation) while the analyzer crystal traversed an angular range twice as large as listed in the Table. Since the longitudinal scans are sensitive only to strains in the GaN layers and not to misorientations or mosaic distribution, these results suggest that the thicker GaN layers are relatively free of inhomogeneous strains. With the 22 μm sample, however, the strain distribution must be larger. This fact is even more obvious in Figure 3, which shows the longitudinal scans themselves. Note that the profile from the 22 μm layer is significantly broader both near the peak (which on the logarithmic intensity scale is close to the FWHM) and in the tails of the curve far from the peak. This behavior may be due to the fact that the 90% absorption depth for CuKα x-rays in GaN is about 21 μm, so it is likely the profile from the thinnest layer is being influenced by strains at the GaN/sapphire interface.

3. Effect of growth rate

Table II shows the effect of GaN growth rate on the breadths of the 0002 longitudinal and transverse profiles. We have performed this analysis on both thin (10 to 15 μm) and thick (100 to 120 μm) GaN layers. In all cases we observed a decrease in the breadth of the transverse profile width as the growth rate was increased; in the thin layers, the change was in the transverse FWHM was about 50%. A small but systematic decrease in the transverse FWHM was also observed in the thick GaN layers. As was the case in the thickness samples discussed above, the breadth of the (0002) longitudinal scan profile was significantly greater in the thinner samples; we again attribute

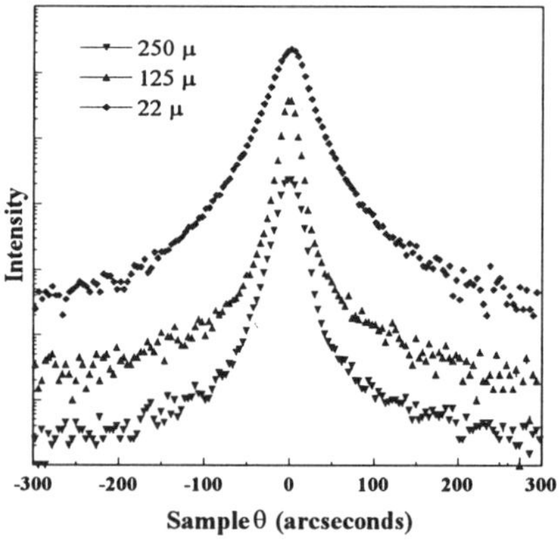

Figure 3. (0002) longitudinal scans from GaN layers of different thicknesses (curves offset vertically for clarity). Top curve: 22 μm; middle curve: 125 μm; bottom curve: 250 μm

Table II. Effect of GaN growth rate on 0002 diffraction peak breadths			
Thickness (μm)	Growth rate (μm/hour)	(0002) transverse scan	(0002) longitudinal scan
16	30	923”	28”
10	30	1023:	44”
14	93	562”	36”
110	42	782”	20”
110	83	682”	16”
125	125	635”	16”

this behavior to the sampling of the strained interface by the CuKα x-rays due to relatively deep 90% absorption depth in GaN.

4. Angular dependence of the transverse scan intensity

A useful method for observing the effects of structural defects is to characterize the dependence of the diffracted intensity on the deviation $\boldsymbol{q}$ from the exact Bragg condition. In the kinematic approximation it is well known that the strain fields associated with specific defects give rise to diffuse scattering with a $\boldsymbol{q}$ dependence that is characteristic of a particular defect. For instance, point defect clusters are known to produce scattering with a $\boldsymbol{q}^{-2}$ dependence for small values of $\boldsymbol{q}$ (Huang scattering) which increases to $\boldsymbol{q}^{-4}$ at large values of $\boldsymbol{q}$ (Stokes-Wilson scattering). Transitions between different regimes have been detected in GaAs [6] and InP [7] by observing the slope of the dependence of the diffuse intensity on $\boldsymbol{q}$ on a log-log plot.

Figure 4 illustrates the dependence of log I versus log $\boldsymbol{q}$ for the 22 μm GaN layer described earlier in Table I. The Figure shows that far from the main Bragg peak the slope of the diffuse scattered intensity reaches a limiting value of about -5.4. A similar analysis of the samples with layer thicknesses of 125 μm and 250 μm revealed limiting slopes of -6.5 and -7.8, respectively. Thus the log I versus log $\boldsymbol{q}$ dependence extracted from the transverse (0002) scans shows a systematic increase in slope with increasing sample thickness.

DISCUSSION

The rocking curve data from samples of different thicknesses show that an intermediate film thickness (125 μm in this case) yields an optimum level of structural perfection as determined from the (0002) transverse scans. A decrease in dislocation density and rotational misorientation is well known in other epitaxial systems [8] and is likely to be occurring in GaN/sapphire. The reason for the observed increase in the FWHM from the layer with the greatest thickness is currently not known. Both intrinsic factors such

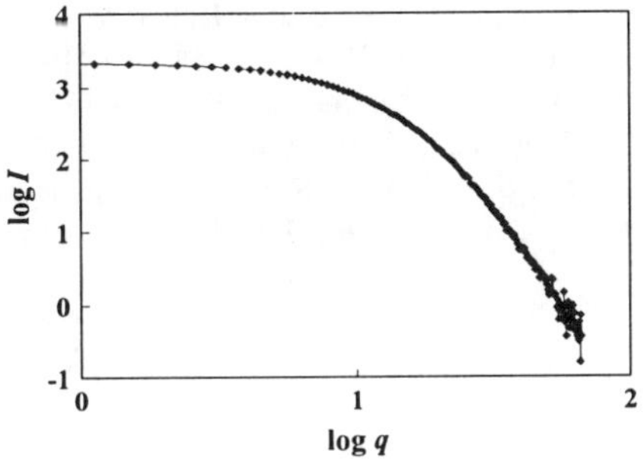

Figure 4. Log I versus log $\boldsymbol{q}$ dependence for 22 μm GaN with limiting slope of -5.4.

as an increase in strain due to thermal expansion mismatch and extrinsic ones such as a degradation of film quality during HVPE growth may be the origin of this behavior.

It is very clear from the x-ray data that the structural perfection of the HVPE-grown GaN increases with increasing growth rate. The fact that this behavior is especially pronounced for the thinner layers suggests the possibility of competition between the kinetics of defect nucleation and propagation versus the kinetics of growth. Further studies of the initial phases of defect generation during the HVPE growth of GaN are clearly warranted.

In all case we observed that the breadth of the longitudinal scans parallel to [0002] were far narrower than the perpendicular transverse scans. Broadening parallel to [0002] would arise from variations in lattice parameter within the layer, particularly from the presence of non-uniform strains. In layers that are thin enough so that appreciable diffraction can occur from regions in close proximity to the (presumably strained) heterointerface, broadening of the longitudinal scans is expected and was indeed observed. In the thicker layers this strain broadening is negligible, and essentially all of the broadening of the diffraction profile can be attributed to a rotational mosaic structure about the [0002] such as that observed in GaN grown by MOCVD [3]. We note that the mosaic structure cannot be due to rotations about axes orthogonal to the [0002], since this would generate scattering off the Bragg peak that would have a $\boldsymbol{q}^{-2}$ dependence.

The systematic changes in the limiting slope of the plots of log I versus log $\boldsymbol{q}$ suggests a monotonic change in the defect structure in the near-surface region of the GaN layer (witching the kinematic absorption depth) with increasing thickness. While it is not possible to ascribe a mechanism to this behavior transmission electron microscopy studies are currently in progress to determine if a simple defect mechanism exists. Modeling of the $\boldsymbol{q}$-dependence of the scattering based on the dislocation model of Krivoglaz and Ryaboshapka [9] is currently in progress and will be reported elsewhere.

ACKNOWLEDGMENTS

Support from the ARPA URI on Visible Light Emitters and the Office of Naval Research is gratefully acknowledged. One of us (RJM) acknowledges the support of the National Science Foundation under grant DMR-9319421.

REFERENCES

1. Lester, F.A. Ponce, M.G. Craford and D.A. Steigerwald, *Appl. Phys. Lett.*, **66**, 1249 (1995).
2. R.J. Matyi, *Rev. Sci. Instr.*, **63**, 5591 (1992).
3. B. Heying, X.H. Wu, S. Keller, Y. Li, D. Kapolnek, B.P.Keller, S.P. DenBaars and J.S. Speck, *Appl. Phys. Lett.*, **68**, 643 (1996).
4. Q. Zhu, A. Botchkarev, W. Kim, Ö. Aktas, A. Salvador, B. Sverdlov, H. Morkoç, S.-C. Y. Tsen and D.J. Smith, **68**, 1141 (1996).
5. See S.A. Safvi, N.R. Perkins, M.R. Horton and T.F. Kuech, this proceedings.
6. R. Bloch, D. Bahr, J. Olde, L. Brügemann and W. Press, *Phys. Rev. B.*, **42**, 5093 (1990).
7. E.L. Garstein, *Z. Phys. B.*, **88**, 327 (1992).
8. T.F. Kuech, A. Segmüller, T.S. Kuan and M.S. Goorsky, *J. Appl. Phys.*, **67**, 6497 (1990).
9. M.A. Krivoglaz and E. Ryaboshapka , *Phys. Metals. Metallog.*, **15**, 18 (1963))

STRUCTURAL CHARACTERIZATION OF THICK GAN FILMS GROWN BY HYDRIDE VAPOR PHASE EPITAXY

L.T. Romano, R.J. Molnar*, B.S. Krusor, G.A. Anderson, D.P. Bour, and P.Maki*
Xerox Palo Alto Research Center, Palo Alto, CA 94304, romano@xerox.parc.com
*Massachusetts Institute of Technology, Lincoln Laboratory, Lexington, MA 02173-9108

ABSTRACT

The structural quality of GaN films grown by hydride vapor phase epitaxy (HVPE) was characterized by transmission electron microscopy (TEM), x-ray diffraction (XRD), and atomic force microscopy (AFM). Films were grown up to 40μm on sapphire with either a GaCl pretreatment prior to growth or on a ZnO buffer layer. Dislocation densities were found to decrease with increasing film thickness. This is attributed to the mixed nature of the defects present in the film which enabled dislocation annihilation. The thickest film had a defect density of $5x10^7$ dislocations/cm^2.

INTRODUCTION

Defect densities in GaN blue LED devices are as high as 10^{10} dislocations/cm^2 and appear not to introduce nonradiative recombination centers as with other III-V semiconductor systems. However the lifetimes of these devices may depend on the defects as shown with other III- V devices [1]. Achieving lower defect densities in the GaN material system has been difficult since substrates that are both lattice and chemically matched, and have the same symmetry as GaN are not available. It would also be desirable for the substrate to cleave in order to obtain facets for edge emitting lasers. Bulk GaN single crystals have been difficult to grow wider than a few millimeters [2]. Currently, sapphire is the most common substrate with a 16% lattice mismatch and thermal expansion difference 1.4 times greater than GaN. Silicon carbide substrates are also used which have a 3.5% lattice mismatch and a thermal expansion difference that is 2.0 times smaller than GaN [3].

In this paper we report the structural properties of thick GaN films grown on sapphire by hydride vapor phase epitaxy (HVPE) which is capable of growing films at a rate of several tens of microns/hour. Room temperature mobilities up to 768cm^2/Vs have been obtained for films with a ZnO buffer layer [4]. These films could then be used as substrates for nitride devices. Homoepitaxial growth of GaN by metal organic chemical vapor deposition (MOCVD) and a AlGaN/GaN/AlGaN double heterostructure (DH) grown by electron-cyclotron-resonance molecular beam epitaxy (ECR-MBE) were made on 15μm thick GaN films.

EXPERIMENT

The details of the HVPE process have been published previously [5]. GaN films were grown to thicknesses ranging from 10 - 40 μm on sapphire at temperatures between 1050 - 1100°C. The sapphire surface either had a GaCl pretreatment at 1050°C or was grown on a sputter deposited ZnO buffer layer. Growth of the GaN/AlGaN DH by ECR-MBE was initiated at 700 - 750 °C without a low temperature buffer layer as discussed elsewhere [6]. Homoepitaxial layers of GaN were grown by MOCVD directly at high temperature without the need of a low temperature buffer layer.

Mat. Res. Soc. Symp. Proc. Vol. 423 © 1996 Materials Research Society

Plane view (PVTEM) and cross sectional TEM (XTEM) samples were prepared by flat polishing to < 10 μm and Ar ion milling to electron transparency with a liquid nitrogen cold stage. Samples were studied in a JEOL 300 kV microscope equipped with energy dispersive x-ray analysis (EDX). X-ray diffraction (XRD) measurements were made both normal to the film surface along the (0001) direction and along the (1-102) direction. Atomic force microscopy was carried out in a Park instrument with a Si tip.

RESULTS AND DISCUSSION

Figure 1 is the XRD spectra of the (002) and (112) reflections for a 15μm thick GaN films grown on sapphire with a GaCl pretreatment. The width of the (002) reflection is associated with mixed and screw dislocations whereas broadening of the (112) reflection is associated with edge dislocations as described by Heying et al. [7]. It can be seen that the full widths half maximum (FWHM) is 11 minutes for the (112) peak and 4.5 minutes for the (002) reflection. These results are similar to GaN films grown on a ZnO buffer layer.

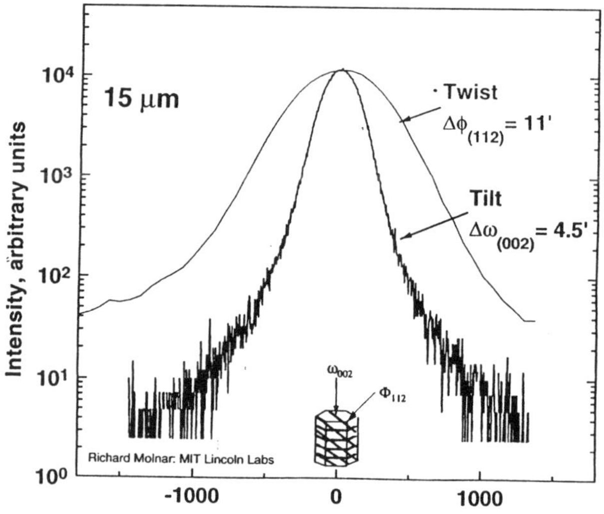

Figure 1. XRD spectra of a 15μm GaN film on sapphire with a GaCl pretreatment.

Figure 2 shows the interface of the 15μm GaN film grown with the GaCl pretreatment. A 200nm layer with a high density of dislocations is observed at the GaN/ sapphire interface. Above this layer, there is a sharp transition into a region with a much lower defect density. The defects visible in this image are either edge <11-20> or mixed <11-23> perfect dislocations. It can be seen that the presence of these mixed defects enables dislocation annihilation as indicated by A, B, C in the micrograph. The interface defect layer was found to contain mostly stacking faults associated with 1/3 [10-10] Shockley partial dislocations and <2-203> and <0001> Frank partial dislocations. Reactions of these partial dislocations with each other and with perfect dislocations can explain the drastic reductions of the defects above the interfacial layer and with film thickness. The possible dislocation reactions are discussed by Ospiyan and Smirnova [8].

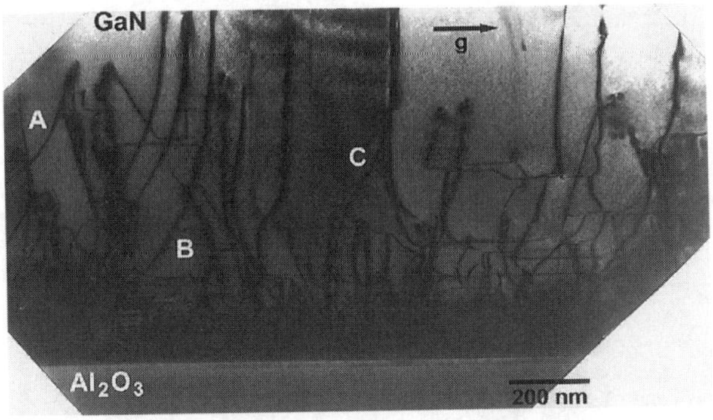

Figure 2 - TEM cross section of a 15μm GaN film on sapphire with a GaCl pretreatment. The image is taken near the [10-10] zone with diffraction vector g = [1-210].

Figure 3 is a PVTEM micrograph for the same film as in Fig. 2. The defects at the surface are both mixed and edge dislocations with a density of 3×10^8 dislocations/cm^2. This corresponds to the AFM image shown in Fig. 4. The defects present in the AFM image are associated with steps which would occur if the dislocations are mixed and therefore contain a screw component. The density of defects is 3×10^8 /cm^2 which agrees with the PVTEM. Both the PVTEM and the AFM show that the dislocations tend to cluster and form cell boundaries.

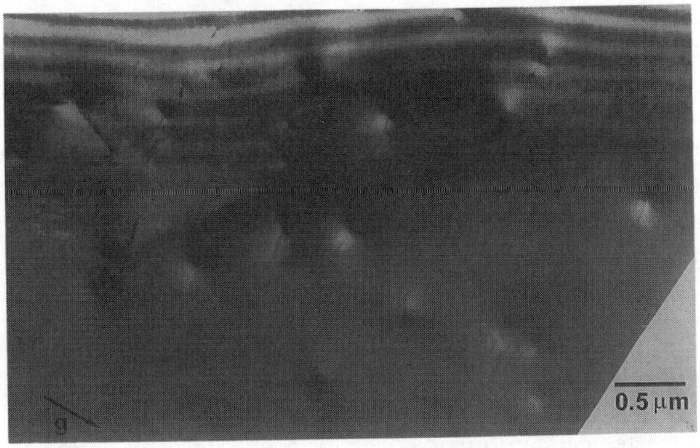

Figure 3 - Plane view TEM of a 15μm film on sapphire with a GaCl pretreatment. The image is taken near the [0001] zone with diffraction vector g = [1-210].

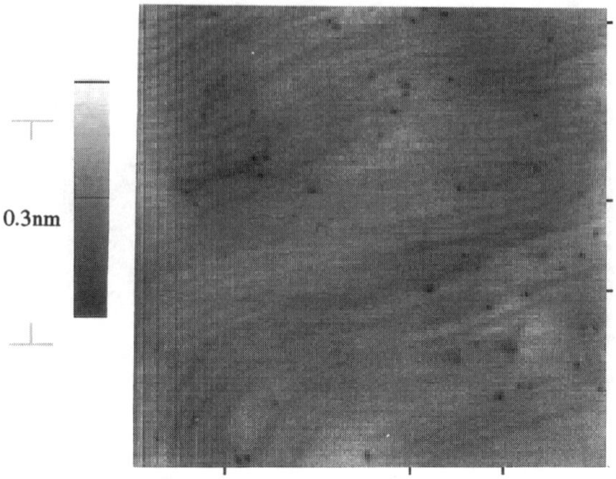

Figure 4 - AFM micrograph of a 5μm x 5μm area of the surface of a 15μm GaN film on sapphire with a GaCl pretreatment. The dark spots are defects associated with 0.3nm steps on the surface.

The dislocation density of GaN films grown with a ZnO buffer layer was measured as a function of film thickness. Figure 5 is a PVTEM of a 40μm film showing the dislocations present at the top surface of the film. The defect density was found to be $5x10^7$ dislocations/cm^2 and formed cells in a similar way to films grown with a GaCl pretreatment (Fig. 3). The defect density was lower by a factor of ten compared to a 15μm film also grown on a ZnO buffer layer.

The HVPE films were used as substrates for MBE and MOCVD growth. Figure 6a is a XTEM image of a 2μm GaN film grown by MOCVD on a 15μm HVPE film. The interface could not be detected between the two layers. Figure 6b is an AlGaN/GaN DH showing that the dislocations of a 15μm HVPE layer penetrate through the MBE layer. However in some regions of the GaN active layer, basal and prism faults were generated from the dislocation loops in the HVPE layer. Laser emission has been reported with well defined cavity modes from this sample [5].

CONCLUSIONS

The defect structure of films grown by HVPE show a mixture of mostly edge and mixed dislocations. Dislocation annihilation was able to occur near the interface due to a defect layer that forms during the high temperature growth. The dislocation density decreased by almost a factor of ten when the film thickness was increased from 15 to 40μm. Both MOCVD and MBE growth of GaN was demonstrated on the HVPE thick layers.

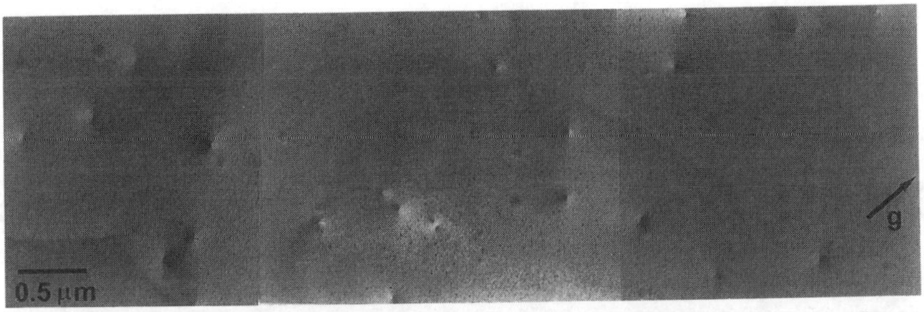

Figure 5 - Plane View TEM of a 40μm film on sapphire with a ZnO buffer layer. The image is taken near the [0001] zone with diffraction vector g = [1-210].

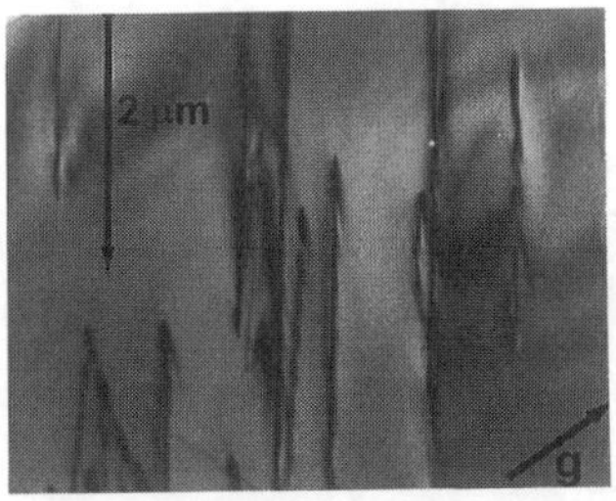

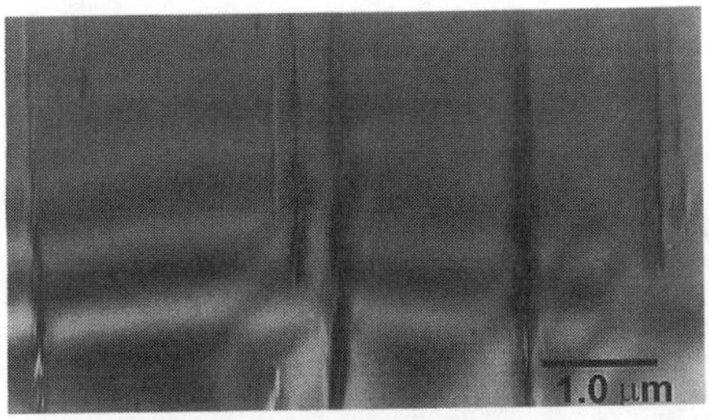

Figure 6a - Cross sectional TEM of a 2 μm GaN layer grown by MOCVD on 15 μm GaN film grown by HVPE. The image is taken near the [10-10] zone with diffraction vector g = [1-212].

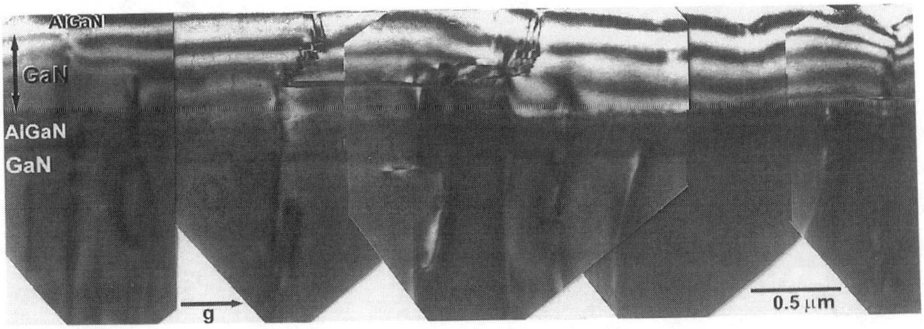

Figure 6b -TEM crossection of an AlGaN/GaN DH grown by MBE on a 15μm HVPE film. The image is taken near the [1-210] zone with diffraction vector g = [10-10].

ACKNOWLEDGEMENTS

The work at Xerox was supported by DARPA (agreement # MDA972-95-3-008). The work at Lincoln Labs was supported by the Department of the Air Force.

REFERENCES

1. S.D. Lester, F.A. Ponce, M.G. Craford, and D.A. Steigerwald, Appl. Phys. Lett. **66**, p.1249 (1995)

2. M. Leszczynski, T. Suski, H. Teisseyre, P. Perlin, I. Grzegory, J. Jun, and S. Porowski, J. Appl. Phys. **76**, p. 4909 (1994)

3. Data in Science and Technology, Semiconductors Group IV Elements and III-V Compounds, edited by O. Madelung, Springer Verlag, Berlin, 1991

4. R. J. Molnar, I. Melngailis, W. Goetz, L. T. Romano, to be published

5. R .J. Molnar, K.B. Nichols, P. Maki, E.R. Brown, and I. Melngailis (Mater. Res. Soc. Proc. 378, Pittsburgh, PA, 1995), p. 479-484

6. P.A. Maki, R.J. Molnar, R.L. Aggarwal, Z-L. Liau, and I. Melngailis (Mater. Res. Soc. Proc. November 1995) to be published

7. B. Heying, X.H. Wu, S. Keller, Y. Li, B. P. Keller, S.P. DenBaars, and J.S. Speck, Appl. Phys. Lett. **68**, p.643 (1996)

8. Y. A. Osipyan and I.S. Smirnova, J. Phys. Chem. Solids, **32**, p.1521 (1971)

Part II

Crystal and Film Growth

Film Growth

SILICON CARBIDE CVD APPROACHES INDUSTRIAL NEEDS

R. RUPP, P. LANIG, J. VÖLKL AND D. STEPHANI
Siemens AG, Corporate Research and Development, Dep. ZFE T EP 6, PO. Box 3220, D-91050 Erlangen, Germany

ABSTRACT

In this paper an overview is given on the current state of epitaxial growth of SiC with special regard to our work at SIEMENS CRD. Problems concerning impurity incorporation and ways to achieve background doping levels as low as 10^{14} cm^{-3} are discussed as well as the influence of high speed wafer rotation on the gas flow in our reactor and related effects on uniformity in thickness and doping. Precise control of the C/Si ratio in the gas phase, which is easily achieved in the described reactor, and the use of reduced pressure lead to a good control of dopant incorporation over more than 3 orders of magnitude while maintaining smooth surface morphology even at growth rates higher than 5 µm/h. Doping variations < ±8 % across 30 mm wafers can routinely be obtained. The quality of the epilayers is proven by electrical breakdown fields as high as $2*10^6$ V/cm at N_A-N_D=$5*10^{-15}$ cm^{-3} and an electron mobility greater than 700 cm^2/Vs at 300 K (4H-SiC).

Finally it is demonstrated that the gas composition at the end of the epitaxial growth process is an important step in order to get oxygen resistant surface properties for subsequent device processing.

INTRODUCTION

Silicon carbide is a material of increasing importance not only for optical applications (blue LED's) but also for high power, high frequency and high temperature electronics. The development of substrates with low defect density and diameters up to 2" within the last few years offers the prospect for the realization of a wide variety of demonstrator devices. But to gain commercial interest for such devices it is necessary to obtain a reasonable yield of operable devices in conjunction with a small spread of their properties. In the field of power electronics these demands are especially harsh, because these devices must be able to handle 1000 V or more in reverse mode and at least several Amps in forward direction. This leads to necessary active device areas in the order of mm^2. Besides the substrate defect density the quality and homogeneity of the epitaxial layer is the dominating issue for fulfilling this economical demands.

In the last ten years significant progress can be observed in the understanding and technical realization of the SiC-VPE process. One very important milestone was the introduction of the step controlled epitaxial growth technique by Matsunami et al. [1,2] in 1986. This was the key to grow epilayers of the same polytype as the substrate by using slightly off oriented surfaces. Nevertheless, the reproducibility of dopant incorporation in the growing layers was a problem until Rottner [3] showed the importance of excess carbon coming from uncoated graphite surfaces and Larkin reported the influence of the carbon to silicon ratio in the source gas on the incorporation of both acceptors and donors in SiC [4,5] two years ago.

But even with all this new knowledge there is an ongoing competition between different epitaxial setups in reaching a point which can be called *production suited* and no final decision can be made up to now, which one will be the most successful. In principal these setups can be classified as follows:

Type 1: Cold wall horizontal atmospheric pressure CVD[6,7,8]; **type 2:** Hot wall horizontal atmospheric pressure CVD[9,10]; **type 3:** Cold wall vertical low pressure CVD[11,12]. All these reactors described in the literature up to now are single wafer machines.

A lot of different groups all over the world have gathered experience with the type 1 reactor design and excellent results have been achieved, but there is one major drawback for this

Mat. Res. Soc. Symp. Proc. Vol. 423 © 1996 Materials Research Society

configuration: the not satisfying homogeneity in doping and thickness of the layer along the main flow direction. Furthermore it is very difficult to avoid gas recirculation in the reactor, preventing the use of this reactor type without a fully SiC-coated susceptor. This usually causes a very thick unintentional deposition of SiC on the back of the wafer. Similar arguments hold for the hot wall configuration as well, but the SiC deposition on the backside is less, due to the very small temperature gradients perpendicular to the wafer surface. By using this reactor type the highest breakdown voltage value (4500 V, [13]) and minority carrier lifetime (2 μs) has been reported up to today, which makes this reactor concept very promising for the future, if the doping homogeneity can be improved and the reactor concept can be transferred to multi-wafer processing.

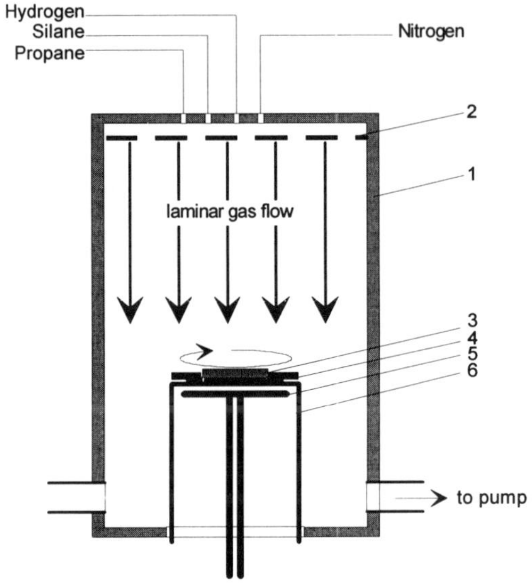

Fig. 1: Sketch of the principal design of the CVD reactor chamber:
(1) double-walled water cooled reactor chamber
(2) gas diffuser
(3) SiC-wafer
(4) substrate holder
(5) rf pancake coil
(6) bell-jar-like graphite susceptor

In a different approach, the type 3 reactor which is in use at the authors location is based on a long tradition of VPE reactors in the III-V epilayer growth. A high speed rotation of the substrate holder and a reduction of pressure are employed to adjust a stationary flow without recirculation. With this method good homogeneity and reproducibility together with low background doping should be achievable even without SiC coating of all graphite parts. The following chapters will introduce this type of a SiC-CVD reactor in detail and give an overview of the results attained in the last two years.

EXPERIMENTAL

CVD-system

The CVD-system was built by EMCORE Corp. (New Jersey) following their experience with III-V epi-equipment. A palladium cell is used to provide high purity hydrogen to the process and a loadlock equipped with a turbo pump is attached to the process chamber for fast batch processing. A pressure control system with a big roughing pump allows accurate pressure adjustment in the growth chamber, even at high flow rates. The process control and data acquisition for all relevant parameters is computer based, a precondition of good reproducibility. The special needs of SiC-CVD, e. g. very high temperature capability, are taken into account for the construction of the growth chamber which is displayed schematically in fig. 1. The reactor allows processing of single wafers with diameters up to 2 ¼". The rf-heated bell-jar-like susceptor and the substrate holder (aid for transferring the wafer from the loadlock into the reactor) are made of high purity graphite. A SiC-coating of the susceptor was not used in order to avoid the above mentioned unintentional SiC-deposition on the back of the wafer during growth. This is not only an advantage for the setup of vertical devices but also allows an easy determination of the thickness of the epilayers by measuring the weight difference before and after growth. The pancake-like rf-coil is stationary and the susceptor is mounted on a rotary ferrofluidic feedthrough sealing both against the coil and the outer reactor wall. Rotation speeds up to 1500 rpm are possible with this setup at a leakage rate < 10^{-7} mbar*l/s. A hydrogen/argon mixture can be used as carrier gas as well. A two color pyrometer is used for temperature control. The point of measurement is on the wafer surface, because the surrounding graphite surfaces change their optical properties by being coated during the run. However, the absolute reliability of the temperature readings is still a problem. It seems that big differences in the doping concentration of the wafer can influence the temperature measurement.

As a dopant source only nitrogen is used, a hydrogen dilution stage is employed to allow an accurate control of the nitrogen flow in the reaction chamber over nearly three orders of magnitude.

4H-SiC-wafers with a diameter of 30 mm purchased from Cree Res. Inc. (Durham NC) were used as substrates for most of our growth experiments. These wafers are oriented in the (0001)-direction (Si-face) with an off-angle of 3.5 and 8 degrees towards $(11\bar{2}0)$. Measurements of the lateral temperature distribution with the pyrometer showed that ΔT is less than 10 K across a 30 mm-wafer.

Typical ranges of the most important parameters for our CVD-growth runs are given in table I.

Table I: Typical growth conditions.

silane flow 2% diluted in H_2	propane flow 5% diluted in H_2	C/Si ratio	temperature	rotation speed	hydrogen shroud flow	growth chamber pressure
500-950 sccm	75-325 sccm	0.5 - 2	1400-1550°C	800 rpm	15-30 slm	50-300 torr

The reliability of a system is a very important industrial issue. With our epi equipment we have made 650 growth runs with an average duration of about 4-5 hours within a time frame of 2¼ years. The average maintenance time in that period including the periodical exchange of the graphite parts was less than 20 % of the process time.

Characterization of the epitaxial layers

The surface quality of the epitaxial layers was examined by means of optical and electron microscopy and with electron diffraction (LEED) together with x-ray photoelectron spectroscopy (XPS) [14]. Secondary ion mass spectroscopy (SIMS) is used to evaluate the thickness homogeneity of the layers.

For the CV-measurement we use metal contacts patterned by means of photolithography and etching, to get a precise contact area at hand for calculating the doping concentration, a significant advantage over the widely used Hg-probers. As Schottky contact material we employ Ti or Ni vapor deposited on the as grown surface. Thus we obtain barrier heights of about 1.1 or 1.6 eV, respectively. In order to determine the lateral and vertical doping distribution, we have installed an automatic wafer mapping system for current-voltage and capacitance-voltage measurements. Typically, we apply an at least 10 by 10 matrix of contacts.

The breakdown behavior of the layers was measured with the same contacts. Without edge termination, the electrical field at the edge of these contacts is more than two times higher than the field below an infinite contact, which can be derived from an electrostatic analysis. To verify this, we have tested pn-structures and Schottky diodes with edge termination as well and can therefore estimate values for the real breakdown field of the epitaxial layers.

Flow visualization:

The gas flow in our reactor configuration can only be stabilized with the help of high speed rotation of the wafer carrier. The adjustment of stable flow conditions is a very important feature for the growth of high quality epitaxial layers. In our reactor configuration we can use the observation of the silane decomposition front for evaluating the actual flow conditions and to estimate whether there is recirculation in the chamber or not. Fig. 2 shows this silane decomposition front for two different flow states in comparison to the numerically determined elemental silicon concentration (simulation work was done at the Institute for Fluid Mechanics at the Univ. of Erlangen, for further details see [15,16]). From the analogy of the pictures 2b,c and from the fact, that the foggy shape in fig. 2a,b is only visible if silane is flowing into the chamber, we conclude that this visible feature is due to light scattering at very small Si droplets formed at a certain temperature range in the process atmosphere.

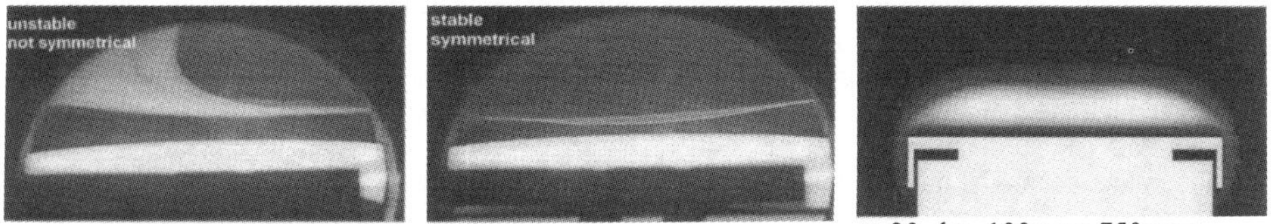

a: 30 slm, 250 torr, 250 rpm **b:** 30 slm, 200 torr, 750 rpm **c:** 30 slm, 100 torr, 750 rpm

Fig. 2: Silane decomposition front (see text, wafer temperature = 1450°C)

a) View inside the reactor chamber through a viewport, the bright plate in the lower half of the picture is the top of the wafer carrier, the gray shape above the carrier is caused by silicon droplet formation due to silane decomposition. The flow is time dependent and not symmetrical
b) Similar to a), but under stable flow conditions.
c) Si concentration field derived from numerical process simulation [16] above the wafer carrier (homogenous bright area on the lower part of the picture). The Si concentration scales with the brightness. The flow is stable.

RESULTS

Flow dynamics

With the method described above, we are able to determine a critical value of the rotation speed for a given set of parameters like total gas flow, pressure and temperature. This critical value marks the onset of time dependent flow patterns, usually accompanied by significant recirculation in the system. This can be explained with the in situ recorded pictures shown in fig. 2a,b. In fig. 2b a sharp and nearly plane silane decomposition front parallel to the hot susceptor is visible, which is typical for stable flow conditions. Conversely, fig. 2a reveals a flow situation, where the rotational symmetry is broken and heavy recirculation occurs. With the help of these observations it is possible to optimize the gas flow regime in the reaction chamber. The observed flow configurations can be reproduced by numerical simulations of the heat and mass transport in our reactor configuration. Due to the fact that the numerical model is strictly rotationally symmetric, we can not precisely calculate flow configurations as shown in fig. 2a, but deduce the onset of recirculation at rotation speeds below the critical value. A further reduction of the rotation then leads to a divergence of the numerical solutions, indicating - in practice - time dependent flow. A comparison between calculated non-recirculating and recirculating flow conditions is given in fig. 3 by displaying both stream functions and isotherms. Evidently, no flat isotherms can be achieved in the recirculating state. Further a significant transport of contaminations, released by the hot graphite parts to the wafer surface, has to be expected. For further details of the numerical modeling of our CVD process including reaction kinetics see [15,16].

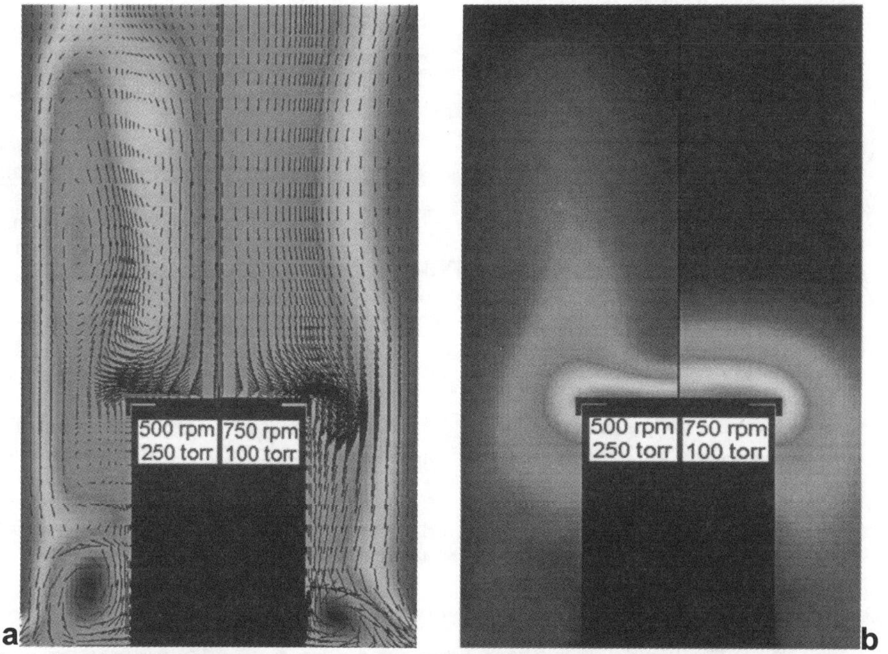

Fig. 3: Results of the numerical process simulation, the black object is the susceptor with the wafer carrier (compare fig. 1), the left part of each picture shows recirculating conditions. Parameters: flow: 30 slm, wafer temperature 1450°C, wall temperature 300°C.

a) Velocity vectors, the gray scale value of the background scales with the stream function [16]
b) temperature field; light gray: high temperature

Growth Behavior

Fig. 4 shows SIMS depth profiles at different radial positions on a 1" wafer with 4 alternately N-doped and undoped epilayers. A total thickness of 1.92 μm was determined by the weight increase of the sample, whereas the SIMS profile reveals a thickness of 1.72 to 1.80 μm. This indicates that on one hand the weight method really is a reasonable tool for a fast determination of layer thickness and that, on the other hand, the thickness inhomogeneity over a 1" wafer is less than 5 %. The comparatively big differences in the absolute N-concentrations of the different profiles is due to the varying nitrogen background of the SIMS-equipment, which is in the order of 10^{17} cm^{-3}.

Growth rate is one of the most important features for an industrial VPE process, because it directly influences the process and accumulated production costs of potential devices. Therefore it is one of the aims of our work, to maximize the growth rates.

At a temperature of 1500 °C growth rates up to 6 μm/h can be achieved on 3.5° off oriented 4H surfaces without any 3C inclusions and degradation of the surface morphology. Under the optical microscope the surface of such a layer looks featureless apart from well known amphitheater like disturbances [17] originated by micropipes or dislocations in the substrate. Using electron microscopy (SEM), we can recognize two different types of steps on these surfaces (fig. 5): large steps with a typical height of 20-30 nm, which can effectively be suppressed by increasing the C/Si ratio in the source gas, and much smaller steps (2-5 nm), which seem to be nearly independent on variations of growth conditions. No steps are visible on 8° off oriented surfaces even at growth rates of 6 μm/h.

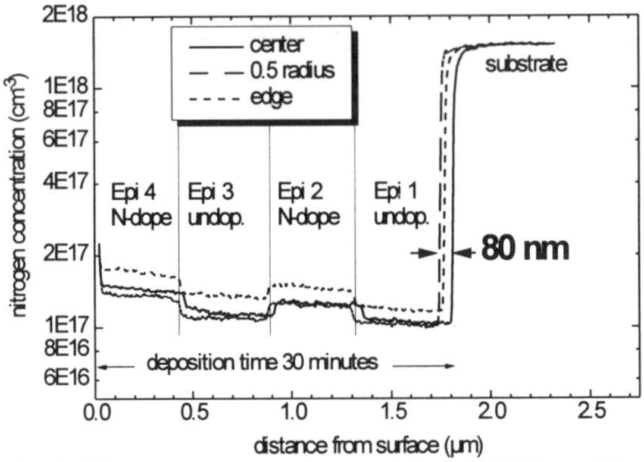

Fig. 4: Nitrogen depth profiles revealed with SIMS on different positions on a SiC-wafer with a sequence of doped and undoped epilayers

Varying both growth rate and C/Si-ratio we found a stability diagram for our process as displayed in fig. 6 for a reactor pressure of 50 torr. Increasing the process pressure leads to a narrowing of the region with good surface quality on the silicon excess side of the diagram, i.e. silicon droplet formation is enhanced by the increasing pressure.

At growth rates exceeding 6 μm/h we frequently observe triangular defects on the epitaxial surfaces, which can be identified by photoluminescence at 4K and by the oxidation behavior to be of the 3C polytype. No differences could be found in the electrical behavior (breakdown field, see above; conductivity) of the layers grown at different growth rates and C/Si-ratios. PL spectra are similar in the near band region and do not show any additional lines pointing to defects generated by the high growth rates.

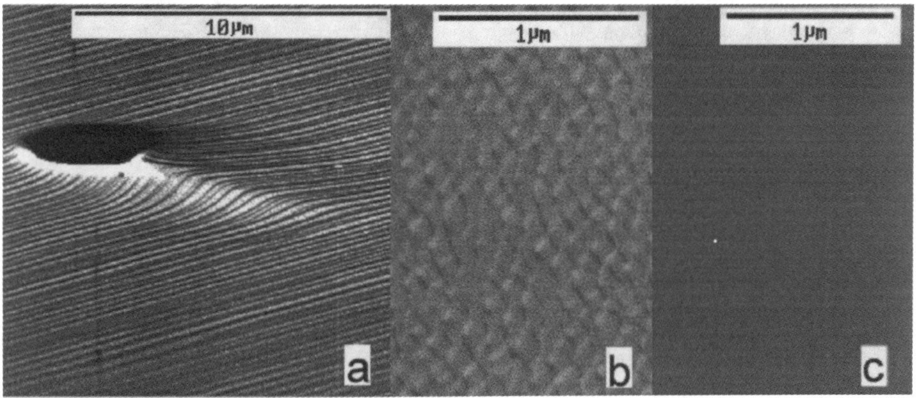

Fig. 5 SEM pictures of epitaxial grown SiC surfaces

a) Macrosteps (height 20-30 nm) in the vicinity of a micropipe, which causes a depression (so called amphitheater, [17]).
b) Small steps (2-5 nm) typical for C/Si ≥ 1 and 3.5 degree off orientation.
c) No steps are visible on 8 degree off oriented surfaces.

First growth experiments with 8° off oriented wafers indicate that in accordance to the predictions made by Kimoto [18], significantly higher growth rates can be achieved without running into problems with the surface morphology and heterogeneous nucleation of 3C.

Background impurities and nitrogen incorporation

In contradiction to many other groups we do not use SiC-coated parts to place our wafer on. This is possible due to the high gas flow velocity generated by the high speed rotation in our system, which counteracts the impurity diffusion from the surrounding parts. Nevertheless, the main impurity in our system, dominating the electrically active background in our layers, is boron (determined by Hall and admittance measurements). A typical distribution of the background acceptor level for a layer grown with C/Si=1 is shown in fig. 7. In the central region of the wafer the background doping is in the order of $2\text{-}4 \cdot 10^{14}$ cm^{-3}. The bowl-like distribution indicates that especially at the wafer edge the diffusion of the boron impurity (released by the dissolution of carbon in the process atmosphere) against the main flow direction still is not negligible. A further reduction of these background impurities would be possible by silicon excess in the source gases, but we do not make use of this for our standard processing in order to avoid increased macrostep formation as described above.

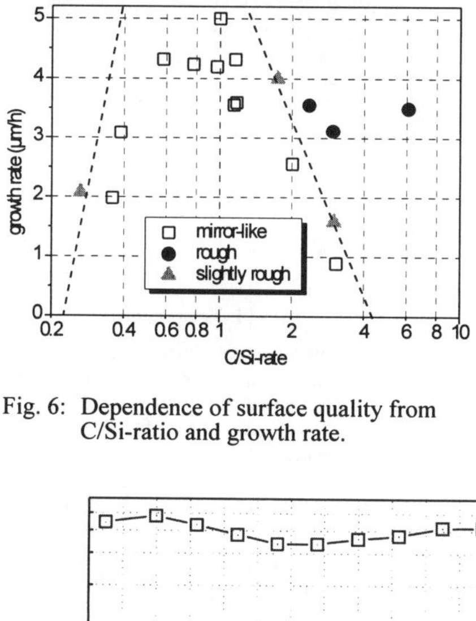

Fig. 6: Dependence of surface quality from C/Si-ratio and growth rate.

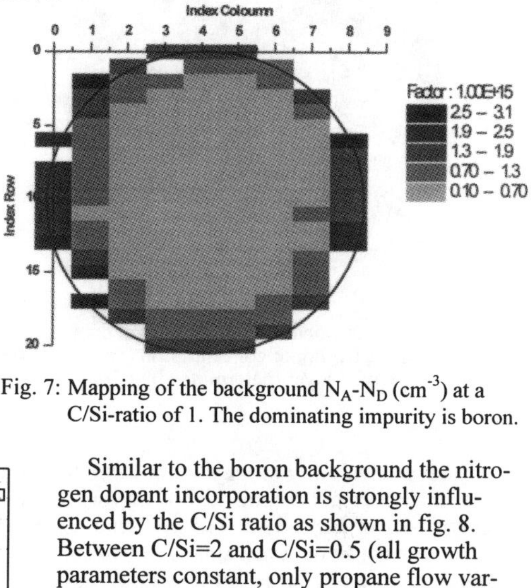

Fig. 7: Mapping of the background N_A-N_D (cm^{-3}) at a C/Si-ratio of 1. The dominating impurity is boron.

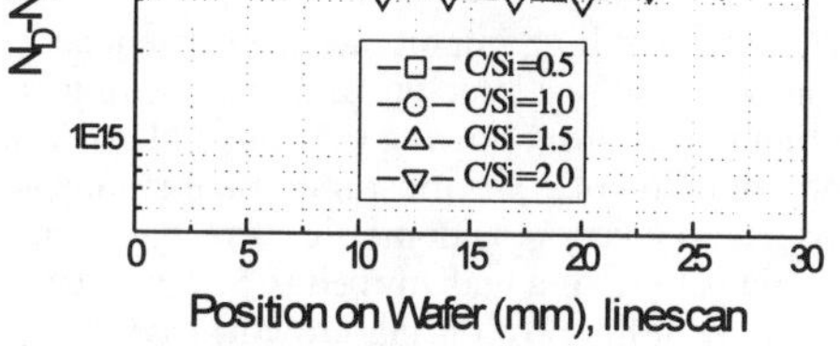

Fig. 8: Dependence of the net donor concentration on the carbon to silicon ratio in the source gas (N_2-flow = 2.5 sccm, chamber pressure = 300 torr, silane flow constant, propane varied)

Similar to the boron background the nitrogen dopant incorporation is strongly influenced by the C/Si ratio as shown in fig. 8. Between C/Si=2 and C/Si=0.5 (all growth parameters constant, only propane flow varied; see figure caption) the average net donor concentration is increased by one order of magnitude. Interestingly, the lateral doping distribution along the linescan across the wafer has nearly the same shape for all 4 samples: There is a minimum in the center and the total variation between minimum and maximum value is about 20 % for this series of growth runs.

This indicates that this typical lateral profile is not due to a shift of the C/Si ratio in the gas phase passing over the wafer (specific for most of the horizontal reactors). In this special case, we measure a temperature linescan across the wafer which has essentially the same shape, exhibiting a total ΔT of about 15

K for these runs. A readjustment of our rf-coil reduces this value to 7 K and leads to an increased doping homogeneity ($\Delta(N_D-N_A)\sim12$ %) as displayed in fig. 9.

The dependence of the nitrogen incorporation on the nitrogen flow was found to be linear from 0.1 sccm to 100 sccm (= available flow range for our nitrogen dilution configuration, see fig. 10). With the help of the site competition effect, we can gain another order of magnitude, leading to a total controllable n-doping range from about $6\text{-}8*10^{14}$ cm^{-3} to $1\text{-}2*10^{18}$ cm^{-3} with a single dopant source. The lower limit is caused by the background impurity concentration (comp. fig. 7).

In the vertical direction the doping concentration is constant within margins of ± 2 % even in the 10^{15} cm^{-3} range, which is about the error of the CV-measurement. This is made possible by an accurate and time stable automatic control of all relevant process parameters.

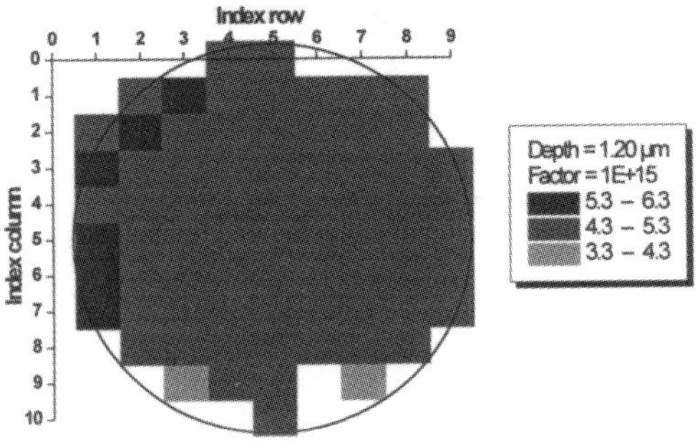

Fig. 9: 2-dimensional map of the lateral distribution of the net donor concentration (cm^{-3}) of a 4H epitaxial layer (determined by CV)

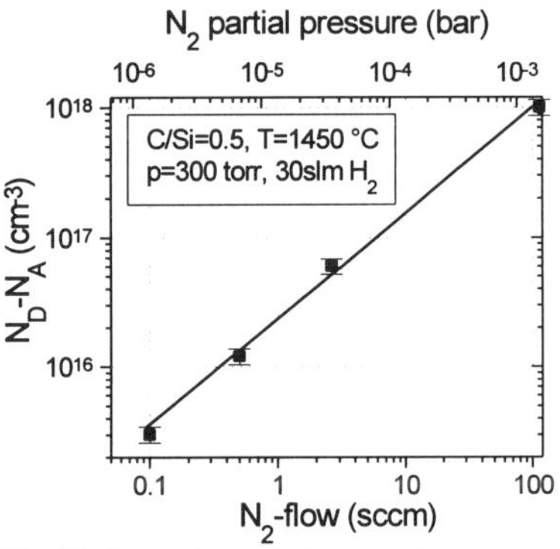

Fig. 10: Dependence between nitrogen flow into the reactor and net donor conc. (the error bars are due to variations on the wafer, comp. fig. 8, 9)

The reproducibility of the doping value from one run to the other is excellent, as long as the wafer thickness and doping are not varying too much: For a series of ten consecutive runs, which can typically be carried out with one set of graphite parts, the average doping value in the layers can be kept constant in margins of ±10%. After replacement of the graphite parts usually 1 or 2 calibration runs are necessary to readjust the desired doping value within these boundaries.

Surface constitution after the epitaxial growth

The way of terminating the VPE growth process can be of high importance for subsequent processing steps. Therefore we have investigated how the surface structure and composition is influenced by the gas composition during the cooling procedure following the epitaxial growth. A detailed description of these experiments can be found in [14]. The most important results are summarized in table II. A hydrogen atmosphere produces a very stochiometric surface, which exhibits the typical bright and sharp hexagonal LEED pattern of a undisturbed α-SiC surface (see fig. 11a), whereas a small silane partial pressure (5-10% of the flow during growth) leads to a nearly amorphous surface with very weak diffraction patterns. In both cases the surface spontaneously forms a natural oxide at room temperature (about 1-2 nm within a few hours). If the cool-down process is carried out in vacuum or inert gas (e. g. Ar), a completely different surface shows up (fig. 11b). The LEED pattern now consists of two simultaneously visible hexagons rotated by 30 degrees against each other. One hexagon is similar to the diffraction pattern of the typical α-SiC surface (periodicity in real space 0.3 nm), while the other hexagon, having a larger unit cell in the reciprocal space, exhibits a real space lattice constant of 0.25 nm, which is very similar to the lattice constant of graphite (0.246 nm). The photoelectron spectrum of this surface shows an additional line at a binding energy of 284.3 eV, which is characteristic for sp^2-bonded graphite species. As a whole, we can conclude from the surface analysis [14] of this sample that the silicon evaporation at the high temperatures in the beginning of the cooling process leads to a carbon enriched surface, which reconstructs itself in a well ordered manner. The thickness of the graphite layer was found to be about 0.8 nm.

The technical importance of this behavior becomes obvious after a closer look to table II and fig. 11: In contradiction to all other samples, the carbon rich surface shows nearly no spontaneous oxidation, even after a exposure to air for several days. This in situ surface passivation can be of high importance especially for the formation of ohmic and rectifying contacts on this surface, where an oxide interfacial layer usually leads to a degradation of the contact properties. Furthermore it has to be expected that silicon or carbon excess at the surface controls the formation of metal carbides or silicides for metals which can form both.

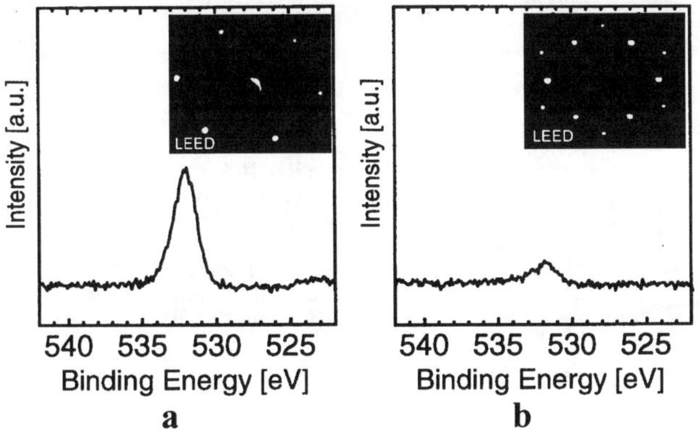

Fig. 11: O1s XP-spectra and LEED patterns (inserts) of epitaxial surfaces cooled in different gas atmospheres (see text and table II)
a) hydrogen
b) argon

Table II: Dependence of the surface properties of epitaxial layers from the atmosphere (p=300 torr) during the cooling process from growth temperature (1500 °C) to room temperature.

		Cooling after epitaxial growth done in:		
Preparation technique	**Wafer as delivered (Cree)**	**H_2 atmosphere**	**H_2/silane atmosphere**	**Ar atmosphere**
Surface composition	nearly stochiometric	nearly stochiometric	Si-rich, nearly amorphous	well ordered graphite top layer 0.8 nm (sp^2),
Natural oxide formation	natural oxide 2 nm	natural oxide 1 nm after few hours	natural oxide 1 nm after few hours	nearly no oxide even after some days of exposure to air

Electrical behavior

The final test for the epitaxial layer is the performance of a device with this layer as an active region. From the analysis of the series resistance of Schottky diodes in the On State we can estimate the electron mobility (‖c-axis, i.e. direction of current flow) in our epitaxial layers to be at least 700 cm^2/Vs (N_D-N_A=5-10*10^{15} cm^{-3}), even at growth rates of 5-6 μm/h. The properties in the reverse direction are given in table III and fig. 12. Both Schottky and pn-diodes have an edge termination to avoid an enhancement of the electrical field at the edge of the devices and are measured in

Flourinert™ oil. Breakdown fields of more than 2 MV/cm can reproducibly be achieved in pn-diodes [19] which is close to the theoretical limit of SiC.

The yield of diodes exhibiting this high breakdown field depends primarily on the size of the active area and on the wafer quality, i. e. micropipe density. For wafers with a very low micropipe density (20-50 cm^2) a yield of about 50 % was attained for devices having an active area of 0.75 mm^2.

Table III: Reverse characteristics of different edge terminated rectifying structures.

	pn-diode	Schottky-diode
structure	top p^+-layer implanted	barrier height ~ 1.1 eV
net donor conc.	$5*10^{15}$ cm^{-3}	$8*10^{15}$ cm^{-3}
epi-thickness	13 μm	10 μm
breakdown voltage	1800 V	> 1200 V
electrical field	2.1 MV/cm at breakdown	1.9 MV/cm at 1200 V
leakage current at 1100 V	< 10^{-7} A/cm^{-2}	~ 10^{-4} A/cm^{-2}

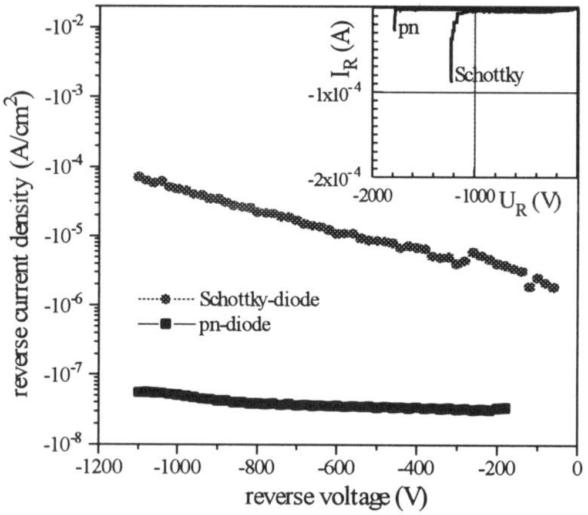

Fig. 12: Typical blocking characteristic of low doped epi-layers (see text and table III). Reverse current measurement done with a Keithley SMU 237 (voltage limit 1100V) in *Fluorinert™* oil. The insert shows the breakdown behavior (taken from a Sony curve tracer 370A, limit 2000V).

CONCLUSIONS

Making use of a type of epitaxial equipment that is more similar to typical III-V epi machines than to the widely used horizontal SiC reactors, within less than 2 years we have developed a stable and reliable process which satisfies many demands for an industrial SiC-VPE-process. These are especially high growth rates (> 5 μm/h) while maintaining quality, accurate and reproducible process control, excellent homogeneity of all relevant properties and reasonable electrical characteristics and yields for a first upcoming commercial SiC device generation.

Nevertheless to make SiC-VPE really cost effective and therefore production suited it is especially necessary, to scale up our process to a multi wafer system and/or to use larger wafers (e. g. 4"). Due to the complicated heat transfer and flow dynamics problems we will try to solve this task with the help of a numerical process simulation to get a fast optimization of the reactor design and the process parameters.

To enlarge the field of SiC applications to real high power devices in the future, a further significant reduction of the background impurity concentration and an increase in growth rate is necessary. For example a 130 μm thick epilayer with a homogenous net donor concentration of $5\text{-}6*10^{14}$ cm^{-3} (non punch through design) and a minority carrier lifetime of about 4-5 μs is needed to build a 10 kV device in SiC. The question for the right method to fulfill this harsh demands is not yet answered. The results recently presented by Kordina et al. [20] with the so called HTCVD-process are a remarkable step towards this direction, and it is our objective now, to show that these goals can be met with our technical approach in the next years.

ACKNOWLEDGMENT

The authors would like to thank H. Behner (SIEMENS, Erlangen) for his contributions concerning the surface characterization of the SiC epitaxial layers. Further thanks go to Y. Makarov and P. Kaufmann (Inst. Fluid Mechanics, Erlangen) for their very helpful numerical simulation of our epitaxial process.

This work was supported by the German Ministry of Education and Research under grant number 03M2746.

REFERENCES

1) H. Matsunami, T. Ueda, H. Nishino 1990 Mater. Res. Soc. Symp. Proc. 162 p 397
2) A. Itoh, H. Akita, T. Kimoto, H. Matsunami 1994 Appl. Phys. Lett. 65 p. 1400
3) K. Rottner, R. Helbig 1994 J. Cryst. Growth 144 p. 258
4) D.J. Larkin, P.G. Neudeck, J.A. Powell, L.G. Matus 1994 Appl. Phys. Lett. 65 p. 1659
5) D.J. Larkin Proc Int. Conf. on Silicon Carbide and Rel. Mat. ICSCRM-95, Kyoto Japan 1995, to be published
6) S. Karmann, C. Haberstroh, F. Engelbrecht, W. Suttrop, A. Schoener, M. Schadt, R. Helbig, G. Pensl, R.A. Stein, S. Leibenzeder 1993 Physica B 185 p. 75
7) T. Kimoto, H. Nishino, W.S. Yoo, H.J. Matsunami 1993 Appl. Phys. 73 p. 726
8) J.A. Powell, D.J. Larkin, L.G. Matus, W.J. Choyke, J.L. Bradshaw, L. Henderson, M. Yoganathan, J. Yang, P. Pirouz 1990 Appl. Phys. Lett. 56 p. 1442
9) O. Kordina, C. Hallin, R.C. Glass, A. Henry, E. Janzén 1994 Inst. Phys. Conf. Ser. 137 p. 41
10) O. Kordina, A. Henry, J.P. Bergman, N.T. Son, W.M. Chen, C. Hallin, E. Janzén 1995 Appl. Phys. Lett. 66 p. 1373
11) R. Rupp, P. Lanig, J. Voelkl, D. Stephani 1995 J. Cryst. Growth 146(1-4), p. 37
12) R. Rupp, P.Lanig, R. Schörner, K.-O. Dohnke, J. Völkl and D. Stephani Proc. Int. Conf. on Silicon Carbide and Rel. Mat. ICSCRM-95, Kyoto Japan 1995, to be published
13) O. Kordina, J.P. Bergman, A. Henry, E. Janzén, S. Savage, J. Andre, L.P. Ramberg, U. Lindefelt, W. Hermansson, K. Bergman 1995 Appl. Phys. Lett. 67 p. 1561
14) H. Behner, R. Rupp 1996 Appl. Surf. Sci, in print
15) S.Yu. Karpov,Yu.N. Makarov, M.S. Ramm Proc. Int. Conf. on Silicon Carbide and Rel. Mat. ICSCRM-95, Kyoto Japan 1995, to be published
16) Yu. N. Makarov Proc. MRS Spring Meeting 1996, Symp. E, to be published
17) J.A. Powell, D.J. Larkin, P.B. Abel 1995 J. Electron. Mater. 24 p. 295
18) T. Kimoto, H. Matsunami 1994 J. Appl. Phys. 75 p. 850
19) K. Hölzlein, H. Mitlehner, R. Rupp, R. Stein, D. Peters, J. Völkl, D. Stephani Proc. Int. Conf. on Silicon Carbide and Rel. Mat. ICSCRM-95, Kyoto Japan 1995, to be published
20) O. Kordina, C. Hallin, A.S. Bakin, I. Ivanov, A. Henry, M. Tuominen, R. Yakimova, A. Vehanen, E. Janzén Proc. Int. Conf. on Silicon Carbide and Rel. Mat. ICSCRM-95, Kyoto Japan 1995, to be published

KINETICS APPROACH TO THE GROWTH OF CUBIC BORON NITRIDE

C. A. TAYLOR II and ROY CLARKE
Randall Laboratory of Physics, University of Michigan, Ann Arbor, MI 48109-1120

ABSTRACT

We have deposited cubic BN films on silicon (100) using a novel ion-assisted RF-sputtering process. Our efforts over the past several years have enabled us to significantly reduce the ion-energy needed to form the cubic phase, with values now substantially less than 100 eV. Through a better understanding of the growth process we have been able to make improvements in film crystallinity and orientation, with an associated reduction of the high film stress which has severely limited film thickness in the past. The combination of low-energy nitrogen ions and high temperature growth has enabled us to deposit cubic BN films as thick as 1.9 μm. A nucleation study, performed using scanning force microscopy, shows that the cubic BN is nucleating as triangular crystallites, indicative of (111) growth. More specifically, the cubic BN appears to be nucleating on the edges of perpendicularly oriented hexagonal planes such that the cubic BN [111] is normal to the hexagonal BN [0002] (c-axis). The results suggest a pathway to "compliant" oriented growth on a variety of substrates.

INTRODUCTION

Significant advances have been made in the growth of cubic BN films as the result of the renewed interest in III-V nitride materials for optical and electronic devices. In the past, forming the metastable cubic phase of BN was a challenge in itself which was met with only limited success. Early films were typically characterized as nanocrystalline materials containing mostly sp^2-bonded phases with only a small fraction of the sp^3-bonded cubic phase.

Today, through the sustained research efforts of several groups, many of the difficulties in forming the cubic phase have been overcome and it is now possible to deposit BN films which are almost entirely cubic. That is not to say that significant problems do not remain; the films, which are generally grown on Si substrates, are polycrystalline with many defects. This may be expected due to the substantial lattice-mismatch to silicon (~40%), the effects of energetic ions which are needed to form the cubic phase (typically 100's of eV), and the relatively low growth temperatures achievable with conventional substrates. Despite these difficulties, significant improvements continue to be made, notably through increased growth temperature and a reduction in ion-energy[1-3], with some research progressing toward alternative substrate materials[4].

In this paper we report advances that we have made in the growth of cubic BN films utilizing a high flux of low-energy nitrogen ions combined with high temperature growth. In particular, we have been able to significantly reduce the ion-energy needed to form the cubic phase, with values now substantially less than 100 eV. Through a better understanding of the growth process we have made improvements in film crystallinity and orientation, with an associated reduction of the high film stress which has severely limited film thickness in the past. Furthermore, recent scanning force microscopy results indicate that highly oriented cubic BN films may be possible on silicon substrates.

EXPERIMENTAL DETAILS

BN films were deposited on silicon (100) substrates in an ultrahigh-vacuum-compatible deposition system by a process described previously[1]. To review, RF-magnetron sputtering of a hexagonal BN target is performed in an Ar/N_2 atmosphere at a total chamber pressure of 1 mTorr. The substrate, which is resistively heated, is immersed in a nitrogen plasma produced by an Astex compact ECR source. The ion flux, measured at the sample with a commercially available Faraday probe, can be varied to a maximum of approximately 1.2 mA/cm^2. Nitrogen ion-energy is precisely controlled by biasing the substrate with respect to ground. Under fixed conditions of nitrogen ion-flux and RF-magnetron power, the growth of cubic BN can be optimized within a

Mat. Res. Soc. Symp. Proc. Vol. 423 © 1996 Materials Research Society

narrow range of ion-energy (substrate-bias). At substrate temperatures of 1100-1200°C and optimized growth parameters, the cubic BN growth rate is approximately 4.5 Å/minute.

Fourier transform infrared (FTIR) absorptance spectroscopy was performed at near-normal incidence and the sample spectra were corrected using a clean silicon wafer as background. Growth parameters were optimized by maximizing the sp^3-bonding percentage (fraction of cubic BN) in the films as determined by IR-spectroscopy. The percentage was determined using a method which has become common in BN related literature[4-8]. We have also probed the surface morphology of the films using a Digital Instruments NanoScope III scanning force microscope (SFM) under ambient (in air) conditions. The microscope was operated in "tapping" mode which is a contact-SFM method where a very sharp stylus, mounted on the tip of a cantilever, is in constant contact with the surface of the sample. The scan rate was 4 lines/second and only the standard planefit (surface-leveling) correction was applied to the data. All images were checked for rotational and scaling variability to insure that the observed features were not simply images of the cantilever geometry.

RESULTS AND DISCUSSION

Figure 1 shows the sp^3-bonding percentage as a function of substrate bias voltage for two series of BN films grown under a high flux of nitrogen ions, 0.65 mA/cm^2 and 1.1 mA/cm^2, respectively. In each case, formation of the cubic BN phase is achieved in a narrow (~30 eV) "window" of ion-energy and optimization occurs in an even narrower range - only several eV in width. The window of experimental conditions where cubic BN forms is affected mainly by the ratio of ion-current to deposition flux[5,6,8]. The low-energy windows we observe are considerably narrower than those reported for growth at several hundreds of eV. The narrowing of the window with increased nitrogen-ion/boron flux ratio is most likely related to the resputtering process since the sputtering rate of the film can quickly exceed the deposition rate as the ion-flux is increased.

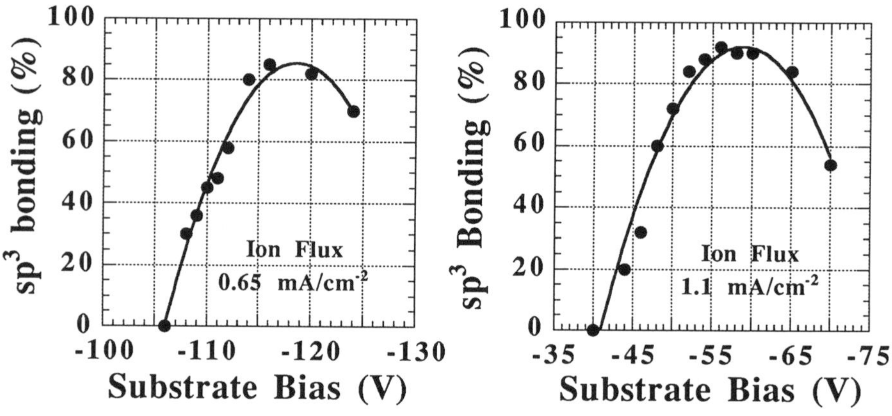

Figure 1. Percent of sp^3-bonding (cubic BN) calculated from the FTIR absorptance spectra of two series of BN films grown with different nitrogen ion currents. The solid line is a polynomial fit to the data. In each case, formation of the cubic phase is achieved in a narrow (~30 eV) window of ion-energy and optimization occurs in an even narrower range - only several eV in width.

Electron diffraction and high resolution imaging have revealed that the well crystallized BN films grow with a layered morphology; a thin amorphous layer (~50Å) at the substrate interface, followed by a region of hexagonal/turbostratic BN oriented with the (0002) planes perpendicular to the substrate, upon which a polycrystalline cubic BN layer forms. This has been well documented for BN films grown by ion-assisted processes on silicon[9-11] as well as other[4,12] substrates. An example of the film morphology can be seen in the high resolution cross-sectional TEM micrograph shown in Figure 2. The film is near the initial nucleation stage of the cubic BN layer and shows a cubic crystallite, ~ 500 Å in length, on top of perpendicularly oriented hexagonal planes. The amorphous and hexagonal layer thicknesses are ~ 50 and 300 Å, respectively.

Figure 2. High resolution cross-sectional TEM micrograph of a BN film near the initial nucleation stage of the cubic BN layer. The micrograph depicts the layered film morphology; an amorphous layer (~50 Å) at the substrate interface, followed by a region of hexagonal/turbostratic BN oriented with the (0002) planes perpendicular to the substrate, upon which a polycrystalline cubic BN layer forms.

Although the interface between the hexagonal and cubic layers is quite rough, once a continuous layer of cubic BN is formed, further deposition leads to essentially single phase cubic material. This is supported by the IR absorptance spectra shown in Figure 3. The sp^2-bonding features associated with the amorphous and hexagonal interface layers remain of similar intensity, while the sp^3-bonding feature of the cubic BN TO phonon increases with film thickness. This demonstrates that for well crystallized films, the cubic BN fraction (or sp^3-bonding percentage) is simply a function of film thickness.

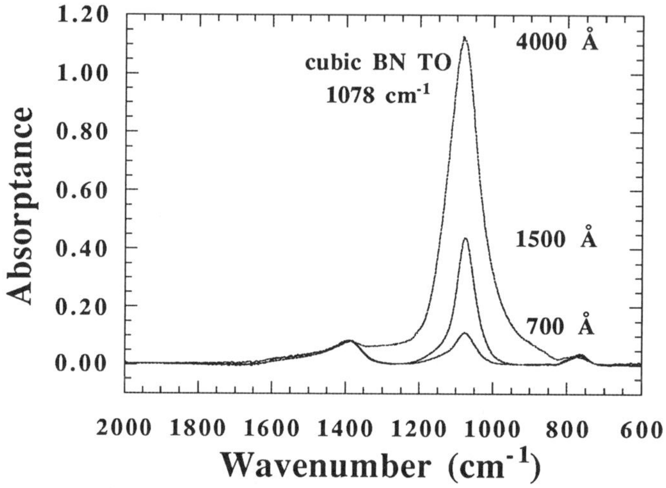

Figure 3. Infrared absorptance spectra from a series of cubic BN films of different thickness. The cubic TO phonon absorptance at 1078 cm^{-1} increases with film thickness while the hexagonal BN features, near 770 cm^{-1} and 1380 cm^{-1}, are of similar intensity in each film. This supports TEM diffraction data from thick cubic BN films which indicate that once the cubic BN layer is formed, continued growth results in nearly single phase cubic material.

The frequency of the zone center TO phonon mode of cubic BN has been shown to be a sensitive measure of internal strain[13,14]. As shown in Figure 3, the cubic TO phonon frequency is observed at 1078 cm^{-1} in this series of films (deposited at 1100°C). This is a large shift from either the 1065 cm^{-1} value measured by infrared reflectivity[15] or the 1056 cm^{-1} value measured by Raman spectroscopy[16] in bulk cubic BN crystallites. Compressive film stress, either inferred by IR spectroscopy or measured directly from substrate curvature, has been frequently reported in cubic BN films[3,11,17,18]. The stress results from the buildup of unannealed defects generated from the ion-assisted process. Many research groups report that the stress limits film thickness, causing delamination at thicknesses greater than several thousand angstroms[9,11].

We have investigated film growth at temperatures ranging from 800-1200°C on silicon substrates and find that compressive film stress can be greatly reduced by growth at high temperature. Figure 4 shows the temperature dependent position of the cubic BN zone center TO phonon as measured by IR absorptance. The infrared spectra show, within experimental error, a linear dependence of peak position with temperature in the measurement range. The solid line in the figure is a least squares fit to the data which has been extrapolated (dashed line) to the bulk TO frequency (1065 cm^{-1}) as measured by infrared reflectivity. We suspect that a large portion of the stress may be relieved through the substrate since the growth temperatures (and extrapolated data) are approaching the melting temperature, T_m, of silicon (1420 °C). The principal conclusion is that high growth temperatures are needed to provide the necessary mobility to anneal defects generated during growth. A general rule established from theoretical studies of epitaxy is that growth temperatures need to be ~5/8 T_m (of the deposited material) to achieve reasonable mobility and grain growth[19].

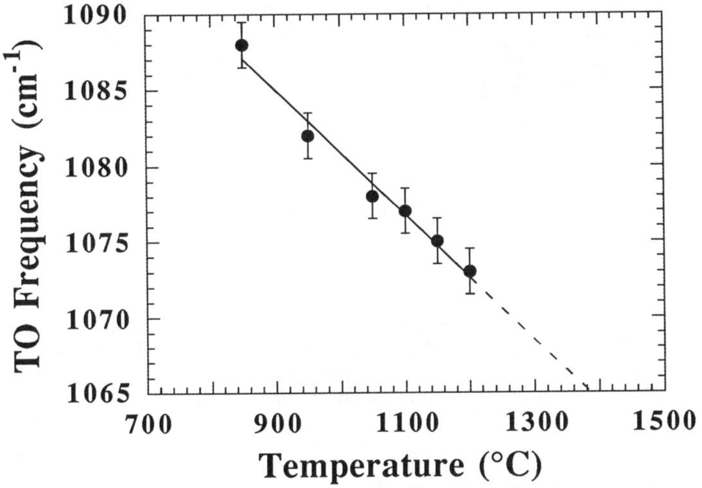

Figure 4. Growth temperature dependence of the cubic BN zone center TO phonon frequency measured by infrared absorptance. The line is a linear fit to the data using the equation $\omega_{TO} = 1122.1 - 0.0417T$.

Although 1200°C is quite low for cubic BN deposition by the 5/8 T_m standard (T_m>2973°C), we have seen significant improvements in film quality at these temperatures. Films deposited at substrate temperatures between 1100 and 1200°C exhibit increased grain size[20] and show no signs of delamination, even at thicknesses up to 1.9 μm. Furthermore, recent scanning force microscopy (SFM) data (discussed in the following paragraph) indicates that a reduction of ion-energy coupled with high-temperature growth may lead to highly oriented cubic BN. Shown in Figure 5 are the IR absorptance spectra from a series of very thin BN films. The weak cubic TO phonon features, which are substantially shifted from the bulk phonon frequency, demonstrate that the films are near the initial nucleation and growth stage of the cubic BN layer. We have recently shown that the large shift in the IR-active cubic BN TO phonon mode, observed during initial nucleation, is based on finite size effects resulting from island nucleation[21]. Specifically, at the initial nucleation stage, the long wavelength infrared radiation polarizes the small separated cubic BN crystallites, leading to a long range depolarization field. The effect of this field is seen as a shift in the TO phonon frequency. As islands coalesce to form a continuous layer, the long range field parallel to the substrate surface rapidly vanishes and the TO mode returns to an equilibrium value. This is an effect well documented in IR spectra of small crystallites[22].

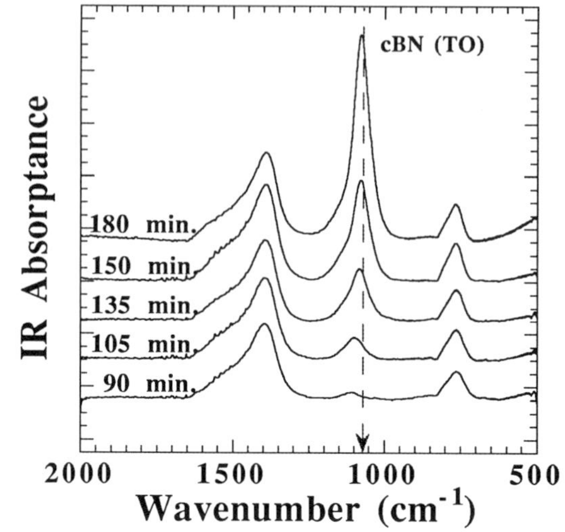

Figure 5. Infrared absorptance spectra from a series of thin BN films. The weak cubic BN TO phonon feature indicates that the films are near the initial nucleation stage of the cubic BN layer. The films are identified by total growth time which includes the growth time for the amorphous and hexagonal BN layers.

SFM images from this same series of films are shown in Figure 6. The scan area in each image is 2x2 μm and the total z-height is 1000 Å. Image (a) shows the first cubic BN islands nucleating on the film surface. The islands are clearly triangular, indicative of (111) cubic growth. A slightly thicker film, panel (b), shows that further nucleation occurs and the average grain size grows to approximately 700 Å. The film grown for 135 minutes, panel (c), is relatively smooth and continues to show triangular crystallites with an average grain size similar to the previous panel. The film grown for 150 minutes, panel (d), shows crystallites of similar size and shape but appears more randomly oriented compared with the thinner films. These results are significant for understanding the mechanism involved in the hexagonal to cubic transition, a subject of current interest in the BN community[10,23-26]. Although by no means a conclusive measure of crystallographic orientation, the images suggest that the cubic BN initially nucleates in an ordered [111] growth direction on the edges of the (0002) planes of the hexagonal BN layer. More specifically, the cubic BN [111] is normal to the hexagonal BN [0002] (c-axis).

The crystallographic orientation suggested by our SFM data differs from that recently reported by Reinke *et al.*[24] and Medlin *et al.*[25] for BN films grown at low temperature (500°C). In both studies, high resolution TEM micrographs show small regions where a cubic BN crystallite is aligned with the {111} fringes parallel to the underlying hexagonal BN (0002) fringes. The authors point out that the relative spacings of the hexagonal BN (0002) (3.33 Å) and cubic BN (111) (2.09 Å) planes approach a 3:2 ratio such that every third cubic BN (111) plane could be closely matched (within 6%) with every alternate hexagonal BN (0002) plane, forming a semicoherent interface. The edges of the highly oriented initial hexagonal BN layer may then serve as preferential nucleation sites for growth of oriented cubic BN. Medlin *et al.* have also suggested[23,26] a direct route to cubic BN formation via a rhombohedral intermediate phase. In this case, transformation to the cubic phase occurs by a diffusionless pathway, whereby stress leads to the compression of (0003) rhombohedral planes into the (111) cubic BN planes. They

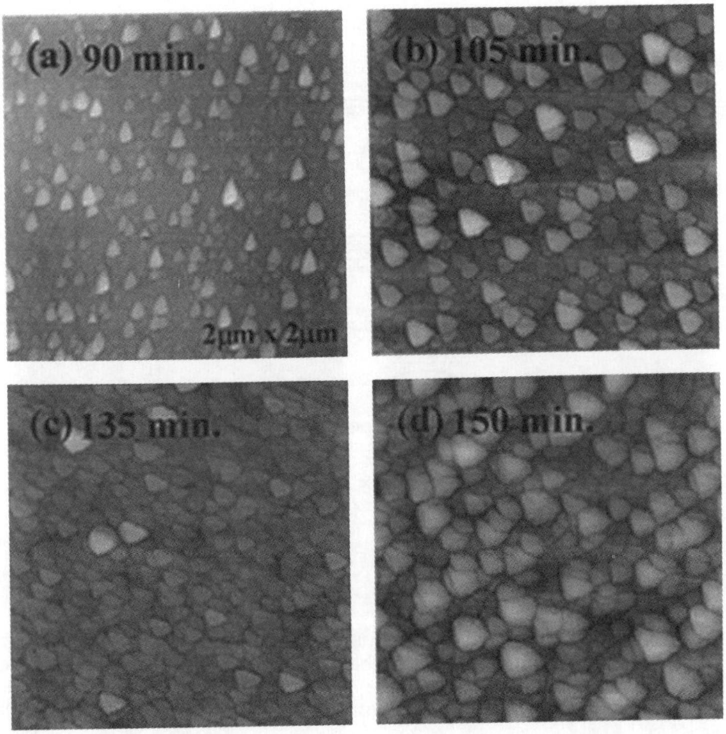

Figure 6. Scanning force microscopy images of a series of thin BN films near the initial nucleation stages of the cubic BN layer. The triangular crystallites, which are indicative of (111) growth, are seen in panels (a) and (b) to first grow in size to approximately 700 Å . The grains then coalesce, panel (c), to form a smooth continuous layer. Thicker films, panel (d), show further nucleation and a more random orientation as growth continues.

have observed[26] rhombohedrally configured domains in BN films but note that further work is needed to determine the predominance of this stacking over the hexagonal arrangement.

We note that another diffusionless pathway exists, via a wurtzitic intermediate, which leads to an epitaxial relationship between the edges of hexagonal BN (0002) basal planes and cubic BN. This mechanism, shown in Figure 7, was first proposed by Wheeler and Lewis[27] for the hexagonal to cubic transition of carbon and leads to (111) cubic planes normal to the (0002) hexagonal basal planes, as suggested by our SFM results. The wurtzitic-cubic structural relationship has actually been observed in bulk BN samples[28]. Although such a transformation has not been directly observed in BN films, the underlying mechanism provides yet another plausible pathway for the hexagonal to cubic transition. We are currently performing transmission electron microscopy on our samples to provide a more detailed and conclusive study of the crystallographic orientation suggested by the SFM images.

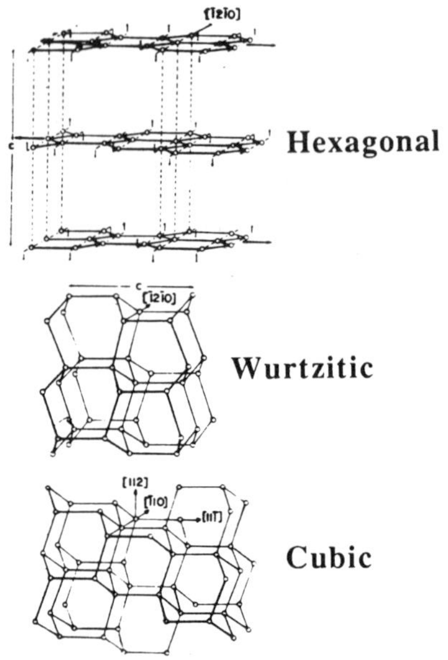

Figure 7. Hexagonal to cubic transformation mechanism first proposed by Wheeler and Lewis[27]. This transformation mechanism provides an epitaxial relationship, via a wurtzitic intermediate, between the edges of hexagonal (0002) basal planes and cubic BN, such that the cubic (111) planes are normal to the hexagonal (0002) planes.

CONCLUSIONS

We have developed a novel ion-assisted RF-sputtering deposition process for the growth of cubic BN films. Our efforts over the past several years have enabled us to significantly reduce the ion-energy needed to form the cubic phase, now substantially less than 100 eV. Through continual process development we have seen improvements in film crystallinity and orientation, with an associated reduction of the high film stress which has severely limited film thickness in the past. The combination of low-energy nitrogen ions and high temperature growth has enabled us to deposit cubic BN films as thick as 1.9 μm. Recent scanning force microscopy images indicate that the cubic BN is nucleating as triangular crystallites, indicative of (111) growth. More specifically, the cubic crystallites appear to be nucleating on the edges of the hexagonal planes such that the cubic BN [111] is normal to the hexagonal BN [0002] (c-axis). This indication of oriented cubic BN growth on silicon substrates not only helps define the mechanism for the hexagonal to cubic transition but also gives promise for the technological development of this material.

Work supported by ONR Grants N00014-91-J-1398 and N00014-94-J-0763

REFERENCES

1. S. Kidner, C. A. Taylor II, and R. Clarke, Appl. Phys. Lett. **64**, 1859 (1994).

2. C. A. Taylor II and R. Clarke, paper presented at *The 6th European Conference on Diamond, Diamond-like and Related Materials*, Barcelona, Spain, Sept. (1995).

3. C. A. Taylor II, P. Encarnacion, and R. Clarke, poster presented at *Materials Research Society Fall Meeting*, Boston, Mass., Nov. (1995).

4. P. B. Mirkarimi *et al.*, Appl. Phys. Lett. **66**, 2813 (1995).

5. P. B. Mirkarimi *et al.*, J. Mater. Res. **9**, 2925 (1994).

6. S. Reinke *et al.*, Diamond Relat. Mater. **4**, 272 (1995).

7. T. Wada and N. Yamashita, J. Vac. Sci. Technol. A **10**, 515 (1992).

8. D. J. Kester and R. Messier, J. Appl. Phys. **72**, 504 (1992).

9. D. J. Kester *et al.*, J. Mater. Res. **8**, 1213 (1993).

10. D. L. Medlin *et al.*, J. Appl. Phys. **76**, 295 (1994).

11. S. Reinke, M. Kuhr, and W. Kulisch, in *Proc. of 4th International Symposium on Diamond Materials*, Reno, Nevada, May 1995.

12. D. J. Kester *et al.*, J. Vac. Sci. Technol. A **12**, 3074 (1994).

13. S. Fahy, Phys. Rev. B **51**, 12873 (1995); erratum (in press).

14. J. A. Sanjurjo *et al.*, Phys. Rev. B **28**, 4579 (1983).

15. P. J. Gielisse *et al.*, Phys. Rev. **155**, 1039 (1967).

16. O. Brafman *et al.*, Solid State Commun. **6**, 523 (1968).

17. D. R. McKenzie *et al.*, Diamond Relat. Mater. **2**, 970 (1993).

18. T. A. Friedmann *et al.*, J. Appl. Phys. **76**, 3088 (1994).

19. C. P. Flynn, J. Phys. F **18**, L195 (1988).

20. S. Kidner, Ph.D. Thesis, The University of Michigan, 1994.

21. C. A. Taylor II, S. Fahy, and Roy Clarke, submitted for publication.

22. R. Ruppin and R. Englman, Rep. Prog. Phys. **33**, 149 (1970).

23. D. L. Medlin *et al.*, Phys. Rev. B **50**, 7884 (1994).

24. S. Reinke, M. Kuhr, and W. Kulisch, in *Proceedings of the 187th Meeting of the Electrochemical Society*, 1995 (unpublished).

25. D. L. Medlin *et al.*, J. Appl. Phys. **79**, 3567 (1996).

26. D. L. Medlin *et al.*, Phys. Rev. B **51**, 10264 (1995).

27. E. J. Wheeler and D. Lewis, Mat. Res. Bull. **10**, 687 (1975).

28. F. R. Corrigan and F. P. Bundy, J. Chem. Phys. **63**, 3812 (1975).

THE IMPACT OF PREGROWTH CONDITIONS AND SUBSTRATE POLYTYPE ON SIC EPITAXIAL LAYER MORPHOLOGY

A.A. BURK, JR., L.B. ROWLAND, G. AUGUSTINE, H.M. HOBGOOD, AND R.H. HOPKINS
Northrop Grumman Science & Technology Center, 1310 Beulah Rd., Pittsburgh PA USA 15235

ABSTRACT

4H and 6H-SiC epitaxial layers exhibit characteristic morphological defects caused by process and substrate interferences with the a-axis directed step-flow growth. 4H-SiC is shown to typically exhibit worse morphology than 6H-SiC for a given off-axis orientation. SiC epitaxial layer defects are significantly reduced by the optimization of growth conditions and substrate surface preparation. The remaining highly variable defects are shown to emanate from the substrate surface with densities of ≥ 1000 cm^{-2}.

INTRODUCTION

Vapor phase epitaxial (VPE) growth is currently the most practical means of SiC active layer formation. While prototype SiC VPE devices have already demonstrated performance exceeding that of Si and GaAs based devices [1,2], SiC materials technology including active layer formation is still immature. In particular, SiC epitaxial layers still exhibit relatively large numbers of morphological defects in comparison with the more mature semiconductors. In this study, epitaxial layer morphology was significantly improved by identifying and minimizing *in situ* pregrowth and substrate related sources of morphological defects.

EXPERIMENT

The epitaxial growths and pretreatments conducted in this study were performed in a horizontal, inductively heated, water-cooled, quartz-walled SiC VPE reactor. The hydrogen carrier gas was purified by diffusion through a Pd-cell membrane. The susceptor temperature was measured with a pyrometer by way of a sapphire light pipe placed in the growth chamber approximately 2 cm from the inductively heated SiC-coated susceptor. Temperature measurements were calibrated by observing the indicated melting point of silicon samples. The substrates used consisted of commercially obtained physical vapor transport (PVT) grown off-axis (3.5° towards the <1210> a-axis direction) 6H and 4H SiC Si(0001) and 8°-off 4H SiC [3]. Substrates and epitaxial layers were measured by optical interference contrast and replica transmission electron microscopy(TEM).

RESULTS

Table I contains a summary of the condition of SiC surfaces after exposure to a variety of pregrowth treatments. Si-droplets were observed to develop on SiC surfaces when exposed to hydrogen at temperatures above the melting point of silicon (1410 °C). These droplets, which interfere with specular epitaxial growth, were effectively inhibited by the presence of propane or HCl overpressures prior to growth. These results are presented in greater detail elsewhere [4]. Despite this significant reduction of droplets afforded by optimization of pregrowth conditions varying numbers of morphological defects were still observed in subsequent epitaxial growths.

Mat. Res. Soc. Symp. Proc. Vol. 423 © 1996 Materials Research Society

Case No.	Experimental Condition	Silicon-droplets formed?	Explanation
1)	30 min. @1450--1520°C under 1 atm H_2 on 4H, 6H, a- and vicinal c-axis PVT SiC and 6H Lely SiC.	Yes on all types	1) $SiC + H_2 \rightarrow$ Si(ℓ)+hydrocarbons 2) Si(ℓ) accumulates on surface
2)	30 min. @1450--1520°C under 1 atm Ar	No	1) SiC is stable under Ar and 2)vapor pressure of Si>>C
3)	30 min. @1450--1520°C under 1 atm H_2 and 140 ppm C_3H_8	No	$SiC + H_2 \leftarrow$ Si(ℓ) + hydrocarbons (C_3H_8 inhibits H_2 etching of SiC)
4)	30 min. @1450--1520°C under 1 atm H_2 with exposed carbon	No	1) $C + H_2 \rightarrow$ hydrocarbons 2) hydrocarbons inhibit SiC etching
5)	30 min. @1450--1520°C under 1 atm H_2 and 3 % HCl	No, but etch pits	1) $SiC + H_2 \rightarrow$ Si(ℓ)+hydrocarbons 2) Si(ℓ) + HCl$\rightarrow$ chlorosilane gas
6)	30 min. @1450--1520°C under 0.1 atm H_2	edges only	Si(ℓ) evaporation rate increased by reduced diffusion layer

Table I. Causes of silicon-droplet formation in SiC epitaxy.

The following experiment was performed to determine the cause and spatial location of the remaining morphological defects in the SiC epitaxial layers. Samples of 4H and 6H-SiC were subjected to an *in situ* etch and growth nucleation procedure that effectively minimized Si-droplets and etch pits (based on the above pregrowth investigation) but then the growth was suddenly halted within a few seconds. Photomicrographs of the resulting surfaces were obtained. They still revealed a few barely visible submicron-sized Si-droplets, impressions where Si-droplets had rested and then evaporated, micropores and other unresolvable "point defects". Before this treatment (not shown) only the pores and some very faint polishing scratches had been discernible. Subsequently the samples were placed back into the growth chamber followed by another identical *in situ* etch and growth nucleation. This time however, the growth was allowed to continue to a thickness of 1 μm, after which the growth was ended with propane overpressure during cool-down to suppress end-growth Si-droplet formation. Photomicrographs of the resulting epitaxial layer were taken and compared to those taken of the same location after the first nucleation.

On the 1 μm thick 4H-SiC sample shown in Figure 1 all the features identified after the first nucleation resulted in morphological defects in the epitaxial layer. While most were amphitheater-shaped shallow depressions, some were larger and deeper tetrahedral shaped pits. There was a rough correlation between tetrahedral pits and the sites that had Si-droplets evident after the first nucleation step (as opposed to those that only had impressions of Si-droplets). In addition, large micropores were observed to initiate tetrahedral pit formation. The dominant source of tetrahedral pits, however, were residual substrate polishing scratches. It is noteworthy that many of the scratches are decorated with numerous tetrahedral pits while others have none. This distribution is consistent with that observed for Si droplets on polishing scratches. Identical experiments performed on 6H-SiC substrates resulted in only the small amphitheater type defect even though the residual polishing scratches appeared to be comparable.

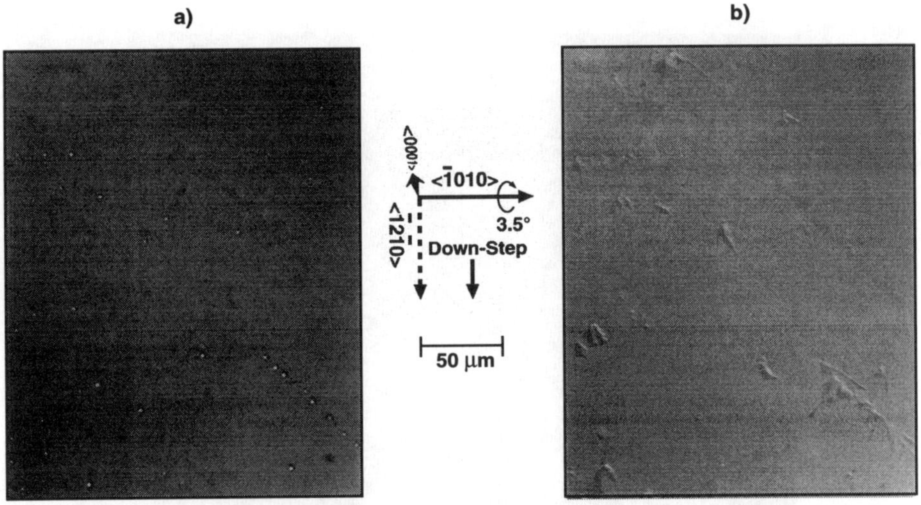

Fig. 1 Interference contrast photomicrographs of a 4H-SiC sample, a) before and b) after epitaxial growth. Every morphological defect in the epitaxial layer can be attributed to a substrate surface imperfection revealed by a previous growth nucleation.

An important result of the above nucleation experiments is that as the first and second etching and nucleation procedures were identical, the lack of a second population of morphological defects in the final epitaxial layer indicates that the substrate was the source for each of the morphological defects in the epitaxial layer. Conversely, this shows that the etching and epitaxial layer nucleation procedure used has effectively eliminated etch pits and Si-droplets generated spontaneously by the VPE process.

While the morphology of 4H epitaxial layers was typically observed to be inferior to that of 6H (particularly on 5 mm square samples), equivalent results were observed on a small percentage of as received whole 4H-SiC wafers. Obvious polishing scratches were observable by interference contrast microscopy on some 4H substrates that yielded highly defective epitaxial layers, while others that showed no prior evidence of significant residual polishing problems also resulted in poor morphology. Closer inspection of one such substrate by replica TEM however revealed a field of very fine polishing scratches (Figure 2). Another substrate (not shown) which resulted in excellent morphology exhibited no discernible polishing scratches by this method. This anecdotal evidence suggests that residual polishing damage is the source of many of the morphological defects in 4H SiC epitaxial layers but does not explain the marked polytype dependence.

Having optimized the pregrowth conditions and thereby obtained specular epitaxial layers, we have begun using a microscope and image analysis software to automatically count

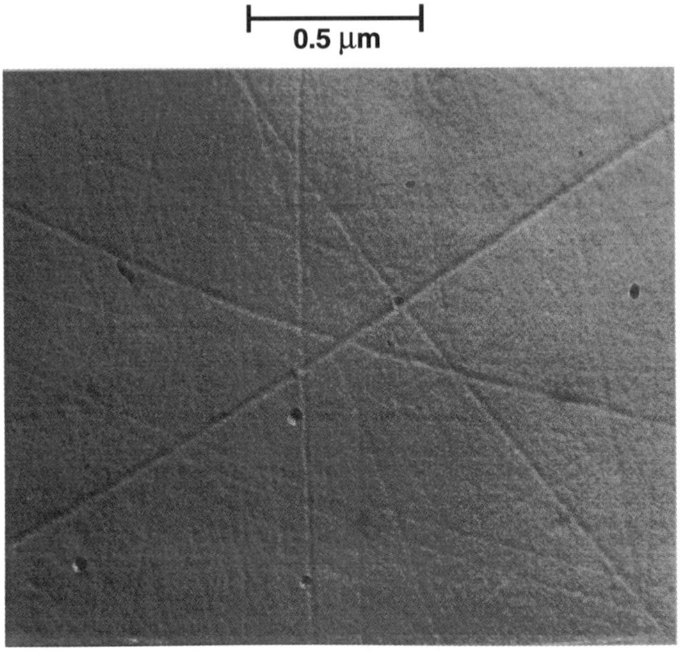

Fig. 2 Replica-TEM photomicrograph of a 4H-SiC substrate before epitaxial growth.

morphological defects. In this way we can determine more quantitatively the impact of substrate preparation (and other conditions) upon epitaxial layer morphology. To obtain a rough figure of merit we multiply the average defect densities and areas to obtain a percentage defect area. Figure 3 contains a compilation of results for one hundred 4H-SiC epitaxial wafers having approximately 5 μm thick layers (except as noted). The data are highly scattered as indicated by the large single standard deviation error bars; however, certain trends are apparent. Improved epitaxial layer morphology is obtained by repolishing incoming substrates, optimizing Si/C ratio, and increasing wafer off-orientation. These enhancements are all consistent with increasing ideality of the operative step-flow epitaxial growth mechanism.

DISCUSSION

With these observations we can better understand the often irreproducible SiC epitaxial layer morphologies reported by numerous researchers using diverse substrates, growth conditions, and reactor constructions. For example our observations of dramatic improvement in 3.5° off-oriented 4H and 6H-SiC epitaxial layer morphology with judicious application of C_3H_8 and HCl overpressures prior to growth can be understood in terms of their impact on SiC etch rates, faceting, and Si-droplet formation [4]. Recently, Powell et al. have reported similar

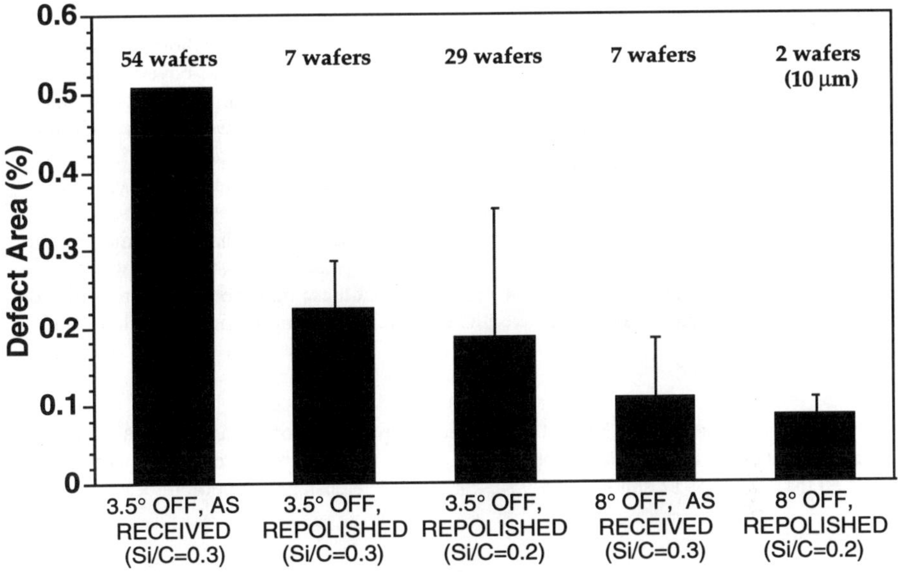

Fig. 3 Effective defect area of a series of ~5 μm thick 4H-SiC epitaxial layers as a function of polishing, orientation, and input Si/C ratio.

improvements by extending the HCl etch up to growth initiation to avoid Si-droplet formation [5]. Hallin et al. has shown that increasing the growth temperature reduced the density of triangular morphological defects [6]. Carter et al. demonstrated that increasing the off-orientation of 4H-SiC substrates from 3.5° to 8° had a similar effect [7]. Also recently, Kimoto indicated that the average terrace widths are larger on 4H-SiC than 6H-SiC surfaces given the same off-axis orientation [8]. This is a likely reason for the large polytype morphology difference. In fact, all of the layer morphology improvements are consistent with an increased ideality for step-flow epitaxial growth.

The two-dimensional step-flow growth model of Frank and van der Merwe is well known in crystal growth. It was first applied to SiC by Kuroda et al. [9]. Our proposed mechanism for morphological defect formation is that the normal step-flow epitaxial growth is interrupted by crystal imperfections such as micropores, polishing scratches, and *in situ* deposits or damage such as Si-droplets or etch pits. In the case of the shallow features the step-flow growth "wraps around" the crystal imperfection, thereby minimizing its impact. In the case of the larger tetrahedral pit, an exposed basal plane (0001) facet generated early in the growth or during etching, propagates by growing principally along its down-step a-axis edge with relatively little three-dimensional growth of the bottom facet. The lateral extent of the tetrahedral pit is consistent with step flow growth; i.e., the projected length in the down-step a-axis direction is approximately equal to the epitaxial layer thickness divided by the tangent of the off-axis tilt angle.

CONCLUSIONS

This work has shown that the large variability in epitaxial layer morphology that continues to be reported, can be explained in part by the variation of the SiC coverage of both initially SiC coated and uncoated graphite susceptors with use, and the large impact that this has on the SiC etch rate, faceting, and Si-droplet formation prior to growth. Even when Si-droplets and etch pits are avoided and growth temperature and substrate off-orientations are increased however SiC epitaxial layers still exhibit highly variable and large densities (~1000 cm^{-2}) of optically detectable morphological defects. The remaining features have been shown to originate in the bulk SiC substrate and its surface. While the SiC vapor phase epitaxial growth process may still be improved so that the decoration of the remaining substrate surface imperfections is further reduced, it would be preferable that these imperfections be minimized. Attempts are currently underway to improve SiC epitaxial layer morphology by both means.

ACKNOWLEDGMENTS

We thank Lin Thomas and Jim Chance for performing epitaxial growths and Jim Haugh for the TEM measurements. This work was supported in part by the Department of the Air Force under contract F33615-92-C-5912 and F33615-95-C-5427 (Tom Kensky contact monitor).

REFERENCES

1. S. Sriram, R. C. Clarke, A. A. Burk, Jr., H. M. Hobgood, P. G. McMullin, P. A. Orphanos, R. R. Siergiej, T. J. Smith, C. D. Brandt, M. C. Driver, and R. H. Hopkins, IEEE Electron Device Lett., EDL-15, 458 (1994).

2. R. R. Siergiej, S. Sriram, R. C. Clarke, A. K. Agarwal, C. D. Brandt, A. A. Burk, Jr., T. J. Smith, A Morse, and P. A. Orphanos, Tech. Digest Int. Conf. SiC and Rel. Mat'95, (Kyoto Japan 1995), p. 321.

3. Cree Research Inc., Durham, North Carolina.

4. A. A. Burk, Jr. and L. B. Rowland, accepted for publication in J. Crystal Growth.

5. J. A. Powell, D. J. Larkin, and P. B. Abel, J. Electron. Mater. 24, 295 (1995).

6. C. Hallin, A. O. Konstantinov, O. Kordina, and E. Janzén, Tech. Digest Int. Conf. SiC and Rel. Mat'95, (Kyoto Japan 1995), p. 35.

7. C. H. Carter, Jr. and V. F. Tsvetkov, Tech. Digest Int. Conf. SiC and Rel. Mat'95, (Kyoto Japan 1995), p. 11.

8. T. Kimoto, A. Itoh, and H. Matsunami, Appl. Phys. Lett. 66, 26 (1995).

9. N. Kuroda, K. Shibahara, W. S. Yoo, S. Nishino, and H. Matsunami, Ext. Abstr. of the 19th Conf. on Solid State Devices and Materials, (Tokyo Japan 1987), p.227.

NUCLEATION AND GROWTH OF ORIENTED DIAMOND FILMS ON NICKEL SUBSTRATES

P.C. YANG, W. LIU, D.A. TUCKER, C.A. WOLDEN, R.F. DAVIS, J.T. GLASS*, J.T. PRATER** AND Z. SITAR.
Department of Materials Science and Engineering, NCSU, Raleigh, NC 27695-7919
* Kobe Steel USA, Inc. RTP, NC 27709.
** Army Research Office, RTP, NC 27709

ABSTRACT

(100) and (111) oriented diamond films were grown on similarly oriented single crystal Ni substrates using a multi-step seeding and growth process. *In-situ* reflection monitoring revealed large surface reflectivity changes upon heating of the seeded substrate in H_2. The reflectivity change was attributed to the surface melting and dissolution of the seeding particles. The presence of atomic hydrogen lowered the eutectic melting point of the Ni-C compound from 1325°C to about 1100°C. It appeared that the molten Ni-C-H surface layer suppressed graphite formation, which is normally observed in diamond growth on Ni, and promoted diamond nucleation. The oriented diamond films were also obtained using non-diamond carbon seeding. Based on experimental observation, a proposed model is described to explain the nucleation mechanism from the molten Ni-C-H surface layer.

INTRODUCTION

The growth of highly oriented heteroepitaxial films represents an important step toward the attainment of large-area, device-quality diamond [1,2,3]. The extremely high surface energy of diamond and the existence of interfacial strain between diamond films and non-diamond substrates are believed to be primary obstacles in formation of oriented, two-dimensional diamond nuclei. Nickel is one of few materials that has a close lattice match with diamond (a=3.52 Å for Ni and a=3.56 Å for diamond). However, its high solubility for carbon and its strong catalytic effect on hydrocarbon decomposition at low pressures have prevented CVD diamond nucleation on the Ni surface without the deposition of an intermediate graphite layer [4]. On the other hand, it has been known for decades that Ni is an effective solvent-catalyst metal for diamond crystallization under high pressure and high temperature(HPHT) conditions [5]. Thus, it is interesting to consider if this same characteristic can be made useful in a low pressure CVD process. Sato et al. reported that both (111) and (100) oriented diamond nuclei could be grown on Ni substrates, but the overall percentage of oriented nuclei was rather low [6]. We have observed that under special conditions oriented diamond nuclei precipitate from supersaturated Ni-C-H solution [7]. A similar phenomenon has been observed for the heteroepitaxial growth of diamond on (111) Pt by Kobashi and Tachibana [8]. Roy also reported on precipitation of diamond particles from different mixtures of diamond and metal powders [9].

In this paper, we report on a multi-step CVD process that results in nucleation of oriented diamond on Ni substrates without a graphite interlayer. The following sections describe in detail experimental conditions, results of analytical characterization, and suggest an explanation of the observed phenomena.

EXPERIMENT

The experiments were carried out in a hot-filament CVD (HFCVD) system. A schematic diagram of the system is shown in Fig.1. The HFCVD chamber was a water-cooled, modified six-way cross. The substrate holder was made of molybdenum or stainless steel. The substrates were heated by radiation from the tungsten filament which was kept at a constant temperature between 2300°C to 2400°C. The substrate temperature was controlled by adjusting the distance between the filament and the substrate, and by water cooling the substrate holder. All cited temperatures were calibrated against the melting points of Ge and Ni-Cu alloys, and measured with thermocouples

Mat. Res. Soc. Symp. Proc. Vol. 423

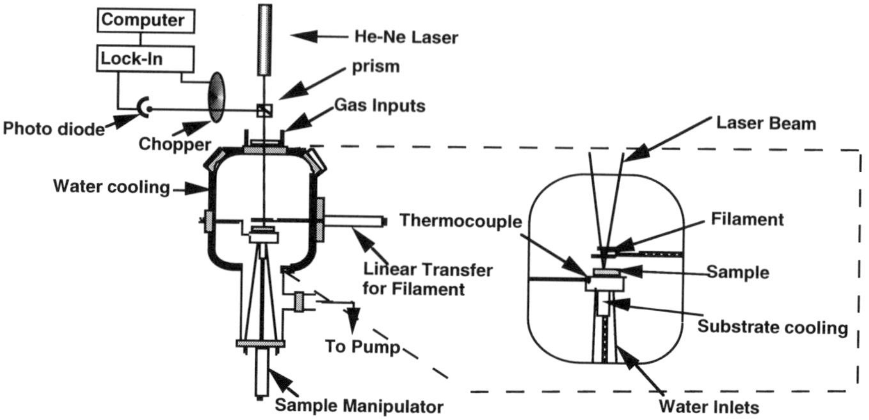

Fig.1 Schematic diagram of the hot HFCVD and Laser reflection systems.

contacting the subsurface of the substrates. A laser reflection system was set up to monitor sample surface reflectivity during the nucleation and growth process.

Both (100) and (111) single crystal Ni were used as substrates. The substrates were polished using standard metal polishing techniques [10] and subsequently cleaned with acetone and DI water. A suspension of 0-2 μm diamond powder in acetone was used for seeding. Samples were simply immersed into suspension and allowed to dry.

As an alternative, non-diamond carbon seeds were also used. They included graphite and fullerene (C_{60}) powders, and graphite layers obtained by decomposition of hydrocarbons. All these seeds were effective in producing oriented diamond nuclei.

The nucleation and growth procedure consisted of a three -step process: (1) deoxidization, (2) high temperature annealing, and (3) diamond deposition. Step 1 involved annealing of the seeded substrates at a temperature of 900°C in a hydrogen atmosphere for 10-30 minutes to remove oxides. In step 2, the substrate holder was raised toward hot filament until a substrate temperature of about 1100°C was reached. In the presence of atomic hydrogen the diamond seeds dissolve rapidly into the Ni lattice at this high temperature. The high temperature annealing time

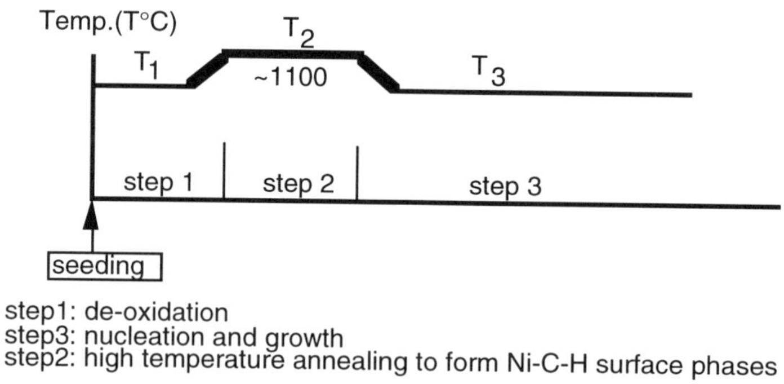

Fig.2 Schematic diagram of the multi-step nucleation and growth process.

Table I. Experimental conditions of the multi-step process.

conditions	substrate temperature (°C)	CH4/H2 flow rate (sccm)	duration time	pressure (Torr)
step 1	900	0/400	10 - 30 minutes	30
step 2	1100	0/400	10 - 60 minutes	30
step 3	900	2/400	4 - 20 hours	30

depended on the seeding conditions and the exact surface temperature. It is believed that the annealing time must be long enough to permit sufficient reaction between the nickel, diamond seeds, and hydrogen to form a molten Ni-C-H intermediate layer which suppresses graphite formation and promotes diamond nucleation. In practice, the duration of the anneal was actually determined by the optical appearance of the seeded substrates. Upon melting, the surface characteristic changed from dark gray to highly reflective or "shiny". A laser reflection measurement was used to detect the surface reflectivity changes during the multi-step process, as shown in Fig.3. In step 3, the substrate temperature was lowered to about 900°C and 0.5% CH_4 was admitted with the hydrogen flow, standard conditions for diamond growth. The experimental conditions of the multi-step process are summarized in TableI.

Morphology of the diamond films after the growth was observed by scanning electron microscopy (SEM). Micro-Raman spectroscopy was performed to assess the quality of the deposit. For this purpose, an argon ion laser beam at a wavelength of 514.5 μm was focused on the diamond surface with a spot size of approximately 5 μm. X-ray diffraction and TEM analyses were carried out to investigate the microstructures and phases involved in the oriented diamond nucleation and growth on Ni and are presented in detail in another paper [11].

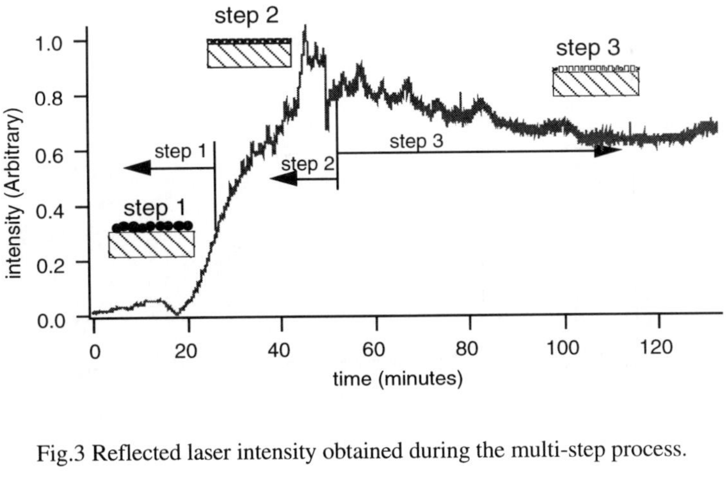

Fig.3 Reflected laser intensity obtained during the multi-step process.

(a) (b)

Fig.4 SEM of (111) (a) and (100) oriented diamond on Ni (b).

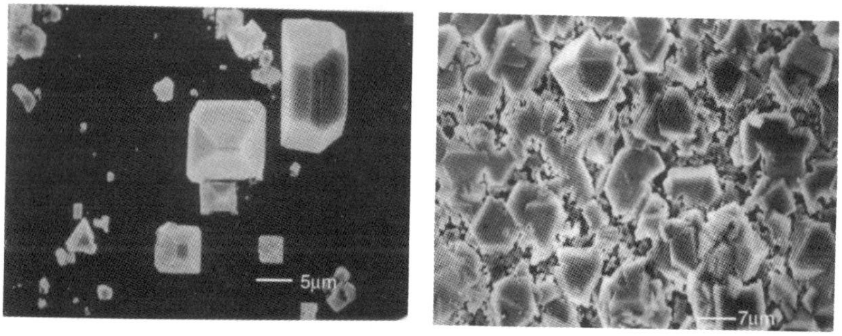

(a) (b)

Fig.5 SEM of the oriented diamond on Ni obtained by (a) graphite seeding and (b) CH_4 seeding.

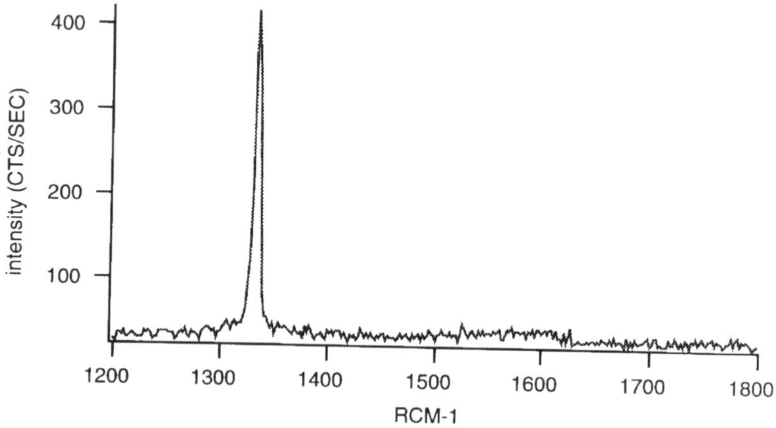

Fig. 6 A typical Raman spectrum obtained from diamond on Ni.

RESULT AND DISCUSSION

Fig.3 shows a typical reflected intensity curve obtained during the multi-step process. Initially, the intensity of reflected light was very low due to scattering by the carbon seeds on the surface. It increased rapidly as the substrate temperature was raised to about 1100°C where it reached a saturated value, indicating the formation of a molten surface layer.

Fig.4 shows scanning electron microscope images of oriented diamond films deposited by HFCVD on single crystal (111) and (100) Ni, respectively. One can see that the diamond facets are oriented to each other both axially and azimuthally. SEM micrographs of oriented diamond films obtained from graphite powder and gas phase CH_4 seeding are shown in Fig.5. The substrate saturated with gaseous carbon species at high temperatures yielded the highest and most uniform nucleation density of diamond. The nucleation density of diamond on the graphite-seeded substrate was low, possibly due to the partial gasification of the powders through reaction with atomic hydrogen, although relatively heavy seeding was employed.

A representative Raman spectrum is shown in Fig.6. The high-quality of the diamond is confirmed by the present of 1-phonon band at 1332 cm^{-1} and the absence of graphitic peaks between 1500 cm^{-1} to 1800 cm^{-1}. A very small amount of graphite (no amorphous sp^2C) is observed on the substrate areas not covered by the diamond nuclei.

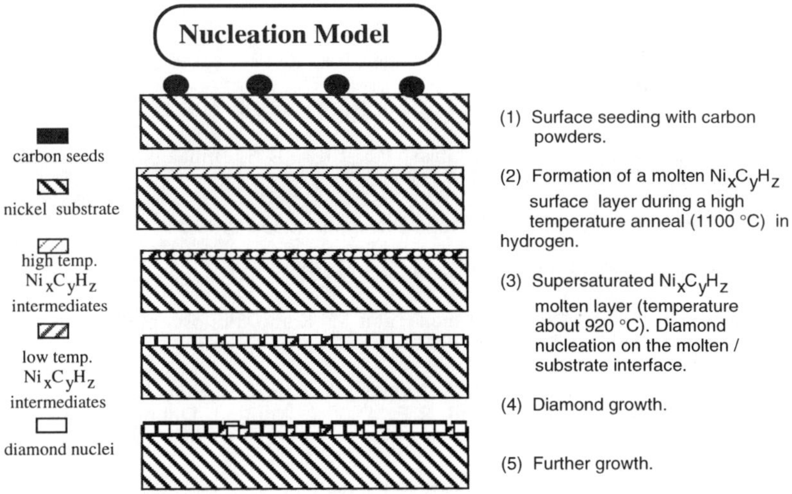

Fig.7 Nucleation model of the seeding and multi-step process.

Based on experimental evidence, the model shown schematically in Fig.7 is proposed to explain the mechanism of oriented nucleation and growth of diamond on nickel substrates. The process starts with seeding a single crystal nickel substrate with carbon species. When the seeded substrate is heated in presence of atomic hydrogen, reactions among nickel, carbon, and hydrogen occur and eventually lead to the formation of a molten Ni-C-H compound on the substrate surface. During the subsequent cooling to a substrate temperature of 900-950°C, diamond nucleates from the molten eutectic compound which is supersaturated with carbon. The diamond nuclei grow with an epitaxial relationship to the substrate which has a very close lattice match to that of diamond. In the continuation of the process, epitaxial diamond is grown on the obtained nuclei, using normal CVD diamond growth conditions, resulting in large area oriented diamond particles. The formation of a molten Ni-C-H surface layer during the high temperature anneal in atomic hydrogen atmosphere is crucial for this process. The presence of atomic hydrogen appears to depress the eutectic point of Ni-C to below 1100°C.

CONCLUSION

Oriented diamond films were grown on both (100) and (111) single crystal Ni substrates using a multi-step method. *In-situ* laser reflectometry was used to monitor and control the process. A model was proposed to explain the nucleation mechanism. The key to the process appears to be a molten surface layer of Ni-C-H. The presence of atomic hydrogen dramatically reduced the melting temperature of this phase, which facilitated the dissolution of seeded carbon into nickel. Diamond nuclei form from the supersaturated solution which are oriented with respect to the original nickel substrate.

ACKNOWLEDGMENTS

The financial support of Army Research Office under grant number DAAH04-93-D-0003 is gratefully acknowledged. Helpful discussion with Dr. R.J. Nemanich is greatly appreciated.

REFERENCES

1. P.C. Yang, W. Zhu, and J.T. Glass, J. Mater. Res. **8**, 1773 (1993).

2. P.C. Yang, W. Zhu, J.T. Glass, U.S. Patent No. 5 298 286 (1994).

3. S.D. Wolter, B.R. Stoner, J.T. Glass, P.J. Ellis, D.S. Buhaenko, C.E. Jenkins, and P. Southworth, Appl. Phys. Lett. **62**, 1215 (1993).

4. D.N. Belton and S.J. Schmieg, J. Appl. Phys. **66,** 4233 (1989).

5. R.H. Wentorf, Jr., Adv. Chem. Phys. **9**, 365 (1965).

6. Y. Sato, H. Fujita, T. Ando, T. Tanaka, and M. Kamo, Philoscophical Transactions of the Royal Society of London **A342**, 225 (1993).

7. P.C. Yang, T.J. Kistenmacher, D.A. Tucker, W. Liu, F.R. Sivazlian, S.P. Bozeman, B.R. Stoner, J.T. Prater, J.T. Glass and R.F. Davis, Applied Diamond Conference 1995, Gaithersburg, MD. P.329.

8. K. Kobashi and T. Tachibana, Private communication.

9. H.S. Dewan, D. Ravichandran, J.P. Cheng, W.R. Drawl, K.A. Cherian and R. Roy, Proc. of the Applied Diamond Conference 1995, Gaithersburg, MD, p.387.

10. Operation Manural of Polishing, Buehler.

11. W.Liu, P.C. Yang, D.A. Tucker, C.A. Wolden, R.F. Davis, J.T. Glass, J.T. Prater and Z. Sitar, to be published in the proceeding of the MRS Spring, 1996, San Francisco, CA.

NUCLEATION AND GROWTH OF GALLIUM NITRIDE FILMS ON Si AND SAPPHIRE SUBSTRATES USING BUFFER LAYERS

N.R. Perkins,† M.N. Horton,† D. Zhi,† R.J. Matyi,¥ Z.Z. Bandic,* T.C. McGill* and T.F. Kuech‡
†University of Wisconsin, Materials Science Program, Madison, WI
‡University of Wisconsin, Dept. of Chemical Engineering, Madison, WI
¥University of Wisconsin, Department of Materials Science and Engineering, Madison, WI
*California Institute of Technology, Department of Applied Physics, Pasadena, CA

ABSTRACT

We have investigated the nucleation and growth of gallium nitride (GaN) films on silicon and sapphire substrates using halide vapor phase epitaxy (HVPE). GaN growth was carried out on bare Si and sapphire surfaces, as well as on MOVPE-grown GaN buffer layers. HVPE growth on MOVPE GaN/AlN buffer layers results in lower defect densities as determined by x-ray than growth directly on sapphire. HVPE GaN films grown directly on sapphire exhibit strong near-edge photoluminescence, a pronounced lack of deep level-based luminescence, and x-ray FWHM values of 16 arcsec by an x-ray θ–2θ scan. The crystallinity of GaN films on sapphire is dominated by the presence of rotational misorientation domains, as measured by x-ray ω-scan diffractometry, which tend to decrease with increasing thickness or with the use of a homoepitaxial MOVPE buffer layer. The effect of increasing film thickness on the defect density of the epilayer was studied. In contrast, the HVPE growth of nitride films directly on silicon is complicated by mechanisms involving the formation of silicon nitrides and oxides at the initial growth front.

INTRODUCTION

Despite recent technological advances in nitride epitaxy, mechanisms controlling the crystallinity of nitride layers on heteroepitaxial substrates are not well understood. The application of low temperature buffer layers has been shown to dramatically improve the crystallinity of epitaxial GaN on sapphire (1,2); however, the current state of technology for nitride epitaxy still lags well behind that of Si and GaAs. There are several sources of defects in GaN grown on sapphire. Defects in the material are generated, such as threading dislocations, due to lattice mismatch at the heterointerface. The threading dislocation density is expected to decrease with increasing thickness of the film, due to occasional defect annihilation. An additional significant type of defect is the boundary between rotationally misoriented domains within the GaN epilayer. The application of a suitable buffer layer acts to decrease the degree of rotational misorientation in the film (3). The nature and extent of development of the microstructure, dominated by these defects as the film thickness is increased, has not been previously investigated in detail. In this work, we utilize two nitride growth techniques, metalorganic vapor phase epitaxy (MOVPE) growth and halide vapor phase epitaxy (HVPE) growth, to quantify differences in epitaxial GaN arising from the presence of rotational misorientation domains and threading dislocations as a function of film thickness and buffer layer.

From a technological viewpoint, a promising route for the development of GaN substrates is the heteroepitaxial growth of GaN films by rapid growth techniques, such as halide vapor

Mat. Res. Soc. Symp. Proc. Vol. 423

phase epitaxy (HVPE), followed by the *in-situ* etch removal of the initial substrate to leave a thick free-standing GaN film. The availability of GaN homoepitaxial substrates would significantly reduce current problems in heteroepitaxial nucleation and eliminate defects arising from differences in the coefficient of thermal expansion between the epilayer and the substrate.

EXPERIMENTAL PROCEDURE

The HVPE technique has been discussed widely in the literature (4,5,6,7). Details concerning our HVPE growth of GaN have been presented elsewhere (8). MOVPE buffer layers and growth studies were carried out in a horizontal MOVPE system, operated at 76 Torr, utilizing trimethylaluminum, trimethylgallium, and ammonia as the reactants. Typical gas velocities were over 30 cm/sec, with cation precursor partial pressures on the order of 0.011 Torr, and a V/III ratio of 1500. Low temperature AlN buffer layers were utilized for MOVPE growths of GaN on sapphire, in accordance with the techniques reported by other researchers (1). For growths of AlN on Si(111) substrates, the Si substrates were initially oxidized, with the oxide removed by HF etch just prior to growth. MOVPE AlN layers were grown on Si(111) using trimethylaluminum and ammonia as the precursors at 1100°C, similar to the method reported by Watanabe *et al* (9).

Both ω-scan (transverse) and θ-2θ scan (longitudinal) x-ray diffraction geometries were used to characterize the GaN films. The x-ray apparatus was equipped with a four-reflection monochrometer and a three-reflection Si analyzer crystal at the detector. In the ω-scan geometry, the detector was held at a fixed position, and the sample was tilted during the scan. In the θ-2θ geometry both the sample and detector were moved, with the detector moving at twice the angular velocity of the sample.

RESULTS AND DISCUSSION

The growth rate of GaN, based on the apparent activation energy for growth and the dependence on GaCl partial pressure, previously reported (8), appears to be limited by mass transport of GaCl to the growth front rather than by surface reactions. The growth thickness uniformity was ±8% over a 2.5 cm sample, with a smooth specular surface. Attention to surface cleaning and higher growth rates were found to improve the growth morphology. Higher GaCl partial pressures may increase the rate of nucleation, leading to a higher density of GaN nuclei on sapphire, and transition to the 2-D growth mode at a lower thickness. The theoretical relationship between the heterogeneous nucleation rate and the supersaturation for the general case is widely discussed in the literature (10).

HVPE GaN layers on sapphire were analyzed using X-ray scans in the ω-rocking (transverse) and θ-2θ rocking geometries (longitudinal). The ω-scan values always exhibit a much broader FWHM to the peak than the θ-2θ scan (i.e., 636 arcsec versus 16 arcsec for a 110 μm thick HVPE GaN/sapphire sample). Broadening in the ω-scan is due to composite broadening from both lattice parameter variation and the presence of large angular misorientation domains (mosaic spread) in the epitaxial film. The θ-2θ scan measures broadening primarily due to variation of the lattice parameter in the material (11). In all samples, regardless of thickness, angular misorientation domains dominate the GaN microstructure. While this conclusion has been reported in the growth of MOVPE GaN, our measurements indicate a similar defect structure in the HVPE-grown materials. The relative magnitude of broadening due to variation in the c-axis lattice parameter as compared to the effect of the in-plane angular distribution of GaN domains is demonstrated in Figure 1, which illustrates a reciprocal space map for this material.

Low temperature (7K) photoluminescence studies of thick GaN/sapphire HVPE films obtained with an excitation laser power density of 5 W/cm^2 possess a negligible amount of yellow luminescence, and very strong near-band transitions at 3.48 eV (FWHM value 10.4 meV). This photoluminescence spectrum has been published elsewhere (8). It is interesting to note that although other HVPE (and MOVPE) samples generally demonstrate strong yellow luminescence (12,13,14), the films in this study demonstrate very little luminescence in the 550 nm region. Thus, the lack of yellow luminescence in these samples is not an inherent feature of the HVPE growth technique. Moreover, the x-ray FWHM values of the HVPE materials in this study are not dramatically different from high quality MOVPE GaN, suggesting that the presence of extended defects (such as low angle grain boundaries) are probably not directly responsible for the presence of yellow luminescence.

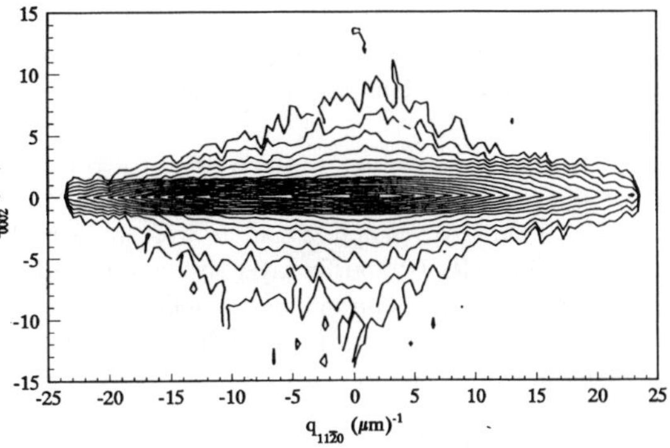

Figure 1: Reciprocal space map for 110 μm GaN/sapphire, T_{gr}=1050°C (min contour=$10^{-0.25}$ counts, max=$10^{3.75}$ counts, 17 contours)

In many epitaxial systems, the density of threading dislocations and rotational misorientation domains decreases with increasing film thickness (15). X-ray measurements in the ω scan compared to the θ-2θ scan geometries were completed for a series of HVPE GaN/sapphire films grown at different thicknesses. The results of this analysis are shown in Figure 2. In the (0002) reflection, the x-ray penetration depth is estimated to be on the order of 20 μm (I/I_o=0.1). These thick HVPE layers are well suited for a study of the change in defect structure with thickness. The dominant change in defect structure in the HVPE GaN/sapphire samples with thickness is the angular misorientation of the GaN domains. The variation in the c-axis lattice parameter is fairly small and changes little with increasing thickness (θ-2θ scan), while the FWHM component attributed to the rotational misorientation

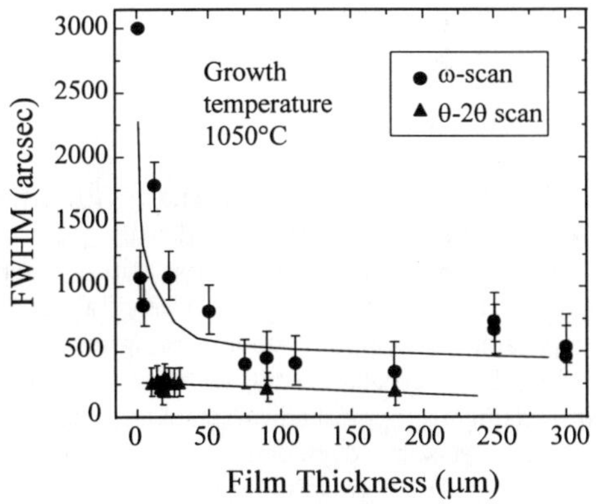

Figure 2: GaN/sapphire crystallinity as a function of HVPE GaN thickness (note detector aperture ~ 0.3 mm on θ-2θ scan)

decreases from several thousands of arcsec (ω scan), at low layer thickness, to 300-500 arcsec for layer thickness above 100 μm (see Figure 2). This result indicates that for all HVPE GaN/sapphire samples the dominant defect structure is the presence of rotational misorientation domains. The contribution to the x-ray peak broadening from these domain or grain boundaries decreases with increasing film thickness, possibly due to ripening growth in the domain size or other routes for annihilation of the extended defects within the domain or comprising the domain boundary.

Based on an analogy to the MOVPE literature, the presence of a suitable buffer layer is expected to significantly reduce the defect density of HVPE GaN layers. To quantify this effect, an MOVPE GaN/AlN/sapphire sample was produced for use as an HVPE buffer layer (starting substrate). The MOVPE buffer layer structure provides a homoepitaxial surface for the initiation of the HVPE growth, eliminating difficulties with the initial nucleation on the sapphire and leading directly to a 2D growth mode. Consequently, samples produced using an MOVPE GaN/AlN/sapphire substrate for the HVPE growth may result in samples with a reduced defect density. Figure 3 illustrates the ω-scan curves for HVPE GaN films grown on MOVPE GaN/AlN/sapphire buffer layers, as compared with the ω-scan curves for HVPE GaN films grown directly on sapphire as well as the initial MOVPE GaN/AlN/sapphire films. The HVPE GaN film grown on the MOVPE buffer layer exhibits a much smaller FWHM than the thicker HVPE film grown without the buffer layer. The MOVPE process, and specifically the application of a low temperature buffer layer, clearly provides a mechanism for the reduction in the rotational spread of growth domains at a low film thickness (1,2). The HVPE growth on the MOVPE buffer layer has a higher FWHM and hence higher defect density compared to the MOVPE buffer layer itself. Several possible mechanisms could give rise to this observation. First, due to the large difference in the thermal expansion coefficient between GaN and sapphire, cracks will appear in both the GaN, and possibly the sapphire, upon cooling for GaN layers above ~20 μm in thickness. Significant cracking has been observed in such films. As a result, inhomogeneities and misalignments in the film arising from mechanical cracking can contribute to the measured FWHM of the film. Additionally, some degradation of the MOVPE buffer surface may be occurring prior to the initiation of HVPE growth, due to contamination of the sample during transfer from the MOVPE reactor, or loss of nitrogen from the MOVPE buffer during the HVPE heatup sequence in ammonia. Optimized pre-growth preparation must be developed to fully utilize these GaN buffer layers.

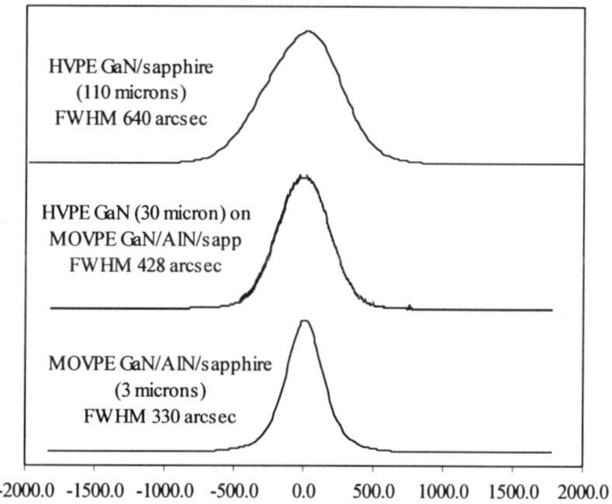

Figure 3: X-ray ω-scan curves for GaN as a function of growth technique and buffer layer

HVPE growths of GaN were also carried out on (111) Si substrates, utilizing a MOVPE AlN buffer or by growth directly on the Si surface. The threefold symmetry of the (111) Si leads to the growth of wurtzite GaN (9). Moreover, Si substrates are of interest because a convenient route exists for removal of the Si substrate *in-situ* to leave a thick free standing GaN film; *e.g.*, sacrificial etch of the backside of the Si using HCl gas. This approach has the potential advantage of producing free-standing GaN substrates free from strain and cracks induced by differences in the heteroepitaxial coefficient of thermal expansion. Initial attempts to deposit GaN directly on (111) Si resulted in polycrystalline deposition, probably due to either direct nitridization of the (111) Si surface during the HVPE heatup in nitrogen (16) or due to reaction of chlorides in the HVPE system with the Si surface, providing volatile silicon chlorides which subsequently react with ammonia to form silicon nitrides at the epitaxy surface. In fact, the formation of silicon nitride through such volatile silicon chlorides is supported by the highly favorable free energy of formation (-130 kJ/mole $SiCl_4$ at 1050°C). Additionally, ammonia can react directly with the solid Si surface to form silicon nitrides (17). To alleviate the reactivity of the Si substrate in the HVPE growth environment and provide an improved surface for nucleation, MOVPE AlN buffer layers were deposited on the Si substrates. A key aspect of the AlN/Si deposition was the growth temperature. AlN buffer layers deposited at temperatures of less than 1050°C in this study were not found to be single crystal (also, see ref (9)). However, as observed by other researchers (9), films deposited at temperatures exceeding 1100°C led to epitaxial single crystal growth. The high growth temperature may lead to reduced oxygen contamination of the Si surface through *in-situ* reduction of the Si surface by hydrogen prior to the nucleation of the AlN buffer. AlN buffer layers exceeding a few tenths of microns were found to exhibit cracks with threefold symmetry. Theoretical calculation of the stress in the AlN/Si film (18) indicate a high tensile stress, on the order of 10^8-10^9 dynes/cm^2 in the AlN film, arising from the difference in the coefficient of thermal expansion between the AlN and Si.

HVPE GaN films deposited on AlN/Si buffers were observed to be smooth and specular except in areas where the AlN buffer layer was cracked. Near the cracked areas, a rough granular morphology was present. Auger analyses confirmed the presence of Ga, N, Si, Al, and O. We postulate that these granular deposits form as a consequence of attack of the exposed Si by HCl, leading to silicon nitrides through the following reactions:

$$Si + xHCl \rightarrow SiCl_x + \frac{x}{2} H_2$$

$$SiCl_x + yNH_3 \rightarrow SiN_y + xHCl + \frac{3y - x}{2} H_2 \cdot$$

CONCLUSIONS

The film properties for GaN grown by the HVPE technique as a function of substrate and buffer layer structure have been examined. The defect structure of the films is found to be dominated by the presence of rotational misorientation domains. Reduction in the defect density is observed with increasing thickness of the HVPE films, primarily due to reduction in the degree of rotational misorientation in the films. Thus, some mechanism for domain ripening or other reduction in the degree of rotational misorientation must be occurring with increasing film thickness. The homoepitaxial nucleation and growth of the HVPE film on a GaN MOVPE buffer layer results in decreased angular spread in the rotational misorientation domains as indicated by the reduced FWHM values in the x-ray ω-scan measurements. The HVPE layers grown on

MOVPE buffers remain structurally inferior to initial MOVPE GaN/AlN/sapphire layers, possibly due to additional complication induced by mechanical fracture in the thick HVPE films. Excellent optical and crystalline characteristics are obtained for the HVPE films on sapphire, with longitudinal x-ray scans of GaN/sapphire indicating a FWHM of 16 arcseconds for a 110 μm thick film by θ-2θ scan. The 7K photoluminescence for GaN/sapphire shows strong near-band emission at 3.48 eV with a FWHM value of 10.4 meV. Moreover, no detectable deep level emission was found in PL measurements at modest pump power densities. For HVPE GaN growth on (111) Si substrates, a high temperature AlN/Si MOVPE buffer layers were required to protect the growth surface from silicon nitride formation.

ACKNOWLEDGMENTS

The authors gratefully acknowledge the support of the ARPA URI on Visible Light Emitters and the Naval Research Laboratory for various aspects of this work.

REFERENCES

1. I. Akasaki, H. Amano, Y. Koide, K. Hiramatsu, and N. Sawaki, J. Crystal Growth **98**, 209-219 (1989).
2. S. Nakamura, Jap. J. Appl. Phys. **30**(10A), L1705-L1707 (1991)
3. Y. Koide, N. Itoh, K. Itoh, N. Sawaki, and I. Akasaki, Jap. J. Appl. Phys. 27(7), 1156-1161 (1988).
4. H. Maruska and J. Tietjen, Appl. Phys. Lett. **15**, 327 (1969)
5. T. Detchprohm, K. Hiramatsu, H. Amano, and I. Akasaki, Appl. Phys. Lett **61**, 2688-2690 (1992).
6. O. Lagerstedt and B. Monemar, J. Appl. Phys. **45,** 2266-2272 (1974).
7. J.I. Pankove, J.E. Berkeyheiser, and E.A. Miller, J. Appl. Phys. **45,** 1280-1286 (1974).
8. N.R. Perkins, M.N. Horton, and T.F. Kuech, (Proc. Mat. Res. Soc., Boston, MA, 1995, paper AAA2.9), to be published.
9. A. Watanabe, T. Takeuchi, K. Hirosawa, H. Amano, K. Hiramatsu, and I. Akasaki, J. Crystal Growth **128**, 391-396 (1993).
10. For example, see A. Milchev, S. Stoyanov, and R. Kaischew, Thin Solid Films, **22**(22), 255 (1974).
11. I. Nikitina and V. Dmitriev, (Inst. Phys. Conf. Ser. 141, San Diego, CA, 1994), 431-436.
12. H.G. Grimmeiss and B. Monemar, J. Appl. Phys. **41**, 4054 (1970).
13. T. Suski, P. Perlin, H. Teisseyre, M. Leszczynzki, I. Grzegory, J. Jun, M. Bockowski, S. Porowski, and T.D. Moustakis, Appl. Phys. Lett **67**(15), 2188-2190 (1995).
14. P. Hacke, A. Maekawa, N. Kiode, K. Hiramatsu, and N. Sawaki, Jpn. J. Appl. Phys **33**(1, 12A), 6443-6447 (1994).
15. T.F. Kuech, A. Segmuller, T.S. Kuan, and M.S. Goorsky, J. Appl. Phys. **67**(10), 6497-6506 (1990).
16. T. Ito, S. Hijiya, T. Nozaki, H. Arawaka, M. Shinoda, and Y. Fukukawa, J. Eectrochem. Soc. **125**(3), 448-452 (1978).
17. C.-Y. Wu, C.-W. King, M.-K. Lee, and C.-T. Chen, J. Electrochem. Soc. **129**(7), 1559-1563 (1982).
18. N. Perkins and T.F. Kuech, manuscript in preparation.

SEEDING WITH A DIAMOND SUSPENSION FOR GROWTH OF SMOOTH POLYCRYSTALLINE DIAMOND SURFACES

I. ST. OMER, T. STACY, E. M. CHARLSON AND E. J. CHARLSON
University of Missouri, Department of Electrical Engineering, Columbia, MO 65211

ABSTRACT

A number of techniques have been used to smooth polycrystalline diamond films. Recent work in substrate seeding with nanocrystalline diamond powder, alone or in a carrier fluid, has shown that diamond seeding improves nucleation density and reduces diamond surface roughness. In this work, silicon substrates were seeded using a commercially available water-based 0.1 micrometer diamond polishing suspension. Growth was achieved using conventional hot-filament chemical vapor deposition (HFCVD). Films were characterized using optical microscopy, scanning electron microscopy (SEM), x-ray diffraction (XRD), and surface profilometry. The resulting diamond films exhibited well-faceted crystals, small grain size and minimal surface roughness. Additionally, the silicon substrate was chemically etched in order to permit examination of the backside of the diamond film. Results show that the diamond surface at the silicon-diamond interface is ultra-smooth. Comparison of the backside of these surfaces with those prepared using conventional diamond grit abrasion indicates that a significant improvement in surface quality is achieved using this diamond seeding technique.

INTRODUCTION

Diamond materials hold great promise for application in microelectronics because of their unique thermal, mechanical, chemical and electronic properties.[1-3] However, progress in this area has been limited by the difficulties associated with doping, patterning, and reproducible growth of smooth, uniform diamond films fabricated using chemical vapor deposition (CVD). Control of nucleation is critical to obtaining the grain size, orientation, surface roughness and adhesion required for optimal application of CVD diamond. Recent studies indicate substrate seeding with nanocrystalline diamond powder enhances nucleation density and surface roughness without any subsequent postdeposition treatment.[4-7] Yang and Aslam reported mean surface roughness of 30 nm and 124 nm over a 3 μm^2 area for 1μm thick films seeded using diamond powder in water and photoresist, respectively.[4] This work demonstrates a seeding technique which uses a commercially available water-based diamond polishing suspension containing 0.1 μm natural diamond particles. The accessibility and relative low-cost of the diamond suspension

Mat. Res. Soc. Symp. Proc. Vol. 423

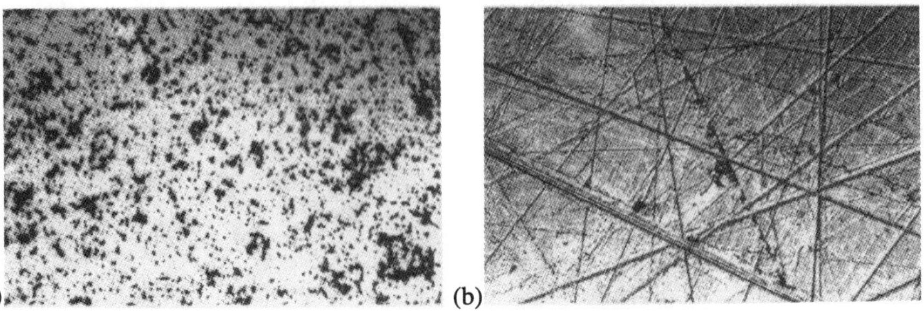

(a) (b)

Figure 1: (a) Seeded silicon substrate (b) Scratched silicon substrate prior to growth. Magnification is 60x.

make this seeding methodology an attractive alternative to diamond powder. Additionally, as evidenced by the exposed silicon-diamond interface, this technique does not cause any damage to the substrate surface.

EXPERIMENT

Boron-doped (100) silicon wafers with resistivities in the range of 14-16 Ω cm were used as substrates. Prior to seeding, the silicon was chemically cleaned and baked at 80 °C in air to insure dehydration. A 2:1 solution of the diamond suspension and isopropyl alcohol was applied to the substrate spinning at approximately 3200 rpm. The seeded samples were then returned to the dehydration oven to evaporate any remaining fluids. The process was then repeated to achieve optimum uniformity. For comparison, additional silicon substrates were prepared by abrading the surface with a 0.25 μm diamond paste. Following abrasion, the samples were rinsed in deionized (DI) water and allowed to air dry. Figure 1 contains optical photographs of both seeded and scratched substrates.

Films were grown on the prepared substrates using the HFCVD method with hydrogen and methane as the reactant gases. Substrates were mounted on a molybdenum base about 7 mm beneath treated tungsten filaments. Filaments were heated to approximately 2000 °C as measured with an optical pyrometer. During the deposition, the substrate temperature was maintained at approximately 765 °C and the pressure at 28-32 Torr. The flow rates were as follows: 148.0 sccm of hydrogen and 2.0 sccm of methane.

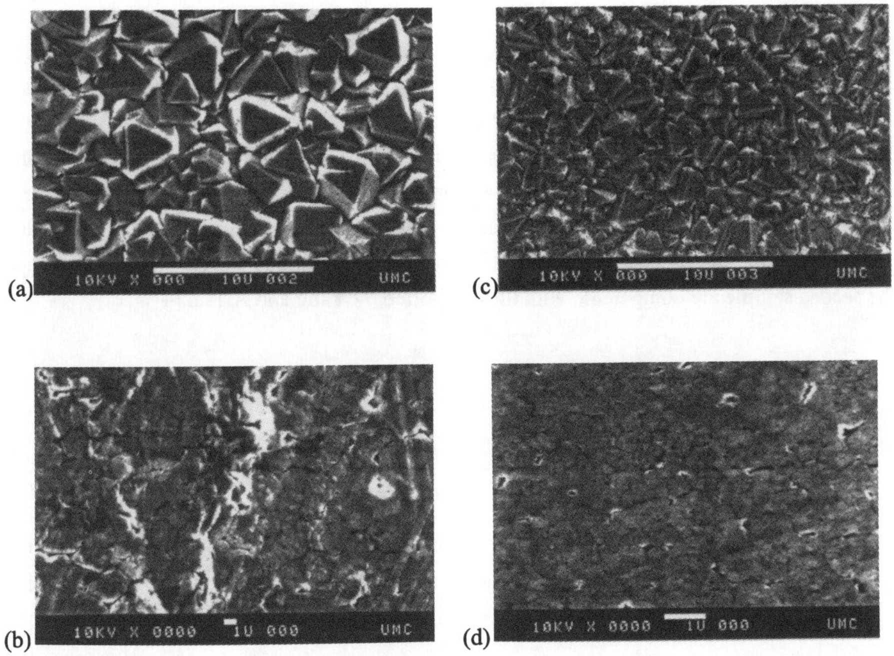

Figure 2: (a) Front of film grown on scratched substrate (b) Back of film grown on scratched substrate (c) Front of film grown on seeded substrate (d) Back of film grown on seeded substrate. The white bar at the bottom represents a scale of 10 µm for (a) and (c), and 1 µm for (c) and (d).

After removal from the growth chamber, the films were then immersed in a saturated solution of CrO_3+H_2SO_4 at 200 °C followed by a rinse in boiling H_2O_2+HN_4OH to remove any nondiamond surface layer. The silicon substrate was then chemically etched using CP4A (3:HF, 5:HNO_3, 3:acetic acid) and a Teflon mask as outlined by Cardinale and Tustison.[8]

RESULTS AND DISCUSSION

Figure 2 contains SEM images of the front and back of the diamond film grown on the seeded and scratched substrates. Both films exhibit well-faceted triangular crystals consistent with (111) diamond growth. The front of the seeded sample has a smoother surface and finer grain size. More significant is the difference in surface roughness observed at the exposed diamond-silicon interface. Scratch profiles are clearly visible in Figure 2(b) whereas the seeded surface in Figure 2(d) reflects that of the polished silicon substrate.

The crystal structure of the diamond films was confirmed using XRD. Figure 3 shows the XRD results for a seeded sample approximately 10 μm thick. As expected from the SEM images, (111) is the dominant peak with smaller (220), (311) and (400) peaks also visible. These peaks correspond to a lattice parameter of 3.566 ± 0.002 Å.

An average roughness for each surface was obtained for a 400 μm profilometer trace on films of approximately 5 μm in thickness using a Tencor Instruments Alpha-step 200. Figure 4 contains traces typical of the front and back surfaces of both the seeded and scratched samples. The average roughness values of 765 Å from the front surface, and 325 Å from the back surface, for the seeded sample are comparable with those reported by Yang and Aslam.[4]

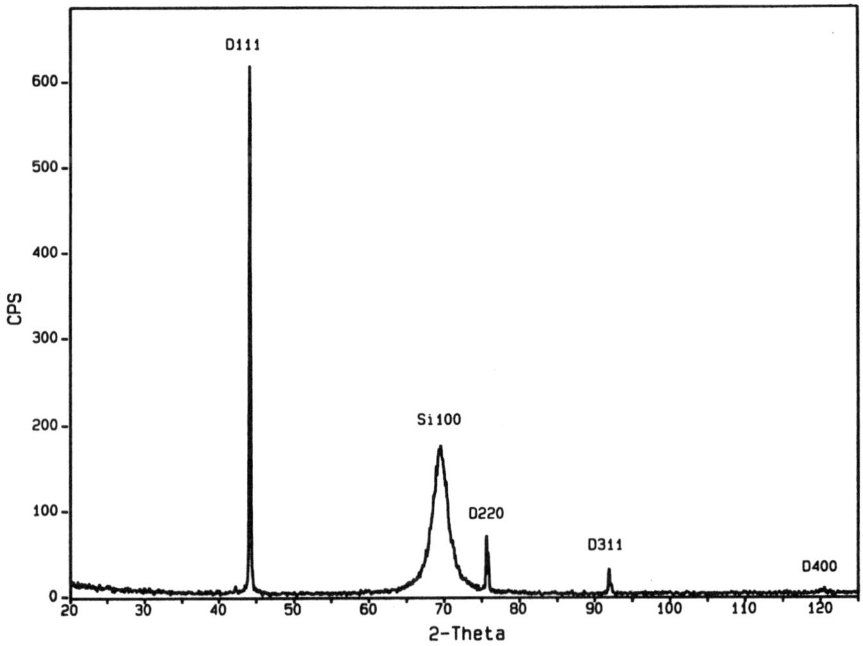

Figure 3: XRD results typical of seeded film approximately 10 μm thick.

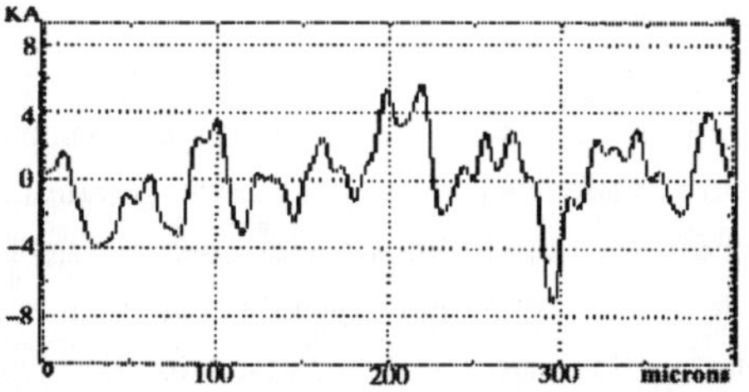

(a) Scratched sample D25 top surface; Average Roughness Ra=1.840 kÅ. Total Indicated Run TIR=12.80 kÅ.

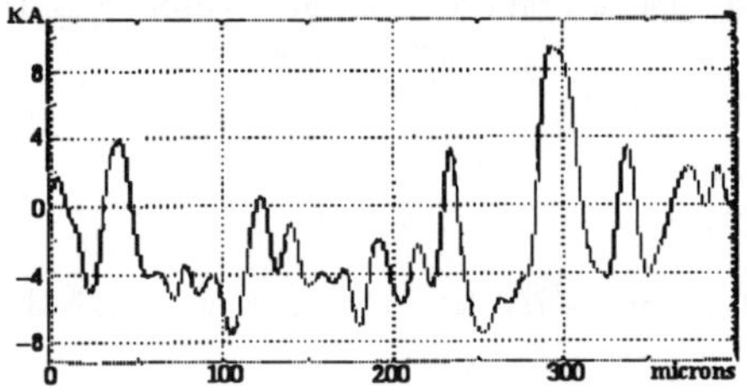

(b) Scratched sample D25 back surface; Average Roughness Ra=2.910 kÅ. Total Indicated Run TIR=16.85 kÅ.

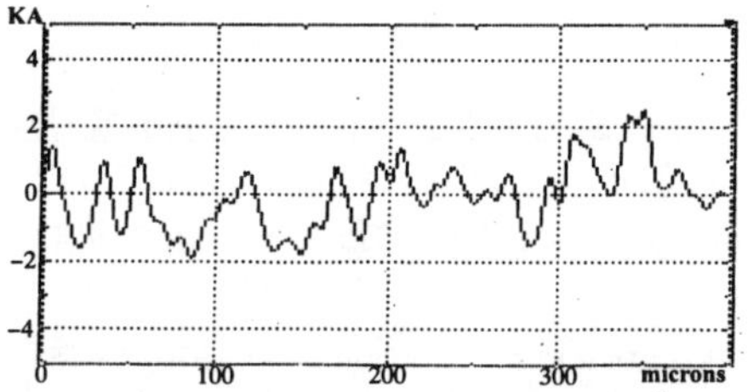

(c) Seeded sample EV2 top surface; Average Roughness Ra=765 Å. Total Indicated Run TIR=4.355 kÅ.

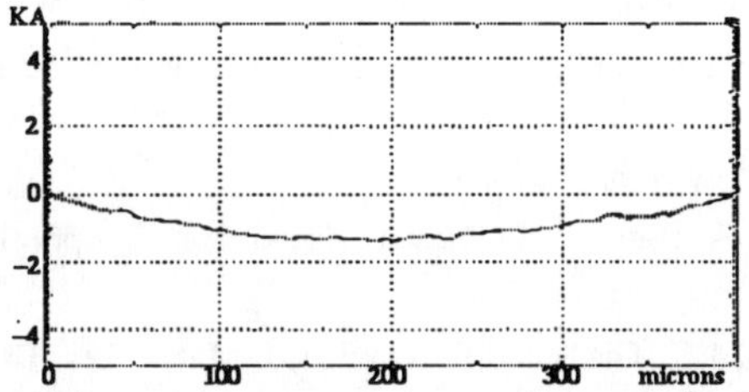

(d) Seeded sample EV2 back surface; Average Roughness Ra=325 Å. Total Indicated Run TIR=1.360 kÅ.

Figure 4: Typical profilometer traces, 400 μm in length, for a scratched and seeded sample on the top and back surfaces. Total Indicated Run is defined as the difference in height between the highest and lowest points in the scan.

CONCLUSION

A diamond seeding technique for growth of CVD diamond has been successfully demonstrated using a commercially available 0.1 μm diamond polishing suspension. Using this methodology, the diamond suspension is thinned with isopropyl alcohol and applied to the substrate prior to growth. The resulting polycrystalline films have well-faceted crystals, good adhesion and minimal surface roughness in comparison to those prepared with conventional diamond abrasion. Additionally, the smooth surface found at the diamond-silicon interface has potential for improved application in standard semiconductor processing such as photolithography.

ACKNOWLEDGMENTS

The authors would like to thank Mr. Louis M. Ross and Mr. Naiyu Zhao for their assistance in the SEM and XRD studies respectively.

REFERENCES

1. M. N. Yoder in Diamond Films and Coatings: Development , Properties, and Applications, edited by R. F. Davis (Noyes Publications, Park Ridge, NJ, 1993) pp. 1-28.
2. J. C. Beam in High-Speed Semiconductor Devices, edited by S. M. Sze (Wiley, New York, 1990) pp. 22-24.
3. G. Sh. Gildenblat, S. A. Grot and A. Badzian, Proc. IEEE **79** (5), 647-668 (1991).
4. G. S. Yang and M. Aslam, Appl. Phys. Lett. **66** (3), 311-313 (1995).
5. H. Makita, T. Yara, N. Jang, K. Nishimura, A. Hatta, T. Ito and A. Hiraki in Applications of Diamond Films and Related Materials: Third International Conference , edited by A. Feldman, Y. Tzeng, W. A. Yarbrough, M. Yoshikawa and M. Murakawa (NIST, Gaithersburg, MD, 1995) pp. 333-336.
6. T. Yara, H. Makita, A. Hatta, T. Ito and A. Hiraki, Jpn. J. Appl. Phys. **34**, L312-L315 (1995).
7. W. Zhu, P. C. Yang, J. T. Glass and F. Arrezzo, J. Mater. Res. **10**, 1455-1459 (1995).
8. G. F. Cardinale and R. W. Tustison in Diamond Optics III: Proceedings of SPIE-The International Society for Optical Engineering edited by A. Feldman and S. Holly (International Society for Optical Engineering **1325**, Bellingham, WA, 1990) pp. 90-97.

PROPERTIES OF TETRAHEDRAL AMORPHOUS CARBON DEPOSITED BY A FILTERED CATHODIC VACUUM ARC

M Chhowalla*, C W Chen, B Kleinsorge, J Robertson, G A J Amaratunga*, W I Milne,
Dept of Engineering, Cambridge University, Cambridge CB2 1PZ, UK.
* present address, Electronic Engineering Dept, Liverpool University, L69 3BX, UK

ABSTRACT

The properties of a highly sp^3 bonded form of amorphous carbon denoted ta-C deposited from a filtered cathodic vacuum arc (FCVA) are described as a function of ion energy and deposition temperature. The sp^3 fraction depends strongly on ion energy and reaches 85% at an ion energy of 100 eV. Other properties such as density and band gap vary in a similar fashion, with the optical gap reaching a maximum of 2.3 eV. These films are very smooth with area roughness of order 1 nm. The sp^3 fraction falls suddenly to almost zero for deposition above about 200°C.

INTRODUCTION

Diamond-like carbon (DLC) is of interest for the formation of transparent, hard films. DLC is a metastable, higher density form of amorphous C (a-C) or hydrogenated amorphous C (a-C:H) containing a significant fraction of tetrahedrally coordinated sp^3 C-C bonding [1]. It is known that a highly sp^3 bonded form of a-C denoted tetrahedral amorphous carbon, or ta-C is obtained by deposition from a highly ionized beam of medium energy, from for example a filtered ion beam [2], the laser ablation of graphite [3] or from a filtered cathodic vacuum arc (FCVA) [4]. The most complete form of ion filtering is found in mass selected ion beam deposition (MSIB). On the other hand the FCVA is a lower cost source of ta-C suitable for laboratory or industrial use. This paper describes the dependence of the properties of ta-C on deposition conditions, in particular the ion energy and deposition temperature.

EXPERIMENTAL DETAILS

In the FCVA, an arc is struck on a graphite target, forming a highly ionised plasma. The plasma beam is led around a 90° solenoid magnetic filter which removes neutral species and particulates. The filter does not provide mass or charge selection. The plasm beam is then condensed on a substrate of Si or quartz held on a temperature controlled copper block. The effective ion energy is varied by applying a DC bias to the substrate, or an RF bias for the quartz substrate.

The sp^2 fraction is found by electron energy loss spectroscopy (EELS), from the C K edge spectrum [5]. The microscopic density is derived from the valence plasmon energy, also measured by EELS. The films can possess a high intrinsic compressive stress, which is derived from the substrate curvature by Stoney's equation. More details of the methods are given elsewhere [6].

RESULTS AND DISCUSSION

The properties of ta-C deposited at room temperature depend strongly on ion

Mat. Res. Soc. Symp. Proc. Vol. 423 © 1996 Materials Research Society

energy. Fig. 1 shows that the sp^3 fraction, density and compressive stress each pass through a maximum as a function of ion energy. The maximum occurs at 60 eV. The maximum sp^3 fraction is 85%, which compares well with that found by others [4,5]. The maximum occurs at an ion energy of 100 eV, slightly lower than found by Fallon [5] by FCVA. A broader maximum is found for ta-C prepared by MSIB by Lifshitz et al [7]. The reason for their broader maximum is not known, but may be related to their slower growth rates. The films are also very smooth with an areal roughness of order 1 nm.

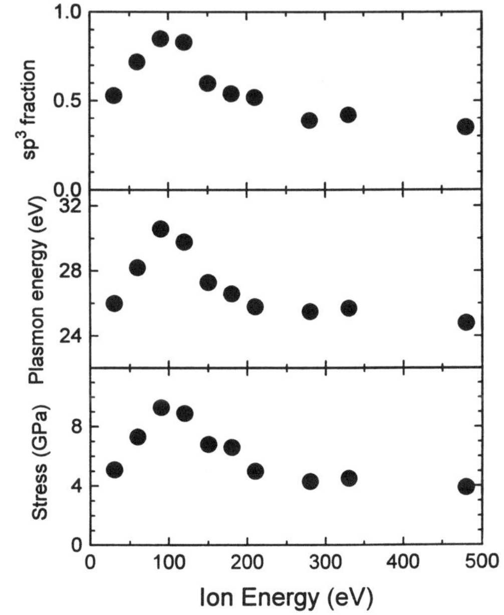

Fig. 1. sp^3 fraction, plasmon energy and compressive stress versus ion energy.

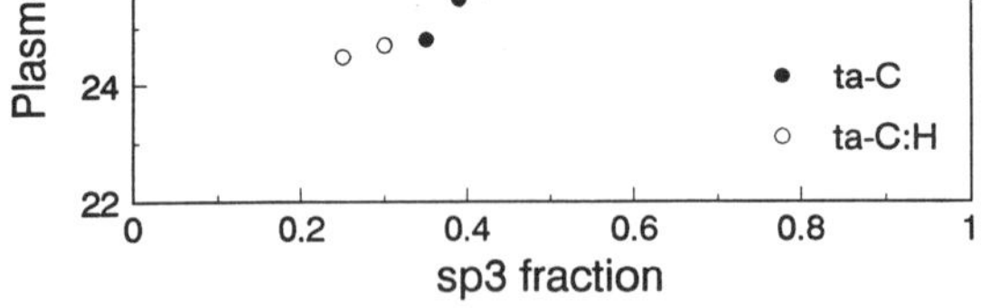

Fig. 2. Correlation of sp^3 fraction and plasmon energy, including data for t-C:H [8].

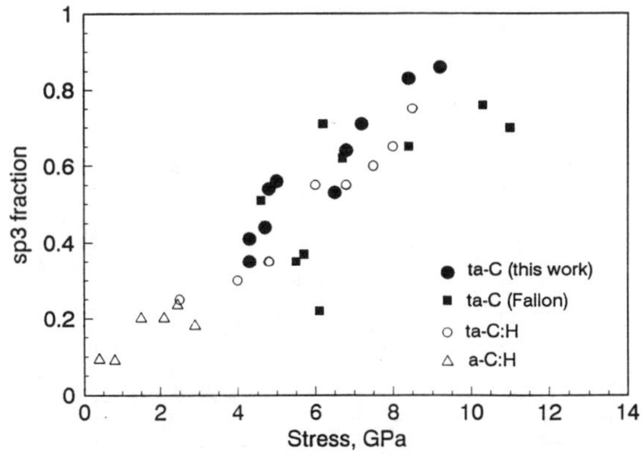

Fig. 3. Correlation of sp^3 fraction and stress, with ta-C, ta-C:H, and a-C:H [5,8,9].

These properties are each strongly correlated to sp^3 fraction. Fig. 2 shows that the sp^3 fraction increases monotonically with the plasmon energy, from which is found an approximately linear variation with the density. This is expected for a random network whose bonding changes continuously from sp^2 to sp^3. Fig. 3 shows the variation of stress with sp^3 fraction. Data for other ta-C samples, ta-C:H deposited from a plasma beam source [8] and for plasma deposited a-C:H [9] are included for comparison. Overall, the stress is seen to vary linearly with sp^3 fraction. Mechanical properties such as hardness also correlate with sp^3 fraction for ta-C [1].

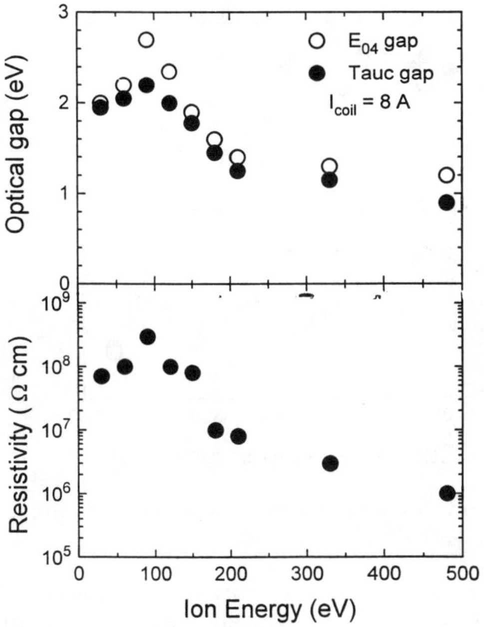

Fig. 4. Variation of optical gap and resistivity with ion energy.

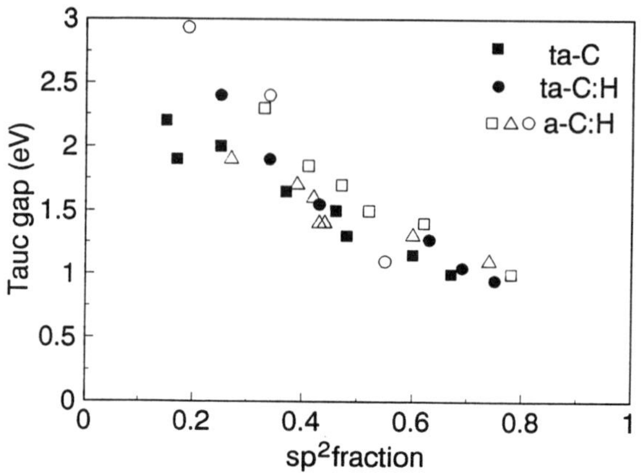

Fig. 5. Correlation of optical gap with sp^2 fraction, including data for ta-C:H [8] and a-C:H [9-11].

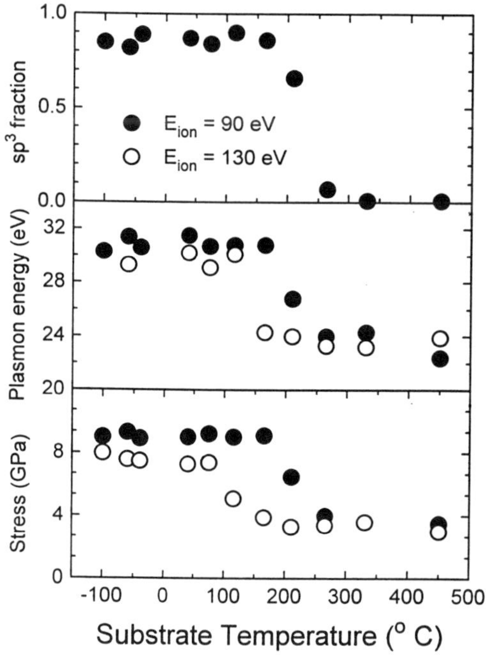

Fig. 6. Variation of sp3 fraction, plasmon energy and stress with deposition temperature for 90 and 130 eV ion energies.

Fig. 4 shows the variation of the resistivity and optical gap (Tauc and E_{04}) for samples deposited on quartz. The resistivity and gap also pass through a maximum with ion energy, with the Tauc gap reaching a maximum of 2.3 eV.

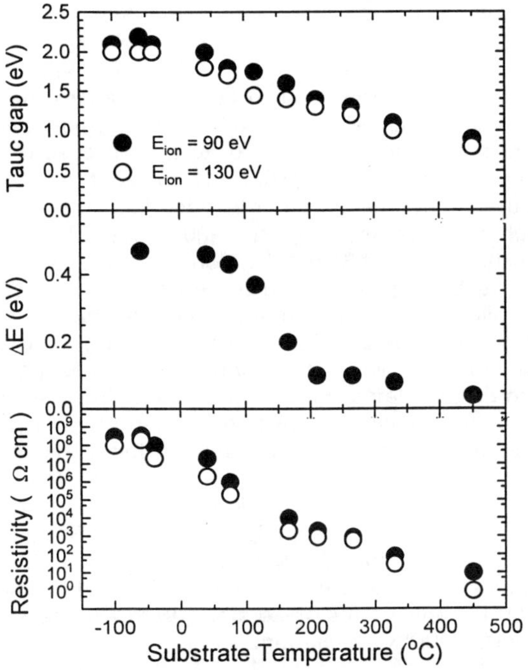

Fig. 7. Variation of optical gap, resistivity and activation energy with deposition temperature.

Fig. 5 shows the variation of the Tauc gap with sp^2 fraction. The gap is seen to increase as the sp^2 fraction decreases. Fig. 5 also includes data for ta-C:H [8] and a-C:H [9-11], which is seen to follow a similar trend, but with a slightly higher gap. This is an important result. It shows that the gap depends basically only on the amount and configuration of sp^2 sites, not directly on the hydrogen content. The band gap in a mixed network of sp^2 and sp^3 sites occurs between the π and π^* states of sp^2 sites. A simplified bonding model suggested that sp^2 sites pair up and may form planar clusters and proposed that the band gap varied inversely with the size of the clusters [12]. The original model is now known to have over-estimated the cluster size. It is now believed that the ion bombardment during deposition limits the cluster size to 2 -10 sp^2 sites, and that the band gap depends on the distortions on the sp^2 configurations as well as the size of the cluster [13]. Fig. 5 confirms that the gap does depend fundamentally on the sp^2 sites, and not directly on the H content.

Fig. 6 shows the variation of the sp^3 fraction, plasmon energy and stress with deposition temperature T_s for an ion energy of 90 eV. The properties are each found to suddenly decrease when T_s exceeds a transition temperature T_1 of about 200°C. Clearly the properties are each strongly correlated. A similar result was found for ta-C deposited by MSIB, but with slightly lower transition temperature [7,14]. The transition temperature T_1 is found to decrease with increasing ion energy. A similar transition to sp^2 bonding was found for ta-C:H [15]. There, a slightly higher value of T_1 was found and it was also found to decrease with ion energy.

Fig. 7 shows the variation of optical gap and conductivity activation energy. These properties are found to decline in a more gradual manner, not showing a sharp transition. The decline starts before T_1. It is possible that the remaining sp^2 sites in ta-C begin to order below the temperature T_1 and this causes the decrease in gap.

Ta-C and all forms of diamond-like carbon are a metastable, higher density form of amorphous C. The deposition mechanism of ta-C is believed to be the subsurface implantation or 'subplantation' of ions [16,2,14]. Incident ions of sufficient energy can penetrate the surface atom layer, enter a subsurface interstitial site where they give rise to a quenched in increase in density and convert the local bonding to sp^3. The penetration threshold is about 30 eV [16]. Ions of lower energy do not penetrate, but stick to the outer surface as sp^2 bonded a-C. Ions of higher energy must lose their excess energy in a collision cascade or a dilute thermal spike of about 10^{-12} s [17], during which time the excess density can relax by thermally activated diffusion. This causes the sp^3 fraction to decline again at higher ion energies. Lifshitz suggests that this decline is due to greater defect formation at higher ion energies [14].

The changes in properties with deposition temperature can also be explained, broadly, within the subplantation model. Higher deposition temperatures increase the rate of relaxation, and a graphitic sp^2 phase suddenly forms, through which interstitials can diffuse with very low migration energy. The deposition temperature has such an effect because it is a significant fraction of the spike temperature. Whereas an idealised thermal spike would start at a single site with a very high temperature, simulations [17] show that in low Z solids at low ion energies, the ion cascade is rather diffuse and so the spike starts with a finite size and rather low temperature of perhaps 4000°K for carbon.

MC wishes to acknowledge the financial support of Multi-Arc Inc.

REFERENCES

1. J. Robertson, Prog Solid State Chem **21** 199 (1991);Pure Applied Chem **66** 1789 (1994)
2. Y. Lifshitz, S. R. Kasi, J. W. Rabalais, Phys Rev Lett **68** 620 (1989)
3. A A Voevodin et al, J Appl Phys **78** 4123 (1995)
4. D R McKenzie, D Muller, B A Pailthorpe, Phys Rev Lett **67** 773 (1991)
5. P J Fallon, V S Veerasamy, C A Davis, J Robertson, G Amaratunga, W I Milne, J Koskinen, Phys Rev B **48** 4777 (1993)
6. M Chhowalla, J Robertson, C W Chen, G A J Amaratunga, J Appl Phys (1996)
7. E Grossman, G Lempert, J Kulik, Y Lifshitz, App Phys Lett **68** 1214 (1996)
8. M Weiler, S Sattel, K Jung, H Ehrhardt, V S Veerasamy, J Robertson, Appl Phys Lett **64** 2797 (1994); Phys Rev B **53** 1594 (1996)
9. R Kleber, et al, Thin Solid Films **205** 274 (1991)
10. R H Jarman, G J Ray, R W Stadley, Appl Phys Lett **58** 592 (1991)
11. M A Tamor, W C Vassell, K R Carduner, Appl Phys Lett **58** 592 (1992)
12. J Robertson, E P O'Reilly, Phys Rev B **35** 2946 (1987)
13. J Robertson, Diamond Related Mats **4** 297 (1995)
14. Y Lifshitz et al, Phys Rev Lett **72** 2753 (1994); Diamond Rel Mats **4** 318 (1995)
15. S Sattel, M Weiler, J Gerber, H Ehrhardt, Diamond Related Mats **4** 333 (1995)
16. J Robertson, Diamond Related Mats **3** 361 (1994)
17. T Diaz de la Rubia, et al, Mat Res Soc Symp Proc **373** 555 (1995)

X-RAY DIFFRACTION ANALYSIS OF STRAIN AND MOSAIC STRUCTURE IN (001) ORIENTED HOMOEPITAXIAL DIAMOND FILMS

W. Brock Alexander*, Pehr E. Pehrsson**, David Black[S], and James E. Butler**

* ASEE/NRL Postdoctoral Research Fellow, Chemistry Division, Naval Research Laboratory, Washington, DC 20375.
** Chemistry Division, Naval Research Laboratory, Washington, DC 20375.
[S] National Institute of Standards and Technology, Gaithersburg, MD 20899.

ABSTRACT

Homoepitaxial diamond films were grown on (001) oriented high pressure, high temperature type Ib diamond by microwave plasma-assisted chemical vapor deposition to thicknesses of 27-48 μm. Substrates were polished off-axis 5.5° ±0.5° in the [100] direction prior to film deposition. Some of the diamond films developed tensile stress sufficiently large to result in cracking on {111} cleavage planes, while other films exhibited compressive stress. The strain and mosaic structure were measured with seven crystal x-ray diffraction. This characterization tool allowed the separation of misorientation effects from those of lattice parameter variation. Films exhibited smaller (~88 ppm) and larger (~27 ppm) perpendicular lattice parameters relative to the HPHT substrates. A cross-sectional approach for probing strain in diamond films with micro-Raman analysis was used to show stress distributions (~100-300 MPa) through the thickness of the film.

INTRODUCTION

Although progress has been made in the growth of homoepitaxial diamond films, these films typically exhibit stress and/or cleavage cracking [1-4]. The development of stress and defects in these films are at least partly due to the crystallographic structure of the substrates. X-ray topography has been used to reveal dislocations, stacking faults, twins, and growth sector boundaries in high pressure-high temperature (HPHT) type Ib diamond substrates [5,6]. Compared with other semiconductor substrates (e.g. Si) the dislocation densities may be orders of magnitude greater in HPHT type 1b diamonds. Further, the lattice parameter varies by 10 ppm due to nitrogen concentration in the different growth sectors [6].

Strain may be relieved by the formation of misfit dislocations in lattice mismatched systems, and these dislocations typically propagate into the material (substrate or film) with the lower shear modulus. With diamond's high shear modulus ($G_{diamond}$=553 GPa) [7] it is resistant to plastic flow and dislocation propagation, thus tensile stress may lead to cleavage cracking.

The purpose of this paper is to report lattice parameter variations in several homoepitaxial films. Nitrogen concentrations were varied in the feed gas in hopes of modifying the film's lattice parameter. This was done in an attempt to modify stress in the films. The tensile stress distribution through the film/substrate interface was also measured with micro-Raman spectroscopy.

EXPERIMENTAL

Homoepitaxial diamond films were grown by microwave plasma-assisted chemical vapor deposition (MPCVD) with 1% CH_4 in hydrogen. Film depositions were interrupted approximately 50% into the total deposition time to check for cleavage cracks with optical microscopy. The substrate temperature was 980°C ±10°C as measured with an optical pyrometer

Mat. Res. Soc. Symp. Proc. Vol. 423 © 1996 Materials Research Society

with a ~1 mm spot size. Nitrogen (at atomic concentrations of 0 ppm, 950 ppm, or 1800 ppm) was added to the feed gas. The C(001) HPHT substrates (~3 mm x 3 mm) were polished off-axis ~5.5°± 0.5° in the [100] direction prior to film deposition. The off-axis angle and orientation were measured with Laue diffraction.

A Philips HR1 x-ray diffractometer was used to perform x-ray analysis on the homoepitaxial diamond films. The x-ray tube generator was operated at 40 kV and 40 mA and produced CuK_α radiation at a wavelength (λ) of 1.54056 Å. The radiation was conditioned by a Bartels [8] 4-crystal monochromator to produce radiation with $\Delta\theta$=12 arcsec divergence and $\Delta\lambda/\lambda=2.3\times10^{-5}$ wavelength spread. The spot size of the x-ray beam at the sample was ~0.5 mm x 5 mm. X-ray analysis performed with the 4-crystal monochromator is generally referred to as high-resolution x-ray diffraction (HRXRD). The x-ray diffractometer had a U-shaped (Bonse-Hart) [9] analyzer crystal in front of the detector to limit the acceptance angle to ~14arcsec, which allowed separation of lattice parameter and misorientation contributions to diffraction peak broadening. The x-ray diffracts off of seven crystals, including the sample, and thus is referred to as seven crystal x-ray diffraction.

Both omega/2theta ($\omega/2\theta$) and omega (ω) rocking curves were recorded. The terminology reflects the configuration geometry of the sample, detector, and the optical axis of the x-ray source. During recording of an $\omega/2\theta$ rocking curve (also referred to as an $\omega/2\theta$ scan), both the detector (2θ) and the sample rock through the Bragg angle (θ), with the detector moving twice the angular velocity of the sample. During an ω scan, only the sample rocks through θ. The $\omega/2\theta$ rocking curve is sensitive to lattice plane spacing [10]. The angular separation of the film and substrate peaks, $\Delta\omega$, are used to calculate the lattice mismatch through Bragg's law. Strain calculations were relative to the substrate lattice parameter. The ω rocking curve is sensitive to lattice plane tilt or mosaic structure [10,11].

Micro-Raman analysis was performed using the 514.5 nm line from an Ar^+ ion laser operating at a power of 100 mW. The objective lens had a magnification of 80x and the spot size at the sample was ~2 µm. Cross-sectional micro-Raman analysis was performed to separate film and substrate contributions. A cross-section was prepared by mechanical polishing.

RESULTS

Figure 1A shows the (004) $\omega/2\theta$ rocking curve from an as received HPHT substrate. The full width at half maximum (FWHM) of this diamond standard was 33 arcsec; slightly broader than the (004) $\omega/2\theta$ rocking curve from a (001) silicon wafer (8 arcsec) [not shown]. The diamond standard rocking curve indicates a $\Delta d_{(004)}/d_{(004)}$ of $\sim6\times10^{-5}$ ($\Delta d_{(004)}/d_{(004)}$ = range of (004) lattice plane spacing). This range of lattice parameters is not surprising due to the variation in nitrogen content for different growth sectors.

Rocking curves from three homoepitaxial films on HPHT substrates are shown in Figures 1B, 1C, and 1D. The experimental data and best fit curve intensities are off-set for clarity. The best fit curve intensity was the summation of two calculated (Gaussian) peak intensities; the peak positions and FWHM's were determined from these calculated peaks. Figures 1B (#09035) and 1C (#09185) demonstrate films with smaller $d_{(004)}$ than the substrate with a peak at higher angles (ω). This is referred to as compressive strain relative to the substrate lattice parameter. In contrast a film exhibiting larger $d_{(004)}$ (tensile) is shown in Figure 1D (#08125). This film also demonstrated {111} cleavage cracks. Cracks were observed in this film after 10 hours of growth (~20 µm). The x-ray diffraction results are summarized in Table 1 along with N_2 feed gas concentration, growth time, film thickness, and the observation of cracks.

The (004) ω rocking curves from the HPHT standard and the three homoepitaxial films are shown in Figure 2. Intensity is shown in logarithmic form to more clearly illustrate the low intensity features. The FWHM of the diamond standard was 10 arcsec versus 8 arcsec for silicon. The cracked film (#08125) exhibits the largest angular spread compared to the other films. The cracked sectors of the film are most likely tilted relative to each other, and the angular spread was ~280 arcsec. The other two films, which did not crack, exhibited angular spreads of ~86 arcsec (#09185) and ~106 arcsec (#09035).

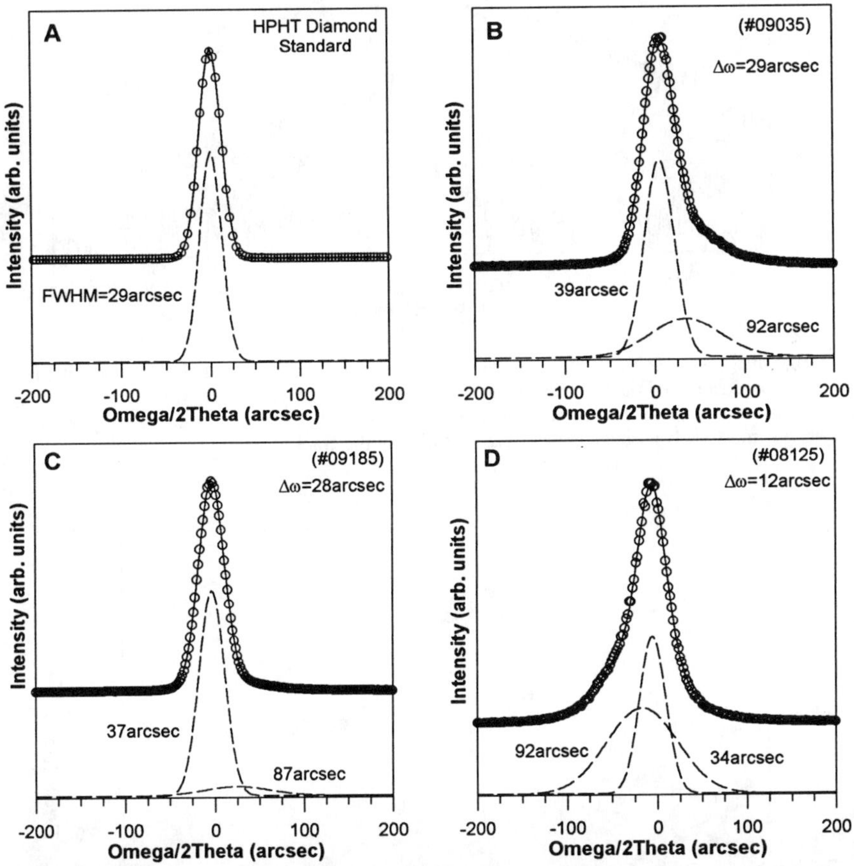

Figure 1. Omega/2theta (ω/2θ) rocking curves from a HPHT diamond standard (A) and three homoepitaxial diamond films (B, C, and D). The actual data (o) and best fit curve (—) are off-set for clarity. The FWHM, peak separation (Δω), and best fit curve were calculated from the Gaussian curves (---).

Table 1. Feed gas N_2 concentration, film growth time, film thickness, cracking, and perpendicular lattice mismatch of three homoepitaxial films.

sample	N_2 (ppm)	growth-time (hrs.)	thickness (μm)	cleavage cracks	X-ray *Δd/d* (ppm)
09035 (B)	0	38	27	no	-88
09185 (C)	1800	15.5	36	no	-85
08125 (D)	950	25	48	yes	+27

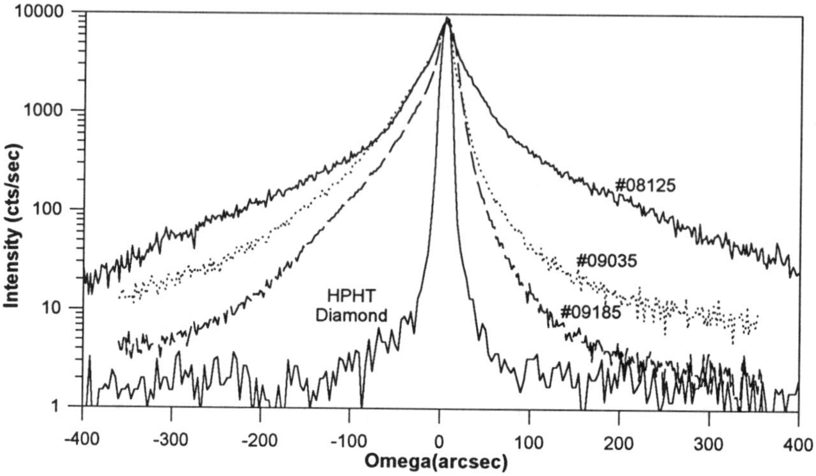

Figure 2. Omega (ω) rocking curves of the HPHT diamond standard and three homoepitaxial films. The broadening of these curves relative to the standard is associated with misorientation and/or dislocation arrays; except for the cracked film (#08125) where the majority of the broadening is most likely associated with relative tilting of the cracked sectors.

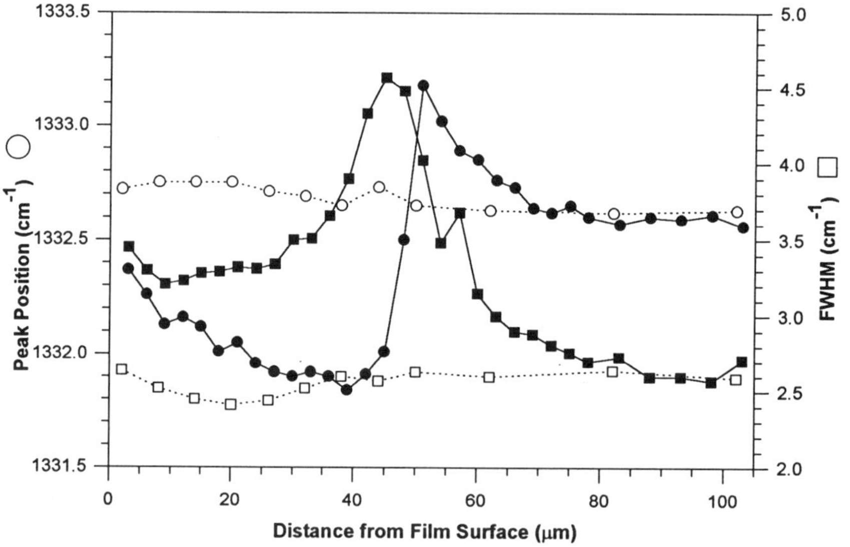

Figure 3. Peak positions (o) and FWHM's (□) of the diamond Raman mode from a cracked (—) and non-cracked (---) homoepitaxial diamond film. The shift to higher wavenumbers is indicative of compressive stress and a shift to lower wavenumbers is indicative of tensile stress. A type IIa natural diamond had a Raman peak position of 1332.70 cm^{-1} and a FWHM of 2.5 cm^{-1}.

The Raman peak position and FWHM as a function of distance from the film surface for cracked (#08125) and non-cracked (#09185) homoepitaxial films and are shown in Figure 3. For comparison, a type IIa natural diamond had a Raman peak position of 1332.70 cm^{-1} and a FWHM of 2.51 cm^{-1}. The cracked film exhibits a decrease in peak position going from the film surface toward the interface. Around the interface, the peak position increases sharply, then decreases going into the substrate. The FWHM also increases around the interface, indicating a larger distribution of strain or possibly a result of film and substrate contributions to the Raman peak. The non-cracked film exhibits only a slight increase in the peak position and a decrease in the FWHM, relative to the substrate.

DISCUSSION

It is unclear why the film grown with 950 ppm N_2 developed tensile stresses and cracked, while the films grown with 0 ppm and 1800 ppm N_2 developed compressive stresses. This may relate to the amount of nitrogen which was incorporated, versus the feed gas concentration. The expectation that stress would not develop since the lattice parameters and thermal expansion coefficients of the film and substrate were nominally the same was unfounded as demonstrated by cracking of one of the films. Mechanisms which generate intrinsic stress include incorporation of impurities, voids, and/or dislocations [12,13]. Substitutional nitrogen may vary from 89 ppm for {111} versus ~1 ppm for {110} growth sectors [14,15] in HPHT diamond, and may approach ~362 ppm in regions [16]. Incorporation of substitutional nitrogen results in expansion of the lattice and has been expressed as $\Delta a/a=(0.14\pm0.02)C_N$ [14], where C_N is the atomic nitrogen concentration. Incorporation of 300 ppm of nitrogen would result in a lattice expansion of 42 ppm relative to nitrogen free lattice. A difference in the thermal expansion coefficients of the film and the substrate may lead to thermal stress ($\sigma_{TH}=(\alpha_{Film}-\alpha_{Sub})\Delta TE_{Film}$) [11]. The values found in Field [12] predict that a 10% difference in thermal expansion coefficients ($\Delta\alpha_{Film\text{-}Sub}$) yields ~120 MPa of stress. Thus, minute differences in thermal expansion coefficients may partially account for stress in the films.

It was observed that films which crack typically exhibit macro- and micro-steps in the <110>, similar to those observed by others [1]. Some non-cracked films exhibited over-grown hillocks as described by van Enckevort et al. [1]. Although substrates were polished off-axis, a small region of the surface was left on-axis (±1.0°). It was observed that for some films the off-axis region would crack while the on-axis region, which was decorated with hillocks or renucleation sites, would not crack. The development of stress and surface defects (hillocks) appear to be interrelated. It may be that the surface defects are associated with dislocations.

Micro-Raman samples a much smaller volume than seven crystal x-ray diffraction, allowing separate analysis of the film and substrate. The cracked film (#08125) exhibited a tensile stress of ~100 MPa ($\Delta\nu$=-0.3 cm^{-1}), using 2.94cm^{-1}/GPa [3], near the film surface which increased to ~300 MPa ($\Delta\nu$=-0.9cm^{-1}) near the interface. Although the film cracked, the stress was not fully relieved. The substrate exhibited a compressive stress of ~170MPa ($\Delta\nu$=+0.5 cm^{-1}) at the interface which decreased (nearly exponentially) with increasing distance from the interface. The Raman peak FWHM was ~3.3cm^{-1} near the surface of the film compared to ~2.6 cm^{-1} deep (~1 mm) within the substrate. The increase in FWHM near the interface is associated with the strain distribution in the sampling volume and individual contributions from the film and substrate.

A mosaic structure acts as discrete diffracting blocks whose diffraction vectors generate broadening or even discrete peaks in the ω rocking curve. At least three Gaussian curves were required to fit the ω rocking curves (not shown). Mosaic structures are generally associated with formation of dislocation arrays. In the homoepitaxial diamond films, the mosaic structure and/or other extended defects broadened the rocking curves relative to the diamond standard. The cracked film exhibited the greatest angular spread, probably due to the relative tilt of the individual sectors separated by the cleavage cracks, which effectively removes the ability to measure mosaic structure. It would have been more appropriate to compare three non-cracked films with more similar thicknesses. Other films, not shown here, have developed cleavage cracks and exhibited tensile stress which was not fully relieved by the cracking. However, once the film cracks it is difficult to measure defect concentrations with the ω rocking curve.

SUMMARY

Both smaller (~85 ppm) and larger (~27 ppm) perpendicular lattice parameters, relative to the HPHT standard, were measured with seven crystal XRD for the homoepitaxial films. Tensile stress caused one film to crack along {111} cleavage planes. Micro-Raman spectroscopy was used to measure the tensile stress distribution through the film/substrate interface and demonstrate that the cracking had not fully relieved the stress. The tensile stress in the cracked film varied from ~100-300 MPa. The films exhibited mosaic and misorientation characteristics associated with dislocation arrays, extended defects, and cracking. The development of stress did not appear to be related with N_2 concentration in the feed gas.

ACKNOWLEDGMENTS

This work was supported in part by the Office of Naval research and the Naval Research Laboratory. WBA gratefully acknowledges the support of the American Society for Engineering Education/Naval Research Laboratory Postdoctoral Fellowship Program.

REFERENCES

1. W.J.P. van Enckevort, G. Jansen, W. Vollenberg, and L.J. Giling, J. of Crystal Growth, **148**, 365 (1995).
2. P.E. Pehrsson, T. McCormick, W.B. Alexander, M. Marchywka, D. Black, J.E. Butler, and S. Prawer, Fall MRS Proc. *Diamond for Electronic Applications*, in press (1995).
3. W.B. Alexander, P.H. Holloway, P. Doering, and R. Linares, Fall MRS Proc. *Diamond for Electronic Applications*, in press (1995).
4. J.J. Schermer, W.J.P.van Enckevort, and L.J. Giling, Diamond and Related Mat., **3**, 408(1994).
5. A.R. Lang, J. Appl. Cryst, **27**, 988 (1994).
6. W. Wierzchowski, M. Moore, A.P.W. Makepeace, and A. Yacoot, J. of Crystal Growth, **114**, 209 (1991).
7. J.E. Field, Eds., *The Properties of Natural and Synthetic Diamond* (Academic Press, London, 1992).
8. W.J. Bartels, J. Vac. Sci. Technol. B, 1, 338 (1983).
9. U. Bonse and M. Hart, Appl. Phys. Lett., 7, 238 (1965).
10. P.F. Fewster and N.L. Andrew, J. Appl. Phys., **74**, (5), 3121 (1993).
11. B.D. Cullity, *Elements of X-Ray Diffraction* p. 103, 292 (Addison-Wesley Publishing Co. Inc., reading, MA. 1978).
12. M. Ohring, *The Materials Science of Thin Films* (Academic Press, Boston, 1992).
13. K.L. Chopra, *Thin Film Phenomena* (McGraw-Hill, Inc., New York, 1969).
14. G.S. Woods, J.A. van Wyk, and A.T. Collins, Phil. Mag., **B62**, 589 (1990).
15. R.C. Burns, V. Cvetkovic, C.N. Dodge, D.J.F. Evans, M.-L.T. Rooney, P.M. Spear, and C.M. Welbourne, J. Cryst. Growth, **104**, 257 (1990).
16. F.C. Frank, A.R. Lang, D.J.F. Evans, M.-L.T. Rooney, P.M. Spear, and C.M. Welbourne, J. Cryst. Growth, **100**, 354 (1990).

TEM and PL characterisation of MBE-grown epitaxial GaN/GaAs

Yan Xin, P.D. Brown, C.B. Boothroyd, A.R. Preston, C.J. Humphreys, T.S. Cheng,* C.T. Foxon,* A.V. Andrianov** and J.W. Orton**
Department of Materials Science and Metallurgy, University of Cambridge, Pembroke Street, Cambridge, CB2 3QZ, UK.
*Department of Physics, University of Nottingham, University Park, Nottingham, NG7 2RD, UK.
**Department of Electrical and Electronic Engineering, University of Nottingham, University Park, Nottingham, NG7 2RD, UK.

Abstract

MBE-grown epitaxial GaN deposited at 700°C on {001}, {111}A and $\{\bar{1}\bar{1}\bar{1}\}$B GaAs has been characterised using the combined techniques of transmission electron microscopy (TEM) and photoluminescence (PL). On both {111}A and $\{\bar{1}\bar{1}\bar{1}\}$B GaAs substrates, single crystal wurtzite GaN was formed, but with very high densities of threading defects. Best epitaxy occurred on $\{\bar{1}\bar{1}\bar{1}\}$B GaAs in accordance with PL measurements. An amorphous phase was identified at the GaN/{111}A GaAs interface and the GaN epilayer evolved in this instance with the same N-terminated growth surface as for the case of growth on $\{\bar{1}\bar{1}\bar{1}\}$B GaAs, as determined by convergent beam electron diffraction (CBED). Growth on {001} GaAs produced highly faulted columnar grains of zincblende GaN. Conversely, growth on {001} GaAs under an additional arsenic flux at 700°C resulted in the deposition of single crystal zincblende GaN with a high density of stacking faults and microtwins. Thus, the microstructure of epitaxial GaN depends very much on the detailed growth conditions and substrate orientations used.

Introduction

Recent progress in the epitaxial growth of group III-nitride semiconductors and the demonstration of intense blue light emission from light-emitting diodes [1] has stimulated world-wide study of these wide-gap semiconductors for short wavelength optoelectronic applications. Such wide-gap devices offer immense associated commercial benefits for displays, flat panel TVs and data storage systems. The favourable thermal and structural properties of (Al,Ga,In)N also make it suitable for high power, high frequency and high temperature applications, with the additional benefit of radiation resistance. One fundamental problem which influences GaN layer growth and device performance is the lack of a suitable substrate material with comparable lattice and thermal match to GaN [2]. Hence, interest has also been directed towards growth on alternative substrates. GaN is a polytypic material having two phases, with the wurtzite structure being more stable than the zincblende phase. The crystal structure adopted by epitaxial GaN is found to depend strongly on the growth conditions, substrate material and substrate orientation [2]. In this paper, the microstructure of epitaxial GaN grown on {001}, {111}A (Ga-terminated) and $\{\bar{1}\bar{1}\bar{1}\}$B (As-terminated) GaAs substrates by molecular beam epitaxy (MBE) are characterised and correlated with photoluminescence measurements.

Experimental

The samples examined here were grown by MBE using elemental solid sources of Ga and As and an activated N plasma source. Samples were grown under similar conditions at 700°C on {001}, {111}A and $\{\bar{1}\bar{1}\bar{1}\}$B GaAs substrates with an active nitrogen arrival rate adjusted to give

Mat. Res. Soc. Symp. Proc. Vol. 423 © 1996 Materials Research Society

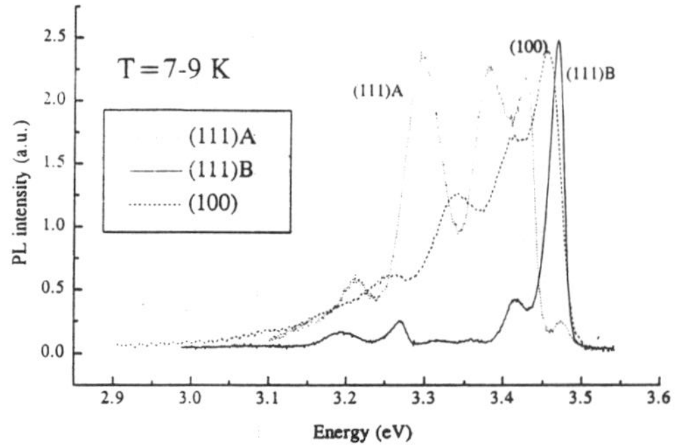

Fig. 1 PL spectra of epitaxial GaN grown on {001}, {111}A and $\{\bar{1}\bar{1}\bar{1}\}$B GaAs substrate.

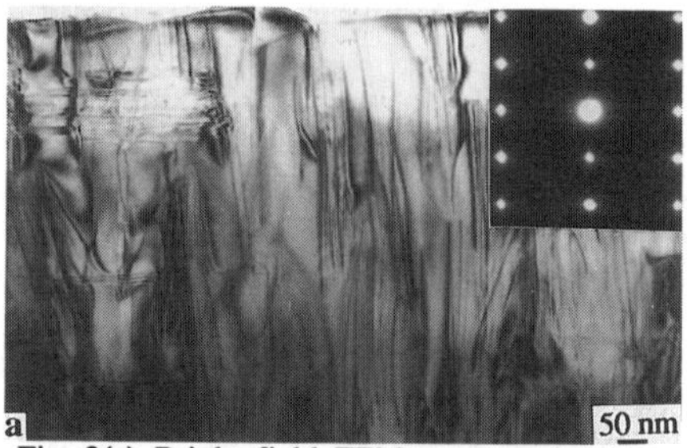

Fig. 2(a) Bright field TEM cross-sectional image of GaN/$\{\bar{1}\bar{1}\bar{1}\}$B GaAs showing high density of threading defects; [11$\bar{2}$0] SAD pattern is inset.

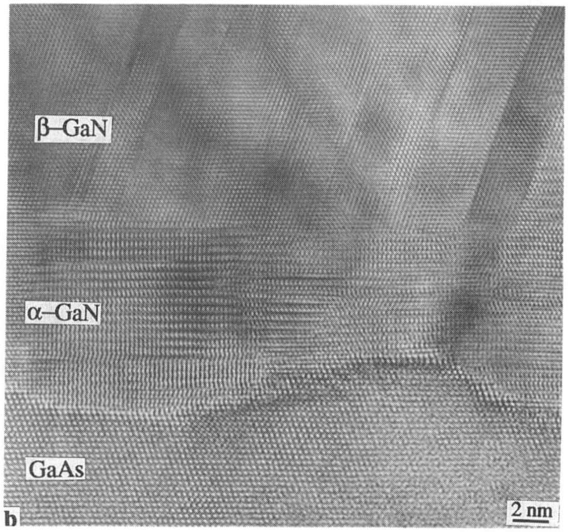

Fig. 2(b) HRTEM image showing the interfacial region. The initial deposit is wurtzite GaN, while small domains of zincblende GaN are present.

Fig. 2(c) Plan view image of GaN epilayer showing mosaic structure.

slightly nitrogen rich growth. Epitaxial GaN was also grown on {001}GaAs under an As flux at the same temperature. In this instance the substrate was cleaned in the presence of N flux at 620°C. Growth with As was then initiated by opening the Ga and N shutters, with the As flux being held constant throughout the growth run. For the photoluminescence (PL) measurements, the samples were mounted on the cold finger of a continuous-flow He cryostat in which the sample temperature could be varied between 7K to 300K. PL from the GaN samples was excited using a Kimmon He-Cd laser at 325nm, dispersed with a 0.75m monochromator and detected by a bi-alkai photomultiplier. The PL resolution was typically in the range of 1-2meV. TEM characterisation was performed in cross-section using a JEOL 4000EX-II for high resolution electron microscopy (HRTEM), and a Philips CM30 microscope for conventional microscopy and Energy Dispersive X-ray (EDX) analysis. The crystal polarity for all of the samples examined was determined using convergent beam electron diffraction (CBED).

Results and Discussion

PL spectra from the three samples grown on {001}, {111}A and $\{\bar{1}\bar{1}\bar{1}\}$B GaAs are shown in Fig. 1. For the GaN grown on the $\{\bar{1}\bar{1}\bar{1}\}$B substrate, the PL is dominated by the recombination of excitons bound to shallow donors, indicative of the good overall quality of the layer. For the other samples, however, additional defect/impurity related features are clearly present. However, it should be noted that we did not observe the typical deep (yellow) emission in any of the three samples.

The microstructure of the epilayer grown on $\{\bar{1}\bar{1}\bar{1}\}$B, illustrated in Fig. 2a, comprises a very high density of threading defects. The [11$\bar{2}$0] selected area diffraction (SAD) pattern inset in Fig. 2a confirms the epilayer to be mainly wurtzite single crystal. The striations parallel to the growth surface in the top left corner of this figure correspond to a small region of zincblende GaN. Fig. 2b shows an HRTEM image of the GaN/$\{\bar{1}\bar{1}\bar{1}\}$B GaAs interface region. It is evident that the substrate surface is rough, with undulations on the scale of several nanometres. Although GaN zincblende domains are present as illustrated, the initial deposit is predominantly of the wurtzite structure. Plan view observations of this sample indicated threading dislocations aligned along low angle grain boundaries to form a mosaic structure (Fig. 2c). However, the density of the threading dislocations in Fig. 2c ($>10^{10}cm^{-2}$) would appear to be significantly lower than the density of the threading defects in Fig. 2a, implying that many of these defects may not be dislocations. Further work is in progress to characterise the nature of all the defects in the epilayer.

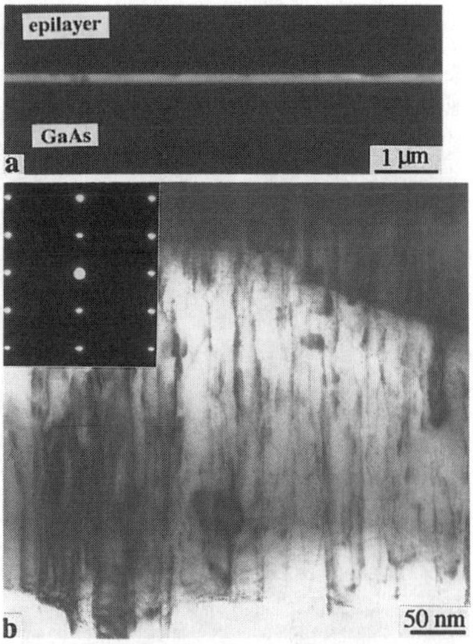

Fig. 3 (a) Low magnification bright field TEM images of GaN/{111}A GaAs showing the presence of an amorphous interlayer. (b) Bright field image of the epilayer showing a high density of threading defects. The [11$\bar{2}$0] SAD pattern is inset.

Fig. 3a is low magnification bright field image of GaN/{111}A GaAs showing a band of light contrast of 100 nm thick between the epilayer and substrate. HRTEM observations and SAD patterns confirmed this interlayer to be amorphous, while EDX analysis demonstrated the presence of Ga, As with additional peaks due to S and Si. Thus, the interlayer is attributed not to interfacial reaction but is merely some amorphous layer remnant on the GaAs substrate prior to growth (although the origin of this layer and the S and Si peaks remains unclear). Nevertheless, the epilayer has still grown single crystal wurtzite on this amorphous layer, as confirmed by the inset SAD pattern, although the epilayer exhibits a very high density of threading defects (Fig. 3b) and is of poorer quality compared with growth on clean $\{\bar{1}\bar{1}\bar{1}\}$B GaAs.

It is of interest to determine the crystal polarity of the epitaxial GaN films. It is noted that X-ray photoemission [3] and Auger measurements [4] on GaN grown on sapphire and SiC present contradictory results. For the case of growth on (0001) and (000$\bar{1}$)SiC, the results indicate that Si-Ga and C-N bonding (in the growth direction) is preferred respectively. For the case of GaN grown on $\{\bar{1}\bar{1}\bar{1}\}$B GaAs substrate, it is to be expected that As must bond with Ga rather than As with N (in the growth direction), such that the GaN growth surface will be N-terminated. Conversely, epitaxial GaN on {111}A GaAs is expected to exhibit a Ga-terminated growth surface.

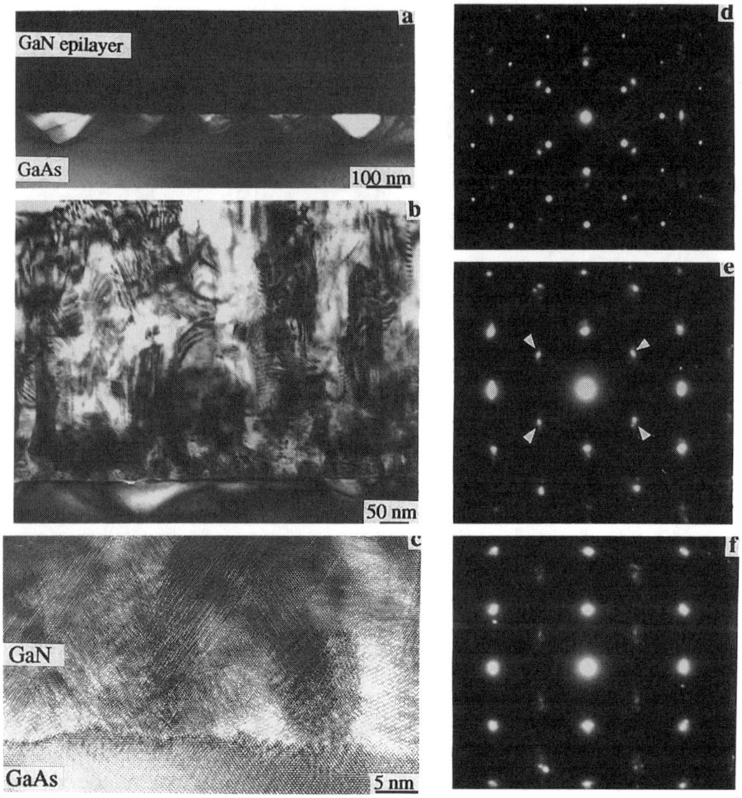

Fig. 4 (a) Low magnification bright field TEM image of GaN/{001}GaAs showing V-pits in the substrate. (b) Bright field image showing columnar grains and a high density of planar defects. (c) HRTEM image of the interfacial region. (d to f) A sequence of SAD patterns recorded through the epilayer showing a phase transition from zincblende to wurtzite.

CBED was used to determine the GaN crystal polarity from these two samples. Precise orientation of a GaN low index zone axis was found to be very difficult in practice. Accordingly, samples were tilted such that a {0002} systematic row was established. Asymmetries of contrast between (0002) and (000$\bar{2}$) diffraction discs when compared with simulations allowed the crystal polarity to be determined. The smallest possible probe size was used in order to avoid the many threading defects present. Comparison with simulations confirmed that the polar growth surface of epitaxial GaN was the same for growth on both {111}A and {$\bar{1}\bar{1}\bar{1}$}B GaAs in our case. Hence, the epilayer on the amorphous layer on the {111}A substrate is also N-terminated. Accordingly, a polarity inversion is attributed to the sample grown on {111}A GaAs since it ought to be Ga-terminated if the epilayer were grown epitaxially on {111}A substrate with a Ga-terminated surface. Thus, the substrate has no effect on the epilayer because of the amorphous interlayer.

Fig. 4a is a low magnification bright field image of GaN grown on {001}GaAs. V-shaped features beneath the epilayer in the substrate are evident and arise from the preferential removal of material during preparation of the TEM sample foil. They are due to regions of material of

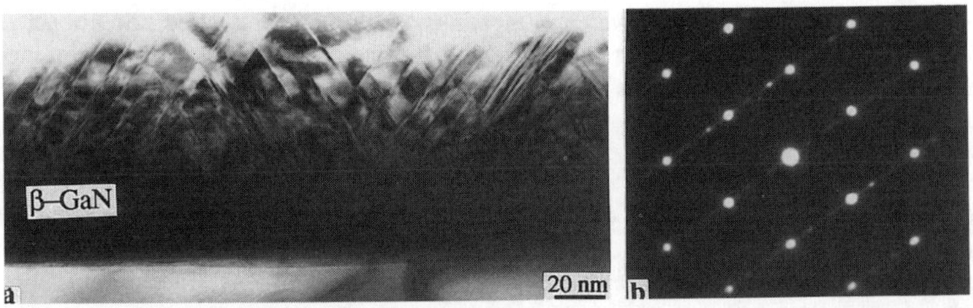

Fig. 5 (a) Bright field TEM image of GaN/{001}GaAs grown under an As flux. The epilayer is single crystal zincblende with a high density of planar defects on inclined {111} planes. (b)<110> SAD pattern from the epilayer shows additional spots and streaks due to microtwins.

differing composition to that of the substrate. Chemical maps of the related epitaxial GaN/{001}GaP system, recorded using a Gatan imaging filter, clearly show that N has diffused into the substrate [5]. The presence of similar pits in the GaAs substrate here is therefore tentatively attributed to the interdiffusion of N into the substrate during growth. The consequent change in composition and associated strain makes these regions susceptible to preferential ion beam milling. This phenomena is common to all of the samples whose substrates were nitrided prior to growth.

Fig. 4b is a bright field image showing the epilayer is comprised of columnar grains within which there are many planar defects parallel to the substrate surface, and also some inclined to it at an angle of about 60°. There are many more fine streaks close to the epilayer/substrate interface, while the HRTEM image of Fig. 4c reveals that this region is zincblende in structure with a high density of planar defects on inclined {111} planes. A sequence of SAD patterns recorded at increasing distance from the epilayer/substrate interface is shown in Figs. 4d to 4f. Fig. 4d corresponds to the interfacial region and demonstrates the presence of zincblende GaN with the same orientation as the substrate. SAD patterns recorded from areas progressively further from the interface into the epilayer such as (Fig. 4e) show changes in relative spot intensities implying a gradual transition from the cubic phase to a higher proportion of hexagonal material. The stronger spots (arrowed) belong to the $[1\bar{1}00]$ zone axis of wurtzite GaN, while the four weaker spots around the transmitted beam are {111} reflections from the cubic phase. The hexagonal phase becomes dominant towards the top of the epilayer (Fig. 4f). The SAD patterns also demonstrate the very strong texture of the columnar grains.

Under the same growth condition, very different GaN microstructures evolve for growth on {111} and {001}GaAs substrates. Thus, the epitaxial growth of GaN depends very much on the substrate orientation used, with TEM results indicating best epilayer quality for growth on $\{\bar{1}\bar{1}\bar{1}\}$B GaAs. Although the hexagonal phase is generally preferred for growth on {111}GaAs, the actual energy difference between the cubic and hexagonal phases of GaN is quite small (approx 10meV [6]). Hence, the presence of a small number of domains of zincblende GaN. It is difficult to compare our results with those of others [7] which indicate better material quality of zincblende GaN on {111}A rather than $\{\bar{1}\bar{1}\bar{1}\}$B GaAs, since we find that the initial deposit on $\{\bar{1}\bar{1}\bar{1}\}$B tends to be of the hexagonal phase (these two observations are not contradictory). More surprising is the fact that single crystal hexagonal GaN forms on an amorphous layer on {111}A GaAs. The many line defects running through the epilayer in this case suggests that they do not arise as a consequence of the large lattice-mismatch between epilayer and the substrate, but instead reflect the process of island coalescence during the first stages of growth [8,9].

Attention must also be given to the effect of threading defects on the electrical and optical properties of devices based on (In,Ga)N, which interestingly show that such defects do not act as efficient nonradiative recombination centres, in marked difference to the case of III-V arsenides and phosphides [10]. Indeed, there is no clear correlation of our microstructural observations with the PL data as yet. Investigations on the effects of the dislocations on electrical properties of the GaN by Electron Energy Loss Spectroscopy (EELS) using a VG HB501 STEM were carried out by acquiring the very low energy loss region of the spectra, which contains information around the band edge feature of ~ 3.4 eV. Preliminary results acquired on and off dislocations down <0001> GaN in plan view geometry shows no difference between the spectra. The same is true for nitrogen K-edge spectra. This is consistant with dislocations not affecting material luminescence.

The microstructure of epitaxial GaN grown on {001}GaAs under an As flux was also characterised. Even though this epilayer was grown at the same temperature of 700°C, the epilayer is single crystal zincblende (Fig. 5a). The SAD pattern (Fig. 5b) shows the presence of extra spots due to the very high density of inclined planar defects on {111} planes. No evidence for the ternary $Ga(As_{1-x}N_x)$phase was found. Thus, arsenic plays a crucial role in the formation of the cubic GaN phase [11]. This nicely demonstrates that parameters other than just growth temperature can dramatically change the microstructures of epitaxially grown GaN.

Summary

In summary, the phase and structural quality of MBE-grown GaN on GaAs depends strongly on the growth conditions and substrate orientations used. While best epitaxy occurs on $\{\bar{1}\bar{1}\bar{1}\}$B GaAs in agreement with PL measurements, all samples are in fact highly faulted. Wurtzite GaN with a high density of threading defects is formed on $\{\bar{1}\bar{1}\bar{1}\}$B GaAs. More interestingly, a similar microstructure evolves for growth on an amorphous layer on {111}A GaAs. Growth columns of mixed phase GaN with preferred orientation are formed on {001}GaAs, however, improved epitaxy is facilitated by growth under an As flux, leading to the production of single cubic crystal material with a high density of planar defects.

Acknowledgements

YX thanks the Cambridge Overseas Trust, the Foreign and Commonwealth Office and the ORS Award scheme for funding. PDB wishes to acknowledge the EPSRC for funding under contract No. GR/J37966.

References

[1] S. Nakamura, T. Mukai, and M. Senoh, Appl. Phys. Lett., **64**, p1687 (1994).

[2] S. Strite and H. Morkoç, J. Vacuum Sci. Technol., **B10**, p1237 (1992).

[3] T. Sasaki, T. Matsuoka and A. Katsui, Appl. Surf. Sci., **41-42**, p504 (1989).

[4] M. A. Khan, J. N. Kuznia, D. T. Olson and R. Kaplan, J. Appl. Phys., **73**, p3108 (1993).

[5] Y. Xin, P.D. Brown, R.E. Dunin-Borkowski, C.J. Humphreys, T. S. Cheng and C.T. Foxon, manuscript in preparation.

[6] C.Y. Yeh, Z.W. Lu, S. Froyen and A. Zunger, Phys. Rev. **B46**, p10086 (1992).

[7] J. W. Yang, J. N. Kuznia, Q. C. Chen, M. Asif Khan, T. George, M. De Graef and S. Mahajan, Appl. Phys. Lett., **67**, p3759 (1995).

[8] B. N. Sverdlov, G. A. Martin, H. Morkoç and D. J. Smith, Appl. Phys. Lett., **67**, p2063 (1995).

[9] D. Kapolnek, X. H. Wu, B. Heying, S. Keller, B. P. Keller, U. K. Mishra, S. P. DenBaars and J. S. Speck, Appl. Phys. Lett., **67**, p1541 (1995).

[10] S. D. Lester, F. A. Ponce, M. G. Craford and D. A. Steigerwald, Appl. Phys. Lett., **66**, p1249 (1995).

[11] T.S. Cheng, L.C. Jenkins, S.E. Hooper, C.T. Foxon, J.W. Orton and D.E. Lacklison, Appl. Phys. Lett., **66**, p1509 (1995).

CHEMICAL BEAM EPITAXY OF GAN_xP_{1-x} USING A N RADICAL BEAM SOURCE

N. Y. LI*, D. H. TOMICH*, W. S. Wong**, J. S. Solomon***, and C. W. TU*
*ECE Department, University of California, San Diego, La Jolla, CA 92093, nli@sdcc3.ucsd.edu
** Present address: MSME Department, University of California, Berkeley, CA 94720
*** Research Institute, University of Dayton, Dayton, OH 45469

ABSTRACT

In this study we report the growth behavior of GaN_xP_{1-x} by chemical beam epitaxy using triethylgallium, tertiarybutylphosphine, and a RF-plasma N radical beam source. We demonstrate that the N radical beam source is an effective N source for the growth of GaN_xP_{1-x}, compared to ammonia (NH_3) with co-injection of phosphine (PH_3) or tertiarybutylphosphine (TBP). At a growth temperature of 640°C, the N composition increases slowly from 2.5 to 2.8% even though the N_2 flow rate is doubled. When the N_2 flow rate is increased further, the reflection high-energy electron diffraction pattern (RHEED) becomes spotty. The N composition, however, shows a strong dependence on the growth temperature. For a fixed N plasma radical beam flux, the lower the substrate temperature is, the higher the N incorporation. The N composition can be adjusted from 0.7 to 10.2% by lowering the growth temperature from 690 to 400°C.

INTRODUCTION

In the last decade much effort has been undertaken to explore the growth of direct-bandgap GaAs and InP on Si substrates [1,2] for the integration of optoelectronic devices with Si circuits. The performance of optoelectronic devices grown on Si, however, is still limited by problem of high-density threading dislocations and stacking faults in the heteroepitaxial layers due to mismatch in lattice constants and thermal expansion coefficients. Therefore, other III-V compounds, such as GaN_xP_{1-x}, GaN_xAs_{1-x}, and $GaN_xAs_yP_{1-x-y}$ alloys, which can be lattice-matched to Si substrates, are of interest. To lattice match to Si, only 2% N is needed for GaN_xP_{1-x}, whereas 20% N is needed for GaN_xAs_{1-x} [3]. For $GaN_xAs_yP_{1-x-y}$, the minimum N composition required for a direct bandgap transition is predicted to be 8% [4].

GaN_xP_{1-x} alloys have been grown using co-injection of NH_3 with PH_3 or TBP [5,6]. The maximum obtainable N composition in GaN_xP_{1-x} using co-injection of NH_3 and PH_3 is 7.6%. It is achieved by maximizing PN molecules in the mixture [5], so the N composition can not be controlled independently by NH_3 alone. We have found previously that the N incorporation efficiency using co-injection NH_3 and TBP is very low due to the strong interaction of TBP with NH_3 in the same cracker [6]. Therefore an alternative efficient and independent N source is highly desirable to increase the N incorporation.

The RF-plasma N radical beam source is highly reactive and has been successfully demonstrated to grow GaN [7], GaN_xAs_{1-x} [3], and p-type doped ZnSe for blue-light-emitting laser diodes [8]. In addition, the nitrogen content in GaN_xAs_{1-x} [3] can be controlled independently by changing the amount of active nitrogen species. In this study we report the growth behavior of GaN_xP_{1-x} using such a source in a chemical beam epitaxy (CBE) system. Our results demonstrate that the N composition in GaN_xP_{1-x} can be achieved up to 10.2%.

Mat. Res. Soc. Symp. Proc. Vol. 423 © 1996 Materials Research Society

EXPERIMENT

The GaN_xP_{1-x} samples were grown by CBE using triethylgallium (TEGa), nitrogen (N_2), and cracked TBP. The temperature of the TBP cracker was 800°C. TEGa and TBP were introduced into the chamber without any carrier gas and the flow rates were controlled by mass-flow controllers. Ultra-high purity nitrogen (>99.999%) was introduced into an Oxford Applied Research RF-activated plasma source with the flow rate controlled by a mass flow controller. The plasma N radical beam source operated at 13.56 MHz with a forward power of 300 W. The N_2 flow rate was varied from 0.38 to 1.50 sccm and the chamber pressure was about 4×10^{-5} Torr during growth. The substrates used in this study were nominally undoped (100) GaP. GaP substrates were cleaned and passivated prior to growth according to the procedures reported previously [6]. Typically, the growth rate and growth temperature were varied from 0.56 to 0.85 monolayer per second (ML/sec) and 470 to 690°C, respectively. The gallium and phosphorus incorporation rates were determined by group III- and group V-induced RHEED intensity oscillations, respectively [9]. The epilayers were characterized by high-resolution x-ray rocking curves.

RESULTS

In this study, the free-standing lattice constant of GaN_xP_{1-x} is obtained from x-ray {511} asymmetric reflection rocking curves by using the elastic theory [10], and then the N composition can be determined using Vegard's law. The following parameters of GaN and GaP were used to obtain the Poisson ratio of GaN_xP_{1-x}. The lattice constants of GaP and cubic GaN are 5.4512 and 4.4820 Å, respectively. The elastic constants, C_{11} and C_{12}, of GaP are 14.12 and 5.25 [11], and those of cubic GaN are 26.40 and 15.30, respectively [12]. All units are 10^{11} dynes/cm^2.

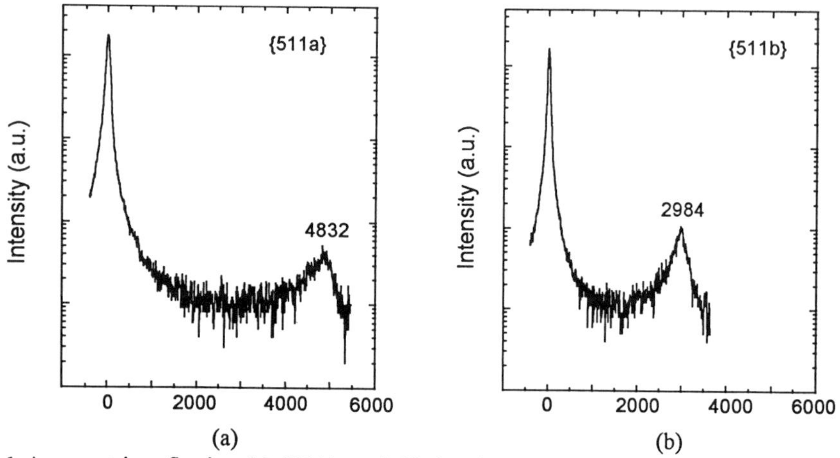

Fig. 1 Asymmetric reflection (a) $\{511\}_a$ and (b) $\{511\}_b$ x-ray rocking curves of a 0.5 μm-thick $GaN_{0.057}P_{0.943}$

Fig. 1(a) and (b) show the asymmetric $\{511\}_a$ and $\{511\}_b$ x-ray rocking curves of $GaN_{0.057}P_{0.943}$. No film tilt was observed by rotating the samples in 4 azimuthal directions of (400) and {511} x-ray rocking curves. The N composition determined from (400) x-ray diffraction is found about a factor of 2 larger than that determined from (511) x-ray diffraction, indicating elastic distortion must be taken into account for a correct N composition. The tensile-strain relaxation in GaN_xP_{1-x} is normally relieved by microcracks, and the full width at half maximum of GaP substrates increases with N composition in GaN_xP_{1-x} due to microcracks propagation into the substrates. The details in strain relaxation via microcrack formations in GaN_xP_{1-x} is reported elsewhere [13].

Fig. 2 shows the N composition of GaN_xP_{1-x} grown at 640°C as a function of TEGa flow rate (growth rate) or V/III (P/Ga) incorporation rate ratio. The phosphorus incorporation rate is fixed at 1.04 ML/sec and the N radical beam flux is also kept constant with a N_2 flow rate of 0.5 sccm and a forward power of 300 W. The TEGa flow rate varies from 0.35 to 0.52 sccm, corresponding to a growth rate of 0.56 to 0.86 ML/sec.

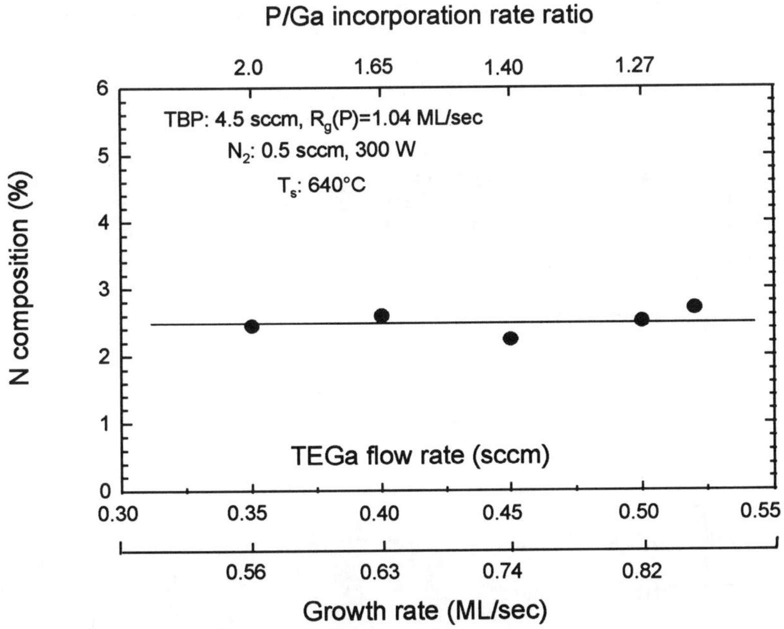

Fig. 2 The dependence of N composition of GaN_xP_{1-x} on the TEGa flow rate and P/Ga incorporation ratio

As shown in Fig. 2, the N composition is independent of the growth rate or P/Ga incorporation rate ratio, and it is about 2.5±0.3%. Generally speaking, the N incorporation in GaP should be increased by lowering the Ga incorporation rate from 0.86 to 0.56 ML/sec at a fixed N flux, but it shows an independence on the growth rate. Two possible reasons may explain this growth behavior. One is that the excess P supply on the GaN_xP_{1-x} growing surface limits the N incorporation. In this case the P incorporation rate is fixed at 1.04 ML/sec, resulting in a high P/Ga incorporation rate ratio of 2.1 at a low growth rate of 0.56 ML/sec. The N incorporation in GaP, therefore, could not be increased significantly under the excess P molecules on the growing surface. The other is that the N incorporation in GaP is limited by the large miscibility gap of GaN_xP_{1-x}.

To understand more about N incorporation in GaP, GaN_xP_{1-x} samples are grown at a fixed P/Ga incorporation rate ratio of 1.3 with different growth rates of 0.56 and 0.82 ML/sec. The N radical beam flux is varied to study the growth behavior of GaN_xP_{1-x}. Fig. 3 shows the dependence of N composition on N radical beam flux. At a forward plasma power of 300 W, the reactive N radical beam flux, estimated from the output of an optical emission detector, can be increased by increasing the N_2 flow rate, but it eventually saturates when the N_2 flow rate is greater than 1.3 sccm. At a growth temperature of 640°C and a growth rate of 0.56 ML/sec, however, the N composition increases slowly from 2.5 to 2.8% as the N_2 flow rate increases from 0.35 to 0.50 sccm. The RHEED pattern becomes spotty as growth proceeds and the surface morphology becomes hazy for further increased N_2 flow rate. For the samples grown at a higher growth rate of 0.82 ML/sec, we find that even though the N_2 flow rate increases from 0.50 to 1.13 sccm, the N composition increases slowly and then saturates at 2.9%. Similar spotty RHEED patterns are observed when the N_2 flow rate is larger than 1.13 sccm. However, the N incorporation in GaP at 640°C can be increased to 5.6% at a N_2 flow rate of 1 sccm by adding a small amount of As (0.4%) as shown in Fig. 3. Compared to 2.8 % N composition in GaN_xP_{1-x}, this result implies that the N incorporation in GaP at 640°C can be increased about a factor of 2 by adding a small amount of As. Based on these results, we believe that the N incorporation in GaP at 640°C is limited by the miscibility gap between GaN and GaP, and it shows a negligible dependence on the growth rate, P/Ga incorporation rate ratio, and N radical beam flux.

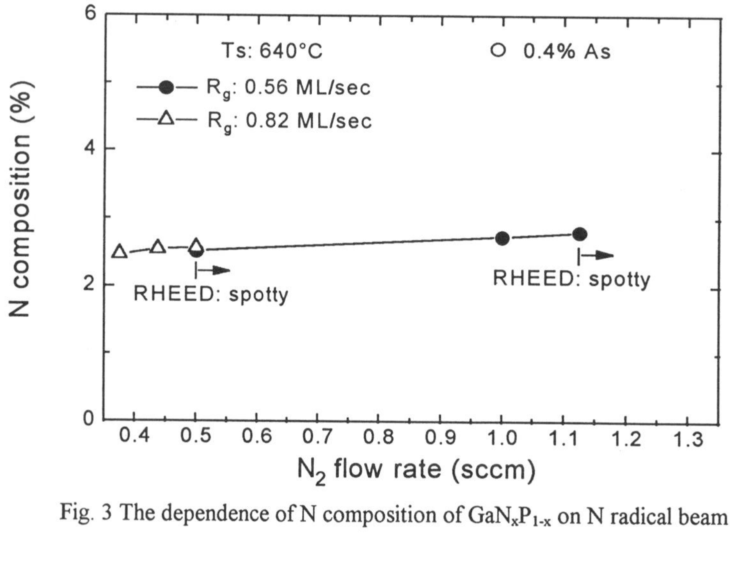

Fig. 3 The dependence of N composition of GaN_xP_{1-x} on N radical beam flux

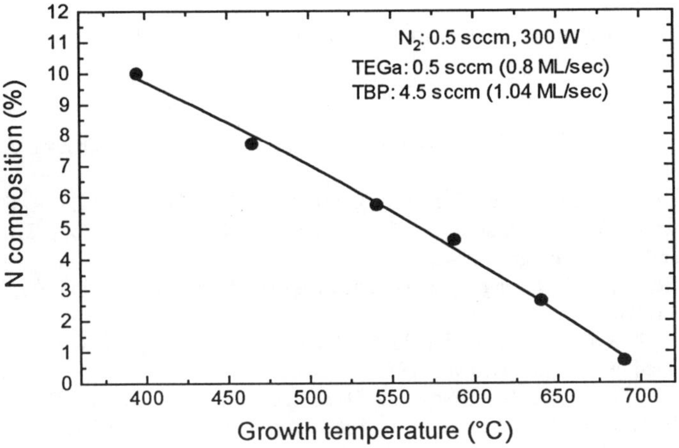

Fig. 4 The dependence of N composition of GaN_xP_{1-x} on the growth temperature

The N composition exhibits a strong dependence on substrate temperature, as shown in Fig. 4. For the same N radical beam flux, the N composition can be increased by lowering the growth temperature. The highest N composition achieved in this work is 10.2% at a growth temperature of 400°C. We believe that the N composition can be increased further by lowering the growth temperature and simultaneously increasing the N radical beam flux. Recently, we have demonstrated that the N composition in GaP, using the same N source, can be incorporated up to 16% by gas-source molecular beam epitaxy [14]. This large N incorporation in GaP can be explained by the enthalpies of formation. The enthalpies of formation of GaP and GaN at 25 °C are 102.5±8.4 and 109.6±9.2 KJ/mole, respectively [15]. These values are very close so that N can bond to Ga as easily as P, but the maximum N incorporation in GaP might still be limited by the miscibility gap.

CONCLUSIONS

CBE of GaN_xP_{1-x} using TEGa, TBP and a N plasma radical beam source has been studied. The N incorporation in GaN_xP_{1-x} is achieved up to 10.2%. The growth behavior of GaN_xP_{1-x} is also investigated. The N composition of GaN_xP_{1-x} samples grown at 640°C is independent of the growth rate, P/Ga incorporation ratio, and N_2 flow rate, but it shows a strong dependence on the growth temperature. The N composition can be adjusted from 0.7 to 10.2% by lowering the growth temperature from 690 to 400°C. It also can be increased about a factor of 2 at a substrate temperature of 640°C by introducing a small amount of As (0.4%).

ACKNOWLEDGMENTS

This work is supported by Texas Instruments System Components Laboratory.

REFERENCES

1. S. F. Fang, K. Adomi, S. Iyer, H. Morkoc, H. Zabl, C. Choi, and N. Otsuka, J. Appl. Phys. **68**, p. R31 (1990).
2. M. Tachikawa, T. Yamada, and T. Sasaki, Jpn. J. Appl. Phys. **34**, p. L657, (1995).
3. M. Kondow, K. Uomi, K. Hosomi, and T. Mozume, Jpn. J. Appl. Phys. **33**, p. L1056 (1994).
4. W. G. Bi, and C. W. Tu, unpublished.
5. J. N. Baillargeon, K. Y. Cheng, G. E. Hofler, P. J. Pearah, and K. C. Hsieh, Appl. Phys. Lett. **60**, p. 2540 (1992).
6. N. Y. Li, W. S. Wong, D. H. Tomich, H. K. Dong, J. S. Solomon , J. T. Grant, and C. W. Tu, to be published in J. Crystal Growth (1996).
7. C. T. Foxon, T. S. Cheng, S. V. Novikov, D. E. Lacklison, L. C. Jenkins, D. Johnston, J. W. Orton, S. E. Hooper, N. Baba-Ali, T. L. Tansley, and V. V. Tret'yakov, J. Crystal Growth **150,** p. 892 (1995).
8. K. Ohkawa, A. Ueno, and T. Mitsuya, J. Crystal Growth **117**, p. 375 (1992).
9. T. P. Chin, B. W. Liang, H. Q. Hou, M. C. Ho, C. E. Chang, and C. W. Tu, Appl. Phys. Lett. **58**, p. 254 (1993).
10. A. T. Macrander, G. P. Schwartz, and G. T. Gualtieri, J. Appl. Phys. **64**, p. 6736 (1988).
11. J. Hornstra, and W. J. Bartels, J. Crystal Growth **44**, p. 513 (1978).
12. M. E. Sherwin, and T. J. Drummond, J. Appl. Phys. **69**, p. 8423 (1991).
13. N. Y. Li, W. S. Wong, D. H. Tomich, K. L. Kavanagh, and C. W. Tu, to be published in J. Vac. Sci. Technol. (1996).
14. W. G. Bi, and C. W. Tu, unpublished.
15. O. Kubaschewski, C. B. Alcock, and P. J. Spencer, *Materials Thermochemistry*, Pergamon Press, New York, 1993, p. 279.

IMPROVED ALUMINUM NITRIDE THIN FILMS GROWN BY MOCVD FROM TRITERTIARYBUTYLALUMINUM AND AMMONIA

T. Metzger*, E. Born*, R. Stömmer**, W. Rieger***, R. Dimitrov***, D. Lentz***, H. Angerer***, O. Ambacher*** and M. Stutzmann***

*Lehrstuhl für Angewandte Mineralogie und Geochemie, Technical University Munich
Lichtenbergstr. 4, 85747 Garching, Germany
**Siemens Corporate Research and Development, Siemens AG
Otto-Hahn-Ring 6, 81739 Munich, Germany
***Walter Schottky Institution, Technical Universiy Munich
Am Coulombwall, 85747 Garching, Germany

ABSTRACT

AlN thin films were grown on c-plane sapphire by metalorganic chemical vapor deposition from tritertiarybutylaluminum and ammonia at 1050°C. These films exhibit a full width at half maximum of the 002 X-ray rocking curve below 200 arcsec indicating high epitaxial quality. By measuring asymmetric reflections, a structural disorder of the lattice mainly due to edge dislocations can be observed. For further investigations, atomic force microscopy and photothermal deflection spectroscopy were performed. In order to study the effect of increasing AlN layer thickness on the optical and structural properties of GaN in an AlN/GaN heterostructure, AlN thin films with increasing thickness ranging from 0.02 to 0.36 μm were used as sublayers for the deposition of 0.75 μm thick GaN layers. Photoluminescence, micro-Raman and X-ray diffraction measurements confirm the relaxation of biaxial compressive stress in the GaN layers due to different thermal expansion coefficients by increasing AlN layer thickness. The pressure dependence of the band gap shift was determined as 24 meV/GPa for biaxial compressive stress. Our results indicate that the growth of AlN with metal organic chemical vapor deposition from tritertiarybutylaluminum and ammonia is a promising method for obtaining high quality epitaxial films.

INTRODUCTION

Among the group-III-nitrides InN, GaN and AlN, AlN is the only material capable of extending the operation of optoelectronic devices into the ultraviolet region of the spectrum. In order to obtain high quality epitaxial films of the group-III-nitrides, different growth methods [1,2] as well as the use of various substrate materials [3,4] are discussed. For AlN the best results are obtained by growing films on c-plane sapphire with MOCVD using either trimethylaluminum (TMA) [5] or triethylaluminum (TEA) [6] and ammonia (NH_3). Here, we describe the use of the alternative precursor tritertiarybutylaluminum (TTBA) as aluminum source for MOCVD growth of AlN. One advantage of TTBA especially in comparison to TMA may be a reduction of the carbon contamination in the grown AlN films because of the β-hydride elimination. This reaction leads to a desorption of the tertiarybutyl-radical from the growth surface without deposition of carbon as it was observed by the deposition of Al from tritertiarybutylaluminum [7]. The good structural quality of the AlN films as analyzed by atomic force microscopy (AFM), high resolution X-ray diffraction (HRXRD) and photothermal deflection spectroscopy (PDS) also encouraged us to grow AlN/GaN hetero-

Mat. Res. Soc. Symp. Proc. Vol. 423 © 1996 Materials Research Society

structures in order to investigate the influence of an AlN sublayer with varying thickness on the optical and structural properties of a GaN layer.

EXPERIMENT

AlN films with a thickness of 0.3 μm were grown by metalorganic chemical vapor deposition (MOCVD) from tritertiarybutylaluminum (TTBA) and ammonia (NH_3) on c-plane sapphire without a buffer-layer in a horizontal cold wall reactor. The growth temperature was 1050 °C and the deposition pressure was 100 mbar. Using hydrogen as transport gas the TTBA flux was maintained at 2.7 μmol/min.

To investigate the structural properties of the AlN thin films, high resolution X-ray diffraction (HRXRD) and atomic force microscopy (AFM) was applied. A modified triple axis diffractometer, which enables the recording of weak reflections, equipped with a graded multilayer mirror (Göbel mirror) [8] and two four-bounce Ge(220) channel cut crystals as incident beam optics and a second Göbel mirror for the diffracted beam optics was used. The Göbel mirror installed in front of the detector limits the angular acceptance of the detector to ~60 arcsec. Both the symmetric 002 reflection and the asymmetric 105 reflection were measured with $CuK\alpha_1$ radiation in ω-2Θ-scan mode as well as in ω-scan mode. For further structural investigations the samples were imaged by atomic force microscopy.

The position of the optical bandgap as well as the sub-bandgap absorption of the thin AlN films was analyzed by photothermal deflection spectroscopy (PDS) in the energy range from 0.9 to 6.5 eV. With this method it is possible to determine absorption coefficients as low as $\alpha = 0.1\ cm^{-1}$ for 1 μm thick films. The experimental details are described elsewhere [9,10].

In order to investigate the influence of an increasing AlN layer thickness on the optical and structural properties of GaN, AlN/GaN-heterostructures were grown. AlN layers with variable thickness of 0.02, 0.06, 0.18, 0.25 and 0.36 μm were deposited on c-plane sapphire without a buffer layer as mentioned above. Subsequently GaN layers with a constant thickness of 0.75 μm were grown using triethylgallium (TEG) and NH_3 at a temperature of 950°C and a pressure of 100 mbar with growth rates of about 1.5 μm/h.

Photoluminescence-(PL)-spectra of GaN were measured with the 333.2 nm line of an Ar^+-laser operating at an excitation power of 0.25 mW. Recording several reflections of GaN, the lattice constant c_0 of the GaN films was determined by XRD with a Philips PW3040 diffractometer using the $CuK\alpha_1$ line. Raman spectra of the GaN layers were recorded with a micro-Raman system operating in back scattering geometry, where the 514.5 nm line of an Ar^+ laser was focused through an orifice and a microscope on the GaN films.

RESULTS

Fig. 1 shows a typical AFM-image of a 0.3 μm thick AlN film grown on c-plane sapphire at 1050 °C without a buffer layer. The surface roughness is of the order of 12 nm (rms-roughness). Clearly visible are individual crystallites with an averaged size of 0.1 μm. The orientation of the crystallites follows the epitaxial relationship (0001)AlN || (0001)Al_2O_3 and $[1\bar{1}00]$AlN || $[1\bar{2}10]Al_2O_3$ as noticeable proven by XRD-measurements, but a slight in-plane rotation between individual grains can be observed. To further investigate the structural features, high resolution X-ray diffraction (HRXRD) was applied using a modified triple-axis diffractometer. Fig. 2 a-d show both the ω scan diffraction profiles and the ω-2Θ scan

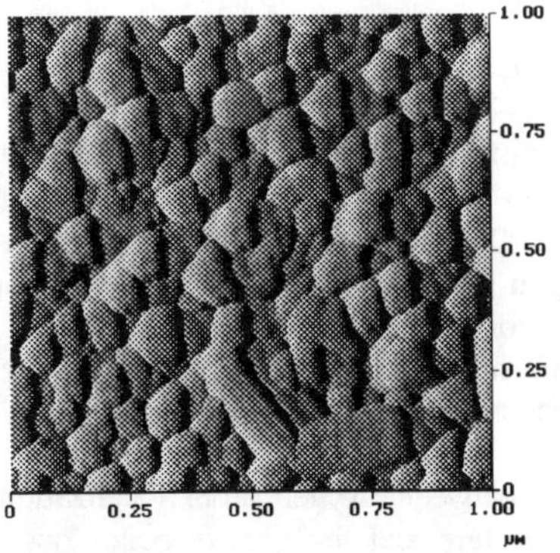

Fig. 1: AFM image of AlN grown on c-plane sapphire at 1050°C

diffraction profiles of the symmetric 002 and the asymmetric 105 reflection of AlN. Compared to the 006 reflection of sapphire, which exhibits only a FWHM of 30 arcsec in both scan directions, the peak broadening of the AlN 002 reflection for the ω-2Θ scan is mainly due to a small coherence length in the direction normal to the substrate surface and due to screw dislocations with Burgers vectors lying parallel to [0001] [11,12]. The low FWHM of the AlN 002 rocking curve (ω scan) indicates a small out-of-plane mosaic spread [13]. The 002 peak broadening for the ω scan is mainly attributed to a small coherence length of the AlN film parallel to the substrate surface which can also be seen in the Lorentzian peak shape. In contrary, the AlN 105 reflection exhibits for both scan directions a distinct peak broadening.This effect can be attributed to the presence of a large number of edge dislocations with Burgers vectors $1/3[11\bar{2}0]$ [12]. This sort of dislocation cannot be seen in the symmetric 00l reflections, because in this case burgers vector and diffraction vector form an angle of 90°. The presence of such edge dislocations can also be seen in a slight in-plane rotation of the individual crystallites with respect to the hexagonal c-axis due to the formation of small angle grain boundaries in the film structure.

Fig. 2: X-ray diffraction peaks of 0.3 µm AlN grown on c-plane sapphire a) ω scan AlN 002 b) ω scan AlN 105 c) ω-2Θ scan AlN 002 d) ω-2Θ scan AlN 105

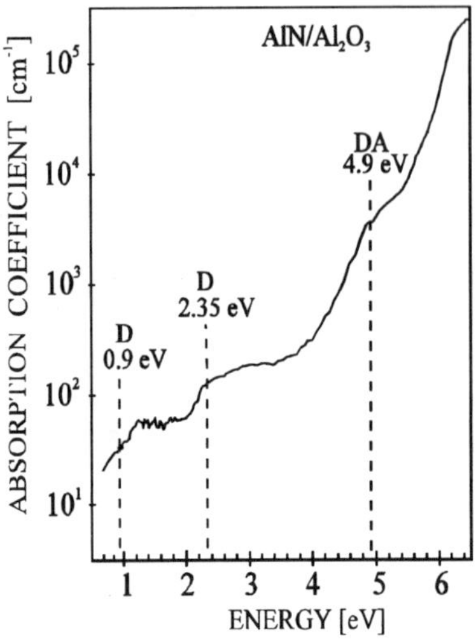

Fig. 3: PDS spectrum of AlN grown on c-plane sapphire

The energetic position of deep defects in the AlN films was analyzed by photothermal deflection spectroscopy (PDS). Fig. 3 shows the absorption coefficient measured by PDS for an AlN sample of 0.4 μm thickness. Three absorption shoulders can be distinguished at energies of 4.90, 2.35 and 0.90 eV, which probably are correlated with oxygen-induced deep donor levels and a donor-acceptor transition [14,15]. Since the oxygen content is known to strongly influence the micro-structure of the grown AlN films [16], the PDS results can be used to investigate the relationship between oxygen atom localization in the AlN structure and the growth conditions and their connection with the change in the film morphology.

In order to study the influence of an increasing AlN layer thickness on the optical and structural properties of GaN layers with constant thickness, AlN/GaN heterostructures were analyzed by PL measurements, Raman spectroscopy and XRD. Fig 4. shows the PL-spectra at room temperature of 0.75 μm thick GaN layers grown on 0.02, 0.06, 0.18, 0.25 and 0.36 μm thick AlN films, where the PL maxima are attributed to excitons bound to neutral donors [17,18].

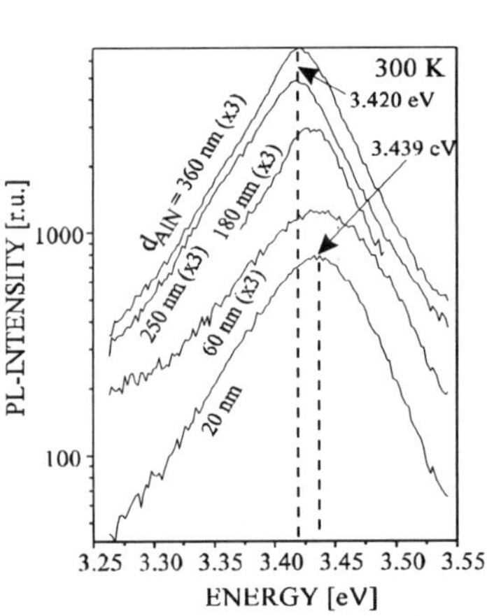

Fig. 4: PL-spectra of 750 nm GaN films grown on AlN sublayers of different thicknesses

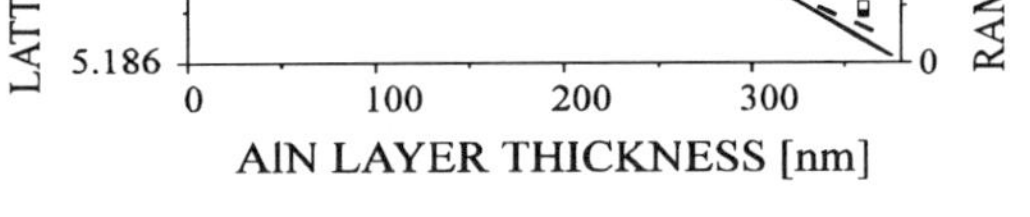

Fig. 5: Shift of the E_2-Raman mode and lattice constant c_0 of GaN versus AlN sublayer thickness

GaN films grown on thin AlN films (0.02 and 0.06 μm) exhibit a broad line shape and a significant shift of the PL-maxima to higher energies (> 3.435 eV), whereas at an AlN-layer thickness of 0.18 μm the PL-maximum is positioned at 3.430 eV with a FWHM of only 75 meV. Samples with thick AlN layers (0.25 and 0.36 μm) emit at only 3.420 eV. The tendency, that the GaN PL-maximum decreases with increasing AlN thickness can also be seen in a significant reduction of the lattice constant c_0 of GaN from 5.194 Å (0.06 μm AlN) to 5.187 Å (0.36 μm AlN).

To exclude the possible influence of an AlGaN alloy formation on the XRD results, micro-Raman measurements were also performed. A shift of the E_2 mode of GaN relative to the intrinsic value of 586 cm^{-1} with increasing AlN thickness was determined. The obtained XRD and micro-Raman results are summarized in Fig. 5.

The value of the lattice constant c_0 of a 0.75 μm thick GaN layer becomes almost equal to the bulk value of 5.185 Å [19,20] and the Raman shift almost vanishes. This behaviour is clearly attributed to a reduction of the biaxial compressive stress in the GaN films with increasing AlN sublayer thickness. In addition, the linear coefficient for the near bandgap luminescence shift due to biaxial compressive stress was calculated as (27 ± 4) meV/GPa for the XRD experiment and (21 ± 5) meV/GPa for micro-Raman spectroscopy [21].

CONCLUSIONS

Epitaxial AlN thin films were grown on c-plane sapphire without a buffer layer by MOCVD using the precursors tritertiarybutylaluminum (TTBA) and NH_3. Structural investigations by atomic force microscopy and high resolution X-ray diffraction indicate a slight in-plane rotation of the individual crystallites resulting in the formation of small angle grain boundaries, where edge dislocations with burgers vectors lying parallel to the hexagonal a-axes are predominant. Photothermal deflection spectroscopy shows an oxygen contamination in the AlN layers, which may influence the film morphology. Using a 0.36 μm thick AlN film as a sublayer in an AlN/GaN heterostructure grown on c-plane sapphire, the biaxial compressive stress in a 0.75 μm thick GaN film due to the lattice mismatch and the large difference in the thermal expansion coefficients between GaN and sapphire almost vanishes. The value of the photoluminescence maximum shift due to the biaxial compressive stress was independently obtained by micro-Raman and X-ray diffraction experiments as about 24 meV/GPa. This value differs significantly from 41 meV/GPa determined by hydrostatic pressure experiments [22]. This difference is attributed to a volume conserving distortion of the unit cell for the case of biaxial compressive stress in contrast to an uniform reduction of the unit cell, when a hydrostatic pressure is applied.

It is intended to improve the structural qualities of the AlN films by growing either AlN or GaN buffer layers between the substrate and the AlN film. Further investigations will show if tritertiarybutylaluminum is a suitable precursor for the deposition of GaAlN alloys using triethylgallium as Ga source.

ACKNOWLEDGEMENTS

This work was supported by the Deutsche Forschungsgemeinschaft (Research Grant No. Stu 193/3) and the Bayerisches Staatsministerium für Unterricht, Kultus, Wissenschaft und Kunst (Research Grant No. IX/2-52 C(12)-61/27/87 107). The authors would like to thank Dr. H.

Göbel and Dr. M. Schuster from Siemens AG Munich for their assistance in the HRXRD-measurements and M. Kelly for critically reading the manuscript.

REFERENCES

1. I. Akasaki, H. Amano, Y. Koide, K. Hiramatsu and N. Sawaki, J. Cryst. Growth **98**, 209 (1989).
2. T. D. Moustakas, T. Lei and R. J. Molnar, Physica B **185**, 36 (1993).
3. A. Kuramata, K. Horino, K. Domen, K. Shinohara and T. Tanahashi, Appl. Phys. Lett. **67**, 2521 (1995).
4. J. Chaudhuri, R. Thokala, J. H. Edgar and B. S. Sywe, J. Appl. Phys. **77**, (1995) 6263.
5. M. Morita, S. Isogai, N. Shimizu, K. Tsubouchi and N. Mikoshiba, Jpn. J. Appl. Phys. **19**, L173 (1981).
6. M. Asif Khan, J. N. Kuznia, R. A. Skogman, D. T. Olson, M. Mac Millan and W. J. Choyke, Appl. Phys. Lett. **61**, 2539 (1992).
7. A. C. Jones, J. Auld, S. A. Rushworth and G. W. Critchlow, J. Cryst. Growth **135**, 285 (1994).
8. M. Schuster and H. Göbel, J. Phys. D: Appl. Phys. **28**, A270 (1995).
9. W. B. Jackson, N. M. Amer, A. C. Boccara and D. Fournier, Appl. Optics **20**, 1333 (1981).
10. O. Ambacher, W. Rieger, P. Ansmann, H. Angerer, T. D. Moustakas and M. Stutzmann, Solid State Communications **97**, 365 (1996).
11. D. Kapolnek, X. H. Wu, B. Heying, S. Keller, B. P. Keller, U. K. Mishra, S. P. DenBaars and J. S. Speck, Appl. Phys. Lett. **67**, 1541 (1995).
12. B. Heying, X. H. Wu, S. Keller, Y. Li, D. Kapolnek, B. P. Keller, S. P. DenBaars and J. S. Speck, Appl. Phys. Lett. **68**, 643 (1996).
13. Q. Zhu, A. Botchkarev, W. Kim, Ö. Aktas, A. Salvador, B. Sverdlov, H. Morkoc, S.-C. Y. Tsen and David J. Smith, Appl. Phys. Lett. **68,** 1141 (1996).
14. S. Pacesova and L. Jastrabik, Czech. J. Phys. B **29**, 913 (1979).
15. J. Pastrnak, S. Pacesova and L. Roskovcova, Czech. J. Phys. B **24**, 1149 (1974).
16. S. Abaidia, V. Serin, G. Zanchi, Y. Kihn and J. Sevely, Philosophical Magazine A **72**, 1657 (1995).
17. O. Lagerstedt and B. Monemar, J. Appl. Phys. **45**, 2266 (1974).
18. B. C. Chung and M. Ghershenzon, J. Appl. Phys. **72**, 651 (1992).
19. H. P. Maruska and J. J. Tietjen, Appl. Phys. Lett. **15**, 327 (1969).
20. T. Detchprom, K. Hiramatsu, K. Itoh and I. Akasaki, Jpn. J. Appl. Phys. **31**, L1454 (1992).
21. W. Rieger, T. Metzger, H. Angerer, R. Dimitrov, O. Ambacher and M. Stutzmann, Appl. Phys. Lett. **68**, 970 (1996).
22. P. Perlin, I. Gorczyca, N. E. Christensen, I. Grzegory, H. Teisseyre and T. Suski, Phys. Rev. B **45**, 13307 (1992).

PROPERTIES OF HOMOEPITAXIALLY MBE-GROWN GaN

T. Suski*,#, J. Krueger*, C. Kisielowski*, P. Phatak+, M.S.H. Leung+, Z. Liliental-Weber+, A. Gassmann*, N. Newman*, M.D. Rubin+, E.R. Weber*, I. Grzegory#, J. Jun#, M. Bockowski#, S. Porowski#, and H.I. Helava++

*Lawrence Berkeley National Laboratory, and Dept. of Materials Science, University of California, Berkeley, CA 94720, USA. tadek@ux5.lbl.gov tadek@iris.unipress.waw.pl;

+Lawrence Berkeley National Laboratory, Berkeley, CA 94720, USA

#UNIPRESS, High Pressure Research Center, 01-142 Warszawa, POLAND

++ American Xtal Technology, Dublin, CA 94568, USA

ABSTRACT

Bulk single crystals of GaN were used for epitaxial growth of GaN films by molecular beam epitaxy. Low temperature photoluminescence yields much higher intensity emission in the near band-edge region for epitaxial films with respect to the situation in bulk crystals. Character of this luminescence changes also. Dominant band-to-band transitions in the bulk crystals are exchanged by bound exciton and/or donor-acceptor pair transitions observed in the epitaxial layers. We will compare the obtained results with the available data on the homoepitaxial samples grown by metalorganic chemical vapor deposition method and discuss the importance of establishing the basic information on energetic positions of excitonic transitions in stress free samples.

INTRODUCTION

Very recently GaN and other III-V nitrides have attracted widespread attention for optoelectronic applications in the blue and ultraviolet spectral range. Despite the successful fabrication highly efficient light emitting diodes [1] and first demonstration of a blue laser diode [2] the p-type doping problems and inadequate structural quality of III-V nitride epitaxial films remain important issues. Large lattice mismatch and differences in thermal expansion between the commonly used substrates (sapphire, SiC) and constituent epitaxial layers leads to a large number of extended defects, e.g. dislocations. Their density reaches the range of 10^9cm^{-2} -$10^{10}cm^{-2}$. Low/high temperature buffer layers made of GaN or AlN films lead to a significant improvement in the quality of heteroepitaxially grown layers of various nitrides [3]. Another idea consists of using new substrates with properties well matched with the nitride compounds (e.g., ZnO, $LiGaO_2$, $LiAlO_2$, $AlMg_2O_4$).

However, the use of GaN single crystals as substrates for the homoepitaxial growth of GaN represents the most natural way to reduce the number of defects in an epitaxial film. This idea has become realistic after wurtzite GaN crystals in the form of platelets with lateral dimensions of a few millimeters had been grown by means of a high-pressure, high-temperature method [4]. Then, a homoepitaxy using metalorganic chemical vapor deposition (MOCVD) has been successfully employed to grow high quality GaN layers [5,6].

The other important aspect of the existence of well-characterized homoepitaxial films of GaN consists in the opportunity of establishing benchmark values for nitride compounds. Stress and strain effects, present in GaN layers grown on lattice mismatched substrates with a different thermal expansion, modify various properties of GaN films. For example, excitonic transitions studied by photoluminescence exhibit a significant shift of their energetic positions (see e.g. [7]).

The purpose of this paper is to present some experimental results on homoepitaxial GaN samples grown by molecular beam epitaxy (MBE). We will concentrate on the optical properties obtained by low temperature photoluminescence measurements. We will compare the obtained results with the available data on the homoepitaxial samples grown by MOCVD method and discuss the importance of establishing the basic information on energetic positions of excitonic transitions in stress free samples.

EXPERIMENT

Bulk GaN crystals, which were used as substrates, were grown from a dilute solution of atomic

Mat. Res. Soc. Symp. Proc. Vol. 423 © 1996 Materials Research Society

nitrogen in liquid gallium at temperatures about 1500°C and a nitrogen pressure of 1.5 GPa [8]. The high-pressure, high-temperature method of GaN synthesis yield crystals which grow in the wurtzite structure, usually as hexagonal platelets perpendicular to the *c*-axis. Their one side is usually very flat [9]. This surface (in crystals with lateral dimensions of 3-4 mm) was used for the GaN film depositions. Background doping in bulk samples is n-type with a carrier concentration above 10^{19} cm^{-3}. This is presumably due to a high concentration of nitrogen vacancies. The width of the X-ray rocking curve for (0004) CuKα reflection ranges between 30-60 arcsec. Prior to the film growth , the GaN substrate was cleaned with HF. The GaN layers were grown in a home-built MBE chamber (sample A) and in the Riber chamber (sample B), using a hollow anode source to provide activated nitrogen species [10]. The nitrogen source is designed to minimize defect formation from contamination and ion damage. The MBE chambers were cryopumped and reached base pressure in the 10^{-9} Torr range. For sample A, we initiated the film growth at 600°C and ramped the substrate temperature up to about 675°C within 5 minutes. This procedure allowed for a minimal decomposition of the bulk GaN substrate prior to deposition. For sample B, the deposition started from 10 min nitradation at 695°C by exposure to nitrogen plasma. The growth took place at 675°C. The films were grown in 8-hour processes, resulting in approximately 0.5 um thick layers. No buffer layer was used. The thin films were then characterized by low temperature photoluminescence (PL) and Raman scattering (T=300K). In addition, a transmission electron microscopy study were performed on Sample A. In the PL studies, which are the main subject of this paper, the samples were excited by the 325 nm line of a 50 mW HeCd laser. The luminescence signal was dispersed by a 0.85 m double monochromator and detected by an UV-sensitive photomultiplier.

RESULTS AND DISCUSSION

Fig. 1a (1b) compares the low-temperature PL (T=6K) in the sample A (sample B) consisting of the bulk GaN crystal substrate and the GaN film deposited on the bulk sample. For the sample A, the thin film shows a sharp intense peak at 3.467 eV with a full width at half maximum (FWHM) of 16 meV. This peak is assigned to a recombination of exciton bound to a shallow neutral donor (EBND). The lower energy shoulder at 3.420 eV and a peak at 3.367 are observed also. They are likely associated with more deeply bound excitons. The yellow luminescence peak at ~2.3 eV which is commonly observed in GaN is almost entirely absent in the spectrum of the deposited film. The intensity ratio, R, of the band-edge luminescence to the midgap luminescence at 2.3 eV for the homoepitaxial film is 600. This compares to a ratio about 0.1 in the bulk substrates of the sample A.

The PL spectra of the both bulk crystals show a weak broad peak near the band edge (FWHM of 0.1 eV-0.2 eV). The observed emission is related to band-to-band transitions. For majority of bulk GaN crystals, the position of this peak exceeds the value of the GaN band gap (3.5 eV at 4.2K). The Burstein-Moss effect seems to be the most likely explanation of this shift.

One can notice easily that the low temperature PL spectrum of the homoepitaxial GaN layers of the sample B (Fig.1b) differs from that characterizing sample A. The near band edge luminescence consists of two peaks at 3,465 eV and 3,423 eV (with FWHM of the order 30-40 meV). We ascribe the higher energy band to EBND transition and the lower energy band to more deeply bound excitons. This double peak structure is accompanied by the complex band caused by donor-acceptor pair (DAP) recombination. The zero phonon line transition occurs at 3.266 eV and it is followed by two phonon replicas. The yellow luminescence band is visible at 2.33 eV and the intensity ratio of the band-edge luminescence to the midgap luminescence, R, for this homo-layer is 4. Comparing this value with R=600 for the film A and taking into account differences in the FWHM of the excitonic bands (two times more narrow in the layer A) show that the film A is of much higher quality. Moreover, one can compare the data of PL measurements performed on samples A and B with results obtained for the homoepitaxial samples grown by MOCVD methods [5,6] (Table 1). Sample 1 grown in Warsaw exhibits excitonic bands originating in free excitonic transitions. This fact as well as the information that the FWHM of the bound excitonic transitions can be below 1 meV show the superior quality of this sample with respect to all others described in the Table 1. The lower quality of the sample 2 can be caused by the mechanical polishing of its surface before the GaN film deposition.

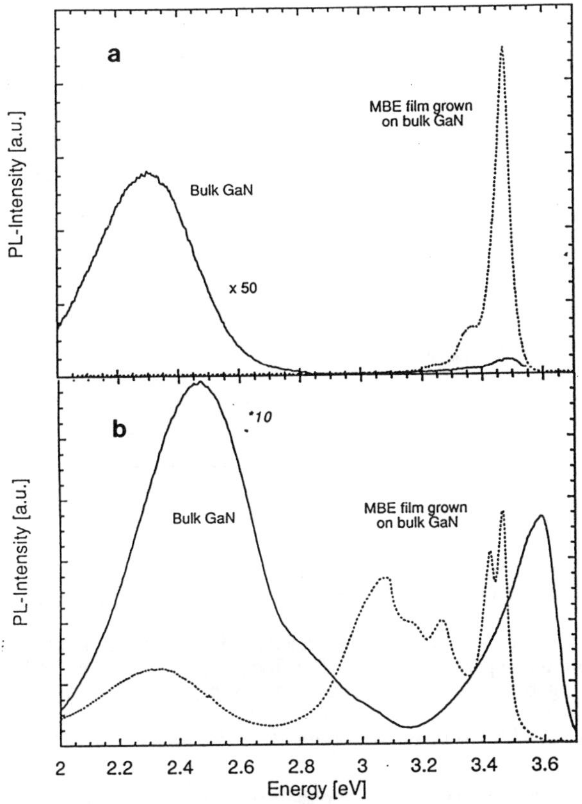

Figure 1. Low temperature (T=6K) photoluminescence of two homoepitaxially grown GaN structures A (a) and B (b). The spectra for epi-layers and bulk GaN substrates are shown.

Now, we will discuss the usefulness of the obtained information about the energetic position of EBND transition an zero phonon line of DAP recombination in homo-layers which we assume are free of stress. Most common structures of GaN consist of heteroepitaxial layers grown on highly lattice mismatched substrates of Al2O3 or SiC. It is an accepted assumption that most of the related stress (mostly biaxial in its nature) is relaxed at a growth temperature by forming misfit dislocations. However, differences in thermal expansion of a layer and a substrate result in the stress incorporation into the grown structure during its cooling (see e.g.[11]). One of the basic questions here concerns properties of stress-free GaN material. Homoepitaxial samples give the best opportunity to determine the required benchmark values. Creation of defect network (mainly dislocations) and a complicated stress field distribution caused by the lattice and thermal expansion misfits, should be eliminated in a homoepitaxial growth.

The most basic parameters characterizing a semiconductor material are its band gap, Eg, and frequencies of phonons. Their changes could be directly related to the stress magnitude present in the material. We used the energetic position of EBND to monitor Eg variation. This band dominates usually low temperature PL spectra. E2 phonon mode which has transverse optical character (E2 is not sensitive to electron concentration) was chosen to trace the response of the crystal lattice onto the stress incorporated in the growth and cooling processes. Fig. 2 represents a graph on which values of E2 phonon frequency are plotted as a function of energetic position of EBND for several epitaxial layers of GaN. The experimental points, though scattered, are arranged along the line corresponding to shift to higher (lower) energies of the both E2 and EBND. The filled and open squares represent the homo-

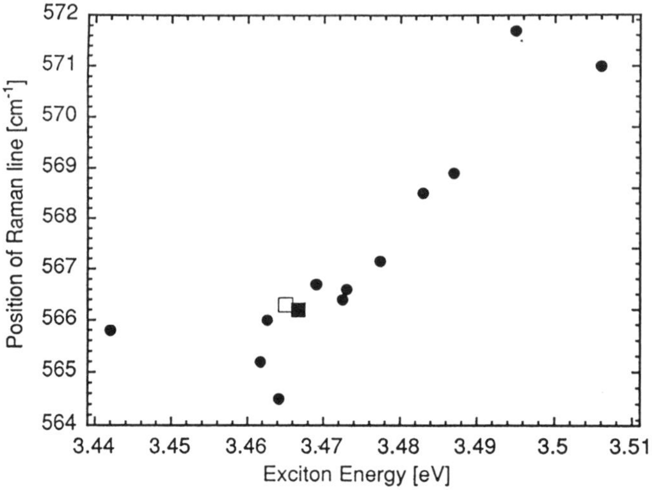

Figure 2.Frequency of E2 phonon mode versus energy of exciton bound to neutral donor for different GaN layers. Squares correspond to homoepitaxially grown films. Circles represent hetero-layers.

layers A and B, respectively, giving a value for stress free material. The position of the corresponding data points is the same within the experimental error. The filled circles correspond to heteroepitaxially grown samples. Points lying below the squares represent samples with the tensile stress. The experimental data points situated above the squares characterized samples with a compressive stress. The magnitude of the incorporated stress could be as high as 1 GPa. The details of this diagram will be described elsewhere [11].

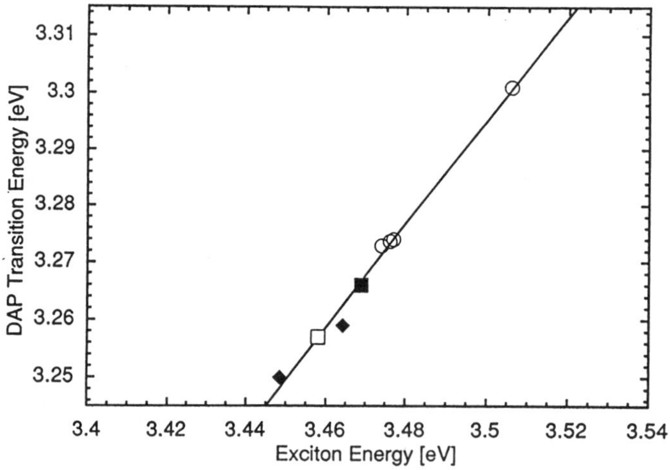

Figure 3. Energy of zero phonon line of donor acceptor pair transition versus the energy of the exciton bound to neutral donor for homo- and hetero-epitaxial GaN samples.

TABLE I. Energetic positions of the optical transitions in the exciton region for homoepitaxially grown GaN samples. Neutral donor bound exciton transition are underlined.

Sample	1 - MOCVD Warsaw, Ref.[5]	2 - MOCVD Xerox, Ref. [6]	A - MBE LBL, this work	B- MBE LBL, this work
Free excitons [eV]	E_A= 3.4780 E_B= 3.4835 E_C= 3.502			
Bound excitons [eV]	E_0= 3.4719 $\underline{E_1= 3.4666}$	$\underline{E_a= 3.458}$	$\underline{E_a= 3.467}$ E_b= 3.420 E_c= 3.367	$\underline{E_a= 3.465}$ E_b= 3.423
Donor-acceptor pair, zero phonon line, [eV]		E_p= 3.257		E_p= 3.266

Fig. 3 represents a plot of the position of zero phonon line of DAP transition versus the energetic position of the EBND for homo- and hetero-epitaxial GaN samples. Since the shallow donor and shallow acceptor impurity states are involved in the DAP transition one can expect that the experimental dependence can be fitted by the line with the slope 1. We found a slope about 0.9 for the discussed dependence. The filled and open squares on Fig. 3 represent the homo-layers A and the sample 2 [6], respectively. We believe that the sample A represents the layer very close to the stress free GaN material. This can be deduced from the similarities in the data for EBND in the samples 1, A, and B. High structural perfection of the sample 1 is important here, whereas, polishing of the sample 2 prior to the growth might introduce some amount of stress and/or strain. The shift in the position of the near band edge PL for the substrate bulk crystal to 3.32 eV (~ 3.5 eV usually observed) supplies an argument supporting this suggestion.

From the TEM studies of samples 1 and A it has become clear [9] that a preparation of bulk crystal surface plays an important role in a drastic reduction of defects present at the interface between homoepitaxial layer and the bulk substrate crystal.

SUMMARY

We report the growth of the homoepitaxial GaN layers by MBE method. In contrast to the heteroepitaxial layers of GaN, they represent structures with very low amount of incorporated stress originating from lattice mismatch and differences in thermal expansion coefficients between the GaN substrate and an GaN epilayer. In this paper we concentrated on the optical properties of the obtained films. Low temperature photoluminescence and Raman scattering studies were employed to establish values of energy characteristic for neutral-donor bound exciton and donor-acceptor pair transitions in stress-free GaN material.

ACKNOWLEDGEMENTS

This work was supported by the Office of the Energy Research, Office of Basic Energy Science, Division of Advanced Energy Project (BES-AEP) and by the Laboratory Technology Transfer Program (ER-LTT) of the U.S. Department of Energy under Contract No. DE-AC03-76SF0098. T. Suski was supported by Fulbright Scholarship and Z. Liliental-Weber by the Director, Office of Energy Research, the U.S. Department of Energy under the Contract No. DE-AC03-76SF00098.

REFERENCES

1. S. Nakamura, M. Senoh, N. Iwasa, S. Nagahama, T. Yamada, and T. Mukai, Jpn. J. Appl. Phys. 34, 797 (1995).
2. S. Nakamura, M. Senoh, S. Nagahama, N. Iwasa, T. Yamada, T. Matsushita, H. Kiyoku, and Y. Sugimoto, Jpn. J. Appl. Phys. **35**, L74 (1996).
3. H. Morkoc, S. Strite, G.B. Gao, M.E. Lin, B. Sverdlov, and M. Burns, J. Appl. Phys. **76**, 1363 (1994).
4. S. Porowski, I. Grzegory, and J. Jun, in High Pressure Chemical Synthesis, edited by B. Baranowski and J. Jurczak (Elsevier Science, Amsterdam, 1989) p. 21.
5. K. Pakula, A. Wysmolek, K.P. Korona, J.M. Baranowski, R. Stepniewski, I. Grzegory, M. Bockowski, J. Jun, S. Krukowski, M. Wroblewski, and S. Porowski, Solid State Commun. **97**, 919 (1996).
6. F.A. Ponce, D.P. Bour, W. Gotz, N.M. Johnson, H.I. Helava, I. Grzegory, J. Jun, and S. Porowski, Appl. Phys. Lett., **68**, 917 (1996).
7. H. Amano, K. Hiramatsu, and I. Akasaki, Jpn. J. Appl. Phys. 27, L1384 (1988).
8. I. Grzegory and S. Krukowski, Physica Scripta, T**39**, 242 (1991).
9. Z. Liliental-Weber et al., Proc. of 1995 MRS Fall Meeting, Boston.
10. A. Anders, N. Newman, M. Rubin, M. Dickinson, A. Thomson, E. Jones, P. Phatak, and A. Gassmann, Rev. Sci. Instr., (in press).
11. Ch. Kisielowski et al. , to be published.

LATTICE-MATCHED InAsN(X=0.38) ON GaAs GROWN BY MOLECULAR BEAM EPITAXY

Y.C. Kao*, T.P.E. Broekaert*, H.Y. Liu*, S. Tang*, I.H. Ho**, and G.B. Stringfellow**
*Texas Instruments, MS147, P.O. Box 655936, Dallas, TX 75265, kao@ti.com
**University of Utah, College of Engineering, Salt Lake City, Utah 94112

ABSTRACT

In this paper, we report the MBE growth of high nitrogen content lattice-matched $InAs_{1-x}N_x$ (x=0.38) single crystal epitaxial films on GaAs. The nitrogen incorporation is about an order higher than previously reported on other mixed group V nitride alloys. These data are consistent with a nitrogen solubility limit calculation in various III-V binary alloys, which predicts orders of magnitude higher nitrogen incorporation in InAs than any other alloys. InAsN growths were obtained using a modified ECR-MBE system with atomic-nitrogen generated by an ECR plasma source. Improved crystal quality was obtained using a "template" growth technique. An x-ray linewidth of 270 arc-s was achieved on a 0.4 μm thick $InAs_{0.62}N_{0.38}$/GaAs multi-layer structure. Hall effect data show these InAsN films are semi-metallic.

INTRODUCTION

Recently, the growth of mixed group V nitride alloy III-V compound semiconductors has gained interest [1-3]. This increased interest in nitride-based semiconductors is primarily related to the potential applications in short wavelength (blue and violet) emitting devices and the capability of growing arsenide nitride and phosphide nitride lattice-matched to silicon [2]. The potential of combining the direct bandgap materials on a Si substrate is attractive for Si-based bandgap engineering applications such as base layers in HBTs, channel layers in FETs, low

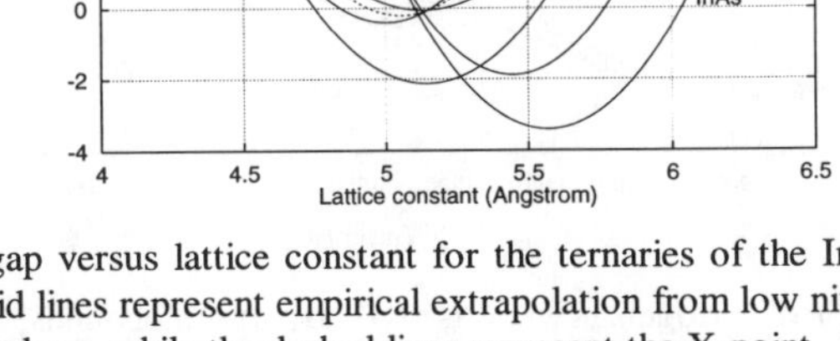

Fig. 1 Estimated bandgap versus lattice constant for the ternaries of the InGaAlAsPN material system. The solid lines represent empirical extrapolation from low nitrogen content alloys for the direct bandgap, while the dashed lines represent the X-point. The point Fam is the *ab initio* theoretical prediction for the bandgap of famatinite $GaAs_{0.75}N_{0.25}$.

Mat. Res. Soc. Symp. Proc. Vol. 423 © 1996 Materials Research Society

ohmic contact layers, and quantum well layers for resonant tunneling devices. Figure 1 shows the estimated bandgap versus lattice constant for the ternaries of the InGaAlAsPN material system [4]. The solid lines represent empirical extrapolation from low nitrogen content alloys for the direct bandgap, while the dashed lines represent the X-point. A bowing coefficient in GaAsN (18 eV) for arsenide nitride and in GaPN (14 eV) for phosphide nitride were used in the calculation. The point labeled Fam represents an *ab initio* estimate from famatinite ordered $GaAs_{0.75}N_{0.25}$with a predicted bandgap of about 1 eV, indicating that some of the intermediate alloys may have finite bandgaps contrary to the low nitrogen quadratic extrapolation.

The major challenge in growing the group III-V-nitride alloy compounds results from the large difference in the atomic radii of nitrogen and the other group V elements. Such material systems with large differences in atomic radii are known to have large miscibility gaps and cannot be grown by equilibrium techniques [5]. However, immiscible material systems can be grown by non-equilibrium techniques such as molecular beam expitaxy (MBE) with some limited success. For example, the previously reported maximum N contents in single-crystal epitaxial GaAsN [1] and GaPN [6] are 1-2% and 3-5%, respectively.

Nitrogen Solubility Limit

To better understand the nitrogen incorporation in various binary III-V systems, we performed a detailed calculation of the solubility of N in InAs, InP, GaAs, GaP, AlAs, and AlP. The calculation, obtained under equilibrium conditions, is based on an assumption that the lattice is relaxed around a nitrogen atom to the 6th neighbor in a zincblende nitride structure [7,8]. The results (Figure 2) show a dramatic spread in solubility. The solubilities of N in both InP and InAs are orders of magnitude higher than for N in GaAs, while the solubilities in AlAs and AlP are orders of magnitude smaller [7]. We believe the major factor affecting solubility may be the force constant. It is relatively weak for the In compounds, as compared with GaAs, and relatively strong for the Al compounds. Although these results were obtained based on equilibrium conditions, we found N incorporation is indeed much higher in InAs when grown by the non-equilibrium MBE technique.

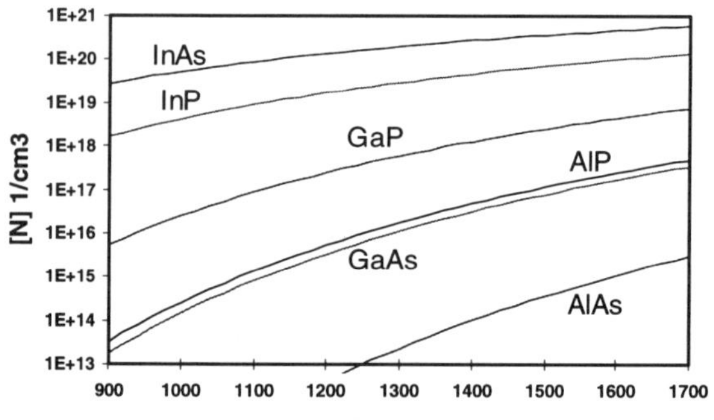

Fig. 2 Solubility limit of nitrogen in various III-V binary compounds. The system having the highest calculated nitrogen solubility isInAsN.

In this paper, we report the experimental results of MBE growth of $InAs_{1-x}N_x$ single-crystal epitaxial films with much higher incorporation of nitrogen (38% to date), which is lattice-matched to GaAs. We expect the lattice matching of $InAs_{1-x}N_x$ on silicon, which requires 60% nitrogen, will also be achievable.

EXPERIMENT

The III-AsN growth was carried out in a Riber MBE system equipped with elemental effusion sources for Ga, In, and Al and an electron cyclotron resonance (ECR) plasma source for atomic-N generation. Arsenic flux was generated and accurately controlled by a valved cell. Figure 3 shows a schematic of the retrofit ECR-MBE system used in this work. The ECR source was mounted at a glancing angle (15 to 25 degrees) with respect to the substrate to reduce ion damage. ECR plasma power was varied from 100 W to 250 W. The distance between ECR plasma chamber and wafer can be adjusted for uniformity and atomic-N intensity control. InAsN films were grown on either 3 in. GaAs(001) on-orientation substrates or on 3 in. Si wafers with (001) orientation tilted toward the [110] direction. Typical growth rate is around 0.7 μm/hr. Different As/N flux ratios are achieved by varying the As flux, while the atomic-N flux is fixed by maintaining a constant plasma power and N_2 flow (6N, ULSI grade) at 10 sccm using a UHV mass flow controller. Typical growth temperature is 400°C to 600 °C.

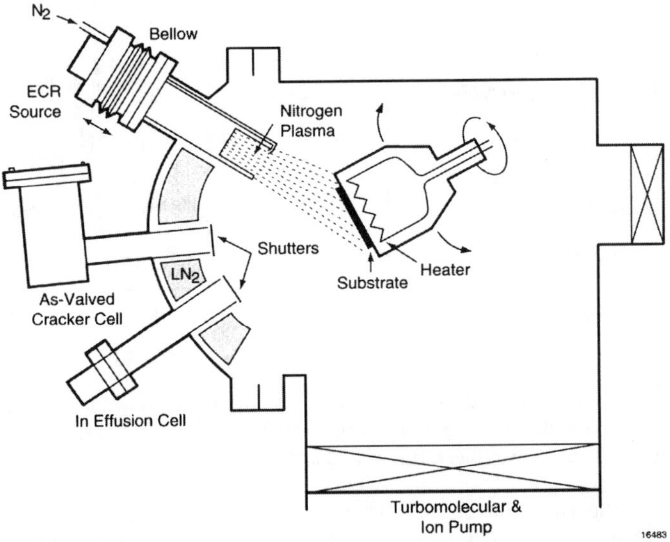

Fig. 3 Schematic of a modified ECR-MBE system. Note both incident angle and the distance of the ECR source to the substrate can be adjusted.

RESULTS AND DISCUSSION

Figure 4(a) shows the double crystal x-ray diffraction spectrum of a 200 nm thick single crystalline $InAs_{1-x}N_x$ (x=0.381) sample grown on a GaAs(001) substrate. The full width at half maximum (FWHM) is 460 arc-s and the lattice mismatch to GaAs is less than 1%. From an x-ray thin-film diffraction (XTFD) spectrum of this sample, we found this single InAsN layer actually contained polycrystalline cubic-InAs and InN grains. For InAsN samples grown at growth temperature over 550 °C, the nitrogen incorporation is significantly lower. However, at lower growth temperatures, nitrogen incorporated as poly-InN instead of single cyrstalline InAsN. For InAsN samples thicker than 200 nm, the x-ray spectra became wide and broad, indicating deteriorated crystalline quality. Using a "template" growth technique to maintain InAsN crystal quality by inserting a GaAs layer between the InAsN layers, we obtained InAsN epitaxial films with better crystalline quality. Figure 4(b) shows the x-ray rocking curve spectrum of a 4 period InAsN/GaAs multi-layer structure (4x100/100 nm). The FWHM of the InAsN peak is 270 arc-s, indicative of improved crystal quality compared to single-layer InAsN. The GaAs layers serve as "template" layers before everyInAsN layer growth.

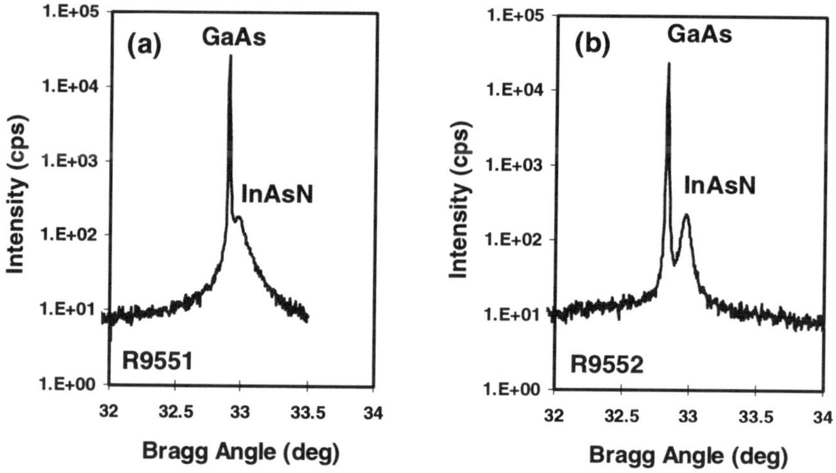

Fig. 4 Double crystal x-ray rocking curve spectra of (a) a 200 nm thick InAsN/GaAs sample showing InAsN linewidth of 460 arc-s, and (b) an InAsN/GaAs 4 period multi-layer structure with an InAsN linewidth of 270 arc-s.

The mobility and sheet charge density of the InAsN films were measured by Hall effect. Typical Hall mobility of the InAsN samples is around 50 to 100 cm^2/Vs with a high carrier density ($1 \times 10^{20} cm^{-3}$), indicating that the material may be semi-metallic. No photoluminescence has been observed from the InAsN material.

The growth of InAsN is not optimized at this moment. Figure 5 shows the cross-sectional scanning electron microscope (X-SEM) micrograph of the multi-layer InAsN/GaAs sample. The surface became roughened after the first InAsN layer growth. Details of the roughening are not clear; possibly some non-single-crystal defects formed on the InAsN surface prevent consequent 2D-layered growth.

Growth of high In containing InAsN compounds is critical for the realization of epitaxial InAsN layers lattice-matched to Si, which requires 58% nitrogen incorporation or around 2-to-3 As to N incorporation ratio. Several growths of InAsN on Si substrates were attempted under various As/N flux ratio. However, no single crystalline peaks were obtained from x-ray. The data

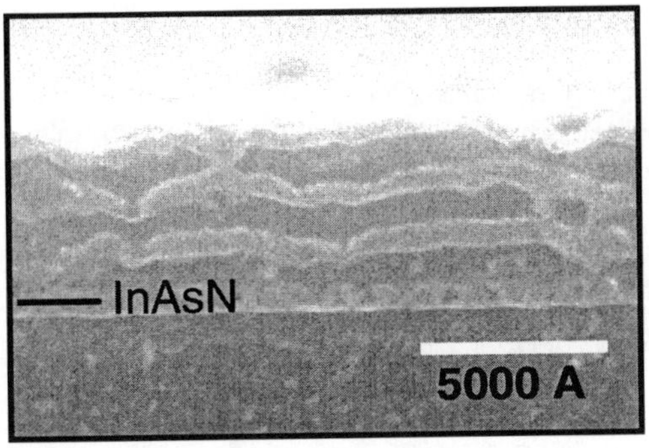

Fig. 5 Cross-sectional GaAs/InAsN multi-layer structure grown on GaAs. Each InAsN layer thickness is 0.1 µm. The growth becomes roughened after the first InAsN film growth.

from XTFD indicate the as-grown InAsN films are polycrystalline with cubic-InAs, cubic-InN, and hexagonal-InN grains observed. Figure 6 shows the TFD spectra of two InAsN/Si samples grown under different As/N flux ratio. For the sample grown with a lower As flux ratio, the InAs peak decreases over 20 times, while the c-InN peak intensity increases slightly.

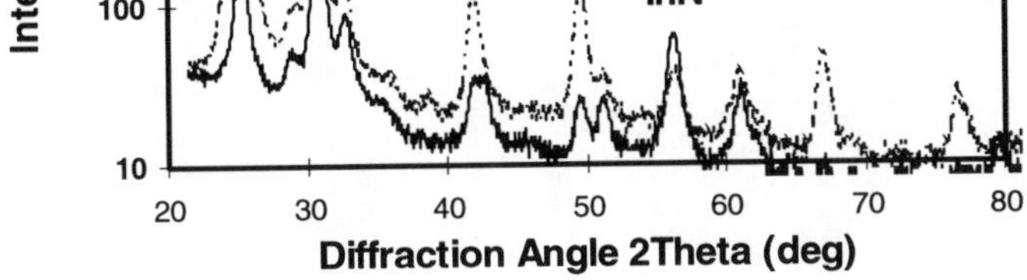

Fig. 6 X-ray thin film diffraction spectra of two InAsN/Si samples grown at different As/N flux ratios. No InAsN x-ray rocking curve peaks were observed in these two samples.

SUMMARY

We report, for the first time, the MBE growth of high N-content single crystalline lattice-matched InAsN (x=0.38) on GaAs. The nitrogen incorporation is about an order higher than previously achieved on other mixed group V nitride alloys. These data are consistent with a nitrogen solubility limit calculation in various III-V binary alloys. Single-crystal InAsN films were obtained using a modified ECR-MBE system with atomic-nitrogen generated by an ECR plasma source. We found thick InAsN films usually contain large amounts of poly-InAs and InN grains. We developed a "template" growth technique to maintain InAsN crystal quality by inserting a GaAs layer between InAsN layers. $InAs_{0.62}N_{0.38}$/GaAs layers with an x-ray linewidth of 270 arc-s (0.4 μm thick film) have been achieved. Hall effect data show these films are semi-metallic. We also attempted InAsN growth on Si and only polycrystalline films were achieved.

REFERENCES

1. M. Weyers, M. Sato, and H. Ando, Jpn. J. Appl. Phys. **31**, L854 (1992).
2. C.T. Foxon, T.S. Cheng, D.E. Lacklison, S.V. Novikov, D. Johnson, N. Baba-Ali, T.L. Tansley, S. Hooper, and L.J. Challis, presented at the 1994 Electronic Materials Conference, Boulder, CO, 1994.
3. M. Kondow K. Uomi, K. Hosomi, and T. Mozume, Jpn. J. Appl. Phys. **33**, L1056 (1994).
4. T.P.E. Broekaert, S. Tang, R.M. Wallace, E.A. Beam III, W.M. Duncan, Y.C. Kao, and H.Y. Liu, presented at the International Conference on Integrated Micro-Nanotechnology for Space Applications, Houston, TX, 1995 (Proceeding to be published).
5. G.B. Stringfellow, J. of Crystal. Growth **58**, 194 (1983).
6. J.N. Baillargeon, K.Y. Cheng, G.E. Holfer, P.J. Pearah, K.C. Hsieh, Appl. Phys. Lett. **60**, 2540 (1992).
7. I. H. Ho and G.B. Stringfellow (unpublished results).
8. P.N. Keating, Phys. Rev. **145**, 637 (1966).

Growth and Characterization of AlInGaN/InGaN Heterostructures

J. C. ROBERTS*, F. G. MCINTOSH*, M. AUMER*, V. JOSHKIN*, K. S. BOUTROS*, E. L. PINER**, Y. W. HE**, N. A. EL-MASRY**, S. M. BEDAIR*
*ECE Dept., North Carolina State University, Raleigh, North Carolina 27695-7911.
**MSE Dept., North Carolina State University, Raleigh, North Carolina 27695-7916

ABSTRACT

The emission wavelength of the $In_xGa_{1-x}N$ ternary system can span from the near ultraviolet through red regions of the visible spectrum. High quality double heterostructures with these $In_xGa_{1-x}N$ active layers are essential in the development of efficient optoelectronic devices such as high performance light emitting diodes and laser diodes. We will report on the MOCVD growth and characterization of thick and thin InGaN films. Thick $In_xGa_{1-x}N$ films with values of x up to 0.40 have been deposited and their photoluminescence (PL) spectra measured. AlGaN/InGaN/AlGaN double heterostructures (DHs) have been grown that exhibit PL emission in the violet, blue, green and yellow spectral regions, depending on the growth conditions of the thin InGaN active layer. Preliminary results of an AlInGaN/InGaN/AlInGaN DH, with the potential of realizing a near-lattice matched structure, will also be presented.

INTRODUCTION

The achievement of device quality InGaN films that led to the realization of blue, green and yellow light emitting devices are distinguished developments that have set Nichia Chemical Industries apart from the rest of the nitride community[1]. There is a paucity of successful activities in the growth of $In_xGa_{1-x}N$ with high values of x due to several difficulties that can be unique for this ternary alloy. The main problem is related to the weak In-N bond that necessitates a high nitrogen vapor pressure during the growth process. Plasmas generated by electron cyclotron resonance (ECR) in the molecular beam epitaxy (MBE) environment do not seem to produce enough of a nitrogen overpressure. For metalorganic chemical vapor deposition (MOCVD), the high temperature growth needed to crack NH_3 will enhance the dissociation of In-N bonds.

This paper will present the growth and characterization of $In_xGa_{1-x}N$ bulk films and $In_xGa_{1-x}N$ based heterostructures that have been deposited between ~750 - 800 °C. There are apparent differences between the properties of the relatively thick (e.g., > ~2000 Å) InGaN films versus the thinner (e.g., < ~200 Å) InGaN active layers of DHs. Although we have grown thick $In_xGa_{1-x}N$ films in our reactor with a value of x up to 0.40, the optical properties of these $In_xGa_{1-x}N$ films with x > ~0.30 begin to degrade rapidly with increasing x, and tend to have In metal segregated either on the surface or in the film, as detected by X-ray diffraction (XRD) analysis. Thinner $In_xGa_{1-x}N$ films that are active layers of double heterostructures can be grown in our reactor with apparently higher values of x as determined by PL measurements. Emission in the yellow region of the visible spectrum with a good value of full width

Mat. Res. Soc. Symp. Proc. Vol. 423 © 1996 Materials Research Society

at half maximum (FWHM), has been observed. Most of the InGaN double heterostructures have been grown with AlGaN upper and lower cladding layers, but we have recently performed a few trials with AlInGaN cladding layers.

EXPERIMENT

Epitaxial growth of $In_xGa_{1-x}N$ was performed in a hybrid ALE/MOCVD growth system, that has been previously described[2]. This hybrid ALE/MOCVD system was previously used for the epitaxial growth of $In_xGa_{1-x}N$ ($0 < x < 0.27$) by ALE[3] and of AlInGaN quaternary alloys by MOCVD[4]. Source gases used were, trimethylgallium (TMG, -10 °C), trimethylaluminum (TMA, +18 °C), ethyldimethylindium (EdMIn, +10 °C) and NH_3; N_2 was used as the carrier gas. Basal plane sapphire, cleaved into ~1.5 cm x 1.5 cm squares, was used as substrate material. After solvent cleaning, substrates were loaded and annealed in N_2 and NH_3 for 15 minutes and 1 minute respectively. ALE growth of AlN/GaN buffer layers[5] was performed at 700 °C, while InGaN films were grown by MOCVD between 750 and 800 °C. The thick InGaN alloys were typically grown on GaN that was deposited by MOCVD at higher temperatures (900 - 950 °C). AlGaN grown at 950 °C was typically used as the cladding layers for the double heterostructures.

RESULTS

InGaN epitaxial layers have been grown with values of InN up to 40% in the ternary alloy. The full width at half maximum (FWHM) of the double crystal X-ray diffraction (DCXRD) data for these films is broad for high values of x, but is

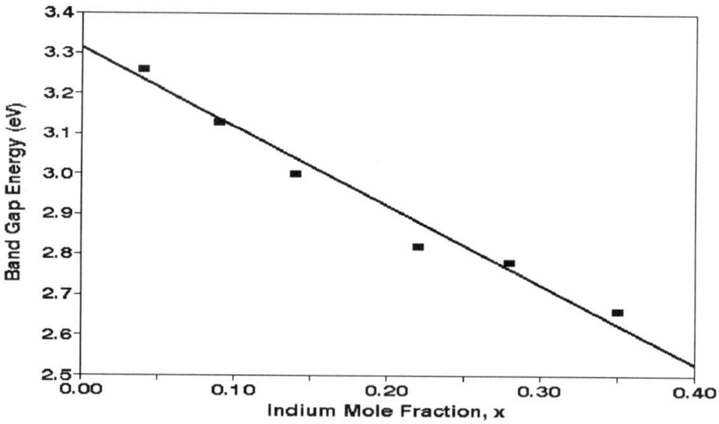

Figure 1. Band gap energy of $In_xGa_{1-x}N$ (as determined by PL) vs. indium mole fraction, x (as determined by XRD).

comparable to that of GaN grown at higher temperatures for low values of InN% in the ternary. For example, DCXRD data for an $In_{0.06}Ga_{0.94}N$ alloy grown at 780 °C on GaN has a FWHM of ~250 arc seconds, one of the lowest reported to date for InGaN. However, Θ - 2Θ X-ray data for the $In_{0.40}Ga_{0.60}N$ film grown at 750 °C on GaN has a broader linewidth, due at least in part to the increased lattice mismatch between the underlying GaN layer and the InGaN alloy and the deterioration of the crystalline quality of the ternary alloy. This composition of 40% InN was achieved by MOCVD with a EdMIn/TMG molar gas phase ratio of only ~2, a relatively low ratio as compared to other published results on InGaN having even lower InN%[6]. Efforts to achieve higher values of x in thick InGaN films usually resulted in fairly broad and weak XRD and PL spectra. The optical properties of these relatively thick InGaN films show intense band edge (BE) emission that is sometimes accompanied by deep levels. Figure 1 shows the variation of the band gap, determined from PL emission, versus the value of x obtained from XRD. Our data is consistent with the predicted variation of E_g vs x by Osamura et. al.[7] and previously reported experimental data by Nakamura et. al[8].

A variety of double heterostructures with $In_xGa_{1-x}N$ as the active layer have been grown with AlGaN cladding layers having ~10% AlN. InGaN active layers were typically grown on InGaN prelayers, which provide a gradual increase in In mole fraction between the AlGaN lower clad layer and the InGaN active layer. A schematic of a typical AlGaN/InGaN/AlGaN heterostructure is illustrated in Figure 2, where the InGaN prelayer is grown over a

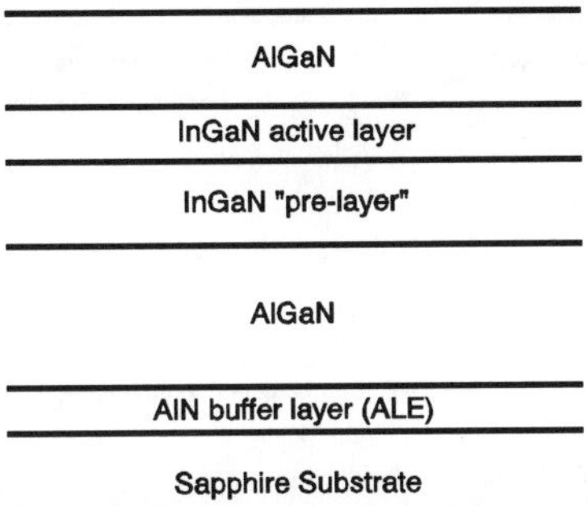

Figure 2. Schematic of AlGaN/InGaN/AlGaN double heterostructure.

temperature range of 780 - 810 °C, and the active layer grown at ~750 °C. PL spectra for some of these heterostructures is presented in Figures 3, 4 and 5. The data displayed in Figure 3 corresponds to a structure having a thicker active layer while the active layers of the structures corresponding to Figures 4 and 5 are each successively thinner. While absolute thickness measurements for these three samples are not currently available, preliminary TEM data from other similar samples coupled with appropriate scaling of the active layer growth times suggest that the thickest layer (Figure 3) should be < ~100 Å. The active layers are deposited on a graded InGaN prelayer of low InN%. The spectral peaks for the samples corresponding to Figures 3, 4 and 5 are 456, 505 and 550 nm respectively, and from these preliminary results there appears to be a gradual red shift in the emission

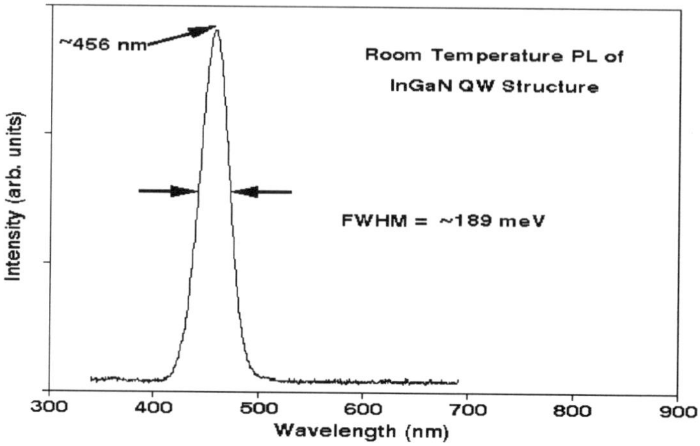

Figure 3. PL of $In_xGa_{1-x}N$ DH with peak emission corresponding to x = ~0.32.

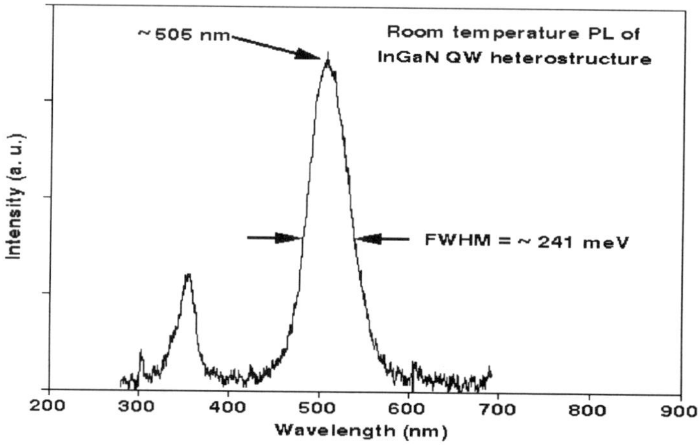

Figure 4. PL of $In_xGa_{1-x}N$ DH with peak emission corresponding to x = 0.48.

spectra as the active layer thickness decreases. However, more studies are required, especially TEM analysis, to confirm this possible relationship between the active layer thickness and the red shift in PL emission. The FWHM of the spectra have values of 189, 241 and 211 meV for Figures 3, 4 and 5 respectively.

It should be possible to obtain a lattice matched heterostructure if an

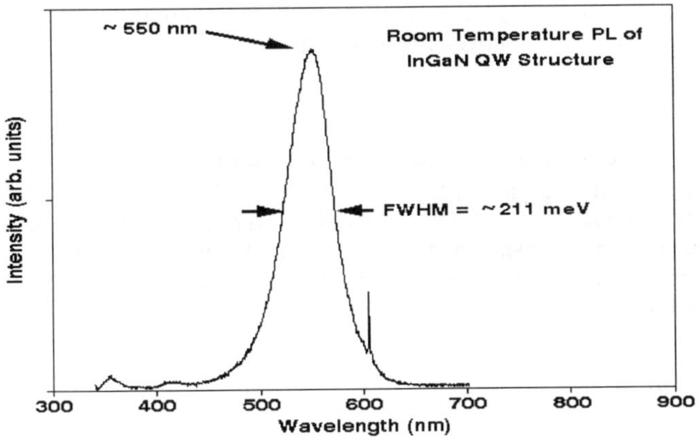

Figure 5. PL of $In_xGa_{1-x}N$ DH with peak emission corresponding to x = 0.63.

$Al_yIn_xGa_{1-x-y}N$ layer can be grown with correct molar compositions x and y to obtain a lattice match to the overlying InGaN active layer. As a first attempt to address this problem, we deposited an $Al_yIn_xGa_{1-x-y}N$ layer on AlGaN and then proceeded to grow the active InGaN layer. The lower AlInGaN cladding layer was grown at 780 °C on AlGaN. This was followed by the growth of the InGaN active layer at 780 °C which in turn was capped by the upper AlInGaN cladding layer also grown at 780 °C. The active layer was grown by simply turning off the TMA flow and increasing the EdMIn

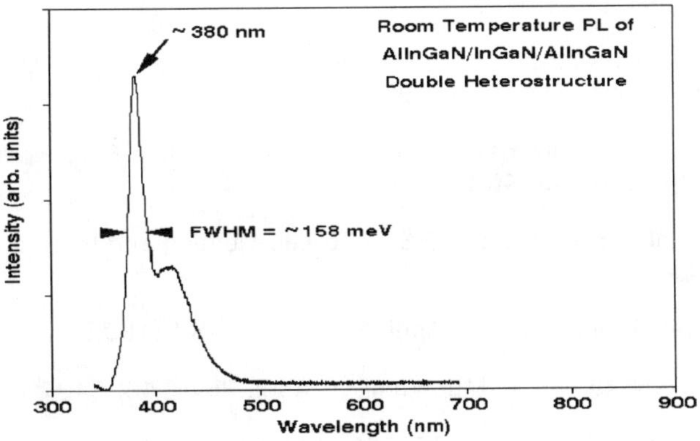

Figure 6. PL of AlInGaN/InGaN/AlInGaN DH.

flow while keeping all other growth conditions constant. The PL spectra for this structure, shown in Figure 6, exhibits apparent sharp band edge emission accompanied by emmission from a deeper level. The weaker deep level emission may be due to the nonoptimized AlInGaN cladding layers.

CONCLUSIONS

Thick $In_xGa_{1-x}N$ films were grown with values of x up to ~0.40. XRD and PL linewidth increase with increasing InN% in the film. AlGaN/InGaN/AlGaN DHs with thin $In_xGa_{1-x}N$ layers were deposited and exhibit intense PL emission from the violet through yellow, which corresponds to an apparent value of x of up to 0.63 in these thin active layers. A preliminary investigation of an AlInGaN/InGaN/AlInGaN DH has been made and PL of this potentially nearly lattice matched structure has been presented.

ACKNOWLEDGEMENTS

This work is supported by the Office of Naval Research (ONR), University Research Initiative (URI) program.

REFERENCES

1. S. Nakamura, M. Senoh, N. Iwasa and S. Nagahama, Jpn. J. Appl. Phys. **34**, L797 (1995).

2. J. C. Roberts, F. G. McIntosh, K. S. Boutros, S. M. Bedair, M. Moussa, E. L. Piner, Y. He and N. A. El-Masry, presented at MRS 1995 Fall Meeting and to be published in MRS proceedings.

3. K. S. Boutros, F. G. McIntosh, J. C. Roberts, S. M. Bedair, E. L. Piner and N. A. El-Masry, Appl. Phys. Lett. **67**, 1856 (1995).

4. F. G. McIntosh, K. S. Boutros, J. C. Roberts, S. M. Bedair, E. L. Piner and N. A. El-Masry, Appl. Phys. Lett. **68**, 40 (1996).

5. E. L. Piner, et. al., presented at MRS 1995 Fall Meeting and to be published in the MRS proceedings.

6. S. Nakamura, T. Mukai, Jpn. J. Appl. Phys. **31**, 1457 (1992).

7. K. Osamura, S. Naka and Y. Murakami, J. Appl. Phys. **46**, 3432 (1975).

8. S. Nakamura, Microelectron. J. **25**, 651 (1994).

GROWTH AND CHARACTERIZATION OF AlGaN/GaN HETEROSTRUCTURES WITH MULTIPLE QUANTUM WELLS BY PAMBE

K. H. Shim,* J. M. Myoung,* O. V. Gluschenkov,** C. Kim,*** K. Kim,** M. C. Yoo,****
S. Kim,** D. A. Turnbull,** and S. G. Bishop**

Departments of Materials Science and Engineering*, Electrical and Computer Engineering**, and Physics***, kevinkim@ux1.cso.uiuc.edu
University of Illinois at Urbana-Champaign, Urbana, IL 61801
On leave from Samsung Advanced Institute of Technology****

ABSTRACT

AlGaN/GaN heterostructures with multiple quantum wells were grown by plasma-assisted molecular beam epitaxy (PAMBE). Structural and optical properties of the heterostructures were analyzed using x-ray diffraction, cathodoluminescence, and photoluminescence. Interband transitions were clearly observed in the GaN quantum wells at both room- and liquid-helium temperatures. The efficiency of the interband recombination due to the confinement effect was greatly enhanced in the thinner quantum wells. The functional dependence of the interband peaks on the well thickness is shown to be in good agreement with the calculated positions of the quantized levels in the wells.

INTRODUCTION

The advances in the material growth and processing of III-V nitrides have led to the demonstration of a current injection laser diode (LD) with InGaN-based multiple quantum wells (MQWs)[1,2]. In the attempt to realize AlGaN/GaN QW LDs, there have been a few reports on the growth and characterization of the AlGaN/GaN single quantum well structures[3,4]. Krishnankutty et al.[3] reported the strain-induced energy gap shift, which was deduced from the position of bound excitons in an AlGaN/GaN strained-layer quantum well grown by metalorganic chemical vapor deposition (MOCVD). Recently, Salvador et al.[4] fabricated a Si-doped AlGaN/GaN 60 Å thick single quantum well using gas source MBE with ammonia gas as a source of active nitrogen. From the observed bound exciton transitions, they estimated heavy hole and electron effective masses as 0.3 m_o and 0.19 m_o, respectively.

In this paper, we present and analyze charcteristics of AlGaN/GaN MQWs grown by plasma-assisted molecular beam epitaxy (PAMBE), which employs inductively coupled nitrogen plasma as a source of active nitrogen. The optical quality of the present quantum wells is shown to be comparable to that for similar heterostructures fabricated with other growth techniques.

EXPERIMENTAL

A detailed description of the PAMBE system developed at the University of Illnois can be found in our earlier publication[5] and in the companion paper[6]. Here we only emphasize the specific features of the system or process relevant to the growth of quantum wells. Figure 1 shows a schematic of the MQW structures studied in the present work. First, a 30 nm AlN

Mat. Res. Soc. Symp. Proc. Vol. 423 © 1996 Materials Research Society

buffer layer was grown on a sapphire substrate. Then, a 0.5 μm thick GaN epilayer was deposited at a substrate temperature of 750 °C. The purpose of this layer was to accommodate stress caused by the sapphire/GaN lattice mismatch. Multiple layers of AlGaN/GaN were grown on top of the GaN layer. In a separate set of experiments, the growth rate and Al mole fraction were determined with sufficient accuracy to precisely control Al concentration and to grow nanolayers of AlGaN and GaN with predetermined thicknesses.

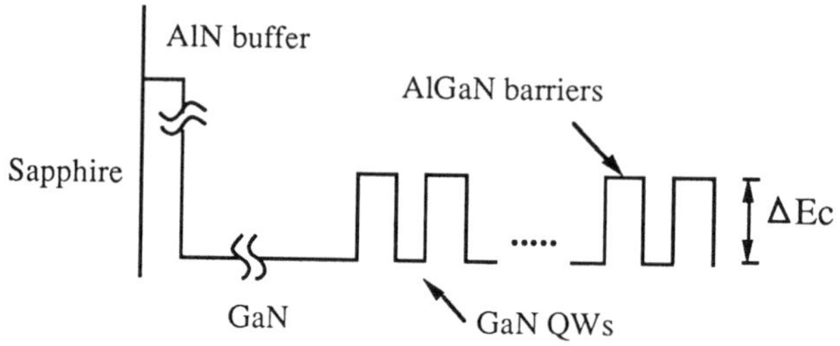

Figure 1. Diagram of conduction energy band of GaN/AlGaN MQW structures.

To avoid formation of dislocations and cracks at the AlGaN - GaN interfaces, the thicknesses of the GaN quantum wells were kept within the pseudomorphic limit[7]. The thicknesses of the GaN QW layers were less than the critical thicknesses (486, 208, and 125 Å for GaN QWs with AlGaN barriers, $x_{Al} = 0.1$, 0.2 and 0.3, respectively). The critical thicknesses were calculated using the Matthew and Blankeslee force balance model[8]. To study radiative transitions in unstrained GaN MQWs, we prepared the samples with different well thicknesses (25 Å, 50 Å, 100 Å) and the same Al mole fraction (0.2) in the barriers. To prevent substantial coupling between the wells, the barrier thicknesses were maintained at 50 Å, 75 Å, and 150 Å, respectively, for the three different GaN QWs.

The crystallinity of the AlGaN/GaN MQWs was examined by x-ray diffractometry (XRD) using a Rigaku D-Max with CuKα radiation. Cathodoluminescence (CL) spectra and images were observed at room temperature using the Oxford monoCL in a Zeiss DSM960 scanning electron microscope. Low-temperature (7 K) photoluminescence (PL) was excited with the 325-nm line of a 25-mW HeCd laser.

RESULTS AND DISCUSSION

Figure 2 is an XRD spectrum obtained from an $Al_{0.2}Ga_{0.8}N$(21 layers of 50 Å)/GaN(20 layers of 25 Å) MQW. The clear separation in Kα1/Kα2 doublets for the (0002) reflection of GaN shows that the structural coherence parallel to the growth direction is good. The measured 2d spacing (5.177 Å) of the GaN (0002) plane is almost the same as that of a 1-μm-thick GaN film (5.179 Å). The measured 2d spacing of $Al_{0.2}Ga_{0.8}N$ (5.144 Å) agrees with the value of 5.14 Å, which was calculated using a linear approximation. Therefore, one may conclude that only a negligible amount of strain (less than 0.04 %) was exerted in the direction of the c-axis in the grown GaN QWs.

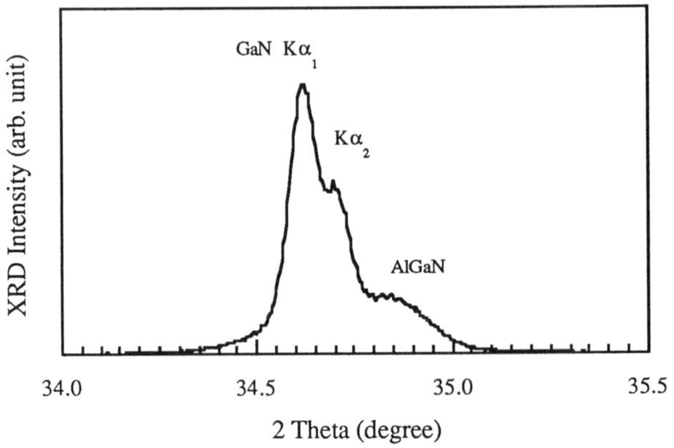

Figure 2. XRD data for the (0002) reflection of $Al_{0.2}Ga_{0.8}N$(21 layers of 50 Å) /GaN(20 layers of 25 Å) MQW structure.

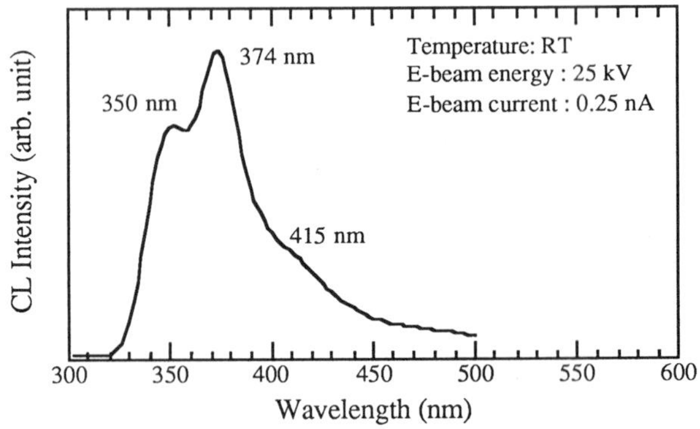

Figure 3. Room-temperature CL spectrum measured from $Al_{0.2}Ga_{0.8}N$(21 layers of 50 Å)/GaN(20 layers of 25 Å) MQWs.

Figure 3 is a room-temperature CL spectrum obtained from an $Al_{0.2}Ga_{0.8}N$ (50 Å)/GaN (25 Å) heterostructure with 20 identical periods. The peak at 350 nm (3.543 eV) is attributed to interband recombination in the GaN QWs. The peaks at 374-nm (3.315 eV) and 415-nm (3.303 eV) are due to the recombinations related to the donor and acceptor impurity levels.

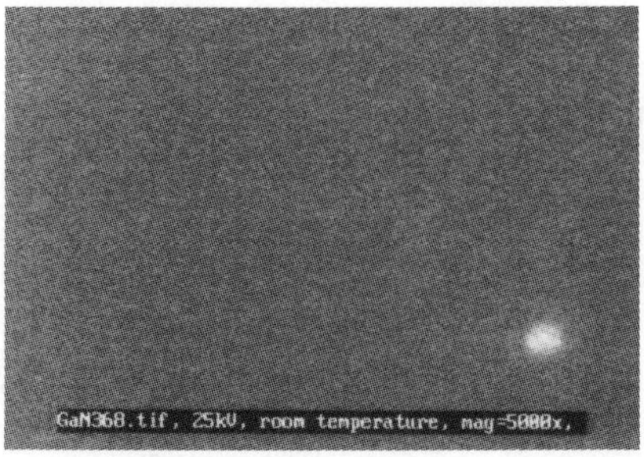

Figure 4. Room-temperature CL image of the 350-nm light emitted from $Al_{0.2}Ga_{0.8}N$(21 layers of 50 Å)/GaN(20 layers of 25 Å) MQWs. The bright spot is for focusing.

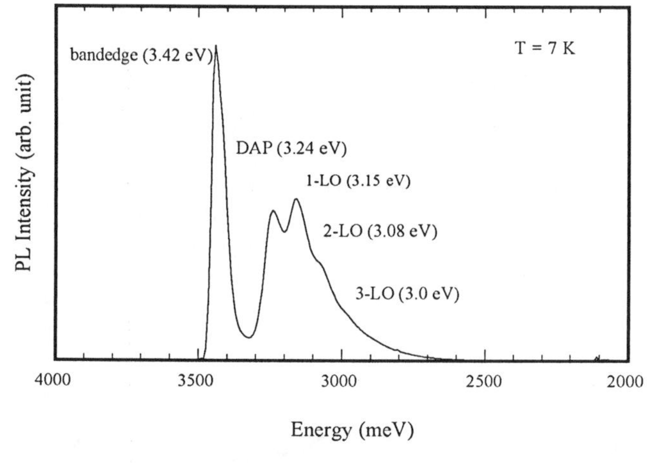

Figure 5. PL spectrum of 1 μm thick GaN epitaxial layer grown on sapphire substrate.

Figure 4 is a 350-nm CL image of the excited GaN MQW structure. It shows an excellent spatial uniformity of the interband luminescence, which indicates the high crystalline quality of the MQWs. This strong light emission observed from the present sample should be an indication that one may expect equally strong light emission at room temperature in both electroluminescence in LEDs and stimulated emission in LDs.

Figure 5 shows a PL spectrum measured from a 1-μm-thick GaN epitaxial layer grown on sapphire. The peak at 3.42 eV, with the full width at half maximum (FWHM) of 45 meV, is attributed to the bandedge transition. The four lower energy peaks at 3.24, 3.15, 3.08 and 2.99

eV correspond to the donor-acceptor pair (DAP) and its phonon replicas (1-LO, 2-LO, and 3-LO replicas with 70-80 meV intervals).

The PL spectra of the $Al_{0.2}Ga_{0.8}N/GaN$ MQWs with different GaN QW widths, (a) 25, (b) 50 and (c) 100 Å, are presented in Figure 6. The high energy peaks with transition energies of (a) 3.58, (b) 3.54, and (c) 3.50 eV are identified as interband transitions in the GaN QWs. The possibility that these peaks could come from the AlGaN barriers is excluded since the exciton energy in the $Al_{0.2}Ga_{0.8}N$ barriers is expected to be 3.87 eV. As expected, these transitions have higher energy than the free exciton energy (3.485 eV) of bulk GaN due to the quantum confinement effect. Thus the confinement energies ($E_{e1} + E_{h1}$) for 25, 50, and 100 Å thick GaN QWs were determined as 95, 55, and 25 meV, respectively, by subtracting the free exciton energy of bulk GaN from the interband transition energies, 3.58, 3.54, and 3.50 eV. Three other well-resolved peaks of the specrum (a), 3.38, 3.34, and 3.31 eV, are attributed to the phonon replica of a donor-bound exciton (DBE-LO), the phonon replica of an acceptor-bound-exciton (ABE-LO), and DAP-2LO, respectively, in bulk GaN. We believe that the same, but broader, peaks are responsible for the low energy spectra of (b) and (c).

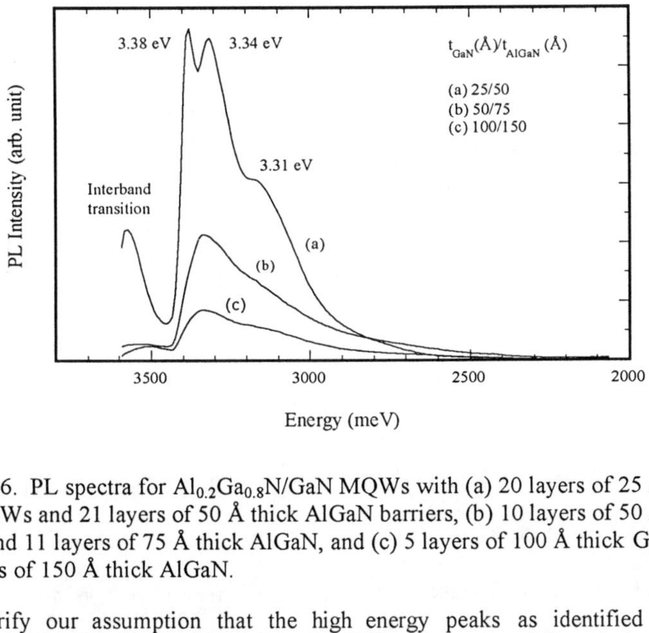

Figure 6. PL spectra for $Al_{0.2}Ga_{0.8}N/GaN$ MQWs with (a) 20 layers of 25 Å thick GaN QWs and 21 layers of 50 Å thick AlGaN barriers, (b) 10 layers of 50 Å thick GaN and 11 layers of 75 Å thick AlGaN, and (c) 5 layers of 100 Å thick GaN and 6 layers of 150 Å thick AlGaN.

To verify our assumption that the high energy peaks as identified above indeed correspond to the interband recombination in the QWs, we compared these transitions with the positions of the quantum levels. The energy of the recombination peak in a QW is given by $E = E_g + E_{en} + E_{hn} + E_{BS}$, where E_g is the bulk energy gap, E_{en} and E_{hn} are the energies of the quantized levels of electrons and holes in the QW, and E_{BS} is the energy gap shift because of applied strain. The E_{BS} is negligible due to the small strain[9]. The confinement energy is defined as a sum of E_{en} and E_{hn}. We assume that the heterojucntion band offset is $\Delta E_C/\Delta E_V = 67/33$[4]. The energy gap of $Al_xGa_{1-x}N$ may then be calculated from its quadratic dependence on the molar fraction[10]: $E(x) = 3.485 + 1.715\,x + x^2$, where the bowing parameter is 1.0 eV[11]. The effective masses of the electrons and the heavy holes are assumed to be independent of the composition and the temperature, and equal to $m_e = 0.236\ m_o$ and $m_{hh} = 0.8\ m_o$, respectively[12]. Thus calculated theoretical values of the confinement energies are 124, 45, and 16 meV which are in good agreement with the experimental values of 95, 55, and 25 meV, respectively, for the three QW structures with different QW thicknesses of 25, 50, and 100 Å.

In summary, high-quality AlGaN/GaN MQW structures have been grown by PAMBE. The structures showed uniform CL radiation characteristics and enhanced efficiency of the interband transitions relative to the bulk samples caused by the confinement of the electrons and holes in the GaN quantum wells. The functional dependence of the PL peaks on the well thickness agrees well with the calculated positions of the quantized levels in the wells.

ACKNOWLEDGMENTS

The authors would like to thank Professor Shun-Lin Chuang for valuable discussions. This work was supported by Samsung Electronics Co. Ltd.

REFERENCES

1. S. Nakamura, M. Senoh, S. Nagahama, N. Iwasa, T. Yamada, T. Matsushita, H. Kiyoku, and Y. Sugimoto, Jpn. J. Appl. Phys. **35**, L74 (1996).
2. H. Amano, K. Hiramatsu and I. Akasaki, Jpn. J. Appl. Phys. **27**, L1384 (1988).
3. S. Krishnankutty, R. M. Kolbas, M. A. Khan, J. N. Kuznia, J. M. Van Hove, and D. T. Olson, J. Electronic Mater. **21**, 609 (1992).
4. A. Salvador, G. Liu, W. Kim, O. Aktas, A. Botchkarev, and H. Morkoc, Appl. Phys. Lett. **67**, 3322 (1995).
5. K. Kim, M. C. Yoo, K. H. Shim, and J. T. Verdeyen, J. Vac. Sci. Technol. **B13**, 796 (1995).
6. J. M. Myoung, C. Kim, K. H. Shim, O. Glusschenkov, K. Kim, and M. C. Yoo, Mater. Res. Soc. Symp. Proc. April 1996 (this volume).
7. C. Kim, I. K. Robinson, J. M. Myoung, K. H. Shim, K. Kim, and M. C. Yoo, Mater. Res. Soc. Symp. Proc. April 1996 (this volume).
8. J. W. Matthew and A. E. Blankeslee, J. Crystal Growth, **27**, 118 (1974).
9. S. L. Chuang and C. S. Chang, Appl. Phys. Lett. **68**, 1657 (1996).
10. W. Shan, T. J. Schmidt, R. J. Haustein, J. J. Song, and B. Goldenberg, Appl. Phys. Lett. **66**, 3492 (1995).
11. Y. Koide, H. Itoh, M. R. Khan, K. Hiramtu, N. Sawaki, and I. Akasaki, J. Appl. Phys. **61**, 4540 (1987).
12. S. Fischer, C. Wetzel, E. E. Haller, and B. K. Meyer, Appl. Phys. Lett. **67**, 1298(1995).

ORIGIN OF HIGH-CONDUCTIVITY LAYER NEAR THE SURFACE IN AS-GROWN DIAMOND FILMS

SADANORI YAMANAKA *, KAZUSHI HAYASHI **, HIDEYO OKUSHI, AND KOJI KAJIMURA *
Electrotechnical Laboratory, 1-1-4, Umezono, Tsukuba, Ibaraki 305, Japan

ABSTRACT

In order to clarify the origin of *p*-type high-conductivity layers (HCL) near the surfaces of as-grown diamond films prepared by chemical vapor deposition, we have investigated the properties of Schottky junctions fabricated from HCL. The Schottky junctions between Al and HCL of undoped homoepitaxial films with step-flow growth showed high-rectification properties. The forward current-voltage characteristics of the junctions in the temperature range between 83 K and 400 K are found to be described by the thermionic-field emission theory. Analysis indicates that thin enough depletion layer is formed at the junction by a high concentration (~$10^{18}/cm^3$) of acceptors existing in HCL due to hydrogenation. The origin of HCL is the acceptors related to incorporated hydrogen near the surface.

INTRODUCTION

It has been known that as-grown diamond films prepared by chemical vapor deposition (CVD) have relatively high-conductivity layers (HCL) near the surfaces both with and without boron (B)-doping [1, 2]. The HCL is of *p*-type conduction [3, 4], which disappears after oxidation and reappears by exposure to a hydrogen plasma. Using HCL, a semiconducting device such as a field effect transistor was actually fabricated [5]. However, the origin of HCL has not yet been clarified.

Several works pointed out the relevance of hydrogenation to HCL [1-12]. Landstrass and Ravi [1] proposed that the origin of HCL is the deep level passivation by terminated hydrogen incorporated near the surface. Kawarada *et al.* [8] suggested that negative charges in the surface states originated from hydrogen termination induce a hole accumulation layer due to surface band bending. It has been reported that HCL is attributed to holes generated by additional acceptors arising from incorporated hydrogen [3, 9, 10]. Based on Hall measurements of HCL and secondary ion mass spectroscopy (SIMS), we supported the last model of the existence of acceptors related to incorporated hydrogen [4].

Recently, we have successfully grown undoped homoepitaxial diamond films with atomically flat region and good crystallinity [11]. The epitaxy has been attained by step-flow growth in which atomic precursors migrate on terraces and incorporate into atomic steps before they form isolated nuclei on the terraces. The current-voltage (*I-V*) characteristics of Schottky junctions between Al and HCL of the films showed excellent high-rectification properties with very low leak currents at reverse bias.

In this study, we have investigated carrier-transport mechanism in the Schottky barrier by the temperature dependence of *I-V* characteristics of the good-quality junctions. We clarify the origin of HCL on the basis of the present experimental results, in addition to the results of Hall measurements and SIMS.

EXPERIMENT

Diamond films were deposited on synthetic Ib diamond (001) substrates using a CVD system made by Applied Science Technology, Inc. (ASTeX). The substrate was 4.0×4.0 mm^2 in area and 0.3 mm in thickness. Source gas was 0.5% CH_4 diluted by H_2. Gas

Mat. Res. Soc. Symp. Proc. Vol. 423 © 1996 Materials Research Society

pressure, total gas flow rate, and microwave power were 25 Torr, 400 sccm, and 750 W, respectively. Substrate temperature was kept at 800 °C measured by a thermocouple attached to the backside of a susceptor. Deposition duration was 6 hrs and resulting film thickness was approximately 2.4 μm.

Ohmic contacts were fabricated on the as-grown diamond surface by evaporating Ti followed by Au to prevent the oxidization of Ti electrode. To fabricate Schottky contacts, Al with diameter of 200 μm was evaporated on the as-grown diamond surface.

In order to avoid adsorption of H_2O onto the sample surface, *I-V* measurements of the Schottky junctions were carried out after the sample was heated up to 400 K for 15 min in vacuum. The temperature dependence of *I-V* characteristics was measured in the range between 83 K and 400 K. The *I-V* measurements were carried out in a vacuum of 10^{-3} Torr using a Hewlett Packard 4140B pico-ampere meter.

RESULTS AND DISCUSSION

The forward *I-V* characteristic of the Schottky junction between Al and as-grown undoped films with step-flow growth at 400 K is shown in Fig. 1. The leak currents at reverse bias were lower than the detection limit 10^{-13}A of the measurement system. The high-rectification properties, in particular, the very low leak currents at reverse bias are much improved than those of previous reports [12], indicating that diamond films with step-flow growth are very useful to obtain high-quality Schottky barriers.

The forward *I-V* characteristics of the junction as a function of temperature from 83 K to 360 K are shown in Fig. 2. *I-V* curves obey conventional forward-characteristics equation for $V \gg kT/q$,

$$I = I_S \exp\left(\frac{qV}{E_0}\right), \qquad (1)$$

where I_S is the saturation current and q the electron charge. The weak temperature dependent slope of *I-V* curves can not be explained by the thermionic emission process ($E_0 = nkT$ with $n = 1$) or the generation-recombination process ($E_0 = nkT$ with $n = 2$) [13] where k is Boltzman's constant.

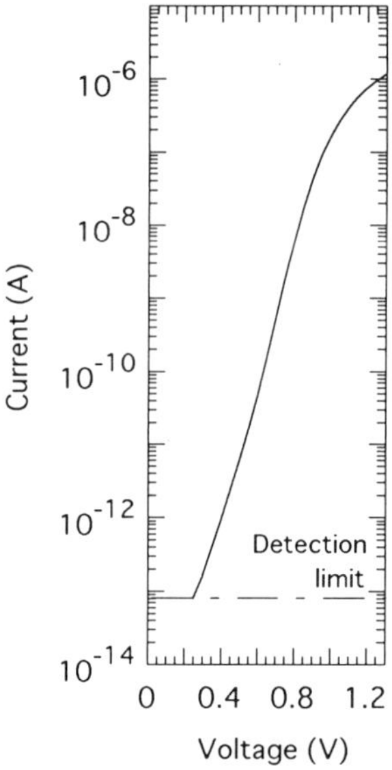

Fig. 1. Forward *I-V* charcteristic at 400 K of Schottky junction between Al and as-grown undoped diamond film with step-flow growth.

As another carrier-transport mechanism through the junction, the tunneling current should be considered. For a Schottky junction with a very thin depletion layer such as that fabricated using a heavily doped semiconductor, carriers can tunnel quantum mechanically through the barrier, whose process, known as field emission, does not depend on temperature. However, in a conventional Schottky junction with relatively high doping, the pure field emission does not occur. Thermally excited electrons with energy where the depletion layer is thin enough to tunnel contribute to the current flow. This thermally assisted tunneling process is known as thermionic-field emission.

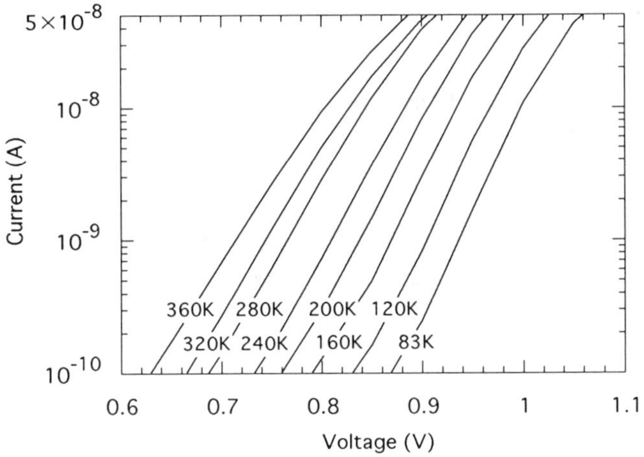

Fig. 2. Forward *I-V* characteristics of the Schottky junction in the temperature range between 83 K and 360 K.

According to the thermionic-field emission theory [13] for a conventional Schottky junction made of a *p*-type semiconductor, E_0 in Eq. (1) can be described by

$$E_0 = E_{00} \coth\left(\frac{E_{00}}{kT}\right), \qquad (2)$$

$$E_{00} = \frac{h}{4\pi}\left(\frac{N_A}{m^* \varepsilon}\right)^{\frac{1}{2}}, \qquad (3)$$

where h is Plank's constant, m^* the effective mass, ε the dielectric constant, and N_A the impurity (acceptor-type) concentration.

Based on Eqs. (1)-(3), we determined E_0 as shown in Fig. 3 from the data of *I-V* characteristics in the temperature range between 83 K and 400K. The solid line in Fig. 3 shows Eq. (2) with E_{00} = 27 meV in excellent agreement with experiment. Using E_{00} = 27 meV, the acceptor concentration was determined by Eq. (3) as $6\times10^{18}/\text{cm}^3$, using $\varepsilon = 5.7\varepsilon_0$ and $m^* = 0.75m_0$ [14]. This result leads to an important conclusion that a high concentration of acceptors exists near the surface of the as-grown film, *i.e.* in HCL. It is worthwhile to note that the model of HCL based on the charged surface states [8] does not provide the space charges which make the depletion layer thin enough.

As shown in Fig. 3, the forward *I-V* characteristics of the Schottky junctions are mostly due to the thermionic-field emission in measurement temperature range less than about 300 K. At higher temperatures around 400 K, however, the field emission can be neglected compared to the thermionic emission, namely, *I-V* characteristics around 400 K can be explained only by the thermionic emission theory [13] whose I-V characteristics for $V>3kT/q$ is given by

$$I = I_s \exp\left(\frac{qV}{nkT}\right), \qquad (4)$$

where n is the ideality factor. In the case of Fig. 1, the estimated n was as low as 1.1, indicating that the present Schottky junction is nearly ideal.

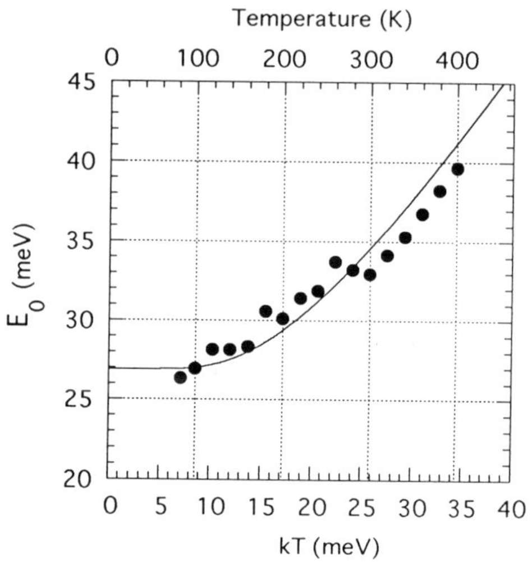

Fig. 3. Experimental value of E_0 as a function of temperature. Solid line is the theoretical value of Eq. (2) with E_{00} = 27 meV.

According to our result of Hall measurements on HCL [4], the carrier density per unit area of HCL in as-grown (hydrogenated) undoped films was approximately $8\times10^{12}/cm^2$ in the temperature range between 150 K and 400 K. The carrier density per unit area was used instead of the carrier concentration because the thickness of HCL cannot be accurately determined at present. The carrier density per unit area of HCL in as-grown (hydrogenated) B-doped films at 300 K was also around $1\times10^{13}/cm^2$, which is 4–5 orders of magnitude lager than that of conventional oxidized (no hydrogen) B-doped films. The atomic B concentration of the film is around $4\times10^{16}/cm^3$. The oxidized B-doped film was obtained by oxidizing as-grown hydrogenated B-doped film and showed an activation energy of 0.38 eV which is consistent with the B-acceptor level reported by Collins and Williams [15]. These facts indicate that the free holes in as-grown hydrogenated films are not due to the B-acceptor but have another origin. Kiyota *et al.* [10] pointed out from capacitance-voltage characteristics that additional acceptors, which are not related to B-acceptors, exist in as-grown B-doped films.

We examined, in the previous work [4], the existence of hydrogen in HCL of as-grown undoped films using SIMS. Incorporated hydrogen distributed spatially in an exponential manner from the surface of the as-grown undoped films. Concentrated hydrogen existed in the region of 20 nm from the surface. In the oxidized undoped films, on the other hand, hydrogen was not detected. If we assume that the thickness of HCL is 20 nm, the carrier concentration at 300 K corresponds to $4 \times 10^{18}/cm^3$. This value is almost the same value of the acceptor concentration obtained from the *I-V* measurements and 2 orders of magnitude greater than atomic B concentration. This experimental result also suggests that acceptors in HCL form a shallow-level.

Due to the high concentration of shallow-acceptors related to incorporated hydrogen, the Fermi level of HCL is close to the valence-band edge in comparison with that expected from the bulk region of undoped diamond films. Figure 4 shows a schematic energy-band diagram of the present as-grown films at thermal equilibrium. In the figure, HCL and the bulk region behave as a p^+ layer and an intrinsic layer, respectively, and then a p^+-i junction is formed through the film. This energy band diagram is consistent with the experimental result of the surface Fermi level position of as-grown films determined by X-ray photoelectron spectroscopy and Kelvin probe method [7].

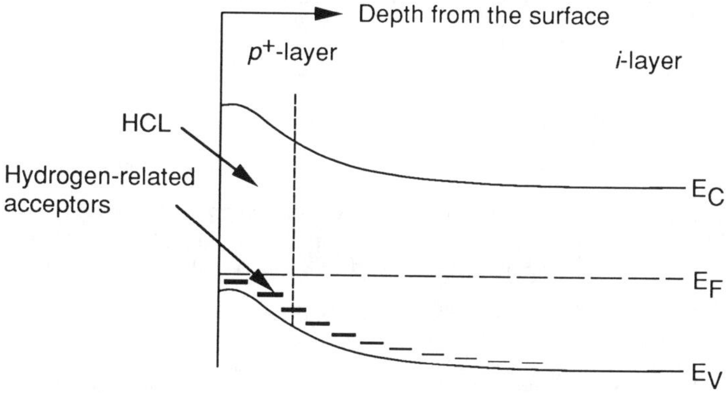

Fig. 4 Proposed model of energy-band diagram.

SUMMARY

We have investigated the properties of Schottky junctions fabricated from HCL. The Schottky junctions between Al and as-grown undoped homoepitaxial films with step-flow growth showed high-rectification properties with very low leak currents at reverse bias. The forward *I-V* characteristics of the junctions in the temperature range between 83 K and 400 K are explained by the thermionic-field emission theory. This means that the depletion layer at the junction is thin enough for the carriers to tunnel through the Schottky barrier. The straightforward analysis leads to an important conclusion that a high concentration ($\sim 10^{18}/cm^3$) of acceptors exists in HCL due to hydrogenation. It is concluded that HCL originates from the acceptors related to incorporated hydrogen near the surface.

ACKNOWLEDGMENTS

The authors would like to acknowledge S. Hara, T. Shimizu, Y. Toyoshima, K. Arai, and P. L. Hacke for their useful discussion. We also thank H. Watanabe and H. Miyoshi of Tokai Univ. for their technical assistance. J. Tanaka of National Institute for Research in Inorganic Materials is also acknowledged for his assistance in Hall measurements.

REFERENCES

* Also with Faculty of Materials Science, University of Tsukuba, 1-1-1, Tennoudai, Tsukuba, Ibaraki 305, Japan

** On leave of absence from Kobe Steel Ltd., Electronics Research Laboratory, 1-5-5, Takatsukadai, Nishi-ku, Kobe 651-22, Japan

1. M. I. Landstrass and K. V. Ravi, Appl. Phys. Lett. **55**, 1391 (1989).
2. S. A. Grot, G. Sh. Gildenblat, C. W. Hatfield, C. R. Wronski, A. R. Badzian, T. Badzian, and R. Messier, IEEE Electron Device Lett. **11**, 100 (1990).
3. T. Maki, A. Shikama, M. Komori, Y. Sakaguchi, K. Sakuta, and T. Kobayashi, Jpn. J. Appl. Phys. **31**, L1446 (1992).
4. K. Hayashi, S. Yamanaka, H. Okushi, and K. Kajimura, Appl. Phys. Lett. **68**, 376 (1996).
5. H. Kawarada, M. Aoki, and M. Ito, Appl. Phys. Lett. **65**, 1563 (1994).
6. S. Albin and L. Watkins, Appl. Phys. Lett. **56**, 1454 (1990).
7. T. Sugino, Y. Sakamoto, A. Furukawa, and J. Shirafuji, Mater. Res. Soc. Symp. Proc. **339**, 45 (1994).
8. H. Kawarada, H. Sasaki, and A. Sato, Phys. Rev. **B52**, 11351 (1995).
9. H. Shiomi, Y. Nishibayashi, and N. Fujimori, Jpn. J. Appl. Phys. **30**, 1363 (1991).
10. H. Kiyota, E. Matsushima, K. Sato, H. Okushi, T. Ando, M. Kamo, Y. Sato, and M. Iida, Appl. Phys. Lett. **67**, 3596 (1995).
11. K. Hayashi, S. Yamanaka, H. Okushi, and K. Kajimura, Appl. Phys. Lett. **68**, 1220 (1996).
12. H. Kawarada, M. Aoki, H. Sasaki, and K. Tsugawa, Diam. Relat. Mater. **3**, 961 (1994).
13. E. H. Rhoderick and R. H. Williams, *Metal-Semiconductor Contacts*, 2nd ed. (Clarendon, Oxford, 1988) p. 89.
14. J. E. Field, *The Properties of Diamond* (Academic Press, London, 1979) p. 84.
15. A. T. Collins and A. M. S. Williams, J. Phys. **C4**, 1789 (1971).

GaN FILM GROWTH BY A SUPERSONIC ARCJET PLASMA

M. A. CAPPELLI,* A. E. KULL,* K. SCHWENDNER,* H. LEE,** S.J. HARRIS, Jr.,** and J. MROCZKOWSKI‡

* Mechanical Engineering Department, Stanford University, Stanford, CA. 94305-3032

** Solid State Electronics Laboratory, Stanford University, Stanford, CA.

‡ Loral Infrared & Imaging Systems, Lexington, MA.

ABSTRACT

This paper reports on preliminary studies of GaN growth obtained using a supersonic nitrogen arcjet plasma expanding into low pressure. A hydrogen-free and carbon-free growth environment was achieved by use of a Ga vapor source positioned downstream of the expanding plume. GaN growth rates of around 8 μm/hour were obtained in these preliminary studies. We believe these growth rates are the highest reported for a non-chemical decomposition process. The growth rates are attributed to the very high nitrogen atom fluxes estimated to be on the order of 10^{19} atoms/cm^2/sec. The initial growth rates are believed to be well below those possible with a fully optimized system. The GaN films are characterized by Raman spectroscopy, visible-IR transmission, and x-ray diffraction. The deposited films appear to be single crystal and epitaxial on basal plane sapphire substrates.

INTRODUCTION

Gallium nitride (GaN) is difficult to grow in thick, high-quality, single-crystal layers. The available epitaxial growth approaches, which can be considered for growing such layers, can be divided into two broad categories. The first growth category encompasses all MBE type approaches which invariably use low pressure electron-cyclotron resonance (ECR) nitrogen plasmas to produce a flux of excited or ionized nitrogen molecules (N_2^+), and neutral nitrogen atoms (N) [1-3]. GaN deposition using these approaches generally exhibit growth rates not exceeding 1 μm/hour (more typically 0.1 - 0.2 μm/hour [2,3]) which is quite inadequate for thick layer growth.

The second broad GaN growth category includes all processes based on some form of chemical decomposition to produce Ga and N -containing radicals either at the substrate surface or in the vapor above the substrate. Processes such as vapor phase epitaxy (VPE) or metalorganic vapor phase epitaxy (MOVPE) are included in this category. The growth rate in these cases is diffusion-limited and rates of up to 100 μm/hour have been reported [4,5]. However, GaN grown by these techniques may contain high concentrations of process by-products such as carbon (i.e., from MOVPE growth). Also, MOVPE and VPE methods seem to lead to GaN films with relatively high background n-type carrier concentrations [2].

We have performed preliminary studies into the possibility of an alternative technique, based on an arcjet plasma supply of a very high flux (10^{19}/cm^2/sec) of nitrogen atoms. The arcjet in our experiments was a direct-current (DC) arc plasma discharge expanded to low densities (~ 10^{16} molecules/cm^3) and high supersonic speeds (Mach numbers of 2-3). This low density, supersonic plasma jet is achieved by appropriately contouring the discharge nozzle and maintaining approximately 0.3 - 1 torr of background pressure in the vacuum chamber into which the jet discharges. Plasma jets of this type, operating on hydrogen, have been previously employed by our group to deposit high quality diamond films [6,7]. Within the context of GaN synthesis, this approach falls into neither of the two general categories identified above. As discussed below, the Ga atoms in this approach are supplied from an electron-bombarded Ga source.

EXPERIMENTAL FACILITY

Basic Configuration

A schematic of the experimental arcjet facility used to grow GaN is shown in Fig.1. Nitrogen is used as the arc source gas, which is dissociated as it passes through the discharge sustained between a central cathode and surrounding anode (which also serves as the nozzle). Gallium vapor is introduced by exposing metallic gallium contained within a pyrolitic boron nitride

Mat. Res. Soc. Symp. Proc. Vol. 423 © 1996 Materials Research Society

fixture to the plasma and suitably biasing the gallium with respect to the plasma potential to attract electron current from the plasma plume. The current is sufficient to heat the gallium by electron-bombardment to temperatures as high as 1000°C, generating a gallium jet that is entrained by the dissociated nitrogen flow. The nitrogen and gallium atoms in the arcjet plume impinge on a resistively-heated sapphire substrate to grow the epitaxial GaN layers.

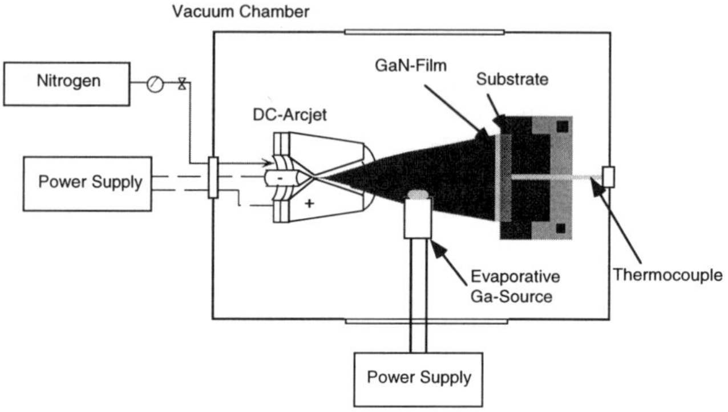

Figure 1. Schematic illustration of low density arcjet facility employed for GaN growth.

Nitrogen Flux and Potential Growth Rate Estimates

We believe that the GaN gowth rate and quality will be limited by the nitrogen flux that can be delivered to the substrate. In order to estimate the fraction of the nitrogen dissociated by the arcjet and the maximum possible nitrogen atom flux that can be delivered to a substrate, we have applied mass and energy conservation to a generic arcjet control volume encompasing the discharge and expansion region of the nozzle. For the purpose of this calculation, we assume uniform flow of velocity, u_e, at the exit of the arcjet nozzle, and that the species comprising the plasma gas at the arcjet exit share a common temperature, T. We assume that the gas introduced at the inlet is molecular nitrogen with a negligible kinetic energy, and that the gas exiting the nozzle is comprised mainly of atomic and molecular nitrogen (a reasonable assumption, since the plasma is found to be only weakly ionized). Under these conditions, an energy balance gives:

$$\eta P_{arc} = \dot{m}^o_{N_2}\left\{\frac{u_e^2}{2} + h_{N_2} - h^o_{N_2} + \theta_D\left(h_N - h_{N_2}\right)\right\}, \tag{1}$$

where η is the arcjet thermal efficiency (typically about 50% for our design), P_{arc} is the operating arcjet power, $\dot{m}^o_{N_2}$ is the total mass flow rate of nitrogen, h_j, and h^o_j represent the specific enthalpies of species j at the exit and inlet to the discharge chamber respectively, and θ_D is the nitrogen dissociation fraction, defined here as:

$$\theta_D = \frac{\chi_N}{\chi_{N_2} + \chi_N} = \frac{\dot{m}_N}{\dot{m}^o_{N_2}}. \tag{2}$$

Here, χ_N and χ_{N2} are the atomic and molecular nitrogen mass fractions at the exit of the nozzle, and $\dot{m}_N$ and $\dot{m}_{N2}$, the corresponding mass flow rates.

In order to demonstrate the potential use of an arcjet as a source of atomic nitrogen, we have computed the dissociation fraction for a generic arc source operating with an exit velocity of 1.4 - 2.7 x 10^5 cm/sec [8], exit temperature of 2000K, thermal efficiency of 50%, nitrogen mass flow rate of 150 mg/sec, and powers of 3kW - 6 kW. These values are not too unlike those used in our experiments, and can often be directly measured or estimated in an experiment. The calculated variation in dissociation fraction with arcjet power for these conditions is shown in Fig. 2.

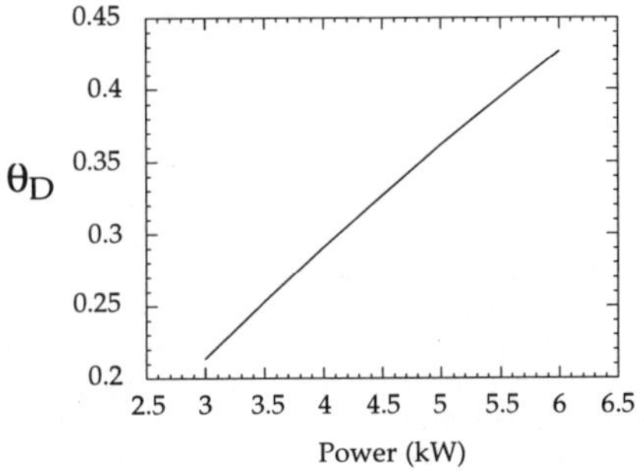

Figure 2. Calculated variation in nitrogen dissociation fraction with arcjet power.

The corresponding nitrogen atom flux at the arcjet exit plane is shown in Fig. 3. This flux is obtained by assuming that the entire partially dissociated nitrogen plume impinges onto a substrate that is 10 cm in diameter, and that the transport of nitrogen to the substrate is not limited by gas-phase diffusion (discussed below).

In a convection-dominated flow, the concentration of minor species (comprising a mixture), at a reactive surface, relative to that outside the boundary layer is given by [9]:

$$\frac{[N]_s}{[N]_o} \approx \frac{1}{1+\gamma\delta/\lambda} \qquad (3)$$

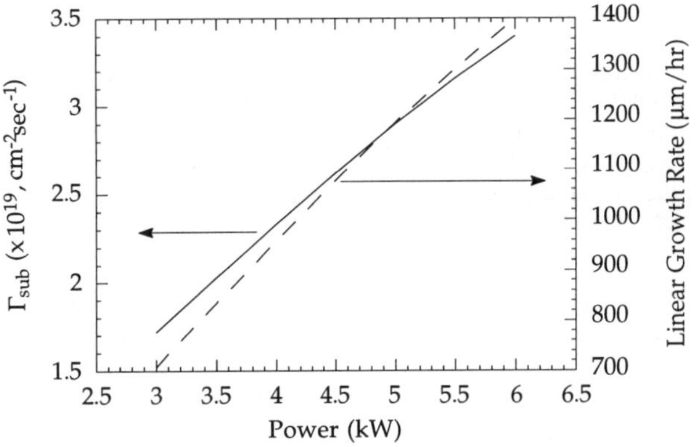

Figure 3. Calculated variation in maximum atom flux and predicted growth rate.

where λ is the mean free path for collisions of the species of interest with the background gas (in this case, atomic nitrogen in molecular nitrogen), δ is the boundary layer thickness, and γ is the surface nitrogen atom reaction probability. The quantity $\gamma\delta/\lambda$ reduces to approximately $10^3\gamma\sqrt{p_e d_s/u_o}$ for atomic nitrogen diffusing in a background gas of mainly molecular nitrogen [9]. Here, p_e, d_s, and u_o represent the stagnation pressure (in Torr), substrate mount diameter (in cm), and arcjet velocity (in cm/s). In order to be in a regime that is free of diffusion transport limitations, it is desirable to operate at lower pressures and with higher jet velocities such that we minimize the transport parameter $10^3\gamma\sqrt{p_e d_s/u_o}$. For conditions typical of our experiments, $p_e \approx$ 10 torr, d_s = 10 cm, and $u_o \approx 2 \times 10^5$ cm/sec. These properties are typical of highly expanded, supersonic, hydrogen arcjet flows, and, we expect similar characteristics for nitrogen flows under similar operating conditions. If we assume $\gamma \approx 0.1$ (a conservative value), then we arrive at $[N]_s/[N]_o \approx 0.3$. Provided $\gamma < 0.05$, the nitrogen atom transport to the substrate in these supersonic dc arcjets is not expected to be diffusion-limited. The maximum growth rates computed on the basis of the computed flux and $\gamma \approx 0.05$ are also shown in Fig. 3. The calculations in Fig. 3 indicate that potentially high linear GaN growth rates can be obtained for operating conditions typical of our supersonic arcjets.

Experimental Procedure

Basal plane oriented sapphire substrates 3.8 cm in diameter were first degreased and cleaned using a final hot 1:1 sulfuric and phosphoric acid soak followed by a water rinse. The substrates were then mounted onto a resistively-heated copper substrate mount. A small hole through the central axis of the copper substrate mount allowed access for a thermocouple to be positioned immediately behind the substrate. A two-color optical pyrometer sighting the front surface of the substrate and mount is also used to estimate the substrate temperature during deposition. The resistive substrate heater was capable of pre-heating the substrates to 400°C prior to turning on the arcjet. After arcjet ignition, the substrate temperature increases to approximately 700-800°C, however, on most occasions, the substrates mounted in this way crack due to thermal stress associated with the substrate clamping mechanism. It is noted that these temperatures are much too low to support rapid growth of optimally smooth crystalline layers. To avoid cracking of the substrates, some growth runs were performed with the substrates removed from the heater environment and held almost parallel to the plasma plume axis. While these substrates did not fracture, it was even more difficult to estimate substrate temperature for this configuration.

The typical growth process could be described by the following steps: (i) the chamber is first evacuated to ~ 10^{-2} torr (while flowing nitrogen at a rate of 20 mg/sec) and the substrate heater is turned on for approximately 45 minutes to establish a substrate temperature of 400°C; (ii) the arcjet is turned on with pure nitrogen flowing at a rate of 146 mg/sec and a power of 0.9 kW, establishing a chamber pressure of 0.3 torr; (iii) the power to the arcjet is increased to approximately 1.3 kW over a period of about 3 minutes; (iv) a bias voltage (~5 V) is applied to the gallium to heat it by electron bombardment, defining the begining of the deposition period; (v) after approximately 5 minutes, the bias and arcjet power are disconnected, and the substrate is allowed to cool to room temperature while partially filling the chamber with nitrogen.

The gallium vapor produced by electron bombardment was clearly observable and discernable from the nitrogen plasma plume as a purple plume arising from the 417 nm atomic gallium fluorescence transition.

RESULTS

Films were generally non-uniform and of varying thickness. This was attributed in part to the non-uniform flux of atomic gallium from the electron bombardment source. In many cases, some regions of the substrate had an array of large (~tens of micron size) particles which were identified to be GaN by micro-Raman spectroscopy. All of the samples had fairly large regions of the substrate which appeared slightly yellow to the eye yet transparent in the visible wavelength range. The samples were also analyzed by scanning electron microscopy (SEM). A typical SEM image of this region of the substrate is shown in Figure 4. The films appear to be island-like, not entirely continuous, but possibly heteroepitaxial. These deposited films were analyzed by visible-infrared transmission spectroscopy, Raman spectroscopy, and x-ray diffraction to further understand the chemical composition and structure.

Figure 4. SEM image of a thin arcjet-deposited GaN layer on sapphire.

A typical visible-IR transmission spectrum for a sample deposited at a temperature of 700 - 800°C is shown in Fig. 5. A Raman spectrum taken with illumination using the 514 nm line from an argon-ion laser is shown in Fig. 6. The sample depicts the sharp drop in transmission due to band-edge absorption at 365 nm, characteristic of αGaN. The measured interference fringes at longer wavelengths indicate a thickness of 0.7 mm, assuming a refractive index of 2.4, typical of bulk GaN. The growth rate for this layer was approximately 8 μm/hr.

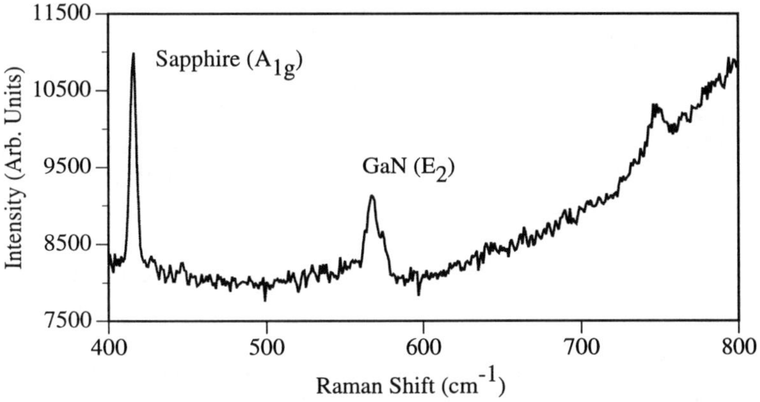

Figure 5 . Typical Raman spectrum of a thin arcjet-deposited GaN layer on sapphire.

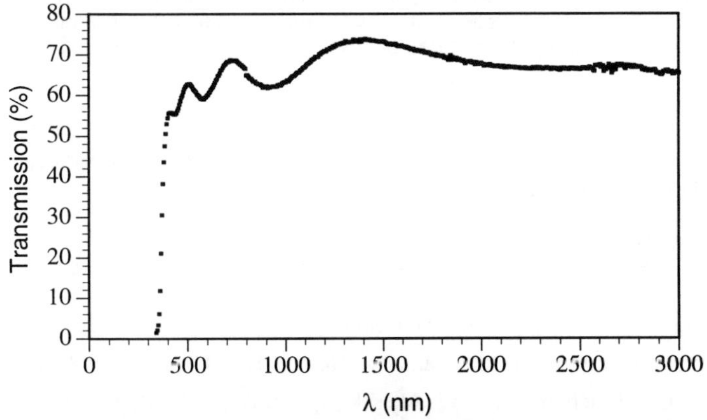

Figure 6 . Typical visible-IR transmission spectrum of a thin GaN layer on sapphire

The Raman spectrum clearly confirms the deposition of GaN through the presence of the characteristic Raman-active 569 cm^{-1} E_2 mode [10]. The feature at approximately 420 cm^{-1} is due to the Raman-active A_{1g} mode of sapphire.

A typical 2θ/ω X-ray diffraction pattern of a GaN film grown on the c- oriented sapphire is shown in Fig. 7. Only (000l) type plane peaks of the wurzitic GaN phase were observed. This means that the GaN film is oriented along the [0001] direction on the basal plane of the sapphire. Fig. 7 also shows the ϕ-scan of both the GaN (1,0,-1,2) plane and sapphire (1,1,-2,3) plane to determine in-plane orientation. Only one set of reflection peaks from the GaN (1,0,-1,2) plane was observed, with 60° spacing. This implies that the GaN film was grown as a single crystal layer. The angular position of the GaN (1,0,-1,2) peaks and sapphire (1,1,-2,3) peaks are coincident, defining the orientation of this deposited layer with respect to the underlying sapphire substrate.

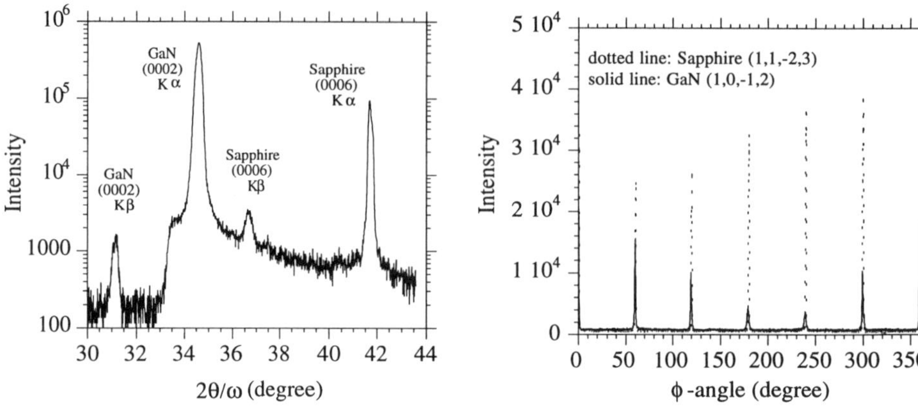

Figure 7. Left: 2θ/ω X-ray diffraction pattern of GaN films grown on the c- oriented sapphire; right: φ-scan of both GaN (1,0,-1,2) plane and sapphire (1,1,-2,3) plane.

SUMMARY

While these preliminary experiments lacked adequate temperature controls and involved no attempts at surface buffer layer growth to improve subsequent nucleation, they did provide the first evidence for the possibility of high GaN growth rates using high nitrogen-atom fluxes without the accompaniment of many of the by-products associated with more typical high growth rate or activated plasma approaches. Single crystal layers approximately 0.5 -0.7 microns thick were deposited on the basal plane of sapphire, corresponding to growth rates of approximately 8 μm/hour.

ACKNOWLEDGEMENTS

This research is supported in part by the National Science Foundation, and by Loral Infrared and Imaging Systems.

REFERENCES

1. T. Lei, T.D. Moustakas, R.J. Graham, Y. He, and S.J. Berkowitz, J. Appl. Phys. 71, 4933, 1992.
2. S. Strite, M.E. Lin, and H. Morkoc, Thin Solid Films **231**, 197, 1993.
3. D.B. Oberman, H. Lee, W.K. Gotz, and J.S. Harris, Jr., J. Crystal Growth **150**, 912, 1995.
4. K. Hiramatsu, T. Detchprohm, and I. Akasaki, Jpn. J. Appl. Phys., **32**, 1528, 1993.
5. T. Detchprohm, K. Hiramatsu, H. Amano, and I. Akasaki, Appl. Phys. Lett., **61**, 2688, 1992.
6. M.H. Loh and M.A. Cappelli, Diamond and Related Materials **2**, pp. 454-461, 1993.
7. M. A. Cappelli, "Arcjet Synthesis of Diamond", in Handbook of Industrial Diamond and Diamond Films, 1st Edition, Edited by M. Prelas, Marcel Dekker, 1996.
8. We are using an empirical relation for the flow velocity based on measurements of average velocity (thrust force/mass flow rate) as a function of specific energy.
9. D.G. Goodwin, J. Appl. Phys. **74**, 6895, 1993.
10. T. Kozawa, T. Kachi, H. Kano, Y. Taga, M. Hashimoto, N. Koide, and K. Manabe, J. Appl. Phys. **75**, 1098, 1994.

PARTICLE-ASSISTED ORIENTED DEPOSITION OF DIAMOND THIN FILMS

DONG-GU LEE AND RAJIV K. SINGH
University of Florida, Department of Materials Science and Engineering, Gainesville, FL 32611

ABSTRACT

We have developed a method for <111> oriented diamond film synthesis using micron-sized diamond particles. Different size of diamond powders were electrophoretically seeded on silicon substrates using diamond suspensions in organic solvents (acetone, methanol, and ethanol). Diamond suspension in acetone was found to be the best for obtaining uniform diamond seeding by electrophoresis. The thickness of diamond seeded films was changed by varying the applied voltage to observe the effect on the orientation of diamond particles. Then diamond films were deposited by the hot filament chemical vapor deposition (HFCVD) process. A preferred orientation with <111> direction normal to the substrate was obtained for monolayer coatings. The surface morphology, crystal orientation, and quality of diamond films were investigated using scanning electron microscopy, x-ray diffractometry, and Raman spectroscopy.

INTRODUCTION

Heteroepitaxial or textured growth of diamond thin films has been one of major issues for semiconducting electronic devices in order to get films with high crystal perfection and uniform doping control. Some studies have been conducted on heteroepitaxial or textured growth of diamond films on silicon [1-3], β-SiC [4],and cubic boron nitride substrates [5]. These techniques employ the use of substrate bias to control the orientation of nuclei.

In this research, we investigate the electrophoretic seeding method for obtaining the preferred orientation of diamond thin films. This seeding technique has several advantages. One advantage is that the electrophoretic seeding does not damage the surface of sensitive device materials, compared to scratching by ultra-sonication in solution or polishing for enhancing nucleation sites. Earlier report using this technique was conducted on selectively seeded silicon regions to fabricate patterned films [6]. Another advantage to electrophoretic seeding is that this processing provides diversity in diamond morphology by controlling nucleation density and particle size. Lastly, this processing also shows the possibility of growth of textured films similar to the case of coatings of superconductors [7]. However, in spite of a lot of potential applications, there has not been much research in this area.

EXPERIMENT

Dispersions containing 0.1-3 g/L of synthetic diamond particles in organic solvent were used for electrophoretic seeding. Diamond particle size (0.25 μm, 1 μm, 5 μm: Warren Diamond Co.) and organic solvents (acetone, methanol, ethanol : Fisher optima grade) were used for this process. Organic liquids are normally preferred as a suspending medium rather than water solution because water is dissociated into gases near the electrode material. Silicon (111) substrates were used as the positive electrode for deposition and aluminum plate was used as the negative electrode because diamond particle always acquires negative surface charge spontaneously when mixed with these solvents. The distance of the electrodes was 3 cm and the applied voltage ranged from 250-1000 V. After diamond particles were seeded on the silicon substrate, diamond thin film was grown by HFCVD process. Substrate temperature and filament temperature of 850 °C and 2100 °C,

Mat. Res. Soc. Symp. Proc. Vol. 423

respectively were employed for the deposition process. The chamber pressure was 30 torr, the gas flow rate was 200 sccm, gas composition corresponding to 1%CH_4 and 99%H_2, and deposition time of 17 hrs were employed in these experiments. Scanning electron microscopy, x-ray diffractometry, and Raman spectroscopy were used to observe surface morphology, crystal orientation, and quality of diamond thin film. For a reference, a scratched sample was prepared by sonication in 10 g/L of diamond particles in acetone for one hour.

RESULTS AND DISCUSSION

Effect of Organic Solvents

Figure 1 (a), (b), and (c) shows the surface morphology of electrophoretically-seeded diamond particles using diamond suspension in acetone, methanol, ethanol, respectively. Dispersions containing 0.1 g/L of 0.25 μm diamond particles were prepared for each organic solvent. Applied voltage and time for electrophoresis were 1000 V and 10 sec, respectively. The deposition rate of diamond particle in electrophoretic seeding was found to vary in different medium. The deposition rate was highest for acetone and least for ethanol suspension. Figure 1 (d) shows the particle distribution of seeding using suspension in acetone at 500 V. It can be clearly seen that the seeded films formed from diamond suspension in acetone were more uniform. In the case of methanol, the seeded films were also fairly good. However, most films seeded using ethanol showed some agglomerations of particles. Therefore, it is assumed that diamond suspension in acetone may be the most stable one with no clustering of particles [8], resulting in uniform seeding.

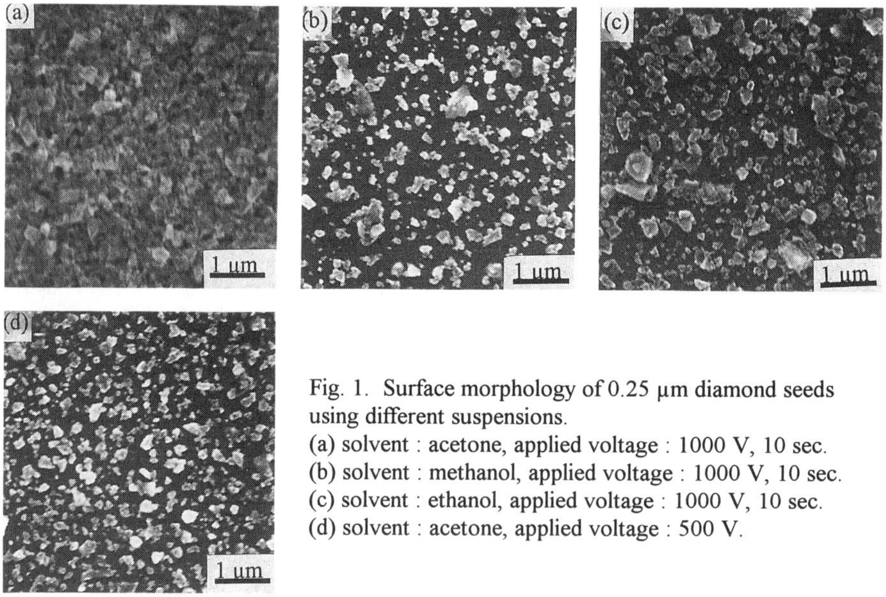

Fig. 1. Surface morphology of 0.25 μm diamond seeds using different suspensions.
(a) solvent : acetone, applied voltage : 1000 V, 10 sec.
(b) solvent : methanol, applied voltage : 1000 V, 10 sec.
(c) solvent : ethanol, applied voltage : 1000 V, 10 sec.
(d) solvent : acetone, applied voltage : 500 V.

Effect of Electrophoretic Seeding on Crystal Orientation

In order to observe change in crystal orientation during electrophoresis and HFCVD diamond films growth, the intensity ratio of certain crystallographic planes of diamond obtained from x-ray measurement was compared as shown in Table 1. The applied voltage was varied from 250 V to 1000 V. Little evidence of preferred orientation for 0.25 µm seeds was observed when compared to a randomly oriented sample due to multiple stacking of diamond particle layers. For 5 µm particle seeding up to around monolayer coverage, highly oriented <111> seeds were obtained. Figure 2 (a) and (b) shows x-ray diffraction patterns for 0.25 µm diamond particles seeded at applied voltage of 750 V and for 5 µm particles seeded at 1000 V, respectively. As shown in Fig. 2(b), peak intensities of (220) , (311), and (400) planes for 5 µm seeds significantly decreases with respect to that of (111) plane. There was little change in intensity ratio between diamond seeds before HFCVD and diamond films after HFCVD, clearly indicating that the initial nucleation density controls the film orientation.

It is speculated that surface of {111} planes tends to have more charges due to the highest number of dangling bonds. Thus, the {111} planes may be more responsive and aligned to the electric field. If multilayers are deposited (as in case of 0.25 µm films), the irregularities in particle shape may prevent {111} plane from aligning vertically to the planar electrode. Figure 3 gives surface morphology of seeded diamond particles and grown films by HFCVD process. In Fig. 3(f), many

Table 1. X-ray intensity ratio changes of diamond seeding under different electrophoretic conditions.

	Seeding before CVD				After CVD for 17 hrs.		
	Applied voltage(V)	Thickness (µm)	I_{111}/I_{220}	I_{111}/I_{400}	Thickness (µm)	I_{111}/I_{220}	I_{111}/I_{400}
Randomly oriented[1]			4	12.5			
Scratched					2.5	9.6	12.5
0.25 µm diamond suspension in acetone (1 g/L)	250	~1	4.2				
	500	~1	5.6				
	750	~2.5	5.0	35.8	4	4.0	57.5
	1000	~5	4.7	38			
5 µm diamond suspension in acetone (3 g/L)	250	~5 (~10%cover)	37.7				
	500	~5 (~50%cover)	28.3				
	1000	~5	24	484	6	27	504
	1000, 3 sec.	~10	8	46			

1, from 6-675 JCPDS files
* distance between electrodes : 3 cm

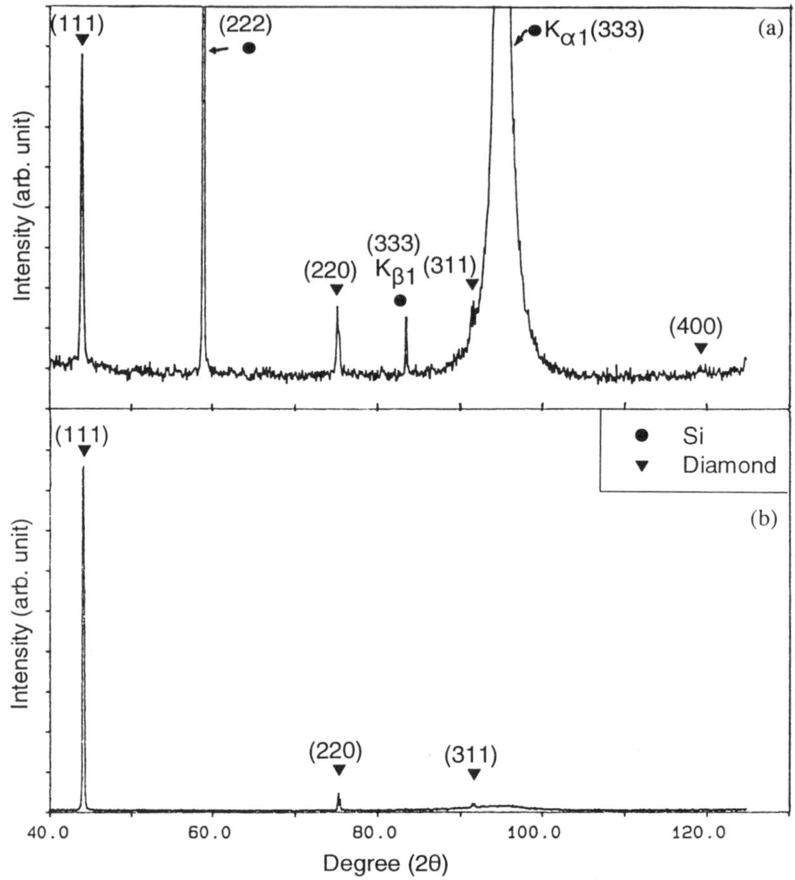

Fig. 2. X-ray diffraction patterns of electrophoretically seeded diamond particles.
(a) 0.25 μm seeds, applied voltage : 750 V, (b) 5 μm seeds, applied voltage : 1000 V.

grains having triangular {111} planes perpendicular to silicon substrate are shown. Figure 4 gives Raman spectra of diamond film grown by HFCVD. Diamond peaks at 1331 cm^{-1} are clearly seen and the broadened peaks of diamond-like carbon or graphite in the range of 1540-1600 cm^{-1} are also observed. However, the impurity peaks was the least for scratched sample. These amount of impurities may be affected by the top surface area exposed to free surface. It has been also observed that the graphitic phases have been found to increase when the nucleation density is increased.

CONCLUSIONS

The crystal orientation of diamond films for controlled seeding of diamond particles has been

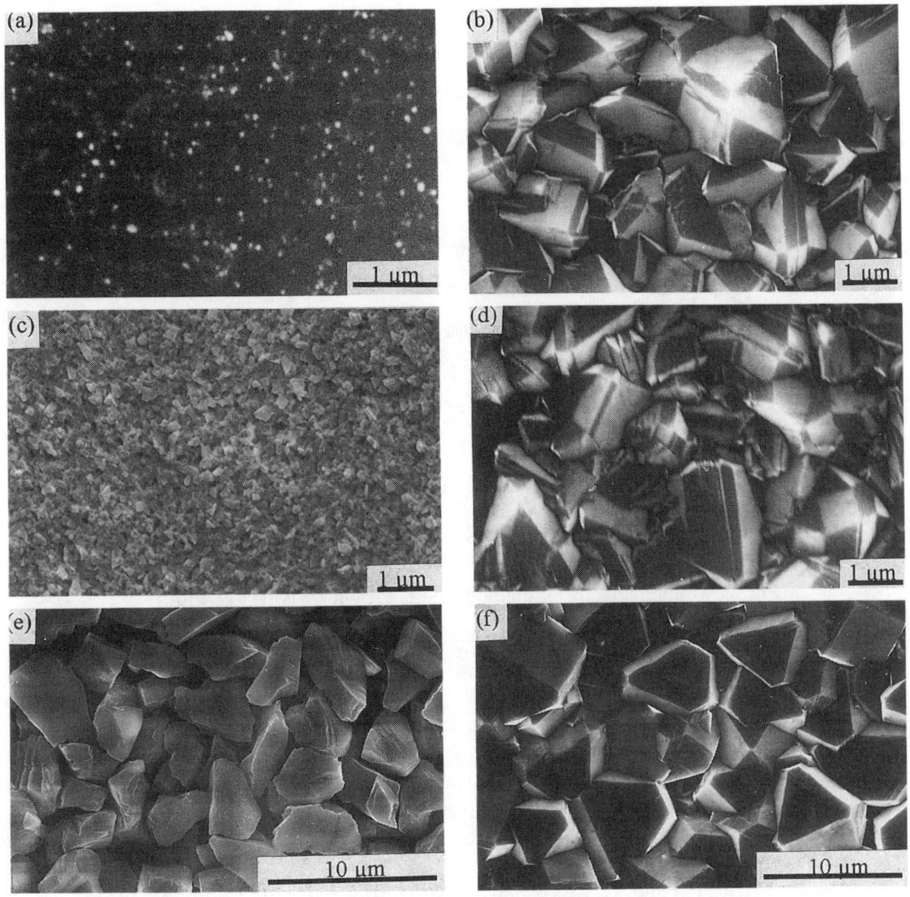

Fig. 3. Surface morphology of diamond films before and after growth by HFCVD.
(a) scratched silicon surface, (b) grown diamond film of (a).
(c) seeded surface with 0.25 μm, applied voltage : 750 V, (d) grown diamond film of (c).
(e) seeded surface with 5 μm, applied voltage : 1000 V, (f) grown diamond film of (e).

investigated. Diamond suspension in acetone was found to be the best for getting uniform diamond seeding by electrophoresis. Preferred orientation of diamond particles to <111> resulted in highly oriented diamond films.

ACKNOWLEDGMENTS

The authors would like to acknowledge the financial support of the University of Florida, National Science Foundation, and The Engineering Research Center for Particle Science and Technology at the University of Florida, NSF grant # EEC 94-02989.

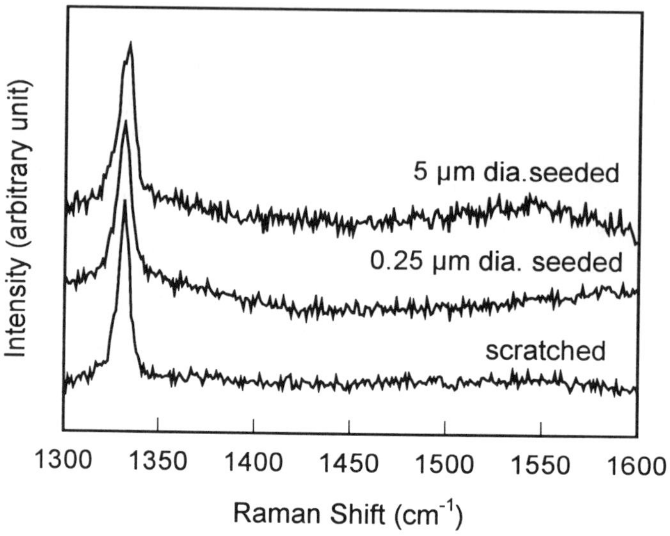

Fig. 4. Raman spectra of diamond films grown by HFCVD with various sample preparation.

REFERENCES

1. C. Wild, P. Koidl, W. Müller-Serbert, H. Walcher, R. Kohl, N. Herres, R. Locher, R. Samlenski, and R. Brenn, *Diamond Relat. Mater.*, **2**, 158 (1993).

2.. R. E. Clausing, L. Heatherly, L. L. Horton, E. D. Specht, G. M. Begun, and Z. L. Wang, *Diamond Relat. Mater.*, **1**, 411 (1992).

3. H. Maeda, M. Irie, T. Hino, K. Kusakabe, and S. Morooka, *J. Mater. Res.*, **10**(1), 158 (1995).

4. B. R. Stoner and J. T. Glass, *Appl. Phys. Lett.*, **60**(6), 698 (1992).

5. S. Koizumi, T. Murakami, T. Inuzuka, and K. Suzuki, *Appl. Phys. Lett.*, **57** (6), 563 (1990).

6. J. L. Valdes, J. W. Mitchel, J. A. Mucha, L. Seibles, and H. Huggins, *J. Eletrochem. Soc.* **138** (2), 635 (1991).

7. C.T. Chu and B. Dunn, *Appl. Phys. Lett.*, **55**, 492 (1989).

8. J. B. Birks, in *Progress in Dielectrics, Vol 1*, edited by J. B. Birks and J. H. Schulman (John Wiley & Sons, New York, 1959), pp 273-312.

9. Michael Seal, in *Synthetic Diamond : Emerging CVD Science and Technology*, edited by K. E. Spear and J. P. Dismukes (John Wiley & Sons, New York, 1994), pp. 524-525.

DEPOSITION OF ADHERENT DIAMOND FILMS ON SAPPHIRE SUBSTRATES

DONALD R. GILBERT, RAJIV K. SINGH
Department of Materials Science and Engineering, University of Florida, Gainesville, FL 32611-2066

ABSTRACT

Sapphire is an important material for infrared-transmission applications and SOI device structures. The use of diamond as a protective coating for sapphire is a desirable, yet elusive process. The primary obstacle to the successful deposition of diamond on sapphire is the large mismatch in thermal expansion characteristics between the two materials. To overcome this problem, we have used a low temperature deposition process involving an electron cyclotron resonance (ECR) assisted plasma CVD system. Continuous, adherent diamond films have been deposited over 1 to 2 cm^2 areas. These films have been characterized for structural quality using Raman spectroscopy and x-ray diffraction.

INTRODUCTION

The chemical vapor deposition of diamond thin films on optically transparent substrates such as sapphire is of immense technological importance for many applications.[1] A diamond coating protects the window from erosion and related environmental degradation effects. Thus far, attempts to grow diamond thin films directly on sapphire and other infrared transparent materials have received limited success.[2] There are two primary obstacles to the successful growth of adherent diamond films on sapphire. First, the large thermal expansion coefficient mismatch between diamond and the substrate results in very high compressive stresses in the thin film.[3-5] This problem is especially acute in conventional diamond chemical-vapor-deposition methods, where typical substrate temperatures are between 800 and 950 °C. Second, the lack of chemical affinity between the film and substrate prevents the formation of an interfacial carbide "glue" layer, which results in poor adherence. Thermodynamically, sapphire has a very stable structure and conversion from Al_2O_3 to an intermediate carbide layer is unfavorable.

To understand the magnitude and nature of the thermal stresses, the cumulative thermal expansion of sapphire, silicon, and diamond as a function of temperature is shown in Figure 1.[6] The cumulative thermal expansion values are calculated by integrating the linear thermal expansion coefficients over the desired temperature range (i.e., room temperature to deposition temperature). This figure shows that at 900 °C, a compressive strain of approximately 0.45% is expected in diamond on sapphire film. The stress induced in the films under these conditions is expected to be approximately 5 GPa, which can easily cause delamination of a weakly bonded diamond film from a sapphire substrate.

In this paper, we demonstrate a method to deposit adherent diamond thin films successfully directly on optically transparent sapphire. To increase the adhesion of the diamond film on sapphire substrates we have adopted a two-step method that involves (i) lower temperature deposition and (ii) growth of diamond from many uniformly dispersed nuclei. To reduce the thermal stresses in the films, a low temperature deposition method (deposition temperature 500 - 550 °C) using a low pressure electron-cyclotron-resonance (ECR) plasma has been developed and optimized. The combination of low pressure and a highly activated gas source, provided by the intense ECR plasma, ensures the deposition of high quality diamond films at low processing temperatures. Under these deposition conditions, the strain in diamond is

Mat. Res. Soc. Symp. Proc. Vol. 423

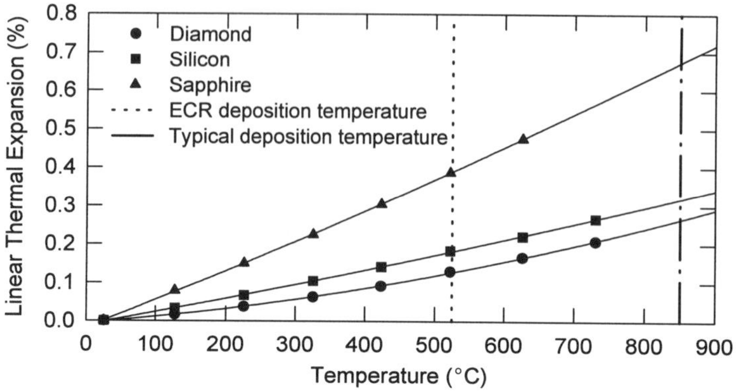

Figure 1. Linear thermal expansion of diamond, silicon and sapphire versus temperature.

expected to be less than 0.25%, resulting in a substantial decrease in the thermal stresses as compared to deposition of diamond films under standard conditions (substrate temperatures ~ 800 to 950 °C). To achieve a high density (~10^9 cm^{-2}) of nuclei on the surface, a colloidally dispersed suspension of diamond powder with average grain size of 0.2 μm was prepared and uniformly dispersed on the surface of the sapphire substrate.

EXPERIMENT

Due to the large energy barrier to diamond nucleation on non-diamond material, it was necessary to provide a preexisting coverage of diamond nuclei on the sapphire surface. To achieve this, a colloid of diamond with an average particle size of less than 0.25 μm was suspended in acetone. The sapphire substrate was then immersed in the stable suspension so that a uniform seeding of diamond was achieved. This seeding step was found to be critical for the formation of continuous films. Diamond films were deposited using an electron-cyclotron-resonance (ECR) enhanced chemical-vapor-deposition system using methanol and hydrogen gas as growth precursors. The ECR plasma system uses an imposed magnetic field to enhance the efficiency of energy transfer from the applied microwaves to the free electrons in the plasma at low pressures. A substrate temperature of approximately 500 - 550 °C was employed during the deposition process, while the total gas pressure was kept at 1.00 Torr. The composition of the process gas was approximately 1% methanol and 99% hydrogen. The relatively low pressure in the plasma chamber results in the formation of a uniform large area plasma. Under these conditions, the lower limit for the deposition of good quality diamond films was approximately 500 °C, and the average growth rate was ~0.2 μm/hour.

RESULTS

Figure 2 shows the surface morphology of a diamond film on a sapphire substrate at two different magnifications. Low magnification (Figure 2(a)) shows that the film is continuous over a relatively large area. Optical evaluation of the diamond-coated sapphire samples showed no visible peeling or cracking over the entire surface (approximate area ~ 2 cm^2), except at the edges of the film. The edge cracking was attributed to the high shear stresses induced by discontinuity of the film at the substrate edge. These results are in sharp contrast to other studies, where the continuous areas in the films were less than 20 μm^2.[2] Figure 2(b) shows the surface morphology of the diamond films at higher magnification (50× of Figure 2(a)). The figure shows faceted particles characteristic of the diamond phase. The small grain-size (<1 μm) is due to the high density of nucleation sites achieved by the colloidal diamond pre-treatment step and the low deposition temperature.

(a) (b)

Figure 2. (a) Low and (b) high magnification SEM of diamond film deposited on sapphire.

The microstructure of the diamond films on sapphire substrates was examined both by X-ray diffraction and Raman spectroscopy. Figure 3 shows the Raman spectra obtained from a diamond film deposited on a (110) sapphire substrate. The Raman spectrum shows a distinct peak at 1338 cm^{-1}, which is characteristic of the diamond phase. However, this peak is shifted from the equilibrium position of 1332 cm^{-1}, primarily because of the high compressive stress in the diamond film. Besides the diamond peak, broad peaks at 1450 cm^{-1} and 1550 cm^{-1} are also observed. These peaks are attributed to a small amount of non-diamond carbon, which has more than a factor of 50 higher Raman-scattering cross-section than diamond. The full width at half maximum (FWHM) of the diamond peak is approximately 12 cm^{-1}, compared to the instrumental resolution of 1 cm^{-1}. To determine the possible influence of the original diamond seeding material on the resulting Raman spectra, a scan was run on the seeded, undeposited substrate surface. The Raman system used was unable to detect the seed layer, indicating that the spectrum detected from the deposited film is entirely produced by grown material. The broadening of this peak can be attributed to high compressive stresses in the diamond film which cause the unstressed diamond peak to be split into two components (doublet at higher frequency and singlet at lower frequency).[9] Other factors contributing to the broadening of the diamond peak are the small domain size of the diamond crystals and microstructural defects in the diamond films. The compressive bi-axial stress (τ) in the film can be determined from the shifting of the diamond peak according to the formula

$$\tau = -0.384(\nu_s - \nu_0)\left(\frac{GPa}{cm^{-1}}\right) \qquad \textbf{(1)}$$

where ν_s is the observed Raman shift for a stressed diamond film and ν_0 corresponds to the equilibrium diamond-peak position in the unstressed state, which corresponds to approximately 1332 cm^{-1}. Using this formula, the compressive stress in the diamond film is estimated to be ~ 2.2 GPa. This value of compressive stress in the film is consistent with the calculated thermal stress expected for diamond deposited at 525 °C.

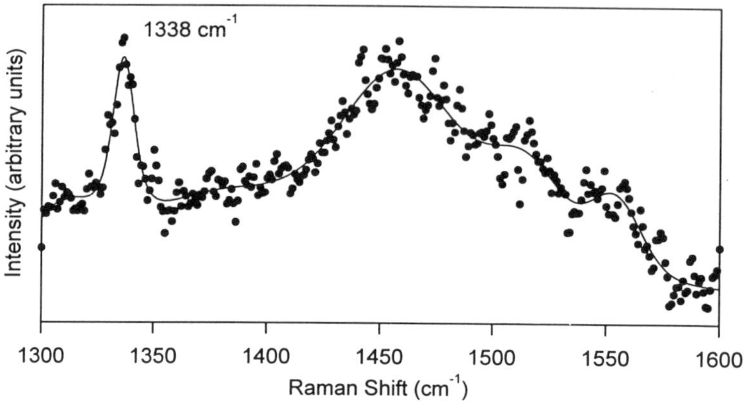

Figure 3. Raman spectra of diamond film deposited on sapphire.

To determine the texture of the diamond films, x-ray diffraction measurements were also conducted. Figure 4 shows the θ-2θ x-ray scan obtained from a diamond film on a sapphire substrate. The x-ray spectra shows only the characteristic peaks for sapphire and diamond. The relative intensities of the diamond peaks reveal the random polycrystalline nature of the diamond film growth on sapphire. Other crystalline phases of carbon were not observed in the X-ray diffraction pattern. However, it should be noted that x-ray is quite limited in its ability to detect non-diamond carbon inclusion in deposited films due to its low atomic number and density. Again, observation of a seeded, undeposited substrate did not show the presence of diamond, thus eliminating any possible influence on post-deposition characterization.

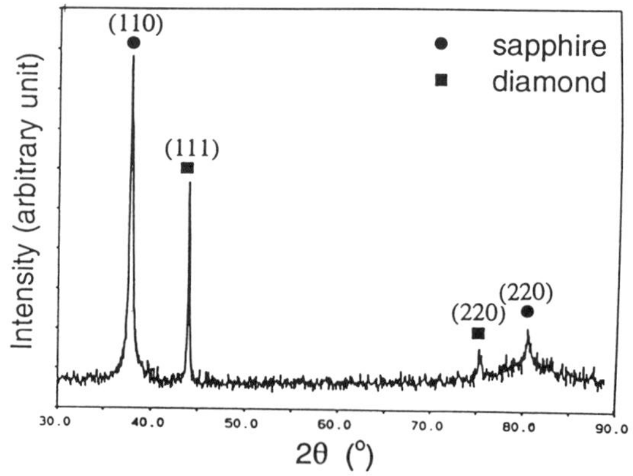

Figure 4. XRD spectrum of diamond film deposited on sapphire substrate.

CONCLUSIONS

We have developed a novel method to deposit diamond films directly on sapphire. This method utilized colloidal diamond combined with a low temperature ECR-CVD process to fabricate continuous, adherent films on sapphire substrates. Raman analysis of the resulting film indicated the presence of residual stress of ~2.2 GPa, which was consistent with the deposition process conditions used. While this process significantly reduces problems due to thermal stress, issues concerning chemical bonding at the film-substrate interface still need to be addressed.

ACKNOWLEDGMENTS

The authors would like to acknowledge the financial support of National Science Foundation, The Engineering Research Center for Particle Science and Technology at the University of Florida, NSF grant # EEC 94-02989.

REFERENCES

1. Jacques I. Pankove and Chang-Hua Qiu, in Synthetic Diamond: Emerging CVD Science and Technology, edited by K. E. Spear and J. P. Dismukes (Wiley, New York, 1994), pp. 401-418

2. P. Y. May, C. A. Rego, C. G. Trevor, E. C. Williamson, M. N. R. Ashfold, K. N. Rosser and N. M. Everitt, Diamond and Related Materials **3**, p. 1375 (1994)

3. M. D. Drory and J. W. Hutchinson, Science **263**, p. 1753 (1994)

4. R. Ramesham and T. Roppel, J. Mater. Res. **7**, p. 1144 (1992)

5. D. K. Snood, W. R. Drawl, R. Messier, Surf. Coat. Technol. **51**, p. 307 (1992)

6. Purdue University, Thermophysical Properties of Matter, Y. S. Touloukian Ed. (IFI/Plenum, New York, 1970), vol. 13

7. R. K. Singh, D. Gilbert, R. Tellshow, P. H. Holloway, R. Ochoa, H. Simmons and R. Koba, Appl. Phys. Lett. **61**, p. 2863 (1992)

8. R. K. Singh and D. G. Lee (unpublished)

9. J. W. Ager and M. D. Drory, Phys. Rev. B **48**, p. 2601 (1993)

GaN GROWTH BY NITROGEN ECR-CVD METHOD

K.H. Chen[1], C.H. Chao[1], T.J. Chuang,[1] Y.J. Yang[2], L.C. Chen[3], C.K. Chen[3], Y. F. Huang[4], C.H. Yang[4], H.Y. Lin[5], I.M. Chang[6], and Y.F. Chen[6]

[1] Institute of Atomic and Molecular Sciences, Academia Sinica, Taiwan.
[2] Department of Electrical Engineering, National Taiwan University, Taiwan.
[3] Center for Condensed Matter Sciences, National Taiwan University, Taiwan.
[4] Department of Physics, Fu-Jen University, Taiwan.
[5] Tar-Tong Institute of Technology, Taipei, Taiwan
[6] Department of Physics, National Taiwan University, Taiwan.

ABSTRACT

A new approach toward GaN growth using electron cyclotron resonance assisted microwave plasma enhanced chemical vapor deposition (ECR-CVD) method has been implemented. This growth technique allows for low- as well as high-temperature deposition, the use of pure nitrogen source, and a wide operating pressure that is between MOCVD and MOMBE. The unique features of this technique enable the growth of the epitaxial layer of GaN on a variety of substrates including sapphire, silicon, and $LiGaO_2$. SEM, XRD, Raman, photoluminescence (PL), and Hall measurement are employed to characterize the deposited films. Highly oriented, (0001) textured films in the expected wurtzite structure with blue emission have been obtained.

INTRODUCTION

The growth and applications of III-Nitride materials has been the focus of many institutes in the field of optoelectronics in the past few years.[1-5] The blue light-emitting materials, such as GaN and AlN, are of high potential toward high intensity LED, laser diode, and high speed, high temperature electronic device.[6] Therefore, tremendous efforts have been applied to the epitaxial growth and doping of III-Nitride materials. Most of the work so far can be divided into two categories: the MOCVD method[7,8] and the low pressure MOMBE method.[9,10] The MOCVD method uses thermal excitation to generate nitrogen atoms from NH_3 precursor. In such a process, high temperature and high NH_3 flux are needed in order to produce enough reaction species at a pressure range between Torrs to atmospheric pressure. However, the choice of substrate material is restricted by the high temperature and the large amount of hydrogen during MOCVD process. In contrast, the MOMBE method requires UHV system and allows lower temperature deposition.[9,10] Despite that the MOMBE method offers some apparent advantage over the MOCVD method, the optical property of the film deposited by the MOMBE method is still unsatisfactory. While high performance film can be obtained by the MOCVD method, the quality of the film is yet inferior compared with most other heteroepitaxial semiconductor systems. Hence, not only there are rooms for further improvement of the film quality but also it remains to be a great challenge to develop novel technique for the growth of high quality III-Nitride materials.

In this work, we present a new approach to the growth of III-Nitride materials by using electron cyclotron resonance (ECR) assisted microwave plasma enhanced CVD method. For brevity, ECR-CVD is used hereafter for the process in the present report. The ECR-CVD process is operated in a pressure range between those for the MOMBE and the MOCVD methods. Furthermore, since the nitrogen atoms and ions are generated

Mat. Res. Soc. Symp. Proc. Vol. 423 © 1996 Materials Research Society

by ECR and microwave source from either nitrogen or NH_3, the deposition can be carried out in a much wider range of gas compositions, e. g., from hydrogen-free to very hydrogen rich environments. The ECR-CVD process also allows a flexible substrate temperature range, which in turn, together with the possibility of a hydrogen-free environment, opens up more choices for the substrate materials. The basic design of our reactor is presented. Preliminary characterization and analyses of the material grown are also reported.

EXPERIMENTAL

The ECR-CVD reactor consists of a 16" diameter chamber equipped with a substrate stage containing a boron nitride heater. An Astex A5000 microwave source and an AX4400 electromagnet comprise the ECR source for the reactor. In the reactor, pure nitrogen with a typical flow rate of 30 sccm is introduced through the ECR plasma by which the atomic and ionic nitrogen are generated for the growth. A trimethyl gallium (TMG) source with a typical flow rate of 1.0 sccm is introduced through a separate shower ring, which bypasses the ECR plasma. The purity of nitrogen and TMG is greater than 5N5 and 6N5, respectively. During deposition, the system can be maintained at a fixed pressure by a turbo molecular pump in a pressure range from 10^{-5} to 10^{-2} Torr. A thermocouple placed underneath the substrate holder is used to monitor the holder temperature. In addition, a two-color pyrometer is also used to monitor the temperature of the 1" diameter Mo holder, on top of which the sapphire, silicon, and $LiGaO_2$ (from Crystal Photonics, Inc.) substrates are placed. The temperature of the Mo holder can be controlled to within -10 K between 500 K and 1300 K. The low temperature range, in particular, is difficult to obtain N-atoms for MOCVD method.

Sample preparation for the optical grade polished (0001)-orientation sapphire first includes degreasing in successive rinses with trichloroethane, acetone, and methyl alcohol. It is then etched in a 10% HF solution followed by a hot (160° C) 3:1 H_2SO_4:H_2PO_3 solution for ten minutes. For silicon substrate, only HF etching is applied. Sample preparation for $LiGaO_2$ substrate includes the same degreasing procedure as for sapphire but followed by a much mild etching condition with 1:1:7 of H_2SO_4:H_2O_2:H_2O for 5 minutes.

After the substrate is loaded into the system, the chamber is pumped down to a base pressure of 10^{-7} Torr before the growth. To remove residuals on the substrate surface, the system is immersed in a nitrogen plasma for 15 minutes before TMG is introduced. For the case of sapphire and Si substrates, a buffer layer[11] is deposited at 500° C prior to the growth of the epitaxial layer to enhance the quality of the GaN films. Typical thickness of the buffer layer is 200 . The growth rate is about 0.8 μm/hr for sapphire and Si substrates and 0.5 μm/hr for $LiGaO_2$ substrates.

The deposited films have been characterized by Ramam,[12,13] SEM, XRD, photoluminescence (PL),[14] and Hall measurement.[15] A Renishaw system 2000 micro-Raman spectrometer with a 25 mW He-Ne laser operating at 632 nm was used for the Raman measurement. With a 5 micron slit, the spectral resolution of the Raman spectrometer is 1 cm^{-1}. Both the X-ray θ-2θ diffraction and rocking curve measurement were performed using a Bede high resolution X-ray spectrometer. The surface morphology and thickness of the film were characterized by SEM. Hall-effect measurement were performed by van der Pauw method[15] at room temperature. Photoluminescence spectra were obtained at temperatures of 10 K and 300 K using a He-Cd laser operating at 325 nm.

RESULTS AND DISCUSSION

Morphology and rocking curves

The SEM micrograph of a highly oriented, (0001) textured film is shown in Figure 1(a) where hexagonal-shaped grains are still visible in a predominantly featureless surface morphology. The corresponding cross section view of the film is shown in Figure 1(b). This film was grown at 1000° C on a (0001) sapphire substrate. The surface roughness of this film is about 1000 while much larger roughness was observed when grown on a Si(100) substrate under identical growth condition. The surface roughness is strongly dependent on the buffer layer thickness and the growth condition. Typically, a buffer layer of around 200 thick is necessary to grow a smooth and continuous film when sapphire and Si(100) substrate are used. Presumably, the need of a buffer layer is to accommodate a large lattice mismatch between the film and the substrate.

Comparing with the 13.6% lattice mismatch between the GaN film and the sapphire substrate, the mismatch can be reduced dramatically by using $LiGaO_2$ substrate. The lattice constants for GaN and $LiGaO_2$ are 3.180 Å and 3.186 Å, respectively, which amounts to only 0.19% mismatch. Therefore, a buffer layer to accommodate the lattice mismatch can be waived when grown on a $LiGaO_2$ substrate. However, the substrate preparation and processing prior to the film growth is critical due to the reactiveness of the material. The low melting point and reactiveness of $LiGaO_2$ also restrict the deposition temperature to below 800° C. Figures 2(a) and 2(b) show the plain view and the cross-section view, respectively, of SEM micrographs taken from a film grown on the $LiGaO_2$ substrate.

Further quantitative measurement of the crystal quality is done by the x-ray rocking curve (θ-scan) and radial peak width (θ-2θ scan). Figure 3(a) and 3(b) show the results for the film grown on sapphire. Both are taken from the (0002) reflection. A FWHM of 1725 arcsec of the rocking curve was measured. The θ-2θ peak width of 263

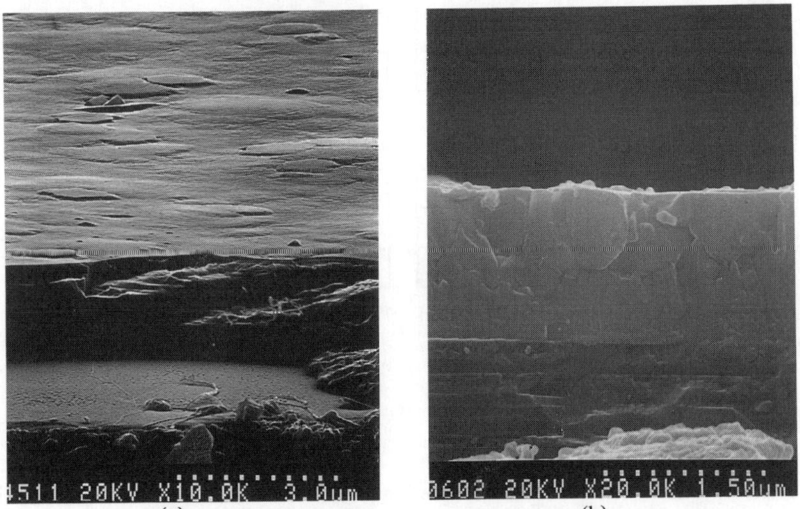

(a) (b)

Figure 1. SEM micrographs for plain view (a) and cross-section view (b) of GaN film grown on a (0001) sapphire substrate with a buffer layer.

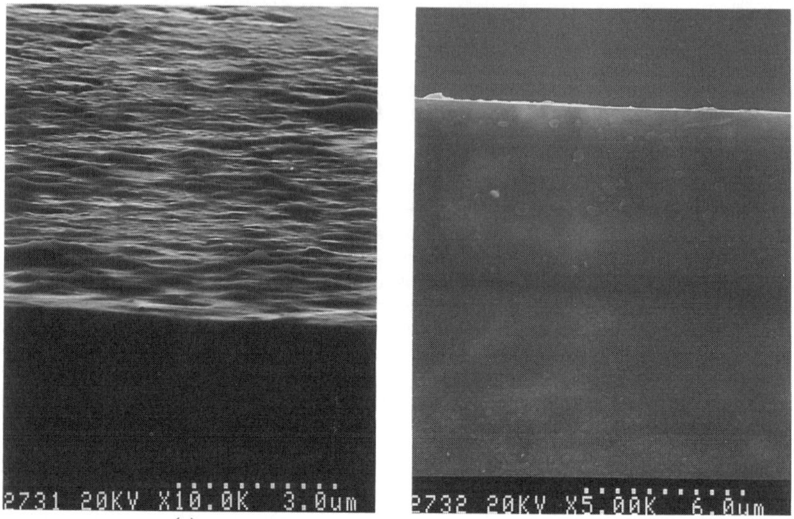

(a) (b)

Figure 2. SEM micrographs for plain view (a) and cross-section view (b) of GaN film grown on a (0001) $LiGaO_2$ substrate without a buffer layer.

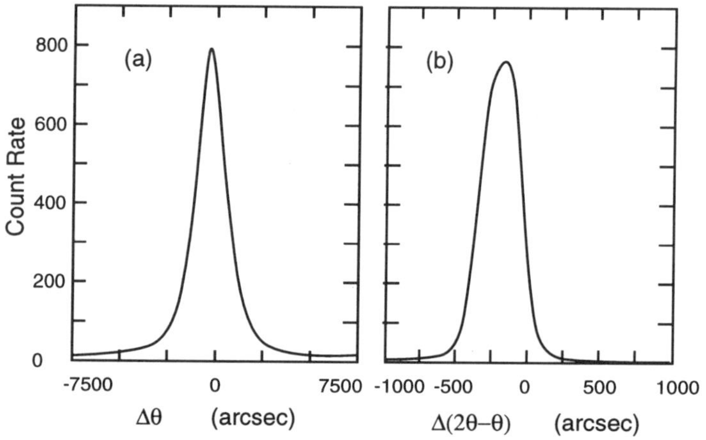

Figure 3. XRD rocking curve i.e., θ-scan (a) and radial peak width i.e., θ-2θ scan (b) of the film in Fig. 1.

arcsec corresponds to a structural coherence length of 1140Å. This coherence length is only about one tenth of the film thickness and is probably limited by the high defect density, presumably originated from the film/substrate interface. The large defect density also causes significant broadening in the rocking curve. Again, the widths of both θ-scan and θ-2θ scan are strongly dependent on the buffer layer thickness and the growth condition. Although our present results of the peak width are still large comparing to the best literature data by other techniques, we believe that, considering the flexible and wide range of growth conditions that the ECR-CVD process can offer, much improvement can be achieved by modifying our process.

Raman spectra

As shown in Figure 4(a), typical Raman spectrum of a good quality, transparent film with Raman shift of 565.5 and 740.2 cm^{-1} were observed. The Raman lines around 420 and 640 cm^{-1} are attributed to the phonon modes of the sapphire substrate. The main peak at 565.5 cm^{-1} can be attributed to three phonon modes of GaN, namely, 534.6 cm^{-1} (A1(TO)), 560.8 cm^{-1}(E1(TO)), and 569.6 cm^{-1}(E2), all of which are indicative of a GaN wurtzite structure. Presumably, the broadening of the Raman lines is due to the high defect density, consistent with our X-ray diffraction results. The Raman spectra of a slightly colored film taken as grown and after post annealing at around 800° C in a nitrogen ambient are shown in Figure 4(b). For this film, the 534.6 cm^{-1} peak and the sapphire peaks are more pronounced than those in a higher quality film. After annealing, the GaN modes show an increase in intensity, suggesting an increase of order of the film.

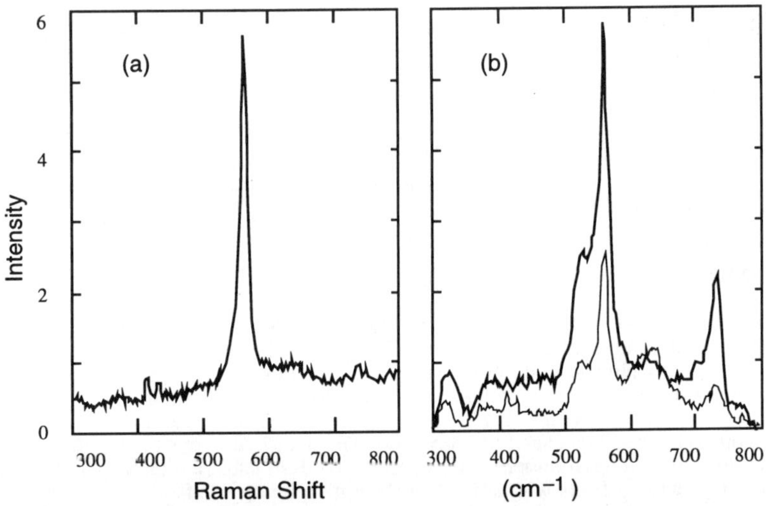

Figure 4. The Raman spectra of a typical transparent film (a) and of a slightly colored film before and post thermal annealing depicted by thin and thick lines, respectively in (b).

Photoluminescence

The PL of the films taken at 10 K are shown in Figure 5(a) and 5(b). The 3.472 eV band edge transition was dramatically quenched in Figure 5(a) and even diminished in Figure 5(b), presumably due to the high intrinsic n-type carriers of the film. An intense donor-to-acceptor emission centered around 3.3 eV and a broad band around 2.2 eV associated with structural defects were observed. The quenching of the band edge transition and the pronounced donor-to-acceptor emission in the PL spectra clearly indicate high background doping level, which is detrimental to further growth of p-type material. This observation is also confirmed by the Hall-effect measurement. The n-type carrier concentration measured by van der Pauw method is typically around $5 \times 10^{18} cm^{-3}$. Further improvement to reduce the background level is needed to achieve p-type doping.

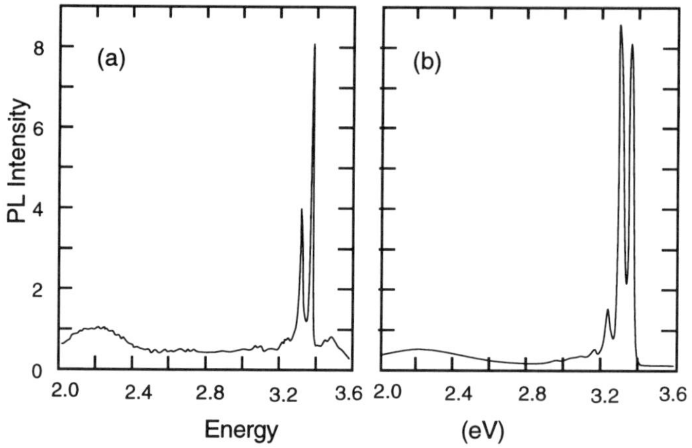

Figure 5. PL spectra of GaN films grown at 1000° C with different degree of intrinsic doping level.

CONCLUSION

An ECR-CVD reactor for the growth of III-Nitride materials has been implemented. Preliminary results show that ECR-CVD method is a promising technique for the growth of III-Nitride materials. The unique features of this growth technique include a wide pressure and temperature range for deposition. In addition, ECR-CVD method offers the flexibility of using pure nitrogen as the source gas instead of other hydrogen containing reactants such as ammonia. The near hydrogen-free together with low temperature growth environment allows more choices of substrate materials. Highly oriented (0001) textured films with smooth surface morphology and a coherence length over 1000 have been successfully grown on a variety of substrates which include sapphire, Si, and $LiGaO_2$ by this novel process. While introducing a buffer layer prior to the growth of GaN on sapphire or Si substrate is crucial to the film quality, a direct growth of epitaxial films without any buffer layer is possible by the use of a lattice matched $LiGaO_2$ substrate. The Raman spectra of our samples clearly show 565.5 cm^{-1} phonon peak indicative of a hexagonal GaN structure, consistent with the SEM and XRD observations. The photoluminescence spectra exhibit a narrow emission band centered at 3.3 eV, which is near but slightly below the band edge, indicating a high intrinsic level of doping. Further optimization of the ECR-CVD process is underway to produce higher quality films for blue light-emitting, high speed/high temperature electronic applications.

ACKNOWLEDGMENT

This work is supported by internal funding from Institute of Atomic and Molecular Sciences, Academia Sinica, Taiwan. One of the authors, C.H. Chao, acknowledges the post-doctoral fellowship from National Science Council, Taiwan.

REFERENCES

1. S. Strite and H. Morkoc, J. Vac. Technol. **B 10**, 1237 (1992).
2. S. Nakamura, T. Mukai, and M. Senoh, Appl. Phys. Lett. **64**, 1687 (1994).
3. M. AsifKhan, S. Krishnankutty, R.A. Skogman, J.N. Kuznia, D.T. Olson, and T. George, Appl. Phys. Lett. **65**, 520 (1994).
4. H. Amano, T. Tanaka, Y. Kunii, S.T. Kim, and I. Akasaki, Appl. Phys. Lett. **64**, 1377 (1994).
5. H Morkoc, S. Strite, G.B. Gao, M.E. Lin, B. Sverdlov, and M. Burns, J. Appl. Phys. **76**, 3 (1994).
6. T.P. Chow and R. Tyagi. IEEE Trans. Electron. Dev. **41**, 1481 (1994).
7. S. Nakamura, Jpn. J. Appl. Phys. **30**, L1705 (1991).
8. S. Nakamura, M. Senoh, and T. Mukai, Jpn. J. Appl. Phys. 62, 2390 (1993).
9. C.R. Abernathy, P. Wisk, F. Ren, and S.J. Pearton, J. Vac. Sci. Technol. **B 11**, 179 (1993).
10. T.D. Moustakas, R.J. Molnar, T. Lei, G. Menon and C.R. Eddy Jr., Mat. Res. Soc. Symp. Proc. **242**, 427 (1992).
11. I. Akasaki, H. Amano, Y. Koide, K. Hiramatsu and N. Sawaki, J. Cryst. Growth **98**, 209 (1989).
12. K.H. Chen, Y.L. Lai, J.C. Lin, K.J. Song, L.C. Chen, amd C.Y. Huang, Diamond and Related Mater. **4**, 460 (1995).
13. K.H. Chen, Y.L. Lai, L.C. Chen, J.Y. Wu, and F.J. Kao, Thin Solid Films **270**, 143 (1995).
14. L. Pavesi, M. Guzzi, J. Appl. Phys. **75**, 15 (1994).
15. L.J. van der Pauw, Philips Res. Repts **13**, 1-9, (1958).

HIGH QUALITY P-TYPE GaN FILMS GROWN BY PLASMA-ASSISTED MOLECULAR BEAM EPITAXY

J. M. Myoung*, C. Kim**, K. H. Shim*, O. Gluschenkov***, K. Kim***, and M. C. Yoo****
Departments of Materials Science and Engineering*, Physics**, and Electrical and Computer Engineering***, k-kim8@uiuc.edu
University of Illinois at Urbana-Champaign, 1406 W. Green St., Urbana, IL 61801
On leave from Samsung Advanced Institute of Technology****

ABSTRACT

p-type GaN films were grown on a (0001) sapphire substrate by the plasma-assisted molecular beam epitaxy. A low-contamination, high-power efficiency inductively coupled radio frequency plasma source was used, which was developed at the University of Illinois. Using an MBE system equipped with this plasma source, high-quality p-type GaN films were grown without post-growth treatment. X-ray rocking curve measurements for (0002) diffraction showed a full width at half maximum of less than 7 arcmin. The highest room-temperature hole concentration obtained was 1.4×10^{20} cm^{-3}, and for the same sample, the mobility was 2.5 $cm^2/V{\cdot}s$. It is believed that the Mott-Anderson transition occurred in this sample resulting in a metallic-type conductivity in the impurity subband. Low-temperature photoluminescence had a blue emission band and no deep-level transitions, indicating the high quality of the grown films. Uniformity of the Mg doping was confirmed by secondary ion mass spectrometry.

INTRODUCTION

III-V nitride semiconductors are considered to be the basis for optical devices in the blue and UV regions of the optical spectrum and for high-temperature, high-power electronic devices.[1,2] They have direct and wide bandgaps and high thermal stabilities[1,2]. Despite these advantages, the progress in the fabrication of III-V nitride devices has been slow for several reasons,[1-3] two of which are the low concentration and low mobility of the holes in the p-doped samples. There are three crucial issues concerning p-doping: a) hydrogen passivation of acceptors; b) high concentration of nitrogen vacancies unintentionally introduced during the film growth; and c) deep acceptor level.

The Mg-doped films grown by metalorganic chemical vapor deposition (MOCVD) with ammonia gas as a nitrogen source typically show high resistivities in the range of 10^6 $\Omega{\cdot}cm$ after the growth.[4,5] The formation of an inactive Mg-H complex during the growth is responsible for this phenomenon. To achieve a p-type conductivity in these samples one has to dissociate this complex and remove hydrogen. Amano and coworkers[6] discovered that by using low energy electron beam irradiation (LEEBI) at room temperature, one can decrease the resistivity of Mg-doped films from above 10^8 to 35 $\Omega{\cdot}cm$, thereby achieving a mobility of 8 $cm^2/V{\cdot}s$. Nakamura et al.[4,5] employed thermal annealing in a nitrogen atmosphere to activate acceptors. They were able to decrease the resistivity to 2 $\Omega{\cdot}cm$ and obtain a mobility of 10 $cm^2/V{\cdot}s$. In contrast to MOCVD, no post-growth treatment is required for plasma-assisted molecular beam epitaxy (PAMBE) since it employs a hydrogen-free nitrogen plasma as a source of active nitrogen. Moustakas and Molnar[7,8] reported on p-type GaN grown by PAMBE using an electron cyclotron resonance (ECR) plasma source. In this work, the room temperature hole mobility and hole concentration of the Mg-doped GaN films were 6 $cm^2/V{\cdot}s$ and 8×10^{18} cm^{-3}, respectively.

Mat. Res. Soc. Symp. Proc. Vol. 423 © 1996 Materials Research Society

Yang et al.[9] also reported on p-type GaN grown by MBE using ammonia and nitrogen gases. In their work, to circumvent the effect of hydrogen passivation on the Mg-doped GaN film, a nitrogen plasma produced by ECR was introduced during the growth in addition to ammonia. Their room-temperature hole mobility and hole concentration were 10 $cm^2/V{\cdot}s$ and $4{\times}10^{17}$ cm^{-3}, respectively.

In spite of the fact that the p-conductivity of their MOCVD-grown nitride films was not high, Nakamura et al.[10] recently demonstrated an InGaN-based multi-quantum well laser diode operating at 417 nm under pulsed conditions at room temperature. A high turn-on voltage (34 V) of the laser, however, caused high power dissipation in the device structure. With high-quality p-type GaN/AlGaN films having higher hole concentrations, one should be able to reduce the threshold voltage and power dissipation in these lasers.

Using an MBE system equipped with a low-contamination, high power efficiency, inductively coupled radio frequency (RF) plasma source developed in our laboratory, we recently grew high-quality p-type GaN films without post-growth treatment. Using Mg as the p-dopant, the hole concentrations achieved were as high as $1.4{\times}10^{20}$ cm^{-3} and the resistivities as low as 10^{-2} $\Omega{\cdot}cm$. To confirm the quality of the Mg-doped samples, double-crystal X-ray and low-temperature photoluminescence measurements were performed. To examine the Mg-distributions in the films, SIMS was employed. To the best of our knowledge, the p-conductivity reported in this work is the highest for the GaN-based material systems. In addition, our results indicate that for the growth of nitride films, the RF plasma source is as good as the ECR plasma source.

EXPERIMENTAL

Figure 1 shows a schematic of the growth system with a plasma source. The flow of

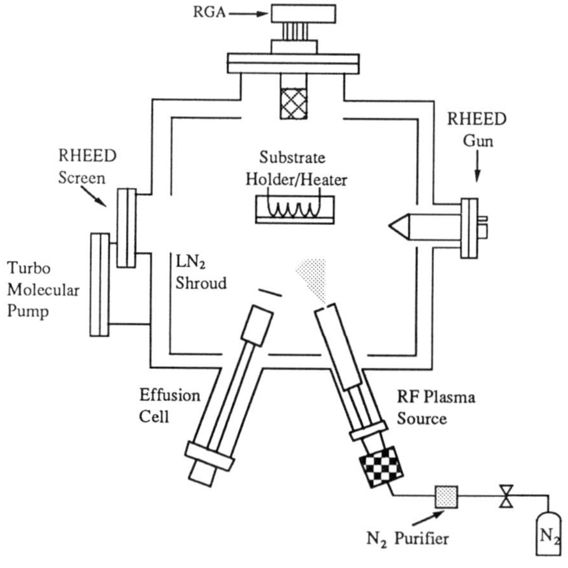

Fig. 1. MBE growth system equipped with a radio frequency plasma source.

ultrahigh-purity nitrogen was controlled and monitored by a mass flow controller. An inductively coupled RF plasma source developed at the University of Illinois was used to produce the active atomic nitrogen needed for the formation of nitrides. In this source we employed high-purity refractory materials and water cooling of various parts to avoid contamination, and the power transmission to the plasma was optimized. We found that overall power efficiency of this source could be very low (10-15%) and that it depended strongly on the pressure inside the plasma tube, the design of the RF coil, and the input power. The detailed analysis of the power efficiency of our plasma source will be published elsewhere[11]. *c*-plane sapphire substrates were used, which were cleaned by the conventional organic solvents (trichloroethane, acetone, and methanol), followed by etching in 3:1 solution of H_2SO_4 and H_3PO_4 at 170 °C. These substrates were mounted on Mo blocks and then loaded into the growth chamber equipped with a reflection high energy electron diffraction (RHEED) assembly. In the growth chamber, the substrates were subjected to thermal desorption for 30 min at 850 °C. Before starting the growth of nitride films on sapphire, the substrates were exposed to a nitrogen plasma for 20 min to form a thin AlN prelayer followed by the growth of a buffer layer at 550 °C. *p*-type GaN films were grown on a 200 Å thick AlN buffer layer at different growth temperatures (650 and 700 °C). During the growth of p-GaN, the pressure of the MBE system was maintained at 9.0×10^{-5} - 1.0×10^{-4}torr.

RESULTS AND DISCUSSION

The crystalline quality of the grown GaN films was monitored in situ by RHEED. Figure 2 (a) shows a RHEED pattern for the AlN buffer layer. Athough it was not perfectly sharp and streaky, the RHEED pattern became sharper as the GaN-film grew thicker (Figure 2 (b)). When the GaN film became thick enough, a sharp (1×1) structure was observed (Figure 2 (c)), which indicated two-dimensional growth. After film growth, X-ray rocking curve measurements using a double-crystal diffractometer with a Mo target were performed for the Mg-doped samples. For the sample grown at 700 °C, the FWHM value of 7 arcmin was measured (Figure 3).

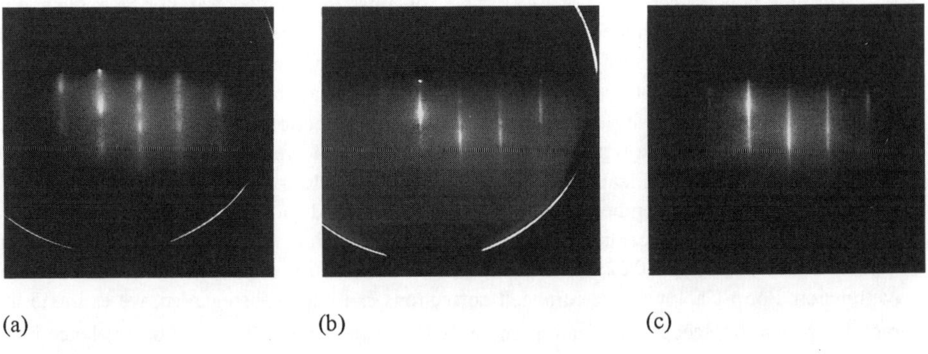

(a) (b) (c)

Fig. 2. RHEED patterns at different stages of the film growth. (a) after 200 Å thick AlN buffer layer, (b) after growth of 0.4 μm thick Mg-doped GaN layer, and (c) after 1.35 μm thick Mg-doped film. Sharp (1×1) structure is observed.

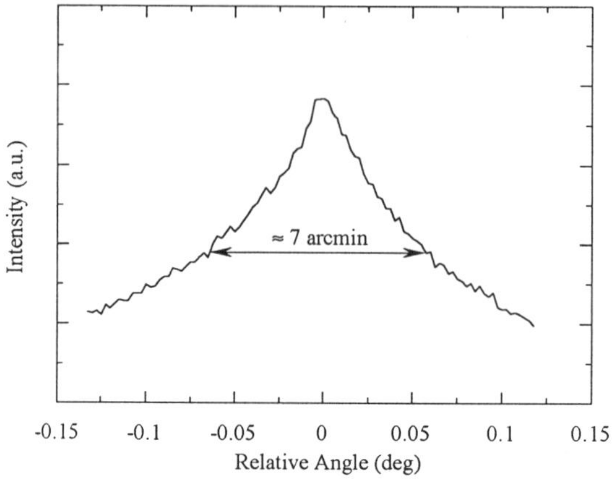

Fig. 3. The result of X-ray rocking curve measurement for (0002) diffraction from Mg-doped GaN film. The FWHM value of 7 arcmin is measured suggesting high-quality GaN film.

To examine the electrical properties of the Mg-doped samples, van der Pauw Hall measurements were performed on 0.7 to 1.35 μm thick films at room temperature. Ohmic contacts were made by soldering indium on the sample surface. Hall measurements showed that these films were p-type with extremely low resistivities in the range of 10^{-1}-10^{-2} Ω·cm without post-growth treatment. The Mg-doped films grown at 700 °C had a hole concentration and a hole mobility of 1.4×10^{20} cm^{-3} and 2.5 cm^2/V·s, respectively. For the samples grown at 650 °C, the room-temperature hole mobility and hole concentration were 13 cm^2/V·s and 2.7×10^{18} cm^{-3}, respectively.

The high hole concentration of $1.4\text{x}10^{20}$ cm^{-3} obtained in this work can not be explained in terms of the temperature-induced ionization of acceptors since it would require at room temperature a concentration of Mg atoms comparable to that of Ga atoms assuming an acceptor ionization energy of 150-170 meV. At a high acceptor concentration one can observe a transition to a metallic-type conductivity where the acceptor wavefunctions overlap substantially and produce a half-filed, metallic-type impurity subband through which the conduction of holes occurs without thermal ionization of acceptors[12]. This Mott-Anderson transition[12,13] was previously observed in a number of n- and p-type semiconductors at dopant concentrations higher than some critical concentrations. The critical concentration, N_M, can be estimated from an empirical formula[12], $N_M a^3=0.02$, where a is a characteristic dimension of the impurity wavefunction. For p-GaN, where central cell corrections can not be disregarded, we estimate the parameter, a, via the acceptor ionization energy[14], E_a, from $a=\hbar/\sqrt{2m_h^* E_a}$. For a valence band effective mass, m_h^*, of $0.8m_o$ and an E_a of 170 meV the Mott transition is expected at the concentration of Mg atoms of $1.3\text{x}10^{20}$ cm^{-3}. Thus, the nature of the presently obtained high conductivity in the GaN:Mg system is likely due to the metallic-type conductivity in the impurity subband. This fact was confirmed by the measurement of the sample resistivity at room and liquid nitrogen temperatures. No temperature dependence of the resistivity was observed in the case of high hole concentration whereas for lower levels of doping the resistivities at 77K were one to three orders of magnitude higher than those at room temperature.

To examine the optical properties of the Mg-doped GaN films, photoluminescence (PL) measurements were conducted at 5 K. The PL spectra were excited with a 325-nm HeCd laser

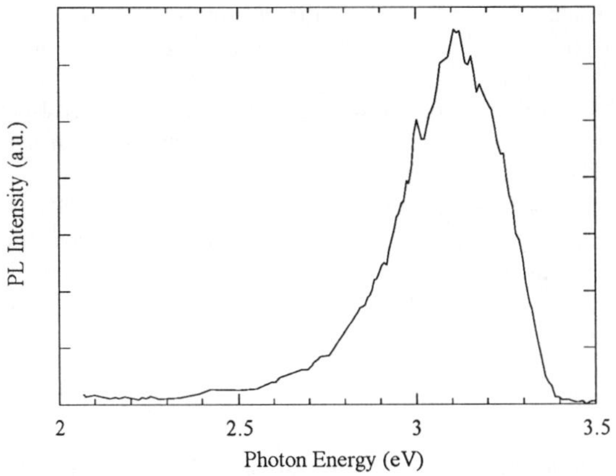

Fig. 4. Photoluminescence spectrum of the sample with a hole concentration of 1.4×10^{20} cm^{-3}. The broadening of the spectrum is due to the heavy doping.

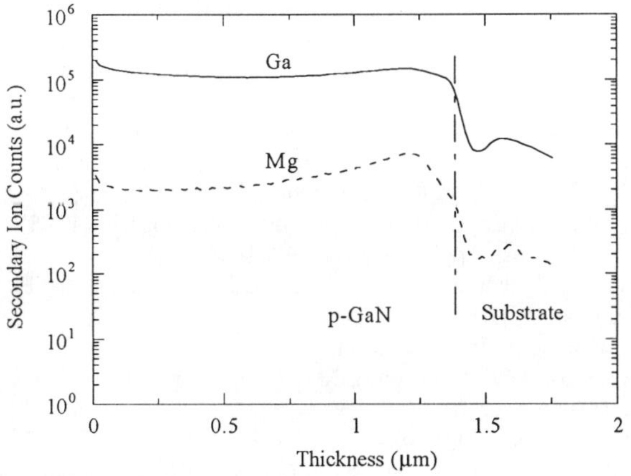

Fig. 5. Depth profile of Mg in p-type GaN film measured by SIMS. Mg distribution is relatively uniform.

and collected with a monochromator and a photomultiplier. Figure 4 shows the PL spectrum for the film grown at 700 °C. It has a blue emission band and shows no deep-level transitions. We believe that this band consists of donor-acceptor (DA) transition and its LO-phonon replicas. Similar PL spectra were previously observed in p-type GaN[8,9]. The observed broadening of the PL spectrum in the present case is believed to be caused by the banding of the impurity states induced by heavy doping.[15]

The Mg distribution in the p-type GaN films was examined by secondary ion mass spectrometry (SIMS) using a 5.5-keV Cs^+ primary ion beam in a Cameca ims 5f instrument. The positive secondary ions were detected with zero voltage offset. The raster size was 175×175 μm^2, and the diameter of the detected area was 30 μm. Figure 5 shows the depth profile of Mg from the film grown at 700 °C as a function of film thickness. The Mg-profile is reasonably uniform and indicates no substantial surface segregation. The films grown at other growth temperatures also showed similar uniformity.

In summary, we have grown high-quality p-type GaN films with high hole concentrations and low resistivities without post-growth treatment using an MBE system equipped with a home-made inductively coupled RF plasma source. It is believed that the Mott-Anderson transition occurred in this sample resulting in a metallic-type conductivity in the impurity subband.

ACKNOWLEDGMENTS

This work was supported by Samsung Electronics Co. Ltd.. The authors would like to thank Dr. S. Kim and Prof. S.G. Bishop for the photoluminescence measurements and valuable discussions.

REFERENCES

[1]S. Strite and H. Morkoç, J. Vac. Sci. Technol. **B 10**, 1237 (1992).
[2]H. Morkoç, S. Strite, G.B. Gao, M. E. Lin, B. Sverdlov, and M. Burns, J. Appl. Phys. **76**, 1363 (1994).
[3]M. E. Lin, G. Xue, G. L. Zhou, J. E. Greene, and H. Morkoç, Appl. Phys. Lett. **63**, 932 (1993).
[4]S. Nakamura, T. Mukai, M. Senoh, and N. Iwasa, Jpn. J. Appl. Phys. **31**, L139 (1992).
[5]S. Nakamura, N. Iwasa, M. Senoh, and T. Mukai, Jpn. J. Appl. Phys. **31**, 1258 (1992).
[6]H. Amano, M. Kito, K. Hiramatsu, and I. Akasaki, Jpn. J. Appl. Phys. **28**, L2112 (1989).
[7]T. D. Moustakas and R. J. Molnar, Mat. Res. Soc. Symp. Proc. **281**, 753 (1993).
[8]R. J. Molnar, R. Singh, and T. D. Moustakas, Appl. Phys. Lett. **66**, 268 (1995).
[9]Z. Yang, L. K. Li, and W. I. Wang, Appl. Phys. Lett. **67**, 1686 (1995).
[10]S. Nakamura, M. Senoh, S. Nagahama, N. Iwasa, T. Yamada, T. Matsushita, H. Kiyoku and Y. Sugimoto, Jpn. J. Appl. Phys. **35**, L74 (1996).
[11]O. Gluschenkov et al., manuscript in preparation.
[12]N. F. Mott, Proc. Phys. Soc. **A62**, 416 (1949); Metal-Insulator Transition (Taylor & Francis, London, 1990), pp. 145-169.
[13]P. W. Anderson, Phys. Rev. **109**, 1492 (1958).
[14]B. I. Shklovskii and A. L. Efros, Electronic Properties of Doped Semiconductors, Springer Series in Solid-State Sciences **45,** (1984).
[15]J. I. Pankove, Optical Processes in Semiconductors (Prentice-Hall, New Jersey, 1971), pp. 132-139.

INFLUENCE OF THE GROWTH ATMOSPHERE ON THE PROPERTIES OF AlN GROWN BY PLASMA - ASSISTED PULSED LASER DEPOSITION

T. OGAWA, M. OKAMOTO, Y. MORI, and T. SASAKI
Department of Electrical Engineering, Osaka University, 2-1 Yamada-Oka, Suita, Osaka 565, Japan

ABSTRACT

We have grown highly oriented aluminum nitride (AlN) thin films on Si (100) substrates by using pulsed laser deposition from sintering AlN targets. Three different growth environments, vacuum, nitrogen gas, and nitrogen plasma, have been used in order to investigate the effect of the ambient on the film quality. Rutherford backscattering spectrometry suggests that the N/Al ratio increases when the AlN film is grown in a nitrogen-contained ambient. Cathodoluminescence study implies the decrease of oxygen content in the film grown in a nitrogen plasma ambient.

INTRODUCTION

Group III nitrides, such as aluminum nitride (AlN) and cubic boron nitride (c-BN), have been widely investigated as candidates for wide-band-gap semiconductors. It is easier to grow AlN thin films than to grow c-BN thin films. Furthermore, because of its capability to achieve n-type doping, AlN is one of candidates for n-type wide gap semiconductors[1]. AlN growth technology should thus be developed in order to obtain semiconducting AlN films.

AlN films have been grown by various methods, such as MOCVD[2-4], reactive sputtering[5], plasma-assisted MBE[6], and pulsed laser deposition (PLD)[7-11]. PLD has advantages of low-temperature growth and less contamination of impurities. In this paper, we report the successful growth of oriented AlN thin films by the PLD technique from a sintering AlN target on Si (100) substrates. Nitrogen plasma has been generated in order to control the Al/N ratio of the AlN thin film which is very important for obtaining high quality AlN films. We have grown AlN films in three following kinds of environments, vacuum, nitrogen gas, and nitrogen plasma ambient and investigated the effect of growth ambient on the film quality.

EXPERIMENT

The deposition chamber was evacuated to a base pressure of $1.2X10^{-7}$ Torr by a turbomolecular pump. The fourth harmonics of a Nd: YAG laser ($\lambda = 266$ nm) was employed to ablate the target. The repetition rate was 10 Hz and the nominal pulse width was 3-5 ns. The incident laser beam was focused by a 250 mm focal length lens into a fluence of ~1.4 J/cm^2 on the target surface. Sintering AlN (Al : 62.8%, N : 32.1%, Y : 3.4 %, O : 1.7%) was used as the target material. The Si (100) substrate was located 3 to 5 cm away from and parallel to the target surface. Before it was introduced into the deposition chamber, the substrate was cleaned with acetone in an ultrasonic bath to remove organic impurities. The Si substrate was resistively heated to 800 °C. An optical pyrometer was used to monitor the

Mat. Res. Soc. Symp. Proc. Vol. 423 © 1996 Materials Research Society

substrate temperature. The deposition of AlN films was carried out in three types of environments: (a) in vacuum (10^{-7} Torr), (b) in nitrogen gas ambient, and, (c) in nitrogen plasma ambient for 10 hours. The AlN deposition in nitrogen gas was performed at 40 mTorr. The dc glow discharge nitrogen plasma was generated by using a ring-shaped electrode biased negatively to about 600 V placed between the target and the substrate as shown in Fig. 1. In this case, the substrate and the target were grounded. The discharge current was adjusted to be less than 1 mA. A faint violet glow could be observed near the ring.

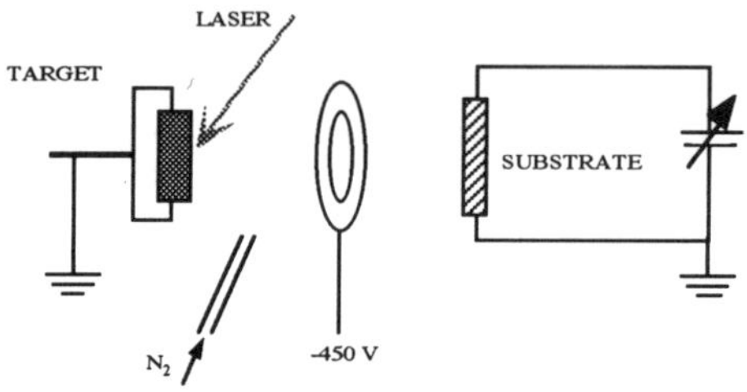

Fig. 1 Schematic diagram of the PLD apparatus.

The AlN thin films were characterized by the X-ray diffraction (XRD) technique, scanning electron microscopy (SEM), Rutherford backscattering spectrometry (RBS), micro-Raman scattering spectroscopy, and cathodoluminescence (CL). The x-ray source is Cu Kα. RBS measurements were performed using 2.1 MeV He^{2+} ions. Raman spectra were excited by an argon-ion laser operated at 514.5 nm with 200 mW. CL spectra were measured at room temperature using a 20 keV electron beam with a current of the order of 10 nA as the excitation source.

RESULTS AND DISCUSSIONS

The XRD measurements were carried out to confirm the growth of AlN thin films. For all types of AlN films, dominant (0002) diffraction peaks were observed, indicating the oriented growth of AlN films on the Si (100) substrate. The XRD pattern of the AlN film grown in high vacuum is shown in Fig. 2.

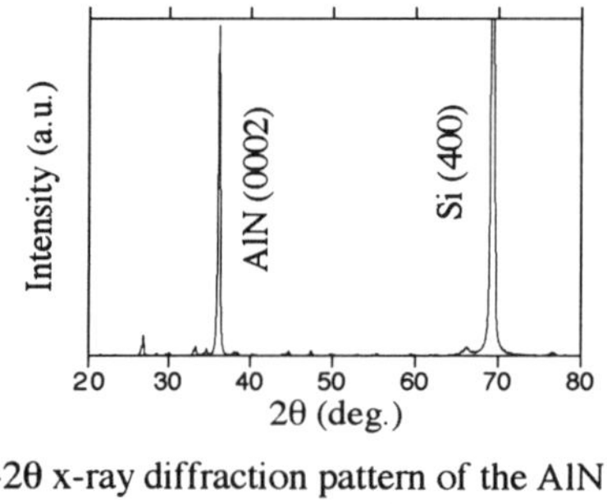

Fig. 2 θ–2θ x-ray diffraction pattern of the AlN film grown in high vacuum (10^{-7} Torr) on Si (100) of 800 °C.

Figures 3 (a), (b), and (c) show the SEM micrographs of the AlN films grown in the vacuum, nitrogen gas ambient and nitrogen plasma ambient, respectively. Comparison of these figures indicates that growth of AlN film in vacuum resulted in a rough surface because large particles could reach the substrate during the deposition. However, the presence of nitrogen gas during deposition seems to prevent large particles from depositing on the substrate as shown in Figs. 3 (b) and (c). Moreover, the generation of nitrogen plasma may enhance the chemical reactivity of nitrogen on the surface. This effect seems to reduce grain size as shown in Fig. 3 (c). This smaller grain may lead to a smoother surface.

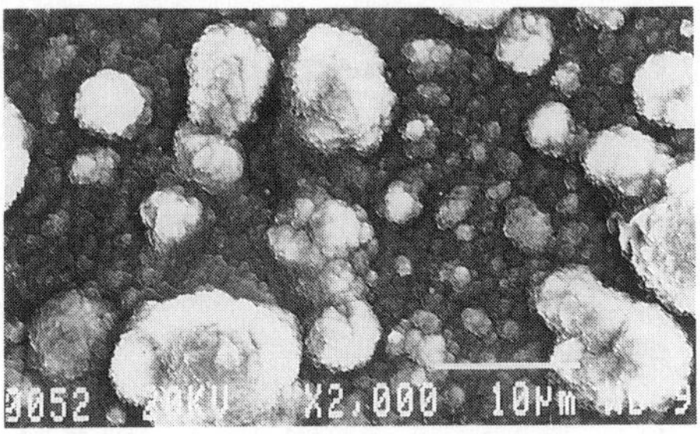

(a)

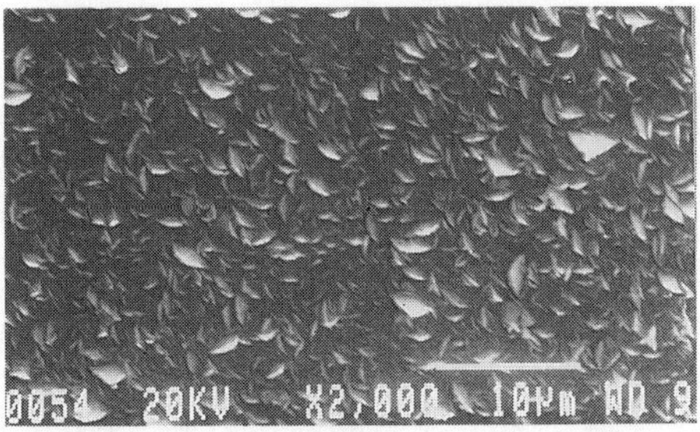

(b)

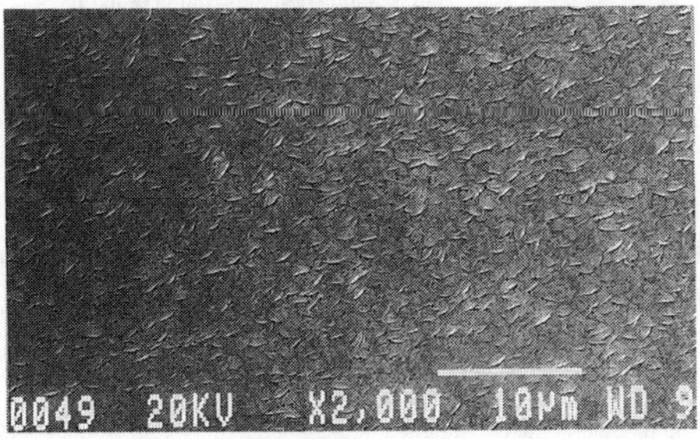

(c)

Fig. 3 SEM micrographs of AlN films deposited (a) in high vacuum, (b) in nitrogen gas, and (c) in nitrogen plasma.

In general, the film fabricated using the PLD technique most nearly maintains the target composition ratio. The target used in this work is deficient in nitrogen and contains yttrium as an impurity (Al : 62.8%, N : 32.1%, Y : 3.4 %, O : 1.7%). Nitrogen gas and nitrogen plasma ambient were used in order to compensate for the shortage of nitrogen. The effects of these ambient were characterized by a RBS study which determined the relative N/Al ratio in the grown films. The results are shown in Fig. 4. Figures 4 (a) and (b) are RBS spectra from the sintering AlN target and the AlN film grown in the vacuum, respectively. As is evident from Fig. 4, yttrium contamination in the film was less than that in the target. The edge of the RBS spectrum for N in Fig. 4 (b) seems to be slightly more obvious in comparison with Fig. 4 (a). This difference means that the N/Al ratio in the film grown in vacuum has been slightly improved as compared to the sintering AlN target.

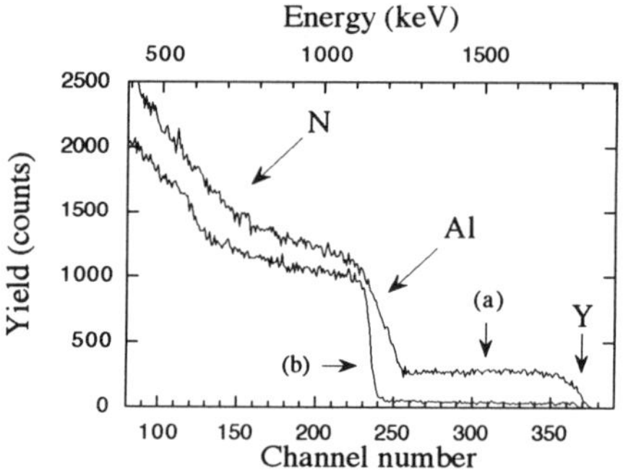

Fig. 4 RBS spectra of (a) sintering AlN target and (b) AlN film grown in high vacuum.

The RBS spectra were also measured for the films grown in nitrogen plasma and nitrogen gas ambients. The N/Al ratios of the films grown in nitrogen plasma and nitrogen gas ambients were estimated to be higher than that of the film grown in high vacuum (1.43 times for nitrogen plasma and 1.37 times for nitrogen gas). This result reveals that the presence of a nitrogen-contained ambient during the growth is effective to increase the nitrogen concentration in the AlN films and is essential for the growth of high-quality AlN films.

Micro-Raman studies were performed to examine the quality of the AlN films obtained. The Raman spectra of AlN films grown in nitrogen gas and nitrogen plasma ambient are shown in Figs. 5 (a) and (b), respectively. The peaks attributed to LO phonons of AlN were seen at 655.6 cm^{-1} in Fig. 5 (a) and at 649.2 cm^{-1} in Fig. 5 (b) (655.0 cm^{-1}[12] and 649.0 cm^{-1}[13]). The peaks were fitted by Lorentzian peak shape functions. The full width at half maximum (FWHM) for the peak due to LO phonons in Fig. 5 (a) was 38.9 cm^{-1}. In contrast, the peak width in Fig. 5 (b) was narrower with a FWHM value of 31.7 cm^{-1}. In addition, though Fig. 5 (a) showed a broad luminescence over 600 cm^{-1} presumably arising from impurities or defects states, no broad luminescence was observed in Fig. 5 (b). These results reveal that the AlN film grown in the nitrogen plasma ambient exhibits better film quality than that obtained in the nitrogen gas. No Raman peak was obtained from the AlN film grown in high vacuum despite the clear XRD peaks as shown in Fig. 2. This indicates the poor crystallinity of this sample.

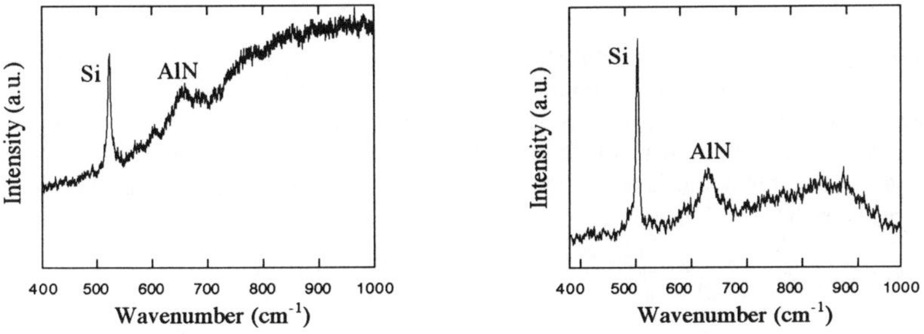

(a) (b)

Fig. 5 Micro-Raman spectra of AlN films grown (a) in nitrogen gas ambient and (b) in nitrogen plasma ambient.

In general, owing to the high gettering efficiency of Al for oxygen, thin films of AlN are prone to oxidation, resulting in significant degradation of the film qualities and properties[14]. It is important to reduce the oxygen content is important for further developments. We characterized the AlN films by CL. CL spectra measured are shown in Fig. 6. The film grown in vacuum has broad band luminescence with a peak at 395 nm (see Fig. 6 (a)). The film grown in nitrogen plasma ambient exhibited two broad bands with peaks at 405 nm and 330 nm (spectrum (c)). The luminescence intensities obtained for these films were almost same. In contrast, as shown in spectrum (b), the film grown in nitrogen gas ambient had one order stronger luminescence intensity with a peak position at 443 nm compared with CL spectra (a) and (c). According to R. A. Youngman, these emission peaks arise from oxygen-related defect states[15]. They reported that the luminescence peak position linearly shifts toward longer wavelength up to around 380 nm with increasing oxygen content.

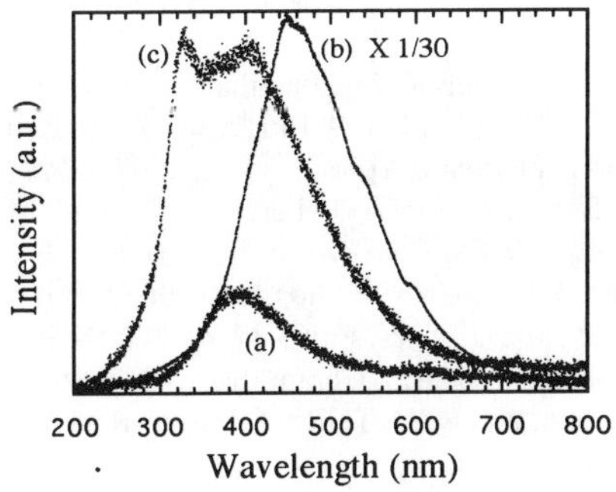

Fig. 6 CL spectra for AlN films grown (a) in high vacuum, (b) in nitrogen gas, and (c) in nitrogen plasma.

The luminescence intensity also depends on the oxygen concentration. The luminescence increases in intensity as oxygen increases in the AlN film. From the discussions above, CL measurements in Fig. 6 suggest that the film grown in nitrogen gas ambient includes a large amount of oxygen. This result is considered to be due to the degassing of oxygen from the chamber wall because of the shut off of the gate valve which was located between the deposition chamber and the turbomolecular pump. CL investigation also suggest a low oxygen content in the film grown in nitrogen plasma ambient. J. H. Harris et al reported oxygen was substituted on nitrogen sites in the AlN lattice[16]. Therefore, a very likely explanation for this result is the higher reactivity of nitrogen plasma than that of residual oxygen gas with Al.

CONCLUSIONS

We have succeeded in growing oriented AlN thin films by the PLD technique from sintering AlN target. We found that the quality of the AlN films depends on the growth ambient. The surface morphology of the films grown in a nitrogen background gas ambient is very smooth. RBS, Raman studies, and CL characterization reveal that larger N/Al ratio, better film quality, and less oxygen contamination are obtained for the AlN film grown in the nitrogen plasma ambient.

REFERENCES

1. S. Strite and H. Morkoç, J. Vac. Sci. Technol. B **10**, 1237 (1992).
2. M. A. Khan, J. N. Kuznia, R. A. Skogman,D. T. Olson, M. M. Millan, and W. J. Choyke, Appl. Phys. Lett. **61**, 2539 (1992).
3. P. Kung, A. Saxler, X. Zhang, D. Walker, T. C. Wang, I. Furguson, and M. Razeghi, Appl. Phys. Lett. **66**, 2958 (1995).
4. J. Chaudhuri, R.Thokala, J. H. Edgar, and B. S. Sywe, J. Appl. Phys. **77**, 6263 (1995).
5. S. Kumar and T. L. Tansley, Jpn. J. Appl. Phys. **34**, 4154 (1995).
6. K. S. Stevens, A. Ohtani, M. Kinniburgh, andR. Beresford, Appl. Phys. Lett. **65**, 321 (1994).
7. M. G. Norton, P. G. Kotula, and C.B. Carter, J. Appl. Phys. **70**, 2871 (1991).
8. K. Seki, X. Xu, H. Okabe, J. M. Frye, and J. B. Halpern, Appl. Phys. Lett. **60**, 2234 (1992).
9. R. D. Vispute, J. Narayan, H. Wu, and K. Jagannadham, J. Appl. Phys. **77**, 4724 (1995).
10. W. T. Lin, L. C. Meng, G. J. Chen, and H. S. Liu, Appl. Phys. Lett. **66**, 2066 (1995).
11. R. D. Vispute, H. Wu, and J. Narayan, Appl. Phys. Lett. **67**, 1549 (1995).
12. O. Brafman, G. Lengyel, S. S. Mitra, P. J. Gielisse, J. N. Plendl, and L. C. Mansur, Solid State Commun. **6**, 523 (1968)
13. W. Changwen and L. Ling, J. Non-Cryst. Solids **112**, 296 (1989)
14. S. Kumar and T. L. Tansley, Jpn. J. Appl. Phys. **34**, 4154 (1995).
15. R. A. Youngman and J. H. Harris, J. Am. Ceram. Soc. **73**, 3238 (1990).
16. J. H. Harris, R. A. Youngman, and R. G. Teller, J. Mater. Res. **5**, 1763 (1990).

LASER WAVELENGTH DEPENDENT PROPERTIES OF BN THIN FILMS DEPOSITED BY LASER ABLATION

Y. Suda*, T. Nakazono*, K. Ebihara**, and K. Baba***
*Department of Electrical Engineering, Sasebo National College of Technology, 1-1 Okishin-machi, Sasebo, Nagasaki 857-11, Japan
**Department of Electrical Engineering and Computer Science, Kumamoto University, 2-39-1 Kurokami, Kumamoto 860, Japan
***Technology Center of Nagasaki, 2-1303-8 Ikeda, Omura, Nagasaki 856, Japan

ABSTRACT

Boron Nitride (BN) thin films are deposited on heated (650 ℃) silicon (100) substrates using Nd:YAG (532 nm) and KrF excimer (248 nm) laser ablation. The laser beam is focused on the hBN targets. The films are grown using a laser repetition rate of 10 Hz at energy density 3.8 J/cm^2 . Argon gas is mixed with the reactant nitrogen gas and the total gas pressure in the chamber during deposition is 20 Pa. Auger Electron Spectroscopy shows that the N/B composition ratio depends on the mixture ratio of nitrogen and the relative emission intensity of B^+ (345.1 nm). The surface morphology of the films prepared by the 532 nm laser is rough with large particulates, whereas much smoother surfaces with fewer and smaller particulates are obtained with the 248 nm laser. Fourier Transform IR measurement shows that the cBN phase in the films is enhanced by applying the negative RF self-bias voltage (~ 200 V) on the substrate electrode.

INTRODUCTION

There has been significant effort and progress made over the past several years in depositing quality Ⅲ - Ⅴ nitride films for a variety of electronic, optical and tribological applications. Despite the successes for most Ⅲ - Ⅴ nitrides, the growth of single-phase cubic boron nitride (cBN), potentially the most valuable of these compounds because of its electronic and mechanical properties, remains unperfected. In addition, hexagonal BN (hBN) and turbostratic BN (tBN) have the desired properties for X-ray transmission windows.

Recently, BN films with an appreciable fraction of the cubic phase have been deposited by a variety of techniques,[1-6] including pulsed laser ablation.[7-13]

In this work, we describe a Nd:YAG laser (532 nm) ablation process and a KrF excimer laser (248 nm) ablation process for the preparation of BN thin films. We investigate the properties of prepared BN films and the spectroscopic characteristics of hBN plasma plume.

EXPERIMENT

The schematic diagram of the experimental apparatus is shown elsewhere,[13] thus only the experimental procedure is briefly described here. A deposition chamber was evacuated by a turbo-molecular pump and a rotary pump. The lasers used in the present study were (i) a Nd:YAG laser (Lumonics YM600, wavelength of 532 nm, pulse duration of 6.5 ns, maximum output energy of 340 mJ) and (ii) a KrF excimer laser (Lambda-Physik LPX-305icc, wavelength of 248 nm, pulse duration of 25 ns, maximum output energy of 850 mJ). The laser beam was focused on the hBN targets at 45° , which were placed in the center of the stainless steel deposition chamber. The energy density of the pulsed laser beam was maintained at 3.8 J/cm^2 and the repetition rate was 10 Hz. The targets were rotated at about 20 rpm in order to prevent crater formation on their surfaces. The Si(100) substrates were located at a distance of 60 mm from the facing target and were heated up to 650 ℃ by an IR lamp. The substrate electrode was coupled to an RF generator through an impedance matching network with a blocking capacitor. The RF power was applied between the grounded chamber and the substrate electrode. A negative DC self-bias voltage was

Mat. Res. Soc. Symp. Proc. Vol. 423 © 1996 Materials Research Society

Table I . Deposition conditions for the BN films.

Laser	Pulsed Nd:YAG laser	KrF excimer laser
Wavelength	λ =532 nm	λ =248 nm
Pulse width	τ =6.5 ns	τ =25 ns
Energy density	Ed=3.8 J/cm ^{2}	
Repetition rate	f=10 Hz	
Target	hBN (purity 99.5%)	
Rotating speed of the target	~ 20 rpm	
Substrate	Si (100)	
Target-substrate distance	d=6.0 cm	
Gas pressure	$P_{N_2}+P_{Ar}$=20.0 Pa	
Substrate temperatures	Ts=650 ℃	
RF bias power (13.56 MHz)	P_{RF}=0 ~ 100W	
Self-bias voltage	Vs=0 ~ -200V	
Deposition time	30 minutes	

generated basically due to the higher mobility of electrons than ions in a megahertz grow discharge. After 18,000 laser pulses, the deposition process was completed. Table I shows the deposition conditions for the preparation of BN films. BN films were characterized by Scanning Electron Microscopy (SEM), Auger Electron Spectroscopy (AES), and Fourier Transform IR measurement (FT-IR). The optical emissions from the plasma plume generated by the pulsed laser irradiation were detected by an optical multichannel analyzer (OMA) with a 1024 photodiode array.

RESULTS

Figure 1(a) and 1(b) show SEM pictures of typical BN films deposited with a Nd:YAG laser (532 nm) and a KrF excimer laser (248 nm), respectively. The density and size of particulates are much higher for the BN film prepared with a Nd:YAG laser (532 nm) [Fig.1(a)] as compared to the film prepared with a KrF excimer laser (248 nm) [Fig.1(b)] even though these films have been deposited using the same number of laser pulses and energy density.

(a) 10 μm

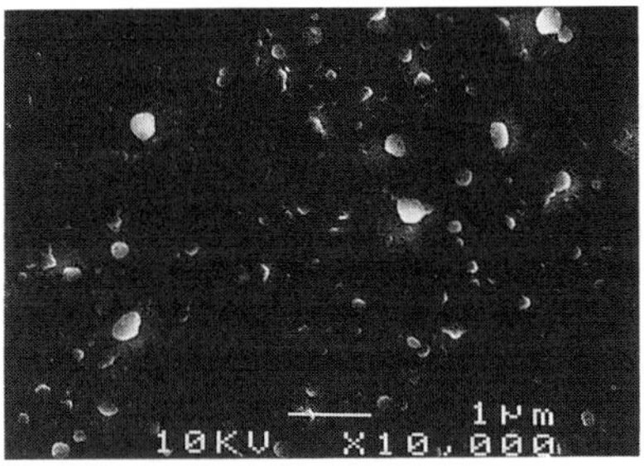

(b) 1 μm

Figure 1. SEM pictures of BN films on Si(100) deposited with (a) a Nd:YAG laser (532 nm) and (b) a KrF excimer laser (248 nm). Other deposition conditions were nitrogen gas pressure; P_{N_2}=20 Pa, energy density; Ed=3.8 J/cm ^{2} , target-substrate distance; d=6.0 cm, substrate temperature; Ts=650 ℃ .

These films have been investigated by AES and the results are shown in Fig. 2(a) and Fig.3(a), respectively. There are B and N peaks present in the spectrum, as well as smaller peaks of C and O. The N/B composition ratio of the film deposited with a 532 nm laser at P_{N_2}=20 Pa, P_{Ar}=0 Pa was 0.86 [Fig.2(a)] and that of the film with a 248 nm laser at P_{N_2}=20 Pa, P_{Ar}=0 Pa was 0.81 [Fig.3(a)]. The film composition ratio (N/B=0.81 ~ 0.86) was 8 ~ 13 % smaller than that (N/B=0.93) of the virgin BN target used here.[14] These films were boron rich like the films (N/B ≈ 0.70) deposited on heated (600 ℃) Si(100) substrate with KrF excimer laser (248 nm) at P_{N_2}=4 ~ 13 Pa.[9] This nitrogen deficiency problem may be linked to the target depletion problem.[14] Ion–assisted treatments are required to fabricate stoichiometric and cubic structure BN films.[7, 10–12] The N/B composition ratio of the film deposited with a 532 nm laser at P_{N_2}=0 Pa, P_{Ar}=20 Pa was 0.09 [Fig.2(b)] and that of the film with a 248 nm laser at P_{N_2}=0 Pa, P_{Ar}=20 Pa was 0.18 [Fig.3(b)]. These films showed an absence of nitrogen and were composed almost entirely of metallic boron.

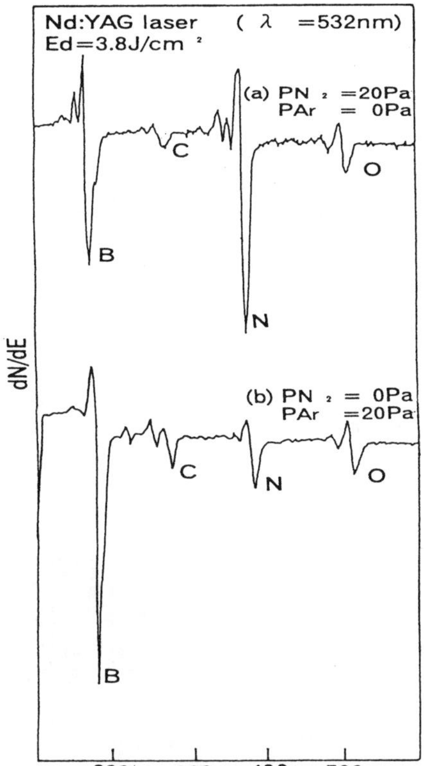

Figure 2. AES spectra of BN films deposited with a Nd:YAG laser (532 nm) at (a) P_{N_2}=20Pa, P_{Ar}=0Pa and (b) P_{N_2}=0 Pa, P_{Ar}=20 Pa.

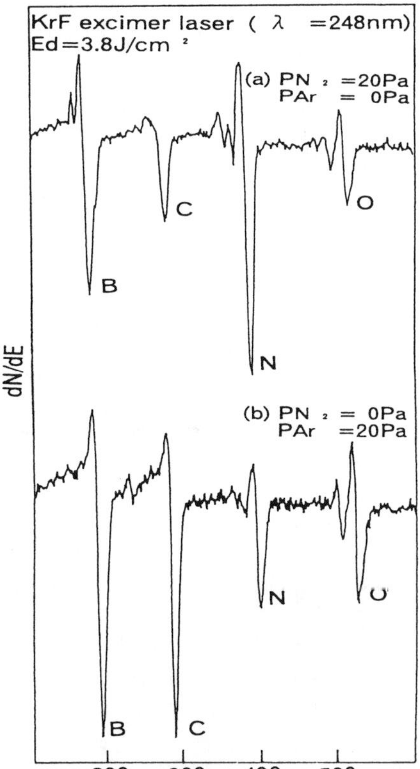

Figure 3. AES spectra of BN films deposited with a KrF excimer laser (248 nm) in (a) P_{N_2}=20Pa,P_{Ar}=0Pa and (b) P_{N_2}=0 Pa, P_{Ar}=20 Pa.

Figure 4 shows optical emission spectra of hBN plasma plume generated by a KrF excimer laser (248 nm) at (a) P_{N_2}=20 Pa, P_{Ar}=0 Pa and (b) P_{N_2}=0 Pa, P_{Ar}=20 Pa . Many spectral lines originated from neutral and ionic species, such as B (208.9, 249.7nm), B^+ (345.1 nm), N (397.6, 417.6 nm), N^+ (413.4 nm) and N_2 (350.1, 353.7, 391.6, 394.4 nm) are identified in Fig.4(a). Emission spectrum of BN (360.03 nm) molecule cannot be observed clearly. Therefore, the hBN target seemed to be mainly ablated as an atomic state. In Fig.4(b), spectral lines from Ar (404.4, 415.9, 419.8 nm) can be observed. The intensities of the emission lines depend on the argon mixture ratio. Figure 5 shows the relative intensity of the emission spectrum of each species and the N/B composition ratio as a function on N_2 content, where each intensity was normalized by their own strongest intensity. The N/B composition ratio depends on the mixture ratio of nitrogen and the relative emission intensity of B^+ (345.1 nm).

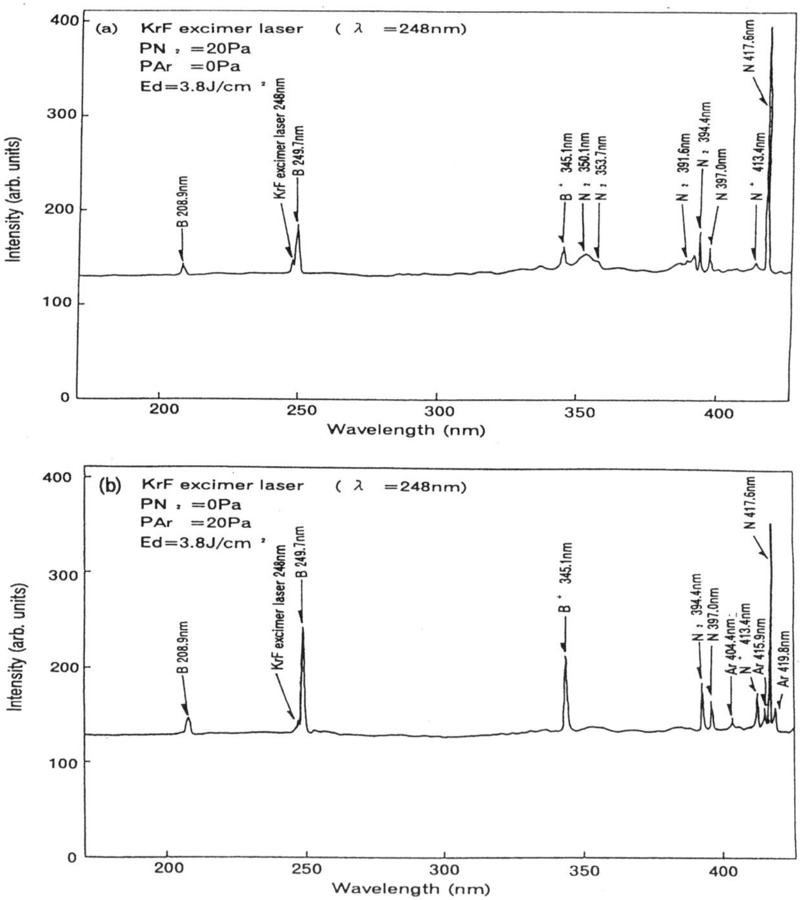

Figure 4. Optical emission spectra obtained with the OMA from the hBN plasma plume generated by the KrF excimer laser (248 nm) at (a) P_{N_2}=20 Pa, P_{Ar}=0 Pa and (b) P_{N_2}=0 Pa, P_{Ar}=20 Pa.

Figure 6 shows the FT–IR absorption spectra of BN films deposited by Nd:YAG laser (532 nm) at various N_2 contents. An infrared spectrum of the cBN crystal shows a strong absorption near 1050 cm^{-1} due to the reststrahlen band (TO mode). The crystalline or amorphous hBN phase results in a distinct and strong, asymmetrical band at around 1380 cm^{-1}, which is attributed to B–N bond stretching and a weaker, sharper band at around 800 cm^{-1} which is attributed to B–N–B bond bending.[15] As can be seen in Fig. 6,

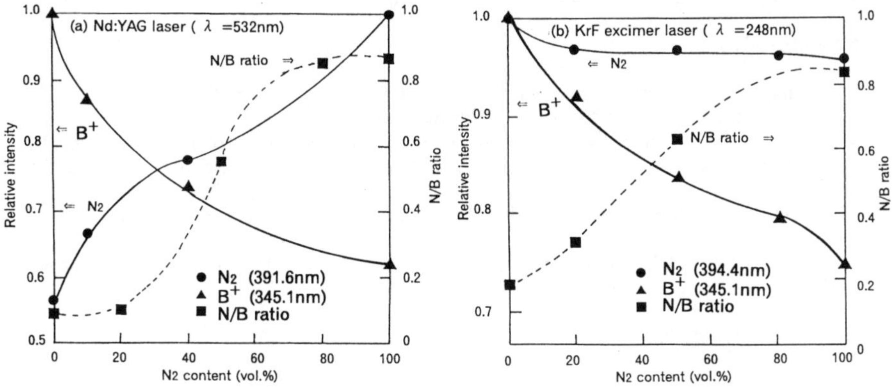

Figure 5. Relative intensity of the emission spectra and the N/B composition ratio as a function on N_2 content; (a) the Nd:YAG laser (532 nm) and (b) the KrF excimer laser (248 nm).

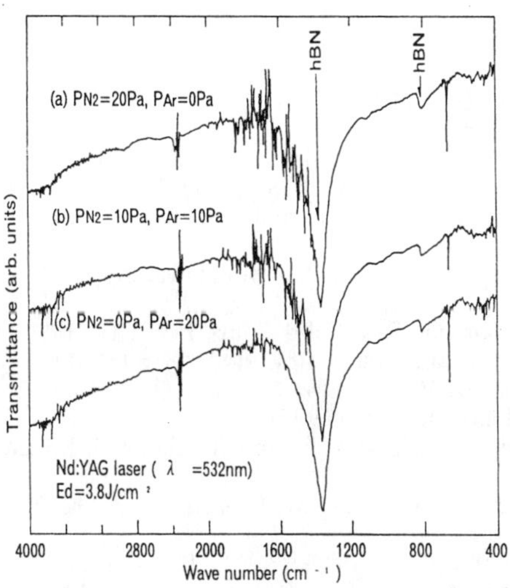

Figure 6. FT–IR spectra measured for BN films deposited by the Nd:YAG laser (532 nm) at various N_2 contents without negative RF self–bias voltage.

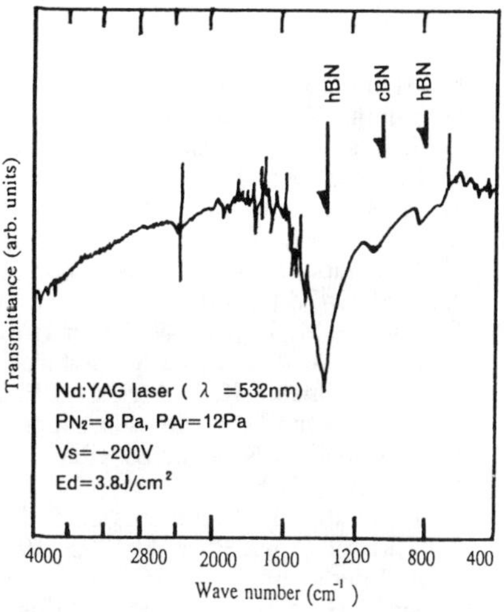

Figure 7. FT–IR spectra measured for BN films deposited by the Nd:YAG laser (532 nm) with negative RF self–bias voltage.

dominant absorption peaks at around 1380 cm^{-1} (B–N stretching) and 800 cm^{-1} (B–N–B bending) are consistent with sp^2 bonds that occurs in the hBN crystal structure. The peak at 1050 cm^{-1} caused by the TO mode of cBN does not appear in these samples.

The effect of negative RF self–bias voltage on the nature on bonding in the film is present in the FT–IR spectrum of Fig. 7. The film deposited by Nd: YAG laser (532 nm) under the conditions (self–bias voltage; Vs=–200 V, Ed=3.8 J/cm^2, d=6.0 cm, Ts=650 ℃, P_{N_2}=8Pa, P_{Ar}= 12Pa) shows the characteristic absorption spectrum at 1050 cm^{-1}, corresponding to the cBN reststrahlen band in addition of hBN.

CONCLUSIONS

BN films were deposited using Nd: YAG laser (532 nm) and KrF excimer laser (248nm) ablation. The following conclusions can be drawn from the results.

(1) The density of the particulates is much lower and the average size is much smaller in the film prepared with the 248 nm laser as compared to the film prepared with the 532 nm laser.

(2) The N/B composition ratio is significantly affected by the relative emission intensity of B^+ (345.1 nm).

(3) The formation of cBN phase in the films is enhanced by applying the negative RF self–bias voltage (~ 200 V) on the substrate electrode.

ACKNOWLEDGMENTS

The authors wish to thank Dr. T. Ikegami and Dr. Y. Yamagata of Kumamoto University for their helpful discussions, and Dr. H. Abe of Ceramic Research Center of Nagasaki and R. Hatada of Technology Center of Nagasaki for their technical assistance with the data. This work was supported in part by a Research Fund of Nagasaki Super Technology Development Association and a Grant–in–Aid for Scientific Research from the Ministry of Education, Science and Culture, Japan.

REFERENCES

1. M. Mieno and T. Yoshida, Jpn. J. Appl. Phys. **29**, L1175 (1990)
2. H. Saitoh, W. A. Yarbrough, Appl. Phys. Lett. **58**, 2482 (1991)
3. D. J. Kester, K. S. Ailey, R. F. Davis and K. L. More, J. Mater. Res. **8**, 1213 (1993)
4. C. R. Eddy, Jr and B. D. Sartwell, J. Vac. Sci. Technol. **13**, 2018 (1994)
5. H. Hofsass, C. Ronning, U. Griesmeier, M. Gross, S. Reinke and M. Kuhr, Appl. Phys. Lett. **67**, 46 (1995)
6. S. Watanabe, S. Miyake, W. Zhou, Y. Ikuhaya, T. Suzuki, M. Murakawa, Appl. Phys. Lett. **66**, 1478 (1995)
7. S. Mineta, M. Kohata, N. Yasunaga and Y Kikuta, Thin Solid Films **189**, 125 (1990)
8. G. L. Doll, J. A. Sell, C. A. Taylor Ⅱ and R. Clarke, Phys. Rev. **B43**, 6816 (1991)
9. T. A. Friedmann, K. F. McCarty, E. J. Klaus, D. Boehme, W. M. Clift, H. A. Johnsen, M. J. Mills and D. K. Ottesen, Appl. Phys. Lett. **61**, 2406 (1992)
10. A. K. Ballal, L. Salamanca–Riba, G. L. Doll, C. A. Taylor Ⅱ and R. Clarke, J. Mater. Res. **7**, 1618 (1992)
11. D. L. Medlin, T. A. Friedmann, P. B. Mirkarimi, P. Rez, M. J. Mills and K. F. McCarty, J. Appl. Phys. **76**, 295 (1994)
12. T. A. Friedmann, P. B. Mirkarimi, D. L. Medlin, K. F. McCarty, E. J. Klaus, D. R. Boehme, H. A. Johnsen, M. J. Mills, D. K. Ottesen and J. C. Barbour, J. Appl. Phys. **76**, 3088 (1994)
13. Y. Suda, T. Nakazono, K. Ebihara and K. Baba, T. IEE of Jpn. **115–A**, 1263 (1995)
14. K. Kaneda and K. Shibata, Jpn. J. Appl. Phys. **33**, 266 (1994)
15. T. Ikeda, Y. Kawate and Y. Hirai, J. Vac. Sci. Technol., **A8**, 3168 (1990)

PREPARATION AND CHARACTERIZATION OF WURTZITIC GaN SINGLE CRYSTALS IN NANO AND MICRO SCALE

SAN YU, HONGDONG LI, HAIBIN YANG, DONGMEI LI, HAIPING SUN, GUANGTIAN ZOU
National Laboratory of Super Hard Materials, Jilin University, Changchun 130023, P.R. China, shtem@mail.jlu.edu.cn

ABSTRACT

Gallium nitride and its alloys are the most promising materials for short wave light emitters. If high quality GaN single crystals can be prepared, the GaN base light emitters should be fabricated directly on the lattice-matched GaN substrate.

In this work, GaN crystals in nano and micro scale with definite faces have been prepared by dc arc discharge using gallium and N_2+NH_3 as starting materials. Transmission electron microscope, selected area diffraction, x-ray microanalysis of energy dispersive spectroscopy, and x-ray diffraction investigation of the as grown GaN crystals show that the well faceted crystals are single crystalline GaN in wurtzite structure having lattice constants a_0=3.18Å and c_0=5.18Å. The crystal size of stoichoimetric GaN in wurtzite structure depends on the partial pressure of nitrogen in the plasma. The maximum crystal size in this work is about several micrometers .

INTRODUCTION

Gallium nitride is a III-V group direct gap semiconductor with enormous potentials for optoelectronic devices from the blue to the near ultraviolet wavelengths which are presently inaccessible to semiconductor technology. GaN and its In and Al alloys span most of the visible region and extend well into the ultraviolet wavelengths[1]. The fabrication of high-quality single crystal GaN has been a formidable challenge to researchers to date[2]. In order to produce GaN, it has been experimentally realized that high temperature, activated nitrogen species and/or high nitrogen pressures are necessary to overcome the large kinetic barriers of formation. For this reason, conventional bulk and thin film growth technologies which are commonly used to produce III-V group semiconductors such as GaAs, and InP can not all be directly transferred to GaN production. Standard bulk growth technologies are not practical alternatives for the growth of GaN because experimental technologies have not been developed which can contain the high nitrogen pressure at the melt temperature. Refractory nitrides such as InN, GaN, and AlN, their solid solutions, and heterojunctions between them are one of the most promising families of electronic materials. All three are direct-band-gap semiconductors with their energy gaps covering the region from 1.95eV(InN) and 3.4eV(GaN) to 6.28eV(AlN). Thus, the growth of high quality crystals of these materials should lead to applications in optoelectronic devices from the visible to the ultraviolet part of the electromagnetic spectrum, as well as for high-power and high-temperature electronics[2,3]. GaN, in particular, was predicted to be the most promising materials for optoelectronic devices .

GaN films have been grown using many growth technologies, including chemical vapor deposition(CVD)[4], metal-organic chemical vapor deposition(MOCVD)[5], molecular beam epitaxy(MBE)[6], and the plasma assisted processes[7,8] on a variety of substrates, such as silicon, spinel, silicon carbide, and various crystallographic orientations of sapphire. Consequently, the satisfied crystal quality of GaN heteroepitaxy layers on foreign substrate has

Mat. Res. Soc. Symp. Proc. Vol. 423 © 1996 Materials Research Society

not been achieved up to date for the lack of a suitable lattice-matching substrate. If high quality bulk GaN can be fabricated, a whole new region of the optical spectrum will be accessible to semiconductor technology in much the same way that the infrared and longer visible wavelengths are presently exploited by GaAs technology.

The fabrication of high-quality single crystalline bulk GaN has been a formidable challenge to researchers to date[9]. The efforts directed to this end by solution growth in a predominately Ga melt under high pressure(~2GPa) and high temperature(~1500°C) conditions have been attempted[10,11]. Consequently, it has been experimentally realized that high temperature, activated nitrogen species and/or high nitrogen pressures are necessary to overcome the large kinetic barriers of GaN lattice configuration.

In this paper, dc arc discharge plasma was employed to activated nitrogen and gallium atoms(ions), wurtzitic GaN(w-GaN) single crystals in nano and micro scale were prepared using metal gallium and the gas mixture of N_2+NH_3 as starting materials The morphology, microstructure and composition of w-GaN have been investigated using transmission electron microscope(TEM), X-ray diffraction(XRD), selected area diffraction(SAD), nano-probe diffraction(NPD), energy dispersive spectroscopy of x-ray(EDS), etc.

GaN nano crystals can be directly used as a short-wavelength optoelectronic ultra-fine-powder(UFP) for its unique properties. Moreover, we can also use it as a starting materials to synthesize large size single crystals by HP/HT technology in the same way as diamond and cubic boron nitride synthesis, which is the final aim of this research.

EXPERIMENT

The samples reported here were all prepared by dc arc discharge evaporating metal gallium in the ambient of N_2+NH_3. The crucible, which acts as an anode, was used to evaporate gallium. The cathode is fixed on the opposite side of the chamber. The experimental conditions are as follows: base pressure 1~20 kPa maintained by control outlet pumping rate and the inlet flow rate of the mixture of N_2+NH3, dc arc current 100~ 500 Ampere stabilized between the cathode and the anode, volume ratio of N_2/NH_3 0.3~0.6. GaN crystals were collected on the water cooled wall. The dc plasma system is illustrated schematically in Figure1.

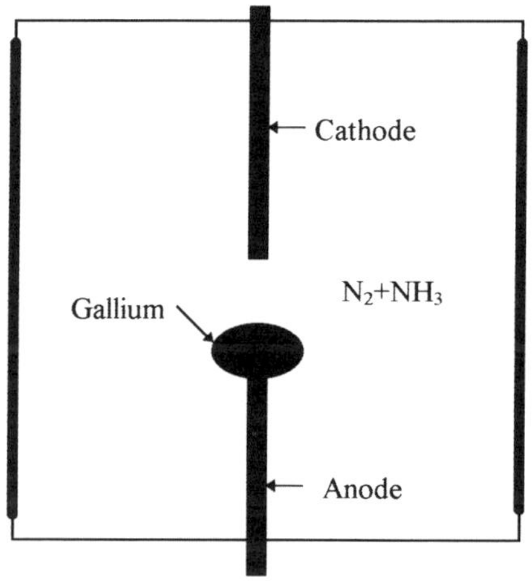

Figure1. Schematic of DC arc plasma system

microanalysis of as-grown GaN crystals were carried out using Hitachi-8100 transmission electron microscope at 200kV and the attachment of Philips DX-4 energy dispersive spectroscopy of x-ray with a super-ultra-thin window (SUTW) capable of detecting element down to boron. Electron nano-probe diffraction and electron nano-probe EDS microanalysis were used to investigation the microstructure and composition of the samples. The microanalysis samples were prepared by two different method: a conventional powder sample preparation by spraying GaN crystals on Formvar support copper grid for nano sized samples and a standard bulk sample preparation by ion milling the in situ compressed GaN plate for microsized samples. The ion milling were performed using model GL-6900 argon ion milling system at 6.0kV accelerating voltage and $80\mu A/cm^2$ beam current. X-ray diffraction data were collected using Cu Kα radiation line as an x-ray source.

RESULTS

The morphology and crystal feature of as-grown powders were investigated by TEM at 200kV. The crystal structure of the sample was determined by XRD. The crystals are well faceted with the size ranging from several nanometers to several micrometers. Figure2 shows

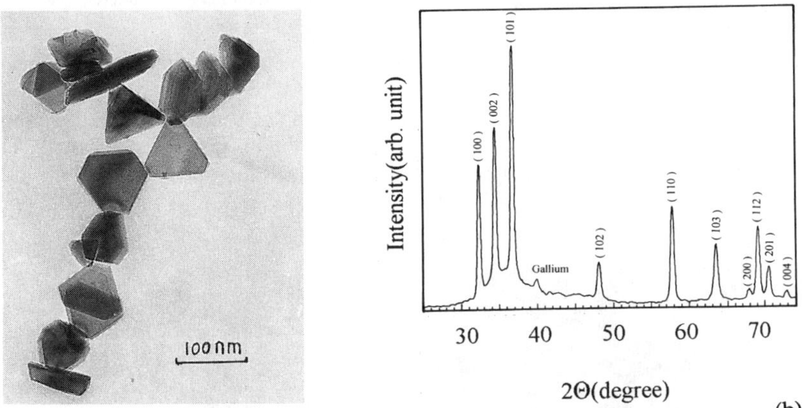

(b)

Figure 2. Bright field TEM micrograph(a) and XRD pattern(b) of W-GaN in nanoscale

the bright field TEM micrograph and XRD pattern of W-GaN crystals in nano scale. According to Figure 2, we can derive that all crystals are well faceted W-GaN having lattice constants of a_0=3.18Å and c_0=5.18Å, and most of the crystals are in triangle or hexagonal morphology.

In contrast to all other III-V semiconductor compounds, fabrication of GaN using conventional techniques occurs under metastable conditions[9]. By employing the DC plasma in this work, more reactive species such as activated nitrogen atoms(or ions) and activated gallium atoms(or ions) were produced with sufficient energy to overcome the kinetic barriers of GaN formation. Since the nitrogen pressure and reactive temperature are much higher in the plasma region, it is believed that nucleation and crystal growth of GaN occur in the gas phase under equilibrium conditions. The quality and size of crystals are determined by the growth rate of W-GaN in the plasma region. The higher the nitrogen pressure, the lager the crystal size. Well faceted W-GaN single crystals can be obtained under lower pressure conditions, In this case, more perfect crystal morphology often appears and SAD pattern is sharply spotty. Figure 3 shows the bright field TEM micrograph of a typical gallium nitride single crystal in nanoscale and its spotty SAD pattern. From Figure3, we can recognize that nano sized single crystalline W-GaN are usually faceted by {001} and {100} crystallographic faces. On the other hand, secondary nucleation on single crystalline W-GaN surfaces often appear while nitrogen pressure is much higher.

Small amount gallium phase detected by x-ray diffraction(as shown in Figure 2b) in the as grown powder is identified by TEM and NPD. The portion of gallium phase is never exceed 5% of the final product. So, we can conclude that the yield of stoichiometrical wurzitic GaN in this method is more than 95%. The result exhibits that it appears as ball-like particles in nano scale(as shown in Figure 4). It is believed that gallium particles were formed in the ununiform plasma regimes where nitrogen pressure and reactive temperature are not enough to configure GaN lattice.

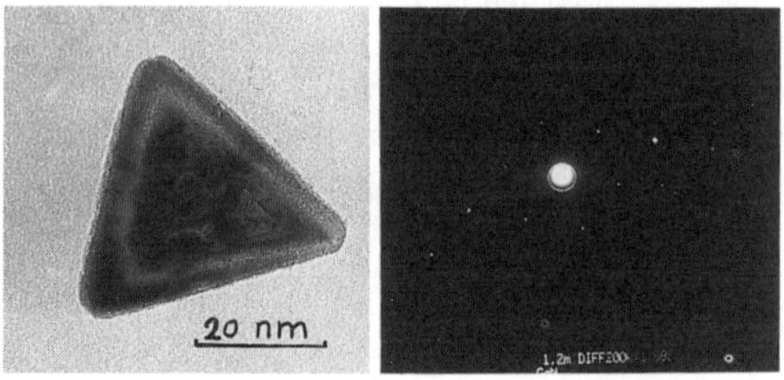

(a) (b)

Figure 3. W-GaN single crystal in nanoscale.
(a) Bright field TEM micrograph
(b) SAD pattern

The particle size of wurtzitic GaN single crystals mainly depends on the partial pressure of the ambient gas. But secondary nucleation often occurs during GaN crystal growth. Figure 5 shows the bright field TEM micrograph of the secondary nucleation.

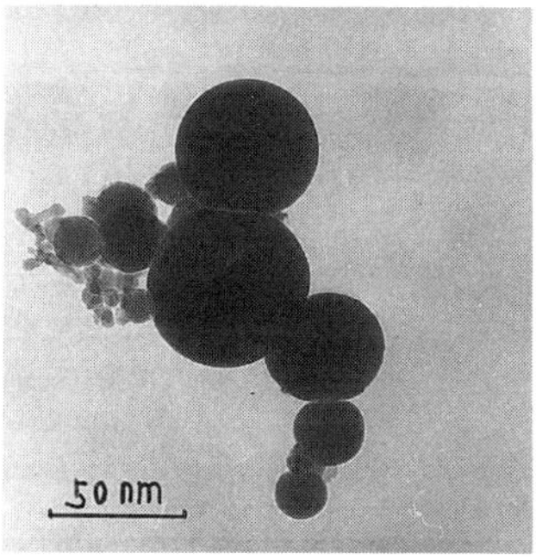

Figure 4. Ball-like Gallium nano particles

Figure 5. Secondary nucleation of GaN

Under the moderate pressure and optimized conditions, comparatively large W-GaN single crystals can be prepared. To investigate the size and structure of the crystals, we obtained the TEM bright field micrograph and NPD pattern on the thin specimens of W-GaN powder, which were prepared by ion milling the in situ compressed plates. Figure 6 shows the the bright field TEM micrograph and NPD pattern of GaN single crystals in microsize. NPD results taken on different analysis point of the same crystal show the same spotty pattern, which proves that the crystal is single crystalline.

The atomic ratio of nitrogen and gallium in the crystals is an important factor as well. To determined it, we carred out the EDS measurement on the thin samples of W-GaN. Analysis data was taken on the ion milled thin sample using nano-probe electron beam. The accurate quantitative result can be a chieved for the sample is very thin after ion milling. The EDS spectrum is shown in Figure 7. Analysis results exhibit that the atomic ratio of nitrogen and gallium is about 1:1. This means that GaN crystals prepared in this work are stoichiomitrical single crystalline.

300nm (a) (b)

Figure 6. TEM micrograph(a) and NPD pattern(b) of W-GaN in micro scale

CONCLUSIONS

In conclusion, wurtzitic GaN crystals in nano and micro scale have been fabricated by employing nitrogen and gallium plasma to activated reactive species. The as grown crystals are well faceted in triangle and hexagonal polyhedron shapes with a stoichiomitric atomic ration of N/Ga.

Further investigation to produce pure-phase and large size wurtzitic and cubic GaN crystals by high pressure/high temperature technology, using wurtzitic GaN powder as starting materials, is being undertaken.

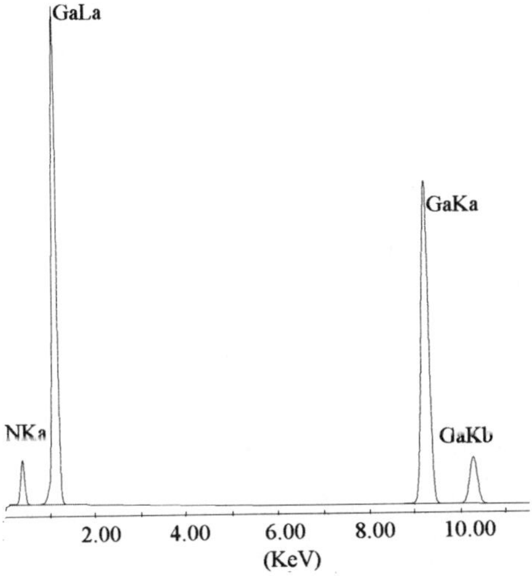

Figure 7. EDS spectrum of wurtzitic GaN single crystal

REFERENCES

1. S. Strike and H. Morocco, J. Va. Sci. Technol., **B10,** p.1237 (1992).

2. R. F. Davis, Proc. IEEE **79,** p. 702 (1991).

3. J. I. Pankove, MRS Symp. Proc. **162,** p. 515 (1990).

4. H. P. Maruska, and J. J. Tietjen, Appl. Phys. Lett. **15, p.** 327 (1969).

5. T. Sasaki and S. Zembutsu, J. Appl. Phys. **61,** p. 2533 (1986).

6. Z. Sitar, M. J. Paisley, B. Yan, and R. F. Davis, MRS Symp. Proc. **162,** p. 537 (1990).

7. S. Zembutsu, T. Sasaki, Appl. Phys. Lett. **48,** p. 870 (1986).

8. T. P. Humphreys, C. A. Sukov, R. J. Nemanich, J. B. Posthil, R. A. Rudder, S. V. Hattangady, and R. J. Markunas, MRS Symp. Proc. **162,** p. 531 (1990).

9. N. Newman, T. C. Fu, X. Liu, Z. Liliental-weber, M. Rubin, J. S. Chan, E. Jones, J. T. Ross, I. Tidswell, K. M. Yu, N. Cheungand E. R. Weber, Mat. Res. Symp. Proc. **339**, p. 483 (1994).

10. P. Perlin, J. Gorczyca, N. E. Christensen, T. Grzegory, H. Teisseyve and T. Suski, Phys. Rev. **B45**, p. 13307 (1992).

11. J. Karpinski, J. Jun and S. Porowaski, J. Crystal Growth, **66**, p. 1 (1984).

GROWTH OF CRYSTALLINE QUALITY SiC ON THIN AND THICK SILICON-ON-INSULATOR STRUCTURES

*F. NAMAVAR, *P. Colter and *A. Cremins-Costa, +C-H. Wu, *E. Gagnon, **D. Perry, and +P. Pirouz
*Spire Corporation, One Patriots Park, Bedford, MA 01730-2396
**Purdue University, West Lafayette, IN
+ Case Western Reserve University, Cleveland, OH

ABSTRACT

We have grown silicon carbide (SiC) on ultrathin Si (about 300Å) and on thick Si (about 2000Å) on commercial SIMOX (from IBIS Corp and SOITEC, Inc.), and bulk Si. Electron diffraction and Rutherford backscattering spectroscopy (RBS)/channeling studies indicate epitaxial growth of single-crystal β-SiC even at growth temperatures as low as 1100°C.

We have already demonstrated the fabrication of ultrathin Si, as thin as 140Å on SiO_2 by using the low-energy SIMOX (LES) (20 to 30 keV) process to produce films of lower cost and excellent integrity compared to thinned commercial SIMOX. Based on these results, ultrathin Si-on-insulator (SOI) substrates appear to have great potential for device quality SiC films. However, the carbonization and/or growth of SiC on ultrathin Si requires further optimization because the processes for surface cleaning and growth of SiC on bulk Si substrates cannot be applied because of the thinness of the substrate layers. Additional carbonization work at higher temperatures has indicated the possibility of converting the entire Si top layer.

INTRODUCTION

Fabrication of SiC-on-Si wafers is desirable for a number of reasons including their use as low-cost, large-area substrates for high-power, high-temperature electronics, and their use as lattice-matched substrates for group III-nitrides. In addition, SiC on SOI substrates may pave the way for monolithic integration of GaN photonic devices with advanced Si technology. An advantage of growing GaN on SiC (or on SiC on Si, or on SiC on SOI) is that the film and the substrate both have the same cleavage planes, enabling formation of cleaved laser facets. This is not the case for GaN on sapphire where different crystal structures cause a 30° misalignment of crystal planes [1], making facet cleaving problematic.

SiC is grown epitaxially on Si either by direct deposition, or by carbonization of the Si surface. The carbonization process has to date, provided the lowest defect SiC on Si [2,3,4,5]. However a lower defect density is needed before SiC on Si is suitable for device applications.

In the past few years, there has been rapid progress with separation by implantation of oxygen (SIMOX) technology, in which a relatively defect-free Si top layer is produced on an oxide layer on a Si substrate [6]. These structures have been used as substrates to grow epitaxial layers such as GeSi [7,8] or silicides [9] or GaAs [10]. Recently, we developed a low-energy SIMOX [11,12,13,14] process (LES) which can produce SIMOX structures with ultrathin (< 500Å) Si top layers. When the ultrathin Si layer of a SIMOX wafer (rather than the surface of a Si wafer) is converted to SiC by carbonization, or used as a substrate for SiC epitaxy, it is expected to have fewer crystal defects, for two reasons: (1) the buried SiO_2 layer will soften at the carbonization or growth temperature and will relieve stresses associated with lattice constant changes, and (2) since it is possible to produce high quality crystalline Si on SiO_2 (which is amorphous), by analogy it should also be possible to produce SiC on SiO_2.

Mat. Res. Soc. Symp. Proc. Vol. 423 © 1996 Materials Research Society

In this paper, we report preliminary results on the epitaxial growth of SiC on a Si thin film with a thickness about 300Å, on Si 2000Å thick, and on bulk Si. Our recent TEM results (for growth at about 1100°C) indicate that good crystalline quality SiC can be obtained on thin Si, even when the thin Si layer is not fully carbonized. Ongoing work (growth above 1300°C) identified the process steps to carbonize the entire Si top layer of SIMOX and produce good quality crystalline SiC. The high temperatures samples are presently being analyzed by TEM.

EXPERIMENTAL PROCEDURE

This work employed a SPI-MOCVD™ reactor typically used for growth of III-V compounds. The growth chamber features a vertical, water-cooled quartz belljar with a three-inch diameter silicon-carbide-coated graphite susceptor. A solid quartz rod supports the susceptor and serves as a light pipe for temperature measurement with a two-color optical pyrometer.

In the standard approach for SiC or Si epi sample preparation, a Si substrate is typically etched by HCl to remove the surface oxide and impurities. However, for a LES wafer (with a thickness under 500Å) special care is needed to avoid etching the Si thin film or reducing the SiO_2 buried layer. We developed a special process to clean the surface layer of LES wafers without damaging or etching the Si thin film or buried oxide [12].

Thin SiC on LES, bulk Si, and standard SIMOX were grown at the same time to study the effect of the thickness of the Si top layer on the initial stage of SiC epitaxial growth. The growth procedure was similar to Powell, *et.al.*[5] except for the differences discussed. Growths were conducted at 1100°C. Research and ULSI-grade gas mixtures were used. Note, that the low Si:C ratio should result in a reduction in the residual nitrogen-doping background by site-competition epitaxy [15,16].

Rutherford backscattering spectroscopy (RBS) was also performed to determine the solid phase Si:C ratio, as well as the thickness of SiC and SIMOX layers. Commonly RBS is performed using an MeV He^+ beam with the sample oriented such that the incident beam is 10° from the normal of the sample. However, measurements were performed with the beam incident 10°, 60° and 75° to the normal of the sample, which increased the effective thickness of the layer up to a factor of four.

A Phillips CM20 transmission electron microscope at Case Western Reserve University was utilized to observe cross-sectional specimens. Cross-sectional transmission electron microscopy (XTEM) samples were prepared by sandwiching two films face to face with M-bond epoxy and curing it. Subsequently, the sandwiched sample was sliced by a diamond wheel and each slice was polished down to a thickness of 20 µm, and mounted on a Cu grid. Mounted specimens were further ion-milled to electron transparency.

EXPERIMENTAL RESULTS

Figure 1 is an XTEM of a typical sample implanted at 30 keV with a dose of 2.5×10^{17} O^+/cm^2 and annealed for six hours at 1300°C in N_2. The Si top layer is only about 310Å thick, and has kept its integrity; no defects were observed in this micrograph. No Si islands can be seen in the SiO_2 buried layer, and no threading dislocations are observed in the Si top layer.

Figure 2 shows RBS spectra of two commercial SIMOX substrates and an LES substrate. Using the RUMP code we determined the thickness of the Si top layer of SIMOX wafers used in this study, *i.e.* LES:400Å, SOITEC:1400Å, and IBIS:2350Å. A typical RBS spectra for both carbonized and CVD grown SiC is shown in Figure 2b. Accidental channeling of He ions clearly shows epitaxial growth of SiC on SIMOX substrates.

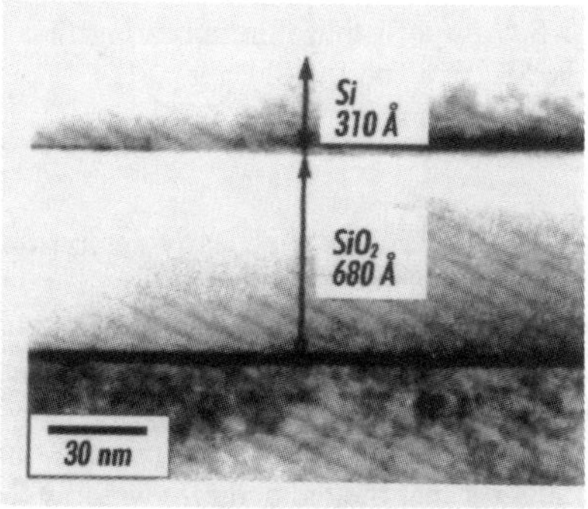

Figure 1 *XTEM of a low-energy SIMOX substrate produced with a total dose of 2.5 x 10^{17} O/cm^2 at 30 keV and high temperature annealing at 1300 °C in N_2 for 6 hours.*

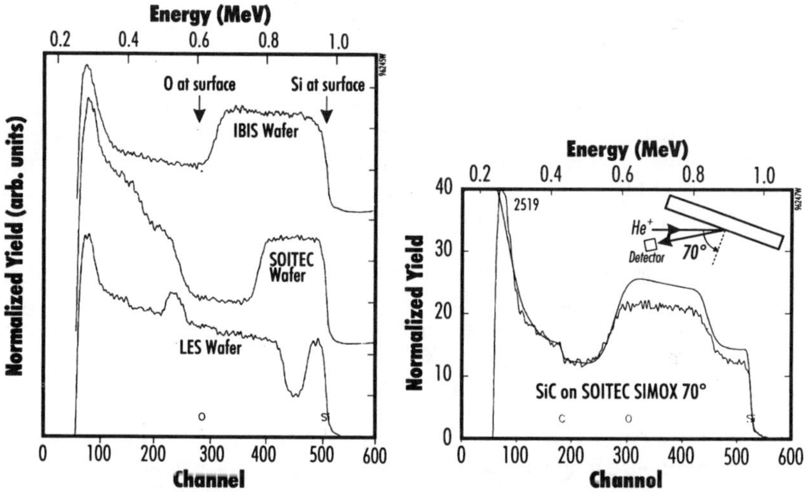

Figure 2 *RBS spectra of a) three SIMOX wafers with different thicknesses used in this study, and b) 570Å SiC grown on a SOITEC SIMOX substrate at 1200 °C. Using the SiO_2 buried layer(amorphous) as a reference, accidental channeling of He ions indicates epitaxial growth of SiC on SIMOX.*

Figure 3 shows a cross-sectional bright-field (BF) TEM micrograph revealing uniform layers with sharp interfaces between them. The SiC film grown on the topmost single-crystalline Si layer was determined to be single crystal by selected area diffraction (Figure 4). The cross-sectional BF micrograph also indicates that the SiC film contains a number of planar defects such as like twins and stacking faults, which show darker contrast in the SiC image. The voids that are usually present in the Si substrate at the interface were absent in the isolated single-crystalline Si film. Some defects were observed in the Si layer which did not appear to affect or change the crystallinity in SiC. Figure 4 shows growth of β-SiC on a Si thin film indicating that a parallel epitaxial orientation relationship exists between the SiC film and the Si layer.

DISCUSSION AND CONCLUSIONS

Our TEM (Figures 3 and 4) and RBS results clearly demonstrate growth of good crystalline quality SiC at 1100°C. The Si:C (0.007) ratio used for growth at 1100°C was considerably lower than the silicon to carbon ratio (0.4) used by others [5]. The reduced cracking efficiency of propane at low temperatures (1100°C) is partially compensated by an increase in the propane concentration. The low concentration of silane (which controls the growth rate) provides the conditions for growth of good quality crystalline SiC at 1100°C, even though the surface mobility is low at this temperature[17]. Our studies suggest that the temperature window for growth of epitaxial SiC may be less dependent on precursor decomposition and more dependent on surface mobility than has been reported [18].

These preliminary results indicate that there is no major difference in the SiC films grown on SIMOX substrates as compared to those grown by others on bulk Si substrates at high temperatures [19]. The high resolution TEM micrograph clearly shows sharp interfaces between the layers in the SOI structure with no sign of predominant interdiffusion having occurred during SiC growth. The voids that are usually present in the SiC/Si interface were absent in our samples. The absence of voids at the SiC/Si interface significantly affects the electrical properties of the SiC film [20]. The reason for the absence of voids in our samples is not clear, however, one possibility is the low growth temperature, which results in higher nucleation density and rapid coverage of Si by an initial SiC film. Consequently, out-diffusion of Si cannot occur and voids cannot form [21]. TEM micrographs of the SiC film grown at 1100°C (which show darker contrast in the SiC image) indicate a number of planar defects such as twin and stacking faults. These defects are typically observed in SiC grown by CVD [19].

In summary, this preliminary work demonstrates that thin (<500Å) Si layers support growth of good quality SiC layers, even when the Si layer is not totally carbonized. This will pave the way for growth of SiC directly on SiO_2 (a compliant substrate) by carbonizing the entire thin Si top layer of SIMOX. Thinned SIMOX is useful for SiGe growth, however, the superior integrity of films produced by the LES process is necessary because growth of SiC involves very high temperature (1100 to 1360°C).

It should be noted that growth at low temperatures can also reduce thermal effects such as diffusion in the substrate during SiC deposition. A combination of low temperature and the buried oxide layer could also prevent slip bands from being generated in the Si substrates and propagating into the SiC films [22]. These results indicate that an optimized SiC growth procedure on thin SIMOX structures is a promising candidate for the production of device quality SiC on Si. SiC on SIMOX structures may facilitate fabrication of MEMS devices because of the additional opportunities provided by SOI structures for selective etching [23].

This work was supported in part by an SBIR from the Office of Naval Research and monitored by Dr. Yoon Park.

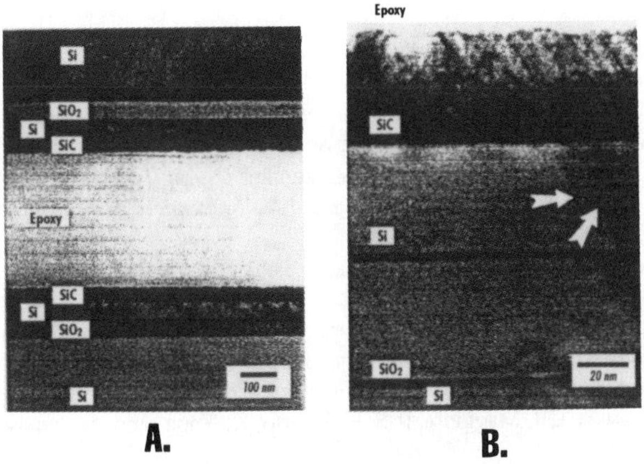

Figure 3 *a) Bright-field TEM micrograph of a SiC film grown on a thin SIMOX substrate epoxied together face-to-face. The Si layer on SiO_2 has been only partially converted to SiC, thus leaving a thin layer of Si between the SiC and the SiO_2. b) A higher magnification BF TEM of the same sample as that shown in Figure 1a. Note the few defects in the Si layer between SiO_2 and SiC film.*

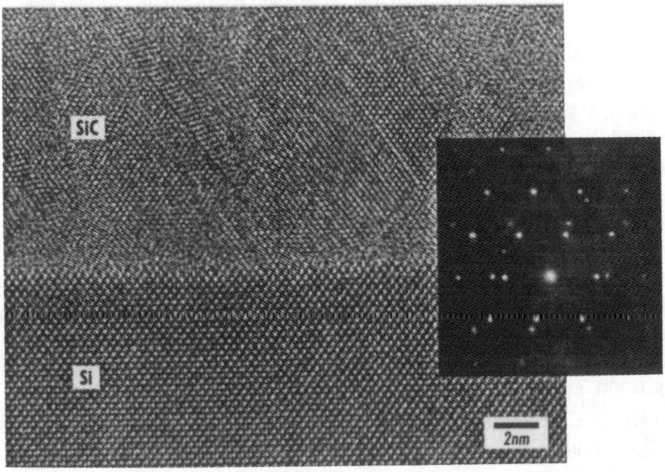

Figure 4 *a) High-resolution cross-sectional TEM micrograph of the SiC/Si interface on the SIMOX substrate and b) electron diffraction of the SiC/Si interface with the incident beam B=[110], indicating single-crystal films with a parallel epitaxial orientation relationship.*

REFERENCES

1. B.G. Levi, "GaN Laser Diode Brightness Hopes for a Long Lived Short-Wavelength Device," *Physics Today*, April 1996.
2. S. Nishino, J.A. Powell, and H.A. Will, *Appl. Phys. Lett.* **42**, 460 (1983).
3. P. Liaw, and R.F. Davis, *J. Electrochem. Soc.* **132**, 642 (1985).
4. S. Nishino, H. Suhara, H. Ono, and H. Matsunami, *J. Appl. Phys.* **61**, 4889 (1987).
5. J.A. Powell, L.G. Matus, M.A. Kuczmarski, *J. Electrochemical Soc.: Solid State Science and Technology,* 1558 (1987).
6. M.K. ElGhor, S.J. Pennycook, F. Namavar, and H.K. Karam, *Appl. Phys. Lett*, **57** 156 (1990).
7. F. Namavar, E. Cortesi, D.L. Perry, E.A. Johnson, N.M. Kalkhoran, J.M. Manke, N.H. Karam, R.F. Pinizzotto, and H. Yang, *MRS Symp. Proc.*, **198**, 503 (1990).
8. A.R. Powell, S.S. Iyer, and F.K. LeGoues, *Appl. Phys. Lett.* **64**, 1856 (1994).
9. L. Luo, M. Nastasi, C.J. Maggiore, R.F. Pinizzotto, H. Yang, and F. Namavar, *J. Appl. Phys.* **73**, 4107 (1993).
10. S.J. Pearton, S.M. Vernon, K.T. Short, J.M. Brown, C.R. Abernathy, R. Caruso, S.N.G. Chu, V.E. Haven, and S.N. Bunker, "Characterization of GaAs Grown by MOCVD on Si-on-Insulator," *Appl. Phys. Lett.* **51**, 1188 (1987).
11. F. Namavar, E. Cortesi, B. Buchanan, and P. Sioshansi, *1989 IEEE SOS/SOI Technology Conference Proceedings,* 117 (1989).
12. F. Namavar, E. Cortesi, N.M. Kalkhoran, J.M. Manke, and B. Buchanan, *1990 IEEE SOS/ SOI Technology Conference Proceedings*, 49 (1990).
13. F. Namavar, "Silicon-on-insulator (SOI) Technical Development," Final Report #RL-TR-91-175, Rome Laboratory, Griffiss Air Force Base, August (1991).
14. F. Namavar, E. Cortesi, B. Buchanan, J.M. Manke, and N.M. Kalkhoran, *MRS Symp. Proc.* **235**, 109 (1992).
15. D.J. Larkin, P.G. Neudeck, J.A. Powell, and L.G. Matus, "Site-Competition Epitaxy for Controlled Doping of CVD Silicon Carbide," *International Conference on Silicon Carbide and Related Materials,* Washington, DC, (November 1993).
16. D.L. Larkin, P.G. Neudeck, J.A. Powell, and L.G. Matus, *Appl. Phys. Lett.* **65**, 1659 (1994).
17. Kimoto, Hironori Nishino, Woo Sik Yoo, and Hiroyuki Matsunami "Growth Mechanisms of 6-HSiC in Step-Controlled Epitaxy," *J. Appl. Phys.* **73**, 726 (1993).
18. A.J. Steckl, C. Yuan, J.P. Li, and M.J. Loboda, "Growth of Crystalline 3C-SiC on Si at Reduced Temperatures by Chemical Vapor Deposition from Silacyclobutane,". *Appl. Phy. Lett.* **63**, 3347 (1993).
19. Chien-Hung Wu, Chacko Jacob, X.J. Ning, S. Nishino, and P. Pirouz, *J. Crystal Growth,* **158**, 480-490, (1996).
20. Chacko Jacob, Chien-Hung Wu, Mehran Mehregany, P. Pirouz. To be presented and High Temp. Electronics Conference, Albuquerque, NM, June 1996.
21. J.P. Li and A.J. Steckl, *J. Electrochem Soc.* **142**, 634, 1995.
22. Christian A. Zorman, *et al. J. Appl. Phys.* **78** (8) 15, 1995.
23. T. Hirano, T. Furuhata, K.J. Gabriel, H. Fujita in Technical Digest, *6th Conf. Solid-State Sensors and Actuators* (San Francisco, June 1991) 873.

GROWTH AND CHARACTERIZATION OF POLYCRYSTALLINE DIAMOND THIN FILMS ON POROUS SILICON BY HOT FILAMENT CHEMICAL VAPOR DEPOSITION

S. MIRZAKUCHAKI*, E. J. CHARLSON*, E. M. CHARLSON*, T. STACY*, F. SHAHEDIPOUR**, H. W. WHITE**,
*Electrical Engineering Department, University of Missouri-Columbia, Columbia, MO. 65211
**Department of Physics and Astronomy, University of Missouri-Columbia, Columbia, MO. 65211

ABSTRACT

Hot filament chemical vapor deposition (HFCVD) was utilized to grow high quality diamond film on porous silicon (PS) substrates to a thickness of 5-6 μm. Boron-doped silicon substrates of <100> orientation and resistivity of 5-15 ohm-cm were anodized by the electrochemical process to form PS. A slurry of diamond paste (1/4 micron average grain size) was rubbed on the samples for a few seconds before introduction into the chamber. Diamond film growth on the PS has the advantages of shorter incubation time and higher nucleation density as evident from scanning electron microscopy (SEM). The results of X-ray diffraction confirm the growth of predominatly (111) oriented high quality diamond film. Electrical properties were also studied by sputtering circular gold contacts on top of diamond film and measuring current-voltage (I-V) characteristics.

INTRODUCTION

The nucleation and growth of polycrystalline diamond thin films on silicon (as the dominant material in semiconductor industry) has been the subject of intense investigation. The effects of Si surface pretreatment on the nucleation enhancement of diamond has been studied by many researchers.[1-2] One of the most useful techniques in promoting nucleation and growth of diamond films with reduced stress is the formation of PS layers on silicon substrates. Porous silicon is quickly becoming an increasingly important and versatile electronic material in today's fabrication technology. Ever since Canham[3-5] demonstrated visible light photoluminescence from PS, much effort has been focused on the possibility of producing optoelectronic devices using this material.[6-9]

Diamond growth on PS by chemical vapor deposition was reported for the first time by Spitzl et. al.[10] These authors demonstrated that PS allows for a reduction of the stress in diamond film as well as enhancing the quality of the diamond structure. Another group of authors[11] have shown that PS (compared to mechanically scratched Si) has a smoother surface and more evenly distributed defects for the growth of diamond film. They also investigated the infrared transmission spectra and concluded that the transmittance of diamond film grown on PS is higher than for film grown on silicon substrates.

Diamond nucleation and growth on PS substrates is the focus of the present work. The study of anodization conditions and their effect on the nucleation density and the quality of grown diamond films was carried out to determine the most suitable set of parameters for formation of porous layers. Electrical characteristics of the deposited diamond film were studied by observing the I-V data on a gold-diamond-PS-Si structure.

Mat. Res. Soc. Symp. Proc. Vol. 423

EXPERIMENT

In this work, several boron-doped Si samples of <100> orientation and resistivity of 5-15 ohm-cm were used as starting substrates. These samples underwent organic cleaning in an ultrasonic bath with trichloroethylene, acetone, methanol, and DI water for five minutes each. Substrates were further cleaned in hot nitric and hydrochloric acids for five minutes each followed by rinsing in DI water. One of these substrates was scratched by mechanical polishing with diamond paste of 0.25 μm grit size.

A thin layer of aluminum, ~500 nm, was then thermally evaporated on the unpolished (back) side of two other substrates in a vacuum system to provide a good electrical contact during the anodization process. PS was formed by anodization in an electrolyte of 1:1 49% HF and ethanol at a constant current density of 50 $mA\text{-}cm^{-2}$ and a reaction time of 60 seconds. During anodization, the back Al contact was protected with a layer of acid-resistant wax. After the formation of PS, a diamond paste slurry (made by dissolving 1/4 micron diamond grit in DI water) was rubbed on this sample to promote diamond nucleation.

Polycrystalline diamond films were deposited on these samples in a HFCVD reactor. Trimethyl phosphite diluted in acetone to a concentration of 400 ppm was used as a source of in situ doping. The deposition conditions used in this study are as follows; hydrogen flow rate of 98 sccm, methane flow rate of 1.5 sccm, a mixture of acetone and trimethyl phosphite at a flow rate of 0.5 sccm, pressure of 25 Torr, filament temperature of 2000 °C, substrate temperature of 720 °C, and filament to substrate distance of 4 mm. The deposition time for one run was 20 minutes in order to measure the nucleation density on scratched and porous silicon. In a separate run, diamond was grown on a different PS sample for 6 hours to a thickness of 5-6 μm.

Circular contacts (0.021 cm^2 in area) were deposited on the phosphorous-doped diamond films by sputtering a thin layer, ~500 Å, of gold through a metal shadow mask in a DC glow discharge system. Prior to sputtering of gold contacts, the surface of the diamond films were cleaned by a saturated solution of $CrO_3 + H_2SO_4$ at 200 °C for 20 minutes followed by rinsing in a 1:1 boiling solution of H_2O_2 and NH_4OH for 10 minutes and then in boiling DI water for 3 minutes. This procedure removes the nondiamond carbonaceous surface layer on top of the diamond film. In order to provide an ohmic contact to the back of the samples, a thin layer of Al, ~650 nm, was thermally evaporated on the unpolished side of the Si substrates.

RESULTS

The size, shape, thickness, and density of the formed pores are determined by the anodization conditions which include HF concentration, temperature, Si dopant type and concentration, anodization time, and current density. Unno et. al.[12] provide an empirical formula for the calculation of the thickness of PS as a function of the current density and the anodization time as follows;

$$T = k \, . \, J^m \, . \, t \qquad (1)$$

where T is the thickness of the PS layer, J is the current density in $mA\text{-}cm^{-2}$, t is the anodization time in seconds, and k and m are constants which are dependent on the HF concentration. For the HF concentration used here k and m are given as $1.01E^{-7}$ and 0.98, respectively. The thickness of the PS layer using the above conditions was calculated by this formula to be ~2.8 μm.

Diamond growth on PS exhibits a higher nucleation density and a shorter incubation time than growth on scratched silicon substrates. Fig. 1 shows the diamond nucleation on the scratched and porous silicon after 20 minutes of deposition. The nucleation density was calculated by counting the number of diamond nuclei on each sample and then dividing by the unit surface area. This was determined to be $2*10^8$ cm^{-2} and $1.5*10^9$ cm^{-2} for scratched and porous silicon, respectively.

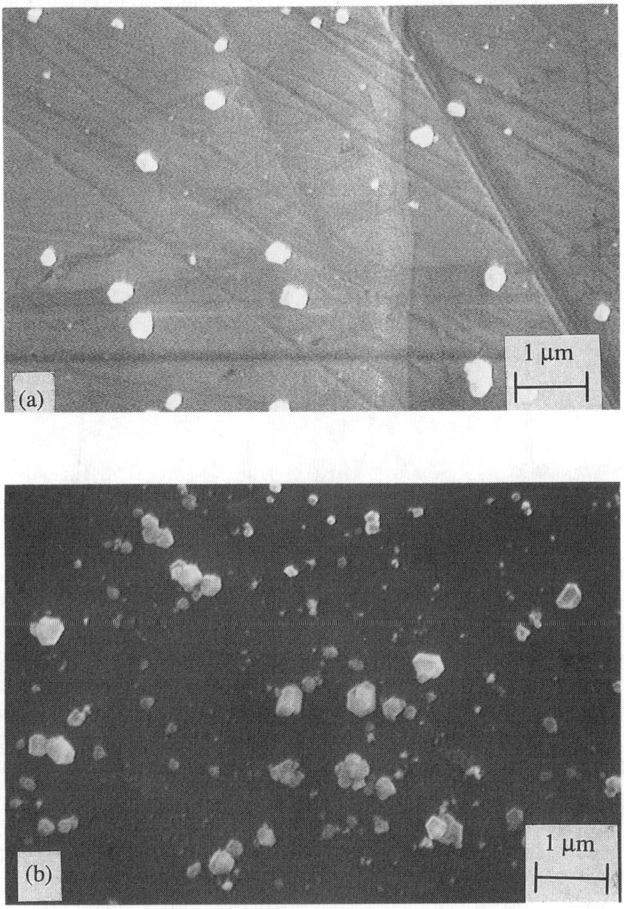

Figure 1. SEM micrographs of diamond on scratched Si (a) and on PS (b).

Heiderhoff et. al.[13] investigated the effect of the PS thickness on the stress induced in diamond films. They used microwave plasma CVD to study the growth of diamond film on a number of PS substrates with various thickness and have presented an empirical curve relating the porous layer thickness to the stress in diamond films. Based on the results presented by this group, a PS layer of 3 μm thickness (approximately the thickness used for this study) would cause a Raman peak shift of 2 cm^{-1} to higher wavenumbers, indicating the presence of some compressive stress. This compares favorably with the Raman shift observed in diamond samples grown here and is smaller than reported values for diamond film grown on monocrystalline Si (3-5 cm^{-1}). The results of the Raman analysis are presented elsewhere.[14]

X-ray diffraction analysis was also performed with the result shown in Fig. 2. Four Bragg diamond peaks are evident in this figure. The X-ray diffraction reveals the predominat orientation of (111) as well as the presence of some (100) texture in this film as the intensity of (400) peak relative to the (111) peak is higher than a powder without preferential orientation.

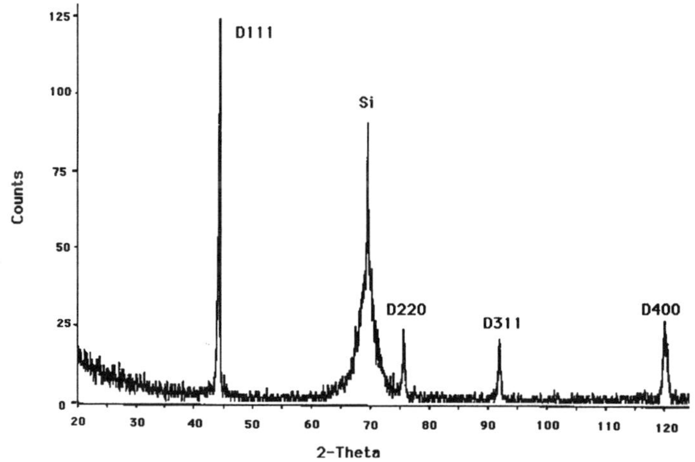

Figure 2. XRD diagram of diamond growth on PS.

It may be concluded that the nanoscale structure and the high surface area of PS provides an effective nucleation site for CVD grown diamond. Recently, a group of researchers[15] demonstrated the HFCVD growth of diamond on PS pretreated with diamond paste in an ultrasonic bath. These authors concluded that the PS layer relaxes the stress due to the large lattice mismatch as well as the large difference in thermal expansion coefficients between the diamond growth layer and the Si substrate.

Fig. 3 shows the I-V curve measured on the diamond film grown on PS. The straight line relationship between current and voltage demonstrates an ohmic contact between gold and P-doped diamond film. The low level of current indicates a high resistivity in the polycrystalline film as well as the PS layer. The high resistivity of the diamond film is attributed to crystal defects such as grain boundaries and the low percentage of electrically activated P atoms.

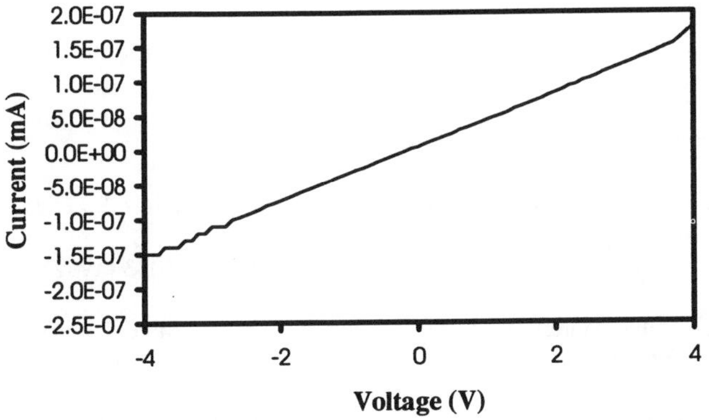

Figure 3. I-V curve measured on gold contact to diamond on PS.

CONCLUSIONS

In conclusion, the results of the growth of CVD diamond thin films on PS substrates has been demonstrated. The diamond film thus grown has a high nucleation density and some (100) texture in addition to the dominant (111) orientation which shows the suitable conditions provided by a PS layer for diamond growth. Results of the electrical measurement suggest the formation of an ohmic contact between gold and diamond grown on porous silicon.

ACKNOWLEDGMENTS

The authors wish to thank Louis M. Ross and Naiyu Zhao for their assistance in the SEM and XRD studies. The Raman work is supported by the U.S. Army Research Office (Research Triangle Park, N.C.) under Grant No. DAAH04-94-G-0305.

REFERENCES

1. P. Ascarelli and S. Fontana, Appl. Surf. Sci. **64**, 307 (1993) .

2. J. H. D. Rebello, D. L. Straub, V. V. Subramaniam, E. K. Tan, S. A. Dregia, B. L. Preppernau, and T. A. Miller, Materials and Manufacturing Processes **6**, 501 (1991).

3. L. T. Canham, Appl. Phys. Lett. **57**, 1046 (1990).

4. L. T. Canham, M. Houlton, W. Leong, C. Pickering, and J. Keen, J. Appl. Phys. **70**, 422 (1991).

5. A. Gullis and L. T. Canham, Nature **353**, 335 (1991).

6. C. Mazzoleni and L. Pavesi, Appl. Phys. Lett. **67**, 2983 (1995).

7. S. Y. Shieh and J. W. Evans, J. Electrochem. Soc. **140**, 1094 (1993).

8. L. Pavesi, C. Mazzoleni, A. Tredicucci, and V. Pellegrini, Appl. Phys. Lett. **67**, 3280 (1995).

9. D. Stievenard and D. Deresmes, Appl. Phys. Lett. **67**, 1570 (1995).

10. R. Spitzl, V. Raiko, and J. Engemann, Diamond and Related Materials **3**, 1256 (1994).

11. G. Q. Ke, Z. J. Xing, X. Yin, K. Chen, and Y. Shen, Vaccum **43**, 1043 (1992).

12. H. Unno, K. Imai, and S. Muramoto, J. Electrochem. Soc. (Solid State Science and Technology), p. 645 (1987).

13. R. Heiderhoff, R. Spitzl, M. Maywald, V. Raiko, L. J. Balk, and J. Engemann, Proc. SPIE, **2151**, (Los Angeles, CA, 1994), p.59.

14. S. Mirzakuchaki, M. Hajsaid, H. Golestanian, R. Roychoudhurry, E. J. Charlson, E. M. Charlson, and T. Stacy, Appl. Phys. Lett. **67**, 3557 (1995).

15. Z. Liu, B. Q. Zong, and Z. Lin, Thin Solid Films **254**, 3 (1995).

RECONSTRUCTION OF NONPOLAR GaN SURFACES

J. Elsner, M. Haugk, Th. Frauenheim *
* Theoretische Physik III, Technische Universität Chemnitz, 09107 Chemnitz, Germany

ABSTRACT

A density–functional theory based nonorthogonal tight–binding scheme is used to investigate the nonpolar surfaces of GaN in the wurtzite and zinc–blende phases. In III-V semiconductors the nonpolar surfaces mainly reconstruct by a bond–length conserving rotation of the surface atoms. In contrast to that for all nonpolar GaN surfaces this bond–length is shortened by $\approx 3-4\%$. We furthermore find the rotation angle to be significantly smaller than could be expected from results for GaAs and GaP.

INTRODUCTION

Due to its interesting optical and electrical properties, GaN has become a promising material for future device applications. At equilibrium GaN condenses in the wurtzite structure, but can also be forced into the metastable zinc–blende phase if grown on a suitable cubic substrate [1]. A detailed knowledge of the reconstructions of cleavage surfaces is important for the investigation of grown crystals, which is possible by experimental techniques as for example cross–sectional scanning–tunneling microscopy. Up to now even advanced growth techniques such as metallo–organic chemical vapor deposition (MOCVD) and molecular–beam epitaxy (MBE) have not yet succeeded in producing crystals with smooth surfaces. As there exists still a considerable portion of cubic phase in the mainly hexagonal crystals and vice versa, no pure single phase material could be grown to a size where cleavage starts to become feasible. However, there is hope that with the continuing advance in growth techniques, GaN crystals might be cleaved within the main cleaving directions terminated by the $(10\bar{1}0)$ and $(11\bar{2}0)$ surfaces in case of the wurtzite and the (110) surface in case of the zinc–blende phase. In this paper we therefore determine the atomic structures of these main nonpolar cleavage surfaces by using a density–functional based tight–binding (DF–TB) molecular–dynamics (MD) scheme.

This scheme has already been very successful in describing C, Si and BN structures [2, 3, 4]. The considered approach produces results of a comparable accuracy to those of self-consistent calculations for a wide range of systems, from clusters to extended structures such as bulk materials and surfaces containing several hundred atoms at a considerably lower computational cost. In contrast to empirical tight–binding schemes, where the matrix elements are fitted to the properties of equilibrium structures, we calculate the matrix elements in a straightforward manner out of an atomic local orbital basis set, which is derived by self–consistent single atom calculations using density–functional theory with the local–density approximation (DFT-LDA). We thus avoid any difficulties arising from an empirical parametrisation and guarantee high transferability.

This paper is organized as follows: We firstly give a brief outline of the method and describe the simulation geometries. Subsequently we go on to determine the surface reconstructions of the nonpolar zinc–blende (110) and wurtzite $(10\bar{1}0)$ and $(11\bar{2}0)$ surfaces. We finally summarize the results.

Mat. Res. Soc. Symp. Proc. Vol. 423 © 1996 Materials Research Society

METHOD

The interatomic potentials and forces for molecular dynamics are derived by a density-functional based non-orthogonal tight-binding scheme. As has been recently described in detail in the literature [2], the Kohn-Sham orbitals of the many-atom structure are represented by a linear combination of atomic orbitals (LCAO)-Ansatz, where we use the localized valence electron orbitals as a minimal basis. Within a two-center approach all Hamilton and overlap matrix elements are derived in a parameter free way via the construction of pseudo-atomic orbitals and potentials by self-consistent single-atom calculations on the LDA level. This two-center approach enables us to tabulate the corresponding Slater-Koster integrals versus interatomic distance. We then perform a non-self consistent solution of the Kohn-Sham equations for the many-atom structure in order to determine the electronic energy and finally derive a universal short-range repulsive pair potential with respect to scf-LDA results. The total energy of the system can then be written in the usual tight-binding form as a sum of the electronic band-structure energy (sum over all occupied Kohn-Sham energies ϵ_i) and the repulsive energy:

$$\begin{aligned} E_{tot}(\{\mathbf{R}_k\}) &= E_{BS}(\{\mathbf{R}_k\}) + E_{rep}(\{|\mathbf{R}_k - \mathbf{R}_l|\}) \\ &= \sum_i^{occ.} \varepsilon_i(\{\mathbf{R}_k\}) + \sum_k \sum_{<l} V_{rep}(|\mathbf{R}_l - \mathbf{R}_k|) \ . \end{aligned} \quad (1)$$

The method is sufficiently accurate in describing the equilibrium geometry and properties of all scale gallium, nitrogen and gallium nitride structures. Thus it can be shown to be transferable and efficient for predictive molecular-dynamic simulations on surface reconstructions. We model our surfaces by a slab with periodic boundary conditions in two dimensions. The slabs contain 8 layers of GaN, each layer consisting of 24 atoms. We saturated the dangling bonds of the atoms on the bottom of the slab with hydrogen and thus prevent any artificial charge transfer between the two surfaces. The coupling to the bulk-substrate was simulated by fixing the hydrogen atoms and the two lowest lying GaN layers during the relaxation. Starting from the unreconstructed cells, we determine the possible surface reconstructions by annealing the slabs at 300 K and finally perform a conjugate gradient relaxation. All calculations make use of a Γ-point approximation in the Brillouin zone, which is justified when using sufficiently large supercells.

RECONSTRUCTIONS OF NONPOLAR SURFACES

It is by now well established that the nonpolar cleavage surfaces of tetrahedrally coordinated III-V and II-VI semiconductors all show a common reconstruction behavior [5]. During the reconstruction process the anion moves away from the bulk towards a pyramidal geometry whereas the cation moves into the bulk towards a planar geometry by a nearly bond-length- conserving rotation [6, 7]. Although the atomic displacements may be considerable the relaxed structures preserve the bulk symmetry. These general results seem to be valid for all III-V semiconductors, however the values of some characteristic quantities, as e.g. the layer rotation by a tilt angle ω, show a strong dependence on the anion component in the compound. Whereas large tilt angles ($27^o \leq \omega \leq 31^o$) have been reported for GaP and GaAs [8] at almost preserved surface bond length, the situation seems to be different in the case of GaN. To our knowledge there have been no experimental measurements and only two

Table I: **GaN zinc–blende (110) surface**

	DF-TB (sp–basis)	DF-TB (sp + Ga 3d–basis)	HF-all electron [11]
$\Delta_{1,\perp}$	0.27 Å	0.24 Å	0.032 Å
$\Delta_{1,y}$	3.49 Å	3.48 Å	3.62 Å
$d_{12,y}$	2.39 Å	2.37 Å	2.45 Å
ΔR (Ga-N)	-3.7%	-3.4%	-6.5% (6.6%)*
ω	15.5^o	13.5^o	2.06^o (6.1^o)*

***Values were obtained with the inclusion of second layer relaxation**

Table II: **GaN wurtzite (10$\bar{1}$0) surface**

	DF-TB (sp–basis)	DF-TB (sp + Ga 3d–basis)	HF-all electron [11]
$\Delta_{1,\perp}$	0.25 Å	0.23 Å	0.046 Å
$\Delta_{1,y}$	3.36 Å	3.35 Å	3.35 Å
$d_{12,y}$	2.69 Å	2.66 Å	2.70 Å
ΔR (Ga-N)	-4.8%	-4.7%	-7.4%
ω	8.0^o	7.0^o	1.45^o

theoretical papers investigating the properties of nonpolar GaN surfaces. In an early work Swarts *et al.* [9] examined the zinc–blende (110) surface for various III-V semiconductor compounds using small cluster Hartree–Fock pseudopotential calculations. They determine the layer rotation angle ω to be significantly smaller for the nitrides (19.4^o for GaN and 19.3^o for AlN) than for GaAs and AlAs, where they report 27.2^o in both cases.
In another very recent work of Jaffe *et al.* [11], reconstructions of the GaN zinc–blende (110) and wurtzite (10$\bar{1}$0) surfaces have been studied by using an all electron Hartree–Fock scheme. The structures were allowed to be optimized by non–bond–length–conserving relaxation of the surface atoms. In both cases they obtain a very small layer rotation angle combined with a considerable shortening of the Ga-N surface bond for the zinc–blende (2.1^o and 6.5%) and for the wurtzite (1.5^o and 7.4%) surface.

In addressing these large discrepancies, we search for the corresponding minimal energy surface configurations by applying the DF–TB MD to these surfaces using unconstrained annealing techniques followed by a conjugate gradient relaxation. In addition, we investigate a second nonpolar cleavage surface of the wurtzite phase, the (11$\bar{2}$0) surface.

The results for the most relevant geometrical parameters, [10, 5] are summarized in Tables I, II and III, where they are compared with data obtained from *ab initio* calculations [11]. Fig. 1 shows the corresponding relaxed structures. Comparing the geometrical parameters of the GaN (110) surface with those of the GaAs (110) surface, see Table IV, we confirm the significantly smaller layer rotation angle ω and a shortened surface bond for the GaN surface.
In [11] the values for the rotation angle and the bond length are found to be even smaller,

Figure 1: Side view of reconstructed nonpolar GaN (110) (*top*), ($10\bar{1}0$) (*middle*) and ($11\bar{2}0$) (*bottom*) surfaces. Ga (N) atoms are represented by large (small) circles. The main reconstruction mechanism consists of an outward rotation of the surface N atoms. All reconstructions show a relatively small rotation of the surface layer.

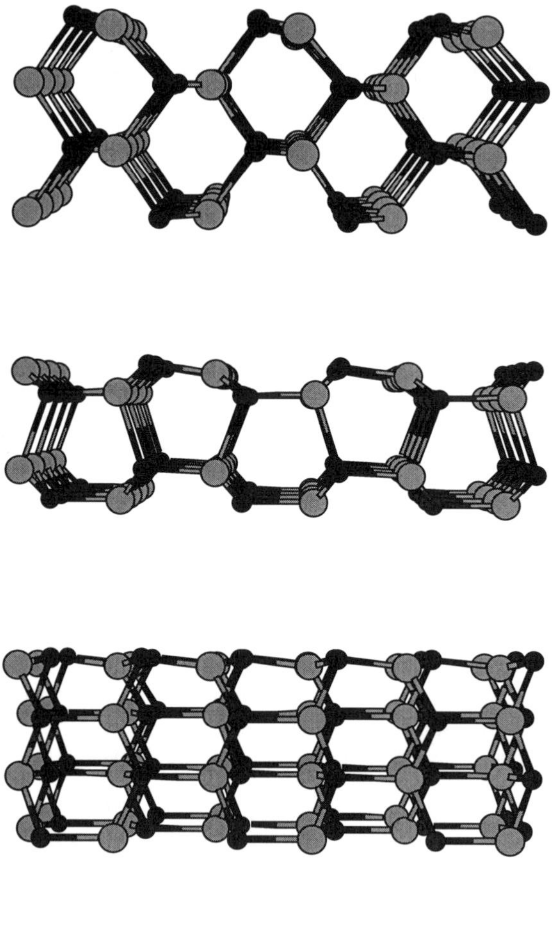

Table III: **GaN wurtzite (11$\bar{2}$0) surface**

	DF-TB (sp–basis)	DF-TB (sp + Ga 3d–basis)
$\Delta_{1,\perp}$	0.24 Å	0.21 Å
$\Delta_{1,y}$	3.35 Å	3.34 Å
$d_{12,y}$	2.70 Å	2.68 Å
$\Delta_{1,x}$	0.14 Å	0.17 Å
ΔR (Ga-N)	-3.3%	-3.2%
ω	7.6^o	6.5^o

Table IV: **GaAs (110) surface**

	DF-TB (sp–basis)	Pseudopotential LDA [8]	LEED [12]
$\Delta_{1,\perp}$	0.62 Å	0.67 Å	0.69 Å
$\Delta_{1,y}$	4.47 Å	4.41 Å	4.52 Å
$d_{12,y}$	3.07 Å	3.34 Å	3.19 Å
$d_{12,\perp}$	1.68 Å	1.42 Å	1.44 Å
ΔR (Ga-N)	-1.8%	–	–
ω	27.6^o	30.2^o	31.1^o

which is attributed to the effects of 3d–electron core relaxation. This means that the Ga 3d and the N 2s levels are energetically very close which may lead to a hybridisation of these levels. We tested the influence of 3d–electron core relaxation effects by taking the Ga 3d orbitals as valence states, but did not find a significant effect on the geometrical parameters of all structures, compare Tables I to III. We therefore postulate that the small layer rotation angle ω is due to a combination of other properties inherent to GaN such as the strong difference in the atomic size and thus electronegativity of Ga and N, which results in a shortening of the Ga–N surface bonds. This may then lead to the observed reduction of ω compared to other III-V semiconductors which are less ionic.

CONCLUSION

We have investigated the main nonpolar cleavage surfaces of GaN in the cubic and hexag onal phase. Although the principal reconstruction mechanisms known from other III-V semiconductors are also established for GaN, we observe some quantitative differences in the reconstructed geometries. These can be summarized in a smaller layer rotation angle and a stronger bond length contraction on the surface.

REFERENCES

1. O. Brandt, H. Yang, B. Jenichen, Y. Suzuki, L. Däweritz and K. H. Ploog, Phys. Rev. B **52** (1995) R2253

2. D. Porezag, Th. Frauenheim and Th. Köhler Phys. Rev. B **51** (1995) 12947.

3. Th. Frauenheim, F. Weich, Th. Köhler, S. Uhlmann, D. Porezag and G. Seifert, Phys. Rev. B **52** (1995) 492.

4. J. Widany, Th. Frauenheim, Th. Köhler, M. Sternberg, D. Porezag, G. Jungnickel and G. Seifert, Phys. Rev. B **53** (1996) 4443.

5. J.P. LaFermina and C.B. Duke, J. Vac. Sci. Technol. **A9** (1991) 1847.

6. C. P. Duke, J. Vac. Sci. Technol. **A6** (1988) 1957.

7. A. Kahn, Surf. Sci. Rept. 3 (1983) 193.

8. A. Alves, J. Hebenstreit and M. Scheffler Phys. Rev. B **44** 6188.

9. C.A. Swarts, W.A. Goddard III and T.C. Mc Gill, J. Vac. Sci. Technol. **17(5)** (1980) 982.

10. Y.R. Wang and C.B. Duke, Surf. Sci. **192** (1987) 309.

11. J. E. Jaffe, R. Pandey and P. Zapol, Phys. Rev. B **53** (1996) R4209.

12. C. Mailhiot, C.B. Duke and D.J. Chadi, Surf. Sci. **149** (1985) 366.

STABILITY AND RECONSTRUCTION OF β-SiC (100) SURFACES

R. Gutierrez, Th. Frauenheim, *
* Theoretische Physik III, Technische Universität, 09107 Chemnitz, Germany

ABSTRACT

We present a detailed investigation of β-SiC (100) surface reconstructions. Minimal energy configurations at both C- and Si-terminations are determined by using a density-functional based molecular-dynamics method. We provide information on the energetic order of various metastable surfaces and describe the geometrical details. In addition, we also study Si- and C-enriched configurations.

INTRODUCTION

During the past decade the properties of cubic and hexagonal polytypes of bulk SiC have been extensively studied by *first principle* investigations [1, 2]. While most studies of surface properties have been performed on a semiempirical level [3, 4, 5], *self-consistent* results by using density-functional (DF) methods became available only recently [6, 7, 8].

On the (110) facets a (1x1) surface symmetry is well established. Wenzien et. al. [6] and Sabisch et. al. [7] report on well relaxed buckled surface cells, which adjust through a top-layer bond-length-contracting rotation moving the Si surface atoms closer towards the substrat while the C-atoms relaxe only parallel to the surface plane. Yan et al. by using the *Car-Parrinello* method, firstly presented an *ab initio* study of the β-SiC (100) surface [8]. They obtain a $c(2\times2)$b reconstruction to be the most stable C-rich surface by bridging triply bonded carbon dimers between neighboring underlying silicon dimers confirming earlier *empirical Tersoff*[9] and *semiempirical* calculations [4, 5]. As most stable structure for a Si-rich (3×2) surface they discuss a 1/3 monolayer of excess silicon in an alternate dimer row structure rather than an added dimer row model proposed by other authors [10].

SIMULATION METHOD

The interatomic potentials and forces for the molecular dynamics simulations are derived by means of a *density-functional* based nonorthogonal *tight-binding* (DF-TB) scheme. The Kohn-Sham orbitals of the system are represented by a *linear combination of atomic orbital* (LCAO)–Ansatz with respect to a minimal basis of the localized valence electron orbitals of all atoms. Within a two-center approach all Hamiltonian and overlap matrix elements are derived in a parameter-free way via the construction of pseudo-atomic orbitals and potentials by self-consistent single-atom calculations using the *local-density approximation* (LDA). Molecular-dynamics simulations for determining the different reconstructions of the SiC surface use the slab method (two-dimensional periodic boundary conditions with oversized periodicty normal to the surface). Each surface layer contains 16 atoms and we employ usually 8 layers for the (100) surface. Since the atoms on the bottom layer have terminating dangling bonds, they were saturated with hydrogen. For simulating approximately the bulk substrate the two deepest lying layers in addition to hydrogen were maintained fixed during the relaxation.

Mat. Res. Soc. Symp. Proc. Vol. 423 © 1996 Materials Research Society

SiC (100) SURFACES

Si-termination

Four possible reconstructions of the (100) surface were investigated: 2×1, 4×1, c(4×2) and p(2×2), as shown in Table **I**. In all cases the building of asymmetric silicon dimers lowers the energy of the system. Since both silicon atoms become chemically inequivalent, charge transfer from the lower to the upper Si-atom takes place, as Mullikan population analysis shows. According to our results the 2×1 reconstruction, see Fig. **1a**, is the most stable one followed by a 4×1-reconstruction, which is only 0.02 eV/dimer higher in energy. At the third place we determine a c(4×2) structure at 0.21 eV/dimer above the ground state and last a p(2×2) reconstruction at 0.33 eV/dimer.

Table I. Energy differences of various reconstructions on the (100) surface.

SiC (100)	Reconstruction	Δ E [eV/Si-C pair]
Si-term.	2x1	0
	4x1	0.021
	c(4x2)	0.21
	p(2x2)	0.33
C-term.	4x2	0
	2x4	0.197
	4x1	0.506
C-rich	2x1-I	0
	c(2x2)b	0.052
	2x1-II	0.34
	c(2x2)s	0.68

The dimer lengths within the different symmetries described above range from 2.4-2.52 Å, which is about 2-6 % larger than the bulk *c*-Si bond length. The obtained r_{Si-Si} = 2.42 Å in the case of the most stable assymmetric 2×1 dimer arrangement with a buckling height of Δz = 0.51 Å almost matches the dimer length and buckling on the corresponding silicon surface (2.41 Å, 0.59 Å) obtained by the present method.[11] With these data we confirm the Tersoff-potential calculations [9, 12] (r_{Si-Si} =2.40 Å 2.46Å) , while ASED-band techniques [4] and SLAB-MINDO results of Craig and Smith,[3] report on smaller dimer lengths, 2.16 Å and 2.33 Å, respectively.

The energy lowering as the result of the 2×1 dimer formation from an unreconstructed 1×1 cell is determined to stabilize the surface by 0.87 eV/dimer compared to 1.31 eV on the corresponding 2×1 Si(100).[11] This value is slightly smaller than the reported ones for symmetric 2×1 dimer reconstructions obtained by applying empirical *Tersoff* potentials (1.02 eV/dimer)[12] (0.89, 1.06 eV/dimer)[9] but gets very close to the 0.90 eV obtained by Yan et al. using the *ab initio* Car-Parrinello method.[8] However, these authors, starting from rows of symmetric dimers in a (2×1) arrangement, after a final relaxation seem to be trapped in a p(2×2) surface cell with very short Si-dimers of 2.26 Å tilt at only Δz = 0.05 Å in a staggered pattern. This structure with a somewhat different geometry is found within our calculations to be 0.33 eV less stable than the 2×1 ground state configuration.

While the c(4×2) (alternating buckled dimers) and the 2×1 reconstructions are confirmed experimentally by LEED, [13] there are no indications, to our knowledge, of a 4×1-structure. Since both 2×1 and 4×1 reconstructions lie very close in energy, changes in the order of stability may occur under real experimental conditions making them indistinguishable at finite temperatures.

In the presence of excess silicon the most stable structures reported on experimentally are 3×2 and 5×2 surface cells[13, 10]. Based on the experimental results, indicating that the excess silicon coverage is 1/3 monolayer on the 3×2 surface, Yan et al. on the *ab initio* level favor an alternate dimer row structure[8] to be by 3.58 eV/ad-dimer significantly more stable than an added dimer configuration. We have checked both structures by using the present method and confirm the proposed energetic order but at a considerably decreased energy difference of 0.54 eV/ad-dimer. For a comparison the most relevant structural parameters are summarized in Table **II** and the structure is shown in Figs. **1**b and c.

The reasons for the obvious deviations in the described geometries and the calculated energy difference for the (3×2) cells from the *ab initio* data can only be speculated. All reconstructions discussed here have been obtained within a complex annealing search for the minimal energy configurations followed by a final conjugate gradient minimization. Using this procedure the ratio in the number of subsurface dimers increases from 2/3 in the originally proposed added dimer reconstruction to 5/6 after relaxation, since three additional dimers develop. This can leads to a further stabilization of the surface and thus to a reduction of the energy difference within our approach.

Table II. Geometrical parameters of (3×2) reconstruction models

(100)-Si-rich surface					
	r^{sur}_{Si-Si}[Å]	r^{sub}_{Si-Si}[Å]	Δz_{sur}[Å]	Δz_{sub}[Å]	method
(3x2)a-alternate dimer row model [8]	2.33	2.45/2.53	0.52	0.52	DF-TB
	2.28	2.60/2.75	0.52	0	*ab initio* [8]
(3x2)b-added row model [10]	2.34	2.60	0.48	0.1	DF-TB
	2.65	2.43	0	0	*ab initio* [8]

C-termination

As in the case of Si-termination we also have to distinguish a stoechiometric (100)-carbon terminated surface and a C-rich (100) surface which have been obtained by either the sublimation of a Si-monolayer[14] or by adsorption of carbon during prolongued exposure to C_2H_2 [15].

Considering first the C-terminated (100) surface, modeled by an even number of Si- and C-layers, we do not have any references available from the literature for comparison. After performing an annealing of a bulk 1×1 structure we obtain a considerable energy lowering by the formation of symmetric carbon dimers. But in contrast to related diamond surfaces an equal partition of two different dimer bond lengths, 1.37 Å and 1.82 Å is established, rather than one as we would have expected. A conjugate gradient relaxation yields finally the three most stable surface cells: 4×2, 2×4 and 4×1, from which the lowest energy 4×2 structure is shown in Figure **1**d, all surfaces developing symmetric carbon dimers at the two different lengths, as shown in Table **III**. The cohesive energy gain for the 4×2-reconstruction amounts 3.16 eV/dimer with respect to the unreconstructed 1×1 surface. The shorter dimers, r^{s}_{C-C} =

1.37 Å are slightly longer than a double bond in ethylene, r_{C-C} = 1.31 Å, getting close to the corresponding dimer length on diamond (100). The other dimers at r^{l}_{C-C}=1.82 Å, however, probably due to the more strained C-C bonding situation on SiC (100), are much longer than the typical single bond, r_{C-C} =1.54 Å, in diamond.

Focusing next on the carbon rich surface, it has been difficult to decide about the nature of the observed c(2×2) structure [16]. Mainly two models have been proposed for explaining the atomic configurations. The first one, c(2×2)s, yields a staggered doubly-bonded C-C dimer structure, where each carbon atom bridges two Si-atoms in the second layer. This model has received support from *semiempirical* SLAB-MINDO calculations by Craig and Smith [17] as well as LEED and Auger electron spectroscopy measurements by Bermudez and Kaplan [16]. In the other c(2×2)b model proposed by Powers et. al. [14] using dynamical calculations of LEED intensities, C-C dimers are located as bridges on top between (2×2) staggered symmetric Si-dimers in the subsurface layer. According to the authors the Si-dimer formation depends on the surface preparation. Yan et. al. [8], by using *ab initio* Car-Parrinello methods, determine the c(2×2)b as being energetically favoured over the staggered carbon dimer model by 0.34 eV/dimer in LDA and 0.68 eV/dimer on the GGA-level. This is in support of Badziag's earlier results obtained within the MNDO-PM3 approach [5].

Table III. Geometrical parameters of (100) C-terminated surfaces.

(100) C-term. surface					
	r^{l}_{C-C}[Å]	r^{s}_{C-C}[Å]	r^{l}_{C-Si}[Å]	r^{s}_{C-Si}[Å]	
4x2	1.819	1.371	1.908	1.894	
(100) C-rich surface					
	r_{C-C}[Å]	r_{Si-Si}[Å]	r_{Si-C}[Å]	∠(Si-C-C) [°]	method
(2x1)-I	1.226	2.296	1.847	136	DF-TB
c(2x2)b	1.227	2.316	1.85	135.5	DF-TB
	1.22	2.381		133	non-scf-TB [18]
	1.21	2.55			MNDO-PM3 [5]
c(2x2)s	1.369	-	1.867	113.2	DF-TB
	1.39	-	1.91	106	DFT-CP [8]
	1.49				Tersoff pot. [12]

In order to verify these results, we have investigated different carbon rich surfaces with c(2×2) and 2×1 symmetry. In modeling these surfaces we were using a slab with 9 atomic layers, i.e. a SiC(100)-Si-terminated surface with an additional carbon monolayer on the top. We establish the c(2×2)b reconstruction with staggered C-dimers in Si-bridging positions, see Fig. **1**e, to be a low energy configuration, confirming the results of Yang et. al. [8] and Badziag [5]. The obtained dimer length of 1.22-1.23 Å clearly indicates developing triple bonds, compare Table **III**. The subsurface Si-dimer length adjusts at 2.31 Å, compared to 2.55 Å(MNDO-PM3)[5] and 2.38 Å(non-scf TB)[18], while dynamical calculations of LEED intensities yield very long 2.71 Å.[14] A 2×1-reconstruction without additional

Si-dimerization in the subsurface layer is about 0.29 eV/dimer higher in energy than the c(2×2)b-structure, supporting the result of Badziag [5]. The C-C dimer length of 1.369 Å in the 2×1 configuration indicates similar double bonding as obtained on the 2×1 (100) diamond surface and on the C-terminated SiC(100) 4×2 surface described above.

Comparing the 2×2 models the c(2×2)s surface cell lies about 0.63 eV/dimer higher in energy than the c(2x2)b configuration, compare Table **III**. While the structural parameters of both models, compare Table **III**, agree quite well with the *ab initio* results of Yan et. al. [8], our determined energy difference between the bridged and staggered configuration is almost the same as obtained on the GGA-level. However, the slight asymmetries in the C dimers and the Si subsurface layer, discussed by Yan et al. cannot be confirmed.

Additionally to the surface reconstructions discussed so far, we have found a new even more stable C-rich 2×1-structure, assigned as 2×1-I, which to our knowledge has not been described earlier and is shown in Fig. **1**f, too. It consists of symmetric triply-bonded C-dimer rows with r_{C-C} =1.226 Å alternating with symmetric Si-Si dimer rows in the subsurface layer with r_{Si-Si} =2.29 Å. However, the energy gain with respect to the c(2×2)b-configuration is only 0.05 eV/dimer.

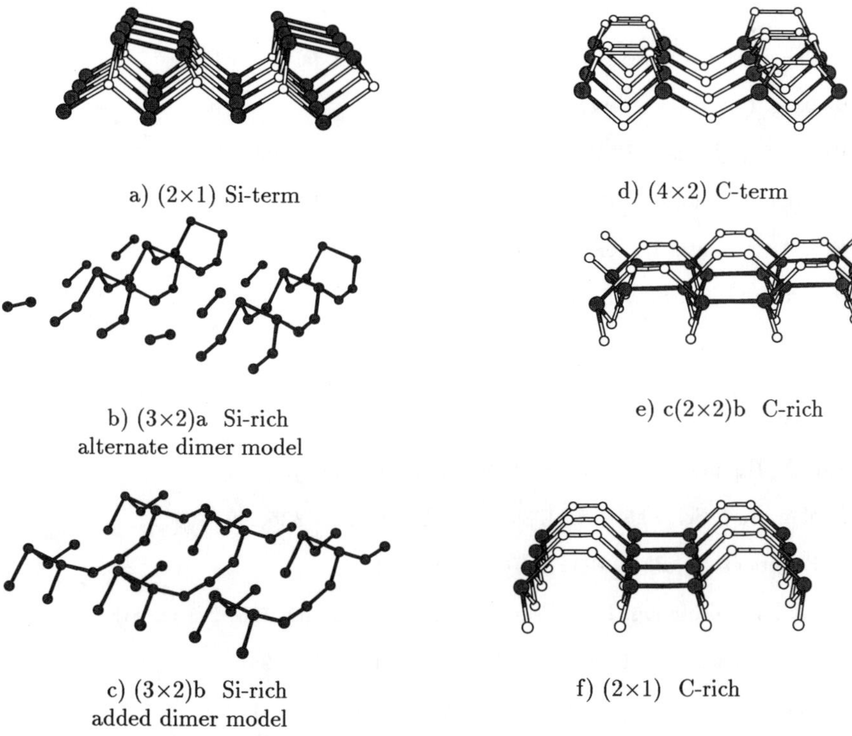

a) (2×1) Si-term

d) (4×2) C-term

b) (3×2)a Si-rich
alternate dimer model

e) c(2×2)b C-rich

c) (3×2)b Si-rich
added dimer model

f) (2×1) C-rich

Fig. 1. Lowest energy configurations of SiC(100).

CONCLUSIONS

Detailed investigations on SiC (100) surface reconstructions has been carried out using a density-functional based tight-binding scheme. The surfaces are stabilized by the formation of asymmetric and symmetric dimer rows at Si- and C-termination, respectively. Concerning the enriched (100) surfaces , our method favours for the Si-rich surface with (3×2) symmetry the alternate dimer row model. For C-rich surfaces the c(2×2)b model is favoured over the c(2×2)s pattern and a new configuration with (2×1) symmetry was found consisting of C-dimer rows alternating with subsurface Si-dimer rows.

REFERENCES

1. C. Cheng, R. J. Needs, V. Heine, J. Phys. C **21**, 1049 (1988).
2. W. R. L. Lambrecht, B. Segall, M. Methfessel, M. van Schilfgaarde, Phys. Rev. B **44**, 3685 (1991).
3. B. I. Craig and P. V. Smith, Surf. Sci. **233**, 255 (1990).
4. S. P. Mehandru and A. B Anderson, Phys. Rev. B **42**, 9040 (1990).
5. P. Badziag, Phys. Rev. B **44**, 11143 (1991).
6. B. Wenzien, P. Käckell, F. Bechstedt, Surf. Sci. **307**, 989 (1994) .
7. M. Sabisch, P. Krüger, J. Pollmann, Phys. Rev. B **51**, 13367 (1995).
8. H. Yan, A. P. Smith, H. Jonsson, Surf. Sci. **330**, 265 (1995).
9. T. Halicioglu, Phys. Rev. B **51**, 7217 (1995) .
10. S. Hara et. al., Surf. Sci. **231**, L196 (1990).
11. Th. Frauenheim, F. Weich, Th. Köhler, S. Uhlmann, D. Porezag, G. Seifert, Phys. Rev. B **52**, 11492 (1995).
12. H. Yan, X. Hu and H. Jonsson, Surf. Sci. **316**, 181 (1994).
13. R. Kaplan, Surf. Sci. **215**, 111 (1989). Thin Solid Films **225**,240 (1993).
14. J. M. Powers et. al., Phys. Rev. **B40**, 11159,(1991).
15. T. Fuyuki, T. Yoshinobu, H. Matsunami, Thin Solid Films **225**, 225 (1993).
16. V. M. Bermudez and R. Kaplan, Phys. Rev. B **44**, 11149 (1991).
17. B. I. Craig and P. V. Smith, Surf. Sci. Lett. **256**, L609 (1991).
18. X. Hu, H. Yan, M. Kohyama and F. S. Ohuchi, J. Phys.: Condens. Matter **7**, 1069,(1995).

A NEW, NEARLY SINGLE-DOMAIN SURFACE STRUCTURE OF HOMOEPITAXIAL DIAMOND (001) FILMS

YALEI KUANG, NAESUNG LEE*, ANDRZEJ BADZIAN*, and TERESA BADZIAN*
Department of Physics and Materials Research Laboratory*, The Pennsylvania State University, University Park, PA 16802

TIEN T. TSONG
Institute of Physics, Academia Sinica, Taipei, Taiwan 11529, Republic of China

ABSTRACT

Boron-doped homoepitaxial diamond films were grown on natural diamond (001) substrates using microwave-assisted plasma chemical vapor deposition techniques. The surface structures were investigated using scanning tunneling microscopy (STM). This showed a dimer-type 2x1 reconstruction structure with single-layer steps where dimer rows on the upper terrace are normal to or parallel to the step edges. We found that dimer rows parallel to the step edges are much longer than those normal to the step edges. The nearly single-domain surface structure observed by STM is in agreement with the low-energy electron diffraction (LEED) patterns from these surfaces. The high atomic resolution STM image showed that the local 1x1 configurations exist.

INTRODUCTION

Recently, great efforts have been made to investigate homoepitaxial diamond (001) films grown using chemical vapor deposition (CVD). These films have smooth surfaces and are nearly mono-crystalline [1-7]. An observation on an atomic scale of the homoepitaxial diamond (001) films is highly desirable for better understanding the growth mechanism. For this purpose, scanning tunneling microscopy (STM) with atomic resolution has been used to investigate the surface structure of CVD-grown diamond (001) films. A 2x1 unit cell has been previously observed in both ultra-high vacuum (UHV) [8] and air [9]. Both single-layer (S_A, S_B) and double-layer (D_A, D_B) steps have been found [9-12]. Antiphase boundaries are also present on diamond (001) surface [13,14]. Earlier STM observations showed a nearly perfect single-B-domain surface which is formed by B-type (1x2) terraces on diamond (001) surface [12]. Similar single-B-domain structures have been observed on the molecular beam epitaxy (MBE) grown and annealed Si(001) surface [15,16]. In this paper, however, we report on the observation of a nearly single-A-domain surface which is formed by A-type (2x1) terraces rather than B-type (1x2) terraces using STM and LEED in UHV. Local 1x1 configurations have been observed.

EXPERIMENT

The homoepitaxial growth was carried out on high-pressure high-temperature (HP/HT) type-Ia natural single crystalline (001) diamond substrates by microwave-assisted plasma CVD. The misorientation of the substrates was determined by x-ray diffraction before deposition. A gas mixture of 1% methane in hydrogen gas was used for deposition. The gas flow rate was 100 sccm. 100 ppm diborane in hydrogen with the gas rate of 5 sccm was used for boron doping during growth. The growth occurred for 1 hour at the pressure of 80 Torr. The substrate temperature was 875 °C, measured by an optical pyrometer. Following the deposition, samples

Mat. Res. Soc. Symp. Proc. Vol. 423

were annealed with pure hydrogen plasma for 5 minutes for etching treatment while the temperature was kept at growth temperature. The samples were then transferred to an UHV chamber equipped with a STM and LEED system from CVD chamber through air. The base pressure of the UHV chamber was maintained at 1.0×10^{-10} Torr while STM and LEED observations were made. STM observations were performed at room temperature in both constant height and constant current modes. After STM observations, the samples were transferred from the sample holder of the STM to that of LEED apparatus for LEED observations.

RESULTS

In order to analyze the morphology of the diamond (001) surface, we acquired STM images of the same film in the different regions in a 1.5 x 1.5 mm^2 area. Figure 1 shows a typical 29.4 x 29.4 nm^2 STM image of the diamond (001) film. It shows a dimer-type 2x1 reconstruction structure with clearly resolved dimer rows. The adjacent dimer rows are spaced 0.50 nm apart. The surface consists of A-type (2x1) terraces and B-type (1x2) terraces separated by single-layer steps, marked A and B. The dimer rows on the alternating terraces change their orientation by 90°. The single-layer steps can be seen, marked S_A and S_B, where dimer rows on the upper terraces are parallel to or normal to the step edges, respectively. The step-down direction is from the lower right to the upper left in the [110] direction. As shown in figure 1, the dimerized surface structures comprise A-type (2x1) terraces with long dimer rows in the [$\bar{1}$10] direction and B-type (1x2) terraces with short dimer rows in the [110] direction. The distribution of the terraces is not uniform. A-type (2x1) terraces dominate the surface, with average width 3.15 nm in the [110] direction, while the average width of B-type (1x2) terraces in the same [110] direction is 1.14 nm. By measurement of the average terrace width, the calculated misorientation of this surface is 2.38° toward [110] direction in agreement of the misorientation of the substrate measured by x-ray diffraction. Most of the STM images we obtained from this surface have the same features as above. However, as shown in figure 2, in some regions, we observed an even distribution of A-type (2x1) and B-type (1x2) terraces. The diamond (001) surface exhibits remarkably irregular domains locally in contrast to the regular atomically flat domains on the Si(001) surface. Since diamond (001) substrates are much rougher than Si(001) surfaces, it is not

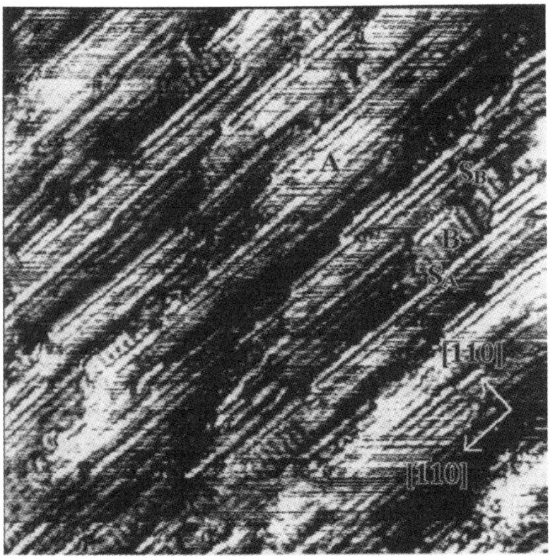

5.0 nm

Fig. 1 A typical STM image of the diamond (001) film grown on the 2.3° off (001) diamond substrate, with 2.3° in the [110] and 0.2° in the [$\bar{1}$10] directions showing nearly single-A-domain structure (sample bias: 860 mV; tunneling current: 0.6 nA).

surprising to observe different features locally on the same surface of the diamond (001) films.

Figure 3 shows a LEED 2x1 pattern of this surface. From the observation of surface

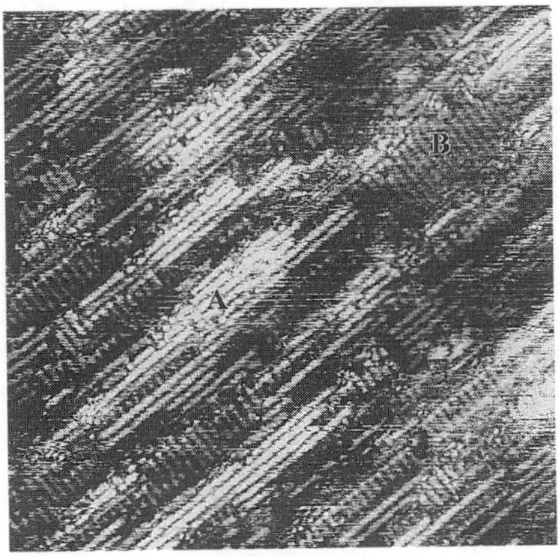

5.0 nm

Fig. 2 STM image of the diamond (001) film showing local even distribution of A-type (2x1) and B-type (1x2) terraces (sample bias: 750 mV; tunneling current: 0.5 nA).

Fig. 3 A LEED pattern of the homoepitaxial (001) film exhibiting 2x1 single-domain surface structure. The electron energy is 170 eV.

structures using STM, we can identify the crystallographic orientations of the sample which can be used to determine the dimerization directions corresponding to the relative A-type (2x1) and B-type (1x2) terraces. It is noted that the half-order spots from A-type (2x1) terraces are much bright while those from B-type (1x2) terraces completely disappear, indicating that the dimerization is mainly parallel to the [110] direction. This result agrees with the observations using STM. However, our typical STM image of figure 1 showed that the steps are not D_A double-layer steps on our nearly single-domain surface where A-type (2x1) terraces dominate the surface. Rather than observation of local D_A steps, there exist short B-type (1x2) terraces between adjacent A-type (2x1) terraces. Our observations are different from the results obtained by Tsuno et al. [12], where the single-domain diamond (001) surface is dominated by B-type (1x2) terraces. In that case, there was no evidence for A-type (2x1) terraces between adjacent B-type (1x2) terraces. From these STM and LEED measurements, we conclude that this surface has a nearly single-A-domain surface structure.

The evidence of anisotropy on the CVD-grown diamond (001) surface have been discussed by Lee and Badzian [17]. They observed different surface structures of H-plasma annealed and as-grown diamond (001) surface using reflection high-energy electron diffraction. The surface structures of the as-grown 3.1° off (001) diamond film annealed in the H-plasma were converted from the single-B-domain to single-A-domain. Tsuno et al. concluded that the formation of the single-B-domain surface on the as-grown 4.3° off (001) diamond film is caused by step flow growth with higher growth rate of S_B steps. Lee and Badzian proposed that step formation energies are very different for the H-plasma annealed vs. the as-grown diamond (001) surface.

Recently, single-A-domain surfaces have been obtained on Si(001) surface annealed by Xe sputtering [18]. Vacancy annihilation is considered to play an important role on the formation of single-A-domain surfaces. But, on the diamond (001) surface, there are far fewer vacancies on the terraces than on the Si (001) surface [9,19]. We propose that the formation of anisotropic domains with dominant A-type (2x1) on the H-plasma annealed diamond (001) films is caused by a different etching rate on the S_A and S_B step edges. The rate of the detachment of hydrocarbons on the SB step edges are higher than those on the S_A step edges. So, retraction of S_B steps results in the formation of the single-A-domain surface.

Figure 4. is a high resolution STM image of a homoepitaxial diamond (001) film. Individual dimers are well resolved, which are about 0.25 nm apart from each other along the dimer rows. It is interesting to note that the region, marked in a box, is not reconstructed. The distance between adjacent bright spots are measured to be 0.25 nm, which is in agreement with theoretical distance between nearest carbon atoms on the unreconstructed diamond (001) surface. Thus, this region is considered to be 1x1 configurations. It has been reported [20] that homoepitaxially grown diamond (001) surface is composed of the monohydride phase with 2x1 structure, where two hydrogen atoms saturate two dangling bonds of a dimer on the surface. In our STM observations, 2x1 structures are dominant on the diamond (001) surface. Since the samples were annealed in hydrogen plasma, it is highly probable that the 2x1 structure is with a monohydride configuration, and the 1x1 structure is with a dihydride configuration Schematics of 2x1 and 1x1 structures on the diamond (001) surface are shown in figure 5.

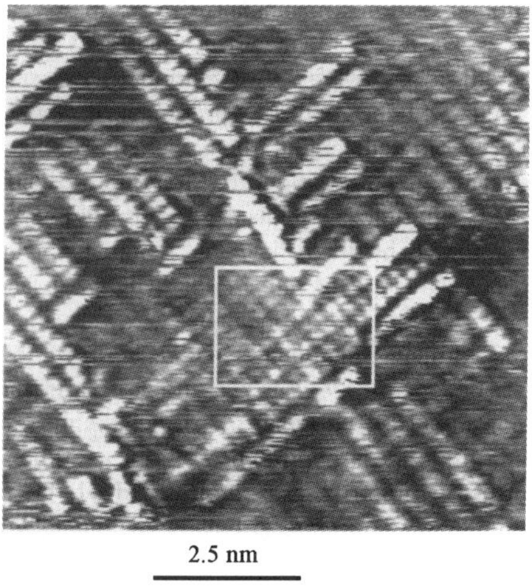

Fig. 4 A high resolution STM image of the diamond (001) film showing local 1x1 structure (sample bias: 800 mV; tunneling current: 0.5 nA).

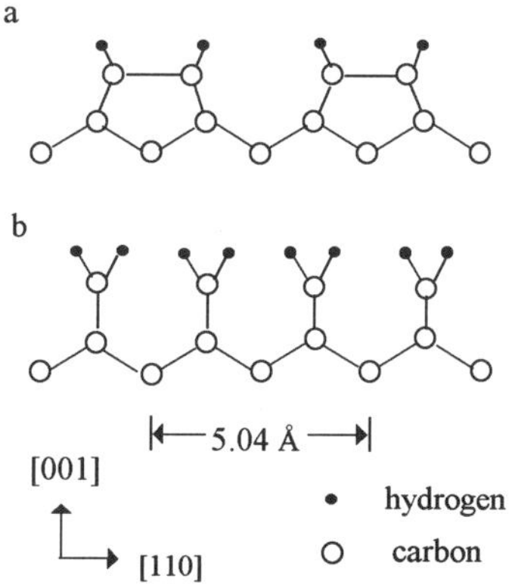

Fig. 5 Structural models of side views of hydrogen-terminated 2x1 and 1x1 structures on diamond (001) surface. (a) 2x1 structure. (b) 1x1 structure.

CONCLUSIONS

In summary, a new, nearly single-A-domain 2x1 surface on homoepitaxial diamond (001) films has been observed using UHV STM and LEED. The dimer-type 2x1 reconstructed surface is composed mainly of A-type (2x1) terraces while very short B-type (1x2) terraces are present between the adjacent A-type (2x1) terraces. We concluded that the different rates of the detachment on the S_A and S_B steps leads to the anisotropy of the diamond (001) surface. The local 1x1 configurations have been found on the mainly 2x1 reconstructed diamond (001) surface.

ACKNOWLEDGMENTS

The authors wish to express their thanks to Dr. R. Diehl for discussion about STM and LEED results. This work was supported in part by National Science Foundation under Grant No. DMR 9117061 and DMR 9522566, Office of Naval Research BMDO Contract N0000-95-I-0905, and Republic of China National Science Council Grant No. 84IA01B032.

REFERENCES

1. M. Kamo, H. Yurimoto, and Y. Sato, Appl. Surf. Sci. **33/34**, 553 (1988).
2. H. Shiomi, K. Tanabe, Y. Nishibayashi, and N. Fujimori, Jpn. J. Appl. Phys. **29**, 34 (1990).
3. W. J. P. van Enckevort, G. Janssen, and L. J. Giling, J. Cryst. Growth **113**, 295 (1991).
4. H.G. Maguire, M. Kamo, H.P. Lang, and H.-J. Güntherodt, Appl. Surf. Sci. **60/61**, 301 (1992).
5. L. F. Sutcu, M. S. Thompson, C. J. Chu, R. H. Hauge, J. L. Margrave, and M. P. D'Evelyn, Appl. Phys. Lett. **60**, 1685 (1992).
6. S. J. Harris and D. G. Goodwin, J. Phys. Chem. **97**, 23 (1993).
7. A. Badzian and T. Badzian, Diamond Relat. Mater. **2**, 147 (1993).
8. R. E. Stallcup, A. F. Aviles, and J. M. Perez, Appl. Phys. Lett. **66**, 2331 (1995).
9. Y. Kuang, Y. Wang, N. Lee, A. Badzian, T. Badzian, and T.T. Tsong, Appl. Phys. Lett. **67**, 2721 (1995).
10. T. Tsuno, T. Imai, Y. Nishibayashi, K. Hamada, and N. Fujimori, Jpn. J. Appl. Phys. **30**, 1063 (1991).
11. H.-G. Busmann, and W. Zimmermann-Edling, H. Sprang, H.-J. Güntherodt, and I.V. Hertel, Diamond Relat. Mater. **1**, 979 (1992).
12. T. Tsuno, T. Tomikawa, S. Shikata, T. Imai, and N. Fujimori, Appl. Phys. Lett. **64**, 572 (1994).
13. Y. Kuang, N. Lee, A. Badzian, T.T. Tsong, T. Badzian, and C. Chen, Diamond Relat. Mater. **4**, 1371 (1995).
14. H. Kawarada, H. Sasaki, and A. Sato, Phys. Rev. B **52**, 11351 (1995).
15. A.J. Hoeven, J.M. Lenssinck, D. Dijkkamp, E.J. van Loenen, and J. Dieleman, Phys. Rev. Lett. **63**, 1830 (1989).
16. J.E. Griffith, G. P. Kochanski, J.A. Kubby, and P.E. Wierenga, J. Vac. Sci. Technol. A7, 1914 (1989).
17. N. Lee and A. Badzian, Phys. Rev. B **53**, 1744 (1996).
18. P. Bedrossian and Tklistner, Phys. Rev. Lett. **68**, 646 (1992).
19. R.J. Hamers, R. M. Tromp, and J.E. Demuth, Phys. Rev. B**34**, 5343 (1986).
20. T. Ando, T. Aizawa, K. Yamamoto, Y. Sato and M. Kamo, Diamond Relat. Mater. **3**, 975 (1994).

COMPUTER SIMULATION OF SURFACE DIFFUSION OF SILICON AND CARBON ADATOMS ON SiC(001)

Q.A. Bhatti, G.J. Moran and C.C. Matthai
Department of Physics and Astronomy,
University of Wales College of Cardiff, P.O. Box 913,
Cardiff CF2 3YB, Wales, UK

ABSTRACT

We have performed molecular dynamics simulations of adatom diffusion on the SiC(001) surface and found that the barriers for carbon adatoms is less than that for silicon adatoms. The diffusion paths were also found to be temperature dependent and at high temperatures the adatom diffusion constant was found to of the order of $10^{-5} cm^2 s^{-1}$.

INTRODUCTION

Silicon Carbide (SiC) has attracted much interest in recent years because it is a wide gap material displaying a high breakdown field and high electron saturation drift velocity. This coupled with the recent advances in growth techniques of SiC has made it a strong candidate material for high-temperature, high-power devices. In order to grow high quality single crystal wafers of SiC, one needs to have a good understanding of the dynamics of deposited atoms on the crystal surface. Molecular dynamics (MD) is a technique that can give detailed information about adsorption and diffusive motions [1]. Theoretical studies to date of the SiC(001) surface have concentrated on the static properties [2, 3] and little or no work has been done on the dynamics of adatoms. However, with new experimental results [4] on surface diffusion in this system, it is important to undertake theoretical studies of quantities which can be related directly to experimental measurements.

CALCULATIONAL PROCEDURE

Because of the technological interest of the (001) face of cubic SiC, we have focussed our attention to this surface. The substrate was constructed with (110), $(1\bar{1}0)$ and (001) block vectors with a lattice parameter of 4.29Å. Periodic boundary conditions were applied in the (110) and $(1\bar{1}0)$ directions and the substrate held by a fixed layer of atoms at the bottom. Both the C- and Si- terminated surfaces reconstruct into a (2×1) structure with the surface atoms forming dimer chains (Figure 1). The adsorption sites of the C and Si adatoms were determined by placing the adatom on the surface and allowing the whole system to relax according to standard MD techniques. The adatom was constrained to move only in the direction perpendicular to the surface plane. By plotting the minimum energies for each lateral position of the adatom, a surface potential energy contour map may be constructed and from this, both the adatom adsorption sites (troughs in the contour plot) and static activation energies for surface diffusion (the heights of the barriers between troughs) may be obtained.

Mat. Res. Soc. Symp. Proc. Vol. 423 © 1996 Materials Research Society

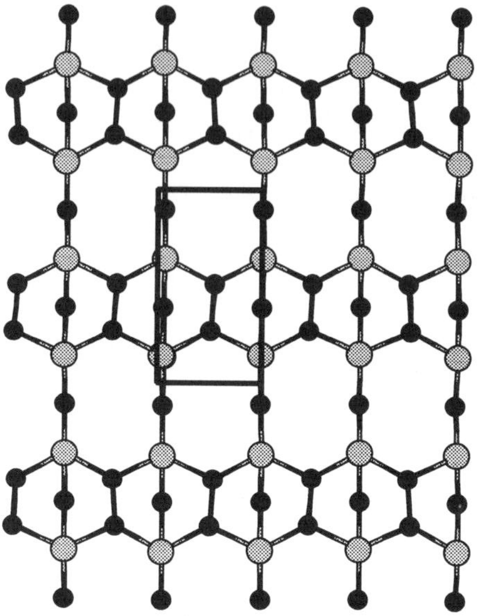

Figure 1: Area scanned on the (2×1) reconstructed SiC(001) surface to create the energy contour map. The light and dark shaded atoms are Si and C atoms respectively.

The MD method requires a knowledge of the forces on each atom. Although it is now possible to perform *ab initio* calculations to extract these forces, such calculations are rather computer intensive. Semi-empirical interatomic potentials have been shown to give a good description of the surface reconstructions found in semiconductor structures. We have therefore chosen to use the coordination dependent potential of Tersoff [5] in performing the calculations described below.

The dynamic surface diffusion properties (activation energies, diffusion paths and diffusion constants) of the adatoms were also obtained by the application of the MD method. In this case, an adatom was initially placed at or near a minimum energy site and the whole system (or just the adatom) given some kinetic energy (temperature). The system was then allowed to evolve according to the equations of motion. The diffusion path is obtained by simply charting the path taken by the adatom on the substrate surface. The activation energy for diffusion can be obtained using the method of Matthai [6] or from a knowledge of the temperature dependence of the diffusion coefficient.

The diffusion coefficient, D, can be calculated from the mean square displacement which is valid at long time (t) scales and is given by [7]

$$D = \frac{1}{6t}||(r(t) - r(0)||^2 \quad (1)$$

Information on the dynamics of the diffusing atom can be also be obtained from the velocity autocorrelation funciton (VACF). One of the advantages of using the MD method is that the VACF is relatively easy to calculate from a knowledge of the velocities of the diffusing atom. The diffusion coefficient is given by

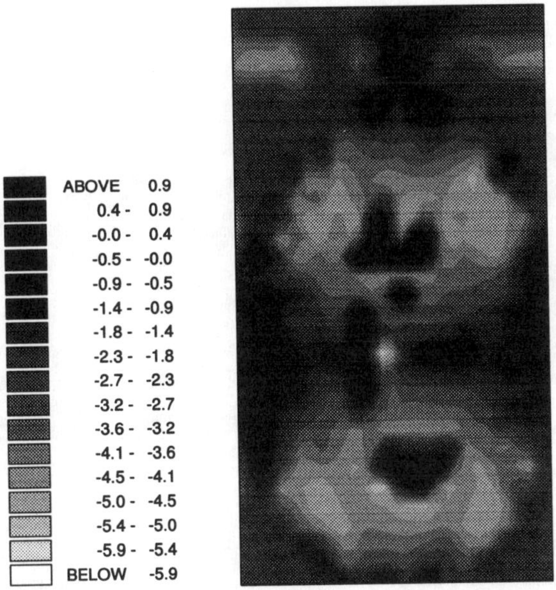

Figure 2: Contour energy map for a carbon adatom on a C-terminated surface over the area scanned in Figure 1. The energy scale in eV is shown on the left. Note that the energy minimum is atop a dimer.

$$D = \frac{1}{3}\int_0^\infty v(\tau)v(\tau + t))dt \qquad (2)$$

It has been shown that the value of D given by these two methods differ by 10-15% for between 1000 and 2000 time steps, but reduces to less than 2% for 10^4 time steps.

RESULTS

Once the relaxed reconstructed surfaces were determined, Si and C adatoms were placed onto the surface and energy contour maps produced. On the C-terminated surface, Si adatoms were found to be in a minimum energy site when it was a bulk continued site (ie, atop a carbon dimer). The C-dimers were found to open up to 2.53Åas compared to their original separation of 1.46Å. Interestingly, this was also the favoured site for the C-adatom (see Figure 2), as the three carbon atoms form a trimer structure. From an analysis of the energy contours, there are two possible diffusion paths, one being along the channel between dimer rows and the other over the dimer rows. As the preferred adsorption site is atop the dimer, the diffusion barrier of the latter path is the relevant one and this was found to be 0.67eV for the C-adatom and 0.98eV for the Si-adatom. Similar results were found for the Si terminated surface.

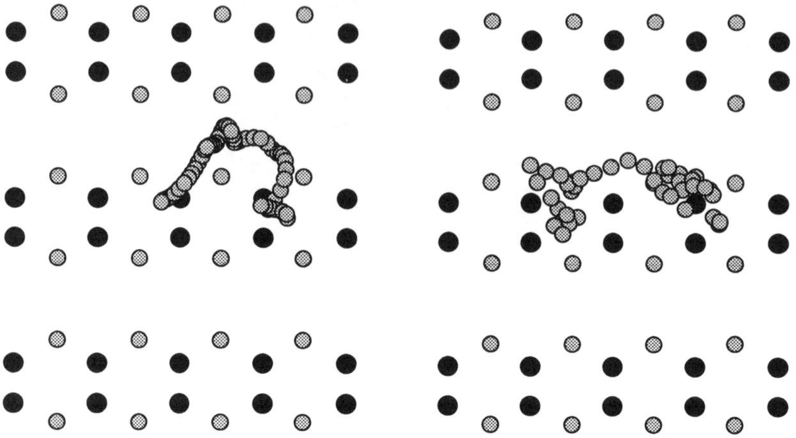

Figure 3: Silicon adatom diffusing on the C-terminated (001) surface at two different temperatures. The light and dark shaded atoms are Si and C atoms respectively. The multiple Si atom images represent the path taken by the diffusing atom.

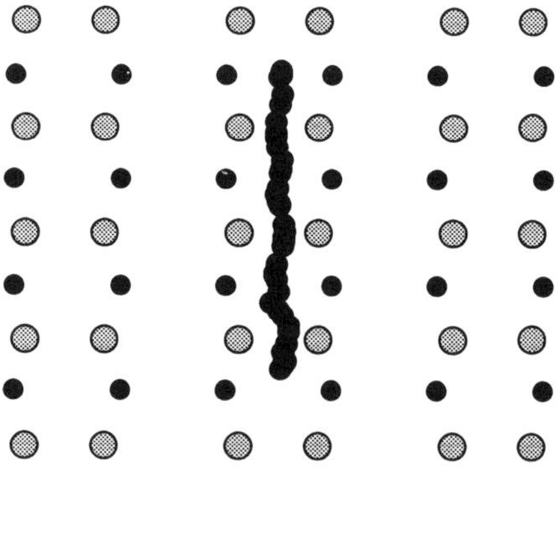

Figure 4: Carbon adatom diffusing on the Si-terminated (001) surface. The symbols are the same as in Figure 3

The dynamic diffusion properties were obtained by giving the adatom a velocity in the surface plane scaled to the kinetic energy of a particle at temperature T. The system was then allowed to evolve dynamically without quenching. The position of the carbon adatom was monitored during the simulation and if the adatom had not moved to another site after a fixed number of time steps, the temperature was increased until the atom had enough energy to overcome the barriers. Because the velocity of the adatom is random in direction, it can move along any one of the available paths, choosing the one that presents the smallest barrier. To ensure that the energy given to the atom did not heat up the substrate and melt it, a dissipating boundary condition at the bottom of the cell was maintained. The simulations were carried out for between 8 and 20 ps.

The diffusion paths obtained were found to be dependent on the temperature and in general agreement with that expected on the basis of the static energy contour maps. At the lower temperatures where diffusion was possible (1000K - 1200K), the adatom was found to move from its initial position atop a dimer into the region between the dimer rows before returning to an adjacent minimum energy site as its energy is lost (the left picture in Figure 3). If the temperature is further increased, the adatom moves along the channel between the dimer rows before settling down to its final position atop a dimer (the right picture in Figure 3). For even higher temperatures (1350K), the adatom diffuses directly along and over the dimer rows from one minimum energy site to another (Figure 4). This temperature dependent behaviour was found for both Si and C adatoms, with the only difference being in the actual temperature when diffusion occured.

From equations 1 and 2, we have been able to estimate the diffusion constant for the high temperature diffusion process. At the lower temperatures, the number of diffusion steps were too few for the results to be meaningful. The diffusion constant at 1350K was found to be $5 \times 10^{-5} cm^2 s^{-1}$ for the silicon diffusion and $1.2 \times 10^{-4} cm^2 s^{-1}$ for carbon adatoms. As the diffusion constant is related to the diffusion activation energy

$$D = D_0 \exp \frac{-E_a}{kT} \tag{3}$$

the values are consistent with the activation barrier being higher for silicon compared to carbon. The value of D is also consistent with the measured surface diffusion lengths of Kimoto and Matsunami [4].

CONCLUSION

We have performed molecular dynamics simulation to detemine the diffusion parameters for silicon and carbon adatoms on the SiC(001) (2×1) reconstructed surface. The diffusion paths were found to along the channels between the dimers at lower temperatures and over the dimers at higher temperatures. The diffusion coefficients thus obtained are consistent with the activation energies for Si diffusion being greater than that for C diffusion which is in agreement with the calculated activation energies.

ACKNOWLEDGEMENTS

We thank Jonathan Crombie for valuable assistance. QAB acknowledges financial support from the Government of Pakistan.

References

[1] D. Srivastava and B. J. Garrison, *Phys. Rev. B*, **47**, 4464 (1993)

[2] S. P. Mehandru and A. P. Anderson, *Phys. Rev. B*, **42**, 9040 (1992)

[3] T. Halicioglu, *Phys. Rev. B*, **51**, 7217 (1995)

[4] T. Kimoto and H. Matsunami, *J. Appl. Phys.*, **78**, 3132 (1995)

[5] J. Tersoff, *Phys. Rev. B*, **39**,5566 (1989)

[6] C. C. Matthai, *Philos. Mag. A*, **52**, 305 (1985)

[7] P. W. M. Jacobs, Z. A. Rycerz and J. Moscinski, *Adv Solid State Chem*, **2**,113 (1991)

HIGH TEMPERATURE GAS PHASES REACTIONS OF TRIMETHYLGALLIUM WITH AMMONIA AND TRIMETHYLAMINE

A. Thon, S.A. Safvi, T.F. Kuech
Department of Chemical Engineering, University of Wisconsin, 1415 Engineering Drive, Madison, WI 53706.

ABSTRACT

The use of trimethylgallium-trimethylamine (TMG:TMN) adduct as alternative cation precursor for MOVPE of GaN was studied by means of *in-situ* mass spectroscopy in an isothermal flow tube reactor. The temperature, pressure and reaction time were chosen to emulate the gas phase environment typical of the metal-organic vapor phase epitaxy (MOVPE) of GaN. Dynamic changes in the mass spectra are reported for the gas phase reactions between trimethylgallium (TMG) and TMN in hydrogen (H_2 / D_2) and ammonia (NH_3 / ND_3) ambients. Evidence is presented for the high temperature TMG:TMN adduct formation, followed by ethane elimination. The strength of the adduct bonding is comparable to that of TMG:NH_3 and thus suppresses TMN displacement by ammonia. The thermal stability of TMG:TMN was found to be higher in ammonia ambient than it is in hydrogen. Kinetic parameters for some of the decomposition processes are given.

INTRODUCTION

The emergence of metal organic vapor phase epitaxy (MOVPE) as the major technique for the growth of GaN for device structures[1], has resulted in an increased activity in understanding the growth process. The growth of device-quality GaN is complicated by gas phase interactions between the conventional Ga source, trimethylgallium (TMG) and the nitrogen source, ammonia [2]. These interactions can lead to changes in the type of and gas phase depletion of growth nutrients, leading to a degradation in the growth uniformity, quality and efficiency[3]. The main gas phase reaction is a strong adduct formation between NH_3 and TMG[4,5]. Other nitrogen sources, such as trimethylamine (TMN), when mixed with TMG, will also form adducts, but the nature of the adduct, specifically the Ga:N adduct bond strength and the molecular structure, will be different [6]. The decomposition temperature and reaction mechanisms of the TMG:TMN adduct may be more suitable for GaN growth by MOVPE. Ammonia would be required, as the primary nitrogen source, to have a high V/III ratio in order to stabilize the growth front. The interaction between NH_3 and the TMG:TMN adduct will determine whether such adduct would be beneficial for the GaN MOVPE process. In this study, we have directly monitored the gas phase reactions between TMG, TMN, and NH_3, in order to better understand their impact on the growth process. These reactions have been monitored over the temperature range of 200-800°C, which is typically encountered in the gas phase environment of the MOVPE reactor as the growth nutrients are transported from the inlet to the growth front.

The pyrolysis of TMG [7,8] and ammonia [9] have been studied and their kinetic parameters are available. To the best of our knowledge, kinetic parameters of homogeneous gas phase decomposition of TMN are not available in literature.

Mat. Res. Soc. Symp. Proc. Vol. 423

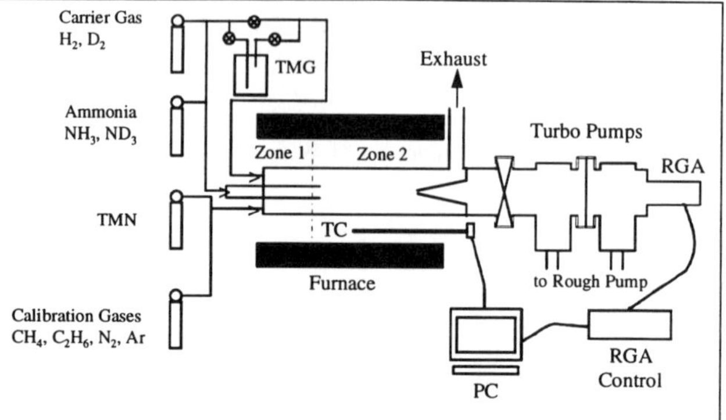

Figure 1: Schematic of the experimental system

At low temperatures, a Lewis acid - Lewis base interactions between TMG (electron acceptor) and NH_3 (electron donor) lead to the intermediate adduct compound, Trimethylgallium monoamine [TMG:NH_3][6,10,11,12]:

$$(1) \qquad (CH_3)_3Ga + NH_3 \Leftrightarrow (CH_3)_3Ga\!:\!NH_3$$

The structure and thermodynamic of this adduct, at low temperatures, have been intensely studied[10,11,12]. Recent studies[11] show that above 90°C one methane molecule per one Ga atom is eliminated, and a six member ring Cyclo(triammido-hexamethyltrigallium) $[(CH_3)_2Ga:NH_2]_3$ is formed in the reaction:

$$(2) \qquad 3\big((CH_3)_3Ga\!:\!NH_3\big) \Rightarrow \big[(CH_3)_2Ga\!:\!NH_2\big]_3 + 3CH_4 .$$

This overall reaction has been suggested to proceed in two steps. The first step is the hydrogen-elimination reaction with the subsequent release of a CH_4. The second step is the oligimerization of this specie into the three member ring.

$$(3) \qquad (CH_3)_3Ga\!:\!NH_3 \Rightarrow (CH_3)_2Ga\!:\!NH_2 + CH_4$$

$$(4) \qquad 3\big((CH_3)_2Ga\!:\!NH_2\big) \Rightarrow \big[(CH_3)_2Ga\!:\!NH_2\big]_3$$

These reactions were recently studied in the high temperature range (200 - 800°C) [5]. A stable species, probably the trimeric compound, decomposes above a temperature of 500 °C with an apparent activation energy of 50.5 kcal/mole [5]. Similar information for the adduct formation between TMG and TMN and the interaction of such an adduct with ammonia is not yet available and is the scope of this study

EXPERIMENT

We have studied the high temperature gas phase reactions between TMG, TMN, and, NH_3 by means of *in situ* mass spectrometry within a flow tube isothermal reactor suitable for kinetic studies [8,13]. A schematic of the experimental system is shown in Figure 1. A two temperature zone furnace is used. TMG and TMN were allowed to mix in the first hot zone of the reactor, that is maintained at a temperature of ~150 °C. This temperature is low enough to prevent thermal decomposition of TMG or the TMG:TMN adduct but is high enough to suppress adduct condensation. This adduct and its initial reaction products are transported to the second temperature zone were they are mixed with NH_3. The temperature of the second zone is varied from 200 - 800°C. The gases are sampled from within the hot isothermal regions of the second zone of the reactor through ~150 μm diameter quartz nozzle. The gas was further expanded in molecular flow and monitored by a residual gas analyzer (RGA). The inlet gas stream contained a small amount (0.5%) of Ar during all experiments, to allow for data normalization and to avoid gas expansion artifacts. Data were obtained as function of temperature through the continuous ramping of the reactor temperature in the second zone. The TMG mole fraction in the inlet of the reactor was 0.015 - 0.05. Experiments utilized both NH_3 as well as deuterated ammonia (ND_3), H_2 and D_2, in order to label the reaction products and to distinguish between possible reaction pathways. The experiments were done in a pressure range of 76 - 100 Torr. Peaks at m/e values of 16, 17, 20, 29, 42, 58, 69, 84 and 99 were monitored corresponding to molecular fragments of CH_4, CH_3D, ND_3, C_2H_6, DMN, TMN, Ga, CH_3Ga and $(CH_3)_2Ga$ respectively. Dilute concentrations of CH_4, C_2H_5, and N_2 in hydrogen were used to calibrate the RGA.

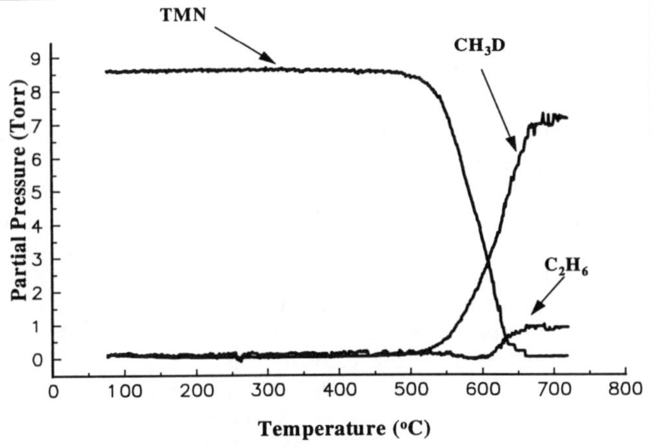

Figure 2: Plot of partial pressure of gas phase species over a range of temperature during homolysis of TMN.

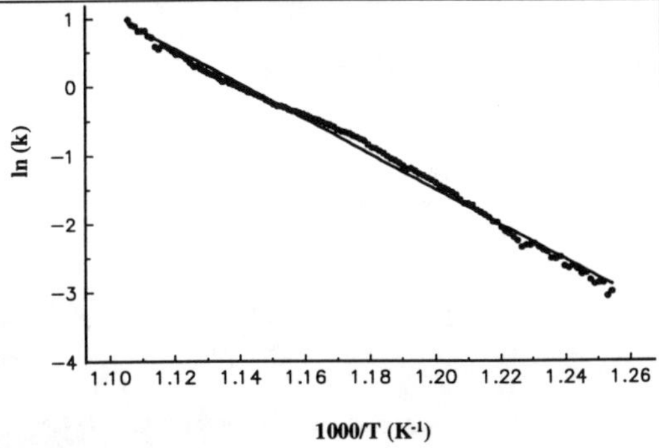

Figure 3: Arrhenius plot for TMN homolysis.

RESULTS

The gas phase homolysis of TMN in D_2 was carried out as a base line for the decomposition characteristics of TMG:TMN adduct in D_2 and in ammonia (NH_3 and ND_3). The decay of the parent peak (m/e = 58) characteristic of TMN, together with the build up of the methane (CH_3D) and ethane, which were found to be the sole decomposition products, are shown in Figure 2. A complete decomposition is observed at 650°C. At this temperature, the amount of methane released (calibrated to pure gas standard) is a little smaller than the initial concentration of TMN introduced into the reactor. The carbon mass balance, with the contribution of ethane that is formed at temperatures above 600°C accounts for a single carbon atom per TMN molecule. The graphs describing the decay of the TMN peak and the rise of the methane peak, are both monotonic and show no changes in the slopes over a range of temperature. These observations suggest that at 650°C only one methyl group was release from the parent TMN molecule, probably to form dimethylamine (MW = 42). The final result of the TMN pyrolysis, beyond the formation methane and ethane, could not be determined from these measurements. As shown in Figure 3 the decomposition process is very well described as a first order Arrhenius reaction with apparent activation energy of 50.8 kcal/mol and pre-exponential factor of $10^{12}s^{-1}$. The statistical error in the measurements is less then 2%. Similar results (47.6 kcal/mol and $10^{-12}s^{-1}$) were deduced for this reaction from the methane formation curve.

The formation and decomposition of TMG:TMN adduct in D_2 were also examined. At this point it should be noted, that currently, work is being done for acquiring a complete quantitative description of the process. The results in hand clearly show that TMG and TMN react rapidly over the whole temperature range with the gas phase evolution of ethane. This is deduced from the change in intensity of the m/e = 58 peak, compared to that of TMN and from the step jump of the ethane peak (m/e = 29). Through calibration of this peak intensity to a pure gas standard it is concluded that the concentration of ethane is close, though a bit smaller, to that of the feed concentration of TMG. The sudden ethane rise, upon mixing, is probably due to the immediate elimination of ethane from the TMG:TMN adduct, in a process similar to that of methane elimination from TMG:NH_3 adduct [5]. As shown in Figure 4, methane (CH_3D) begins to evolve only above 425°C. The evolution of methane follows a two step process, where saturation is approached at about 700°C. The decay of the m/e = 58 peak (not calibrated to a reactant partial pressure) also takes place in a stepwise manner. The first reaction saturates slightly above 500°C, the same temperature in which the m/e = 99 disappears and corresponds to a plateau in the

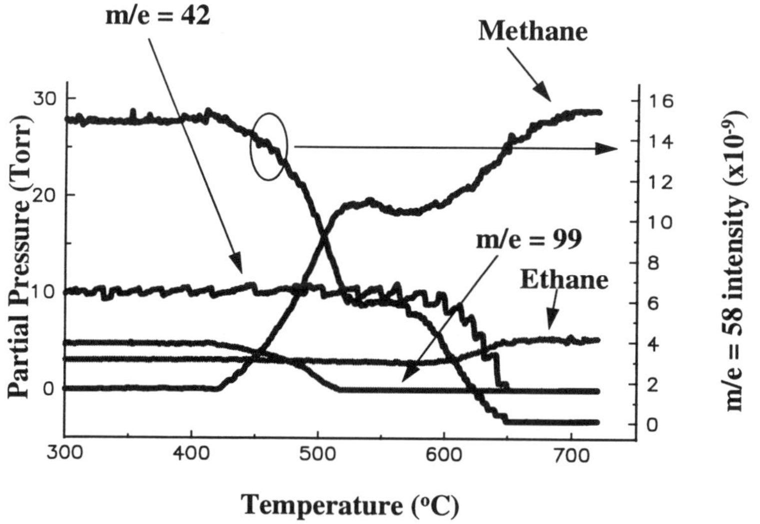

Figure 4: Temperature dependence of TMG:TMN decomposition in H_2. Partial pressures of gas phase species present in the system are indicated.

methane peak. This temperature is close to that of TMG decomposition in H_2 [7,8]. The intensity of the m/e = 42 peak contains contributions from the parent TMN moleculeas well as from any DMN formed in the flow tube reactor. This peak decays to zero at 650°C, the same temperature of TMN decomposition in Figure 2. Since at this temperature, the m/e = 58 peak also diminishes, it is suggested, that the decay of these three peaks represents the homolysis of the excess TMN over TMG. (In this experiment the feed ratio of TMN:TMG was ~3:1). The activation energy for the first step was measured to be 46.4 kcal/mol and the pre-exponential factor 10^{-13} sec^{-1}. The measured activation energy for the second step was unreasonable low, and the decay of the m/e = 42 peak does not obey an Arrhenius process within our experimental accuracy.

The temperature dependence of the reaction of TMG:TMN adduct with ammonia (both NH_3 and ND_3) was next examined. The adduct was first formed in the first zone of the reactor, at 150°C and only then was allowed to mix with the ammonia. The same peaks, characteristic of the TMG:TMN adduct decomposition, were observed in this case, plus those of ammonia. The behavior of the chemical system was different from that of TMG:TMN decomposition in hydrogen. In ND_3 ambient, a small amount of deuterated methane is immediately released upon mixing of the adduct with the ammonia and is observed over the entire temperature range. This methane is most probably a result of TMG:ND_3 adduct formation, followed by the fast methane elimination reaction, which was observed previously [5]. Subsequent increases in the methane peak with the temperature is due to the reaction of methyl groups with the carrier gas. Additionally the m/e = 42 peak disappears at 700°C which is higher than that of the TMG:TMN decomposition in hydrogen (650°C). The activation energies for this process as measured from the decay of the m/e = 99 peak and the raise of the methane peak, were very high, 70 - 72 kcal/mol, while the pre-exponential factor was 10^{21} sec^{-1} to 10^{23} sec^{-1}.

The implication of these studies for the design and operation of MOVPE reactors for GaN growth is several fold. In most or all MOVPE growth systems operating under conventional pressures (1-760 Torr) and flow regimes, little, if any, $(CH_3)_3Ga$ exists in the growth environment during NH_3 - based GaN growth. In our previous study, we suggested that the dominant gas phase species within the reactor is $((CH_3)_2GaNH_2)_x$. It was further deduced that the reactions of this specie in the gas phase, through decomposition or further oligimerization, should be the principal mechanism by which the growth rate and growth rate uniformity is affected by the specific reactor design. It was further stated that decomposition of this species can be suppressed by an appropriate design of the MOVPE reactor, and that only by controlling the gas phase reaction high quality growth will be achieved. The results of the current study show that the TMG:TMN adduct is very rapidly formed in an analogous mechanism to that of the TMG:NH_3 formation, followed by ethane elimination. The higher adduct decomposition temperature in ammonia ambient, compared to that in hydrogen, and the higher activation energy required for its decomposition reveal same role for ammonia in thermally stabilizing this adduct derived compound. The mechanism for this process is yet not understood and is currently being further studied.

ACKNOWLEDGMENTS

The authors would like to thank Prof. D. Gaines and Dr. D. Saulys from the Chemistry Department, the University of Wisconsin-Madison, Prof. K.F. Jensen and Mr. H. Simka from the Department of Chemical Engineering at the MIT for their helpful comments and discussions. Funding from ARPA-URI on visible light emitters and NSF-MRG on CVD are acknowledged.

REFERENCES

1. I. Akasaki and H. Amano, J. Crystal Growth **146**, 455 (1995).
2. D. A. Neumayer and J. G. Ekerdt, Chem. Mater. **8**, 9 (1996).
3. R. H. Moss, J. Crystal Growth **68**, 78 (1984).
4. B. S. Sywe, J. R. Schlup and J. H. Edgar, Chem. Mater. **3**(4), 737 (1991).
5. A. Thon and T.F. Kuech, to be published in APL.
6. A. Zaouk, E. Salvetat, J. Sakaya, F. Maury and G. Constant, J. Crystal Growth **55**, 135 (1981).
7. S. P. DenBaars, B. Y. Maa, P.D. Dapkus, A. D. Danner and H. C. Lee, J. Crystal Growth **77**, 188 (1986).
8. Q. Chen and P.D. Dapkus, J. Electrochem. Soc. **138**(9), 2821 (1991).
9. J. E. Dove and U. S. Nip, Canadian Journal of Chemistry **57**, 689 (1979).
10. J. R. Durig, C. B. Bradly and J. D. Odom, Inorg. Chem. **21**, 1466 (1982).
11. M. J. Almond, C. E. Jenkins, D. A. Rice and K. Hagen, J. Organomet. Chem. **439**, 251 (1992).
12. G. E. Coates, J. Chem. Soc. , 2003 (1951).
13. J.M. Redwing, T.F. Kuech, D. Saulys and D.F. Gaines, J. Crystal Growth **135**, 423 (1994).

ATOMIC FORCE MICROSCOPY STUDY OF GaN-BUFFER LAYERS ON SiC(0001) BY MOCVD

DONGSUP LIM[1], DONGJIN BYUN[2], GYEUNGHO KIM[2], OK-HYUN NAM[2], IN-HOON CHOI[1], DALKEUN PARK[4] and DONG-WHA KUM[2]

1 Korea University, Department of Materials Science and Engineering, Seoul, Korea.

2 KIST, Division of Metals, Seoul, Korea.

3 KIST, Division of Ceramics, Seoul, Korea.

4 KIST, Division of Chemical Engineering, Seoul, Korea.

ABSTRACT

Buffer layers promote lateral growth of films due to a decrease in interfacial free energy between the film and substrate, and large 2-dimensional nucleation. Smooth surfaces of the buffer layers are desired. Optimum conditions for GaN-buffer growth on the vicinal surface of 6H-SiC(0001) were determined by atomic force microscope (AFM). AFM analysis of the GaN nucleation layers led to an optimum growth conditions of the GaN-buffer layer which was confirmed by cross-sectional transmission electron microscopy, Hall measurements and photoluminescence spectra. Optimum growth conditions for GaN-buffer layer on SiC(0001) was determined to be 1 minute growing at 550°C.

INTRODUCTION

GaN is an attractive material with applications in blue and ultraviolet light emitting diodes (LEDs) and laser diodes (LDs). MOCVD technique has greatly improved the quality of GaN over the past few years. There are still remaining problems in the III-V nitride growth like high n-type carrier concentration [1], difficulty of p-type doping [2,3] and lack of suitable substrate materials [4].

Low temperature grown AlN or GaN thin buffer layers on sapphire has improved the quality of GaN films grown by MOCVD at temperatures of about 1000°C [5,6]. The small lattice mismatch of about -3.5% between GaN and SiC makes the SiC a good candidate for the GaN film growth [7]. In this study, a thin GaN-buffer layer is introduced on the vicinal 6H-SiC(0001) substrate to improve the quality of the MOCVD grown GaN.

Vertical layer thickness variations in the nanometer range can be easily detected by AFM, making this technique an ideal tool to study the formation of the initial nuclei and subsequent growth. Pavlidis *et al.* employed AFM technique with Monte Carlo simulation of the nucleation analysis, and reported successful results for the buffer layer on GaAs [8]. It is our intention to find the optimum growth conditions for the GaN-buffer layer on SiC(0001) utilizing AFM analysis.

EXPERIMENTAL

GaN-buffer layers were grown on the vicinal 6H-SiC(0001) substrate (3.5° off toward $<11\bar{2}0>$) using an IR lamp heated horizontal reactor at atmospheric pressure. SiC substrates

Mat. Res. Soc. Symp. Proc. Vol. 423 © 1996 Materials Research Society

were degreased, and dipped into 10% HF solution for 10 minutes to remove the protective oxide layer and rinsed in deionized water. GaN-buffer growth was carried out at 500~650 °C. GaN film depositions on the buffer layers were carried out at 1020 °C with the GaN film thickness of about 0.8 μm. Source gases for gallium and nitrogen were trimethylgallium (TMG) and ammonia (NH_3), respectively. N_2 was used as a diluent and carrier gas for TMG. Flow rates of TMG, NH_3 and N_2 were 4.5 μmole/min., 1 slm and 1 slm, respectively.

Surface morphologies of the GaN-buffer layers were observed by AFM (Park Scientific Instrument STM-SU2-210) with scanning frequency of 4 Hz. Cross-sectional TEM samples were prepared to less than 30 μm thick. A Philips CM30 TEM equipped was used at 200 kV. Both bright-field (BF) images and selected area diffraction patterns were recorded at the $<01\bar{1}0>$ zone axis orientation. Room temperature Hall measurements were carried out using Van der Pauw technique for the electrical properties of specimens. Optical properties of samples were measured by low temperature photoluminescence (PL) at 10 K using a He-Cd laser (λ=325nm). Power used for PL analysis was 40 mW.

RESULTS AND DISCUSSIONS

The essential role of buffer layers is to supply nucleation centers having the same orientation as the substrate, and to promote lateral growth by decreasing the interfacial free energy between the film and substrate. For the promotion of lateral growth, large 2-dimensional (2-D) nucleation and smooth buffer layer surfaces are desired.

AFM images of GaN-buffer grown for 1 minute at various temperatures are shown in Figure 1. At 500°C and 550°C, 2-dimensional (2-D) islands are observed to be dominant. The specimen grown at 550°C shows relatively larger 2-D nucleation islands than that of the sample grown at 500°C. As the GaN-buffer growth temperature increases, 3-dimensional (3-D) growth becomes dominant as expected. Above 600°C, mixed features of 2-D and 3-D islands are observed. AFM results in Figure 1 suggests that the optimum temperature for GaN-buffer growth on SiC(0001) is 550°C.

To find the optimum GaN-buffer growth time, more depositions were carried out at 550° C with various growing times. AFM analysis was again performed on these specimens and Figure 2 shows the root mean square (RMS) roughness and the average roughness of these samples as a function of growth times. Among these specimens, samples grown for 1 and 2 minutes exhibited the minimum roughness. AFM results show that the optimum GaN buffer layer growth conditions are growth at 550°C and for 1 or 2 minutes, since large 2-D nucleation and smooth buffer layer surface are desired for optimized buffer layer. Figure 2 also shows the tendency of increasing roughness with the increase of buffer layer thickness. After the substrate is covered with the nucleation layer, tendency of 3-D growth increases significantly. However the surface diffusion of the reactants are relatively slow due to the low growth temperature and 3-D growth becomes dominant, and results in rough surface.

Cross-sectional TEM was used to understand the effect of buffer layer, surface morphology and defect density of the GaN film. TEM images of GaN films are presented in Figure 3. All images are taken at $<01\bar{1}0>$ zone axis orientation. When a buffer layer is not used, rough surface and many growth defects in GaN film is apparent as shown in Figure 3(a). Two types of lattice defects are observed, i.e. threading dislocations in the GaN film and stacking faults lying parallel to the GaN/SiC interface or basal plane of GaN. From the contrast of the TEM image, these stacking faults appear to be terminated at the threading dislocations.

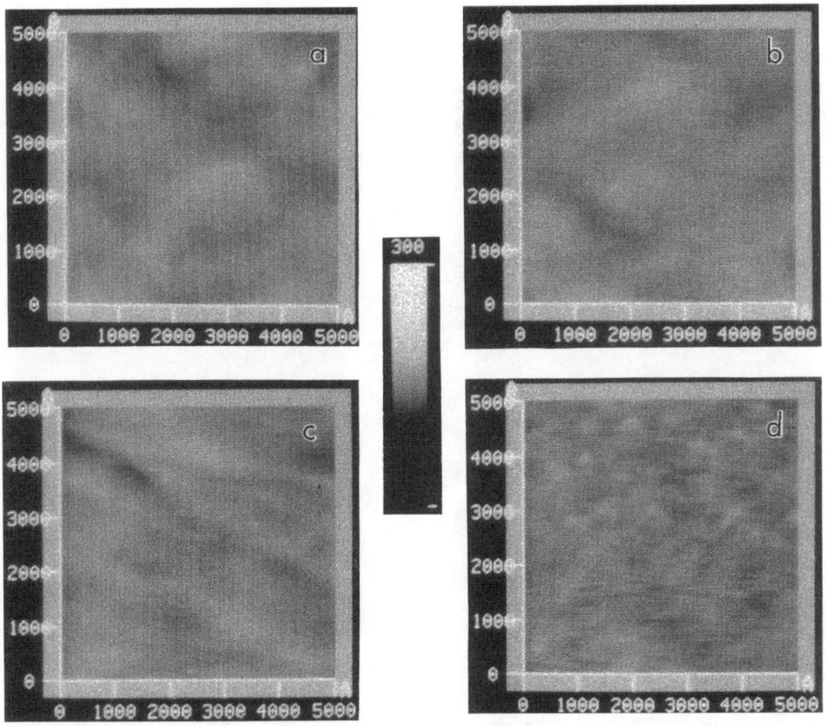

Figure 1. AFM images of the GaN buffer layers grown on 6H-SiC(0001) for 1 minute at (a) 500°C, (b) 550°C, (c) 600°C and (d) 650°C. Units for all AFM measurements are in Å.

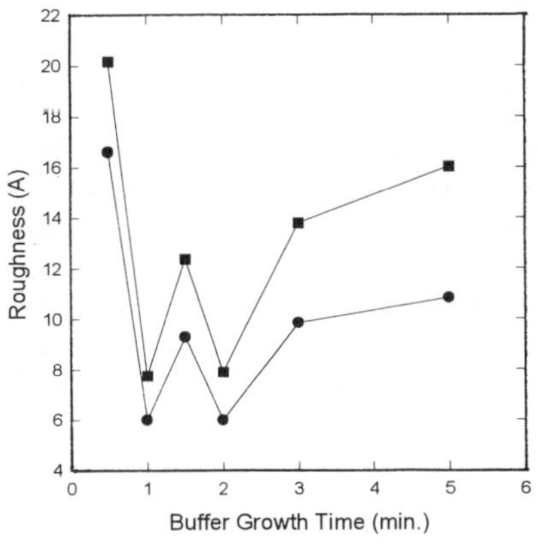

Figure 2. RMS(■) and average (●) roughness of GaN buffer layers grown at 550°C measured by AFM as a function of buffer layer growth time

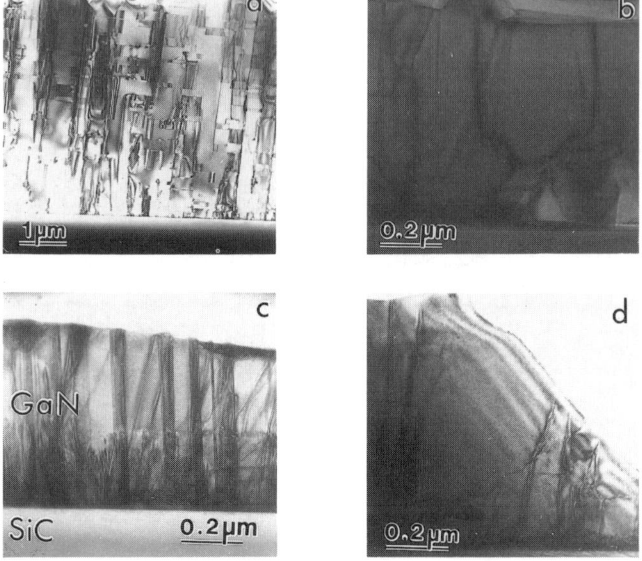

Figure 3. Cross-sectional TEM bright field images of GaN/SiC(0001) (a) without buffer layer, and with buffer layers grown (b) at 550°C for 1 minute, (c) at 550°C for 2 minutes and (d) at 600°C for 2 minutes.

When buffer layer is grown at 550°C for 1 minute, significant reduction in dislocation density is apparent as shown in Figure 3(b). The density of threading dislocations near GaN/SiC interface is about $5x10^9/cm^2$, which is comparable to those found in GaN film grown on sapphire using AlN-buffer layer [9].

As growth time for the buffer layer is increased to 2 minutes at 550°C (Figure 3(c)), more threading dislocations are observed. The defect density near interface is approximately of $2x10^{10}/cm^2$. Both surface roughness and dislocation density become higher at longer buffer growth time. When the GaN-buffer was grown at 600°C for 1 minute (Figure 3(d)), the dislocation density at GaN/SiC interface region is estimated to be $1.4x10^{10}/cm^2$. From the cross-sectional TEM observation, optimum buffer growth time at 550°C is 1 minute.

There are several differences in the distribution of lattice defects among these samples. It may reflect an effect of buffer layer on the quality of GaN epilayer grown subsequently. First, by introducing thin GaN-buffer layer, it is possible to decrease stacking faults in the GaN film. Second, threading dislocation plays a critical role in controlling the surface morphology of the grown film. In Figure 3(c) and (d), the protrusion or intrusion of surface is made when threading dislocations intersect the free surface of GaN film.

Room temperature carrier concentrations and mobility of the GaN/SiC(0001) are shown with respect to the buffer growth temperature and time in Figure 4. The specimen without GaN-

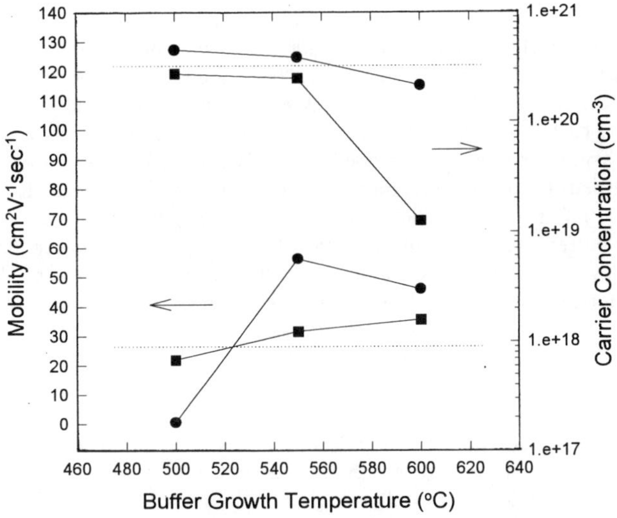

Figure 4.
Room temperature Carrier concentration an mobility of GaN/SiC(0001) as a function of buffer growth temperature; (••••) without buffer layer, (●) buffer layer grown for 1 minute; (■) buffer layer grown for 2 minutes.

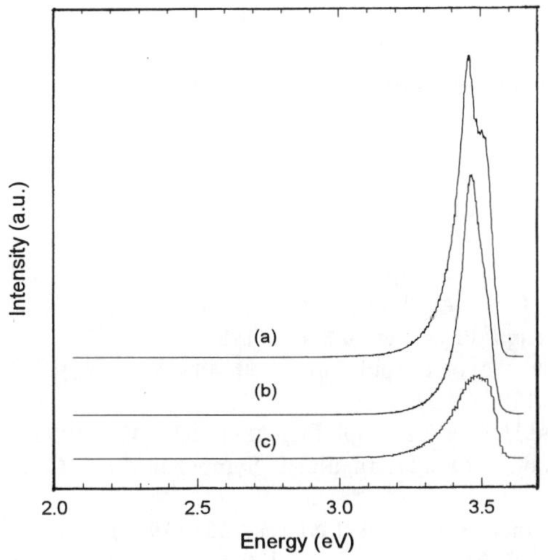

Figure 5.
Low temperature (10K) PL spectra of GaN/SiC(0001) with buffer layer grown (a) at 550°C for 1 minute, (b) at 500°C for 2 minutes and (c) without buffer layer are presented.

buffer layer are also shown in Figure 4. It can be seen that most of the samples with GaN-buffer exhibit better electrical properties than that without the GaN-buffer layer. It is consistent with the TEM observations that the specimens with buffer layers exhibit less defect density and higher mobility. The highest mobility is obtained from the specimen when the buffer layer was grown at 550°C for 1 minute. Nevertheless, there is little variation in carrier concentrations.

Low temperature (=10K) PL spectra of the GaN/SiC(0001) samples with and without buffer layers are compared in Figure 5. All samples show donor bound exciton peak around 3.47 eV as reported previously [10-12]. The specimen with the buffer layer grown at 550°C for 1 minute, which is suggested as the optimum growth conditions of GaN-buffer layer by AFM analysis, shows lowest defect density, highest mobility and the most intense PL peak. Even though the specimen with GaN-buffer grown at 500°C for 2 minutes shows the lower mobility than the specimen without the buffer layer, this sample exhibits the better optical property than the specimen without buffer layer. It should be mentioned that the specimen with the buffer layer grown at 550°C for 1 minute shows the presence of a peak around 3.5 eV which is not clear currently.

CONCLUSIONS

AFM analysis on the GaN nucleation layers on SiC(0001) leads to an optimized buffer layer growth condition, that is, 1 minute growth at 550°C. GaN/SiC(0001) with optimized buffer layer shows the minimum defect density as well as highest mobility and the most intense PL intensity. It is obvious that the use of AFM to find the optimum condition for buffer growth is very successful.

ACKNOWLEDGMENT

This work was supported through KIST-2000 Program (Grant No. BSV00020-002-4).

REFERENCES

1] W. Seifert, R. Franzheld, E. Butter, H. Subotta and V. Riede, Cryst. Res. Technol. **18**, 383 (1983).
2] H. Amano, M. Kito, K. Hiramatsu and I. Akasaki, Jpn. J. Appl. Phys. **28**, L2112 (1989).
3] S. Nakamura, M. Senoh and T. Mukai, Jpn. J. Appl. Phys. **30**, 1708 (1991).
4] S. Yoshida, S. Misawa and S. Gonda, Appl. Phys. Lett. **42**, 427 (1983).
5] H. Amano, N. Sawaki, I. Akasaki and Y. Toyoda, Appl. Phys. Lett. **48**, 353 (1986).
6] S. Nakamura, Jpn. J. Appl. Phys. **30**, L1705 (1991).
7] M. E. Lin, B. Sverdlov, G. L. Zhou and H. Morkoc, Appl. Phys. Lett. **62**, 3479 (1993).
8] K. Wang, D. Pavlidis and J. Sigh, 22nd International Symposium on Compound Semiconductors Proceedings, in press (1995).
9] W. Qian, M. Skowronski, and M. De Graef, Appl. Phys. Lett. **66**, 1252 (1995).
10] K. Maier, J. Schneider, I. Akasaki and H. Amano, Jpn. J. Appl. Phys. **32**, L846 (1993).
11] S. Strite and H. Morkoc, J. Vac. Sci. Technol. **B10**, 1237 (1992).
12] B-C. Chung and M. Gershenzon, J. Appl. Phys. **72**, 651 (1992).

TEM ANALYSIS OF THE OBSERVED PHASES DURING THE GROWTH OF ORIENTED DIAMOND ON NICKEL SUBSTRATES

W. Liu, P.C. Yang, D.A. Tucker, C.A. Wolden, R.F. Davis, J.T. Glass*, J.T. Prater** and Z. Sitar
Department of Materials Science and Engineering, NCSU, Raleigh, NC 27695.
*Kobe Steel USA, Inc. RTP, Durham, NC 27709.
** Army Research Office, RTP, Durham, NC 27709.

ABSTRACT

Transmission electron microscopy (TEM) was used to investigate the interfacial microstructures and phases involved in the nucleation and growth of the oriented diamond on Ni substrates by a multi-step growth process. A molten surface layer is formed during the process, which appears to be critical for both promotion of the diamond nucleation and suppression of graphite formation. Cross-section TEM analysis revealed that a polycrystalline nickel carbide interfacial structure exists between the diamond particles and the single crystal Ni substrate. X-ray diffraction analysis (XRD) identified the carbide phase as Ni_4C. It is suggested that the Ni_4C is formed in the molten layer and stabilizes sp^3C precursor for diamond nucleation.

INTRODUCTION

The heteroepitaxy of diamond films on nickel substrate by chemical vapor deposition (CVD) has been the subject of intensive research both experimentally and theoretically[1-3]. Nickel is one of the few materials that has a close lattice-parameter match with diamond (a=3.52Å for Ni vs. a=3.56Å for diamond). However, its high solubility for carbon and its strong catalytic effect on hydrocarbon decomposition at low pressures have prevented CVD diamond nucleation on the Ni surface without the deposition of an intermediate graphite layer [4]. The graphite interlayer generally forms immediately when Ni substrates are placed in a methane-hydrogen CVD environment. Belton and Schmieg reported on a study of filament-assisted diamond film growth on single crystalline Ni (100) substrates. Graphite islands with poor azimuthal orientation were observed after two minutes of growth. At longer times, the graphite became disordered. This precludes the possiblity of an orientational relationship between the diamond film and the Ni substrate, even though diamond may eventually nucleate and grow on the graphitic interlayer.

On the other hand, it has been known for decades that Ni is an effective solvent-catalyst metal for diamond crystallization under high pressure and high temperature (HPHT) conditions [5]. HPHT synthetic diamond often containes crystalline nickel inclusions. Early XRD studies of this material showed additional reflections corresponding to a lattice spacing close to that of diamond (200) planes. It was found that they arose from a Ni-rich face-centered cubic phase which was identified as Ni_4C carbide [6]. Sato et al. [7] reported that both (111) and (100) oriented diamond nuclei could be grown on Ni substrates, but the overall percentage of oriented nuclei was rather low. Yang et al. [1-3] reported a novel hot filament CVD (HFCVD) process for nucleation of oriented diamond films on both single crystal and polycrystalline nickel substrates. Kobashi [8] reported on highly oriented (111) diamond grown on Pt substrate using similar conditions. Although the process is very successful, there are still many questions associated with the mechanism of the oriented diamond nucleation on Ni and Pt substrates. In this paper we report for the first time direct observation and identification of the interfacial structure of oriented diamond on Ni using cross-sectional TEM and XRD analysis. Based on these observations, the nucleation mechanism is discussed.

EXPERIMENT

Briefly, the deposition process started by seeding the Ni surface with carbon powder. The seeded substrates were then annealed at a temperature of 900°C in a hydrogen atmosphere to allow for deoxidation of the substrate surface. After annealing, the substrate temperature was raised to about 1100°C. At this high temperature, the substrate surface became molten and carbon rapidly

Mat. Res. Soc. Symp. Proc. Vol. 423

dissolved into the Ni lattice. The temperature was then lowered to about 900°C to allow diamond precipitation. Finally, the nuclei are grown using standard diamond CVD conditions of 0.5% CH_4. A full description of the process can be found elsewhere[1].

Plan-view and cross-section TEM samples were prepared from oriented diamond on (100) single crystal Ni, that has been grown for 4 hours. The plan-view sample was prepared by standard methods [9]. For the cross-section samples difficulties arose due to extreme differences in hardness between diamond and nickel. The following technique was developed to overcome some of the difficulties. Two samples were glued together with diamond sides face to face, and mounted into a molybdenum tube with G-1 epoxy. The tube was cut into slices 400 μm thick that were flattened, dimpled, and then polished on one side to a depth of 60 μm. Then the sample was turned over and further ground and polished to a thickness of less than 10 μm. Finally, the sample was ion milled using "sector rotation speed control", which rotates the specimen slowly while the ion beam is perpendicular to the specimen's interfaces, and rapidly through the remaining sectors. In addition, low incident angle and low voltages were used to thin the interfacial region. Although time-consuming, this procedure resulted in good samples from which TEM images could be obtained. The microstructural studies were performed with a TOPCON EM-002B operated at 200 kV.

X-ray diffraction analysis was undertaken [10] using the wide-film Debye-Scherrer XRD technique developed by Read[11]. In a typical experiment, the angle between the incident X-ray beam and diamond film surface was 15°. An excitation potential of 26 kV and tube current of 40 mA gave an adequate single-to-noise ratio after 2 hours of exposure.

RESULTS AND DISCUSSION

Figure 1 shows an SEM image of a highly oriented diamond nuclei grown on (100) Ni. The micro-Raman spectrum in Figure 2 shows a strong diamond peak and absence of any graphitic phase. These two results present a strong evidence that the developed multi-step deposition process yielded high quality diamond deposits on nickel substrates.

Figure 1. SEM micrographs of oriented diamond nuclei formed on the <100> Ni substrate.

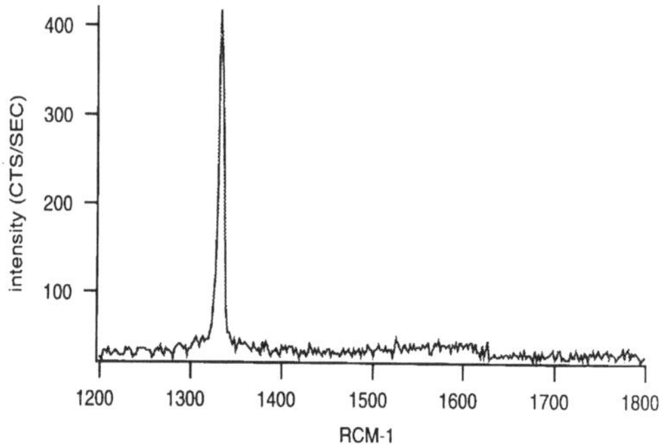

Figure 2. A typical Raman spectrum from the diamond nuclei.

In order to obtain information on structural relationships and intermediate phases, extensive TEM and XRD studies were performed. Figure 3 shows a TEM micrograph of two oriented diamond particles which was attained after the sample preparation discussed above. Three distinct regions could be observed: (i) the diamond particles, (ii) an irregular approximately 0.5 μm thick interfaces, and (iii) the Ni substrate. The selected area diffraction (SAD) pattern analysis (shown in Figure 3) clearly identifies each phase. Nickel and diamond exhibited single crystalline diffraction patterns while the interfacial region showed a finely-grained, polycrystalline character.

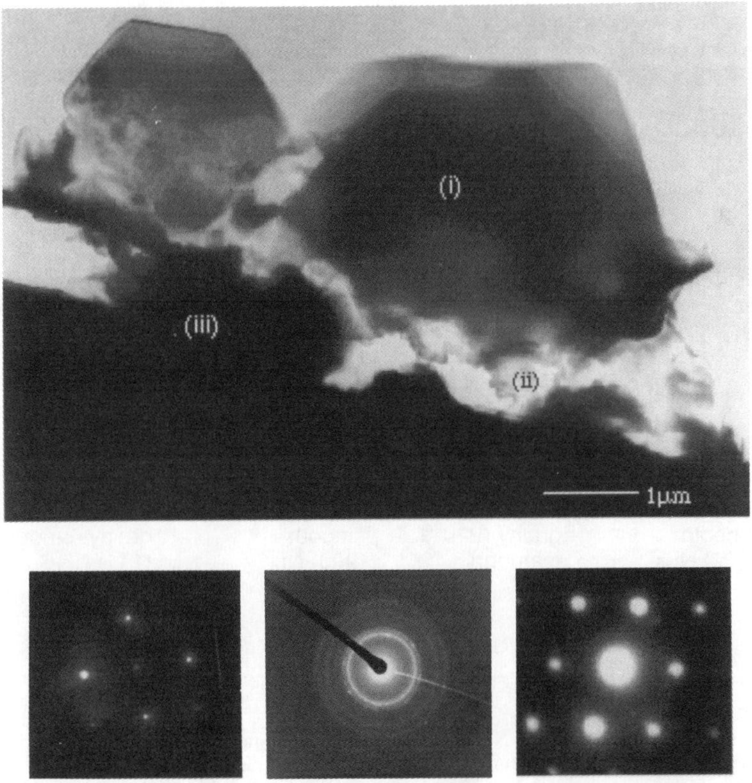

(i) diamond [100] (ii) Ni_4C ring pattern (iii) Ni [110]

Figure 3. Cross-section TEM analysis of the interfacial Ni_4C phase involved in the nucleation and growth of oriented diamond on Ni.

The (200) ring observed in the interfacial region could not be assigned to polycrystalline Ni or diamond. For diamond, the (200) reflection is forbidden, although it may be allowed if the particles are either defective or very small. However, the SEM micrographs show relatively large crystallites and there are no signs of twinning. Ni could not have contributed to this pattern either. The minimum observed grain size of Ni on the substrate surface solidified from the molten phase was about 80 µm. This is much too large to form a continuous ring pattern in SAD. As such, the (200) ring was attributed to the presence of Ni_4C phase.

The interfacial nickel carbide phase was confirmed also by XRD. Figure 4 shows the diffraction pattern from the diamond thin film deposited on (100) single crystal Ni substrate. The X-ray diffraction pattern indicated the presence of three distinct phases in the thin film, which is consistent with the TEM analysis. Diffraction from relatively large diamond crystallites dominate the pattern. However, weaker and smoother lines indicative of a fine randomly-oriented polycrystalline structure were readily detected. These lines were consistent with the presence of metastable Ni_4C phase which is normally very difficult to detect due to the overlap of polycrystalline X-ray pattern of Ni, diamond and Ni_4C.

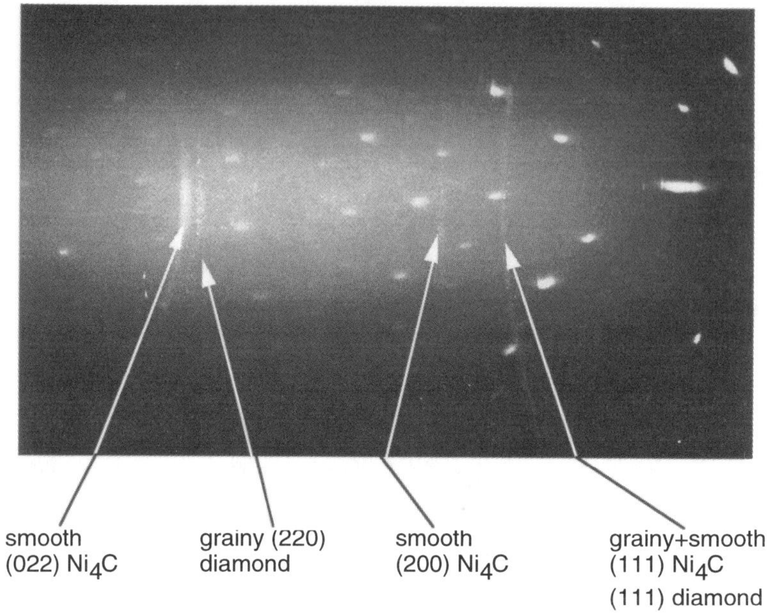

Figure 4. XRD photograph of the high density (100) oriented diamond film on (100) single crystal Ni substrate.

TEM and XRD analyses of the Ni_4C phase show that the metastable Ni_4C carbide phase can be characterized as a NaCl-type structure with carbon atoms in the ordered octahedral interstitial sites of Ni lattice [12]. Based on this structure, the following mechanism is proposed to explain diamond nucleation on Ni_4C. Oriented diamond was obtained on both (111) and (100) single crystal Ni substrates [1]. The proposed relationship between Ni, diamond, graphite, and Ni_4C is shown using the plane was for clarity. Figure 5 (a) depicts the (111) planes of Ni_4C. The filled-in circles are Ni atoms positioned on the lattice points, and the small circles are carbon atoms in the 3-fold interstitial position 1.04 Å beneath the surface Ni layer. The bold triangle shows an area where 6 nickel atoms lie on the surface and one carbon atom (*) resides in the octahedral interstitial position.

Now consider a group of carbon atoms is placed on top of each hollow position of this triangle as shown in Figure 5 (b). The presence of the interstitial C (*) attracts the carbon atom in position 1, pulling it nearer to the surface than atoms 2, 3 or 4. This "puckered" network would favor tetrahedrally-bonded carbon. Alternatively, if Ni_4C is not formed, the carbon atoms would simply lie in a plane, which is the structure of sp^2-bonded graphite. It is argued that the surface melting step is critical, as it allows the dissolved carbon to move as free atoms and arrange themselves properly. Undoubtedly, atomic hydrogen is central to stabilizing the electron configuraton, though its exact role is unclear. In addition, the cause of the observed registry between the underlying Ni substrate and diamond particles remains unknown.

To the best of our knowledge, this is the first direct observation of Ni_4C during the low pressure growth of diamond on nickel. Since the Ni_4C carbide is also found in the process of HPHT catalytic diamond synthesis, these results indicate a similarity between the CVD and HPHT processes in the sense of using Ni as a catalyst to nucleate diamond. The presence of atomic H is central to the multi-step process and is believed to play a role analogous to high pressure in favoring the formation of Ni_4C. It is not known whether H is incorporated into the Ni_4C lattice. It is suggested that the Ni_4C provides the sites and template for the subsequent nucleation of the diamond particles as a result of the close lattice matching of the lattice vectors of both phases.

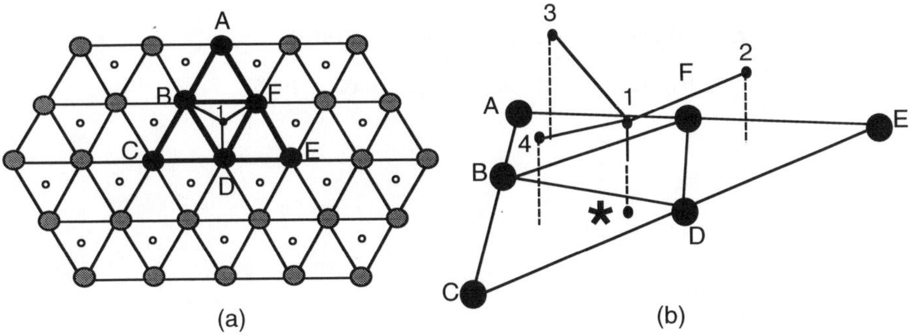

large shaded circle--Ni atoms
small light circle--C atoms

Figure 5. (a) Ni (111) planes with the carbon atoms in the interstitial sites 1.04Å beneath the Ni layer; (b) sp^3C precursor originated from the Ni_4C structure

CONCLUSIONS

For the first time, a polycrystalline Ni_4C interfacial structure was identified between the oriented diamond particles and Ni substrate by cross-section TEM and XRD analyses. The interfacial layer was about 0.5 μm thick and finely grained. Its presence appears to be critical for the nucleation of the oriented diamond on nickel.

ACKNOWLEDGMENTS

This work was supported by Army Research Office, and by BMDO/IST through ONR.

REFERENCES

1. P. C. Yang, W. Zhu and J. T. Glass, J. Mater. Res., **8**, 1773 (1993).

2. W. Zhu, P. C Yang and J. T. Glass, Appl. Phys. Lett., **63**, 1640 (1993).

3. P. C. Yang, W. Zhu and J. T. Glass, J. Mater. Res., **9**, 1063 (1994).

4. D. N. Belton and S. J. Schmieg, J. Appl. Phys., **66**, 4233 (1989).

5. R. H. Wentorf, Jr., Adv. Chem. Phys., **9**, 365 (1965).

6. H. Li. D. Pugh, J. Lees and J. A. Bland, Nature, **4791**, 865 (1961).

7. Y. Sato, H. Fujita, T. Ando, T. Tanaka and M. Kamo, Philosophical Transactions of the Royal Society of London A **343**, 225 (1993).

8. Kobashi and Tachibana, Private communication.

9. P. J. Goodhew, Thin foil preparation for electron microscopy, Elsevier, New York, 1985.

10. P. C. Yang, T. J. Kistenmacher, D. A. Tucker, W. Liu, F. R. Sivazlian, S. P. Bozeman, B. R. Stoner, J. T. Prater, J. T. Glass and R. F. Davis, Applied Diamond Conference 329 (1995).

11. M. H. Read and C. Altmen, Appl. Phys. Letter, **7**, 51 (1965).

12. P. C. Yang, Master thesis, North Carolina State University (1995)

Part III

Defects, Dopants and Characterization

Defects and Structural Properties

THEORY OF DEFECTS, DOPING, SURFACES AND INTERFACES IN WIDE GAP NITRIDES

J. BERNHOLC, P. BOGUSŁAWSKI,[1] E. L. BRIGGS, M. BUONGIORNO NARDELLI, B. CHEN, K. RAPCEWICZ, Z. ZHANG

Department of Physics, North Carolina State University
Raleigh, NC 27695, Bernholc@ncsu.edu

ABSTRACT

The results of extensive theoretical studies of group IV impurities and surface and interface properties of nitrides are presented and compared with available experimental data. Among the impurities, we have considered substitutional C, Si, and Ge. C_N is a very shallow acceptor, and thus a promising *p*-type dopant. Both Si and Ge are excellent donors in GaN. However, in AlGaN alloys the DX configurations are stable for a sufficiently high Al content, which quenches the doping efficiency. At high concentrations, it is energetically favorable for group IV impurities to form nearest-neighbor X_{cation}-X_N pairs. Turning to surfaces, AlN is known to exhibit NEA. We find that the NEA property depends sensitively on surface reconstruction and termination. At interfaces, the strain effects on the band offsets range from 20% to 40%, depending on the substrate. The AlN/GaN/InN interfaces are all of type I, while the $Al_{0.5}Ga_{0.5}N$/AlN zinc-blende (001) interface may be of type II. Further, the calculated bulk polarizations in wurtzite AlN and GaN are -1.2 and -0.45 $\mu C/cm^2$, respectively, and the interface contribution to the polarization in the GaN/AlN wurtzite multi-quantum-well is small.

INTRODUCTION AND SUMMARY

Current interest in wide band-gap nitrides is due to possible applications in blue/UV light-emitting diodes and lasers, and in high-temperature electronics [1, 2]. The recent demonstration of stimulated emission in the blue region has served to further increase awareness of the potential of nitride-based devices [3] and elicited the interest of a number of theoretical groups [4].

In this paper we describe several of our recent results [5, 6, 7, 8, 9, 10] concerning substitutional group-IV impurities in wurtzite GaN and AlN, electron affinity of AlN surfaces in the wurtzite structure, and the zinc-blende (001) interfaces of AlN/GaN/InN.

The group IV impurities are potentially important dopants (*e.g.*, Si is frequently used as a *n*-type dopant of epitaxial GaN). Both C and Si may also be unintentionally incorporated as contaminants during growth. In general, a group-IV atom is likely to become a donor when incorporated on the cation site, and an acceptor on the anion site. This possibility of the amphoteric behavior critically depends on the conditions of growth. We have recently shown [7] that C is preferentially incorporated on the N site under Ga-rich conditions of growth, while both Si and Ge occupy cation sites under N-rich conditions. However, at high concentrations self-compensation, *i.e.*, a simultaneous incorporation of the dopant on both cation and anion sublattice, is expected, which would limit the doping efficiency. We therefore also discuss the electronic structure of X_{cation}, X_N, and X_{cation}-X_N nearest-neighbor pairs. We further consider a possible transition of X_{cation} from the substitutional

[1]also at Institute of Physics, PAN, 02-668 Warsaw, Poland.

Mat. Res. Soc. Symp. Proc. Vol. 423 © 1996 Materials Research Society

to the DX-like configuration. Such a transition is commonly accompanied by a capture of electron on a stable (or metastable) state, which quenches the doping efficiency. Some of the results for C were published in [6].

Motivated by the experimental discovery of negative electron affinity at AlN surfaces [11], we have investigated the electron affinity [8]. We find that the 2×2 aluminium-vacancy reconstruction on the aluminium-terminated face has a negative electron affinity and that the 1×1 hydrogen-passivated nitrogen-terminated surface has a very small electron affinity. The electronegativity difference between the adatom and the atomic species comprising the surface is found to be a guide for anticipating the presence of NEA. In particular, NEA may occur when the adatom is much more electropositive than the surface atoms.

We have also investigated the zinc-blende (001) interfaces of AlN/GaN/InN [9]. The elastic energy of GaN grown on AlN is lower than AlN grown on GaN, indicating that the quality of pseudomorphic growth should be higher when AlN is used as the substrate. Further, the effects of strain on the valence-band offset are significant, as is illustrated by the calculated values of the valence-band offset for AlN/GaN (-0.58 eV) and GaN/AlN (-0.70 eV). The band offsets of AlN/GaN/InN were computed using the AlN in-plane lattice constant and including strain effects. They are all of type I and the transitivity rule is satisfied. No interface states were found in the gap. The $Al_{0.2}Ga_{0.8}N/In_{0.1}Ga_{0.9}N$ interface was studied using the virtual crystal approximation. It is of type I and there are no interface states in the gap.

CALCULATIONS

The calculations were performed using either quantum molecular dynamics [12], or standard plane wave codes. Technical details for the defect and impurity calculations were given in [5]. Soft pseudopotentials for N and C were used [13], while the pseudopotential of Ge was generated according to Ref. [14].

The calculations of the electron affinity used Γ-point sampling and a plane-wave energy cut-off of 30 Ry. Norm-conserving pseudopotentials [15, 16, 17] were used for aluminium (p-local), while for nitrogen a soft-core p-local pseudopotential was used [18].

The plane-wave energy cut-off for the interface calculations was 50 Ry. The equivalent of 10 k-points for bulk and superlattice calculations in the zinc-blende structure [19] were chosen while 6-kpts were used for calculations in the wurtzite structure. For all species, d-local norm-conserving pseudopotentials [15, 16, 17] were used. Further, the nonlinear core correction was included for the cation species [20]. The calculated bulk properties of the zinc-blende nitrides obtained using these pseudopotentials are shown in Table 1. The agreement with experiment is excellent. For interface calculations, we studied 4+4 superlattices (16 atoms in the supercell) along (001).

GROUP IV IMPURITIES

We first briefly discuss configurations of substitutional impurities and strain effects. A substitutional impurity in a wurtzite crystal has four nearest neighbors. One of them, located along the *c*-axis relative to the impurity (called here type-1 neighbor), is non-equivalent to the remaining three neighbors (called here type-2 neighbors). For the group-IV atoms considered here, this non-equivalence is small, since the bond lengths with type-1 and type-2 neighbors are equal to within 1 per cent. In all cases, we find that breathing mode distortions preserve the local hexagonal symmetry. The calculated changes in bond lengths are given in Table 2, together with the energy gain E_{rel} due to the relaxation

Table 1: Calculated bulk properties of zinc-blende nitride semiconductors. The values of the gap at the Γ-point (E_Γ) and of the valence-band width (ΔE_{vbw}) are the LDA results. Experimental values are in brackets and follow Ref. [25].

	AlN	GaN	InN
a_0 (Å)	4.37 (4.38)	4.52 (4.5)	5.01 (4.98)
B_0 (MBar)	2.02 (2.02)	1.70 (1.90)	1.58 (1.37)
E_Γ (eV)	4.09	2.24 (3.45)	0.16
ΔE_{vbw} (eV)	14.86	15.48	14.01
c_{11} (10^{11} dyne/cm^2)	24.85	23.74	20.24
c_{12} (10^{11} dyne/cm^2)	13.37	11.23	12.96

from the ideal substitutional configuration to the final one. As follows from Table 2, the inclusion of relaxation effects is necessary for a proper description of C_{cation}, Si_N, and Ge_N, due to the large differences between the atomic radii of the impurity and host atoms. The calculated ΔE_{rel} are systematically greater for AlN than for GaN, reflecting the higher stiffness of AlN.

As follows from Table 2, C, Si, and Ge occupying the cation site in both GaN and AlN are effective-mass donors. The only exception is C_{Al} in AlN, where the C-induced level is at about 0.4 eV below the bottom of the conduction band. In GaN, because of its lower band gap, the C-derived level is a resonance situated at 0.9 eV above the bottom of the conduction band.

In addition to the substitutional configuration of X_{cation} donors, we have investigated the stability of the DX-like configuration [21]. In this configuration one bond between the impurity and its first neighbor is broken, and one of these atoms (or both) are shifted to an interstitial site. We have analyzed only one possible DX state, namely that with the broken bond between the impurity and the type-1 neighbor; the configuration with the broken bond to the type-2 N atom should have very similar properties.

We first consider GaN. We find that C^*_{Ga} is metastable not only in the negative but also in the neutral charge state. The energy of C^*_{Ga} is higher than that of the substitutional C_{Ga} by 0.54 and 0.35 eV for the neutral and the negative charge state, respectively. In the DX state both the host N atom and the impurity are significantly displaced along the c-axis (see Ref. [6] for details). In contrast, Si is unstable in the DX state, since there is no energy barrier for the transition from the initial DX configuration to the substitutional one. This holds for both the neutral and the negatively charged Si. Finally, a DX state is metastable for the negatively charged Ge, and its energy is higher by 0.3 eV than that of Ge^-_{Ga}. Unlike for C^*_{Ga}, Ge atom is located at the lattice site, and the nearest-neighbor N atom is displaced along the *c*-axis. The Ge-N distance is 2.61 Å, to be compared with 1.93 Å for the substitutional configuration. $Ge^*_{Ga}(-)$ introduces a singlet at about 0.4 eV below the bottom of the conduction band. For the neutral charge state, the DX configuration is unstable.

Turning to AlN, we find that in the negative charge state C^*_{Al} is more stable than C^-_{Al} by 0.2 eV. In the neutral charge state C^*_{Al} is metastable, as its energy is higher by 0.48 eV than that of the ground state. Unlike in GaN, the DX configurations are *stable* for both Si and Ge in the negative charge state. Their energies are lower by 0.15 and 1.2 eV than those of the respective substitutional cases. In both cases the impurity remains on

Table 2: Effects of atomic relaxations around impurities: Δb is the change of the bond length, ΔE_{rel} is the relaxation energy, E_{imp} is the position of the impurity level, with e.m. denoting the effective-mass state, and ΔE_{imp} is the relaxation-induced shift of the impurity level.

	Δb (%)	ΔE_{rel} (eV)	E_{imp} (eV)	ΔE_{imp} (eV)
		GaN		
C:Ga	-18.1	1.65	e.m.	–
Si:Ga	- 5.6	0.65	e.m.	–
Ge:Ga	- 1.4	0.25	e.m.	–
C:N	-2.0	0.1	E_v+0.2	–
Si:N	13.6	3.9	E_v+1.2	0.8
Ge:N	13.5	4.1	E_v+1.35	0.8
		AlN		
C:Al	-17.2	2.6	E_c-0.4	1.2
Si:Al	- 7.0	0.7	e.m.	–
Ge:Al	- 2.9	0.3	e.m.	–
C:N	2.0	0.4	E_v+0.4	0.1
Si:N	16.5	6.75	E_v+1.7	1.0
Ge:N	16.7	6.95	E_v+1.75	1.05

the substitutional site, and the N atom is strongly displaced. For Si^*_{Al} the distance to the N atom is 2.95 Å compared to 1.78 Å for Si_{Al}, and for Ge the respective values are 2.83 and 1.86 Å. $Si^*_{Al}(-)$ introduces a level at –1.5 eV below the bottom of the conduction band. For Ge, the gap state is even deeper, at –2.0 eV below the bottom of the conduction band. Finally, the DX configurations are unstable for both Si and Ge in the neutral charge state.

The different DX configurations found for C, as opposed to Si and Ge, are in part due to atomic size effects. For example, when C substitutes for the much bigger Ga atom, the C-N bonds are highly stretched, because they are shorter than the equilibrium Ga-N bonds by about 15 %. In the DX state one C-N bond is broken, and C is free to relax and to shorten the three remaining C-N bonds. Consequently, the C^*_{Ga}-N bonds are shorter by about 7 % than C_{Ga}-N bonds in the substitutional case. At equilibrium, C^* forms a nearly planar configuration with the type-2 neighbors. Such a geometry is additionally stabilized by the tendency of C to form planar sp^2 bonds. Both factors may contribute to the stabilization of C^*_{Ga} in the neutral charge state. In contrast, these factors are not operative for Si and Ge impurities, which remain on the substitutional site.

The predicted stability of DX^- states for C, Si, and Ge in AlN implies that the reaction $2d^0 \rightarrow d^+ + DX^-$ is exothermic, and the electrons are captured by the deep DX-derived levels. Consequently, the doping efficiency is quenched. In $Al_{1-x}Ga_xN$ alloys the doping efficiency with C, Si, and Ge should thus strongly depend on the alloy composition. These dopants should be efficient donors up to a crossover composition given by the stability of the DX state.

We will now consider the acceptors X_N. In zinc-blende crystals, acceptor levels of group-IV atoms are three-fold degenerate. Due to the wurtzite symmetry of GaN and AlN, the triplets are split into doublets and singlets by the energy E_{split}. In all cases considered here, the energies of doublets (given in Table 2), E_D, are higher than those of singlets, E_S. Both the acceptor energies and the splittings strongly depend on the impurity. We find that C is a shallow acceptor with E_D=0.2 eV in GaN, which is in an excellent agreement with the recent experimental data [22]. Thus, C is a promising *p*-type dopant, since it is a shallower acceptor than the commonly used Mg [2]. In AlN, C is deeper (E_D=0.4 eV) and more localized. The doublet-singlet splitting E_{split} is about 0.2 eV in both materials. In contrast, both Si and Ge are deep acceptors. For GaN:Si E_D=1.2 eV and E_{split}=0.6 eV, and for GaN:Ge E_D=1.35 eV and E_{split}=0.6 eV. In AlN the binding energies and the splittings are even higher, and we find E_D=1.7 (1.75) eV and E_{split}=0.7 (0.7) eV for Si_N and Ge_N, respectively).

Finally, we turn to the electronic structure of the nearest-neighbor X^+_{cation}-X^-_N pairs. Compared to the case of distant X_{cation} and X_N impurities, the main modification is a substantial increase of the doublet-singlet splittings. In the case of GaN, E_{split} rises from 0.6 to 1.0 eV for Si-Si pairs, and from 0.6 to 1.05 eV for Ge-Ge pairs. For AlN, the increase is from 0.7 to 1.4 eV for Si-Si pairs, and from 0.7 to 1.35 eV for Ge-Ge pairs. This effect is due to the close proximity of the X_{cation} donor, located along the *c*-axis, relative to the acceptor X_N.

NEGATIVE ELECTRON AFFINITY AT AlN SURFACES

The electron affinity is defined as the difference between the bottom of the conduction band and the vacuum level. Negative electron affinity occurs when the conduction band bottom is below the vacuum level. It has potential use in electron emitters, cold cathodes and other optoelectronic devices. The recent experimental observation of negative electron affinity at AlN surfaces [11] provides the motivation to study the electron affinity at AlN surfaces.

The value of the electron affinity is modified by the surface dipole moment and the space charge due to the occupation of surface states by free carriers from the bulk. The presence of free carriers leads to band bending near the surface; however, this effect is significant only in doped materials and need not be considered at present. Rather, it is the effect of chemisorption and surface reconstructions on the the surface dipole that constitutes our present interest.

The electron affinity of a semiconductor is defined as

$$\chi = D - (E_c - V_{bulk}) \tag{1}$$

where D, the surface dipole strength, determines the relative positions of the bulk electron states and the vacuum level [28], while $(E_c - V_{bulk})$ is the position of the conduction band minimum relative to the bulk-averaged electrostatic potential. The latter is purely a bulk property and can be obtained in a calculation of the AlN bulk. It is well-known that the local density functional theory consistently underestimates the size of semiconductor band-gaps. For this reason, the conduction-band minimum was rigidly shifted to the experimental value of 6.2 eV using the so-called scissors operator. The surface dipole strength was obtained from the difference of the electrostatic potential across the surface.

The 1×1 (0001) Al- and N-terminated surfaces (with neither adatoms nor vacancies) were studied. Both surfaces exhibit some inward relaxation, while the geometries in the surface plane did not significantly change. The electron affinities are 0.85 eV and 0.30 eV for the Al- and N-terminated surfaces, respectively. These 1×1 structures are expected to have higher surface energies than the 2×2 reconstructed structures. As the 2×2 vacancy reconstructions were among the lower energy 2×2 reconstructions of (0001) GaN, the 2×2 vacancy reconstructions of the (0001) surfaces of AlN were next investigated. With one-quarter of a monolayer of the Al surface atoms removed, the surface can accomodate a much larger inward relaxation than the 1×1 structure. In this case, the electron affinity is -0.70 eV. For the same reconstruction on the N-terminated face, the electron affinity is 1.40 eV. The effect of hydrogen chemisorption on the electron affinity of the 1×1 (0001) Al- and N-terminated surfaces was also examined. Hydrogen adsorption reduced the electron affinity of the N-terminated surface to 0.05 eV; however, the electron affinity of Al-terminated surface increased to 1.60 eV.

The changes in the electron affinity reflect modification of the net surface dipole. During chemisorption, the adsorbates saturate the dangling bonds of the surface atoms resulting in charge transfer between the adsorbates and surface atoms. This transfer manifests itself as an additional dipole field, the strength and direction of which determines the change of the electron affinity. The diamond (111) surface provides another example of this effect. The bare surface has a positive electron affinity; however, it has a negative electron affinity when it has been passivated by hydrogen. Since carbon is more electronegative than hydrogen, charge tranfers from the hydrogen adatom to the surface carbon atom. This induces an additional dipole which points outwards and reduces the electron affinity to a negative value [29]. As nitrogen is more electronegative than hydrogen, this also occurs when hydrogen is chemisorbed on the N-terminated AlN surface. In contrast, aluminium has a smaller electronegativity than hydrogen and consequently we expect the opposite to occur when hydrogen adatoms are chemisorbed on the Al-terminated surface. This is indeed the case. The electronegativity difference between the surface species and the adsorbed species provides insight into manner in which the adatoms modify the surface charge density and hence the electron affinity. Of course, bonding properties will be affected by the detailed nature of the surface, but nonetheless strongly electropositive elements, such as Li and Be, may result in negative electron affinity when adsorbed on the surface.

INTERFACES OF WIDE-GAP NITRIDES

Because of the lattice-mismatch between AlN and GaN (2.5%) and between InN and AlN (12.1%), superlattices of these materials will be strained. We have studied the energetics of two different interfaces, namely the strained GaN/AlN (with AlN in-plane lattice constant) and the strained AlN/GaN (with GaN in-plane lattice constant). The elastic energy for these interfaces, which is the difference of the strained and unstrained energy per pair, is 46 meV/pair for the GaN/AlN interface, while it is 63 meV/pair for AlN/GaN. The interface energy, defined as the excess or deficit energy due to the presence of the interface, is extremely small, namely $\sim$ 1 meV/atom, which is of the order of the precision of the calculations. Thus the interfaces show similar bonding characteristics. These results suggest that the critical thickness for pseudomorphic AlN/GaN is significantly smaller than for GaN/AlN, in agreement with the observation that epitaxially grown GaN/AlN results in higher quality samples [30].

The band offsets of the strained heterojunctions have also been studied. Following the conclusions of the previous paragraph, AlN is taken to be the substrate in all of the cases presented below. The valence-band offsets are -0.70 eV for GaN/AlN and -0.44 eV for AlN/GaN (001) strained interfaces, indicating that the effect of strain on the value of the valence-band offset is significant. The interfaces are all of type I and there are no interface states in the gap. We then investigated the band offsets of the (001) GaN/AlN, InN/AlN and InN/GaN strained heterojunctions. The band offsets are shown in Fig. 1.

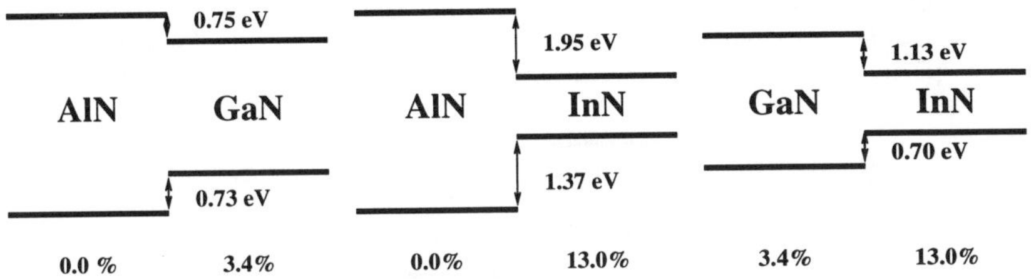

Figure 1: Calculated band offsets for the three interfaces described in the text (an AlN substrate is assumed). The lattice mismatch shown is with respect to AlN.

Table 3: Calculated bulk properties of zinc-blende $Al_{0.2}Ga_{0.8}N$ and $In_{0.1}Ga_{0.9}N$ semiconductors within the virtual crystal approximation. The values of the gap at the Γ-point (E_Γ) and of the valence-band width (ΔE_{vbw}) are the LDA results. The AlN in-plane lattice constant is used (the mismatch is with respect to the AlN lattice constant).

	$Al_{0.2}Ga_{0.8}N$	$In_{0.1}Ga_{0.9}N$
a_0 (Å)	4.51	4.61
B_0 (MBar)	1.67	1.78
E_Γ (eV)	2.56	1.68
ΔE_{vbw} (eV)	15.10	15.31
c_{11} (10^{11} dyne/cm^2)	23.46	20.75
c_{12} (10^{11} dyne/cm^2)	10.88	12.02
Vegard's Rule (Å)	4.49	4.57
Lattice Mismatch	3.1	5.2

In determining the conduction-band offsets, the conduction band minima were shifted to their experimental values using the so-called scissors operator. Each of the interfaces is of type I and the transitivity rule is satisfied. Further, states at the top of the valence band are mostly confined to the epilayer with the smaller gap and there are no interfaces states in the gap of the superlattice.

The possible sources of systematic error of the calculation of the band offsets include the neglect of the anion p- and cation d-state repulsion [31] and the well-known neglect of many-body effects in the local density approximation. In GaN/AlN interfaces, the inclusion of the $3d$-electrons as valence electrons results in a constant shift of 0.2 eV which is less than the experimental error [32] and does not change the character of the interface. Incorporating this shift gives results in agreement with previous estimates using a d-valence pseudopotential [33] and an all-electron calculation [34]. The importance of many-body effects on the band offsets is not known and awaits a future GW calculation.

We also studied an alloy interface based upon the nitride-based multi-quantum-well structure that Nakumura et al. [3] used to demonstate stimulated emission in the blue region of the spectrum. The bulk properties of the two zinc-blende alloys comprising the epilayers of the multi-quantum-well, calculated using the virtual crystal approximation, are shown in Table 3. In the virtual crystal approximation, the pseudopotentials of the constituent species are averaged by their fractional composition [35]. Vegard's rule is seen to be satisfied.

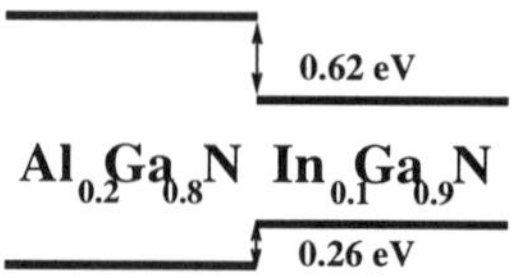

Figure 2: Calculated band offsets for the zinc-blende (001) $Al_{0.2}Ga_{0.8}N/In_{0.1}Ga_{0.9}N$ interface described in the text assuming an AlN substrate.

The zinc-blende (001) interface of these alloys was investigated. Strain effects were explicitly included using macroscopic elasticity theory (using the theoretically determined elastic constants shown in Table 2) and a total energy minimization. The values for scissors corrections of the band gap of the alloys were estimated using a Vegard-type rule, namely the gap of the alloy was taken to be the average of the band gap of the pure phase weighted by its fractional composition. The band offsets are shown in Fig. 2. The interface is of type I and there are no interface states in the gap.

Because of its low symmetry, the wurtzite system may display pyroelectric and piezoelectric behavior [36]. These effects, if present, will manifest themselves macroscopically in multi-quantum-wells along those directions that do not have a perpendicular mirror-plane [37, 38]. Thus, for instance, macroscopic fields are not observed in the strained (001) interfaces in zinc-blende structure, while they are observed in strained (111) interfaces [37]. Indeed, a polarization will be induced only if off-diagonal components of the strain are present. In (0001) strained GaN/AlN, we observe a substantial electric field as has been previously noted by Satta and coworkers [39]. In order to distinguish the bulk pyroelectric and piezoelectric contributions to this field from that induced by the interface, we have calculated the spontaneous bulk polarization of unstrained AlN and the strain induced polarization for the GaN epilayer [40]. The spontaneous polarization (P_3) of AlN and GaN in equilibrium are -1.227 $\mu C/cm^2$ and -0.448 $\mu C/cm^2$ respectively; the polarization of the strained GaN is -0.454 $\mu C/cm^2$. These values are comparable to the computed bulk polarization in BeO [40]. The effect of the interface dipole is small; the polarization in the GaN/AlN multi-quantum-well is well described by the superposition of the polarizations of the constituent epilayers of the multi-quantum-well. The estimated contribution of the interface dipole (which includes the response of one epilayer to the field of the other) is 0.057 $\mu C/cm^2$, which is of opposite sign and an order of magnitude smaller than the bulk polarizations. The computed value of the polarization in the superlattice agrees with that estimated from experiment by Martin *et al.* [32].

This work was supported in part by grants NSF DMR 9408437, ONR N00014-92-J-1477, and KBN 2-P03B-178-10.

REFERENCES

[1] R. F. Davis, Physica B **185**, 1 (1993).

[2] H. Morkoc, S. Strite, G. B. Gao, M. E. Lin, B. Sverdlov, and M. Burns, J. Appl. Phys. **76**, 1363 (1994).

[3] S. Nakumura et al., Jpn. J. Appl. Phys. **35**, L74 (1996).

[4] See, e.g., B. Min, C. Chan and K. Ho, Phys. Rev. B **45**, 1159 (1992), V. Fiorentini, M. Methfessel and M. Scheffler, Phys. Rev. B **47**, 13353 (1993), E. Albanesi, W. Lambrecht and B. Segall, J. Vac. Sci. Technol. B **12**, 2470 (1994), A. Rubio, J. Corkill and M. L. Cohen, Phys. Rev. B **49**, 1952 (1994), C. Yeh, S. Wei and Z. Zunger, Phys. Rev. B **50**, 2715 (1994), A. F. Wright and J. S. Nelson, Phys. Rev. B **51**, 7866 (1995) and M. Palummo, L. Reining, R.W. Godby, C.M. Bertoni and N. Bornsen, Europhys. Lett. **26**, 607 (1995). For a general reference, see W. Lambrecht and B. Segall, in *Properties of Group III Nitrides*, edited by J. Edgar, EMIS Data Series (IEE, London, 1994), chapt. 5.

[5] P. Bogusławski, E. M. Briggs, and J. Bernholc, Phys. Rev. B **51**, 17255 (1995).

[6] P. Bogusławski, E. M. Briggs, and J. Bernholc, Appl. Phys. Lett., July (1996).

[7] P. Bogusławski and J. Bernholc, to be published.

[8] B. Chen, K. Rapcewicz, Z. Zhang and J. Bernholc, to be published (1996).

[9] M. Buongiorno Nardelli, K. Rapcewicz and J. Bernholc, to be published (1996).

[10] K. Rapcewicz, B. Chen, B. Yakobson and J. Bernholc, to be published (1996).

[11] M. C. Benjamin, C. Wang, R. F. Davis and R. J. Nemanich, Appl. Phys. Lett. **64**, 3288 (1994).

[12] R. Car and M. Parrinello, Phys. Rev. Lett. **55**, 2471 (1985).

[13] G. Li and S. Rabii, unpublished (1992).

[14] X. Gonze, R. Stumpf, and M. Scheffler, Phys. Rev. B **44**, 8503 (1991).

[15] G. Bachelet, D. R. Hamann and M. Schlüter, Phys. Rev. B **26**, 4199 (1982).

[16] D. R. Hamann, M. Schlüter and C. Chiang, Phys. Rev. Lett. **43**, 1494 (1979).

[17] D. R. Hamann, Phys. Rev. B **40**, 2980 (1989).

[18] G. Li, and S. Rabii, unpublished (1992).

[19] S. Froyen, Phys. Rev. B, **39**, 3168 (1989).

[20] S. Louie, S. Froyen and M.L. Cohen, Phys. Rev. B **26**, 1738 (1982).

[21] D. J. Chadi and K. J. Chang, Phys. Rev. Lett. **61**, 873 (1988).

[22] S. Fisher, C. Wetzel, E. E. Haller, and B. K. Meyer, Appl. Phys. Lett. **67**, 1298 (1995).

[23] R. Bauer and G. Margaritondo, Physics Today, **40**, 27 (1987).

[24] M.-E. Lin, B. N. Sverdlov, S. Strite, H. Morkoç and A.E. Drakin, Electron. Lett. **29**, 1759 (1993).

[25] A. F. Wright and J. S. Nelson, Phys. Rev. B **50**, 2159 (1994).

[26] J. Lee, H. Aaronson and K. Russel, Surf. Sci. **51** 302 (1975) and E. Arbel and J. Cahn, Surf. Sci. **51**, 305 (1975).

[27] N. Chetty and R.M. Martin, Phys. Rev. B **45**, 6074 (1992), *ibid.* **45**, 6089 (1992) and *ibid.* **44**, 5568 (1991).

[28] A. Zangwill, Physics at Surfaces, (Cambridge University Press, Cambridge, England, 1988).

[29] Z. Zhang, M. Wensell, and J. Bernholc, Phys. Rev. B. **51**, 5291 (1995).

[30] See, for instance, H. Morkoç et al., J. Appl. Phys. **76**, 1363 (1994), M. Paisley and R.F Davis, J. Cryst. Growth **127**, 136 (1992).

[31] S. Wei and A. Zunger, Phys. Rev. Lett. **59**, 144 (1987).

[32] G. Martin et al., Appl. Phys. Lett. **65**, 610 (1994).

[33] V. Fiorentini, M. Methfessel and M. Scheffler, Phys. Rev. B **47**, 13353 (1993).

[34] E. Albanesi, W. Lambrecht and B. Segall, J. Vac. Sci. Technol. B **12**, 2470 (1994).

[35] M. Peressi, S. Baroni, A. Baldereschi and R. Resta, Phys. Rev. B **41**, 12106 (1990).

[36] N. W. Ashcroft and N. D. Mermin, *Solid State Physics* (Saunder College, Philadelphia 1976). Ch. 27.

[37] D. Smith, Solid State Commun. **57**, 919 (1986).

[38] A. Bykhovski, B. Gelmont and M. Shur, Appl. Phys. Lett. **63**, 2243 (1993); J. Appl. Phys. **74**, 6734 (1993).

[39] A. Satta, V. Fiorentini, A. Bosin, F. Meloni and D. Vanderbilt, preprint (1996).

[40] M. Posternak, A. Baldereschi, A. Catellani and R. Resta, Phys. Rev. Lett. **64**, 1777 (1990).

STRUCTURAL DEFECTS AND THEIR RELATIONSHIP TO NUCLEATION OF GaN THIN FILMS

WEIDA QIAN*, MAREK SKOWRONSKI, AND GREG S. ROHRER
Department of Materials Science & Engineering
Carnegie Mellon University, Pittsburgh, PA 15213

*current address: Department of Materials Science, Northwestern University, Evanston, IL 60208

ABSTRACT

Microstructure and extended defects in α-GaN films grown by organometallic vapor phase epitaxy on sapphire substrates using low temperature AlN (or GaN) buffer layers have been studied using transmission electron microscopy. The types and distribution of extended defects were correlated with the film growth mode and the layer nucleation mechanism which was characterized by scanning force microscopy. The nature of the extended defects was directly related to the initial three-dimensional growth. It was found that inhomogeneous nucleation leads to a grain-like structure in the buffer; the GaN films then have a columnar structure with a high density of straight edge dislocations at grain boundaries which are less likely to be suppressed by common annihilation mechanisms. Layer-by-layer growth proceeds in many individual islands which is evidenced by the observation of hexagonal growth hillocks. Each growth hillock has an open-core screw dislocation at its center which emits monolayer-height spiral steps.

INTRODUCTION

Gallium nitride and its related alloys (AlGaN and InGaN) are important wide band-gap semiconductors that have potential applications in both short wavelength optoelectronic and high power/high frequency devices [1-4]. Epitaxial films of GaN have been grown by organometallic vapor phase epitaxy (OMVPE) [5] and molecular beam epitaxy (MBE) [6,7], on a number of substrates including Si [8,9], GaAs [10], SiC[11], α-Al_2O_3 (sapphire) [12,13] and MgO [14]. Among them, sapphire is the most common substrate and generally yields the best film. However, nitride films deposited on sapphire, which is poorly matched to GaN both in terms of lattice parameter and thermal expansion coefficient (see Table I), are known to contain a high density of extended defects (mainly threading dislocations) [15,16]. These defects affect both the electrical and optical properties of the material and are known to be detrimental to its device applications. For example, dislocation cores can form non-radiative recombination centers which limit the efficiency of light emitting diodes and often lead to injection failures in lasers. By low temperature deposition of an AlN (or GaN) buffer layer, the crystal quality of the GaN epilayers can be considerably improved [13,17]. However, such films typically have dislocation densities in the 10^9 -10^{10} cm^{-2} range [15,16].

It is widely accepted that the crystal structure, the nature and relative concentration of extended defects in group-III nitride films are strongly influenced by the growth techniques and conditions used, and by the type and orientation of the substrates. The dominant defects in OMVPE grown GaN films on sapphire are straight threading dislocations which penetrate the entire GaN film [16]. Both the structural and electron optical properties of the GaN film depends on the thickness, the deposition temperature of the low temperature buffer, and the orientation of the sapphire substrate [18-20]. In this paper, we report studies of OMVPE growth of GaN on sapphire using a low temperature AlN (or GaN) buffer layer, with particular emphasis on its initial nucleation and its relation to the resulting microstructure. GaN films grown on either (0001) (c plane) or $(11\bar{2}0)$ (a plane) sapphire substrates under different conditions were studied by transmission electron microscopy (TEM). The types and distribution of extended defects were correlated with the film

Mat. Res. Soc. Symp. Proc. Vol. 423 © 1996 Materials Research Society

growth mode and the layer nucleation mechanism which was characterized by scanning force microscopy.

Table I. Lattice parameters (in Å) and thermal expansion coefficients (in K^{-1}) of wurtzite GaN, AlN and hexagonal α-Al_2O_3

		GaN	AlN	α-Al_2O_3
Lattice	a	3.189	3.111	4.758
parameter	c	5.178	4.979	12.991
Thermal exp.	a	5.59×10^{-6}	4.2×10^{-6}	7.5×10^{-6}
coefficient	c	3.17×10^{-6}	5.3×10^{-6}	8.5×10^{-6}

EXPERIMENTAL

The GaN samples used in this study were grown on both (0001) and $(11\bar{2}0)$ sapphire substrates using an inductively-heated, water cooled, vertical OMVPE reactor operated at 57 torr [19,20]. An AlN (or GaN) buffer layer, 200 to 500 Å in thickness, was deposited at 450 °C to 550 °C before the high temperature deposition of GaN. TEM specimens were prepared by first lapping with diamond paste on a titanium plate down to a thickness of about 100 µm, followed by polishing with consecutively smaller size diamond pastes and/or alumina (Al_2O_3) in a Gatan dimpler down to a thickness of about 20 µm. The specimen was then sputtered to electron transparency by Ar^+ at liquid nitrogen temperature. Transmission electron microscopy observations were carried out on a Philips EM420-TEM and a JEOL 4000 EX-TEM, operated at 120 keV and 400 keV, respectively. For surface observations, the as-grown films were examined in air with a Park Scientific Instruments scanning force microscope. The 5 µm scanner was operated at 2 Hz. All images were acquired in the constant force mode using 3 to 12 nN of contact force.

RESULTS

Crystalline Orientation and the Interface

GaN epitaxial layers grown on both c-plane and a-plane α-Al_2O_3 substrates crystallize with the wurtzite structure and the growth is along the c-axis. The crystal orientation relationships of GaN and AlN grown on sapphire substrates are summarized in Table II. Selected area electron diffraction (SAD) and high resolution electron microscopy reveal that both GaN and AlN layers grow epitaxially on their substrates. For growth on (0001) α-Al_2O_3, the epitaxial relationships were determined to be: $(0001)_{GaN}//(0001)_{AlN}//(0001)_{\alpha\text{-}Al_2O_3}$, with in-plane orientation relations of $[1\bar{1}00]_{GaN}//[1\bar{1}00]_{AlN}//[11\bar{2}0]_{\alpha\text{-}Al_2O_3}$, in agreement with previously published results [15, 21]. For growth on $(11\bar{2}0)$ α-Al_2O_3, both GaN (0001) and AlN (0001) planes were found to be parallel to the $(11\bar{2}0)$ substrate plane, with in-plane orientation relations of $[1\bar{1}00]_{GaN}//[1\bar{1}00]_{AlN}//[0001]_{\alpha\text{-}Al_2O_3}$, in accordance with observations by other groups [12, 22,23].

Table II. Epitaxial relationships during growth of GaN, AlN on c- and a-plane sapphire

Growth direction		In-plane orientation relationship	
Epilayer	Substrate	Epilayer	Substrate
(0001)GaN (0001)AlN	(0001) α-Al_2O_3	$[1\bar{1}00]$ GaN $[1\bar{1}00]$ AlN	$[11\bar{2}0]$ α-Al_2O_3
(0001) GaN (0001) AlN	$(11\bar{2}0)$ α-Al_2O_3	$[1\bar{1}00]$ GaN $[1\bar{1}00]$ AlN	[0001] α-Al_2O_3

The interface between the AlN (or GaN) buffer layer and the sapphire substrate was found to be locally coherent and the misfit dislocation spacing is almost uniform. Figure 1 shows the TEM micrographs of the AlN/(0001) α-Al_2O_3 interface. The measured misfit dislocation spacing shown in figure 1a is 2.1 nm, which is close to the expected spacing of 2.2 nm corresponding to a fully relaxed layer (about 14 % lattice constant mismatch between AlN and sapphire). Similar

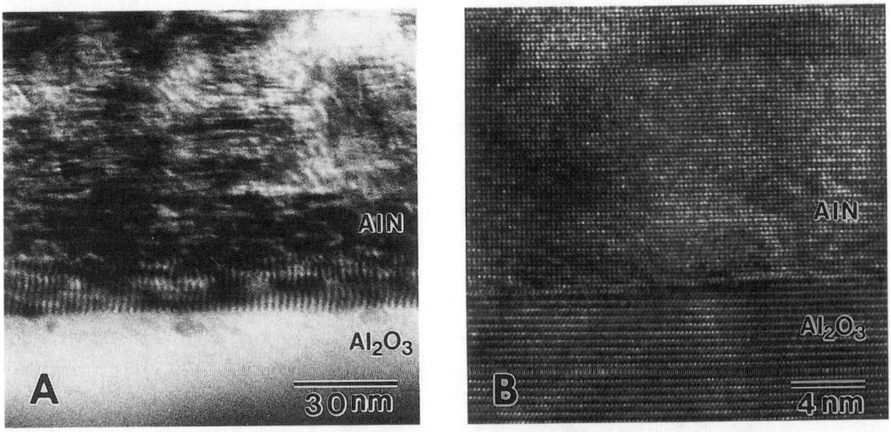

Figure 1. Transmission electron micrographs of the AlN/(0001) α-Al_2O_3 interface: (a). bright field image shows misfit dislocations at the AlN/sapphire interface; (b). Lattice image along $[1\bar{1}00]$ of sapphire shows the crystalline structure.

results have been reported during AlGaN epitaxial growth on sapphire [24]. The high resolution TEM micrograph shown in figure 1b reveals that the interface of AlN/(0001) α-Al_2O_3 is sharp, with the $\{1\bar{1}00\}$ planes of AlN parallel to the $\{11\bar{2}0\}$ planes of sapphire and the basal (0002) planes of AlN parallel to those of sapphire. Surface steps on the sapphire side are most likely due to the interaction of ammonia (NH_3) during surface treatments before deposition. However, extensive observation along the interface with high resolution TEM reveals that, along with many extended defects, the AlN buffer layer is typically composed of columnar fine crystals [21]. Lateral dimension of columnar grains is on the order of 10 nm, and the misorientation among neighboring grains is a few degrees. The GaN/sapphire interface and the GaN buffer layer exhibits similar features. The columnar structures were also reported in GaN films grown on 6H SiC [25].

Microstructure and the Extended Defects

Figure 2a shows the cross-sectional weak beam TEM micrograph of a GaN film grown on $(11\bar{2}0)$ α-Al_2O_3 using a 375 Å GaN buffer layer deposited at 450 °C, imaged near the $[1\bar{1}00]$ orientation with the active beam of **g**= $11\bar{2}0$. The dominant defects in the GaN film are dislocations resulting from the misfit strain introduced by the lattice mismatch between the epilayer and the substrate, and low angle grain boundaries. A high density of dislocation half loops and stacking faults along {0002} GaN planes were observed in regions close to the buffer. A defect density reduction of about an order of magnitude was observed within the initial 0.4 µm of the GaN film. In the immediate vicinity of the low temperature buffer layer, the defect density was so high that we were unable to resolve the defects individually using conventional TEM. At small film thickness, the dislocation lines in the GaN epilayer orient themselves irregularly. However, at film thickness larger than about 0.6 µm, most remaining threading dislocations lie close to the [0001] growth direction [16].

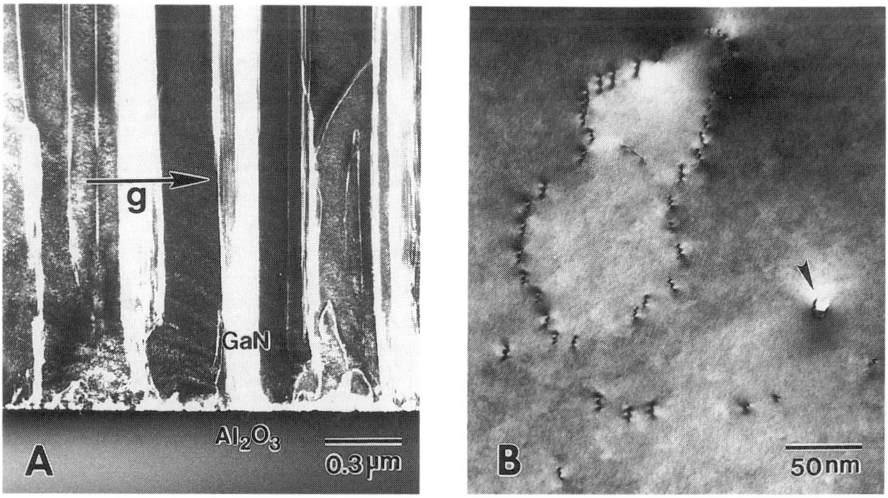

Figure 2. TEM micrographs of a GaN film grown on a-plane sapphire with a 375 Å GaN buffer layer deposited at 450 °C. (a). Weak beam cross-sectional view; (b). Bright field plan-view.

When viewed from plan-view TEM micrographs (Fig. 2b), the extended defects in relatively thick (thicker than about 1 µm) GaN films are basically of two types: threading dislocations and nanopipes (labeled with an arrow) [26]. The threading dislocation density near the top surface of a 3 µm thick GaN layer is typically on the order of $10^9 \sim 10^{10}/cm^2$, as measured using plan-view TEM. It has been reported that other extended defects, such as microtwins and double-positioning boundaries were frequently observed at the vicinity of the GaN/substrate interface using cross sectional high resolution TEM [9,25,27].

Threading Dislocations

A typical plan-view bright field TEM micrograph ($\mathbf{g}=11\bar{2}0$) of an α-GaN film grown on sapphire is shown in Fig. 3a. Since the TEM specimen was polished only from the substrate side, Fig. 3a is representative of GaN films that are thicker than about 2 μm. Except for a relatively small density of nanopipes (not shown), the defects are almost exclusively threading dislocations. All the threading dislocation lines form short segments, indicating that they are lying close to the film growth direction (c-axis). The majority of them are pure edge, with Burgers vectors of the $1/3\langle 11\bar{2}0\rangle$ type, and are arranged to form well defined low angle grain boundaries. The grain size ranges from 50 nm to 500 nm, while the dislocation spacing at the boundaries ranges from 5 to 50 nm, corresponding to in-plane misorientations of less than 3° [16]. Through TEM observations carried out on large number of MOVPE GaN films grown under different conditions, we found that low angle grain boundaries are the main source of threading dislocations. Threading density in the GaN film shown in figure 3b, which has less grain-like structure, is about one order of magnitude lower than that in the film which is shown in figure 3a ($10^9/cm^2$ compared to $10^{10}/cm^2$).

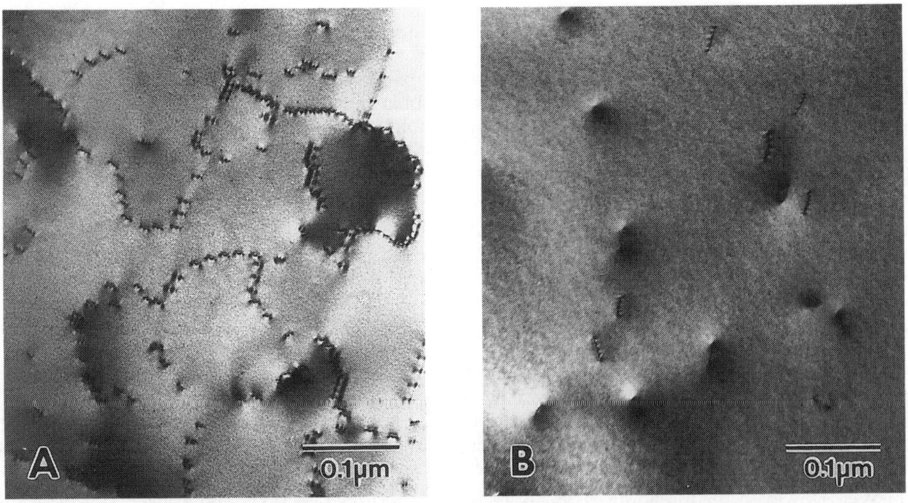

Figure 3. Plan-view TEM micrographs of GaN films grown on sapphire. (a). A typical micrograph taken from samples with high threading densities. The majority of threading dislocations form well defined low angle grain boundaries. (b). A typical micrograph taken from samples with low threading densities, which shows less grain-like structure.

The grain-like structure can also be clearly observed in weak beam cross-sectional TEM micrographs (figure 2a). The image contrast is produced by differing diffraction conditions in neighboring grains. Except near the vicinity of the buffer layer, the dimension of each grain keeps an almost constant value throughout the GaN film. The individual dislocations at grain boundaries can hardly be resolved in thicker cross-sectional specimens since many of them overlap along the beam direction. Although high resolution TEM micrograph often reveals some local tilting of c-planes in some regions, grain misorientation along the c-axis is much smaller and could not be detected by electron diffraction experiments. In fact, x-ray rocking curves [(0004) reflection] taken from these samples have a typical width of 200 - 350 arcsecs [19,20].

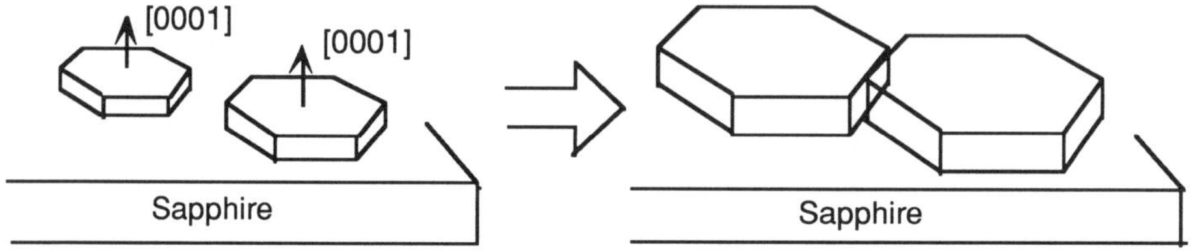

Figure 4. Schematic representation of the initial growth of GaN on sapphire and the formation of grain-like structures in the film.

Low angle grain boundaries are believed to be formed during coalescence of islands at the initial stages of GaN growth, based on the nucleation mechanism first proposed by Akasaki et al. [13]. The buffer layer is thought to have an amorphous-like structure at its deposition temperature (450 - 550 °C). It then recrystallizes and forms columnar fine crystals during annealing [21]. The initial growth of GaN is also highly three dimensional. Since these islands are more likely to preserve a hexagonal shape, and often have small misorientations with their neighbors, dislocation arrays are created during coalescence of these islands, as illustrated schematically in Fig. 4.

The reduction of the defect density within the initial 0.4 μm of the GaN film is most likely due to annihilation and coalescence through mutual interactions and the formation of half loops, as is evidenced in figure 2a. In these regions, both the defect density and the percentage of randomly oriented dislocations are high, and therefore, there is a high probability of defect interactions. However, the remaining quasi-parallel dislocations which thread along the growth direction are apparently not annihilated appreciably during further film growth because of their invariant mutual spacing along the growth direction.

It is well known that a strained layer supperlattice (SLS) can be used to suppress threading dislocations in zinc-blende semiconductor epilayers [28]. The stress/strain field in the epilayer introduced by the SLS is used to initiate dislocation glide on the {111} slip planes, causing the dislocation lines to bend either towards another threading line so that they annihilate one another, or normal to the growth direction along the interface so that they do not propagate further into the epilayer. However, the threading dislocations in α-GaN films described above have slip planes of the $\{1\bar{1}00\}$ type, and generally require larger resolved shear stress to initiate glide. Furthermore, the stress/strain field introduced by the SLS is parallel to the (0001) growth plane, which does not provide driving force for dislocation glide on the $\{1\bar{1}00\}$ slip planes. Therefore, it can be expected that, to suppress, or eventually eliminate, this type of threading dislocations with a SLS could be more difficult. This argument is partially supported by our observations on several samples, each of which has a SLS composed of different periods of GaN/AlN layers placed at different positions in the GaN film. Figure 5 shows the cross-sectional TEM micrograph of a sample which has 5 periods of (6 nm GaN/6 nm AlN) placed after 2 μm of GaN film growth. An additional 0.5 μm GaN film was grown on top of the SLS. Although most dislocation lines were slightly bent as they passed through the SLS, only about 10% of threading dislocations were actually suppressed.

Basal plane dislocations were sometimes observed in regions far away from the buffer layer in a number of GaN films. Figure 6 shows a cross-sectional TEM micrograph taken from one of these samples. Detailed TEM analysis indicates that these are Frank partial loops containing stacking faults. Though more complete work is needed to develop a satisfactory explanation for the formation of these dislocations, an unstable growth condition or the thermal stresses [29] might be responsible for their formation.

Figure 5. Cross-sectional TEM micro-graph showing the ineffectiveness of a SLS for threading dislocation suppression in the GaN films.

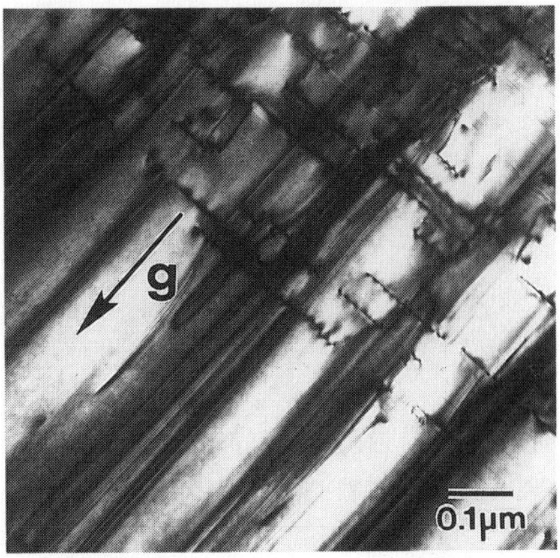

Figure 6. Cross-sectional TEM micro-graph showing GaN basal plane dislocations in regions near the inserted AlN thin layer.

Nanopipes

In addition to threading dislocations, another major type of extended defect in GaN films which is potentially detrimental to device applications is the nanopipes which has an estimated density of 10^5 10^7 /cm^2. These defects are long, faceted empty tubes that penetrate through the GaN epilayer [26,30]. The radii of the pipes are in the 3 - 50 nm range and they appear to propagate along the c-axis of the film. Figure 7 shows plan-view bright field images of GaN film from a region containing a nanopipe (marked with letter A). In figure 7a, the micrograph was taken along the [0001] zone axis, in which the pipe is a perfect hexagon with a diameter of about 9 nm. Two dislocations (marked with letters B and C) are visible along this orientation. After the specimen was tilted about 30° until the $(02\bar{2}1)$ beam near the $[1\bar{2}16]$ zone axis was excited (figure 7b), a wavy contrast can be observed within the pipe which is similar to the contrast from dislocation lines, in addition to the contrast from the pipe walls. TEM analysis indicates that the nanopipe image contrast is in agreement with the $\mathbf{g}\cdot\mathbf{b}$ criteria for a screw dislocation of Burgers vector $\mathbf{b}$ = nx1/3[0003] (where n is an integer and c=5.18 Å for α-GaN). Figure 8 shows the lattice image of a similar nanopipe taken with the [0001] orientation. The internal surfaces of the pipe are formed by six close-packed $\{1\bar{1}00\}$ prism planes (lattice spacing of 2.76 Å), and a closed circuit drawn around the pipe shows no net atomic displacement. This was also the case for several other nanopipes examined in detail by HRTEM. Based on a combination of diffraction contrast analysis,

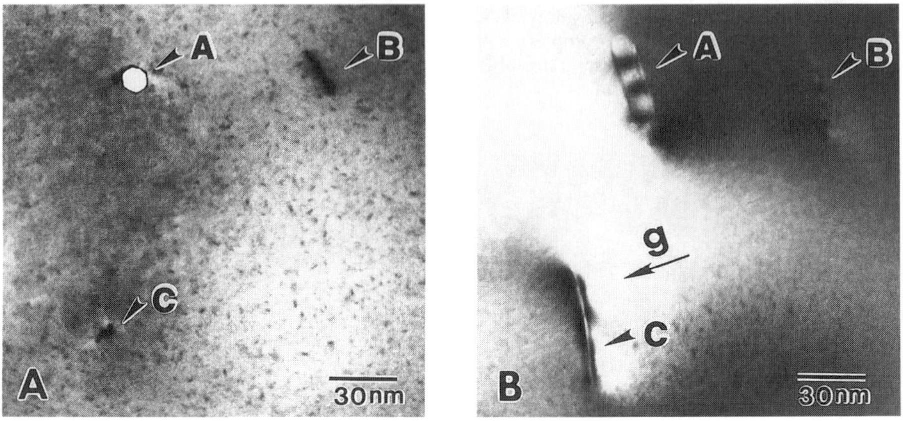

Figure 7. Bright field TEM micrographs of a nanopipe taken under different diffraction conditions. (a). Along the [0001] zone axis; (b). With **g**=$(02\bar{2}1)$.

Figure 8. Lattice image of a nanopipe obtained with the [0001] zone axis. The internal surfaces of the pipe are formed by six close-packed $\{1\bar{1}00\}$ prism planes (lattice spacing of 2.76 Å).

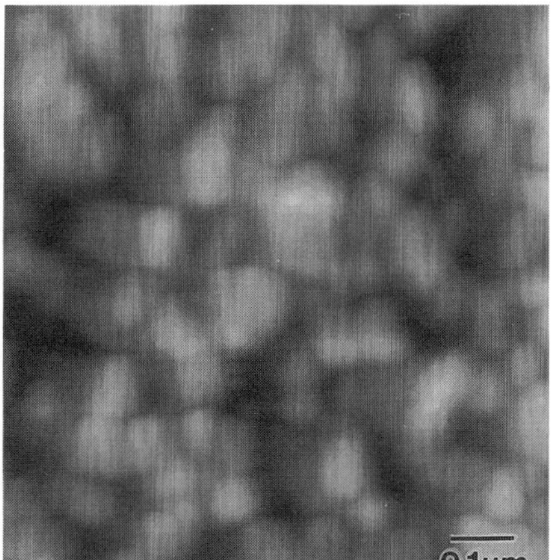

Figure 9. Scanning force microscopy image of a low temperature GaN buffer layer on c-plane α-Al_2O_3. The buffer forms grain-like structure with surface roughness of about 20 nm, and the grain size of 100 nm.

high resolution TEM and scanning force microscopy (SFM) study (shown below), we conclude that these nanopipes are the open cores of screw dislocations with Burgers vectors of a single or multiple c parameters [26,30]. Another nanometer dimension defect along the growth direction has been reported in GaN films grown on sapphire, and has been characterized as microtwins due to local lattice twists or a difference in stoichiometry [23].

Similar, but orders of magnitude larger, open-core dislocations have been reported in a number of crystals including SiC [31,32] and some flux grown garnets [33]. The SiC micropipes have received the greatest attention because they are known to be the defects that limit the breakdown voltage of high power devices [34]. The Burgers vector associated with the spiral steps in SiC is usually greater than 100 Å and the defect is, therefore, visible in conventional optical microscopes.

The existence of open-core dislocations has been predicted by Frank [35], who argued that a state of local equilibrium exists in which a dislocation should have less energy when its core is hollow than when its core is filled with strained lattice. In other words, the open-core is the consequence of the balance between the surface free energy and the lattice strain energy in a dislocation. The lattice strain energy associated with a dislocation provides a driving force for preferential evaporation in the vicinity of the dislocation core [36]. For crystals with relatively large shear modulus and small surface free energy, an open-core should thus be established for dislocations with relatively large Burgers vectors. In fact, all the previously observed open-core dislocations have Burgers vector on the order of, or greater than, 10 Å. The smaller radii of the open-cores in GaN crystals by comparison to those previously observed in other crystals can be explained by its smaller lattice parameter (c=5.2Å). According to Frank's theory [35], the radius of the hole is proportional to the square of its Burgers vector. However, our observation do not quantitatively agree with Frank's theory [30].

GaN Surface and the Film Nucleation

Optical microscopy, scanning force microscopy (SFM) and scanning electron microscopy have been used to study the surface structure of the sapphire substrate, the buffer layer and the GaN free surface. Figure 9 shows a SFM image of a 100 nm thick low temperature GaN buffer layer deposited on c-plane α-Al_2O_3. After annealing at 1025 °C, the buffer forms a grain-like structure with surface roughness of about 20 nm, and an island size of about 100 nm. However, our study revealed that surface morphology of the buffer layer is strongly influenced by its deposition and annealing temperatures, and by the substrate surface cleaning procedures. Lateral growth and smooth coalescence of islands could be promoted through carefully optimizing the buffer conditions. It has been recently reported that, before annealing, the low temperature GaN buffer layers consist of predominantly cubic phase with a high density of stacking faults and twins [37]. The average grain size increases with increasing layer thickness and annealing temperature. At temperatures necessary for the growth of high quality GaN (over 1000 °C), the buffer layer partially converts to hexagonal GaN [37].

The free surfaces of most GaN films typically exhibit cone-like and hexagonal mounds and ledge-facets that are only incompletely bounded by visible steps [38]. These mounds and ledge-facets, which have a typical dimension on the order of 10 µm and can be viewed under an optical microscope, are believed to be growth hillocks which are directly related to the three dimensional nature of the initial GaN growth. The ledge-facets are formed during coalescence of neighboring islands which have slight c-axis misorientations. These misorientations would result in increments in the x-ray rocking curve widths. The apparently large dimension of the growth hillocks compared to those of islands in the buffer can be explained by the mechanism proposed by Hiramatsu et al [21]. The initial growth of GaN is also columnar on top of the buffer fine crystals. During further growth, some fine crystals become larger size islands as the result of geometric selection [39]. The front area of each hillock increases as the merge continues, and the number of growth hillocks emerging at the front gradually decreases. However, there is no direct correlation between sizes of the growth hillocks and those of the small grains observed by TEM. The misoriented grains, bounded by arrays of edge dislocations, once formed, are not likely to expand.

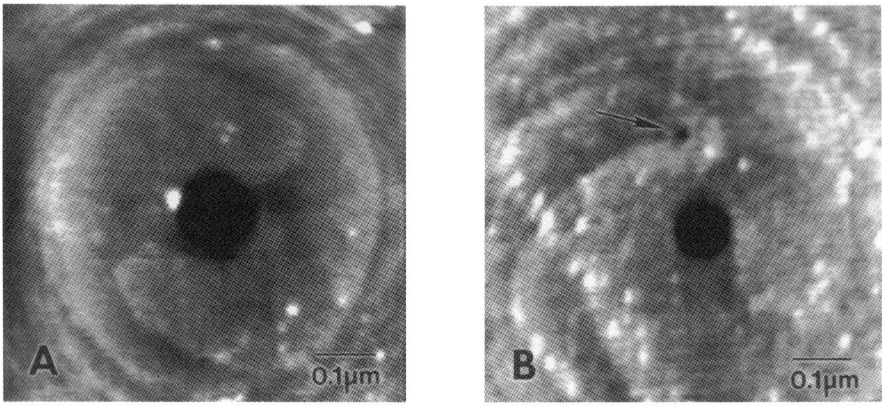

Figure 10. High resolution SFM images of open-core screw dislocations at the hillock centers. (a). A pair of monolayer spiral steps (3.1 ± 0.8Å) originate from the open core and propagate outwards; (b). A single spiral step in addition to a pair of monolayer spiral steps from the open core.

SFM observation reveals single atomic layer steps in flat regions between the hillocks and spiral single atomic layer steps on the top of faceted hillocks [30,37]. This indicates that at the later stages, GaN grows by both a layer-by-layer and a spiral mechanism [37]. Nanopipes were also observed during SFM observations [30,37]. These are located at the centers of faceted small hillocks and were accompanied by a pair or two pairs of single atomic layer spiral steps. Figure 10 shows high resolution SFM images of nanopipes with a pair of spiral steps (Fig.10a) and with an additional single spiral step (Fig. 10b). A pair of spiral steps, each 3.1± 0.8 Å high, originates at the big hole. The measured step height is consistent with the expected dimension of a single diatomic layer of GaN which is one half the length of the c-lattice parameter (2.6 Å). Thus, the two steps form an additional complete GaN unit cell. The origin of these extra steps at the center of the hillock indicates that there is a screw dislocation in the center of the hole with a Burgers vector (**b** = 1/3[0003]) equal in length to the c lattice parameter (5.2 Å). Numerous hillocks were examined with different probe tips and similar observations were made; each has a hole with an approximately 600 Å radius at the center. In some cases, four single steps, each 1/2 c high, emerged from the same hole which had a larger radius (approximately 925 Å). This corresponds to a "giant" dislocation with Burgers vector **b**=2/3[0003]. After spiraling away from the hole on the flat top of the hillock, the steps bunch together and interfere with the step emerged from other sources. In many regions, very small holes which emit single spiral steps were observed (marked with an arrow in Fig.10b). These are believed to be the mixed dislocations with a screw component of c/2, rather than sessile Frank partials. Since stacking faults were not observed in plan-view TEM. The apparent discrepancy between the nanopipe sizes determined from the SFM and TEM images is resolved by recognizing that the SFM image actually measures a crater on the growth surface that is always wider than the nanopipe itself. The crater can be easily observed in plan-view TEM specimens which preserve the GaN growth surface [26] and is a predicted structural component of an open-core dislocation [35].

CONCLUSIONS

Microstructure, extended defects and the growth mechanism of GaN films grown by organometallic vapor phase epitaxy on sapphire substrates using low temperature AlN (or GaN) buffer layers have been studied using a number of microscopy techniques. GaN layers grown on both a-plane and c-plane sapphire substrates using both GaN and AlN buffer layers crystallize with the same wurtzite structure and have similar microstructures. Although a number of other extended defects are often observed in the GaN films, such as stacking faults, basal plane dislocations and microtwins, the major extended defects in these films are threading dislocations and nanopipes. Low angle grain boundaries are the main source of threading dislocations. These dislocations are mostly of pure edge type, running along the c-axis. The nature of these threading dislocations suggests that there should be no appreciable annihilation with the further increase of the GaN layer thickness, and the suppression of dislocations by commonly used buffer layer schemes could be more difficult. Nanopipes are open-cores of screw dislocations formed under local thermodynamic equilibrium. They are about 3 - 50 nm in radii, propagate along the c axis of the film and are potentially detrimental to device applications of GaN. The nanopipes are located at the centers of faceted small growth hillocks and were often accompanied by a pair or two pairs of single atomic layer spiral steps. The grain-like structure in the GaN film and the growth hillocks on the GaN free surface are related to the three dimensional growth of the buffer and the initial GaN growth. At the later stages, GaN grows by both a layer-by-layer and a spiral mechanism. It is, therefore, important to carefully optimize the buffer deposition conditions to minimize or avoid the formation of columnar structures in the GaN film.

ACKNOWLEDGMENTS

The authors acknowledge Drs. K. Doverspike, L. B. Rowland and D. K. Gaskill of Naval Research Laboratory for providing the MOVPE GaN samples used for this study. M. Shin and A. Y. Polyakov of Carnegie Mellon University for providing low temperature buffer samples used in surface studies (figure 9). W. Q. and M. S. acknowledge support under AFOSR Grant No. F29620.94.1.0392 and G.S.R. acknowledges support from the NSF under YIA Grant No. DMR-9458005.

REFERENCES

[1] H. Morkoç, S. Strite, G. B. Gao, M. F. Lin, B. Sverdlov and M. Burns, J. Appl. Phys. **76**, 1363 (1994).
[2] H. Morkoc and S.N. Mohammad, Science, **267**, 51 (1995).
[3] J. I. Pankove, Mater. Res. Soc. Symp. Proc. **97**, 409 (1987).
[4] S. Strite, M.E. Lin, and H. Morkoç, Thin Solid Films, **231**, 197 (1993).
[5] H. Amano, N. Sawaki, I. Akasaki, and Y. Toyoda, Appl. Phys. Letters, **48**, 353 (1986).
[6] H. Gotoh, T. Suga, H. Suzuki, and M. Kimata, Japan. J. Appl. Phys. **20**, L545 (1981).
[7] M.J. Paisley and R.F. Davis, J. Crystal Growth, **127**, 136 (1993).
[8] T. Lei, T.D. Moustakas, R.J. Graham, Y. He and S.J. Berkowitz, J. Appl. Phys. **71**, 4933 (1992).
[9] S.N. Basu, T. Lei and T.D. Moustakas, J. Mater. Res., **9**, 2370 (1994).
[10] S. Strite, J. Ruan, Z. Li, N. Manning, A. Salvador, H. Chen, D.J. Smith, W.J. Choyke, H. Morkoc, J. Vac. Sci. Technol. B, **9**, 1924 (1991).
[11] M.E. Lin, S. Strite, A. Agarwal, A. Salvador, G.L. Zhou, N.Teraguchi, A. Rockett and H. Morkoç, Appl. Phys. Lett. **62**, 702 (1993).
[12] D.K. Wickenden, K.R. Faulkner, R. W. Brander, and B.J. Isherwood, J. Crystal Growth, **9**, 158 (1971).

[13] I.Akasaki, H.Amano, Y.Koide, K.Hiramatsu, and N.Sawaki, J. Crystal Growth, **98**, 209 (1989).
[14] R.C. Powell, G.A. Tomasch, Y.W. Kim, J.A. Thornton, J.E. Greene, Mater. Res. Soc. Symp. Proc. vol. **162**, 525 (1990).
[15] R. C. Powell, N.-E. Lee, Y.-W. Kim and J. E. Greene, J. Appl. Phys., **73**, 189 (1993).
[16] W. Qian, M. Skowronski, M. D. Graef, K. Doverspike, L. B. Rowland and D. K. Gaskill, Appl. Phys. Lett., **66**, 1252 (1995).
[17] S. Yoshida, S. Misawa, and S. Gonda, J. Vac. Sci. Technol. B, **1**, 250 (1983).
[18] J. N. Kuznia, M. A. Khan, D. T. Olson, R. Kaplan, and J.A. Freitas, Jr., J. Appl. Phys., **73**, 4700 (1993).
[19] L. B. Rowland, K. Doverspike, A. Giordana, M. Fatemi, D. K. Gaskill, M. Skowronski, and J. A. Freitas Jr., Inst. Phys. Conf. Ser., **137**, 429 (1993).
[20] K. Doverspike, L. B. Rowland, D. K. Gaskill, S. C. Binari and J. J.A. Freitas, J. Electron. Mater., **24**, 269 (1995).
[21] K. Hiramatsu, S. Itoh, H. Amano, I. Akasaki, N. Kuwano, T. Shiraishi and K. Oki, J. Crystal Growth **115**, 628 (1991).
[22] T. Lei, K.F. Ludwig, and T.D. Moustakis, J. Appl. Phys. **74**, 4430 (1993).
[23] Z. Liliental-Weber, H. Sohn, N. Newman and J. Washburn, J. Vac. Sci. technol. B **13**, 1578 (1995).
[24] F.A. Ponce, J.S. Major, W.E. Plano and D.F. Welch, Appl. Phys. Lett. **65**, 2302 (1994).
[25] D.J. Smith, D. Chandrasekhar, B. Sverdlov, A. Botchkarev, A. Salvador and H. Morkoç, Appl. Phys. Lett. **67** 1830 (1995).
[26] W. Qian, M. Skowronski, K. Doverspike, L. B. Rowland and D. K. Gaskill, J. Crystal Growth, **151**, 396 (1995).
[27] D. Chandrasekhar, D.J. Smith, S. Strite, M.E. Lin and H. Morkoc, J. Crystal Growth **152**, 135 (1995).
[28] S. F. Fang, K. Adomi, S. Iyer, H. Morkoç, H. Zabel, C. Choi, and N. Otsuka, J. Appl. Phys. **68**, R31 (1990).
[29] K. Hiramatsu, T. Detchprohm and I. Akasaki, Lpn. J. Appl. phys. 32, (1993) 1528.
[30] W. Qian, G.S. Rohrer, M. Skowronski, K. Doverspike, L. B. Rowland and D. K. Gaskill, Appl. Phys. Lett. **67**, 2284 (1995).
[31] V. G. Bhide, Physica, **24**, 817 (1958).
[32] H.M. Hobgood, D.L. Barrett, J.P. McHugh, R.C. Clarke, S. Sriram, A.A. Burk, J. Greggi, C.D. Brandt, R. H. Hopkins and W. J. Choyke, J. Crystal Growth, **137**, 181 (1994).
[33] S. Takasu and S. Shimanuki, J. Crystal Growth, **24/25**, 641 (1974).
[34] J. A. Powell, P. G. Neudeck, D. J. Larkin, J. W. Yang and P. Pirouz, Inst. Phys. Conf. Ser., **137**, 161 (1994).
[35] F.C. Frank, Acta Cryst., **4**, 497 (1951).
[36] J.P. Hirth and G.M. Pound, in: Progress in Materials Science 11, Condensation and Evaporation, Ed. B. Chalmers (Pergamon Press, New York, 1963).
[37] X.H. Wu, D. Kapolnek, E.J. Tarsa, B. Heying, S. Keller, B.P. Keller, U.K. Mishra, S.P. DenBaars and J.S. Speck, Appl. Phys. Lett., **68**, 1371 (1996).
[38] G.S. Rohrer, J. Payne, W. Qian, M. Skowronski, K. Doverspike, L. B. Rowland, and D. K. Gaskill, Mater. Res. Soc. Symp. Proc. (in press).
[39] A.A. Chernov, Modern Crystallography III: Crystal Growth (Springer, Berlin, 1984).

EFFECT OF Si DOPING ON THE STRUCTURE OF GaN

Zuzanna Liliental-Weber, S. Ruvimov, T. Suski,* J.W. Ager III, W. Swider, Y. Chen, Ch. Kisielowski, J. Washburn, I. Akasaki,** H. Amano,** C. Kuo, and W. Imler, ***

Materials Sciences Division, E.O. Lawrence Berkeley National Laboratory 62/203
Berkeley, CA 94720
* High Pressure Research Center, UNIPRESS, Polish Academy of Sciences, Warsaw, Poland
** Meijo University, Nagoya, Japan, *** Optoelectr. Division, Hewlett Packard, San Jose, CA

ABSTRACT

The influence of Si doping on the structure of GaN grown by metal-organic chemical-vapor deposition (MOCVD) has been studied using transmission electron microscopy (TEM), x-ray diffraction and Raman spectroscopy. Undoped and low Si doped samples were compared with samples of increased dopant concentration. In addition, defect reduction due to different buffer layers (AlN and GaN) is discussed. Silicon doping improves surface morphology and influences threading dislocation arrangement. High doping leads to a more random distribution of dislocations. Based on this study it appears (for the same dopant concentration) that an AlN buffer layer can significantly reduce the number of threading dislocations, leaving the samples more strained. However, no significant reduction of threading dislocations could be observed in the samples with GaN buffer layer. These samples are the least strained.

INTRODUCTION

In the last few years III-V nitrides (GaN, AlN, InGaN..) have attracted considerable interest due to their potential applications including blue light emitting devices and high temperature electronics [1-4]. Two main difficulties prevent their further development. First is the lack of suitable substrates with low lattice and thermal mismatch to GaN and the second is a difficulty of achieving useful doping, in particular p-doping in high concentration [3]. High lattice and thermal mismatch between popular substrates such as SiC and sapphire leads to high stress in the layer and eventual layer cracking. To obtain high-brightness blue-light-emitting diodes high reliability and high speed are necessary. High defect density can preclude potential applications. It is not clear at this point how particular doping influences the type of defects and their distribution, and very little is known about how a particular buffer layer helps defect reduction and strain relaxation in the layer. In this paper the influence of Si doping on the structural properties of GaN is described. Undoped and Si doped samples grown by MOCVD with different buffer layers have been studied using transmission electron microscopy (TEM), x-ray diffraction and Raman spectroscopy.

MOCVD GROWN GaN

Two sets of GaN layers were grown by MOCVD on sapphire substrates. In one case $(11\bar{2}0)$ Al_2O_3 substrates (α plane) were used for the growth of GaN layers with AlN as a buffer layer. These samples will be called "A." Undoped and Si-doped samples were studied. Undoped samples grown with the low-temperature AlN buffer layer on the α plane of sapphire were 3.5 μm thick and for Si-doped samples (3×10^{18} cm^{-3}) were 5 μm thick. The second set of GaN layers were grown on (0001) Al_2O_3 substrates using low-temperature GaN as a buffer

Mat. Res. Soc. Symp. Proc. Vol. 423 © 1996 Materials Research Society

layer. These samples will be called "B." These samples were 1.8 μm thick and all were Si doped with doping levels $9x10^{16}$ cm^{-3}, $3x10^{18}$ cm^{-3} and $1x10^{19}$ cm^{-3}. X-ray double-crystal diffractometry was used for general characterization of the structural quality of these layers, while TEM techniques were applied for defect investigation. Both plan-view and cross-section specimens were studied by TEM. Raman spectroscopy was used for determination of the strain in the layer.

Different amount of broadening of the FWHMs (3.3-8 arc min) of the rocking curves taken for Θ– and Θ/2Θ scans were observed for the "A" samples. More broadening was observed for the undoped samples (Table I). Optical micrographs taken from these two samples show completely different surface roughness. The doped samples had smoother surface. Some large defects could be seen on optical micrographs for both doped and undoped "A" samples. Atomic force microscopy shows that micro-tubes (voids or pin-holes) are formed on the sample surfaces with a diameter of 0.2-0.5 μm (Fig. 1a).

Some micro-tubes (pinholes or voids) were observed as well, in plan-view TEM samples (Fig. 1b). These micro-tubes (about 3 nm diameter) were associated with dislocations with edge components. It is not clear at this point if larger pin-holes observed by AFM have the same nature as these seen in our plan-view samples, since the size of larger pinholes (0.2-0.5 μm) and distribution practically precludes their observation in high resolution.

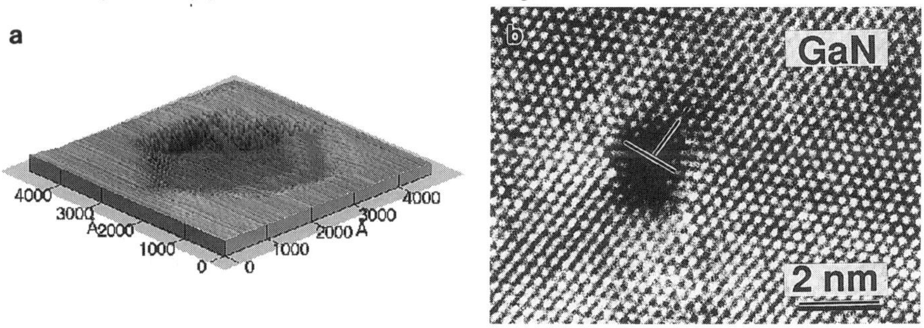

Fig. 1. (a) AFM image of a pinhole formed in Si-doped sample "A;" (b) Much smaller pinholes seen in plan-view samples.

Transmission electron microscopy from plan-view images clearly showed differences in densities of threading dislocations and their distribution for Si-doped compared to undoped GaN layers. Dislocation densities at the top of the layers were about 5×10^9 and 7×10^8 cm^{-2} for undoped and Si-doped GaN, respectively. While the dislocations in the Si-doped layer were distributed homogeneously (Fig. 2a), they were arranged in small-angle boundaries in the undoped GaN layers (Fig. 2b). In both cases, threading dislocations near the layer surface were close to edge orientation with Burgers vector of $1/3\langle\bar{1}2\bar{1}0\rangle$ and $\{10\bar{1}0\}$ glide plane (prism plane). This explains why these dislocations are not very mobile. Dislocation boundaries in undoped samples divide the layer into grains of 0.3-0.5 μm in size, slightly misoriented with respect to each other. The presence of these subgrains in the undoped GaN layer explains its rough surface morphology, in comparison with the rather smooth morphology of the Si-doped layers. In general, misorientation between subgrains has two components, parallel and perpendicular to the layer surface, depending on the type of dislocations at the boundaries. In-plane misorientation is associated with tilt boundaries, while perpendicular misorientation is associated with twist boundaries. A pure twist boundary contains a crossed grid of screw dislocations with Burgers vectors parallel to the dislocation lines. A pure tilt boundary is formed by edge dislocations with Burgers vectors perpendicular to the boundary plane. Grain boundaries in our samples were more complex and had both twist and tilt components. The type of dislocations at the boundaries will influence the broadening of the x-ray rocking curve FWHM.

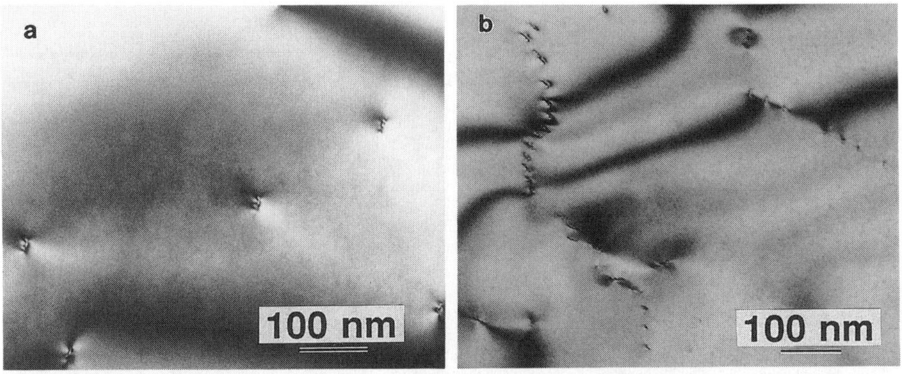

Fig. 2. Plan-view micrographs of the "A" GaN layers grown on the AlN buffer layer: (a) with Si doping and (b) undoped layer. Note the difference in the threading dislocation distribution in both samples.

Electron diffraction patterns from cross-sections of both samples showed that the GaN grew in the **c** direction, with the orientation relationship to the Al_2O_3 substrate of (0001)GaN//(11$\bar{2}$0)Al_2O_3 and [1$\bar{1}$00]GaN//[0001]Al_2O_3. A high resolution image from the AlN/Al_2O_3 interface (Fig. 3a) showed a high density of misfit dislocations. They were more visible when image filtering was applied (Fig. 3b). It was noticed that eight (1$\bar{1}$00) AlN planes fit five (0003) Al_2O_3 planes. No difference in misfit dislocation arrangement at the AlN/Al_2O_3 interface could be detected in undoped and doped samples. A high density of parallel defects, most probably dislocation loops, was formed in the AlN buffer layer. Such defects appeared not to propagate to the sample surface, but they did promote threading dislocation interactions and the formation of half-loops near the AlN/GaN interface instead of the straight dislocations, that are usually formed when GaN grows on sapphire without a buffer layer. Such half loops formed at the AlN/GaN interface had mixed character; and, depending on the type of dislocation formed, these half loops remained near the AlN/GaN interface or propagated to the surface.

The distribution of threading dislocations seen in cross-sectioned samples appears to be similar in both types of samples. This indicates that based exclusively on cross-section projections one cannot conclude if dislocations are randomly distributed or if they are arranged on subgrain boundaries. The highest defect density was observed near the interface of the GaN layer and the AlN buffer, but it decreased sharply as the GaN layer thickness reached 0.5 μm. It was surprising to see defects along the c axis of GaN (Fig.3c) distributed layer by layer with almost equal distances parallel to the interface (dislocation loops or bent threading dislocations). The reason for the appearance of these parallel defects and their distribution is not understood; however, they were similar to those observed in the AlN layer and in bulk GaN [4]. Whenever threading dislocations interacted with such a loop parallel to the sample surface, they bent over and did not simply extend directly to the sample surface (Fig.3c). This interaction of the parallel defects with dislocations was more effective in the doped samples. Formation of these parallel defects could explain the Raman shift of E2 phonon. This shift is related to the strain distribution in the layer. These Raman measurements were obtained on cross-section samples with about 1μm decrement along the c-axis (Fig. 4a). They indicate that in the undoped samples the highest strain could be detected at the interface and only a small decrease could be observed with increase of the sample thickness. However, a large gradient was observed in the Si doped sample. When compared to homoepitaxial GaN layers grown on bulk GaN substrates (566.2 cm^{-2}) the average

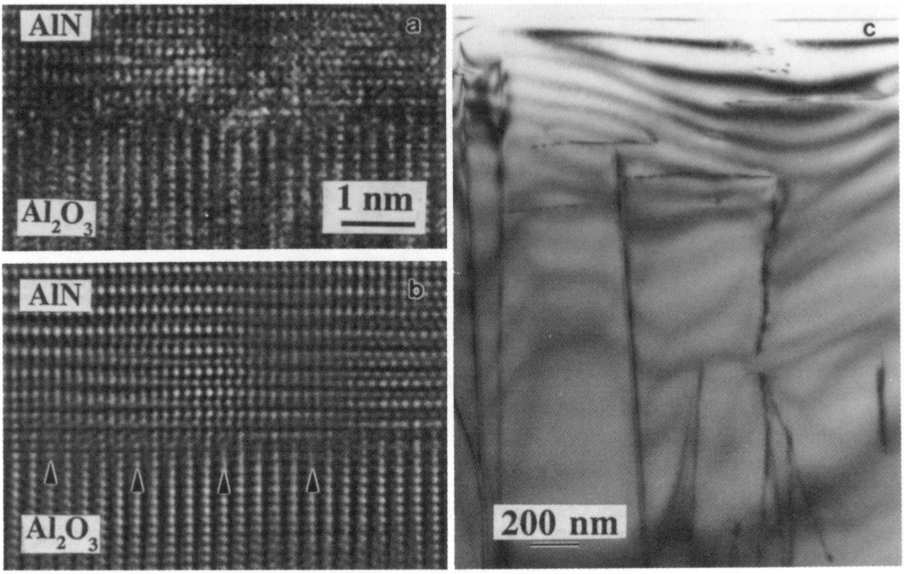

Fig. 3. (a) High resolution image of AlN/Al_2O_3 interface; (b) the same interface after noise filtering. Misfit dislocations are marked by arrows; (c) Formation of parallel defects almost equally distanced through the GaN layer and their interaction with the threading dislocations.

strain in the undoped samples was estimated as $\Delta\omega = 3.0$ cm^{-1} which is equivalent to 0.97 GPa. A similar value (0.99 GPa) was obtained based on x-ray measurements (Table I). The average strain estimated in the Si-doped layer based on the Raman shift of the E2 phonon was 0.81 GPa and based on x-ray measurements 0.77 GPa.

Table 1. TEM, x-ray and Raman data

	Samples "A" undoped	Si-doped	Samples "B Si-doped		
$N_{Si}(cm^{-3})$	----	$3x10^{18}$	$9x10^{16}$	$3x10^{18}$	$1x10^{19}$
t (μm)	3.5	5	1.8	1.9	1.8
$\rho_D(cm^{-2})$	$5x10^9$	$7x10^8$	$1x10^{10}$	$6x10^9$	$3x10^9$
$FWHM_\Theta$	8.0	4.2	12	10	10
c (Å)	5.1914	5.1900	5.1852	5.1857	5.1869
$\Delta c/c$ (%)	0.123	0.96	0.02	0.07	0.19
$\Delta\omega$ (cm^{-1})	3.0	2.5	0.15	0.50	0.90
$\delta_{\Delta\omega}$(GPa)	0.97	0.81	0.05	0.16	0.29
δ_c(GPa)	0.99	0.77	0.04	0.11	0.2

The c parameter of strain-free GaN is c=5.1850 Å [5].

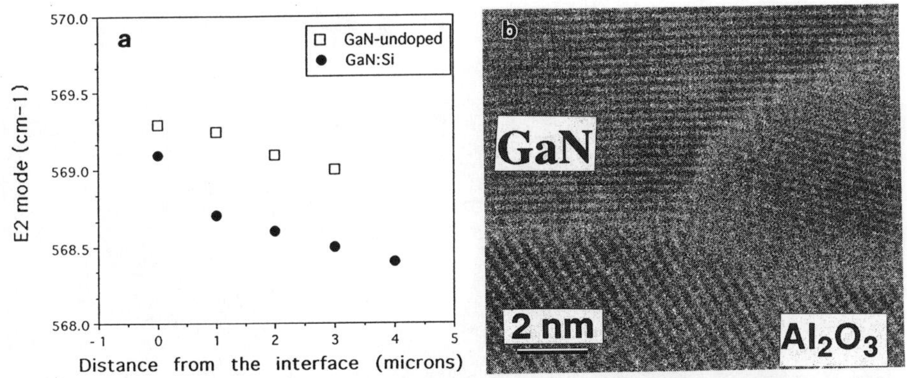

Fig.4. (a) Strain distribution in "A" samples measured by Raman shift; (b) High-resolution image of the substrate/buffer layer interface of "B" sample with Si concentration of $3x10^{18}$ $cm.^{-3}$ Note formation of a differently oriented grain at the interface with the substrate.

Doped GaN "B" samples grown on (0001) sapphire with the GaN buffer layer also showed threading dislocations mostly arranged in subgrain boundaries, similarly to undoped samples "A." However, for the highest doping ($1x10^{19}cm^{-3}$) a random dislocation arrangement could be observed with an average separation distance of a few hundred nm, much smaller than the grain diameter in the lower doped samples (Fig. 5a). The dislocation density in this highest doped sample was about $3x10^{9}cm^{-2}$ which is much higher than the average dislocation density in doped "A" samples. Since the grain diameter increases with doping to some extent, and for higher doping a more random dislocation distribution is observed, the movement of misfit dislocations appears to be larger in the samples with lower doping, since they move across the grain diameter. In the samples with the highest doping the misfit dislocations are almost immobile, so they quickly produce a high density of threading dislocations. This explains their random distribution in the layer.

In addition to threading dislocations small pinholes (6 nm in diameter) were observed in all three "B" samples. In plan-view these pin-holes appear as faceted slightly elongated hexagons. High resolution images from these pin-holes show that they are formed within the GaN layer since lattice images of $\{1\bar{1}00\}$ planes could be seen inside the hole. Some small particles about 1 nm in diameter (see Fig.5b) were always visible in all high resolution images of these pin-holes. No dislocation Burgers vector can be detected around such particles. Studying the contrast changes with defocus suggests that these small particles are probably foreign atoms, most probably small dopant clusters. The reason for the formation of these pinholes is not clear. They might be formed around a screw dislocation as suggested earlier [6], or around dopant inclusions since the density of these pinholes increases with dopant concentration. Association of a screw dislocation with such pinholes is unlikely, since no contrast typical for screw dislocations could be detected on any of the observed images of the pinholes.

The GaN buffer layer in "B" samples looks completely different in comparison with the AlN buffer layer in "A" samples. Dislocation loops parallel to the interface were not formed. Occasionally differently oriented GaN grains were formed at the interface (Fig. 4b). Half-loops were formed at the buffer/layer interface. Starting from this interface a steady decrease of dislocation density was visible due to the formation of parallel defects as was observed in samples "A."

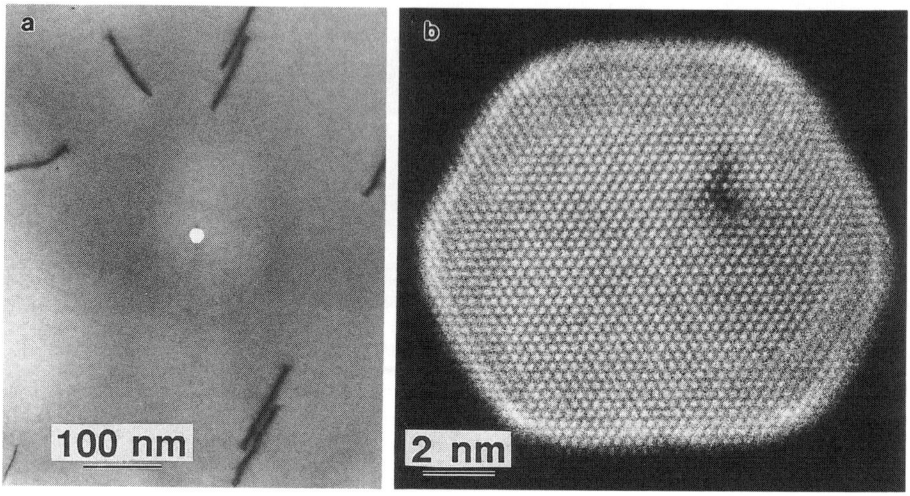

Fig.5. (a) Plan-view of the "B" sample with Si concentration of $1x10^{19}$ cm.$^{-3}$ Note randomly distributed dislocations and a pinhole in the center of the micrograph; (b) High resolution micrograph from a pinhole. Note black contrast of the defect (dopant cluster) within the pinhole.

The Raman shift observed in "B" samples indicated a much lower strain in these samples compared to "A" samples grown with the AlN buffer layer. The stress in the low-doped samples ($9x10^{16}$ cm^{-2}), is very small, close to that observed in homoepitaxial layers. Surprisingly, the stress increases with doping level in "B" samples, (whereas in "A" samples doping caused a stress decrease). However, on average the samples with the highest dopant concentration have only a small fraction (1/3) of the stress compared to the undoped "A" samples (Table I). This would suggest that the role of Si doping in strain relaxation can be rather small. The buffer layer appears to determine strain level in these samples.

CONCLUSIONS

Si doping of GaN layers grown by MOCVD improves the structural quality of the layer: the top surface is smoother and the density of threading dislocations is smallest. This is in agreement with the observation of a decrease of dislocation density with increase of dopant concentration (S, Te, and Zn) in GaAs and (InP) [7].

It was evident that strain relaxation was related to dislocation density and therefore, was directly influenced by the buffer layers. If one compares two samples with the same concentration of dopant ($3x10^{18}$cm^{-3}) it appears clear that the density of dislocations is much smaller in "A" samples. Based on early work of Amano et al [3] one can almost neglect the fact that samples "A" were grown on α-planes of sapphire and samples "B" were grown on c-planes. Therefore, the AlN buffer layer causes a reduction of the threading dislocation density in the layer. However, their smaller density causes an increase of strain. At this point it is clear that the buffer layer plays a much more important role in strain relaxation than does the Si doping.

Comparing both sets of samples one can see that the FWHM of x-ray rocking curves is much larger in the samples where dislocations are arranged on subgrain boundaries (see Table I). These dislocations do not interact with the parallel defects as effectively as do the randomly distributed dislocations. This finding confirms Raman studies that stress distribution in the samples with dislocations arranged on grain boundaries is equally distributed through the layer thickness, whereas a gradient in strain (smaller toward the sample surface) was found in the samples with random dislocation arrangement.

Different types of micro-tubes (pinholes or voids) were found in the two sets of samples. Samples "A" grown on the α-plane of sapphire with an AlN buffer had a high density of these defects always connected with a threading dislocation with an edge component, whereas the defects found in "B" samples were not associated with a dislocation Burgers vector. Based on the study of contrast around these defects it is also unlikely that screw dislocations could be associated with them. Therefore, it appears that formation of these pinholes in "B" samples is rather related to the dopant clusters, since the density of pinholes increases with dopant concentration. It is not clear at this point if the large diameter micro-tubes seen by AFM are of the same nature as the pinholes with small diameter diameter seen by TEM in these plan-view samples.

ACKNOWLEDGEMENT

This work was supported by the Director, Office of Basic Science, Materials Science Division, U.S. Department of Energy under the Contract No. DE-AC03-76FS00098. Use of the microscopes in the National Center of Electron Microscopy in LBNL is greatly appreciated.

REFERENCES:

1. T.D. Moustakas and R.J. Molnar, Mat. Res. Soc. Conf. Proc. **281**, 753 (1993).
2. S. Nakamura, T. Mukai, and M. Senoh, Appl. Phys. Lett., 64, 1687 (1994).
3. H. Amano, K. Hiramatsu, and I. Akasaki, Jpn. J. Appl. Phys. 27, L1384 (1988).
4. Z. Liliental-Weber, C. Kisielowski, S. Ruvimov, Y. Chen, J. Washburn, I. Grzegory, M. Bockowski, J. Jun, and S. Porowski, J. Electr. Mat. (1996) in print.
5. M. Leszczynski, T. Suski, H. Teisseyre, P. Perlin, I. Grzegory, J. Jun, S. Porowski, and T.D. Moustakas, J. Appl. Phys. 76, 4909 (1994).
6. W. Qian, M. Skowronski, K. Doverspike, L.B. Rowland, and D.K. Gaskill, J. Cryst. Growth 151, 396 (1995).
7. T. Kamejima, S. Matsui, and H. Watanabe, J. Appl. Phys. 50, 3312 (1979)

ESR AND LESR STUDIES IN CVD DIAMOND

C.F.O. Graeff*, E. Rohrer*, C.E. Nebel*, M. Stutzmann*, H. Güttler**, R. Zachai**
*Walter Schottky Institut, TU-München, D-85748 Garching, Germany,
carlos.graeff @wsi.physik.tu-muenchen.de
**Daimler-Benz AG, Forschung und Technik, D-89013 Ulm, Germany

ABSTRACT

CVD diamond films with nitrogen content varying from 10 ppm to 132 ppm have been studied by electron spin resonance (ESR), light-induced ESR (LESR) as well as spin-dependent conductivity (SDC). Two characteristic signals have been observed. A carbon-related defect line with $g = 2.0029 \pm 0.0002$ and width 4 ± 1 G, is observed in ESR, LESR and SDC. The intensity of this line measured by ESR increases linearly with nitrogen content. For low-defect-density samples, or after illuminating the high-defect-density samples with UV light, a second signal is observed both in ESR and LESR, but not in SDC, with a central line at $g = 2.0024 \pm 0.001$ and width 0.2 ± 0.1 G and related hyperfine satellites ≈ 30 G away from the central line. This line is assigned to isolated substitutional nitrogen, the so-called P1 center. The density of N-related paramagnetic states is strongly affected by illumination and heat treatments. Spin-dependent conductivity measurements show that the dark conductivity at room temperature in CVD-diamond is dominated by hopping at the $g = 2.0029$ defects.

INTRODUCTION

The unique properties of diamond have attracted considerable interest, especially after the discovery that low cost and large area diamond films can be grown by the chemical vapor deposition (CVD) technique [1]. The structural properties of diamond grown by CVD have significantly improved in recent years. High quality polycrystalline films can be grown up to several hundreds of micrometers in thickness. The addition of small amounts of nitrogen to the process gas gives rise to faceted growth with (100) columnar texture, making this material a promising candidate for active electronic devices [1]. Successful applications require, however, the optimization of material properties with respect to carrier mobility, doping, and defect densities.

From studies on natural and synthetic diamond, it is known that nitrogen is the dominant impurity, with many possible configurations from isolated substitutional to aggregate [1,2]. The substitutional dispersed form is a deep donor with a level 1.7 eV below the bottom of the conduction bands [3], giving rise to a well-established electron spin resonance (ESR) signal, known as the P1 center. On the other hand, very little is known about nitrogen in CVD-diamond films and the data available in the literature are not consistent. For example, despite the fact that nitrogen is a deep donor, it has been argued that „intrinsic" diamond, which contains typically 10 to 80 ppm nitrogen, is p-type [4].

Here we report on electron spin resonance (ESR), light-induced (LESR) and spin-dependent conductivity (SDC) experiments on CVD-diamond films with nitrogen content varying from 10 to 132 ppm, with the aim of better understanding the effects of nitrogen on the film properties.

EXPERIMENT

The diamond films investigated in the present experiments have been grown by a standard microwave plasma-assisted CVD described in ref. [5], using methane and hydrogen as the main gas sources. Two series of samples have been investigated. In the first series, the nitrogen content in the gas phase has been varied from 0 to 4%, while other deposition parameters were kept constant. In the second series, an optimization of the deposition parameters was performed (substrate temperature and substrate bias voltage, while the N content was kept at around 10 ppm). The substrate was silicon. The absolute nitrogen content has been determined by elastic recoil detection [6]. During growth, the substrate temperature for the first series was fixed at 750 °C, and in the second series was varied between 800 and 900 °C.

Mat. Res. Soc. Symp. Proc. Vol. 423 © 1996 Materials Research Society

For the ESR and LESR experiments, free-standing samples were placed in the TE_{102} cavity of a standard X-band ESR spectrometer. All experiments have been performed at room temperature using 0.07-100 kHz magnetic field modulation, with phase sensitive detection. Samples were illuminated with a Hg/Xe arc lamp for the LESR experiments. Low-pass cut-off optical filters were used to control the photon energy in the LESR experiments. Thermal annealing from 200 to 300 °C was performed with typical heating and cooling times of 60 s in a thermally stabilized oven under air. For the SDC measurements, samples with coplanar contacts were used. The spin-dependent change of the conductivity was also measured by modulating the magnetic field.

RESULTS

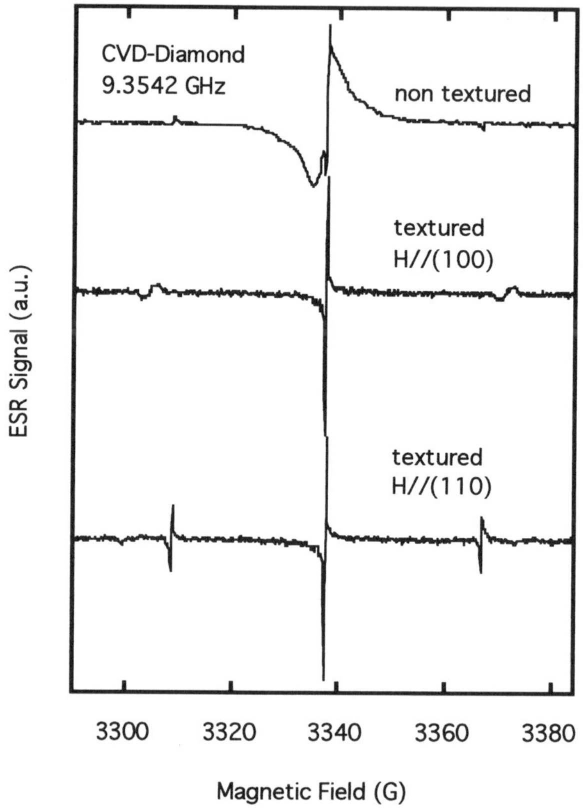

Fig.1: Typical ESR and LESR spectra for two different types of CVD-diamond films: highly oriented (100) textured (ESR), and non textured (LESR). Note that for the former, the anisotropy in the hyperfine constants are clearly observed.

In Fig. 1, typical ESR and LESR spectra are shown for two types of CVD-diamond films: highly oriented, (100) textured, and non textured (35 ppm N content). The signal consists of two (narrow and broad) central lines and satellite peaks. The broader line has a g-factor of 2.0029 ± 0.0002 and a peak-to-peak width ΔH_{pp} of 4 ± 1 G. The narrow line is a triplet composed of a central line with g = 2.0024 ± 0.0001 and ΔH_{pp} = 0.2 ± 0.1 G, and satellite peaks coming from the hyperfine interaction of the electron spin with the ^{14}N nuclei.

The broader (g=2.0029) line is present in all samples, independent of nitrogen content, both in the dark (ESR) and under illumination (LESR). Note that this signal consists of a single central line with unresolved shoulders, attributed recently to carbon dangling bonds, hyperfine-coupled to nearby hydrogen nuclei [7,8].

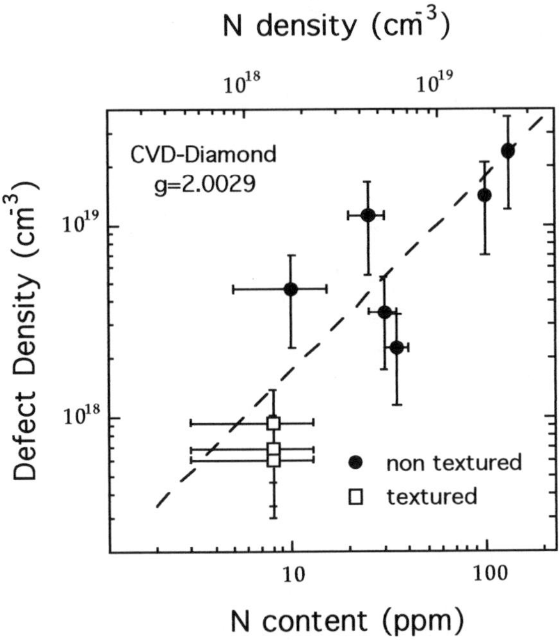

Fig. 2: Defect density of the C-related defect as measured by ESR, for textured and non textured samples, as a function of N content. The dashed line corresponds to the density of N in the film.

In Fig. 2 the density of the g=2.0029 defects is plotted against the N content, for textured and non textured samples. As can be seen, optimized deposition conditions that lead to highly oriented, (100) textured samples are related to a decrease by a factor of 10 in the C-related defect density. Note that the density of C-related defects is to a first approximation equal to the density of N in the film (dashed line). This would indicate that N incorporation induces the creation of C-related deep defects, and thus to achieve a material with low defect density, the incorporated N content must lie below 10 ppm.

The detrimental effects of the C-related defects are clearly observed from the spin-dependent conductivity results presented in Fig. 3. The dark conductivity of the present samples, at room temperature, is dominated by a hopping process, most probably through states in the grain boundaries [11]. Typical dark conductivity activation energies at room temperature are very small, of the order of 100 meV. As can be seen in Fig. 3, the SDC signal is very similar to the g=2.0029 line measured with ESR.

The narrower (g=2.0024) line, attributed to paramagnetic nitrogen, is found to be strongly influenced by deposition conditions, thermal, and illumination treatments. In Fig.1, the sample identified as non textured corresponds to a sample grown under non optimized conditions, with 35 ppm incorporated N. For this sample, the g=2.0024 line can only be detected after illumination with photon energies hν greater than 4 eV. An isotropic set of two satellite peaks is found at ± 29.2 G relative to the central line. In the case of the highly oriented, (100) textured samples, the anisotropy of the hyperfine splitting is clearly observed. From the curves, the anisotropic hyperfine coupling parameters are easily obtained, $A_{//}$ = 114.0 and $A_{\perp}$ = 81.3 MHz, in perfect agreement with the well known P1 center in natural diamond [2]. In fact the non-textured line shape corresponds to the powder, or randomly oriented spectrum using these hyperfine parameters [9]. Note that the distortions in the satellite line shapes for the textured samples are due to imperfect orientation of the crystallites. In fact, from the line shape it is possible to quantify the degree of disorientation of these films. With a simple geometrical model that takes into account the presence of a distribution of crystallite orientations, it is possible to simulate the observed ESR spectra. Using such a procedure, it is that for these highly oriented (100) textured samples, deviations in the crystallite orientation are of the order of 10°, in good quantitative agreement with results obtained from X-ray diffraction in the same samples.

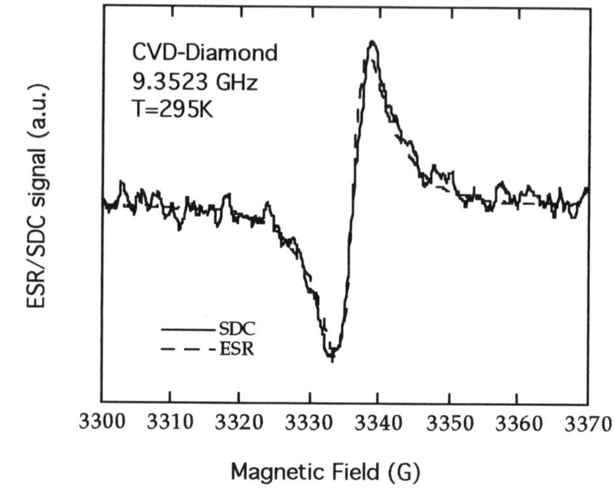

Fig. 3: Typical SDC and ESR spectra for CVD-diamond films. The line shape and g-factor for SDC and ESR spectra are practically identical.

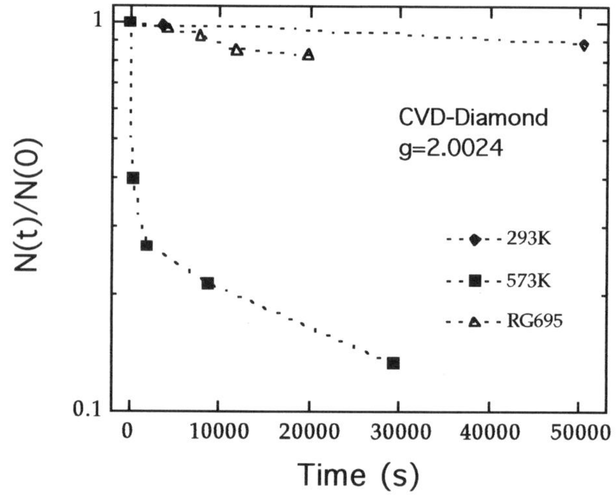

Fig. 4: Normalized N-related signal intensity (N(t)/N(0)) decay as a function of time after UV illumination, plotted for typical thermal and illumination treatments. The illumination treatment (RG695 in the legend) was performed by illuminating the sample with low band pass filter cutting at 695 nm. The dashed lines are just guides to the eye.

In the textured samples, the N-related signal is detected in the as-grown state, but can be enhanced by a factor of 3 by light exposure with photons of hν > 4 eV. In both textured and non textured samples, the increase in the N-related defect is accompanied by the same increase of the C-related defect spin density. Note however, that the density of C-related defects is at least a factor 20 higher than the N-related P1 center. The light-induced changes in the occupation of the P1 center is weakly dependent on light intensity, varied between 1 and 100 mW/cm^2, and stable when the sample is left in the dark at room temperature.

Accelerated quenching of the light-enhanced signal can be achieved by applying higher temperatures or by illuminating the sample at room temperature with photon energies smaller than 3 eV. Typical results are plotted in Fig. 4 for the normalized signal intensity (N(t)/N(0)) as a function of time, for different experimental conditions. The activation energy of the characteristic decay time τ, measured from the dynamics of the process at different temperatures, or using different monochromatic light sources, is found to be around 1.5 eV.

The results presented above are in good agreement with photoconductivity experiments on synthetic diamonds [10] and with theoretical calculations of the electronic structure [3] of crystalline diamond, where isolated N is found to be a deep donor with a level of about 1.7 eV below the top of the conduction band. The energetic position of the C-related defect is not known, but must lie well bellow the conduction band. The density of P1 centers after illumination is found to be about 5% of the incorporated N (or the C-related defect). Thus N can be either incorporated substitutionaly (paramagnetic), or in other forms. 95 % of the incorporated N is not paramagnetic, and apparently responsible for the creation of a C-related defect (g=2.0029), which then acts as an efficient compensation center.

For the SDC, the g=2.0024 line was not observed, even after illumination with UV light. This result is expected since the SDC is selectively sensitive only to states that leads to the transport of charge, in this case dominated by hopping in C-related defects near the Fermi level.

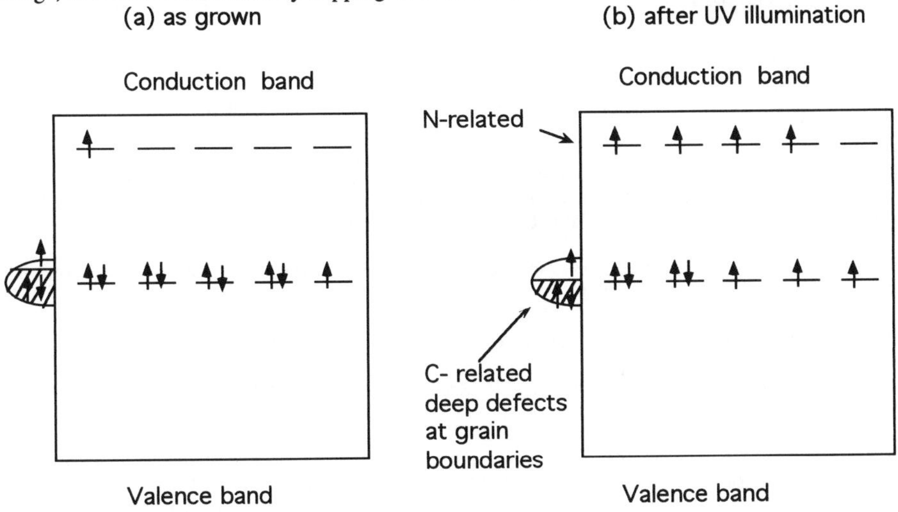

Fig. 5: Schematic diagram showing the changes in occupation of the N-related defects, and C-related deep defects, before (a) and after UV-illumination (b).

A simple model to explain such effects is presented schematically in Fig. 5. Note that it is still not clear whether the C-related defects are mainly in the grain boundaries, or in the bulk of the crystal, and the same is true in the assignment of their energetic position in the band gap. In the as-grown condition, the N-donor states are practically empty due to the presence of the C-related defects with a density at least a factor 20 higher than the former. The C-related defects must lie deep in the band gap (at least deeper than the N-donor state), either in the bulk or at the interface (grain boundaries), and are predominantly diamagnetic. With UV illumination, part of the electrons in the C-related states are transferred to the N-donor states, increasing the density of both paramagnetic N-donor states (P1 center), and C-related defects.

CONCLUSIONS

In CVD-diamond, two lines can be observed with ESR, LESR, and SDC. One with a g=2.0029, related to a C defect, and the second with g=2.0024, attributed to dispersed substitutional N. Both lines are strongly dependent on growth conditions, and their intensity can be manipulated with illumination and heat treatments. Our results indicate that N has a donor level at about 1.5 eV below the conduction band, but is also responsible for the creation of deep defects in a relation of roughly 20 defects for each donor state. Thus N will hardly be useful as a dopant, and it is important to keep N-concentration levels in the film below 10 ppm in order to achieve materials with low deep defect densities. SDC indicates that hopping through deep defects, most probably in the grain boundaries, is the dominant conductivity path at room temperature in our films.

ACKNOWLEDGMENTS

This work has been supported by the Bundesministerium für Bildung, Forschung, Wissenshaft und Technologie under contract number 3N1001G4. We thank Dr. Michael Kelly for a critical reading of the manuscript. C.F.O.G. is pleased to acknowledge support from the Alexander von Humbolt Stiftung (Germany), and FAPESP (Brazil).

REFERENCES

1. J.I.B. Wilson and W. Kulisch, special issue of physica status solidi (a) **154**: Diamond Thin Films, (1996).

2. W.V. Smith, P.P. Sorokin, I.L. Gelles and G.J. Lasher, Phys. Rev. **115**, 1546 (1959). For a review on electron spin resonance in natural and high pressure synthetic diamond see, J.H.N. Loubser and J.A. van Wyk, Rep. Prog. Phys. **41**, 1201 (1978).

3. S.A. Kajihara, A. Antonelli, B. Bernholconkin, and K. Okumura, Appl. Phys. Lett. **59**, 3148 (1991).

4. J. Mort, M.A. Mach, R. Car, Phys. Rev. Lett. **66**, 2010 (1991).

5. H.-J. Füsser, M. Rosler, M. Hartweg, R. Zachai, X. Jiang and C.-P. Klages, Electrochem. Soc. **93** (17), 102 (1993).

6. A. Bergmeier, G. Dollinger, T. Faestermann, C.M. Frey, M. Ferguson, H. Guehler, G. Schulz, H. Willerscheid, Diamond and Related Materials, 1996 (unpublished).

7. H. Jia, J. Shinar, D.P. Lang and M. Pruski, Phys. Rev. B **48**, 17595 (1993).

8. S.L. Holder, L.G. Rowan and J.J. Krebs, Appl. Phys. Lett. **64**, 1091 (1994).

9. M. Hoinkis, E.R. Weber, M.I. Landstrass, M.A. Plano, S. Han, and D.R. Kania, Appl. Phys. Lett. **59**, 1870 (1991).

10. R.G. Farrer, Solid State Commun. **7**, 685 (1969).

11. E. Rohrer, C.F.O. Graeff, R. Jansen, C.E. Nebel, M. Stutzmann, H. Güttler, and R. Zachai Phys. Rev. B, 1996 (unpublished).

ELECTRON PARAMAGNETIC RESONANCE STUDIES OF HF-CVD DIAMOND FILMS

B. RAMAKRISHNAN*, D. J. KEEBLE **, H. RODRIGO †, A. KULKARNI ††

*Department of Physics, Michigan Technological University, Houghton, MI 49931.

**Carnegie Laboratory of Physics, Department of APEME, University of Dundee, Dundee, DD1 4DH, UK.

†Department of Electrical Engineering, Royal Melbourne Institute of Technology, Melbourne, Australia.

††Department of Electrical Engineering, Michigan Technological University, Houghton, MI 49931.

ABSTRACT

Diamond thin films have been deposited on silicon substrates by hot-filament chemical vapor deposition (HF-CVD). Substrate temperature and methane concentration have been varied and the resulting structural properties of the deposited films studied. Raman spectroscopy, scanning electron microscopy and electron paramagnetic resonance (EPR) measurements were performed. The EPR measurements showed a single spectrum at g = 2.0027(2). The bulk concentration of the paramagnetic species, as determined from the total EPR absorption were found to vary in the range 10^{17} to 10^{19} cm^{-3}. Low paramagnetic defect concentrations were found for samples exhibiting a low non-diamond carbon contribution to the Raman spectrum. These samples were those grown with a methane concentration of 1 % or less.

INTRODUCTION

The growth of diamond at low temperatures and pressures by chemical vapor deposition has now developed so that films of high quality, as characterized by Raman spectroscopy, can be routinely fabricated. Extensive studies have been made on diamond films using various characterization techniques. The technological important thermal[1,2] and electrical properties[3,4] of these films are dependent on the incorporation of point and extended defects and non-diamond phases. An approximate relation has been shown between the Raman full width at half maximum (FWHM) and the electrical properties, and has been attributed to differences in the defect density[3]. Electron paramagnetic resonance (EPR) spectroscopy has high sensitivity to paramagnetic charge states of point defects in carbon phases. A number of EPR studies on diamond films have been reported[5-10], but a complete understanding of the origin of the paramagnetic defects observed has yet to be achieved.

Mat. Res. Soc. Symp. Proc. Vol. 423 © 1996 Materials Research Society

Previous studies[5-10] have reported the observation of an intense EPR spectrum from as-grown diamond films. The reported g-values range from 2.0021 to 2.0028. Variations in lineshape and linewidth have also been observed. Typically the spectrum exhibits partially resolved structure. Jia *et. al.*[8] have studied the EPR spectra from diamond thin films grown with gas mixtures H_2/CH_4, D_2/CH_4 and $H_2/^{13}CH_4$ from which they were able to conclude that the dominant contribution to the spectrum involved a paramagnetic center localized on a carbon site. The satellite lines were assigned to an interaction with hydrogen nuclei. In a 9.5 and 35 GHz EPR study of diamond films Holder *et.al.*[7] observed that the satellite lines were at ~ $\pm g_N \beta_N B_0$ with respect to the central line showing that they could not be due to hyperfine interaction but likely resulted from weak magnetic dipole coupling between the electronic spin and neighboring hydrogen nuclei[11] and were so called 'spin-flip' satellites. Earlier cross-polarization (CP) combined dynamic nuclear polarization (DNP) magic angle spinning (MAS) experiments[12] on diamond films had detected proton polarization transferred to the ^{13}C spin-system of a paramagnetic center. Only ^{13}C atoms associated with a paramagnetic center and with protons in the immediate vicinity can contribute to such a signal. However, in a recent multi-frequency EPR study[10] the splitting of the satellite lines was studied and found to deviate from theory. The satellite lines were attributed to a biradical center. The overall EPR lineshape was fitted using two lorentzian lines in combination with the doublet spectrum for the biradical center[10].

There is a similarity between the diamond thin film spectrum and that observed from amorphous carbon [13] (a-C) thin films. It has been suggested that the diamond thin film spectrum is due to dangling bond defects present in non-diamond carbon phases[14]. An alternative interpretation has been given by Fanciulli and Moustakas[5] who have noted the similarity with the spectrum assigned to the negative charge state of the carbon vacancy in single crystal diamond.

A limited number of studies on the influence of growth conditions on the EPR spectrum have been made. Watanabe and Sugata[9] studied films synthesized at different pressures and also at various concentrations of different hydrocarbon gases. Fanciulli and Moustakas[5] used the variation of substrate temperature and, in particular, system pressure to study the defects in the diamond thin films. In this study, we report EPR measurements on diamond films fabricated with the variation of methane concentration and substrate temperature. The results from EPR are compared to the results from Raman Spectroscopy.

EXPERIMENTAL DETAILS

The diamond films in this study were grown by hot-filament chemical vapor deposition (HFCVD), from a gas mixture of CH_4 and H_2. The samples were held on a molybdenum substrate holder heated by tungsten filaments. Growths were performed using tungsten filaments at 1900-2100 ^{o}C with a mixture of high purity methane and hydrogen giving a total flow rate of 200 sccm at a pressure of 40 Torr. Growths were typical of 8 hrs duration but several longer 12-16 hrs growths were also made. The films were deposited on a *p*-type (100) silicon (>100 Ωcm) scratched with diamond paste and ultrasonically cleaned in distilled water and acetone. A series of films were grown with the substrate temperature (T_S) varied between 825 -950 °C with a CH_4/H_2 of 1%. Films were also grown using T_s = 850 °C but with different CH_4/H_2 ratios in the range 0.25 to 5%. The data from twenty nine growths are presented here, at least three growths for each methane concentration were performed.

The films were characterized by Raman spectroscopy and Scanning Electron Microscopy (SEM). Raman spectroscopy was carried out using a ISA JOBIN-YVON U1000 spectrometer with the 514.5nm line from an argon ion laser. A JOEL JSM-35 scanning electron microscope was used to observe the morphology and measure grain size and film thickness. EPR measurements were carried out on a Varian E109 X-band spectrometer at room temperature with 100-kHz field modulation. Spectra were recorded using modulation amplitude of 0.05 mT and with a microwave power of 12 mW. Power saturation studies were carried on a selection of the samples. The N_S^0 spectrum from a synthetic diamond Ib crystal was used as a reference to calculate the g-value. However, measurements were also performed on a Bruker ESP300 X-band spectrometer where the magnetic field was calibrated using a NMR magnetometer and the microwave frequency was measured using a frequency counter. The area under the absorption was determined from a calibration using a series of α, α'-diphenyl-β-picarylhydrazyl (DPPH) samples of known weight.

RESULTS AND DISCUSSION

Typical Raman spectra for samples fabricated with the variation of substrate temperature and methane concentration are shown in figures 1a and 1b, respectively. It can be seen from figure 1a that all the films exhibit the characteristic first order Raman peak centered at 1332 cm^{-1}. The signal from the non-diamond phases (in the region around 1550 cm^{-1}) are negligibly small except for the samples grown at substrate temperatures of 825 and 950 °C. As expected, the increase in methane concentration did produce films with significantly increasing amount of non-diamond carbon content as shown in figure 1b. The surface morphology indicated a polycrystalline film with a mixture of (100) and (111) facets for all the samples grown with a methane concentration less than 3.5 % (see figure 2a). The samples grown with a methane concentration of 3.5 % and 5 % indicated a ball-like structure (see figure 2b) characteristic of diamond/non-diamond codeposition. The grain size was found to vary between 2-9 μm and the film thickness ranged between 2-15 μm.

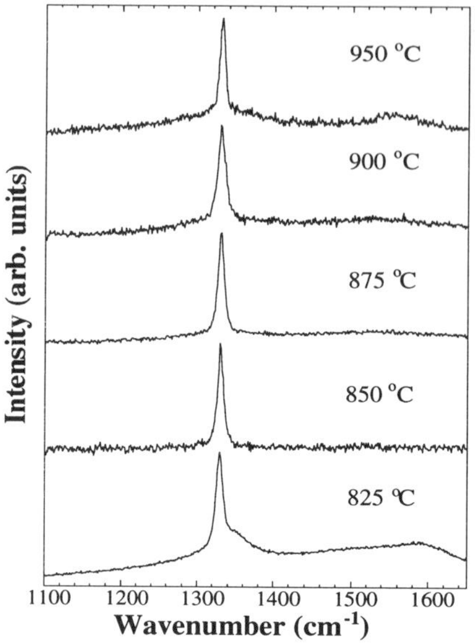

Figure 1a - Raman spectra of diamond films fabricated with the variation of substrate temperature.

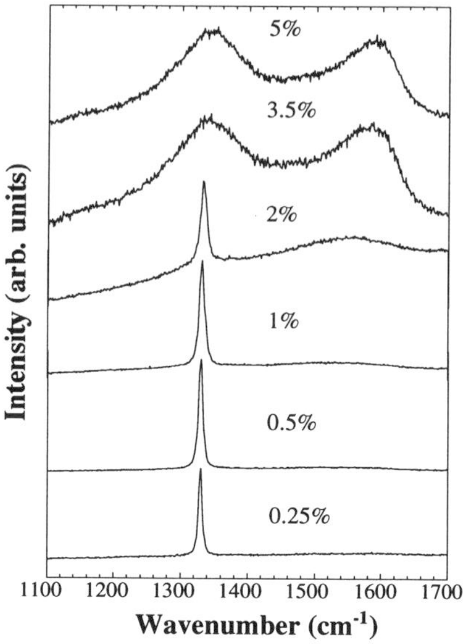

Figure 1b - Raman spectra of diamond films fabricated with the variation of methane concentration.

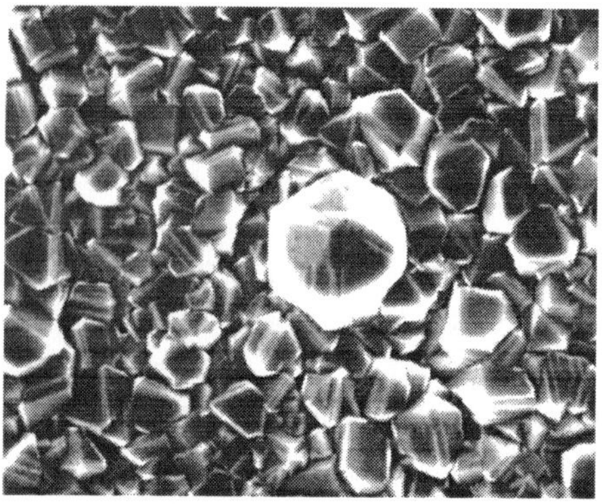

Figure 2a - SEM picture of a typical poly-crystalline film with a mixture of (100) and (111) facets.

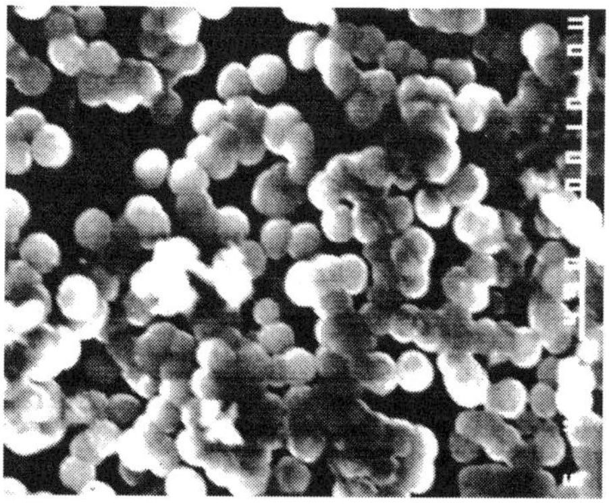

Figure 2b - SEM picture indicating the ball-like morphology.

All samples studied exhibited an EPR spectrum at $g = 2.0027(2)$ with an intensity greater than that observed for the silicon damage signal from the scratched substrate. The variation of spin concentration with respect to substrate temperature and methane concentration are shown in figures

3a and 3b, respectively. As seen from figure 3a, EPR spin concentration does not change significantly for the samples grown with various substrate temperatures except for the sample grown at a substrate temperature of 950 °C. A substantial change in the EPR spin concentration is observed for samples grown with varying methane concentration. It can be seen from figure 3b that EPR spin concentration increases with increasing CH_4/H_2 ratio and is minimum for sample grown with a methane concentration of 0.5%. A comparison of figures 1b and 3b suggest an increase in EPR spin concentration with the increase in amorphous carbon content of the films. The amorphous carbon content could not be evaluated quantitatively because of the different Raman scattering efficiencies of diamond and non-diamond carbon phases.

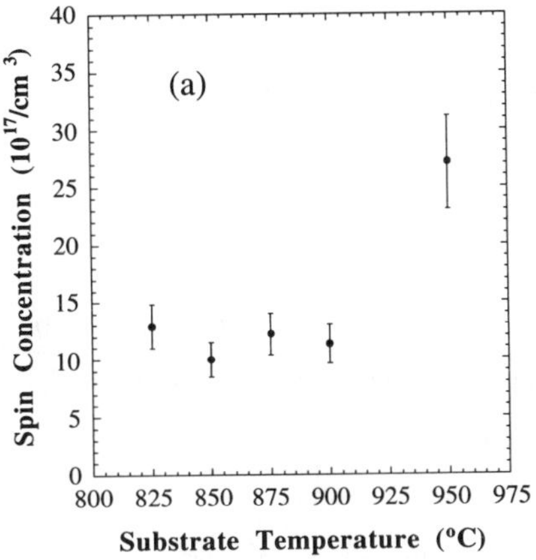

Figure 3a - Substrate temperature vs. EPR spin concentration.

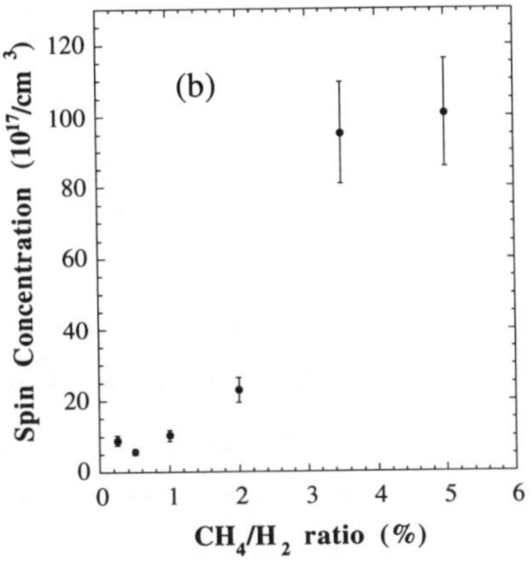

Figure 3b - Methane concentration vs. EPR spin concentration.

CONCLUDING REMARKS

Optimal growth conditions for diamond thin films were found to be in the range 0.25-1 % methane with a substrate temperature of 850 °C. Increasing methane concentration results in an increase in the amount of non-diamond phases. Electron paramagnetic resonance studies show that the concentration of paramagnetic defects increases systematically with increasing methane concentration. These results are consistent with a model that includes a non-diamond phase paramagnetic center contribution to the observed EPR spectrum.

ACKNOWLEDGMENTS

The authors wish to thank Mr. K. Paxton, Mr. T. Gasperich and Dr. J. Hwang, Institute of Materials Processing, Michigan Technological University for their hospitality. BR would like to thank Mr. D. Hembree and Mr. C. T. Olmstead at ORNL for some of the Raman spectra. DJK also wishes to thank Dr. E.H. Poindexter, ARL, Fort Monmouth for his hospitality. This work was funded in part by the Michigan Research Excellence Fund.

REFERENCES

[1]. C. Uher and D. T. Morelli, J. Appl. Phys. **76** (3), 1515-7 (1994).

[2]. J. E. Graebner, S. Jin, J. A. Herb *et al.*, J. Appl. Phys. **76** (3), 1552-6 (1994).

[3]. M. A. Plano, M. I. Landstrass, L. S. Pan *et al.*, Science **260**, pp. 1310-12, (1993).

[4]. M. I. Landstrass and K. V. Ravi, Appl. Phys. Lett. **55** (14), 1391-3 (1989).

[5]. M. Fanciulli and T. D. Moustakas, Phys. Rev. B **48** (20), 14982-88 (1993).

[6]. M. Hoinkis, E. R. Weber, M. I. Landstrass *et al.*, Appl. Phys. Lett. **59** (15), 1870-1 (1991).

[7]. S. L. Holder, G. L. Rowan, and J. J. Krebs, Appl. Phys. Lett. **64** (9), 1091-93 (1994).

[8]. H. Jia, J. Shinar, D. P. Lang *et al.*, Phys. Rev. B **48** (23), 17595-8 (1993).

[9]. I. Watanabe and K. Sugata, Jpn. J. Appl. Phys. **27**, 1808-11 (1988).

[10]. D. F. Talbot-Ponsonby, M. E. Newton, J. M. Baker *et al.*, J. Phys.: Condens. Matter **8**, 837-849 (1996).

[11]. G. T. Trammell, H. Zeldes, and R. Livingston, Phys. Rev. **110** (3), 630-4 (1958).

[12]. Herman Lock and Gary E. Maciel, J. Mater. Res. **7** (10), 2791-7 (1992).

[13]. M. Hoinkis, E. D. Tober, R. L. White *et al.*, Appl. Phys. Lett. **61** (22), 2653-6 (1992).

[14]. I. Watanabe, K. Haruta, and Y. Shimamura, Jpn. J. Appl. Phys. **33**, 2035-40 (1994).

Electrical and optical investigation of the position of vanadium related defects in the 4H and 6H SiC bandgaps

J.R. Jenny, M. Skowronski, W.C. Mitchel*, S. R. Smith*, A. O. Evwaraye*, H.M. Hobgood**, G. Augustine**, and R.H. Hopkins**
Department of Materials Science and Engineering, Carnegie-Mellon University, Pittsburgh, PA 15213
*Wright Laboratories, Materials Laboratory (WL/MLPO) Wright-Patterson Air Force Base, OH
**Northrop Grumman Science and Technology Center, Pittsburgh, PA 15235

Abstract

Hall effect, deep level transient spectroscopy, optical absorption, and optical admittance spectroscopy were employed to determine the position of the vanadium acceptor and vanadium-nitrogen complex in vanadium- and nitrogen-doped 4H and 6H SiC. Hall effect results indicate that the acceptor position in 4H(6H) SiC is 0.80(0.66) eV beneath the conduction band edge. The DLTS signature of the defect in the 4H polytype showed an ionization energy of 806 meV and a capture cross section of 1.8×10^{-16} cm^{-2} The optical absorption measurements proved that the acceptor level investigated is related to isolated vanadium, and therefore the vanadium acceptor level. Based upon DLTS and SIMS measurements, the maximum solubility of vanadium in SiC was determined to the 3×10^{17} cm^{-3}. An examination of polarized light experiments indicates that vanadium also complexes with another element to form electronic(at 5000 cm^{-1}) and vibrational absorption(at 683 cm^{-1}) bands. While the other constituent cannot be identified, evidence suggests that nitrogen is a likely candidate. This complex introduces a deep level at E_c-0.78 eV as determined using optical admittance spectroscopy.

I. Introduction

Due to its wide band gap and inherent properties, silicon carbide has been sought out as a material which can be employed in unique conditions, where other semiconductors would fail. Silicon carbide is specifically well suited to both high temperature and high power applications. However, due to the extreme conditions associated with the production of suitable wafer-sized boules of this material, the incorporation of impurities is required to facilitate its utility. Deep levels are required to elucidate the desired properties which would otherwise be present in pure silicon carbide. One element which has been identified as having the potential of producing such a level is vanadium. It has been studied with electron paramagnetic resonance (EPR)[1], photo-EPR[2], and infrared absorption[1]. From these measurements, vanadium has been found to occupy silicon substitutional sites in the three following charge states: positive ($3d^0$), neutral ($3d^1$), and negative ($3d^2$), and therefore produces two levels in the SiC bandgap. The donor level (0/+) has been shown[3] to reside near the middle of the band gap (at E_c-1.35 eV).

In the absence of all other considerations, vanadium appears to be the most promising element for the production of high resistivity silicon carbide. However, one must not only look at isolated defects which are created when an element is incorporated, but also the different complexes formed as well as the viability for their formation when examining a potential deep dopant. We report on the discernment of a complex which can be attributed to the pairing of vanadium and nitrogen. Prior to this examination of this vanadium complex, an discussion of the vanadium acceptor level is undertaken.

Mat. Res. Soc. Symp. Proc. Vol. 423 © 1996 Materials Research Society

II. Sample Preparation/Experimental Setup

For this study, SiC samples were sectioned from a vanadium-doped SiC c-axis oriented boule grown using a physical vapor transport technique which has been previously described.[4, 5] The boules were first sliced into 40 mm diameter wafers which were then cut into 1.0 cm x 1.0 cm x 0.5 mm samples for optical absorption studies and 0.5 cm x 0.5 cm x 0.5 mm for hall effect experiments. The sample surfaces were prepared by lapping and subsequent polishing with progressively finer diamond pastes. The absorption samples were employed without further preparation, but the Hall effect samples required contact deposition.

5000 Å layers of tungsten were sputtered onto SiC samples in a van der Pauw configuration.[6] The samples were then annealed for 60 seconds at 550°C in a reducing ambient consisting of 90%-N_2 and 10%-H_2. Hall effect experiments were performed from 20 to 500°C.

III. Results/Discussion: A. Vanadium Acceptor Level

The very first examination of vanadium in SiC yielded the existence of the acceptor level at a position above the donor level.[1] In the past 2 years two articles have sought to establish the activation energy of this level at ~0.62[7] and <0.25 eV[8]. Both estimates relied upon inference, and neither provided any direct experimental evidence. As will be evinced in this section, these values are too low.

Shown in figure 1 are two absorption spectra observed at 13K from a vanadium-doped 4H sample. Clearly visible are the characteristic $^2E \rightarrow {}^2T_2$ intracenter transition absorption bands from the vanadium $3d^1$ center in the 4H polytype at 0.929 and 0.970 eV. The solid line represents the spectrum taken under normal operating conditions, while the dashed spectrum was taken immediately following a 10 minute illumination with a trans-band gap halogen source. The initial spectrum is assumed to be in equilibrium, and therefore the Fermi level in the sample is pinned to either the donor or the acceptor level.

Plotted in figure 2 is the resistivity of two vanadium-doped samples as a function of inverse temperature. The circular symbols represent the resistivity of a 6H sample , while the squares represent that of a 4H sample. The lines in the graph are a least squares fit of the Ahhrenius

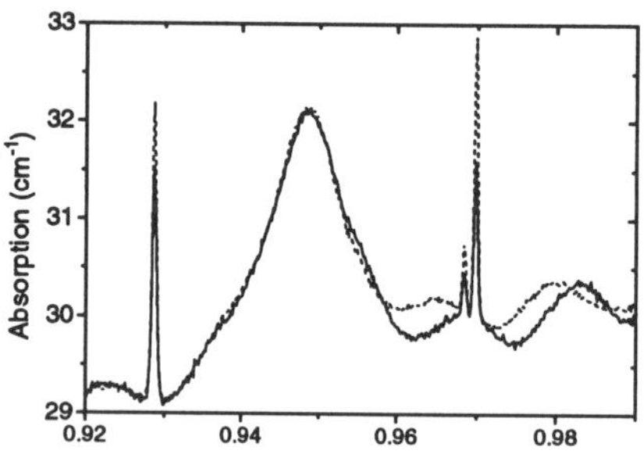

Figure 1-NIR absorption spectrum of a 4H V-doped SiC sample both before (solid line) and after (dashed line) white light illumination.

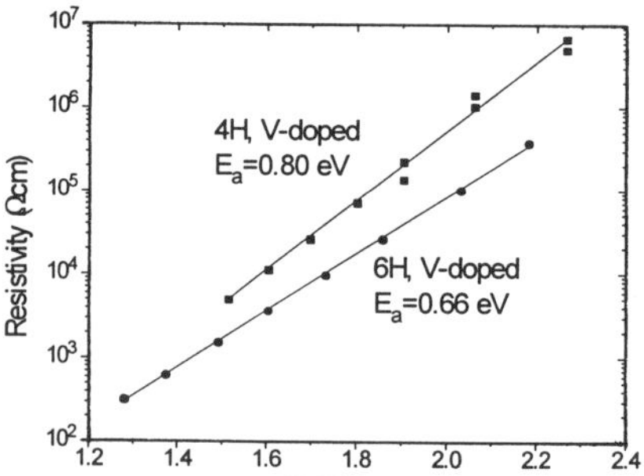

Figure 2-Temperature dependent resistivity measurements on both V-doped 4H (square) and 6H (round symbols) SiC.

relationship between temperature and resistivity (modified for the conduction band density of states). The fitting parameter in each case is the activation energy of the defect. The ionization energy of the 4H sample was found to be 0.80 eV, or the same value as was found using DLTS. The position of the acceptor level in the 6H material was found to reside 0.66 eV beneath the conduction band edge.

In order to determine the activation energy of the defect responsible for the DLTS signal, a plot of e/T^2 versus reciprocal temperature was created, and is shown in figure 3. The activation energy was extracted from this plot using the following equation:

$$e_n=\sigma N_c V_{th} \exp((E_c-E_a)/kT) \qquad (1)$$

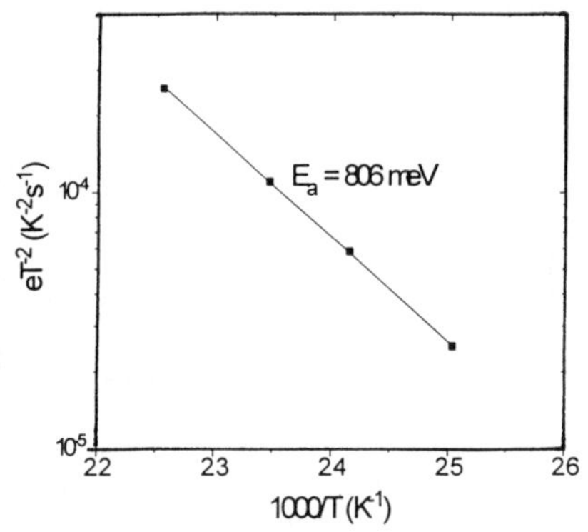

Figure 3-DLTS activation energy plot of a V-doped 4H sample. The capture cross section was found to be 1.79×10^{16} cm^{-2}.

where σ is the capture cross section, N_c is the conduction band density of states, V_{th} is the thermal velocity of electrons, E_c-E_a is the depth of the level from the conduction band (in the case of electrons), k is the Boltzmann constant, and T is the temperature. The activation energy of the defect was calculated to be 806(20) meV. The thermal velocity of electrons in this sample was calculated using a method devised by Green.[9] With this information, the temperature independent capture cross section was determined by examining a single point from figure 3 and was found to be 1.8×10^{-16} cm^{-2}. No other levels were detected in these samples throughout the temperature range examined (90 to 450K). This fact coupled with the knowledge that these samples were intentionally doped with vanadium, indicate that this may be the vanadium acceptor level. The maximum trap concentration of vanadium acceptors was found to be 3×10^{17} cm^{-3}. Since the boules were doped with vanadium and since SIMS results indicated a higher total concentration of vanadium, this is believed to be the maximum concentration of isolated, substitutional vanadium in SiC.

B. The Vanadium-Nitrogen Complex

With the amounts of unintentional impurities present in the PVT-grown SiC, there is a real likelihood that the vanadium will form complexes with these elements. A priori, there is no way of determining the effect of such a complex on the properties structure of silicon carbide. During the course of this investigation, a vanadium complex was found. The fact that vanadium is present in this complex has been established, but the second component cannot be ascertained unequivocally, but nitrogen is forwarded as the most likely candidate.

The discovery of this vanadium complex stemmed from an examination of an electronic transition observed in the near infrared spectrum of silicon carbide samples. Shown in figure 4 is this electronic transition observed in a typical 6H sample. The fine structure is defined by two no-phonon doublets at 5000 cm^{-1} and 4850 cm^{-1}. These bands are ubiquitous in n-type silicon carbide, and therefore most likely related to a common contaminant.

In 1995, Kunzer et. al.[8] discussed the magnetic circular dichroism of the absorption (MCDA)-detected ESR of the absorption band at 5000 cm^{-1}. From this, they determined that the element

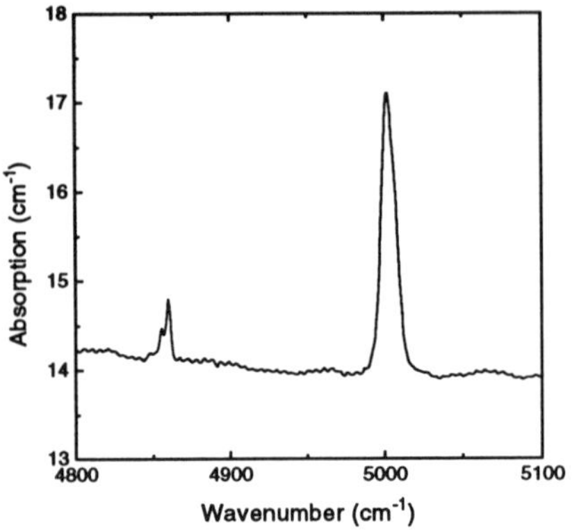

Figure 4 - Electronic transition of the V-N complex

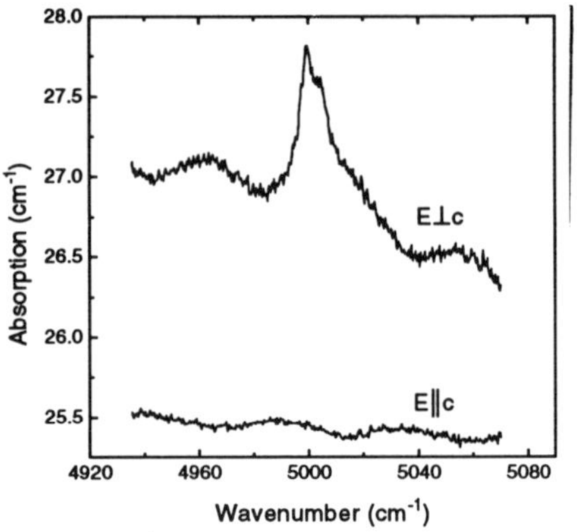

Figure 5-Absorption of the 5000 cm^{-1} band with the light polarized both: a) perpendicular and b)parallel to the c-axis.

responsible for the absorption band was vanadium. The only element lacking from the report of Kunzer et al.[7] was that they did not perform angular dependent measurements, which would prove whether or not this was a isolated or a complex defect.[10]

A polarization study of the 5000 cm^{-1} band is the first indication that the defect responsible is a complex(figure 5). Shown are the experimental spectrum of light polarized both perpendicular(figure 5a) and parallel(figure 5b) to the c-axis. In the latter condition, either no absorption is observed, or the absorption is within the S/N ratio of the experiment. For defects possessing cubic symmetry, the selection rules necessitate that light absorption experiments excite defects identically whether the light is polarized along the x, y or z (or any combination of x, y, or z). This statement stems from the fact that the polarization vector is triply degenerate under cubic symmetry, and therefore the triple product of the initial symmetry, the polarization vector, and the final symmetry is identical for each polarization. From this significant change of the absorption in the different polarizing conditions, some anisotropy of the defect must be present. The first likely symmetry is C_{3v}. This symmetry would imply a complex whose adjacent constituent species resides on a substitutional lattice site.

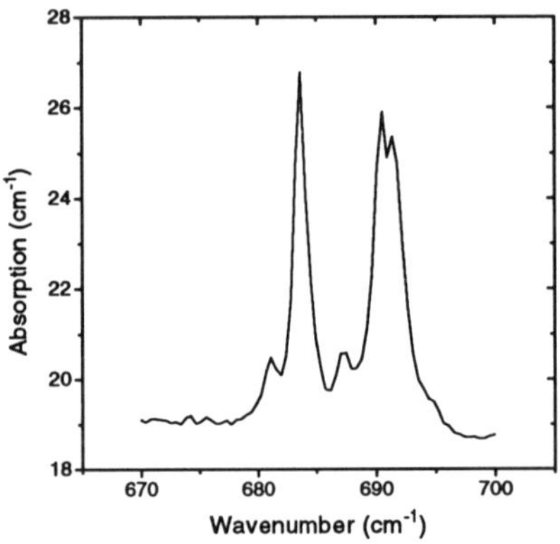

Figure 6-MIR absorption bands attributable to the vibrational modes of the V-N complex

Shown in figure 6 is a middle infrared absorption band structure observed in nearly all SiC spectra. These bands are intricate, and cannot, as yet, be interpreted. They were correlated to the 5000cm^{-1} absorption band (figure 7).

This correlation, coupled with the presence of a phonon replica (at 683 cm^{-1}) observed near the electronic transition, indicates that the bands at 683 cm^{-1} are related to the vanadium complex. The cause of this absorption band must be the other constituent of the complex. The structure of the vibrational absorption band is not yet understood, but may be related to the presence of the three inequivalent lattice sites in the 6H polytype, coupled with the low symmetry of the complex.

With the assumption that this complex forms a center with C_{3v} symmetry, the ground state of the complex is the $^3A_2$ state, while the first excited state is the ^3E. From this, the transition involved in the observed absorption band is the $^3A_2 \rightarrow {}^3E$. In order to gauge whether the transition is orbitally allowed, the cross product of the initial symmetry, the electric dipole, and the final symmetry must contain the totally symmetric representation, A_1. This is illustrated by the following relationship:[12]

$$\begin{matrix} \text{x, y direction} \\ \text{z direction} \end{matrix} \quad A_2 \times \begin{pmatrix} E \\ A_1 \end{pmatrix} \times E = \begin{pmatrix} A_1+A_2+E \\ E \end{pmatrix} \begin{matrix} \text{(contains } A_1) \\ \text{(does not contain } A_1) \end{matrix} \qquad (2)$$

From this, the cross product of symmetries for light polarized along the x and y directions contain the totally symmetric representation (A_1), and therefore, the transition is orbitally allowed. The same is not true for light polarized along the z direction. These arguments suggest that the C_{3v} symmetry is most likely a correct descriptor of the defect.

The other possible transitions that this symmetry would allow include: $^3A_2 \rightarrow {}^3A_2$, $^3A_2 \rightarrow {}^3A_1$, and $^3A_2 \rightarrow {}^3E(2)$. The first two are orbitally forbidden. A subsequent correlation (figure 8) will be used to show that the bottom-most $^3A_2$ state is 0.78 eV beneath the conduction band edge. The energy of the transition to the first ^3E state is 0.625 eV, and therefore, the upper state of the third transition ($^3E(2)$) is most likely degenerate with the conduction band. This degeneracy would significantly shorten the time the defect would be allowed to exist in this ^2E configuration, which would broaden the transition so that it would not be observed.

At this juncture, speculation as to the chemical nature of the second element is in order. In order to isolate the species, the two primary restrictions are employed: that the total defect possesses two electrons and the second element is ubiquitous in silicon carbide. The first consideration is a result of research that Kunzer et. al.[7] performed which found that in n-type material, the complex contains two electrons.

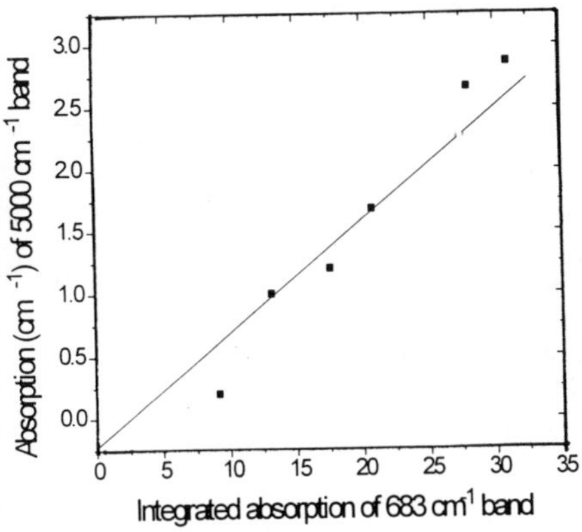

Figure 7-Correlation between the absorption intensities of the 683 cm^{-1} and the 5000 cm^{-1} bands

The vanadium atom substitutes for silicon, and, therefore, the other element must substitute on the carbon sub-lattice. Since the band lies close to the optic phonon branch, and neglecting the significant modification of the force between the impurity and the matrix, the impurity examined must be slightly heavier than the carbon atom it replaces. Additionally, the number of electrons in a vanadium-nitrogen complex (2, 1 from V and 1 from N) is exactly the same as was observed in the MCDA-ESR study of the 5000 cm^{-1} band.

As a result of these arguments, the most likely candidate is nitrogen. Another issue which must be addressed is the preferential orientation of the complex along the c-axis. This is not difficult to rationalize because such a phenomenon is observed in another transition element complex (Ti-N), which preferentially orients perpendicular to the c-axis.[12] The reason behind this is not well understood, but if the Ti-N complex has a preferred orientation, not much rationalization is required to substantiate the preferred orientation of the vanadium-nitrogen complex. As was previously mentioned, one reason behind the study of this complex was to determine whether it would detract from the high-resistivity nature of vanadium doped silicon carbide boules. This is partly established through the determination of the depth of the complex's level in band gap. Optical admittance spectroscopic (OAS) measurements were employed to correlate the 683 cm^{-1} absorption band to a deep level(at E_c-0.78 eV) observed in the OAS spectrum(figure 8). This figure was constructed from five samples (including two at the origin). An examination of this correlation shows that the vanadium-nitrogen complex creates a deep level 0.78 eV from the conduction band edge.

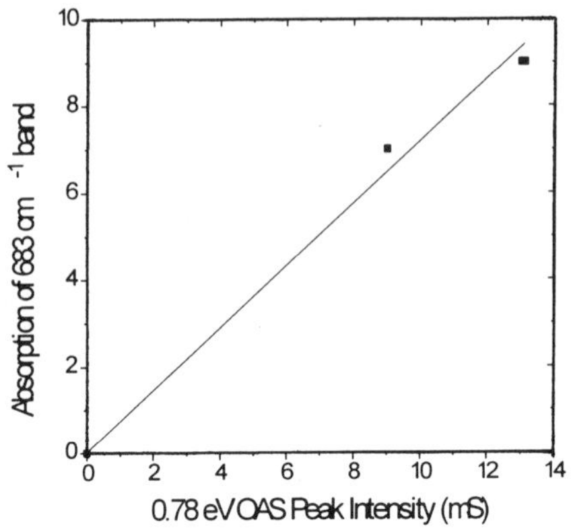

Figure 8-Correlation between the 683 cm^{-1} band and the intensity of an OAS peak at E_c-0.78 eV.

IV. Conclusions

Using Hall effect, DLTS, optical absorption spectroscopy, and OAS, the activation energies of the vanadium acceptor and vanadium nitrogen complex were established. For the acceptor level, the activation energy was determined to be 0.80 (0.66) eV for the 4H (6H) SiC modification. Using DLTS, the concentration of isolated vanadium centers was found to be $3x10^{17}$ cm^{-3}. The complex introduced a level slightly deeper at E_c-0.78 eV for the 6H polytype. Optical absorption spectroscopy was used to characterize the structure of the vanadium complex as C_{3v}. The vanadium complex is responsible for absorption bands in the MIR(683 cm^{-1}, vibrational transition) and NIR(5000 cm^{-1}, electrical transition).

V. References

1. Schneider, J., *et al.*, Appl. Phys. Lett. **56**, 1184-1186 (1990)
2. Maier, K., *et al.*, Mat. Sci. & Engr. B **11**, 27-30 (1992)
3. Jenny, J.R., *et al.*, J. Appl. Phys. **78**, 3839 (1995)
4. Tairov, Y.M. and V.F. Tsvetkov, J. Cryst. Growth **52**, 146-150 (1981)
5. Hobgood, H.M., *et al.*, J. Crystal Growth **137**, 181-186 (1994)
6. van der Pauw, L.J., Philips Technical Review **20**, 220-224 (1958)
7. Kunzer, M., *et al.*, Mat. Sci. & Engr. B **29**, 118-121 (1995)
8. Jenny, J.R., *et al.*, J. Appl. Phys. **78**, 3160-3163 (1995)
9. Green, M.A., J. Appl. Phys. **67**, 29442954 (1990)
10. Private communication with Dr. M. Kunzer, 1995.
11. Harris, D.C. and M.D. Bertolucci, *Symmetry and Spectroscopy: An Introduction to Vibrational and Electronic Spectroscopy* (1989, New York: Dover Publications) p. 550
12. Vainer, V.S., *et al.*, Sov. Phys. Solid State **28**, 201 (1986)

ELECTRON SCATTERING BY NATIVE DEFECTS IN III-V NITRIDES AND THEIR ALLOYS

L. HSU[1][2], W. WALUKIEWICZ[2]
[1]Department of Physics, University of California, Berkeley, CA 94720
[2]Materials Science Division, Lawrence Berkeley National Lab, Berkeley CA 94720

ABSTRACT

We have calculated the electron mobilities in GaN and InN taking into consideration scattering by short range potentials, in addition to all standard scattering mechanisms. These potentials are produced by the native defects which are responsible for the high electron concentrations in nominally undoped nitrides. Comparison of the calculated mobilities with experimental data shows that scattering by short range potentials is the dominant mechanism limiting the electron mobilities in unintentionally doped nitrides with large electron concentrations. In the case of $Al_xGa_{1-x}N$ alloys, the reduction in the electron concentration due to the upward shift of the conduction band relative to the native defect level can account for the experimentally measured mobilities. Resonant scattering is shown to be important when the defect and Fermi levels are close in energy.

INTRODUCTION

Much research currently focuses on the growth of high quality epitaxial films of III-V nitrdes. Because of their wide direct bandgap, these materials are considered prime candidates for the manufacture of visible to ultraviolet optoelectronic devices [1] and the III-V compatible dielectrics [2]. In recent years there have been several reports on calculations of electron mobilities in III-V nitrides [3-5]. Those calculations were aimed at an assessment of the relative contributions of the standard scattering mechanisms and the establishment of the phonon mobility limits in these materials. Comparison of the calculations with experimental results was rather difficult as undoped GaN and InN very often exhibit large electron concentrations.

It has been argued before and there is an increasingly convincing body of evidence that the electrons found in nominally undoped nitrides originate from nitrogen vacancies with the first ionization level resonant with the conduction band [6,7]. It has been found that by controlling the crystal stoichiometry one can change the electron concentrations by many orders of magnitude with the highest concentrations exceeding 10^{20} cm^{-3}. The highly localized nature of the native donors raises the question on how the short range potential associated with the donors affects the electron mobilities and whether there is any difference in the electron mobilities in intentionally and unintentionally doped materials.

MOBILITIES IN BINARY III-V NITRIDES

Our calculations of the electron mobilities are based on a variational approach used previously to calculate mobilities in other III-V compounds [8]. We incorporate all standard scattering mechanisms; optical phonon, acoustic deformation potential, acoustic phonon piezoelectric and Coulombic ionized-impurity scattering. We also include the scattering arising from the short range potential of the native defects. This type of scattering has been previously considered in narrow gap semiconductors [9].

The total potential of an ionized and highly localized defect can be separated into two parts,

$$V(r) = V_{Coul}(r) + V_{sr}(r) \tag{1}$$

where V_{coul} is the long range screened Coulomb potential and V_{sr} is the short range potential. Both potentials are localized on the same center. The electron scattering rate is proportional to the square matrix element of the potential (1) and consists of three terms; the standard Coulomb

Mat. Res. Soc. Symp. Proc. Vol. 423 © 1996 Materials Research Society

term, a short range term proportional to the square of the matrix element $\langle S|V_{sr}|S\rangle$ and an interference term proportional to $\langle S|V_{sr}|S\rangle \langle S|V_{Coul}|S\rangle$, where $|S\rangle$ is the conduction band Bloch amplitude. Therefore, incorporation of the short range potential results in two additional terms in the total scattering rate,

$$\tau_{sr} = \frac{\hbar^3 \pi}{N\, m^*\, k\, A^2} \tag{2}$$

and

$$\tau_{int} = \frac{\hbar^3 \varepsilon_0\, k}{N\, e^2\, m^*\, A\, F_{int}} \tag{3}$$

where A is the matrix element of the short range potential, N is the concentration of native defects, k is the electron wavevector, m* is the effective mass in units of the free electron mass, and ε_0 is the dielectric constant.

The screening factor is

$$F_{int} = 4\left[1 - \frac{1}{\xi_0}\ln\left(1 + \xi_0\right)\right] \tag{4}$$

where $\xi_0 = (2k\lambda)^2$ and the screening length λ is [8]

$$\lambda = 1.72 \times 10^{-14} \frac{\varepsilon_0}{\left(m^*/m_0\right)^{3/2}\, T^{1/2}} \frac{1}{F_{-1/2}(x)} \tag{5}$$

where $F_{-1/2}(x)$ is the Fermi integral of -1/2 order. It is important to note the different dependence of the scattering rates on electron energy. For a highly degenerate electron gas with concentration n, $k \propto n^{1/3}$, $\lambda \propto n^{-1/6}$ and to a good approximation $F_{int} = 4$. Assuming that in the unintentionally doped samples the electron concentration is proportional to the defect concentration, one finds from Eqs. (2) and (3), that $1/\tau_{sr} \propto n^{4/3}$ and $1/\tau_{int} \propto n^{2/3}$. It can also be shown for scattering by the coulomb centers that $1/\tau_{Coul} \propto n^{1/3}$. Hence, the short range scattering is expected to play a more important role in samples with a high free carrier concentration. The results of our calculations for unintentionally doped GaN along with the best experimental data found in the literature [10-13] can be seen in Fig. 1. Here we assume that the concentration of electrons is equal to the concentration of the native defects. Shown are the calculated mobilities assuming that the sources of the electrons are all extrinsic Coulombic impurities (dotted line) or are native defects with a short range potential (solid line). Also plotted are the individual contributions to the solid line mobility from the short range and interference terms. As one can see, the experimental mobilities at very high carrier concentrations show a sudden drop far below the dotted line values, which can be understood by the incorporation of scattering by short range potentials associated with highly localized native defects. Although the dotted line mobilities were calculated assuming zero compensation, it is easily demonstrated that even with compensating acceptors, the mobility at large carrier concentrations would not show the sudden drop present in the experimental data, as Coulomb scattering has a relatively weak dependence on the carrier concentration. Although not shown, experimental data for mobilities in GaN intentionally doped with Si or Ge show only the expected weak dependence on electron concentration [14-17], even at very high carrier concentrations. The material parameters used in the calculations are listed in Table 1. We find that a value of $A = 3.7\times10^{-34}$ erg cm^3 for the short range scattering strength is required to account for the experimental data. Assuming that the short range potential is restricted to a

Table 1

	GaN	InN
density (g/cm^3)	6.1	6.81
ε_0	9.5	15.3
ε_∞	5.35	8.4
LO phonon energy (meV)	90.5	76
m^*/m_0 (at band minimum)	0.2	0.11
lattice parameter a_0 (Å)	4.52	4.98
acoustic phonon vel. (cm/s)	6.6 x 10^5	5.07 x 10^5
piezoelectric constant (V^2/dyn)	1326[a]	1326[a]
deformation potential (eV)	9.6	2.5
$\langle S \vert V_{sr} \vert S \rangle$	3.7×10^{-34}	2.5×10^{-34}

[a]Estimated from ref. 22. It is assumed that the value is similar for the two nitrides. Piezoelectric scattering is not an important mechanism in any concentration range.

single unit cell this value corresponds to a potential well with a depth of about 10 eV. As there are no discussions of the magnitude of this parameter in the literature, the experimental mobilities were our only guide to estimating a value.

Another III-V compound that exhibits large concentrations of free electrons is InN. In as grown, undoped InN, electron concentrations are as high as 10^{21} cm^{-3}. It is generally accepted that in this case, the electrons also originate from highly localized native defects which have an energy level located well above the conduction band edge. One can therefore expect that scattering by short range potentials should play a significant role in this material as well.

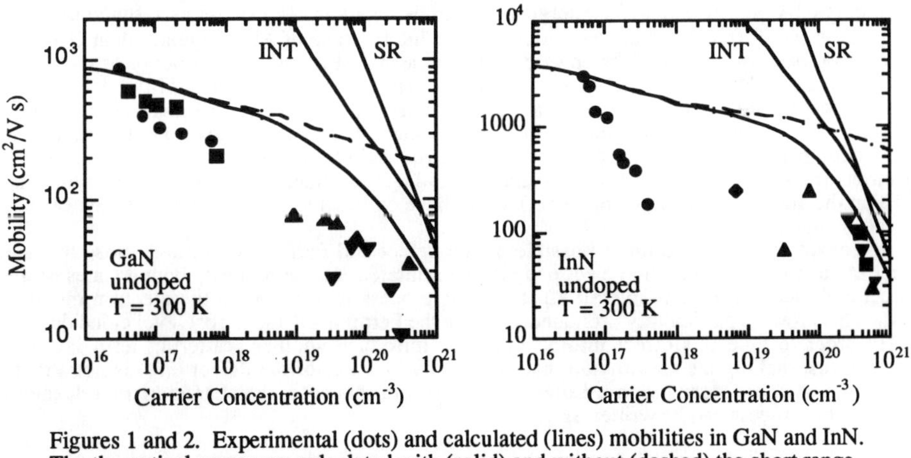

Figures 1 and 2. Experimental (dots) and calculated (lines) mobilities in GaN and InN. The theoretical curves are calculated with (solid) and without (dashed) the short range potential. INT and SR show the contributions from the interference and short range terms, respectively.

Calculated room temperature electron mobilities in unintentionally doped InN are shown in Fig. 2 along with the best available experimental data [17-21]. Because of the relatively small energy gap of E_G =1.9 eV, at large carrier concentrations (>10^{19} cm^{-3}) the Fermi energy becomes comparable to the energy gap. In such cases it is necessary to incorporate the effects of nonparabolicity on the density of states effective mass. In a simple approximation, the energy dependent effective mass is given by,

$$m^* = m_0^*\left(1 + 2\frac{E}{E_G}\right) \tag{6}$$

where m_0^* is the conduction band edge effective mass and E is the electron energy.

The results in Fig. 2 show that, as in GaN, the mobility is entirely determined by the short range and interference scattering at very high carrier concentrations. Although there is only a very limited amount of experimental data on electron mobilities in InN, it is quite clear that the strong concentration dependence observed at very high electron densities cannot be explained by Coulomb scattering. The rapid decrease of the mobility for n larger than 10^{20} cm^{-3} can be explained by scattering by native defects with the parameter A = 2.5×10^{-34} erg cm^3, which is close to it's value in GaN.

Unfortunately, there is no data available on InN which has been doped to such a high level. From the calculations presented in Fig. 2 one finds that at high electron concentrations the mobility in InN doped with hydrogenic donors should be at least one order of magnitude higher than in the unintentionally doped material. Therefore it is expected that doping with shallow hydrogenic donors should greatly improve the electrical characteristics of InN.

MOBILITIES IN THE ALLOYS OF III-V NITRIDES

In addition to the binary III-Nitrides, we have also examined mobilities in $Al_xGa_{1-x}N$. One noteworthy feature of the experimental mobility data in this material is that the electron mobilities remain relatively constant as a function of alloy composition [23,24]. In order to try to reproduce this trend, we introduce two more terms in our mobility calculations, alloy disorder [25], which is a well established scattering mechanism, and resonant scattering [26].

Figures 3A and 3B show the energies of various levels in $Al_xGa_{1-x}N$ as a function of Al content for two different defect concentrations. As the bandgap of AlN is greater than GaN, the conduction band minimum shifts upward with increasing Al fraction. The energy of the defect level remains relatively constant as it is localized. This decrease in energy of the defect level with respect to the conduction band leads to a "freeze out" of carriers originating from these defects. The calculations presented in Figs 3A and 3B include a downward shift of the conduction band edge resulting from electron-electron and electron-ion interaction [27]. As the electrons freeze out, the effect becomes smaller, leading to a superlinear dependence of the total shift on the alloy composition.Another effect we have considered is the influence of the carriers on the conduction band.

Resonant scattering becomes possible as the defect and Fermi levels move close to each other. At this point, conduction electrons may be scattered into temporarily bound states of the defect and vice versa. The overall effect of this process is to decrease the electron mobility [28]. This effect is particularly pronounced where the Fermi level and defect level coincide.

In order to take this effect into account, we introduce another scattering term into our calculations, making the assumption that the density of states of the defect level is not a delta function, but is broadened with a Lorentzian shape. In this case, the characteristic relaxation time for this process can be written as

$$\tau_{res} = \frac{N m^* k}{2\pi\hbar}\left[\frac{\Gamma^2}{(E_{def} - E)^2 + \Gamma^2}\right]^{-1} \tag{7}$$

where E is the electron energy and Γ is the broadening of the defect level [29]. A value of 25 meV was adopted for Γ in these calculations.

The results of our calculations for alloys are shown in Figures 4A and 4B along with the corresponding experimental data. The general agreement between the two sets of data is quite good. The sudden decrease in the calculated mobility occurs where the Fermi and defect levels cross, suggesting that the experimentally measured dips are not due to experimental error or poor growth, but are in fact signatures of resonant scattering. We expect that this effect could also be seen if hydrostatic pressure were used to bring the defect level down in energy with respect to the conduction band. Such a measurement could be used to determine the pressure coefficients of the conduction band and defect level more accurately.

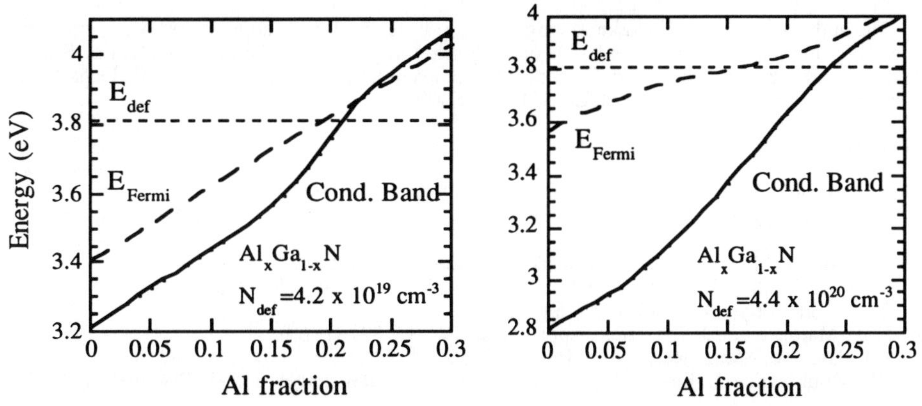

Figure 3A/B. Plot of energy levels v. Al content in $Al_xGa_{1-x}N$ for two defect concentrations showing Fermi and defect levels and conduction band.

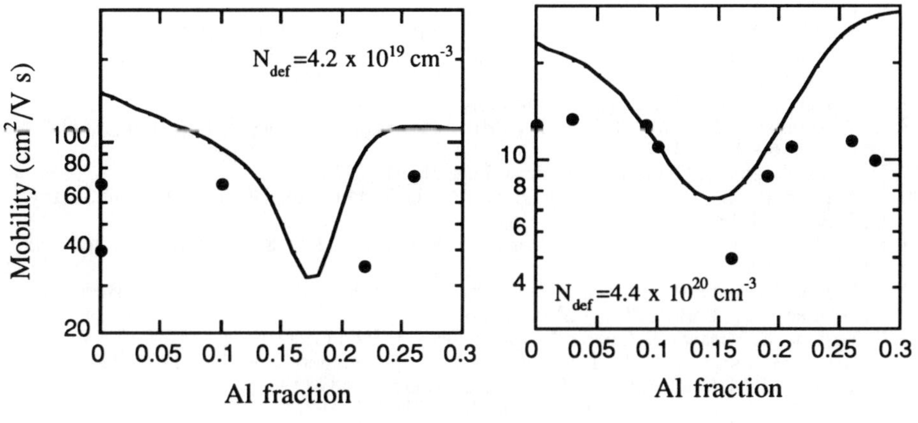

Figure 4A/B. Plot of experimental (dots) and theoretical (line) mobility in $Al_xGa_{1-x}N$ as a function of Al content for two defect concentrations.

CONCLUSIONS

We have presented the results of theoretical calculations of the electron mobility in GaN and InN. We have shown that scattering by the short range potential of highly localized native defects is responsible for the mobility drop observed at high electron concentrations in unintentionally doped materials and that resonant scattering can be seen by measuring mobilities in the case where the defect and Fermi levels are close in energy. The matrix element of the short range potential has been determined by comparison of the theoretical calculations with the experimental mobilities in unintentionally doped materials with high electron concentrations.

ACKNOWLEDGMENTS

The authors would like to acknowledge Eugene E. Haller for his support and his suggestions in editing this paper. This work has been supported in part by the Director, Office of Energy Research, Office of Basic Energy Sciences, Materials Science Division of the U. S. Department of Energy under contract number DE-AC03-76SF00098.

REFERENCES

1. S. Strite and H. Morkoc, J. Vac. Sci. Technol. B **10**, 1237 (1992).
2. X. J. Li, Z. Xu, Z. J. He, H. Z. Cao, W. Su, Z. C. Chen, F. Zon, and E. G. Wang, Thin Solid Films **139**, 261 (1986).
3. R. J. Molnar, T. Lei and T. D. Moustakas, Appl. Phys. Lett. **62**, 72 (1993).
4. V. W. L. Chin, T. L. Tansley, and T. Osotchan, J. Appl. Phys. **75**, 7365 (1994).
5. D. L. Rode, Phys. Stat. Sol. B **55**, 687 (1973).
6. H. P. Maruska and J. J. Tietjen, Appl. Phys. Lett. **15**, 327 (1969).
7. J. I. Pankove, Mater. Res. Symp. Proc. 162, 515 (1990).
8. W. Walukiewicz, L. Lagowski, L. Jastrzebski, M. Lichtensteiger, and H. C. Gatos, J. Appl. Phys. **50**, 899 (1979).
9. E. Litwin-Staszewska, S. Porowski, and A. A. Filipchenko, Phys. Stat. Sol. B **48**, 525 (1971).
10. T. Sasaki and S. Zembutsu, J. Appl. Phys. **61**, 2533 (1987).
11. C. J. Sun and M. Razeghi, Appl. Phys. Lett. **63**, 973 (1993).
12. S. Nakamura, Jpn. J. Appl. Phys. **30**, L1705 (1991).
13. D. K. Gaskill, K. Doverspike, L. Rowland and D. L. Rode, Inst. Phys. Conf. Ser. No. 141, 425 (1995).
14. D. K. Wickenden, W. A. Bryden, Inst. Phys. Conf. Ser. No. 137, 381 (1993).
15. L. B. Rowland, K. Doverspike, D. K. Gaskill, Appl. Phys. Lett. **66**, 1495 (1995).
16. S. Nakamura, T. Mukai, M. Senoh, Jpn. J. Appl. Phys. **31**, 2883 (1992).
17. T. L. Tansley and C. P. Foley, Electron. Lett. **20**, 1066 (1984).
18. C. R. Abernathy, J. D. MacKenzie, S. R. Bharatan, K. S. Jones, and S. J. Pearton, Appl. Phys. Lett. **66**, 1632 (1995).
19. H. J. Hovel and J. J. Cuomo, Appl. Phys. Lett. **20**, 71 (1972).
20. A. Yamamoto, M. Tsujino, M. Ohkubo, and A. Hashimoto, Solar Energy Mat. and Solar Cells **35**, 53 (1994).
21. P. W. Wisk, C. R. Abernathy, S. J. Pearton, F. Ren, Mater. Res. Soc. Symp. Proc. **282**, 599 (1993).
22. D. L. Rode in *Semiconductors and Semimetals*, eds. R. K. Willardson and A. C. Beer (Academic, New York, 1975), Vol 10, Chap. 1.
23. S. Yoshida, S. Misawa, S. Gonda, J. Appl. Phys. **53**, 6844 (1982).
24. Y. Koide, H. Itoh, N. Sawaki, I. Akasaki, et al., J. Electrochem. Soc. **133**, 1956 (1986).
25. L. Makowski, M. Glicksman, J. Phys. Chem. Solids **34**, 487 (1973).
26. M. E. Raikh, A. L. Efros, Sov. Phys. Solid State **28**, 735 (1986).
27. W. Walukiewicz, Phys. Rev. B 41, 10218 (1990).
28. O. F. Sankey, J. D. Dow, K. Hess, Appl. Phys. Lett. 41, 664 (1982).
29. Z. Wilamowski, K. Swiatek, T. Dietl, and J. Kossut, Solid State Comm. **74**, 833 (1990).

CHARACTERIZATION OF DEEP LEVEL DEFECTS IN 4H AND 6H SIC VIA DLTS, SIMS AND MEV E-BEAM IRRADIATION

J.P. Doyle, M.O. Aboelfotoh*, M.K. Linnarsson, and B.G. Svensson
Royal Institute of Technology, Solid State Electronics
Electrum 229, S-164 40, Kista-Stockholm, Sweden

A. Schöner, N. Nordell, and C. Harris
Industrial Microelectronics Center
P.O. Box 1084, S-164 21, Kista-Stockholm, Sweden

J.L. Lindström
National Defense Research Institute
P.O. Box 1165, S-581 11 Linköping, Sweden

E. Janzén and C. Hemmingsson
Department of Physics and Measurement Technology
Linköping University, S-581 83 Linkoping, Sweden

Abstract

Electrically active defects in both 4H and 6H polytypes of SiC have been observed through the use of deep level transient spectroscopy (DLTS). Schottky contacts were grown by vapor phase epitaxy (VPE) with doping concentrations, the epitaxial layer having a doping concentration in the range of 10^{14} cm^{-3} to 10^{17} cm^{-3}. Numerous levels have been found in the as-grown n-type 6H-SiC samples and secondary ion mass spectrometry (SIMS) and MeV electron irradiation have been employed to corrrelate the defect levels to impurities or structural defects. In contrast, only a single level is observed in the as-grown 4H-SiC samples.

Introduction

The use of silicon carbide as a high temperature, power, and frequency semiconducting material has been recognized for quite some time, but advances in the fabrication of devices have been limited by the quality of both substrates and epitaxially grown layers [1]. Rapid advancements in both have resulted in the current production of high quality material. In order to fully exploit the material, a knowledge of the electrically active deep levels is required as their effects as efficient recombination centers or traps may limit the minority carrier lifetime, or, conversely, result in a method for the fabrication of semi-insulating substrates. The electrically active deep levels may be the result of impurity atoms [2-6] or structural defects such as vacancies or interstitials. The most common technique that has been employed for the identification of the deep levels has been deep level transient spectroscopy (DLTS). To date, there are numerous reports in the literature describing deep levels in the 6H polytype , although in most cases, defect levels are found to be the result of electron, ion, or neutron irradiation [1,7,8]. There are few reports of defects in as-grown epitaxial films [9].

Experimental

The epitaxial layers were grown by vapor phase epitaxy (VPE) in a horizontal reactor reported elsewhere [10]. The epitaxial layers were grown at about 1600 °C with SiH_4 and C_3H_8 used as the precursor gases while nitrogen is used the n-type dopant. The samples were de-greased with organic solvents (trichloroethylene, acetone, and ethanol) and dipped into a 10% HF solution prior to loading into an electron beam evaporation system which was pumped to a pressure below 1×10^{-7} Torr. Schottky contacts were then formed by depositing either copper, titanium, or nickel films of 200 nm thickness through a metal mask with a diameter of 1 mm. Deep level transient spectroscopy was

Mat. Res. Soc. Symp. Proc. Vol. 423 © 1996 Materials Research Society

typically performed using six to eight rate windows ranging from (20 ms)$^{-1}$ to (3.2 s)$^{-1}$. Secondary ion mass spectrometry (SIMS) was performed using a Cameca IMS 4f microanalyzer utilizing either a cesium or oxygen ion beam. Electron irradiations were conducted at 2 MeV with the dose varying from $1x10^{15}$ e$^-$/cm^2 up to $1x10^{17}$ e$^-$/cm^2. The dose being selected according to the doping concentration of the samples.

Results

Figure 1 shows a DLTS spectrum for an as-grown n-type SiC-6H sample which displays a number of deep levels which are more clearly observed after electron irradiation as shown in figure 2. In the as-grown case we observe a strong peak at a position 0.41 eV below the conduction band edge (E_c) with evidence of another closely spaced deep level indicated by the slight shoulder on the low temperature side of the peak.

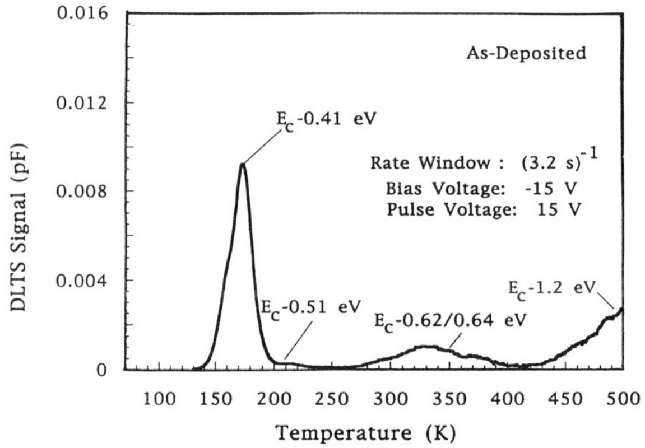

Figure 1. DLTS spectra of an as-grown layer of n-type VPE grown SiC-6H doped with nitrogen to a level of $1x10^{15}$ cm^{-3}. A nickel Schottky contact was used for the measurement.

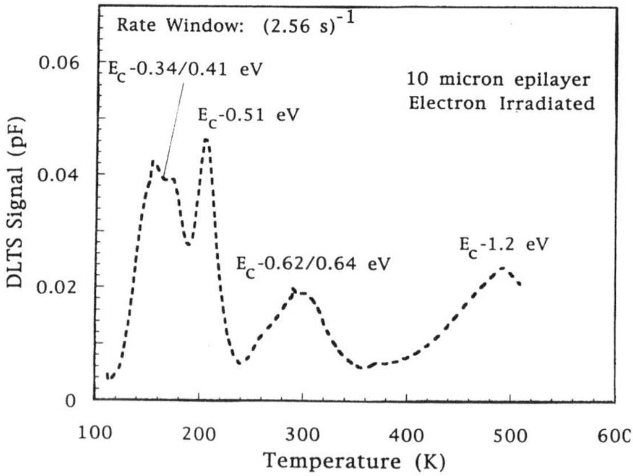

Figure 2. Results from electron irradiation of a portion of the sample shown in figure 1. A dose of $1x10^{15}$ e$^-$ /cm^2 at an energy of 2 MeV. The E_c-0.51 eV level is found to have a concentration of $1x10^{13}$ cm^{-3}.

Additionally, we observe evidence for the Z_1/Z_2 center (E_c-0.62/0.64 eV) in the as-grown state which has previously been observed in Lely grown crystals or after electron irradiation of epitaxial films [1,8]. With the aid of the electron irradiated sample we have also been able to assign positions to levels corresponding to E_c-1.2 eV and E_c-0.51 eV. Additionally, the shoulder observed in the case of the E_c-0.41 eV peak can be ascribed to the E_c-0.34 eV level.

In order to assess the nature of the defect level via electron irradiation, a sample doped to a level of $1x10^{17}$ cm^{-3} was electron irradiated to a dose of $1x10^{17}$ e^- $/cm^2$ with the resulting DLTS spectra shown in figure 3.

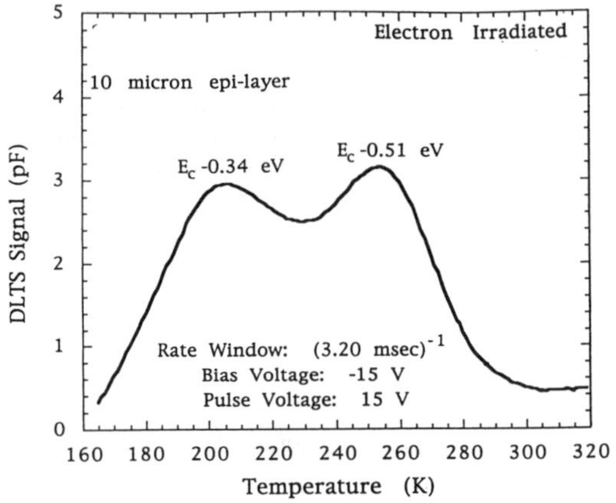

Figure 3. N-type SiC-6H sample doped to a level of $1x10^{17}$ cm^{-3} and irradiated with 2 MeV electrons with a dose of dose of $1x10^{17}$ e^- $/cm^2$. The concentration of the E_c-0.34 and E_c-0.51 eV levels is found to approach 10^{16} cm^{-3}. The rate window is shorter than in figures 1 and 2 in order to avoid the freezeout of the carriers ar lower temperatures.

Note that the rate window chosen was three orders of magnitude shorter than the previously discussed samples in order to avoid carrier freezeout at lower temperatures which has been observed after high dose electron irradiation. We find that the E_c-0.51 eV peak continues to increase indicative of either an intrinsic defect or the interaction of a radiation induced defect with an intrinsic impurity whose concentration is at least 10^{16} cm^{-3}, which corresponds to the concentration of the observed defect level. Analysis of the lower temperature peak indicates a position at E_c-0.34 eV peak, although the presence of the E_c-0.41 eV peak cannot be ruled out due to the overlapping signals.

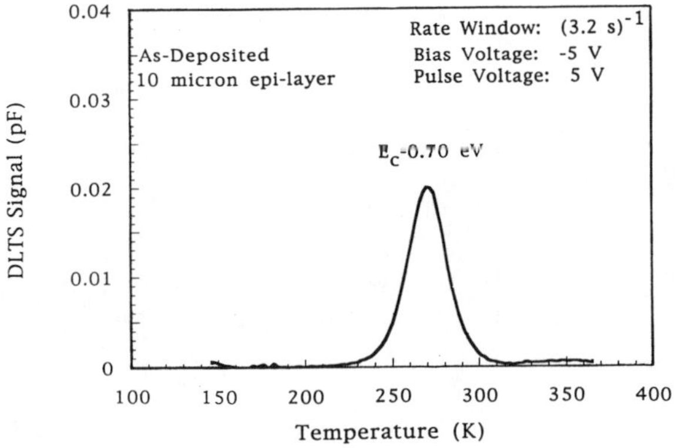

Figure 4. N-type SiC-4H as-grown with a nitrogen doping level of $1x10^{15}$ cm^3.The concentration of the deep level observed at E_c-0.70 eV is found to be $1x10^{13}$ cm^{-3}.

DLTS spectra for 4H polytypes of n-type SiC, both as-grown and after electron irradiation, are shown in figures 4 and 5, respectively. In contrast to the shoulder observed on the E_c-0.41 eV peak in the 6H polytype, only a single symmetrical peak at E_c-0.70 eVwas observed in the as-

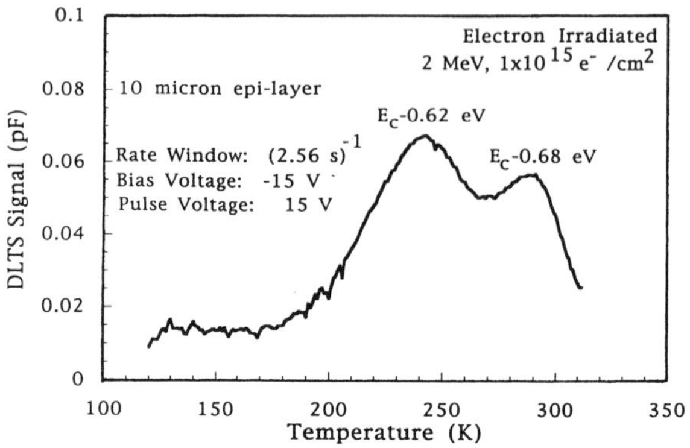

Figure 5. A portion of the sample in figure 4 after rradiation with a dose of $1x10^{17}$ e^-/cm^2 at 2 MeV.

deposited case. After electron irradiation, we observe that a double peak has formed with energy levels assigned at E_c-0.62 eV and E_c-0.68 eV, respectively.

Metal impurities were assessed using SIMS analysis. Among the most widely discussed intrinsic impurities in silicon carbide, titanium has attracted the most attention since the identification of the species in SiC by Choyke and Patrick [4,5]. As shown in figure 6, a uniform titanium concentration of $1x10^{15}$ cm^{-3} is found in the substrate; however, the concentration of Ti in the epitaxial layer is found to be inhomogeneously distributed in depth with higher concentrations near the front surface as well as the interface with the substrate. In the bulk of the epitaxial film, the Ti concentration is below the limit of detectability. Lateral variations in the Ti concentration across the sample have also been observed. The limit of detectability for titanium via SIMS is on the order of $1x10^{13}$ cm^{-3} or below. In figure 7, a thicker epi-layer of SiC-6H is also found to display a higher concentration of Ti near the front surface.

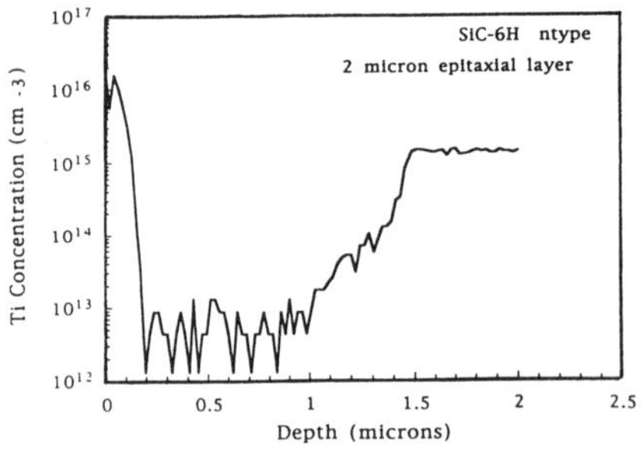

Figure 6. Ti concentration of a SiC-6H via SIMS profiling. A portion of the substrate is also shown.

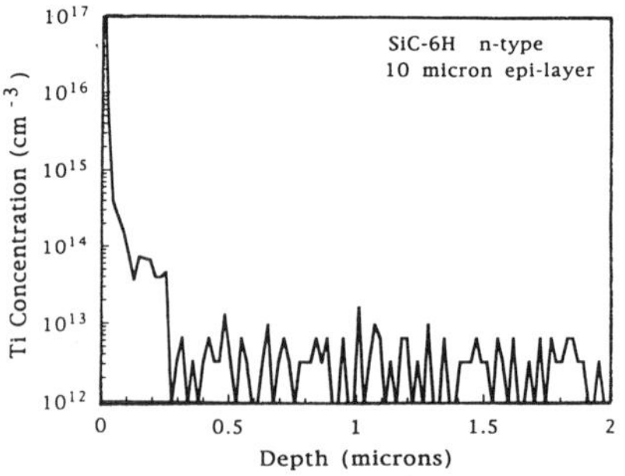

Figure 7 Ti concentration of a SiC-6H epitaxial layer via SIMS profiling.

Discussion

It is clear from these results that in as-grown epitaxial layers of the 6H polytype, electrically active deep levels are present. When compared to previous studies [1,2], the low doping levels used in this work allow for the observation of low concentrations of deep levels. Numerous levels are observed, with the E_c-0.41 eV level displaying the highest concentration on the order of $5x10^{12}$ cm^{-3} After electron irradiation, we observe the growth of all levels observed in the as-grown case. It should be noted that after the irradiation with a $1x10^{15}$ e^-/cm^2 dose, the concentration of the peaks does not exceed approximately $2x10^{13}$ cm^{-3}. Thus we cannot rule out the interaction of defects with metallic impurities, such as titanium, which has been observed with concentrations in some areas of the epitaxial layers above this amount.

Under higher electron doses, the growth of the single peak at E_c-0.51 eV suggests that the level at this energy position is due to a purely intrinsic defect (i.e. a vacancy or interstitial type of defect). On the other hand, it is possible that a defect-impurity complex has formed. With a concentration measured at approximately $1x10^{16}$ cm^{-3}, the impurity would be of relatively high concentration as compared to the doping concentration. One potential candidate for this would interstitial carbon but this remains speculative at this point. Additionally, the increase in the E_c-0.34 eV level, but not the corresponding E_c-0.41 eV level that was observed in both the as-grown case (as a low temperatrure shoulder) and after lower dose irradiation (as shown in figures 1 and 2, respectively) is not seen. De-convolution of the peak is presently inconclusive, which makes it difficult to determine whether or not the E_c-0.41 eV peak has saturated or has continued to increase.

In contrast, the 4H polytype only displays a single peak at E_c-0.70 eV. The symmetry of the peak and the lack of any other peaks in the spectra of figure 4, indicates that either a defect or an impurity resides preferentially on a single site. This is supported by the electron irradiated spectra of figure 5 where two peaks are observed, which is consistent with the picture of a defect residing on two inequivalent sites as is expected due to the nature of the SiC lattice structure [1]. The low concentration of the level in figure 4, at $1x10^{13}$ cm^{-3} makes the determination of the presence of metallic species, such as titanium, by SIMS profiling, inconclusive at the present time.

Conclusion

Electrically active deep levels are found to occur in both 6H and 4H as-grown low doped epitaxial films. A similarity in the levels obtained after electron irradiation is observed in the case of the 6H polytype, but not in the 4H. The concentration of the levels in the 6H are found to increase with electron irradiation. In particular the
E_c-0.34 eV and the E_c-0.51 eV levels are found to increase in concentration approaching levels of $1x10^{16}$ cm^{-3} after electron irradiation of $1x10^{17}$ e^-/cm^2. In the case of the 4H polytype, the single level observed in the as-grown state suggests that the presence of an impurity or defect that occupies, preferentially, a single site.

Acknowledgements

Financial support from the Swedish Board for Technical Development (NUTEK, SiC Consortium) for this project is gratefully acknowledged.

* Permanent address: North Carolina State University, Department of Materials Science and Engineering, Raleigh North Carolina

References

1. G. Pensl and W.J. Choyke, Physica B, 185, 264, (1993)

2. W.J. Choyke and L. Patrick, in *Silicon Carbide 1973*, (University of South Carolina Press, Columbia, 1974)

3. L. Patrick and W.J. Choyke, Phys. Rev. B 10, 5091, (1974)

4. K.M. Lee, L.S. Dang, G.D. Watkins, and C.J. Choyke, Phys. Rev. B, 32 (4), 2273, (1985)

5. V.S. Vainer, V.A. Il'in, V.A. Karachinov, and Y.M. Tairov, Sov. Phys. Solid State, 28(2), 201, (1986)

6 K. Maier, H.D. Muller, and J. Schneider, Mat. Sci. Forum, 83-87, 1183, (1992)

7. G. Pensl and R. Helbig, Festkorperprobleme 30 (1990)

8. H. Zhang, PhD. Thesis (1990), Erlangen University, Germany

9. S.E. Saddow, M. Lang, T. Dalibor, G. Pensl, and P.G. Neudeck, Appl. Phys. Lett., 66(26), 3612, (1995)

10. N. Nordell, S.G. Andersson, and A. Schoner, International Conference on SiC and Related Materials, Kyoto, Japan, (1995)

DEEP LEVEL DEFECTS IN Mg-DOPED GaN

GYU-CHUL YI and BRUCE W. WESSELS
Department of Materials Science and Engineering and Materials Research Center, Northwestern University, Evanston, Illinois 60208

ABSTRACT

Deep level defects in Mg compensated GaN grown by metal-organic vapor phase epitaxy were investigated using photocapacitance spectroscopy measurements on Schottky barrier diodes. Addition of magnesium resulted in the formation of a series of deep centers with optical threshold energies of 1.0, 1.2, 1.8, and 3.1 eV. Upon annealing the epitaxial GaN in nitrogen at 850°C the mid-gap levels disappeared and only the trapping level at 3.1 eV remained. The mid-gap levels are ascribed to Mg dopant complexes which may in part be responsible for low doping efficiency of Mg in the as-grown, doped GaN. The deep level at 3.1 eV commonly observed from all Mg-doped GaN most likely involves the Mg acceptor. The photo-excited state of the 3.1 eV level had relaxation times of the order of 10^3 sec at 295 K.

INTRODUCTION

As-grown, epitaxial Mg-doped GaN is usually highly resistive [1,2]. However, after post-deposition nitrogen-ambient thermal annealing the *p*-type conductivity is improved [2]. The nature of the defect centers responsible for compensating the *p*-type material is not well understood [3–6]. Recently photoconductivity spectroscopy of Mg-doped GaN has been studied and deep tail states have been observed [7]. We have recently used photocapacitance spectroscopy to study deep level defects in Mg-doped GaN, a technique that is especially applicable for the study of deep levels in wide band-gap semiconductors [8,9]. In this study, both steady state and transient photocapacitance measurements were carried out on deliberately Mg doped GaN prepared by metalorganic vapor phase epitaxy to determine the deep level defects. The properties of the Mg related level at E_c-3.1 eV were of special interest.

EXPERIMENT

The layers were prepared by atmospheric pressure metal-organic vapor phase epitaxy [10,11]. The films compensated by the addition of Mg during growth were all *n*-type as determined by Hall effect measurements. Schottky barrier diodes were used for photocapacitance measurements on both undoped and Mg-doped epitaxial GaN layers. They were fabricated by evaporating either gold or aluminum thin films on the GaN films through a shadow mask. The typical thickness of the thin metal film is of the order of 500 Å. The diode area was 0.6–1.0 mm^2. Good rectifying current-voltage (I-V) characteristics were observed for the diodes. Carrier concentrations were 1×10^{17}–2×10^{18} cm^{-3} in the autodoped, *n*-type GaN and 5×10^{15}–2×10^{17} cm^{-3} in the Mg-doped GaN as determined by capacitance-voltage measurements.

Steady state photocapacitance measurements were performed over the spectral range of 0.5–3.4 eV at room temperature. Photocapacitance spectra were measured by illuminating from low energy to high energy with monochromatic light from a Zeiss MM12 double prism monochromator equipped with a 250 W quartz-halogen lamp [12]. Transient photocapacitance was measured by illuminating with light at a fixed energy. The energy resolution was better than

Mat. Res. Soc. Symp. Proc. Vol. 423 © 1996 Materials Research Society

0.04 eV in all experiments. The photocapacitance signal was measured using a Boonton 72B capacitance meter and recorded via an analog-to-digital converter and a personal computer. With this system capacitance changes of the order of 0.01 pF could be measured. The typical capacitance of a junction was in the range of 10 to 400 pF. An applied reverse bias voltage of -2 V was used. The barrier layer capacitance was not influenced by the internal photoemission from the metal.

RESULTS

The photocapacitance spectrum of a Mg-doped, *n*-type GaN Schottky barrier diode at 295 K is shown in Fig. 1 (dotted line). For comparison the spectrum of an undoped layer is also included. The photocapacitance spectrum of the autodoped GaN (with a carrier concentration of 1.6×10^{17} cm^{-3}) exhibited defect levels with optical threshold energies of 1.0, 1.4, and 3.25 eV. The thresholds at 1.0 and 1.4 eV are due to the electron emission from the deep centers to the conduction band. As for the deep traps in as-grown Mg-doped GaN (grown using a cp_2Mg flow rate of 0.2 μmol/min), the photocapacitance spectrum is shown in Fig. 1 (dotted line). The addition of Mg resulted in the elimination of the deep level at 1.4 eV and the formation of a series of trapping levels at 1.0, 1.2, and 3.1 eV. For a net donor concentration of $N_D=1.2\times10^{17}$ cm^{-3}, the concentrations of the traps at 1.0 and 1.2 eV are 4.6×10^{13} cm^{-3} and 2.8×10^{14} cm^{-3}, respectively. The concentration of the deep level at 3.1 eV is greater than 1.2×10^{15} cm^{-3}. The observed trap energies are summarized in Table I.

Upon annealing the epitaxial Mg-doped GaN in nitrogen at 850°C for 10 min, the mid-gap levels at 1.0 and 1.2 eV disappeared and only the trapping level at 3.1 eV remained (Fig. 2(a)). Moreover the measured carrier concentration decreased to 1.0×10^{16} cm^{-3}. For the Mg-doped GaN grown using the higher cp_2Mg flow rate of 1.0 μmol/min, the photocapacitance spectra of the as-grown (dotted line) and the post thermal annealed (solid line) Mg-doped GaN are shown in Fig. 2(b). More Mg doping resulted in a higher concentration of traps at 3.1 eV and the formation of a new deep center at 1.8 eV whose concentration is 1.8×10^{15} cm^{-3}. Upon annealing

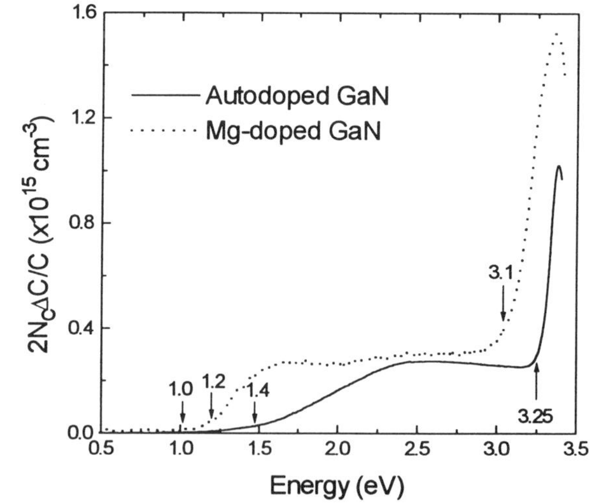

Fig. 1. Photocapacitance spectra of autodoped GaN (solid line) and intentionally Mg-doped GaN (dotted line). The photocapacitance spectrum of the autodoped GaN shows the deep traps at the optical threshold energies of 1.0, 1.4, and 3.25 eV. Addition of magnesium resulted in the formation of a series of deep centers at 1.0, 1.2, and 3.1 eV.

Table I. Summary of the optical threshold energies of defects observed in epitaxial GaN films.

Sample number	cp_2Mg flow rate (μmol/min)	Post-annealing	Carrier concentration (10^{16} cm^{-3})	Observed defect levels (eV)
Y209	0.0	No	16	1.0, 1.4, 3.25
Y174	0.2	No	12	1.0, 1.2, 3.1
Y177	0.2	Yes	1.0	3.1
Y172	1.0	No	1.2	1.8, 3.1
Y191	1.0	Yes	0.8	2.2, 3.1

in nitrogen at 850°C for 10 min, the mid-gap level at 1.8 eV disappeared and another trapping level is observed at 2.2 eV. Moreover the carrier concentration also decreased from 1.2×10^{16} to 0.8×10^{16} cm^{-3}. The traps at 1.8 eV and 2.2 eV in highly Mg-doped GaN might be the origin of the deep tail states observed in the photoconductivity spectra of the Mg-doped GaN [7].

The levels at 1.0 and 1.2 eV are unidentified, but the level at 3.1 eV most likely results from the Mg acceptor. In order to determine the spectral dependence of a photoionization cross section of Mg acceptor level at 3.1 eV, the transient photocapacitance was measured at photon energies ($h\nu$) between 2.75 and 3.3 eV. Fig. 3 shows transient photocapacitance measured at the photon

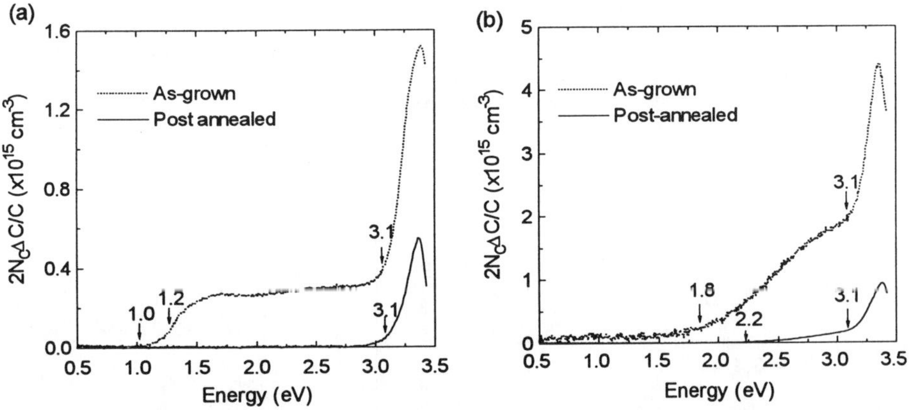

Fig. 2. (a) Photocapacitance spectra of the as-grown (dotted line) and the post thermal annealed (solid line) Mg-doped GaN grown using the cp_2Mg flow rate of 0.2 μmol/min. Upon annealing the Mg-doped GaN in nitrogen at 850°C the mid-gap levels disappeared and only the trapping level at 3.1 eV remained. (b) Photocapacitance spectra of the as-grown (dotted line) and the post thermal annealed (solid line) Mg-doped GaN grown using the cp_2Mg flow rate of 1.0 μmol/min. Heavy Mg doping resulted in the formation of a center at 1.8 eV and an increase in the concentration of the trap at 3.1 eV. Annealing in nitrogen resulted in the removal of the deep level at 1.8 eV and the formation of the deep level at 2.2 eV.

energies of 3.1 eV and 3.35 eV. After commencement of the light, the photocapacitance signal ΔC increases according to the relationship given by [13]:

$$\Delta C(t) = C_o[1-\exp(-t/\tau_R)] \tag{1}$$

where C_o is the steady-state capacitance and τ_R is the rise time constant. The rise time constant is approximately the inverse of optical emission rate. The time constant in this case is 160 and 0.3 seconds for 3.1 and 3.35 eV illumination, respectively.

Rise time constants at each photon energy were obtained from the transient photocapacitances measurements. The photoionization cross section is inversely proportional to the time constant for a negligible thermal emission rate. The spectral dependence of the photoionization cross section was obtained (Fig. 4) and fitted using the Lucovsky expression [14]:

$$\sigma(h\nu) \propto \nu^{-3}E_i^{1/2}(h\nu-E_i)^{3/2} \tag{2}$$

with E_i = 3.15 eV which is the ionization energy of the defect level. From E_i = 3.15 eV and energy band gap of GaN at room temperature ($E_g \cong 3.43$ eV), the photoionization energy of the Mg acceptor is estimated to be 280 meV. This is in contrast to a value of 160 meV for thermal ionization energy measured by Hall effect on *p*-type GaN [15]. The large difference between the thermal and optical ionization energies indicates that the Mg acceptor undergoes lattice relaxation.

To examine the thermal emission processes, the thermal decay of the photocapacitance was measured. After cessation of the optical pulse the photocapacitance decays by thermal emission of the trapped carriers. The photocapacitance decays exponentially with time according to the relationship:

$$\Delta C(t) = C_o\exp(-t/\tau_D) \tag{3}$$

where in this case the decay constant is given by τ_D the decay time. For the case of levels near the valence band the thermal emission involves holes. The decay constant is inversely proportional to the thermal emission rate constant for holes. As shown in Fig. 3, the observed

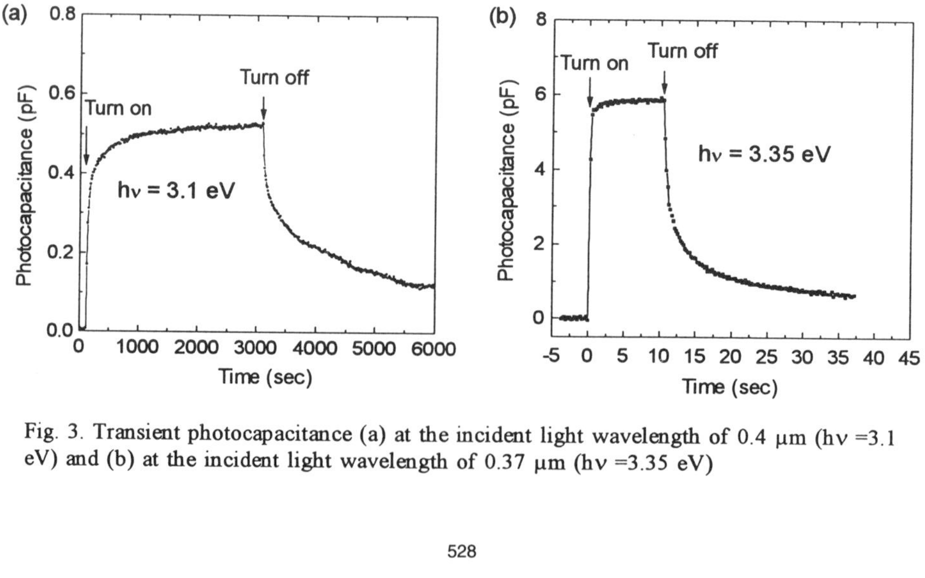

Fig. 3. Transient photocapacitance (a) at the incident light wavelength of 0.4 μm (hν =3.1 eV) and (b) at the incident light wavelength of 0.37 μm (hν =3.35 eV)

time constant is long ranging from 10–1000 seconds. This suggests that there is a barrier to emission and that lattice relaxation is occurring. This phenomenon is the subject of further study.

These results indicate that Mg is not a well-behaved shallow acceptor. Doping with Mg introduces mid-gap levels into the GaN in addition to the Mg related acceptor level. These midgap levels are thermally unstable since nitrogen-ambient thermal post-annealing decreases their concentrations. This result suggests that the Mg may form complexes (Mg–X) that are electrically active and are deep. Formation of these complexes would result in a lowered doping efficiency for the Mg acceptors. Moreover they could serve to pin the Fermi level near midgap leading to high resistivity material. Presumably upon heating, the complexes disassociate to form Mg acceptors. In the case of *n*-type material a lower electron concentration would result upon heating as observed in this study. For *p*-type material, disassociation would lead to a higher hole concentration upon thermally treating.

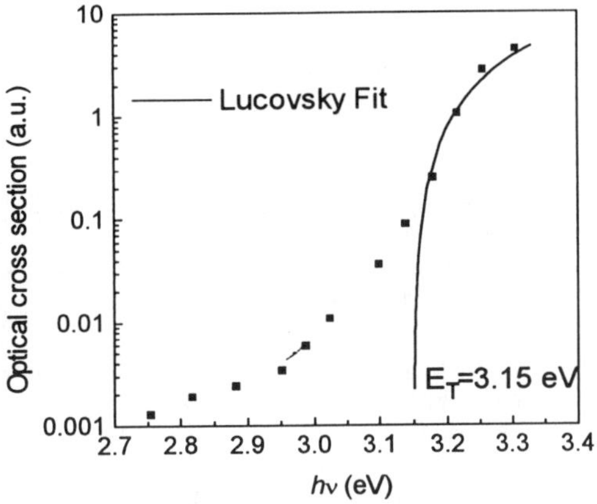

Fig. 4. Transient photocapacitance time constants vs photon energy $h\nu$ at room temperature. From the Lucovsky fitting curve (solid line), the defect level is shown to be 3.15 eV.

The Mg center itself may undergo lattice relaxation as suggested by the anomalously long time constant associated with thermal relaxation of the associated photocapacitance.

CONCLUSIONS

In summary, deep level defects in Mg-doped GaN grown using MOVPE were observed using photocapacitance spectroscopy on Schottky barrier diodes. Addition of magnesium resulted in the formation of deep centers at 1.0, 1.2, 1.8, and 3.1 eV. Upon annealing the epitaxial GaN in nitrogen at 850°C the mid-gap levels disappeared and only the trapping level at 3.1 eV remained with another trapping level at 2.2 eV for highly Mg-doped GaN. Mg complex formation may be responsible for the observed low doping efficiency of Mg in the as-grown, doped GaN as well as its semi-insulating behavior. The deep level at 3.1 eV commonly observed in the photocapacitance spectra of Mg-doped GaN most likely involves the Mg acceptor. The large difference between thermal and optical ionization energies as well as the anomalously long thermal emission time indicates that the Mg acceptor undergoes lattice relaxation.

ACKNOWLEDGEMENTS

This work was supported by the National Science Foundation Ceramics and Electronic Materials Program under grant number DMR-9302357. Extensive use of the facilities of the Materials Research Center at Northwestern University supported in part by the NSF under grant number DMR-9120521 is acknowledged.

REFERENCES

1. H. Amano, M. Kito, K. Hiramatsu, and I. Akasaki, Jpn. J. Appl. Phys. **28**, L2112 (1989).
2. S. Nakamura, N. Isawa, M. Senoh, and T. Mukai, Jpn. J. Appl. Phys. **31**, 1258 (1992).
3. J. A. Van Vechten, J. D. Zook, R. D. Horning, and B. Goldenberg, Jpn. J. Appl. Phys. **31**, L139 (1989).
4. M. S. Brandt, N. M. Johnson, R. J. Molnar, G. Singh, and T. D. Moustakas, Appl. Phys. Lett. **64**, 2264 (1994).
5. J. Neugebauer and C. G. Van de Walle, Phys. Rev. Lett. **75**, 4452 (1995).
6. W. Götz, N. M. Johnson, J. Walker, D. P. Bour, and R. A. Street, Appl. Phys. Lett. **68**, 667 (1996).
7. C. H. Qiu, C. Hoggatt, W. Melton, M. W. Leksono, and J. I. Pankove, Appl. Phys. Lett. **66**, 2712 (1995).
8. W. Götz, N. M. Johnson, R. A. Street, H. Amano, and I. Akasaki, Appl. Phys. Lett. **66**, 1340 (1995).
9. G.-C. Yi and B. W. Wessels, submitted.
10. G.-C. Yi and B. W. Wessels, in Abstracts of the 37th TMS Electronic Materials Conference, Charlottesville VA, June 21-23, 1995.
11. G.-C. Yi and B. W. Wessels, Mater. Sci. Forum **196-201**, 49 (1995).
12. E. J. Bawolek and B. W. Wessels, Thin Solid Films **102**, 251 (1983).
13. H. G. Grimmeiss, Ann. Rev. Mater. Sci. **7**, 341 (1977).
14. G. Lucovsky, Solid State Commun. **3**, 299 (1965).
15. T. Tanaka, A. Watanabe, H. Amano, Y. Kobayashi, I. Akasaki, S. Yamazaki, and M. Koike, Appl. Phys. Lett. **65** 593 (1994).

DLTS AND CV ANALYSIS OF DOPED AND N-IMPLANTED GaN

D. HAASE, M. SCHMID, A. DÖRNEN, V. HÄRLE, H. BOLAY, F. SCHOLZ, M. BURKARD, and H. SCHWEIZER
4. Physikalisches Institut, Universität Stuttgart, 70550 Stuttgart
Federal Republic of Germany
e-mail: dh4@pi4wap0.physik.uni-stuttgart.de

ABSTRACT

We studied by deep level transient spectroscopy (DLTS) and capacitance-voltage (CV) measurements the effects of doping (Zn, S), nitrogen implantation and annealing of n-type GaN grown on sapphire by MOVPE. The DLTS spectra of the as grown samples show two defect levels which are assumed to be identical with recently reported levels [10, 11]. In N-implanted GaN a third level is introduced which is not detectable in our as grown samples. This levels concentration follows the increasing N-implantation density. The depth profiles of its concentration correlate with the distribution of implantation defects expected from Monte-Carlo simulation. After annealing at 900°C for 60s the additional defect level vanishes. The DLTS spectrum then resembles those of annealed as grown samples. The n-type carrier concentration (CV measurements) increases in samples with low N-implantation dose. This implantation effect can be removed also with the RTA step. The increasing carrier concentration provides evidence that the N vacancy is a donor in GaN. For Zn and S doped GaN deep defect levels has been found, which are reported here.

INTRODUCTION

Gallium nitride has recovered great interest in research and development of opto-electronic devices and materials since Akasaki et al. [1] showed that *p*-type doping is achievable by low-energy–electron-beam irradiation. The direct bandgap energy of $E_G = 3.4$ eV at room temperature allows the realization of blue-light-emitting devices. Nakamura et al. have realized LED and laser diode technology [3, 4]. To optimize devices based on GaN it is necessary to understand the doping mechanisms of this material.

In recent studies Binari et al. [7] and Pearton et al. [6] have shown that implantation damage, which produces vacancies in large concentration, leads to a highly resistive material. On the other hand the nitrogen vacancy (V_N) has been discussed for a long time to be responsible for the intrinsic *n*-type doping of GaN [8]. To achieve high *n*-type carrier concentration intentionally doping is necessary. As a group-V element sulfur is a candidate for this purpose. For this reason we investigated the effect of S doping. The doping with Zn is of interest due to the blue luminescence center [2] and the acceptor states which are assigned to Zn [5].

EXPERIMENT

The analysis of shallow donor concentration was done by using the conventional capacitance voltage (CV) technique which uses the fundamental relationship between the dynamic capacity of the space-charge region of a Schottky contact and the net carrier concentration. The measurement of the capacitance was done with a BOONTON 72B capacitance meter. The test signal frequency was 1MHz and held at a level of 100mV. Deep level transient spectroscopy (DLTS) was used to investigate ionization energies and concentrations of deep levels between $E_C - 100$meV and $E_C - 900$meV. DLTS was performed

Mat. Res. Soc. Symp. Proc. Vol. 423 © 1996 Materials Research Society

with a commercial POLARON DL 4600 equipment. All DLTS measurements shown in this paper are measured with a rate window of 200 s^{-1} and a pulse width of 1 ms.

The samples were grown by metal-organic vapor-phase epitaxy (MOVPE) on sapphire substrate oriented along the [0001] direction using an AlN buffer. The growth parameters were $T_G = 1000°C$ and $P_G = 100$ mbar. The as-grown samples were characterized by X-ray diffraction and photoluminescence (PL) measurements. Both indicate a good crystal quality. Full width at half maximum (FWHM) of X-ray diffraction was 60 arcsec. The intense PL signal of the dominant excitonic luminescence at room temperature showed a FWHM of 3.6 meV.

Doping of GaN with Zn and S was performed during MOVPE growth with diethylzinc and H_2S respectively. Schottky and ohmic contacts were fabricated as published by Hacke et al. [10]. To ensure a reliable contact to the measurement equipment all samples were fixed on a submount and the contacts were bonded with Au wires. The performance of the Schottky contacts (max. reverse bias $U_R = -30$ V, reverse current density $j_R = 7 \cdot 10^{-4}$ A/cm^2, threshold voltage $U_{th} = +0.5$ V) was almost identical for all samples.

NITROGEN IMPLANTATION

To investigate the doping effect of the V_N in n-type GaN we studied N-ion implantation at low doses. Implantation of N^{2+} ions with 270 keV was performed with ion doses between $I = 1 \cdot 10^{10}$ cm^{-2} to $1 \cdot 10^{13}$ cm^{-2}. The beam direction was tilted 7° off the [0001] direction to avoid channeling. To study the annealing behavior of the centers rapid-thermal annealing (RTA) was performed at a temperature of $T = 900°C$ for 60 s. During RTA the samples were covered by a silicon wafer to form proximity caps. The Monte Carlo simulation [15] of the implantation defects is shown in Fig. 1. The depth profile of the point defects for the implantation parameters listed above shows for the N and Ga vacancy an almost equal concentration. The Ga and N interstitial defect is estimated to be in the same order of magnitude as the vacancies. A raw estimation for the antisite defect of N, N_{Ga}, extracted from the replacement calculation of the simulation gives a ratio between the V_N and N_{Ga} of 6 to 1.

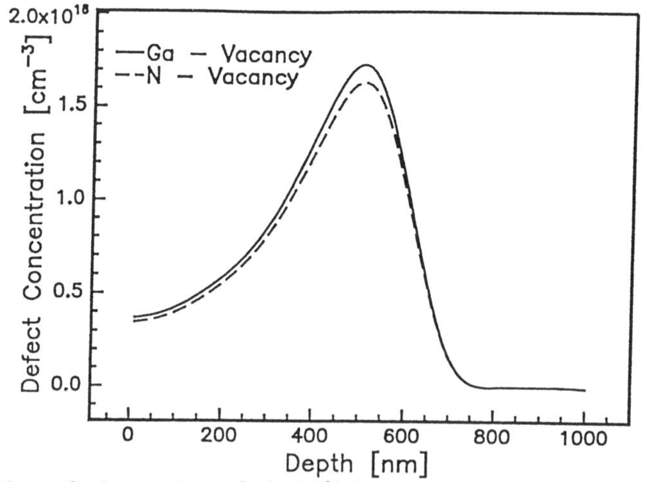

Figure 1: *Monte-Carlo simulation [15] of N^{2+}-implantation defects for $E_{impl.}$=270keV and a dose I=1 $\cdot$ 10^{10} cm^{-2}. Shown is the concentration depth profile of N and Ga vacancies.*

CAPACITANCE VOLTAGE MEASUREMENTS

The CV measurements of the as-grown sample as shown in [Fig. 2(a)] has a flat electron concentration profile with $n = 2 \cdot 10^{16}$ cm^{-3}. The results for the low implantation doses $I_0 = 1 \cdot 10^{10}$ cm^{-2} and $I_1 = 1 \cdot 10^{11}$ cm^{-2} are depicted in Fig. 2(b) and 2(c). After implantation of I_0 the electron concentration shows a profile with a maximum of $n = 6 \cdot 10^{16}$ cm^{-3}. This maximum increases up to a value of $n = 5 \cdot 10^{17}$ cm^{-3} with increasing the N^{2+}-implantation dose to I_1. After the annealing step at $T = 900°$C for 60s for both of the implanted samples the results are shown in Fig. 2(d): the n-type-carrier concentration has decreased and shows a flat profile with $n = 5 \cdot 10^{16}$ cm^{-3}, which is close to the conditions measured for the as-grown sample. Figure 2(e) shows the development of the maximum electron concentration corresponding to the sample treatment. For the implantation dose I_1 the value is averaged in the region of the maximum due to the apparent noise of the measurement.

DLTS MEASUREMENTS

The DLTS spectra for the as-grown, low-dose implanted, and annealed samples are shown in Fig. 3. The measurements were done with the identical samples which were used for the CV-measurements.

Figure 3(a) shows the spectrum of the as-grown sample. Two defect levels labeled D1 and D2 were found with ionization energies of $E_{D1} = 270 \pm 20$ meV and $E_{D2} = 598 \pm 20$ meV respectively. They correspond within the measurement uncertainty to the defect levels E1 and E2 found by Hacke et al. [10].

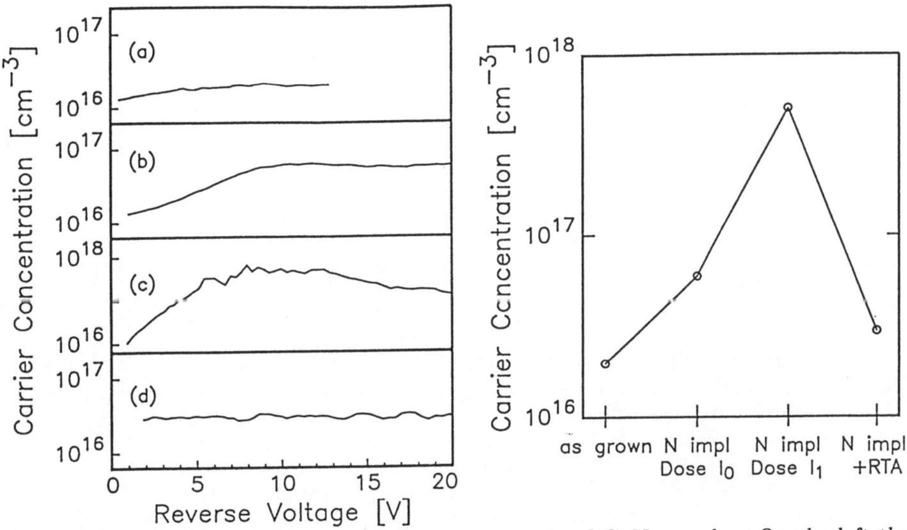

Figure 2: *Results of capacitance-voltage measurements of GaN samples. On the left the free electron concentration versus applied reverse bias @ $T = 300$ K is shown: (a) as grown, (b) after N implantation with $I_0 = 1 \cdot 10^{10}$ cm^{-2} @ 270 keV, (c) after N implantation with $I_1 = 10 \cdot I_0$, (d) sample with same treatment as (c) after RTA 60 s @ 900°C. On the right: the relationship between carrier concentration and sample treatment for the maxima of the measurements (a) to (d).*

Implantation of N^{2+} with 270 keV and an the ion dose $I_0 = 1 \cdot 10^{10}$ cm^{-2} introduces another deep level, labeled D3 in Fig. 3(b) . This level has an ionization energy of $E_{D3} = 670 \pm 20$ meV and is therefore comparable to the defect E3 found in GaN grown by hydride-vapor-phase epitaxy published in [10]. After increasing the implantation dose to $I_1 = 1 \cdot 10^{11}$ cm^{-2} the DLTS spectrum in Fig. 3(c) is recorded. In this spectrum the defect D3 is dominant. The effect of the annealing step is shown in Fig. 3(d). After RTA at $T = 900^oC$ for 60 s the level D3 has decreased whereas the level D2 shows a much smaller change in concentration. This result is shown more clearly in Fig. 3(e), where the concentrations of the defect levels D2 and D3 corresponding to the sample treatment is depicted.

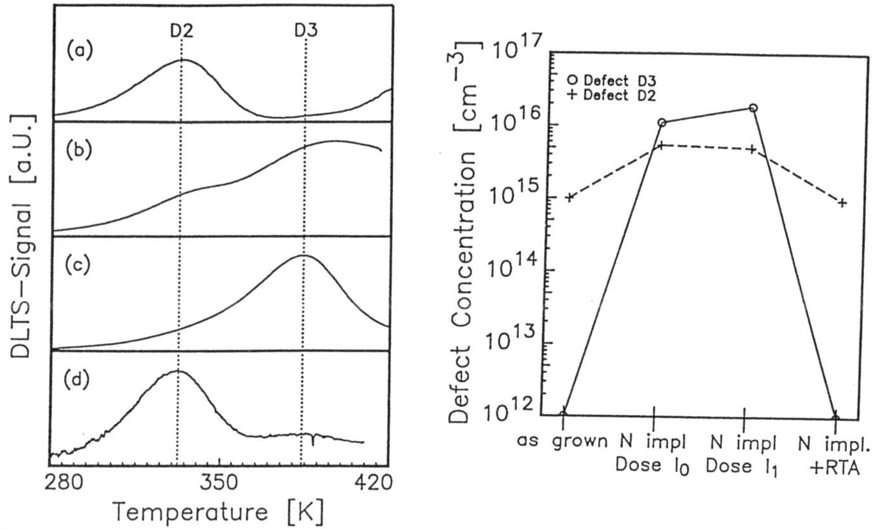

Figure 3: *DLTS signal of the samples of Fig. 2, measured with a rate window of 200* s^{-1}, *1 ms pulse width: (a) as grown, (b) after N implantation with* $I_0 = 1 \cdot 10^{10}$ cm^{-2} *@ 270 keV, (c) after N implantation with* $I_1 = 10 \cdot I_0$, *(d) sample with same treatment as (c) after RTA 60 s @ 900°C. On the right: concentration of levels D2 and D3 in dependence of sample treatment.*

To investigate the effect of a higher dose N implantation was done with $I_2 = 1 \cdot 10^{13}$ cm^{-2}. For this sample a reliable determination of the carrier concentration with CV measurements was not possible. Although CV measurements of this sample are questionable, a DLTS spectrum given in Fig. 4 gives insight in a change of implantation effect. Defect level D3 has for this sample a lower FWHM than the minimum FWHM (fit in Fig. 4:$\Delta E = 0$, the fit for D3 after I_0:$\Delta E = 20meV$, not shown) which the theory for the usual DLTS line shape based on the assumption that $N_T \ll n$ [9] allows. The extreme sharp DLTS line shape might indicate that this assumption is no more valid.

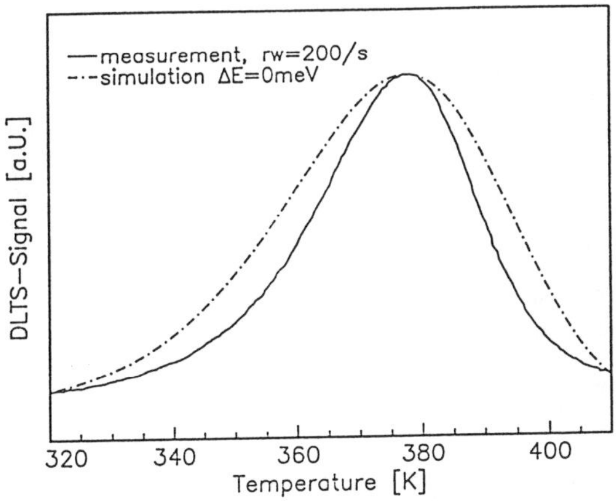

Figure 4: *DLTS spectrum (rate window=200 s^{-1}, 1 ms pulse width) of a GaN sample after N implantation with $I_2 = 1 \cdot 10^{13}$ cm^{-2} @ 270 keV. The fact that the simulation with $\Delta E_T = 0$ is the broader line might indicate that the assumption $N_T \ll n$, which is basic for the usual model [9] is not valid for this spectrum.*

DISCUSSION

Comparing our results referring to the increasing carrier concentration after N implantation at low doses with published data concerning N implantation in GaN the increasing free electron concentration after N implantation seems to contravene the results reported by Binari et al. [7] who observed an increasing sheet resistance after N implantation. They concluded on a lower n-type-carrier concentration due to implantation-induced higher deep-level concentration. We observed the generation of a deep level, too, but we assume that its effect on the free carrier concentration is overcompensated for low implantation doses by additionally generated shallow donors. For higher implantation doses the deep level concentration is not necessarily much lower than the carrier concentration, as the results above suggest (Fig. 4). These might be a hint that the deep level generation overtakes the generation of shallow donors with high implantation doses which would explain the results of [7] and [6] under the light of deep level concentration.

The results of tight binding calculations by Dow and Jenkins [12], [13] predict the ground state of the V_N close to the conduction band. Neugebauer and Van de Walle used first-principles total-energy calculations to analyze native defects [14]. Their results showed that the donor level of the V_N should lay in the conduction band, but the formation energy is too high to stabilize the V_N under n-type conditions. They mentioned that this does not exclude clusters out of them to form stable donors. Our results show that the N implantation causes doping levels which result from forming N vacancies or clusters out of them. No other point defect of intrinsic nature is, up to our knowledge, predicted by theoretical means to have a donor level close enough to the conduction band.

To interpret the presented data referring to defect level D3, which obviously shows strong dependence on N implantation, we have to consider the possible implantation defects. Nitrogen implantation produces the intrinsic point defects as vacancies and interstitials in a first step. The Monte-Carlo simulation [15] of the distribution of implantation induced defects results in a comparable concentration for both, N and Ga vacancies as shown in Fig. 1 . For the gallium vacancies Neugebauer and Van de Walle [14] predicted a level close to the valence band, so we consider this defect not as a probable explanation for defect level D3.

It is tempting to identify level D3 with the N interstitial. First, level D3 is generated in a high concentration by N implantation and anneals out as expected for interstitials, second level D3 lies in the upper part of the band gap which may correspond to the mid-gap level predicted in [14] for the N interstitial. The DLTS-spectrum of the as-grown sample shows level D2 which is common to all presented DLTS-spectra. As mentioned above D2 is comparable to level E2 published in [10]. The concentration of D2 changes little with N implantation and annealing. For that reason an intrinsic point defect, which is not mainly affected by implantation, could be identified with defect D2. Hacke et al. [10] identified level E2 with the N antisite. From the weak reaction to N implantation and annealing we follow also this identification.

Zn AND S DOPED GaN

The intention for Zn doping was to investigate the recombination center responsible for the blue luminescence in LEDs [2]. The DLTS spectra depicted in Fig. 5 is completely different to that one of the nominally undoped, as grown sample shown in Fig. 2. We obtain a defect level concentration of $1 \cdot 10^{15}$ cm^{-3} ($n = 1 \cdot 10^{18}$ cm^{-3}) and a ionization energy of $E_{Zn} = 350$meV . This value is clearly different from the deep acceptor levels of Zn obtained from D-A transitions in photoluminescence spectroscopy [5]. .

Doping with sulfur was done with the intention of increasing n-type carrier concentration. In fact the carrier concentration determined by CV measurements was $n = 2 \cdot 10^{17}$ cm^{-3}. The DLTS spectra shows one defect level depicted in Fig. 5 . This level has comparable parameters as level D2 in Fig. 3. The defect level concentration for the doped sample is with $N_T = 6 \cdot 10^{14}$ almost two orders of magnitude smaller than in the undoped sample.

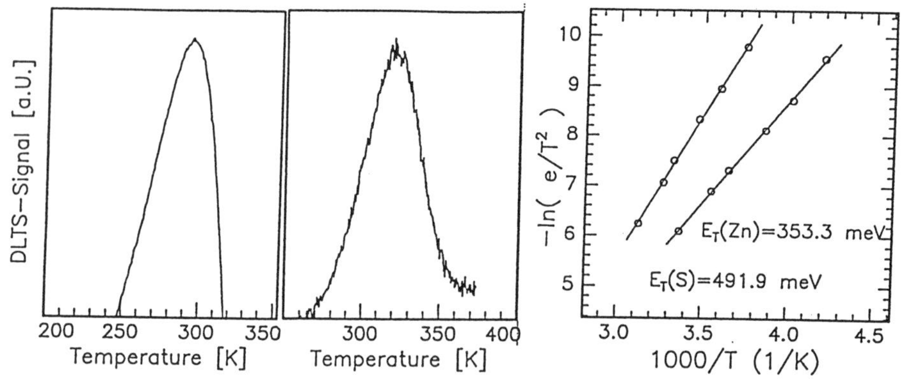

Figure 5: *DLTS spectra for zinc (left) and sulfur (middle) doped GaN. The Arrhenius plot (right) indicates the levels parameters.*

ACKNOWLEDGEMENTS

We thank M. H. Pilkuhn for his steady interest in our work.

REFERENCES

1. H. Amano, M. Kito, K. Hiramatsu, I. Akasaki, Jpn. J. Appl. Phys. **28** L2112-14 (1989)
2. I. Akasaki, K. Hiramatsu, H. Amano, Memories of the Faculty of Engineering, Nagoya University, Vol. 43, No. 2. March 1992, Nagoya, Japan
3. S. Nakamura, M. Senoh, T. Mukai, Jpn. J. Appl. Phys. **32** L8-L11 (1993)
4. S. Nakamura, M. Senoh, S. Nagarama, N. Iwasa, T. Yamada, T. Matsushita, H. Kiyoku, Y. Sugimoto, Jpn. J. Appl. Phys. **35** L74 (1996)
5. B. Monemar, O. Lagerstedt, H. P. Gislason, J. Appl. Phys. **51**, 625 (1980)
6. S. J. Pearton, C. B. Vartuli, J. C. Zolper, C. Yuan, R. A. Stall, Appl. Phys. Lett. **67**, 1435 (1995)
7. S. C. Binari, H. B. Dietrich, G. Kelner, L. B. Rowland, K. Doverspike, D. K. Wickenden, J. Appl. Phys. **78**, 3008 (1995)
8. H. P. Maruska and J. J. Tietjen, Appl. Phys. Lett. **15**, 327 (1969)
9. G. Pensl in Landolt Börnstein: Data Tables on Impurity and Defects, Series III, Vol 22, Part b, edited by O. Madelung and M. Schulz (Springer, Berlin/Heidelberg, 1989), Chap. 3.
10. P. Hacke, T. Detchprohm, K. Hiramatsu, N. Sawaki, J. Appl. Phys. **76**, 304 (1994)
11. W. Götz, N. M. Johnson, H. Amano, I. Akasaki, Appl. Phys. Lett. **65**, 463 (1994)
12. D. W. Jenkins and J. D. Dow, Phys. Rev. B **39**, 3317 (1989)
13. D. W. Jenkins, J. D. Dow, M. H. Tsai, J. Appl. Phys. **72**, 4130 (1992)
14. J. Neugebauer, C. G. Van de Walle, Phys. Rev. B **50**, 8067 (1994)
15. We used the software TRIM for this calculations

THE RELATIONSHIP BETWEEN MICROPIPES AND SCREW DISLOCATIONS IN PVT GROWN 6H-SiC

JENNIFER GIOCONDI, GREGORY S. ROHRER, MAREK SKOWRONSKI
Carnegie Mellon University, Department of Materials Science and Engineering
Pittsburgh PA 15213, USA

V. BALAKRISHNA, G. AUGUSTINE, H.M. HOBGOOD, AND R.H. HOPKINS
Northrop Grumman Science and Technology Center
Pittsburgh, PA 15235, USA

ABSTRACT

The growth surface of a 6H-SiC boule, grown by physical vapor transport, was examined using scanning force microscopy. The dimensions of surface/micropipe intersections and screw dislocation Burgers vectors have been determined from topographic data. All micropipes are positioned along the lines of super screw dislocations with a Burgers vectors of at least 4 times the c-axis repeat distance (15.2 Å). Perfect c-axis screw dislocations with Burgers vectors of only 15.2 Å are stable and do not have open cores. Measurements show that micropipe core radii, determined indirectly from the width of the craters formed at the surface/micropipe intersections, increase with the square of the dislocation Burgers vector.

INTRODUCTION

Micropipes are cylindrical voids with diameters in the 1-10 μm range that are found oriented along the c-axis of SiC crystals grown from the vapor phase. For example, in crystals grown by the physical vapor transport (PVT) process, these defects occur with a density of 50-10^3 cm^{-2} and extend throughout most of the boule volume [1,2]. Because micropipes limit the operating range of high voltage SiC devices, their elimination has important practical implications [3]. The objective of this paper is to specify the nature of this extended defect.

A possible mechanism for the stabilization of such a long, narrow cylindrical void was proposed by Frank in 1951 [4]. If a dislocation with a sufficiently large Burgers vector threads the length of the crystal, it can be energetically favorable to replace the most highly strained part of the crystal in the vicinity of the dislocation line with an empty cylinder. Frank demonstrated that a state of local equilibrium can be achieved by balancing the elastic energy of the dislocation against the surface energy of facets bounding a narrow cylinder. One of the central predictions of Frank's theory is that the pipe radius (r_0) should be proportional to the square of the Burgers vector (b); the relevant physical parameters that determine the proportionality are the surface energy (γ) and the shear modulus (G):

$$\frac{b^2}{r_0} = \frac{8\pi^2\gamma}{G} \tag{1}$$

There are several pieces of evidence supporting the idea that micropipes are, indeed, empty core screw dislocations. First, Verma [5] noted that micropipes intersect the growth surface at the origins of optically visible spiral steps. Also, measurements based on optical micrographs of the growth surface have been used to show that the core radius increases with

Mat. Res. Soc. Symp. Proc. Vol. 423 © 1996 Materials Research Society

the height of the spiral step [6]. More recently, white beam synchrotron topography was used to show that micropipes in PVT grown 6H-SiC crystals were empty core screw dislocations with Burgers vectors 3 to 7 times the c lattice constant (15.2 Å) [7]. In the present paper, we present quantitative scanning force microscopy (SFM) measurements of the dislocation Burgers vectors and micropipe radii based on topographic images of the growth surface of a 6H SiC crystal produced by the PVT method. Our results show that single, perfect c-axis oriented screw dislocations with Burgers vectors of 15.2 Å = b_0 are stable and do not have open cores. On the other hand, all micropipes are found to lie along the lines of super-dislocations with Burgers vectors of nb_0, where n is an integer greater than or equal to 4.

EXPERIMENTAL

All measurements described in this paper were conducted on the same 6H-SiC growth surface. A wafer containing the growth surface was removed from the top of a large single crystal boule grown by the PVT process and imaged without further treatment [1]. SFM analysis was carried out using a Park Scientific Instruments (PSI) scanning probe microscope. Because the scanner field of view was 5 microns, only surface/micropipe intersections smaller than this size were measured. An optical microscope was used to identify the points where the micropipes emerged on the surface and to position the scanning probe. While measurements of more than 25 spirals were made in the course of this study, we compare only a subset of these in the current paper. This subset of measurements was made using exactly the same probe tip (a high aspect ratio PSI ultralever) and instrumental parameters (stage tilt, scan rate, and feedback parameters). While the remaining measurements are consistent with those presented here, limiting our analysis to only this subset allows us to conclude that scatter in the results is not due to differences in the details of the probe shape or the feedback response.

When a dislocation line intersects a surface, the component of the Burgers vector normal to the surface is equal to the height of the step created by the intersection. In this case, we take the screw component of the dislocation's Burgers vector to be equal to the height of the step that emerges from the micropipe. Note that only the component of the Burgers vector normal to (0001) is determined; topographic measurements can not be used to distinguish pure screw dislocations from those of mixed character. Figure 1a shows a typical topographic profile from which a step height was determined. This is a single scan line taken from the image in Fig. 2b, along the fast-scan direction. In order for rotating spiral steps to produce the 6H polytype, they must occur in integer multiples of the c-axis repeat distance, which is the length one perfect screw dislocation along [0001], b_0=15.2 Å. In fact, measured step heights seldom deviated from these intervals by more than a few angstroms (a few percent of the total height). For the final analysis, step heights were assigned to the closest integer multiple of b_0. For example, the height of the step shown in Fig. 1a (136 Å) is very near $9b_0$.

One difficulty associated with determining the core radius from any microscopic examination of the growth surface is that the micropipe radius at the surface is much larger than the radius in the bulk. Because of capillarity, the surface/micropipe intersection forms a sloping crater which is widest at the surface and slowly tapers to the stable radius in the bulk, r_0. A typical SFM line profile extracted from an image of a surface/micropipe intersection is shown in Fig. 1b. The flat part in the center of the depression is where the probe has extended as far as possible into the void (the maximum vertical depth of field is limited by the piezoelectric scanner design and the high voltage amplifiers used in the microscope). Clearly, it is not possible to directly observe r_0 using surface images. However, in Frank's original theoretical description of the empty-core dislocation, he predicted this effect and determined the relationship between the initial slope of the crater, which is determined accurately by SFM, and the ultimate radius [4]. For example, at the point where the slope of the crater wall (dz/dr) is equal to 1/4, the crater radius is 10.5 times r_0. In each case, the crater diameter (d_c) was determined using the points where the slope equaled 1/4 and this value was used to calculate r_0, as illustrated in Fig. 1b.

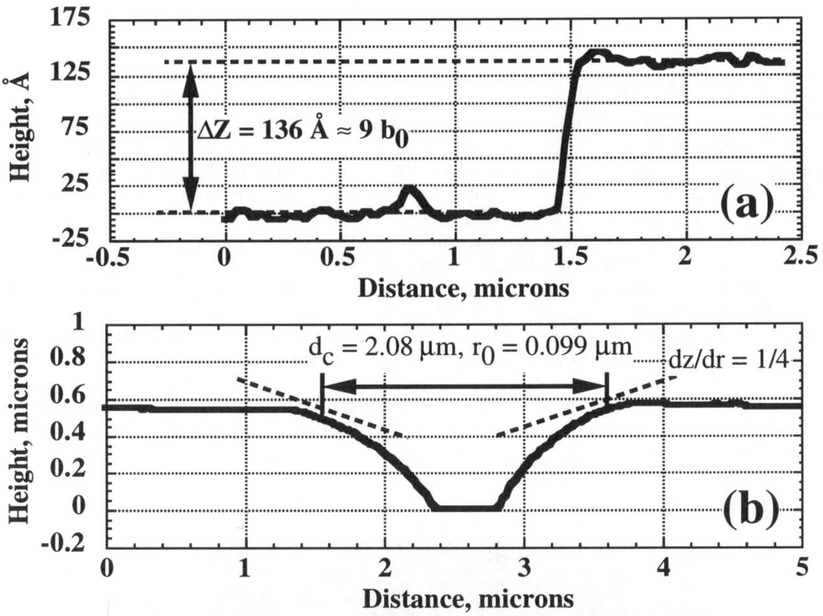

Figure 1. Topographic profiles used to measure the sizes of (a) the Burgers vectors associated with micropipes, b, and (b) the micropipe core radius, r_0.

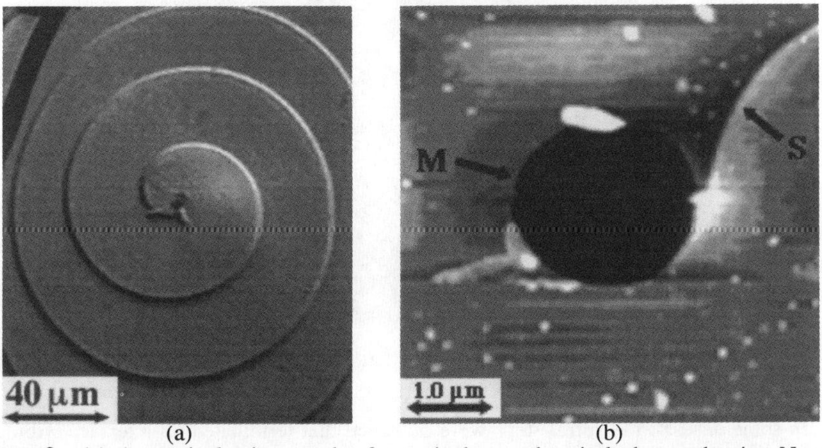

Figure 2. (a) An optical micrograph of a typical growth spiral observed using Nomarski contrast. The small black spot at the end of the spiral step is a surface/micropipe intersection. (b) A SFM image of a different surface/micropipe intersection. The micropipe is the circular black spot in the center, labeled M. Due to the small field of view, only a small portion of the spiral step (labeled S) is shown. A profile taken over this step is shown in Fig. 1a.

RESULTS

The SiC (0001) growth surface was comprised of flat terraces, separated by large, circular growth steps. The flat terraces were decorated with growth spirals such as the one shown in Fig. 2a. A typical SFM image, recorded in the center of a similar spiral, is shown in Fig. 2b. This topographic image is shaded so that the lowest regions appear dark and the highest ones are light. Thus, the roughly circular black spot in the center of the micrograph (labeled M) is the intersection of the micropipe with the surface and the abrupt change in contrast in the upper right hand corner of the image is a step (labeled S) emerging from the pipe. The regions of roughly constant contrast are atomically flat terraces and the white spots are small, ubiquitous, island-like inhomogeneities. A Burger circuit taken around the pipe shows that the height of this step must equal the Burgers vector of a dislocation whose line emerges from the crystal at this point. In this case, the step height (see the profile in Fig. 1a) is 136 Å, so the Burgers vector of this super screw dislocation has a magnitude 9 times that of a single perfect dislocation. In other cases, multiple spiral steps emerged from the same pipe and the sum of the heights was used to determine the total Burgers vector. When the steps are less than 45 Å, or equal to only one or two times b_0, they are not optically visible. In some cases, micropipes were located in the optical microscope without visible spiral steps. However, upon closer examination in the SFM, multiple spiral steps, all with heights of approximately 15 Å, were found to emerge from the defect. Every micropipe examined as part of this study occurred at the point where a super-dislocation emerged from the crystal.

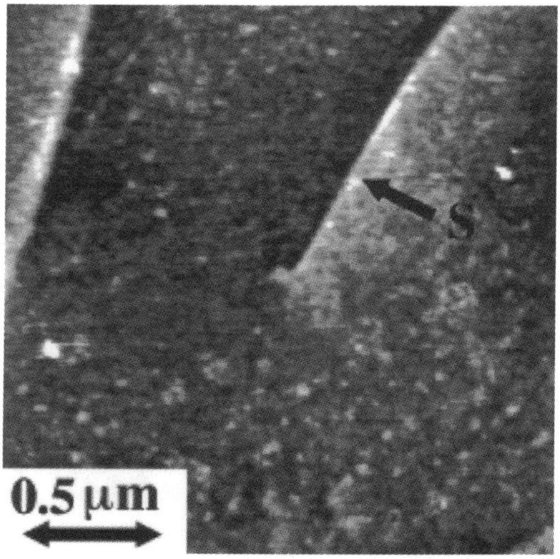

Figure 3. SFM image of a 15 Å step (labeled S) that ends abruptly on a flat terrace without a depression or micropipe.

In some cases, 15 Å spiral steps were observed that terminated on a terrace without a micropipe. One such example is shown in Fig. 3, where a step (labeled S) terminates near the center of the micrograph without an observable depression. Based on studies of surface/dislocation intersections on the surfaces of GaN films using the same instrumentation, we know that it is possible to detect depressions as small as 200 Å in diameter (this is similar to the probe diameter) [8-9]. If there is an even smaller, undetected crater at the end of the step in Fig. 3, the pipe radius associated with such a crater would be less than 10 Å, a physically unrealistic value. Therefore, we conclude that single perfect dislocations with Burgers vectors of 15 Å do not form micropipes.

Based on eqn 1, the micropipe radius should increase with the square of the Burgers vector. Figure 4 shows a plot of b^2 v. r_0 for 9 carefully measured micropipes. Assuming that the shear modulus of SiC is 200 GPa [10] and the surface energy is 4 J/m^2 [11], the slope of this line should be equal to approximately 16 Å; the dashed line on the graph shows this "ideal" slope. Although there is a considerable amount of scatter, the results are consistent with the predicted slope.

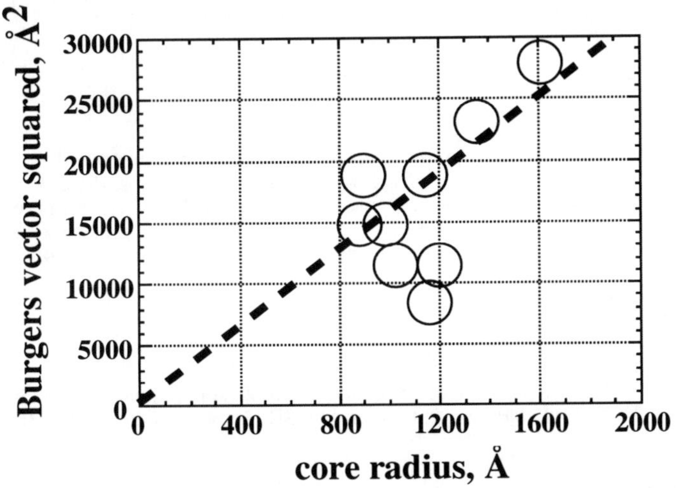

Figure 4. The core radius v. the square of the Burgers vector for 9 carefully measured micropipes. The size of the circle is used to indicate the maximum estimated uncertainties in the determination of b and r_0. The dashed line has a slope of 16 Å.

DISCUSSION

Based on our observations, Frank's theory for the open core screw dislocation seems to provide an appropriate description of the micropipe defect in SiC. The smallest possible perfect dislocations, with Burgers vectors of only 15.2 Å, have stable cores and do not form micropipes. The larger super-dislocations, however, which have more elastic strain, do form the cylindrical voids and the radii of these defects are correlated to the square of the dislocation Burgers vector.

It is, however, important to note that the observed deviations from the predicted behavior are larger than the estimated uncertainties involved in the measurement. We can identify two possible sources for this error. First, Frank's analysis predicts the dimensions of equilibrium defects [4]. The structures described here were formed on a growing crystal. In this case, the core radii should be influenced by the supersaturation of the vapor above the growing crystal surface. In particular, the kinetic radii should be systematically smaller than the predicted equilibrium radii and this contraction can be computed with knowledge of the critical radius for two dimensional nucleation, r_c [12]. We can use the observed spiral terrace width to estimate that r_c is on the order of 1000 Å [13] and, based on this estimate, use the results in ref. 12 to find that the kinetic core radius is about 10% less than the equilibrium core radius. Since this is comparable to other experimental uncertainties, we conclude that kinetic effects can not account for the observed deviations.

A second possible cause for the deviations is that the shapes of the craters are influenced by second phase inhomogeneities that are frequently found on the growth surface and always found on the inner wall of the surface crater at the micropipe. For example, in Fig. 2b, a white elongated feature is seen near the top of the micropipe. Similar features are found near all micropipes and because some of them have influenced the shape of growing steps, we can be certain that they were present on the surface during growth. Such inhomogeneities might pin the walls of the craters and make them artificially large, an effect that would account for deviations below the ideal line on Fig. 4. The role that these inhomogeneities play in micropipe nucleation will be described in a forthcoming paper.

CONCLUSION

Scanning force microscopy measurements have been used to demonstrate that micropipes form in SiC to relieve strain at the cores of dislocations with large screw dislocation components. Single complete screw dislocations are stable and micropipes have been observed to occur only when the dislocation Burgers vector is at least 4 times that of the single dislocation. Preventing the formation of these super-dislocations should suppress micropipe formation in SiC.

ACKNOWLEDGMENTS

J.G. and G.S.R. acknowledge support from an NSF REU supplement to YIA Grant No. DMR-9458005. M. S. acknowledges support under AFOSR Grant No. F29620.94.1.0392.

REFERENCES

[1] D.L. Barrett, J.P. McHugh, H.M. Hobgood, R.H. Hopkins, P.G. McMullin, R.C. Clarke, W.J. Choyke, J. Cryst. Growth **128**, 358 (1993).
[2] H.M. Hobgood, D.L. Barrett, J.P. McHugh, R.C. Clarke, S. Sriram, A.A. Burk, J. Greggi, C.D. Brandt, R.H. Hopkins, W.J. Choyke, J. Cryst. Growth **137**, 181 (1994).
[3] J. A. Powell, P. G. Neudeck, D. J. Larkin, J. W. Yang and P. Pirouz, Inst. Phys. Conf. Ser., **137**, 161 (1994).
[4] F. C. Frank, Acta Cryst. **4**, 497 (1951).
[5] A.R. Verma, Crystal Growth and Dislocations, (Butterworths, London, 1953).
[6] H. Tanaka, Y. Uemura, Y. Inomata, J. Cryst. Growth **53**, 630 (1981).
[7] M. Dudley, S. Wang, W. Huang, C.H. Carter, Jr., V.F. Tsvetkov, and C. Fazi, J. Phys. D: Appl. Phys. **28**, A63 (1995).
[8] W. Qian, G. S. Rohrer, M. Skowronski, K. Doverspike, L. B. Rowland, and D. K. Gaskill, Applied Physics Letters **67**, 2284 (1995).
[9] G. S. Rohrer, J. Payne, W. Qian, M. Skowronski, K. Doverspike, L. B. Rowland, and D. K. Gaskill, in GaN and Related Materials edited by R.D. Dupuis, F.A. Ponce, J.A. Edmond, and S. Nakamura (Mater. Res. Soc. Proc., Pittsburgh, PA, 1996), in press.
[10] W.R.L. Lambrect, B. Segall, M. Methfessel, and M.v. Schilfgaarde, Phys. Rev. **B44**, 3685 (1991).
[11] B. Wenzien, P. Käckell, F. Bechftebt, Surface Science **307-9**, 989 (1994).
[12] H. Müller-Krumbhaar, T. Burkhardt, and D. Kroll, J. Crystal Growth **38**, 13 (1977).
[13] N. Cabrera, M.M. Levine, and J.S. Plaskett, Phys. Rev. **96**, 1153 (1954).

CHARACTERIZATION OF DEFECT STRUCTURES IN 3C-SiC SINGLE CRYSTALS USING SYNCHROTRON WHITE BEAM X-RAY TOPOGRAPHY

W. Huang,* M. Dudley,* and C. Fazi**
* Dept. of Materials Science and Engineering, SUNY at SB, Stony Brook, NY11794-2275
**U.S. Army Research Laboratory, 2800 Powder Mill Road, Adelphi, MD 20783, USA

ABSTRACT

Defect structures in (111) 3C-SiC single crystals, grown using the Baikov technique, have been studied using Synchrotron White Beam X-ray Topography (SWBXT). The major types of defects include complex growth sector boundary structures, double positioning twins, stacking faults on {111} planes, inclusions and dislocations (including growth dislocations and partial dislocations bounding stacking faults). Detailed stacking fault and double positioning twin configurations are determined using a combination of Nomarski interference microscopy, SEM and white beam x-ray topography in both transmission and reflection geometries. Possible defect generation phenomena are discussed.

INTRODUCTION

SiC has attracted intensive attention due to its potential as a material for the manufacture of electronic devices[1-11]. Because of its unique features of wide bandgap, high chemical stability and high resistance to radiation damage, etc., SiC is expected to have broad applicability in high-temperature, high power and high frequency devices. Among all its polytypes including 6H, 4H, 15R, etc., the 3C-SiC shows the highest electron mobility. Its intrinsic carrier concentration is five orders of magnitude larger than that of 6H- or 4H-SiC and it offers a better chance to support electron impact ionization with minimum anisotropy [1-3]. Therefore 3C-SiC is one of the most promising of the polytypes for application to active devices.

Some efforts have been made to demonstrate the feasibility of fabricating inversion-type MOSFETs, Schottky-barrier FETs and PN junction diodes using cubic-SiC [4-7]. However, because of the lack of large, high quality 3C-SiC substrates, all these devices are fabricated by growing 3C-SiC epitaxial layers on Si or 6H-SiC substrates. The lattice mismatch between 3C-SiC and Si is 20% and highly strained epitaxy is unavoidable. Even when lattice mismatch is not significant, for instance 3C- on 6H-SiC, the 3C-SiC epilayers can still have poor structural quality [8].

Obviously, if large, high quality 3C-SiC crystals were available for use as substrates for 3C-SiC epitaxy, the expected improvement in epilayer quality should lead to improved device performance. Unfortunately, the growth of large, high quality 3C-SiC substrates has remained elusive. Since the 1960's, various techniques have been tried including solution growth, sublimation, and the Van Arkel method [7, 9, 10]. However, reports of crystals grown by these techniques indicate very poor quality. Recently, the Baikov method has been used to grow relatively large size 3C-SiC crystals. In order to examine the crystal quality, Synchrotron White Beam X-ray Topography (SWBXT) is employed to provide detailed characterization of defect content in these crystals. Due to its unique capabilities, SWBXT can reveal the overall uniformity of crystal structure as well as local defect configurations. Since the electrical properties of a device are inextricably related to the nature of defects and their distribution within the active region of the device, the information obtained from SWBXT is valuable from both the crystal growth and device fabrication points of view.

Mat. Res. Soc. Symp. Proc. Vol. 423 © 1996 Materials Research Society

EXPERIMENTAL

The crystals analyzed in this work were supplied by the U.S. Army Research Laboratory. They are as-grown, (111) 3C-SiC platelets with sizes of around 3-5 mm across and 1 mm thick. These crystals are transparent and light yellow in color. This color indicates that the crystal is either lightly doped or undoped, since the presence of any metallic impurities or nitrogen introduces a green cast. As the impurity concentration increases the coloration will rapidly approach black [1]. Optical microscopy, in both transmission and reflection modes, shows that these crystals are generally quite uniform, except for occasional inclusions. Scanning Electron Microscope (SEM) observation revealed faceted surfaces, as shown in figure 1.

Synchrotron white beam X-ray topography experiments were carried out at the Stony Brook Synchrotron Topography Facility, Beamline X-19C, at the National Synchrotron Light Source (NSLS), Brookhaven National Laboratory. The transmission, or Laue geometry, was employed, with the detector (Kodak SR-5 film) oriented perpendicular to the direction of the area-filling incident beam. Using the synchrotron white beam a single exposure can provide several useful topographic images with linear resolution better than 5 μm. In addition, a new experimental arrangement, involving the use of a narrow beam in reflection geometry, is applied to identify crystal layer structures. As shown schematically in figure 2, the crystal is set parallel to the incident beam and the beam height is adjusted to be smaller than the crystal thickness. The incident beam is controlled to hit only part of the crystal. This way, the diffracted image only shows the structure of the material covered by incident and diffracted beams. Through the precise control of the beam position, a sequence of topographs can be recorded by moving the beam from top to bottom of the crystal. Comparison of these topographs enables one to locate the structural configuration and defect distribution with the crystal. In this study, the beam height is typically 0.2 mm.

Figure.1 SEM micrograph of 3C-SiC platelet

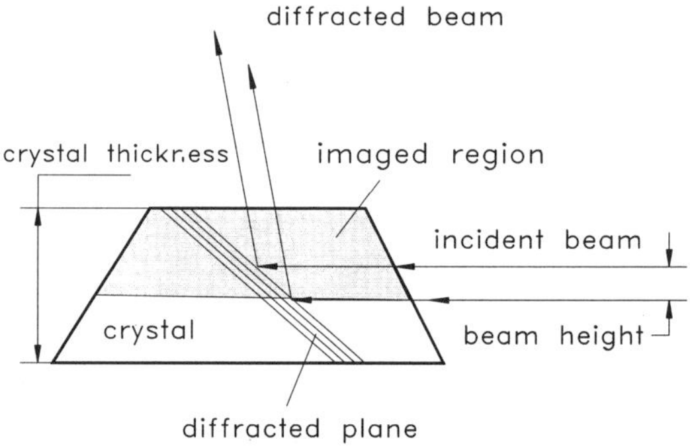

Figure 2. Schematic diagram showing the narrow beam reflection geometry.

RESULTS AND DISCUSSION

About a dozen crystals were examined. Results from three of them are presented below. Figure 3 shows an X-ray transmission topograph recorded from crystal 1, showing the overall defect distribution. The presence of Pendellosung fringes, labeled **F**, at the top left edge of the crystal indicates the high quality of this crystal. Stacking faults are found in the middle of the crystal, as indicated by **S**. Dislocations, indicated by **D**, are also observed. Partial dislocations bounding the stacking faults are indicated by **P** and inclusions by **I**. Some evidence for the existence of growth sector boundaries, **G**, is also visible. Comparing with figure 1, it is clear that the lines labeled by **L** in figure 3 are related to the surface step morphology. Since the bottom of the crystal is attached to a carbon rod during the growth process, this region is covered by carbon and does not show a clear facet feature [11]. Also, the lines indicated by **B** correspond to the lines of intersection between facets. It is found that inclusion density is higher on these edges than elsewhere.

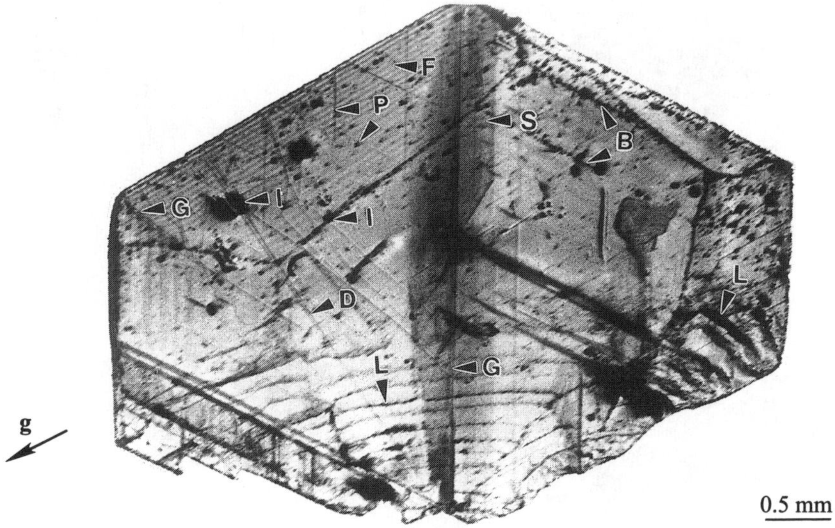

Figure 3. SWBXT transmission topograph (**g**=$20\bar{2}$, λ=0.690Å) recorded from crystal 1.

A transmission topograph recorded from crystal 2, shown in figure 4(a) displays a similar defect configuration to that in figure 3. Again the Pendellosung fringes (**F**), show the high crystal quality. Stacking faults (**S**), dislocations (**D**), growth sector boundaries (**G**), and inclusions (**I**) are the major defects in this crystal. The dislocation density in the crystal is determined to be less than $10\text{-}10^2$ cm/cm^3. Most of these dislocations are partial dislocations bounding stacking faults. The others are growth dislocations since they originate from the left part, where the crystal started to grow, and follow the crystal growth directions propagating towards the crystal facets. Some of them have been forced to glide or climb during cooling after growth forming complicated dislocation configuration.

More details of the stacking fault configuration in this crystal are shown in the (111) image shown in figure 4(b). It is found that these stacking faults are similar to those previously observed by TEM (Transmission Electron Microscopy) in 3C-SiC epitaxial layers grown by CVD (Chemical Vapor Deposition) method [12]. By applying the **g**·**u**=0 contrast extinction criterion of stacking faults, in which **g** is the diffraction vector and **u** is the fault vector of the stacking faults, the stacking faults observed are determined to be of {111} type with fault vectors of 1/3<111> or 1/6<112>. Stacking faults appear as equally spaced parallel fringes on topograph 4(b) due to the fact that their fault planes, $(\bar{1}11)$, $(1\bar{1}1)$ and $(11\bar{1})$, are inclined to the crystal surface. Stacking faults with fault vector $1/3[1\bar{1}1]$ vanish in $(\bar{2}02)$ and $(20\bar{2})$ images but appear in

$(1\bar{1}1)$ and (202) images. Similarly, stacking faults with fault vector 1/6[112] are not observed in $(2\bar{2}0)$ and $(\bar{2}20)$ images but are visible in (111) and (202) images. As an example, a stacking fault with fault vector $1/3[1\bar{1}1]$, as indicated by **S1** in figure 4(b), vanishes in figure 4(a). Since the stacking of the Si-C elementary layer along <111> is random and the stacking fault energy on {111} is low, the probability of occurrence of stacking fault is enhanced [14]. Thus {111} type stacking faults are the dominant defects in these crystals as expected.

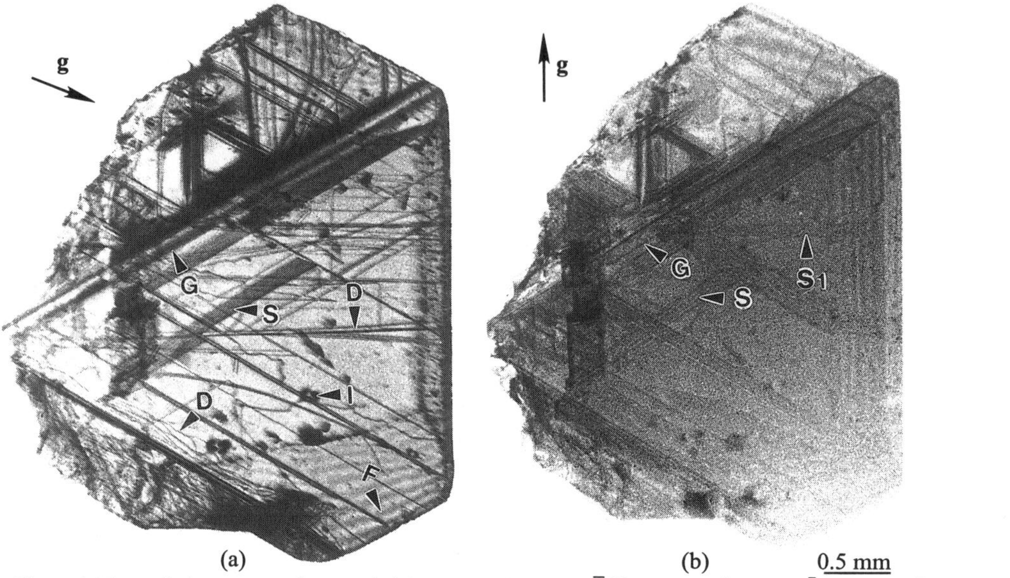

Figure 4. Transmission topographs recorded from crystal 2. (a) **g**=$\bar{2}02$, λ=0.400Å. (b) **g**=$1\bar{1}1$, λ=0.558Å

Double positioning twins are found in most of the crystals examined here. The stacking sequences of ABCABC and ACBACB are related by a 180° twinning operation, which is often referred to as double positioning twinning [15]. This kind of twin structure is frequently observed in 3C-SiC epilayers [4, 8, 12] and can be easily identified by SWBXT [8]. In this paper, the two twin related configurations are referred to as 3C(I) and 3C(II), respectively. When both 3C(I) and 3C(II) co-exist, the corresponding diffraction patterns superimpose as shown later. A typical twin structure is shown in figures 5(a) and (b), which were recorded from crystal 3. Both are $(11\bar{3})$ images of two related regions 3C(I) and 3C(II), respectively. The quality of the 3C(II) structure, as shown in figure 5(b) is similar to that in the 3C-SiC crystals grown by the Van Arkel method[14]. Growth sector boundaries, indicated by **G**, and a high density of inclusions, indicated by **I**, are clearly visible. This kind of growth sector boundary structure is usually associated with doping or impurity segregation [16, 17]. Since these crystals are either low doped or undoped, the growth sector boundary structure may be formed due to carbon segregation. The quality of the 3C(II) region, as shown in figure 5(b), is not as good as that in 3C(I) region, as shown in figure 5(a), due to the large strains induced by impurity. Growth sector boundaries become visible due to the strain between growth sectors, which originates in differences in the rates of impurity segregation in the different growth directions [18]. Thus the presence of growth sector boundaries is likely to be indicative of difference in impurity segregation rates between adjacent growth sectors. Such differences may have a detrimental influence on device performance.

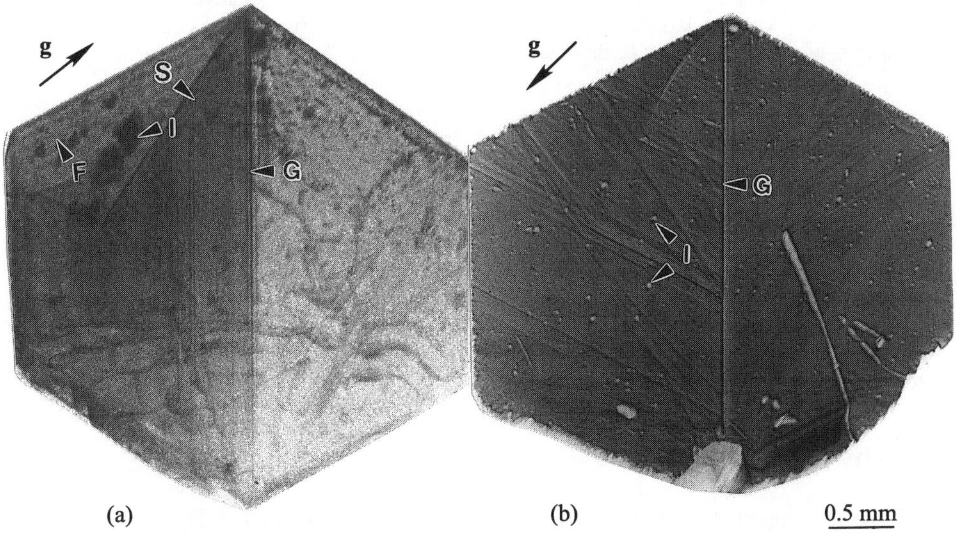

Figure 5. Transmission topographs recorded from crystal 3 show double positioning twins. (a) 3C(I) **g**=$11\bar{3}$, λ=0.459Å. (b) 3C(II) **g**=$11\bar{3}$, λ=0.459Å

By using a narrow incident beam, the 3C(II) region is determined to be at the bottom of the crystal. Figure 6(a) is the diffraction pattern produced when the incident beam is 0.2 mm above the crystal bottom whereas figure 6(b) is the one when the incident beam hits the very bottom of the crystal. Diffraction from 3C(II) appears only in figure 6(b). This proves that 3C(II) structure exists within a layer thinner than 0.2 mm at the bottom of the crystal. There is experimental evidence that the growth rate of the (111) face in the platelet is infinitesimally small compared with the growth rates of other faces, including the face parallel to it ($\bar{1}\bar{1}\bar{1}$ face) [11]. Therefore the (111) faces are always covered with a thin film of carbon formed by the decomposition of

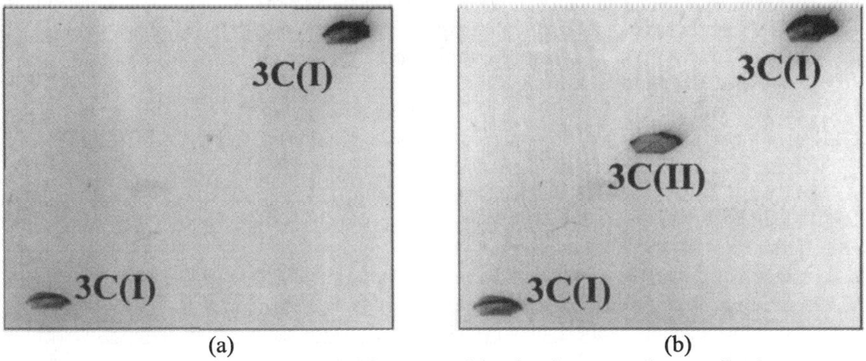

Figure 6. Diffraction patterns recorded from crystal 3 using the narrow beam reflection geometry.

CH_4 which is the reaction product of hydrogen with the graphite substrate. The further growth on (111) faces is expected bring about a twin structure. The difference between growth rates on (111) and $(\bar{1}\bar{1}\bar{1})$ faces determines that the thickness of 3C(II) is much smaller than that of 3C(I). Apparently this kind of structural preference eliminates the influence of double positioning boundaries, which benefits device fabrication.

CONCLUSIONS

Synchrotron white beam X-ray topography has been used to reveal the defect structures in (111) 3C-SiC single crystals grown by the Baikov method. The major types of defects include complex growth sector boundary structures, double positioning twins, and stacking faults on {111} planes, inclusions and dislocations (including growth dislocations and partial dislocations bounding stacking faults). The dislocation density in these crystals is determined to be less than $10\text{-}10^2$ cm/cm^3. Layer structures of double positioning twins, 3C(I) and 3C(II), are determined by using a narrow beam in reflection geometry. A thin layer of 3C(II) is located at the bottom of the crystal and its quality is relatively poor compared with that of the 3C(I) region. The presence of growth sector boundaries may have a detrimental influence on device performance.

ACKNOWLEDGMENTS

Research supported by U.S. Army Research Laboratory, Adelphi, MD under grant No. DAAL02-92-R-9147/2FE425, and U.S. Army Research Office (Grant numbers DAAH04-94-G-0091 and DAAH04-94-G-0121, contract Monitor Dr. John T. Prater). Topography carried out at the Stony Brook Synchrotron Topography Station, Beamline X-19C, at the National Synchrotron Light Source in Brookhaven National Laboratory, which is supported by DOE.

REFERENCES

1. E. Nelson, F. A. Halden and A. Rosengreen, J. Appl. Phys., **37**, 333 (1966)
2. C. Fazi, M. Dudley, S. Wang and M. Ghezzo, Inst. Phys. Conf. Ser., **137**, 487 (1994)
3. G. L. Harris, K. H. Jackson et al, Materials Letters, **4**, 77 (1986)
4. K. Shibahara, T. Saito et al, IEEE Elcetron Devices Letters, **DEL-7**, 692 (1986)
5. S. Yoshida, H. Daimon et al, J. Appl. Phys., **60**, 2989 (1986)
6. S. E. Saddow, M. Lang et al, Appl. Phys. Lett., **66**, 3612 (1995)
7. V. B. Shields, K. Fekade and G. Spencer, Appl. Phys. Lett., **62**, 1919 (1993)
8. M. Dudley, W. Huang et al, J. Phys. D: Appl. Phys., **28**, A56 (1995)
9. V. V. Babovets, Inorganic Materials, **11**,1623 (1975)
10. R. W. Bartlett and G. W. Martin, J. Appl. Phys., **39**, 2324 (1968)
11. S. I. Gorin, et al, in Growth of Crystals, edited by N. N. Sheftal, (New York 1968), P. 25
12. T. D. Gulden, J. of the American Ceramic Society, **54**, 498 (1971)
13. M. Benaissa, J. Werckmann et al, J. of Crysal Growth, **131**, 5 (1993)
14. J. A. Power, D. J. Larkin et al, in 1992 Amorphors and Crystalline Silicon Carbide IV, edited by C. Y. Yang et al, (Berlin, springer), P. 23
15. L. S. Aivazova, L. G. Nikolaeva et al, Inorgainc Materials, **9**, 133 (1973)
16. A. S. Aivazova, S. N. Gorin et al, Inorganic Materials, **9**, 1201 (1973)
17. H. Klapper, in Charaterization of Crystal Growth Defects by X-ray Methods, (Durham, England, 1979)

STRUCTURE OF THE 2H-AlN/6H-SiC INTERFACE

P. VERMAUT, P. RUTERANA, G. NOUET, and H. MORKOÇ*

LERMAT URA CNRS N° 1317, ISMRA, 6 Blvd du Maréchal Juin, 14050 Caen Cedex, France, philippe@leriris1.ismra.fr
* University of Illinois, Materials Research Laboratory, Urbana, Illinois, IL 61801 USA

ABSTRACT

Steps at the AlN/SiC interface have been investigated by HREM. Some of them do not introduce any defects. Others steps present a dislocation character which can contribute to relieve the strain due to the lattice mismatch. A third kind of steps gives rise to a prismatic planar fault for which a model is given. A reaction between prismatic and basal stacking faults is analyzed.

INTRODUCTION

Due to their large direct band gap, III-nitride materials are very interesting materials for optoelectronic applications such as light emitting diodes and lasers in the deep ultraviolet region [1]. Since bulk crystals or wafers are not available, AlN and GaN have to be epitaxially grown on a large variety of substrates. Device quality films have been obtained on sapphire and SiC substrates. However, SiC is still the most promising one as it has good lattice and thermal match with AlN and GaN [2]. Previous results have shown that growth of an AlN buffer layer strongly enhances the crystallinity, so improving the optical and electrical properties of the deposited films [3]. The buffer layer thickness and the growth conditions were reported to be critical parameters. Unfortunately, the epitaxial layers still contain high density of defects [4]. Three kinds of defects are mainly reported. Dislocation half loops and threading dislocations with a density of 10^{10} cm^{-2} have been observed, showing **a** and **a+c** Burgers vectors [4]. Numerous Stacking Faults (SFs) close to the substrate surface are also observed. Their density decreases with the distance from the substrate[5]. Planar defects identified as Double Positioning Boundaries (DPB) have been reported by Tanaka et. al [6] in AlN films grown on SiC. Similar defects were called Stacking Mismatch Boundaries (SMB) in GaN grown on SiC, by Sverdlov et al.[7]. In the two cases, the defects limit domains in which the stacking sequences are different. However, the structure of such defects and their configuration are still unclear. In the present work, the structure of the AlN/SiC interface has been studied by High Resolution Electron Microscopy, and the structure of prismatic faults was determined.

EXPERIMENT

SiC-6H substrates were cut 3.5 off the basal plane toward <11$\bar{2}$0>. The (0001)Si surface is cleaned using the classical way followed by a hydrogen plasma step in order to reduce the amount of oxygen-carbon bonding down to below the X-ray photoemission detection limit. The details of this procedure were reported by Lin et al.[8]. Depositions started by an AlN buffer layer which has intermediate lattice parameters between GaN and SiC. The thickness of the buffer layer was 50 nm. Growth was performed by an Electron Cyclotron Resonance plasma enhanced Molecular Beam Epitaxy at a rate 40 nm/h with a substrate temperature between 600 and 650 °C.

Mat. Res. Soc. Symp. Proc. Vol. 423 © 1996 Materials Research Society

TEM cross section samples were thinned down to 100 μm by mechanical grinding and dimpled down to 10 μm. Electron transparency was achieved by ion milling with a LN_2 cold stage at 5 kV. A final step at 3 kV was used to remove possible ion damage. HREM observations were made along the <11$\bar{2}$0> direction on a Topcon 002B electron microscope operating at 200 kV with a point to point resolution of 0.18 nm (Cs=0.4 mm). Image simulations were carried out using the multislice package of the EMS software [9].

RESULTS

The deposited AlN film and the SiC substrates present hexagonal symmetry structures with different stacking sequences. AlN exhibits the wurtzite structure, or 2H polytype in the Ramsdel notation, with a ...AaBbAaBb... stacking sequence, whereas SiC has the 6H polytype with the ...AaBbCcAaCcBbAa... sequence. Although the two materials have tetrahedrally coordinated structures with a good lattice match in the basal plane, their stacking is very different along the **c** direction.

Due to the misoriention of the substrates, numerous steps are present at the interface. The displacement vector relating two terraces of the substrate separated by an interface step does not always correspond to a translation operation of the wurtzite structure. Therefore, the two crystals which grow on the two different terraces can present two different stacking sequences e.g.: ..AaBb.. and ..AaCc.. A simple procedure [10] has been devised in order to characterize all the possible interface steps, and at the AlN/SiC interface, only three types of steps were shown to exist.

Firstly, there are steps for which the displacement vector corresponds to a translation operation of the wurtzite structure. An example is shown fig. 1. In this case, the step height is of 6 atomic bilayers (atoms are black on the micrograph).

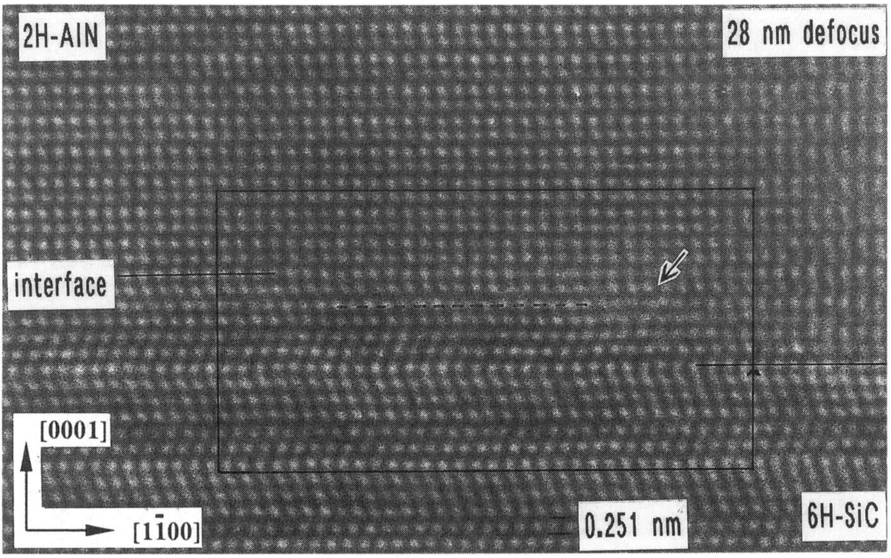

Fig. 1 HREM image of a 6 atomic bilayers step height. The Burgers circuit drawn on the micrograph shows that such a step does not present any dislocation character.

The displacement vector introduced by the step is then 1 **c** translation of the SiC and corresponds to 3 **c** translations of the AlN structure, regardless the small lattice mismatch (~1%). The Burgers circuit drawn on the micrograph does not point out any dislocation character for the step. A close examination of the step shows that it can be divided into two parts. The leftern part corresponds to the core (arrow fig. 1); the step face is vertical and extends along 4 atomic bilayers. Then, along a distance of 4.5 nm (dashed lines fig. 1), the step is lifted up another 2 bilayers, along the stacking misfit appears to be compensated. At this place, the coherency of the interface seems to be loss.

The second kind of interface step is illustrated on figure 2. The step is 2 atomic bilayers high. The displacement vector $\mathbf{d} = \mathbf{c}_{AlN} + 1/3<10\bar{1}0>$ introduced by the step was illustrated by the closure failure of the Burgers circuit drawn on the micrograph. The $1/3<10\bar{1}0>$ component corresponds to the fault vector of the I_2 type SF of the wurtzite structure [11]. In the core of this step the 6H stacking which can be considered as a 2H faulted sequence, is equivalent to a I_2 type SF bounded by a Shockley partial dislocation.

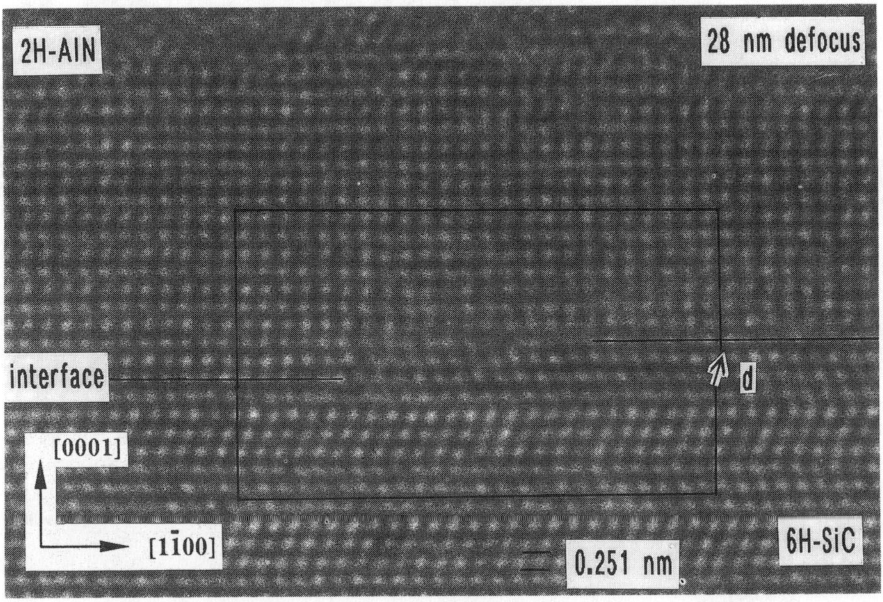

Fig. 2 HREM image of a 2 atomic bilayer height step. The closure failure of the Burgers circuit underlines the displacement vector $\mathbf{d}= 1/3<10\bar{1}0> + \mathbf{c}_{AlN}$ of the step.

A typical example of the third type of interface step is shown on figure 3. In contrast to the I_2 type step, this one gives rise to a defect which extends in the AlN layer and links the step to a SF arrowed on the micrograph. This limits domains with different stacking sequences. The tetrahedra drawn on the micrograph shows evidence of a 1/2 **c** translation vector between the two crystals. It has been extensively studied in a previous work [10] and was identified as due to prismatic SFs already observed by Conventional TEM by Drum [12].

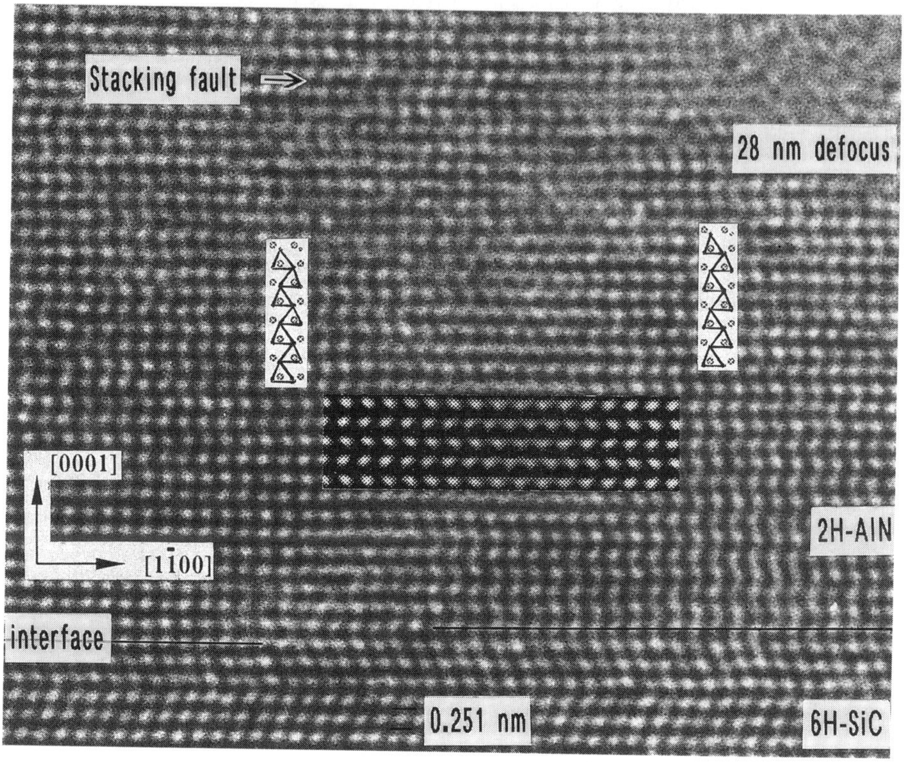

Fig. 3 HREM image of a defect which links a one atomic bilayer high interface step to a I_1 type SF at 8 nm inside the AlN film. A simulated image inset in the micrograph has been obtained with a model shown in fig. 4. The specimen thickness is 6 nm.

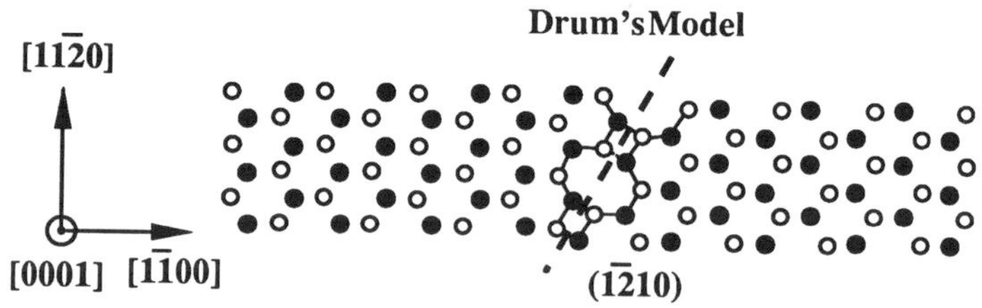

Fig. 4 [0001] projection of Drum's model of the prismatic defect. Black atoms lie above the plane of the white atoms.

The proposed model suggests a fault contained in the $(11\bar{2}0)$ plane as well as its 1/2 $[10\bar{1}1]$ fault vector. A projection of such a defect along the [0001] direction is shown figure 4, it results in 4 and 8 atoms cycles. Such a defect, observed along the $[11\bar{2}0]$ direction is not seen edge on, it is at 30° of the electron beam direction. The contrast observed on the micrograph is then a projection of the defect. A HREM image simulation obtained with Drum's model (inset fig. 3) is in good agreement with the experiment. Such defects are known to be growth defects and to link $1/6\langle 20\bar{2}3\rangle$ SFs named I_1 type by Hirth and Lothe [11], in hexagonal closed packed materials. This type of interface step has the same displacement vector than a I_1 type SF and then will named I_1 type interface step.

Figure 5 shows a configuration of such defects. Two adjacent I_1 type interface steps which have generated prismatic faults are indicated by vertical arrows. The two faults are joined by a I_1 type SF indicated by the horizontal arrows. This results in a closed domain with a different stacking sequence than the surrounding crystal. The closure failure of the Burgers circuit drawn on the micrograph is d=1/3 $[\bar{2}203]$, which corresponds to the sum of the displacement vectors of the two I_1 type steps (1/6 $[\bar{2}203]$).

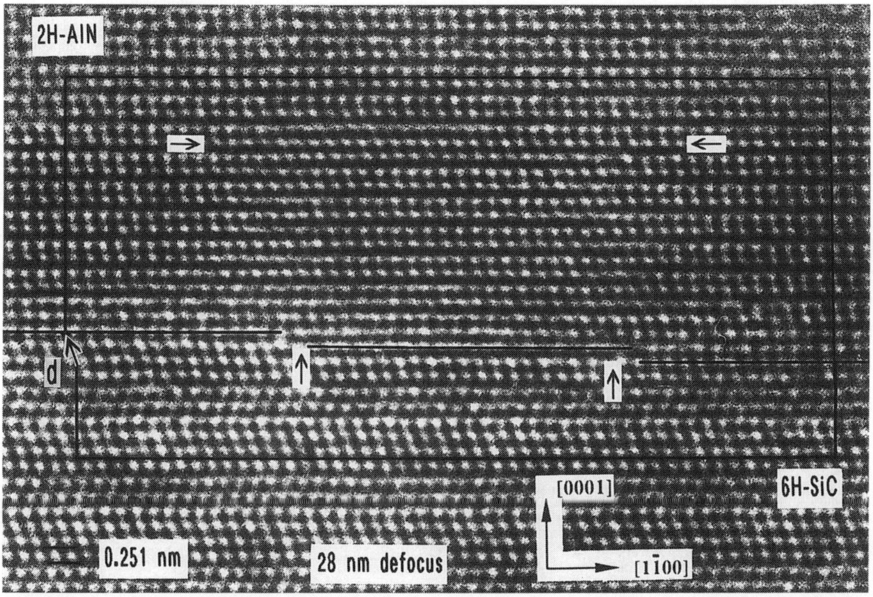

Fig. 5 Two prismatic defects generated by two interface steps (vertical arrows) and joined by a I_1 type SF (horizontal arrows).

The result is equivalent to a I_2 type step with the dislocation character of a Shockley partial, meaning that the SF corresponds to one of the two prismatic faults which have fold from the prismatic to the basal plane and reacted with the other. As the fault vector 1/2 $[10\bar{1}1]$ of the prismatic defect and that of the SF 1/6 $[20\bar{2}3]$ are different; stair rod dislocations with $1/6[10\bar{1}0]$ Burgers vectors are expected at the intersection lines.

DISCUSSION AND CONCLUSION

This study of the AlN/SiC interface has shown that although some interface steps do not present a dislocation character, a number of them introduce defects in the deposited film. Two kinds of steps are observed to generate defects. The first one only gives rises to a dislocation equivalent to a Shockley partial. Such steps are convenient as they can contribute to the relief of the strain due to the lattice mismatch.

The second kind of interface steps generates prismatic defects. They may cross the epitaxial layer and are observed to end close to the interface in two ways:

- bounded by a Frank partial dislocation [10],
- react with an other prismatic fault which have fold from the prismatic to the basal plane.

These defects may be related to DPBs and SMBs which seem to originate also at interface steps. However, in our case, the prismatic faults are contained in a $(11\bar{2}0)$ plane seen at 30° from the observation direction, and not edge on.

REFERENCES

1. Morkoç H., Strite S., Gao G. B., Lin M. E., Sverdlov B., and Burns M., *J. appl. Phys.*, **76**, 1363 (1994).
2. Lin M. E., Sverdlov B., Zhou G. L., and Morkoç H., *Appl. Phys. Lett.* **62**, 3479 (1993).
3. Kuznia J. N., Asif Khan M., and Olson D. T., *J. appl. Phys*, **73**, 4700 (1993).
4. Vermaut P., Ruterana P., Nouet G., Salvador A., Botchkarev A., Sverdlov B., and Morkoç H., Proceeding of the IXth Conference on Microscopy of Semiconducting Materials, Inst. of Phys. Conf. Series N° **146**, 289 (1995).
5. Ponce F. A., Krusor B. S., Major J. S., Plano W. E., and Welch D. F., *Appl. Phys. Lett.* **67**, 410 (1995).
6. Tanaka S., Kern R. S., Davis R. F., *Appl. Phys. Lett.* **66,** 37 (1994).
7. Sverdlov B. N., Martin G. A., Morkoç H., and Smith D. J., *Appl. Phys. Lett.* **67,** 2063 (1995).
8. Lin M. E., Strite S., Agarwal A., Salvador A., Zhou G. L., Teraguchi N., Rockett A., and Morkoç H., *Appl. Phys. Lett.* **62,** 702 (1993).
9. Stadelmann P. A., *Ultramicroscopy*, **21**, 131 (1987).
10. Vermaut P., Ruterana P., Nouet G., and Morkoç H.,submitted for publication in *Phil. Mag. A*.
11. Hirth J. P. and Lothe J., Theory of dislocations, Second Edition, edited by Wiley Interscience, p. 354 (1982).
12. Drum C. M., *Phil. Mag. A*, **11**, 313 (1965).

Extrapolation of critical thickness of GaN thin films from lattice constant data using synchrotron X-ray

Chinkyo Kim[*], I. K. Robinson[*], Jaemin Myoung[**], Kyuhwan Shim[**], Kyekyoon Kim[***], and Myung-Cheol Yoo[****]
[*]Department of Physics, University of Illinois at Urbana-Champaign, 1110 W.Green St., Urbana, IL 61801
[**]Department of Material Sciences and Engineering, University of Illinois at Urbana-Champaign
[***]Department of Electrical and Computer Engineering, University of Illinois at Urbana-Champaign, 1406 W.Green St., Urbana, IL 61801, kevinkim@ux1.cso.uiuc.edu
[****]Samsung Advanced Institute of Technology, Suwon, Korea

ABSTRACT

In some materials, Van der Merwe's equilibrium theory of strain relief is believed to explain the sudden transition from pseudomorphic growth of a thin film to a progressively relaxed state. We show, for the first time for GaN, how an accurate estimate of the critical thickness of a thin film can be extrapolated from suitable measurements of lattice constants as a function of film thickness using synchrotron X-ray. We do this both for an elementary elastic energy function, in which the interactions between the dislocations are ignored, and for a more realistic energy estimate due to Kasper. The method is found to work quantitatively for thin films of GaN on AlN. The critical thickness is determined to be 29 ± 4 Å.

INTRODUCTION

Ever since molecular beam epitaxy has been used to grow materials, many attempts have been made to grow pseudomorphic epitaxial films on substrates with different lattice constants for many different combinations. However, due to the lattice mismatch it is not possible to grow pseudomorphic films with arbitrary thickness. In the initial stage of growth, the misfit can be accommodated by the strained film alone, but above a certain thickness, the so-called critical thickness, a spontaneous emergence of misfit dislocations cannot be avoided. Equilibrium theory generally predicts extremely small values for the critical thickness, which are sometimes almost impossible to measure directly. So we demonstrate here a method which allows us to extrapolate the critical thickness from a series of lattice constant measurements.

In thermodynamic equilibrium, misfit dislocations appear at the interface of strained layer heterostructures when the strained layer is thick enough that it is energetically favorable for the mismatch to be accommodated by a combination of elastic strain and interfacial misfit dislocations, rather than by elastic strain alone.[1] The equilibrium assumption does not always apply: in some well-studied systems, such as Ge_xSi_{1-x}, kinetic effects are more important.[2] In this work we show that the equilibrium theory is appropriate for thin GaN films on AlN buffer layer on sapphire (0001). Therefore, in the initial stage of growth, when the film is below the critical thickness, GaN is expected to grow pseudomorphically on AlN. Above the critical thickness, the strain begins to relax by spontaneous creation of dislocations at the GaN/AlN interface. Subsequently the lattice constant begins to approach the bulk value as the thickness of GaN increases further.

Mat. Res. Soc. Symp. Proc. Vol. 423 © 1996 Materials Research Society

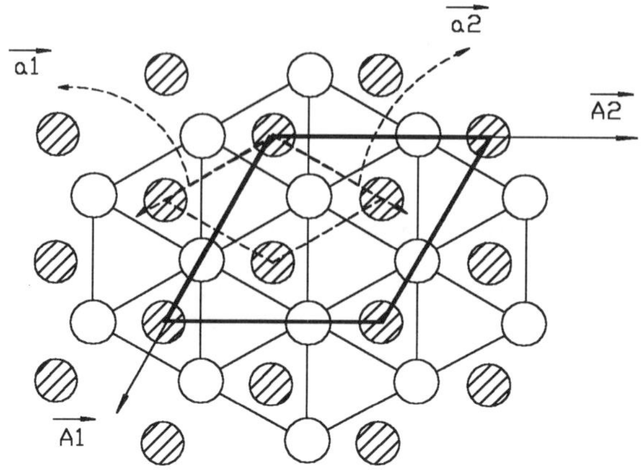

Fig. 1 Unit cells of sapphire (0001) and AlN on top of it. Solid line is for sapphire and dashed line for AlN. Open circles and hatched ones represent O in sapphire and Al in AlN, respectively.

EXPERIMENT

To test those ideas which will be explained in the next section we examined measurements of lattice constants of GaN thin film grown on AlN buffer layers on sapphire (0001).[3] The AlN buffer layer between the sapphire and the GaN was used because its in-plane lattice constant is intermediate, as used previously[4], which redistributes the substantial mismatch between two interfaces. GaN samples with thicknesses in the range from 50 Å to 1 μm were grown by MBE (Molecular Beam Epitaxy) on sapphire (0001) using a 32 Å AlN buffer layer. In order to have a workable signal from the thinnest films, X-ray beamline X16C at the National Synchrotron Light Source (NSLS) was employed to determine the lattice constant a of the GaN films using a least-squares fit method.[3] Fig.1 shows the known epitaxial arrangement of AlN on sapphire (0001)[5,6] and table I[7] lists the known bulk hexagonal lattice constants of the relevant materials. If there were no lattice mismatch, the in-plane lattice constant of AlN would be expected to be $4.758/\sqrt{3} = 2.747$ Å, considerably less than the bulk value of 3.112 Å. Instead, we found a = 3.084 Å indicating a partial compression of the buffer layer.

THEORY

First we consider the relationship between the film thickness and the lattice constant a using Van der Merwe's energy minimization theory.[8] From elementary elasticity theory the strain energy per unit area is proportional to $h(a_o - a)^2$, where h is a film thickness, a_o is the lattice constant of the (completely relaxed) material of the film and a is the actual lattice constant of the film. Assuming that the effective range of the dislocation field is constant, the energy due to

Table I Physical parameters of GaN, AlN and Sapphire

	a (Å)	c (Å)
GaN	3.189	5.185
AlN	3.112	4.982
Sapphire	4.758	12.991

dislocations depend on their density alone, which is proportional to $(a - a_s)$. Then, the total energy of the strained film is given by

$$E = \alpha h(a - a_o)^2 + \beta(a - a_s) \quad (1)$$

where α and β are constants. For the system to be in equilibrium, this energy should be a minimum: $\partial E/\partial a = 0$. Then, to obtain the value of h_c, the critical thickness, we simply need to consider $a = a_s$. Setting $\partial E/\partial a = 0$ also gives a relation between h and a so long as $a > a_s$. At $a = a_s$, h is equal to h_c below which the film is pseudomorphic and a will have the same lattice constant as the substrate, a_s. The derived relation between a and h from Eq.(1) is

$$a = a_o + \frac{h_c}{h}(a_s - a_o) \quad (2)$$

The method we have used to determine h_c experimentally is to fit lattice constant data for thicker films, and use Eq.(2) as an extrapolation formula.

Now we consider the total energy more rigorously. The elastic strain ε depends on the mismatch between a film and its substrate, m, as well as the average number of dislocations present at the interface.[9] The relationship is given by

$$m = |\varepsilon| + \frac{b'}{p} \quad (3)$$

where $m=(a_o-a_s)/a_s$, $|\varepsilon| = (a_o-a)/a_o$, a_s is the lattice constant of the substrate, b' is the active component of the Burgers vector and p is the average distance between dislocations. Assuming ideal misfit dislocations we obtain the following expression.

$$\frac{b}{p} \approx m - |\varepsilon| \approx \frac{(a - a_s)}{a_o} \quad (4)$$

Using the expressions from Kasper[2,9], a better estimate of the energy of the dislocations per unit length is given by

$$E_{ds} = \frac{\mu b^2}{4\pi(1-\nu)p}\left[1 + \ln\left(\frac{Q}{q}\right)\right] \quad (5)$$

where q is the inner cut-off radius, given by[2] $q = \pi b/(2\sqrt{2}(1-\nu))$, μ is the shear modulus, ν is the Poisson's ratio, b is a Burgers vector. Q is the effective range of the misfit dislocation field, and has two limiting values that depend on the density of dislocations, p

$$Q = p/2 \qquad \text{when } h \gg p/2 \quad (6)$$

$$= \frac{p(4hp)}{2(p^2 + 4h^2)} \approx 2h \qquad \text{when } h \ll p/2 \quad (7)$$

We will use another approximation, Q = hp/(2h + p/2), to interpolate smoothly between the two limits of Kasper. This then applies over the entire domain of h. With this approximation, the total energy of a film with thickness h is given by

$$E \approx \alpha(a - a_o)^2 h + \beta(a - a_s)\left(1 + \ln(\frac{2hp}{q(p+4h)})\right) \tag{8}$$

Following the same calculation we used to obtain Eq.(2), we find,

$$(a - a_o)h = \left[1 + \ln\left(\frac{2hp}{q(p+4h)}\right) - \frac{1}{p^2} + \frac{1}{p(p+4h)}\right]\left[\frac{(a_s - a_o)h_c}{1 + \ln(2h_c / q)}\right] \tag{9}$$

where $p=ba_o/(a-a_s)$ as before. Since Eq.(9) is an implicit function, we must solve it numerically to obtain a as a function of h.

RESULTS

Our results for the lattice constant a of the GaN films are plotted in Fig. 2 as a function of their thickness. They show a progressive trend away from the bulk GaN lattice constant for thinner films. The in-plane lattice constants are found to lie between those of bulk GaN and the AlN buffer layer. It is clear that if the trend towards thinner films were extrapolated slightly, the GaN lattice constant would cross that of the AlN buffer layer. This is the situation that occurs at

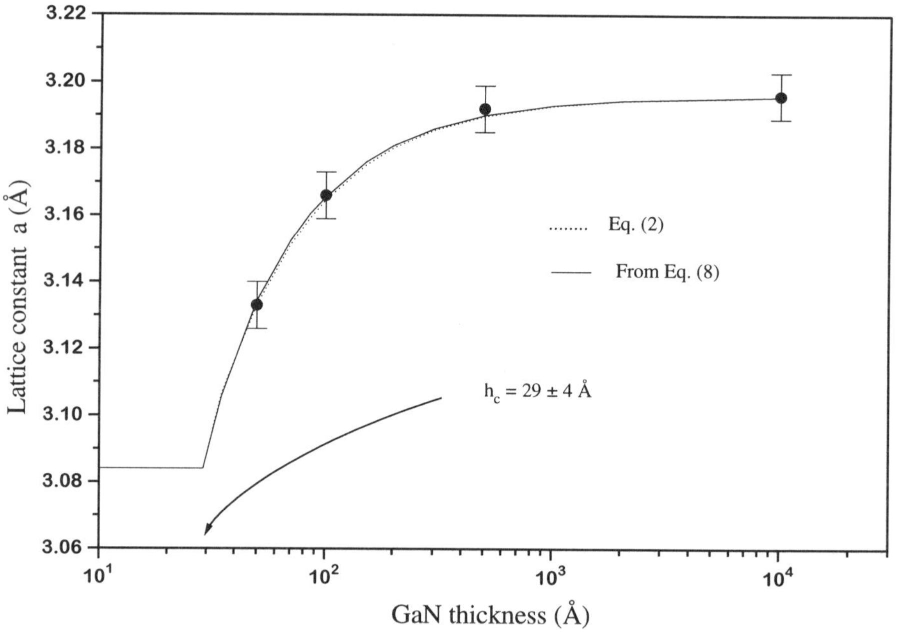

Fig. 2 Fit of lattice constant a as a function of GaN layer thickness

the critical thickness, h_c of GaN; thinner films would simply have a pseudomorphic epitaxial relationship with the AlN substrate. In order to establish a quantitative estimate of h_c, it is necessary to make a theoretical fit to the data for the trend of lattice constant with thickness, which we explained above. We fitted our data to our extrapolation formula, Eq.(2) and obtained the best fit over the entire region of film thickness with $h_c = 28 \pm 4$ Å.

With this initial knowledge of the relevant parameters we find that the thickness crossover between the two limits of Kasper (Eqs.(6) and (7)), h=p/2, falls at h= 72Å, right in the middle of the range of our data. We are therefore justified in using the more accurate treatment of Eq. (9). We therefore fit the data again using Eq.(9) with q = 5.525 Å, obtained from $\nu = 0.38$. This gave the second curve in Fig.2 and $h_c = 29 \pm 4$ Å.

SUMMARY AND CONCLUSIONS

We extrapolated the lattice constants of GaN film and obtained the critical thickness of GaN on AlN buffer layer with a sapphire substrate. The lattice constant a was observed in the region between 3.133 Å and 3.196 Å. The critical thickness of GaN was determined to be 29 ± 4 Å. This value agrees extremely well with the value of h_c predicted by the theory of Fischer et al,[3,10] 31.5 Å Since this theory is also based on energetic equilibrium arguments, it appears to support the conclusion that the behavior of GaN thin films, unlike Ge_xSi_{1-x} films for example, can indeed be described by equilibrium models.

Our second conclusion is a practical one. The logarithmic correction for the interaction between dislocations, which was used to obtain the more accurate interpolation formula, Eq.(9), is apparently not very important for the determination of h_c. This can be seen by comparing the two fit curves in Fig.2, which are almost identical. The simple analytic form of Eq.(2) is apparently sufficient to describe the thickness trend of the lattice constant and thereby allow the accurate determination of the critical thickness from its fit.

ACKNOWLEDGMENTS

We thank Dr. Mauro Sardela for useful discussions. This work was supported by the United States Department of Energy under grant DEFG02-91ER45439 and by Samsung Electronics Co. Ltd.. NSLS (National Synchrotron Light Source) is supported by the U. S. DOE under grant DE-AC012-76CH00016.

REFERENCES

[1]G. C. Osbourn, IEEE J. Quantum Electron. QE-22, 1677 (1986)

[2]S.C. Jain, J. R. Willis, and R. Bullough, Adv. in Physics **39**, 127 (1990)

[3]C. Kim, I. K. Robinson, J. Myoung , K. Shim, K. Kim, M. Yoo, to be published

[4]H. Amano, K. Hiramatsu and I. Akasaki, Jpn. J. Appl. Phys. **27**, L1384 (1988)

[5]R. C. Powell, N. -E. Lee, Y. -W. Kim, and J. E. Greene, J. Appl. Phys. **73**, 189 (1993)

[6]P. Kung, C. J. Sun, A. Saxler, H. Ohsato, and M. Razeghi, J. Appl. Phys. **75**, 4515 (1994)

[7]H. Morkoc, S. Strite, G. B. Gao, M. E. Lin, B. Sverdlov, and M. Burns, J. Appl. Phys. **76**, 1363 (1994)

[8]J. H. van der Merwe, Surf. Sci. **31**, 198 (1972)

[9]E. Kasper and H. -J. Herzog, Thin Solid Films **44**, 357 (1977); E. Kasper, Surf. Sci. **174**, 630 (1986)

[10]A. Fischer, H. Kuhne, and H. Richter, Phys.Rev. Lett. **73**, 2712 (1994)

Ex Situ and In Situ Methods for Complete Oxygen and Non-Carbidic Carbon Removal from $(0001)_{Si}$ 6H-SiC Surfaces

Sean W. King, Mark C. Benjamin, Richard S. Kern, Robert J. Nemanich, and Robert F. Davis,
Department of Materials Science and Engineering, and Department of Physics
North Carolina State University, Raleigh, NC 27695.

ABSTRACT

Comparisons between the wetting characteristics of $(0001)_{Si}$ 6H-SiC and (111) Si surfaces in various acids and bases were made. It was found that 10:1 HF dipped Si (111) surfaces were hydrophobic where as the $(0001)_{Si}$ 6H-SiC surfaces were hydrophilic. $(0001)_{Si}$ 6H-SiC surfaces capped with a 20Å Si layer, however, were hydrophobic after HF dipping and exhibited outgassing levels on annealing which were several orders of magnitude lower than SiC wafers dipped in HF without the capping layer. Annealing the Si capped $(0001)_{Si}$ 6H-SiC surfaces in UHV at 1100°C for 5 min. was found to be sufficient to thermally desorb the Si capping layer and produce a (3x3) Si rich, oxygen free $(0001)_{Si}$ 6H-SiC surface.

INTRODUCTION

For SiC to succeed as the substrate/semiconductor of choice for high frequency/ high temperature, high power devices and III-N heteroepitaxy, a considerable reduction in defects (line, planar, point, etc.) must be achieved. Following Si technology, where surface cleaning and preparation are critical first step in all processes [1], a continued reduction in defects in SiC should be expected as a result of improved SiC wafer surface cleaning techniques. In Si technology for example, improper removal of surface contamination and oxides prior to Si homoepitaxy has been shown to result in an increase in the density of line and planar defects in epitaxial films from < $10^4/cm^2$ to > $10^{10}/cm^2$ [2-5] and an associated drop in device yield [2].

Typically SiC ex situ cleaning consists of solvent degreasing and RCA cleaning with the last step usually a 5-10 min. dip in an HF solution (composition ranging from 0.1-50%) [6-8]. The HF dip is intended to remove any native or intentionally grown (dry or thermal) oxides formed on the SiC surface. For silicon, the HF etch has been found to be beneficial in that it not only removes oxides (SiO_2 or SiO_x) from the surface but also passivates the surface by terminating all the dangling bonds with hydrogen [9-11]. The hydrogen termination inhibits re-oxidation of the silicon surface on removal from the HF solution and produces a hydrophobic surface [11]. For $(0001)_{Si}$ 6H-SiC this has not been found to be the case. Surface analysis by the authors and others (see Figure 1a) has revealed that the SiC surface is still terminated with an ≈ 10 Å layer of carbonaceous material (consisting or C-C, C-O, C=O, and C-H bonding) after HF dipping [7-9]. Thermal desorption of this non-carbidic carbon can be easily achieved by annealing at temperatures as low as 500-700°C. However, unlike Si, approximately 1/2 to 3/4 monolayer of oxide remains at the surface (see Figure 2a-b), and thermal desorption of this oxide in UHV requires temperatures in excess of 1000°C [12]. As shown in Figure 1c, removal of the oxide by this method can result in the formation of some non-carbidic/graphitic carbon at the SiC surface. To combat this, others have followed the pioneering work of Kaplan [13] and employed in situ techniques in which the SiC surface is annealed in a flux of Si (solid or gas source) which allows thermal desorption of the oxide at lower temperatures (< 900°C) while maintaining a Si rich surface [12,14]. In this paper, we report on our efforts to use a Si capping layer to provide both a hydrophobic surface during ex situ processing and a Si rich surface during in situ processing.

EXPERIMENTAL

The vicinal n-type (typically $N_d = 10^{18}/cm^3$) $(0001)_{Si}$ 6H-SiC wafers used in these experiments contained an n type epilayer (typically $N_d = 10^{17}/cm^3$) and 500-1000Å of thermally grown oxide. They were supplied by Cree Research, Inc. The thermal oxide was removed using a 5-10 min. dip

Mat. Res. Soc. Symp. Proc. Vol. 423

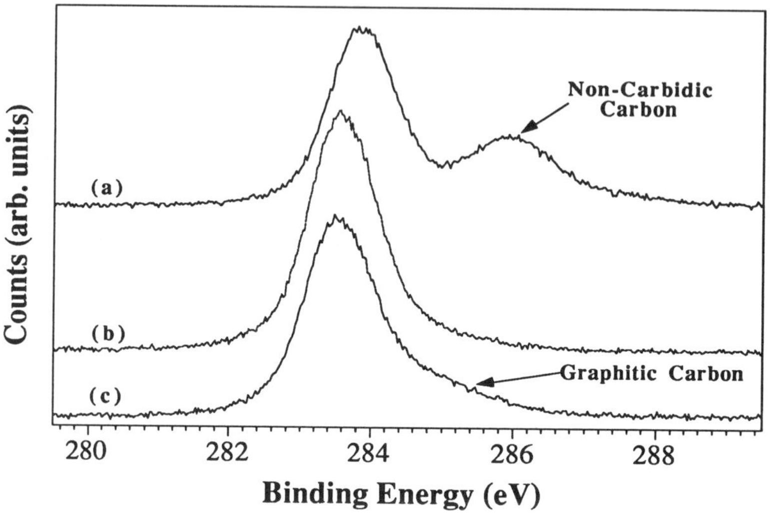

Figure 1. XPS of C 1s from $(0001)_{Si}$ 6H-SiC surface after: (a) 5 min. 10:1 HF dip, (b) after thermal annealing at 800°C, and (c) after annealing at 1200°C in UHV.[8]

in a 10:1 HF dip (CMOS grade, J.T. Baker). Further cleaning of this surface was then investigated by immersion in other acid/base solutions or by reoxidizing the SiC surface using a UV/O_3 treatment followed by a wet chemical treatment. The UV/O_3 treatments described in this study employed a box in which was positioned a high intensity Hg lamp in close proximity to the SiC wafer. The details of this process have been described previously [15]. The wet chemistries examined included 10:1 HF, 10:1 buffered HF (7:1 NH_4F:HF), 40% NH_4F, HCl:HF, and NH_3OH:HF solutions, HNO_3, H_2SO_4, acetic, and lactic acid. Except where noted, after all wet chemical cleans the samples were rinsed in DI water (18 MΩ) and blown dry with N_2 (UHP). All wet chemicals were of CMOS grade purity (J.T. Baker).

The in situ cleaning and the surface analyses of the samples subjected to ex situ and in situ cleaning were conducted in a unique ultra-high vacuum (UHV) system consisting of a 36 ft. long UHV transfer line to which were connected several surface analysis and thin film deposition units. The details of each and the transfer line have been described elsewhere [15]. Surfaces prepared in the above manner were then subsequently mounted to a molybdenum sample holder and loaded into the loadlock for subsequent analysis by AES, XPS, EELS, and LEED. XPS analysis was performed using the Al anode (hυ = 1486.6 eV) at 20 mA and 10kV. AES spectra were obtained using a beam voltage of 3 keV and an emission current of 1 mA. LEED was performed using rear view optics, a beam voltage of approximately 100 eV, and an emission current of 1 mA. Calibration of the XPS binding energy scale was performed by measuring the position of the Au $4f_{7/2}$ and shifting the spectra such that the peak position occurred at 83.98 eV. All sample temperatures quoted here were measured using an optical pyrometer and an emissivity of 0.5.

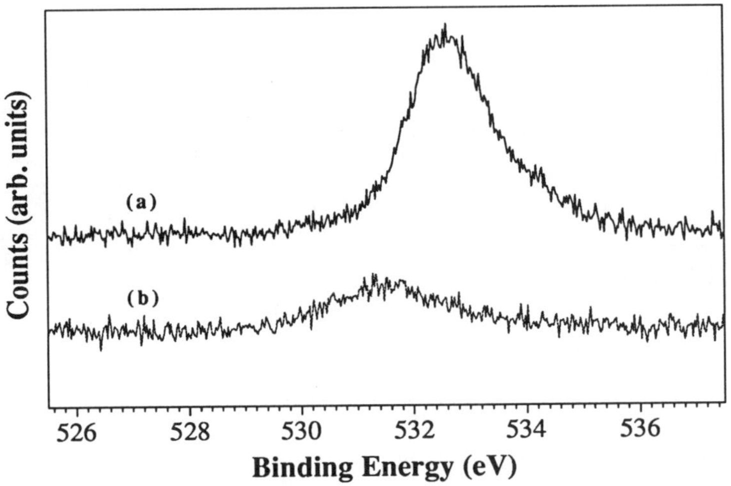

Figure 2. XPS of O 1s from (a) $(0001)_{Si}$ 6H-SiC and (b) Si (111) wafers after dipping in 10:1 HF for 5 min.[12]

RESULTS

Si and SiC wetting experiments. As previously mentioned, oxide removal from (0001) 6H-SiC surfaces using HF leaves a hydrophilic surface containing significant amounts of oxide surface (see Fig. 3a). To determine a wet chemistry which produces a more hydrophobic SiC surface, several $(0001)_{Si}$ 6H-SiC wafers were dipped in HF, NH_4F, NH_3OH, HCl, HNO_3, H_2O_2, H_2SO_4, acetic acid, and lactic acid and the wetting characteristics of these surfaces in these acids and bases and de-ionized water monitored visually. For comparison purposes, Si (111) and Si (100) wafers were also dipped simultaneously in each acid with the SiC wafer. All wafers (Si or SiC) were initially dipped in 10% HF to remove any native oxides from the surfaces before dipping in the acids and bases of interest. For Si, the surfaces remained hydrophobic when dipped in NH_3OH, HCl, or H_2O_2. Dipping the Si wafers in HNO_3 or H_2SO_4 removed the hydrophobic nature of the surface. Dipping Si in the organic acids resulted in a strongly adhering thin film of the acid to the silicon surface which could be removed in DI H_2O leaving a hydrophobic surface. For SiC, all acids and bases wetted the surface and none were found to produce a hydrophobic SiC surface. 10% HF solutions with pH's adjusted from strongly acidic, neutral, and to strongly basic using HCl, NH_4F, and NH_3OH, respectively, were also examined as they have been reported to produce better hydrogen termination of Si (111) surfaces [17,18]. In these experiments, the dipping times were held constant at 10 min. No changes in the wetting characteristics of the $(0001)_{Si}$ SiC surface in HF solutions with the different pH's were found. In an additional experiment, the dipping time was varied from 5 min. to 1 hr.; however, no difference was found. Additionally, neither XPS nor AES indicated a significant change (within AES and XPS experimental accuracy) in the amount of surface oxygen on $(0001)_{Si}$ SiC wafers after dipping in HF solutions of various pH. No differences were observed between on axis and vicinal surfaces, as well.

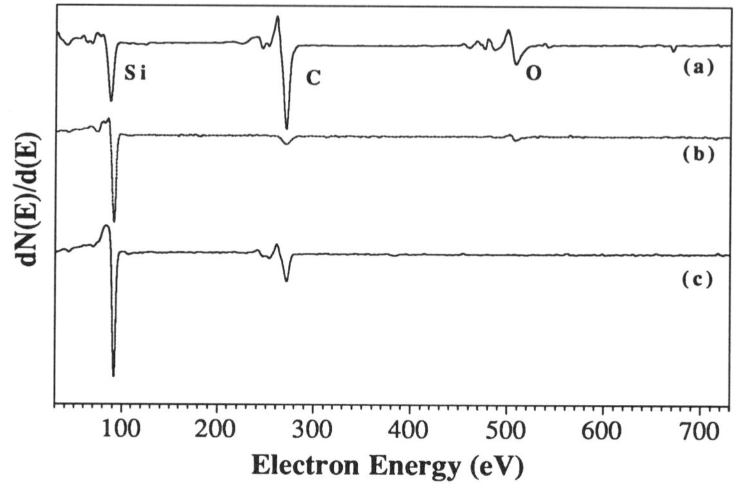

Figure 3. (a) AES of (0001)Si 6H-SiC surface after dipping in 10:1 HF for 5 min., (b) $(0001)_{Si}$ 6H-SiC with a 20Å Si capping layer after dipping in 10:1 HF, 5min., (c) $(0001)_{Si}$ 6H-SiC after hermal desorption of the Si capping layer at 1100°C for 5 min.

Si capping layer. A new technique for in situ oxide removal rapidly gaining acceptance consists of annealing SiC in a flux of silicon at temperatures > 900°C [13,14]. Since wet chemical cleaning of Si was found to more readily produce hydrophobic and cleaner surfaces, the use of a 20Å Si capping layer on $(0001)_{Si}$ 6H-SiC was investigated. The following procedure was used to prepare the Si capping layer. Firstly, the (0001) 6H-SiC wafer was given a 5 min. dip in 10:1 HF, DI rinsed, N_2 blow dried and loaded into vacuum. The 6H-SiC was then degassed and annealed in a SiH_4 flux (10^{-6} Torr) at 950°C for 20 min. in the GSMBE. This produced an oxygen free Si rich (1x1) SiC surface. Next, a Si-Ge electron beam MBE was used to deposit 200Å of Si on the SiC at room temperature in situ. The Si/SiC sample was then given two UV/O_3 treatments followed by dips in 10:1 HF to thin the Si capping layer down to 20Å. After each HF dip, the polished face of the Si/SiC wafer was found to be hydrophobic as with Si wafers. AES of this surface after HF dipping showed only small amounts of oxygen and carbon contamination (see Figure 3b). XPS of this surface after HF exposure showed a Si 2p peak at 99.5 eV with a shoulder at 101.5 eV indicating that the surface consisted of an ≈ 20 Å Si film on top of the SiC. On annealing this surface to desorb the Si capping layer, it was observed that the outgassing from the wafer was several orders of magnitude lower than that typical for other HF dipped SiC wafers (10^{-7} Torr ax vs. 10^{-5} Torr typical). In fact, the outgassing levels were typical for those observed from Si (111) wafers. The reason's for these higher and lower outgassing rates will be discussed later.

Following annealing of the Si/SiC wafer at 1100°C for 5 min., LEED showed a sharp (3x3) reconstructed surface commonly observed for surfaces prepared via annealing in Si fluxes [12]. XPS of this surface, now shows a Si 2p peak located at 101.5 eV with a shoulder at 99.5 eV indicating that the 3x3 surface is due to the presence of a Si bilayer. Continued annealing at 1100°C resulted in further desorption of Si which resulted in first a (1x1) and then a (√3x√3)R30° LEED pattern. AES of the (3x3) surface showed no oxygen within the detection limits of AES and

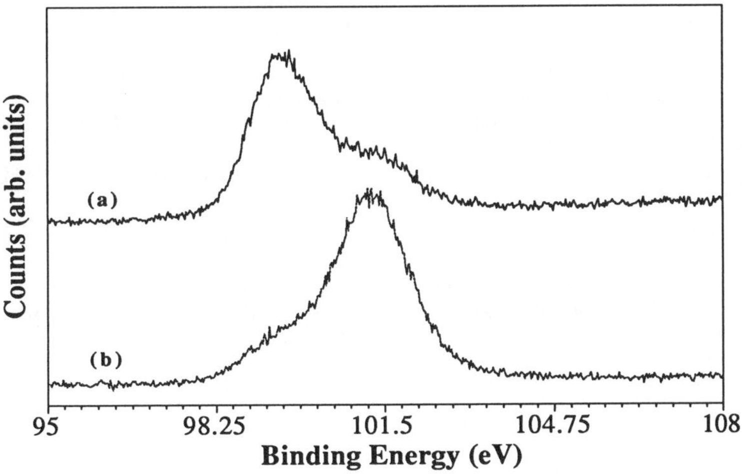

Figure 4. XPS of Si 2p from a $(0001)_{Si}$ 6H-SiC with a 20Å Si capping layer after: (a) 5 min. dip in 10:1 HF, and (b) after thermal desorption of the Si capping layer at 1100°C for 5 min.

a Si/C ratio of 3/1. It should be pointed out that (3x3) surfaces have been previously prepared, removed from vacuum, dipped in HF, and the wetting characteristics observed. In this case, the surfaces were hydrophilic indicating that the Si bilayer is readily removed by the combination of air exposure and HF dipping. As such, 15-20 Å probably represents the optimum thickness for the Si capping layer.

DISCUSSION

One frequently noted problem with (0001) 6H-SiC wafers is the existence of internal micro-pipes. Aside from being potential device killers, the authors have frequently experienced other problems during processing of SiC wafers which can be attributed to the micro-pipes as well. The authors have frequently observed that these entities can harbor HF and H_2O after HF dipping and N_2 drying. Presumably, this is due to capillary action. This conclusion is based primarily on the observation that during annealing/thermal desorption in UHV of HF dipped SiC wafers phenomenally high outgassing rates have been observed (i.e. 10^{-5} Torr). RGA analysis of the background vacuum during thermal annealing of the SiC wafer reveals that HF and H_2O are the two main constituents outgassing from the wafers. As outgassing rates for Si wafers which have experienced an HF dip & N_2 dry are several orders of magnitude lower (10^{-8} to 10^{-7} Torr), we suggest that HF/H_2O trapped in the micro-pipes due to capillary action are the source of the extremely high outgassing. The importance of this observation is that the high outgassing can lead to difficulties in completely removing the oxide from the SiC surface as well as creating a large background of H_2O and HF in the growth system. The H_2O/HF levels can subsequently result in an increase of background oxygen or fluorine in epitaxial films. Owing to the observed change from a hydrophilic to a hydrophobic surface with the addition of the Si capping layer, it is not surprising that lower outgassing rates were observed for the SiC surface with the Si capping layer.

In addition to the oxide removal and the lower outgassing rates, another advantage of the use of the Si capping layer is the potential for its incorporation into already existing processing routes. Potentially, the Si capping layer could be deposited during cooling from SiC thin film CVD epitaxy. Currently, the SiC wafer/film assemblies are cooled in H_2 which produces a carbon rich layer which must be removed by a second processing step of thermal oxidation [9,18]. Unfortunately, high quality SiC CVD epitaxial deposition typically occurs at temperatures higher than the Si melting point (1440°C) and hence results in the deposition of Si droplets instead of a continuous film of Si [19]. However, the authors have initially found that this approach can still eliminate the need for oxidation as the Si droplets can be etched away (15 HNO_3: 2 HF: 5 CH_3COOH) leaving a surface concentration equal (by XPS and AES) to that produced by thermal oxidation and oxide removal by HF. Perhaps, the Si capping layer can best be utilized with 3C-SiC surfaces where epitaxy can occur at lower temperatures.

CONCLUSIONS

The wetting characteristics of (0001) 6H-SiC surfaces in HF and other acids were found to be hydrophilic. Hydrophobic surfaces could be achieved by use of a thin 20 Å Si capping layer. The use of the capping layer also resulted in lower outgassing rates during thermal annealing during in situ processing. Annealing the Si capping layer/SiC wafer at 1100°C for 5 min. resulted in the desorbtion of the excess silicon and a (3x3) reconstructed SiC surface.

ACKNOWLEDGMENTS

The authors would like to express appreciation to Dr. Ja-Hum Ku for assistance with the deposition of the Si capping layer and input into these experiments. Appreciation is also expressed to Cree Research, Inc. for the 6H-SiC wafers. The work described herein was supported by the Office of Naval Research under contract N00014-95-1-1080.

REFERENCES

1. W. Kern, J. Electrochem. Soc., **137** (6) 1887 (1990).
2. R. Williams, Modern GaAs Processing Methods, 2nd ed. (Artech House, Inc., New York, 1990), pp. 81-114.
3. G.R. Srinivasan and B.S. Meyerson, J. Electrochem. Soc., **134** (6) 1518 (1987).
4. B.S. Meyerson, E. Ganin, D.A. Smith, and T.N. Nguyen, J. Electrochem. Soc., **133** (6) 1232 (1986).
5. M.K. Sanganeria, M.C. Ozturk, G. Harris, K.E. Violette, I. Ban, C.A. Lee, and D.M Maher, J. Electrochem. Soc. **142** (11), 3961 (1995).
6. Y. Mizokawa, S. Nakanishi, O. Komoda, S. Miyase, H.S. Diang, C. Wang, N. Li, and C. Jiang, J. Appl. Phys., **67** (1) 264 (1990).
7. H. Tsuchida, I. Kamata, and K. Izumi, Jpn. J. Appl. Phys., **34** 6003 (1995).
8. L.M. Porter, R.F. Davis, J.S. Bow, M.J. Kim, R.W. Carpenter, R.C. Glass, J. Mater. Res. **10** (3) 668 (1995).
9. B.S. Meyerson, F.J. Himpsel, and K.J. Uram, Appl. Phys. Lett. **57**, 1034 (1990).
10. M. Grundner and H. Jacob, Appl. Phys. A **39**, 73 (1986).
11. Y.J. Chabal, G.S. Higashi, K. Raghavachari, and V.A. Burrows, J. Vac. Sci. Technol. A, **7** (3), 2104 (1989).
12. R.S. Kern, S.W. King, and R.F Davis, to be published.
13. R. Kaplan, Surf. Sci. **215**, 111 (1989).
14. A. Fissel, B. Schroter, E. Ducke, B. Schroter, and W. Richter, J. Cryst. Growth **154**, 72 (1995).
15. J.A. McClintock, R.A. Wilson, and N.E. Byer, J. Vac. Sci. Technol. **20**, 241 (1982.)
16. J. van der Weide and R.J. Nemanich, Appl. Phys. Lett. **62** 1878 (1985).
17. G.S. Higashi, R.S. Becker, Y.J. Chabal, A.J. Becker, Appl. Phys. Lett. **58**, 1656 (1991).
18. G.S. Higashi, Y.J. Chabal, G.W. Trucks, and K. Raghavachari, Appl. Phys. Lett. **56**, 656 (1990).
19. L. Rowland, Northrop-Grumman. Private communication.

HIGH TEMPERATURE SURFACE DEGRADATION OF III-V NITRIDES

C.B. Vartuli[1], S.J. Pearton[1], C.R. Abernathy[1], J.D. MacKenzie[1], J.C. Zolper[2] and E.S. Lambers[1]
[1]Department of Materials Science and Engineering,University of Florida, Gainesville FL 32611
[2]Sandia National Laboratories, Albuquerque NM 87185-0603

ABSRACT

The surface stoichiometry, surface morphology and electrical conductivity of AlN, GaN, InN, InGaN and InAlN was examined at rapid thermal annealing temperatures up to 1150 °C. The sheet resistance of the AlN dropped steadily with annealing, but the surface showed signs of roughening only above 1000 °C. Auger Electron Spectroscopy (AES) analysis showed little change in the surface stoichiometry even at 1150 °C. GaN root mean square (RMS) surface roughness showed an overall improvement with annealing, but the surface became pitted at 1000 °C, at which point the sheet resistance also dropped by several orders of magnitude, and AES confirmed a loss of N from the surface. The InN surface had roughened considerably even at 650 °C, and scanning electron microscopy (SEM) showed significant degradation. In contrast to the binary nitrides the sheet resistance of InAlN was found to increase by ~ 10^2 from the as grown value after annealing at 800 °C and then remain constant up to 1000 °C, while that of InGaN increased rapidly above 700 °C. The RMS roughness increased above 800 °C and 700 °C respectively for InAlN and InGaN samples. In droplets began to form on the surface at 900 °C for InAlN and at 800 °C for InGaN, and then evaporate at 1000 °C leaving pits. AES analysis showed a decrease in the N concentration in the top 500 Å of the sample for annealing ≥800 °C in both materials.

INTRODUCTION

Much progress has recently been made in the areas of growth, dry etching and implant isolation and doping of the III-V nitrides and their ternary alloys. This has resulted in nitride-based blue/UV light emitting and electronic devices.[1-3]. High temperature annealing is necessary in many of the processing steps for these devices, including activation of implanted ions,[4] maximization of implant isolated regions[5,6] or high temperature alloying of metal contacts.[7] A key issue is the question of surface degradation of the III-V nitrides during these high temperature anneals.[8]

In all of this previous work the equilibrium N_2 pressures above the solid (or solid plus liquid) have been the focus. In many process steps rapid thermal annealing (RTA) using the proximity geometry is employed, and this is a non-equilibrium situation. Zolper et. al.[9] observed that the luminescence and surface morphologies of GaN annealed in flowing N_2 actually improved for RTA temperatures up to 1100 °C. Similar results were obtained at lower temperatures by Lin et. al.[10]

In this paper we report on an investigation of the thermal stability of AlN, GaN, InN, InAlN and InGaN during rapid thermal annealing. the electrical conductivity, surface morphology and surface stoichiometry have all been measured as a function of annealing temperature. The In-containing materials are found to be substantially less

Mat. Res. Soc. Symp. Proc. Vol. 423 © 1996 Materials Research Society

thermally stable than either GaN or AlN, and loss of nitrogen creates a thin n-type surface layer in all three binary nitrides.

EXPERIMENTAL

The GaN, AlN, InN, InGaN and InAlN samples were grown using Metal Organic Molecular Beam Epitaxy (MO-MBE) on semi-insulating, (100) GaAs substrates or Al_2O_3 c-plane substrates in an Intevac Gen II system as described previously.[11] The group-III sources were triethylgallium, trimethylamine alane and trimethylindium, respectively, and the atomic nitrogen was derived from an ECR Wavemat source operating at 200 W forward power. The layers were single crystal with a high density (10^{11} - 10^{12} cm^{-2}) of stacking faults and microtwins. The GaN and AlN were resistive as-grown, and the InN was highly autodoped n-type (>10^{20} cm^{-3}) due to the presence of native defects. InAlN and InGaN were found to contain both hexagonal and cubic forms. The $In_{0.75}Al_{0.25}N$ and $In_{0.5}Ga_{0.5}N$ were conducting n-type as grown (~ 10^{19} cm^{-3}) due to residual autodoping by native defects. The samples were annealed in a rapid thermal anneal (RTA) system (AG 410T) face down on a GaAs substrate for 10 s at temperatures between 650 - 1150 °C in a N_2 atmosphere.

RESULTS AND DISCUSSION

The sheet resistance normalized to the as grown value for all of the nitride samples is shown in Fig. 1 as a function of annealing temperature. The values for the GaN, AlN and InN are found to drop by approximately three orders of magnitude with annealing, up to 900 °C. The material becomes strongly n-type in all cases, even in the GaN and AlN which were resistive as grown. The sheet resistance for AlN continues drop steadily with anneal temperature until 1100 °C. AlN shows only a small loss of N from the surface as determined by AES. However the electrical measurements are more sensitive to small changes in the composition than the Auger. Here we believe the N vacancies created by the loss of N from the surface are creating shallow donors. This agrees with the theoretical prediction of Maruska and Tietjen.[12] The actual values of sheet resistance for AlN are much higher than the GaN up to 900 °C, and significantly higher than InN at all temperatures. The data in Figure 1 is in agreement with the melting point and vapor pressure curves for these materials. AlN is predicted to be stable under N_2 gas up to ~2500 °C,[8] and to melt at ~ 3700 °C at atmospheric pressure[8]. GaN is predicted to melt at ~3000 °C and InN at only ~2400 °C[8] and to degrade at 600 °C AES has confirmed loss of N from the annealed GaN sample which would suggest that N vacancies are contributing to the conductivity. At 1150 °C the sheet resistance for the GaN drops sharply indicating that N is being lost at a much greater rate than the Ga. Groh et. al.[13] showed loss of nitrogen beginning at 710 °C in vacuum annealed GaN, with significant loss at ≥ 980 °C. The sheet resistance for the InN drops steadily over the temperature range, which correlates to the problems of nonstochiometry in InN. The large size difference between the N and In make the material less stable.

The sheet resistance for InGaN and InAlN on the other hand, increases with annealing. The InAlN sheet resistance increases by 10^2 from the value for the as grown material when annealed at 800 °C. Its resistance then remains constant to 900 °C, and then decreases slightly at 1000 °C. For InGaN the sheet resistance remains constant up to

700 °C and then increases rapidly with increasing temperature. This suggests that simple N vacancies are not the cause of the residual n-type conductivity in these samples since at the highest temperatures we are losing N from the surface, as described below. However these samples become less conducting, suggesting creation of compensating acceptors or annealing of the native donors is occurring.

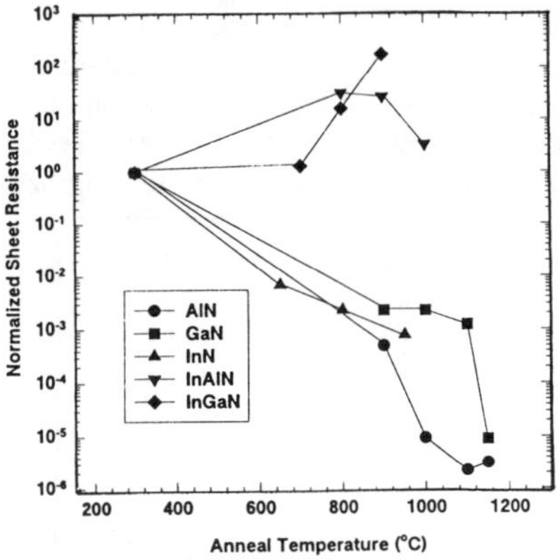

Figure 1. Sheet resistance normalized to the as grown value for AlN, GaN, InN, InAlN and InGaN as a function of annealing temperature.

The RMS data normalized to the as grown roughness as a function of anneal temperature is shown in Fig. 2 for AlN, GaN and InN. The AlN is still smooth at 900 °C, but becomes quite rough at 1000°C. Further surface reconstruction continues at higher annealing temperatures. At 1150 °C the sample becomes smooth again- in fact slightly smoother than the as grown sample. GaN shows no roughening, becoming smoother with annealing due to defect annealing and surface reconstruction. InN on the other hand is a factor of two rougher than for as grown samples at 650 °C, indicating the weaker bond strength of this material.

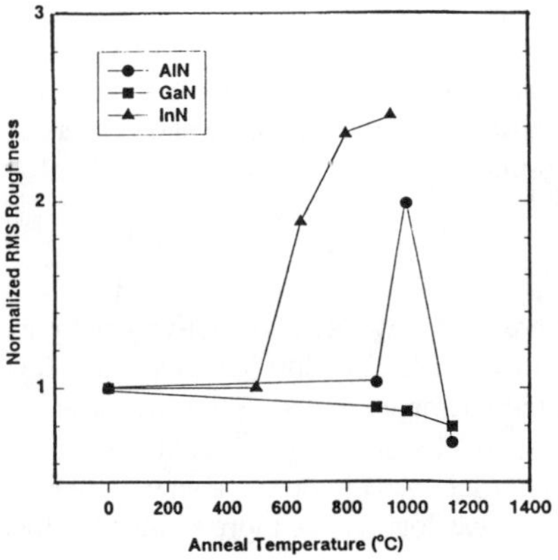

Figure 2. The RMS data normalized to the as grown roughness as a function of anneal temperature for AlN, GaN and InN.

In Fig. 3 the RMS roughness for InAlN and InGaN are shown as a function of rapid thermal anneal temperature. We see that the InAlN remained smooth until 800 °C, and at 900 °C has increased an order of magnitude in roughness. At 1000 °C the RMS roughness returns to a value close to that of the value for the as grown material. We found this to be a result of In droplets forming on the surface above 800 °C and then evaporating above 900 °C. The InGaN surface was unchanged at 700 °C, with the roughness increasing above that temperature. In Fig. 4 the individual AFM scans are shown for the ternaries. We see the surface of the samples becoming coarser above 800 °C, with large droplets forming. In the case of InAlN these droplets are removed by annealing at 1000 °C, where the surface evaporation is more congruent.

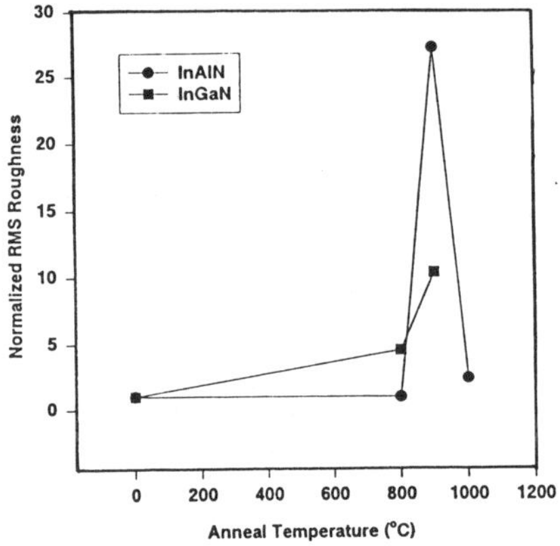

Figure 3. The RMS data normalized to the as grown roughness as a function of anneal temperature for InAlN and InGaN.

For GaN, the Ga/N ratio measured by raw AES counts increased from 1.73 on the as-grown samples to 2.34 after 1150 °C annealing, indicating that nitrogen was indeed lost from the surface. Similar results were obtained for for InAlN and InGaN (Fig.5) as grown and annealed at 1000 °C and 800 °C respectively. Both materials showed a definite decrease in the amount of N at the surface of the samples after anneal. In the case of InGaN there was also a reduction in In, which could be related to the changes in the electrical properties. The surfaces of all the samples show a loss of N, consistent with the SEM, AFM and EDAX results.

CONCLUSION

The III-V nitrides are thermally stable to relatively high temperatures. AlN and GaN remain smooth and stochiometric at 1000 °C, InAlN and InGaN up to 800 °C, and InN up to 600 °C. Above these temperature capping is necessary to prevent the loss of N and, sometimes, ın. Consistent with the predicted melting temperatures and thermal stabilities of the nitrides, we found AlN to be somewhat more stable than GaN, and much more stable than InN. InAlN was found to be more stable than InGaN, as expected from a consideration of the binary component N_2 vapor pressures. AlN may prove to be a good capping material for the other nitrides, because of its high stability and the fact that it can be selectively removed by wet etching in KOH based solutions.

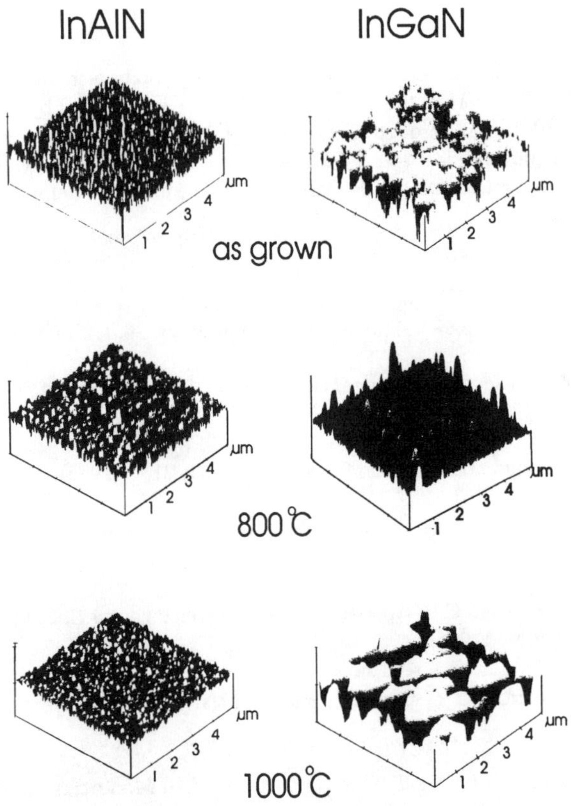

Figure 4. Individual AFM scans of InAlN and InGaN. Vertical scale is 100 nm per division.

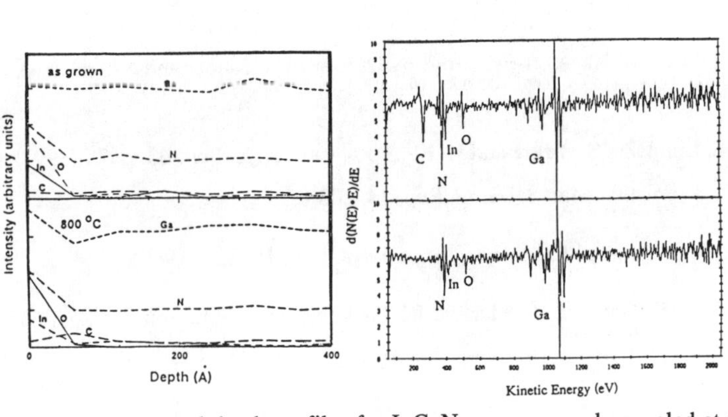

Figure 5. AES surface scans and depth profiles for InGaN as grown and annealed at 800 °C.

ACKNOWLEDGMENTS

The authors would like to thank the staff of the Microfabritech Facility for their help with this work. The work at the University of Florida is supported by NSF (DMR-9421109), an AASERT grant through ARO (Dr. J. M. Zavada), and a University Research Initiative grant #N00014-92-J-1895 administered by ONR. The work at Sandia is supported by DOE contract DE-AC04-94AL85000. The technical help of J. Escobedo at SNL is appreciated.

REFERENCES

1. S. Nakamura, M. Senoh, and T. Mukai, Jpn. J. Appl. Phys. **30**, L1708 (1991).

2. S.C. Binari, L.B. Rowland, W. Kruppa, G. Kelner, K. Doverspike, and D.K. Gaskill, Electron. Lett. **30**, 1248 (1994).

3. M.A. Khan, M.S. Shur, and Q. Chen, Electron. Lett. **31**, 2130 (1995).

4. J.C. Zolper, A.G. Baca, R.J. Shul, R.G. Wilson, S.J. Pearton and R.A. Stall, Appl.Phys. Lett. **68**, 166 (1996).

5. C.B. Vartuli, S.J. Pearton, C.R. Abernathy, J.D. MacKenzie and J.C. Zolper, J. Vac. Sci. Technol. B, **13**, 2293 (1995).

6. J.C. Zolper, S.J. Pearton, C.R. Abernathy and C.B. Vartuli, Appl. Phys. Lett. **66**, 3043 (1995).

7. M.E. Lin, Z. Ma, F.Y. Huang, Z.F. Fan, L.H. Allen and H. Morkoc, Appl. Phys. Lett. **64**, 1003 (1994).

8. S. Porowski and I. Grzegory, in Properties of Group III Nitrides, ed. J.H. Edgar (INSPEC, Stevenage, UK 1994).

9. J.C. Zolper, M. Hagerott-Crawford, A.J. Howard, J. Rainer and S.D. Hersee, Appl. Phys. Lett. **68**, 200 (1996).

10. M.E. Lin, B.N. Sverdlov and H. Morkoc, Appl. Phys. Lett. **63**, 3625 (1993).

11. C.R. Abernathy, Mat. Sci. Eng. Rep. **14**, 203 (1995).

12. H.P. Marsuka, and J.J. Tietjen, Appl. Phys. Lett. **15**, 327 (1969).

13. R. Groh, G. Gerey, L. Bartha and J.I. Pankove, Phys. Stat. Solidi A **26**, 363 (1974).

MORPHOLOGY OF OPTICALLY TRANSPARENT CUBIC SILICON CARBIDE PREPARED BY CHEMICAL VAPOR DEPOSITION

Michael W. Russell[§,*], Jaime A. Freitas, Jr.[¥] and James E. Butler[*]

§ NRC Postdoctoral Research Associate
* Naval Research Laboratory, Washington, DC 20375
¥ Sachs/Freeman Associates, Landover, MD 20785

Abstract

Crystals of cubic silicon carbide (3C-SiC) were grown in an RF-induction furnace on graphite substrates by atmospheric pressure chemical vapor deposition (APCVD) from a single precursor, methyltrichlorosilane (MTS) in hydrogen. The deposits were characterized by micro-Raman spectroscopy, x-ray diffraction, and atomic force and scanning electron microscopies. Above ~1600°C preferential <110> growth directions were identified for the majority of the crystals. At intermediate deposition temperatures (1600-1700°C) the dominant morphology consisted of yellow prismatic crystals heavily twinned along {111} and $\{11\bar{1}\}$. At substrate temperatures exceeding ~1750°C hexagonally-shaped {111} oriented 3C-SiC platelets were formed with alternating {001}/{101} edges. The dependence of nucleation density, film morphology and film orientation on deposition conditions will be discussed with emphasis on the growth of high quality single crystals of 3C-SiC.

Introduction

In addition to its superior mechanical properties, the wide band gap (2.3 eV), high breakdown field (~5×10^6 Vcm^{-1}), and high saturated electron velocity (~1000 $Vcm^{-2}s^{-1}$) of cubic silicon carbide (3C-SiC) make it an attractive candidate for elevated-temperature, high-frequency, and high-power electronic devices [1]. Chemical vapor deposition has been demonstrated to yield dense, high purity 3C-SiC which has found applications as wear and corrosion-resistant coatings and as a semiconductor device material particularly in aggressive environments [2].

Previous work at the A.A. Baikov Institute in Moscow demonstrated that high-quality, high-purity 3C-SiC single crystals (~3 mm) could be grown on resistively heated graphite rods by APCVD using MTS in H_2 [3]. Earlier results from NRL have also demonstrated that high quality 3C-SiC was deposited at ~1650°C under conditions of low MTS concentration in H_2 [4]. An important objective of this study was to examine the effect of deposition parameters on the microstructure of the deposited layers. Deposition conditions which favored platelet growth were of particular interest.

Experimental Procedure

The growth of 3C-SiC was carried out at atmospheric pressure by the thermal decomposition of methyltrichlorosilane (MTS) in excess hydrogen in a vertical, cold-walled quartz reaction chamber (2.54 cm ID x 30.5 cm). A 5 kW RF induction power supply was used to heat a cylindrical graphite substrate (1.83 cm Φ x 2.54 cm) which was supported in the quartz tube by a molybdenum rod. Prior to each deposition run the reaction chamber was evacuated to 0.1 torr and then purged with hydrogen. The substrate was preheated for 30 minutes in flowing hydrogen at the deposition temperature prior to initiating the MTS flow.

The hydrogen flow rate was controlled with a mass flow controller. The MTS precursor was maintained in a constant temperature bath to control its vapor pressure. The vapor pressure of MTS is ~170 torr at room temperature and decreases to ~89 torr and ~54 torr

Mat. Res. Soc. Symp. Proc. Vol. 423

at 10°C and 0 °C, respectively [5]. The H_2/MTS [l/mol] ratio to the reactor was controlled by adjusting the MTS bath temperature, the H_2 flow through the bath, and an additional H_2 flow added before the reactor. The H_2/MTS ratio was varied between 100 and 5000; the total hydrogen flow was 10-750 sccm.

The substrate temperature was varied between 1500°C and 2000°C for various deposition times and was monitored with a Mo-sheathed W/Re (type C) thermocouple fitted into a cavity in the backside of the substrate. The surface temperature of the graphite was monitored additionally with a two-color optical pyrometer. Following each deposition the substrate was maintained at temperature for 15 minutes and then cooled to room temperature in flowing H_2.

Surface morphology of the silicon carbide layers was examined by scanning electron microscopy (SEM). Micro-Raman scattering measurements were performed at room temperature with 514.5 nm excitation using an Ar^+ ion laser.

Results and Discussion

Cubic silicon carbide crystallizes in the zincblende structure (a_0=0.436 nm) with two atoms per formula unit, and has three allowable optical phonon modes. As an isotropic polar crystal, the optical modes are split into one longitudinal optical (LO) phonon (972 cm^{-1}) and two degenerate transverse optical (TO) phonons (796 cm^{-1}) [6]. Anisotropy in each of the other polytypes of silicon carbide (4H, 6H, ...) makes the Raman spectrum of 3C-SiC readily distinguishable. The Raman spectra of 3C-SiC crystals grown at substrate temperatures of ~1650°C and ~1750°C with an H_2/MTS ratio of 700 are shown in Figure 1a and 1b, respectively. Each spectra shows an LO band (971 cm^{-1}) and a single TO band (796 cm^{-1}) in good agreement with published values. The low intensity peak at 728.5 cm^{-1} in Figure 1a is as yet unidentified, but may due to defect-related scattering (structural or impurity) or a multiple-phonon process [7]. For comparison, Figure 1c shows the Raman pattern for a Lely 6H-SiC crystal. In this spectrum, an LO band (966.5 cm^{-1}), two TO bands located at 787.6 and 796 cm^{-1}, and a planar E_1 mode (766.1 cm^{-1}) are visible [8-9]. X-ray diffraction patterns for the

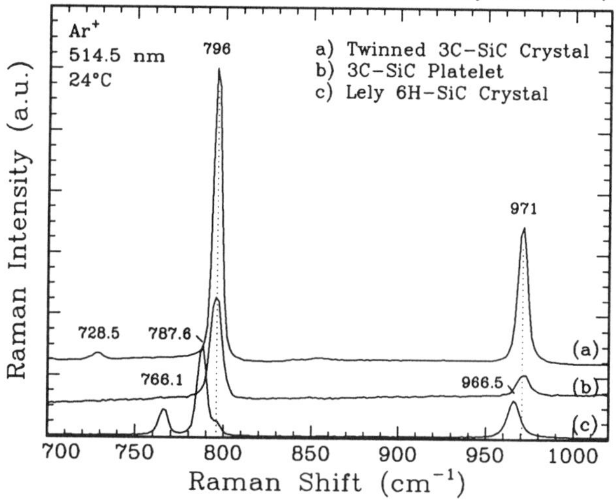

Figure 1. Raman spectra of 3C-SiC crystals (H_2/MTS=700) grown at a) 1650°C; and b) 1750°C; c) Lely 6H-SiC.

various deposits grown in this study were consistent with 3C-SiC. The only additional phase detected was graphite, most likely from the substrate.

As-grown, the crystals ranged in appearance from transparent yellow to yellow-green. Variation in color was most likely due to different doping concentrations of unintentional nitrogen donors. Nearly all of the deposits exhibited a darkened carbon-rich region at the interface with the substrate. As reported previously, the yellow appearance of CVD 3C-SiC became increasingly brilliant with decreasing MTS concentrations at higher H_2 flow rates. It was suggested that excess hydrogen may directly or indirectly improve the etching of free carbon from the growing SiC surface, or may assist in the formation of stable compounds which facilitate the removal of free carbon from the reactor. Free-standing crystals showed well-defined nitrogen bound exiton photoluminescence spectra with single- and multiple-phonon modes indicating that high-quality crystallites had been grown [4].

Normalized for the deposited area, an MTS flux greater than ~0.15 $gcm^{-2}hr^{-1}$ produced dense, thick (ex. 1-2 mm in 4 hours for H_2/MTS=200) polycrystalline deposits which could be separated from the substrate to yield free-standing layers. Lower nucleation rates at lower concentrations of MTS resulted in the formation of individual, well-defined crystallites.

The morphology of the SiC deposits was a strong function of the deposition temperature, concentration of MTS in H_2, and the total gas flow. The effect of deposition parameters on morphology has been considered in detail by a number of other authors [10-11]. For example, Pampuch et al. [12] described the formation of silicon carbide using $SiCl_4$+CCl_4 in hydrogen. They observed the formation of twinned skeletal grains and prismatic grains at substrate temperatures ranging from 1600-1700°C, consistent with the findings of this work. They also reported the growth of tabular, hexagonally-shaped crystals above 1800°C although the morphology of these deposits was not shown. Based in part on IR absorption data, they identified these high-temperature crystals as 6H-SiC.

Examination of SEM images for individual crystals revealed that the dominant growth direction was usually <110>. In many of the faceted crystallites, a common feature was the occurrence of microtwin lamellae along {111} planes. The presence of these microtwins is likely due to the formation of stacking faults during growth. Variable growth rates in a particular plane will favor twinning. A twin plane re-entrant edge mechanism, proposed initially to account for growth of germanium crystals [13-14], has been used to describe the development of 3C-SiC facets [15].

In this study, the effect of temperature on the morphology of 3C-SiC deposits was in good agreement with the literature for temperatures up to ~1750°C. At ~1750°C the formation of hexagonally-shaped platelets was observed for low MTS concentrations in H_2. SEM micrographs of twinned columnar and platelet crystals, grown at 1650°C and 1750°C respectively, are shown in Figure 2. The Raman spectra from these samples are illustrated in Figure 1.

The platelets were {111} oriented and also exhibited a preferred <110> growth direction. Atomic force microscopy revealed the surface roughness of the {111} faces to be better than 2 nm over 2x2 mm^2 with no discernible surface structure away from the near-edge region. The observation of square pyramidal and rectangular faceting on alternate faces adjacent to the {111} surfaces suggested a sequential {110}/{001} habit for the edges of the platelets. SEM observations confirmed the expected 35.3° {111}∠{110} and 54.7° {111}∠{001} angular relationships between the faces. Typical surface features on the {110} and {001} faces are shown in Figure 3. The morphology of an individual 3C-SiC platelet can be modelled as a thin (111) slice from the corner of a cubo-octahedron. A schematic illustration of this relationship is shown in Figure 4.

Platelets grew radially and axially in six-fold clusters about a central <110> axis, spaced by 60° along the <110> projection. Together, several clustered platelets formed the backbone for the twinned columnar crystals observed at the lower growth temperatures. A

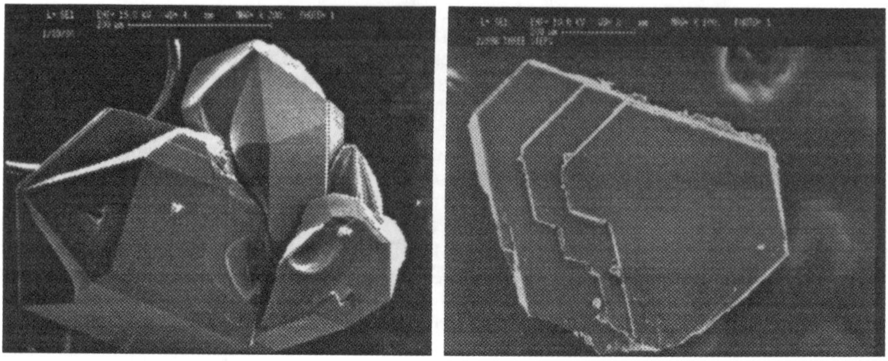

a) b)

Figure 2. SEM micrographs of 3C-SiC crystals (H_2/MTS=700) grown at a) 1650°C; and b) 1750°C. Bar = 200 μm.

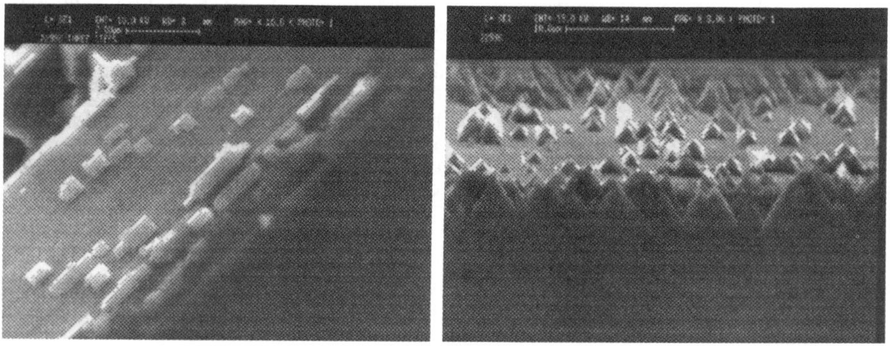

a) b)

Figure 3. SEM micrographs of surface faceting on a) {101}; and b) {001} faces of a {111} oriented 3C-SiC platelet. Bars = 2 μm and 10 μm, respectively.

semiquantitative calculation of the nucleation rate for the various low-index faces of 3C-SiC assuming two-dimensional nucleation and layer growth revealed that the preferential growth direction shifts from <111> to <101> with increasing temperature [11]. According to the model, growth in the <114> direction should dominate at the highest temperatures, although temperature values for each growth regime were not reported. The model does anticipate a growth rate anisotropy parallel to {111} between $<\bar{2}11>$ and $<2\bar{1}\bar{1}>$ for 3C-SiC which would favor twinning in the cubic structure relative to the hexagonal polytypes. In the CVD of 3C-SiC platelets, the apparently flux-limited condition of the higher substrate temperature results in a growth rate normal to {111} sufficiently low so as to not facilitate the filling of the gap between neighboring platelets.

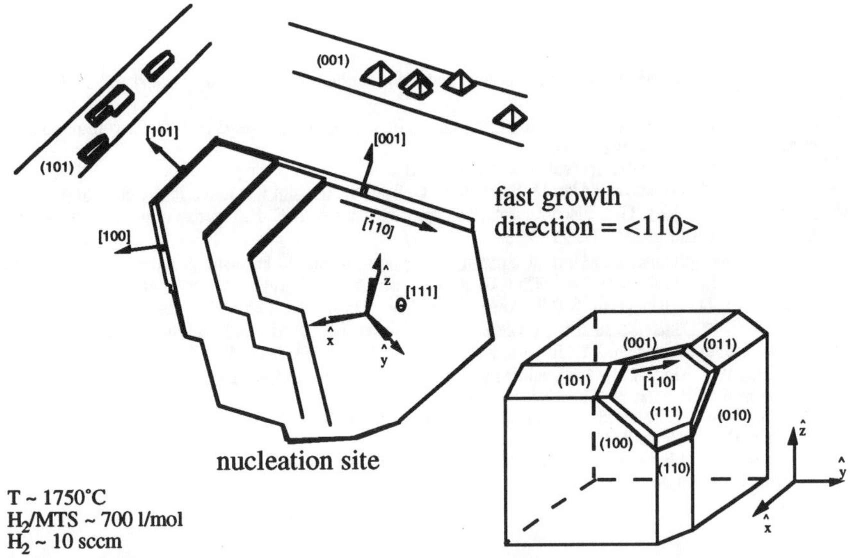

Figure 4. Schematic illustration of a (111) oriented 3C-SiC platelet. Rectangular facets and square pyramidal facets appeared on alternating {101} and {001} edges, respectively.

Conclusions

Optically transparent 3C-SiC has been grown on graphite substrates from MTS in excess hydrogen. The microstructure of CVD silicon carbide is strongly depended on temperature, MTS flux, and carrier gas flow rate. Faceted crystals were deposited at ~1650°C under conditions of low MTS supersaturation. These crystals were transparent yellow, though heavily twinned along {111}. For substrate temperatures of ~1750°C hexagonally-shaped {111} oriented platelets were grown. The platelets grown in this work were also transparent yellow and were unambiguously identified by Raman spectroscopy as 3C-SiC. Platelet edges displayed an alternating {101}/{001} character that was manifested in the geometry of the respective surface features. The morphological transition from twinned columnar growth to platelet growth with increasing temperature was described and related to a semiquantitative nucleation and growth model for 3C-SiC. Future work will focus on increasing the size of the platelets as well as on determining the nature (Si or C) of the {111} 3C-SiC surfaces.

Acknowledgments

T.L. McCormick is acknowledged for her assistance with the SEM and micro-Raman spectroscopy. T.W. Mercer performed the Atomic Force Microscopy. The authors wish to thank S.N. Gorin and L.M. Ivanova of the Baikov Institute in Moscow as well as B.L. Mackey at the Naval Research Laboratory for valuable discussions on this work. MWR is grateful for the support of an NRC/NRL postdoctoral research associateship. This work was supported in part by the Office of Naval Research and the Naval Research Laboratory.

References

1. R.F. Davis, G. Kelner, M. Shur, J.W. Palmour, and J.A. Edmond, *Proceedings of the IEEE*, **79** (5) 677 (1991).
2. T. Hirai and M. Sasaki in Silicon Carbide Ceramics-1, Eds. Shigeyuki Somiya and Yoshizo Inomata, Elsevier Applied Science, 87-95 (1991).
3. S.N. Gorin and L.M. Ivanova [private communication].
4. M.W. Russell, J.A. Freitas, Jr., Alan D. Berry and J.E. Butler, Covalent Ceramics III-Science and Technology of Non-Oxides, G.S. Fischman, A.F. Hepp, P.N. Kumta, A.E. Kaloyeros, and J.J. Sullivan, Eds., MRS Proc., 410, Pittsburgh, PA (1995).
5. Howard W. Post, Silicones and Organic Silicon Compounds, Reinhold Publishing, New York, 57 (1949).
6. D.W.Feldman, J.H. Parker, Jr., W.J. Choyke, and L. Patrick, *Phys. Rev.*, **173** 787 (1968).
7. C.A. Arguello, D.L. Rousseau, S.P.S. Porto, *Phys. Rev. (USA)*, **181** 1351-63 (1969).
8. J.A. Freitas, Jr. in Properties of Silicon Carbide, G.L. Harris, Ed., INSPEC, London, 21 (1995).
9. H. Harima, S. Nakashima, and Y Uemura, *J. Appl. Phys.*, **78** (3) 1996 (1995).
10. D.J. Cheng, W.J. Shyy, D.H. Kuo, and M.H. Hon, *J. Electrochem. Soc.*, **134** (12) 3145 (1987).
11. C.H. Chu and M.H. Hon, *J. Cer. Soc. Jpn.*, **101** (1) 95 (1993).
12. P. Pampuch and L. Stobierski, *Ceramurgia Int.*, **3** (2) 43 (1977).
13. R.S. Wagner, *Acta Metall.*, **8** 57 (1960).
14. D.R. Hamilton and R.G. Seidensticker, *J. Appl. Phys.*, **31** 1165 (1960).
15. C.H. Chu, Y.M. Lu and M.H. Hon, *J. Mater. Sci.*, **27** 3883 (1992).

Part III

Defects, Dopants and Characterization

Doping and Electrical Properties

Dopant incorporation efficiencies of SiC crystals grown on $\{1\bar{1}00\}$-face

Naohiro Sugiyama, Atsuto Okamoto, Toshihiko Tani and Nobuo Kamiya
Toyota Central Research & Development Laboratories, Inc.
Nagakute Aichi, 480-11, Japan, e0881@mosk.tytlabs.co.jp

ABSTRACT

The dopant incorporation efficiencies of SiC bulk single crystals were investigated by the measurement of optical absorption coefficients and chemical analysis by Glow Discharge Mass Spectroscopy. The SiC crystals were grown on the seed with the $\{1\bar{1}00\}$ face by a sublimation method. The grown ingot consisted of approximately three parts, which were the regions grown with the $\{1\bar{1}00\}$ face (I), the $\{1\bar{1}0n\}$ face (II) and the $\{1\bar{1}0\bar{n}\}$ face (III) as a growing front. The nitrogen concentrations in these parts were in the order of II>I>III. The wafers exhibited the color distributions in the c-plane due to the difference of the growing facet. The nitrogen incorporation efficiencies were discussed in terms of atomic structure on the surface. The aluminum concentration in the crystal grown on a $\{1\bar{1}00\}$-faced seed was in the medium between those in crystals grown on Si-face and C-face. On the other hand, the boron concentration was similar to that of Si-face grown crystal, and higher than that of C-face grown crystal.

INTRODUCTION

Silicon Carbide (SiC) is an attractive semiconductor material for high-temperature, high-power and/or high-frequency electronic devices due to its wide band gap, high saturated drift velocity and high field breakdown strength [1]. The feasibility of these devices is dependent on the quality of SiC substrates. SiC crystals are usually grown by a sublimation method (the modified Lely method) [2]. There exist, however, large defects called micropipes in the SiC crystals which are grown on {0001}-faced seed crystals by this method. Micropipe defects must be eliminated because they could give fatal damages to device functions. Carter et al. recently reported that micropipes densities were diminished to 2-3/cm^2, but not eliminated completely [3] Takahashi et al. reported that micropipes were not observed when SiC crystal grew on the face of $\{1\bar{1}00\}$ or $\{11\bar{2}0\}$, which were perpendicular to the {0001} face [4].

For SiC crystals as semiconductor substrates, the control of electric properties, which is related to dopant incorporation, is required as well as the diminution of structural defects. We have reported the dependence of the dopant incorporation efficiencies in bulk SiC crystals on the growing directions [5]. It suggests that SiC crystals preferentially incorporate nitrogen atoms on the $(000\bar{1})$ C-face and aluminum and boron atoms on the (0001) Si-face. Moreover, it was clarified that nitrogen incorporation efficiency was determined by the orientation of the growing front. On the other hand, the distribution in aluminum and boron concentrations exhibited a double ring pattern in the c-cut wafer plane grown on Si face. Subsequently, we investigated the dopant incorporation efficiency of SiC crystals grown on the $\{1\bar{1}00\}$ faced seed in the present

Mat. Res. Soc. Symp. Proc. Vol. 423 © 1996 Materials Research Society

work, so that the incorporation efficiencies are discussed by comparing the dopant concentrations in the crystals grown with different front facets.

EXPERIMENTAL PROCEDURE

Crystal Growth

SiC crystals were grown in a graphite crucible with a lid. SiC source powder was laid at the bottom of the crucible and the seed crystal attached on the lid. SiC crystals grow by vapor transport from a source powder (2295℃) to a seed (2230℃) under Ar ambient pressure of 1 Torr. Abrasive grade SiC powder (Showa Denko) was used as a source powder. The 6H-SiC seeds with a $\{1\bar{1}00\}$ face were fabricated by cutting from an ingot grown on $(000\bar{1})$ faced seed by a sublimation method, since Acheson crystals are not thick enough in the c-direction. The crystals were grown on the $\{1\bar{1}00\}$ oriented seed with 10 mm in diameter. The growth run for 24 hrs. The impurities were supposed to derive from the source material and the graphite crucible. Intentional doping was not carried out in this experiment.

Characterization of Crystals

The grown ingot was sliced parallel to the $\{1\bar{1}00\}$ face at a distance of 2 mm from the top of the crystal (ref. fig.1). The sliced wafer was polished to a thickness of 1 mm (wafer-A). The concentrations of impurities were determined in the central area (ϕ 8mm) of the wafer-A by Glow Discharge Mass Spectroscopy (GDMS). The results were compared with those of wafers grown on (0001) Si-face and $(000\bar{1})$ C-face.

Additionally, three c-cut wafers were sliced out of the grown ingot to investigate the distributions of impurities in the crystal. They were sampled from the parts where the SiC grew with a $\{1\bar{1}00\}$ facet (wafer-I), with a $\{1\bar{1}0n\}$ facet which is inclined from $\{1\bar{1}00\}$ face to Si-face (wafer-II) and with a $\{1\bar{1}0\bar{n}\}$ facet which inclined to C-face (wafer-III). (n : natural numbers) Absorption coefficients were measured in the visible light range for the qualitative estimation of nitrogen concentration on these wafers.

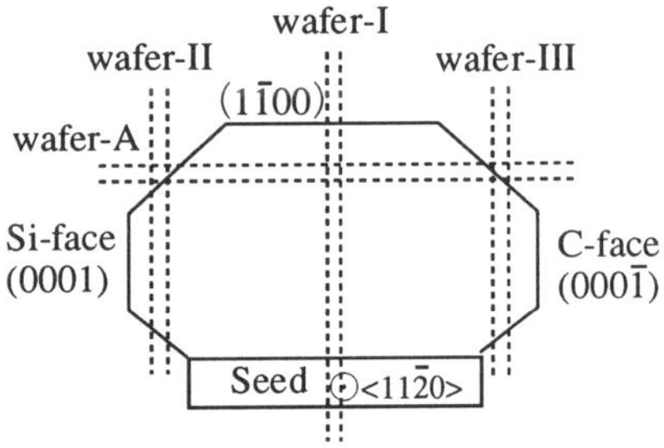

Fig. 1 Schematic view of grown ingot and sampled wafers.

Fig. 2 Outview of the SiC ingot grown on $\{1\bar{1}00\}$ faced seed. The maximum diameter was about 17 mm.

RESULTS AND DISCUSSIONS

Grown Crystal

Figure 2 shows the SiC ingot grown on the $\{1\bar{1}00\}$ faced seed. The ingot had growing facets of three $\{1\bar{1}00\}$ faces and their linking $\{1\bar{1}0n\}$ faces and $\{\bar{1}\bar{1}0n\}$ faces. The growth rate was approximately 8mm / 24hrs in height. Figure 3 shows the wafer-A horizontally sliced with the $\{1\bar{1}00\}$ face from the ingot. This figure indicates that the wafer-A consists of three regions with different color contrast, which are central area, Si-face side area and C-face side area of the wafer. Si-face side area was greener. On the other hand, C-face side area was colorless and striped due to high stacking fault densities. This suggests that the dopant incorporation efficiencies are different on these three areas. The c-cut wafers (wafer-I, II, III) underwent more detailed examination to see the difference of the dopant incorporation efficiencies on these regions.

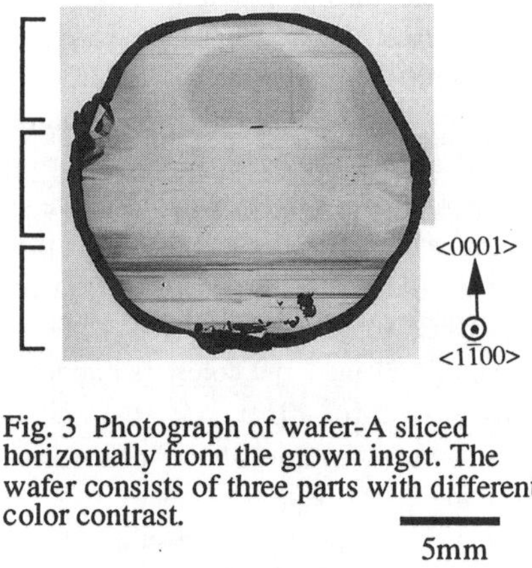

Fig. 3 Photograph of wafer-A sliced horizontally from the grown ingot. The wafer consists of three parts with different color contrast.

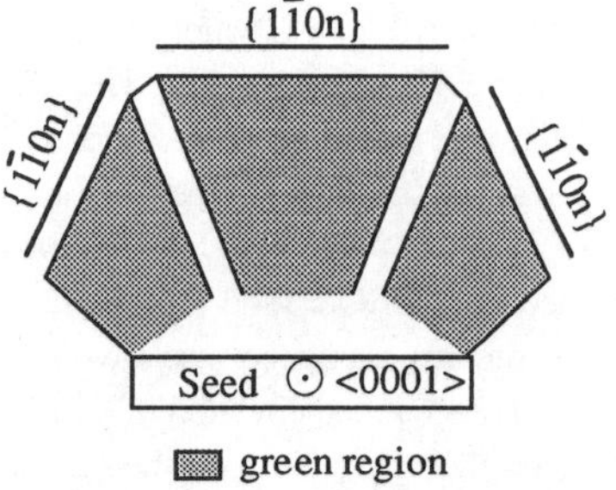

Fig. 4 Schematic color contrast in the wafer-II. The nitrogen concentration was higher in the areas where SiC had grown with $\{1\bar{1}0n\}$ facets.

Nitrogen Incorporation Efficiency

Biedermann has reported that nitrogen incorporation gives 6H-SiC crystals green color [6]. It suggests that the nitrogen concentration can be estimated qualitatively by optical absorption measurement. The color contrast in wafer-II shows an inhomogeneous distribution in nitrogen concentration. Greener regions extended widely, where crystals would have grown with $\{1\bar{1}0n\}$ facets. Colorless region is observed at the narrow area where the growth would have been performed between two $\{1\bar{1}0n\}$ facets (ref. fig.4). This results agree with the color distribution of the wafer grown on Si-faced seed (fig.5-a). Greener region was located at the fringe area where the crystal would have grown on $\{1\bar{1}0n\}$ facets. Briefly, nitrogen

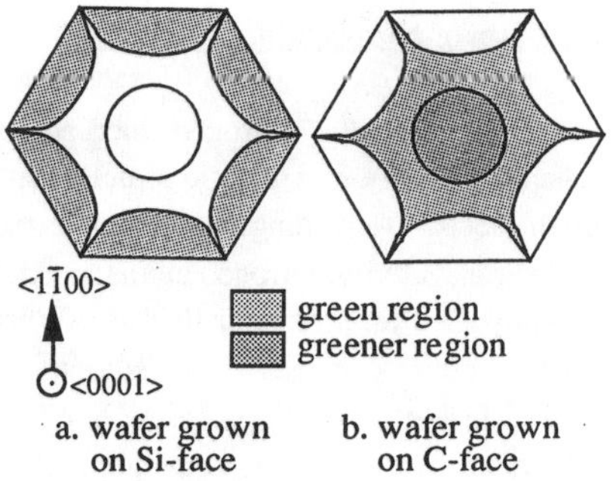

Fig. 5 Distributions of color contrast in the {0001} wafer grown on the Si-faced seed (a) and on the C-faced seed (b).

atoms were incorporated effectively onto the $\{1\bar{1}0n\}$-faceted surface, which was inclined from $\{1\bar{1}00\}$ face to Si-face.

We speculate that the nitrogen concentration in the wafer-III could correspond to that of the crystal grown on C-face in the directions of $<\bar{1}\bar{1}0n>$ or $<11\bar{2}n>$, as is the same for wafer-II. The wafer grown on C-faced seed shows hexagonal color contrast with greener region in the center and colorless regions in the fringe (fig.5-b). That is, the nitrogen concentration was low at the area where SiC grew on $\{1\bar{1}0\bar{n}\}$ face.

Figure 6 shows the results of optical absorption measurement of wafer-I, II, III. The absorption coefficient of wafer-I lay between those of wafer-II and wafer-III. This indicates that the nitrogen incorporation efficiency of the wafer-I was in the medium between wafer-II and wafer-III. J. Yang et al. have reported that the surface of the SiC crystal grown on $\{1\bar{1}00\}$ substrate contains elongated facets with $\{1\bar{1}04\}$ (or $\{1\bar{1}0\bar{4}\}$) face in the direction of $<11\bar{2}0>$ [7]. This means that the growth on the $\{1\bar{1}00\}$ face can be considered as a combination of growths on the $\{1\bar{1}04\}$ and $\{1\bar{1}0\bar{4}\}$ faces. The nitrogen concentrations in the grown crystal are summarized in fig. 7.

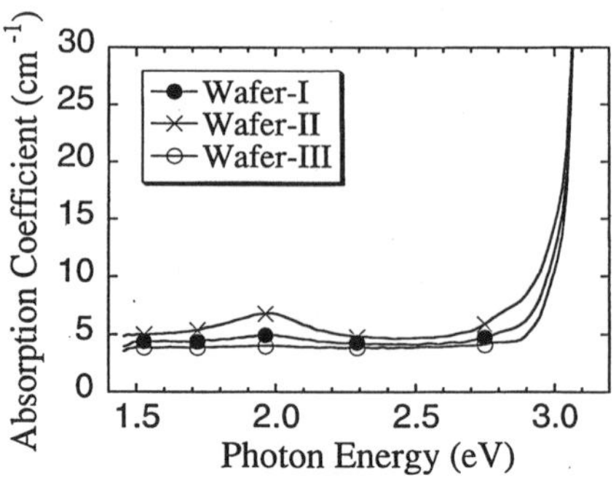

Fig. 6 Optical absorption coefficients of the wafer-I, II and III. The peaks near 2.0 eV are related to incorporated nitrogen.

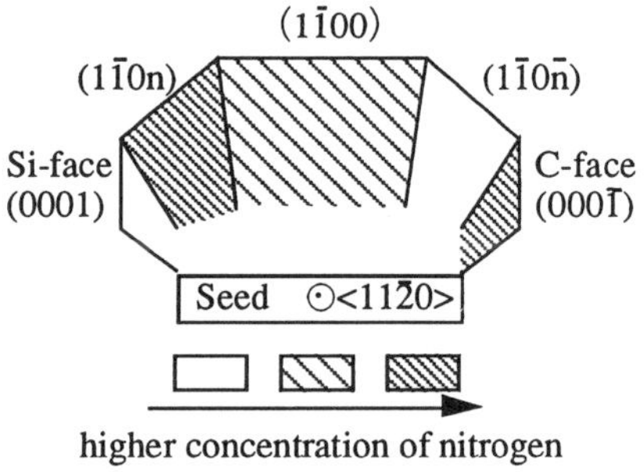

Fig. 7 Schematic diagram for the distribution of nitrogen concentration in the vertical section of a grown ingot. The color contrast was not clear near the seed.

Mechanism of Nitrogen Incorporation

We interpret the nitrogen incorporation mechanism in terms of atomic structure at the growth facets. According to the atomic radius and electronegativity, nitrogen atoms are known to occupy C-sites preferentially [8]. When a nitrogen atom, occupies a C-site in the $(000\bar{1})$C-face growth, no dangling bond exists on the atom by forming a nonbonding electron pair. This state is considered relatively stable. This is the reason why the nitrogen atoms are effectively incorporated on the C-face. Figure 8

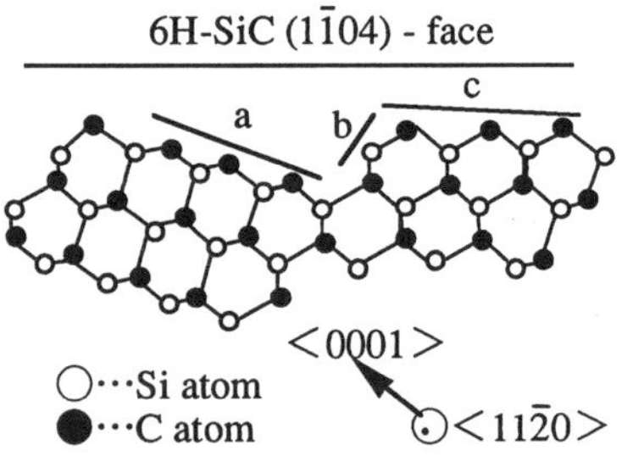

Fig. 8 surface structure of $(1\bar{1}04)$-face. a- and b-plane are equivalent to C-face and Si-face, respectively. c-plane consists of carbon atoms with two dangling bonds.

shows the atomic structure of the $\{1\bar{1}04\}$ face. This surface partially includes the same type of structure as C-face. Therefore, the nitrogen concentration was high in the region where SiC grew on the $\{1\bar{1}04\}$ face. In general, $\{1\bar{1}0n\}$ face contains three types of surface structure, which are the surface structure of C-face (a in fig. 8), Si-face (b) and the structure consisting of Carbon atoms with two dangling bonds (c). Although the nitrogen incorporation efficiency of type-c is not clear, high efficiency is expected in type-a. This explains the high nitrogen concentration in the part grown with $\{1\bar{1}0n\}$ facets. On the other hand, there is no surface structure of C-face on the surfaces of the (0001)Si-face and little on the $\{\bar{1}10\bar{n}\}$ faces. So, the nitrogen atoms are difficult to be incorporated in the crystal on these faces. The nitrogen incorporation efficiencies can be qualitatively estimated by the density of C-face surface structure on the growing front.

Aluminum and Boron Incorporation Efficiencies

Table I shows the result of chemical analysis on the wafer-A by GDMS, compared to those of the wafers grown on Si-face and C-face. The aluminum concentrations of the wafers were in the order of growth on Si-face > on $\{1\bar{1}00\}$ face > on C-face. This tendency is opposite to that for the nitrogen concentration. It is agreeable if aluminum atoms could preferentially occupy Si-sites [8]. It is noted, however, that the distribution of aluminum concentration shows a double ring pattern, which is quite different from that of nitrogen.

The concentration of boron in the wafer-A was as high as that in the Si-face grown wafer, which was higher than that in the C-face grown wafer. It is not clear why boron atoms showed equally effective incorporation in the $\{1\bar{1}00\}$ face to the Si-face, but it might be related to the possibility that boron atoms could occupy both Si- and C-sites [8].

CONCLUSIONS

The dopant incorporation efficiencies were examined for SiC bulk single crystals grown on the $\{1\bar{1}00\}$ faced seed. The results obtained are as follows. The grown ingot consists of three parts with different color contrasts, which were the regions grown with the $\{1\bar{1}0n\}$ face, the $\{\bar{1}10\bar{n}\}$ face and the $\{1\bar{1}00\}$ face as a growing front. The nitrogen concentrations are different in each part and exhibited distributions in the c-plane. The aluminum concentration in the crystal grown on a $\{1\bar{1}00\}$ oriented seed was in the medium between those of crystals grown on Si-face and C-face.

Table I. Impurity concentrations in the wafers grown on different faces and in source powder.

Growing Surface	Impurities [ppm]				
	Al	B	Fe	Ti	Ca
$(1\bar{1}00)$ -face	2.5	3.1	0.15	0.02	ND
(0001) Si-face	4.1	2.7	0.25	0.05	ND
$(000\bar{1})$ C-face	1.2	0.9	0.28	0.05	ND
source powder	21	5	89	38	27

On the other hand, the boron concentration is similar to that in Si-face grown crystal and higher than that in C-face grown crystal.

ACKNOWLEDGEMENTS

The authors wish to thank Dr. A. Fukumoto and Dr. H. Tadano in Toyota Central R&D Labs., Inc. for valuable discussions on the results.

REFERENCES

1. M. Bhatnagar and B. J. Baliga, IEEE Trans. Electron Devices 40, 645 (1993)
2. Y. M. Tairov and V. F. Tsvetkov, J. Cryst. Growth 43, 209 (1978)
3. C. H. Carter Jr. and V. F. Tsvetkov, the 6th International Conf. on Silicon Carbide and Related Mareials (Kyoto) Techenical Digest (1995) p.11
4. J. Takahashi, M. Kanaya and Y. Fujiwara, J. Cryst. Growth 135, 61 (1994)
5. N. Sugiyama, A. Okamoto and T. Tani, the 6th International Conf. on Silicon Carbide and Related Materials (Kyoto) Techenical Digest (1995) p.22
6. Biedermann E. Solid State Comm. 3, 343, (1965)
7. J. Yang, S. Nishino, M. Mehragany and P. Pirouz, Inst. Phys. Conf. Ser. No.137, Silicon Carbide and Related Materials, (IOP, 1994) p.17
8. J. A. Freitas Jr., W. E. Carlos and S. G. Bishop, "Amorphous and Crystalline Silicon Carbide III", (Springer, 1990) p.135

LOW VOLUME RESISTIVITY CHEMICAL VAPOR DEPOSITED BORON DOPED POLYCRYSTALLINE THIN DIAMOND FILM GROWTH ON SAPPHIRE

Hassan Golestanian, S. Mirzakuchaki, E. J. Charlson, T. Stacy, and E. M. Charlson
University of Missouri-Columbia, Department of Electrical Engineering, Columbia, MO 65211

ABSTRACT

Hot-filament chemical vapor deposited (HFCVD) boron doped polycrystalline diamond thin films having low volume resistivity were grown on sapphire. The films were characterized using scanning electron microscope (SEM), X-ray diffraction, and current-voltage measurements. SEM micrographs show good crystalline structure with preferred (100) orientation normal to the surface of the film. X-ray diffraction pattern revealed diamond characteristics with the four typical diamond peaks present. Finally, the obtained I-V characteristics indicated that the film's volume resistivity is at least two orders of magnitude lower than those of HFCVD polycrystalline diamond thin films grown on silicon under similar growth conditions.

INTRODUCTION

Because of its unique properties, diamond has been explored for diverse applications in electronics, optics, and material coatings.[1-3] The wide energy gap, surface stability at high temperatures, and low relative dielectric constant of diamond makes it an ideal candidate for use in high speed, high power, high temperature, and high frequency applications.[4] Unfortunately, the use of bulk diamond crystals severely limits the semiconductor applications due to difficulty in doping, device integration, high cost, and small size.[2] In the hope of alleviating these problems, a great deal of effort has been undertaken for study of heteroepitaxial diamond synthesis by chemical vapor deposition (CVD). Since Spitsyn et. al. reported the growth of diamond films on a foreign substrate under metastable conditions,[5] heteroepitaxial diamond synthesis using CVD has become a reality. It has been suggested that for reproducibly growing diamond thin films, substrates of carbides and materials capable of forming interfacial carbides, e.g. Si, Mo, SiC, $Si_2 N_4$, Al_2O_3, and WC are desirable.[1] For substrates having large thermal expansion coefficient, i.e. steel, or exhibiting adhesion problems, an intermediate layer like SiC has been suggested.[1]

CVD diamond on insulating high temperature substrate is desirable for various electronic and optical applications. Sapphire substrates are one of the most promising possibilities. Unfortunately, diamond thin films grown on sapphire exhibit very poor adhesion.[1] CVD diamond growth on sapphire is severely affected by the difference in thermal expansion coefficients between the two materials. Different thermal expansion coefficients of the layer and the substrate results in interfacial stress and thus poor adhesion. Therefore, the grown films tend to crack and to peel off from the substrate both during sample cool down and when the sample is exposed to atmospheric pressure. For an Al_2O_3 substrate, it has been suggested that tension rather than compression is needed for good adhesion of diamond.[6]

EXPERIMENT

Synthetic boron doped HFCVD polycrystalline thin diamond films were grown using

Mat. Res. Soc. Symp. Proc. Vol. 423 © 1996 Materials Research Society

in-situ doping on sapphire. In this study, 1″ diameter α-Al_2O_3 discs were used as substrates. The HFCVD reactor consisted of straight tungsten filaments placed horizontally over the substrate with a filament-to-filament distance of 7 mm. In order to keep the level of tungsten contamination low, the filaments were pre-treated using a gas mixture of hydrogen and methane of 85 and 15 sccm, respectively. Since the CVD diamond films grown on sapphire are expected to have poor adhesion, the substrate underwent a two-step surface treatment first using 15 µm grit diamond paste for 20 minutes followed by a smoothing period of 10 minutes with 0.25 µm grit diamond paste. The first abrasion was found necessary to improve the adhesion while the smoothing step was utilized to promote the growth of smaller size crystals for growth of a smooth film surface. Prior to introduction to the CVD system , the substrate was rinsed with DI water and air dried. To further reduce cracking and peeling, the growth process was divided into two to three timed growth periods.

The substrate was placed about 7 mm beneath the treated tungsten filaments on a Mo plate. Its temperature was monitored using a K-type thermocouple placed against the substrate backside through the Mo plate. Reacting gases were H_2 and CH_4 with flow rates of 98 and 1.5 sccm, respectively. The dopant source used was trimethyl borate, $B(COCH_3)_3$, in acetone. Flow of the dopant gas was controlled at 0.5 sccm to provide the desired ratio of boron to carbon in the chamber (200 to 1000 ppm). Reacting constituents were introduced into the chamber using a gas feed through with a 1/4" copper tube positioned above the filaments. Chamber pressure was kept at 29-31 Torr with a substrate temperature of 765-775°C.

RESULTS

Mechanical abrasion using diamond paste and a timed growth process resulted in a significant reduction in cracking and peeling of the films. Extra care was also taken during the cool down period and subsequent sample introduction to atmospheric pressure to further reduce the flaking. To show the area covered by the diamond film, a photograph of sample S_{11} is given in Fig. 1. Viewing the figure, one can see that the grown diamond film has completely adhered to the substrate. This film has been stable for several months. It must be noted that the non-uniformity present on the picture is due to the residue of the silver thick film conductors that were applied for I-V measurements.

In order to characterize the as-grown diamond films on sapphire, SEM, X-ray diffraction, and I-V measurements on the films were obtained. Crystal structure of the films were examined using scanning electron microscope. A micrograph of sample S_7 is presented in Fig.2. Well faceted crystals having good crystal structure with preferred (100) orientation normal to the surface of the film is observed in the figure.

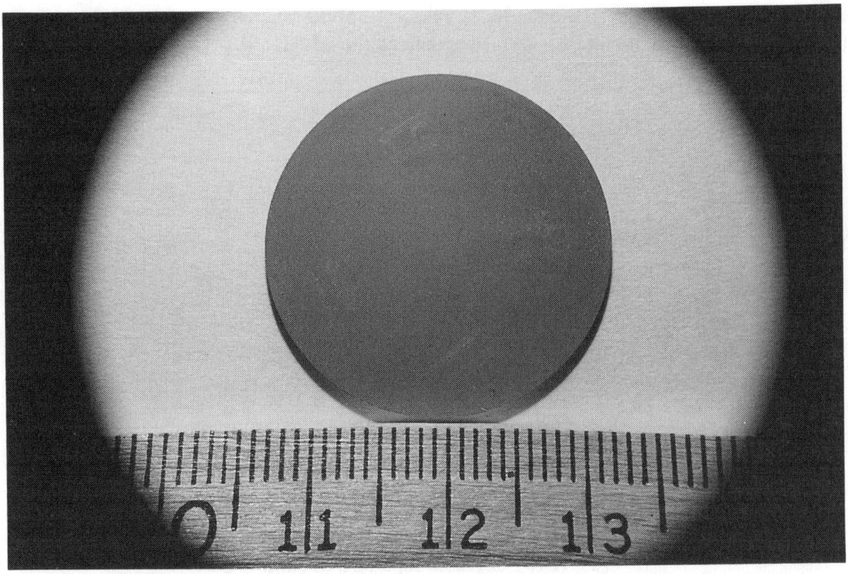

Fig. 1: Photograph picture of sample S_{11} showing area covered by the diamond film

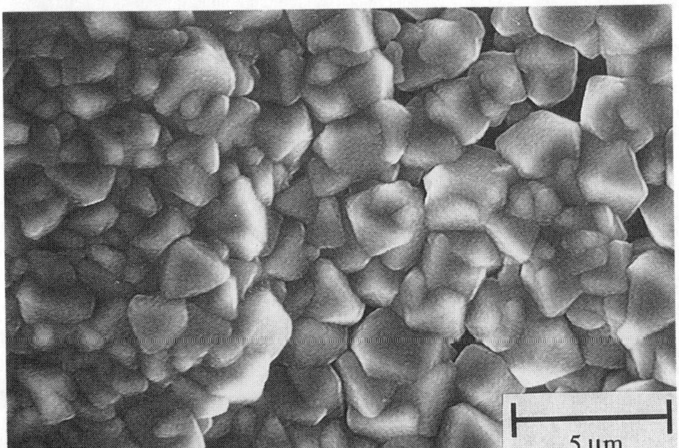

Fig. 2: SEM micrograph of sample S_7

The X-ray diffraction pattern of the samples indicated the existence of the four conventional diamond peaks. Fig. 3 shows the X-ray diffraction pattern of sample S_9. In order to quantitatively determine the crystalline plane orientation and possible texture, the relative intensities of the peaks were compared to the standard ASTM file for carbon. To approximate the degree of texture, the Harris texture index of the diffracting planes were then calculated. It was found that for the (400) plane the index ($T_{(400)}$=1.34) dominated. A preferred orientation of (h_i k_i l_i)-plane parallel to the film surface is indicated by T_i>1.[7]

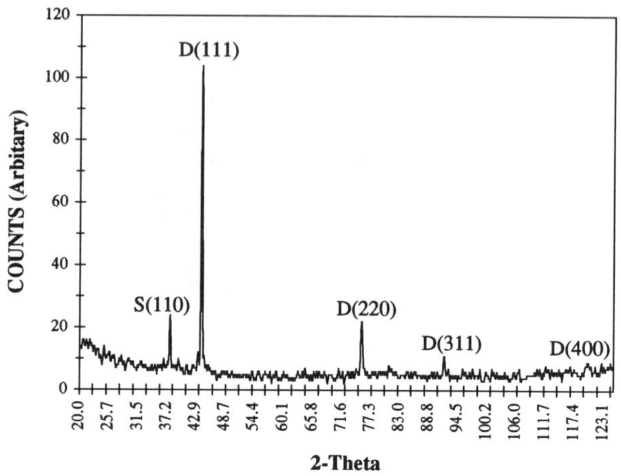

Fig. 3: X-ray diffraction pattern of sample S_9

Conductivity type of the carriers in sample S_7 was verified using a carefully isolated thermal probe station. The current variations due to a heat cycle and subsequent cool down is shown in Fig. 4. The hot probe was grounded to the free standing sample with a cold ohmic contact connected to the input of the pico ammeter. The positive current indicates p-type conductivity.

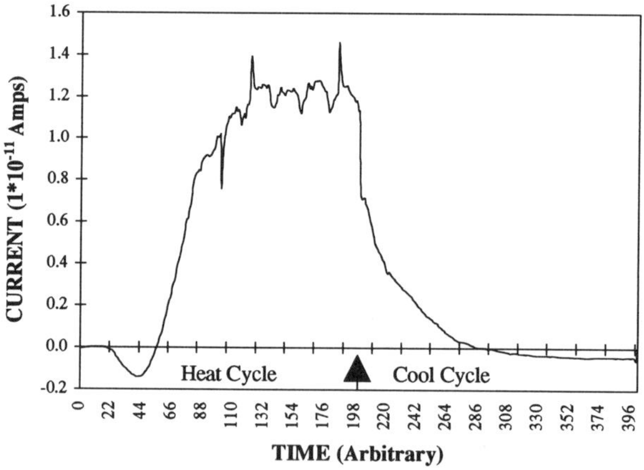

Fig. 4: The current variations due to a heat cycle and subsequent cool down

Two stripes of silver thick film conductors were applied on top of the films about 0.1cm apart and 0.6 cm long. Length to distance ratios of the stripes were kept large to minimize possible sources of error in the I-V measurements. The measured values of current vs. applied voltage for two samples are plotted in Fig. 5.

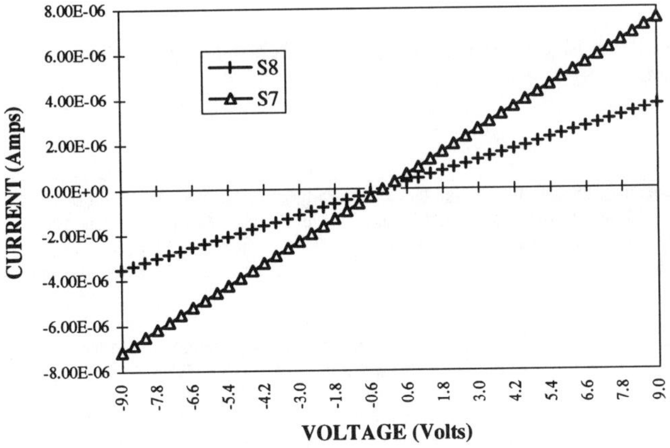

Fig. 5: I-V obtained for CVD diamond films grown on sapphire

Using resistance values obtained from the figures and the corresponding physical data, volume resistivity of the films were then estimated. Table I contains dopant concentration in the chamber, growth process length, and calculated volume resistivity for several samples studied in this work.

Table I: Dopant concentration, growth process length, and calculated volume resistivity.

Sample	B/C (ppm)	Growth Process (Hours)	Resistivity (Ω-cm)
S_7	400*	4 hrs. and 15 min.	1.5×10^3
S_8	1000	One 2 hr. and a 1 hr.	3.1×10^3
S_9	200**	Two 2 hr. runs	4.8×10^5
S_{11}	200	One 2.5 hr. and a 1.5 hr.	3.41×10^4

* A delta doping period of 15 minutes was used
** First two hours was undoped

CONCLUSION

Hot-filament chemical vapor deposited boron doped polycrystalline diamond thin films having low volume resistivity were grown on sapphire in spite of the significant difference between thermal expansion coefficient of sapphire and CVD diamond. The calculated film volume resistivity ($\rho=1.5 \times 10^3$ Ω-cm) was found to be less than two orders of magnitude lower than that for films grown on silicon ($\rho=8.42 \times 10^5$ Ω-cm) under similar growth conditions. Adhesion problem was eliminated by vigorous mechanical treatment of the substrate, timed growth periods, and a slow cool down process.

ACKNOWLEDGMENTS

The authors would like to thank Mr. L. M. Ross and Mr. Naiyu Zhao for their assistance in SEM and XRD studies.

References

[1] M. N. Yoder, in Applications of Diamond Films and Related Materials, edited by Y. Tzeng, M. YoshiKawa, M. Murakawa, and A. Feldman (Elsevier Science Publishers, New York, 1991) pp. 287-294; Satoru Hosomi and Isao Yoshida, ibid. , pp. 15-24; L. Schafer, M. Sattler, and C. P. Klages, ibid., pp. 453-459; Walter A. Yarbrough, ibid. , pp. 25-34; R. Ramesham, C. Ellis, and T. Roppel, ibid. , pp. 411-416.

[2] G. SH. Gildenblat, " The Electrical Properties and Device Applications of Homoepitaxial and Polycrystalline Diamond Films," IEEE Proceedings, Vol. 79, No. 5, pp. 647-668, 1991.

[3] G. Zhao, T. Stacy, E. J. Charlson, E. M. Charlson, C. H. Chao, M. Hajsaid, J. M. Meese, G. Popovici, and M. Prelas, " Silver on Diamond Schottky Diodes Formed on Boron Doped Hot-Filament Chemical Vapor Deposited Polycrystalline Diamond Films," J. Appl. Phys. Lett. Vol. 61 No. 9, pp. 1119-1121, 1992.

[4] R. F. Davis, Z. Sitar, B. E. Williams, H. S. Kong, H. J. Kim, J. W. Palmour, J. A. Edmond, J. Ryu, J. T. Glass, and C. H. Carter, Jr., " Critical Evaluation of the Status of the Areas for Future Research Regarding the Wide Band Gap Semiconductors Diamond, Gallium Nitride, and Silicon Carbide, " Materials Science and Engineering, Bl. pp. 77-104, 1988.

[5] B. V. Spitsyn, L. L. Bouilov, and B. V. Derjaguin, " Vapor Growth of Diamond on Diamond and other Surfaces," J. of Crystal Growth, Vol. 52, pp. 219-226, 1981.

[6] Benno Lux and Roland Houbner, in Diamond Films and Coatings Development, Properties, and Applications, edited by Robert F. Davis (Noyes Publications, Park Ridge, New Jersey, 1993) pp. 215-219.

[7] J. B. Cohen and J. E. Hilliard, in Local Atomic Arrangement Studied by X-Ray Diffraction, (Gordon and Breach, New York, 1966).

ACTIVATION OF ACCEPTORS IN Mg-DOPED, p-TYPE GaN

W. GÖTZ, N.M. JOHNSON, J. WALKER, D.P. BOUR
Xerox Palo Alto Research Center, Palo Alto, California 94304, USA

ABSTRACT

The activation of acceptors was investigated for Mg-doped, heteroepitaxial layers of GaN grown by metalorganic chemical vapor deposition. After growth the samples were exposed to isochronal rapid thermal anneals in the temperature range from 500°C to 775°C. The samples were studied by variable temperature Hall effect measurements and photoluminescence (PL) spectroscopy in the as-grown condition and after each temperature step. The thermal treatment leads to the formation of acceptors which are characterized by an activation energy for ionization in the range between 165 meV and 182 meV. These acceptors are attributed to Mg atoms substituting for Ga in the GaN lattice. The experimental results for the acceptor activation are consistent with the dissociation of electrically inactive acceptor-hydrogen complexes. The reversibility of this process is investigated by the exposure of the activated, p-type GaN samples to atomic hydrogen in a remote-plasma hydrogenation system at 600°C and reactivating at 850°C.

INTRODUCTION

Mg-doped, p-type GaN exhibits a shallow acceptor state, the ionization energy of which ranges between ~160 meV and ~180 meV depending on the concentration of acceptors and compensating donors [1]. This acceptor state is believed to be due to Mg atoms substituting for Ga (Mg_{Ga}) in the GaN lattice. However, as-grown, Mg-doped GaN is semi-insulating when grown with growth techniques that furnish a hydrogen rich ambient and a post-growth treatment (e.g., furnace annealing) is required to activate the acceptor dopant [2,3]. Results from *ab-initio* total energy calculations [4] suggest, that hydrogen in its positive charge state (H^+) compensates acceptors at growth temperature to enable charge neutrality [5]. Thus, the formation of donors is suppressed, since H^+ has a lower formation energy than other potential donors (e.g. the nitrogen vacancy). After growth, acceptor-hydrogen complexes form which are dissolvable in a post-growth thermal activation process at temperatures below the growth temperature.

Conclusive experimental evidence for the existence of acceptor-hydrogen complexes in GaN from vibrational mode spectroscopy has yet not been achieved [6]. However, it has been shown that the exposure of Mg-doped, p-type GaN to monoatomic hydrogen at elevated temperatures leads to a reduction of the acceptor concentration and to an increase of the hole mobility [7]. This increase is indicative of the removal of ionized scattering centers and consistent with the formation of acceptor-hydrogen complexes upon introduction of hydrogen into the GaN sample.

In this study we investigate the activation of acceptors in Mg-doped GaN with isochronal (5 min) RTA annealing at temperatures in the range between 500°C and 775°C. The GaN material was characterized by variable temperature Hall effect measurements and low temperature (2 K) photoluminescence (PL) in the as-grown state and after each annealing step. To test whether the activation process is reversible, the activated samples were exposed to atomic hydrogen at 600°C.

EXPERIMENTAL

The GaN material used in this study was grown by MOCVD and doped with Mg during growth. The films were grown heteroepitaxially on sapphire substrates to a thickness of 2 μm

Mat. Res. Soc. Symp. Proc. Vol. 423 © 1996 Materials Research Society

with a low temperature GaN buffer layer. After growth, these films were semi-insulating with resistivities of $>10^7$ Ω cm at 500 K. The samples were not initially exposed to any post-growth acceptor activation processes. The incorporation of Mg was verified by secondary ion mass spectrometry (SIMS). The SIMS depth profiles revealed a uniform distribution of Mg throughout the film at a concentration of $\sim 6\times10^{19}$ cm^{-3}.

The GaN films were annealed in a RTA system under flowing N_2. The heat-up time to the annealing temperatures was ~5 s and temperature overshoots were minimized (< 3 K). The hydrogenation was performed with a remote microwave plasma operating at 2 torr [7].

The Hall measurements were conducted in the temperature range from 80 K to 500 K. The magnetic field was 17.4 kG. Samples of 5 × 5 mm^2 size were cut and metal dots were vacuum evaporated in the four corners to obtain ohmic contacts in the Van der Pauw geometry.

For the PL measurements, the samples were mounted in a cryostat and immersed in liquid He to achieve a sample temperature of ~2 K. The PL spectra were excited with the 325 nm line of 30 mW HeCd laser with a pumping power of ~1 W / cm^2. The resolution of the spectra was 0.05 nm.

ACTIVATION OF ACCEPTORS

The as-grown, Mg-doped GaN films exhibited high resistivity, and reliable resistivity measurements could only be obtained at elevated temperatures [8]. After annealing at 500°C, the resistivity of the sample remained basically unchanged (~10^7 Ωcm at 500 K). After the annealing temperature was raised to 775°C, the resistivity reaches a value of ~ 3 Ωcm at 300 K.

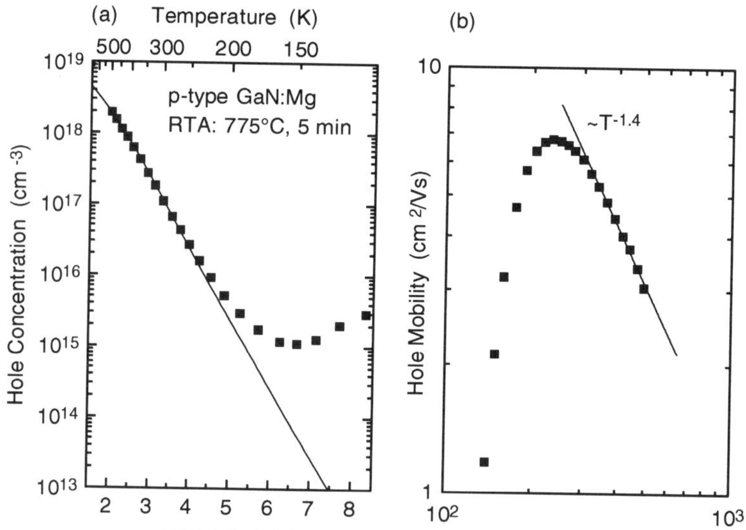

Fig. 1. Hole concentration (a) and hole mobility vs temperature (b) as determined from Hall effect measurements. The symbols refer to experimental data and the solid line in Fig. 1a is calculated using the charge neutrality condition. The solid line in Fig. 1b indicates a temperature dependence of ~$T^{-1.4}$ of the hole mobilities at high sample temperatures.

Hall effect results after annealing at 775°C are shown in Fig. 1. Figure 1a and b show the hole concentration and the hole mobility, respectively, as functions of the sample temperature. The hole concentrations p were obtained from the experimental Hall constants R_H with $p = r_H/qR_H$ (q = electronic charge); the Hall scattering factor was assumed to be isotropic, temperature independent and of unity value ($r_H \equiv 1$). The temperature dependence of the electron concentration shown in Fig. 1a reveals activation by a shallow acceptor only for temperatures above 200 K with impurity band conduction at lower temperatures. The solid line in Fig. 1a is calculated from the charge neutrality condition. The model assumes a single acceptor species (Mg_{Ga}), specified by its thermal activation energy and uniform bulk concentration, and partial compensation by an unidentified donor species, specified by its bulk concentration. The onset of impurity band conduction severely restricts the temperature interval over which the three-parameter model is applicable. Therefore, the data were analyzed by fixing the donor compensation at 3×10^{18} cm^{-3} [8] and performing a two-parameter fit to the high-temperature (i.e., >200 K) portion of the data. After the 775°C anneal, an acceptor concentration of ~2.3×10^{19} cm^{-3} and an activation energy for ionization of ~170 meV were determined.

The room temperature (300 K) hole mobilitiy (Fig. 1b) after annealing at 775°C is 6.7 cm^2/Vs and the peak mobilitiy appears at 235 K with a value of 6.1 cm^2/Vs. The high temperature branch of the mobility follows a temperature dependence of ~$T^{-1.4}$ which is indicative of lattice scattering and close to the ideal temperature dependence of this scattering mechanism, ~$T^{-1.5}$. At lower sample temperatures, the hole mobility is affected by impurity band conduction and drops rapidly.

Results from PL spectroscopy (2 K) taken from the Mg-doped GaN film after growth and after each annealing step have been reported elsewhere [8] and are summarized as follows. The PL spectrum of as-grown, Mg-doped GaN is dominated by a set of lines emitting in the blue region of the electromagnetic spectrum. The zero phonon line (ZPL) is located at 382 nm (3.25 eV) and LO phonon (0.09 eV) replicas are visible at 393 nm (3.16 eV) and 403 nm (3.08 eV). After annealing at 500°C these lines "red-shift". The ZPL is now positioned at 414 nm (3.00 eV) with the first LO phonon replica at 425 nm (2.92 eV). The intensity of the PL increases by a factor of two after annealing at 500°C. The spectra remain unchanged after annealing at 600°C. After annealing at 700°C the intensity of the PL lines decreases and the lines broaden. The ZPL line appears only as a shoulder and its position cannot be exactly determined. However, the line of the first phonon replica indicates further red-shift of ~0.01 eV. After annealing at 775°C a weak blue band centered at 435 nm (2.85 eV) with a full width at half maximum of ~0.23 eV appears in the spectrum. In addition to the blue band, a set of sharp PL lines located at 369.55 nm (3.358 eV), 376.25 nm (3.298 eV) and 377.55 nm (3.286 eV) appears.

REMOTE PLASMA HYDROGENATION OF p-TYPE, Mg-DOPED GAN

After annealing at 850°C (5 min), the p-type, Mg-doped GaN sample was exposed to atomic deuterium in a remote-plasma deuteration system at 600°C for one hour. This treatment rendered the sample semi-insulating and no Hall effect measurement was attempted. However, after further annealing at 850°C (5 min), the sample exhibited p-type conduction. Results from variable temperature Hall effect measurements are shown in Fig. 2. Figure 2 demonstrates hole concentrations (a) and hole mobilities (b) after annealing at 850°C and after deuteration and annealing at 850°C. The solid lines in Fig. 2a were calculated from the charge neutrality condition by assuming a single acceptor and donor compensation. The activation energy for ionization of the acceptor is ~170 meV and the concentration of the compensation is 3×10^{18} cm^{-3} for both data

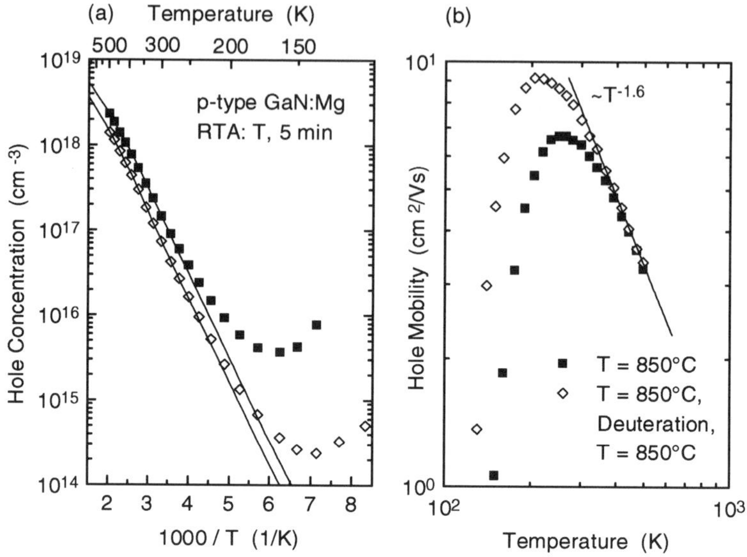

Fig. 2. Hole concentration (a) and hole mobility vs temperature (b) as determined from Hall effect measurements. The symbols refer to experimental data and the solid lines in Fig. 2a are calculated using the charge neutrality condition. The solid line in Fig. 2b indicates a temperature dependence of ~$T^{-1.6}$ of the hole mobilities at high sample temperatures.

sets displayed in Fig. 2a. The concentration of the acceptors is 3.0×10^{19} cm^{-3} after annealing at 850°C and 2.0×10^{19} cm^{-3} after deuteration and annealing.

The hole mobilities significantly increase after the sample was exposed to the deuterium plasma and reactivation as demonstrated in Fig. 2b. The room temperature and peak mobilities prior to deuteration are 6.4 cm^2/Vs and 6.7 cm^2/Vs (265 K), respectively, and after deuteration and reactivation the corresponding values 7.3 cm^2/Vs and 9.2 cm^2/Vs (205 K), respectively. The high temperature branches of the mobilities exhibit a temperature dependence of ~$T^{-1.6}$ and at lower sample temperatures the mobilities are affected by impurity band conduction.

In Fig. 3, we show PL spectra taken at 2 K for Mg-doped GaN after annealing at 850°C, after remote plasma deuteration at 600°C for 1 h, and after annealing, deuteration and acceptor reactivation at 850°C. We also show the PL spectrum that was obtained from the same film in the as-grown state. This spectrum exhibits a set of PL lines consisting of the ZPL line at 382 nm (3.25 eV) and LO phonon replicas [8]. The spectrum taken after annealing at 850°C is almost identical to the spectrum after annealing at 775°C [8] and exhibits the "red-shifted" blue band centered at 435 nm (2.85 eV) and set of sharp lines closer to the band edge [8]. After the remote plasma deuteration, these PL lines disappear and the broad line which includes the ZPL line L2 increases in intensity. After reactivation of the acceptor dopant (RTA: 850°C, 5 min), the broad PL line decreases in intensity but its energy position remains almost unchanged.

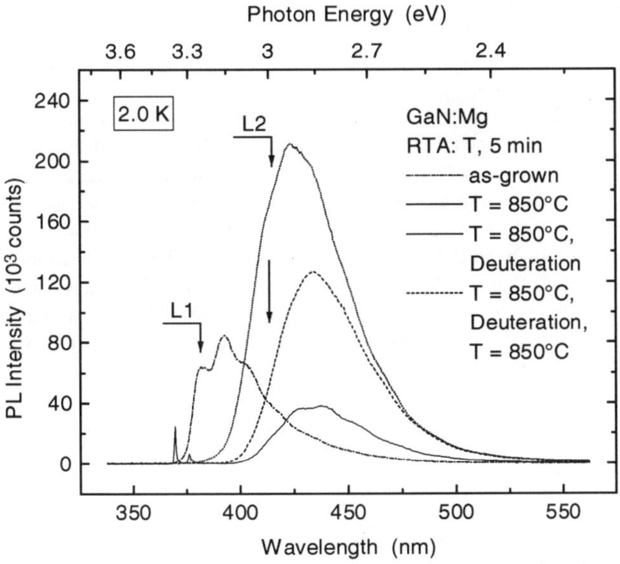

Fig. 3. PL spectra of Mg-doped GaN in the as-grown state and after annealing at 850°C, subsequent exposure to a remote-plasma deuteration at 600°C for 1h, and subsequent activation of the acceptor dopant at 850°C. Zero phonon lines (ZPL) are labeled as L1 (382 nm) and L2 (414 nm).

DISCUSSION

An activation energy of ~ 170 meV is typically observed for Mg-doped, p-type GaN. The acceptor level is likely due to Mg atoms residing on Ga substitutional lattice sites. The reduction of the activation energy with increasing annealing temperature is consistent with increasing acceptor concentrations and impurity band conduction. A hole activation energy of 0.16 eV has been determined for MBE-grown, Mg-doped GaN by Hall effect measurements [9], and 0.15 - 0.16 eV were determined from the temperature dependence of donor-acceptor pair recombination of MOCVD-grown, Mg-doped, p-type GaN observed by PL [10]]. The Hall effect data presented in Fig. 1 are consistent with the picture of an increasing number of acceptors being generated by the thermal treatments as opposed to the removal of compensating donors. After growth, only a fraction of Mg atoms acts as acceptors to compensate the presence of native or impurity related donors. A combination of acceptors and donors is an energetically favorable condition for crystal formation in wide bandgap semiconductors under thermodynamic equilibrium [5]. In Mg-doped GaN compensation is enabled by the presence of H^+ during growth and the formation of Mg-H complexes during cool-down. As a consequence, as-grown, Mg-doped GaN is semi-insulating. The presence of high donor concentrations ($3x10^{18}$ cm^{-3}) is consistent with low formation energies for native donors and donor impurities in semi-insulating material as compared to n-type material. Furthermore, the appearance of impurity band conduction in our Mg-doped, p-type GaN material after annealing at 775°C (Fig. 1a) clearly indicates that the activation process of Mg acceptors is due to the generation Mg-related acceptor states in the bandgap of GaN and therefore strongly supports the concept of Mg-H complexes.

The shift in the position of the PL line series has previously been used to suggest that the Mg-related acceptor level is located 250 meV above the valence band edge [11]. However, our data show no evidence that the set of PL lines with a ZPL at 382 nm (L1) is directly related to Mg or that its "red-shift" after annealing at 500°C is related to the activation of Mg acceptors (Fig. 3). First, the red-shift of 0.25 eV is too large to be accounted for by the generation of Mg states into the bandgap unless large lattice relaxation is associated with the dissociation of Mg-H complexes. Second, the red-shift occurs after annealing at 500°C [8], which appears to be grossly out of proportion to the negligible amount of electrical activation of acceptors occurring at that temperature. Third, remote plasma hydrogenation at 600°C renders the Mg-doped GaN semi-insulating, however, the PL band centered at 435 nm does not "blue-shift" to its original position with the ZPL at 382 nm (Fig. 3). To identify the origin of these PL lines will require further investigations. The complex pattern of defect generation and removal revealed by PL strongly indicates that acceptor activation involves more that simply the dissociation of Mg-H complexes, although this process is clearly implicated.

SUMMARY

Experimental results from variable temperature Hall effect measurements and PL obtained for the activation of acceptors in Mg-doped GaN during post-growth thermal anneals in the temperature range between 500°C and 850°C suggest that hydrogen passivates acceptors by the formation of electrically inactive acceptor-hydrogen complexes. A "red-shift" (0.25 eV) of a set of PL lines with a ZPL at 385 nm to a broad band centered at 435 nm which was thought to be indicative of the activation of acceptors in Mg-doped GaN is shown to occur at an annealing temperature of 500°C and, therefore, is likely to be unrelated to the p-type conductivity. This finding is supported by the observation that after remote-plasma hydrogenation which renders Mg-doped, p-type GaN semi-insulating the position of the PL band at 435 nm remains unchanged.

ACKNOWLEDGEMENT

The authors are pleased to thank C.G. Van de Walle and J. Neugebauer for helpful discussions. The work reported was supported by ARPA, agreement # MDA972-95-3-008.

REFERENCES

1. W. Götz, N.M. Johnson, D. P. Bour, C. Chen, H. Liu, C. Kuo, and W. Imler, Mater. Res. Soc. Symp. Proc. **395** (1996)
2. H. Amano, M. Kito, K. Hiramatsu, and I. Akasaki, Jpn. J. Appl. Phys. **28**, L2112 (1989)
3. S. Nakamura, T. Mukai, and M. Senoh and N. Iwasa, Jpn. J. Appl. Phys. **31**, L139 (1992)
4. J. Neugebauer and C.G. Van de Walle, Appl. Phys. Lett. **68**, (1996)
5. J.A. van Vechten, J.D. Zook, R.D. Hornig, and B. Goldenberg, Jpn. J. Appl. Phys. **31**, 3662 (1992)
6. M.S. Brandt, J.W. Ager III, W. Götz, N.M. Johnson, J.S. Harris, R.J. Molnar, and T.D. Moustakas, Phys. Rev. B **49**, 14758 (1994)
7. W. Götz, N.M. Johnson, J. Walker, D.P. Bour, H. Amano and I. Akasaki, Appl. Phys. Lett. **67**, 2666 (1995)
8. W. Götz, N.M. Johnson, J. Walker, D.P. Bour, and R.A. Street, Appl. Phys. Lett. **68**, 667 (1996)
9. R.J. Molnar and T.D. Moustakas, Bull. Am. Phys. Soc. **38**, 445 (1993)
10. I. Akasaki, H. Amano, M. Kito, and K. Hiramatsu, J. Lumin. **48 & 49**, 666 (1991)
11. S.S. Liu, T.R. Cass, and D.A. Stevenson, J. Electron. Mater. **6**, 237 (1977)

ELECTRICAL CHARACTERIZATION OF MAGNESIUM-DOPED GALLIUM NITRIDE GROWN BY METALORGANIC VAPOR PHASE EPITAXY

J. W. HUANG*, H. LU**, J. G. CEDERBERG*, I. BHAT**, T. F. KUECH*
*Chemical Engineering Department, University of Wisconsin, Madison, WI 53706
**Electrical, Computer and Systems Engineering Department, Rensselaer Polytechnic Institute, Troy, NY 12180

ABSTRACT

We have applied frequency-dependent capacitance measurements and admittance spectroscopy on metalorganic vapor phase epitaxy GaN:Mg to study the electronic states associated with Mg doping. Samples with different Mg doping levels were grown and annealed in nitrogen. Lateral dot-and-ring Schottky diodes using Au/Ti were fabricated. After a 800 °C anneal, frequency-dependent measurements show that the capacitance is reduced at a higher frequency, most likely due to the inability of a deep center to maintain an equilibrium ionization state under a high frequency modulation. The net ionized acceptor concentrations was found to be greater at a higher Mg doping level. Admittance spectroscopy, in which the conductance is monitored as a function of temperature, verifies the existence of at least one impurity-related acceptor level with an activation energy of ~ 140 meV. A reduction in the annealing temperature was found to lead to a lower net ionized acceptor concentration, as well as a higher activation energy.

INTRODUCTION

The successful p-type doping in metalorganic vapor phase epitaxy (MOVPE) GaN using Mg has allowed the realization of blue light emitting p-n junction diodes. Treatments, such as electron beam irradiation [1] or thermal annealing [2], are usually required to activate the p-type conduction. Electrical characteristics of these Mg-related acceptors are, however, not well understood. The Mg acceptor binding energy has been reported to be 250 and ~ 160 meV from the donor-acceptor pair transition and from temperature-dependent photoluminescence (PL) experiments respectively [3]. The temperature-dependent Hall measurements revealed Mg-related activation energies of 125 and 157 meV for two samples from the same growth run but annealed for 60 and 25 seconds respectively [4]. The 60-second annealed sample had more electrically active acceptors. In this paper, we report the electrical characterization of GaN:Mg, using frequency-dependent capacitance measurements and admittance spectroscopy, in order to study the electronic states associated with Mg doping. Both the effects of Mg doping levels and thermal annealing temperatures were investigated. It is well known that deep traps greatly affect the free carrier profiles, as determined by capacitance-voltage (C-V) measurements [5]. Given the aforementioned energy level depths associated with Mg, as well as the fact that Mg-doping is responsible for most or all of the observed p-type conductivity, the small signal capacitance of Mg-doped GaN is expected to be dependent on measurement frequency. We had also found that deep level transient spectroscopy (DLTS) could not be successfully used to characterize these Mg-related deep levels, principally due to the low diode capacitance level at low measurement temperatures. In contrast, admittance spectroscopy is better-suited for the study of these relatively shallow and/or fast levels [6].

Mat. Res. Soc. Symp. Proc. Vol. 423 © 1996 Materials Research Society

EXPERIMENTS

Mg-doped GaN samples were grown in a horizontal MOVPE reactor at 100 Torr using trimethyl gallium (TMGa), bis-cyclopentadienyl magnesium (Cp_2Mg), and ammonia (NH_3) on (0001) sapphire substrates [7]. The flow rates of TMGa and NH_3 were 16 μmole/min and 2 standard liters/min (slpm) respectively. The carrier gas was H_2, and the total gas flow rate was 5 slpm. A thin, undoped buffer GaN layer was grown first at 600 °C, followed by the growth at 1050 °C until the thickness reached ~ 0.8 μm. A Mg-doped layer, ~ 2 μm in thickness, was then grown at 1050 °C at a growth rate of ~ 2 μm/hour. Samples with three different Cp_2Mg flow rates at 0.72 (sample A), 0.6 (sample B), and 0.84 (sample C) μmole/min were grown separately. The undoped GaN layer grown at 1050 °C is n-type with a free carrier concentration at ~ $1x10^{17}$ cm^{-3}. After growth, all samples were annealed at 800 °C in a 600 Torr N_2 ambient for 30 min. Additional pieces of sample C were also annealed at 500, 600, 700, and 900 °C. Lateral dot-and-ring Schottky diodes using Ti/Au were fabricated. The dot electrode has a diameter of 560 μm, surrounding by a ring electrode with 40 μm gap. The area of the ring electrode was greater than 10 times that of the dot. C-V and frequency-dependent capacitance measurements were performed at room temperature, in the dark, using a Hewlett-Packard 4284A LCR meter in series mode with an AC modulation level at 15 mV and a frequency, f, ranging from 100 Hz to 1 MHz. In admittance spectroscopy, the conductance (G) as well as the susceptance ($j2\pi fC = j\omega C$) were monitored over the temperature range of 80 - 400 K.

RESULTS

Effect of Mg Doping Levels

We have measured the sample capacitance at a zero DC bias voltage as a function of frequency, and found that the capacitance value was strongly frequency-dependent. The results of samples A and B were shown in fig. 1. The observed variation in capacitance could be due to either the high resistivity of the film or the well-known dispersion effect which occurs when a deep level is unable to follow the high frequency voltage modulation and contribute to the net space-charge in the depletion region [8]. The former effect is relatively unimportant since the capacitance values in fig. 1 were obtained from the series mode measurement of the LCR meter, in which the capacitance is determined independently of any series resistance. The low frequency capacitance should be dependent on the sum of deep and shallow impurities, while the high frequency capacitance is given by the free carrier density [9]. We have performed C-V measurements at

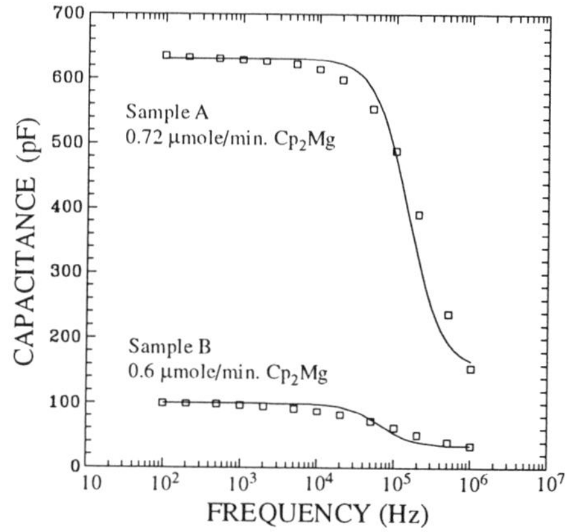

Fig. 1 Measured (□) and predicted (-) zero DC bias capacitance as a function of measurement frequency for samples A and B. The experimental values were obtained using a 15 mV modulation voltage.

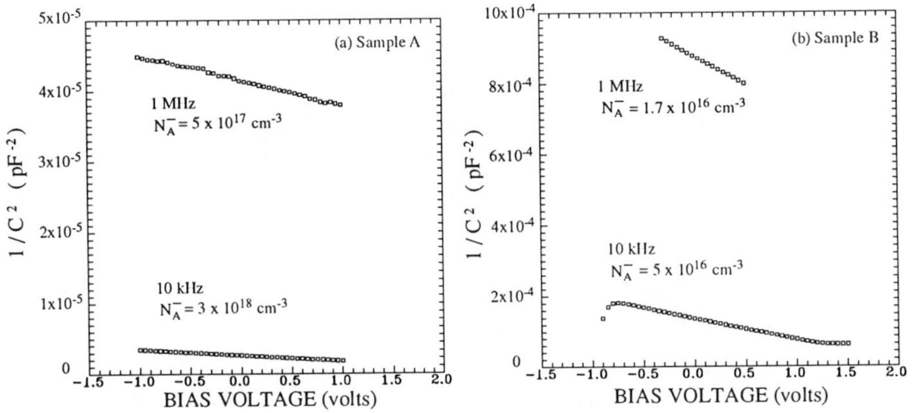

Fig. 2 The room temperature capacitance-voltage characteristics of Ti/Au/GaN:Mg Schottky contact for (a) sample A, and (b) sample B of fig. 1.

high (1 MHz) and low (10 kHz) frequencies to determine the net ionized impurity concentrations (N_a-N_d) as shown in fig. 2 for both samples in fig. 1. N_a-N_d for sample C was found to be 1×10^{19} cm^{-3} at 10 kHz, indicating an increasing N_a-N_d with Mg doping level.

Samples A, B, and C were also analyzed using admittance spectroscopy [10, 11] in order to investigate the deep levels associated with Mg-doping. The admittance as a function of the measurement frequency, ω, and temperature, T, is the sum of G and $j\omega C$, where [10]

$$G(\omega, T) = \frac{\omega_t (C_o - C_\infty)}{1 + (\frac{\omega_t}{\omega})^2} \quad (1) \qquad \text{and} \qquad C(\omega, T) = C_\infty + \frac{C_o - C_\infty}{1 + (\frac{\omega}{\omega_t})^2} \quad (2)$$

In eqns. (1) and (2), C_o and C_∞ are low and high frequency capacitance values respectively. The characteristic frequency, ω_t, is proportional to the emission rate (e_p) of free carriers from the deep level to a band edge state. At a fixed ω, a peak in G(ω, T) as a function of T is found to occur at $\omega_t = \omega$, where C(ω, T) is at the midpoint of a step change [11]. Since

$$\omega_t = \omega = 2\pi f \quad \propto \quad e_p = c_p N_V \exp(-\Delta E / kT) \qquad (3)$$

where c_p is the hole capture coefficient of the deep level, N_V is the valence band density of states, and ΔE is the depth of the deep level from the valence band edge. ΔE can be determined from the slope of log (T^2/f) vs. 1/T derived from measurements at several different ω to acquire the corresponding temperatures where G is maximized at a given ω. The admittance spectra of samples A and B are shown in fig. 3 at 10 kHz, with insets showing the determination of ΔE. The spectrum of sample C is very similar to that of sample A with ΔE being about the same. For sample A, only one deep state was observed with ΔE at 136 meV, and the characteristic frequency associated with Mg acceptor at room temperature (294 K) was determined to be ~ 148.4 kHz from the log (T^2/f) vs. 1/T relationship. By using eqn. (2), the frequency dependence

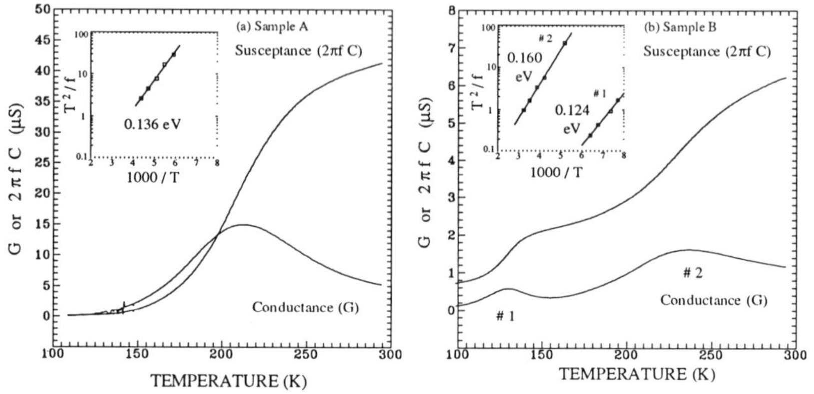

Fig. 3 Measured conductance and susceptance as a function of temperature for (a) sample A, and (b) sample B at 10 kHz, 15 mV modulation, and a zero DC bias voltage. The inset graph shows the determination of the activation energy.

of capacitance at 294 K can be predicted and found to be in reasonably good agreement with the measured data as shown in fig. 1. The 136 meV energy level is very close to the reported Mg acceptor binding energy from Hall measurement or PL [3, 4], and would contribute to the low net ionized impurity concentration at $5x10^{17}$ cm^{-3} as determined by C-V at 1 MHz (fig. 2(a)). In this high frequency case, not all Mg acceptors can follow the fast voltage modulation used in the C-V measurement and these Mg acceptors do not contribute to the measured change in the net space charge. The total Mg concentration can be calculated from the charge neutrality equation given by

$$N_A^- = \frac{N_A}{1 + \frac{g\,p}{N_V}\exp(\Delta E / kT)} = p \qquad (4)$$

A total Mg concentration of $N_A = 6.67x10^{18}$ cm^{-3} was determined by assuming $p = 5x10^{17}$ cm^{-3} as measured in fig. 2(a), $N_V = 1.74x10^{19}$ cm^{-3} [4], $\Delta E = 136$ meV, T = 294 K, degeneracy factor g = 2 [4], and neglecting the existence of any other additional shallow impurities. This N_A is about a factor of 2 greater than the $3x10^{18}$ cm^{-3} as determined by C-V at 10 kHz (fig. 2(a)).

At the lower Mg doping concentration, two deep levels appeared in sample B at 160 and 124 meV. The characteristic frequency associated with the 160 meV level at 294 K is ~ 64.7 kHz from the log (T^2/f) vs. 1/T relationship (fig. 3(b)). The agreement between the predicted and the measured 294 K frequency dependence of the capacitance, as shown in fig. 1, is quite good through the use of this single characteristic frequency in eqn. (2). An extension of eqn. (2) to include both deep levels should lead to an even better agreement [10]. N_A, calculated through eqn. (4) using $\Delta E = 160$ meV and $p = 1.7x10^{16}$ cm^{-3} from fig. 2(b) at a frequency of 1 MHz, is $3.5x10^{16}$ cm^{-3}, which compares favorably with the measured 10 kHz value of $5x10^{16}$ cm^{-3}. The calculated values of N_A in samples A and B are in relatively good agreement with experimental values in light of the simple model used to derive these quantities.

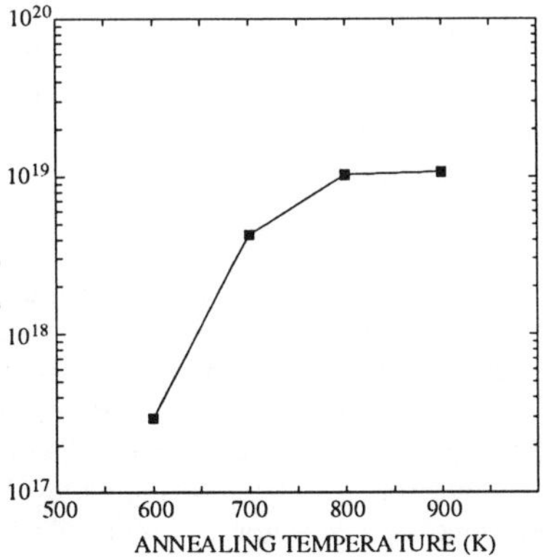

Fig. 4 Measured C-V N_a-N_d of sample C at 1 kHz as function of annealing temperatures.

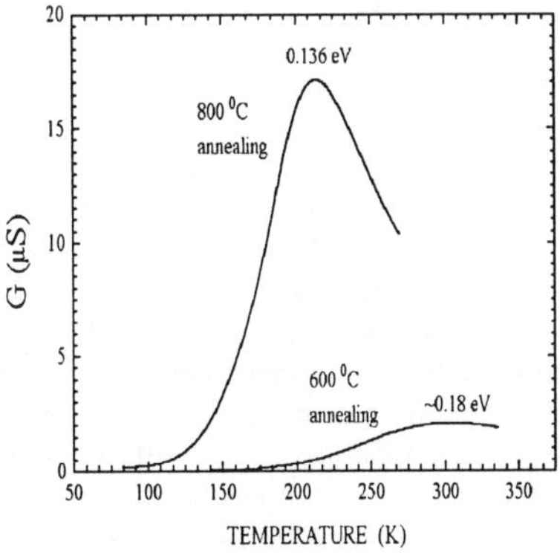

Fig. 5 Admittance spectroscopy spectra of sample C annealed at 600 vs. 800 °C.

Effect of Thermal Annealing Temperatures

We have performed C-V measurements on several pieces of sample C with annealing temperatures ranging from 500 to 900 °C. A temperature of ~ 800 °C or higher was found to be needed to fully activate the Mg acceptors, as shown in fig. 4. On the other hand, no electrical activity was observed for temperatures lower than 600 °C. Admittance spectroscopy spectra of these samples (fig. 5) indicate that a higher annealing temperature reduces the activation energy of Mg-related acceptors, leading to a greater N_a-N_d. These observations are consistent with the reported results using temperature-dependent Hall measurements [4].

DISCUSSION

The energy levels of samples A, B, and C, derived from our admittance spectroscopy measurements, appear to be in close agreement to the results of a Hall measurement study by Tanaka et al [4]. While only one peak was observed in the spectra of samples A and C, two spectral features were present in the spectrum of sample B. We have performed secondary ion mass spectroscopy (SIMS) analysis on samples A and B. Due to the lack of a GaN:Mg standard, only relative counts were available for the concentration of Mg. Sample A has about a factor of 6 greater Mg concentration than sample B. This observation is quite different from the ratio of electrically active species in these two samples of $3x10^{18}/5x10^{16} = 60$ as determined by C-V measurements at low frequency (fig. 2 with f = 10 kHz). This result suggests that there are more electrically active Mg acceptors in sample A than in B. It has been proposed that hydrogen passivation is related to the hole compensation mechanism [12]. In previous studies, the amount of hydrogen was found to increase linearly with the Mg concentration in the as-grown MOVPE GaN:Mg samples [13]. After annealing, the hydrogen density would decrease in general, but the exact relationship between the amount of hydrogen extraction, the Mg doping level, and the annealing time is not presently clear [14]. A recent Raman and infrared absorption spectroscopic study by Brandt et al. [15] has demonstrated the existence of four local vibrational modes for samples of GaN:Mg possessing a high concentration of hydrogen. One pair of the modes, with room temperature frequencies of 2168 and 2219 cm^{-1}, is assigned to Mg-H complexes in the c-plane and parallel to the c axis. A reduced hydrogen content may play a role in the presence of the two Mg-related deep levels, as well as the lower level of electrically active Mg acceptor in our sample B.

CONCLUSIONS

Frequency-dependent capacitance measurements and admittance spectroscopy were applied on GaN:Mg to study the electronic states associated with Mg doping. Both the effects of Mg doping levels and thermal annealing temperatures were investigated. Frequency-dependent measurements show that the capacitance is reduced at a higher frequency, due to the inability of the deep centers to maintain an equilibrium ionization state under a high frequency modulation. The net ionized acceptor concentrations was found to be greater at a higher Mg doping level. Admittance spectroscopy verifies the existence of at least one deep acceptor level in GaN:Mg with an activation energy ~ 140 meV. A reduction in the annealing temperature was found to lead to a lower net ionized acceptor concentration, as well as a higher activation energy. Admittance spectroscopy has also been shown to be applicable for the direct observation of 'shallow' levels in GaN and other wide bandgap materials.

ACKNOWLEDGMENTS

The authors would like to thank K. L. Bray for several useful discussions. The financial support of this work at the University of Wisconsin was provided by the Army Research Office and the National Science Foundation. The work at Rensselaer Polytechnic Institute was partially supported by Philips Laboratories, Briarcliff Manor, NY, and by Texas Instruments.

REFERENCES

1. H. Amano, M. Kito, K. Hiramatsu, and I. Akasaki, Jpn. J. Appl. Phys. **28**, L2112 (1989).
2. S. Nakamura, T. Mukai, M. Senoh, and N. Iwasa, Jpn. J. Appl. Phys. **31**, L139 (1992).
3. S. Fischer, C. Wetzel, E. E. Haller, and B. K. Meyer, Appl. Phys. Lett. **67**, 1298 (1995).
4. T. Tanaka, A. Watanabe, H. Amano, Y. Kobayashi, I. Akasaki, S. Yamazaki, and M. Koike, Appl. Phys. Lett. **65**, 593 (1994).
5. L. C. Kimerling, J. Appl. Phys. **45**, 1839 (1974).
6. D. L. Losee, J. Appl. Phys. **46**, 2204 (1975).
7. H. Lu and I. Bhat, to be published in Mat. Res. Soc. Symp. Proc. vol. 395.
8. G. H. Glover, Solid-State Electron. **16**, 973 (1973).
9. E. Schibli and A. G. Milnes, Solid-State Electron. **11**, 323 (1968).
10. W. G. Oldham and S. S. Naik, Solid-State Electron. **15**, 1085 (1972).
11. H. C. Casey, A. Y. Cho, D. V. Lang, E. H. Nicollian, and P. W. Foy, J. Appl. Phys. **50**, 3484 (1979).
12. S. Nakamura, N. Iwasa, M. Senoh, and T. Mukai, Jpn. J. Appl. Phys. **31**, 1258 (1992).
13. Y. Ohba and A. Hatano, Jpn. J. Appl. Phys. **33**, L1367 (1994).
14. W. Götz, N. M. Johnson, J. Walker, D. P. Bour, and R. A. Street, Appl. Phys. Lett. **68**, 667 (1996).
15. M. S. Brandt, J. W. Ager, W. Götz, N. M. Johnson, J. S. Harris, R. J. Molnar, and T. D. Moustakas, Phys. Rev. B **49**, 14758 (1994).

THE INFLUENCE OF HYDROGEN PLASMA PASSIVATION ON ELECTRICAL AND OPTICAL PROPERTIES OF AlGaN SAMPLES GROWN ON SAPPHIRE

A.Y. POLYAKOV, M. SHIN, S.J. PEARTON*, M. SKOWRONSKI, D.W. GREVE**, J.A. FREITAS***
Department of Materials Science & Engineering, Carnegie Mellon University, Pittsburgh, PA 15213-3890, polyakov@andrew.cmu.edu
* Department of Materials Science and Engineering, University of Florida, 132 Rhines Hall, Gainesville, Florida, FL 32611-6400
** Department of Electronic and Computer Engineering, Carnegie Mellon University, Pittsburgh, PA 15213-3890
*** Sach/Freeman Assoc. Inc. Landover, MD 20785, Naval Research Lab.: Contract # N00014-93-C-2227

ABSTRACT

Hydrogen plasma passivation effects are studied for undoped AlGaN layers grown by MOCVD. Hydrogen treatment at 200°C for 1h led to a substantial decrease in carrier concentration accompanied by an increase in electron mobility. The magnitude of the near-band edge absorption tails also dramatically decreased after hydrogen passivation. The effect is explained by pairing of negatively charged hydrogen acceptors with positively charged native donors commonly believed to be related to nitrogen vacancies. The bond strength in such hydrogen complexes increases as the composition of AlGaN moves towards AlN, as revealed by results of post hydrogenation annealing.

INTRODUCTION

The material system of AlGaN/GaN/InGaN has some very interesting properties, such as wide bandgap, high saturation velocity, and high thermal conductivity, which make it very attractive for applications in optoelectronics and high frequency/high power/high temperature electronics. Nowadays most of the groups are growing AlGaN films by metalorganic chemical vapor deposition (MOCVD), predominantly on lattice mismatched sapphire substrates. Due to some relatively recent advances in growth involving deposition of low temperature AlN or GaN thin buffer layers the crystalline and electronic quality of GaN films grown by MOCVD has remarkably improved. The electron concentration decreased from 10^{20} cm^{-3} to about 10^{16} cm^{-3} or below, the electron mobilities increased from 10 cm^2/V·s to about 600-800 cm^2/V·s (a comprehensive review and relevant references can be found in Gaskill et al.[1]). Improvement of the quality of AlGaN layers has also been achieved, and several groups reported having been able to grow films with low carrier concentrations of about 10^{17} cm^{-3} [2,3] (or even semi-insulating [4]) and respectable mobilities on the order of 100 cm^2/V·s. However, frequently undoped $Al_xGa_{1-x}N$ films with x<0.4 show electron concentrations in the 10^{18}-10^{19} cm^{-3} range and mobilities on the order of 10cm^2/V·s [5-10]. In addition, many groups observed very strong tails at optical absorption near the band edge. High carrier concentration and strong band tailing in AlGaN make it difficult to use such layers in many interesting applications and it would be of great interest to find a means to alleviate this problem. Hydrogen passivation (see e.g. [11] and references therein) could be such a means. It is known that hydrogen coming from MOCVD growth ambient can effectively passivate Mg acceptors in GaN (see e.g. [12,13]). Hydrogen treatment has been shown to dramatically decrease carrier concentration in undoped InGaN and InAlN layers [14] and in p-GaN films [15]. However,Brandt *et al* [15] did not report any passivation of Si donors or native donors in their n-GaN samples grown by molecular beam epitaxy (MBE). It would be interesting to see if the passivation can be achieved by using a different source of hydrogen (we use direct high frequency plasma exposure instead of a remote microwave plasma source) and also to explore the passivation effect in AlGaN Alloys.

Mat. Res. Soc. Symp. Proc. Vol. 423 © 1996 Materials Research Society

EXPERIMENTAL

$Al_xGa_{1-x}N$ layers with x varying in the whole composition range were grown by low pressure (76 Torr) MOCVD, using trimethylgallium and trimethylaluminum for group III precursors and ammonia as a source of nitrogen. Details of growth will be discussed elsewhere [4]. When grown at 1000°C with growth rate of 1μm/h, $Al_xGa_{1-x}N$ layers had high electron concentration of 10^{18}-10^{19} cm^{-3} for all x-values below 0.5. The carrier concentration was greatly reduced in samples grown at 1050°C with a lower growth rate of 0.7 μm/h. Electrical characterization of the layers before and after hydrogen plasma treatment was performed at room temperature by van der Pauw technique (in some cases, for highly resistive samples, these measurements were extended to higher temperatures of 140°C to estimate the depth of the energy levels for predominant defects). For these measurements indium contacts were soldered to the samples before and after the hydrogen plasma treatment.
Optical transmission measurements before and after plasma treatment were made at room temperature using a UV-vis HP-8452 spectrometer, with a sapphire substrate as a reference. Transmission coefficient was converted to absorption coefficient using the known spectral dependence of reflectivity of GaN and AlN [16,17] and extrapolating between them for solid solutions. Photoluminescence measurements were done at 6K, using excitation from a He-Cd laser. Hydrogen plasma treatment was done in a parallel plate reactor at 30 kHz, at 200°C, for 1h. Annealing of the samples was done in nitrogen, in the temperature range 400-800°C, with 50°C steps in temperature and with 10 min annealing at each step.

RESULTS AND DISCUSSION

The effect of hydrogen plasma treatment on electron concentrations and mobilities of AlGaN samples is shown in Table 1. Several features can be observed. First, hydrogen plasma treatment does decrease the carrier concentrations significantly, and since this decrease in carrier concentration is accompanied by an increase in electron mobility, the effect should be due to true passivation rather than compensation. Secondly, the hydrogen passivation efficiency for AlGaN appears to be higher than for pure GaN. Thirdly, there is hardly any change in electron concentration and mobility for the closely compensated GaN sample with initial concentration of $8 \cdot 10^{14}$ cm^{-3} (sample #3 in Table 1). At the same time, for AlGaN changes in carrier concentration still occur even when the starting concentration is very low. One possible explanation would be that passivation of native donors in AlGaN proceeds via pairing with negatively charged hydrogen ions [18] and that the efficiency of passivation drops down rapidly as the Fermi level crosses the hydrogen acceptor level. That would place the H^- level in GaN somewhere above E_C-0.2 eV, and much deeper than that in AlGaN (for example in $Al_{0.12}Ga_{0.88}N$ it should be close to E_C-0.5 eV). Measurements of the temperature dependence of carrier concentration in high resistivity AlGaN samples after hydrogen treatment yield the same activation energies as before treatment (about 0.3 eV in $Al_{0.12}Ga_{0.88}N$ as opposed to 0.27 eV before treatment; 0.22 eV in $Al_{0.67}Ga_{0.33}N$ both before and after treatment) which indicates that no deeper compensating centers have been introduced and the decreased carrier concentration is a result of passivation of the existing electrically active centers. Our results seem to contradict the observations of Bradt *et al* [15]. The reason for such a difference could be attributed to the difference in passivation efficiency when using direct plasma as opposed to remote microwave plasma sources.

TABLE 1

Electrical properties of AlGaN layers before and after hydrogen plasma passivation

Sample number	Al content	Before treatment n, cm^{-3}	Before treatment μ, $cm^2/V{\cdot}s$	After treatment n, cm^{-3}	After treatment μ, $cm^2/V{\cdot}s$
1	0	8×10^{17}	250	1.6×10^{17}	303
2	0	1.3×10^{18}	9.9	2.5×10^{16}	182
3	0	8×10^{14}	43	6.4×10^{14}	47.2
4	0.12	1.5×10^{18}	2.1	3.3×10^{10}	115.4
5	0.12	1.6×10^{11}	110.8	1.6×10^{10}	115.8
6	0.2	1.1×10^{19}	2.2	7.8×10^{17}	32.4
7	0.31	3.6×10^{18}	5.2	2.3×10^{17}	18.2
8	0.38	1.1×10^{19}	1.9	2.9×10^{18}	5
9	0.67	1.9×10^{14}	19.1	2.9×10^{13}	59.6

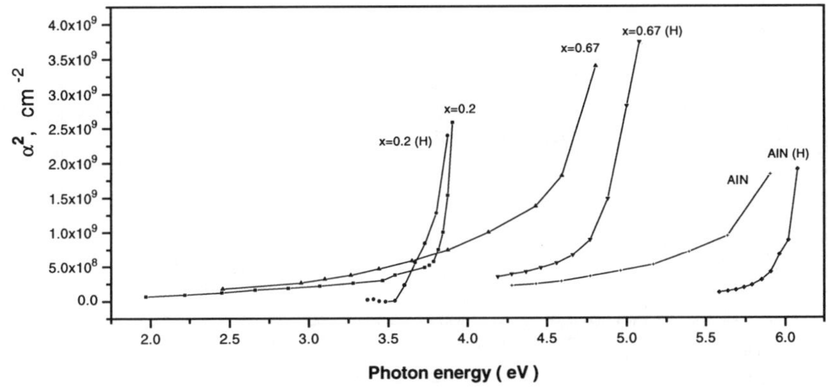

Fig. 1. Dependence of squar of absorption coefficient (α^2) on photon energy before and after hydrogen plasma treatment of AlGaN samples with various Al content. Spectra after hydrogen treatment are marked with H in brackets. Compositions are shown near each spectrum

The effect of hydrogen plasma passivation on the absorption spectra near the fundamental absorption edge in AlGaN and AlN samples are shown in Fig1. It is apparent that before passivation the magnitude of the band tailing (manifested in deviation of the squared absorption coefficient versus photon energy from a straight line) is quite considerable in all the samples measured. This band tailing is strongly suppressed after hydrogen plasma treatment. One could argue that the origin of the band tails is related to fluctuations of local electric fields due to fluctuations of the density of charged defects. As the concentration of such defects is diminished after hydrogen plasma exposure one could expect to also see the decrease in the magnitude of the near band edge absorption, which is exactly the case. For AlN no measurements of electrical properties could be performed because of the very high resistivity of the layers. The near band edge absorption observed in that case could come from the local fields associated with deep levels, and suppression of such absorption could be related to the suppression of electrical activity of these deep centers.

Changes observed in PL spectra of AlGaN samples after hydrogen treatment depend on initial density of deep centers . In GaN sample #1 (see Table 1) only the near band edge luminescence at 3.47 eV was detected and the intensity of this band stayed practically unchanged after hydrogen passivation. In GaN sample #2 the intensity of this 3.47 eV band was lower by about order of magnitude, and in addition to that band two defect realated bands at 2.3 eV and 3.1 eV were observed (ratio of respective intensities was 12:2:2). After exposure to hydrogen plasma the intensity of the 2.3 eV band decreased by about 4 times and the intensity of the band edge PL increased by about a factor of two.

In $Al_{0.12}Ga_{0.88}N$ three defect bands were observed in sample #4 (bands at 3.53 eV, 3.1 eV and 2.3 eV, ratio of intensities 18:5:30) and two defect related bands (3.65 eV and 2.3 eV, ratio 250:5) in sample#5. After hydrogen treatment the 2.3 eV band was strongly suppressed in both samples (corresponding intensities decreased by about 5 times), and the 3.65 eV band intensity decreased by more than two orders of magnitude in sample #5. However, the intensity of the 3.5 eV band increased by about 3 times in sample #4. A strong suppression of the 2.3 eV photoluminescence band was also observed by the authors of [15] in GaN and now it has also been demonstrated to occur in AlGaN.

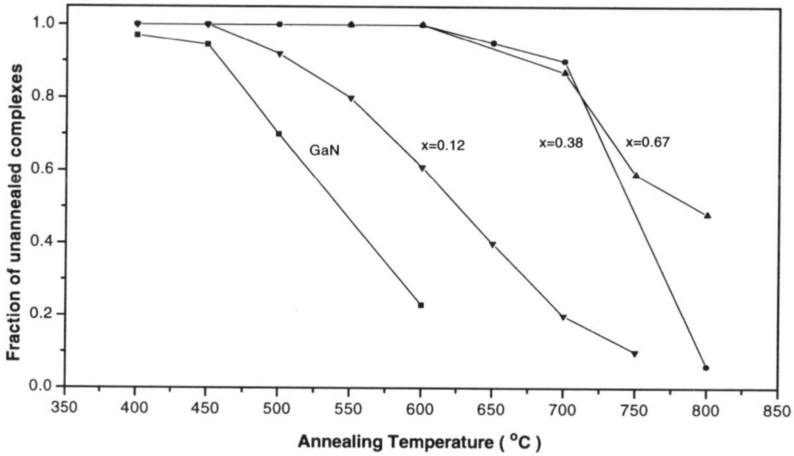

Fig. 2. Dependence of the fraction of unannealed native donor-hydrogen complexes in AlGaN layers.

The thermal stability of native donor-hydrogen complexes was studied by isochronal annealing experiments (Fig 2)[19]. (This fraction of unannealed complexes, f, is defined as $f=(N_{in}-N_T)/(N_{in}-N_{after})$, where N_{in}, N_{after} and N_T are respectively electron concentrations before hydrogen treatment, after hydrogen treatment and after annealing at temperature T.) It can be clearly seen that the thermal stability of such complexes gradually increases with the increase in AlN mole fraction. Such increased thermal stability of hydrogen complexes with native donors could in part be responsible for frequently observed decrease in carrier concentration in as-grown AlGaN layers with high Al mole fractions [2-7].

CONCLUSIONS

Hydrogen plasma passivation was shown to significantly improve electrical and optical properties of undoped AlGaN layers. The effect is most likely due to pairing of native defect donors with negatively charged hydrogen ions. The ioninzation energy of the H^-level seems to increase in AlGaN as the Al content increases along with the bond strength in donor-hydrogen complexes in AlGaN. The temperatures to which the hydrogen passivation effect persists appear to be high enough for the hydrogen passivation to be looked upon as a plausible technique to be used in processing of AlGaN based devices.

ACKNOWLEDGMENTS

The work at CMU was supported in part by AFOSR Grant F49520-95-1-0087.

REFERENCES

[1] D.K. Gaskill, L.B. Rowland and K. Doverspike, Electrical Transport Properties of AlN, GaN and AlGaN, in Properties of Group III Nitrides, ed. J.H. Edgar (INSPEC Publications, London, 1994)pp 101-116

[2] I. Akasaki, H. Amano, Y. Koide, K. Hiramatsu and N. Sawaki, J. Cryst. Growth, **98**, 209 (1989)

[3] X. Zhang, P. Kung, A, Saxler, D. Walker, T.C. Wang and M. Razhegi, Appl. Phys. Lett., **67**, 1745 (1995)

[4] M. Shin, A.Y. Polyakov, M. Skowronski, D.W. Greve, R.G. Wilson, J.A. Freitas, to be published in the Proceedings of the MRS Spring Meeting, 1996.

[5] Koide, H. Itoh, N. Sawaki and I. Akasaki, J. Electrochem. Soc., **133**, 1956 (1986)

[6] Y Koide, H. Itoh, M.R.H. Khan, K. Hiramatsu, N. Sawaki and I. Akasaki, J. Appl. Phys., **61**, 4540 (1987)

[7] Yoshida, S. Misawa and S. Gonda, J.Appl. Phys., **53**, 6844 (1982)

[8] M.A. Khan, A. Bhattarai, J.N. Kuznia and T. Olson, Appl. Phys. Lett., **63**, 1214 (1993)

[9] M.A. Khan, J.N. Kuznia, J.M. Van Hove, N. Pan and J. Carter, Appl. Phys. Lett., **60**, 3027 (1992)

[10] M.A. Khan, Q. Chen, C.J. Sun, J.M. Yang, M. Blasingame, M.S. Shur and H. Park, Appl. Phys. Lett., **68**, 514 (1996)

[11] Hydrogen in Compound Semiconductors, ed. S.J. Pearton (Trans. Tech. Publications, Switzerland, 1994)

[12] S. Nakamura, M. Senoh, T. Mukai, Jpn. J. Appl. Phys., **30**, L1708 (1991)

[13] B. Goldenberg, J.D. Zook, R.J. Ulmer, Appl. Phys. Lett., **62**, 381 (1993)

[14] S.J. Pearton, C.R. Abernathy, P.W. Wisk, W.S. Hobson and F. Ren, Appl. Phys. Lett., **63**, 1143 (1993)

[15] M.S. Brandt, N.M. Johnson, R.J. Molnar, R. Singh and T. Moustakas, Appl. Phys. Lett., **64**, 2264 (1994)

[16] S. Loughlin and R.H. French, Optical Functions of AlN, in Properties of Group III Nitrides, ed. J.H. Edgar (INSPEC Publications, London, 1994) pp 175-189

[17] J.A. Miragliotta, Optical Functions of GaN, ibid., pp 190-194

[18] A.Y. Polyakov, Hydrogen Passivation of Ternary and Quaternary III-V Materials, in Hydrogen in Compound Semiconductors, ed. S.J. Pearton (Trans. Tech Publications, Switzerland, 1994) pp. 295-320

[19] S.J. Pearton, J.W. Corbett and M. Stavola, Hydrogen in Crystalline Semiconductors (Springer-Verlag, Heidelberg, 1992)

THEORETICAL STUDY OF HYDROGEN IN CUBIC GaN

Stefan K. ESTREICHER* and Djordje M. MARIC**
*Physics Department, Texas Tech University, Lubbock, TX 79409 – USA
**Swiss Center for Scientific Computing, 6928 Manno – Switzerland

ABSTRACT

Preliminary results of theoretical studies of hydrogen and hydrogen-related defects in cubic GaN are reported. Our calculations contrast with those of other authors in that the host crystal is represented by molecular clusters rather than periodic supercells, and that they are obtained using an all-electron methodology rather than the single effective-particle approach of density-functional theory. Our results confirm some predictions of other authors but conflict with others.

INTRODUCTION

All GaN crystals contain large concentrations of defects and impurities, as shown by the *n*-type character of as-grown samples and the yellow luminescence they exhibit.[1,2] The former implies that native shallow donors are present, and the latter that deep levels are in the gap. FTIR spectra and SIMS profiles show the presence of defect centers containing H, O, Si, C, and others impurities, in concentrations ranging from 10^{17} to as much as $10^{21}\ cm^{-3}$. Hydrogen or deuterium are also easily introduced in high concentrations by exposure to an ECR plasma, proton implantation, etching with H-containing chemicals, or even by boiling in water.[2–6]

Hydrogen is one of the most common and probably the most versatile impurity in all semiconductors.[7–10] In addition to any surface reactions, H diffuses through the bulk, seeks out stretched or dangling bonds at impurities and defects, where it forms covalent bonds. This results in geometrical changes and shifts the energy levels of the original defect center. Thus, hydrogenation profoundly affects the optical and electrical properties of the material. The optical changes are new or different IR and Raman local modes and/or new photoluminescence lines. The electrical changes may be *passivation* (a level moves from the gap to a band), *activation* (a level moves from a band to the gap), or *'level shifting'* (a level shifts within the gap, e.g. a shallow level becomes deep). The latter situation may be common in wide bandgap materials because energy levels may have to shift a lot to disappear from the gap. These reactions are reversible. Annealing (usually a few hundred degrees Celsius) breaks the bond involving H and restores the sample to its pre-hydrogenation condition. In some cases, the injection of minority charge carriers greatly enhances the reactivation reactions. Note that the temperatures required to reactivate H-passivated centers are lower than those required to force H out of the material.

In GaN, it is believed that H is abundant as an isolated interstitial and in various complexes. In *p*-type GaN, the data suggest that H passivates (or compensates) the shallow Mg acceptor, and has a shallow donor level in the gap.[3,5,6,11,12] The formation of {H,Mg} pairs is consistent with the observation that the amount of incorporated H increases linearly with the concentration of Mg.[13] Further, the temperature dependence of the processes is

Mat. Res. Soc. Symp. Proc. Vol. 423 © 1996 Materials Research Society

similar to that for H-acceptor pairs in other semiconductors.[3,5]

Direct experimental evidence of the formation and structure of H-impurity pairs is lacking. Local vibrational mode (LVM) studies of H-passivated, Mg-doped 2H-GaN have revealed[14] IR and Raman lines at 2168 and 2219 cm^{-1}. They were tentatively assigned to two inequivalent configurations of the {H,Mg} pair. As noted by the authors, these frequencies are typical of Si–H bonds rather than of Mg–H ($\sim$ 1450 cm^{-1}) or N–H ($\sim$ 3400 cm^{-1}) bonds. However, the Si concentration in these samples is too low for Si–H vibrations to account for the IR spectrum, and the LVMs are independent of the doping efficiency.

Reports of H affecting defects in n-type GaN have been published[6] and discussed.[15] The data suggest that H influences defect generation rather than forms H-shallow donors pairs. However, other experimental studies[5,14] have failed to produce evidence of substantial penetration of H into n-type material.

In GaN, the complexes involving H are much more stable than in other semiconductors, and temperatures of the order of 900 $°C$ are needed to expel H from the material.[16] Therefore, H may be technologically more important in GaN than in other semiconductors. However, there is little information on the reactions taking place, and all or almost all of the microscopic details are provided by theory.

THEORETICAL APPROACH

A material such as GaN presents special challenges for theorists.[10] It exists in two structures: cubic (zincblende or 3C) and hexagonal (wurtzite or 2H). The lattice sites are similar, and the properties of substitutional impurities or vacancies should be nearly identical in both polytypes. However, the interstitial sites are different, which could affect the equilibrium sites, diffusion paths, activation energies, and perhaps energy levels of interstitial impurities. Then, the two host atoms have very different covalent radii (1.26 Å for Ga and 0.75 Å for N) and electronegativities (1.8 for Ga and 3.0 for N). The latter results in a large ionic character. It is currently not known how much the contribution of the long-ranged Madelung potential affects the energetics of impurities and defects.

The theoretical techniques that have been used so far to study H-related defect centers in GaN are based on the density-functional (DF) theory, and the host crystal is approximated by periodic supercells. In this description, a long-ranged electrostatic contribution to the energy is included, but it is that of a periodic distribution of defect centers rather than that of one defect in an otherwise perfect crystal.

The most complete predictions for isolated H and {H,Mg} pairs have been published by Neugebauer and Van de Walle.[17] They performed DF calculations in 32-atom supercells of 3C-GaN, with plane-wave basis sets (60 Ry cutoff), with and without $3d$'s on the Ga atoms. In their calculations, the entire cell is allowed to relax fully for every position of the H atom. The results can be summarized as follows. *Isolated* interstitial H has a very large negative-U energy, $\sim$ 2.5 eV, implying that H^0 is unstable and should not be found in GaN. The donor level is very near the CB and the acceptor level quite deep ($\sim E_v + 0.9$ eV). For $E_F < 2$ eV, the stable state is H^+, and it is H^- for E_F above that value. Close to $E_F = 2$ eV, H^+ and H^- may coexist, which should result in the formation of H_2 molecules (however, the authors predict a large formation energy for the molecule). H^+ is at the AB site to N (AB_N) with N–H $\sim$1 Å, but the BC configuration is only one-tenth of an eV higher. The calculated activation energy for diffusion of H^+ is 0.7 eV. H^0 is unstable, but has a rather flat potential energy surface. Its lowest-energy configuration is at the AB site of a Ga atom, close to the T_{Ga} site. Finally, H^- is also at the AB_{Ga} $\sim T_{Ga}$ site, with an activation energy of the order of 3.4 eV. The {H,Mg} pair has H at the AB_N site (Mg···N–H). The calculated N–H stretching frequency is 3360 cm^{-1}, and the pair dissociation energy is 1.5 eV.

Bosin et al.[18] performed DF calculations in 16 and 32 atom supercells of 3C- and 2H-GaN. The plane-wave basis set had a 25 Ry cutoff and the Ga 3*d*'s were included explicitly. Some configurations were optimized only in C_{3v} symmetry. They find that hydrogen incorporation is favored in *p*-type GaN (the H-acceptor pair formation is always exothermic) as well as in *n*-type material. Their calculations also predict that H should have negative-U properties, but with a smaller U ($\sim$ 1.2 eV) than predicted by Neugebauer et al.[17] (their calculated gap is only 2.1 eV). The donor and acceptor levels are predicted to be at 2.03 and 0.80 eV, respectively. For E_F near 1.5 eV, the formation energies of all three charge states are about the same. The stable configuration of H^0 is at the BC site, 0.5 eV below the AB_N site. H^+ prefers the AB_N site, while H^- is at the AB_{Ga} site. The predictions regarding the {H,Mg} pair agree with those of Neugebauer et al.[17] in that its formation is always energetically favorable, and the H-stretching frequency is high. In 2H-GaN, the lowest-energy configuration has H at the AB_N site with C_s symmetry (not along the *c*-axis), with a H stretching vibration at 2939 cm^{-1}.

Okamoto et al.[19] also performed DF calculations in 16 atom supercells of 3C-GaN with a 50 Ry cutoff. The Ga 3*d*'s were included in the pseudopotential. They also find that in the {H,Mg} pair, H forms a strong H bond with N, but with H near a BC site: Mg···H–N. The calculated frequencies in 2H-GaN are 2468 cm^{-1} (along the c axis) and 2440 cm^{-1} (roughly perpendicular to it). These frequencies are surprisingly low considering that the calculated N–H distance is only 1.02 Å, but are very close to the LVMs reported by Brandt.[14] Such low values imply a considerable overlap between Mg and the N–H bond, weakening the latter.

PRESENT RESULTS

We are performing approximate ab-initio Hartree-Fock (HF), ab-initio HF, and DF calculations in molecular clusters of 3C-GaN containing 44 host atoms. Most of these calculations are still under way, and we report here only the results of preliminary calculations at the approximate ab-initio HF level (PRDDO/M[20]). The basis sets are Slater atomic orbitals, with 3*d*'s on Ga, with and without frozen-core potentials. Ab-initio HF and DF calculations, with gradient corrections to the local density approximation, will follow. We also intend to include the appropriate Madelung energy corrections. A summary of the results currently available is as follows.

Interstitial H can be found in three charge sates in 3C-GaN. The calculated lowest-energy configurations are shown in Fig. 1. We did not yet calculate formation energies, and therefore have no prediction regarding the negative-U behavior of H in GaN.

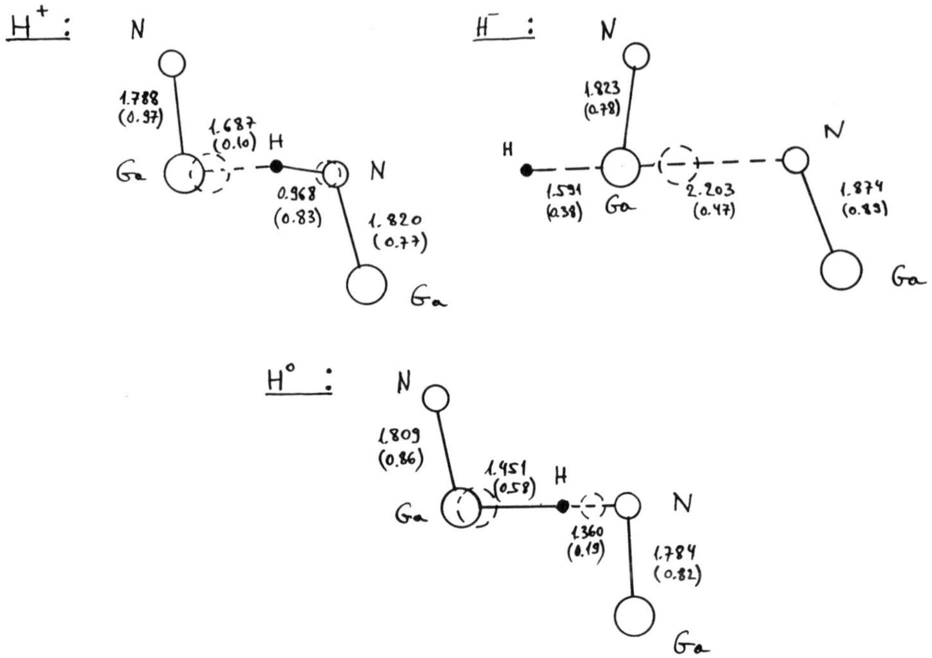

FIGURE 1: Calculated lowest-energy configurations of H^+, H^0, and H^- in 3C-GaN. Note that the configuration with H^0 at the AB_{Ga} site is only slightly higher in energy.

H^+ is bound to N in a BC-like position. Gradient optimizations with no symmetry restrictions place H slightly off the trigonal axis, with a N–H–Ga angle of 178°. The N-H bond length (0.97 Å) and degree of bonding[21] (0.83) indicate a strong N-H bond. The nearest Ga atom moves off its perfect substitutional site almost to the plane of its three N nearest neighbors (nns), where it is 3-fold coordinated. The AB_N site is higher in energy, but also has H strongly bound to N.

The lowest-energy configuration for H^0 is at the BC site, with H bound to Ga and most of the odd electron localized on N. Another configuration is nearly energetically degenerate (+0.04 eV, an insignificant number at the present level of theory). It has H^0 is on the trigonal axis, very near the T_{Ga} site, with only a little overlap with the adjacent Ga atom (Ga–H=1.85 Å, degree of bonding 0.10). There is very little lattice relaxation for this position and charge state of H. H^0 has a donor level in the gap. An overestimate for the activation energy for diffusion of H^0 is 1 eV, with the AB_N site about 0.7 eV above the AB_{Ga} site.

H^- is at the AB_{Ga} site, but a large relaxation of the Ga atom. The Ga–H internuclear distance is 1.59 Å, with degree of bonding 0.38. This covalent overlap comes at the expense of the Ga–N bond, which stretches from 1.89 to 2.20 Å while its degree of bonding drops from 0.75 to 0.47. The activation energy for H^- is about 1.5 eV, with a saddle point near the T_N site.

Figure 2 shows two complexes involving H. The lowest-energy dimer (molecular hydrogen at the T_{Ga} site) and the {H,Mg} pair. Note that dimers should not form in materials containing only H^+ or only H^-, because the long-range Coulomb repulsion will prevent pair formation.

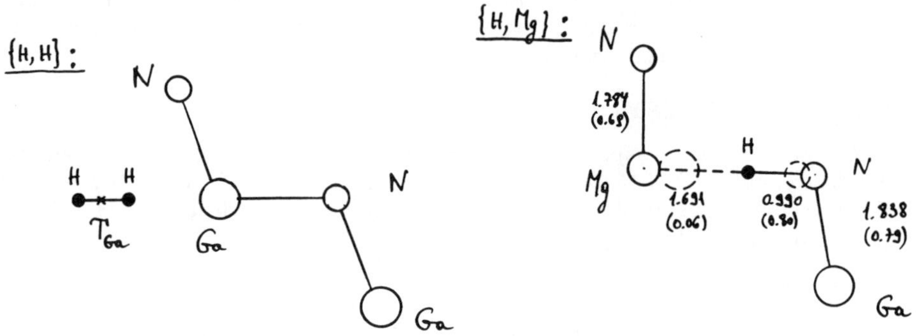

FIGURE 2: Calculated lowest-energy configurations of the H_2 molecule and the {H,Mg} pair.

Five possible dimer states were considered: the molecule at a T site (two possibilities), BC–AB pairs or H_2^*'s (two possibilities), or a complex with both H's near the same BC site but off the trigonal axis, one H bound to N, the other to Ga, in C_1 symmetry. The lowest-energy state is the molecule at the T_{Ga} site, which has no levels in the gap.

The {H,Mg} pair has H in a BC position, strongly bound to N (H–N=0.99 Å, with degree of bonding 0.86). The Mg atom moves all the way to the plane of its three N nns, and shows negligible overlap with H. Although no vibrational frequencies have been calculated at this time, it is clear from the electronic structure that it must be close to an unperturbed N–H one.

DISCUSSION

The theoretical predictions of various authors for the behavior of H in GaN differ in many respects. This is not surprising in view of the theoretical complexities associated with GaN. In particular, nobody yet has correctly included the long-ranged Madelung energy term. The magnitude of the energy correction could be of the order of several tenths of an eV. It may be that this correction could be large and have different signs for H^+ and H^-. It is therefore significant that there is agreement on some properties of H in GaN.

There is **qualitative agreement** on the following points. *(i)* H^- is at the AB_{Ga} site, with a large activation energy for diffusion, *(ii)* H^+ is within 1 Å or so of N, and *(iii)* in the {H,Mg} pair, H is strongly bound to the N atom. Two authors also predict that H has a large negative-U energy.

There is **qualitative disagreement** on the following points. *(i)* H^0 is at the AB_{Ga} site in Ref. 17 and in the present work, but at the BC site in Ref. 18. *(ii)* H^+ is at the AB_N site in Refs. 17,18 but at the BC site, slightly off-center, in the present work. Note that in Ref. 17, the AB_N and BC sites are off by only 0.1 eV. *(iii)* In the {H,Mg} pair, H is at the AB_N site of N in Refs. 17,18 with a very high N–H stretching vibrational frequency, while in Ref. 19, H is at the BC site, with a much lower frequency. In the present work, H is also near the BC site but strongly bound to N, suggesting a high frequency.

As each author is more confident in his/her own results than in those of other theorists, it would be most useful to have microscopic experimental data such as FTIR or Raman spectra, including uniaxial stress studies to analyze the symmetry, and isotope substitutions to determine which species are involved.

ACKNOWLEDGEMENTS

The work of SKE is supported by the grant D-1126 from the R.A. Welch Foundation and the contract XAX-5-15230-01 from the National Renewable Energy Laboratory. Most of the computer time required for the present research was provided by the Swiss Center for Scientific Computing.

REFERENCES

1. For a review of the history and key properties of GaN, see J.I. Pankove, MRS Proc. **162**, 515 (1990).
2. *'GaN and Related Materials'*, ed. S.J. Pearton (Gordon & Breach, New York, in print).
3. J.M. Zavada, R.G. Wilson, C.R. Abernathy, and S.J. Pearton, Appl. Phys. Lett. **64**, 2724 (1994).
4. S.J. Pearton, C.R. Abernathy, and F. Ren, Electrochem. Lett. **30**, 527 (1994).
5. M.S. Brandt, N.M. Johnson, R.J. Molnar, R. Singh, and T.D. Moustakas, Appl. Phys. Lett. **64**, 2264 (1994).
6. B. Molnar, C.R. Eddy, Jr., and K. Doverspike, ECS Proc. **95-21**, 236 (1995).
7. S.K. Estreicher, Mat. Sci. Engr. Reports **14**, 319 (1995).
8. S.J. Pearton, J.W. Corbett, and M.J. Stavola, *'Hydrogen in Crystalline Semiconductors'* (Springer-Verlag, Berlin, 1992).
9. *'Hydrogen in Compound Semiconductors'*, ed. S.J. Pearton, Mat. Sci. Forum **148-149** (Trans Tech, Aedermannsdorf, Switzerland, 1994).
10. See the chapter by S.K. Estreicher and D.E. Boucher in Ref. 2.
11. S. Nakamura, N. Iwasa, M. Senoh, and T. Mukai, Jpn. J. Appl. Phys. **31**, 1258 (1992).
12. J.A. Van Vechten, J.D. Zook, R.D. Horning, and B. Goldenberg, Jpn. J. Appl. Phys. **31**, 3662 (1992).
13. Y. Ohba and A. Hatano, Jpn. J. Appl. Phys. **33**, L1367 (1994).
14. M.S. Brandt, J.W. Ager III, W. Götz, N.M. Johnson, J.S. Harris, Jr., R.J. Molnar, and T.D. Moustakas, Phys. Rev. B **49**, 14758 (1994).
15. S.J. Pearton, private communication.
16. S.J. Pearton, R.J. Shul, R.G. Wilson, F. Ren, J.M. Zavada, C.R. Abernathy, C.B. Vartuli, J.W. Lee, J.R. Mileham, and J.D. Mackenzie, ECS Proc. **95-21**, 178 (1995).
17. J. Neugebauer and C.G. Van de Walle, Phys. Rev. Lett. **75**, 4452 (1995).
18. A. Bosin, V. Fiorentini, and D. Vanderbilt, MRS Proc. **395** (1996) in print.
19. T. Ogino and M. Aoki, Jpn. J. Appl. Phys. **19**, 2395 (1980).
20. See the discussion in Ref. 7 and L. Throckmorton and D.S. Marynick, J. Comp. Chem. **6**, 652 (1985).
21. D.R. Armstrong, P.G. Perkins, and J.J.P. Stewart, J. Chem. Soc. Dalton Trans. (1973), p. 838.

ROLE OF HYDROGEN AND HYDROGEN COMPLEXES IN DOPING OF GaN

JÖRG NEUGEBAUER[1,2] AND CHRIS G. VAN DE WALLE[1]
[1]Xerox Palo Alto Research Center, 3333 Coyote Hill Road, Palo Alto, California 94304
[2]Fritz-Haber-Institut der Max-Planck-Gesellschaft, Faradayweg 4–6, D-14195 Berlin, Germany

ABSTRACT

We have calculated electronic structure, energetics and migration for hydrogen and hydrogen complexes in GaN employing state-of-the-art first-principles calculations. Using these results in combination with previous detailed investigations about native defects we have calculated the concentration of hydrogen and dopants for different growth conditions. Our results reveal a fundamental difference in the behavior of hydrogen in *p*-type and *n*-type material. In particular, we explain why hydrogen has little effect on donor impurities and why H concentrations are low in *n*-type GaN. We discuss why hydrogen is beneficial for acceptor incorporation in GaN, and identify the limitations of this process.

INTRODUCTION

Hydrogen is a well known impurity which plays an important role in semiconductors by compensating or passivating native defects and impurities, saturating dangling bonds or forming complexes. In high-temperature growth techniques [such as metalorganic vapor deposition (MOCVD) and hydride vapor phase epitaxy (HVPE)] which are commonly employed for GaN growth hydrogen is highly abundant. Several experimental investigations indicate that hydrogen plays an important role in the doping of GaN: Nakamura *et al.* demonstrated that Mg-doped and initially semiinsulating GaN can be made *p*-type by thermal annealing in vacuum and nitrogen ambient but not in NH_3 or H ambient.[1] Furthermore, Nakamura *et al.* showed that the activation of the Mg acceptors can be reversed by annealing in a hydrogen ambient,[1] revealing the crucial role played by hydrogen. Based on these observations Van Vechten *et al.* suggested that hydrogen enables *p*-type doping by suppressing compensation by native defects.[2] These authors went on to propose the incorporation (and subsequent removal) of hydrogen as a general method for improving *p*-type as well as *n*-type doping of wide-band-gap semiconductors. The Van Vechten model highlights the important role of hydrogen, but leaves various issues unexplained, such as the lack of hydrogen incorporation in *n*-type GaN, and the success of *p*-type doping (without post-growth treatments) in MBE (molecular-beam epitaxy).

In the present paper we summarize the properties H exhibits in GaN and discuss how H affects doping in GaN. In particular, we explain why H is beneficial for *p*-type doping but has little effect on donor impurities. Based on these results we identify the specific conditions under which H is beneficial and discuss the limitations of this process.

Mat. Res. Soc. Symp. Proc. Vol. 423

METHOD

Employing first-principles density-functional theory we have calculated the total energy surface, electronic structure and atomic geometry for hydrogen in all relevant charge states (H^+, H^0, H^-).[3] 32-atom supercells were used and atomic relaxation is fully taken into account. The calculations were performed for cubic GaN which has a higher symmetry than the wurtzite structure. As shown in Ref. [4] the wurtzite and the cubic phase show nearly equivalent formation energies and electronic structure for defects. We expect therefore that the results are valid both for H in zincblende and wurtzite GaN. The electron-ion interaction is described by soft Troullier-Martins pseudopotentials.[5] Details of the computational approach can be found elsewhere.[6, 7, 8]

The key to describing doping issues is the calculation of the equilibrium concentrations of dopants, impurities and native defects:

$$c = N_{\text{sites}} \exp^{-E^f/k_B T} \tag{1}$$

where N_{sites} is the number of sites the defect or impurity can be built in, k_B the Boltzmann constant, T the temperature, and E^f the formation energy. The formation energy is not a constant but depends on the various growth parameters. To be more specific, the formation energy of a Mg acceptor is determined by the relative abundance of Mg, Ga, and N atoms. If the Mg acceptor is charged, the formation energy depends further on the Fermi level (E_F), which acts as a reservoir for electrons. Forming a substitutional Mg acceptor requires the removal of one Ga atom and the addition of one Mg atom; the formation energy is therefore:

$$E^f(\text{GaN:Mg}_{\text{Ga}}^q) = E_{\text{tot}}(\text{GaN:Mg}_{\text{Ga}}^q) - \mu_{\text{Mg}} + \mu_{\text{Ga}} + qE_F \tag{2}$$

where $E_{\text{tot}}(\text{GaN:Mg}_{\text{Ga}}^q)$ is the total energy derived from a calculation for substitutional Mg, and q is the charge state of the Mg acceptor. Similar expressions apply to the hydrogen impurity, and to the various native defects.

For the following discussion we will fix the chemical potentials. For the Ga chemical potential we assume Ga-rich conditions (which appear to be common in experimental growth conditions). For Mg we are interested in the highest possible concentration. We therefore set the Mg-chemical potential to its upper limit which is given by the formation of Mg_3N_2 ($\mu_{\text{Mg}} = \mu_{\text{Mg}_3\text{N}_2}$); increasing the Mg-chemical potential further makes the system thermodynamically unstable against the formation of Mg_3N_2 droplets or precipitates. Using the same arguments we set the H chemical potential to that of H_2 molecules and the Si chemical potential to that of Si_3N_4. Using these values for the chemical potentials the formation energy is solely a function of the Fermi energy.

ROLE OF H IN DOPING

Before discussing the role of H in *p*-type doping let us briefly focus on the H-free case, i.e., only the acceptor (Mg) and native defects are present. Figure 1(a) shows the corresponding formation energies as a function of the Fermi energy. The dominant native defect under *p*-type conditions is the N vacancy; all other defects are higher in energy. The slope of the formation energies characterizes the charge state; a positive slope (as found for the N vacancy) indicates a positive charge state, corresponding to a donor. For Mg the kink in the formation energy describes a change in the charge state from neutral to negative as characteristic for an acceptor. The position of the kink at $E_F \approx 0.2\,\text{eV}$ gives

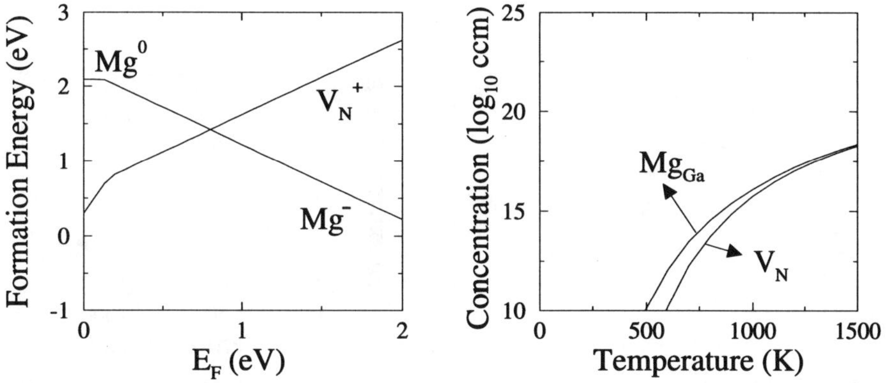

Figure 1: Formation energy vs. Fermi level for the Mg_{Ga} acceptor and the native defects. The corresponding equilibrium concentrations are given in (b). Ga-rich conditions are assumed.

the position of the calculated acceptor level, which is close to the experimental value of 0.16 eV.[9]

Using the calculated formation energies, and taking into account that the Fermi energy is fixed by the condition of charge neutrality, the equilibrium concentration [Eq. (1)] for each defect can be calculated as a function of temperature. The results are shown in Fig. 1(b). As expected, the Mg concentration increases with increasing temperature. However, with increasing temperature the N vacancy concentration also increases; at temperatures exceeding 1000 K the Mg acceptors become increasingly compensated by N vacancies. Native defect compensation is therefore potentially an important concern for high-temperature growth techniques. Low temperature growth techniques such as MBE should suffer less from this problem. This conclusion would be consistent with the fact that only in MBE-grown GaN can *p*-type conductivity be achieved without any post-growth processing.[10, 11]

We will now consider H-rich conditions which are characteristic for many of the high-temperature growth techniques such as MOCVD and HVPE. Figure 2(a) shows that under these conditions H becomes the dominant donor; the formation energy of H is in the considered interval of Fermi energies always lower than the dominant native defect, the N vacancy. The calculated equilibrium concentrations are displayed in Fig. 2(b). The Mg and H concentration are for all temperatures virtually identical indicating that H completely compensates the Mg acceptors. Further, compared to the H-free case the concentration of Mg acceptors is *increased* and the defect concentration is *decreased.* Both effects are crucial to increase doping levels.

What is the mechanism by which H changes the acceptor and defect concentration? In a plot such as Fig. 1(a) and 2(a), the Fermi level position can be roughly estimated to be near the crossing point between the acceptor and the dominant donor species. At this point their formation energies (and hence there concentrations) are equal, ensuring charge neutrality.[12] By going from H-free conditions (only Mg and N-vacancies are present) to H-rich conditions the crossing point shifts to higher Fermi energies (from ≈ 0.8 eV to ≈ 1.2 eV). An increase in the Fermi energy *generally* decreases the formation energy

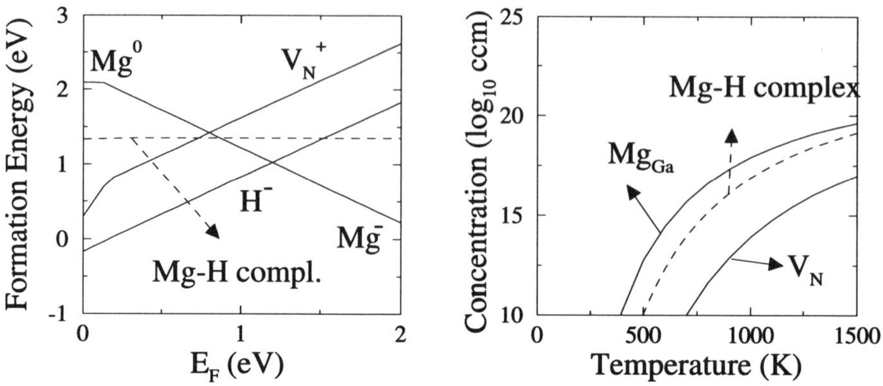

Figure 2: Formation energy vs. Fermi level for the Mg_{Ga} acceptor and the neutral Mg-H complex. Also included are the native defects and interstitial H. The corresponding equilibrium concentrations are given in (b) as a function of the growth temperature. H rich conditions are assumed.

of acceptors and increases the formation energy of donors (defects), thus resulting in a lowered defect concentration and an increased acceptor concentration. We note that this mechanism works only if H is able to significantly shift the Fermi energy which is the case if (i) H is the dominant donor (i.e., its formation energy must be lower than that of all native defects) and (ii) its formation energy must be comparable to that of the dopant impurity (a crossing point must exist in the band gap). It is interesting to note that condition (ii) is not valid for n-type doping with, e.g., Si.[13] The reason is that H under n-type conditions has a higher formation energy than under n-type conditions;[3] the Si donor has for *all* Fermi energies a lower formation energy.

ACTIVATION MECHANISM OF THE DOPANTS

Despite the fact that growing under H-rich conditions improves acceptor and defect concentrations, Fig. 2(a) shows that the Mg acceptors are almost completely compensated by the H impurities. In order to activate the Mg acceptors, post-growth treatments are necessary to eliminate the compensation by H.

The H donors and Mg acceptors can actually form electrically neutral complexes with a binding energy of $\approx 0.7\,\text{eV}$ [see Fig. 2(a)].[3] For the specific choice of chemical potentials made here, this binding energy is low enough for the complexes to be dissociated at the growth temperature; however, the Mg and H will form pairs when the sample is cooled to room temperature, consistent with experimental observations.[14]

The first step in the activation process is the dissociation of the Mg-H complex (see Fig. 3). Our estimated dissociation barrier for the complex is 1.5 eV, calculated by considering a jump to a nearest-neighbor site; the total barrier may be slightly higher.[3] This barrier should be low enough to be overcome at modest annealing temperatures (around 300°C). Experimental results show, however, that activation has to be carried out at much higher temperatures (> 600°C) [1]. The reason is that dissociation alone is insufficient; in

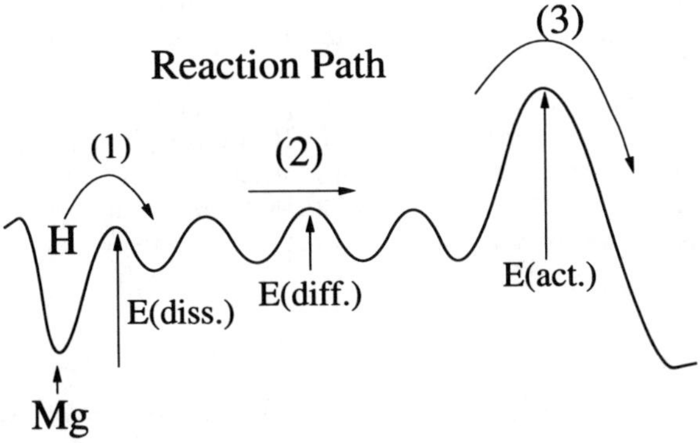

Figure 3: Schematic picture of the reaction path and characteristic energy barriers for the activation of the Mg acceptors. At low temperatures the H forms a neutral complex with the Mg atom. With increasing temperature the Mg-H complex dissociates (1), the positively charged H can then easily migrate through the GaN crystal (2). Increasing further the temperature allows the H to overcome the activation barrier (3) and become electrically inactive.

order to prevent the H from compensating the Mg acceptor it has to be either removed (to the surface or into the substrate) or neutralized (e.g., at an extended defect).

The calculated diffusion barrier for H^+ in GaN is low ($\approx 0.7\,\mathrm{eV}$[3]) indicating that H^+ is highly mobile and can easily migrate to the surface or extended defects. The high temperature necessary to activate the Mg acceptors therefore reflects an activation barrier for eliminating H as compensating center by incorporating it at extended defects (which typically occur in high concentrations in GaN[15]) or by removal of H through desorption at surfaces.

GENERAL CRITERIA FOR HYDROGEN TO ENHANCE DOPING

The mechanism by which hydrogen enhances doping works only under specific conditions. First of all, hydrogen must be the dominant compensating defect (i.e., its formation energy must be lower than that of all native defects, and comparable to the formation energy of the dopant impurity). Second, it must be possible to remove hydrogen from the doped layer after growth; an anneal at high temperature will work only in a specific temperature window. On the one hand, the temperature must be *high* enough to enable all three processes necessary to eliminate H as compensating center: complex dissociation, migration, and overcoming of the activation barrier (see Fig. 3). We call the lowest temperature where this mechanism occurs T_1. On the other hand the temperature must be *low* enough ($< T_2$) to keep the dopant impurity immobile, and to prevent the formation of other compensating defects; in the case of Mg doping of GaN, an anneal at too high a temperature would lead to Mg compensation by N vacancies, as shown in Fig. 1. Thus, a necessary condition for this mechanism to work is $T_1 < T_2$.

This condition is met in *p*-type GaN, judging by the success of thermal annealing procedures.[1] We note that the above analysis is *general* and applies to dopants in any semiconductor. Whether the condition is realized, however, depends on the specific parameters of a system (migration barriers of impurities and defects, activation energies, etc.) and must therefore be addressed independently for each system. It is interesting to note that even for GaN (were the method works well for acceptor doping) it does not work for donor impurities. First of all, the formation energy of H is too high in n-type GaN for it to be incorporated in significant concentrations. And even if H *were* incorporated, the high diffusion barrier of H in n-type GaN ($\approx 3.4\,\mathrm{eV}$ [3]) would render it immobile even at very high annealing temperatures, preventing the H from reaching the zones where it can be neutralized or removed.

ACKNOWLEDGMENTS

This work was supported in part by ARPA under agreement no. MDA972-95-3-0008, and by the DAAD (German Academic Exchange Service). One of the authors (J.N.) acknowledges financial support by the DAAD (German Academic Exchange Service).

References

[1] S. Nakamura, N. Iwasa, M. Senoh, and T. Mukai, Jpn. J. Appl. Phys. **31**, 1258 (1992).

[2] J. A. Van Vechten, J. D. Zook, R. D. Hornig, and B. Goldenberg, Jpn. J. Appl. Phys. **31**, 3662 (1992).

[3] J. Neugebauer and C. G. Van de Walle, Phys. Rev. Lett. **75**, 4452 (1995).

[4] J. Neugebauer and C. G. Van de Walle, Proc. Mater. Res. Soc. Symp. **339**, 687 (1994).

[5] N. Troullier and J. L. Martins, Phys. Rev. B **43**, 1993 (1991).

[6] R. Stumpf and M. Scheffler, Comp. Phys. Commun. **79**, 447 (1994).

[7] J. Neugebauer and C. G. Van de Walle, Phys. Rev. B **50**, 8067 (1994).

[8] J. Neugebauer and C. G. Van de Walle, Proc. Mater. Res. Soc. Symp. **408** (1996).

[9] I. Akasaki, H. Amano, M. Kito, and K. Hiramatsu, J. Lumin. **48&49**, 666 (1991).

[10] R. J. Molnar, T. Lei, and T. D. Moustakas, Proc. Mater. Res. Soc. Symp. **281**, 753 (1993).

[11] M. E. Lin, C. Xue, G. L. Zhou, J. E. Greene, and H. Morkoc, Appl. Phys. Lett. **63**, 932 (1993).

[12] Strictly speaking, this assumption is valid only if we ignore free carriers, a reasonable approximation if the Fermi level is far enough from the valence-band edge.

[13] J. Neugebauer and C. G. Van de Walle, Appl. Phys. Lett. **68**, 1829 (1996).

[14] W. Götz, N. Johnson, J. Walker, D. P. Bour, H. Amano, and I. Akasaki, Appl. Phys. Lett. **67**, 2666 (1995).

[15] S. D. Lester, F. A. Ponce, M. G. Craford, and D. A. Steigerwald, Appl. Phys. Lett. **66**, 1249 (1994).

DIFFUSION OF HYDROGEN IN 6H SILICON CARBIDE

M.K. Linnarsson, J.P. Doyle and B.G. Svensson
Royal Institute of Technology, Solid State Electronics, P.O. Box E229, S-164 40 Kista-Stockholm, Sweden

ABSTRACT

6H polytype silicon carbide (SiC) samples of n-type have been implanted with 50 keV H^+ ions and subsequently annealed at temperatures between 200 °C and 1150 °C. Using depth profiling by secondary ion mass spectrometry motion of hydrogen is observed in the implanted region for temperatures above 700 °C. A diffusion coefficient of $\sim 10^{-14}$ cm^2/s is extracted at 800°C with an approximate activation energy of ~3.5 eV. Hydrogen displays strong interaction with the implantation-induced defects and stable hydrogen-defect complexes are formed. These complexes anneal out at temperatures in excess of 900°C and are tentatively identified as Carbon-Hydrogen centers at a Si vacancy.

INTRODUCTION

SiC was recognized as a candidate material for semiconductor devices operating at high temperatures, powers and frequencies already in the 1950's, for a review see ref.[1]. The application of SiC has, however, been hampered by poor quality of the material; for example, improvements of the bulk growth of SiC are required for elimination of device limiting micropipes[2]. During the last years, considerable progress has been made and today, SiC wafers of reasonable quality are available and can be used as substrates for growth of epitaxial layers by chemical vapour deposition (CVD). If so-called site-competition is utilized in the CVD process incorporation of dopants for both p-type and n-type layers can be well controlled, which results in increased doping range and reproducibility[3].

During CVD growth of SiC films H_2 is normally employed as the carrier gas and hydrogen is anticipated to be incorporated in the SiC film at concentrations exceeding the solubility limit (supersaturation) at the growth temperature. The properties of hydrogen in crystalline semiconductors have been the subject of numerous investigations and for an extensive review, see ref.[4]. Technologically, the principal interest in hydrogen occurs because of its ability to passivate shallow acceptor and donor impurities as well as deep level defects in the two most commonly used semiconductors, Si and GaAs. A similar effect may also be anticipated in SiC, and indeed, this has recently been confirmed by experimental results from different groups[3,5]; in particular, a strong interaction between hydrogen and p-type dopants (B and Al) is revealed.

In order to control the influence of hydrogen, a key issue is the mobility. Because of its pronounced tendency to interact with, and be trapped by, other impurities/defects and to occur in different states (ionic, atomic, molecular), hydrogen is expected to display a complex diffusion behaviour, as has been observed in Si[4]. In this contribution, we have studied the diffusion of hydrogen in 6H SiC samples implanted with 50 keV H^+ ions to doses of $\leq 5 \times 10^{15}$ cm^{-2}. Hydrogen starts to migrate at temperatures above ~700 °C and decorates gradually the defect distribution generated by the ion implantation. The process is modelled by computer simulations

Mat. Res. Soc. Symp. Proc. Vol. 423

where the diffusing hydrogen atoms are assumed to be trapped by the implantation-induced defects, resulting in the formation of stable hydrogen-defect complexes. A diffusion coefficient of ~10^{-14} cm^2/s is obtained at 800 °C with an approximate activation energy of ~3.5 eV. The hydrogen-defect complexes are stable up to ~900 °C where they start to dissociate and rapid out-diffusion of hydrogen takes place.

EXPERIMENT

Bulk SiC samples of 6H polytype with a nitrogen concentration of 5×10^{17} cm^{-3} (n-type) were used as starting material and implanted at room temperature with 50 keV H^+ ions to a dose of 2×10^{15} or 5×10^{15} cm^{-2}. According to Monte Carlo simulations using the transport of ions in matter code (TRIM, version -90)[6], implantation of 50 keV H^+ ions gives a hydrogen concentration versus depth profile with a peak at ~3500 Å, as shown in fig.1. The implantation profile is relatively well confined to the peak region while the distribution of generated vacancies exhibits a large tail towards the surface. After implantation the samples were heat treated in vacuum according to a time and temperature schedule that varied from 0.5 h to 55 h in duration and 200 °C to 1150 °C.

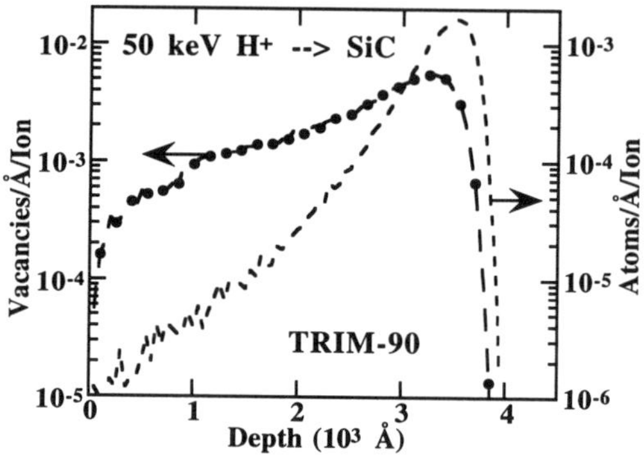

Fig. 1 Calculated profiles by TRIM-90 for the distributions of implanted hydrogen and generated vacancies in SiC bombarded with 50 keV H^+ ions. In the calculations a displacement energy threshold of 20 eV was assumed.

Analysis of the hydrogen depth distribution was performed via secondary ion mass spectrometry (SIMS) using a Cameca IMS 4f microanalyser. A primary sputtering beam of $^{133}Cs^+$ ions was rastered over an area of 75x75 μm^2, and secondary ions of $^1H^-$ were collected from the central region of the sputtered crater (diameter of analysed area ~8 μm). The erosion rate was typically 25 Å/s. In order to minimize the effect of sample charging, calibration of the sample high voltage (4.5 kV) with respect to maximum intensity of the $^{13}C^-$ signal was cyclically performed during depth profiling. The shift in sample voltage was small and never exceeded 15 V. On selected samples electron flooding was also employed for charge compensation but was found to reduce the dynamic range of the $^1H^-$ signal.

EXPERIMENTAL RESULTS

Figure 2 shows hydrogen profiles after annealing at 750 °C for 4 h, 19 h and 55 h; a small shift of the peak position occurs towards shallow depths compared to the as-implanted profile. Furthermore, for long annealing times the tail of the profile is gradually broadened with a rather distinct diffusion front moving towards the surface. Above the detection background of ~10^{18} cm^{-3}, no diffusion of hydrogen into the substrate is observed and the integrated hydrogen dose stays constant within less than 15 % of the implanted dose.

In fig.3 the annealing time is kept constant at 4 h while the temperature is varied. The profile after treatment at 650 °C is identical with the as-implanted one and at higher temperatures the surface tail broadens in a similar way as found for long annealing times at 750 °C. At 850 °C the diffusion front has reached the surface and no further broadening takes place. For longer times at 850 °C or an increase of the temperature, the shape of the hydrogen profile remains the same and only a gradual decrease of the absolute concentration values occurs, as illustrated in fig.4.

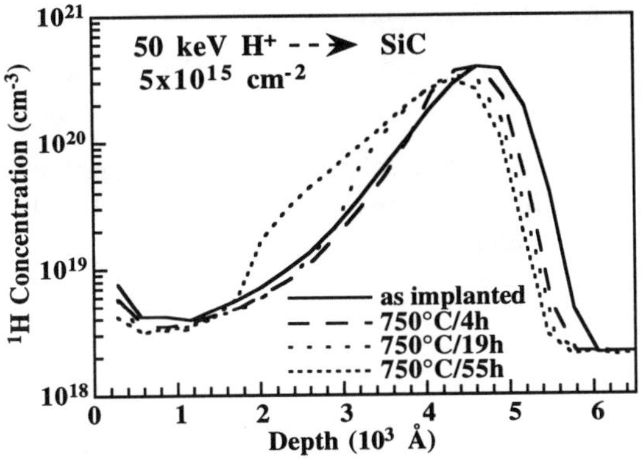

Fig. 2 Hydrogen concentration versus depth profile for four 6H SiC samples measured by SIMS. The samples are implanted with 50 keV H^+ ions using a dose of $5x10^{15}$ cm^{-2}. Three samples are subsequently annealed at 750°C for 4h, 19h, and 55h, respectively.

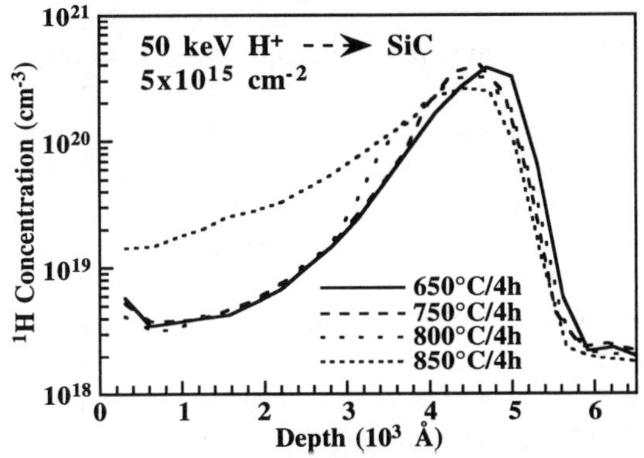

Fig. 3 SIMS measurements of the hydrogen concentration versus depth after implantation with 50 keV H^+ ions, dose $5x10^{15}$ cm^{-2}. The samples are annealed 4h at 650 °C, 750 °C, 800 °C and 850°C.

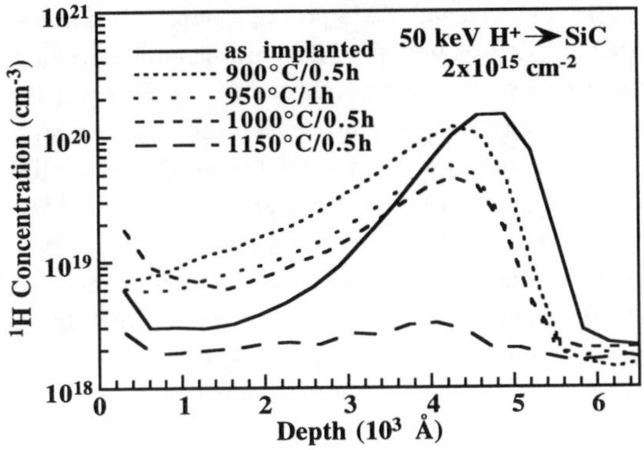

Fig. 4 SIMS measurements of the hydrogen concentration versus depth for samples implanted with 50 keV H^+ ions, dose of $2x10^{15}$ cm^{-2}. Five samples are analysed, one as implanted and four are annealed at 900°C/0.5h, 950°C/1h, 1000°C/0.5h and 1150°C/0.5h.

EVALUATION OF DATA AND DISCUSSION

The SIMS results reveal migration and trapping of hydrogen in the implanted region, and at temperatures below 850 °C, only a minor fraction (≤15 %) diffuses into the bulk of the wafer or is lost by out-diffusion. On the basis of a qualitative comparison with the calculated profiles in fig.1, the experimentally observed broadening of the hydrogen profile towards the surface provides strong evidence for trapping of migrating hydrogen by the implantation-induced defects. In order to model such a process, the following reaction is assumed

$$V + H \text{-->} VH \qquad (1)$$

where H is the mobile species and V represents the trapping defects, which are presumably of vacancy-type[7]. Applying the theory for diffusion-limited reaction processes[8], the corresponding set of coupled differential equations becomes

$$\begin{aligned} d[H]/dt &= D_H d^2[H]/dx^2 - 4\pi R D_H[V][H] \\ d[V]/dt &= -4\pi R D_H[V][H] \\ d[VH]/dt &= 4\pi R D_H[V][H] \end{aligned} \qquad (2)$$

where brackets denote concentration values, t is the annealing time, x is the sample depth, D_H is the diffusion coefficient for hydrogen, and R is a capture radius put equal to 5 Å. As a first approximation, the distributions of [H] and [V] at t=0 are taken from the TRIM calculations shown in fig.1. Vacancies and self-interstitials generated during ion bombardment exhibit correlated recombination and only a small fraction (~10 %) of the initial defects survives[9]. In order to account for this annihilation process in the simulations, the input distribution of vacancies predicted by TRIM is multiplied by a factor of 0.1. Results from the simulations are displayed in fig.5 with $D_H=3\times10^{-13}$ cm^2/s and a hydrogen dose of 5×10^{15} cm^{-2}. Similar to the experimental observations, a gradual broadening of the hydrogen profile takes place towards the surface with increasing time, and a rather abrupt transition occurs between the low and high concentration regimes of the surface tail. The total amount of retained hydrogen atoms in VH-complexes is ~80% of the initial dose, close to that found experimentally.

Applying equations (1) and (2) and using the as-implanted SIMS profile as input distribution for [H] and representing the distribution of trapping defects with the SIMS profile after annealing at 850 °C for 4 h (fig.3), D_H has been quantitatively evaluated. For fitting the simulated profiles to the measured ones, emphasis is put on the depth of transition between the low and high concentration regimes for the surface tail. Figure 6 shows the extracted values of D_H versus the reciprocal absolute annealing temperature. An activation energy of ~3.5 eV is obtained for D_H with a pre-exponential factor of ~300 cm^2/s. In order to increase the sensitivity of the SIMS measurements and to investigate if the same value of D_H is valid for diffusion into the undamaged bulk of the wafer further studies using deuterium ions are being pursued[10]. In an undamaged layer of crystalline 6H SiC the concentration of traps retarding the migration of hydrogen is low, and a more rapid diffusion than predicted by fig.6. may be anticipated.

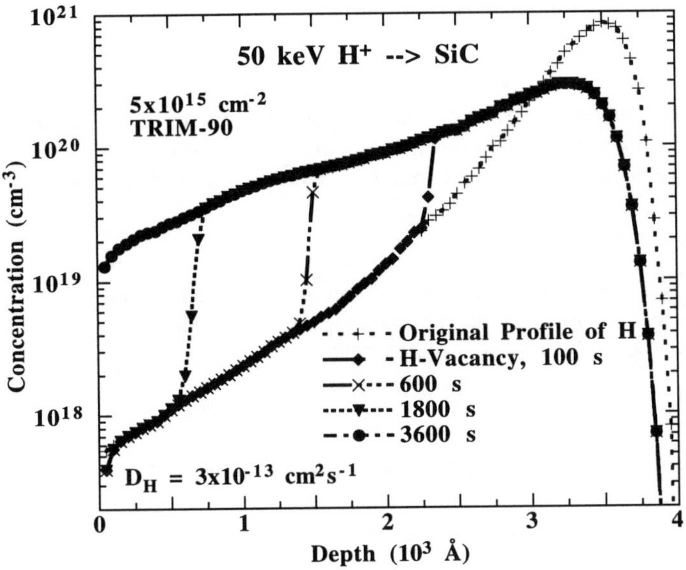

Fig. 5 Simulated profiles of the H-vacancy distribution obtained from the model described by equations (1) and (2) with $D_H = 3x10^{-13}$ cm^2/s and using TRIM profiles as input distributions.

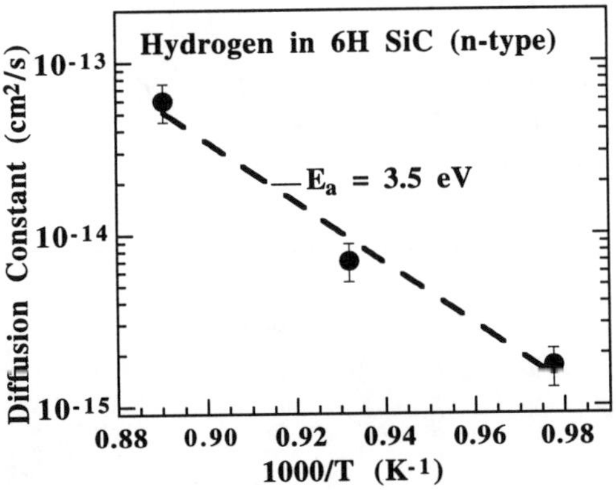

Fig. 6 Arrhenius plot of the extracted diffusion coefficient for hydrogen versus the reciprocal absolute annealing temperature. The slope corresponds to an activation energy of ~3.5 eV. Error bars indicate a relative accuracy of ±25%.

Finally, the gradual decrease of the hydrogen distribution at temperatures above ~900 °C occurs without changing the profile shape and is attributed to dissociation of the hydrogen-defect complexes with a rapid out-diffusion of hydrogen from the implanted region. It is interesting to note that after annealing in the same temperature range (~1000 °C) Patrick and Choyke[7] reported a decrease of the photoluminescence intensity of lines assigned to a center where a H atom is bonded to a C atom at a Si vacancy.

SUMMARY

After ion implantation hydrogen is found to be mobile in n-type 6H SiC at temperatures above ~700 °C with an activation energy of ~3.5 eV for the diffusion coefficient. Defects generated by the ion bombardment are efficient traps for migrating hydrogen and a gradual hydrogen decoration of the damage profile occurs. Based on the thermal stability it is very tempting to identify the hydrogen-defect complexes formed as CH-centers at a Si vacancy.

ACKNOWLEDGMENTS

Discussions with Drs C.I. Harris, N. Nordell, K. Rottner and A. Schöner at the Industrial Microelectronic Center in Sweden are gratefully acknowledged. Financial support was received from the Swedish Board for Technical Development (NUTEK, SiC consortium).

REFERENCES

1. G. Pensl and R. Helbig, Advances in Solid State Physics **30,** 133 (1990).
2. E. Janzén, O. Kordina, A. Henry, W.M. Chen, N. T. Son, B. Monemar, E. Sörman, P. Bergman, C.I. Harris, R. Yakimova, M. Tuominen, A.O. Konstantinov, C. Hallén, and C. Hemmingson, Physica Scripta **T54**, 283 (1994).
3. D.J. Larkin, S.G. Sridhara, R.P. Devaty, and W.J. Choyke, J. Electronic Materials **24**, 289 (1995).
4. S.J. Pearton, J.W. Corbett and T.S. Shi, Appl. Phys. **A43**, 153 (1987), and references therein.
5. F. Gendron, L.M. Porter, C. Porte, and E. Bringnier, Appl. Phys. **67**, 1253 (1995).
6. J.P. Biersack and I.G. Haggmark, Nucl. Instr. Meth. **174**, 257 (1980).
7. L. Patrick and W.J. Choyke, Phys. Rev. **B8**, 1660 (1973).
8. T.R. Waite, Phys. Rev. **107**, 463 (1957).
9. See for example, B.G. Svensson, C. Jagadish, A. Hallén and J. Lalita, Nucl. Instr. Meth. **B106**, 183 (1995).
10. M.K. Linnarsson, J.P. Doyle, C.I. Harris, and B.G. Svensson, to be published.

ELECTRONIC STRUCTURE OF BERYLLIUM, MAGNESIUM AND SILICON IMPURITY IN CUBIC GALLIUM NITRIDE

V. A. GUBANOV*, E. PENTALERI*, C. Y. FONG**, B. M. KLEIN**
*Physics Department, San Jose State University, San Jose, CA 95192, vgubanov@isc.sjsu.edu
** Department of Physics, University of California, Davis, CA 95616

ABSTRACT

Beryllium (Be), magnesium (Mg) and silicon (Si) impurities in zinc-blende gallium-nitride (c-GaN) are investigated by the tight binding-linear combination muffin-tin orbitals (TB-LMTO) method using a 64-atom supercell. Be and Mg impurities at a Ga site, respectively induce partially empty acceptor-like bands at the valence band edge, which result in p-type conductivity of doped c-GaN. Si impurity in the Ga sublattice creates a partially occupied impurity subband overlapping with the conduction band edge and is responsible for the measured n-type conductivity. The impurity levels of a Si at a N site are located deep in the gap and do not influence much the conductivity of c-GaN. The shell-projected, total and partial densities of states and the charge density maps are used to elucidate the energy and spatial localizations of the impurity states

INTRODUCTION

Reproducible and well controllable dopings of both n- and p-types are vital for succcessful production of GaN based blue and ultraviolet light emitting devices and lasers developed recently [1-3]. While the n-type doping of GaN has been rather straightforward using either Si or Ge [4,5], reliably high doping densities of Mg have been achieved only recently in either molecular beam epitaxy (MBE) or metalorganic chemical vapor deposition (MOCVD) [6,7] growth process. However, the doping center is still not clearly identified [8]. In spite of excellent thermal stability for many implanted acceptors into GaN [9], numerous attempts to use such dopants as Be, Si and Cd were largely unsuccessful [3,7,10], though Be has been a commonly used p-type dopant in MBE growth of GaAs based materials [11,12]. At the same time, Zn-doped InGaN/AlGaN double-heterostructures [13,14] revealed superbright blue and green emissions and showed that successful p-type Mg doping is not the only one possible.

In the present paper, we investigate the electronic structure, chemical bonding, the nature and energy and spatial localization of Be, Mg and Si impurity levels in c-GaN using the tight binding-linear combination muffin tin orbitals (TB-LMTO) method.

MODEL AND METHOD CALCULATIONS

The electronic structure calculations of doped c-GaN have been performed by the TB-LMTO method [15] making use a 64-atom supercell model with a single impurity placed at the center of the supercell. Be, Mg dopings at the cation sublattice and Si doping at both cation and anion sublattices were investigated, respectively. Thus, the systems considered are $M_{[0.016]}Ga_{[0.984]}N$ {M=Be, Mg, Si} and $GaSi_{[0.016]}N_{[0.984]}$ doped crystals. The crystal lattice constant, a, of 4.49Å found for epitaxially grown c-GaN [16] has been used in the calculations. Ga 3d states were treated as valence states. No lattice relaxation near the impurity site was taken into account. We added 64 empty spheres placed at standard positions of the unit cell to improve the description of the potentials, so that the supercell consists of 128 sites. The calculations were carried out with the Hedin-Lundqvist exchange-correlation potential [17]. Neither the self-interaction nor the quasi-particle approach was used to improve the gap value. As has been

Mat. Res. Soc. Symp. Proc. Vol. 423 © 1996 Materials Research Society

shown in our previous paper [18] for the ideal c-GaN, the above model provides the energy band structure in excellent agreement with results of more accurate but time consuming methods, such as the full potential augmented plane wave and full potential LMTO methods.

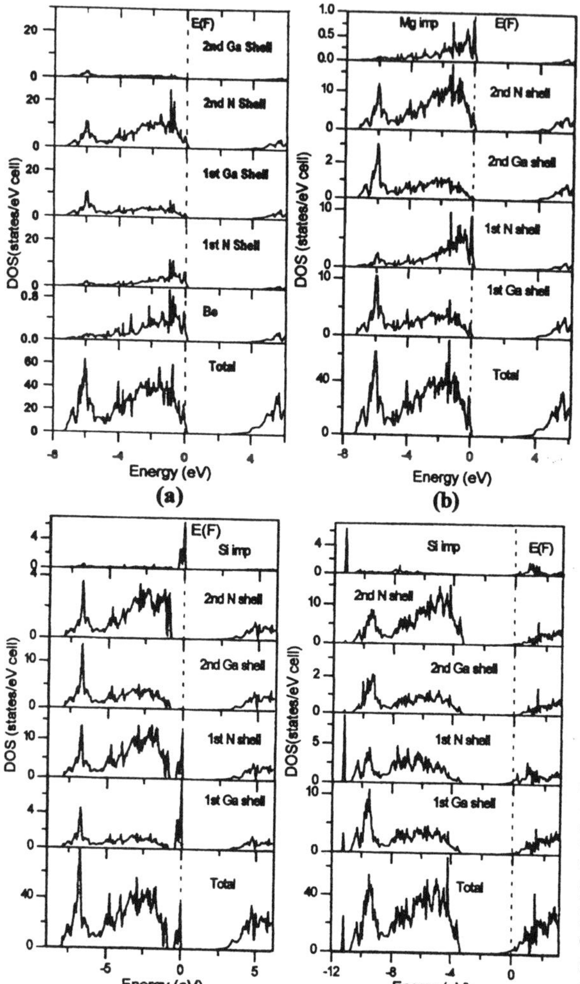

Fig.1. Total and shell-projected densities of states for Be (a), Mg (b) impurities at Ga sites, and Si impurity at Ga (c) and N (d) sites. Shell-projected densities are given for four coordinational spheres around impurity.

RESULTS AND DISCUSSION

The TB-LMTO calculations of the ideal c-GaN (see [18]) give the direct LDA gap of 1.92eV, with the upper part of the valence band (VB) composed of N2p states hybridized with Ga4p states. Both Ga4s and N2p states contribute to the states near the conduction band (CB) edge.

Figs. 1(a) and (b) show total and shell-projected density of states (DOS) obtained in the 64-atom supercell TB-LMTO calculations for the Be and Mg impurities substituting a Ga atom respectively. A narrow impurity subband merges with the VB edge for both cases. The Fermi level crosses the maximum of each impurity band and results in the creation of holes in the VB. Both Be and Mg dopings may result in the p-type conductivity of c-GaN. But the energy and spatial localization of Be and Mg states are very different. Be2p states are hybridized with the whole VB in rather uniform way (Fig. 1(a)), mix noticeably with the states of distant shell of atoms from the impurity (up to 4th N and Ga). Mg states (Fig. 1(b)) are essentially shifted to the upper part of the VB and form the sharp partially occupied band at the VB edge. This is in line with the recent photoluminescence data [7], from which the Mg acceptor binding energy was estimated to be 0.25eV (wurtzite phase). The analysis of shell-projected DOS showed that Mg acceptor states (of Mg2p character) are much more localized in space than the ones of Be

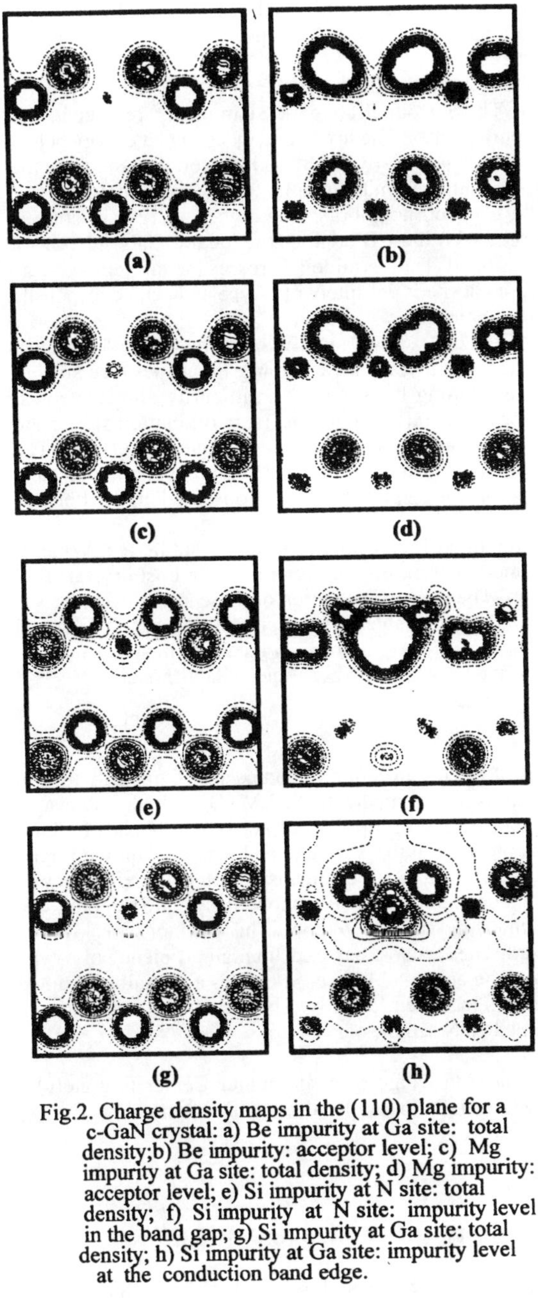

Fig.2. Charge density maps in the (110) plane for a c-GaN crystal: a) Be impurity at Ga site: total density;b) Be impurity: acceptor level; c) Mg impurity at Ga site: total density; d) Mg impurity: acceptor level; e) Si impurity at N site: total density; f) Si impurity at N site: impurity level in the band gap; g) Si impurity at Ga site: total density; h) Si impurity at Ga site: impurity level at the conduction band edge.

impurity: for doping Mg, atoms of the 4-th shell sphere practically do not feel the presence of the impurity.

The spatial extensions of impurity states can be seen by calculating the charge density maps. In Fig. 2, the maps of the total charge density for the c-GaN with a Be (Fig. 2(a)) and a Mg (Fig. 2(c)) impurities are presented. All the charge distributions are given in the (110) plane. As is seen, the chemical bonding of both impurities in the host crystal is of highly ionic character. The charge density maps for impurity levels at the VB edge (Figs. 2(b) and (d)) show that both the Be and the Mg atoms polarize strongly the states of many shells of neighboring atoms. Such a polarization is especially strong in the Be impurity case. These results together with calculated shell-projected DOS (see above) show that Be and Mg acceptor impurities in c-GaN are more complex than the ones accepted in the point-like tight-binding models (see, e.g. [19]). These results confirm that when Mg atoms substitute Ga atoms no deel impurity levels are formed in the band gap, and the deep level luminescence in Mg doped c-GaN observed experimentally (see [8]) cannot be attributed to a completely isolated Mg impurity substituting a cation.

Si is the other widely used and effective doping agent [13]. It can be incorporated into GaN at concentrations above 10^{19} cm^{-3} and, in principle, can be amphoteric either on substitutional Ga or substitutional N sites. Experiments show, however, that Si is a very efficient donor, but does not appear to act as an acceptor. Using the supercell TB-LMTO calculations outlined above, we have investigated both possible types of Si dopings: an isolated substitutional Si impurity at a Ga site, and an isolated Si impurity at a N site in the supercell.

Fig. 1(c) presents the total and shell-projected DOS for the Si at a Ga site case with atoms up to the 4-th neighboring shell. The introduction of a Si impurity results in the appearance of two very different set of impurity states. The lower energy one at an energy below the upper part of the VB has a narrow width and has mostly the Si3s character. As seen from the shell-projected DOS, its spatially extension should be limited mainly to the nearest neighbors of the impurity, that is, the 1st N and the 1st Ga shell neighbors. The higher energy structure is predominantly of Si3p character and merges with the CB edge of the ideal c-GaN. It forms a broad, shallow and highly delocalized donor band. The Fermi level crosses the tail region of the merged impurity band. Thus, Si impurity at a Ga site is definitely of n-type and can be responsile for high n-type conductivity of the GaN:Si samples [5].

The total VB charge density map (Fig. 2(g)) shows mainly ionic character of chemical bonding of the Si impurity at a Ga site. But, in distinct from Be and Mg impurities (see above), there is some covalent contribution to the bonding between the Si impurity and its nearest neighbor N atoms. The charge densities for atoms farther than the 1st coordination sphere are practically unchanged as compared with the charge distribution in the undoped crystal. The charge density calculated for the n-type Si states in the CB (Fig. 2(h)) shows the donor band highly delocalized character in space which may be responsible for the observed high efficiency of n-type doping of GaN by Si impurities.

A completely different picture is obtained for Si doping in the N sublattice. When it happens, a narrow and strong band is formed deep in the gap region of the host crystal (Fig. 1(d)). The Fermi level crosees its maximum. The energy separation of these states from the VB edge is about 1.5eV, and they cannot result in any sizable p-type conductivity of the Si doped GaN crystals. Fig. 2(f) shows that this "Si acceptor" level is strongly localized in space, and has the shape close to the one of the s-like N vacancy states in the N deficient GaN crystals.

CONCLUSIONS

The electronic structure calculations of Be, Mg in cation sublattice and Si impurity in both cation and anion sublattices were carried out using the supercell TB-LMTO method. The results show the appearence of strongly delocalized partially occupied acceptor-type subbands for both the Be and the Mg dopings. The subbands merge into the VB edge and may explain the p-type conductivities of Be and Mg doped crystals. Some qualitative conclusions on better efficiency of the Mg doping has been made from the calculated results. We show also that Si doping can produce only n-type conductivity when Si enters the cation sublattice. The Si donor states overlap with the CB edge and are highly delocalized. These states may explain high efficiency of n-type dopings in c-GaN. When the Si atom enters the anion sublattice, Si creates a partially occupied narrow band deep in the gap region with states localized spatially. These states cannot influence much the electric conductivity of the Si-doped GaN samples.

ACKNOWLEDGMENTS. A grant from the San Diego Supercomputer Center is gratefully acknowledged. C. Y. Fong acknowledges partial support by the US Department of Energy under contract #DE-AC04-95AL85000.

REFERENCES

1. S. Nakamura, T. Mukai, and M. Senoh, Appl. Phys. Lett., **64,** p.1687 (1994).

2. I. Akasaki, H. Amno, N. Koide, M. Kotaki, and K. Manabe, Physica B, **185,** p.428 (1993).

3. R. Davis, J. Cryst. Growth., **137,** p.161 (1994).

4. S. Nakamura, T. Mukai, M. Senoh, Jpn. J. Appl. Phys., **31,** p.2883 (1993).

5. L. B. Rowland, K. Doverspike, D. K. Gaskill, Appl. Phys. Lett., **66,** p. 1495 (1995).

6. H. Amano, M. Kito, K. Hiramatsu, I. Akasaki, Jpn.J.Appl.Phys., **28,** p. L2112 (1989).

7. S. Fischer, C. Wetzel, E. E. Haller, Appl. Phys. Lett., **67,** p. 1. (1995).

8. P. Haske, H. Nakayama, T. Detchprohm, K. Hiramatsu, N. Sawaki, Appl. Phys. Lett., **68,** p.1362 (1996).

9. R. G. Wilson, S. J. Pearton, C. R. Abernathy, J. M. Zavada, Appl. Phys. Lett., **66,** p.2238 (1995).

10. S. Strite, H. Morkoc, J. Vac. Sci. Technol. B, **10,** p.1237 (1992).

11. T. J. Kuo, T. H. Chiu, J. E. Cunningham, K.W. Goossen, C. G. Fonstad, F. Ren, Electron. Lett., **26,** p.1260 (1990).

12. D. N. Talwar, Phys. Rev. B., **52,** p. 8121 (1995).

13. S. Nakamura, M. Senoh, N. Iwasa, S. Nagahama, T. Yamada, T. Mukai, Jpn. J. Appl. Phys., **34,** p.L1332, p.797 (1995).

14. G. Mohs, B. Fluegel, H. Giessen, H. Tajalli, N. Peyghambarian, P. Chiu, B. S. Phillips, M. Osinski, Appl. Phys. Lett., **67,** p.1515 (1995).

15. M. van Schilfgaarde, T. A. Paxton, O. Jepsen, O.K. Andersen, TB-LMTO Program - Version 44, Max Plank Institute for Solid State Physics, 1994.

16. T. Leim, T. D. Moustakas, R. J. Graham, Y. He, S. Berkowitz, J. Appl. Phys., **71,** p.4933 (1992).

17. K. Hedin, B. I. Lundqvist, J. Phys., C **4,** p.2.064 (1971).

18. V. A. Gubanov, A. F. Wright, J. S. Nelson, C. Y. Fong, B. M. Klein, Proceedings of Fall 1995 MRS Meeting, Boston, 1995 (in press).

19. D. W. Jenkins, J.D. Dow, Phys. Rev. B, **39,** p.3317 (1989).

IMPURITY CONDUCTION IN N-TYPE 4H-SIC

A. O. Evwaraye†, S. R. Smith††, W. C. Mitchel and M. D. Roth
Wright Laboratory, Materials Directorate, MLPO
Wright-Patterson Air Force Base, Ohio 45433-7707

ABSTRACT

Impurity conduction (or hopping conduction) has been observed in the more heavily n-type 4H-SiC samples by both temperature dependent resistivity measurements and thermal admittance spectroscopy. The measured activation energies ε_3 for hopping were 4-5 meV and 2.3-3.0 meV respectively. No evidence of hopping conduction was seen by either method in the sample where N_D-N_A < 10^{18} cm^{-3}. The thermal admittance spectrum of the lightly n-type sample showed the two nitrogen levels at 53 and 100 meV.

INTRODUCTION

The conductivity in a semiconductor can in general be expressed by[1]

$$\sigma(T) = \sigma_1 e^{-\varepsilon_1/kT} + \sigma_2 e^{-\varepsilon_2/kT} + \sigma_3 e^{-\varepsilon_3/kT} \qquad (1)$$

where ε_1 , ε_2 and ε_3 are activation energies for conduction in different regimes. The activation energy ε_1 is necessary to excite an electron from a donor in to the conduction band. The activation energy ε_3 is required for jumping from an occupied donor site to an unoccupied donor site without an excursion into the conduction band. When the doping concentration in an n-type semiconductor is high enough, there is an overlap of the electron wave-functions of neighboring sites. As the temperature is reduced, the conduction electrons are frozen on to the donors. Because of compensation and the overlap of the electron wave-functions, there can be movement of charge carriers from occupied donor sites to unoccupied donor sites. This movement constitutes the hopping or impurity conduction. The impurity conduction dominates the conductivity at low temperatures even though the mobility of the electrons moving in the impurity levels is small. As the impurity concentration increases, the temperature at which impurity conduction becomes experimentally observable shifts to higher temperature.

Impurity conduction has been observed in many semiconductors (such as germanium, silicon, silicon carbide, cadmium sulfide, and indium antimonide) at low temperatures[2-7]. Temperature dependent Hall effect and resistivity measurements were the main tools used in the investigations. The anomalies observed in the electrical measurements were adequately explained by the two band model introduced by Hung[8].

Mat. Res. Soc. Symp. Proc. Vol. 423

Busch and Labhart [9] observed impurity conduction in silicon carbide in 1946. But more recently [10, 11] impurity conduction has been observed in n-type 4H-SiC.

In this paper, we shall present the details of temperature dependent resistivity measurements of n-type 4H-SiC samples and of thermal admittance spectroscopy(TAS). The activation energy ε_3 obtained from the resistivity measurements is compared with that obtained from TAS. A satisfactory agreement between the techniques is found.

The samples studied were n-type 4H-SiC grown by the physical vapor transport method. The samples are labeled W1, W2 and W3. Temperature dependent resistivity measurements were made using van der Pauw[12] structures with either Ni, Ti, or Al contacts. Schottky diodes for the thermal admittance spectroscopy were fabricated as described in reference [13] . Capacitance-voltage (C-V) measurements on the diodes were made with 1 MHz capacitance meter. N_D-N_A for W1, W2 and W3 are

1.5x 10^{18} cm^{-3}, 3 x 10^{18} and 1.5 x 10^{17} cm^{-3} respectively. The samples were not intentionally doped.

RESULTS.

Figure 1 shows the resistivity of samples W1 and W2 as functions of temperature.

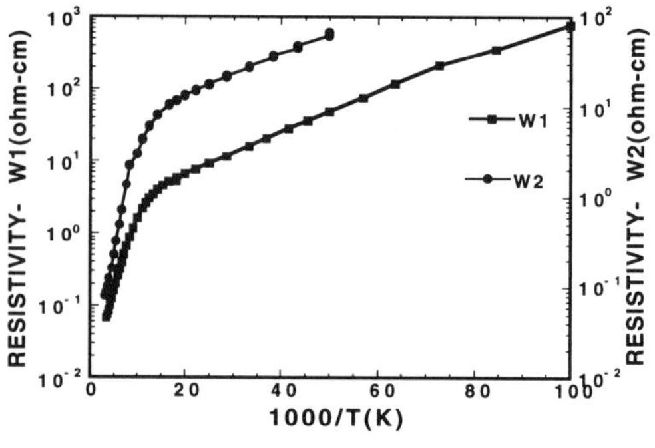

Figure 1. Resistivity measurements as a function of temperature of samples W1 and W2.

The striking feature of the resistivity curves is the change of the slope. This change in slope corresponds to a change in conduction mechanism from ε_1 to ε_3. Figure 2 shows the plot of the Hall coefficient as well as the resistivity for sample W1. The maximum in the Hall coefficient curve occurs at about the same temperature at which the conduction mechanism changes (i.e. at the knee of the resistivity curve). The shape of the Hall coefficient curve can be explained by the so called two-band model [8,14].

The observed Hall coefficient can be expressed as:

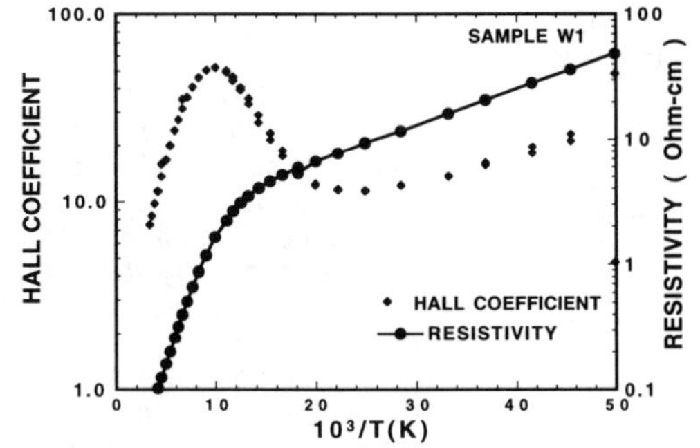

Figure 2. Hall coefficient reaches a maximum at about the temperature the conduction mechanism changes,

$$R=\frac{R_c\sigma_c^2+R_h\sigma_h^2}{(\sigma_c+\sigma_h)^2} \quad (2)$$

using the notations of reference [1]. See also reference [14]. The maximum in the Hall coefficient occurs at $\sigma_c \approx \sigma_h$ where the conductivity mechanism changes from ε_1 to ε_3. To the left of the maximum $\sigma_c > \sigma_h$. the Hall coefficient is proportional to $\exp(\varepsilon_1/kT)$ and to the right of the maximum $\sigma_h > \sigma_c$; R is proportional to $\exp(-\varepsilon_1/kT)$. It is clear from Fig. 2 that the curves have similar slopes on both sides of the maximum. This therefore is a clear confirmation of impurity conduction in 4H-SiC.

The activation energy ε_3 is obtained from the plots of ln(ρ) vs 1/kT for samples W1 and W2. ε_3=4.5 meV for W2(N_D-N_A=3 x 10^{18} cm^{-3}) and is equal to 5.3 meV for W1(N_D-N_A=1.5 x 10^{18} cm^{-3}). It appears that the sample having larger N_D-N_A has a smaller activation energy for hopping. The resistivity versus temperature curve for sample W3 increases exponentially as the temperature is reduced; there is no inflection or change of slope as observed in samples W1 and W2 . The N_D-N_A for W3 is 1.5 x 10^{17} cm^{-3}; it does appear that the occupied donors are sufficiently far away from unoccupied donor sites that there is no overlap of electron wave functions to promote hopping at low temperatures.

Thermal admittance spectroscopy (TAS) has been described in detail elsewhere[15].

The additional conductance that arises as the junction is modulated with a small sinusoidal voltage can be expressed as

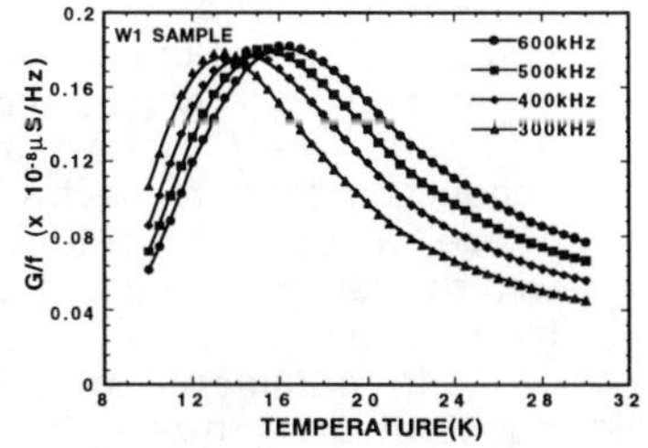

Figure 3. Thermal admittance spectrum of the impurity conduction. The maximum occurs when the hopping frequency equals the measuring frequency.

$$G_p = A\left(\frac{e_n\omega^2}{e_n^2+\omega^2}\right)\frac{N_T}{n_d}\left(\frac{q\varepsilon N_+}{2(V_R+V_D)}\right)^{\frac{1}{2}} \quad (3)$$

the notation is as explained in reference[15]except that n_d is the concentration of carriers in the impurity band and e_n is the hopping rate. At low temperatures, the ε_1 conduction mechanism is negligible. Therefore the additional conductance is due to hopping of carriers from occupied donor to unoccupied donor sites in the impurity band. This changes N_+ in the depletion region. The ε_3 conduction mechanism is temperature activated. When the hopping rate in the impurity band is equal to the measuring frequency ω, a peak occurs in the conductance spectrum.

Figure 3 shows representative normalized conductance spectrum (G/ω) for sample W1 obtained at different measuring frequencies.

A similar spectrum is obtained when sample W2 is used.

The plot of $\ln(\omega/T^2)$ vs 1/kT is shown in Fig. 4. The solid line through the data points is the least square fit to the data. The slope of the line yields the activation energy for hopping which is responsible for the observed changes in the conductance of the diode. The measured activation energy ranges from 2.3 to 3.0 meV for the various samples. These values obtained for ε_3 are in general agreement with the values obtained from resistivity measurements.

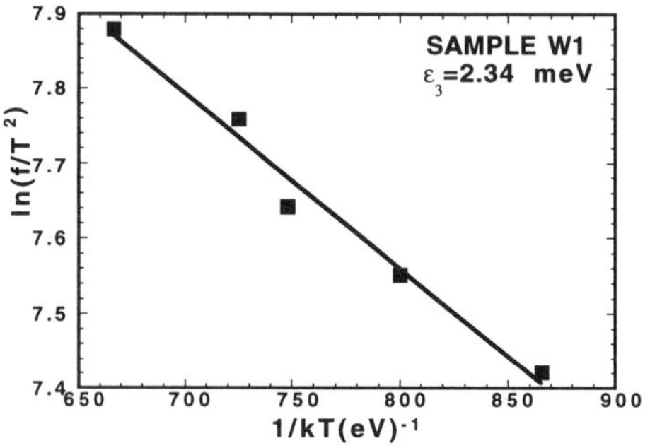

Figure 4. A plot of $\ln(\omega/T^2)$ vs 1/kT. The slope of the straight line through the data points gives the activation energy ε_3

TAS of W3 does not show the impurity band at 2.3 meV which is consistent with the behavior of the resistivity curve of this sample. However, the TAS of W3 clearly reveals the two nitrogen ionization levels. Figure 5 shows representative normalized conductance spectrum (G/ω) for sample W3 obtained for different measuring frequencies.

Two conductance maxima for each frequency are clearly visible. These conductance maxima are attributed to nitrogen sitting on the hexagonal(h) site and on the cubic(k) site. The measured activation energies for nitrogen at the hexagonal and at the cubic site are 53 meV and 100 meV below the conduction band, respectively. These values agree quite well with the reported values of 52 meV and 91 meV.

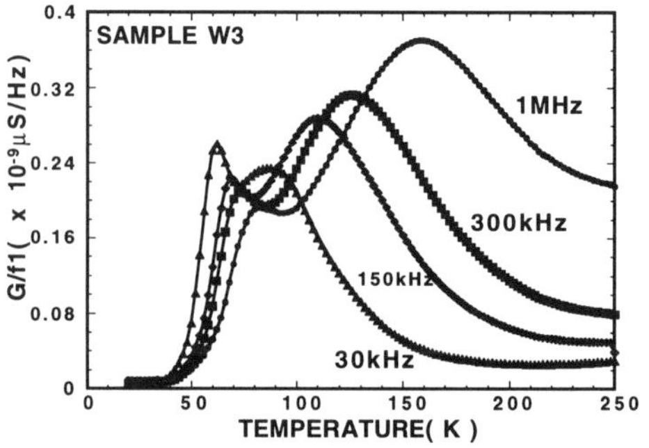

Figure 5. Thermal admittance spectrum of sample W3 where there is no evidence of hopping.

In summary, impurity conduction has been observed in n-type 4H-SiC. The resistivity and Hall coefficient curves can only be explained by postulating the existence of an impurity band. The activation energy obtained from temperature dependent resistivity measurements ranges from 4-5 meV. Thermal admittance spectroscopy was used, for the first time, to study impurity conduction in n-type 4H-SiC. The activation energy, ε_3, of 2.3-3.0 meV, obtained by TAS is in agreement with that obtained from the resistivity measurements.

The two nitrogen ionization levels obtained for sample W3 agree with the reported values obtained by IR measurements.

ACKNOWLEDGMENTS

The authors would like to thank Dr. P. M. Hemenger for his support and encouragement. The work of one of us (SRS) was carried out under Air Force Contract No. F33615-91-C-5603. The authors would like to acknowledge the technical assistance of Paul Von Richter (deceased), Gerald Landis, and Robert V. Bertke

† Department of Physics, University of Dayton, 300 College Park, Dayton, Ohio 45469-2314
†† University of Dayton Research Institute, 300 College Park, Dayton, Ohio 45469-0178

REFERENCES

1. B. I. Shklovskii and A. L. Efros, **Electronic Properties of Doped Semiconductors** (Springer-Verlag Berlin 1984)

2. C. S. Hung and J. R. Gliessman, Phys. Rev. **19**, 726 (1950)

3. R. O. Carlson, Phys. Rev. **100**, 1075 (1955)

4. H. Fritzsche and M. Cuevas, Phys. Rev. **119**, 1238 (1960)

5. F. J. Morion and J. P. Maita, Phys. Rev. **96**, 28 (1954)

6. H. Fritzsche and K. Lark-Horovitz, Phys. Rev. **113**, 999 (1959)

7. H. Fritzsche, Phys. Rev. **99**, 406 (1955)

8 C. S. Hung, Phys. Rev. **79**, 727 (1950)

9. G. Busch and H. Labhart,Helv. Phys. Acta **19**, 463 (1946)

10. W. Götz, A. Schöner, G. Pensl, W. Suttrop, W. J. Choyke, R. Stein and S. Leibenzeder, J. Appl. Phys. **73**, 3332(1993)

11 A. O. Evwaraye, S. R. Smith, W. C. Mitchel and M. D. Roth. Appl. Phys. Lett. (To be published May 1996)

12. L. J. van der Pauw, Philips Research Report Vol. **13**, 1 (1958)

13. A. O. Evwaraye, S. R. Smith, M. Skowronski and W. C. Mitchel, J. Appl. Phys. **74**, 5269 (1993)

14. N. F. Mott and W. D. Twose, **Advances in Physics**, edited by N. F. Mott (Taylor and Francis, Ltd. London), **Vol. 10**, 107 (1961)

15. A. O. Evwaraye, S. R. Smith and W. C. Mitchel, J. Appl. Phys. **75**, 3472 (1994)

FACTORS INFLUENCING THE ELECTRICAL AND OPTICAL PROPERTIES OF AlGaN LAYERS ON SAPPHIRE

M. SHIN, A.Y. POLYAKOV, M. SKOWRONSKI, D.W. GREVE*, R.G. WILSON**, J.A. FREITAS***
Department of Materials Science & Engineering, Carnegie Mellon University, Pittsburgh, PA, 15213-3890, mscy@andrew.cmu.edu
*Department of Electrical and Computer Engineering, Carnegie Mellon University, Pittsburgh, PA 15213-3890
** Hughes Research Laboratories, 3011 Malibu Canyon Road, Malibu, CA 90265
*** Sachs/Freeman Assoc. Inc., Landover, MD 20785, Naval Research Laboratory: Contract N00014-93-C-2227

ABSTRACT

Epitaxial layers of $Al_xGa_{1-x}N$ were grown by MOCVD on sapphire substrates and characterized by temperature dependent Hall effect, optical absorption, SIMS and photoluminescence. The concentration of native donors was found to decrease with increase of growth temperature. Deep levels found in AlGaN layers were interpreted as A_1 double donor state and T_2 donor state of nitrogen vacancy.

INTRODUCTION

AlGaN/GaN material system is of great interest for applications in optoelectronic devices operating in the blue-ultraviolet wavelength region and for high temperature/high frequency/high power electronic devices [1]. Epitaxial layers of AlGaN and GaN are often grown by metalorganic chemical vapor deposition (MOCVD) on sapphire substrates and a remarkable improvement in the crystalline quality of the layers has been achieved recently due to the use of low temperature AlN or GaN buffers (see e.g. a review in [2] and references therein). For GaN, careful optimization of the growth conditions allowed to decrease the electron concentration in undoped layers from about 10^{20} cm^{-3} to below 10^{16} cm^{-3}, with corresponding mobility increase from 10 cm^2/V·s to 600-800 cm^2/V·s at room temperature (see a review in [3]). Much less is known about the impact of growth conditions on the properties of $Al_xGa_{1-x}N$ alloys. Many groups report that the electron concentration in $Al_xGa_{1-x}N$ samples with $0<x<0.2$-0.4 is high, on the order of 10^{18}-10^{19} cm^{-3}, and that in optical absorption a strong near band edge absorption related to a high concentration of charged defects is observed [4,5]. On the other hand, it has been reported that a material with much lower carrier concentration of about 10^{17} cm^{-3} and with virtually no near band edge absorption can also be grown [5,6]. At present it is not quite clear what differences in the growth procedure cause such a difference in the crystal quality. In this paper we report how changes in the growth temperature and in growth rate affect electrical and optical properties of AlGaN films.

EXPERIMENTAL

Films of $Al_xGa_{1-x}N$ with x in the whole composition range were deposited by MOCVD, on basal plane sapphire substrates, using trimethylgallium (TMG), trimethylaluminum (TMA) and ammonia as precursors. Growth was done in a horizontal two-inlet reactor, at 76 Torr, with hydrogen as a carrier gas. Substrate preparation procedure consisted in degreasing in organic solvents, etching in hot H_2SO_4:H_3PO_4, and annealing in hydrogen at 1100°C for 10 minutes and in hydrogen and ammonia for 3 min.

Low temperature AlN buffer layers, about 700A thick, were deposited at 500°C and crystallized by ramping the temperature up to the AlGaN films growth temperature of 1000-1050°C with subsequent annealing under ammonia flow for 1 min. at these temperatures. Two sets of AlGaN

Mat. Res. Soc. Symp. Proc. Vol. 423 © 1996 Materials Research Society

samples were grown: one at a growth temperature of 1000 °C, with a growth rate of 1μm/h, and another at 1050 °C, with a growth rate of about 0.7 μm/h (for the sake of brevity those are henceforth called, respectively, samples of the A series and the B series). Composition of the layers was calculated from the values of lattice parameter c measured by X-ray diffraction. The thickness of the layers was in all cases close to 1 μm. Electron concentrations and mobilities were obtained from conventional van der Pauw measurements in the temperature range from room temperature to 140°C, and ohmic contacts for these measurements were prepared by soldering indium. The depth of the dominant levels was calculated from activation energies in the temperature dependence of electron concentration. It was assumed that the measured activation energies directly correspond to the depth of the energy levels in question.

Optical absorption spectra at room temperature were calculated from optical transmission spectra using the known reflectivity data for AlN and GaN [7,8]. Plots of square of absorption coefficient (α^2) versus photon energy were drawn to obtain the values of the bandgap energies in our AlGaN layers, and deviations from these linear plots for photon energies below the bandgap were used as a measure of band tail absorption [4]. For the sample of the A series with composition x=0.38 we could not achieve as good a morphology as in other layers, despite several attempts, and in this particular sample the high apparent magnitude of band tail absorption could be due to, in part, scattering of light. Photoluminescence (PL) spectra were measured with He-Cd laser excitation, at 6K. Concentrations of C, Si, H, O, S and Se in our $Al_xGa_{1-x}N$ samples with x=0,0.12 and 1 were measured by secondary ion mass spectrometry (SIMS) [9].

RESULTS AND DISCUSSION

The effect of growth temperature on electron concentration in AlGaN is shown in Fig 1. The mobilities of electrons in GaN were about 200 $cm^2/V{\cdot}s$ in both series, but for $Al_xGa_{1-x}N$ layers the mobilities dropped to 2-5 $cm^2/V{\cdot}s$ for the series A (grown at 1000°C), while they remained close to 100-200 $cm^2/V{\cdot}s$ for the series B. The composition dependence of electron concentration for the series A of AlGaN samples is close to what has already been reported. That is, the carrier concentration becomes very high for x values between 0 and 0.5 and then decreases rapidly. In the samples of the series B the trend is somewhat different. For compositions $0<x<0.12$ the resistivities were too high to measure them reliably. For x=0.12 the electron concentration was still quite low, about 10^{11} cm^{-3}. It increased for x>0.12 before going down steeply again for $Al_xGa_{1-x}N$ layers with x higher than 0.5.

Appreciable changes of carrier concentration with temperature in the samples of the series A were only observed for x>0.5 (activation energies measured were 0.11 eV for x=0.55 and 0.22 eV for x=0.67).

The activation energies obtained from Hall effect measurements on AlGaN samples with various Al content are shown in Fig. 2. In highly resistive samples of the series B with x=0.12 the measured activation energy was close to 0.3 eV. For x=0.225 it decreased to 0.08 eV and then increased gradually with x reaching the value of 0.7 eV for x=0.64. The energies in Fig. 2 are measured in respect to the valence band maximum to allow comparison with the composition dependence of the conduction band edge. Also shown in Fig. 2 are the T_2 and A_1 donor states of the nitrogen vacancy (V_N) calculated for various AlGaN compositions by Jenkins *et al* [10]. According to this model if the V_N concentration in AlGaN is much higher than the concentration of compensating acceptors, T_2 states would be the dominant level pinning Fermi energy. For x values lower than 0.5, T_2 levels lie within the conduction band and hydrogen-like shallow donors are expected to be formed as a result of Coulombic interaction of the ionized T_2 states with free electrons at the bottom of conduction band. For x>0.5 the T_2 level of the nitrogen vacancy enters the bandgap and the level becomes deeper as the Al content increases. Such behavior of the T_2 levels of the nitrogen vacancy was invoked by Jenkins *et al* [10] to explain the frequently observed transition from very high conductivity to very low conductivity in AlGaN near the middle of the composition range, as we also see in our AlGaN samples of the series A. If, on the other hand, the concentration of compensating acceptors is higher than the concentration of nitrogen vacancies, the Fermi level will be either pinned by the A_1 double donor level associated with nitrogen vacancies (if the concentration of acceptors N_a is less than $2 \cdot V_N$) or, for even higher N_A (or lower V_N), the Fermi level will be pinned by other traps deeper in the bandgap. Such a behavior is very close to what was observed in the samples of the series B (Fig.1). From Fig. 2 it is apparent that there is a qualitative agreement between the composition dependence of the depth of T_2 and A_1 states in AlGaN and the experimental energy levels in our AlGaN layers of the series A and series B.

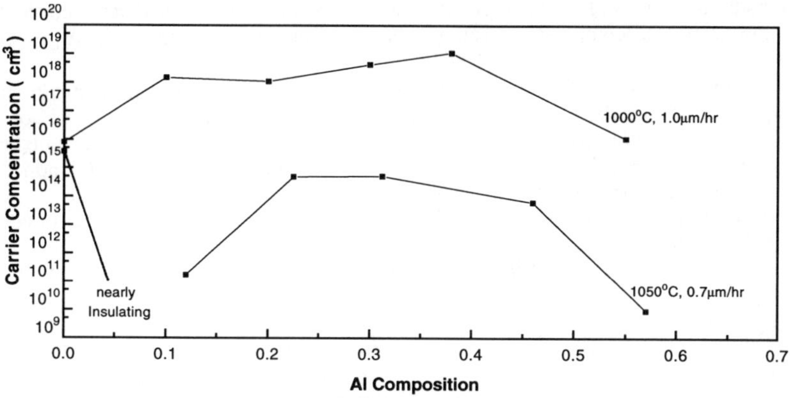

Fig. 1. Composition dependence of the room temperature electron concentration in AlGaN layers grown at 1000 °C (A series) and at 1050 °C (B series).

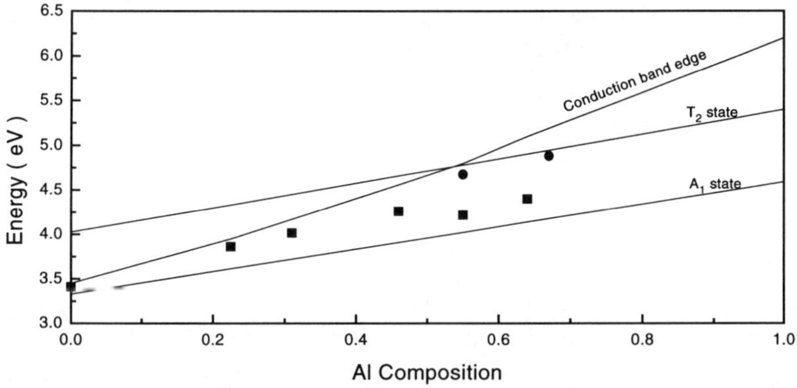

Fig.2. Composition dependence of the energy of the dominant centers in AlGaN samples of the series A (circles) and of the series B(squares). The energies are measured in respect to the valence band maximum. Also shown are the energies of the conduction band edge and of the T_2 and A_1 states of nitrogen vacancies [10].

Two questions could be asked at this point. First, is the lower electron concentration in series B due to higher concentration of compensating acceptors or to a lower concentration of donor centers. Second, are there any impurities that could account for the observed difference in electrical properties of our two series of AlGaN samples?

It has already been mentioned that the mobilities in the samples of the series B are much higher than in the samples of the series A. Such behavior indicates lower total concentration of ionized centers in samples from series B, and therefore favors a decrease of native donor defects at elevated growth temperature. An additional experimental evidence supporting the above was obtained from optical measurements. The near band edge absorption tail is much less pronounced in series B (Fig. 3a.) indicating lower ionized defect concentration rather than increased compensation ratio. Finally, in Table 1 we show the measured concentrations of some common contaminant impurities in MOCVD growth. It can be seen from the table that the concentration of main compensating acceptor, carbon, is in fact higher in the samples of the series A grown at 1000 ^{o}C.

From the same table it also follows that the concentration of most common unintentional donor-type impurities, Si, are the same in all AlGaN samples with low x, although these donors could play some role in $Al_xGa_{1-x}N$ layers of the series A with x>0.1 (Si concentration is higher in AlN grown at 1000 ^{o}C than at 1050 ^{o}C; concentrations of other possible background donors, S and Se, were below the detection limit of SIMS in all the samples).

Among other impurities in $Al_{0.12}Ga_{0.88}N$ only the concentrations of oxygen and hydrogen are substantially higher in the samples of the series B. Hydrogen has been shown to passivate donors in AlGaN [11], but complexes of hydrogen with donors do not survive annealing at temperatures higher than 800 ^{o}C, while the electrical properties of our AlGaN samples did not change after annealing in nitrogen at 800 ^{o}C for 1h. Oxygen is believed to be a donor in AlGaN [12], and, therefore, higher oxygen concentration in the samples of the series B does not easily explain why the carrier concentration is lower in these samples (and more statistics seems to be necessary in this point: the trend in oxygen concentration dependence on growth temperature observed in AlGaN in table 1 goes against the trend in GaN and AlN in the same table) .

The conclusion of the above arguments is that the difference in electrical and optical properties between the AlGaN samples of the two series stems from lower concentration of donor-type native defects in series B. The composition dependence of the activation energies qualitatively agrees with theoretical predictions for the T_2 (V_N) levels for the samples of the series A, and A_1 (V_N) levels for the samples of the series B.

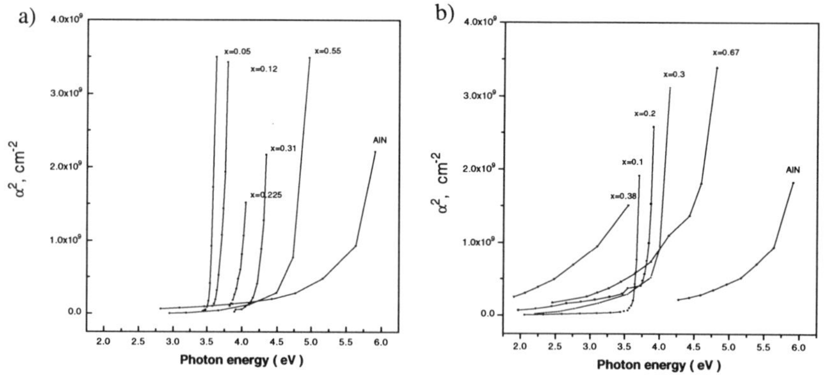

Fig.3a,b. Square of absorption coefficient (α^2) versus photon energy plots for several AlGaN samples of the series A (a) and of the series B (b).

In addition to deep defects discussed above some other deep centers have been detected by PL measurements. Four defect bands were observed in PL spectra of our AlGaN layers: 3.65-3.7 eV, 3.5-3.55 eV, 3.0-3.1 eV and 2.1-2.3 eV. The energy of the bands did not change significantly with composition. The 3.65 eV band was dominant in the $Al_xGa_{1-x}N$ samples of the series B with x<0.2 (the intensity of the 3.1 eV and 2.1 eV bands constituted no more than 5% of the intensity of the 3.65 eV band). The 3.5 eV band first emerged for x=0.22 and totally replaced the 3.65 eV band for higher Al mole fractions. In the samples of the series A with x=0.12, the 3.5 eV band intensity was about an order of magnitude higher than the intensity of the 3.65 eV band, but the most intense band of all was the 2.3 eV band (the ratio of intensities for the 3.65, 3.5, 3.1 and 2.3 eV bands was 2:18:5:30). Limited space does not allow us to discuss the possible nature of these defects bands.

Table 1. Concentrations of some impurities in GaN, $Al_{0.12}Ga_{0.88}N$ and AlN grown at 1000°C and at 1050°C.

Impurity Atoms	GaN 1000 °C	GaN 1050 °C	AlGaN 1000 °C	AlGaN 1050 °C	AlN 1000 °C	AlN 1050 °C
C	2.5×10^{18}	1.5×10^{18}	5×10^{18}	1.5×10^{18}	$2-3\times10^{19}$	1.5×10^{18}
H	3.5×10^{18}	1.5×10^{18}	$2-5\times10^{18}$	10^{19}	$2-3\times10^{19}$	10^{19}
O	6×10^{17}	5×10^{17}	10^{19}	3×10^{19}	$2-3\times10^{20}$	3×10^{19}
Si	2×10^{18}	2×10^{18}	2×10^{18}	2×10^{18}	10^{19}	2×10^{18}

Finally it would be interesting to know which of the two factors, the increased growth temperature or the reduced growth rate, play a major role in improvement of electrical and optical properties of our AlGaN layers. For that purpose we grew $Al_xGa_{1-x}N$ layers with x=0.12 and 0.22 at 1050 °C with a growth rate of 1 µm/h and at 1000°C with a growth rate of 0.7 µm/h. In both cases the electron concentrations were close to 10^{16} cm^{-3}, with mobilities of about 100 $cm^2/V\cdot s$. Thus the concentration of defects in these two cases was lower than for the samples of the first series, but higher than for the samples of the second series.

CONCLUSIONS

Hall effect, near band-edge optical absorption, and SIMS measurements were used to characterize concentrations of native donors and impurities in AlGaN alloys. The native donor concentration decreases with increase of growth temperature and decrease of growth rate. Thermal activation energies of dominant electron traps agree with the A_1 double donor and T_2 donor state of nitrogen vacancy.

ACKNOWLEDGMENTS

The work at CMU was supported in part by AFOSR Grant F49620-95-1-0087.

REFERENCES

[1]. H. Morkoc, S. Strite, G.B. Gao, M.E. Lin, B. Sverdlov and M. Burns, J. Appl. Phys, **76**, 1363 (1994)

[2]. S. strite and H. Morkoc, J. Vac. Sci. Technol., **B10**, 1237 (1992)

[3]. D.K. Gaskill, L.B. Rowland and K. Doverspike, Electrical Properties of AlN, GaN and AlGaN, in Properties of Group III Nitrides, edited by J.H. Edgar (INSPEC Publications, London, 1994) pp. 101-116

[4]. S. Yoshida, S. Misawa and S. Gonda, J. Appl. Phys., **53**, 6844 (1982)

[5]. X. Zhang, P. Kung, A. Saxler, D. Walker, T.C. Wang and M. Razhegi, Appl. Phys. Lett., **67**, 1745 (1995)

[6]. H. Murakami, T. Asahi, H. Amano, K. Hiramatsu, N. Sawaki and I. Akasaki, J. Cryst. Growth, **115**, 648 (1991)

[7]. J.A. Miragliotta, Optical Functions of GaN, in Properties Of Group III Nitrides, edited by J.H. Edgar (INSPEC Publications, London, 1994) pp. 190-194
[8]. S. Loughin and R.H. French, Optical Functions of AlN, ibid., pp. 175-189
[9]. R.G. Wilson, F.A. Stevie and C.W. Magee, Secondary Ion Mass Spectrometry (Wiley, New York, 1989)
[10]. D.W. Jenkins, J.D. Dow and M-H. Tsai, J. Appl. Phys., **72**, 4130 (1992)
[11]. A.Y. Polyakov, M. Shin, M. Skowronski, D.W. Greve, S.J. Pearton and J.A. Freitas, this volume
[12]. S.C. Strite, Impurities and Dopants in Group III Nitrides, in Properties of Group III Nitrides, edited by J.H. Edgar (INSPEC Publications, London, 1994) pp. 272-278

Influence of Forced Diffusion of Boron on Electrical Conductivity of Diamond Films

T. Sung[1], G. Popovici[1], M. A. Prelas[1], R. G. Wilson[2], Kim Bigelow[3], J. Chacon[1], and S. K. Loyalka[1].

[1]Nuclear Engineering Dept., University of Missouri-Columbia, Columbia, MO, 65211.
[2]Hughes Research Laboratories, Malibu, CA 90265.
[3]Norton Diamond Film, Northgboro, MA, 01532.

ABSTRACT:

Four CVD diamond films grown on tungsten carbide were used for diffusion. Diffusion was performed in a hydrogen atmosphere. Three methods of diffusion were used: conventional diffusion due to concentration gradient, forced diffusion under a dc electric bias with thermal ionization and forced diffusion with optical and thermal ionization of boron acceptor level in diamond. The temperature dependence of the electrical conductivity of the diffused samples was measured in the temperature range 300-600 K. The increase of approximately six order of magnitude in conductivity of the films doped by forced diffusion was obtained. The samples diffused with boron by conventional diffusion due to concentration gradient showed only one order of magnitude increase in electrical conductivity.

INTRODUCTION:

Boron doping changes an insulating diamond to a semiconducting diamond. Common methods of doping such as during growth[1,2], ion implantation[3-4], solid state diffusion[5] and forced diffusion[6] have been used for diamond doping. In paper, a study of electrical properties of diamond films is performed. A comparison of electrical properties due to conventional diffusion and forced diffusion is made.

EXPERIMENT:

The four diamond films supplied by Norton Company were investigated. Before diffusion, the samples were cleaned in a hot CrO_3 + H_2SO_4 solution for 20 minutes and rinsed with deionized water to eliminate the non-diamond carbon on the surface. Then, the samples were cleaned in a hot HNO_3 + HCl + H_2O solution for 20 minutes and rinsed with deionized water to purge the surface of metal impurities. Boron powder was used for dopant source and mixed with carbon paste to assure good electric contact. These samples were placed on a graphite support which was heated by tungsten wire. The temperature of support was monitored by a K-type thermal couple. The general arrangement for forced diffusion with optical and thermal ionization is presented in Figure 1. The chamber was flushed with hydrogen gas at one atmosphere. After diffusion, the samples were cleaned again by the same cleaning procedure described above.

The resistance measurements were performed from room temperature to 350 °C in air. The two-electrode scheme was used for measurements. Two parallel carbon paste strips, 8 mm in length, 1 mm in width and separated by 1.5-2 mm, were applied on the surface of the samples. Figure 2 presents the apparatus of conductance measurements.

Mat. Res. Soc. Symp. Proc. Vol. 423 © 1996 Materials Research Society

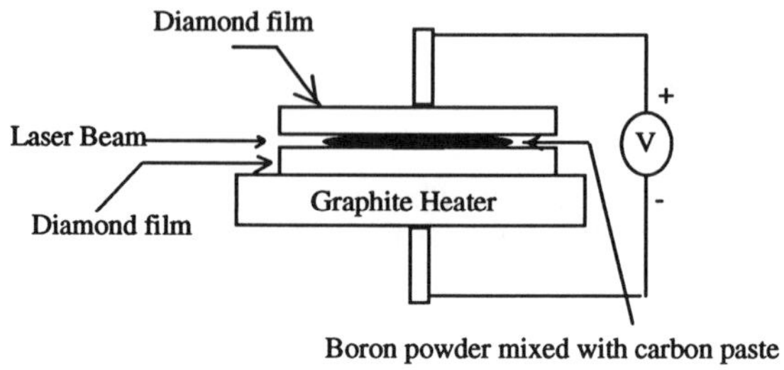

Figure 1. The arrangement of forced diffusion with optical and thermal ionization.

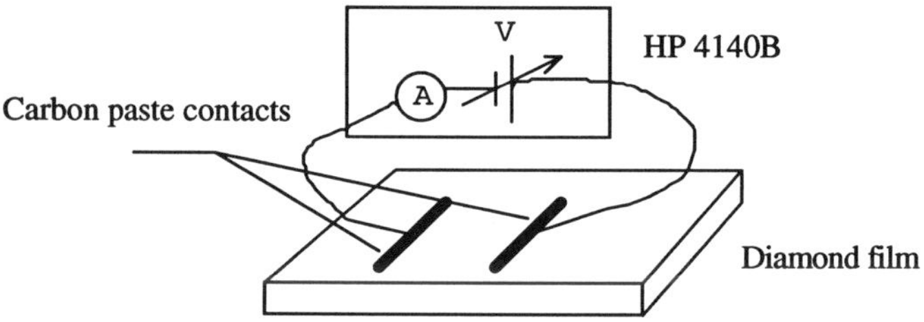

Figure 2. The arrangement of conductance measurements

A computerized Hewlett-Parkard picoammeter with build-in dc voltage source, model 4140B, was used. In the range of bias from -1 V to 1 V, the contact had an ohmic behavior over all temperature range. The conductivity measurements were performed while cooling the sample, in order to minimize the influence of contact change with temperature.

RESULTS AND DISCUSSIONS:

Raman spectrum of all these samples shows the diamond peaks at 1341 cm^{-1} and the broad graphite peaks at 1530 cm^{-1}. The summary of diffusion conditions and electrical measurements are shown in Table 1. The sign of bias is on the surface which has the dopant source. Figure 3. presents the plots of conductance vs. temperature for 4 samples. It is found that the activation energy of the samples before diffusion is about 1.0 eV and the resistance is larger than 4×10^{11} Ω (limit of the measurement system). These values are characteristic of undoped diamond[7,8]. The activation energy ΔE was determined using the formula $G = G_0 \exp(-\Delta E/kT)$, where G is conductance; G_0 is a constant; k is Boltzman constant; T is temperature.

The diffusion of boron due to concentration gradient (for simplicity, named "diffusion" later on) does not produce any change in activation energy. After diffusion in the sample 1 and 2 (step 1) the activation energy remained 1.0 eV, the same as the non-diffusion sample. The change of the electrical conductivity approximately one order of magnitude might be due to thermal annealing of the samples during the diffusion process. We assume that boron did not enter substitutionally the diamond lattice during the diffusion process. The change in activation energy of the sample 1 and 2 was obtained only after the second step of diffusion when forced diffusion was used. A higher change in electrical conductivity was obtained for the sample #2, that have sign minus on the diffusion surface.

Table 1. The summary of the diffusion conditions and electrical measurements

Sample	1	2	3	4
Thickness(μm)	24	17	28	15
Resistance (Ω) at room temperature	>4x10^{11}	>4x10^{11}	>4x10^{11}	> 4x10^{11}
Activation Energy (eV)	1.0	1.0	1.0	0.7
Step 1	800 °C, 22hr, no laser	800 °C, 22hr, no laser	800 °C, 22hr, -15V, -45mA, no laser	800 °C, 22hr, 15V, 45mA, no laser
Activation Energy (eV)	1.0	1.0	0.47	0.43
Step 2	1000 °C, 4.5hr, 300V, <1 mA, no laser	1000 °C, 4.5hr, -300V, <1 mA, no laser	1000 °C, 4.5hr, 300V, <1 mA, laser on	1000 °C, 4.5hr, -300V, <1 mA, laser on
Activation Energy (eV)	0.44, 1.0	0.44, 1.0	0.33, 0.5	0.44, 0.7
Step 3	1000 °C, 8hr, no laser	1000 °C, 8hr, no laser	1000 °C, 8hr, no laser	1000 °C, 8hr, -5V, -40 mA, no laser
Activation Energy (eV)	0.44, 1.0	0.35, 1.0	0.33, 0.5	0.35, 0.7
Step 4	1000C, 4hr, -11V, -40mA, no laser	1000C, 4hr, 11V, 40mA, no laser	1000C, 4hr, no laser	1000C, 4hr, no laser
Activation Energy (eV)	0.44, 1.0	0.23, 1.0	0.21, 0.5	0.35, 0.7

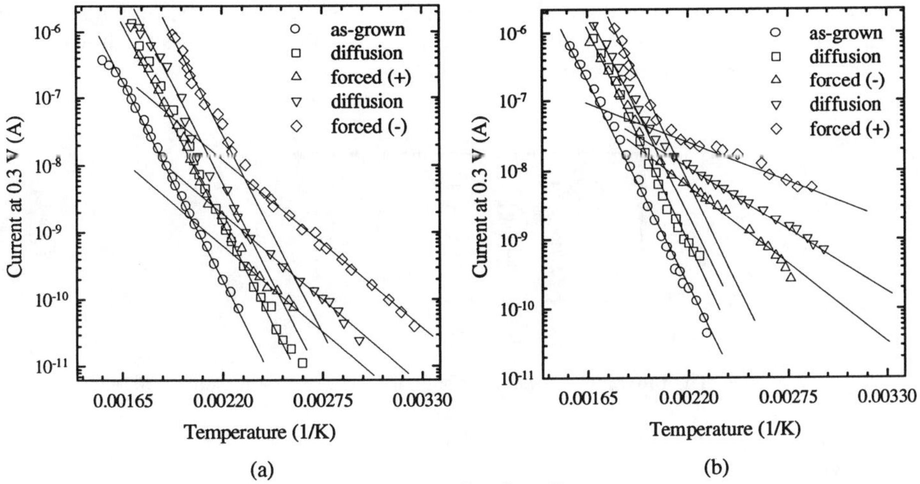

Figure 3. (Continued)

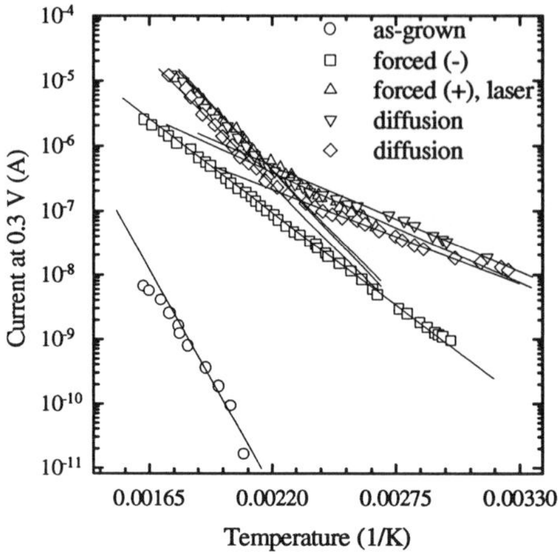

(c)

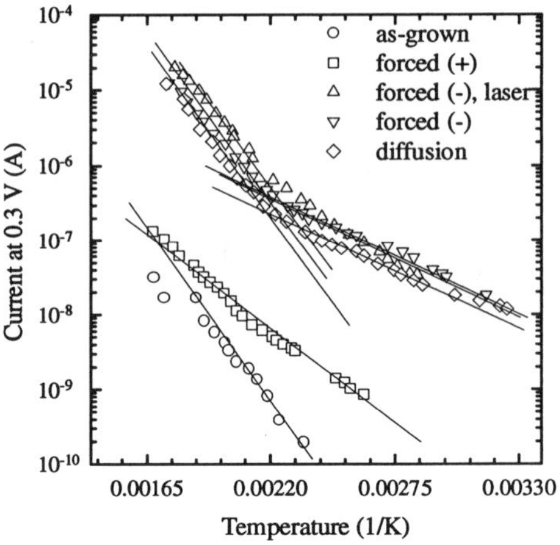

(d)

Figure 3. Temperature vs. Conductance plots : (a) Sample #1, (b) Sample #2 (c) Sample #3 (d) Sample #4

The change of activation energy of the sample 3 and 4 was obtained after the first step (forced diffusion). The activation energy changes from 1.0 eV to 0.44 eV, that is close to the reported activation energy of boron level in diamond ΔE_g = 0.37 eV. The error in the determination of activation energy approximately 20% might be due to ohmic contacts.

The forced diffusion results showed that the voltage sign on diffusion surface is very important. The large change in electrical conductivity was obtained in the step 1 for sample 3 and in the step 2 for sample 4, when a negative bias was applied to the surface. This can be explained that boron acted as negative ions in the diamond lattice. The negative bias has a strong enhancement on boron movement into the diamond lattice. Figure 4. shows the mechanism of moving of boron ions in the diamond lattice under dc bias.

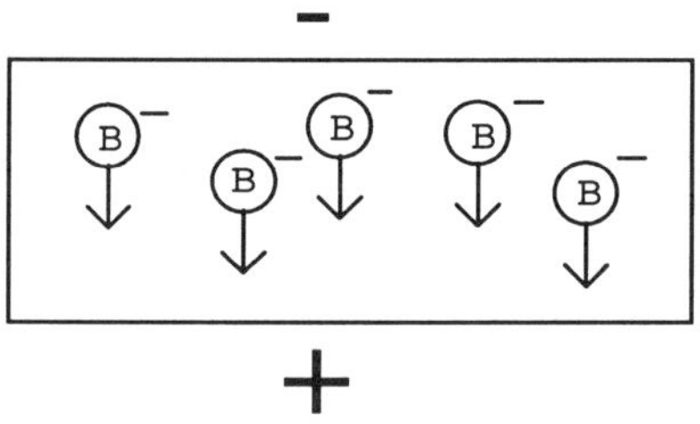

Figure 4. The movement of boron ions in the diamond lattice under dc bias

Diffusion performed after forced diffusion had a little effect on the electrical conductivity of the samples 2 and 4.

The activation energy of conductivity slightly decreased at higher diffusion steps. That might be explained by formation of the boron impurity band at higher concentration of boron.

CONCLUSIONS:

Diffusion due to concentration gradient and forced diffusion of boron in CVD diamond films has been performed. Diffusion due to concentration gradient has little effect on electrical conductivity of the diffused samples. Forced diffusion caused a change of activation energy of the electrical conductivity from 1 eV to ~0.4 eV, indicating that the diffused boron, at least partially, entered the diamond lattice substitutionally.

ACKNOWLEDGMENTS:

The authors gratefully acknowledge the support of NSF foundation DMI #9460657. The authors would also like to thank Dr. Henry White and Fatemeh Shahedipour for the Raman measurements.

REFERENCES:

1. B. V. Spitsyn and A. E. Alexenko, in: Proc. 2nd Int. Symp. on Diamond Materials, v. **91-6**, eds. A. J. Purdes, J. C. Angus, R. F. Davis, B. M. Meyerson, K. E. Spear and M. Yoder (Electrochemical Society Proceedings, 1991, Pennington), 597-603.

2. K. Okano, Y. Akiba, T. Kurosu, M. Iida, and T. Nakamura, Journal of Crystal Growth,1990, v. **99**, 1192-1195.

3. V. S. Vavilov, M. A. Gukasyan, M. I. Guseva and E. A. Konorva, Soviet Phys. Semicond. 1972, **6**, 741.

4. A. V. Densienko, A. A. Melnikov, A. M. Zaitsev, V. I. Kurganski, A. J. Shilov, Ju. P. Gorban and V. S. Varichenko, Materials science and Engineering, 1992, B 11, 273-277.

5. W. Tasi, M. Delifino, D. Hodul, M. Riaziat, L. Y. Ching, G. Reynold, and C. B. Cooper, IEEE Electron Device Letters, 1991, v. **12**, n. 4, 157-159.

6. Galina Popovici, T. Sung, S. Khasawinah, M. A. Prelas and W. G. Wilson, J. Appl. Phys., 1995, v. **77**, 5625-5629.

7. Y. Muto, T. Sugino, and J. Shirafuji and K. Kobashi, Appl. Phys. Lett. 1991, **59**, (7), 843-845

8. Bohr-ran Huang and D. K. Reinhard, Appl. Phys. Lett. 1991, **59**, (12) 1494-1496.

MEASUREMENT OF THE ACTIVATION ENERGY IN PHOSPHOROUS DOPED POLYCRYSTALLINE DIAMOND THIN FILMS GROWN ON SILICON SUBSTRATES BY HOT FILAMENT CHEMICAL VAPOR DEPOSITION

S. MIRZAKUCHAKI, H. GOLESTANIAN, E. J. CHARLSON, T. STACY
Electrical Engineering Department, University of Missouri, Columbia, MO 65211

ABSTRACT

Although many researchers have studied boron-doped diamond thin films in the past several years, there have been few reports on the effects of doping CVD-grown diamond films with phosphorous. For this work, polycrystalline diamond thin films were grown by hot filament chemical vapor deposition (HFCVD) on p-type silicon substrates. Phosphorous was introduced into the reaction chamber as an in situ dopant during the growth. The quality and orientation of the diamond thin films were monitored by X-ray diffraction (XRD) and scanning electron microscopy (SEM). Current-voltage (I-V) data as a function of temperature for gold-diamond film-silicon-aluminum structures were measured. The activation energy of the phosphorous dopants was calculated to be approximately 0.29 eV.

INTRODUCTION

In order for polycrystalline diamond film to be considered for electronic applications, an effective and controllable method for both p- and n-type doping must be found. There are many reports on the formation of prototype devices on diamond doped with boron.[1-4] The difficulty in doping of diamond arises from the fact that C atoms have a small tetrahedral radius of 0.77 Å, and almost all impurity atoms (except B) are larger and deform the lattice.[1]

The solubility of a dopant atom in diamond is related to its formation energy (E_f) which is calculated based on the bulk cohesive energy of diamond and bulk cohesive energy of the impurity. The B atom has a negative E_f which explains its ease of incorporation into the diamond lattice. Phosphorous, on the other hand, has a low equilibrium solubility in diamond as a result of its high formation energy ($E_f = 10.7$ eV), which explains the difficulty of n-type doping of diamond with phosphorous. Okano et. al.[5,6] used phosphorous pentoxide (P_2O_5) as a source of n-type doping. They have reported the formation of a diamond p-n junction diode, but since the electron mobility is extremely low (only 50 cm^2/v.s), n-type conductivity may be due to lattice damage and not P incorporation into the diamond lattice. The main problem with doping diamond film n-type seems to be the low degree of ionization of the P atoms at room temperature.

Other dopant sources such as Li and Na as well as the effect of N impurity (as a shallow donor) have been investigated.[7] Ramesham et. al.[8] found an increase in the resistivity of their diamond films as a result of P incorporation.

As a general rule, it is often difficult to make ohmic contacts to large bandgap semiconducting materials such as diamond. Most of the methods utilized to form ohmic contacts to bulk semiconducting diamond have involved roughening, graphitizing the diamond surface, or carbide formation under the contact.[1] Ohmic contact to p-type homoepitaxially grown diamond has been created by intense argon-fluoride excimer laser radiation.[9]

In order to measure the activation energy of the P atoms incorporated in a diamond film,

Mat. Res. Soc. Symp. Proc. Vol. 423 © 1996 Materials Research Society

it is necessary to study the effect of temperature on the conductivity of the films. In this work we report on the growth and in situ doping of diamond thin film with phosphorous by HFCVD as well as formation of ohmic contacts and the measurement of the activation energy of dopant atoms.

To the best of the authors' knowledge, there are few reports on the activation energy measurements on P doped diamond films. Activation energies of 0.1 to 0.37 eV have been reported for B doped diamond films.[1] Okano et. al.[6] report an activation energy value of 0.05 eV for their diamond film doped with P_2O_5. Moreover, they reported the resistivity as independent of the P concentration in the reactant gas, which puts the n-type conductivity due to P dopant atoms in question. Bernholc et. al.[7] have calculated a shallow level of 0.2 eV below the bottom of the conduction band for phosphorous, which is in close agreement to the result found here.

EXPERIMENT

Boron doped Si wafers of (100) orientation and resistivity 5-15 ohm-cm were cut into various sizes (maximum of 1 inch on each side). Prior to surface pretreatment, these substrates were organically cleaned in an ultrasonic bath by trichloroethylene, acetone, methanol, and DI water for five minutes each. They were further cleaned by submersion into hot nitric and hydrochloric acids (five minutes each) and were finally rinsed by hot DI water. In order to promote nucleation, these samples were scratched by mechanical polishing with diamond paste of 0.25 μm grit size.

Polycrystalline diamond films were deposited on Si samples in a HFCVD reactor. Trimethyl phosphite diluted in acetone to a concentration of 400 ppm was used as a source of in situ doping. The deposition conditions used in this study are as follows; hydrogen flow rate of 98 sccm, methane flow rate of 1.5 sccm, a mixture of acetone and trimethyl phosphite at a flow rate of 0.5 sccm, pressure of 25 Torr, filament temperature of 2000 °C, substrate temperature of 720 °C, and filament to substrate distance of 4 mm.

Circular contacts (0.021 cm^2 in area) were deposited on the phosphorous-doped diamond films by sputtering a thin layer, ~500 Å, of gold through a metal shadow mask in a DC glow discharge system. Prior to the sputtering of gold contacts, the surfaces of the diamond films were cleaned by a saturated solution of $CrO_3 + H_2SO_4$ at 200 °C for 20 minutes followed by rinsing in a 1:1 boiling solution of H_2O_2 and NH_4OH for 10 minutes and then in boiling DI water for 3 minutes. This procedure removes the nondiamond carbonaceous surface layer on top of diamond film, the presence of which has been shown to alter the electrical properties of metal-diamond junctions.[10] In order to provide an ohmic contact to the back of the samples, a thin layer of Al, ~650 nm, was thermally evaporated onto the unpolished side of the Si substrates.

To remove the ambiguity of the possible effects of the diamond-Si interface, a circular diaphragm was etched (by a hydrofluoric solution) on the back Si substrate to fabricate a free standing area of diamond film. Gold contacts designated as C2 and C3 were sputtered over the diamond diaphragm area and contact C1 was sputtered over the entire structure that includes Si substrate.

This sample was then placed in an oven (at atmospheric pressure) and the temperature was gradually increased from 300 K to 600 K and was monitored by a K-type thermocouple placed close to the sample. The I-V characteristics were taken between C1 and the back Al contact at 25 degree intervals using a computer controlled Hewlett Packard model 4140B.

RESULTS

Fig. 1 shows the SEM view of the surface and the cross section of diamond film grown here. As evident in these pictures, the diamond film consists of randomly oriented crystallites with (100) and (111) growth orientation. The cross sectional view reveals the diamond film's smooth surface and a thickness of 5-6 μm. A typical XRD pattern taken in the diffraction angle (2θ) range of 20° - 130° and shown in Figure 2 contains four diamond Bragg peaks in addition to the Si peak.

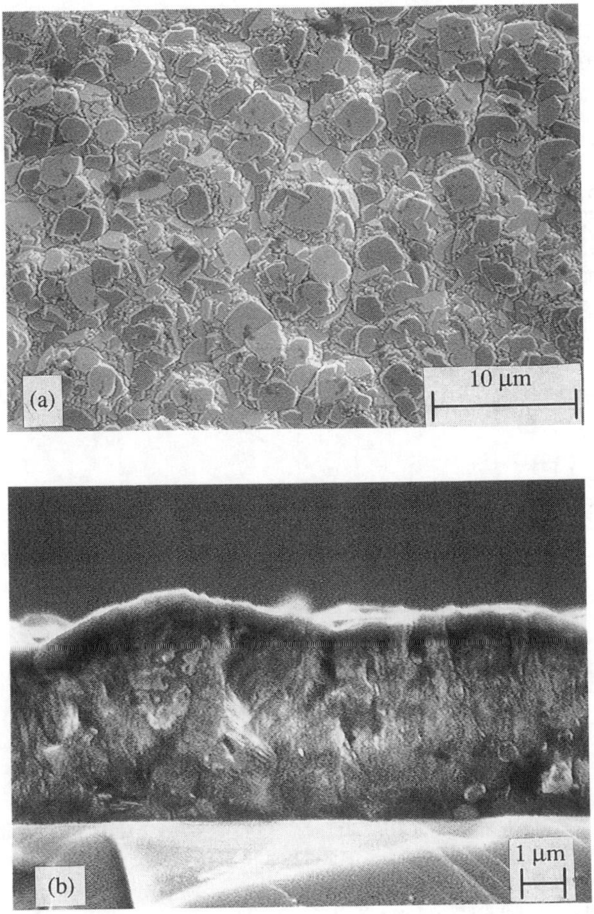

Figure 1. Surface (a) and cross section (b) of the diamond film grown on silicon.

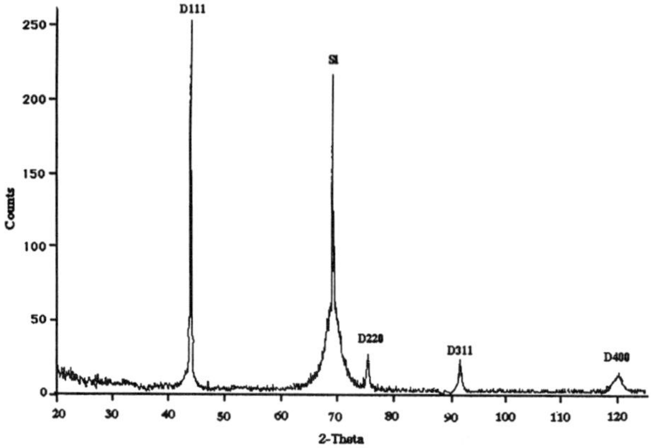

Figure 2. XRD pattern of the P-doped diamond film.

Fig. 3 demonstrates the I-V curves between gold contacts sputtered over the diamond diaphragm (C2 to C3) and also between one such contact and another one on the diamond film supported by the Si substrate (C1 to C2). Both of the curves in this figure show the ohmic behavior of gold contact to self-supporting diamond film.

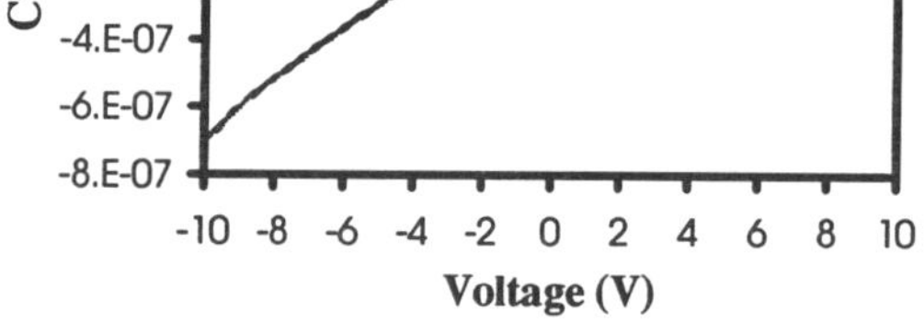

Figure 3. Ohmic I-V curves on the gold-P doped diamond film.

The resistivity of thin film diamond can be estimated from I-V curves because the contact radius (0.8 mm) is much larger than the thickness of the diamond film (6 μm) and therefore the current is confined to the contact area. On the other hand, the diamond bulk resistance is much larger than the Si substrate and hence the voltage is assumed to be mainly dropped in the diamond film.

Figure 4 shows the change in the I-V curve of the diamond film due to exposure to the mentioned range of temperature. The exponential dependence of bulk resistivity on the temperature is depicted in Fig. 5(a) and can be formulated by the Arrhenius law;[11]

$$\rho \propto \exp (Ea / kT) \qquad (1)$$

where ρ is the bulk resistivity, Ea the activation energy, k the Boltzmann's constant, and T is the absolute temperature in K.

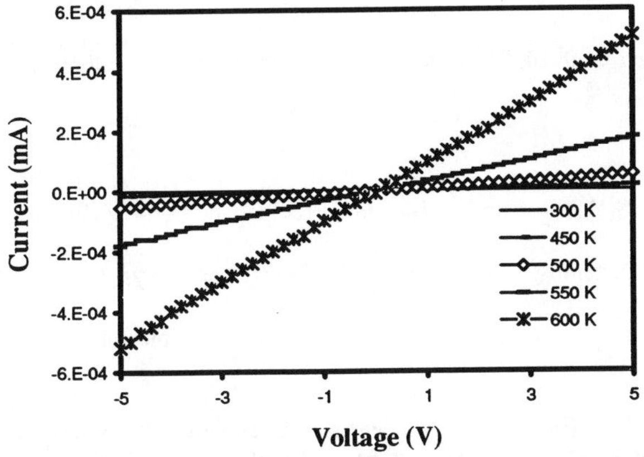

Figure 4. Change of current vs. temperature shown here.

From the Arrhenius equation, the natural log of the resistivity value vs. the inverse of the temperature should be a straight line, as shown in Fig. 5(b), the slope of which is equal to the activation energy divided by the Boltzmann's constant. The activation energy was calculated to be 0.29 eV. The high value of the activation energy demonstrates the low degree of ionization of the P atoms at room temperature and hence the difficulty of effectively doping diamond n-type.

Fig. 5. Resistivity vs. temp. (a), and the natural log of resistivity vs. inverse temp. (b).

CONCLUSIONS

In this paper we have reported on the temperature measurement studies on phosphorous doped polycrystalline diamond thin film. The results of the temperature measurement in the range of 300-600 K suggest an activation energy of 0.29 eV for the dopant atoms.

ACKNOWLEDGMENTS

The authors wish to thank Louis M. Ross and Naiyu Zhao for their assistance in the SEM and XRD studies.

REFERENCES

1. G. Sh. Gildenblat, S. A. Grot, and A. Badzian, Proc. IEEE **79**, 647 (1991).

2. M. W. Geis, N. N. Efremow, and J. A. von Windheim, Appl. Phys. Lett. **63**, 952 (1993).

3. V. Venkatesan, J. von Windheim, and K. Das, IEEE Transactions on Electron Devices **40**, 1556 (1993).

4. G. Zhao, T. Stacy, E. J. Charlson, E. M. Charlson, C. H. Chao, M. Hajsaid, J. Meese, G. popovici, and M. Prelas, Appl. Phys. Lett. **61**, 1119 (1992).

5. K. Okano, H. Kiyota, T. Iwasaki, T. Kurosu, M. Iida, and T. Nakamura in Proceedings of the 2nd International Conference on New Diamond Science and Technology, edited by R. Messier et. al., (Washington, DC, 1990) pp. 917-922.

6. K. Okano, H. Kiyota, T. Iwasaki, Y. Nakamura, Y. Akiba, T. Kurosu, M. Iida, and T. Nakamura, Appl. Phys. A **51**, 344 (1990).

7. J. Bernholc, S. Kajihara, and A. Antonelli in Proceedings of the 2nd International Conference on New Diamond Science and Technology, edited by R. Messier et. al., (Washington, DC, 1990) pp. 923-928.

8. R. Ramesham, C. Ellis, and T. Roppel in Proceedings of the First International Conference on the Applications of Diamond Films and Related Materials, edited by Y. Tzeng, M. Yoshikawa, M. Murakawa, and A. Feldman, (Elsevier, NY, 1991) pp. 411-416.

9. M. Geis, Proc. IEEE **79**, 669 (1991).

10. S. Grot, G. Gildenblat, C. Hatfield, C. Wronski, A. Badzian, T. Badzian, and R. Messier, IEEE Electron Device Letters **11**, 100 (1990).

11. S. Grot, C. Hatfield, G. Gildenblat, A. Badzian, and T. Badzian, Appl. Phys. Lett. **50**, 1542 (1991).

THEORY AND REALIZATION OF A TWO - LAYER HALL EFFECT MEASUREMENT CONCEPT FOR CHARACTERIZATION OF EPITAXIAL AND IMPLANTED LAYERS OF SiC

Adolf Schöner, Kurt Rottner, and Nils Nordell
Industrial Microelectronics Center (IMC)
P.O Box 1084, S-164 21 Kista, Sweden

ABSTRACT

Epitaxial and implanted layers are generally characterized by Hall effect measurements using a pn-junction as electrical insulation of the layer from the substrate. Due to defects, low doping concentrations or thin layers the resistivity of epitaxial or implanted layers is comparable to the resistivity of the pn-junction and the substrate. This results in inefficient electrical insulation between both regions. To be able to determine the properties of epitaxial or implanted layers even in the case of substantial leakage current we developed a two-layer Hall effect measurement concept. This concept is based on the conventional van der Pauw technique applied to the layer and the substrate separately. In addition the current-voltage characteristic of the pn-junction is measured and modeled in the analysis as an ohmic resistor.
This two-layer concept is applied to epitaxial grown SiC and the results are compared with conventional van der Pauw technique. In addition both techniques are compared with the results of capacitance-voltage (CV) measurements and secondary ion mass spectroscopy (SIMS).

INTRODUCTION

The importance of SiC as a material for high temperature and high power devices is steadily increasing. Among other things, this is accomplished by an improved doping control in both epitaxial crystal growth and in ion implantation processes. By this development the demands on the accuracy in techniques for characterization of doped layers is increased.
A powerful and well established technique for determining fundamental electrical transport properties is the measurement of the Hall effect. From this technique resistivity, free carrier concentration and carrier mobility can be deducted. Ionization energy and concentration of dopants governing the conductivity type, as well as the degree of compensation can be determined from the temperature dependence of the carrier concentration by a least squares fit of the neutrality equation to the experimental data. However, this Hall effect measurements are only accurate, if they are performed at homogeneously doped layers. To achieve this conditions when measuring epitaxial grown or implanted layers on a substrate, the use of semi-insulating substrates or the insulation by a pn-junction formed between the substrate and the measured layer are generally be sufficient. In the case of SiC, semi-insulating substrates do not yet commercially exist, and pn-junctions formed by ion implantation or epitaxial growth block high voltages within small areas, but they become leaky, when they are extended over large areas, which is the case in Hall effect measurements. One main reason for this leakage currents is a high concentration of crystal defects, e.g. micropipes.
We have developed a technique to cope with the influence of undesired currents through the substrate on the Hall effect results. Our method makes it possible to separate two layers, e.g. a substrate and an epitaxial layer, from each other. This is accomplished by an accurate measurement of the resistivity of the pn-junction, and subsequently by calculating the carrier concentration and mobility in the substrate and epitaxial layer separately.

Mat. Res. Soc. Symp. Proc. Vol. 423 © 1996 Materials Research Society

In this paper we present the theory of a two-layer Hall effect measurement concept. We show results from measurements on p- and n- doped epitaxial layers grown on 6H-SiC, and compare them with results obtained with the conventional van der Pauw technique, as well as SIMS and CV-measurements.

SIMULATION OF THE TWO-LAYER HALL EFFECT MEASUREMENT

In the conventional van der Pauw technique resistivity, carrier concentration and mobility are calculated for a single layer [1]. For the two-layer Hall effect concept these basic equations have to be extended by taking into account the interaction between the two layers. A detailed description of a possible equivalent circuit for this extension and the equations for calculating Hall coefficients and resistivities of the two layers are summarized in ref. [2]. In this paper we describe a method to simulate Hall effect measurements on two-layer pn-structures.
Assuming two homogeneously doped layers of different resistivity and opposite conductivity type, the free carrier concentration in each layer can be described by the neutrality equation. For the calculation of the neutrality equation values for the ionization energy and concentration of dopants, and for the compensation are needed. Knowing the carrier concentrations the Hall coefficients and the resistivities of the n- and the p-layer can then be calculated by using a Hall scattering factor of 1.0. For the calculation of the resistivity the mobilities of electrons and holes as a function of the temperature are described by a combination of the mobilities for phonon and impurity scattering mechanisms. Using the standard van der Pauw equations and knowing the thicknesses of the layers, the internal Hall voltages and the voltage drops during resistivity measurements can be derived for each layer separately.
If these independently modeled layers of opposite conductivity type are brought into contact, a pn-junction is formed between them. By keeping the voltage across the pn-junction well below its turn on voltage, the current-voltage characteristic of the pn-junction can be linear approximated and the pn-junction can be treated as an ohmic resistor. The parallel equivalent circuit obtained is then the basis for further calculations of current and voltage distributions in the composite structure. In the following it is assumed, that the current is applied only to one of the layers (e.g. to the epitaxially grown or implanted layer) to be able to compare the simulation results with results of conventional van der Pauw measurements on epitaxial layers. The pn-junction at the interface is assumed to be perfectly uniform.

EXPERIMENT

For Hall effect measurements samples were epitaxially grown on commercially available substrates in a horizontal reactor for vapor phase epitaxy in detail presented in ref. [3]. Silane and propane were used as growth precursors, while trimethylaluminium (TMAl) and nitrogen were used to achieve p- and n-type conductivity, respectively. The growth temperature was about 1600°C, the reactor pressure 800 mbar, and the growth rate 2.5 μm/h, with palladium diffused hydrogen as carrier gas.
Four ohmic contacts with a thickness of 100nm were formed both on the substrate and on the epitaxial layer by electron beam evaporation of either nickel or titanium followed by a subsequent anneal at 950°C. Standard van der Pauw Hall effect measurements were performed in a temperature range of 100K to 450K on two-layer structures consisting of a thin epitaxial layer on top of a substrate of opposite conductivity type. The van der Pauw technique was extended for measurements both at the substrate and at the epitaxial layer site. In addition the current-voltage characteristic and the resistance of the pn-junction was recorded.

The atomic concentration of dopants was measured with secondary ion mass spectroscopy (SIMS) using a CAMECA ims 4f instrument. The Al profiles were obtained with O_2^+ primary ion beam and detection of $^{27}Al^+$. Capacitance-voltage (CV) measurements on evaporated titanium Schottky contacts were used to determine the net doping concentration.

RESULTS AND DISCUSSION

Fig. 1 shows the Al concentration determined by different measurement techniques as a function of the Al content in the gas phase during growth. In the concentration range below $10^{18}cm^{-3}$ we found a good agreement of the net doping concentration determined by CV-measurements and the atomic concentration measured by SIMS, indicating an almost complete incorporation of Al on electrically active lattice sites.

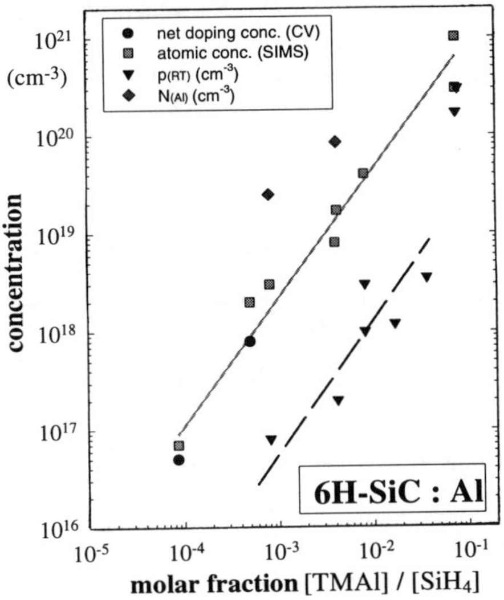

Fig.1 Al concentration measured by different electrical and atomic methods as a function of the Al content in the gas phase of the VPE reactor.

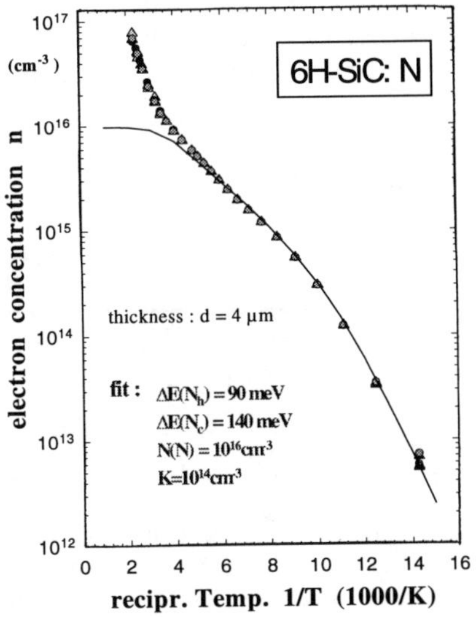

Fig.2 Temperature dependence of the electron concentration of an epitaxially grown n-type layer measured by the van der Pauw Hall effect method.

Compared to the linear dependence of the atomic Al concentration on the TMAl flow (SIMS) the Hall effect analysis gives a different result. Although the free hole concentration at room temperature in the Al-doped layers has the same dependency, resulting in a degree of ionization of around 1%, the Al doping concentration calculated by a fit of the neutrality equation to the experimental data is almost an order of magnitude higher than the atomic concentration. This discrepancy is less pronounced at higher Al concentrations, but is still too large to be explained by inaccurate material parameters (e.g. effective masses) or their temperature dependence.

In the case of N-doped n-type layers we found a similar behavior, where CV-measurements gave a net doping concentration a factor of 10 lower than the Hall effect analysis (fig. 2). To explain in this case the disagreement of CV and Hall effect by compensation is unrealistic as a degree of compensation of around 90% must be assumed.

In addition the Hall effect measurement on a n-layer, as shown in fig. 2, gives a temperature behavior at temperatures above 300K, which could be interpreted as an additional impurity with an ionization energy of about 250 meV. However, deep-level-transient-spectroscopy (DLTS) showed no indication of its presence.

The substantial difference in doping concentrations estimated from Hall effect measurements as compared to other standard characterization techniques could most probably be due to a low pn-junction resistivity. In order to investigate the influence of the resistance of the built-in pn-junction on the measured Hall coefficients and the corresponding free carrier concentrations by simulation, we assumed a two-layer structure consisting of a n-type thin layer on top of a thick p-type substrate. To make results of the simulation comparable with the experimental results (fig. 2) material parameters of 6H-SiC were used and a single impurity level for each layer was modeled by the neutrality equation. The ionization energy of N as n-type dopant and Al as p-type dopant in 6H-SiC were taken from ref. [4] and [5], and the doping concentrations were chosen in that way, that the results of the simulation is comparable to the result of the van der Pauw measurement. The compensation was assumed to be negligible. The parameters used for the simulation are summarized in tab. I.

Tab. I Material and impurity parameters used in the simulations of the two-layer Hall effect measurements

material: 6H-SiC			impurity		
layer	conductivity	eff. mass	symbol	ionization energy	concentration
1	n-type	$0.27\ m_0$	N	100 meV	$10^{16}\ cm^{-3}$
2	p-type	$1.0\ m_0$	Al	240 meV	$5\cdot10^{18}\ cm^{-3}$

Fig. 3 shows the dependence of the electron concentration in the n-layer and the hole concentration in the p-substrate on the pn-junction resistance for three different temperatures. The solid lines are the results of a simulation based on the two-layer concept, which represents a measurement on the combined structure. The dashed lines show for comparison the carrier concentrations calculated by the neutrality equation for each layer separately (i.e., at infinite pn-junction resistance).

For a temperature of 100K the model with separated layers and the combined model give the same result independent of the pn-junction resistance, indicating that at low temperatures the influence of the p-substrate is negligible due to its low hole concentration and therefore high resistivity. The conventional van der Pauw technique gives then accurate results.

At temperatures above room temperature the electron concentration calculated by the neutrality equation saturates as expected at the given N-concentration. The hole concentration in the substrate is strongly increasing, leading to a lower substrate resistance. The substrate is then influencing the measurement of the n-layer. At room temperature and a pn-junction resistance of 500Ω the influence of the substrate is expressed in an increase of the estimated electron concentration in the n-layer by almost a factor of 10. This factor is even higher when the temperature is increased to around 1000K and the hole concentration in the substrate has saturated at the level of complete ionization of the Al-acceptor.

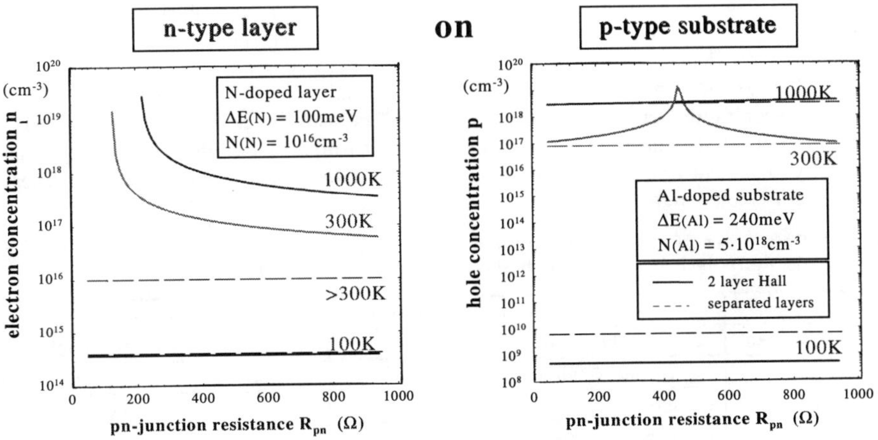

Fig.3 Calculated electron and hole concentrations of a two-layer structure consisting of a thin n-layer on a thick p-substrate as a function of the resistance of the pn-junction between the layers.

Performing a simulated temperature scan using the parameters of tab. I and assuming a linear temperature dependence of the resistance of the pn-junction, a qualitative comparison of the simulation with the real measurement (fig. 2) can be made. In fig. 4 the carrier concentrations are shown as a function of the temperature. The dashed curves are again representing the solution of the neutrality equation and the solid curves are the results of the two-layer Hall effect simulation. In the temperature range below 150K both calculations of the n-layer show the same concentration of free electrons. The concentration of holes in the substrate is much lower than the electron concentration. Therefore the resistance in the substrate is much higher than the n-layer resistance and no influence of the substrate on the electron concentration is seen. This changes, when the temperature is higher than 150K. The hole concentration in the substrate is steeply increasing and when it becomes comparable with the electron concentration in the n-layer the two-layer concept gives a value for the electron concentration which is higher than that estimated from the solution of the neutrality equation. The simulation of the two-layer structure above room temperature agrees quantitatively well with the experimental data on n-6H-SiC epitaxial layers. This clearly shows that the interpretation of the existence of an additional impurity could be falsified by the two-layer Hall effect simulations.

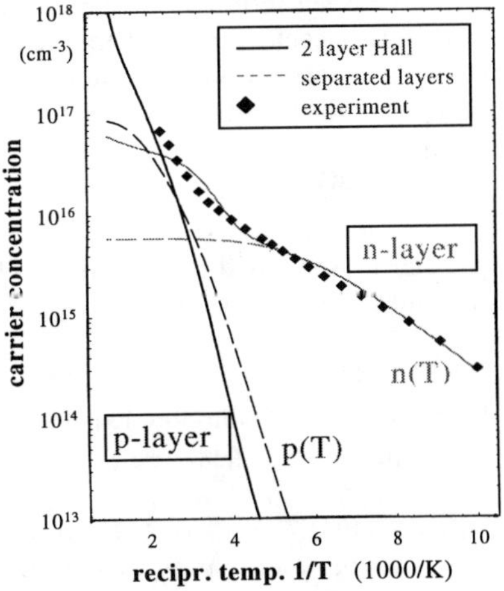

Fig.4 Simulated temperature dependence of the carrier concentrations in both the n- and p-layer compared to experimental data.

The two-layer Hall effect concept was realized in a computer controlled equipment. Room temperature measurements were carried out and are summarized in tab. II. The dopant concentration is calculated from the neutrality equation for comparison with SIMS and CV results, and with results from conventional Hall effect technique. Tab. II shows that the difference in doping concentration derived from standard Hall effect analysis for thin epitaxial or implanted layers on top of a thick substrate, and CV and SIMS measurements on the same samples can be resolved by taken into account the influence of a non perfect insulation on the layer transport properties as assumed in the presented two-layer Hall effect concept.

Tab. II Comparison of results from the two-layer Hall effect measurements with the standard van der Pauw method as well as with SIMS and CV results for Al- and N-doped Layers

		N doped layers	Al doped layers	
Measurement	concentration (cm^{-3})		low doping	high doping
two-layer Hall effect	free carriers (300K)	$1.6 \cdot 10^{15}$	$1.1 \cdot 10^{16}$	$5 \cdot 10^{16}$
	dopants	$2 \cdot 10^{15}$	$1.4 \cdot 10^{17}$	$7 \cdot 10^{18}$
standard Hall effect	free carriers (300K)	$1.2 \cdot 10^{16}$	$3.4 \cdot 10^{17}$	$7 \cdot 10^{17}$
	dopants	$2 \cdot 10^{16}$	$7 \cdot 10^{17}$	$5 \cdot 10^{20}$
CV	net doping	$2 \cdot 10^{15}$	$8 \cdot 10^{16}$	$6 \cdot 10^{18}$
SIMS	atomic	-	$1 \cdot 10^{17}$	$7 \cdot 10^{18}$

CONCLUSION

In conclusion, we have investigated epitaxially grown 6H-SiC films by Hall effect measurements. From a two-layer simulation it was shown, that the substantial leakage current over the pn-junction results in an overestimation of the carrier and impurity concentrations. This explains the difference between results from standard Hall effect measurements and other measurements of the impurity concentration. Furthermore artifacts in the conventional Hall effect measurements, which could be interpreted as additional impurities, can be explained by the two-layer concept as an influence of the substrate due to non-ideal pn-junction.

ACKNOWLEDGMENTS

The authors gratefully acknowledge M.Linnarsson for the SIMS measurements. The investigation was financially supported by Asea Brown Boveri and the Swedish National Board for Industrial and Technical Development (NUTEK).

REFERENCES

1. L.J. van der Pauw, Philips Res. Reports 13[1], 1958
2. R.D. Larrabee and W.R. Thurber, IEEE Trans. Elect. Dev., Vol. ED-27, 1980
3. N. Nordell, S.G. Andersson, and A. Schöner in Proc. 6th Int. Conf. on Silicon Carbide and Related Materials - 1995, Sept.18-21, 1995, Kyoto, Japan
4. A. Schöner, N. Nordell, K. Rottner, R. Helbig, and G. Pensl in Proc. 6th Int. Conf. on Silicon Carbide and Related Materials - 1995, Sept.18-21, 1995, Kyoto, Japan
5. W. Suttrop,G. Pensl,W.J. Choyke, S. Leibenzeder, R.Stein, J. Appl. Phys. 72, 1992

FORMATION AND HIGH FREQUENCY CV-MEASUREMENTS OF ALUMINUM / ALUMINUM NITRIDE / 6H SILICON CARBIDE STRUCTURES

C.-M. ZETTERLING[1,2], K. WONGCHOTIGUL[3], M. G. SPENCER[3], C. I. HARRIS[4], S. S. WONG[2] AND M. ÖSTLING[1]
[1] KTH, Royal Institute of Technology, Dept. of Electronics, Kista, Sweden
[2] Center for Integrated Systems, Stanford University, Stanford, CA 94305
[3] Materials Science Research Center of Excellence, Howard University, Washington, DC 20059
[4] IMC, Industrial Microelectronics Center, Kista, Sweden

ABSTRACT

Undoped single crystalline aluminum nitride films were grown by metal organic chemical vapor deposition (MOCVD) at 1200 °C. The precursors used were trimethylaluminium (TMA) and ammonia (NH_3) in a hydrogen carrier flow, at a pressure of 10 Torr. Silicon carbide substrates of the 4H or the 6H polytype with an epilayer on the silicon face, were used to grow the 200 nm thick AlN films. Aluminum was evaporated and subsequently patterned to form MIS capacitors for high frequency (400 kHz) capacitance voltage measurements at room temperature. It was possible to measure the structure and characterize accumulation, depletion and deep depletion. However, it was not possible to invert the low doped SiC epilayer at room temperature. From an independent optical thickness measurement the relative dielectric constant of aluminum nitride was calculated to be 8.4. The films were stressed up to 50 Volts (2.5 MV/cm) without breakdown or excessive leakage currents. These results indicate the possibility to replace silicon dioxide with aluminum nitride in SiC field effect transistors.

INTRODUCTION

Silicon carbide (SiC) has been suggested as a material for high power and high temperature electronic devices, such as power MOSFETs [1]. The progress that has been achieved in materials and process technology has enabled further device research [2, 3] using either the 4H or the 6H polytype in commercially available wafers. At temperatures around 1200 °C in an oxidising ambient silicon dioxide is readily formed on SiC with satisfactory characteristics, although oxides grown on p-type materials sometimes display disturbingly large flatband voltage shifts [4, 5]. When oxides are used to passivate high voltage junctions the scaling of electric fields with the dielectric constant has to be taken into account. At or around the critical field for breakdown in SiC, 4 MV/cm, the electric field in the oxide would be 10 MV/cm, which is several times higher than what would be wanted for reliable operations. This effectively displaces the advantage of using SiC in the first place, unless novel device structures are found which limit the electrical field in the dielectric, or other materials are found to replace silicon dioxide. Another reason for replacing silicon dioxide is that at the elevated temperatures that SiC devices are proposed to operate, charged impurities can easily diffuse through the oxide and thereby shift the flatband voltage of MIS devices.

Aluminum nitride (AlN) has been proposed as a potential replacement for silicon dioxide in high temperature MIS devices using SiC. The lattice mismatch between the two crystal structures is only 0.97 %, which enables single crystalline growth of AlN films on SiC substrates [6].

Mat. Res. Soc. Symp. Proc. Vol. 423 © 1996 Materials Research Society

Although AlN is a wide bandgap semiconductor (6.2 eV), if left undoped its properties are more like that of an insulator. The band offset is believed to be evenly split between the conductance and the valence bands. The relative dielectric constant of about 8 to 9 alleviates the problem of high fields in the dielectric when a high voltage is applied to the blocking junction even though the breakdown electric field of AlN is not well determined. One reason for believing AlN to be better at elevated temperatures is that it could be an effective diffusion barrier, not allowing charged impurities to diffuse at applied bias and thereby shift the flatband voltage of devices, although this needs to be shown in practice. This paper is a preliminary study in which MIS structures using AlN are fabricated and characterised mainly by capacitance-voltage method to determine possible deposition parameters for further work.

EXPERIMENTAL

The starting material for the deposition experiments were single crystalline silicon carbide wafers of the 4H or the 6H polytype with the silicon face polished slightly off-axis. All substrates but one had an epitaxial layer of 3 - 4 μm low doped SiC (< 10^{16} cm^{-3}), whereas the bulk was doped in the 10^{18} cm^{-3} range. The epitaxial layers were of the same doping polarity as the substrates (i. e. n on n, or p on p) and were supplied by CREE Research Inc. One substrate had a p-type epitaxial layer doped at about 10^{18} cm^{-3}, which was grown at IMC in Sweden.

Prior to deposition the wafers were cut into smaller pieces (1/6 sector of a 30 mm wafer) and cleaned in solvents and acids in the following sequence: 2 minutes of ultrasonic clean in trichloroethylene (TCE), followed by acetone and then methanol. The samples were then cleaned in a 1:1:5 mixture of NH_4OH:H_2O_2:H_2O heated to about 70 °C for 15 minutes followed by a 1:1:5 mixture of HCl:H_2O_2:H_2O, also 70 °C, for 15 minutes. The samples were rinsed in deionized water between each step. The final step before loading into the chamber was a 30 second dip in 1:50 HF:H_2O and a nitrogen blow dry.

The AlN growth was performed in a low pressure, vertically configured, metal organic chemical vapor deposition system. The system has two chambers separated by a load lock, which allows the growth to be performed in one chamber and reflection high energy electron diffraction (RHEED) in the other without breaking the vacuum. Precursors for the growth were trimethylaluminum (TMA) and ammonia (NH_3), and hydrogen was used as a carrier. The growth temperature was 1200 °C, at a pressure of 10 Torr. Typical growth times for 200 nm thick films were 8 minutes. For some of the samples, silane was introduced during ramp up of the temperature, but shut off before growth was initiated. This was done in an attempt to improve the crystallinity of the AlN films and the interface between AlN and SiC, by avoiding silicon depletion of the surface.

After the film thickness was determined, the samples were coated with about 400 nm aluminum in an e-gun evaporator. Some samples were set aside for materials characterisation. Standard optical lithography and wet etching was used to pattern the aluminum into circular dots with a diameter of 400 μm (smaller sizes were also made to check that the capacitance would scale with the area correctly). No backside contact was made for capacitance voltage measurements, instead a large front side capacitive contact was used to connect the substrate (the area is at least 60 times larger than the device area) [7]. For current-voltage measurements, the reverse side was scratched with a diamond scriber and contacted with an InGa eutectic.

MATERIALS CHARACTERISATION

Reflection high energy electron diffraction (RHEED) was used to determine the crystallinity of the AlN films immediately after growth. The patterns obtained had clear dots with slightly varying intensity, and streak marks but no arcs, which showed that the material grown was single crystal rather than amorphous, with a slight surface roughness. The samples where silane was used during ramp up had slightly better RHEED pictures, with hints of Kikuchi lines.

The thickness of each film was determined using interference fringes in optical reflection data. For the same growth condition, but different substrates, there was a thickness variation of 10 percent (film thicknesses were between 180 and 220 nm).

Compositional analysis was performed using sputter profiling Auger spectroscopy. The impurities of concern were carbon, oxygen and silicon, which are all potential dopants in AlN. The amounts were below the detection limit of Auger, less than 1-3 percent.

ELECTRICAL CHARACTERISATION

High frequency capacitance voltage measurements were performed using an HP 4275A multi-frequency LCR meter connected to a computer for simultaneous collection of capacitance and conductance data. Dual direction voltage sweeps were performed at room temperature with a measuring frequency of 400 kHz in a parallel configuration. The samples were checked in accumulation in the series configuration and at other frequencies to ensure that the true capacitance was not distorted by high substrate series resistance. The voltage sweep was started from inversion, with optical exitation of carriers from a focused microscope lamp, and swept to accumulation, and then back to deep depletion.

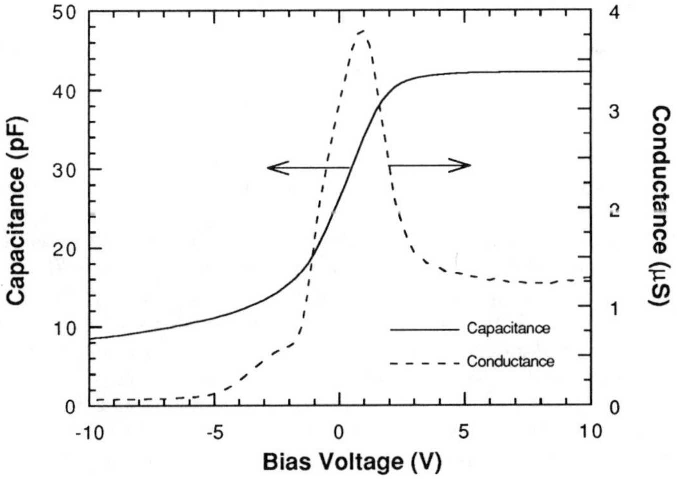

Fig 1a. Simultaneous capacitance and conductance versus voltage characteristics for a 220 nm thick AlN film on a n-type 6H SiC substrate, measured at room temperature and at 400 kHz. The device area is 0.00126 cm^2 and the relative dielectric constant was 8.4

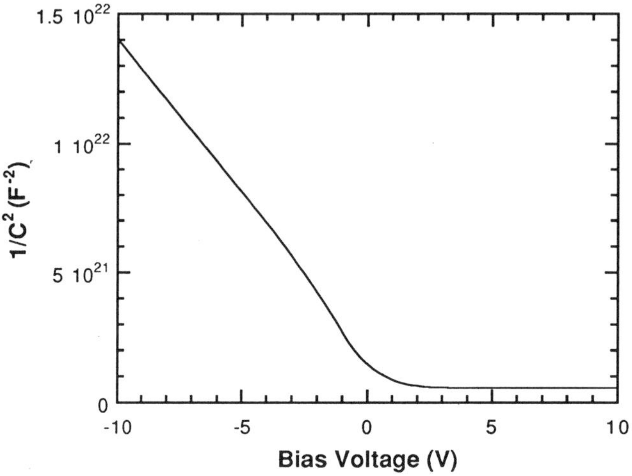

Fig 1b. $1 / C^2$ versus bias voltage plot of the same device as in fig 1a. The slope of the curve between -3 V and -10 V was used to confirm the epilayer doping of $8 \cdot 10^{15}$ cm^{-3}.

Typical capacitance and conductance data for the reverse sweep is shown in Figure 1a for a 400 μm diameter dot on a n-type 6H substrate. Accumulation, depletion and deep depletion can clearly be seen in both the capacitance and the conductance data. The accumulation capacitance was found to scale with the area of the contacts, and along with the thickness determined by optical measurement, the relative dielectric constant was found to be 8.4. From the $1/C^2$ versus applied voltage plot as shown in Figure 1b, the slope in the deep depletion portion of the curve (-3 V to -10 V) was used to determine the epitaxial doping. The results were consistent with data from CREE. Inversion was not seen at room temperature. The 6H n-type sample was grown without silane. In comparison, the 4H n-type was grown with silane. The RHEED was improved, but the CV characteristic was worse, displaying more interface states. Although a complete characterisation of interface states is yet to be done, the slope in the depletion region and the shift of the flatband voltage show that the amount is about the same order as silicon dioxide on SiC.

The three p-type samples (one 4H and two on 6H) were grown with silane during ramp up, and had good RHEED photos. The capacitance voltage measurements at 400 kHz made at room temperature are shown in Figure 2. Note that the scales are different for the two curves. The 6H sample had a heavily doped epi-layer, which can be seen from the small difference in capacitance between accumulation and depletion. It has a relative dielectric constant of about 8. The 4H sample on the other hand shows an accumulation capacitance that is more than a factor two smaller than as expected from optical thickness measurements and theoretical calculations. Similar results were achieved on the low doped 6H sample. The reason for this apparent difference in dielectric constant is not known. Note that the flatband voltage shift for both curves are close to the theoretically expected value of -3 V, and not as much as results obtained from silicon dioxide MIS structures on SiC [4, 5].

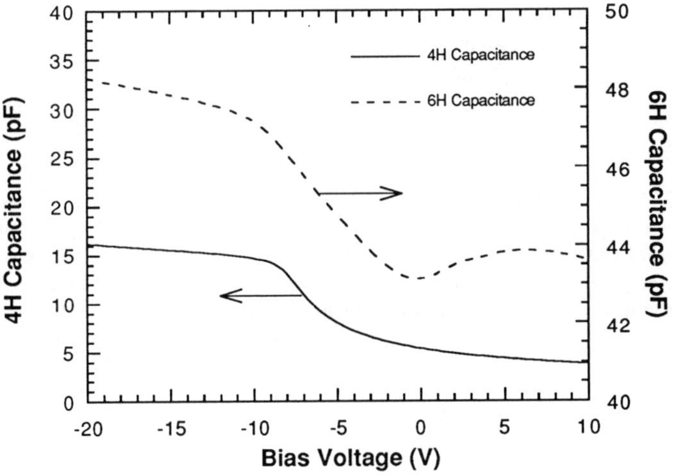

Fig 2. Capacitance versus voltage measurements of 200 nm thick AlN films grown on 4H and 6H SiC p-type substrates. Note the different scales used.

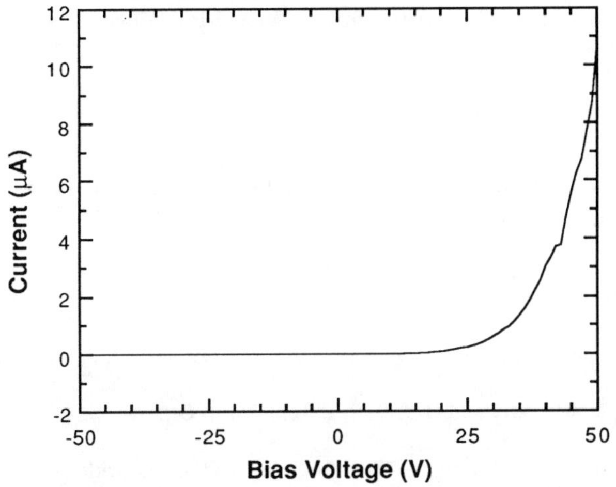

Fig 3. Current versus voltage characteristics of a 220 nm thick AlN film grown on a 6H n-type SiC substrate.

CV-measurements were performed with biases up to +/- 35 V, constrained by the instrument. Even at these voltages the capacitance and conductance data were as expected. To test the high voltage leakage characteristics of the material, current versus voltage measurements were made at room temperature. A typical curve for the 6H n-type sample is seen in Figure 3. In depletion the films could be biased to 100 Volts without breakdown because most of the voltage is dropped in the depletion region. In accumulation, the current increased quickly above 50 Volts bias, and the capacitors became permanently leaky if currents larger than 100 μA were conducted (corresponding to 0.08 A/cm^2). The 50 Volts applied over 200 nm of AlN is equivalent to a field of 2.5 MV / cm in the film. Less than 11 μA of leakage current is conducted. No evidence could be found for a Fowler-Nordheim type of conduction, possibly because the AlN / SiC behaves more like a heterojunction than a dielectric on a semiconductor.

CONCLUSION

Using MOCVD, AlN MIS atructures have been fabricated on various SiC substrates. Both the dc and ac leakage was found to be low, and electric fields up to 2.5 MV / cm could be sustained. Capacitance voltage measurements agree with theoretical calculations, except for low doped p-type substrates. The density of interface states was low. These results indicate the potential of replacing silicon dioxide with aluminum nitride in SiC field effect transistors. More work is needed in optimising the growth conditions and resolving the question of substrate influence on device characteristics.

ACKNOWLEDGEMENTS

The authors acknowledge partial support from the following organisations: the Stanford Nanofabrication Facility (CIS Industrial Grant, prop. 95-38), the Swedish National Board for Industrial and Technical Development (NUTEK), the John and Karin Engblom stipend fund, the National Science Foundation (agr. #HRD-9255 378) and the Office of Naval Research (grant #N00014-94-1-0274).

REFERENCES

1. M. Bhatnagar, D. Alok and B. J. Baliga, "5th International Conference on Silicon Carbide and Related Materials," Washington D.C., Inst. Phys. Conf. Ser., Vol. 137, p. 703, 1993.

2. R. F. Davis, G. Kelner, M. Shur, J. W. Palmour and J. A. Edmond, *Proc. IEEE*, **79**, p. 677, 1991.

3. P. A. Ivanov and V. E. Chelnokov, *Semicond. Sci. Technol.*, **7**, p. 863, 1992.

4. D. Alok, P. K. McLarty and B. J. Baliga, *Appl. Phys. Lett.*, **65**, p. 2177, 1994.

5. C.-M. Zetterling and M. Östling, "Diamond, SiC and nitride wide-bandgap semiconductors," San Francisco, Materials Research Society Proceedings, Vol. 339, p. 209, 1994.

6. K. Wongchotigul, N. Chen, D. P. Zhang, X. Tang and M. G. Spencer, To be published in the proceedings of ICSCRM 95, Kyoto, Japan, 1995.

7. J. N. Shenoy, G. L. Chindalore, M. R. Melloch, J. A. Cooper, J. W. Palmour and K. G. Irvine, *J. Elec. Mater.*, **24**, p. 303, 1995.

Part III

Defects, Dopants and Characterization

Optical and Field Emission Properties

ELECTRONIC STRUCTURE AND TEMPERATURE DEPENDENCE OF EXCITONS IN GaN

I. A. BUYANOVA*, J. P. BERGMAN*, W. LI*, B. MONEMAR*, H. AMANO** AND I. AKASAKI**

* Dept of Physics and Measurement Technology, Linköping University, S-581 83 Linköping, Sweden, irb@ifm.liu.se

** Department of Electrical Engineering, Meijo University, Tempaku-ku, Nagoya 468, Japan

ABSTRACT

We present an optical study of the excitonic properties of epitaxial GaN using reflectivity and photoluminescence (PL) measurements. The values for the intrinsic exciton energies are found to be dependent on the built in strain developed due to the difference in thermal expansion coefficients between the GaN epilayer and the foreign substrate. For GaN on sapphire the compressive biaxial strain causes an upshift of A and B excitons by typically 15 meV relative to the strain free sample, in accordance with previous data. For GaN on SiC, on the other hand, a downshift ~ 8 meV in the free exciton energies is observed at 2K. Only two excitonic peaks about 18 meV apart, are resolved in reflectivity spectra for GaN on SiC, probably due to the overlapping of A and B excitons. The suggested explanation implies the reduction of the bandgap energy and of the valence band splitting under the action of a biaxial tensional strain in the GaN layer. For all structures the strain-induced shifts of the bandgap energy are much smaller at elevated temperatures, presumably reflecting the temperature dependence of the accumulated strain energy. The exciton-polariton structure of the GaN is also discussed. The enhanced intensity of the no-phonon (NP) A line compared to its longitudinal (LO) phonon replica is suggested to be partially attributed to strong defect scattering.

INTRODUCTION

The increasing interest in group-III nitrides is partly motivated by the need for optical materials functioning in the blue and ultraviolet spectral region, and partly by their potential application in high power and high frequency electronic devices capable of operating in hostile environments and at high temperatures. These devices recently became closer to reality due to advances in crystal growth techniques, in particular after the discovery of growth conditions providing p-type doping of the material [1].

The epitaxial growth of III-V nitride structures is mostly performed using foreign substrates due to the lack of high quality bulk GaN. The lattice mismatch and the difference in the thermal expansion coefficient (TEC) between the film and substrate in such structures cause an appearance of built in mechanical strain, which influences the epilayer properties. The effect of the internal stress on the fundamental optical properties of GaN epilayers has been studied only for the layers grown on sapphire substrates [2-4]. The compressive biaxial stress field in the plane of the layer was found to be responsible for the increased bandgap and increased A-B-C exciton splitting compared to the case of unstrained bulk material. A different behaviour is expected for GaN grown on SiC substrates, where a tensional biaxial strain caused by the difference in TEC is revealed by the measurements of lattice parameters [5,6]. However, the influence of this built in strain on the GaN electronic structure has not so far been experimentally established.

In this paper we study the influence of built in strain on the intrinsic optical properties of GaN epilayers using photoluminescence (PL) and reflectivity measurements. The excitonic bandgap in strain free and biaxially strained GaN epilayers is compared by studying bulk samples, GaN/SiC and GaN/Al_2O_3, respectively. A decrease (about 8 meV at 2K) of the bandgap energy is observed for GaN thin films grown on SiC substrate in comparison with the bulk sample. The opposite behaviour is detected for GaN/Al_2O_3 structures, in agreement with previously reported data [2-4]. The effect of the built in strain is found to be more pronounced at low temperatures, suggesting an elastic accommodation of the strain energy under the cooling

Mat. Res. Soc. Symp. Proc. Vol. 423

towards liquid helium temperatures. Exciton-photon coupling responsible for GaN polariton properties is also discussed.

EXPERIMENTAL

In this study we used GaN epilayers grown without buffer layer by metal organic vapour phase epitaxy (MOVPE) on optical grade polished 6H SiC substrates. Similar layers were grown on sapphire substrates, but with a thin AlN buffer layer [7]. The growth temperature was about 1000°C. More than 500 μm thick undoped GaN layers grown by hydride VPE technique at 1000°C were used as a strain-free reference samples.

The PL and reflectance measurements were performed at a temperature range from 2 K up to 300 K in a variable temperature cryostat. The 334.5 nm line of an Ar^+-ion laser was used as an excitation source for PL measurements, whereas reflectance measurements were performed with a quartz halogen lamp. Optical spectra were dispersed with a 0.5-m double-grating monochromator and detected by a GaAs photomultiplier tube.

EFFECT OF THE BUILT-IN STRAIN ON THE OPTICAL SPECTRA OF GaN EPILAYERS.

Fig. 1 shows typical PL spectra recorded from a GaN/SiC (curve 1), a bulk GaN (curve 2) and a GaN/Al_2O_3 (curve 3) samples. All spectra contain PL lines related to bound exciton (BE) and free exciton (FE) recombination. Such an assignment of the PL transitions was done based on the PL temperature dependence, as well as after comparison of the PL and reflectance spectra, as will be demonstrated below at the example of the GaN/SiC epilayer. The excitonic lines detected in strained GaN epilayers are shifted in energy in comparison with bulk samples (Fig. 1). The direction of the shift is opposite for the GaN/SiC and GaN/Al_2O_3 structures. Such a behaviour is consistent with the expected change of the GaN bandgap due to the influence of built in strain field, known to be caused by the difference in TEC for GaN (5.6 10^{-6} K^{-1}), 6H SiC (4.2 10^{-6} K^{-1}) and sapphire (7.5 10^{-6} K^{-1}). At cooldown after the growth on SiC, the GaN epilayer will experience a tensional biaxial strain field in the plane of the layer. This will cause an increase of the c/a ratio of the GaN lattice, as was detected by lattice parameter measurements [5,6]. Consequently, the bandgap energy and valence band splitting (and thus A-B-C exciton splitting) are expected [8] to be decreased in GaN/SiC samples. The opposite behaviour is observed for GaN/Al_2O_3, where a considerable (more than 10 kbar) biaxial compression under (0001)

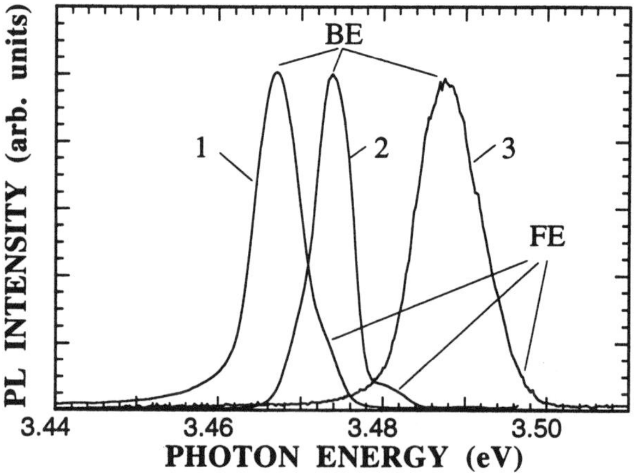

Fig.1. Typical Pl spectra measured at 1.5 K from a GaN/SiC (curve 1), a bulk GaN (curve 2) and a GaN/Al_2O_3 (curve 3).

direction was demonstrated to be responsible for the upshift of the excitonic bandgap and an increase of the splitting between the A, B, and C excitons [2].

The temperature dependence of the PL recorded from GaN/SiC epilayers is shown in Fig. 2. With increasing temperature the rapid quenching of the low energy excitonic emission, dominating the PL spectrum at low temperatures, is observed. Such a behaviour is typical for BE recombination. The high energy component at 3.472 eV (2 - 15 K), denoted as FE_1 in Fig. 2, governs the PL spectrum within a wide temperature range from 30 to 300K. For the temperatures higher than 100K another excitonic emission (denoted as FE_2 in Fig. 2) appears in the spectrum. The maximum position of this new PL band is shifted by about 18 meV towards higher energies. Based on the results of the reflectivity measurements (see below), we believe that this PL is of intrinsic origin and is caused by the recombination of FEs related to the higher valence subband (B or C exciton). The corresponding emission should increase in intensity at higher temperatures due to a thermal population of the higher energy excitonic states. Temperature dependent measurements were also performed for bulk GaN and the GaN epilayers grown on sapphire substrates, revealing a similar PL behaviour. For all structures free excitons dominate PL spectra above 50 K and even at room temperature.

Further evidence of the intrinsic character of the excitonic PL was obtained from the reflectivity measurements. Excitonic-like reflectance spectra can be observed only for intrinsic transitions and, thus, are free from the contribution of bound excitons. The derivative temperature dependent reflectance spectra measured from GaN/SiC samples are shown in Fig.3. All spectra contain two excitonic resonances separated by about 17 meV. Their energy position coincides at all temperatures with the energy position of FE PL lines - Fig. 3, confirming the intrinsic origin of the observed PL.

The valence band in the wurtzite GaN is split into 3 subbands by the combined action of the

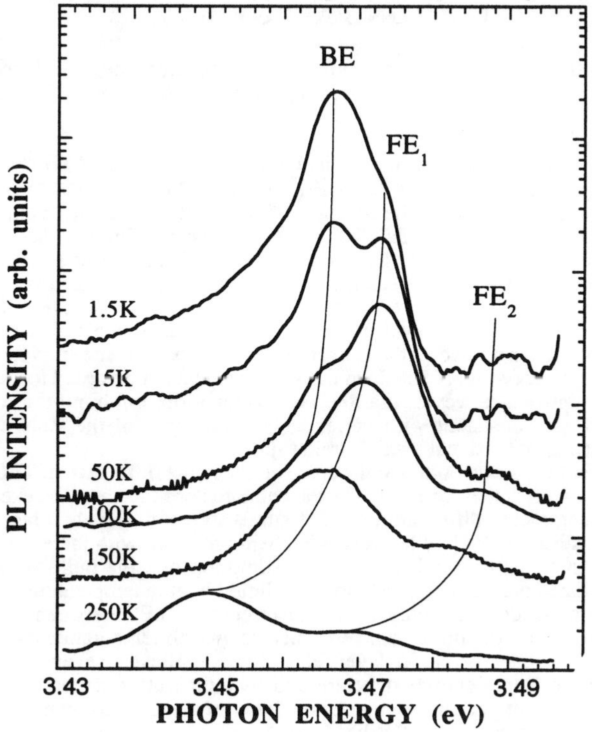

Fig. 2. Temperature dependent PL spectra measured from GaN/SiC epilayers.

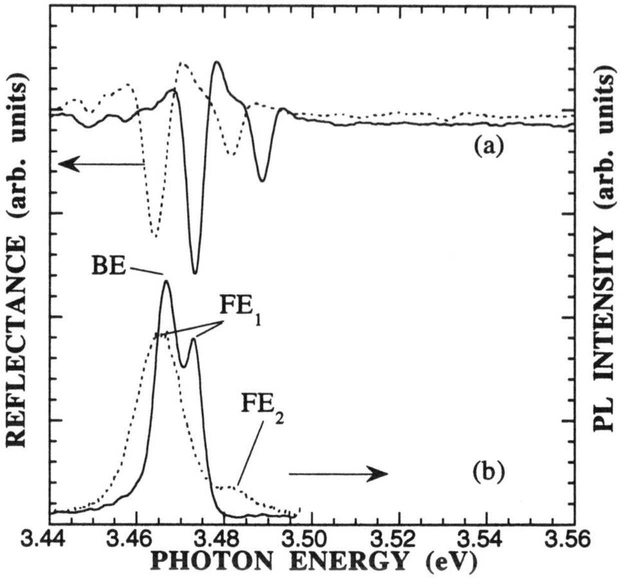

Fig.3. Derivative reflectance (a) and PL (b) spectra measured from GaN/SiC heterostructures. Solid and dashed curves represent spectra measured at 15K and 150K, respectively.

crystal filed and the spin-orbit interaction. Thus, three excitonic states A, B and C are usually detected in reflectance spectra of bulk GaN [9] and GaN/Al_2O_3 epitaxial films [2-4, 10]. However, there is no evidence (within the accuracy of at least 2%) of a third excitonic resonance in the reflectance spectra measured for GaN/SiC epilayers. One possibility is that the C resonance is too weak to be resolved in the spectra. Then, the two detected resonances should be assigned to A and B excitons, respectively, as was recently proposed in Ref.3. An alternative explanation is that the separation between the A and B excitons is smaller than 5 meV and, thus, they can not be separately resolved. In this case the low energy line represents the combined A-B exciton, while the high energy resonance is related to the C exciton. Obviously further experimental studies on samples with a narrower linewidth (~5 meV in our measurements) are needed to clarify this issue. We would like to point out, however, that the last explanation, implying the reduced A-B-C splitting in GaN/SiC, is consistent with the expected behaviour of the GaN exciton energy structure under the action of tensional biaxial strain [8].

The magnitude of the strain built into the epilayer due to the difference in TEC could also be dependent on the measurement temperature. The temperature dependence of the excitonic bandgap in GaN epilayers experiencing different internal strain, is shown in Fig. 4. A large difference in excitonic bandgap is observed at low temperatures, being reduced with increasing temperature. Such a behaviour can be understood within the assumption that the built-in strain energy is elastically accommodated during the cooling towards liquid helium temperatures. The magnitude of internal strain is dependent not only on the difference in TEC between the epilayer and substrate but also on other structure parameters, such as growth temperature, layer sequence in multilayer structure, presence of buffer layer, etc. This complicates severely the quantitative analysis of the GaN excitonic energy structure and its temperature dependence for different heterostructures. Consequently, each heteroepitaxial layer of GaN has its own excitonic bandgap and its own temperature dependence of this bandgap.

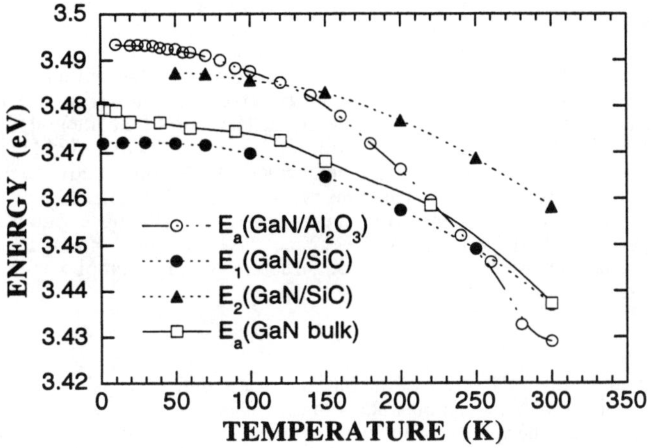

Fig. 4. Temperature dependent energies of intrinsic excitonic transitions determined for bulk GaN and GaN/Al_2O_3 utilizing PL spectroscopy and for GaN/SiC from PL and reflectance measurements.

POLARITON PROPERTIES

Excitonic properties of direct wide bandgap semiconductors are usually discussed within

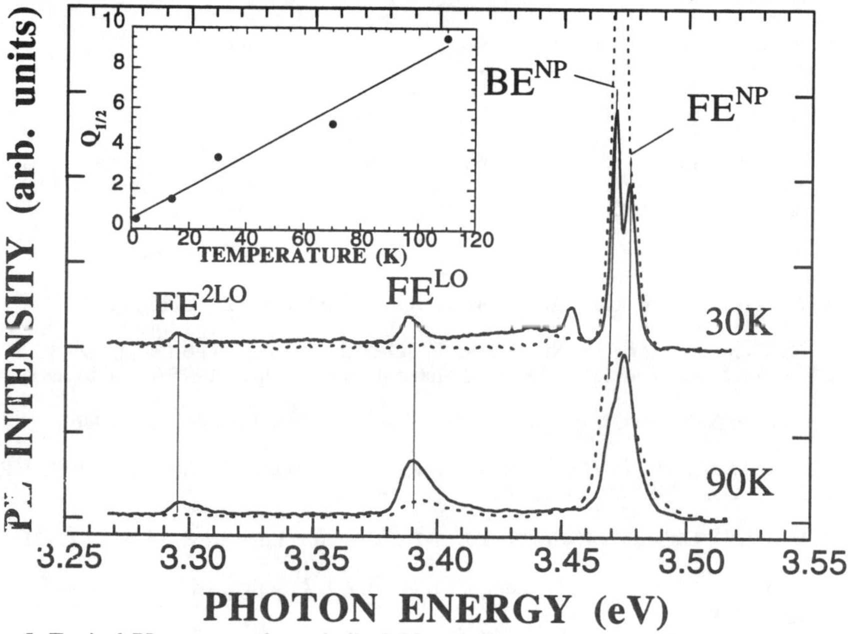

Fig. 5. Typical PL spectra of two bulk GaN samples (denoted by solid and dashed curves, respectively) monitored at 30 K and 110 K. The insert shows the ratio of the first and second LO-phonon replicas as a function of temperature.

the polariton concept, which takes into account strong exciton-photon coupling, characteristic for these materials [11]. As a consequence, FE recombination requires scattering of the FE from the exciton-like to the photon-like branch of the polariton dispersion curve. At low exciton densities this scattering is usually mediated by LO-phonon, leading to an enhanced intensity of the LO-replicas compared to the NP FE line. This is not observed for GaN (Fig.5), where the NP-emission is much stronger than its LO-phonon replicas. The ratio of the integrated intensities of the first and second LO phonon replica $Q_{1/2}$ follows, however, the expected [11] temperature dependence $Q_{1/2} \sim T$ - see insert in Fig. 5. The intensity of the NP line relative to its LO-phonon replicas is found to be higher for the structures with strong BE PL, i.e. higher defect and/or impurity density (compare dashed and solid curves in Fig. 5). This observation suggests that defect scattering is partially responsible for the enhanced NP FE emission in GaN. Evidently, to further study the true intrinsic exciton polariton structure of GaN, samples with lower defect density are needed.

SUMMARY

We have demonstrated the effect of the internal strain on the intrinsic optical properties of GaN epilayers. The internal tensional biaxial strain (in GaN/SiC structures) was found to cause a reduction of the bandgap energy and of the valence band splitting, while compressive stress (in GaN/Al_2O_3 structures) leads to the upshift of the excitonic bandgap of GaN. Temperature dependent built-in strain is also proposed to be responsible for the remarkable difference in temperature dependence of the excitonic bandgap detected for GaN/SiC, GaN/Al_2O_3 and bulk GaN. The exciton-polariton structure of GaN is also discussed. The enhanced intensity of NP FE PL compared to its LO phonon replica is suggested to be partially attributed to strong defect scattering.

REFERENCES

1. H. Amano, M. Kito, K. Hiramatsu, and I. Akasaki, Jpn. J. Appl. Phys. **28,** L2112 (1989).

2. B. Gil, O. Briot, R. L. Aulombard. Phys. Rev. **B52**, R17028 (1995).

3. N. V. Edwards, M. D. Bremser, T. W. Weeks, R. S. Kern, H. Liu, R. A. Stall, A. E. Wickenden, K. Doverspike, D. K. Gaskill, J. A. Freitas, U. Rossow, R. F. Davis, and D. E. Aspnes, presented at the 1995 MRS Fall Meeting, Boston, 1995 (unpublished).

4. K. Kim, W. R. L. Lambrecht, and B. Segall. Phys. Rev. **B50**, 1502 (1994).

5. H. Amano, K. Hiramatsu, and I. Akasaki. Jpn. J. Appl. Phys. **27**, L1384 (1988).

6. A. M. Tsaregorodtsev, I. P. Nikitina, M. P. Scheglov, and A. S. Zubrilov, presented at 6th Int. Conf. on Silicon Carbide and Related Materials, Kyoto, Japan, 1995 (unpublished).

7. H. Amano, N. Sawaki, I. Akasaki, Y. Toyoda, Appl. Phys. Lett. **48**, 353 (1986).

8. W. R. L. Lambrecht, K. Kim, S. N. Rashkeev, and B. Segall, presented at the 1995 MRS Fall Meeting, Boston, 1995 (unpublished).

9. R. Dingle, D. D. Sell, S. E. Stokowski, and M. Ilegems. Phys. Rev. **B4**, 1211 (1971).

10. W. Shan, T. J. Schmidt, X. H. Yang, S. J. Hwang, J. J. Song, and B. Goldenberg. Appl. Phys. Lett. **66**, 985 (1995).

11. S. Permogorov in Excitons, edited by E. I. Rashba, and M. D. Sturge (North Holland Publishing Company, 1982) p.177-203.

OPTICAL PROPERTIES AND MORPHOLOGY OF GaN GROWN BY MBE ON SAPPHIRE SUBSTRATES

E. TUNCEL, D. B. OBERMAN, H. LEE, T. UEDA, J. S. HARRIS, Jr.
Solid State Electronics Laboratory, CIS 329, Stanford University, CA 94305-4075

ABSTRACT

A series of GaN films grown by MBE on sapphire substrates with different nitrogen sources are characterized by optical transmission, spectroscopic ellipsometry (SE), photoluminescence (PL) and cross-sectional atomic force microscopy (AFM). The film thicknesses determined from broad spectral range transmission measurements and the AFM images are used in the analysis of the SE spectra. Interface roughnesses between the constituent layers, such as the substrate, the buffer and GaN layers are included in the modelling of the SE spectra and are also imaged by cross-sectional AFM. An effective medium type model is used for the modelling of interface and surface roughnesses in the SE spectra. The optical constants of the films in the band edge spectral range are determined in such a way as to simultaneously satisfy the transmission and the ellipsometry data.

INTRODUCTION

High crystalline quality nitride films have been difficult to grow because of a lack of suitable substrate materials. The growth, at lower temperatures, of a thin AlN buffer layer was shown to produce GaN films exhibiting crack free surfaces[1]. Although high defect concentrations are found to propagate from the substrate/film interface, a blue laser using InGaN quantum wells has been demonstrated[2]. Recent cathodoluminescence measurements have associated the luminescence emission from GaN films to the presence of crystallites in the layers[3]. Both hexagonal and cubic phases were found to coexist, producing emission at energies characteristic of each crystallographic structure. Imaging of the GaN surface by AFM, combined with cathodoluminescence measurements have also indicated the presence of hexagonal crystallites[4]. We report results obtained by multiple characterization techniques used to correlate the morphology of the GaN layers, as measured by cross-sectional AFM, with optical properties, such as the width of the absorption band edge and luminescence emission efficiency. In addition, spectroscopic ellipsometry (SE) and optical transmission measurements are combined to estimate the film optical constants. While optical transmission measurements provide information on the bulk sample properties, ellipsometry has great sensitivity to the quality of the reflecting surface and the epitaxial layer-substrate interface. Accurate reduction of ellipsometric data to the physical quantities of interest requires knowledge of the thicknesses and optical constants of the constituent layers. We have thus used the imaging of interfaces and surfaces by AFM together with optical transmission results in our SE analysis, with the aim of improving the modelling of the substrate/film and film/ambient interfaces.

EXPERIMENT

The GaN films are grown in our laboratory by molecular beam epitaxy using a solid source for gallium and various nitrogen sources. A thin AlN layer was grown by nitridizing the sapphire surface prior to the GaN growth. The substrates were (0001) oriented sapphire wafers. An ECR

Mat. Res. Soc. Symp. Proc. Vol. 423 © 1996 Materials Research Society

plasma and hydrogen azide (HN_3) sources were used for the growth of samples #42 and 72, respectively. The growth conditions are described in Ref[5]. Samples #83 and 100 were grown under similar conditions using dimethylhydrazine (DMHy) and ammonia (NH_3) gases respectively. Typical growth rates are of the order of 0.15μm/hr. The layers are nominally undoped; samples #42 and 83 are found to be highly resistive.
The photoluminescence (PL) measurements were carried out using the 325nm line of a 25mW He-Cd laser. The spectral emission was analysed by lock-in techniques using a GaAs photomultiplier tube and a 0.5m monochromator. The samples were mounted in a cold finger cryostat which was cooled to 77K for the low temperature data.
The transmission measurements were performed using a spectrophotometer operating between 180nm and 3200nm. In the 250nm to 850nm spectral range, the refractive index, n, and extinction coefficient, k, of the same films were determined from spectroscopic ellipsometry (SE) measurements performed with a SOPRA instrument model GESP, which is of rotating polarizer type. The spectra consist of the ellipsometric angles (ψ,Δ) measured at about 200points. Multiple angles of incidence around Brewster's angle were used in order to increase confidence in the modelling of the results.
A scanning probe microscope was used in contact mode for imaging the cleaved edges of the samples. The topography of the samples is measured by sensing the deflection of the probe which is part of a flexible Si_3N_4 cantilever. A nominally 115μm tip with wide legs corresponding to a spring constant of 0.5k(N/m) was used. The areas of the frames presented here are of the order of 1-3μm^2.

RESULTS

Fig. 1(a)-(c) shows cross-sectional AFM images of samples #42, 72, 100. The substrate is located on the right hand side of the images. In Fig. 1(a), the layer thickness is

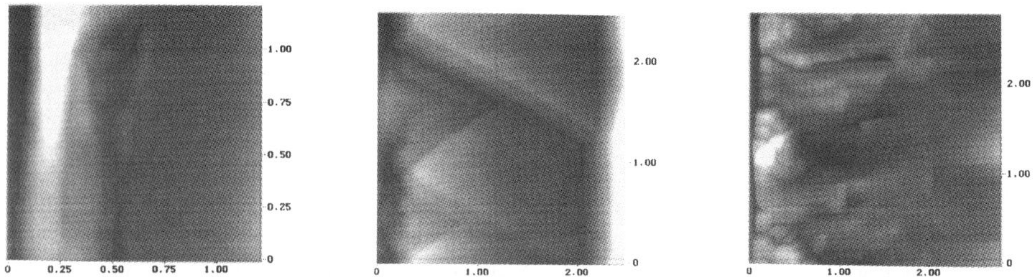

Fig. 1 Cross-sectional AFM images of samples (a) #42 grown using an ECR plasma, (b) #72 grown using hydrogen azide, and (c) #100 grown using ammonia. The horizontal and vertical scales are in μm.

measured to be 0.44μm. In measurements performed on different regions of this sample, all in the vicinity of the piece used for the optical measurements, large variations in the thickness were found. Although this cannot be seen in Fig. 1(a), on a 1μm^2 area, the film is composed of columns separated by empty regions. As can be seen in Fig 1(a), the substrate/layer region is not continuous and we do not observe dislocations in the GaN layer. The AlN buffer layer cannot be distinguished. In contrast, in Fig. 1(b) the HN_3 grown sample, #72, consists of well defined triangular crystallites, corresponding to the side view of hexagonal crystals, with typical lateral dimensions on the order of 1.5μm. The layer thickness is ≈1.69μm. The morphology of the NH_3 grown sample shown in Fig. 1(c) is of a more classical appearance and exhibits columnar growth. The film does not appear to be homogeneous and contains a lower density of vacancies than sample #42 shown in Fig. 1(a). The image of sample #83, which is not shown here, exhibits a cross-section that is similar to that of sample #42 shown in Fig. 1(a) with a smoother film/substrate interface. Fluctuations in the layer thickness from 0.242μm to 0.42μm can be observed over a 1μm lateral range. Layers #42 and 83 are transparent and yellow in color, #72 is dark yellow and opaque, while #100 is white with transparent and milky regions.

Fig. 2 shows the PL spectra measured at 77K for samples (a)#72, and (b)#100. Sample #72 exhibits the highest quality emission with strong band edge luminescence at 3.435eV with a FWHM≈200meV. A broad background is found to extend down to the weaker "yellow band" emission which peaks at 2.145eV. For sample #100, although the band edge emission with a FWHM≈86meV is still observed , it is much weaker than the impurity emission at 2.2eV.

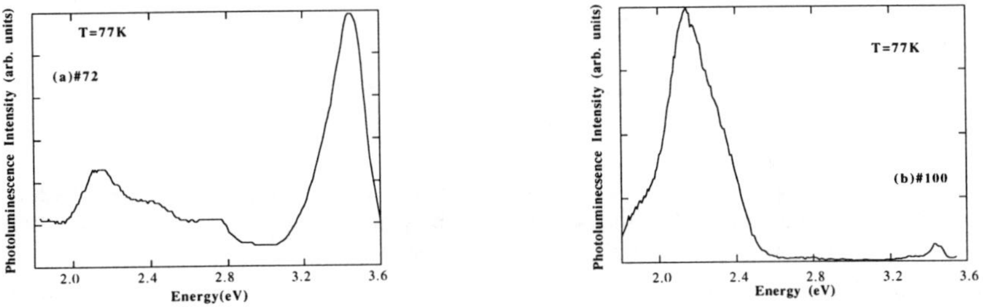

Fig. 2 Photoluminescence spectra taken at T=77K. The sample numbers and nitorgen sources used for their growths are a) #72(HN_3), b)#100 (NH_3).

Fig. 3 shows the transmission spectra, referenced to sapphire, of the four samples limited to the band edge spectral range. The spectra extending out to 3200nm are shown in the inset. Above

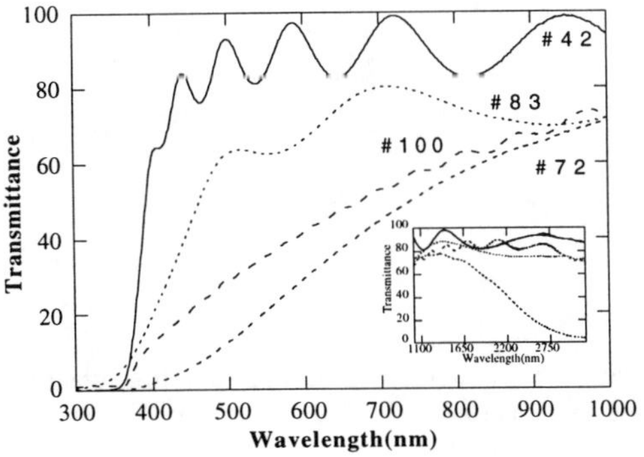

Fig. 3 Transmission spectra in the vicinity of the band gap. The sample numbers are indicated to the left-hand side of the figure. The nitrogen sources used for their growths are from top to bottom, are #42 (ECR plasma), #83(DMHy), #100 (NH_3), #72(HN_3). The inset shows the same spectra extended up to 3200nm.

800nm the crystals become transparent and, providing the films are sufficiently thick, interference fringes resulting from reflections at the air/film and film/substrate interfaces are observed. The film thicknesses, d, and the fringe period, Δ, can be related through the relationship d=1/2nΔ, where n is the refractive index. Taking n=2.35, we measured d=0.62μm, 1.61μm, 0.29μm, 1.99μm for samples #42, 72, 83 and 100 respectively. In the figure inset, which shows the farther infra-red, the effect of free carriers, in the form of increased absorption, can be seen for sample #72. The characteristic dependence of the absorption coefficient on wavelength $\approx\lambda^s$, with s=3.15±0.02, indicates the importance of impurity scattering expected to occur at high defect levels[6]. Fitting our data with the model used for the analysis of the ellipsometric data, which will be described below, provides us with a band gap of E_0=3.3eV for sample #42 and E_0=3.26eV, for #83. These values are smaller than the generally accepted hexagonal GaN band-edge of E_0=3.4eV[3]. Since layer characteristics, such as optical constants and thickness are correlated, these differences in band-edge energies might result from differences in the absorbing film thicknesses. The broadening of the band-edge for sample #72 was attributed to band-tailing resulting from the high electron concentration present in this sample. Samples #42 and 83, which are highly resistive, have the sharpest absorption edges of the series, although they appear to be inhomogeneous in morphology and thickness in the AFM images presented in Fig. 1. Consequently, we attribute the sharpness of the band edge to the absence of free carriers, rather than to crystallographic quality. The shape of the absorption edge for sample 100 is different from the $(E\text{-}Eo)^{1/2}$ behaviour which characterizes the other three samples.

Fig. 4 shows the in-plane refractive index, n, and extinction coefficient, k, of samples #42, and 83. Their values are in agreement with previously published values[7,8]. These constants are computed from SE data, using a modelling and fitting procedure that we will now outline. The SE data consist of the ellipsometric angles (ψ,Δ) which are defined in terms of the complex reflectance ratio $\rho = r_p/r_s = tg\psi e^{i\Delta}$, where r_p, r_s are the complex reflection coefficients for light polarized parallel and perpendicular to the plane of incidence[9]. At the macroscopic level, the reflection coefficientsare related to n and k through Fresnel's equations, with a model

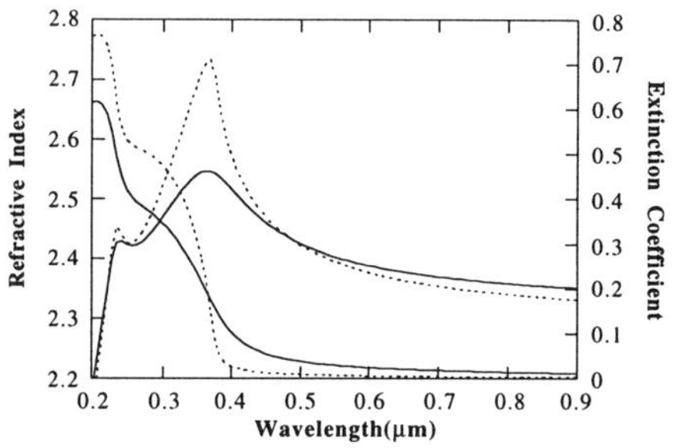

Fig. 4 Refractive index, n, and extinction coefficient, k, for samples #42 (dotted line), 83 (full line).

taking into account the presence of multiple layers. Our model is built from a sapphire substrate, a main GaN layer and anticipated imperfections, such as rough interfaces, surfaces. The surface roughness is described using the Bruggeman effective medium approximation, and is characterized by a layer thickness and a void fraction[10]. It is assumed that the inhomogeneities producing the roughness are spherical inclusions having dimensions smaller than the wavelength of light. The same type of model is used to represent the mixed layers consisting of GaN and voids. The macroscopic optical constants are related to the microscopic properties of the materials through the dielectric function $n + ik = (\varepsilon_1 + i\varepsilon_2)^{1/2}$. The real and imaginary parts of the dielectric function are computed using a model proposed by Adachi[11]. This model takes into account interband transitions at high symmetry points of the Brillouin zome, and uses the Kramers-Kronig relationship to compute $\varepsilon_1(\omega)$ from $\varepsilon_2(\omega)$. Although the ellipsometry data only extends up to about 5eV, the joint density of states includes not only the fundamental band edge, E_0, but also a higher energy transition which has been observed in reflectivity experiments and occurs at E_1=5.3eV[12]. The energy dependence at the fundamental band edge is of the form $\varepsilon_2(\omega) = Af(E, E_0, \Gamma)(E - E_0)^{1/2}$, where f is a function describing the band edge broadening and A contains the oscillator strength of the transition. Spectroscopic or angular data is analysed using a linear regression based on the Levenberg-Marquardt algorithm with the purpose of minimizing an error function consisting of the difference between calculated and measured values of the quantities (ψ,Δ). The starting value for the overall GaN thickness, including rough surfaces and interfaces, is taken from the transmission data shown in Fig. 3. The parameters A, Eo and Γ are fit to the transmission spectra shown in Fig. 3 for each of the samples. These parameters are then input into the model used for the fitting of the ellipsometric data. The fitting parameters for the regression include the thicknesses of the various layers, and their compositions of the main GaN layer. In order to check the validity of our model, in a second iteration, it is used to fit the transmission spectra presented in Fig. 3.

We now discuss the results obtained for our samples. For sample #42, the analysis indicates the presence of an interface region between the substrate and the main GaN layer, modeled by a 352Å±29Å mixture of GaN/sapphire/voids, with a fraction of 42%±4% of voids and -0.8%%±9.2% of sapphire. The main GaN layer was found to have a thickness of 5095Å±41Å. A surface region consisting of 3 layers of mixed material consisting of GaN and voids improved the fitting. These layers were found to have decreasing thicknesses and decreasing concentrations of GaN towards the surface from 556Å±27Å with 92%±0.5% of GaN, to 312Å±3Å with 71%±2% of GaN and finally 287Å±109Å thick with 7%±4.2% of GaN at the surface. In agreement with the transmission result, the total GaN thickness thus determined comes to 6250Å. Only three layers were required to model the data for sample #83. Starting from the sapphire substrate, a GaN/voids mixed layer of 450Å±61Å containing 96%±0.8% of GaN was followed by 2332Å±68Å of GaN and 243Å±15Å of GaN/void with 13%%±0.2% described the surface region. The differences in the modelling of these two samples reflect the differences observed in the cross-sectional AFM images, where sampe #42 was observed to be much more inhomogeneous both locally and on a larger scale. Because sample #72 exhibits roughness on a scale comparable to the wavelength of light, we did not attempt to model its optical constants. For both samples #72 and 100, the shape of the band edge needs additional modelling. Using known values for n and k, it is possible to fit the SE data for these samples using multiple layers and obtain layer thicknesses in agreement with transmission data.

CONCLUSIONS

We have used a combination of optical techniques and cross-sectional AFM measurements on the same samples to characterize their optical properties. There was good agreement between thicknesses measured by AFM and optically. In agreement with recent results in the literature, we have also found that our hydrogen azide grown sample was composed of crystallites, and exhibited the highest band edge luminescence emission intensity.

The AFM measurements, by probing areas of the samples a few μm^2, are able to distinguish inhomogeneities in film compositions and thicknesses that are not achievable by optical techniques, where typical spot sizes can be in the mm^2 range. While the optical transmission measurements provide the overall thickness of the absorbing material, spectroscopic ellipsometry also provides information on the presence of thin surface and interface layers. Once the presence of these additional layers has been included in the model, the optical constants of the materials involved can be extracted. The combined use of these techniques for the study of nitride alloys is promising.

ACKNOWLEDGMENTS

E. Tuncel gratefully acknowledges support from the Swiss National Science Foundation; J. L. Stehle from SOPRA is acknowledged for helpful discussions. ARPA support through the Center for Nonlinear Optical Materials, N00014-92-J-1903, and the Optoelectronic Materials Center, MDA972-94-1-0003, and facilities provided by NSF through the Stanford Center for Materials Research are acknowledged.

REFERENCES

1. H. Amano, N. Watanabe, N. Koide, I. Akasaki, Jpn. J. Appl. Phys. **32**, L1000 (1993).

2. S. Nakamura, et al., Jpn. J. Appl. Phys. **35**, L74 (1996).

3. J. Menniger, U. Jahn, O. Brandt, H. Yang, K. Ploog, Phys. Rev. B **53**, 1881 (1996).

4. C. Trager-Cowan, K. P. O'Donnell, S. E. Hooper, C. T. Foxon, Appl. Phys. Lett. **68**, 355 (1996).

5. D. B. Oberman, H. Lee, W. K. Gotz, J. S. J. Harris, J. Cryst. Growth **150**, 912 (1995).

6. R. D. Cunningham, R. W. Brander, N. D. Knee, D. K. Wickenden, J. Lumin. **5**, 21 (1972).

7. E. Ejder, Phys. Stat. Sol. (a) **6**, 445 (1971).

8. S. Logothetidis, J. Petalas, M. Cardona, T. D. Moustakas, Phys. Rev. B **50**, 18017 (1994).

9. R. M. A. Azzam, N. M. Bashara, *Ellipsometry and Polarized Light* (North Holland, Amsterdam, 1977).

10. D. E. Aspnes, J. B. Theeten, F. Hottier, Phys. Rev. B **20**, 3292 (1979).

11. S. Adachi, Phys. Rev. B **38**, 12345 (1988).

12. S. Bloom, G. Harbeke, E. Meier, I. B. Ortenburger, Phys. Stat. Sol. (b) **66**, 161 (1974).

A STUDY OF ELECTROLUMINESCENT EMISSION FROM CVD DIAMOND

C. Manfredotti*, F. Fizzotti*, E. Vittone*, P. Polesello*, F. Wang*[†], R. Schwarz‡, S. Grebner‡
*Dipartimento di Fisica Sperimentale, Università di Torino, Via P. Giuria 1, 10125 Torino, Italy, INFN, Sezione di Torino and INFM, UdR Torino, Italy, e-mail: manfredotti@to.infn.it
*[†]present address: Physics Department, North Carolina State University, P.O. Box 8202 Raleigh, NC 27695-8202, USA
‡Physik Department E16, Technische Universität München, D-85747, Garching, Germany

ABSTRACT

CVD diamond could be an electroluminescent material. While its photoluminescent and cathodoluminescent properties are relatively well known, the electroluminescent ones have not been intensively investigated. The electroluminescent emission of CVD diamond is reminescent of the so-called "band A"; it is centered at 2.8 eV and extends from 2.6 to 3 eV. The luminescence intensity increases linearly with the electric current and corresponds to a superquadratic regime of I-V characteristics in the dark and to a quadratic regime under a λ = 442 nm illumination. The I-V characteristics can be related to space charge limited current (SCLC) phenomena. Besides, the temperature behaviour of electroluminescence and of photoexcited current can be put into relation with the presence of a deep state in the bandgap where injected carriers from both electrodes recombine. The electroluminescent emission, even in a wavelength region far away from the photopic response, is clearly and strongly visible at naked eye. The luminescence efficiency is being discussed in order to consider CVD diamond as a material for optoelectronic devices.

INTRODUCTION

Visible-ultraviolet light emitting devices and detectors have recently attracted the attention of several researchers due to their potential applications both in the scientific and in the technological field. Among the materials which are object of study for these aims, diamond has earned a certain consideration due to its superior electronic properties [1-4]. Its wide band gap (5.5 eV) lets us look forward to its application as an optoelectronic device in the ultraviolet. Moreover, the recent development of growth techniques such as chemical vapour deposition (CVD) has shown the possibility to get good quality material at a reasonable price. The main problem in the application of diamond in optoelectronics lies in its indirect band gap, which lowers its efficiency in the emission of light. However, both in natural and in synthetic diamond intense luminescence can be observed in the energy range from near infrared to ultraviolet [5]. This can be ascribed to the presence of states in band gap generated by structural imperfections, like dislocations or grain boundaries, or impurities. As a matter of fact luminescence, either excited by photons or by electrons, has been widely used in order to study the nature and behaviour of levels in diamond. [5] Closely linked and complementary to luminescence studies are photocurrent measurements. They can give another deep insight into the transport properties and the electronic structure of diamond films [6], while from the point of view of applications, the analysis of current excited by light can give a hint towards the development of visible-UV radiation detectors.
In this work electroluminescence (EL) and photocurrent (PC) study of CVD diamond is presented. The importance of these techniques both from the applicative and from the fundamental point of view is evident. Particular attention is put in the dependence of EL and PC on temperature, which can give some hints towards the understanding of the electronic properties

Mat. Res. Soc. Symp. Proc. Vol. 423 © 1996 Materials Research Society

of diamond films. We give here a tentative explanation of the electronic behaviour of the diamond film in terms of the presence of a deep state in band gap which rules its main properties.

EXPERIMENT

Electroluminescence (EL) and photocurrent (PC) measurements have been performed on a CVD diamond sample supplied by Norton, Co., Northboro, MA, USA, prepared by the dc arc-jet technique and not intentionally doped. The film is about 400 μm thick with an area of 1×1 cm^2. The substrate and a part of the film (on the bottom) which contains graphite, amorphous carbon and very small crystallites have been cut off. Electric contacts with an area of about 8×8 mm^2 on the polished surfaces have been deposited by subsequent evaporation of Ti/Pt/Au [7]

EL has been excited by the application of voltages ranging from 850 to 1950 V. Emitted light, coming from one side of the sample, has been collected by a system consisting of a standard luminescence apparatus. Spectra have been taken ranging from 340 to 800 nm. The stability of the EL has been checked by operating the sample for one day continuously, and no changes in both intensity and spectra have been observed during operation. In order to study the dependence of EL signal from temperature, the sample has been placed in a cryostat at a pressure lower than 10^{-6} mbar. The temperature has been varied from 346 to 110 K. All the temperature-dependent measurements have been performed under a 800 V bias. Electronic properties have been studied by means of I-V characteristics and subgap PC. PC measurements have been performed under the excitation of the 442 nm (25 mW/cm^2 power per unit area) and 325 nm (12 mW/cm^2) lines of a HeCd laser under a bias varying between 10 and 500 V; temperature has been varied from 110 to 346 K, in order to study the dependence of PC with respect to temperature.

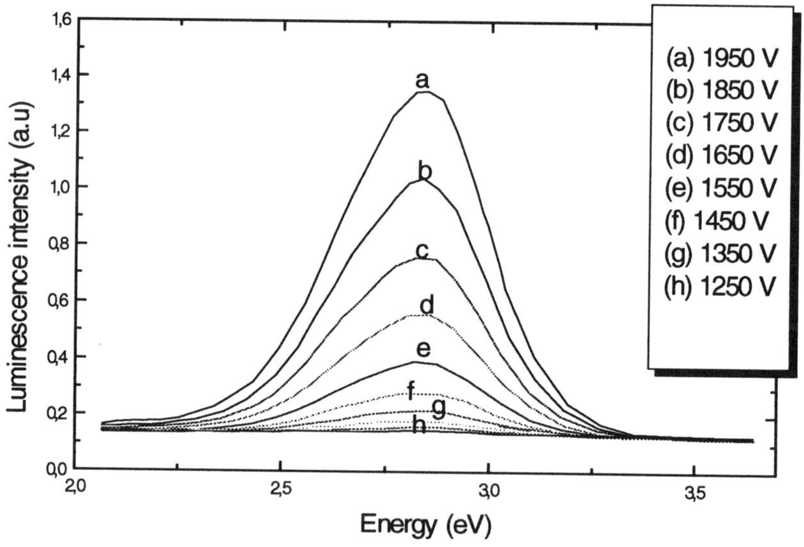

Fig. 1: Electroluminescence spectra at different voltages at room temperature

RESULTS AND DISCUSSION

In Fig. 1 EL spectra at voltages between 850 and 1950 V, taken at room temperature, are shown. They present the characteristic band A feature [5] with a peak at a photon energy of 2.8 eV: all the bands show an asymmetric tail at low energies and a FWHM of about 0.4 eV. Fig. 2 shows I-V characteristics measured at different temperatures (110-346 K): a log I-log V plot shows a power law dependence of current on voltage ($I \sim V^n$). The exponent n, evaluated from a linear fit of log I vs. log V curves, ranges from 6.88 to 9.75; for $T \geq 167$ K it scales linearly with the inverse of temperature. For this range of temperatures this is consistent with a model based on the formation of space charge limited current (SCLC) explaining the behaviour of our film under an applied voltage [8].

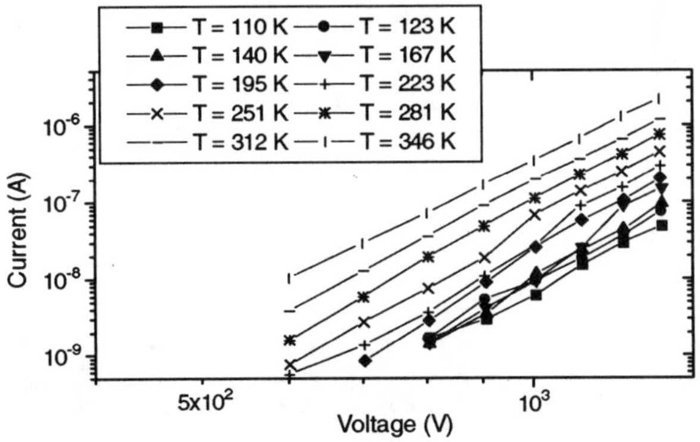

Fig. 2: The current-voltage characteristics of the sample at different temperatures

The linear dependence of EL integrated intensity with respect to current is shown in Fig. 3: we see that the linear coefficient decreases with increasing temperature, which is consistent with a quenching of light emission with temperature. The linear dependence of EL integrated intensity with current confirms our previous hypothesis, that the radiative recombination is linear and could be due to the presence of only one recombination center [4].

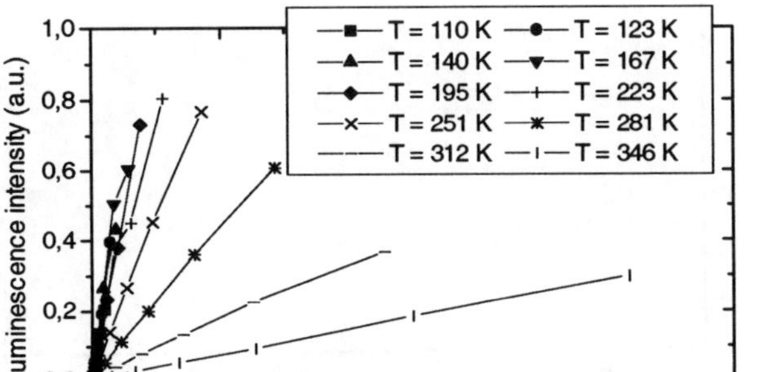

Fig. 3: The integrated EL intensity as a function of the current at different temperatures

In Fig. 4 the relationship between EL integrated intensity per unit current versus temperature at an 800 V bias voltage is presented: we needed to normalize EL intensity with current because of the different I-V characteristics at different temperatures. The shape of the curve is very similar to the one obtained by Khong and Collins [9], which was explained in terms of the presence of a trap center: unfortunately we could not go to temperatures lower than 110 K, which does not allow us to see the low temperature plateau observed by these authors. Nevertheless, a maximum at T=165 K and a long exponential tail at T > 200 K is easily observed, which reproduces the curve shown in [9].

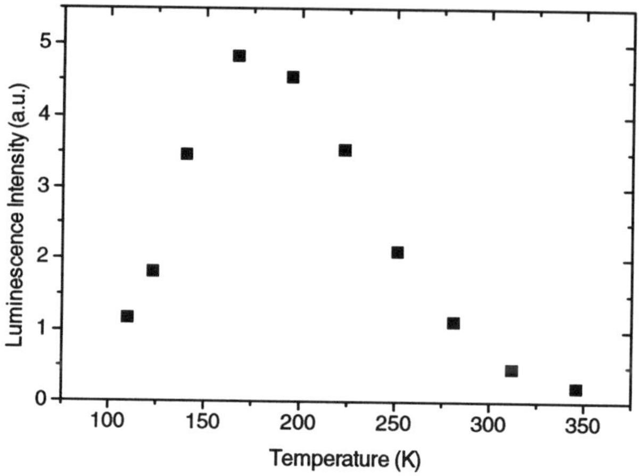

Fig. 4: EL integrated intensity as a function of temperature at an 800 V bias voltage

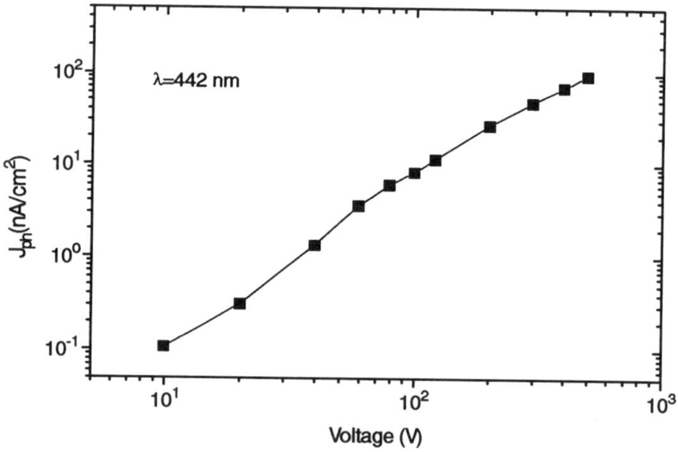

Fig. 5: Photocurrent intensity per unit area as a function of bias voltage at $\lambda = 442$ nm

Another way to understand the origin of the optoelectronic properties of the film is by means of PC measurements. In Fig. 5 PC intensity per unit area (J_{PC})as a function of bias voltage applied at room temperature is shown: on a log-log scale we note that PC intensity grows with the square of voltage. This is a behaviour which closely reminds that of SCLC and is consistent with I-V characteristics shown in Fig. 2 [8, 10]. Moreover, PC intensity is nearly independent of temperature, as shown in Fig. 6, for excitation wavelengths λ=442 nm and λ=325 nm: this could give us a hint towards the presence of a deep level located in the bandgap. In fact, a shallow one would have given a steep Arrhenius plot, which is not clearly our case [8].

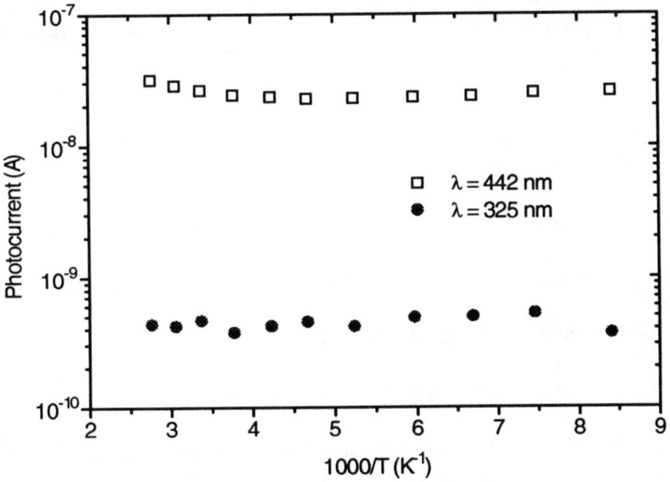

Fig. 6: Photocurrent intensity as a function of temperature with a blue- (open squares) and a UV- (filled circles) light excitation.

Although we have not yet a complete understanding of the mechanisms underlying the luminescence excited by an applied voltage, the above results can be tentatively explained in terms of the presence of a deep recombination level whose origin is still subject of discussion. It is possible that the same centre limits also the transport properties of CVD [4, 7]: as a matter of fact, the charge collection length and, as a consequence, the trapping time are strongly influenced by light in the wavelength range from 350 to 550 nm. The presence of a space charge seems also to strongly affect the transport properties.

The linear relationship between the electroluminescence intensity and the electrical current, the temperature independence of both electroluminescence and photoconductivity, are all a clear indication that recombination is monomolecolar and that electron and holes, both injected through the contacts [3], recombine at a center located near the middle of the gap [9].

CONCLUSIONS

In this work a study of EL and PC properties of a CVD diamond sample is presented, and a tentative explanation is given. The phenomenon has been recognized to be a bulk one, and it has been related to charge injection and recombination at a deep gap level.

The fact that the luminescence emission is visible also at naked eye lets us hope in a future development of CVD diamond technology in the field of optoelectronics: a comparison with analogous EL measurements made on *c*-SiC and *a*-SiC devices lets us estimate the luminescence efficiency of our sample to be about 10^{-4} in magnitude [11]; even if this value is not large, we believe that further studies in this direction towards a deeper understanding of optical and transport processes in CVD diamond can give us hints for an improvements of its device characteristics. In particular, injection efficiancy and, as a consequence, electroluminescence efficiency can be improved by fabricating injecting contacts on both sides by ion implantation.

REFERENCES

[1] M.D. Whitfield, S.S.M.Chan, R.B. Jackman, Appl. Phys. Lett. **68**, 290 (1996)

[2] Y. Taniguchi, K. Hirabayashi, K. Ikoma, N.I. Kurihara, M. Matsuhi, Jpn. J. Appl. Phys. **28**, L1848 (1989)

[3] B. Burchard, A.M. Zaitsev, W.R. Fahrner, A.A. Melnikov, A.V. Denisenko, V.S. Varichenko, Diamond Relat. Mater. **3**, 947 (1994)

[4] C. Manfredotti, F. Wang, P. Polesello, E. Vittone, F. Fizzotti, A. Scacco, Appl. Phys. Lett. **67**, 3376 (1995)

[5] A.T. Collins, Diamond Relat. Mater. **1**, 457 (1992)

[6] P. Gonon, A. Deneuville, F. Fontaine, E. Gheeraert, J. Appl. Phys. **78**, 6633 (1995)

[7] C. Manfredotti, F. Fizzotti, E. Vittone, S. Bistolfi, M. Boero, P. Polesello, Nucl. Instr. Methods B **93**, 516 (1994)

[8] M.A. Lampert, P. Mark, *Current Injection in Solids* (Academic Press, New York, 1970)

[9] Y.L. Khong, A.T. Collins, Diamond Relat. Mater. **2**, 1 (1993)

[10] P. Gonon, S. Prawer, D. Jamieson, Appl. Phys. Lett. **68**, 1238 (1996)

[11] R. Schwarz, private communication

CATHODOLUMINESCENCE STUDIES OF BOUND EXCITONS AND NEAR BAND GAP EMISSION LINES IN BORON- AND PHOSPHORUS-DOPED CVD-DIAMONDS

H. STERNSCHULTE*, T. ALBRECHT*, K. THONKE*, R. SAUER*, M. GRIEßER**, AND M. GRASSERBAUER**
*Abt. Halbleiterphysik, Universität Ulm, D-89069 Ulm, Germany
**Institut für Analytische Chemie, TU Wien, Getreidemarkt 9/151, A-1060 Wien, Austria

ABSTRACT

Cathodoluminescence measurements at cryogenic temperatures are reported on boron- and phosphorus-doped CVD-diamond films grown on silicon substrates. Boron and phosphorus concentrations were determined by SIMS measurements; for boron, they reached from unintentional background doping levels up to 3500 ppm. At increasing boron concentrations, the radiative recombination of boron bound excitons (BE^{TO}) at 5.22 eV photon energy systematically broadens and shifts down to 4.99 eV whereas the free exciton emission (FE^{TO}) disappears for 40 ppm and higher. In the phosphorus-doped films we observe new lines at 5.16 eV and 4.99 eV which we ascribe to TO- and ($TO+O^{\Gamma}$)-phonon assisted transitions of an exciton bound to a shallow impurity other than boron, possibly phosphorus or a phosphorus-related shallow complex.

INTRODUCTION

Synthetic CVD-grown diamond layers are considered an attractive material for a variety of applications including UV transparent optical windows, hard coatings, effective heat conductors, and active electronic devices. The latter application requires knowledge - and finally control - of the electronic properties of the material, i. e. in particular energy levels and concentrations of impurities and defects. An important aim towards electronic devices is p- and n-doping of diamond for p-n junctions. While boron is known to act as an acceptor and boron doping is possible in CVD-growth in a wide range of concentrations no relatively shallow donor is known to date. Phosphorus is a potential candidate following theoretical calculations. A survey on the state of the art is given in Ref. [1]. The present work aims at investigating CVD-layers grown on silicon substrates with boron- or phosphorus-doping at largely varying concentrations by low temperature-cathodoluminescence (CL).

EXPERIMENTAL DETAILS

Cathodoluminescence was excited by a conventional RHEED-electron gun operated with a typical accelerating voltage of (3-4) kV. The electrons impinge on the sample surface under an angle of 45°, and CL is observed under 90° from the electron beam. The spot size is approximately 1 mm in diameter. As most of the CVD-diamond layers are polycrystalline with typical crystallite sizes of a few micrometers, all present measurements integrate over a large number of crystallites.

The samples were mounted with silver paste on the coldfinger copper sample holder of a continous-flow cryostat which can be cooled either with liquid nitrogen or with liquid helium. The lowest temperature obtained on the sample surface is 15 K as estimated from free exciton line shape fits in silicon samples. Based on such temperature adjustments measurements between 15 and 300 K have been made. The luminescence signal was focussed with two Al-coated off-axis parabolic mirrors on the entrance slit of a monochromator and detected either with a Peltier cooled photomultiplier using lock-in technique or with a LN_2-cooled UV-enhanced CCD-camera.

Mat. Res. Soc. Symp. Proc. Vol. 423 © 1996 Materials Research Society

RESULTS AND DISCUSSION

a) Boron-doped CVD-diamond films

Fig. 1 shows a selection of CL spectra demonstrating the general trends upon increasing boron concentrations. The topmost spectrum exhibits dominantly free exciton radiation as TO-, TA-, and TO- plus O^{Γ}-phonon assisted recombination lines (FE^{TO}: 5.27 eV, FE^{TA}: 5.32 eV, and $FE^{TO+O^{\Gamma}}$: 5.105 eV). Weakly observed is radiation from excitons localized at isolated boron atoms ("bound exciton", BE^{TO}: 5.215 eV). These line positions are in accord with the basic luminescence data in Ref. [2]. This sample was not intentionally boron doped but contaminated in the growth apparatus.

All other samples in Fig. 1 were intentionally boron-doped during hot filament CVD-growth. The boron concentrations were determined by SIMS and increase from top to bottom. They are listed in Table I for all samples investigated. In Fig. 1, for sample TD 219A with [B] = 40 ppm, the FE lines have almost disappeared; the BE^{TO} line has largely increased and broadened yet maintaining its energy position. As a new feature, the $BE^{TO+O^{\Gamma}}$ transition is visible, and towards lower energies, there is a rise in intensity up to the boron-related emission band peaking at 4.5 eV [3]. For the next concentration, [B] = 825 ppm (sample B113), the bound exciton lines have shifted and broadened even more. Finally, for [B] = 3500 ppm (sample B110), this trend is continued: The BE^{TO} and $BE^{TO+O^{\Gamma}}$ lines merge due to their largely enhanced widths, however, the two components remain distinguishable with energy spacing of around 165 meV = $\hbar\Omega(O^{\Gamma})$. At the same time, the lines are further shifted down in energy significantly. A lineshape analysis demonstrates that the BE spectra of all samples can be generated progressively from the low-doping case by simple Gaussian-broadening

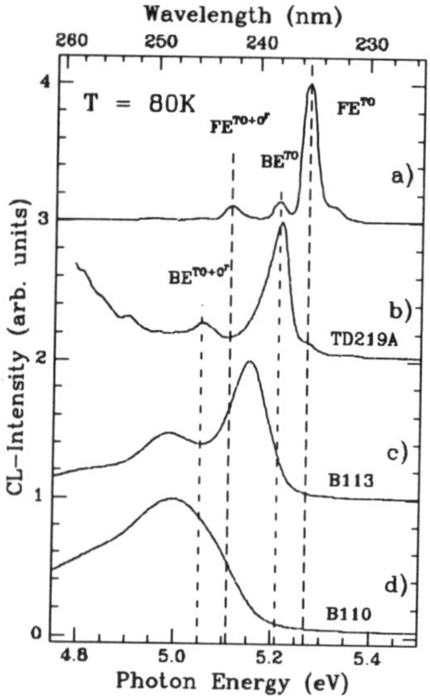

Fig. 1:
Near-band edge boron-doped diamond CL-spectra at 80 K. (a) Spectrum from a nominally undoped MW-CVD film grown on Si:B exhibiting TO-phonon and $(TO+O^{\Gamma})$-phonon assisted free exciton radiation (FE^{TO}, $FE^{TO+O^{\Gamma}}$) along with boron bound exciton radiation (BE^{TO}). (b - d) Spectra from hot filament grown films with intentional boron doping of 40 ppm, 825 ppm, and 3500 ppm, respectively (see Table I).

Table I:
List of the boron-doped diamond films indicating the total boron and phosphorus concentrations as determined by SIMS measurements. The substrates were in all cases Si:B. Sensitivity limit of SIMS ≤ 0.01 ppm.

Sample	B-concentration (ppm)	P-concentration (ppm)
Nr 3 S.2	20	-
TD219A	40	< 0.01
Nr 5 S.2	81	-
WD 13	140	-
Nr 7 S.2	225	-
WD 12	340	-
Nr 8 S.2	500	-
Nr 9 S.2	630	-
Nr 10 S.2	715	-
B113	825	< 1
B110	3500	< 0.1

and spectral shifts essentially equal to the broadening. This implies that the relative coupling strengths of the TO- and (TO+O^{Γ})-phonons remain constant for all boron concentrations. Energy-versus-concentration plots are shown in Fig. 2 for all samples studied. Three samples with boron concentrations above 3500 ppm exhibited spectra virtually identical to that at 3500 ppm. There was no further energy shift.

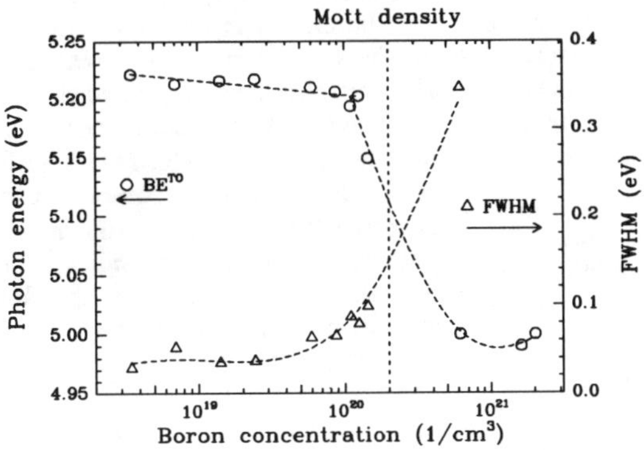

Fig. 2:
Energy positions and halfwidths of the BE^{TO} tranistions as a funktion of the boron concentrations. Indicated is the Mott density which is found to be 2×10^{20} cm^{-3} [11] or 3×10^{20} cm^{-3} [2].

BE line shifts and broadenings at high doping levels have only in silicon been studied to some extent [4]; they have not been reported so far in diamond. We assume that the BE line positions remain constant as long as the acceptors and bound excitons are isolated. Wavefunction overlap results in level broadening and down-shifts which both can be subject to local variations due to impurity concentration fluctuations. The critical boron doping level where the line shift essentially sets in, is approximately $N_{B,crit} \approx 1 \times 10^{20}$ boron atoms/cm^3. In a model where the bound excitons are assumed to have spherical wavefunctions which begin to overlap at $N_{B,crit}$, the excitonic radii are $\approx$ 13 Å. This is roughly four times the acceptor radius of a_B = 3.6 Å. The latter value is calculated from the Effective-Mass-Theory (EMT) using $\varepsilon = 5.7$ and $m^* = 0.84\ m_0$ with the Bohr-radius of 0.53 Å. It is problematic to apply EMT in the present case - although it reproduces the experimental boron ionization energy, 370 meV, quite well - since the acceptor radius, comparable to the diamond lattice constant a_0 = 3.57 Å, is not consistent with the basic requirement of EMT in order to work, $a_B >> a_0$ [5]. Nevertheless, our rough estimate of the exciton extension seems to support a very low value of the acceptor radius a_B. We note that our present observations and interpretation replace our earlier suggestion (based on fewer samples) that the spectrum with the transitions at 5.16 eV and 4.99 eV in sample B113 may be due to a new shallow impurity [6]. Instead, the present sequence of spectra indicates that the spectrum fits fully into the boron bound exciton concept for higher doping concentrations.

b) Phosphorus-doped CVD-diamond films

Eight different phosphorus CVD-diamond films were studied. Data are compiled in Table II. All films were grown by the hot filament technique on Si:B substrates and are therefore contaminated with boron. An exception is TD297C which was deposited on Si:P. Here, the boron doping probably originates from a contamination of the growth chamber. In the other samples, the phosphorus doping was achieved by adding PH_3 to the CH_4/H_2 gas mixture during the growth. Incorporation of boron is optically apparent either by observation of the boron bound exciton or the boron-related broad band at $\approx$ 4.5 eV [3]. As a novel feature, all eight phosphorus doped samples exhibit two new lines at 5.16 eV and 4.99 eV, respectively, which we ascribe to a phosphorus-related shallow complex emitting a TO-phonon and a (TO+O^{Γ}) phonon-assisted transition. CL spectra of two of the eight phosphorus-doped samples are shown in Fig 3 (TD297C and P112). The top spectrum is the same as on top of Fig. 1 for comparison. Sample TD297C shows incompletely resolved boron bound exciton emission and the new lines. Sample P110 shows only the latter two lines. There is no simple correlation between the CL-intensity of the new lines and the phosphorus doping. The samples P112 and P109 of the highest phosphorus concentration exhibit only weak intensities of this line. Assuming our assignment is correct, we conclude that only a fraction of the total phosphorus concentration is in optically active states. SEM measurements show that the surface morphology of the samples is very different reaching from cubo-octahedral structures to ballas. It is very probable that also the microscopic structure is far from being perfect on the length scale of the exciton wavefunction. Raman measurements probing a much smaller length scale, of the order of the lattice constant, reveal in all cases the 1332 cm^{-1} (165 meV) O^{Γ}-phonon line with narrow widths but there are also contributions in the Raman spectra indicating amorphous or graphitic structures in the layers.

Table II:
List of the phosphorus-doped diamond films indicating the total boron and phosphorus concentrations as determined by SIMS measurements. The substrates were in all cases Si:B except for sample TD 297C which was grown on Si:P.

Sample	B-concentration (ppm)	P-concentration (ppm)
TD297C	180	< 0.06
P146	-	SIMS not possible
P126	-	SIMS not possible
P130	0.1 - 0.4	1
P110	0.1 - 0.4	1
P128	-	10 - 15
P112	-	35 - 50
P109	-	70

The new CL line at ≈ 5.16 eV is accompanied by a satellite at 4.99 eV, lower in energy by approximately the O^{Γ}-phonon energy. Hence, it must itself be the TO-phonon replica of the phosphorus-related bound exciton radiation which is not observed as a no-phonon transition. The localization energy of the exciton is then approximately $h\nu(FE^{TO})$ - 5.16 eV ≈ 110 meV. For comparison, the localization energy of the boron bound exciton amounts to 55 meV. In terms of bound exciton data known very well from silicon a larger exciton localization energy at a shallow impurity would imply a larger impurity ionization energy ("Haynes' rule", [7]). Theoretically, substitutional phosphorus is expected to form a shallow donor [8], and recent conductivity measurements on phosphorus implanted high purity natural, type IIa diamonds revealed n-type conductivity with 0.2 - 0.21 eV activation energy [9]. From these considerations it is rather probable that the shallow binding center is not isolated phosphorus but a shallow complex incorporating phosphorus. As boron is a nearly omnipresent contamination, P-B pairs (on nearest neighbor sites) are possible candidates. They could act as an isoelectronic center. Isoelectronic complexes are known from many semiconductors to be highly optically active [10].

Our conclusion that the new line at ≈ 5.16 eV is due to a shallow bound exciton is supported by temperature controlled measurements (Fig. 4). In Fig. 4a, we plot CL-intensities as a function of sample temperature for P126 as an Arrhenius plot. When the 5.16 eV line is thermally deactivated, the free exciton luminescence increases, and the sum of both intensities remains constant up to ≈ 60-70 K. (From there, also the FE is thermally dissociated and all CL-intensities rapidly drop.) Hence, excitons thermally released from the localized state contribute to the free exciton radiation. The bound exciton decay has been fitted theoretically and compared to corresponding data of the samples P109 and P110 in Fig. 4b. We use an expression

$$I = I_0 * \left(1 + a \times \exp(\Delta E_1 / k_B T) + bT^{3/2} \times \exp(\Delta E_2 / k_B T)\right)^{-1} \quad (1)$$

based on the assumption that all excitons are thermally partitioned among three states, namely the bound exciton ground state and two excited states with excess energy of ΔE_1 and ΔE_2, respectively. The second state with ΔE_2 is the free exciton state into which the localized excitons are thermally released. This introduces the $T^{3/2}$ term, and a and b are constants including density of states and lifetime ratios. In all three cases, ΔE_1 is obtained as (10 ... 20) meV. These low excited bound exciton states are not optically observed because they fall below the luminescence widths of the bound exciton ground state-to-ground state transition. The values for ΔE_2 are (112 ... 150) meV, reasonably close together with an average value of ≈ 130 meV. This is in turn fairly consistent with the measured spectroscopic localization energy of ≈ 110 meV. These measurements confirm our conclusion that the new line at ≈ 5.16 eV originates from an exciton that is bound to a phosphorus impurity or a shallow complex incorporating phosphorus.

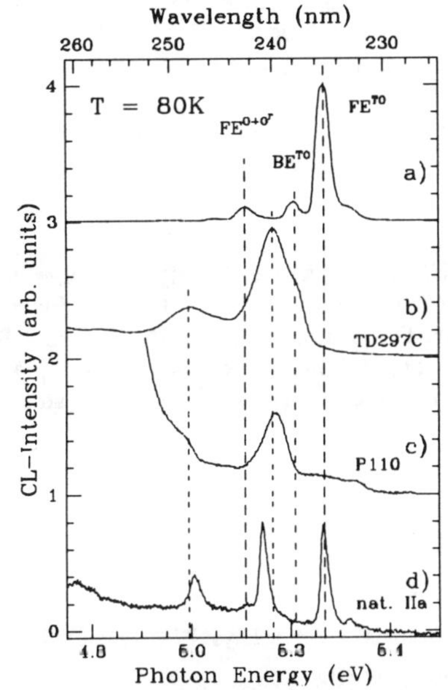

Fig. 3:
Near-band edge phosphorus-doped diamond CL-spectra at 80 K. (a) The spectrum is identical to Fig. 1(a) and shown for comparison. (b, c) Spectra from films phosphorus-doped to < 0.06 ppm and 1 ppm, respectively (see Table II). Sample P110 shows the strong rise in intensity due to the boron-related CL band at 4.5 eV peak energy [3]. (d) Spectrum from a natural, type IIa diamond showing emission lines unrelated to boron or phosphorus.

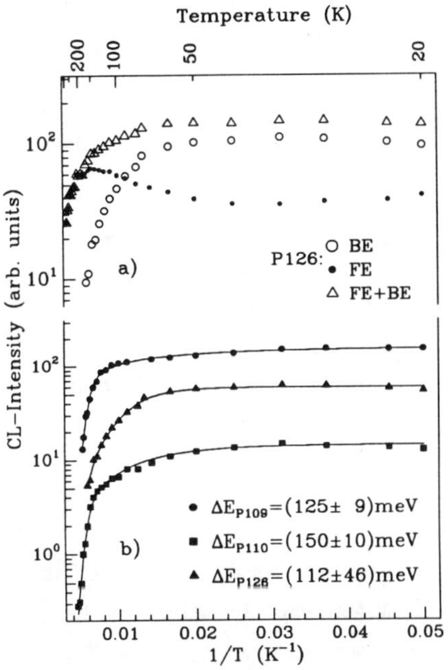

Fig. 4:
Arrhenius-plots from temperature controlled CL-measurements. (a) Temperature dependence of the novel bound exciton line (BE, open circles) at hν ≈ 5.16 eV), of the free exciton line (FE, dots), and of their integrated intensity (FE + BE, triangles). (b) Temperature dependence of the novel bound exciton line at hν ≈ 5.16 eV for three phosphorus-doped samples with high temperature de-activation energies from fits to the data (see text).

We note that the above discussed excitonic lines have to be distinguished from other CL spectra as shown on the bottom of Fig. 3. The latter spectrum was obtained from a natural, type IIa bulk diamond and reveals free exciton emission as narrow FE^{TO} and FE^{TA} lines. In addition, two lower energy, fairly narrow lines emerge, with an energy spacing of ≈ 140 meV ≈ $\hbar\omega(TO)$ and temperature-independent intensity ratio. Thus, the higher energy line at ≈ 5.14 eV is very probable a no-phonon exciton transition at a deeper center. It is not known which elemental species are contained in our particular natural diamond, hence even a speculative assignment of this deeper center cannot be made. However, the accidental coincidence of the line position (≈ 5.16 eV) for the new shallow bound exciton and the latter deeper bound exciton demands great caution in assigning experimental spectra.

CONCLUSION

We have presented CL-studies of boron- and phosphorus-doped CVD-diamond films. In heavily boron-doped layers the boron bound exciton lines shift drastically to lower photon energies and broaden due to wavefunction overlap of the excitons and/or acceptors, to acceptor band formation and to acceptor potential fluctuations. Phosphorus-doped CVD-films show a new excitonic line at ≈ 5.16 eV ascribed to a TO-replica transition with no observable NP-transition. The exciton localization energy amounts to ≈ 110 meV defining the binding center still as a shallow impurity. This is possibly isolated phosphorus or a phosphorus-related complex center, e. g. of isolectronic character, such as a P-B pair on nearest-neighbor sites.

ACKNOWLEDGEMENTS

The authors are grateful to S. Bohr (TU Wien), W. Hänni (CSEM Neuchâtel), W. Dötter and R. Erz (Universität Kaiserslautern), and R. Klarmann (Universität Augsburg) for growing the diamond samples. The financial support by the Deutsche Forschungsgemeinschaft (Contract Sa 520) carried out under the auspices of the D-A-CH-cooperation of Germany, Austria, and Switzerland on the "Synthesis of superhard materials" is gratefully acknowledged. All studied samples are contributed from this cooperation.

REFERENCES

1. G. S. Gildenblatt, S. A. Grot, and A. Badzian, Proc. IEEE **79** (1991) 647.
2. P. J. Dean, E. C. Lightowlers, and D. R. Wight, Phys. Rev. **140A** (1965) 352.
3. S. Lawson, H. Kanda, H. Kiyota, T. Tsutsumi, and H. Kawarada, J. Appl. Phys. **77** (1995) 1729.
4. J. Wagner, Phys. Rev. **B29** (1984) 2002; Solid State Electron. **28** (1985) 25.
5. W. Kohn, Solid State Physics, ed. F. Seitz and D. Turnbull (Academic, 1957), Vol. **5**, p. 257.
6. H. Sternschulte, J. Horseling, T. Albrecht, K. Thonke, and R. Sauer, Diamond Rel. Mater. (in press).
7. J. R. Haynes, Phys. Rev. Lett. **4** (1960) 361.
8. S. Kajihara, A. Antonelli, and J. Bernholc, MRS Symposia, Proc. **162** (1989) 315.
9. J. F. Prins, Diamond Rel. Mater. **4** (1995) 580.
10. D. G. Thomas and J. J. Hopfield, Phys. Rev. **150** (1966) 680; P. J. Dean and D. C. Herbert, in "Excitons", ed. K. Cho (Springer, 1979), p. 55; Landolt-Börnstein Vol. 22b, (Springer, 1989), p. 340
11. A. W. S. Williams, E. C. Lightowlers, and A. T. Collins, J. Phys. C: Solid State Phys. **3** (1970) 1727
12. H. Shiomi, Y. Nishibayashi, and N. Fujimori, Jpn. J. Appl. Phys. **30** (1991) 1363

Characterisation of Diamond-like Carbon by Raman Spectroscopy and Optical Constants

C. Mößner**, P. Grant, H. Tran, G. Clarke, D. J. Lockwood, H. J. Labbé, B. Mason, R. Berriche*
Institute for Microstructural Sciences, * Institute for Aerospace Research,
National Research Council of Canada, Ottawa, Ont., K1A 0R6, Canada
**present address: Swiss Federal Institute of Technology, Lausanne, Switzerland,
moessner@dmx.epfl.ch

ABSTRACT

Crystalline diamond coatings and, increasingly, diamond like amorphous carbon (DLC) films are used for tribological and protective layers for their hardness and chemical inertness. They are also under investigation for their electron emitting properties, with possible applications in field emission displays. DLC films were deposited by laser ablation using a KrF excimer laser and fluences between 0.5 and 2 J/cm^2. FTIR measurements did not show the presence of hydrogen in the films. Raman spectra allowed for the determination of the nature of the graphitic and diamond bonds (sp^2 and sp^3) as well as information about the disorder and short range order in the films. For a better determination of the sp^3-content, which is often hidden in the Raman spectra, a correlation with optical properties in the near IR to near UV region was established. These values depended strongly on the substrate temperature and the laser fluence. DLC formation could be demonstrated even at substrate temperatures close to room temperature. Vickers hardness values and first measurements on the electron emissivity of the films can be correlated to the diamond character and the preparation method of the films.

INTRODUCTION

Diamond like amorphous carbon (DLC) values for hardness, chemical inertness, electrical and optical properties are intermediate to those of graphite and diamond, and, therefore cover a wide range [1-3]. In addition to various sputtering and vapor deposition methods, pulsed laser deposition (PLD) has been shown to be an effective method to generate DLC thin films [4-7]. One major advantage of using PLD instead of other deposition methods is the lack of hydrogen, thus enhancing the density of the films and the index of refraction, which is preferable in some applications. DLC thin films produced by PLD are generally categorized diamond-like, but the degree of diamond-like character varies considerably, and is strongly dependent on the deposition conditions.

In addition to the continued lasting interest in diamond as a protective coating, recent interest has also focused on possible applications as a field emitter for displays and other vacuum microelectronic devices [8-11]. Crystalline diamond exhibits a negative electron affinity (NEA) for certain crystal surface planes via restructuring [12-14], which can under appropriate conditions, lead to a low work function, and thereby supposedly allow high current densities in field emission setups. A high electron emission, which is due to high micro roughness or low work function, is also observed for amorphous films [8], whereas the exact mechanism for high field emission current is still under discussion. Interest in these films also exists due to their possible enhancement of the quality of already existing microtip emitter displays [15-17]. In that case, DLC thin films would protect the emitting tips as well as enhance and stabilize the Fowler-Nordheim current.

Part of the work done to investigate the possibilities and characteristics of PLD deposited DLC thin films as flat field emitters or additional layers in microtip configurations is reported in this paper. The variation of electrical, optical and mechanical properties as

Mat. Res. Soc. Symp. Proc. Vol. 423

a function of deposition parameters within the limit of the experimental setup was investigated. Only the combination of several characterization methods gave useful information on the film. As deposited films were also checked for electron emissivity, which was found to be relatively low, and tests on etching the films with reactive hydrogen for an improvement in the emissivity properties were performed.

EXPERIMENTAL

The films were deposited using a Lumonics Hyperex 400 KrF excimer laser (wavelength 248nm) in a high vacuum setup with a base pressure of $5 \cdot 10^{-7}$mbar, as previously used for the preparation of high temperature superconductor ($YBa_2Cu_3O_{7-x}$) thin films [18]. For the DLC experiments typical conditions were 18ns pulslength, 10Hz, intensities between 7.4 and $1.8 \cdot 10^8$ W/cm^2 corresponding to fluences from 1.3 to 3.3 J/cm^2, and deposition temperatures T_s between 20 °C and 400 °C. A pyrolitic graphite disc was hit with an angle of incidence of 45°, and sputtered material collected on various substrates placed face to face to the target at a distance from 26 to 52mm. Results for deposition on Si(100) are reported.

Raman spectroscopy was performed in a quasibackscattering configuration at 295 K. The spectra were excited with 80 mW of 457.9 nm Ar laser light, analyzed with a Spex triple monochromator, and detected with a cooled Photometrics CCD900 array detector. The incident light was focused to a spot size of roughly 1mm^2 and polarized in the plane of incidence, while the backscattered light was not polarization analyzed [19].

A Woolam VASE Ellipsometer was used for the optical measurements. The films were investigated in the wavelength range between 275 to 1700 nm, and at angles of incidence of 50°, 60° and 70°. The optical constants were determined with the Fourhi-Bloomer model for transparent regions, and directly in the opaque ones.

Hardness measurements were undertaken by nano-indentation using the nanomechanical probe developed at NRC [20].

RESULTS

Several films were deposited with varying parameters like target-substrate distance, additional nitrogen, and film thickness, but the only parameters which affected their properties were the laser intensity resp. fluence and the substrate temperature. All films were optically absorbing, and showed considerable compressive stress. The stress built up at the substrate-film interface due to the incoming ions plays a considerable role in enhancing the formation of sp^3 bonds [21]. Ion energies from 15 to 70 eV have been reported to be optimal for tetrahedral configuration in the films. The energy of emitted carbon particles under the conditions used in our experiments is about 30 eV for C^+ [22], or between 10 and 120 eV following, as reported in [23-24], and two or three times as much for dimers and trimers. This indicates good conditions for the growth of DLC by supporting the formation of sp^3 bonds.

Raman spectra shown in Fig. 1 exhibit an overall pattern, which is generally attributed to DLC [25]. For T_s = 20 °C only a broad peak at 1550 cm^{-1} is visible. For raising substrate temperature a shoulder at about 1390cm^{-1} (for T_s = 400°C) becomes visible, and the peak around 1550 cm^{-1} shifts to 1580 cm^{-1}, which is also the wavenumber for the main peak of crystalline graphite. A similar behavior is observed for increasing laser fluence.

The two peaks can be attributed to the graphitic D and G bands, respectively [25]. The G band itself is due to E_{eg} C=C stretching vibrations of graphite, whereas the broadening is due to increased disorder in the carbon sheets. Peak D becomes active when small graphite crystallites are present, which leads to a loss of translational symmetry and a

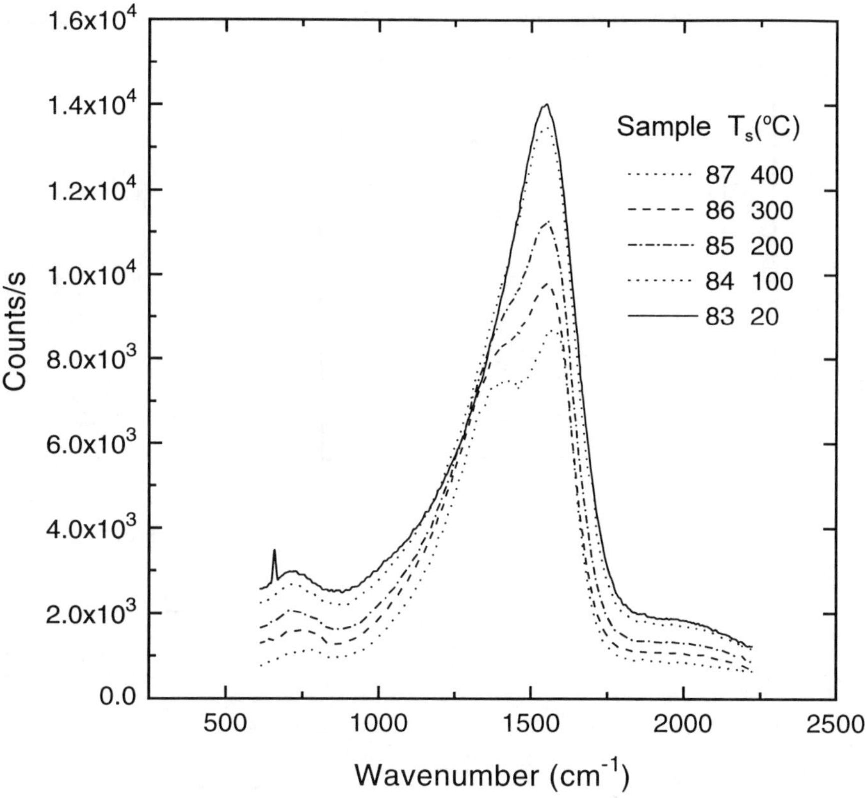

Fig. 1: Raman spectra of five DLC samples deposited at various substrate temperatures T_s.

breakdown of the momentum conservation rule [26-27]. A sharp peak at 1332cm^{-1} would be expected for crystalline diamond, but due to the fact that the sensitivity for this Raman line is 55 times lower than the one for graphite, no conclusion on the presence of any sp^3 -bonds can be drawn from the measured spectra alone. XPS measurements on the same samples strongly indicate, that no crystalline diamon is present, due to the absence of significant peaks below 25eV [28]. The Raman spectra show increased graphitic behavior with larger amounts of small graphite particles for raising substrate temperature (and increasing fluence). This correlates very well with atomic force microscope (AFM) measurements performed on some of the samples, where larger particles for higher T_s and higher laser fluences could be seen.

To obtain further information on possible sp^3-bond content in the film, ellipsometry was performed to determine the optical gap E_g and n_{eff}. n_{eff} is defined as the number of effective valence electrons taking part in optical transitions in the interval from zero to some cutoff energy. By determination of the dielectric constant ϵ_e both values can be extracted [29]. Figure 2 includes the extracted n_{eff} for the temperature dependence of the film properties.

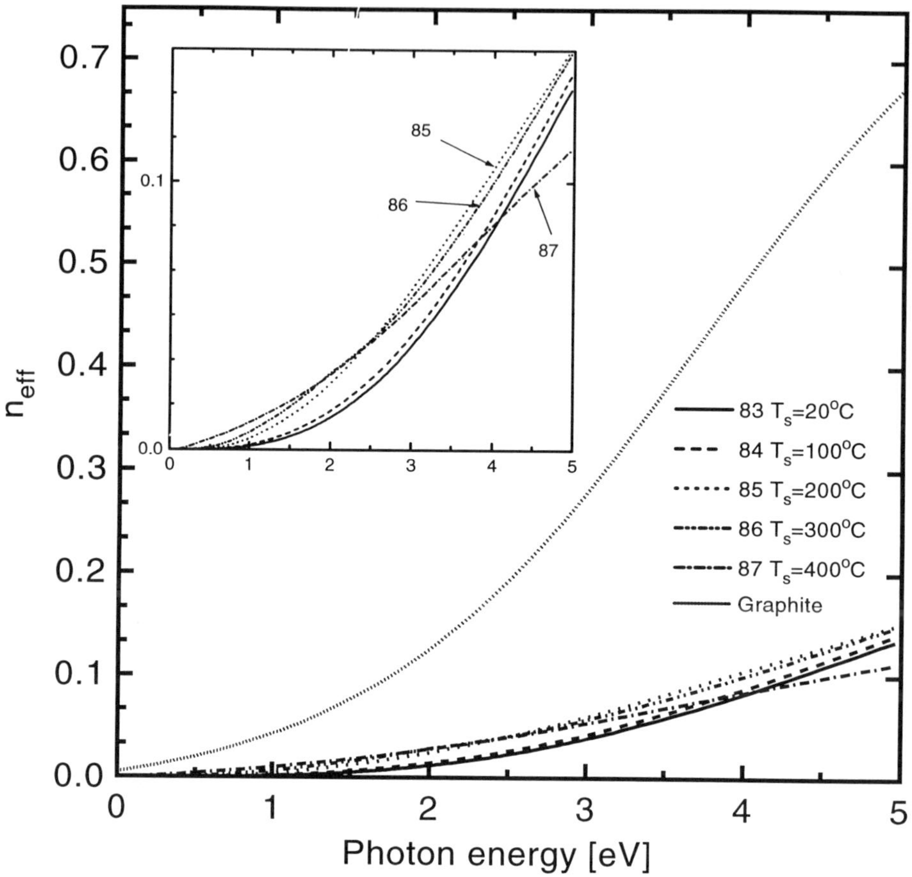

Fig. 2.: Effecitve number of valence electrons per carbon atom n_{eff} contributing to optical interband transitions below photon energy hv. The graphite data show the π-band contributions to n_{eff}. The inset shows the measured data in detail.

The ratio of n_{eff} of the film to n_{eff} of graphite gives, under certain assumptions, the relative concentration of carbon atoms with sp^2 - configuration (C_{sp2}), and is included as the sp^2/sp^3 ratio in Tables 1 and 2. There is a tendency of increasing sp^2 - concentrations with increasing temperature and power, but there are some deviations for samples 86 and 87. Figure 2 also shows the limitations of the model used to fit n_{eff}. As the material gets more graphitic, it will no longer be a semiconductor, but a semimetal, which makes the determination of ϵ_2 problematic, because the values for energies close to zero are not known and have to be interpolated. Therefore, the values in Table 1 for samples 87 and 86 should be treated with some suspicion.

The values determined for the optical gap E_g show a clear correlation with the deposition parameters. A wider bandgap is correlated with more localized π bonding. The values of n_{eff} can be lowered by a depletion of π-bonds (the assumption used for the evaluation presented in the tables) although they may also be affected by the localization of π-bonds. The hardness measurements as included in Table 1 strongly indicate diamond like behavior. From this we assume that the values of C_{sp2} are correct in tendency (except 86 and 87), but the degree of sp^2 bonding may be somewhat larger due to π bond localization.

I-V measurements of the films show very poor electron emission. This is probably related to the high sample purity and a comparably smooth surface. On the other hand, small changes to the surface of the samples, changing either the microstructure or the bonding states, are probably the key for enhancing the emissivity. Some samples treated in an RF-hydrogen plasma (100 W at 330 V bias) exhibit a tripling of the emission current (2.8m A/cm^2 at 400 KV/cm). The enhancement seems to be independent of the film properties before the treatment, and the currents are fluctuating as in the case of uncovered tip-emitters.

Table 1. Dependence of film properties on substrate temperature; laser fluence at 3.2J/cm^2 (n.d.: data not available).

Film number	Substrate temperature [oC]	Optical gap Eg [eV]	Ratio sp2/sp3	Roughness/ size of particles [nm]	hardness HV2 [GPa]
83	20	0.52	0.23	2/ 10-20	58
84	100	0.32	0.24	n. d.	n. d.
85	200	0.12	0.28	n. d.	36.7
86	300	-	(.0.30)	n. d.	n. d.
87	400	-	(0.22)	40/ 20-200	14.8

Table 2. Dependence of film properties on laser fluence; substrate temperature at 200°C.

Film number	Laser Fluence [J/cm2]	Optical gap Eg [eV]	Ratio sp2/sp3	Roughness and size of particles [nm]	hardness HV2 [GPa]
76	1.3	-	0.37	3/10-30	18.0
74	2.3	-	0.35	n. d.	23.1
85	3.2	0.12	0.28	n. d.	36.7

CONCLUSIONS

Amorphous carbon films with diamond character could be deposited using laser ablation with relatively low intensities of 10^8 W/cm^2. Graphitic components could be clearly correlated with the deposition parameters. Temperatures used in other processes to produce polycrystalline diamond already lead to a graphitization of the films. Therefore deposition at room temperature (or even lower) at high laser fluences is recommended for further investigations of inert film deposition. To improve the electron emissivity, the very pure films have to be doped and their surfaces further modified. The combination of the further developed microtip technology with thin diamond coatings is also promising and currently under investigation [17].

ACKNOWLEDGEMENTS

One of the authors (C.M.) would like to thank Irwin Sproule, Jeff Frazer, John Phillips, Jennifer Bardwell, Richard Barber, Bryce Baites and Alexander Blais for additional measurements and helpful discussions.

REFERENCES

1. O.S. Panar, D. Sarangi, S. Kumar, P. Dixit, R. Bhattacharyya, J. Vac. Sci. Technol. A **13**, 2519 (1995)
2. D.L. Pappas, J. Hopwood, J. Vac. Sci. Technol. **12**, 1576 (1994)
3. A.A. Goruppa, N.St.J.Braithwaite, Diamond and Related Materials **3**, 1223 (1994)
4. L. Ganapathi, S. Giles, R. Rao, Appl. Phys. Lett. **63**, 993 (1993)
5. R.T. Demers, D.G. Harris, SPIE Vol. **1146**, 48 (1989)
6. F. Davanloo, E.M. Juengermann, D.R. Jander, T.E. Lee, C.B. Collins, J. Mater. Res. **5**, 2398 (1990)
7. D.L. Pappas, K.L. Saenger, J. Bruley, W. Krakow, J.J. Cuomo, J. Apll. Phys. **71**, 5675 (1992)
8. N. Kumar, H. K. Schmidt, M. H. Clark, A. Ross, B. Lin, L. Fredin, B. Baker, D, Patterson, W. Brookover, C. Xie, C. Hilnert, R.L. Fink, C.N. Potter, A. Krishnan, D. Eichman, SID94 Digest, p.94 (1994)
9. K. Okano, K. Hoshina, M. Iida, S. Koizumi, T. Inuzuka, Appl. Phys. Lett. **64**, 2742 (1994)
10. K. Okano, K.K. Gleason, Electr. Lett. **31**, 74 (1995)
11. Z. Feng, I.G. Brown, J.W.Ager III, J. Mater. Res. **10**, 1585 (1995)
12. F.J. Himpsel, J.A.Knapp, J.V.VanVechten, D.E.Eastman, Phys. Rev. B **20**, 624 (1979)
13. Z. Zhang, M. Wensell, J. Bernholc, Phys. Rev. B **51**, 5291 (1995)
14. T. Yamada, T.J.Chuang, H.Seki, Y.Mitsuda, Molecular Physics **76**, 887 (1991)
15. K.V.Ravi, Mat. Sci. Eng. B **19**, 203 (1993)
16. E. I. Givargizov, J. Vac. Sci. Technol B **13**, 414 (1995)
17. P. Laou, I. Shi, C. Py, C. Mößner, P.Grant, to be published
18. P.D. Grant, M.W. Denhoff, H. Tran, Physica C **185-189**, 2099 (1991)
19. J.-M. Jin, M.W.C. Dharma-wardana, D.J. Lockwood, G.C. Aers, Z.H. Lu, L.J. Lewis, Phys. Rev. Lett. **75,** 878 (1995)
20. R. Berriche, Scripta Metall. Mat. **32**, 617 (1995)
21. D.R McKenzie, D. Muller, B.A. Pailthorpe, Phys. Rev. Lett **67**, 773 (1991)
22. D.J. Krajnovich, J. Cham Phys. **102**, 726 (1995)
23. C. Germain, C. Girault, R. Gisbert, J. Aubreton, A. Catherinot, Diamond and related Materials **3**, 598 (1994)
24. K. Mann, F. Müller, SPIE Vol. **1835**, 13 (1992)
25. D.S. Knight, W.B. White, J. Mater. Res. **4**, 385 (1989)
26. L.C. Nistor, J. VanLanduyt, V.G. Ralchenko, T.V. Konoenko, E.D. Obrazstova, V.E. Strelnitsky, Apll. Phys. Lett. A **58**, 137 (1994)
27. M.L. Terranova, V. Sessa, V. Rigato, F. Caccavale, M. Braglia, G. Cocito, Thin Solid Films **232**, 21
28. C. Mößner, I. Sproule, unpublished results
28. H.R. Philipp, H. Ehrenreich, Phys. Rev. **129**, 1550 (1963)
29. N. Savides, J. Appl. Phys. **58**, 518 (1985)

THERMAL AND OPTICAL ADMITTANCE SPECTROSCOPY STUDIES OF DEFECTS IN 15R-SIC

S.R. SMITH†, **A.O. EVWARAYE**††, **AND W.C. MITCHEL**, Wright Laboratory, Materials Directorate, WL/MLPO, 3005 P St., Wright-Patterson AFB, Ohio 45433-7707

ABSTRACT

Nitrogen is the common *n*-type dopant of the various polytypes of silicon carbide. The nitrogen levels in 4H-SiC (at E_C-53 meV and E_C-100 meV) and in 6H-SiC (at E_C-89 meV, E_C-100 meV, and E_C-125 meV) have been studied in detail by temperature dependent Hall effect measurements, electron spin resonance (ESR), and thermal admittance spectroscopy. Until now, such detailed studies of the nitrogen levels in 15R-SiC have not been carried out.

Lely-grown 15R samples were used in these studies. The net carrier concentrations (N_D-N_A), determined by room temperature CV measurements, ranged from 1 x 10^{18} to 3 x 10^{18} cm^{-3}. The nitrogen levels in 15R-SiC were studied using thermal admittance spectroscopy. Optical admittance spectroscopy (OAdS) was used to study the deeper defects in this polytype. It was found that optical transitions to the conduction band were inhibited in the heavily doped material.

INTRODUCTION

Because of its many superior properties, including a wide bandgap and high breakdown field, silicon carbide is being investigated for a variety of electronic applications including high temperature, high power, microwave, and radiation hard, devices. Significant advances have been made in recent years in both materials and devices[1]. Now that improved material is becoming available, both in bulk and epitaxial form, researchers are investigating the deep levels present in the material. Deep levels can limit carrier lifetime and adversely affect device performance. In addition, there is a need, particularly in the area of high power microwave devices, for insulating substrates. Compensation of residual impurities by intentionally doping with deep level impurities is required to produce this material.

Deep levels in SiC have been studied for some time by a variety of techniques but perhaps the most powerful are those based on capacitance spectroscopy. Deep level transient spectroscopy (DLTS) is the most widely used technique and a number of levels have been detected in 6H-SiC. In particular, a level near 0.70 eV has been reported by Uddin and coworkers[2,3], and Jang et al.[4], and by us[5]. However, no one has been able to identify the defect responsible. Unless special high-temperature facilities are used, 0.7 eV is about as deep a level as can be observed by conventional DLTS. Other techniques have been used to investigate deeper levels, and photoluminescence[6] and electron paramagnetic resonance[7] have proven useful, particularly for the study of transition metal impurities such as vanadium and titanium.

The samples studied were 15R-SiC grown by the Lely method. The crystals were all *n*-type due to the predominance of the nitrogen donor levels. Wafer preparation consisted of oxidation and etching to remove polishing damage[9]. To create the capacitance specimens, nickel was annealed on the Si face to create the ohmic contact and unannealed sputtered aluminum formed the Schottky contacts, which ranged in size from 100 to 600 μm in diameter.

Mat. Res. Soc. Symp. Proc. Vol. 423 © 1996 Materials Research Society

Admittance Spectroscopy

Thermal admittance spectroscopy (TAS)[10], involves the measurement of the capacitance and conductance of a diode as a function of both temperature and frequency. This technique takes advantage of the band bending present in the depletion region of a Schottky diode. When the Fermi level crosses the defect level a peak in the conductance occurs whenever $\omega_k<\tau_i>=1$, where ω_k is the measurement frequency and $<\tau_i>$ is an average time constant for the ith defect level. For each peak in the conductance the logarithm of ωT^{-2} may be plotted versus 1/kT to yield a straight line with slope $-E_a$. An advantage of TAS over the more commonly used deep level transient spectroscopy (DLTS) is that the shallowest uncompensated level can be measured by TAS but not by DLTS. The system at our laboratory is capable of measuring both TAS and DLTS from 4K to 400C, which permits measurement of activation energies up to about 0.8 eV. Due to the wide bandgap of SiC, many levels are not observable using TAS or DLTS, so we have established an optical admittance spectroscopy (OAS) capability. In this variation of admittance spectroscopy[10] carriers are photo excited to the conduction band (in n-type material) and when there change the conductance. With this technique activation energies up to the bandgap of SiC can be measured without resorting to the very high temperatures that would be necessary with thermal admittance or deep level transient spectroscopy.

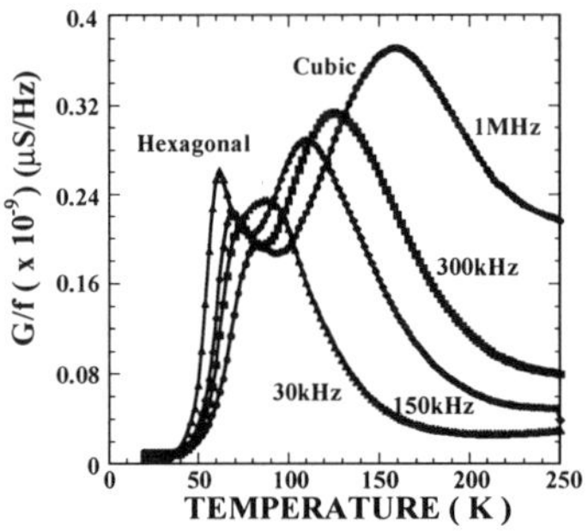

Figure 1. The typical thermal admittance spectrum comprises several frequencies, with peaks where the measurement frequency is equal to the emission rate of the defect. This graph shows peaks for nitrogen shallow donors in 4H-SiC.

RESULTS

A typical thermal admittance spectrum for 4H-SiC is shown in figure 1.[11] The energy of the defect is determined from an Arrhenius plot of $\ln(f/T^2)$. Nitrogen is the principal donor in SiC and occupies the carbon lattice sites giving rise to an energy levels at E_C-0.053 and E_C-0.100 eV. Figure 2 shows the thermal admittance spectrum for a specimen of nitrogen doped Lely SiC. The most striking difference is the temperature at which the peaks occur. Usually, activation energy is proportional to temperature, and this specimen is no exception.

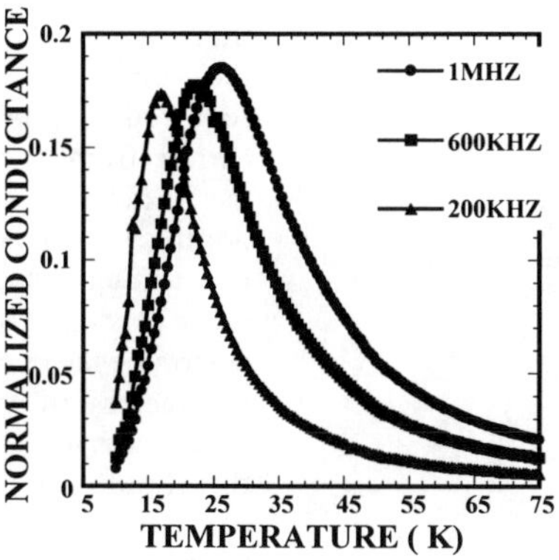

Figure 2. Admittance spectrum for *n*-type 15R Lely SiC.

Figure 3 is the Arrhenius plot of the peak position versus inverse temperature. The slope of the curve gives the activation energy of the defect. In this case the activation energy is calculated to be ~ 4 meV. Since this energy is much smaller than the AC signal used to detect the admittance, this can not be from a transition to the conduction band. It turns out that the signal is due to hopping or impurity conduction. This is verified by the curve shown in figure 4, which is a composite curve of the Hall coefficient and the bulk resistivity as a function of temperature. The curves are characteristic of impurity conduction. the slope of the resistivity curve at low temperature yields an activation energy of ~ 4 meV. In excellent agreement with the results from the admittance measurement.

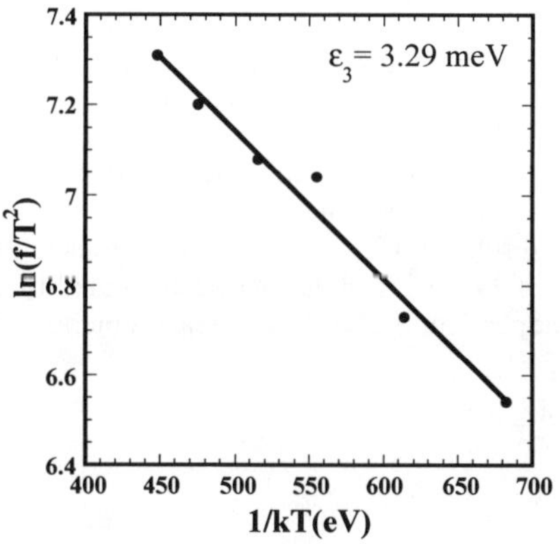

Figure 3. Arrhenius plot of the admittance data for the Lely specimen of 15R-SiC. Activation energy is 3.29 meV.

Hopping conduction, or more formally, impurity conduction, is the mechanism whereby charge carriers 'hop' from an occupied site (in this case, donor sites) to an unoccupied site. This is possible only when the donor concentration is high enough that the charge carrier wave functions overlap, and compensation of the shallow donor provides ionized sites for the carriers to hop to. The hopping takes place under the influence of the applied electric field, in this case, the built in voltage of the Schottky diode used in the experiment and the applied AC signal used to

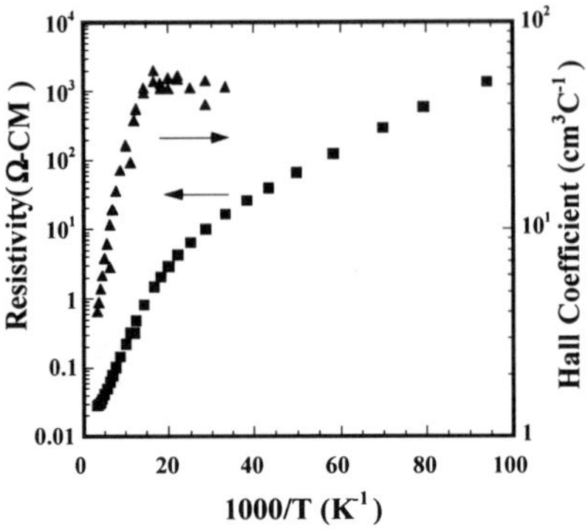

Figure 4. Temperature dependent Hall effect measurements reveal hopping conduction with an activation energy of ~4 meV.

measure the admittance of the diode. The energy that is measured is, therefore, the energy required to move the carrier from one site to the next. This energy is doping dependent, but not enough specimens were available to determine the dependence parameters in 15R-SiC. The carrier density threshold for hopping conduction appears to be around 2×10^{18} cm^{-3}.

Attempts to detect deep levels in the 15R-SiC band gap were unsuccessful. A broad photo response was seen, but no peaks in the admittance spectrum were observed. This is believed to be the result of photo-excited carriers dropping from the conduction band into the impurity band, thus being lost to the conduction process.

CONCLUSIONS

This paper has reported our studies of electronic levels in 15R-SiC. The emphasis has been placed on those results which utilized admittance spectroscopic techniques. We have pointed out the complementarity of thermal admittance spectroscopy with DLTS, and the advantage of being able to characterize shallow levels as well. The observation of levels attributable to N in n-type material was discussed. Nitrogen shallow donor levels were not observed in this heavily doped material. The results of using admittance spectroscopy were to observe a level with an activation energy of 3.29 eV. This level was attributed to hopping conduction and verified by temperature dependent Hall effect measurements. Optical admittance measurements were unsuccessful.

ACKNOWLEDGMENT

The authors wish to acknowledge the assistance of G. Landis, P. von Richter (deceased), and R. Bertke for technical assistance in preparation of the samples. The work of SRS was supported by Air Force Contract F33615-91-C-5603.

†University of Dayton Research Institute, 300 College Park, Dayton, OH 45469-0178.

††University of Dayton, Department of Physics, 300 College Park , Dayton, OH 45469-2314.

REFERENCES

1. "Proceedings of the Fifth Conference on Silicon Carbide and Related Materials," eds. M. G. Spencer, R. P. Devaty, J. A. Edmond, M. Asif Khan, and M. Rahman, Inst. Phys. Conf. Ser. **137** (1994).
2. A. Uddin and T. Uemoto, Jpn. J. Appl. Phys. **32**, L1670 (1993).
3. A. Uddin, H. Mitsuhara, and T. Uemoto, Jpn. J. Appl. Phys. **33**, L908 (1994).
4. S. Jang, T. Kimoto and H. Matsunami, Appl. Phys. Lett. **65**, 581 (1994).
5. A.O. Evwaraye, S.R. Smith, and W.C. Mitchel, J. Appl. Phys. **76**, 5769 (1994).
6. W. Choyke and I. Linkov, "Proceedings of the Fifth Conference on Silicon Carbide and Related Materials," eds. M. G. Spencer, R. P. Devaty, J. A. Edmond, M. Asif Khan, and M. Rahman, Inst. Phys. Conf. Ser. **137** (1994), pg. 141.
7. J. Schneider and K. Maier, Physica B, **185**, 199 (1993).8. H. McD. Hobgood, R. C. Glass, G. Augustine, R. H. Hopkins, J. Jenny, M. Skowronski W. C. Mitchel, and M. Roth, Appl. Phys. Lett. **66**, 1364 (1995).
9. A.O. Evwaraye, S.R. Smith, and W.C. Mitchel, J. Appl. Phys. **74**, 5269 (1993)
10. P. Blood and J. W. Orton, The Electrical Characterization of Semiconductors: Majority Carriers and Electron States (Academic Press, London, 1992).
11. A.O. Evwaraye, S.R. Smith, and W.C. Mitchel, J. Appl. Phys. **75**, 3472 (1994)
12. C. Raynaud, F. Ducroquet, G. Guillot, L. M. Porter, and R. F. Davis, J. Appl. Phys. **76**, 1956 (1994)
13. A.O. Evwaraye, S.R. Smith, and W.C. Mitchel (unpublished).
14. A.O. Evwaraye, S.R. Smith, and W.C. Mitchel, Mater. Res. Soc. Symp. Proc. **325**, 353 (1994)
15. A.O. Evwaraye, S.R. Smith, and W.C. Mitchel, Mater. Res. Soc. Symp. Proc. **339**, 353 (1994)
16. W.C. Mitchel, Matthew Roth, S.R. Smith, A.O. Evwaraye, and J. Solomon, Inst. Phys. Conf. Ser. No. **141**: Chapter 4, p.411, 1995
17. Th. Stiasny, R. Helbig, and R.A. Stein, *Amorphous and Crystalline Silicon Carbide IV, Springer Proceedings in Physics 71*, edited by C.Y. Yang, M.M. Rahman, and G.L. Harris (Springer, Berlin, 1992)
18. A.O. Evwaraye, S.R. Smith, and W.C. Mitchel, J. Appl. Phys. **77**, 4477 (1995)
19. A.O. Evwaraye, S.R. Smith, and W.C. Mitchel, Appl. Phys. Lett. **66**, 2691 (1995)

LOCAL–FIELD AND EXCHANGE–CORRELATION EFFECTS IN OPTICAL SPECTRA OF WIDE–BAND–GAP SEMICONDUCTORS

V.I. GAVRILENKO, F. BECHSTEDT
Friedrich-Schiller-Universität, IFTO, Max-Wien-Platz 1, D–07743 Jena, Germany

ABSTRACT

The density-functional theory with *ab initio* pseudopotentials has been used to study the linear optical response of semiconductors. We present results for optical spectra where the effects of the macroscopic local-field and microscopic exchange-correlation interaction are included beyond diagonal and random-phase approximation. Quasiparticle corrections to the single-particle energies have been added in the polarization function. Numerical calculations are performed for the column-IV materials Si, SiC, and diamond as model substances.

INTRODUCTION

Highly accurate calculations of optical properties of semiconductors without using any empirical parameters are of continuous interest [1, 2, 3, 4]. Most of the theoretical studies are based on the independent-particle approximation [5] [often called the random-phase aproximation (RPA)] and a first-principles description of the electronic and atomic structure in the framework of the density-functional theory (DFT) in the local-density approximation (LDA). The independent-quasiparticle (QP) approximation has been used in Refs. 1 and 6. First attempts [1, 2] have been made to go beyond RPA considering exchange–correlation (XC) corrections. Local-field (LF) effects [5] due to the atomic structure of the matter influence the resulting optical spectra. However, they have been studied only in few papers [2, 3]. These papers supplement earlier studies in the field [7, 8, 9, 10] (and references therein), in which LF and excitonic effects have been discussed in the framework of the empirical-pseudopotential method (EPM) or expansions of the eigenfunctions in terms of localized orbitals. However, a clear and detailed picture is still missing. Moreover, full calculations of the dynamical dielectric function in silicon [2] seems to give rise to strange sharp features at the low-energy side of the absorption spectrum not found experimentally.

In this work the influence of local fields as well as that of exchange-correlation effects on optical spectra of semiconductors is systematically studied beyond RPA in the framework of *ab initio* DFT–LDA by using the equilibrium atomic structure arising within the same description. In addition XC self-energy effects [4, 11], i.e. quasiparticle (QP) shifts of the electron and hole DFT-LDA energies, are discussed. As model substances the column-IV materials diamond (C), silicon (Si), and silicon carbide (SiC) are considered.

MACROSCOPIC DIELECTRIC FUNCTION

According to Adler and Wiser [5] the macroscopic dielectric function, that governs the optical properties of a crystal, may be directly related to the zeroth element of the inverse of the microscopic dielectric matrix $\varepsilon(\mathbf{q}+\mathbf{G},\mathbf{q}+\mathbf{G}';\omega)$, where $\mathbf{q}$ denotes a vanishing wave vector and $\mathbf{G},\mathbf{G}'$ represent elements of the reciprocal Bravais lattice of the crystal. The

Mat. Res. Soc. Symp. Proc. Vol. 423

longitudinal macroscopic function is defined as

$$\varepsilon^M(\hat{\mathbf{q}};\omega) = \lim_{\mathbf{q}\to 0} \frac{1}{\varepsilon^{-1}(\mathbf{q}+\mathbf{G},\mathbf{q}+\mathbf{G}';\omega)}\bigg|_{\mathbf{G}=\mathbf{G}'=0}, \tag{1}$$

where $\hat{\mathbf{q}} = \mathbf{q}/|\mathbf{q}|$ represents the direction of $\mathbf{q}$. The zeroth element of the inverse dielectric matrix is influenced by the off-diagonal elements of the dielectric matrix, which are due to the lattice periodicity and generate "umklapp" processes in the dielectric response. They are generally referred to as "local-field effects" [5]. We define these LF effects more exactly as the discrepancy between $\varepsilon^M(\hat{\mathbf{q}};\omega)$ and the zeroth element of the dielectric matrix, $\lim_{\mathbf{q}\to 0}\varepsilon(\mathbf{q},\mathbf{q};\omega)$.

The microscopic dielectric matrix

$$\varepsilon(\mathbf{q}+\mathbf{G},\mathbf{q}+\mathbf{G}';\omega) = \delta_{\mathbf{G}\mathbf{G}'} - v(\mathbf{q}+\mathbf{G})P(\mathbf{q}+\mathbf{G},\mathbf{q}+\mathbf{G}';\omega) \tag{2}$$

with $v(\mathbf{q}+\mathbf{G}) = 4\pi e^2/|\mathbf{q}+\mathbf{G}|^2$ is directly related to the polarization function P of the system under consideration. It contains the irreducible diagrams of the proper part of the two-particle Green function. Neglecting XC corrections to the two-particle Green function, i.e., applying the independent-particle approximation, P has to be replaced by the polarization function P_0 of independent particles (or quasiparticles). If XC effects on the longitudinal response are not neglected, e.g. within the DFT or DFT-LDA, the polarization function P appears instead that of independent particles P_0. It can be easily proved [2, 3] that the expression for interacting particles takes the form

$$P(\mathbf{q}+\mathbf{G},\mathbf{q}+\mathbf{G}';\omega) = \sum_{\mathbf{G}''}\Gamma(\mathbf{q}+\mathbf{G},\mathbf{q}+\mathbf{G}'';\omega)P_0(\mathbf{q}+\mathbf{G}'',\mathbf{q}+\mathbf{G}';\omega), \tag{3}$$

where the matrix Γ arises from the vertex function of the system. Its inverse matrix is directly related to a kernel K_{XC} that describes the XC effects in the two-particle function beyond RPA. Within the DFT it may be represented in real space by $K_{XC} = \delta^2 E_{XC}/(\delta n \delta n')$ with the electron density n and the total XC energy E_{XC}. Within the DFT-LDA the kernel is local in real space and does not depend on the frequency. As a result, the reciprocal-space representation for K_{XC} is independent of $\mathbf{q}$ and ω. It holds $K_{XC}(\mathbf{q}+\mathbf{G},\mathbf{q}+\mathbf{G}';\omega) = K_{XC}(\mathbf{G}-\mathbf{G}')$.

The $\mathbf{q}\to 0$ limit of expression (1) is required to obtain the optical response functions. This limit has to be taken with care to keep the correct analytical properties of this function and the underlying inverse dielectric matrix [12]. In the limit of vanishing wave vectors the macroscopic dielectric function (1) transforms into

$$\varepsilon^M(\hat{\mathbf{q}};\omega) = \hat{\mathbf{q}}\cdot\hat{\varepsilon}^M(\omega)\cdot\hat{\mathbf{q}}, \tag{4}$$

where the macroscopic optical tensor [12] $(i,j = x,y,z)$

$$\varepsilon^M_{ij}(\omega) = \varepsilon_{ij}(\omega) - \sum_{\mathbf{G},\mathbf{G}'(\neq 0)} \frac{|\mathbf{G}|}{|\mathbf{G}'|} W_i(\mathbf{G};\omega)S^{-1}(\mathbf{G},\mathbf{G}';\omega)W_j^*(\mathbf{G}';-\omega) \tag{5}$$

is related to the corresponding microscopic one following from the zeroth element ("head") of the dielectric matrix (2)

$$\varepsilon_{ij}(\omega) = \delta_{ij} + \frac{16\pi e^2\hbar^2}{V}\sum_{\mathbf{k}}\sum_{c,v}\frac{\langle c\mathbf{k}\,|\mathrm{v}_i|\,v\mathbf{k}\rangle\,\langle v\mathbf{k}\,|\mathrm{v}_j|\,c\mathbf{k}\rangle}{[\epsilon_c(\mathbf{k})-\epsilon_v(\mathbf{k})]\left\{[\epsilon_c(\mathbf{k})-\epsilon_v(\mathbf{k})]^2 - \hbar^2(\omega+i\eta)^2\right\}} \tag{6}$$

and functions

$$W_j(\mathbf{G};\omega) = \frac{16\pi e^2\hbar}{|\mathbf{G}|V}\sum_{\mathbf{k}}\sum_{c,v}\frac{\left\langle c\mathbf{k}\left|e^{i\mathbf{G}\mathbf{x}}\right|v\mathbf{k}\right\rangle\left\langle v\mathbf{k}\left|\mathrm{v}_j\right|c\mathbf{k}\right\rangle}{\left[\epsilon_c(\mathbf{k})-\epsilon_v(\mathbf{k})\right]^2-\hbar^2(\omega+i\eta)^2} \tag{7}$$

arising from "wing" elements $\varepsilon(\mathbf{q},\mathbf{G};\omega)$. Here matrix elements of the velocity operator $\mathbf{v}$ and exponential functions with the Bloch eigenfunctions $|n\mathbf{k}>$ belonging to the bandindex n, the wave vector $\mathbf{k}$ from the Brillouin zone (BZ), and the single-particle energy $\epsilon_n(\mathbf{k})$ are introduced. S^{-1} is the inverse of the lower-right submatrix of $\varepsilon(\mathbf{q}+\mathbf{G},\mathbf{q}+\mathbf{G}';\omega)$ corresponding to nonzero reciprocal lattice vectors $\mathbf{G}$ and $\mathbf{G}' \neq 0$, the so-called "body" of the dielectric matrix.

The electronic-structure calculations underlying the computations of the optical properties are based on the DFT-LDA [13]. The electron-ion interaction is treated by norm-conserving, *ab initio*, fully separable pseudopotentials in the Kleinman-Bylander form. As model systems we consider silicon- and carbon-based crystals. The C-potentials are softened by careful choosing of the core radii [14]. The electronic wave functions are expanded in terms of plane waves. The energy cutoffs for the plane-wave expansion are chosen to 15, 34, and 42 Ry for silicon, silicon carbide, and diamond. The total-energy optimizations give rise to theoretical cubic lattice constants of a = 10.227 a.u. for Si, a = 8.109 a.u. for SiC, and a = 6.681 a.u. for C. We also study the influence of many-body QP effects. Thereby we overcome the scissors-operator approximation. The QP corrections to the DFT-LDA eigenvalues are computed within the GW approximation for the XC self-energy [11] according to a simplified scheme developed by Cappellini et al. [15]. Using the numerical input described above corresponding shift values have been published for Si and diamond in Ref. 4 and for SiC in Ref. 15.

RESULTS

The imaginary parts of the macroscopic dielectric functions resulting for C, SiC, and Si are plotted versus photon energy in Fig. 1 within different approximations: without LF and XC effects, with LF effects, with LF and XC effects, and with LF and XC effects but using QP eigenvalues instead of DFT-LDA ones. The changes of LF contributions due to inclusion of XC kernel are plotted separately. The nonlocality LF and many-body XC effects have practically no influence on the peak positions but give rise to remarkable renormalizations of the oscillator strengths. Compared with the $\mathbf{G} = \mathbf{G}' = 0$ element of the dielectric function, the LF effects reduce the oscillator strength of $\varepsilon^M(\omega)$ in the spectral region below the main absorption peaks. On the other hand, the inclusion of the XC kernel reduces the LF effects in this region. The corresponding curves lie inbetween those for $\varepsilon(\omega)$ and $\varepsilon^M(\omega)$ including only LF effects. The neteffect which incorporates LF and XC amounts roughly 60 % of the pure LF influence using the RPA expression for the polarization function. Sharp spurious peaks as have been observed for Si by Farid and Engel [2] do not appear in our spectra. We believe that their finding is related to a noncoverged continued-fraction expansion but has nothing to do with the neglect of self-energy corrections to the DFT-LDA transition energies. In the region of the main absorption peaks and above them the situation is not unique. Positive and negative variations of the oscillator strengths occur. However strongest theoretical absorption peakes, i.e. E_2 for C and Si, E_o, E_1, E_o', and E_1' for SiC, and E_1' for Si are increased by LF but somewhat reduced by XC effects. In the high–energy region above these peaks local fields slightly enhance the response whenever XC effect in this region is negligible.

The comparison of the spectra for C, SiC, and Si make evident that the LF and XC effects show only weakly pronounced chemical trends with the averaged size of the atoms, the localization of the wave functions, or the averaged density of the electrons. The strongest LF and XC effects appear for silicon whereas their relative influence is slightly reduced changing to diamond. This result is in agreement with earlier semiempirical calculations for Si and C [7, 8]. A simple explanation of this trend arises from expression (5). Assuming that the magnitude of the wing and body elements of the dielectric matrix scales with the electron density according to the generalized f-sum rule, the local-field corrections are proportional to this density. However, the density is weighted by the inverse average gap of the system. As a consequence of the interplay of the two factors the effect of the local fields (or LF combined with XC effects) is absolutely and relatively weaker comparing to the materials with the stronger bonds. The band-index- and wave-vector-dependent QP corrections of the DFT-LDA eigenvalues essentially cause a shift of the spectra to higher photon energies. They roughly act like a scissors operator although there is a small broadening of the spectra [4].

In Fig. 2 the macroscopic dielectric functions of C, SiC, and Si are calculated within the DFT-LDA but including LF, XC, and QP effects and are compared with experimental data [16, 17, 18]. The theoretical spectra are shifted towards lower energies to bring the zeros of the experimental and theoretical real part together. For this purpose the calculated QP shifts $\Delta_n(\mathbf{k})$ are replaced by $A \cdot \Delta_n(\mathbf{k})$ with scaling factors $A = 0.2$ (C), 0.45 (SiC), and 0.5 (Si). This scaling reduces the effect of the wave-vector- and band-index-dependent quasiparticle shifts calculated for Si, SiC, and C [4, 15]. Considering the comparison of theoretical and experimental optical spectra over a wide range of photon energies, one can conclude that the QP effect is overestimated for the most pronounced optical transition, e.g. E_1 and E_2 in Si and C, in contrast to electron-hole pair excitations near the fundamental indirect energy gap. As a consequence the scaling factors $A < 1$ have been introduced. Using the positions of the zero in the real parts of the macroscopic dielectric function in order to define averaged scissors operators Δ, one derives from the wave-vector- and band-index-dependent QP corrections values $\Delta = 0.95$ (Si), 1.65 (SiC), and 2.65 eV (C). However, averaged scissors operators being necessary only amount to $\Delta = 0.47$ (Si), 0.84 (SiC), and $\Delta = 0.40$ eV (C). The reason for this observation is not very clear. One posssible reason could be related to excitonic effects which increase with the localization of the electronic states in the considered material. We mention that similarly small scissors operators have been found to bring the calcuations of ε_∞ in agreement with experiment [1]. Whereas the partial inclusion of QP effects shifts the theoretical peaks towards position of experimental ones.

CONCLUSIONS

In conclusion, we have studied the influence of local-field effects and exchange-correlation corrections beyond RPA on the optical properties of column-IV materials in the framework of an *ab initio* density-functional method. We find that (i) these effects do not shift the prominent peak positions in $Im\varepsilon^M(\omega)$ and that (ii) the agreement of theory and experiment remains still insufficient after inclusion of these effects. We found a weakly pronounced chemical trend. With rising electron localization the influence of LF and XC decreases

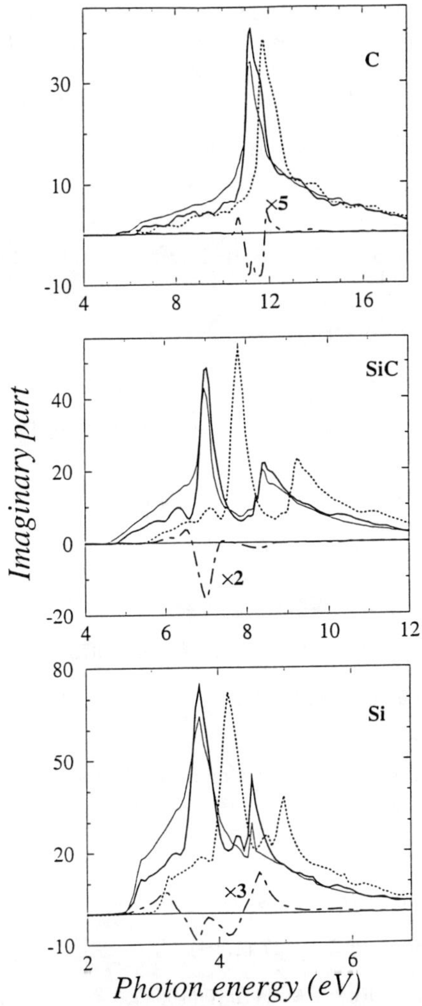

Fig. 1 Imaginary part of macroscopic dielectric function of diamond, silicon carbide, and silicon versus photon energy. Thin solid line: without LF and XC effects (only $\mathbf{G} = \mathbf{G}' = 0$ element), bold solid line: with LF and XC effects, dotted line: with LF and XC effects but shifted by wave-vector- and band-index-dependent QP corrections, dashed–dotted line: change of LF contribution due to inclusion of XC kernel in expr. (3).

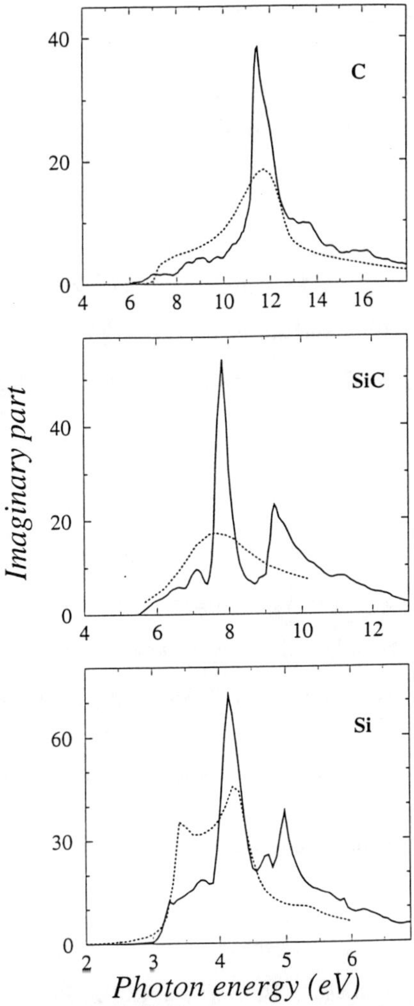

Fig. 2 Imaginary part of macroscopic dielectric function of diamond, silicon carbide, and silicon versus photon energy. Solid lines: calculated results including LF, XC, and QP effects, dotted lines: experimental results for Si [16], SiC [17] and diamond [18].

slightly. The influence of quasiparticle corrections is also checked taking into account wave-vector- and band-index-dependent self-energy shifts. We observe an obvious remarkable overestimation of these shifts. In order to bring theoretical absorption spectra also closer together with measured lineshapes we suggest that excitonic effects have to be also included in the *ab initio* calculations.

References

[1] Z.H. Levine, and D.C. Allan, Phys. Rev. Lett. **63**, 1719 (1989); Phys. Rev. B**43**, 4187 (1991); **42**, 3567 (1990); **44**, 12781 (1991).

[2] B. Engel and F. Farid, Phys. Rev. B**46**, 15812 (1992).

[3] R. Daling, W. van Haeringen, and B. Farid, Phys. Rev. B**45**, 8970 (1992).

[4] B. Adolph, V.I. Gavrilenko, K. Tenelsen, F. Bechstedt, R. Del Sole, Phys. Rev. B**53**, No. 11 (1996).

[5] S.L. Adler, Phys. Rev. **126**, 413 (1962); N. Wiser, Phys. Rev. **129**, 62 (1963).

[6] R. Del Sole and R. Girlanda, Phys. Rev. B**48**, 11789 (1993).

[7] J.A. Van Vechten and R.M. Martin, Phys. Rev. Lett. **28**, 446 (1972).

[8] S.G. Louie, J.R. Chelikowsky, and M.L. Cohen, Phys. Rev. Lett. **34**, 155 (1975).

[9] W. Hanke and L.J. Sham, Phys. Rev. B**21**, 4656 (1980); W. Hanke, Adv. Phys. **27**, 287 (1978).

[10] H. Bross, O. Belhachemi, B. Mekki, and A. Seoud, J. Phys.: Condens. Matter **2**, 3919 (1990).

[11] M.S. Hybertsen and S.G. Louie, Phys. Rev. B**34**, 5390 (1986).

[12] R.M. Pick, M.H. Cohen, and R.M. Martin, Phys. Rev. B**1**, 910 (1970).

[13] R. Stumpf and M. Scheffler, Comp. Phys. Commun. **79**, 447 (1994).

[14] P. Käckell, B. Wenzien, and F. Bechstedt, Phys. Rev. B**50**, 17037 (1994).

[15] B. Wenzien, P. Käckell, F. Bechstedt, and G. Cappellini, Phys. Rev. B**52**, 10897 (1995).

[16] D.E. Aspnes and A.A. Studna, Phys. Rev. B**27**, 985 (1983).

[17] S. Logothetidis, H.M. Polatoglou, J. Petalas, D. Fuchs, and R.L. Johnson, Physica B **185**, 389 (1993).

[18] A.D. Papadopoulos and E. Anastassakis, Phys. Rev. B**43**, 5090 (1991).

PHOTOLUMINESCENCE PROPERTIES OF CVD DIAMOND EXCITED BY ULTRA-VIOLET SYNCHROTRON RADIATION

Jaihyung Won, Akimitsu Hatta, Toshimichi Ito, Takatomo Sasaki. and Akio Hiraki
Department of Electrical engineering, Osaka University, 2-1 Yamada-oka, Suita, Osaka 565, Japan

ABSTRACT

Photoluminescence (PL) properties of microwave-assisted chemical-vapor- deposition (CVD) diamond have been studied using ultraviolet synchrotron radiation. The defect-related 5RL PL feature, which was not detected in cathodoluminescence (CL), was observed for both undoped and boron-doped (200ppm) CVD diamond. The defect formation in the thin near-surface layer is discussed in relation to dependence of PL spectra on incident photon energy. In the case of boron-doped diamond, the boron-related peak was detected at 2.3eV while the band-A feature was not observed at 2.9eV. PL excitation (PLE) spectra associated with these emissions are also discussed in relation to absorption coefficients at energies above the band gap.

INTRODUCTION

Luminescence spectroscopy has been proven to be a sensitive technique for identification of impurities and other defects in semiconducting materials. Luminescence induced by ultra-violet radiation was used for many years in study of natural and synthetic diamonds [1]. The band structure of the ideal diamond is well understood from a theoretical point of view [2,3]. Recently, a number of works were conducted to characterize the growth process and to measure the properties of chemical-vapor-deposition (CVD) diamond [4-7]. However, photoluminescence (PL) experiments have been limited to measurements using excitation lights below the diamond band-gap energy diamond (5.5eV) simply because high intensity and high-energy excitation sources were not conventionally available.

In this work, low-temperature cathodoluminescence (CL) and PL were used to investigate radiative recombination processes in undoped and boron-doped CVD diamonds since both PL and CL are techniques sensitive to the near surface region of diamond. Because of large diamond absorption of excitation lights in the direct-transition energy range (≥7.0eV) [8], PL can be used for characterization of defect formation in the thin near surface layer as well as for study of diamond band structure using excitation spectra.

EXPERIMENT

Diamond films were grown on p^+-Si substrates by means of microwave plasma CVD method. All the substrates used were subjected to ultrasonic pretreatment with diamond powders. The source gas used in the growth experiment consisted of 100 sccm CO (10%)-H_2 (90%) mixture. In order to dope boron into diamond films, H_2 gas was replaced by B_2H_6-diluted H_2 gas. The growth conditions employed were as follows: total pressure 45 Torr, microwave power 250W and substrate temperature 930℃. The concentration of boron dopant in the reaction gases, estimated from boron-to-carbon (B/C) ratio , was 200 ppm in all the doping experiments. PL measurements were carried out in a chamber with pressures below 10^{-8}Torr using synchrotron

Mat. Res. Soc. Symp. Proc. Vol. 423

radiation from the 750-MeV storage ring located in the Institute for Molecular Science, Okazaki, Japan. A toroidal-grating monochromator was used to select photon energy ranging from 6.0eV to 13.8eV. For PL measurements a monochromator was used, with a photomultiplier tube (PMT) and a focusing lens placed in front of the specimen. For CL measurements, a parabolic mirror and a PMT were attached to a scanning electron microscope (SEM), and the electron-beam acceleration voltage employed was 5keV. For both cases of PL and CL measurements the specimens were cooled down to 83K using liquid nitrogen.

RESULTS AND DISCUSSION

CL spectra of undoped CVD diamond, as typically shown in Fig. 1 (a), contain a peak centered at wavelength of 4250 Å (2.9eV), called band A. Also free-exciton recombination radiation peaks were detected, indicating good crystallinity of the undoped specimen [9]. In the case of B-doped specimens, similar spectra were observed, except for boron-related feature located around 2.3eV, as shown in Fig. 1 (b).

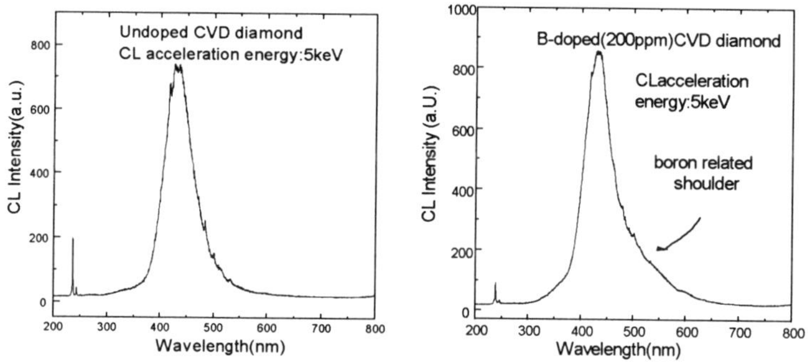

Fig. 1. Typical cathodoluminescence (CL) spectra of (a) undoped and (b) B-doped CVD diamond. The acceleration voltage used was 5kV.

In order to study the diamond properties in the near surface layers, 5-keV incident electrons was employed in the present CL measurements. The penetration depth L (μm) of the electrons used is estimated to be 3400 Å in diamond, using the following formula:

$$L=0.018\times V^{1.825}, \qquad (1)$$

where V is the acceleration voltage in keV [10].

PL spectra of the undoped and boron-doped CVD diamonds are shown in Figs. 2(a) and 2(b), respectively. The photon energy of incident light used was 11.3 eV. The PL spectrum of undoped specimen has two relatively broad peaks (Fig. 2(a)). One is the same band-A peak as in the case of CL spectra. The other peak, called the 5RL center, suggests the presence of intrinsic defects in the undoped diamond [11-14]. The 5RL center contains four narrow peaks located at every 0.234eV. On the other hand, the 5RL center was not observed in the 5-keV CL spectra as in the ordinary case. We found that the 5RL center appears in the CL spectra only when intrinsic defects are induced by ion implantation or high energy electron radiation. For example, Fig. 3 shows a typical CL spectrum of undoped CVD diamond after $^{4}He^{2+}$ ion implantation. It can be compared to PL with the defect-related feature.

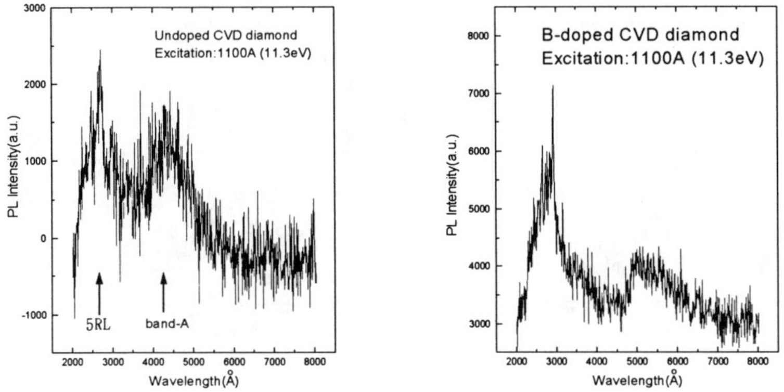

Fig. 2 Photoluminescence (PL) spectra of (a) undoped and (b) boron doped CVD diamonds, respectively. The incident photon energy was 11.3eV.

In the case of B-doped CVD diamond (Fig. 2(b)), the intrinsic-defect peak was observed together with the B-related peak (2.3eV) [15] while band-A (2.9eV) was not observed.

In the PL spectroscopy, the effect of light absorption must be considered. The light absorption coefficient μ is 2.5×10^6 cm^{-1} at a photon energy of 11.3eV [8]. Thus, since the intensity of the incident light, I, at depth X, from the surface is represented by the equation.

$$I = I_0\, e^{-\mu X}, \qquad (2)$$

where I_0 is the incident intensity, the penetration depth X, defined as $1/\mu$, is estimated to be X=40 Å. From the above experimental results follows that the surface region of CVD diamond contains substantially higher density of the intrinsic defects than the bulk region of that does.

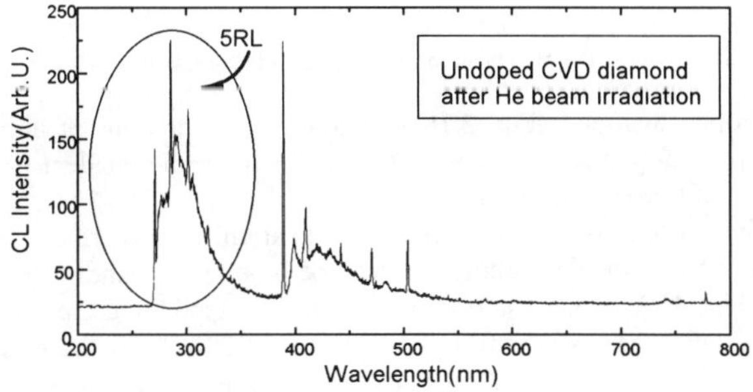

Fig. 3. The intrinsic defect-related 5RL feature is induced in cathodoluminescence (CL) after 2.1-MeV $^4He^{2+}$ ion irradiation.

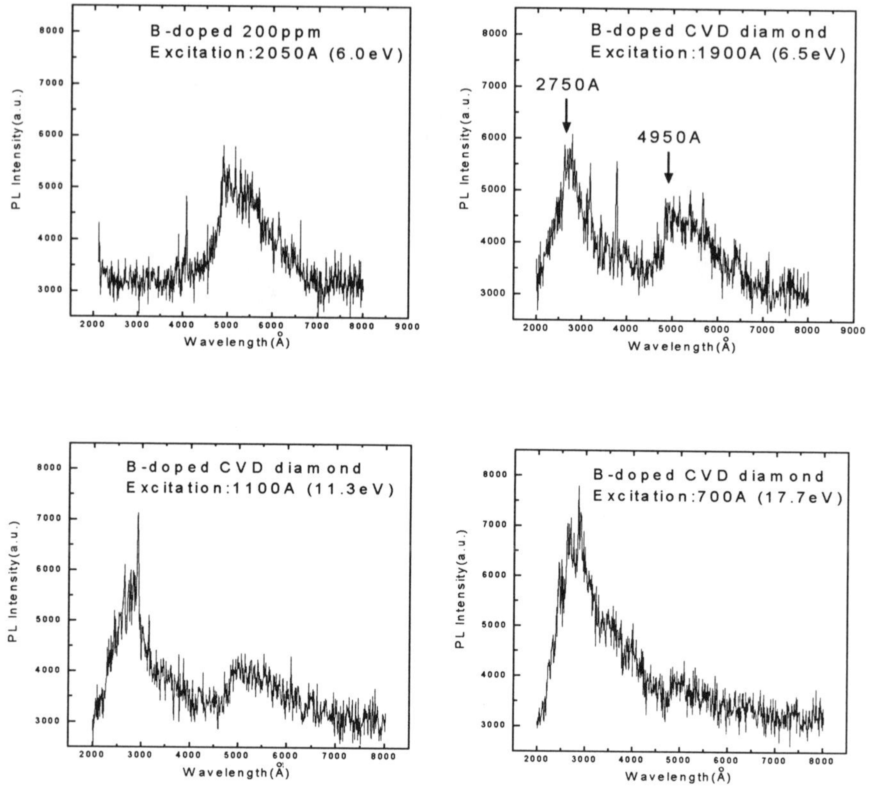

Fig. 4 (a)- (d). Photoluminescence (PL) spectra of the B-doped diamonds measured for different incident photon energies. With increasing photon energy the 5RL peak located at 2750Å becomes dominant.

In the case of B-doped diamond (Fig. 2 (b)), slightly higher amount of intrinsic defects, compared to the case of undoped sample, was observed for the same incident photon energy. This is probably because of boron incorporation.

Various spectra of the B-doped diamonds were measured at different energies of incident light (Fig. 4(a) - Fig. 4(d)). At low incident photon energy of 6.0eV, the band-A feature was not observed as shown in Fig. 4(a), while was observed in the corresponding CL spectra using 5-keV incident electrons which can penetrate to a depth of 3,400Å from the surface. For the incident photons of 6.0eV having μ of 1×10^4 cm^{-1} [16], the photon penetration depth X is 10,000Å.

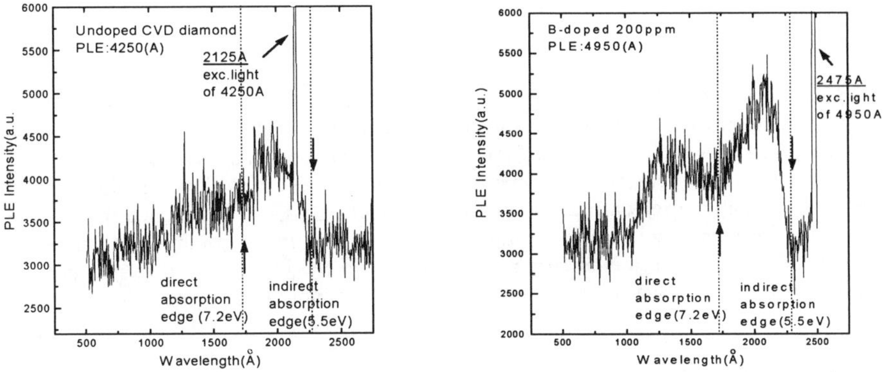

Fig. 5 Photoluminescence excitation spectra for (a) band-A (4250 Å) of undoped CVD and (b) B-related excitation (4950 Å) of B-doped CVD diamond. Sharp intense peaks in (a) and (b) are due to excitation lights passing through the monochrometer.

Similarly, X is estimated to be 2,500Å using μ of 4×10^4 cm^{-1} at 6.5eV (Fig. 4(b)) [16]. This leads to the conclusion that the density of intrinsic defects related to the 5RL peak becomes larger in shallower depth region below the specimen surface. Moreover, increase of photon energy resulted in increase of the 5RL peak amplitude although the photon penetration depths estimated are almost the same for 11.3-eV and 17.7-eV photons. This is because photo-excited high-energy carriers may be more and less sensitive to the 5RL center and the B-related center, respectively, probably due to difference in charge states between both centers.

One may wonder why the CL spectrum (Fig.1(b)) for the electron penetration depth L=3,400Å at 5keV obviously differs from the PL spectrum (Fig. 4(b)) for the photon penetration depth X=2,500Å at 6.5eV, although the penetration depths of the incident probing particles are not very different. The reasons for this are that the creation of electron-hole pairs by the incident particles takes place mainly near the deep region in the case of the electron excitation while it occurs more in the shallower region in the case of the photon excitation, and that diffusion effect of the excited carriers is not important due to their short mean-free-paths in the CVD diamond.

In the measurements of PLE spectra, band-A (4250 Å) signal was monitored as a function of the incident light energy for undoped CVD diamond while the B-related peaks (4950 Å) was used for B-doped CVD diamond. The results were shown in Fig. 5 (a) and Fig 5 (b) for the former and latter, respectively. Down and up arrows correspond to the energy edges for the onsets of the indirect and direct transitions in diamond. In both Fig. 5 (a) and Fig. 5 (b) the indirect transition edge of diamond band structure can be clearly seen at 5.5 eV, while the direct transition edge appears as dip at 7.2 eV. In the PLE case, one should also consider the energy-dependent absorption effect of the incident light, although the PLE spectra partly reflect the joint density-of-states (JDOS) of the specimen. In other words, the signals are reduced more at higher photon energies due to lager absorption effect. Therefore, the observed peaks in the PLE spectra may not corresponds to the peaks of JDOS.

Presented experimental results indicate that both the band-A and boron-related centers in PL spectra can trap conduction electrons and holes created in the conduction and balance band, respectively, and give light emissions.

CONCLUSION

Unlike CL spectra, PL spectra exhibit increase in the amplitude of the defect-related 5RL peak with increasing excitation photon energies above the diamond band gap. PL spectra using above-band-gap light are more suitable for characterization of optical physics in thin near surface layer than CL spectra because the former is more sensitive to the shallower region while the latter probe mainly the region near the penetration depth. Thus, we have found that B-doped CVD specimens contain substantial amounts of intrinsic defects in the specimen subsurface.

Photoluminescence excitation spectra associated with the band A or B-related peak can provide the absorption edges for the direct and indirect transitions of diamond.

ACKNOWLEDGEMENTS

The authors wish to thank H.Yagyu, H.Makita, H.Yagi, N.Jiang, S.Sonoda, and K.Ogawa for their assistance.

This work was supported partly by Joint Studies Program (1995-1996) of the Institute for Molecular Science and partly by a Grant-in-Aid for Scientific Research from the Ministry of Education, Science, Sports and Culture of Japan.

REFERENCES

[1] P.J.Dean and J.C. Male, J. Phys. Chem. Solids, 25, 1369 (1964).
[2] L.Kleinman and J. C. Phillips, Phys. Rev. 116, 880 (1959); F. Herman, *idid.* 93, 1214 (1954).
[3] L.Kleinman and J. C. Phillips, Phys. Rev. 117, 406 (1960).
[4] M.Kamo, Y.Sato, and A.T.Collins, SPIE-Diamond Opt. II 1146, 180 (1989).
[5] A.T.Collins, M.Kamo, and Y.Sato, J. Mater. Res. 5, 2507 (1990).
[6] A.T.Collins, M.Kamo, and Y.Sato, J. Phys. Condens. Mater. 1, 4029 (1989).
[7] P.R.Chalker, in diamond and Diamond-like Films and Coatings, p.127 (1990).
[8] H.R.Philipp and E.A.Taft, Phys. Rev. 127, 159 (1962).
[9] H.Kawarada, Y.Yokota, T.Sogi, H.Matsuyama, and A.Hiraki, SPIE Vol. 1325 Diamond Optics III (1990).
[10] G.Davies, The Properties of Diamond. ed. J.E.Field (Academic press, London,1979) Chap.5.
[11] J.Mazzaschi, J.C.Brabant, M.Brousseau, and F.Viollot, Rev. Phys. Allp. 15, 9 (1980).
[12] A.T.Collins and P.M.Spear, J.Phys. C. 19, 6845 (1986).
[13] A.T.Collins, G.Davies, H.Kanda, and G.S.Woods, J. Phys. C1363 (1988).
[14] J.Ruan and W.J.Choyke, J. Appl. Phys. 69, 9, (1991).
[15] J.Ruan, K.Koba, and W.J.Choyke, Appl. Phys. Letters 60, 22 (1992).
[16] L.H.Robins, E.N.Farabaugh and Albert Feldman, Diamond Films and Tech. 5, 4,199 (1995)

SURFACE CHEMICAL EFFECTS ON THE OPTICAL PROPERTIES OF THIN NANOCRYSTALLINE DIAMOND FILMS

A.V. KHOMICH, V.I. POLYAKOV, P.I. PEROV, V.P. VARNIN*,I.G. TEREMETSKAYA*, V.G. BALAKIREV**, and E.D. OBRAZTSOVA***
Institute of Radio Eng. & Electronics, RAS, Moscow, Russia, vip197@ire216.msk.su
* Institute of Physical Chemistry, RAS, Moscow, Russia
** Institute SMS, Aleksandrov, Vladimirskaya obl.,Russia
***General Physics Institute, RAS , 117942 Moscow,Russia

ABSTRACT

The effect of annealing in air on internal structure and optical properties of hot filament CVD nanocrystalline diamond films was investigated. Oxidation of the films lead to selective removal of intercrystallite layers with formation of highly porous structure with characteristic dimensions of several nanometers. Dramatic changes in optical transmission and Raman spectra were also observed. The origin of the two Raman spectrum maxima at 1140 and 500 cm^{-1} is discussed. Hydrogen absorption and desorption processes in porous diamond were studied.

INTRODUCTION

For many applications, thin diamond films with the thickness in the range from about tenths to several microns can be useful. Optical applications require both low absorption and low surface scattering of light in diamond films. The optical losses due to the absorption and due to growth surface light scattering in such samples are low due to small thickness of the films. Thin CVD diamond films can have some other applications for electronic devices, such as light-to-electricity converters [1], UV-photodetectors and cold emitter devices [2], therefore it is important to understand the surface structure of CVD diamond films.

The post-deposition treatments of CVD diamond films at elevated temperatures are a valuable tool for both modification of their electrical, optical, and photoelectric properties, and investigating the film structure, the thermal stability and for obtaining information about the location of structure defects and impurities. The time-dependent oxygen etching behavior of the structure, Raman and photoluminescence spectra for the differently prepared microcrystalline diamond films were investigated in detail in [3] (see also references herein). In this paper we explore this method for so called "nanocrystalline" diamond films and present the results of an investigation of the hot filament (HF) CVD nanocrystalline diamond film's internal structure and optical properties as influenced by the different post-deposition treatments.

EXPERIMENT

The diamond films with thickness varied from 0.5 to 20 μm were grown on silicon substrates by hot filament (HF) CVD from mixtures of hydrogen and methane (or acetone). Ultrafine diamond powder of 5 nm average particle size was used for Si substrate pre-treatment in ultrasonic bath with the powder suspension in alcohol-acetone-water mixture. This enabled us to increase the diamond nucleation density on the Si surface up to 10^{10} cm^{-2}.

Optical spectroscopy measurements were carried out using an IR and UV-VIS "Specord" spectrometers with a range's 0.185-0.9 μm and 2.5-50 μm. The films were examined also with Raman spectroscopy (488 nm excitation wavelength, 2 μm laser beam spot size). The annealing

Mat. Res. Soc. Symp. Proc. Vol. 423

and oxidation experiments were performed at 450 - 800 C (for 0.5 or 1.0 hour for each temperature) in air, nitrogen or hydrogen atmosphere respectively. The same samples were examined by Raman and optical spectroscopy and by electron microscopic study as well.

RESULTS

Optical properties.

For all the as-grown films, the transmission in the middle and far IR was consistent with the value expected for a film with a smooth surface (with root mean square surface roughness ranging from 20 to 150 nm) and refractive index of 2.35-2.37. The spectra of nanocrystalline diamond films were close to that of the type Iba diamonds with the extra absorption near the indirect band edge in the UV-VIS and in the defect-induced one-phonon diamond band in IR. The concentration of carbon-hydrogen groups in our films was of the order of 1-2 %, as estimated from the integral intensity of the corresponding stretch mode in IR absorption spectra, where three sharp components are seen at 2920 cm^{-1}, 2850 cm^{-1} and at 2836 cm^{-1}. The peak frequencies and shapes of two higher components are quite close to the well-known spectrum of CH_2-groups (antisymmetric and symmetric stretches) involving sp^3-bonded carbon in a-C:H and CVD diamond films. Deconvolution of this band shows also the presence of CH_3 and CH sp^3-groups. The interpretation of the nature of the absorption band with the peak at near 2836 cm^{-1} (which dominates in the CH_x spectra of 10-20 μm thick nanocrystalline films) is less obvious. This band was never observed in the IR spectra of diamond-like carbon films, so it should originate from hydrogen at diamond grain surfaces or associated with crystal structure defects. Some authors assigned this band to hydrogen located at (111) diamond grain surfaces, while others assigned this strong C-H stretch band to nitrogen incorporation in CVD diamond films. We analyzed UV-vis spectra and IR one-phonon defect-induced band in films under study and calculated the maximum possible nitrogen content, like in [4], which occured to be several times lower, than the concentration of the centers, responsible for the 2836 cm^{-1} band. Therefore the incorporation of nitrogen alone cannot be responsible for the whole 2836 cm^{-1} band in our films.

The oxidation in air of diamond films at 590-630 C resulted in gradually decreasing of the diamond film's transmission in the whole spectral range except for the far IR [5]. At these and higher temperatures, oxygen etching results in selective, non-uniform removal of the film material at near surface regions, which leads to formation of surface porosity and hence to increased surface and bulk light scattering. During this stage of oxidation, hydrogen begins to leave the diamond films. As a result, the dangling carbon bonds are occupied by oxygen forming C=O and C-O-C groups, which were registrated in the IR spectra. At the temperatures higher than 630 C the most of the non-diamond carbon is removed from samples investigated and the remaining diamond crystallites in the film structure are attacked by the oxygen. During the annealing in air at such temperatures, oxygen removes not only non-diamond carbon, but diamond with high density of defects also, which results in the significant reducing of the absorption near the band gap edge region. The absorption bands due to CH_x -groups and defect-induced one-phonon band almost disappears after the oxidation at these temperatures [5].

The interference fringes were well observed as before as after oxidation at as high temperatures as 650 C indicating that the surface of the film remained sufficiently smooth. It is noteworthy that the mean transmission of diamond film windows in the far IR grew from 70% up to 85-92 % as a result of oxidation, and the effective refractive index decreased down to 1.5-1.7. We used an effective medium approximation to model the optical constants of such oxidized films, in which we assume that the film is composite heterogeneous medium consisting of diamond and void components. A calculation of the dielectric functions based on this model showed that the volume fraction of the diamond in such porous films lowered down to 60-40 % [5], in agreement with the estimations of the electron microscopy observations.

Nanocrystalline film's structure

The fractographic and transmission electron diffraction study showed that as grown nanocrystalline diamond films consisted of separate dense, without visible voids and sharp interfaces, 100 - 200 nm size bodies (Fig. 1, **a**) that turned out to be in some cases the single crystals built with 10 - 15 nm blocks, but mainly - polycrystalline conglomerates. The stacking faults and twin's interfaces were seen in some crystallites. Along the <111> plane, the superperiodic structure was observed with 1 - 1.5 nm period. Defect-free regions in the diamond single crystals were about 10 nm in size.

After oxidation in air at 530 C the surface morphology of the film became more distinct, some particles with sizes of 30 - 50 nm were seen. The internal structure of the films did not suffer any visible changes. The structure was changed, however, by oxidation at 627 C. Polycrystalline conglomerates with the same electron diffraction image as in as-grown films, appear as broom-like units (Fig. 1, **b**) consisting of stacks of highly flattened crystallites with maximum linear size of about 10 nm. We consider that the visualization of the separate crystallites became possible due to the oxidation removal of intercrystallite layers with the same electron-optical properties as for the crystallite themselves.

Raman spectra

The Raman spectrum of as-deposited diamond films (Fig.2) consisted of the bands of sp^3-bonded and sp^2-bonded carbon phases. The narrow weak band with maximum near 1332 cm^{-1} belonged to the well-crystallized diamond phase. The wide bands at 1350 and 1580 cm^{-1} are known for the nanocrystalline (glassy) carbon. The questionable features are the wide bands near 1140 and 1500 cm^{-1}. This bands are usually absent in the micron-size polycrystalline diamond film, so they are often called as the "fingerprint" of fine-grained diamond films. The position of the wide band at 1140 cm^{-1} coincided with the maximum of TO-like phonon diamond band. An additional wide Raman mode at 500 cm^{-1} (in the range of the acoustical phonons of diamond) also was observed. The appearance of two bands in Raman spectra (with maxims at 500 cm^{-1} and at 1140 cm^{-1}) appropriate to amorphous diamond, indicates the presence of a ultrafine-grained (or even amorphous) diamond phase with the characteristic dimension less than 5 nm, for which all the density of one-phonon states of diamond is displayed [6].

We have observed a non-monotonous modification of the Raman spectrum with the annealing temperature. For the temperatures lower than 620 C the only change is a minor intensity decrease of all spectral bands. For higher temperatures the trace of a "nanodiamond" phase (at 1140 cm^{-1}) has disappeared abruptly, while the peak of a well-crystallized diamond has become more pronounced and its contribution increased with respect to that of the graphite bands. Such type of a behavior may be interpreted as removing of the ultrafine-grained diamond phase. Indirectly it may clarify the position of that phase in the bulk of the film. The IR measurements indicate the spreading of the etching process from the surface to the bulk of the sample at temperatures exceeding 620 C. Since the intensity of the signal of a well- crystallized diamond begins to increase at this temperature, it is plausible to suppose that the coarse diamond grains remain unchanged, while the intergrain phase is removed. It corresponds argument in favor of

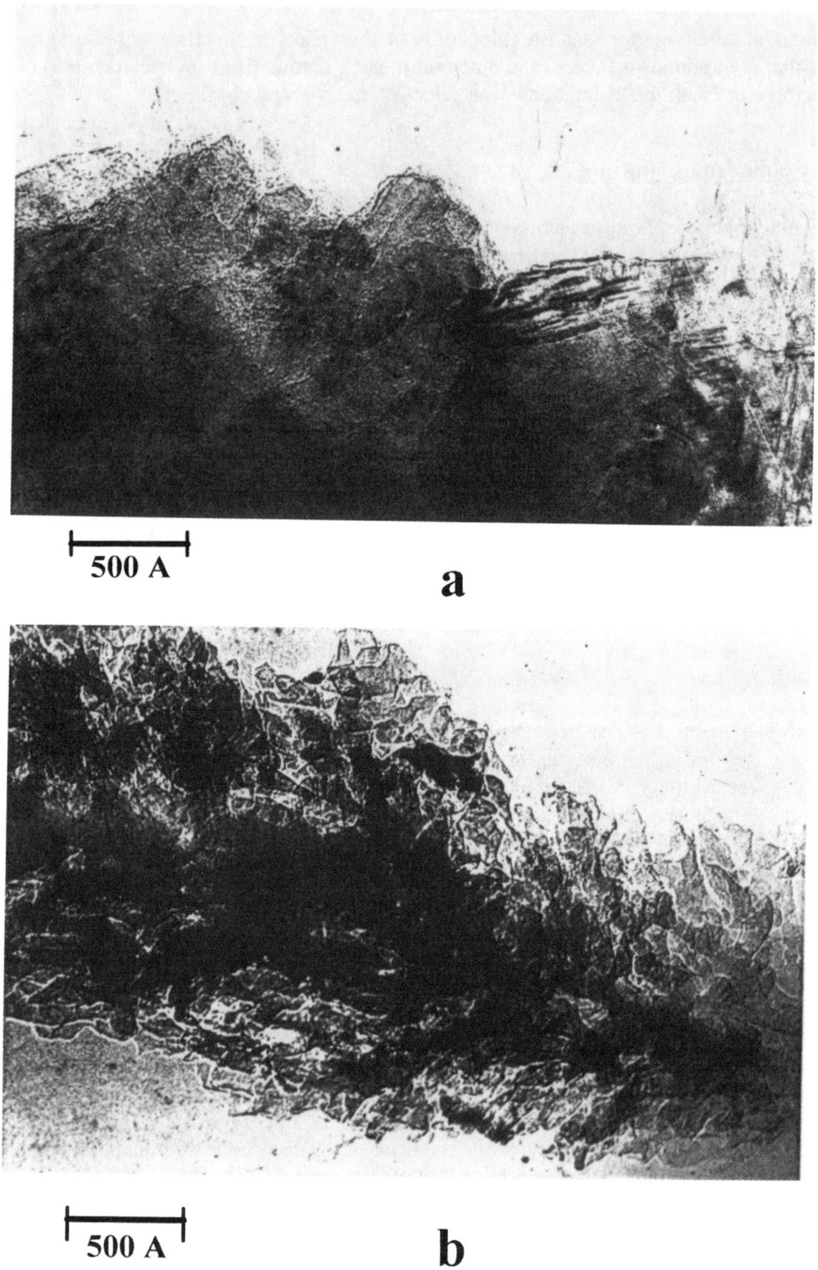

Fig.1. The TEM image of the nanocrystalline diamond film fragments before (**a**) and after (**b**) annealing on air.

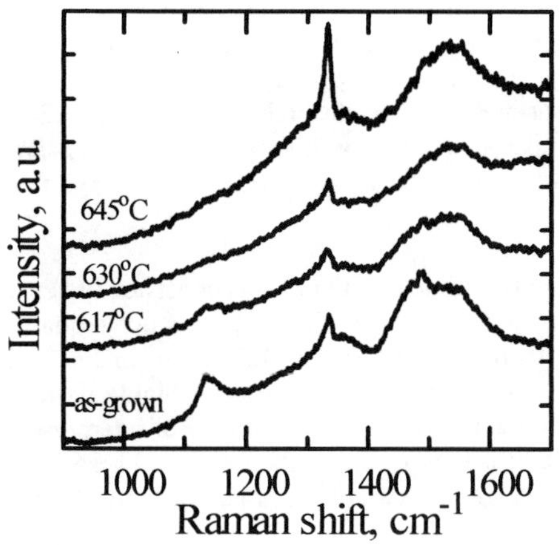

Fig. 2. Raman spectrum transformations due to sequential oxidations of diamond film.

situating nanodispersive phase in the intergrain space, where this phase is stabilized by mechanical stresses. The indirect confirmation of these stresses existence is the high-frequency shift of the fundamental mode of the coarser-grained crystallites, while after oxidation this mode has returned to the value characteristic to the well-crystallized diamond. The annealing of as-grown films in high-purity nitrogen at 650 C didn't influence on Raman, IR and UV-visible spectra. The annealing of oxidized porous films in hydrogen at 850 C during 30 minutes restores the CH_x stretching bands, while the Raman spectra remained unchanged and the 1140 cm^{-1} Raman band didn't appear. This rules out the possibility that this band is caused by C-H groups. The same conclusions were done in [7], where the excitation energy dependencies of the Raman 1140 cm^{-1} and 1470 cm^{-1} peaks were study. The apparent frequency down-shift observed when the incident photon energy increased. This shift is opposite to the up-shift reported for polymeric chains like polyenes or transpolyacetylene. It seems that this band in nanocrystalline diamond films is connected with the nondiamond carbon phase (with the same electron-optical properties as for the diamond), which located between the highly flattened diamond crystallites.

Activated diamond (?)

It is well known that activated carbon possess high capture capability in respect to many of gases and liquids. This effect originates mainly from high surface area of activated carbon. Analogous effect was observed in present work for another substance with highly developed surface - porous diamond films described above. We observed the long-time processes of the sp^3 CH_x bond concentration growth in such porous samples during their storage in air after the oxidation treatment. Characteristic time of these processes was about hundreds of hours, that evidenced probably in favor of multi-step surface chemical processes. The theoretical study of the energetics of (100) diamond surfaces showed, that the surfaces covered by hydroxyl groups were determined to have the lowest potential energy [8], so adsorption of the water vapor is one of possible mechanisms for CH_x -bond formation. The total content of CH_x bonds after long storage prevailed the content of these bonds in the as-grown film by several times. The CH_x-

stretching band shape differed significantly from one for the as-grown films due to higher part of the sp^3 CH_3- groups and to the absence of 2836 cm^{-1} band. The decrease of CH_x-bands' intensity during the heating sample in air began at 150-180 C and heating up to 250-280 C was sufficient to remove all the adsorbed hydrogen. The dehydrogenation processes via oxygen molecules of hydrogenated diamond powder surfaces occurred at higher temperatures (300-380 C) [9], while for samples under study these temperatures are closed to that for diamond-like films. The growth of CH_x-stretching band didn't lead to any changes of the Raman spectra of the films under study.

CONCLUSION

It was shown that as grown nanocrystalline diamond films consisted of separate dense, without visible voids and sharp interfaces, 100 - 200 nm size bodies. After oxidation in air at 627 C, polycrystalline conglomerates with the same electron diffraction image as in as-grown films, were transformed into broom-like units consisting of stacks of highly flattened crystallites with maximum linear size about 10 nm due to the oxidation removal of intercrystallite layers with the same electron-optical properties as for the crystallite themselves. These layers consisted of the amorphous diamond phase, that appeared in Raman spectra as wide bands at 500 and 1140 cm^{-1}.

ACKNOWLEDGMENTS

This work was partly supported by grants No of the 96-02-18373 of the Russian Fund of Fundamental Research and by grant NQ4000 from the International Science Foundation.

REFERENCES

1. V.I. Polyakov, N.M. Rossukanyi, P.I. Perov, A.I. Rukovishnikov, A.V. Khomich, V.P. Varnin, I.G. Teremetskaya: Diamond and Related Materials **4** ,1061 (1995).

2. A.V. Karabutov, V.I. Konov, S.M. Pimenov, V.G. Ralchenko, V.D. Frolov, M.S. Nunuparov, E.D. Obraztsova, V.I. Polyakov, N.M. Rossukanyi, I.A. Leontiev : paper No. E7.12 presented on MRS'96 Spring Meeting.

3. P.K. Bachmann, D. Leers and D.U. Wiechert, *Diamond Relat. Mater.*, **2,** 683 (1993).

4. C.D. Clark and S.T. Davey: J. Phys. C: Solid State Phys. **17**, 1127 (1984).

5. P.I. Perov, V.I. Polyakov, A.V. Khomich, N.M. Rossukanyi, A.I. Rukovishnikov, V.P. Varnin, I.G. Teremetskaya,in Wide Band Gap Electronic Materials, NATO ASI, ser.3,edited by M.A. Prelas, P. Gielisse, G. Popovici, B.V. Spitsyn , T. Stacy (Kluwer Academic Publishers, Dordrecht/Boston/London , 1995), p.171

6. E.D. Obraztsova. K.G. Korotushenko, S.M. Pimenov, V.G. Ralchenko, A.A. Smolin, V.I. Konov, E.N. Loubnin : Nanostructured Materials **6** , 827 (1995).

7. B. Marcus. L. Fayette, M. Mermoux, L. Abello, G. Lucazeau : J. Appl. Phys. **76**, 3463 (1994).

8. S. Skokov, B. Weiner, M. Frenklach, paper No. DD5.5 presented on MRS 1995 Fall Meeting.

9. T. Ando, K. Yamamoto, M. Ishii, M. Kamo, Y. Sato, in Advances in New Diamond Sciences and Technology, edited by S. Saito, N. Fujimori et al., MYU, Tokio, 1994, p.431.

RAMAN SPECTROSCOPY INVESTIGATION OF $(SiC)_{1-x}(AlN)_x$ LAYERS FORMED BY ION IMPLANTATION IN 6H-SiC

D.R.T. ZAHN*, T. WERNINGHAUS*, M. THÜMER*, J. PEZOLDT**, AND V. HEERA***
* TU Chemnitz-Zwickau, Professur für Halbleiterphysik, D-09107 Chemnitz, Germany, werni@physikus.physik.tu-chemnitz.de
** TU Ilmenau, Institut für Festkörperelektronik, D-98684 Ilmenau, Germany
*** FRZ Rossendorf e.V., Institut für Ionenstrahlphysik und Materialforschung, D-01314 Dresden, Germany

ABSTRACT

6H silicon carbide (SiC) substrates were implanted with nitrogen and aluminum at different doses and annealed in the temperature range from 1300°C-1700°C. Micro-Raman Spectroscopy (μ-RS) measurements were performed in two sample geometries (conventional plane-view and cross-sectional). Changes of the polytype from 6H- to a cubic $(SiC)_{1-x}(AlN)_x$ and influences in the 6H-SiC wafer up to depths of 2μm were detected. The results obtained by cross-sectional μ-RS are discussed in comparison to other results from Reflection High Electron Energy Diffraction (RHEED), Rutherford Backscattering (RBS), Auger Electron Spectroscopy (AES), Transmission Electron Microscopy (TEM), and Positron Annihilation Spectroscopy (PAS) measurements.

INTRODUCTION

Wide band semiconductors such as SiC and AlN as well as solid solutions of both compounds recently attracted growing attention due to their excellent physical and chemical properties [1]. Many polytype structures of these compounds, which have different band gap energies, are known. Therefore, a band gap tailoring should be possible in a wide energy range (3C-SiC: 2.4eV, 2H-AlN: 6.2eV) by variation of the AlN content in the solid solution $(SiC)_{1-x}(AlN)_x$. An open question is which kind of polytype is formed after ion beam synthesis of $(SiC)_{1-x}(AlN)_x$ layers in 6H-SiC. Raman Spectroscopy (RS) is a widely used investigation technique for SiC to distinguish between the polytypes of SiC [2,3]. In order to improve the lateral resolution a microscope is used to obtain focus diameters of about 1μm (Micro-RS = μ-RS). A cross-sectional geometry was chosen to provide additional depth-resolved information about sample parameters, e.g. stress, crystallinity, buried layers, polytype variations, etc [4,5]. In this paper the detection of polytype transitions in the region of $(SiC)_{1-x}(AlN)_x$ formation by cross-section μ-RS is demonstrated.

EXPERIMENT

The samples used in this investigation were produced by co-implantation of aluminum (Al) and nitrogen (N) into CREE Research Inc. 6H-SiC (0001) wafers. To produce buried $(SiC)_{1-x}(AlN)_x$ layers with compositions of about x=0.04 and x=0.20 the following implantation sequences were carried out at room temperature: (i) 100keV Al^+ ($1\cdot10^{16}cm^{-2}$), 160keV Al^+ ($2.5\cdot10^{16}cm^{-2}$), 65keV N^+ ($1\cdot10^{16}cm^{-2}$), 120keV N^+ ($2.5\cdot10^{16}cm^{-2}$); (ii) as (i) but with fluences 5 times higher. After the implantation procedure an annealing procedure in the temperature range from 1300°C to 1700°C were carried out to recrystallize the implanted samples. For further

Mat. Res. Soc. Symp. Proc. Vol. 423

details on the implantation procedure, the TRIM simulations used to determine the implantation profile, and the sample preparation see Ref. [6].

The Raman spectra were measured at room temperature in the spectral range from 100 to $1100cm^{-1}$ using a Dilor XY Raman spectrometer equipped with a CCD camera for multichannel detection, an optical microscope providing a laser beam focus diameter of about $1\mu m$, and an automated x-y-stage with a minimum step width of 100nm. All measurements shown in this paper were taken using the 2,47eV (501.7nm) Ar^+ line. The laser power on the sample surface was approximately 5mW and the monochromator slits were set for a spectral resolution of $2.5cm^{-1}$ as measured by the full width at half maximum (FWHM) of the Rayleigh scattered light. In a conventional plane-view sample geometry spectra were taken in the virgin as well as in the implanted regions of as-implanted and of annealed samples. According to the selection rules the overlapping structures of the 3C-SiC transversal optical (TO) phonon and the 6H-SiC E1(TO) phonon can be distinguished [3]. For that purpose a polarisation configuration -z(yx)z with x=[1-100], y=[11-20], and z=[0001] was chosen. Thereafter one-dimensional scans in steps of 100nm were taken in the cross-sectional sample geometry. These cross-sectional scans were taken for all three types of samples with a polarisation configuration of -x(zy,zy)x.

A subsequent fitting of each single spectrum with lorentzian lineshapes provides the phonon parameters, i.e. intensity, FWHM, and frequency position, depending on the location of the laser beam focus. Using a convolution of a gaussian lineshape representing the laser beam in the sample and several box functions - describing the distinct layers in the sample heterostructure - the variation of the phonon parameters can be simulated. The simulations are sensitive to the FWHM of the laser beam. A value of $\approx 1.6\mu m$ was determined by the simulation of several scans across a vacuum/virgin 6H-SiC wafer interface. A more detailed description of the sample geometries used and the fitting and simulation procedure is given in Refs. [4,5]

RESULTS

Plane-View Configuration

Typical Raman spectra taken in the plane-view geometry are displayed in Fig. 1 with part a) the spectrum of a virgin 6H-SiC wafer, b) two spectra of as-implanted samples with a low and a high dose implantation, and c) the spectra of the samples implanted with two different doses and annealed at three different temperatures. Considering the spectrum of the virgin sample (Fig.1a)) all phonon features allowed in the chosen polarisation configuration can be observed: The TO phonon E_2 ($768,0cm^{-1}$), E_2 ($789,0cm^{-1}$), and the longitudinal optical (LO) phonon A_1 ($967,5cm^{-1}$). In contrast to the selection rules an additional weak structure can be observed at $795,5cm^{-1}$ (E_1 (TO)). This is related to the large aperture of the objective lense resulting in a partial lifting of the selection rules.

In Fig. 1b) only very weak (low dose) or no sharp phonon features (high dose) are observable while two broad bands appear at about $500cm^{-1}$ and $750cm^{-1}$. An additional broad structure appearing at about $1250cm^{-1}$ is not displayed. These changes can be explained by an amorphization of the 6H-SiC in the surface region and a strongly reduced light penetration depth. The broad features appearing in the spectra of 1b) are correlated to Si-Si/Si-Al/Si-N ($\approx 500cm^{-1}$), Si-C/Al-N ($\approx 750cm^{-1}$), and C-C/C-N vibrations ($1250cm^{-1}$) in the amorphous layer [7]. As the amorphous layers have similar thicknesses for low and high dose implantation (see Ref. [8]) the different light penetration depths are likely to be related to a difference in the density of the amorphous layers.

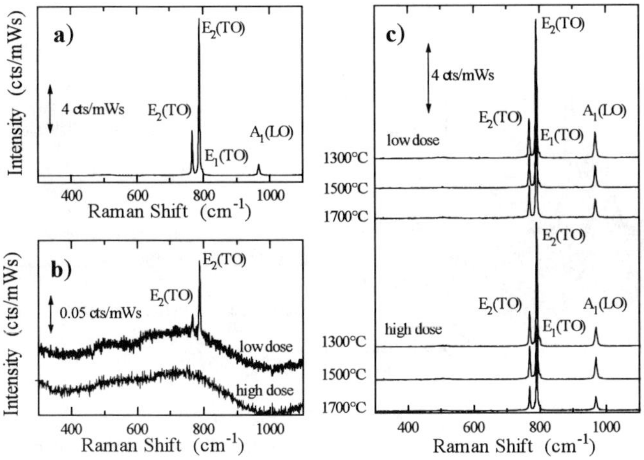

Fig. 1: The Raman spectra of the virgin a), the as-implanted b), and the implanted and annealed c) samples measured in the plane-view geometry.

Considering the spectra in Fig. 1c) only minor differences of the phonon intensities between the spectra of samples annealed at different temperatures are visible. In addition the spectra are similar to that in Fig. 1a). This can be explained with a high transparency of the recrystallized material. As a result the changes in the $(SiC)_{1-x}(AlN)_x$ region can not be distinguished and cross-sectional one-dimensional scans were performed to get more information about the implanted layers.

Cross-section Configuration

In contrast to the plane-view geometry the scattering volume in high transparent materials can be enhanced using the cross-sectional geometry. Considering an implanted layer thickness of 300nm, a laser beam focus diameter of 1.6μm, and a penetration depth of about 500μm an enhancement factor of about 150 for the detection of the implanted layer can be estimated.

Fig. 2 displays the cross-sectional micro-Raman scan taken in steps of 100nm in a 3D representation of a low dose implanted samples annealed at 1300°C (left hand side). The spectra in the front display zero signal due to the laser beam being entirely in the ambient atmosphere while the rear spectra are taken in the unaffected 6H-SiC wafer. At the right hand side the intensity of the three phonon features appearing in the 3D plot are given: A_1(TO)+E_2(TO) (up triangles), E_2(TO) (circles), and the superposition of the E_1(TO) 6H-SiC and TO 3C-SiC (down triangles). The straight lines give the simulation obtained using a model for the distinct layers in the 6H-SiC wafer produced by the implantation and the following recrystallization. Since the A_1(TO)+E_2(TO) (up triangles) and the E_2(TO) phonon (circles) only stem from the 6H-SiC their phonon intensity variation was used to examine the 6H-SiC content in implanted sample. In order to obtain any cubic content , e.g. 3C-SiC, the resulting simulated intensity variation of the E_1(TO) was substracted from the measurement. The resulting variation (given by the diamonds) can be related to the implanted layer $(SiC)_{1-x}(AlN)_x$.

Fig.3 displays the intensity variations depending on the laser beam position obtained from scans across low dose (left part) and high dose (right part) implanted samples annealed at temperatures between 1300°C and 1700°C. Considering the low dose implanted sample set the

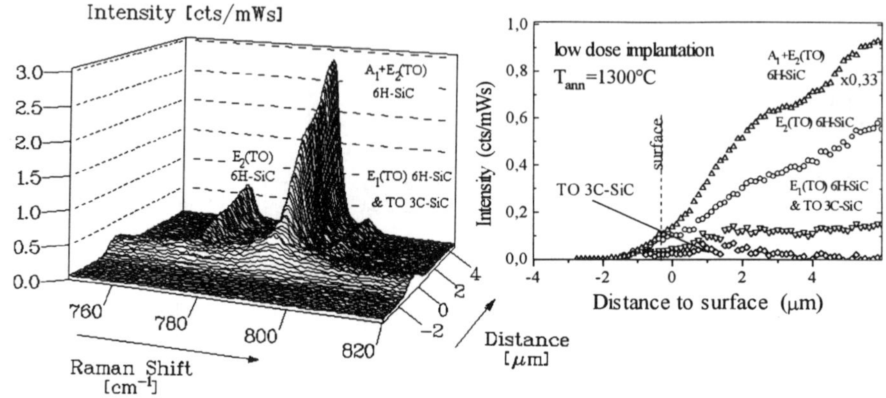

Fig. 2: Left: 3D representation of the cross-sectional scan across a low dose implanted sample annealed at 1300°C in steps of 100nm. Right: Intensity variation of the various phonon contributions obtained from the curve fitting of the individual spectra.

intensity variation of the sample annealed at 1300°C can be simulated using four subsequent 6H-SiC layers. In view of their different participation in the intensity variation these layers can be interpreted as 6H-SiC with different defect densities with no cubic content.

The assumption of a high defect density in the top 6H-SiC layer and a subsequent reduction for the deeper layers is corroborated by the high background intensity as well as by the enlarged FWHM of the phonon structures near the surface which can be seen in the 3D plot in Fig.2. The FWHMs are decreasing with increasing distance from the surface indicating a decreasing defect density to a value of 2.9cm^{-1} in the bulk typical for single crystals. Y. Sasaki *et al.* [8] determined the FWHM in dependence of the correlation length in small SiC particles. Using their results the correlation length in the SiC increases from 1nm in the top layer to ≥50nm in two subsequent layers. Such a layered system, namely top layer/two intermediate layers/bulk was found to be appropriate to simulate the ion implantation induced modification of the 6H-SiC crystals.

Thicknesses of ≈120nm for the top layer and about 950nm and 630nm for the intermediate layers were evaluated by the simulations. In the case of 1500°C and 1700°C the thicknesses are 100nm/870nm/900nm and 20nm/700nm/1200nm, respectively. Obviously for the last two cases a cubic content due to the implanted layer can be observed. The shaded bar gives the region in which this implanted layer is located. For the sample annealed at 1500°C a thickness of ≈300nm starting in a distance of about 110nm from the surface and for the 1700°C sample a thickness of ≈330nm starting in a depth of about 20nm were estimated. The thicknesses and positions of the implanted layer correspond to results obtained by Reflection High Electron Energy Diffraction (RHEED), Rutherford Backscattering (RBS), Auger Electron Spectroscopy (AES), and Transmission Electron Microscopy (TEM) but are somewhat larger [6,9-11]. This point is still under investigation. In addition the modified depth in the 6H-SiC wafers up to ≈2μm is in contrast to the findings of these methods. On the other hand, results obtained by Positron Annihilation Spectroscopy for Ge^+ implanted 6H-SiC reveal modifications up to depths of 2.5μm correlated to vacancies [12].

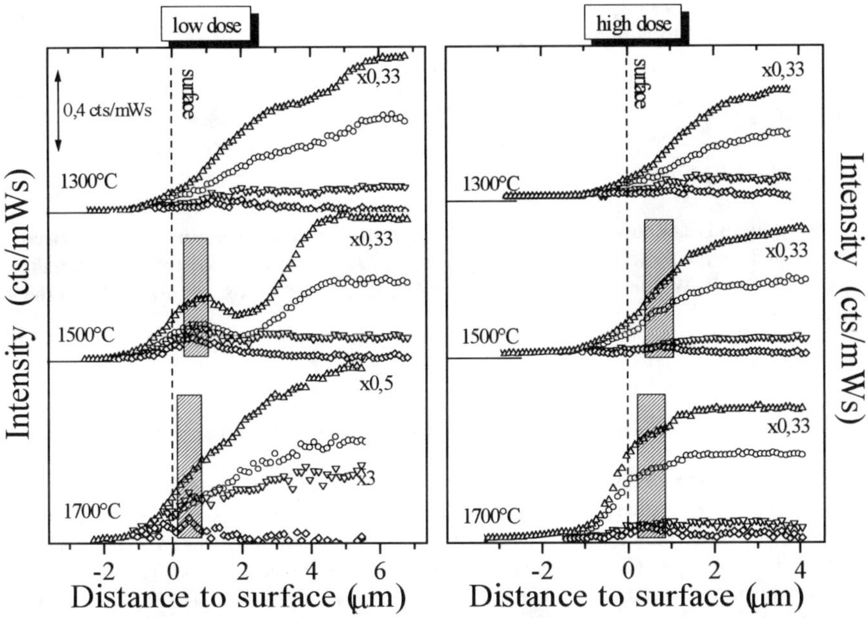

Fig. 3: The variation of the phonon intensities for low dose and high dose implanted samples annealed at temperatures between 1300°C and 1700°C. The circles represent the E_2(TO)-phonon mode, the up triangles the A_1+E_2(TO), and the down triangles the combination of the E_1 (6H-SiC) and the TO (3C-SiC). The shaded bar denote the implanted region where a $(SiC)_{1-x}(AlN)_x$ layer was formed.

For the high dose implanted samples similar results were obtained: 190nm/870nm/860nm (1300°C), 110nm/510nm/1250nm (1500°C), and 20nm/350nm/1540nm (1700°C), respectively. As in the case of low dose implantation for the 1300°C sample no cubic content is detectable in contrast to RHEED and TEM investigations (see Ref. [9-11]) while for the 1500°C and the 1700°C samples implanted layers with thicknesses of ≈290nm and ≈330nm starting in depths of about 130nm and 50nm from the surface, respectively, were detected. This can be explained by a high defect density in the implanted 6H-SiC as well as in the transformed regions. Further investigation are necessary for an improvement for the cubic content detection.

For the interpretation of the phonon parameter variation one has to take into account that the assumption of distinct layers is a simplification because of the gradual implantation profiles [6]. In spite of this simplification the model discribes the measured variations quite well.

CONCLUSION

Micro-Raman Spectroscopy in plane-view as well as in a cross-section sample geometry was used to investigate 6H-SiC wafers low dose and high dose implanted and subsequently annealed at temperatures between 1300°C and 1700°C. In contrast to the plane-view sample geometry one-dimensional scans across the sample heterostructure in the cross-sectional sample geometry yield information about the polytype formation in the $(SiC)_{1-x}(AlN)_x$ layers in dependence on the annealing temperature. A layered system, namely top layer/two intermediate layers/bulk was found to sufficiently describe the measured phonon intensity variations. The non-porportional intensity increase of the E_1(TO) 6H-SiC phonon mode superimposed by the TO 3C-SiC phonon mode can be attributed to the formation of the 3C polytype in the implanted region near the surface.

ACKNOWLEDGEMENTS

We gratefully acknowledge the financial support of this work by the DFG under contract Pe 624/1-1 and by the Graduiertenkolleg "Dünne Schichten und nicht-kristalline Materialien". In addition we thank P. Werner from the Max Planck Institut für Mikrostrukturphysik for the TEM investigations.

REFERENCES

1. H. Morkoc, S. Strite, G.B. Bao, M.E. Lin, B. Sverdlov, and M. Burns, J. Appl. Phys. **76**, 1363 (1994)
2. Z.C. Feng, A.J. Mascarenhas, W.J. Choyke, and J.A. Powell, J. Appl. Phys. **64**, 3176 (1988)
3. D.W. Feldman, J.H. Parker, Jr., W.J. Choyke, and L. Patrick, Phys. Rev. **173**, 787 (1968)
4. T. Werninghaus, S. Laufer, H.-J. Hinneberg, and D.R.T. Zahn, in Diamond Materials IV, K.V. Ravi and J.P. Dismukes (Eds.) Electrochem. Soc. Proc. **4**, Pennington, NJ, (1995) p. 151
5. T. Werninghaus, M. Friedrich, and D.R.T. Zahn, physica status solidi (a) **154**, 269 (1996)
6. V. Heera, J. Pezoldt, X.J. Xing, and P. Piroux, Proc. of ICSCRM-95, Kyoto, Japan, to be published
7. E.A. Fagen in Silicon Carbide, R.C. Marshall, J.W. Faust, Jr., C.E. Ryan (Eds.) Univ. of South Carolina, Columbia, (1974), p. 476
8. Y. Sasaki, Y. Nishina, M. Sato, and K. Okamura, Phys. Rev. **B40**, 1762 (1989)
9. V. Heera, J. Stoemenos, R. Kögler, and W. Skorupa, J. Appl. Phys. 77, 2999 (1995)
10. J. Pezoldt, A.A. Kalnin, D.R. Moskwina, and W.D. Savelyev, Nucl. Instr. and Meth. **B80/81**, 943 (1993)
11. J. Pezoldt, A.A. Kalnin, and W.D. Savelyev, Nucl. Instr. and Meth. **B65**, 361 (1992)
12. Y. Pacaud, W. Skorupa, A. Perez-Rodriguez, G. Brauer, J. Stoemenos, and R.C. Barklie, to be published in Nucl. Instr. and Meth. **B** (1995)

PHOTOREFLECTANCE STUDY OF GaN FILM GROWN BY METALORGANIC CHEMICAL VAPOR DEPOSITION

K. Yang*, R. Zhang*, Y.D. Zheng*, L.H. Qin*, B. Shen*, H.T. Shi*, Z.C. Huang** and J.C. Chen**
*Department of Physics, Nanjing University, Nanjing 210093, People's Republic of China
**Department of Electrical Engineering, University of Maryland Baltimore County, Baltimore, MD 21228-5398, USA

ABSTRACT

Photoreflectance was used to study the optical properties of single crystal hexagonal GaN film on (0001) sapphire substrate grown by metalorganic chemical vapor deposition. The energy gap of GaN was determined as 3.400 eV, and the possible origin of the PR signal was attributed to the modulation of the surface field and lineshape broadening of defects. Optical absorption and cathodoluminescence of the GaN sample were measured, and the optical absorption edge of 3.39 eV and the cathodoluminescence emission peak of 3.461 eV at low temperature (15.6K) confirmed the results of Photoreflectance.

INTRODUCTION

Recently, GaN-based III-V nitride semiconductors attract extensive attention for their potential device applications[1,2]. Among them, gallium nitride (GaN) is one of the most interesting materials which have a wurtzite structure in natural form, and have a wide direct band gap, which provides efficient radiation recombination. The attractive optical properties, together with the outstanding thermal and chemical stability of GaN, make it not only ideally suitable for fabricating the light emitting diodes and detectors operating in the blue and ultraviolet wavelength range, but also for application at high temperature and in hostile environments[3,4]. For most optoelectronic applications, the optical properties of GaN film is the key factor to influence the features of devices. Many studies have been performed on the theoretical calculation of the energy band structures of GaN, and on the experimental characterization of its optical properties. The energy gap of hexagonal GaN at room temperature was determined as ~3.4 eV experimentally, while the theoretical calculation values ranged from 2.3 to 3.6 eV[5,6]. However, most experimental studies were still concentrated on the photoluminescence (PL) and absorption of GaN[7-9]. Currently, with the substantial improvement in epitaxy growth technology, high quality GaN films on sapphire have been fabricated by metalorganic chemical vapor deposition (MOCVD)[3,7], vapor-phase epitaxy (VPE)[10] and molecular-beam epitaxy (MBE)[8]. This makes it possible to use additional optical characterization techniques, such as photoreflectance (PR)[10], to study the optical properties of GaN. The derivative nature of PR suppresses uninteresting background effects and emphasizes structure localized in the energy region of interband transitions at critical points. The high sensitivity of PR even at room temperature and nondestructivity make it a powerful tool to study and characterize semiconductor energy band structures.

In this paper, photoreflectance was used to study the optical properties near the bandgap of hexagonal GaN. The energy band gap of hexagonal GaN was determined as 3.39 and

Mat. Res. Soc. Symp. Proc. Vol. 423

3.400 eV by optical transmission and photoreflectance respectively, and cathodoluminescence (CL) was measured to interpret the results of PR. Moreover, the possible origin of the PR signal was discussed and attributed to the modulation of the surface field and lineshape broadening of defects.

EXPERIMENTS

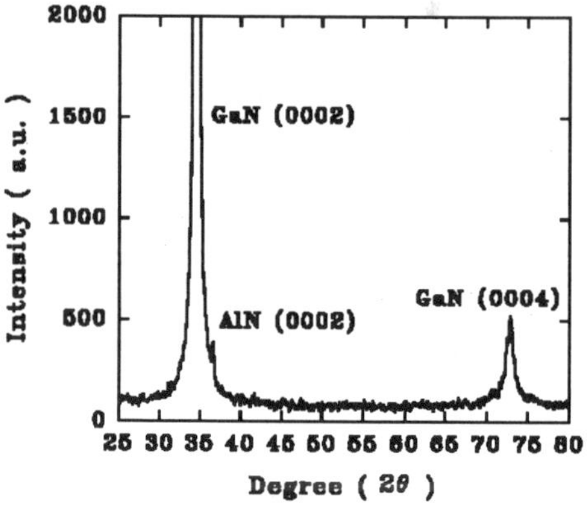

Fig.1 X-ray diffraction spectrum of the GaN film grown by MOCVD.

The GaN sample used in this work was a nominally undoped single-crystal film which was grown on (0001) sapphire substrate with an AlN buffer layer by MOCVD. The Ga, Al (for the buffer layer) and N source gases were trimethylgallium (TMGa), trimethylaluminum (TMAl), and ammonia (NH_3), respectively. H_2 was used as carrier gas. The thin AlN buffer layer with a thickness of 50 nm was deposited at 550°C and followed by the growth of GaN at 720°C. The growth rate of GaN was about 1μm/hour, and all samples were unintentionally n-type doped. The X-ray diffraction (shown in Fig.1) confirmed high quality of the single crystal of the GaN epitaxial film, and the lattice constants in the direction of c-axis were 0.5188 and 0.4948 nm for GaN and AlN, respectively. Further studies showed that the full width at half-maximum (FWHM) of the (0002) GaN diffraction peak determined from the rocking curves of X-ray double-crystal diffractometer $\theta/2\theta$ scan was less than 170".

The optical experiments, such as optical absorption, optical reflectivity and PR, were performed at room temperature on a self-developed optical measurement system controlled by a computer[11]. In the case of PR measurement, an air-cooled 500W Xe lamp filtered by a monochromator with a 2400 lines/mm grating, was used as the probe beam. The modulating beam came from the 488 nm line of a 50 mW Ar^+ laser mechanically chopped at 20 Hz. The light was detected by a photomultiplier, before which another monochromator was used to filter the modulating beam. The output was fed into an EG&G lock-in amplifier. The data were collected and stored in the computer, which also controlled the scan speed of the monochromator.

RESULTS AND DISCUSSIONS

Fig. 2 showed a typical PR spectrum of single crystal GaN on sapphire substrate at room temperature. According to the equation

$$\Delta R/R = Re\{ A\,\Gamma^{-n} e^{i\theta} (E - E_g - i\Gamma)^{-n} \} \qquad (1)$$

we fitted the experimental curve by changing the values of the amplitude A, the linewidth Γ, the phase θ, and the transition energy E_g. The exponent n was fixed at the value of 2.5, which is often used for band to band transition in the bulk samples[12]. Thus we obtained the E_g of 3.400 eV and the linewidth of 0.124 eV. This is well agreed the reflectivity peak of 3.30 eV in the optical reflection spectrum of the GaN sample measured at room temperature, which is the shortest band-to-band transition of E_0: Γ_6 -Γ_7 in GaN, named the fundamental energy gap transition.

The lineshape of PR is a direct result of the phase value in Eq.(1), but embodying physical mechanism. Giordana et al. observed a predominant feature below the band gap, which they thought, was the contribution of the interface states. To clarify this confusion, we performed optical absorption and cathodoluminescence measurements.

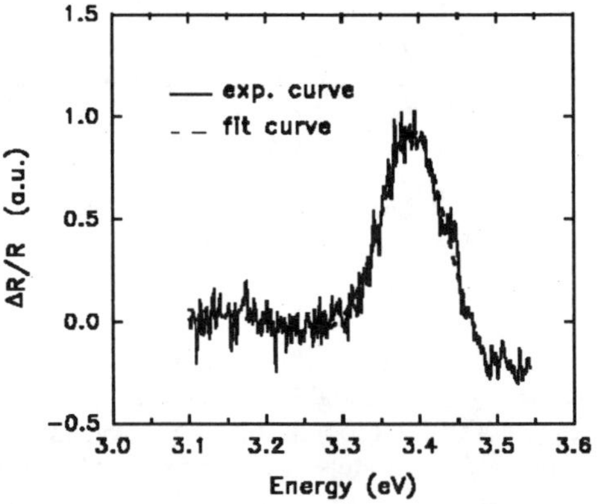

Fig.2 Photoreflectance spectrum of GaN film on sapphire at RT (the solid line). The curve was fitted by the dash line.

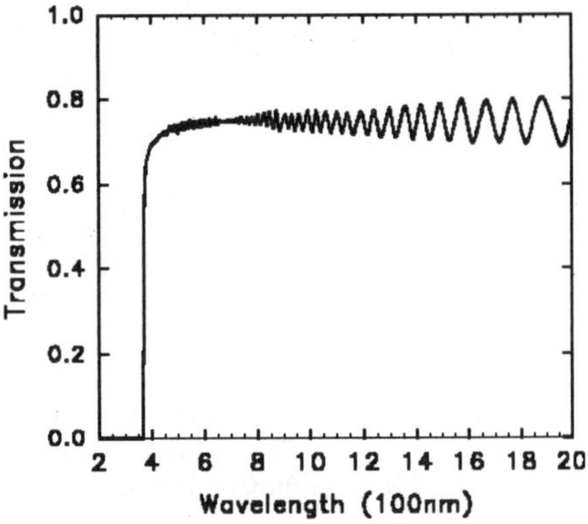

Fig.3 Room-temperature optical transmission spectrum of the GaN film grown on (0001) sapphire substrate.

The optical transmission spectrum of GaN was performed at room temperature from 200 to 2000 nm as shown in Fig.3. By drawing the absorption coefficient square vs. the spectral energy curve, $\alpha^2 \propto h\nu$, shown in Fig.4, it was observed that the absorption coefficient is about 7×10^3 cm^{-1} above the band gap, and almost zero below the gap, with a sharp line shape around the band gap. As seen from the data in Fig.4, we obtained the energy bandgap of 3.39 eV, which was consistence with the results of PR. From the interference features on the transmission data, we obtained the film thickness as about 6 μm.

CL spectra were obtained at room temperature and low temperature in the energy range of 1.5 and 4.0 eV, shown in Fig.5. In room temperature, we didn't observe a band-edge emission peak but a broad emission band in the yellow spectral region (yellow band) with the maxima around 2.2 eV. While at low temperature (15.6K) the CL spectrum of the MOCVD GaN sample was dominated by a narrow peak of 3.461 eV near band-edge, and a broad yellow emission band with the maxima around 2.3 eV. It was found that the band-edge spectral feature and the broadband emission in the yellow spectral region shifted towards higher energy as the GaN band-gap increases at low temperature. Considering the blue shift of band gap at low temperature, the emission peak of 3.461 eV in CL spectrum confirmed the results of PR measurement.

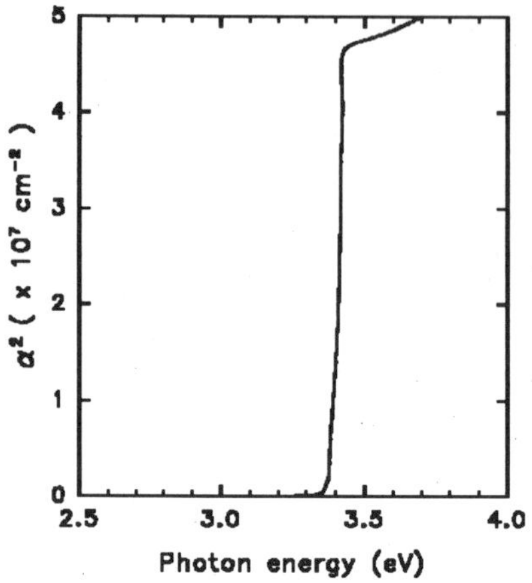

Fig.4 Absorption coefficient square as a function of photon energy for GaN film.

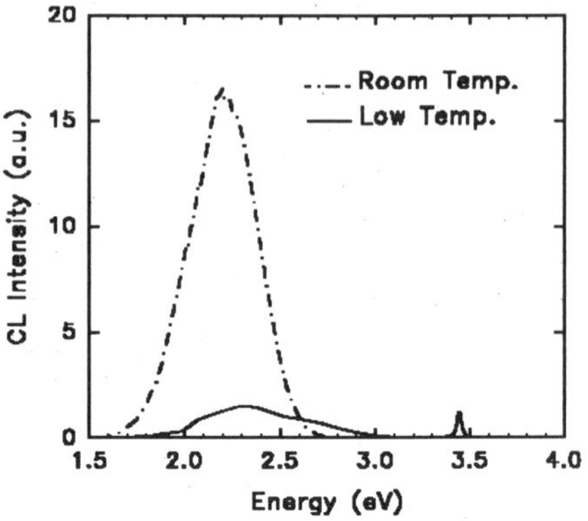

Fig.5 Cathodoluminescence spectra of GaN film at room temperature and low temperature (15.6K).

Though the lineshape of PR is dominant around the band-gap, the PR-below-E_g modulation data provided different information from conventional PR spectra. Just like the case in Ref[10], the PR signal increased with the decreasing modulation frequency, and no detectable PR modulation was achieved by using the 633nm HeNe line(1.96eV), while the Ar^+ laser lines produced a signal. We think the modulation is impurities or defects related, but don't think the PR features are dominated by interface states. Because the penetration depth of photons with energy around 3.4eV is about 1.4μm, and the GaN epilayer used in this study is as thick as 6μm, it is less likely that the PR signal is the contribution of interface states. It must be a surface or bulk related mechanism that requires a more than 2eV transition process to dominate the process, for example, the "yellow band" deep levels in the range of 1.9 to 2.5 eV which can be easily observed in room-temperature CL spectrum, shown in Fig.5, or photoluminescence (PL) spectrum[13], are the possible dominant effect in producing surface field modulation and the PR signal. As in electroreflectance (ER), scattering possibilities, and therefore the linewidth Γ, increase with the photon energy. This should lead to broaden and flatten PR structure toward higher energies[14]. We may say, in our experiment, the PR signal comes from the modulation of the surface field, not the interface, but the defects broaden the lineshape.

CONCLUSIONS

In this paper, we studied the optical properties of hexagonal GaN films on sapphire grown by metalorganic chemical vapor deposition. Photoreflectance near the fundamental band gap was investigated and the possible origin of the signal was discussed. The energy gap and linewidth were determined as 3.400 and 0.124 eV respectively. Optical absorption and cathodoluminescence confirmed the results of PR.

REFERENCES

1. S. Strite and H. Morkoc, J. Vac. Sci. Technol. B, 10, 1237(1992)
2. H. Morkoc, S. Strite, G.B. Gao, M.E. Lin, B. Sverdlov and M. burns, J. Appl. Phys., 76, 1363(1994)
3. S. Nakamura, T. Mukai and M. Senoh, Appl. Phys. Lett., 64, 1687(1994)
4. M.A. Khan, J.N. Kuznia, A.R. Bhattarai and D.T. Olson, Appl. Phys. Lett., 62, 1786(1993)
5. A. Rubio, J.L. Corkill, M.L. Cohen, E.L. Shirley and S.G. Louie, Phys. Rev. B, 48, 11810(1993)
6. S. Bloom, G. Harbeke, E. Meier and I.B. Ortenburger, Phys. Stat. Sol.(b), 66, 161(1974)
7. K.G. Fertitta, A.L. Holmes, J.G. Neff, F.J. Ciuba and R.D. Dupuis, Appl. Phys. Lett., 65, 1823(1994)
8. R. Singh, R.J. Molnar, M.S. Unlu and T.D. Moustakas, Appl. Phys. Lett., 64, 336(1994)
9. K. Yung, J. Yee, J. Too, M. Rubin, N. Newman and J.Ross, Appl. Phys. Lett., 64, 1135(1994)
10. A. Giordana, D.K. Gaskill, D.K. Wickenden and A.Estes Wickenden, J. Electronic Materials, 23, 509(1994)
11. L.H. Qin, K. Yang, Y.D.Zheng, R.Zhang, X.J. Dai, D. Feng, Z.C. Huang and J.C. Chen, Chinese Phys. Lett., 13(2), 153(1996)
12. D.E. Aspnes, Surf. Sci., 37, 418(1973)
13. W. Shan, T.J. Schmidt, R.J. Hauenstein, J.J Song and B. Goldenberg, Appl. Phys. Lett., 66(25), 3492(1995)
14. B.O. Seraphin, Semiconductors and Semimetals, Vol.9, (Academic Press, New York, 1972), p. 29.

THERMOMODULATED REFLECTIVITY SPECTRA OF GAN/SAPPHIRE EPILAYER

Y. Li[a,b], Y. Lu[b], H. Shen[c], M. Wraback[c], M. Schurman[a], L. Koszi[a], R. A. Stall[a]
[a]Emcore Corporation, Somerset, New Jersey 08873
[b]The State University of New Jersey, Rutgers University, Piscataway, New Jersey 08855-0909
[c]Army Research Laboratory at Fort Monmouth, New Jersey 07703-5601

ABSTRACT

Thermomodulation spectra from the metalorganic chemical vapor deposition (MOCVD) grown GaN have been measured in the temperature range 20 K - 310 K. A theoretical model is established to explain the modulation spectrum by considering the modulation of epilayer thickness and dielectric constant. By performing the lineshape analysis, the bandgap energy and broadening parameter were determined in the temperature range. The nonlinear temperature coefficient (Varshni coefficient) of energy gap is measured to be 5.9×10^{-4} eV/K. The temperature dependence of broadening parameter is also measured for the first time.

INTRODUCTION

Wide bandgap group III nitrides are strong candidates for blue light-emitting diodes (LED) and lasers intended for high density optical storage and display technologies. GaN has a wurtzite structure in natural form, and has a direct bandgap of 3.4 eV at room temperature[1]. In the past several years, significant progress has been made on GaN film quality, p-type doping control and growth methods. High efficiency blue LEDs are commercially available[2]. Many optical characterization techniques have been used to determine the optical properties of GaN[3-13]. The temperature dependence of the energy gap in GaN has been previously evaluated experimentally by absorption[3-5], luminescence[6-10], and photoreflectance (PR) studies[12,13].

Modulation spectroscopy has proven to be a powerful experimental technique for studying and characterizing the properties of group IV and III-V semiconductors, both bulk and reduced dimensional systems[14-21]. Recently, W. Shan et al. [13] have reported on a PR study of GaN thin film. A. Giordana et al.[24] have performed PR and electrolyte electroreflectance (EER) measurements on GaN. These measurements were made at room temperature. In this paper, we report the first thermomodulation spectroscopy characterization of GaN. The temperature dependence of the energy gap in GaN thin films was obtained. The band gap temperature coefficient in the 20-310 K range is determined. The broadening parameter and its temperature dependence are also determined.

EXPERIMENTAL

The GaN samples used in this work were grown on c-Al_2O_3 substrates by MOCVD in an EMCORE multi-wafer rotating disk reactor. A detail of growth process can be found elsewhere[25]. The epitaxial GaN layers were Wurtzite in crystal structure. The room temperature electron concentration was 2×10^{16} cm^{-3}. The mobility was about 350 $cm^2 v^{-1} s^{-1}$. The

Mat. Res. Soc. Symp. Proc. Vol. 423 © 1996 Materials Research Society

thickness was estimated to be about 1μm from the cross section view of SEM. An interdigital finger pattern with a Ti/Au metallization layer was deposited on the surface of the sample by standard lift-off process. The finger width and spacing were 8μm, respectively. The area of the pattern was about 4 mm^2. During the characterization, a square voltage was applied to the electrodes. The modulation of input voltage from 0 to 8 V resulted in a current 0 to 40 mA at room temperature. A quasimonochromatic light dispersed by a 0.25 m monochromator from a halogen tungsten lamp was focused on the sample, and the reflection signal was detected by an UV-enhanced Si photodiode connecting to a lock-in amplifier and data acquisition system. The modulation frequency was 280 Hz.

THE RESULTS AND DISCUSSIONS

In Fig.1 a and b the dashed lines represent the typical derivative reflective spectrum at 200 K, and 310 K, respectively. Spectra at different temperatures between 20 K and 310 K have similar lineshapes (not shown). The oscillation below the bandgap is due to the modulated optical interference from the layered epitaxy structure, which will be discussed later. The strength oscillation gradually increases with photon energy and reaches the maximum around the band gap, then rapidly decreases and disappears above the bandgap. The spacing between the peak and the valley decreases when the photon energy approaches the GaN bandgap. As the temperature increases, the structure related to the bandgap shifts towards the lower energy, while other features remain the same. We also found that the normalized modulation signal ($\Delta R/R$) was proportional to the square of modulation voltage.

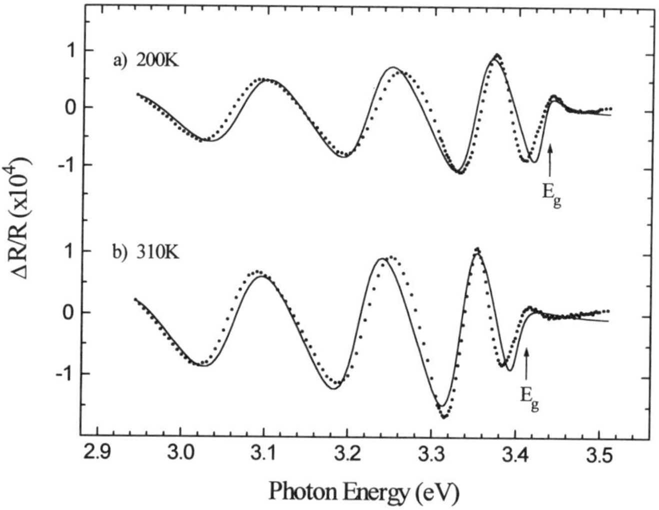

Fig.1. Experimental thermoreflectance spectra (dots) of GaN at 200K and 310K. The solid lines are least-square fit to the theory.

It is important to note that the modulation mechanism is NOT electromodulation (EM), especially for those features below the bandgap. The EM spectrum has a third derivative nature and produces very sharp, derivative-like features usually localized near the critical points of the band structures. Although the structures below the band gap have been previously reported, they were attributed to impurities[26,27] and/or back surface reflection (BSR) effect[28,29]. These features were usually in the vicinity of the band gap within the binding energy of the related impurity and the broadening parameters of the transition.

It is generally understood that the change of reflectance in a modulation spectroscopy is due to the modulation of the dielectric constant[14]

$$\frac{\Delta R}{R} = \alpha\Delta\varepsilon_1 + \beta\Delta\varepsilon_2 \qquad (1)$$

where α, β are Seraphin coefficients, $\Delta\varepsilon_1$ (or $\Delta\varepsilon_2$) is the change of real part (or imaginary part) of dielectric constant. Since the values of α and β are typically of order one, a $\frac{\Delta R}{R} \approx 10^{-4}$ (see fig. 1) implies a $\Delta\varepsilon_1$ (or $\Delta\varepsilon_2$) in the order of 10^{-4}, which is too large for a delocalized spectrum covering a region of 500 meV. However, this delocalization can be understood as follows: Below the bandgap, the GaN epilayer is transparent, the reflection (r_b) from the epilayer (GaN) and the substrate (Sapphire) is high due to the large difference in the refraction indices (n_{GaN}=2.4 and $n_{Al_2O_3} = 1.78$). When the thickness (d) of the epilayer is modulated by some means, the interference between the reflection from Air-GaN surface (r_f) and that from GaN-Sapphire interface, contributes a modulation to the total modulated reflectance $\frac{\Delta R}{R}$ in the order of ($r_b \frac{\Delta d}{d}$). In our case, r_b=0.15; therefore, a $\frac{\Delta d}{d} \approx 10^{-4}$ contribute to $\frac{\Delta R}{R} \approx 10^{-4} \sim 10^{-5}$. The contribution from the modulation of the thickness cannot be neglected. The possible mechanism for modulating the thickness of the sample is, a) thermal effect (thermal expansion); b) piezoelectric effect. By symmetry consideration, piezomodulation in the growth direction (c-axis) with an electric field applied to x-y direction (in plane) does not exist. This was further proven by increasing the modulation frequency to 100, kHz and the signal disappeared. We have also performed a standard thermomodulation (TR) experiment. In the experiment, a silver paint wire was pasted on a GaN sample. The resistance of the silver paint wire was about 2Ω. The wire acted as a heater when a current passed through. A similar spectrum was obtained when the same power was applied to the wire, which proved that the spectroscopy mechanism of our experiment was indeed temperature modulation.

Based on the above considerations, a theoretical model was established. In this model, both the interference and absorption were considered. For the nearly normal light, the reflection coefficient from the GaN epilayer can be written as[30]

$$r = \frac{r_f - r_b e^{2i\beta}}{r_f - r_f r_b e^{2i\beta}}, \qquad (2)$$

where $r_f=(1-n_{GaN})/(1+n_{GaN})$, $r_b=(n_{GaN}-n_{Al_2O_3})/(n_{GaN}+n_{Al_2O_3})$, $\beta=2\pi n_{GaN}d/\lambda$, d is the thickness of the film, and λ is the wavelength of the light. The modulated reflectance is derived as

$$\frac{\Delta R}{R}=2\,\mathrm{Re}\left[\left(\frac{1}{r_f+r_b e^{2i\beta}}-\frac{r_f}{1+r_f r_b e^{2i\beta}}\right)\frac{4\pi n_2 di}{\lambda}r_b e^{2i\beta}\frac{\Delta d}{d}+(\alpha-i\beta)e^{i\varphi}\Delta\varepsilon\right] \quad (3)$$

Here φ is a phase factor due to multiple reflection[16] where $\Delta d=\alpha_T\Delta T$, α_T is the thermal expansion coefficient of GaN, which is equal to 3.17×10^{-6} /K[1]. Because the interested energy range is near the absorption edge of GaN, the index of refraction of GaN, n_{GaN}, is a complex number. For a three dimensional M_0 critical point, the dielectric function can be written as[15]

$$\varepsilon(E,\Gamma)=\varepsilon_0+iA\left(E-E_g+i\Gamma\right)^{1/2}, \quad (4)$$

where E_g is the bandgap energy of GaN; Γ is the broadening parameter due to carrier scattering processes; A is a parameter that depends on the dipole moment matrix element, the energy, and the carrier effective masses; ε_0 is a constant related to the contribution from other transitions and $n_{GaN}^2=\varepsilon_{GaN}$.

By using formula (3) and (4), a theoretical simulation was pursued. The least-square curve fit to the experimental results is shown in Fig.1 (solid line). The simulation results indicate that the contribution of dielectric modulation (second term in equation (4)) is small. The thickness of the film was determined from the least-square curve fit. The value is 1.15 μm and is in agreement with our SEM results. The bandgap energy, E_g, and the broadening parameter, Γ, were also obtained from the least-square curve fit. It was found that the fitting curve was very sensitive to the bandgap energy, E_g, and the broadening parameter, Γ. We estimated the thickness modulation of the film to be about 1 Å, corresponding to a temperature change of about 1 ^{0}C.

The temperature dependence of energy gap in the temperature range 20 K-310 K in our GaN sample was measured, and is shown in Fig. 2 (dots). The solid line is the least-square fit to the Varshni relation[31]:

$$E(T)=E(0)-\frac{\alpha T^2}{T+\beta} \quad (5)$$

where E(0) is the energy gap at 0 K, and β, α are two constants referred to as Debye temperature and Varshni coefficients, respectively. We found E(0)=(3.470 ±0.003) eV, $\alpha=(5.9\pm0.5)\times10^{-4}$ eV/K and β=600 K.

Displayed in Fig. 2 by triangles are the experimental values of Γ as a function of temperature. The solid line is a best fit to Bose-Einstein type expression[32]:

$$\Gamma(T) = \Gamma(0) + \frac{\Gamma_{ep}}{e^{E_p/KT} - 1} \tag{6}$$

where Γ_0 is the broadening at 0 K due to non-uniformity, electron-electron interaction, impurity, and dislocation scattering effects, Γ_{ep} is an electron-phonon coupling coefficient, and E_p is the relevant phonon energy. The solid line in Fig. 2 is a least-square fit to equation (6). We obtained $\Gamma(0) = (8.6 \pm 0.4)$ meV, $\Gamma_{ep} = 179$ meV and $E_p = 93$ meV. To the best of our knowledge, this is the first report on the measurement of temperature dependence of broadening parameter for GaN.

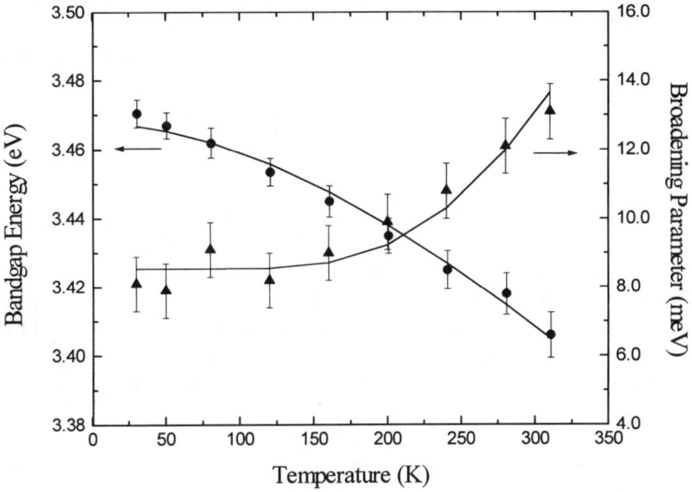

Fig.2. Temperature dependence of the energy gap (dots) and broadening parameter (triangles) of GaN. The solid lines are least-square fits to the Varshni expression and the Bose-Einstein type expression for bandgap and broadening parameter, respectively.

In summary, the thermomodulation spectroscopy technique has been used to characterize the GaN. A theoretic model is developed to explain the obtained spectra by considering the modulation of epilayer thickness and the dielectric constant. The temperature coefficient of energy gap in our GaN film is 5.9×10^{-4} eV/K. The temperature dependence of broadening parameter was measured with $\Gamma(0) = (8.6 \pm 0.4)$ meV, $\Gamma_{ep} = 179$ meV and $E_p = 93$ meV.

REFERENCES

1. H. Morkoc, S. Strite, G.B. Gao, M. E. Lin, B. Sverdlov, and M. Burns, J. Appl. Phys. 76, 1363 (1994), and references therein.
2. S. Nakamura, M. Senoh, T. Mukai, Jpn. J. Appl. Phys., 32, L8 (1993).

3. J. I. Pankove, J. Lumin. 7. 114 (1973).
4. D. L. Camphausen and G. A. N. Connell, J. Appl. Phys. 42, 4438 (1972).
5. H. Teisseyre, P. Perlin, T. Suski, I. Grzegory, S. Porowski, J. Jun, A. Pietraszko, T.D. Moustakas, J. Appl. Phys. 76, 2429 (1994).
6. K. Osamura, S. Naka, Y. Murakami, J. Appl. Phys., 46, 3432 (1975).
7. B. Monemar, Phys. Rev, B 10, 676 (1974).
8. M. Ilegems, R. Dingle, and R. A. Cogan, J. Appl. Phys. 43, 3797 (1972).
9. T. Matsumoto and M. Aoki, Jpn. J. Appl. Phys. 13, 1804 (1974).
10. A M. Zykov and G. K. Gaido. Sov. Phys. Semicond. 6. 154 (1972).
11. R. Dingle and M. Ilegems, Solid State Commun. 9. 175 (1971).
12. G. Ramirez-Flores. H. Navarro-Contreras, A. Lastras-Martinez, R. C. Powell, and J. E. Greene, Phys. Rev. B50, 8433 (1994).
13. W. Shan, T. J. Schmidt, X. H. Yang, S. J. Hwang, and J. J. Song, Appl. Phys. Lett., 66, 987 (1994).
14. F. H. Pollak, Proc. Soc. Photo-Optical Instrum. Eng., 276, 142 (19981).
15. F. H. Shen, M. Dutta, J. Appl. Phys., 78, 2151 (1995).
16. B. Batz, in R. K. Willardson, A. C. Beer (eds.), Semiconductors and Semimetals, Vol. 9 Academic Press, New York, 315 (1972).
17. O. J. Glembocki, B. V. Shanabrook, in D. G. Seiler and C. L. Litter (eds.), Semiconductors and Semimetals, Vol.67, Academic Press, New York, 222 (1992).
18. Sydor, J. Angelo, W. Mitchel, T. W. Haas, M. Y. Yen, J. Appl. Phys., 66, 1989 (1989).
19. D. K. Gaskill, N. Bottka, L. Aina, M. Mattingly, Appl. Phys. Lett., 56 1269 (1990).
20. A. Ksendzov, H. Shen, F. H. Pollak, D. P. Bour, Solid State Commun., 73, 11 (1990).
21. J.S. Foresi, T. Moustakas, Appl. Phys. Letts. 62, 2859 (1993).
22. M. Misora, T. Moustakas, R. P. Vaudo, R. Simgh, J. S. Shah, Proceeding of SPIE, 2519, 78 (1995).
23. A. Giordana, D. K. Gaskill, J. Electro. Mat., 23, 509 (1994).
24. C. Yuan, T. Salagai, R. A. Stall, Y. Li, M. Schurman, C. Y. Hwang, W. E. Mayo, Y. Lu, J. Electrochem. Soc., 142, L163 (1995).
25. O. J. Glembocki, N. Bottka, J. E. Furneaux, J. Appl. Phys., 57, 432 (1985).
26. H. Shen, Z. Hang, S.H. Pan, F. H. Pollak, J. M. Woodall, Appl. Phys. Lett., 52, 2058 (1988).
27. R. L. Tober, J.D. Bruno, J. Appl. Phys. 68, 6388 (1990).
28. V. Behn, H. Roppischer, Phys. State Solidi, 141(b), 325 (1987).
29. H. Shen, S. H. Pan, F. H. Pollak, M. Dutta, T. AuCoin, Phys. Rev., B36, 9384 (1987).
30. M. Born, E. Wolf, *Principles of Optics*, Pergamon, New York, 328 (1965).
31. Y. P Varshni, Physica (Utrecht), 34, 149 (1967).
32. P. Lautenschlager, M. Carriga, M. Cardona, Phys. Rev. B36, 4813 (1987).

OPTICAL CHARACTERIZATION OF GaN FILMS GROWN ON (0001) SAPPHIRE SUBSTRATE

K. Yang*, R. Zhang*, Y.D. Zheng*, L.H. Qin*, B. Shen*, H.T. Shi*, Z.C. Huang** and J.C. Chen**
*Department of Physics, Nanjing University, Nanjing 210093, P.R.China
**Department of Electrical Engineering, University of Maryland Baltimore County, Baltimore, MD 21228-5398, USA

ABSTRACT

The optical properties of single crystal hexagonal GaN films grown on (0001) sapphire substrate by metalorganic chemical vapor deposition were investigated. The energy gap of hexagonal GaN was determined as 3.39 and 3.400 eV by optical transmission and photoreflectance, respectively. The refractive index of GaN as a function of photon energy was drawn from the transmission spectrum. Furthermore, Raman scattering spectra were employed to study the phonon modes of the GaN film. The properties of LO phonon-plasmon coupled modes were further studied, and the carrier concentration and damping constant were determined by line-shape fitting of the coupled modes.

INTRODUCTION

The wide-gap semiconductors of the nitrides belonging to the III-V family, such as GaN, AlN and their alloys, have received enormous attention for their practical use in both optoelectronic and electronic devices[1,2]. GaN has a direct wide energy band gap as well as notable thermal and chemical stability, and these attractive properties make it not only ideally suitable for fabricating blue and ultraviolet (UV) light emitting diodes and detectors, but also for application in harsh environments, such as at high temperature[3,4]. With the substantial improvement in epitaxy growth technology, high quality GaN films on sapphire substrate have been fabricated by metalorganic chemical vapor deposition (MOCVD)[4-6], and other techniques[7,8]. Therefore, high performance devices, including blue light emitting diodes and UV detectors, have been demonstrated[2].

In spite of impressive technological achievements of the last three years, there are still some problems concerning basic physical properties of this compound which have to be solved. Many studies have been reported on the theoretical calculation of the energy band structure of GaN and on the experimental characterization of its optical properties. However, most experimental studies were still concentrated on the photoluminescence and absorption of GaN[5,7,9]. In this paper, we studied the room temperature structural and optical properties of single crystal hexagonal GaN films grown on (0001) sapphire substrate by MOCVD. Photoreflectance was employed to determine the bandgap of GaN film as an additional optical technique, and the optical absorption edge of 3.39 eV confirmed it. For studying the phonon modes of the GaN film, Raman scattering was performed, and the properties of LO phonon-plasmon coupled modes were further studied. As a result, the carrier concentration and damping constant were determined by line-shape fitting of the coupled modes with considering the dominant scattering mechanisms of deformation-potential and electro-optic interaction.

Mat. Res. Soc. Symp. Proc. Vol. 423 © 1996 Materials Research Society

EXPERIMENT

In our work, single crystal hexagonal GaN films were grown on (0001) sapphire substrate with an AlN buffer layer by MOCVD. The Ga, Al (for the AlN buffer layer) and N source gases were trimethylgallium (TMGa), trimethylaluminum (TMAl), and ammonia (NH_3), respectively. The epitaxial layer was grown using TMGa and high purity NH_3 as sources and H_2 as carrier gas. The thin AlN layer (~50nm) was deposited at 550°C as a buffer layer on (0001)-oriented sapphire substrate just before the growth of the GaN at 720°C. The growth rate of GaN in the experiment was about 1μm/hour and all samples were unintentional doped n-type.

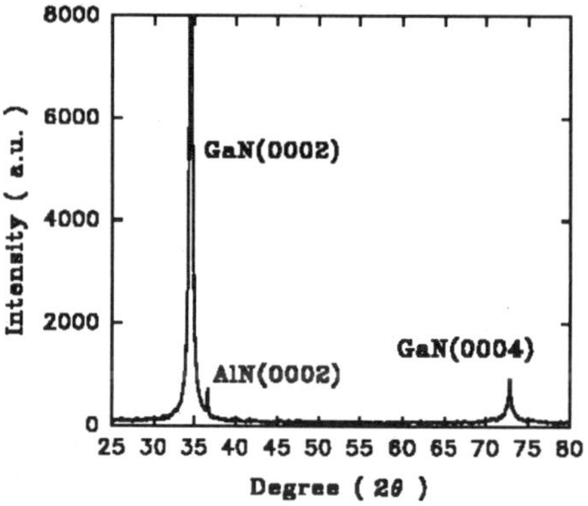

Fig.1 X-ray diffraction spectrum of the GaN film grown on (0001) sapphire substrate.

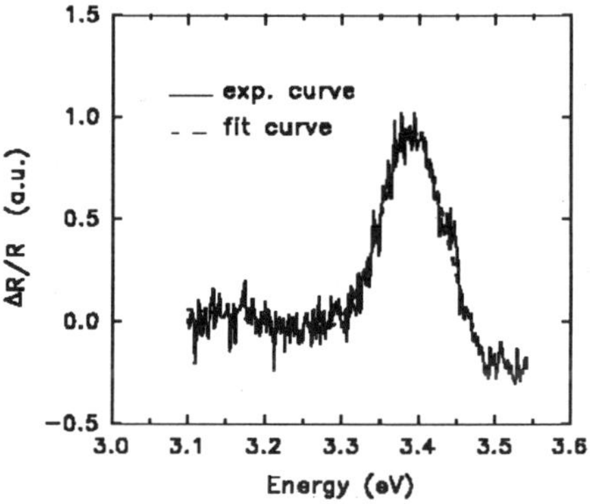

Fig.2 Photoreflectance spectrum of the GaN film at RT (the solid line). The curve was fitted by the dash line.

X-ray diffraction was employed to analyze the structural quality of GaN epitaxial films. Fig.1 presented the typical XRD pattern for the GaN film deposited on (0001) sapphire substrate. From the (0002) and (0004) peak positions of the GaN film, we obtained the lattice constant in the direction of c-axis of 5.188Å. As seen from the data in Fig.1, the (0002) AlN peak of buffer layer with the lattice constant c of 4.948Å was also observed. In further study, it was showed that the full width at half-maximum (FWHM) of the (0002) GaN diffraction peak determined from the rocking curves of X-ray double-crystal diffractometer $\theta/2\theta$ scan was less than 170", which confirmed high quality of the single crystal of the epitaxial GaN film.

RESULTS AND DISCUSSIONS

In order to investigate the band structure of GaN film, PR spectra was measured at room temperature on a self-developed optical measurement system[10]. Figure 2 showed a typical PR spectrum of single crystal GaN on sapphire substrate at room temperature. According to the equation

$$\frac{\Delta R}{R} = Re\{A\,\Gamma^{-n}\,e^{i\theta}\,(E - E_g - i\Gamma)^{-n}\} \qquad (1)$$

we fitted the experimental curve by changing the values of the amplitude A, the linewidth Γ, the phase θ, and the transition energy E_g. The exponent n was fixed at the value of 2.5[11], which is often used for band to band transition in the bulk samples. Thus we obtained the E_g of 3.400 eV and the linewidth of 0.124 eV.

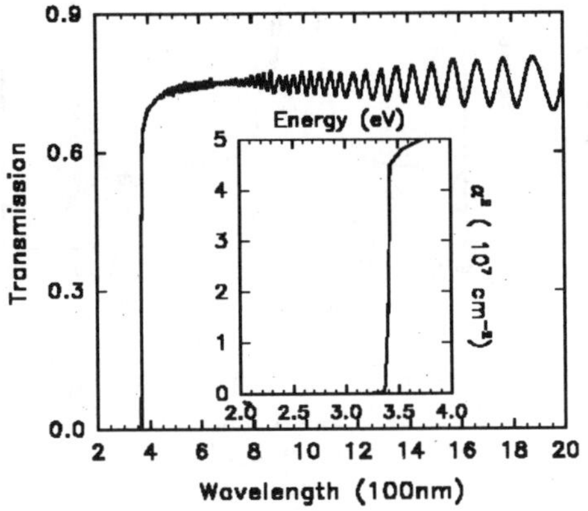

Fig.3 Optical transmission spectrum of the GaN film on sapphire at RT. The bandgap of GaN was determined as 3.39 eV.

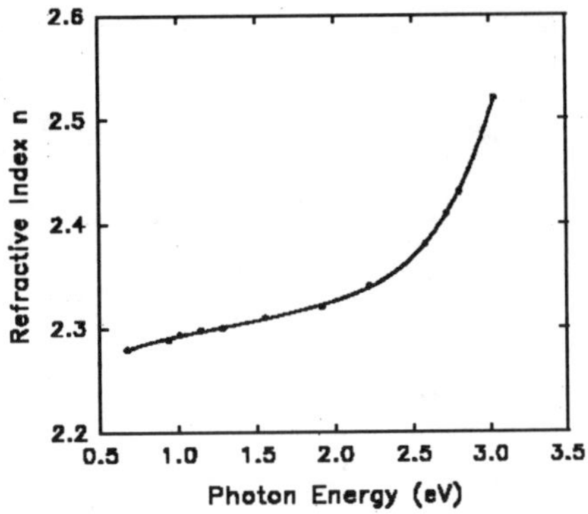

Fig.4 Refractive index n of the GaN film vs. photon energy at room temperature.

The optical transmission spectrum, shown in Fig.3, was performed at room temperature. The absorption coefficient was about 7×10^3 cm^{-1} above the bandgap, and almost zero below the gap with a sharp line shape around the band gap. By drawing the absorption coefficient square vs. the spectral energy curve (shown in Fig.3), $\alpha^2 \sim h\nu$, we obtained the direct energy bandgap of 3.39 eV, which was consistent with the results of PR measurement. From the interference features on the transmission data, the film thickness was determined as $\sim 6\ \mu m$ agreed well with the designed value. Moreover, the refractive index n was drawn from the transmission data[12] as a function of photon energy as shown in Fig.4.

Raman scattering was used to studied the phonon modes of the GaN film. In the measurements of the Raman scattering spectra, the 488 nm light of an Ar^+ ion laser was focused on the sample at room temperature. For GaN has a hexagonal Wurtzite structure and belongs to the C_{6v} symmetry group, there are six Raman active phonons: two E_2 , E_1(TO), E_1(LO), A_1(TO) and A_1(LO). Raman spectra from the GaN film grown on sapphire substrate, shown in Fig.5, were obtained with blackscattering configuration, and A_1(TO), E_1(TO), A_1(LO), E_1(LO), and high frequency E_2 modes, with phonon frequencies as 545 cm^{-1}, 569 cm^{-1}, 735 cm^{-1}, 750 cm^{-1}, and 578 cm^{-1} respectively, were observed together with three modes of 418 cm^{-1}, 430 cm^{-1} and 449 cm^{-1}, come from the sapphire substrate.

Raman scattering technique is a sensitive probe to investigate not only phonon structure, but also to electronic concentration. The position and the shape of the phonon-plasmon coupled mode are sensitive to plasmon frequency, therefore, to density of electron or nitrogen vacancy. In Fig.6, polarized light from the 488 nm Ar^+ ion laser was focused on the sample, and a backscattering geometry, denoted as Z(X-)-Z, with the Z direction parallel to the (0001) axis of GaN, was employed. The coupled modes of LO phonon and

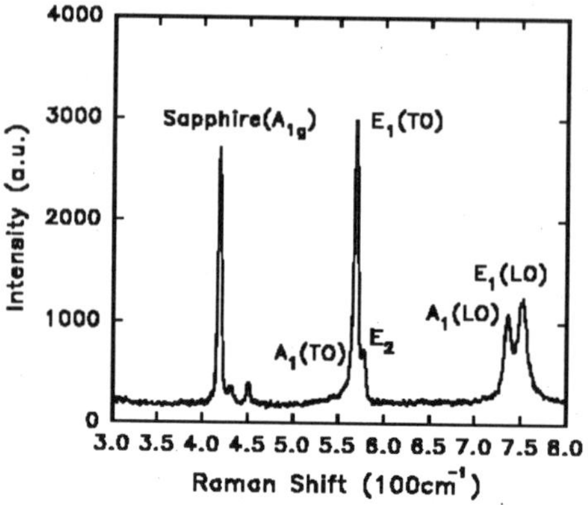

Fig.5 Raman spectrum of the GaN film at room temperature under 488 nm argon laser excitation in backscattering configuration.

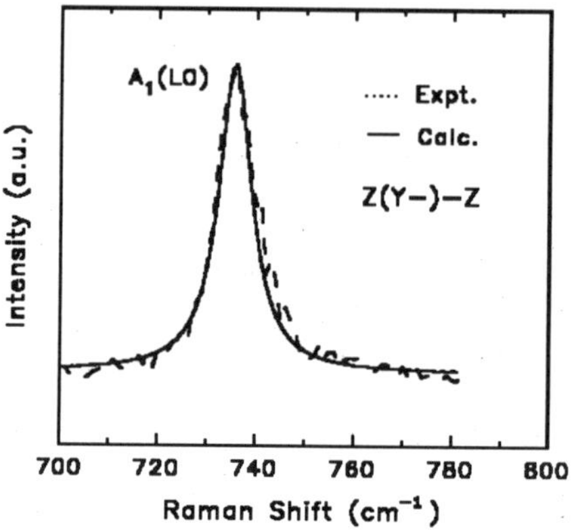

Fig.6 Comparison between the experimental and calculated line shape of the LO phonon-plasmon coupled mode.

overdamped plasmon in GaN have been studied. As the results of T.Kozawa et al.[13], the contribution of the deformation-potential and electro-optic mechanism is dominant in GaN. Thus, the cross section can be expressed by

$$I_B = \frac{d^2S}{d\omega\, d\Omega}\Big|_B = \frac{16\pi h n_2}{V_0^2 n_1}\frac{\omega_2^4}{C^4}\left(\frac{d\alpha}{dE}\right)^2 (n_\omega+1)\, B\, Im\left(-\frac{1}{\epsilon}\right) \qquad (2)$$

and,

$$B = 1 + 2C\frac{\omega_t^2}{\Delta}\left[\omega_p^2\gamma(\omega_t^2-\omega^2) - \omega^2\eta(\omega^2+\gamma^2-\omega_p^2)\right]$$
$$+C^2\left(\frac{\omega_t^4}{\Delta(\omega_l^2-\omega_t^2)}\right)\{\omega_p^2[\gamma(\omega_l^2-\omega_t^2)+\eta(\omega_p^2-2\omega^2)] + \omega^2\eta(\omega^2+\gamma^2)\} \qquad (3)$$

$$\Delta = \omega_p^2\gamma[(\omega_t^2-\omega^2)^2+(\omega\eta)^2] + \omega^2\eta(\omega_l^2-\omega_t^2)(\omega^2+\gamma^2) \qquad (4)$$

Here, ω_t and ω_l are the frequencies of TO and LO phonons, γ is the plasmon damping constant, η is the phonon damping constant, $\omega_{1,2}$ are the incident and scattered photon frequencies, V_0 is the volume of the unit cell, $n_{1,2}$ are the refractive indexes at $\omega_{1,2}$, E is macroscopic electrical field, α is polarizability, n_ω is the Bose-Einstein factor, C is the so-called Faust-Henry coefficient. The dielectric function ϵ is given by a sum of the contribution from phonons and plasmons

$$\epsilon = \epsilon_{\infty}\left(1 + \frac{\omega_l^2 - \omega_t^2}{\omega_t^2 - \omega^2 - i\,\omega\eta} - \frac{\omega_p^2}{\omega(\omega + i\,\gamma)}\right) \quad (5)$$

where ω_p is the plasmon frequency

$$\omega_p^2 = \frac{4\pi n e^2}{\epsilon_{\infty} m^*} \quad (6)$$

and ϵ_∞ is the high frequency dielectric constant, n is the free carrier concentration, and m^* is the effective mass.

By taking ω_p, γ, η and C in Eq.(2) as fitting parameters, we fitted the calculated line shape to observed A_1(LO) phonon-plasmon coupled modes by means of the least squares method (shown in Fig.6). The agreement between the experimental and calculated line shapes of the coupled mode was excellent, and the values of fitting parameters were: $\omega_p = 39\ cm^{-1}$, $\Gamma = 7.7\ cm^{-1}$ and $\gamma = 260\ cm^{-1}$. In the fitting, we used a fixed value of parameter C (=0.4), which was given in Ref.13. Using Eq.(6), we determined the free-carrier concentration n from the value of the plasma frequency derived from the line-shape fitting. Because the data discussed here concerned coupled modes of A_1 symmetry where the electric and lattice displacements were along the c axis of GaN, we used $\epsilon_\infty = 5.35$ and $m^* = 0.19m_0$, thus we obtained $n = 1.6\times10^{16}\ cm^{-3}$. It shows that the GaN films have high quality by using an AlN buffer layer on sapphire substrate.

CONCLUSIONS

In this paper, we have studied the optical properties of hexagonal GaN films on (0001) sapphire substrate grown by MOCVD. XRD spectrum showed high quality single crystal of the GaN epitaxial film. We obtained the energy gap of GaN film as 3.400 eV by photoreflectance, and the optical absorption edge of 3.39 eV in optical transmission confirmed it. Moreover, all of Raman active phonon modes were observed, and the A_1(LO) mode was further discussed. we obtained the free-carrier concentration n as low as $1.6\times10^{16}\ cm^{-3}$, which shows the GaN films have high quality by using an AlN buffer layer on sapphire substrate.

REFERENCES

1. S.Strite and H.Morkoc, J. Vac. Sci. Technol. B, 10 (4), 1237 (1992)
2. H. Morkoc, S. Strite, G.B. Gao, M.E. Lin, B.Sverdlov and M. Burns, J. Appl. Phys., 76, 1363(1994), and references therein.
3. M.A. Khan, J.N. Kuznia, A.R. Bhattarai and D.T. Olson, Appl. Phys. Lett., 62(15), 1786(1993)
4. S.Nakamura, T.Mukai and M.Senoh, Appl. Phys. Lett., 64 (13), 1687(1994)
5. K.G.Fertitta, A.L.Holmes, et al., Appl. Phys. Lett., 65 (14), 1823(1994)
6. H. Amano, M. Kito, K. Hiramatsu and I. Akasaki, Jpn. J. Appl. Phys., 28, L2112(1989)

7. R.Singh, R.J.Molnar, M.S.Unlu and T.D.Moustakas, Appl. Phys. Lett., 64(3), 336 (1994)
8. A.Giordana, D.K.Gaskill, D.K.Wickenden, A.Estes Wickenden, J. Electronic Materials, 23(6), 509(1994)
9. K. Yung, J. Yee, J. Too, M. Rubin, N. Newman and J. Ross, Appl. Phys. Lett., 64(9), 1135(1994)
10. L.H. Qin, K. Yang, Y.D. Zheng, R. Zhang, X.J. Dai, D. Fen, Z.C. Huang and J.C. Chen, Chinese Phys. Lett., 13(2), 153(1996)
11. D.E. Aspnes, Surf. Sci., 37, 418(1973)
12. R. Swanepoel, J. Phys. E: Sci. Instrum., 16, 1214(1983)
13. T.Kozawa, N.Koide et al., J. Appl. Phys., 75 (2), 1098(1994)

OPTICAL AND STRUCTURAL PROPERTIES OF α-$Si_{1-x}C_x$ FILMS

Zhizhong Chen, Kai Yang, Rong Zhang, Hongtao Shi, Youdou Zheng
Department of Physics, Nanjing university, Nanjing 210093, P.R China

ABSTRACT

In this paper, we reported experimental results about optical and structural properties of amorphous silicon carbide (α-$Si_{1-x}C_x$). The films of α-$Si_{1-x}C_x$ were grown by CVD on substrate of quartz glass. Optical constants (n-refractive index, α-absorption coefficient, Eg-optical energy band gap) of these films were determined by transmission spectra. The radial distribution functions (RDFs) of α-$Si_{1-x}C_x$ films were drawn out from the data of x-ray diffraction spectra. According to the RDFs, we imagined the statistic scene from which we could obtain the information of atomic radial distribution. The bond lengths and bond numbers of Si-Si, Si-C, and C-C could be also determined by RDFs. From the analysis of Raman spectra, we obtained the information of their vibration state density, and discerned the peaks of bond vibration, which agreed well with the results of α-$Si_{1-x}C_x$ RDF.

INTRODUCTION

Amorphous silicon carbide (α-$Si_{1-x}C_x$) is an attractive material from a technological point of view: its electronic and optical properties make it potentially useful as solar selective coating and as a blue light material. In addition, because of high temperature resistant, wide band gap and high electron saturation velocity of its semiconducting properties, α-$Si_{1-x}C_x$ alloy is a good candidate for stable high-T semiconductor and high-power device[1].

In recent years, a widespread experimental effort has been devoted to the study of α-SiC. However, the knowledge of its microscopic structure, essential prerequisite to a complete understanding of its physical properties, is still lacking. The short distance order of noncrystal determines its energy band structure, electroconductivity, thermoconductivity and optical properties etc. The method of RDF is comparatively successful in predicting some properties of disordered systems[2]. Through its Raman spectra, bond information and vibration state density is easily acquired[3,4].

In this paper, we present the experiment results of two samples of α-$Si_{1-x}C_x$ about microscopic structure and optical energy gap, and discuss the relationship between macroscopic physics properties and microscopic structure.

EXPERIMENTAL AND DATA ANALYSIS

Experiments have been done on α-$Si_{1-x}C_x$ films of various compositions. In this

Mat. Res. Soc. Symp. Proc. Vol. 423

paper we present the results of two samples, which x values are 0.5 (#081) and 0.4 (#09) respectively. The two samples were grown by PECVD on substrate of quartz glass with similar depositional conditions. In the experiment, the reaction chamber was evacuated with a mechanical pump, then with a diffusion pump prior to the sample growth. The total growth pressure is about 1 Torr. Reactive source gases are H_2, hydrogen-diluted CH_4 and SiH_4. During the process of growth, the quartz substrate was kept at about 200 ℃.

X-ray diffraction (XRD) measurements were performed using a conventional Cu x-ray tube. Fig.1 shows the diffraction intensity (I(k)) of x-ray. The maximum value of the diffraction wave vector (k) was 80 nm^{-1}, and the minimum value is 7 nm^{-1}. Ref[1] has explained that the data extending to 0 nm^{-1} doesn't provide more corrigenda . The

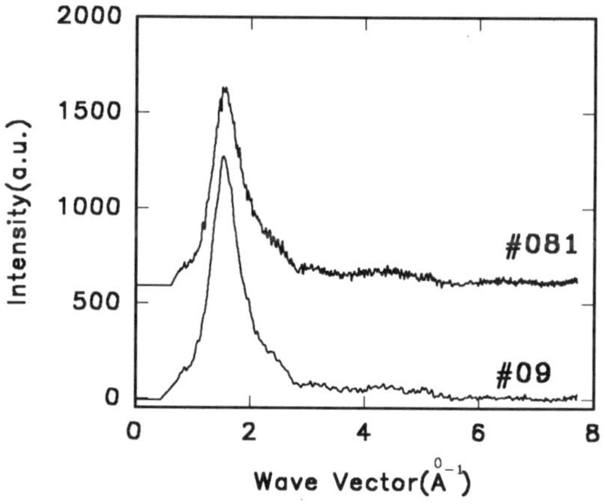

Fig1. X-ray diffraction functions (#09:x=0.4 #081:x=0.5)

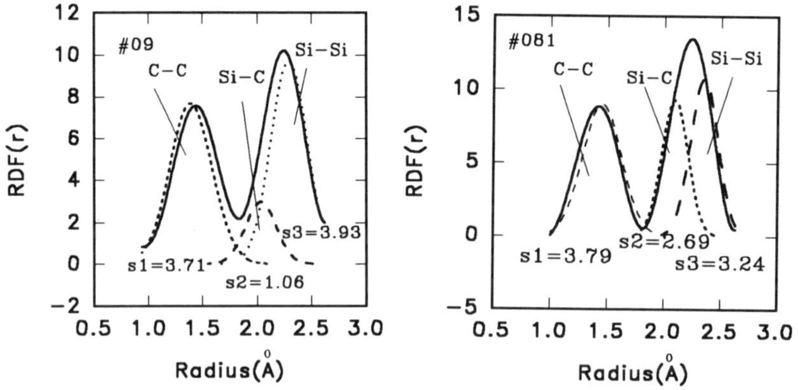

Fig.2 Radial distribution functions

maximum value of k is so small that we have to use a new method to calculate the samples' radial distribution functions, which is called Gaussial cut-and-trial[5~8]. In this case, the factors of air scattering and samples' absorption which affected the diffraction intensity little is neglected. We acquire the total RDFs of the two samples, as shown in Fig.2 (the first shell distribution). The bond length and bond number are obtained by fitting the RDF's first peak with Gaussial distribution function. The area under Gaussial peak gives the bond number, and the peak's abscissa identifies the bond length. More accurate results will be acquired by partial distribution function in which neutron-diffraction spectra is required.

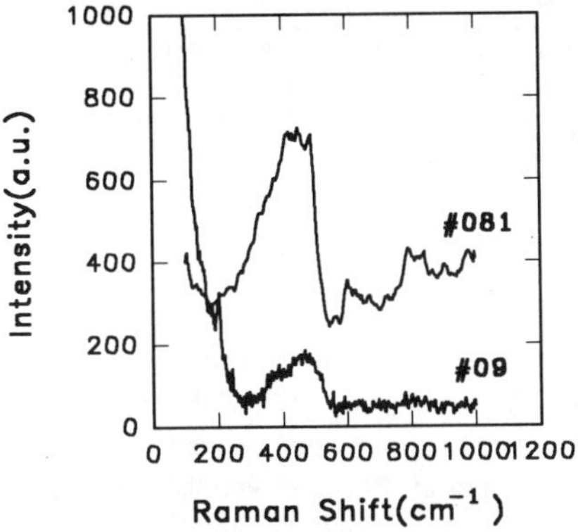

Fig.3 Raman spectra

We analyze Raman spectra (fig.3.) in order to justify the results of RDFs. The range of wave number (v) is from 200cm^{-1} to 1000 cm^{-1}, and the step length is 2 cm^{-1}. With the methods of references[12~16], we can discern the peaks of bond vibration. From the intensity of vibration we also can conclude the bonds information.

We measured the transmission spectra of the samples, as shown in fig.4. We try to find how the optical energy gap, Eg(opt), varies with the microscopic structure. Eg(opt), as shown in fig.5, is calculated by the envelope of the transmission spectra[9,10]. The refractive index and absorption coefficient which agree with the results of RDFs can also be easily got from the envelope.

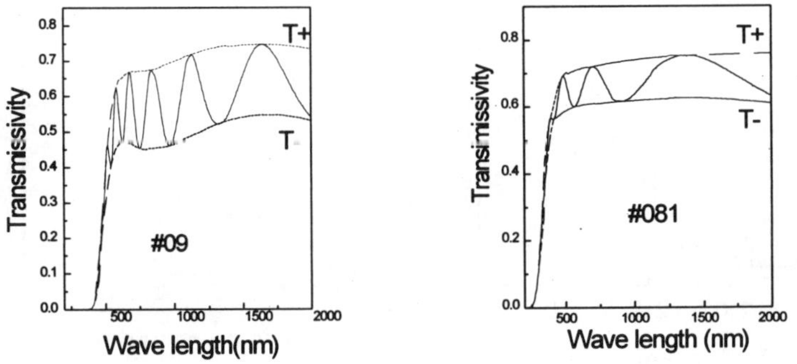

Fig.4. transmission spectra

RESULTS AND DISCUSSION

In fig.2 we show the total RDFs of the two samples. The total RDFs are characterized by the well defined peaks of the first shell[3]. After fitting for the first peaks of RDFs, we tabulate the bonds lengths (Si-Si, Si-C, C-C) and bonds numbers in Tab.1. In crystalline SiC, one atom of carbon is surrounded by four silicon atoms. α-$Si_{1-x}C_x$'s atoms are chemically disordered. Because the bonds numbers of Si-C is small, it seems that there are some small particles of Si, C, or compounds of Si and C. Both the bonds lengths and the bonds numbers of the samples vary with x obviously. All the bonds of the samples studied in this work are shorter than those of crystalline SiC, mainly because non-crystalline bonds are twisted.

	Si-Si		Si-C		C-C	
sample	length (pm)	number	length (pm)	number	length (pm)	number
#09	235	3.92	197	1.06	145	3.93
#081	235	3.24	209	2.69	145	3.79

Tab.1. Samples' bonds length and number

In fig.3 we show the Raman spectra of the samples. According to ref[12~16], the vibration peaks at 330 cm^{-1}, 492 cm^{-1}, 602 cm^{-1}, 798 cm^{-1}, 842 cm^{-1}, 910 cm^{-1} can be discerned. It's sure that the peaks at 330 cm^{-1}, 490 cm^{-1} (slightly crystallized), 602 cm^{-1}, 910 cm^{-1} are amorphous Si-Si peaks, and the peaks at 798 cm^{-1} and 842 cm^{-1} are amorphous Si-C peaks. We'll find C-C peaks out of 1000 cm^{-1}. The total area under amorphous Si-Si peak is much more than that of amorphous Si-C, which confirms the conclusion of RDFs.

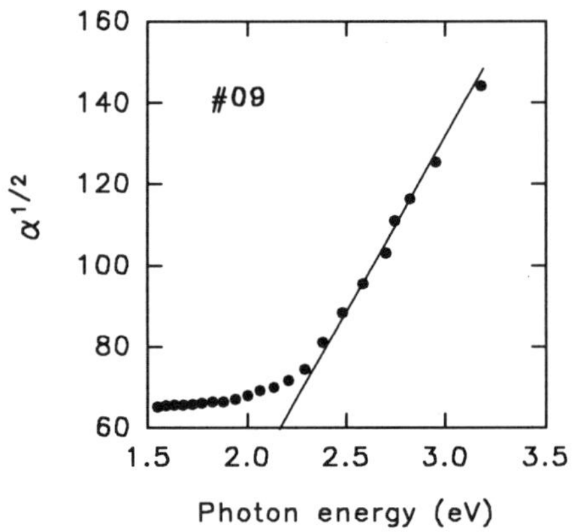

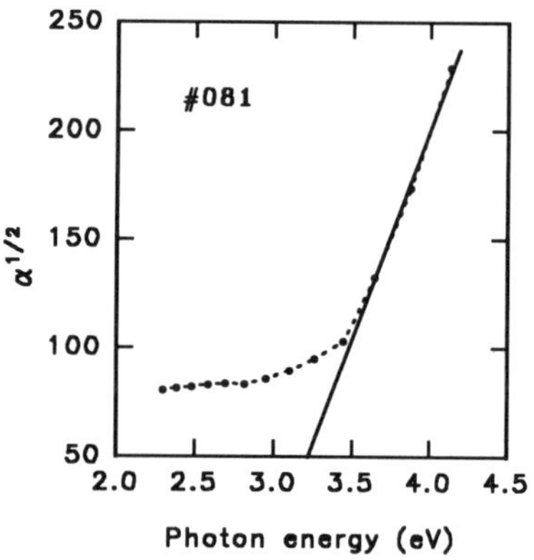

Fig.5 curves of $\alpha^{1/2}$ versus photon energy($h\omega$)

In fig.5, we show the curves which illustrate the relationship of $(\alpha h\omega)^{1/2}$ and $(h\omega)$. The optical energy band gaps are obtained by drawing the straight line at high energy region to the horizontal axis. And the Eg(opt) value of sample #081 is determined as

3.2 eV, which is larger than the 2.2 eV of sample #09. We think that more C-C bonds can result in it, because the bond energy of C-C is more than that of others. The refractive index (n) of the two samples is also obtained from the envelope (the dot line), shown in fig.4. The refractive indices of sample #09 and #081 are 2.34 and 2.1 respectively. We think that the refractive index may be also influenced by the microstructure of the samples, but we still don't know the details.

REFERENCES

1. J.Bullot and M.P.Schmidt, Phys.Stat.Sol., 143, (1987), 345.
2. B.E.Warren, X-ray Diffraction(Addison-Wesley, Reading, MA,1969).
3. R.Shuker, R.Gamon, In Proc., 2nd intern, conf. Light Scattering in Solids, Paris, 1971(Flammarion Science, Paris, 1971).
4. M.Cardona, Light Scattering in Solids (1982).
5. S.T.Huang, X-ray Scattering in Solids (1990).
6. A.H.Yu, Basis of X-ray Diffraction Technology (in Chinese).
7. H.P.Hlug, L.E.Alexander, X-ray Diffraction Technology (Chinese version), X-ray Diffraction Procedurestor Polycrystalline and Amorphous materials, Ied, Wiley, 1974.
8. International Tables for X-ray Crystallography.
9. H.Demiryont, J.R.Site.and K,M,Geib,Appl.Opt,1985,24:490
10. W.X.Liang, Z.Q.Yin, Thin Film Science and Technology(in Chinese), Vol.7. No., Jun., 1994(129-134).
11. W.Schulke, Phil.Mag.B. 43 (1981) 451.
12. M.Gorman, S.A.Solin, Solid State Comm. 15 (1974) 711.
13. Diego.Olego.and M.Cardona, Phys. Rev.B,vol.(25), No.6, (1982) 378-88
14. G.H Li, M.D Tao, Chinese Journal of Semiconductors, Vol.8, No.2., Mar,1987 (182-185).
15. Y.Inoue, S.Nakashina and A.mitsuishi, Solid State Comm.,48,p.1071(1983).
16. P.K. Banerjee, J.M.T. Pereira and S.S Mitra, J.Non.Crys.Solid, Vol.87-88, No.1-4.,p.1-21(1982).

SURFACE PHOTOVOLTAGE EFFECTS IN PHOTOEMISSION FROM DIAMOND SURFACES

C. BANDIS AND B. B. PATE
Department of Physics, Washington State University, Pullman, WA 99164-2814

ABSTRACT

Photovoltaic effects in ultraviolet photoemission spectroscopy of the in-situ "re-hydrogenated", and reconstructed (111) diamond surfaces are evaluated. We show that photovoltaic charging effects during photoemission studies of the in-situ "re-hydrogenated" (111)-(1x1):H diamond surface are significant at temperatures as high as room temperature. In contrast, experiments on the reconstructed (111)-(2x1) diamond surface find no photovoltaic charging which suggests that the surface exhibits relatively high conductivity (effectively grounded surface). These results demonstrate that extra care should be taken in determining the Fermi level pinning position relative to the bands at diamond surfaces and interfaces. The assumption that the UPS photoelectron spectra reflect the equilibrium energy band arrangement should be experimentally confirmed in each case, especially when wide band gap materials are involved.

INTRODUCTION

A significant number of the experiments performed to study diamond interfaces involve photoemission experiments. These are analyzed in order to obtain information about the energy bands (band bending) of the semiconductor near the interface, as well as their evolution with various surface treatments [1-5]. In contrast to other semiconductors (e.g. GaAs), most of the photoemission experiments on diamond have been performed at room temperature, and little is known about low temperature photoemission from diamond. In this paper, we present UV ($h\nu$=21.2 eV) photoemission measurements from both the in-situ "rehydrogenated" negative electron affinity (111)-(1x1):H, and the positive electron affinity (111)-(2x1) diamond surfaces performed at temperatures ranging from 170 to 425 K. Photoelectron energy distributions demonstrate that, in contrast to the reconstructed (111)-(2x1) surface, the in-situ "rehydrogenated" diamond surface exhibits an *apparent* temperature dependent Fermi level pinning position. These results can be understood within Hecht's model for photovoltaic charging during photoemission [6-8]. Furthermore, we find that photovoltaic charging in diamond can be significant even at room temperature. Therefore, the effects of photovoltaic charging must be addressed in photoelectron spectroscopy (PES) studies of wide band gap semiconductors such as diamond, cBN, AlN, etc..

Over the past several decades photoelectron emission has been used to study the electronic structure of semiconductor interfaces. Results from such photoemission studies were generally thought to represent the equilibrium band structure at the interface. Recently however, it has been shown that this is not always a well justified assumption, and that the photon flux used during the photoemission experiments can cause significant deviation from equilibrium even at room temperature due to photovoltaic charging [7, 9, 10]. Hecht [6-8] has calculated the effect of the surface photovoltage (ΔV) resulting from separation of electron-hole pairs in the band bending region for GaAs, and found that, at low temperatures, photovoltaic charging can be significant. After Hecht, in Fig. 1a we show the surface (no illumination) of a p-type semiconductor with an arbitrary density of surface states which exhibits downward band bending, E_F-E_{VBM}=V_b (E_F is the energy of the Fermi level, and E_{VBM} is the energy of the valence band maximum, VBM, at the surface). Under illumination with band gap radiation (Fig. 1b), the photovoltage (ΔV) due to separation of the electron-hole pairs in the band bending region opposes the voltage associated

Mat. Res. Soc. Symp. Proc. Vol. 423

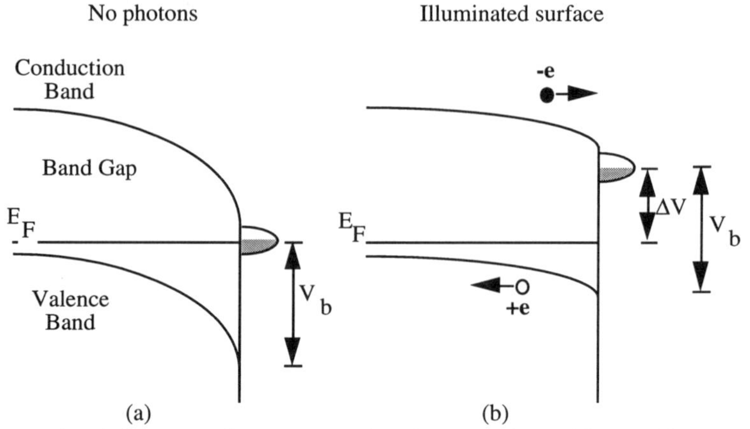

FIG. 1 Energy band diagrams of a p-type semiconductor surface with an arbitrary, partially filled, density of states localized at the semiconductor–vacuum interface, (a) in the dark, and (b) with photon illumination.

with the band bending. This, in the case of an electrically isolated surface, results in flattening of the bands (Fig. 1b) and an *apparent* E_F-E_{VBM}=V_d (=V_b-ΔV) which is determined by the temperature dependent restoring current in the semiconductor. Hecht derived this restoring current in terms of thermionic emission over the surface barrier plus field emission through the barrier and found that it depends exponentially on temperature [6, 7].

EXPERIMENT

Our photoemission experiments were performed on both the in-situ "re-hydrogenated" and reconstructed (111) surfaces of a 4 carat type IIb (p-type, boron doped) natural single crystal diamond (sample D5). Before each study the diamond surface was mechanically polished on a ten-inch cast iron wheel using 1μm grit and olive oil, under a load of several pounds followed by ultrasonic rinse in methyl-ethyl-ketone (MEK), acetone and alcohol ("as-polished" surface). With "re-hydrogenated" we refer to the surface which results after: (1) hydrogen desorption from the as-polished surface by in-vacuo heating to about 1000°C (reconstructed surface), and (2) subsequent re-hydrogenation of the clean surface by in-situ dosing of atomic hydrogen [2] (1 x 10^{-5} Torr of hydrogen). Al Kα x-ray photoelectron spectroscopy (XPS) detected only carbon and oxygen (less than 10% of a monolayer). No surface roughness characterization was performed.

The sample was mounted, using platinum foil and wire, onto a liquid nitrogen cooled sample holder, and positioned at the center of an all-metal ion-pumped ultra-high-vacuum (UHV) chamber (base pressure $\leq 4 \times 10^{-10}$ Torr). A tungsten filament mounted behind the sample was used for e-beam heating. Sample temperatures from 170 K up to 1400 K were routinely achieved. The sample temperature was recorded using a chromel-alumel thermocouple, spot-welded to the platinum support and in contact with the back side of diamond. The energy position of the valence band maximum at the surface relative to the Fermi level was determined using ultraviolet photoemission spectroscopy (UPS). The electron energy distribution curves were taken with a hemispherical electron analyzer (ΔE=0.08 eV).

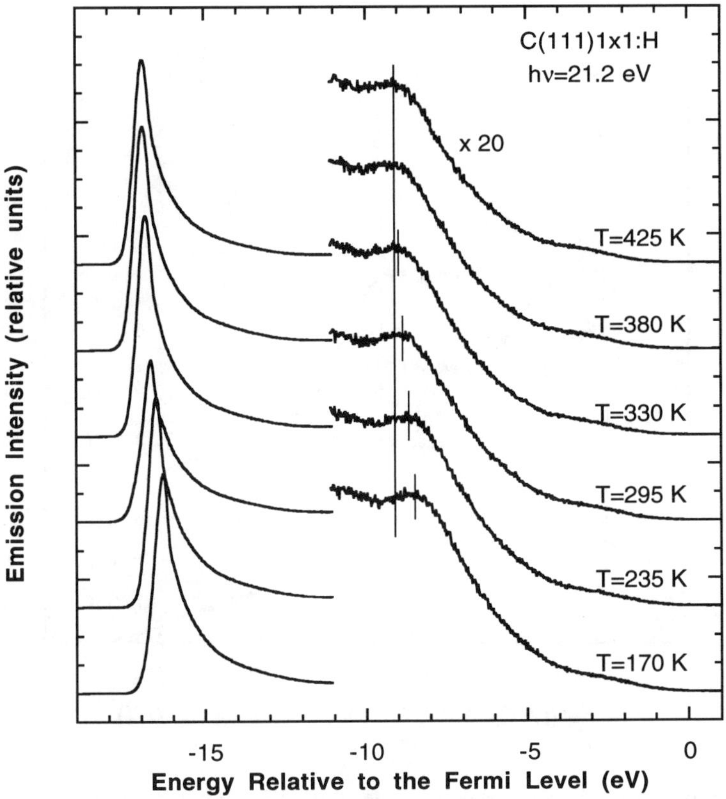

FIG. 2 Electron energy distributions photoemitted from the re-hydrogenated (111)-(1x1):H diamond surface, at a variety of temperatures (T). The apparent shift in energy of the emission is due to photovoltaic charging, which becomes more severe at lower temperatures.

RESULTS AND DISCUSSION

The "as-polished" surface is known to be hydrogen terminated and exhibit a 1x1 low-energy electron diffraction (LEED) pattern [2, 11] as well as negative electron affinity [12]. Upon annealing above 1000°C, the LEED changes to a 2x1 pattern ("reconstructed" surface). This reconstructed diamond surface is hydrogen free and has positive electron affinity [2]. While molecular hydrogen has no effect on the clean (111)-(2x1) diamond surface, room temperature exposure to atomic hydrogen restores the surface symmetry to the original 1x1 (in-situ "re-hydrogenated" (111)-(1x1):H diamond surface) [13]. Similar to the as-polished surface, no surface state emission is observed from the re-hydrogenated diamond surface.

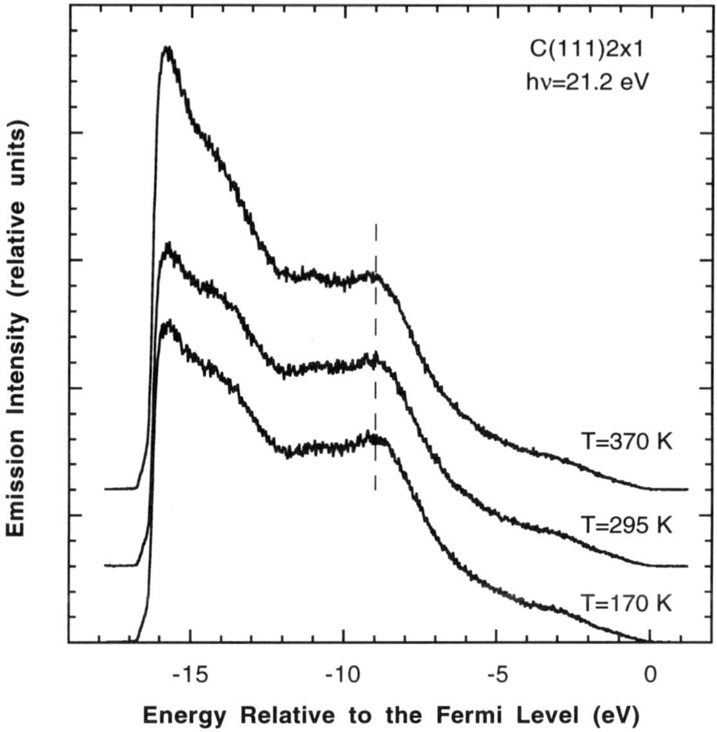

FIG. 3 Electron energy distributions photoemitted (hν=21.2 eV) from the reconstructed (111)-(2x1) diamond surface, at a variety of temperatures (T). Note that there is no energy shift of the distributions as a function of temperature. This suggests that the surface conductivity dominates and that the surface is effectively grounded.

PES from the in-situ "re-hydrogenated" (111)-(1x1):H diamond surface

The amount of band bending (i.e., band bending potential) depends upon the band alignment of the Fermi level, both at the surface and in the bulk. Of course, the alignment of the Fermi level in the bulk is an extrinsic property and varies with doping. Although the surface chemistry is fundamentally the same [2], the re-hydrogenated (111)-(1x1):H diamond surface has additional downward band bending in comparison to the "as-polished" (111)-(1x1):H surface [5]. The difference in band bending may be due to surface sites that remain unsaturated after atomic H exposure [2, 14]. At room temperature (T=295 K), PES measurements have determined an overall downward band bending (≈0.6 eV) of the re-hydrogenated (111)-(1x1):H type IIb diamond surface (see Fig. 2 and Ref. [5]).

In Fig. 2, in addition to the room temperature results, we show electron energy distribution curves (EDC) from the re-hydrogenated (111)-(1x1):H diamond surface for a variety of sample temperatures between 170 K and 425 K. Although independent of temperature for T≥350 K the electron energy distributions shift to higher energies as we lower the sample temperature below 350 K. Note in Fig. 2 that characteristic initial state features (valence band emission) and final state features (emission near the conduction band minimum) both shift together to higher final state energy as the temperature is decreased. This behavior is characteristic of band flattening due to photovoltaic charging during PES as discussed by Hecht [6-8]. As it can be seen (Fig. 2), photovoltaic charging effects are generally more important at low temperature; this is due to the fact that the restoring currents in the semiconductor become smaller as we lower the temperature. Furthermore, we see that the photovoltaic charging is significant even at room temperature (~0.2 eV). Consistent with being due to photovoltage, the temperature dependence of the *apparent* E_F-E_{VBM} of the re-hydrogenated (111)-(1x1):H diamond surface is completely reversible.

The PES measurements (Fig. 2) find a temperature independent E_F-E_{VBM} = 1.2 eV above T=350 K. We assume that this is the actual (no illumination) Fermi level pinning position, E_F-E_{VBM}, at the surface. Quantitative analysis of the *apparent* E_F-E_{VBM} for T<350 K, is in excellent agreement with calculations performed using the theory as developed by Hecht, assuming that the surface conductivity is zero (electrically isolated surface) [15].

PES from the reconstructed (111)-(2x1) diamond surface

In Fig. 3 we show 21.2 eV photoemission spectra from the hydrogen free, reconstructed (111)-(2x1) diamond surface at 170, 295, and 370 K. Consistent with previous room temperature observations [2], compared to the re-hydrogenated surface (Fig. 2), the low kinetic energy threshold of the EDCs moved to higher energies indicating a positive electron affinity surface, and emission due to surface states appeared just below the Fermi level (Fig. 3). The valence band maximum of the reconstructed surface is found to be 1.2 eV below the Fermi level. In contrast to the re-hydrogenated surface, we see no energy shift of the energy distributions as a function of temperature (Fig. 3). Although satisfactory understanding of the surface conductivity requires detailed characterization of the surface morphology, the absence of photovoltaic charging suggests that the surface conductivity dominates and that the illuminated surface is effectively grounded. A possible conduction mechanism could be based on the π-bonded chains at the reconstructed (111)-(2x1) diamond surface. In general agreement to our conclusions, study of the role of hydrogen in the surface conductivity of the (110) diamond surface, by Mackey *et al.* [16, 17], finds that exposure to hydrogen of the bare reconstructed (110) diamond surface significantly increases the sheet resistance of the surface.

CONCLUSION

We have shown that photovoltaic charging effects in photoemission studies of the in-situ "re-hydrogenated" (111)-(1x1):H diamond surface can be significant at temperatures as high as room temperature. In contrast, similar experiments on the reconstructed (111)-(2x1) diamond surface find no photovoltaic charging which suggests that the surface exhibits relatively high conductivity (effectively grounded surface). These results demonstrate that extra care should be taken in determining the Fermi level pinning position relative to the bands at diamond surfaces and interfaces. The assumption that the UPS photoelectron spectra reflect the equilibrium energy band arrangement should be experimentally confirmed in each case, especially when wide band gap materials are involved. Alternative approaches, such as study of the temperature dependence of the Fermi level pinning position, or analysis of the time dependence of the photovoltage must then be used to determine the true E_F-E_{VBM} [8].

ACKNOWLEDGMENTS

The authors acknowledge the research support of the National Science Foundation under grant number ECS-9222368.

REFERENCES

1. F. J. Himpsel, P.Heimann, and D. E. Eastman, Solid State Commun. **36**, 631 (1980).
2. B. B. Pate, Surf. Sci. **165**, 83 (1986).
3. T. Tachibana, J. T. Glass, and R. J. Nemanich, J. Appl. Phys. **73** (2), 835 (1993).
4. J. van der Weide and R. J. Nemanich, Phys. Rev. B **49** (19), 13629 (1994).
5. C. Bandis and B. B. Pate, Phys. Rev. B **52** (16), 12056 (1995).
6. M. H. Hecht, J. Vac. Sci. Technol. B **8** (4), 1018 (1990).
7. M. H. Hecht, Phys. Rev. B **41** (11), 7918 (1990).
8. M. H. Hecht, Phys. Rev. B **43** (14), 12102 (1991).
9. M. Alonso, R. Cimino, and K. Horn, Phys. Rev. Lett. **64** (16), 1947 (1990).
10. K. Horn, M. Alonso, and R. Cimino, Appl. Surf. Sci. **56-58**, 271 (1992).
11. B. J. Waclawski *et al.*, J. Vac. Sci. Technol. **21** (2), 368 (1982).
12. F. J. Himpsel *et al.*, Phys. Rev. B **20** (2), 624 (1979).
13. B. B. Pate *et al.*, J. Vac. Sci. Technol. **19** (3), 349 (1981).
14. B. D. Thoms *et al.*, J. Chem. Phys. **100** (11), 8425 (1994).
15. C. Bandis and B. B. Pate, Surf. Sci. Lett. **345**, L23 (1996).
16. B. L. Mackey *et al.*, in *Diamond Materials IV*, edited by K. V. Ravi and J. P. Dismukes, Electrochemical Society Proceedings, Vol. 95-4 (The Electrochemical Society, Pennington, 1995).
17. B. L. Mackey *et al.*, Phys. Rev. B **52** (24), R17009 (1995).

Investigation of the field emission current from polycrystalline diamond films

J.W. Glesener and A.A. Morrish
Naval Research Laboratory
4555 Overlook Ave.
Washington, DC 20375

Abstract:

The field emission current from boron doped polycrystalline diamond films was characterized as a function of voltage and temperature. The motivation for the current-temperature measurements was to assess the thermal stability of the diamond emitters and gain some insight into a possible emission mechanism.

Results from the current-temperature (I-T) measurements found that the field emission current appeared independent of temperature. The best characterization of the results implied a temperature independent electron tunneling mechanism if not electron emission from the valence band of diamond.

The existence of negative electron affinity (NEA) on the hydrogen terminated single crystal (111) and (100) [1,2] diamond surfaces has been a motivation for the practical exploitation of this phenomenum in polycrystalline diamond (PCD) films. NEA has been observed in PCD films [3], and electron emission at low values of the electric field has been measured [4]. The origin of the electrons observed in field emission from p-type diamond is currently a topic of active research. In order to investigate from what states electron emission was occurring measurements on the temperature variation of the field emission current were carried out on boron doped PCD diamond films

The diamond films characterized in this work were grown using a hot filament chemical vapor deposition system with a methane-hydrogen ratio of 1/120 at a process pressure of 15 torr. Diborane or nitrogen was used as the dopant source. The nominal filament temperature was kept at 2100C and the substrate temperature during growth was 900C. Tantalum was employed as the filament material. Prior to initiation of the growth cycle, several gas purges of the growth chamber were carried out in order to reduce any residual oxygen or nitrogen. Following completion of the growth procedure, the samples were allowed to cool to room temperature in the chamber before being exposed to the atmosphere. Sample A during growth was doped using a boron/carbon ratio of 200 ppm. Sample B during growth was doped with a boron/carbon ratio of 100 ppm. The resultant boron concentration for both films was $\approx 10^{18}/cm^3$.

The field emission measurements were performed under UHV conditions at a pressure of 2.6×10^{-6} Pa in a turbomolecular pumped chamber. The pressure was monitored using a residual gas analyzer. Low levels of oxygen were of particular concern because it has been shown in the literature that oxygen reacts with the diamond surface at a temperature of around 500C [5]. The measurement of the

Mat. Res. Soc. Symp. Proc. Vol. 423 © 1996 Materials Research Society

emission current was carried out using two distinct configurations. The setup shown in figure 1a was used to measure the field emission current as a function of temperature. This particular arrangement was used in order to minimize the influence of temperature effects on the Ta probe-diamond film distance and limit any leakage current. The temperature was measured using a chromel-alumel thermocouple placed in physical contact with the diamond surface adjacent to the emitting area. A Keithley 2001 voltmeter was used to measure the thermocouple voltage. The physical contact of the thermocouple on the diamond film had no effect on the measured emission current. A typical measurement cycle was initiated by heating a sample up to 350C. Upon removal of power from the substrate heater, the field emission current was sampled every 2 seconds. Using this method, the maximum rate of sample cooling at any point on the current-temperature curve was limited to less than 1 K/s. Limiting the sample temperature to below 350C was done to minimize the reactivity of the diamond surface to any residual oxygen in the chamber. When mounted in the vacuum chamber, PCD films were routinely cycled between room temperature and 350C. No degradation of the field emission current was seen as a consequence of this repeated temperature cycling.

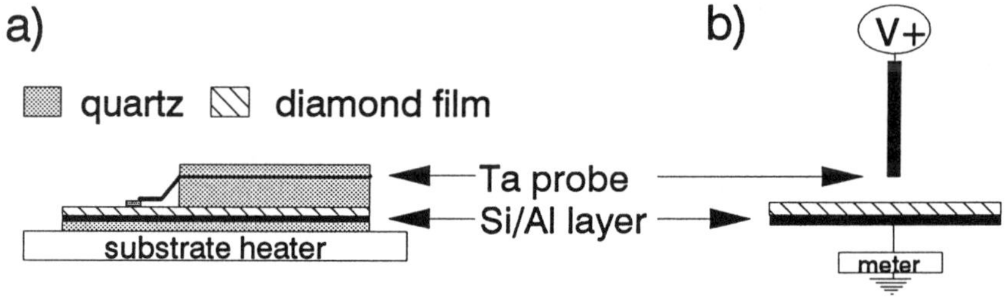

Figure 1. Schematic diagrams of the in vacuo field emission characterization setup.

All the diamond samples were grown on the same batch of p-type Si wafer material and ohmic contact to the Si wafer was made by using an Al foil backing. Repeated heating would bond the uncoated silicon wafer to the foil and diffuse Al into the surface. The Al foil was connected to an electrometer via a vacuum feedthrough. A Keithley 486 electrometer was used to measure the current and the data acquired by computer.

A semilog plot of the field emission current (I) versus inverse temperature (T^{-1}) is shown in figure 2 for the two boron doped samples. Electron emission from a semiconductor surface can occur from the conduction band, valence band, and/or surface states [6,7]. Electron emission from each type of electronic state has a particular theoretical temperature dependence. By examining the current-temperature results shown in figure 2 it was thought possible to isolate a particular emission mechanism.

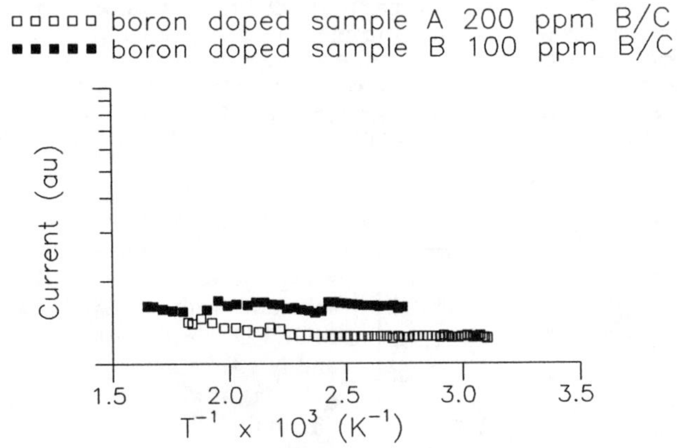

Figure 2. Field emission current versus T^{-1} for samples A and B. Every tenth point is shown.

Because surface states have been only found on the reconstructed (111) diamond surface [8,9] and the reconstructed surface has a positive electron affinity (0.5 eV) [10], electron emission from surface states was not considered. On theoretical grounds, field emission from the valence band is predicted to be independent of temperature [6,7], thus emission from the valence band was indicated.

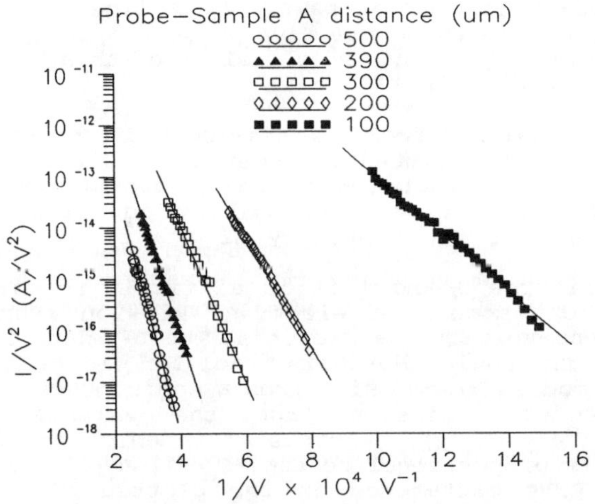

Figure 3. Fowler-Nordheim plot of the room temperature I-V characteristics of sample A. A linear best fit is shown for each of the five probe-sample distances by the straight line and the slope of this line is used in the plot in figure 5.

I-V measurements of the field emission current were carried out on sample A. The data obtained from the experimental setup diagrammed in figure 1b is shown in figure 3 and 4 for five different probe distances. Figure 3 is plotted in the Fowler-Nordheim format and in figure 4 the current-voltage data is graphed according to equation 1. The average increase in the slope of the I-V plots in figure 4 is 10% over the best fit of the same line in figure 3.

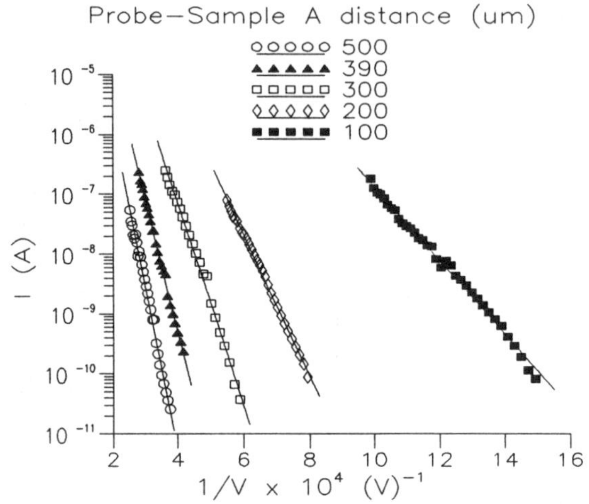

Figure 4. A current versus V^{-1} plot of the room temperature I-V characteristics of sample A. A linear best fit is shown for each of the five probe-sample distances by the straight line and the slope of this line is used in the plot in figure 5.

Figure 5 compares the slopes from a linear best fit of the Fowler-Nordheim data and the data fitted to equation 1. The purpose of this graph is to confirm that the slope is proportional to the probe substrate distance d as predicted by equation 1 and to examine the influence of the constancy of the field enhancement factor β.

As shown in figures 3 and 4 it is difficult to distinguish between two different models of electron emission from diamond based on just current-voltage measurements. The I-V data could just as easily be fit by the Fowler-Nordheim model for electron emission by a metal as by a model for emission from a semiconductor. The I-V behavior for electron emission from the valence band is proportional to $V^2\exp(-C_v/V)$, whereas for emission from the conduction band I is proportional to $\exp(-C_c/V)$, C_v,C_c are constant in both cases[6]. The difference in the plotted slope between either equation is only ≈10%, while the difference in the value for the tunneling barrier which is subsumed into C_v,C_c can be large. Practically, it makes little difference on which format is selected to plot the I-V characteristics as long as the correct barrier height is understood.

Current-temperature measurements on the field emission current from boron doped diamond films appear to indicate at least a temperature independent electron tunneling mechanism, if not electron emission from the valence band. The theoretical difficulty with electron emission from the valence band has been demonstrated by Huang et al.[11], who pointed out that in p-type diamond a field of 5V/nm is required in order to hope to have a measurable current.

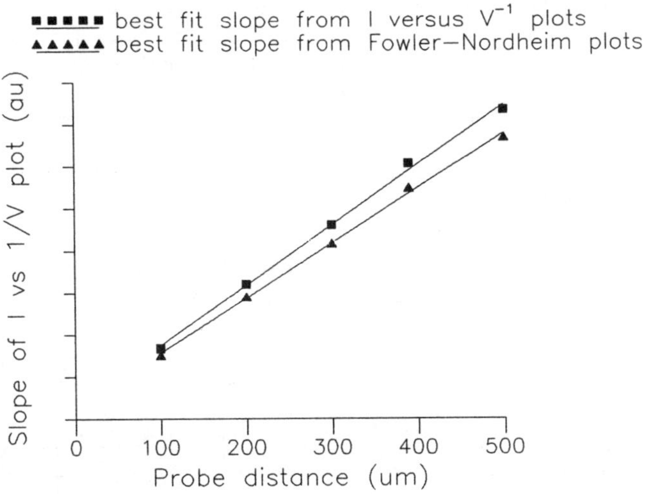

Figure 5. Slopes from the best fit lines from figures 3 and 4 plotted as a function of distance.

References

1. F.J. Himpsel, J.A. Knapp, J.A. VanVechten, and D.E. Eastman, Phys. Rev., B 20 (1979) 624.
2. J. van der Weide and R.J. Nemanich, J. Vac. Sci. Technol., B12 (1994) 2475.
3. N. Eimori, Y. Mori, A. Hatta, T. Ito, and A. Hiraki, Jpn. J. Appl. Phys. 33 (1994) 6315.
4. C. Wang, A. Garcia, D.C. Ingram, M. Lake, and M.E. Kordesch, Electron. Lett., 27 (1991) 1459.
5. H. Nakahata, T. Imai, and N. Fujimori, Proc. 2nd Int. Conf. on New Diamond Science and Technology, 1991 pp. 487-493.
6. A. Modinos, Field, Thermionic, and Secondary Electron Emission Spectroscopy, Plenum, New York, 1980, chap. 8.
7. R. Stratton, Phys. Rev. 125 (1962) 67.
8. B.B. Pate, P.M. Stefan, C.Binns, P.J. Jupiter, M.L. Shek, I. Lindau, and W.E. Spicer, J. Vac. Sci. Technol. 17 (1980) 1087.
9. G.D. Kubiak and K.W. Kolasinski, Phys. Rev. B 39 (1989) 1381.
10. C. Bandis and B.B. Pate, Phys. Rev. B 52 (1995) 12056.
11. Z.-H. Huang, P.H. Cutler, N.M. Miskovsky, and T.E. Sullivan, J. Vac. Sci. Technol. B13 (1995) 526.

FIELD EMISSION AND BAND BENDING CONSIDERATIONS FROM HIGH-QUALITY NEA DIAMOND

C. BANDIS[1], B. B. PATE[1], W. PHILLIPS[2], M. A. PLANO[2], M. D. MOYER[2], AND M. A. MORENO[2]
[1]Department of Physics, Washington State University, Pullman, WA 99164-2814
[2]Crystallume, 3506 Basset Street, Santa Clara, CA 94054

ABSTRACT

The near band gap photoelectric emission and field emission properties of diamond are investigated. Our results find three characteristic photoelectric yield spectra which have been observed from both polycrystalline CVD diamond films and single crystal diamond. The categories correspond to differences in bulk doping/surface preparation and illustrate the importance of band bending at the surface. Field emission experiments also find that the same three categories have distinct field emission properties. Our field emission observations are discussed in terms of electron transport properties from the bulk to the surface.

INTRODUCTION

Field emission from semiconductor surfaces can be due to electron tunneling from the conduction band, valence band, and/or surface states (for a review see Ref. [1]). However, because diamond is a wide band gap semiconductor (E_g=5.47 eV) with no known n-type shallow donor, and the occupied surface states of the (111)-(2x1) surface are centered ~1 eV below the valence band maximum (VBM) [2], emission from neither the conduction band nor the surface states can justify the observed emission properties. On the other hand, emission from the VBM would normally require an unexpectedly low electron affinity. In order to understand the origin and properties of the observed emission from both natural single crystal and CVD diamond films, a number of possible mechanisms have been recently proposed and studied. These possible mechanisms include emission from mid-gap defect sub-bands [3, 4], back contact injection into the conduction band [5], creation of electroformed conducting channels [6], and emission due to sharp asperities (created during dielectric breakdown) with high field enhancement factors [7]. Furthermore, recent study of the electron energy distributions from simultaneous field emission and photoemission finds that in the case of type IIb diamond the field emission originates from the valence band [8]. Nevertheless, independent of the origin of the field emission, one would expect that the transport properties of the electrons would play a significant role in the observed field emission I-V data.

Since photoelectric yield from negative electron affinity (NEA) surfaces is a useful technique to study bulk-to-surface transport and electron escape phenomena, we performed in addition to field emission experiments, total electron yield measurements on NEA diamond surfaces with different band bending conditions. Electron photoexcitation from the valence band to the conduction band for near band gap excitation photon energies is a process that produces an initial carrier distribution spread from the surface to deep into the bulk. For excitation photon energies less than 6.0 eV, the penetration depth exceeds 1 μm [9]. Thus, significant electron emission can only arise with transport of the photoexcited carries to the surface followed by escape into vacuum, and therefore study of the total electron yield spectra can yield significant information about the electron transport properties.

Mat. Res. Soc. Symp. Proc. Vol. 423 © 1996 Materials Research Society

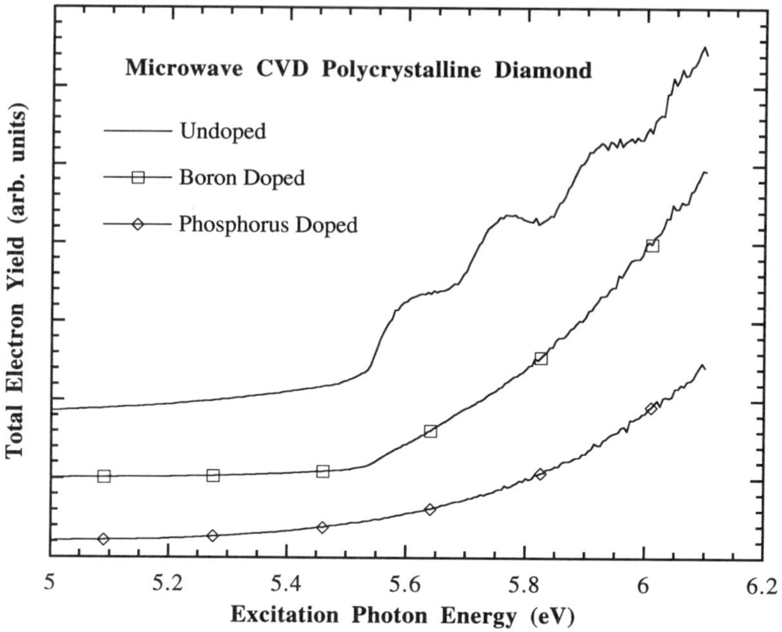

FIG. 1 Photoelectric yield from microwave CVD polycrystalline diamond for near band gap excitation photon energies.

EXPERIMENT

In addition to three polycrystalline diamond films, a type IIb (boron doped, p-type) and a type Ib (nitrogen doped) single crystal diamonds were used in this study. The films were grown using microwave plasma enhanced CVD on silicon substrates. Undoped, boron doped, and phosphorous-doped diamond films were investigated.

The experiments were performed in an ultra high vacuum system with the sample at room temperature. A Hg-Xe arc-lamp was employed as a source for near band-gap radiation. The photon energy was selected with the use of a 1 m normal incidence monochromator and refocused at the sample position into a 1mm x 2mm rectangular spot with an energy resolution of 50 meV or less. The excitation spectra are normalized with respect to the photo-yield of sodium salicylate [10]. The field emission I-V measurements were performed using a flat 1 mm in diameter tungsten anode, positioned at about 100 μm from the sample surface.

RESULTS AND DISCUSSION

In Fig. 1 we show photoelectric yield measurements of the three high quality microwave CVD diamond films as a function of the excitation photon energy (normalized to the incident light intensity). In Fig. 2 we show the photoelectric yield from single crystal diamond under the conditions of nearly flat bands, downwards band bending, and (we suspect) upwards band

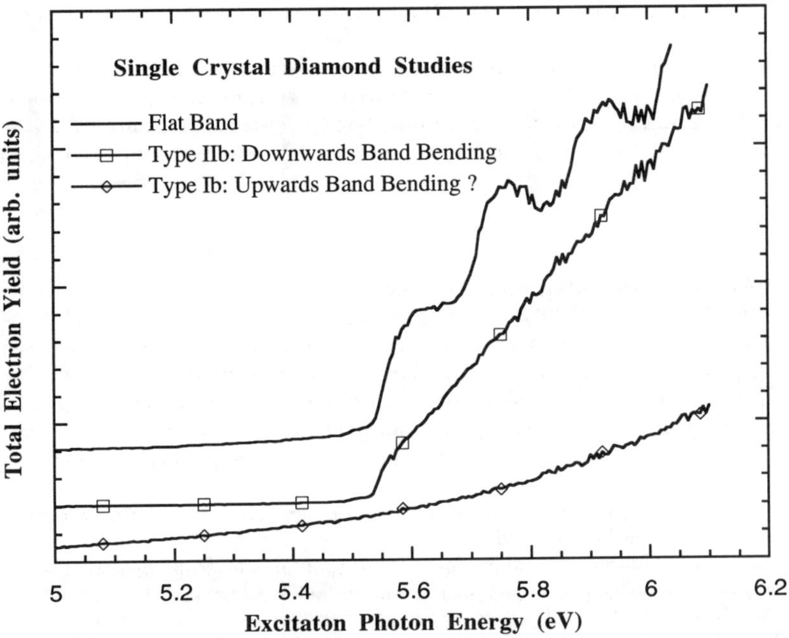

FIG. 2 Photoelectric yield from single crystal diamond for near band gap excitation photon energies.

bending. There is a strong correspondence between the form of the photoelectric yield from the three CVD polycrystalline diamond films and the three single crystal measurements. Single crystal studies [9, 11, 12], based upon the three-step model, have understood the photoelectric yield characteristics of single crystal type IIb diamond (Fig. 2) in terms of the observed changes in band bending (built-in electric fields) at the surface. Furthermore, empirical examination has found [13] that the photoelectric yield from hydrogenated type Ib diamond material (Fig. 2) varies as the excess energy above threshold to the fourth power (threshold energy =4.3 eV).

Near the surface, the band alignment of a semiconductor to the Fermi level is usually determined by the density of surface (and near-surface) states rather than bulk doping. Hydrogenated diamond surfaces are characterized by valence band maximum (VBM) to Fermi level energy separations of from 0.5 to 1.0 eV (see Table II in Ref. [9]). Doping determines the Fermi level alignment in the bulk. Therefore, band bending at the surface can depend upon both bulk doping and surface treatment. With identical surface treatments (all surfaces as grown via microwave CVD) the bulk Fermi level position will determine band bending. If the valence band maximum (VBM) lies closer to the Fermi level in the bulk (p-type doping) than at the surface, then there is downwards band bending. If the conduction band minimum (CBM) in the bulk lies closer to the Fermi level than it does at the surface, then there is upwards band bending.

Two of the films shown in Fig. 1 (undoped and boron doped) exhibit a sharp threshold at 5.53 eV which is similar to that observed (Fig. 2) from natural single crystal NEA diamond. The third CVD diamond film (phosphorus doped) bears strong similarity to the emission measured from hydrogenated type Ib, high pressure high temperature (HPHT) grown diamond (Sumitomo

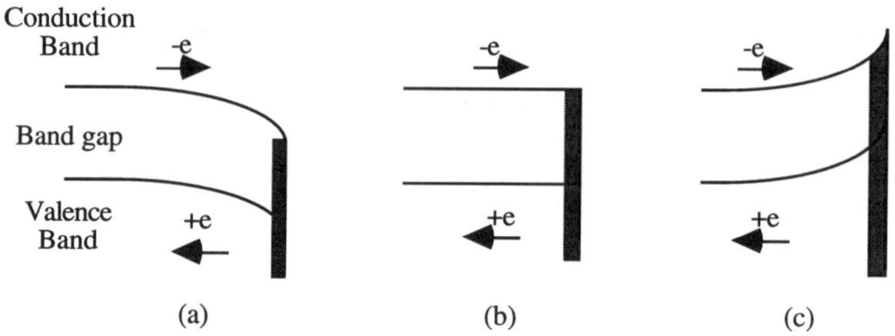

FIG. 3 Typical semiconductor surfaces with (a) downward band bending, (b) flat bands, and (c) upward band bending.

Sumicrystal™). Sumicrystal™ diamond contains substitutional nitrogen at concentrations of 10^{18} to 10^{19} cm^{-3}.

Significant boron doping in diamond drops the bulk Fermi 0.4 eV or less above the valence band maximum. This leads to downwards band bending at the surface (Fig. 3a). In this case, photoelectric emission results from both bulk photoexcitation of excitons and electrons. The form of the emission reflects [9] the presence of conduction band electrons which are being driven towards the surface by the built-in field in the band bending region.

Truly intrinsic (undoped) material is characterized by a mid-gap Fermi level (in the bulk) and by extremely long characteristic band bending lengths. This would lead to nearly flat bands close to the surface (Fig. 3b). In fact, the observation of the oscillations in "as-polished" (111) single crystal diamond is associated with nearly flat bands at the surface and yield dominated by exciton transport [9].

Substitutional phosphorus or nitrogen, should lead to n-type bulk material (although the levels are deep). Diamond material with an n-type bulk will produce upward band bending (Fig. 3c). One would expect that strong upward band bending to the surface would lead to drastically reduced photoelectric emission, even with a true NEA surface. We have found that the yield of the polycrystalline diamond film varies as the excess energy above threshold to the fourth power (threshold energy = 4.8 eV). As mentioned above, this is the same dependence on photon energy (except for a different threshold value) as found for Sumitomo HPHT diamond. The fundamental mechanism responsible for this form of the photoelectric yield has not been determined.

In Fig. 4 we show data from field emission I-V experiments performed on the same samples whose total electron yield is discussed above. The distance between the anode and the emitting surfaces was always ~100 μm, except for the HPHT, type Ib single crystal for which the anode was positioned ~25 μm from the emitting surface. As can be seen the most efficient field emitter is the p-type sample with downward band bending. The two samples, which based on their photoyield properties are thought to exhibit nearly flat bands, are the second best with similar field emission strength. The lowest field emission current density was observed by the phosphorus doped diamond film and the nitrogen doped HPHT single crystal diamond which are thought to exhibit upward band bending.

Based on both the photoemission and field emission data discussed above it becomes apparent that we can categorize our observations in three different groups based on the band bending near the emission surface. The most efficient emitters (via both photoemission and field emission) found to be the surfaces that exhibit downward band bending. This can be understood if we assume that the surface conductivity is low (typical of hydrogenated surfaces [14, 15]), and that electron transport from the bulk to the surface is necessary for the observed emission. In such a case, as it can be seen in Fig. 3, electrons experience a force towards the surface in the presence of downward band bending, and a force away from the surface in the presence of upward band

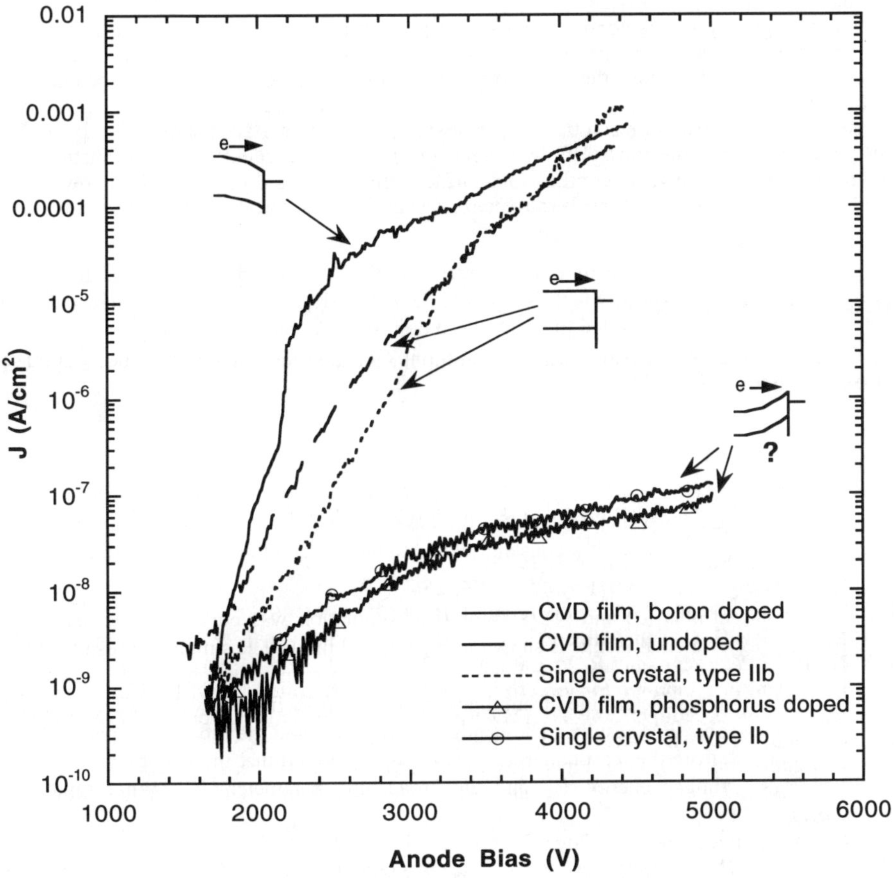

FIG. 4 Field emission as a function of field from CVD diamond thin films and single crystal diamonds.

bending. Therefore, both photoelectrons emitted from the conduction band and field emitted electrons from the valence and/or the conduction band would have to overcome the same kind of barrier in order to reach the surface. Electron transport mechanisms over barriers due to band bending have been studied in the case of metal-semiconductor interfaces [16]. Based on such an analysis, in the case of upward band bending the current density versus field dependence is expected to be temperature dependent. Further experiments such as temperature depended field emission are needed to verify the validity of our arguments.

CONCLUSION

We examined both the photoelectron emission and field emission properties of microwave CVD diamond films and single crystal diamonds. Our results find three characteristic photoelectric yield spectra which have been observed from both polycrystalline CVD diamond films and single crystal diamonds. The categories correspond to differences in bulk doping and illustrate the importance of band bending at the surface. In addition, field emission experiments find that the same three categories have distinct field emission properties. This can be understood by taking into account that electrons from the valence and/or conduction band have to overcome the same kind of barrier to reach the surface, assuming that the surface conductivity is relatively low. In this study, the most efficient electron emitters are found to be the surfaces with downward band bending. In contrast to a surface with downward band bending, upward band bending results in a force that repels the electrons away from the surface and therefore much lower current densities.

ACKNOWLEDGMENTS

Work supported by the Rome Lab at Hanscom AFB and the National Science Foundation under grant number ECS-9222368.

REFERENCES

1. A. Modinos, *Field, Thermionic, and Secondary Electron Emission Spectroscopy* (Plenum, New York, 1984).
2. B. B. Pate, Surf. Sci. **165**, 83 (1986).
3. Z. H. Huang *et al.*, Appl. Phys. Lett. **65**, 2562 (1994).
4. Z. H. Huang *et al.*, J. Vac. Sci. Technol. B **13** (2), 526 (1995).
5. M. W. Geis, J. C. Twichell, and T. M. Lyszezrz, J. Vac. Sci. Technol. B, in press (1996).
6. N. S. Xu, Y. Tseng, and R. V. Latham, J. Phys. D **27**, 1988 (1994).
7. L. S. Pan, in *Diamond for Electronic Applications*, edited by D. Dreifus *et al.*, MRS Symposia Proceedings, Vol. 416 (Materials Research Society, Pittsburgh, 1996), in press.
8. C. Bandis and B. B. Pate, submitted for publication.
9. C. Bandis and B. B. Pate, Phys. Rev. B **52** (16), 12056 (1995).
10. J. Samson, *Techniques of Vacuum Ultraviolet Spectroscopy* (John Wiley, New York, 1967), p. 214.
11. C. Bandis and B. B. Pate, Phys. Rev. Lett. **74** (5), 777 (1995).
12. C. Bandis, D. Haggerty, and B. B. Pate, in *Diamond, SiC and Nitride Wide Bandgap Semiconductors*, edited by C. H. Carter Jr., G. Gildenblat, S. Nakamura, and R. J. Nemanich, MRS Symposia Proceedings, Vol. 339 (Materials Research Society, Pittsburgh, 1994), p. 75.
13. C. Bandis and B. B. Pate, unpublished results.
14. B. L. Mackey *et al.*, Phys. Rev. B **52** (24), R17009 (1995).
15. C. Bandis and B. B. Pate, Surf. Sci. Lett. **345**, L23 (1996).
16. E. H. Rhoderick and R. H. Williams, *Metal-Semiconductor Contacts* (Clarendon Press, Oxford, 1988).

ELECTRON FIELD EMISSION FROM DIAMOND-LIKE CARBON

J Robertson*, S R P Silva**, G A J Amaratunga***, and W I Milne*
* Engineering Dept, Cambridge University, Cambridge CB2 1PZ, UK,
** Electronic Engineering Dept, Surrey University, Guildford, GU2 5XH, UK
*** Electronic Engineering Dept, Liverpool University, Liverpool L69 3BX, UK

ABSTRACT

The low electron affinity of a-C:H is related to that of diamond surfaces and is studied using a chemical bonding model. The electron field emission from hydrogenated amorphous carbon (a-C:H) and nitrogen modified a-C:H showing low turn on fields are described. Nitrogen improves the field emission, apparently by raising the Fermi level.

INTRODUCTION

Flat panel displays with a picture quality comparable to a cathode ray tube can, in principle, be obtained from field emission displays (FEDs). An FED uses the field emission of electrons from a matrix addressed array of cathodes to excite phosphor pixels. The first design of FEDs used of 'Spindt' tips of Mo or Si to provide the high electric fields needed for field emission [1]. Tips are needed because the low emissivity of these high affinity materials. An alternative is to use smooth thin film microcathodes of low affinity materials, such as diamond or diamond-like carbon (DLC) [2]. Diamond is a candidate for field emission cathodes because of its negative electron affinity and chemical inertness. Its emission properties have been widely studied [3-8]. DLC is a semiconducting form of amorphous or hydrogenated amorphous carbon (a-C:H) containing a significant fraction of sp^3 bonding [9]. DLC is a more attractive candidate than diamond because it can be deposited by PECVD at room temperature and so it is compatible with glass substrates. It is therefore of interest to consider the electron emission properties of DLC [10].

Another problem for diamond based emitters is how to provide electrons to its conduction band. There are no active, shallow n-type dopants for diamond; for example nitrogen forms deep levels about 1.4 eV below the conduction band edge. Thus, defective diamond has better emission [7], presumably due to grain boundary conduction. In contrast, a-C:H has a second advantage that it can be doped n-type because N now forms shallow levels in the narrower gap of a-C:H.

THEORY

The hydrogenated diamond surfaces along with other wide gap semiconductors like cubic boron nitride have a negative electron affinity (NEA)[3-5,11]. The most familiar cases of NEA are Cs coated surfaces of p-type semiconductors [12]. These could be called effective NEA surfaces, in which a semiconductor of positive electron affinity becomes NEA because of a surface dipole and charge transfer from the Cs layer. In contrast diamond and c-BN are intrinsic NEA surfaces, in which the NEA property arises from the bonding of the bulk semiconductor itself.

The origin of the NEA can be understood by considering bonding, as in Fig. 1.

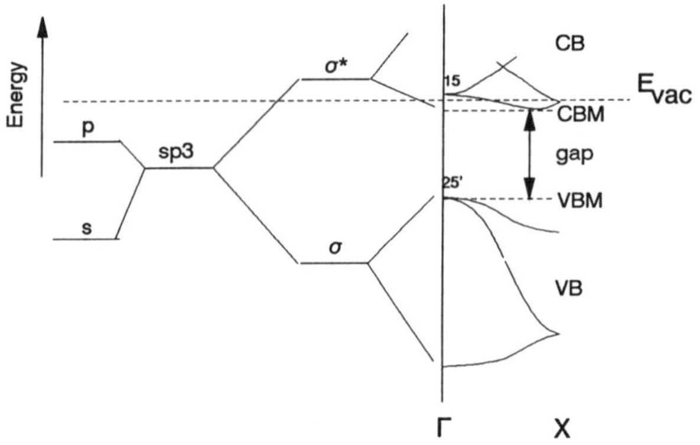

Fig. 1. Schematic diagram of the development of atomic orbitals, bonds and bands in a covalent solid, showing band edge energies with respect to the vacuum level.

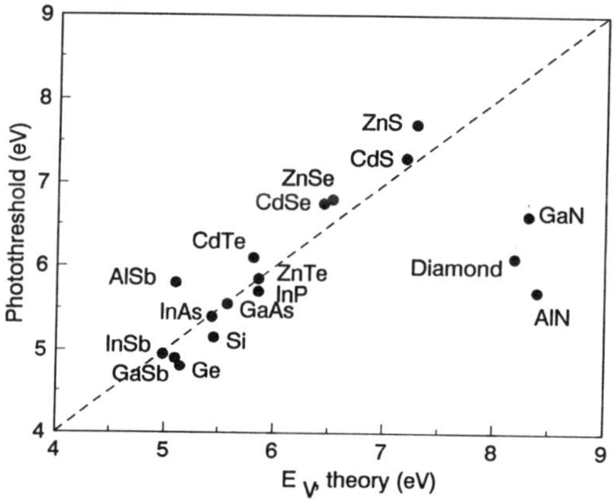

Fig. 2. Comparison of the calculated valence band maximum energy and the measured photoelectric threshold for diamond and zincblende semicnoductors.

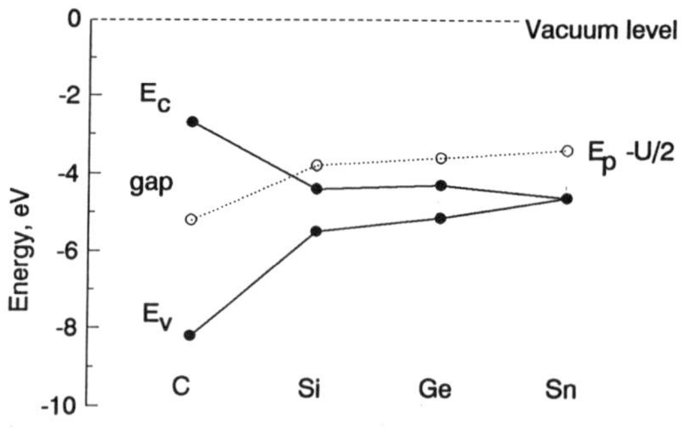

Fig. 3 Variation of band energies for group IV elements.

Here, the 2s and 2p valence orbitals of carbon hybridise to produce the sp^3 hybrids on each atom which interact to form the bonding (σ) and antibonding (σ^*) states. The σ state broadens into the valence band while the σ^* state broadens into the conduction band of the solid. Now, the valence states must lie below the vacuum level E_{vac}. However, there is no such requirement on the σ^* or conduction states. For example, the σ^* states of methane lie above E_{vac}. The conducton band minimum (CBM) E_c can lie below or above E_{vac}.

A simple analysis of the EA can be given in terms of atomic orbitals using tight-binding theory [13,14]. We express E_c as lying at the band gap energy E_g above the valence band maximum (CBM) E_v. This valence state is a bonding p state of Γ_{25}' symmetry in all diamond/zincblende semiconductor crystals. Its energy is given by

$$E(\Gamma_{25}) = V_0 + (V_2^2 + V_3^2)^{1/2}$$

where $V_0 = \frac{1}{2}(E(p,c)+E(p,a))$ the average of the p orbital energy on the cation and anion sites for a zincblende semiconductor like c-BN, $V_3 = \frac{1}{2}(E(p,c)-E(p,a))$ is the ionic energy. V_2 is the covalent energy, which is given empirically by $V_2 = V_2^0/d^2$ where d is the bond length and $V_2^0 = 18.6$ eV/Å^2 [14]. The orbital energies E(p) etc are the orbital energies in the solid which are given in terms of the ionisation potentials of the free atom and the intra-atom correlation energy U by [13-15]

$$E(p) = E - \frac{1}{2}U$$

This treatment has been found to give a reasonably successful accurate value of E_v below E_{vac} which experimentally is the photoelectric threshold (Fig. 2)[13,14]. However, the model is still incomplete as the predictions are less successful for first row systems like diamond.

We now apply this model to the case of diamond and amorphous C in particular. The low EA of diamond is actaully rather surprising, if one considers the work function of metals varies linearly with the electronegativity [16], and that C is the most electronegative of the group IV elements. To understand this further, we plot in Fig. 3 the values of E_v and E_c for the group IV elements. We see that E_v follows the p orbital energy E_p which does decline from Sn to C. In contrast, E_c rises sharply from Si to diamond because of diamond's very wide band gap. Thus, diamond has the lowest electron affinity of the group IV semiconductors, despite having the largest electronegativity, because it has the widest gap.

A second question concerns the 1 to 2 eV decrease in electron affinity of a diamond surface from a positive value for clean, unhydrogenated surface to a negative value for hydrogenated surfaces [5]. This decrease is likely to arise from the addition of hydrogen to the surface rather than the reconstruction of the unhydrgenated surface. The change in affinity has been attributed to the charge rearrangement within a C-H bond which creates a small dipole moment which opposes the potential step at the surface [13]. Thus, changes in affinity of diamond surfaces are related directly to changes in the local bonding.

EXPERIMENTAL

We now describe measurements of electron emission from a-C:H and N-doped a-C:H films. The a-C:H and a-C:H:N films were deposited from CH_4/He/N_2 gas

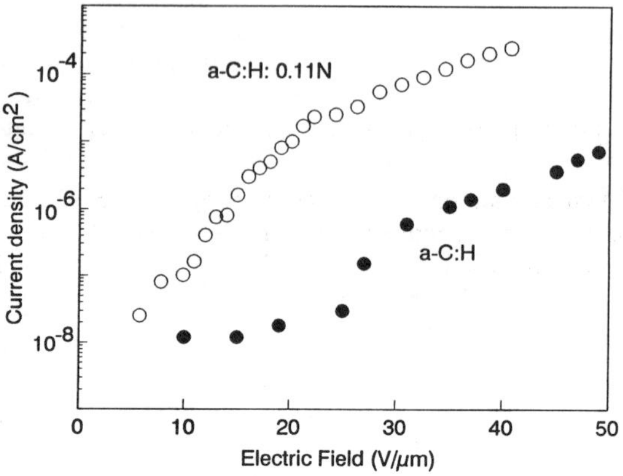

Fig. 4. Electron emission characteristics of an a-C:H film and an a-C:H:N film containing 11% of nitrogen.

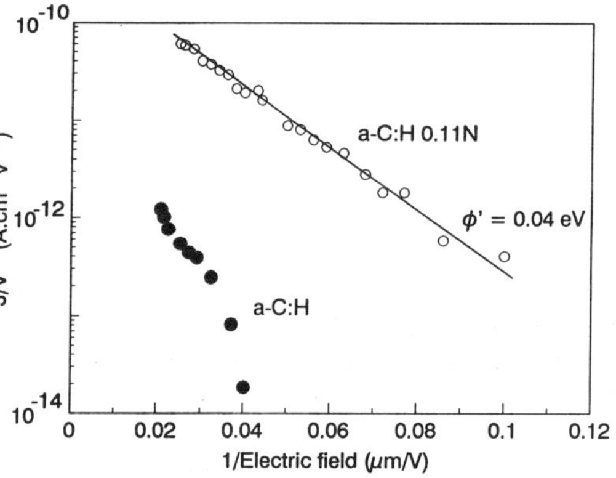

Fig. 5. Electron emission of Fig. 1 plotted according to the Fowler-Nordheim relation, eqn (1).

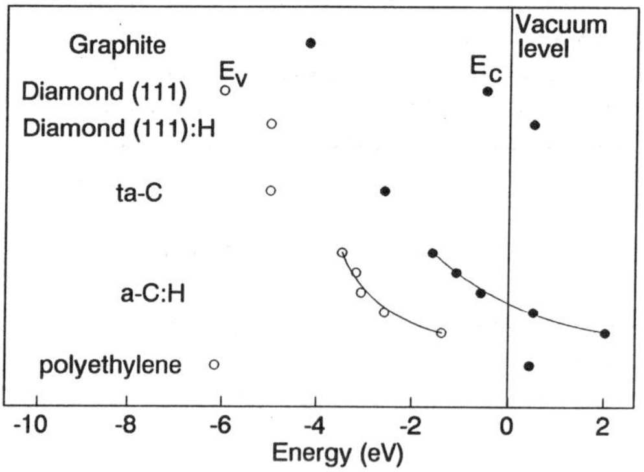

Fig 6 Estimated experimental band edge energies with respect to the vacuum level, for diamond [4], ta-C and a-C:H [20].

mixtures using a capacitively coupled RF PECVD system with magnetic confinement. The films were deposited onto highly doped n-Si substrates at room temperature and had a thickness of 0.3 μm. The characterisation of the structure, bonding and electrical properties is described elsewhere [17,18]. The optical gap is 1.7 eV for the a-C:H and 2.0 eV for the a-C:H:N film discussed.

Field emission measurements were carried out using the a-C:H:N films as a cathode and an ITO coated glass slide as an anode [19]. The anode was separated from the film by 50 μm glass fiber spacers and the emission are was set at 1 cm^2. The field emission measurements were carried out at a pressure of $5x10^{-7}$ Torr. The current density is calculated assuming emission form the total area.

Fig. 4 shows the current density versus electric field for an a-C:H and an a-C:H:N doped film containing 11 at% N. The emission from the a-C:H film is seen to increase rapidly at about 26 V/μm. In the film a-C:H:N film, this onset is lowered to 12 V/μm and the current density is seen to be two orders of magnitude higher.

The current densities have been fitted to the Fowler-Nordheim relationship expected for field emission

$$J=bE^2\exp(-a\phi^{3/2}\frac{\beta}{E}) \qquad (1)$$

where J is the current density in A/m^2, E is the electric field in V/m, ϕ is the emission barrier height in eV, a and b are constants with $a=6.8.10^9$ in these units. β is the field enhancement factor for rough surfaces. Fig. 5 shows the current densities plotted according to eqn (1). The data for the a-C:H film gives a poor fit to eqn (1) while the a-C:H:N film gives a much better fit. The data for a-C:H and a-C:H:N have a slope of corresponding to an effective barrier height ϕ' of 0.05 eV and 0.04 eV, respectively, with $\beta=1$. These are very low effective barrier heights and are consistent with the rather good emission of a-C:H:N.

DISCUSSION

Fig. 6 summarises the experimentally measured band edge energies of various C:H phases with respect to the vacuum level [13,5,20]. As the electron affinity of a semiconductor is determined by its local chemical bonding as argued from the tight-binding model, it does not depend on the crystalline order. We therefore expect amorphous C:H phases should have similar band energies with respect to the vacuum level. This is confirmed by the data in Fig. 4. Here we plot E_v values for a-C:H films derived from experimental photoemission partial yield data [20], with E_c derived by adding the optical gap to E_v. E_c and E_v of a-C:H is seen to lie midway between the positions in diamond. E_c and E_v then gradually rise in energy towards the vacuum level, as the gap E_c-E_v increases towards the bottom of the diagram. Indeed E_c of the wider gap,'polymeric' type of a-C:H has risen above the vacuum level, corresponding to a negative electron affinity. This therefore accounts for the general observation of easy electron emission from a-C:H in Figs 4 and 5.

Electron emission occurs from the Fermi level E_f, not E_c. Emission is therefore easier if E_f is raised by n-type doping, assuming no band bending at the surface. This is basically the cause of the higher electron emission from a-C:H:N films in Fig. 4. There are, however, some questions in the details. Nitrogen doping of amorphous C has been studied by various groups, with conflicting conclusions. N also tends to induce sp^2 bonding, so an increase in conductivity of a-C:H:N could arise instead from a closing of the gap due to graphitisation, rather than true doping [21]. True doping by N has been found in ta-C, a highly sp^3 bonded unhydrogenated form of a-C [22,23]. There, N addition was observed to cause the conductivity to first increase and then decrease as E_f passed from near E_v to near E_c, while the optical gap remained constant. A similar but much weaker doping response is found in the present a-C:H system [18], prepared under magnetically confined PECVD conditions. We therefore believe that the increased electron emission in Fig. 4 does arise from the raising of E_f towards E_c with N addition, by at least 0.3 eV from the change in the activation energy of conduction.

The change in effective barrier height, from the slope of the Fowler-Nordheim plots, was of order 0.01 eV, much less than the change in E_f. This suggests that the electric field in eqn (1) must be corrected by a finite value of the field-enhancement factor β of at least 10. The factor β is expected for traditional cases of field emission, where pointed shapes are used [1]. It was not expected for the case of a-C:H, which forms very smooth films, often with surface roughness under 10 nm, as these can also be used as low wear tribological coatings.

REFERENCES

1. I Brodie, Proc IEEE **82** 1006 (1994)
2. N Kumar, H K Schmidt, C Xie, Solid State Technol **38** p71 (May 1995)
3. F J Himpsel, J A Knapp, J A vanVechten, D E Eastman, Phys Rev B **20** 624 (1979)
4. J van der Weide, Z Zhang, P K Baumann, M G Wensell, R J Nemanich, Phys Rev B **50** 5803 (1994)
5. C Bandis, B B Pate, Phys Rev B **52** 12056 (1995)
6. N Xu, R V Latham, Y Tzeng, Electronics Lett **29** 1596 (1993)
7. W Zhu, G P Kochanski, S Jin, L Seibles, J Appl Phys **78** 2707 (1995)
8. M W Geis, J C Twichell, J Maculey, K Okano, Appl Phys Letts **68** 1328 (1995)
9. J Robertson, Pure Appl Chem **66** 1789 (1994)
10. F Zeng, I G Brown, J W Ager, J Mater Res **10** 1585 (1995)
11. M C Benjamin, C Wang, R F Davis, R J Nemanich, App Phys Lett **64** 3288 (1994)
12. F Bell, 'Negative Electron Affinity Devices' (Oxford University Press, 1973)
13. J Robertson, Diamond Related Mats **5** xxx (1996)
14. W A Harrison, Phys Rev B **31** 2121 (1985)
15. I Lefebvre, M Lannoo, G Allan, Phys Rev B **39** 13518 (1989)
16. M Cardona, L Ley, 'Photoemission in Solids I' (Springer, Berlin, 1978) Ch 1.
17. S R P Silva, et al, submitted to Philos Mag B
18. S R P Silva, et al, submitted to J Appl Phys
19. G A J Amaratunga, S R P Silva, Appl Phys Letts **68** 2529 (1996)
20. J Schafer, J Ristein, L Ley, J Non-Cryst Solids **164** 1123 (1993)
21. A Hembold, P Hammer, J U Thiele, K Rohwer, Philos Mag B **72** 335 (1995)
22. V S Veerasamy, J Yuan, G A J Amaratunga, W I Milne, K Gilkes, M Weiler, L M Brown, Phys Rev B **48** 17954 (1993)
23. J Robertson, C A Davis, Diamond Related Materials **4** 441 (1995)

AUTHOR INDEX

Abernathy, C.R., 163, 569
Aboelfotoh, M.O., 519
Agarwal, A.K., 87
Ager, III, J.W., 487
Aggarwal, R., 221
Akasaki, I., 487, 675
Albrecht, T., 693
Alexander, W. Brock, 305
Amano, H., 487, 675
Amaratunga, G.A.J., 299, 777
Ambacher, O., 323
Anderson, G.A., 245
Andrianov, A.V., 311
Angerer, H., 323
Augustine, G., 275, 507, 539
Aumer, M., 341

Baba, K., 397
Badzian, Andrzej, 433
Badzian, Teresa, 433
Balakirev, V.G., 723
Balakrishna, V., 539
Bandic, Z.Z., 287
Bandis, C., 759, 771
Baumann, P.K., 33,143
Bechstedt, F., 711
Bedair, S.M., 157, 341
Benjamin, Mark C., 563
Bergman, J.P., 675
Bernholc, J., 465
Berriche, R., 699
Bhat, I., 601
Bhatti, Q.A., 439
Bigelow, Kim, 649
Bilbro, G.L., 33
Bishop, S.G., 347
Black, David, 305
Bockowski, M., 329
Boguslawski, P., 465
Bolay, H., 531
Boothroyd, C.B., 311
Born, E., 323
Borst, T.H., 63
Botchkarev, A.E., 201
Bour, D.P., 245, 595
Boutros, K.S., 341
Bozack, M.J., 119
Bozeman, S.P., 143
Brandt, C.D., 87
Brennan, K.F., 45
Briggs, E.L., 465
Broekaert, T.P.E., 335
Brown. P.D., 311
Brown, E.R., 221
Buchwald, W.R., 93
Burk, Jr., A.A., 87, 275
Burkard, M., 531
Burke, T., 93
Butler, James E., 305, 575
Buyanova, I.A., 675
Byun, Dongjin, 451
Bykhovski, A.D., 75

Cappelli, M.A., 359
Casady, J.B., 105, 111
Cederberg, J.G., 601
Chacon, J., 649
Chang, I.M., 377
Chao, C.H., 377
Charlson, E.J., 293, 415, 589, 655
Charlson, E.M., 293, 415, 589
Chen, B., 465
Chen, C.K., 377
Chen, C.W., 299
Chen, J.C., 735, 747
Chen, K.H., 377
Chen, L.C., 377
Chen, Naiqun, 175
Chen, Y., 487
Chen, Y.F., 377
Chen, Zhizhong, 753
Chen, Q.C., 75
Cheng, T.S., 311
Chhowalla, M., 299
Choi, In-Hoon, 451
Chow, T.P., 9
Chuang, T.J., 377
Clarke, G., 699
Clarke, Roy, 265
Clough, F.J., 39
Coleman, J.C., 207
Collins, D.M., 175
Colter, P., 409
Cremins-Costa, A., 409
Crofton, J., 119

Davis, R.F., 281, 457
Davis, Robert F., 563
Delucca, J.M., 137
DenBaars, S.P., 23
Dillard, W.C., 105
Dimitrov, R., 323
Dmitriev, A.V., 69
Donovan, S.M., 163
Dörnen, A., 531
Doyle, J.P., 519, 625
Dudley, M., 545
Dunn, M., 57
Duxstad, K.J., 201

Ebert, W., 63
Ebihara, K., 397
Eckstein, R., 215
Eizenberg, M., 125
El-Masry, N.A., 341
Elsner, J., 421

Estreicher, Stefan K., 613
Evwaraye, A.O., 507, 637, 705

Fan, Z.-F., 201
Fazi, C., 545
Fizzotti, F., 81, 687
Flemish, J.R., 93
Fong, C.Y., 631
Forchel, A., 169
Foxon, C.T., 311
Frauenheim, Th., 421, 427
Freitas, Jr., Jaime A., 575, 607, 643
Frischholz, M., 51

Gagnon, E., 409
Galetto, M., 81
Gassmann, A., 329
Gavrilenko, V.I., 711
Ghezzo, M., 9
Gilbert, Donald R., 371
Giocondi, Jennifer, 539
Glass, J.T., 281, 457
Glesener, J.W., 765
Gluche, P., 63
Gluschenkov, O.V., 347, 385
Golestanian, Hassan, 589, 655
Götz, W., 221, 595
Gradinaru, G., 99, 111
Graeff, C.F.O., 495
Grant, P., 699
Grasserbauer, M., 693
Grebner, S., 687
Greve, D.W., 607, 643
Grießer, M., 693
Grimberg, I., 125
Grzegory, I., 329
Gubanov, V.A., 631
Gutierrez, R., 427
Güttler, H., 495

Haase, D., 531
Haller, E.E., 201
Hara, Shiro, 149
Härle, V., 531
Harris, C.I,, 519, 667
Harris, G.L., 175, 207
Harris, Jr., James S., 233, 681
Harris, Jr., S.J., 359
Hatfield, C.W., 33
Hatta, Akimitsu, 717
Haugk, M., 421
Hayashi, Kazushi, 353
He, Y.W., 341
Heera, V., 729
Helava, H.I., 329
Helbig, R., 51
Hemmingsson, C., 519
Hengehold, R., 57
Hiraki, Akio, 717
Ho, I.H., 335
Hobgood, H.M., 275, 507, 539

Hofmann, D., 215
Hopkins, R.H., 87, 275, 507, 539
Horton, M.N., 227, 239, 287
Hsu, L., 513
Hu, R., 99, 111
Huang, J.W., 601
Huang, W., 545
Huang, Y.F., 377
Huang, Z.C., 735, 747
Humphreys, C.J., 311

Imler, W., 487
Isaacs-Smith, T., 99, 111, 119
Ito, Toshimichi, 717

Janzén, E., 519
Jenny, J.R., 507
Johnson, N.M., 221, 595
Johnson, R.W., 105, 111
Joshkin, V., 341
Jun, J., 329

Kajimura, Koji, 149, 353
Kaminski, V.V., 75
Kamiya, Nobuo, 583
Kao, Y.C., 335
Kapolnek, D., 23
Keeble, D.J., 501
Keller, B.P., 23
Keller, S., 23
Kern, Richard S., 563
Khan, M.A., 75
Khomich, A.V., 723
Kieseling, F., 169
Kim, Chinkyo, 347, 385, 557
Kim, Gyeungho, 451
Kim, Kyekyoon, 131, 347, 385, 557
Kim, S., 347
Kim, W., 201
King, Sean W., 563
Kisielowski, Ch., 329, 487
Klein, B.M., 631
Kleinsorge, B., 39, 299
Koeth, J., 169
Kohn, E., 63
Kolnik, J., 45
Koszi, L., 741
Krueger, J., 329
Krusor, B.S., 245
Kuang, Yalei, 433
Kuech, T.F., 227, 239, 287, 445, 601
Kulkarni, A., 501
Kull, A.E., 359
Kum, Dong-Wha, 451
Kuo, C., 487

Labbé, H.J., 699
Lambers, E.S., 569
Lanig, P., 253
Lee, Dong-Gu, 365

Lee, Heon, 233, 359, 681
Lee, J.W., 131
Lee, J.W., 163
Lee, Naesung, 433
Lentz, D., 323
Leonard, R.T., 157
Leung, M.S.H., 329
Levit, M., 125
Li, Dongmei, 403
Li, Hongdong, 403
Li, N.Y., 317
Li, W., 675
Li, Y., 741
Liau, Z.L., 221
Liliental-Weber, Zuzanna, 201, 329, 487
Lim, Dongsup, 451
Lin, H.Y., 377
Lindstrom, J.L., 519
Linnarsson, M.K., 519, 625
Liu, H.Y., 335
Liu, W., 281, 457
Lo Giudice, A., 81
Lockwood, D.J., 699
Lothian, J.R., 163
Loyalka, S.K., 649
Lu, H., 601
Lu, Y., 741
Luckowski, E.D., 111, 119

MacKenzie, J.D., 163, 569
Madangarli, V., 99, 111
Makarov, Y., 215
Maki, P., 221, 245
Manfredotti, C., 81, 687
Margherita, D., 81
Maric, Djordje M., 613
Mason, B., 699
Matthai, C.C., 439
Matyi, R.J., 239, 287
Matz, W., 195
McGill, T.C., 287
McIntosh, F.G., 341
McMullin, P.G., 87
Melngailis, I., 221
Metzger, T., 323
Milne, W.I., 39, 299, 777
Mirzakuchaki, S., 415, 589, 655
Mishra, U.K., 23
Mitchel, W.C., 507, 637, 705
Moßner, C., 699
Mohammad, S.N., 201
Mohney, S.E., 137
Molnar, B., 183
Molnar, R.J., 221, 245
Monemar, B., 675
Moran, G.J., 439
Moreno, M.A., 771
Mori, Y., 391
Morkoç, H., 201, 551
Morris, A.S., 33
Morrish, A.A., 765
Moyer, M.D., 771
Mroczkowski, J., 359
Müller, J., 169
Müller, St. G., 215
Myoung, Jaemin, 131, 347, 385, 557

Nakazono, T., 397
Nam, Ok-Hyun, 451
Namavar, F., 409
Nardelli, M. Buongiorno, 465
Nebel, C.E., 495
Nemanich, Robert J., 33, 143, 563
Neugebauer, Jörg, 619
Newman, N., 329
Nordell, Nils, 519, 661
Nouet, G., 551

Oberman, D.B., 681
Obraztsova, E.D., 723
Ogawa, T., 391
Oguzman, I.H., 45
Okamoto, Atsuto, 583
Okamoto, M., 391
Okushi, Hideyo, 149, 353
Omer, I. St., 293
Ongaro, C., 81
Orton, J.W., 311
Oruzheinikov, A.L., 69
Östling, M., 667

Park, Dalkeun, 451
Pate, B.B., 759, 771
Pearton, S.J., 163, 189, 569, 607
Pehrsson, Pehr E., 305
Pensl, G., 215
Pentaleri, E., 631
Perkins, N.R., 227, 239, 287
Perry, D., 409
Perov, P.I., 723
Pezoldt, J., 729
Phatak, P., 329
Phillips, W., 771
Piner, E.L., 341
Pirouz, P., 409
Plano, M.A., 771
Poker, D.B., 207
Polesello P., 81, 687
Polyakov, V.I., 723
Polyakov, A.Y., 607, 643
Popovici, G., 649
Porowski, S., 329
Prater, J.T., 281, 457
Prelas, M.A., 649
Preston, A.R., 311

Qian, Weida, 475
Qin, L.H., 735, 747

Ramakrishnan, B., 501
Rao, M.V., 183

Rapcewicz, K., 465
Rea, Laura S., 3
Reinhardt, K., 57
Ren, F., 163
Rieger, W., 323
Roberts, J.C., 341
Robertson, J., 39, 299, 777
Robinson, I.K., 557
Rodrigo, H., 501
Rohrer, E., 495
Rohrer, Gregory S., 475, 539
Romano, L.T., 221, 245
Roth, M.D., 637
Rottner, Kurt, 51, 661
Rowland, L.B., 87, 275
Rubin, M.D., 329
Ruden, P.P., 45
Rupp, R., 253
Russell, Michael W., 575
Ruterana, P., 551
Ruvimov, S., 201, 487

Safvi, S.A., 227, 445
Sasaki, Takatomo, 391, 717
Sauer, R., 693
Schmid, M., 531
Schmitt, E., 215
Scholz, F., 531
Schöner, Adolf, 51, 519, 661
Schurman, M., 741
Schwarz, R., 687
Schweizer, H., 531
Schwendner, K., 359
Scofield, J., 57
Seshadri, S., 87
Shahedipour, F., 415
Shen, B., 735, 747
Shen, H., 741
Sheridan, D.C., 105
Shi, Hongtao, 735, 747, 753
Shim, K.H., 131, 347, 385
Shim, Kyuhwan, 557
Shin, M., 607, 643
Shul, R.J., 163
Shur, M.S., 75
Siergiej, R.R., 87
Silva, S.R.P., 777
Singh, Rajiv K., 365, 371
Sitar, Z., 281, 457
Skorupa, W., 195
Skowronski, Marek, 475, 507, 539, 607, 643
Smith, S.R., 507, 637, 705
Solomon, J.S., 317
Spencer, M.G., 667
Stacy, T., 293, 415, 589, 655
Stall, R.A., 189, 741
Stephani, D., 253
Sternschulte, H., 693
Stömmer, R., 323
Stringfellow, G.B., 335
Stutzmann, M., 323, 495
Suda, Y., 397
Sudarshan, T.S., 99, 111
Sugiyama, Naohiro, 583
Sun, Haiping, 403
Sung, T., 649
Suski, T., 329, 487
Svensson, B.G., 519, 625
Swider, W., 487

Tani, Toshihiko, 583
Tang, S., 335
Taylor, II, C.A., 265
Teraji, Tokuyuki, 149
Teremetskaya, I.G., 723
Thon, A., 227, 445
Thonke, K., 693
Thümer, M., 729
Tin, C.C., 99, 111
Tomich, D.H., 317
Tran, H., 699
Tsong, Tien T., 433
Tu, C.W., 317
Tucker, D.A., 281, 457
Tuncel, E., 681
Turnbull, D.A., 347

Ueda, Tetsuzo, 233, 681

Van de Walle, Chris G., 619
Varnin, V.P., 723
Vartuli, C.B., 163, 569
Vermaut, P., 551
Vescan, A., 63
Vittone, E., 81, 687
Völkl, J., 253

Walker, J., 595
Walukiewicz, W., 513
Wang, F., 687
Wang, R., 45
Ward, B.L., 33, 143
Washburn, J., 201, 487
Weber, E.R., 329
Weishart, Hannes, 195
Weiss, B.-Z., 125
Werninghaus, T., 729
Wessels, Bruce W., 525
White, H.W., 415
White, M.H., 87
Wickenden, A.E., 183
Williams, J.R., 119
Wilson, R.G., 189, 643, 649
Winnacker, A., 215
Wolden, C.A., 281, 457
Wolter, S.D., 63
Won, Jaihyung, 717
Wong, S.S., 667
Wong, W.S., 317
Wongchotigul, K., 667
Wraback, M., 741

Wu, C–H., 409
Wu, Y.F., 23

Xie, K., 93
Xin, Yan, 311

Yamanaka, Sadanori, 353
Yang, C.H., 377
Yang, Haibin, 403
Yang, Kai, 735, 747, 753
Yang, P.C., 281, 457
Yang, Y.J., 377
Yeo, Y.K., 57
Yi, Gyu-Chul, 525
Yoo, Myung-Cheol, 131, 347, 385, 557
Yu, San, 403
Yuri, Masaaki, 233

Zachai, R., 495
Zahn, D.R.T., 729
Zetterling, C.-M., 667
Zhang, Rong, 735, 747, 753
Zhang, Z., 465
Zhao, J.H., 93
Zheng, Youdou, 735, 747, 753
Zhi, D., 227, 239, 287
Zolper, J.C., 189, 569
Zou, Guangtian, 403
Zull, H., 169

SUBJECT INDEX

{1100} face, 583
15R-SiC, 705
3C-SiC, 207, 427, 545, 575
4H-SiC, 105, 111, 119
4H-SiC p^+/n diodes, 57
4H-SiC power UMOSFET, 87
6H-SiC, 51, 119, 125, 195, 451, 539, 563, 729
6H-SiC thyristor, 93

acceptors, 595
activation, 569, 619
activation energy, 595, 601, 637, 655
admittance spectroscopy, 601, 705
AFM, 451
Al doping, 661
AlGaN, 607, 643
AlGaN/GaN, 347
AlInGaN, 341
AlN, 323, 391, 569, 667
AlN buffer, 487
AlN, recombination rate, 69
AlN/GaN heterostructure, 323
AlN/GaN/InN interfaces, 465
AlN/SiC interface, 551
aluminum, 583
amorphous carbon, 39
amorphous silicon carbide, 753
analytical modeling, 9
annealing, 183, 601
annealing temperatures, 195
arcjet, 359
atomically-flat surfaces, 149
Auger Electron Spectroscopy, 175, 397, 569
Au/NiCr contacts and stability, 119
avalanche photodiode, 45

band bending, 771
band gap shift, 323
Be, 631
blue emission band, 385
Bn, cubic, 265
BN thin films, 397
breakdown voltage, 9
broadening parameter, 741
buffer layer(s), 221, 287
bulk 329

C, 465
Ca, 189
capacitance voltage measurements, 667
carbon
    amorphous, diamond like, 699
    diamond like, 299
    tetrahedral amorphous, 299
carbonization, 409
cathodoluminescence, 391, 693, 717
chemical beam epitaxy, 539
chemical bonding model, 777
chloride VPE, 233
compensation, 637
compliant buffer layers, 265
compound semiconductors, 335
contacts, 125, 137
critical thickness, 557
cross-sectional atomic force microscopy, 681
current-temperature, 765
CV, 353, 415, 531, 655, 705
CVD, 377, 433, 575

decomposition of GaN, 183
deep depletion, 99
deep level transient spectroscopy (DLTS), 57
defect(s), 501, 519, 551, 705
    deep level, 525
    structures, 545
density functional theory, 421, 427
deposition temperature, 299
device, 51
    high temperature, 3, 105
    JFET, 105
    microwave, 3
    optoelectronic, 341
    p-n junction diodes, 111
    power, 3
    Schottky diodes, 51, 63
diamond, 33, 365, 415, 433, 457, 501, 589, 759, 765, 771
        B doped, 693, 717
        coatings, 699
        CVD, 495, 687
        electrical conductivity, 649
        film(s), 63, 353, 655
            nanocrystalline, 723
            polycrystalline, 293
        oriented, 281
        P doped, 693
        porous, 723
diffusion, 625
diodes, 207
dislocations, 245, 545, 557
DLTS, 507, 519, 531
donor, 531
doping, 619, 655
    (Zn,S), 531
double positioning twins, 545

ECR, 335, 371, 377
    source, 265
electrical and optical properties, 643
electrical properties, 253
electric field breakdown, 111
electroluminescence, 687

electron
field emission, 777
ionization rate 45
optics, 33
electron cyclotron resonance, 163
assisted MBE, 169
electronic structure, 675
electrophoresis, 365
energy gap, 741
epitaxial, 359
epitaxy, 253
EPR, 501
ESR and LESR, 495
etching, 175
exchange-correlation, 711

Fermi level, pinning, 149
field emission, 143, 765, 771
figures of merit, 9
first-principles calculations, 619
forced diffusion, 649
Fowler-Nordheim, 87
frequency-dependent capacitance
measurements, 601

$GaCl_3$, 233
GaCl pretreatment, 221
GaN(-), 75, 157, 189, 221, 227, 233, 245, 287, 311, 329, 359, 377, 409, 451, 475, 525, 531, 557, 569, 595, 613, 619, 675, 681, 741
buffer layer, 451, 487
cubic, 631
film(s), 735, 747
MBE, 329
MOCVD, 487
nanostructures, 169
n-type, 201
p-type, 385
recombination rate, 69
reconstruction, 421
wire gratings, 169
wurtzitic, 403
characterization of, 403
preparation of, 403
zinc blende, 45
GaN:Mg, 601
GaNP, 317
gas flow, 253
gas phase reactions, 227
gate insulator, 87
Ge, 465
group IV impurities, 465
growth sector boundary structures, 645

H-acceptor pair, 613
H pairs, 613
halide vapor phase epitaxy, 239
Hall effect, 377, 507, 637, 661, 705
two layer simulation, 661
hardness values, 699
HCl, 157
heating, 75
heterostructure(s), 23
double, 341
near lattice-matched, 341
HFCVD, 365, 415, 501, 655
HFETS, 23
high(-)
concentration of acceptors, 353
dose implantation, 195
field properties, 99
power, 93
resistivity and n-type GaN, 183
temperature, 87, 93
growth, 265
hole ionizaton rate, 45
homoepitaxial
diamond, 305
overgrowths, 221
hopping
conduction, 705
or impurity conduction, 637
Hot Filament Chemical Vapor
Deposition, 81
HREM, 551
HVPE, 227, 287
hydride vapor phase epitaxy, 221, 245
hydrogen, 613, 619, 625
plasma passivation, 607

I-V characteristics, 687
I-V-T, 57
III-nitride semiconductors, 311
III-V nitride(s), 163, 569
materials, 265
implant activation, 183
implantation temperature, 195
implanted with nitrogen and
aluminum, 729
InAlN, 569
InAsN, 335
inclusions, 545
incorporated hydrogen, 353
incorporation efficiencies, 583
InGaN, 341, 569
InN, 569
recombination rate, 69
insulator reliability, 87
interfacial microstructures, 457
internal structure and optical
properties, 723
intrinsic exciton, 675
ion energy, 299
ion implantation, 189
ionization, 189

JFET, 105
junction field effect, 189

KrF excimer, 397

laser
ablation, 397, 699
diode (LD), 347
lattice match, Si substrates, 317
least strained, 487
LED, 23
local-field, 711
low
electron affinity, 777
temperature, 371
volume resistivity, 589

mapping, 51
MBE, 31, 335
metal-diamond wire, 81
metallic-type, 385
metalorganic chemical vapor deposition, 735, 747
methyltrichlorosilane (MTS), 575
Mg, 631
Micro-Raman Spectroscopy, 729
micropipe, 539
microwave circuit, 33
MIS capacitors, 667
MODFET, 23
modulation spectroscopy, 741
molecular
beam epitaxy, 335
dynamics, 427, 439
Monte Carlo, 45
morphology, 275, 575
mosaic structure, 305
Mott(-)
Anderson transition, 385
transition, 693
MOVPE, 23, 667
adducts, 445
gas phase reactions, 445
multi-step, 281
multiple quantum wells (MQWs), 347

N, 495, 583, 777
composition, 317
doping, 661
implantation, 531
plasma ambient, 391
radical beam source, 317
vacancy, 531
Nd:YAG, 397
NEA, 33, 143, 771
negative electron affinity, 465
Ni, 281, 457
Ni-C-H, 281
Ni-Ti thin films, 125
Ni_4C, 457
nichrome, 119
nitride, 335
non-chemical decomposition, 359
nonpolar surfaces, 421

nucleation, 475
density, 415
layers, 451
numerical model, 227

O-implantation, 189
ohmic contacts, 119, 149, 201
optical, 711
absorption, 507
properties, 699, 735, 747
transmission, 681
origin of high-conductivity, 353
oxidation, 723
oxide quality, 99

P, 655
particle, 365
photoassisted dry etching, 157
photocapacitance spectroscopy, 525
photocurrent(s), 51, 175
photoemission, 759
photoluminescence spectroscopy, 169, 233, 311, 377, 675, 717
photoreflectance, 735
photothermal deflection spectroscopy, 323
photovoltage, 759
physical vapor transport, 539
plasma(-)
assisted molecular beam epitaxy, 347, 385
CVD, 371
polarization, 465
polytype, 275
change, 729
preferred orientation, 365
pregrowth conditions, 275
prismatic planar fault, 551
protective coating, 371
proton isolation, 207
pulsed laser deposition, 391
pyroeffect, 75

radial distribution functions, 753
radiative recombination rates, 69
Raman spectroscopy, 377, 501, 699, 723, 753
ramp response, 99
rapid thermal annealing, 569
reactive ion etching, 163
reactor geometry, 227
reflectivity, 675
resistivity measurements, 637
rotational misorientation, 287
Rutherford backscattering spectrometry, 391

sapphire, 359, 371, 385, 589, 607, 643, 735, 747
scanning tunneling microscopy, 433

Schottky
barrier height, 143
junction, 353
screw dislocations, empty core, 539
seeding, 293
sensitivity, 81
seven crystal x-ray diffraction, 305
Si, 465, 631
(100), 265
doping, 487
porous, 415
on-insulator, 409
SiC, 3, 63, 137, 175, 253, 409, 439, 519, 625, 667, 705
bulk crystal growth, 215
bulk single crystal, 583
epitaxial, 275
epitaxy, 661
MOS, 99
power devices, 9
SIMS, 519, 625
single(-)
crystalline, 403
domain, 433
space charge limited current, 687
spectroscopic ellipsometry, 681
spiral steps, 539
stacking faults, 545
steps, 551
strain, 75, 557
stress, 305
structural defects, 475
structure, 201
sublimation growth, 215
substrates, 221
surface, 253
charge, 51
diffusion, 439
impurities, O and C, 563
layer, 717
melting, 281
morphology, 397, 487
preparation, 275
reconstruction, 421, 427
roughness, 293
structure, 433
Synchrotron White Beam X-ray Topography, 545
synchrotron X-ray, 557
system efficiency, 9

ta-C, 299
TB-LMTO method, 631
TEM, 201, 245, 311, 403, 457
high resolution, 265
termination design, 9
textured, 365
theory, 613
thermal admittance spectroscopy, 637
thermal expansion, 75
thermionic(-)
emission, 149
field emission, 353
thermodynamic
analysis, 233
equilibrium, 137
thermomodulation, 741
thin films, 475
threading dislocation, 487
Ti/Al, 201
Ti/Al/Ni/Au, 201
transition metal, 137
transmission spectra, 753
transport, 57
trapping, 625
trimethylgallium-trimethylamine adduct, 445
tungsten
carbide, 195
silicide, 195
tunneling, 57

ultrathin Si, 409
UMOS FET, 9
UMOS IGBT, 9
uv photoemission spectroscopy, 143

vacuum microelectronic devices, 33
vanadium
acceptor, 507
nitrogen complex, 507

wide-band-gap, 711

x-ray
diffraction spectra, 753
diffractometry, high resolution, 239, 323
dosimeters, 81